B.L. THE
A.K. THE

A TEXTBOOK OF ELECTRICAL TECHNOLOGY

IN

S.I. SYSTEM OF UNITS

(Including rationalized M.K.S.A. System)

for the Examinations of B.Sc. (Egg.) ; Sec. A & B of A.M.I.E. (I) ; A.M.I.E.E. (London); City & Guilds (London) ; I.E.R.E. (London) ; Grad. I.E.T.E. (I)

VOLUME I

2002

S. CHAND & COMPANY LTD.

RAM NAGAR, NEW DELHI-110 055

S. CHAND & COMPANY LTD.

(An ISO 9002 Company)

Head Office : 7361, RAM NAGAR, NEW DELHI - 110 055

Phones : 3672080-81-82; Fax : 91-11-3677446

Shop at: **schandgroup.com**

E-mail: **schand@vsnl.com**

Branches :

No. 6, Ahuja Chambers, 1st Cross, Kumara Krupa Road, **Bangalore**-560 001. Ph : 2268048
152, Anna Salai, **Chennai**-600 002. Ph : 8460026
Pan Bazar, **Guwahati**-781 001. Ph : 522155
Sultan Bazar, **Hyderabad**-500 195. Ph : 4651135, 4744815
Mai Hiran Gate, **Jalandhar** - 144008. Ph. 401630
613-7, M.G. Road, Ernakulam, **Kochi**-682 035. Ph : 381740
285/J, Bipin Bihari Ganguli Street, **Kolkata**-700 012. Ph : 2367459, 2373914
Mahabeer Market, 25 Gwynne Road, Aminabad, **Lucknow**-226 018. Ph : 226801, 284815
Blackie House, 103/5, Walchand Hirachand Marg , Opp. G.P.O., **Mumbai**-400 001. Ph : 2690881, 2610885
3, Gandhi Sagar East, **Nagpur**-440 002. Ph : 723901
104, Citicentre Ashok, Govind Mitra Road, **Patna**-800 004. Ph : 671366

First Edition 1959
Subsequent Editions and Reprints 1960, 61, 62, 64, 65, 66, 67, 68, 69, 70, 71, 72, 1973 (Twice), 74, 75, 76, 77, 78, 79, 80, 81, 82, 83, 84 (Twice), 85, 86, 87, 88 (Twice), 89, 90, 91, 92, 93, 94, 95, 97, 98 (Twice), 99, 2000, 2001
Reprint 2002

Other Parts Available :
Volume I Basic Electrical Engineering
Volume II A.C. & D.C. Machines
Volume III Transmission, Distribution & Utilization
Volume IV Electronic Devices & Circuits

ISBN : 81-219-0290-8

PRINTED IN INDIA

By Rajendra Ravindra Printers (Pvt.) Ltd., 7361, Ram Nagar, New Delhi-110 055 and published by S. Chand & Company Ltd., 7361, Ram Nagar, New Delhi-110 055.

1 ELECTRIC CURRENT AND OHM'S LAW

1.1. Electron Drift Velocity

Suppose that in a conductor, the number of free electrons available per m^3 of the conductor material is n and let their axial drift velocity be v metres/second. In time dt, distance travelled would be $v \times dt$. If A is area of cross-section of the conductor, then the volume is $vAdt$ and the number of electrons contained in this volume is $vA\,dt$. Obviously, all these electrons will cross the conductor cross-section in time dt. If e is the charge of each electron, then total charge which crosses the section in time dt is $dq = nAev\,dt$.

Since current is the rate of flow of charge, it is given as

$$i = \frac{dq}{dt} = \frac{nAev\,dt}{dt} \qquad \therefore\ i = nAev$$

Current density $\quad J = i/A = ne\,v$ ampere/metre^2

Assuming a normal current density $J = 1.55 \times 10^6$ A/m^2, $n = 10^{29}$ for a copper conductor and $e = 1.6 \times 10^{-19}$ coulomb, we get

$$1.55 \times 10^6 = 10^{29} \times 1.6 \times 10^{-19} \times v \quad \therefore\ v = 9.7 \times 10^{-5} \text{ m/s} = 0.58 \text{ cm/min}$$

It is seen that contrary to the common but mistaken view, the electron drift velocity is rather very slow and is independent of the current flowing and the area of the conductor.

N.B. Current density *i.e.*, the current per unit area, is a vector quantity. It is denoted by the symbol $\vec{J}$.

Therefore, in vector notation, the relationship between current I and $\vec{J}$ is :

$$I = \vec{J} \cdot \vec{a} \qquad [\text{where } \vec{a} \text{ is the vector notation for area 'a'}]$$

For extending the scope of the above relationship, so that it becomes applicable for area of any shape, we write :

$$I = \int \vec{J} . d\vec{a}$$

The magnitude of the current density can, therefore, be written as $J \cdot \alpha$.

Example 1.1. *A conductor material has a free-electron density of 10^{24} electrons per metre^3. When a voltage is applied, a constant drift velocity of 1.5×10^{-2} metre/second is attained by the electrons. If the cross-sectional area of the material is 1 cm^2, calculate the magnitude of the current. Electronic charge is 1.6×10^{-19} coulomb.* **(Electrical Engg. Aligarh Muslim University 1981)**

Solution. The magnitude of the current is

$$i = nAev \text{ amperes}$$

Here, $\quad n = 10^{24}$; $A = 1$ cm$^2 = 10^{-4}$ m^2

$$e = 1.6 \times 10^{-19} \text{ C} ; \ v = 1.5 \times 10^{-2} \text{ m/s}$$

$$\therefore \quad i = 10^{24} \times 10^{-4} \times 1.6 \times 10^{-19} \times 1.5 \times 10^{-2} = \mathbf{0.24\ A}$$

1.2. Charge Velocity and Velocity of Field Propagation

The speed with which charge drifts in a conductor is called the *velocity of charge*. As seen from above, its value is quite low, typically fraction of a metre per second.

However, the *speed* with which the effect of e.m.f. is experienced at all parts of the conductor resulting in the flow of current is called the *velocity of propagation of electrical field*. It is independent of current and voltage and has high but constant value of nearly 3×10^8 m/s.

Example 1.2. *Find the velocity of charge leading to 1 A current which flows in a copper conductor of cross-section 1 cm² and length 10 km. Free electron density of copper = 8.5×10^{28} per m³. How long will it take the electric charge to travel from one end of the conductor to the other.*

Solution. $i = neA\text{v}$ or $\text{v} = i/neA$

$\therefore\ \text{v} = 1/(8.5 \times 10^{28}) \times 1.6 \times 10^{-19} \times (1 \times 10^{-4}) = 7.35 \times 10^{-7}$ m/s = **0.735 μm/s**

Time taken by the charge to travel conductor length of 10 km is

$$t = \frac{\text{distance}}{\text{velocity}} = \frac{10 \times 10^3}{7.35 \times 10^{-7}} = 1.36 \times 10^{10} \text{ s}$$

Now, 1 year = 365 × 24 × 3600 = 31,536,000 s

$t = 1.36 \times 10^{10}/31,536,000$ = **431 years**

1.3. The Idea of Electric Potential

In Fig. 1.1 is shown a simple voltaic cell. It consists of copper plate (known as anode) and a zinc rod (*i.e.* cathode) immersed in dilute sulphuric acid (H_2SO_4) contained in a suitable vessel. The chemical action taking place within the cell causes the electrons to be removed from Cu plate and to be deposited on the zinc rod at the same time. This transfer of electrons is accomplished through the agency of the diluted H_2SO_4 which is known as the electrolyte. The result is that zinc rod becomes negative due to the deposition of electrons on it and the Cu plate becomes positive due to the removal of electrons from it. The large number of electrons collected on the zinc rod is being attracted by anode but is prevented from returning to it by the force set up by the chemical action within the cell.

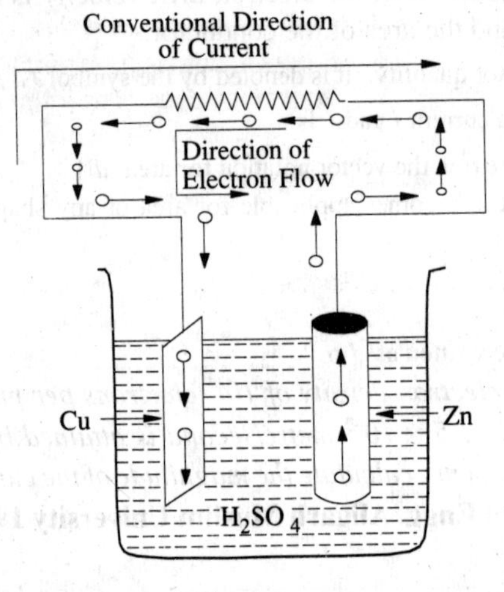

Fig. 1.1

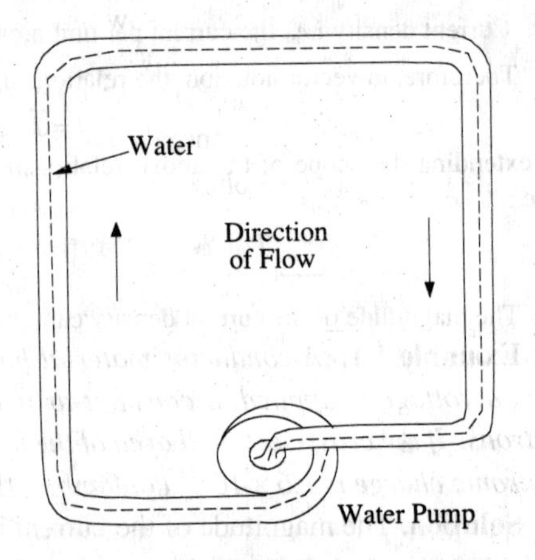

Fig. 1.2

But if the two electrodes are joined by a wire *externally*, then electrons rush to the anode thereby equalizing the charges of the two electrodes. However, due to the continuity of chemical action, a continuous difference in the number of electrons on the two electrodes is maintained which keeps up a continuous flow of current through the external circuit. The action of an electric cell is similar to that of a water pump which, while working, maintains a continuous flow of water *i.e.* water current through the pipe (Fig. 1.2).

It should be particularly noted that the direction of *electronic* current is from zinc to copper in the external circuit. However, the direction of *conventional* current (which is given by the direction of flow of positive charge) is from Cu to zinc. In the present case, there is no flow of positive charge as such from one electrode to another. But we can look upon the arrival of electrons on copper plate (with subsequent decrease in its positive charge) as equivalent to an actual departure of positive charge from it.

When zinc is negatively charged, it is said to be at negative potential with respect to the electrolyte, whereas anode is said to be at positive potential relative to the electrolyte. Between themselves, Cu plate is assumed to be at a higher potential than the zinc rod. The difference in potential is continuously maintained by the chemical action going on in the cell which supplies energy to establish this potential difference.

1.4. Resistance

It may be defined as the property of a substance due to which it opposes (or restricts) the flow of electricity (*i.e.*, electrons) through it.

Metals (as a class), acids and salt solutions are good conductors of electricity. Amongst pure metals, silver, copper and aluminium are very good conductors in the given order.* This, as discussed earlier, is due to the presence of a large number of free or loosely-attached electrons in their atoms. These vagrant electrons assume a directed motion on the application of an electric potential difference. These electrons while flowing pass *through* the molecules or the atoms of the conductor, collide with other atoms and electrons, thereby producing heat.

Those substances which offer relatively greater difficulty or hindrance to the passage of these electrons are said to be relatively poor conductors of electricity like bakelite, mica, glass, rubber, p.v.c. (poylvinyl chloride) and dry wood etc. Amongst good insulators can be included fibrous substances such as paper and cotton when dry, mineral oils free from acids and water, ceramics like hard porcelain and asbestos and many other plastics besides p.v.c. It is helpful to remember that electric friction is similar to friction in Mechanics.

1.5. The Unit of Resistance

The practical unit of resistance is ohm.** A conductor is said to have a resistance of one ohm if it permits one ampere current to flow through it when one volt is impressed across its terminals.

For insulators whose resistances are very high, a much bigger unit is used *i.e.* megaohm = 10^6 ohm (the prefix 'mega' or mego meaning a million) or kilohm = 10^3 ohm (kilo means thousand). In the case of very small resistances, smaller units like milli-ohm = 10^{-3} ohm or microhm = 10^{-6} ohm are used. The symbol for ohm is Ω.

Table 1.1. Multiples and Sub-multiples of Ohm

Prefix	*Its meaning*	*Abbreviation*	*Equal to*
Mega-	One million	M Ω	10^6 Ω
Kilo-	One thousand	k Ω	10^3 Ω
Centi-	One hundredth	–	–
Milli-	One thousandth	m Ω	10^{-3} Ω
Micro-	One millionth	μ Ω	10^{-6} Ω

*However, for the same resistance per unit length, cross-sectional area of aluminium conductor has to be 1.6 times that of the copper conductor but it weighs only half as much. Hence, it is used where economy of weight is more important than economy of space.

**After George Simon Ohm (1787-1854), a German mathematician who in about 1827 formulated the law known after his name as Ohm's Law.

1.6. Laws of Resistance

The resistance R offered by a conductor depends on the following factors:

(*i*) It varies directly as its length, l.

(*ii*) It varies inversely as the cross-section A of the conductor.

(*iii*) It depends on the nature of the material.

(*iv*) It also depends on the temperature of the conductor.

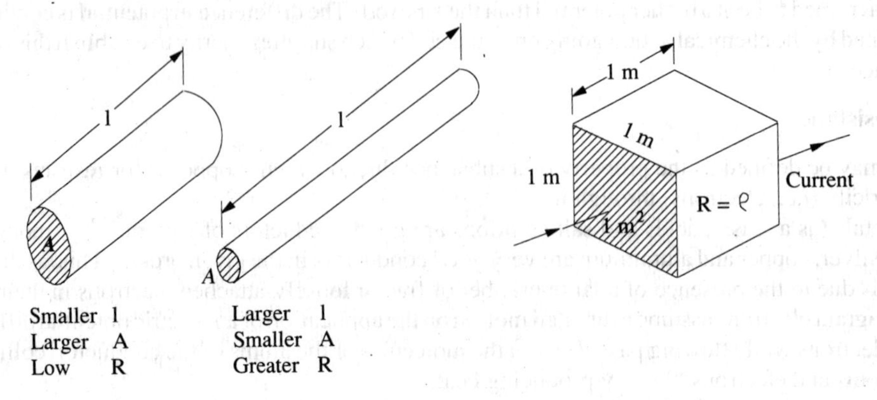

Fig. 1.3

Fig. 1.4

Neglecting the last factor for the time being, we can say that

$$R \propto \frac{l}{A} \quad \text{or} \quad R = \rho \frac{l}{A} \qquad ...(i)$$

where ρ is a constant depending on the nature of the material of the conductor and is known as its *specific resistance* or *resistivity*.

If in Eq. (*i*), we put

l = 1 metre and A = 1 metre², then R = ρ (Fig. 1.4)

Hence, specific resistance of a material may be defined as

the resistance between the opposite faces of a metre cube of that material.

1.7. Units of Resistivity

From Eq. (*i*), we have $\rho = \dfrac{AR}{l}$

In the S.I. system of units,

$$\rho = \frac{A \text{ metre}^2 \times R \text{ ohm}}{l \text{ metre}} = \frac{AR}{l} \text{ ohm-metre}$$

Hence, the unit of resistivity is ohm-metre (Ω-m)

It may, however, be noted that resistivity is sometimes expressed as so many ohm per m^3. Although, it is incorrect to say so but it means the same thing as ohm-metre.

If l is in centimetres and A in cm^2, then ρ is in ohm-centimetre (Ω-cm).

Values of resistivity and temperature coefficients for various materials are given in Table 1.2. The resistivities of commercial materials may differ by several per cent due to impurities etc.

Table 1.2. Resistivities and Temperature Coefficients

Material	*Resistivity in ohm-metre at 20°C($\times 10^{-8}$)*	*Temperature coefficient at 20 °C($\times 10^{-4}$)*
Aluminium, commercial	2.8	40.3
Brass	6 – 8	20
Carbon	3000 – 7000	–5
Constantan or Eureka	49	+0.1 to –0.4
Copper (annealed)	1.72	39.3
German Silver (84% Cu; 12% Ni; 4% Zn)	20.2	2.7
Gold	2.44	36.5
Iron	9.8	65
Manganin (84% Cu; 12% Mn; 4% Ni)	44 – 48	0.15
Mercury	95.8	8.9
Nichrome (60% Cu; 25% Fe; 15% Cr)	108.5	1.5
Nickel	7.8	54
Platinum	9 – 15.5	36.7
Silver	1.64	38
Tungsten	5.5	47
Amber	5×10^{14}	
Bakelite	10^{10}	
Glass	$10^{10} - 10^{12}$	
Mica	10^{15}	
Rubber	10^{16}	
Shellac	10^{14}	
Sulphur	10^{15}	

Example 1.3. *A coil consists of 2000 turns of copper wire having a cross-sectional area of 0.8 mm². The mean length per turn is 80 cm. and the resistivity of copper is 0.02 μ Ω–m. Find the resistance of the coil and power absorbed by the coil when connected across 110 V d.c. supply.*

(F.Y. Engg. Pune Univ. May 1990)

Solution. Length of the coil, $l = 0.8 \times 2000 = 1600$ m; $A = 0.8$ mm² $= 0.8 \times 10^{-6}$ m².

$$R = \rho \frac{l}{A} = 0.02 \times 10^{-6} \times 1600/0.8 \times 10^{-6} = \mathbf{40\Omega}$$

Power absorbed $= V^2/R = 110^2/40 =$ **302.5 W**

Example 1.4. *An aluminium wire 7.5 m long is connected in a parallel with a copper wire 6 m long. When a current of 5 A is passed through the combination, it is found that the current in the aluminium wire is 3 A. The diameter of the aluminium wire is 1 mm. Determine the diameter of the copper wire. Resistivity of copper is 0.017 μ Ω-m; that of the aluminium is 0.028 μ Ω-m.*

(F.Y. Engg. Pune Univ. May 1991)

Solution. Let the subscript 1 represent aluminium and subscript 2 represent copper.

$$R_1 = \rho_1 \frac{l_1}{a_1} \text{ and } R_2 = \rho_2 \frac{l_2}{a_2} \quad \therefore \frac{R_2}{R_1} = \frac{\rho_2}{\rho_1} \cdot \frac{l_2}{l_1} \cdot \frac{a_1}{a_2}$$

$$\therefore \quad a_2 = a_1 \cdot \frac{R_1}{R_2} \cdot \frac{\rho_2}{\rho_1} \cdot \frac{l_2}{l_1} \quad \ldots(i)$$

Now $I_1 = 3$ A; $I_2 = 5 - 3 = 2$ A.

If V is the common voltage across the parallel combination of aluminium and copper wires, then

$$V = I_1R_1 = I_2R_2; \quad \therefore R_1/R_2 = I_2/I_1 = 2/3$$

$$a_1 = \frac{\pi d^2}{4} = \frac{\pi \times 1^2}{4} = \frac{\pi}{4} \text{ mm}^2$$

Substituting the given values in Eq. (*i*), we get

$$a_2 = \frac{\pi}{4} \times \frac{2}{3} \times \frac{0.017}{0.028} \times \frac{6}{7.5} = 0.2544 \text{ m}^2$$

$\therefore \quad \pi \times d_2^2/4 = 0.2544$ or $d_2 = \mathbf{0.569\ mm}$

Example 1.5. *(a) A rectangular carbon block has dimensions 1.0 cm × 1.0 cm × 50 cm. (i) What is the resistance measured between the two square ends ? (ii) between two opposing rectangular faces ? Resistivity of carbon at 20°C is 3.5 × 10⁻⁵ Ω-m.*

(b) A current of 5 A exists in a 10-Ω resistance for 4 minutes (i) how many coulombs and (ii) how many electrons pass through any section of the resistor in this time ? Charge of the electron = 1.6 × 10⁻¹⁹C. **(M.S. Univ. Baroda 1989)**

Solution

(*a*) (*i*) $R = \rho\, l/A$

Here, $A = 1 \times 1 = 1 \text{ cm}^2 = 10^{-4} \text{ m}^2 ;\ l = 0.5 \text{ m}$

$\therefore \quad R = 3.5 \times 10^{-5} \times 0.5/10^{-4} = \mathbf{0.175\ \Omega}$

(*ii*) Here, $l = 1 \text{ cm} ;\ A = 1 \times 50 = 50 \text{ cm}^2 = 5 \times 10^{-3} \text{ m}^2$

$R = 3.5 \times 10^{-5} \times 10^{-2}/5 \times 10^{-3} = \mathbf{7 \times 10^{-5}\ \Omega}$

(*b*) (*i*) $Q = It = 5 \times (4 \times 60) = \mathbf{1200\ C}$

(*ii*) $$n = \frac{Q}{e} = \frac{1200}{1.6 \times 10^{-19}} = \mathbf{75 \times 10^{20}}$$

Example 1.6. *Calculate the resistance of 1 km long cable composed of 19 strands of similar copper conductors, each strand being 1.32 mm in diameter. Allow 5% increase in length for the 'lay' (twist) of each strand in completed cable. Resistivity of copper may be taken as 1.72 × 10⁻⁸ Ω-m.*

Solution. Allowing for twist, the length of the strands.

= 1000 m + 5% of 1000 m = 1050 m

Area of cross-section of 19 strands of copper conductors is

$$19 \times \pi \times d^2/4 = 19\, \pi \times (1.32 \times 10^{-3})^2/4 \text{ m}^2$$

Now, $$R = \rho \frac{l}{A} = \frac{1.72 \times 10^{-8} \times 1050 \times 4}{19\pi \times 1.32^2 \times 10^{-6}} = \mathbf{0.694\ \Omega}$$

Example 1.7. *A lead wire and an iron wire are connected in parallel. Their respective specific resistances are in the ratio 49 : 24. The former carries 80 percent more current than the latter and the latter is 47 percent longer than the former. Determine the ratio of their cross sectional areas.* **(Elect. Engg. Nagpur Univ. 1993)**

Solution. Let suffix 1 represent lead and suffix 2 represent iron. We are given that

$$\rho_1/\rho_2 = 49/24 ;\ \text{if } i_2 = 1 ,\ i_1 = 1.8 ;\ \text{if } l_1 = 1,\ l_2 = 1.47$$

Now, $$R_1 = \rho_1 \frac{l_1}{A_1} \quad \text{and} \quad R_2 = \rho_2 \frac{l_2}{A_2}$$

Since the two wires are in parallel, $i_1 = V/R_1$ and $i_2 = V/R_2$

$$\therefore \quad \frac{i_2}{i_1} = \frac{R_1}{R_2} = \frac{\rho_1 l_1}{A_1} \times \frac{A_2}{\rho_2 l_2}$$

$$\therefore \quad \frac{A_2}{A_1} = \frac{i_2}{i_1} \times \frac{\rho_2 l_2}{\rho_1 l_1} = \frac{1}{1.8} \times \frac{24}{49} \times 1.47 = \mathbf{0.4}$$

Example 1.8. *A piece of silver wire has a resistance of 1 Ω. What will be the resistance of manganin wire of one-third the length and one-third the diameter, if the specific resistance of manganin is 30 times that of silver.* **(Electrical Engineering-I, Delhi Univ. 1978)**

Solution. For silver wire : $R_1 = \rho_1 \dfrac{l_1}{A_1}$; For manganin wire, $R_2 = \rho_2 \dfrac{l_2}{A_2}$

$$\therefore \quad \frac{R_2}{R_1} = \frac{\rho_2}{\rho_1} \times \frac{l_2}{l_1} \times \frac{A_1}{A_2}$$

Now $\quad A_1 = \pi d_1^2/4$ and $A_2 = \pi d_2^2/4 \quad \therefore A_1/A_2 = d_1^2/d_2^2$

$$\therefore \quad \frac{R_2}{R_1} = \frac{\rho_2}{\rho_1} \times \frac{l_2}{l_1} \times \left(\frac{d_1}{d_2}\right)^2$$

$$R_1 = 1\ \Omega;\ l_2/l_1 = 1/3,\ (d_1/d_2)^2 = (3/1)^2 = 9;\ \rho_2/\rho_1 = 30$$

$$\therefore \quad R_2 = 1 \times 30 \times (1/3) \times 9 = \mathbf{90\ \Omega}$$

Example 1.9. *The resistivity of a ferric-chromium-aluminium alloy is 51×10^{-8} Ω-m. A sheet of the material is 15 cm long, 6 cm wide and 0.014 cm thick. Determine resistance between (a) opposite ends and (b) opposite sides.* **(Electric Circuits, Allahabad Univ. 1983)**

Solution. (*a*) As seen from Fig. 1.5 (*a*) in this case,

$l = 15$ cm $= 0.15$ m

$A = 6 \times 0.014 = 0.084$ cm^2

$= 0.084 \times 10^{-4}$ m^2

$$R = \rho \frac{l}{A} = \frac{51 \times 10^{-8} \times 0.15}{0.084 \times 10^{-4}}$$

$= \mathbf{9.1 \times 10^{-3}\ \Omega}$

(*b*) As seen from Fig. 1.5(*b*) here

$l = 0.014$ cm $= 14 \times 10^{-5}$ m

$A = 15 \times 6 = 90$ cm$^2 = 9 \times 10^{-3}$ m^2

$\therefore \quad R = 51 \times 10^{-8} \times 14 \times 10^{-5} / 9 \times 10^{-3}$

$= \mathbf{79.3 \times 10^{-10}\ \Omega}$

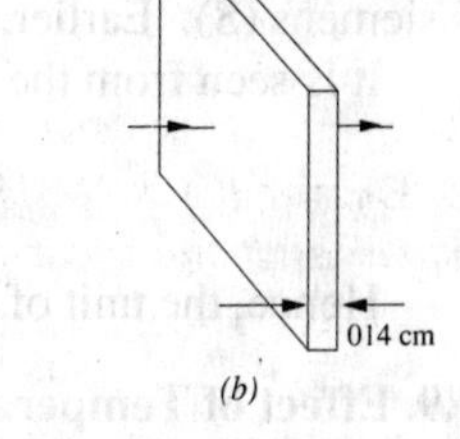

Fig. 1.5

Example 1.10. *The resistance of the wire used for telephone line is 35 Ω per kilometre when the weight of the wire is 5 kg per kilometre. If the specific resistance of the material is 1.95×10^{-8} Ω-m, what is the cross-sectional area of the wire ? What will be the resistance of a loop to a subscriber 8 km from the exchange if wire of the same material but weighing 20 kg per kilometre is used ?*

Solution. Here $R = 35\ \Omega$; $l = 1$ km $= 1000$ m ; $\rho = 1.95 \times 10^{-8}$ Ω-m

Now, $\quad R = \rho \dfrac{l}{A}$ or $A = \dfrac{\rho l}{R} \therefore A = \dfrac{1.95 \times 10^{-8} \times 1000}{35} = \mathbf{55.7 \times 10^{-8}\ m^2}$

In the second case, if the wire is of the same material but weighs 20 kg/km, then its cross-section must be greater than that in the first case.

Cross-section in the second case $= \dfrac{20}{5} \times 55.7 \times 10^{-8} = 222.8 \times 10^{-8}$ m^2

Length of wire $= 2 \times 8 = 16$ km $= 16000$ m $\therefore R = \rho \dfrac{l}{A} = \dfrac{1.95 \times 10^{-8} \times 16000}{222.8 \times 10^{-8}} = \mathbf{140.1\ \Omega}$

Tutorial Problems No. 1.1

1. Calculate the resistance of 100 m length of a wire having a uniform cross-sectional area of 0.1 mm^2 if the wire is made of manganin having a resistivity of 50×10^{-8} Ω -m.

If the wire is drawn out to three times its original length, by how many times would you expect its resistance to be increased ? **[500 Ω ; 9 times]**

2. A cube of a material of side 1 cm has a resistance of 0.001 Ω between its opposite faces. If the same volume of the material has a length of 8 cm and a uniform cross-section, what will be the resistance of this length ? **[0.064 Ω]**

3. A lead wire and an iron wire are connected in parallel. Their respective specific resistances are in the ratio 49 : 24. The former carries 80 per cent more current than the latter and the latter is 47 per cent longer than the former. Determine the ratio of their cross-sectional areas. **[2.5 : 1]**

4. A rectangular metal strip has the following dimensions :

$$x = 10 \text{ cm}, y = 0.5 \text{ cm}, z = 0.2 \text{ cm}$$

Determine the ratio of resistances R_x, R_y and R_z between the respective pairs of opposite faces.

[$R_x : R_y : R_z$ :: 10,000 : 25 : 4] [Elect. Engg. A.M.Ae. S.I. June 1987]

5. The resistance of a conductor 1 mm^2 in cross-section and 20 m long is 0.346 Ω. Determine the specific resistance of the conducting material. **[1.73×10^{-8} Ω-m] [Elect. Circuits-1, Bangalore Univ. 1991]**

1.8. Conductance and Conductivity

Conductance (*G*) is reciprocal of resistance*. Whereas resistance of a conductor measures the *opposition* which it offers to the flow of current, the conductance measures the *inducement* which it offers to its flow.

From Eq. (*i*) of Art. 1.6, $R = \rho \frac{l}{A}$ or $G = \frac{1}{\rho} \cdot \frac{A}{l} = \frac{\sigma A}{l}$

where σ is called the *conductivity* or *specific conductance* of a conductor. The unit of conductance is siemens (S). Earlier, this unit was called mho.

It is seen from the above equation that the conductivity of a material is given by

$$\sigma = G \frac{l}{A} = \frac{G \text{ siemens} \times l \text{ metre}}{A \text{ metre}^2} = G \frac{l}{A} \text{ siemens/metre}$$

Hence, the unit of conductivity is siemens/metre (S/m).

1.9. Effect of Temperature on Resistance

The effect of rise in temperature is :

(*i*) to *increase* the resistance of pure metals. The increase is large and fairly regular for normal ranges of temperature. The temperature/resistance graph is a straight line (Fig. 1.6). As would be presently clarified, metals have a positive temperature co-efficient of resistance.

(*ii*) to *increase* the resistance of alloys, though, in their case, the increase is relatively small and irregular. For some high-resistance alloys like Eureka (60% Cu and 40% Ni) and manganin, the increase in resistance is (or can be made) negligible over a considerable range of temperature.

(*iii*) to *decrease* the resistance of electrolytes, insulators (such as paper, rubber, glass, mica etc.) and partial conductors such as carbon. Hence, insulators are said to possess a *negative* temperature-coefficient of resistance.

1.10. Temperature Coefficient of Resistance

Let a metallic conductor having a resistance of R_0 at 0°C be heated to t°C and let its resistance at this temperature be R_t. Then, considering normal ranges of temperature, it is found that the increase in resistance $\Delta R = R_t - R_0$ depends

(*i*) directly on its initial resistance
(*ii*) directly on the rise in temperature
(*iii*) on the nature of the material of the conductor.

*In a.c. circuits, it has a slightly different meaning.

or $$R_t - R_0 \propto R \times t \text{ or } R_t - R_0 = \alpha R_0 t \qquad ...(i)$$

where α (alpha) is a constant and is known as the *temperature coefficient of resistance* of the conductor.

Rearranging Eq. (*i*), we get $\alpha = \dfrac{R_t - R_0}{R_0 \times t} = \dfrac{\Delta R}{R_0 \times t}$

If $R_0 = 1\ \Omega$, $t = 1°C$, then $\alpha = \Delta R = R_t - R_0$

Hence, the temperature-coefficient of a material may be defined as :

the increase in resistance per ohm original resistance per °C rise in temperature.

From Eq. (*i*), we find that $$R_t = R_0 (1 + \alpha t) \qquad ...(ii)$$

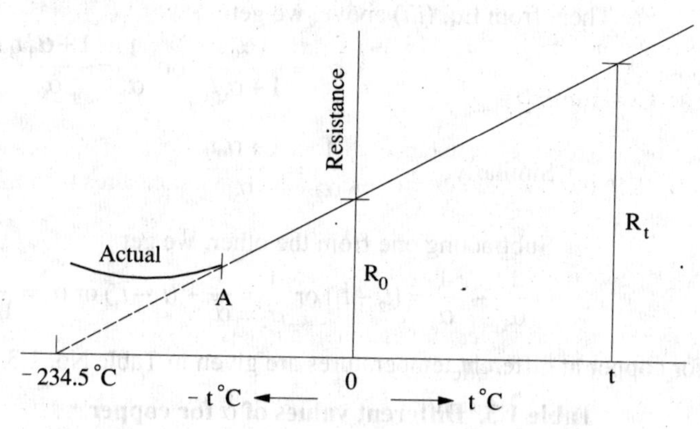

Fig. 1.6

It should be remembered that the above equation holds good for both rise as well as fall in temperature. As temperature of a conductor is decreased, its resistance is also decreased. In Fig. 1.6 is shown the temperature/resistance graph for copper and is practically a straight line. If this line is extended backwards, it would cut the temperature axis at a point where temperature is −234.5°C (a number quite easy to remember). It means that theoretically, the resistance of copper conductor will become zero at this point though as shown by solid line, in practice, the curve departs from a straight line at very low temperatures. From the two similar triangles of Fig. 1.6 it is seen that :

$$\frac{R_t}{R_0} = \frac{t + 234.5}{234.5} = \left(1 + \frac{t}{234.5}\right)$$

$$\therefore \quad R_t = R_0\left(1 + \frac{t}{234.5}\right) \text{ or } R_t = R_0 (1 + \alpha t) \text{ where } \alpha = 1/234.5 \text{ for copper.}$$

1.11. Value of α at Different Temperatures

So far we did not make any distinction between values of α at different temperatures. But it is found that value of α itself is not constant but depends on the initial temperature on which the increment in resistance is based. When the increment is based on the resistance measured at 0°C, then α has the value of α_0. At any other initial temperature *t*°C, value of α is α_t and so on. It should be remembered that, for any conductor, α_0 has the maximum value.

Suppose a conductor of resistance R_0 at 0°C (point *A* in Fig. 1.7) is heated to *t*°C (point *B*). Its resistance R_t after heating is given by

$$R_t = R_0 (1 + \alpha_0 t) \qquad ...(i)$$

where α_0 is the temperature-coefficient at 0°C.

Now, suppose that we have a conductor of resistance R_t at temperature *t*°C. Let this conductor be *cooled* from *t*°C to 0°C. Obviously, now the initial point is *B* and the final point is *A*.

The final resistance R_0 is given in terms of the initial resistance by the following equation

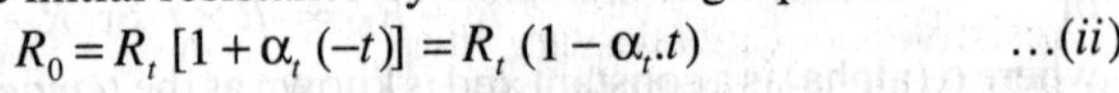

$$R_0 = R_t\,[1 + \alpha_t\,(-t)] = R_t\,(1 - \alpha_t . t) \qquad ...(ii)$$

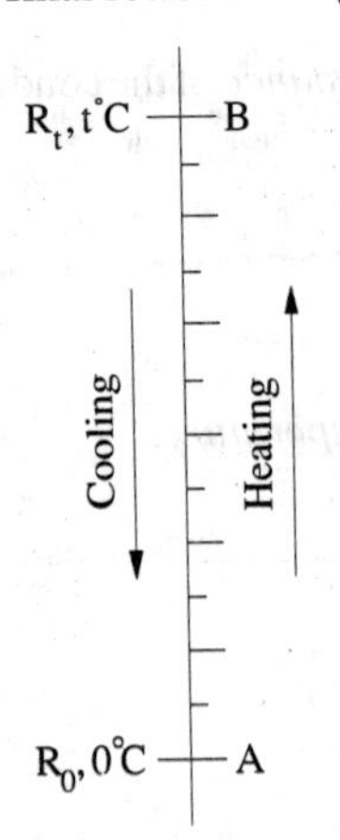

Fig. 1.7

From Eq. (*ii*) above, we have $\alpha_t = \dfrac{R_t - R_0}{R_t \times t}$

Substituting the value of R_t from Eq. (*i*), we get

$$\alpha_t = \frac{R_0(1 + \alpha_0\,t) - R_0}{R_0(1 + \alpha_0\,t) \times t} = \frac{\alpha_0}{1 + \alpha_0\,t} \quad \therefore \alpha_t = \frac{\alpha_0}{1 + \alpha_0\,t} \qquad ...(iii)$$

In general, let α_1 = tempt. coeff. at t_1°C ; α_2 = tempt. coeff. at t_2°C.

Then, from Eq. (*iii*) above, we get

$$\alpha_1 = \frac{\alpha_0}{1 + \alpha_0 t_1} \quad \text{or} \quad \frac{1}{\alpha_1} = \frac{1 + \alpha_0\,t_1}{\alpha_0}$$

Similarly,
$$\frac{1}{\alpha_2} = \frac{1 + \alpha_0 t_2}{\alpha_0}$$

Subtracting one from the other, we get

$$\frac{1}{\alpha_2} - \frac{1}{\alpha_1} = (t_2 - t_1) \text{ or } \frac{1}{\alpha_2} = \frac{1}{\alpha_1} + (t_2 - t_1) \text{ or } \alpha_2 = \frac{1}{1/\alpha_1 + (t_2 - t_1)}$$

Values of α for copper at different temperatures are given in Table No. 1.3.

Table 1.3. Different values of α for copper

Tempt. in °C	0	5	10	20	30	40	50
α	0.00427	0.00418	0.00409	0.00393	0.00378	0.00364	0.00352

In view of the dependence of α on the initial temperature, we may define *the temperature coefficient of resistance at a given temperature as the change in resistance per ohm per degree centigrade change in temperature from the given temperature.*

In case R_0 is not given, the relation between the known resistance R_1 at t_1°C and the unknown resistance R_2 at t_2°C can be found as follows :

$$R_2 = R_0\,(1 + \alpha_0\,t_2) \quad \text{and} \quad R_1 = R_0\,(1 + \alpha_0\,t_1)$$

$$\therefore \quad \frac{R_2}{R_1} = \frac{1 + \alpha_0\,t_2}{1 + \alpha_0\,t_1} \qquad ...(iv)$$

The above expression can be simplified by a little approximation as follows :

$$\frac{R_2}{R_1} = (1 + \alpha_0\,t_2)\,(1 + \alpha_0\,t_1)^{-1}$$

$$= (1 + \alpha_0\,t_2)\,(1 - \alpha_0\,t_1)$$ [Using Binomial Theorem for expansion and neglecting squares and higher powers of $(\alpha_0\,t_1)$]

$$= 1 + \alpha_0(t_2 - t_1)$$ [Neglecting product $(\alpha_0^2\,t_1\,t_2)$]

$$\therefore \quad R_2 = R_1[1 + \alpha_0\,(t_2 - t_1)]$$

For more accurate calculations, Eq. (*iv*) should, however, be used.

1.12. Variations of Resistivity with Temperature

Not only resistance but specific resistance or resistivity of metallic conductors also increases with rise in temperature and *vice versa.*

As seen from Fig. 1.8 the resistivities of metals vary linearly with temperature over a significant range of temperature–the variation becoming non-linear both at very high and at very low temperatures. Let, for any metallic conductor,

ρ_1 = resistivity at t_1°C

ρ_2 = resistivity at t_2°C

m = slope of the linear part of the curve

Then, it is seen that

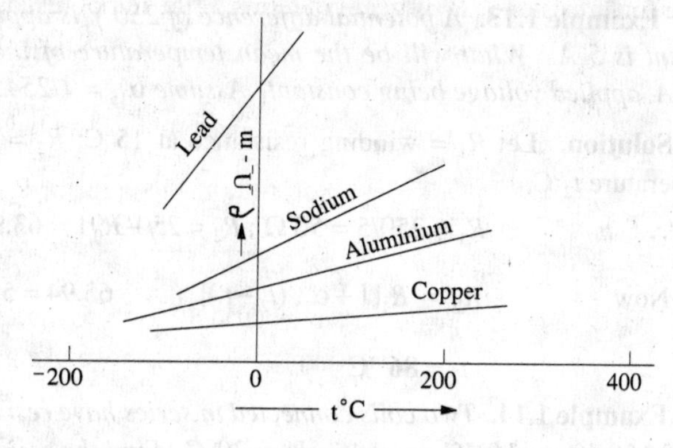

Fig. 1.8

$$m = \frac{\rho_2 - \rho_1}{t_2 - t_1} \quad \text{or} \quad \rho_2 = \rho_1 + m(t_2 - t_1)$$

or
$$\rho_2 = \rho_1 \left[1 + \frac{m}{\rho_1} (t_2 - t_1) \right]$$

The ratio of m/ρ_1 is called the *temperature coefficient of resistivity* at temperature t_1°C. It may be defined as numerically equal to the fractional change in ρ_1 per °C change in the temperature from t_1°C. It is almost equal to the temperature-coefficient of resistance α_1 Hence, putting $\alpha_1 = m/\rho_1$, we get

$$\rho_2 = \rho_1[1 + \alpha_1 (t_2 - t_1)] \quad \text{or simply as} \quad \rho_t = \rho_0 (1 + \alpha_0 t)$$

Note. It has been found that although temperature is the most significant factor influencing the resistivity of metals, other factors like pressure and tension also affect resistivity to some extent. For most metals except lithium and calcium, increase in pressure leads to decrease in resistivity. However, resistivity increases with increase in tension.

Example 1.11. *A copper conductor has its specific resistance of 1.6 × 10⁻⁶ ohm-cm at 0°C and a resistance temperature coefficient of 1/254.5 per °C at 20°C. Find (i) the specific resistance and (ii) the resistance - temperature coefficient at 60°C.*

(F.Y. Engg. Pune Univ. Nov. 1988)

Solution. $\alpha_{20} = \frac{\alpha_0}{1 + \alpha_0 \times 20}$ or $\frac{1}{254.5} = \frac{\alpha_0}{1 + \alpha_0 \times 20}$ $\therefore \alpha_0 = \frac{1}{234.5}$ per °C

(*i*) $\rho_{60} = \rho_0 (1 + \alpha_0 \times 60) = 1.6 \times 10^{-6} (1 + 60/234.5) = \mathbf{2.01 \times 10^{-6}\ \Omega - cm}$

(*ii*) $\alpha_{60} = \frac{\alpha_0}{1 + \alpha_0 \times 60} = \frac{1/234.5}{1 + (60/234.5)} = \mathbf{\frac{1}{294.5}}$ **per°C**

Example 1.12. *A platinum coil has a resistance of 3.146 Ω at 40°C and 3.767 Ω at 100°C. Find the resistance at 0°C and the temperature-coefficient of resistance at 40°C.*

(Electrical Science-II, Allahabad Univ. 1993)

Solution. $R_{100} = R_0 (1 + 100\ \alpha_0)$ ...(*i*)

$R_{40} = R_0 (1 + 40\ \alpha_0)$ ...(*ii*)

$\therefore \frac{3.767}{3.146} = \frac{1 + 100\ \alpha_0}{1 + 40\ \alpha_0}$ or $\alpha_0 = 0.00379$ or 1|264 per°C

From (*i*), we have $3.767 = R_0 (1 + 100 \times 0.00379)$ $\therefore R_0 = \mathbf{2.732\ \Omega}$

Now, $\alpha_{40} = \frac{\alpha_0}{1 + 40\,\alpha_0} = \frac{0.00379}{1 + 40 \times 0.00379} = \mathbf{\frac{1}{304}}$ **per °C**

Example 1.13. *A potential difference of 250 V is applied to a field winding at 15°C and the current is 5 A. What will be the mean temperature of the winding when current has fallen to 3.91 A, applied voltage being constant. Assume α_{15} = 1/254.5.* **(Elect. Engg. Pune Univ. 1991)**

Solution. Let R_1 = winding resistance at 15°C; R_2 = winding resistance at unknown mean temperature t_2°C.

$\therefore \quad R_1 = 250/5 = 50\ \Omega\ ;\ R_2 = 250/3.91 = 63.94\ \Omega.$

Now $\quad R_2 = R_1[1 + \alpha_{15}(t_2 - t_1)] \quad \therefore \quad 63.94 = 50\left[1 + \dfrac{1}{254.5}(t_2 - 15)\right]$

$\therefore \quad t_2 = \mathbf{86\ °C}$

Example 1.14. *Two coils connected in series have resistances of 600 Ω and 300 Ω with tempt. coeff. of 0.1% and 0.4% respectively at 20°C. Find the resistance of the combination at a tempt. of 50°C. What is the effective tempt. coeff. of combination ?*

Solution. Resistance of 600 Ω resistor at 50°C is = 600 [1 + 0.001 (50–20)] = 618 Ω
Similarly, resistance of 300 Ω resistor at 50°C is = 300 [1 + 0.004 (50–20)] = 336 Ω
Hence, total resistance of combination at 50°C is = 618 + 336 = **954 Ω**
Let β = resistance-temperature coefficient at 20°C
Now, combination resistance at 20°C = 900 Ω
Combination resistance at 50°C = 954 Ω

$\therefore \quad 954 = 900\,[1 + \beta(50-20)] \quad \therefore \beta = \mathbf{0.002}$

Example 1.15. *Two wires A and B are connected in series at 0°C and resistance of B is 3.5 times that of A. The resistance temperature coefficient of A is 0.4% and that of the combination is 0.1%. Find the resistance temperature coefficient of B.* **(Elect. Technology, Hyderabad Univ. 1992)**

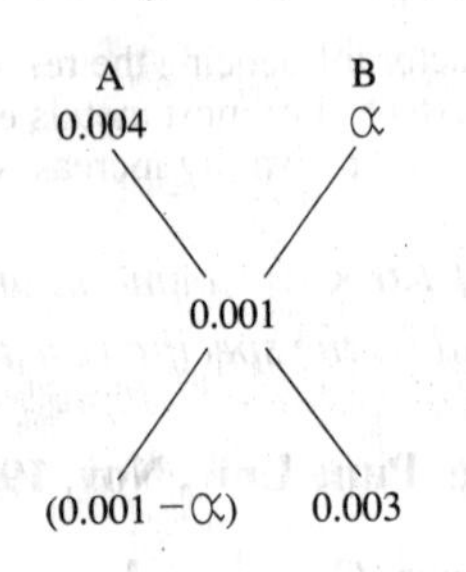

Fig. 1.9

Solution. A simple technique which gives quick results in such questions is illustrated by the diagram of Fig. 1.9. It is seen that

$R_B/R_A = 0.003/(0.001 - \alpha)$ or $3.5 = 0.003/(0.001 - \alpha)$

or $\alpha = 0.000143\ °C^{-1}$ or **0.0143%**

Example 1.16. *Two materials A and B have resistance temperature coefficients of 0.004 and 0.0004 respectively at a given temperature. In what proportion must A and B be joined in series to produce a circuit having a temperature coefficient of 0.001 ?* **(Elect. Technology, Indore Univ. April 1981)**

Solution. Let R_A and R_B be the resistances of the two wires of materials *A* and *B* which are to be connected in series. Their ratio may be found by the simple technique shown in Fig. 1.10.

$$\frac{R_B}{R_A} = \frac{0.003}{0.0006} = 5$$

Hence, R_B must be **5 times R_A.**

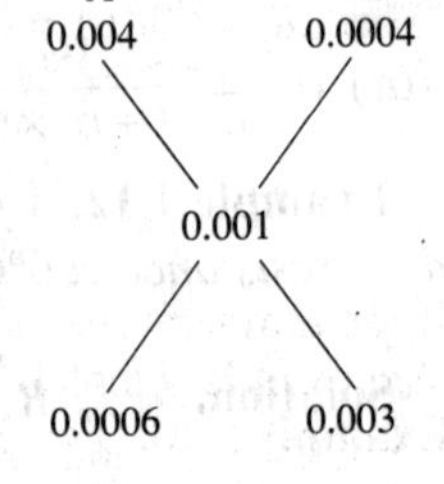

Fig. 1.10

Example 1.17. *A resistor of 80 Ω resistance, having a temperature coefficient of 0.0021 per degree C is to be constructed. Wires of two materials of suitable cross-sectional area are available. For material A, the resistance is 80 ohm per 100 metres and the temperature coefficient is 0.003 per degree C. For material B, the corresponding figures are 60 ohm per metre and 0.0015 per degree C. Calculate suitable lengths of wires of materials A and B to be connected in series to construct the required resistor. All data are referred to the same temperature.*

Solution. Let R_a and R_b be the resistances of suitable lengths of materials *A* and *B* respectively which when joined in series will have a combined temperature coeff. of 0.0021. Hence, combination resistance at any given temperature is ($R_a + R_b$). Suppose we heat these materials through t°C.

When heated, resistance of A increases from R_a to R_a $(1 + 0.003\ t)$. Similarly, resistance of B increases from R_b to R_b $(1 + 0.0015\ t)$.

$\therefore$ combination resistance after being heated through t°C

$$= R_a\ (1 + 0.003\ t) + R_b\ (1 + 0.0015\ t)$$

The combination α being given, value of combination resistance can be also found directly as

$$= (R_a + R_b)(1 + 0.0021\ t)$$

$$\therefore \quad (R_a + R_b)(1 + 0.0021\ t) = R_a(1 + 0.003\ t) + R_b(1 + 0.0015\ t)$$

Simplyfing the above, we get $\dfrac{R_b}{R_a} = \dfrac{3}{2}$ ...(*i*)

Now $R_a + R_b = 80\ \Omega$ ...(*ii*)

Substituting the value of R_b from (*i*) into (*ii*) we get

$$R_a + \frac{3}{2} R_a = 80 \text{ or } R_a = 32\ \Omega \text{ and } R_b = 48\ \Omega$$

If L_a and L_b are the required lengths in metres, then

$L_a = (100/80) \times 32 =$ **40 m** and $L_b = (100/60) \times 48 =$ **80 m**

Example 1.18. *A coil has a resistance of 18 Ω when its mean temperature is 20°C and of 20 Ω when its mean temperature is 50°C. Find its mean temperature rise when its resistance is 21 Ω and the surrounding temperature is 15°C.* **(Elect. Technology, Allahabad Univ. 1992)**

Solution. Let R_0 be the resistance of the coil and α_0 its tempt. coefficient at 0°C.

Then, $18 = R_0\ (1 + \alpha_0 \times 20)$ and $20 = R_0\ (1 + 50\ \alpha_0)$

Dividing one by the other, we get

$$\frac{20}{18} = \frac{1 + 50\ \alpha_0}{1 + 20\ \alpha_0} \quad \therefore \ \alpha_0 = \frac{1}{250} \text{ per °C}$$

If t°C is the temperature of the coil when its resistance is 21 Ω, then,

$$21 = R_0\ (1 + t/250)$$

Dividing this equation by the above equation, we have

$$\frac{21}{18} = \frac{R_0(1 + t/250)}{R_0(1 + 20\ \alpha_0)}\ ;\ t = 65°\text{C}\ ;\ \text{temp. rise} = 65 - 15 = \mathbf{50°C}$$

Example 1.19. *The coil of a relay takes a current of 0.12 A when it is at the room temperature of 15°C and connected across a 60-V supply. If the minimum operating current of the relay is 0.1 A, calculate the temperature above which the relay will fail to operate when connected to the same supply. Resistance-temperature coefficient of the coil material is 0.0043 per °C at 0°C.*

Solution. Resistance of the relay coil at 15°C is $R_{15} = 60/0.12 = 500\ \Omega$.

Let t°C be the temperature at which the minimum operating current of 0.1 A flows in the relay coil. Then, $R_t = 60/0.1 = 600\ \Omega$

Now $R_{15} = R_0\ (1 + 15\ \alpha_0) = R_0\ (1 + 15 \times 0.0043)$ and $R_t = R_0\ (1 + 0.0043\ t)$

$$\therefore \quad \frac{R_t}{R_{15}} = \frac{1 + 0.0043\ t}{1.0654} \text{ or } \frac{600}{500} = \frac{1 + 0.0043\ t}{1.0645} \quad \therefore t = \mathbf{64.5°C}$$

If the temperature rises above this value, then due to increase in resistance, the relay coil will draw a current less than 0.1 A and, therefore, will fail to operate.

Example 1.20. *Two conductors, one of copper and the other of iron, are connected in parallel and carry equal currents at 25°C. What proportion of current will pass through each if the temperature is raised to 100°C ? The temperature coefficients of resistance at 0°C are 0.0043/°C and 0.0063/°C for copper and iron respectively.* **(Principles of Elect. Engg. Delhi Univ. June 1985)**

Solution. Since the copper and iron conductors carry equal currents at 25° C, their resistances are the same at that temperature. Let each be R ohm.

For copper, $R_{100} = R_1 = R\ [1 + 0.0043\ (100 - 25)] = 1.3225\ R$

For iron, $R_{100} = R_2 = R\ [1 + 0.0063\ (100 - 25)] = 1.4725\ R$

If I is the current at 100°C, then as per current divider rule, current in the copper conductor is

$$I_1 = I\frac{R_2}{R_1+R_2} = I\frac{1.4725\,R}{1.3225\,R+1.4725\,R} = 0.5268\,I$$

$$I_2 = I\frac{R_1}{R_1+R_2} = I\frac{1.3225\,R}{2.795\,R} = 0.4732\,I$$

Hence, copper conductor will carry **52.68%** of the total current and the iron conductor will carry the balance *i.e.* **47.32%**

Example 1.21. *The filament of a 240 V metal-filament lamp is to be constructed from a wire having a diameter of 0.02 mm and a resistivity at 20°C of 4.3 μ Ω–cm. If α = 0.005/°C, what length of filament is necessary if the lamp is to dissipate 60 watts at a filament tempt. of 2420°C ?*

Solution. Electric power generated $= I^2R$ watts $= V^2/R$ watts

$\therefore$ $V^2/R = 60$ or $240^2/R = 60$

Resistance at 2420°C $\quad R_{2420} = \dfrac{240\times240}{60} = 960\ \Omega$

Now $\quad R_{2420} = R_{20}\,[1+(2420-20)\times0.005]$

or $\quad 960 = R_{20}(1+12)$

$\therefore$ $\quad R_{20} = 960/13\ \Omega$

Now $\quad \rho_{20} = 4.3\times10^{-6}\ \Omega\text{-cm}$ and $A = \dfrac{\pi(0.002)^2}{4}\ \text{cm}^2$

$\therefore$ $\quad l = \dfrac{A\times R_{20}}{\rho_{20}} = \dfrac{\pi(0.002)^2\times960}{4\times13\times4.3\times10^{-6}} = \mathbf{54\ cm}$

Example 1.22. *A semi-circular ring of copper has an inner radius of 6 cm, radial thickness 3 cm and an axial thickness 4 cm. Find the resistance of the ring at 50°C between its two end-faces. Assume specific resistance of Cu at 20° = 1.724 × 10⁻⁶ ohm-cm and resistance tempt. coeff. of Cu at 0°C = 0.0043/°C.*

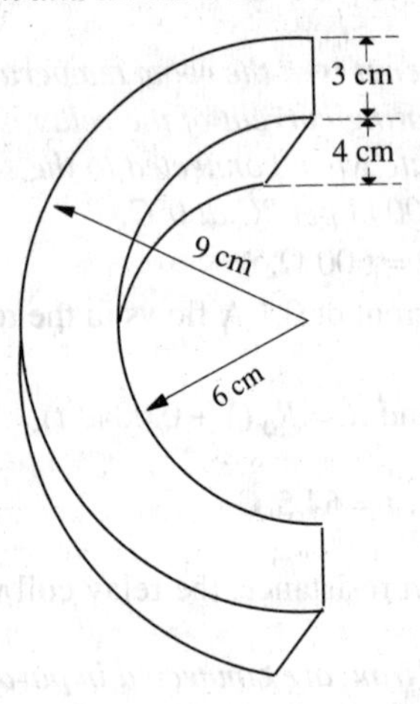

Fig. 1.11

Solution. The semi-circular ring is shown in Fig. 1.11.

Mean radius of ring $= (6+9)/2 = 7.5$ cm

Mean length between end faces $= 7.5\,\pi$ cm $= 23.56$ cm

Cross-section of the ring $= 3\times4 = 12\ \text{cm}^2$

Now $\alpha_0 = 0.0043/°C$; $\alpha_{20} = \dfrac{0.0043}{1+20\times0.0043} = 0.00396$

$$\rho_{50} = \rho_{20}[1+\alpha_0(50-20)]$$

$$= 1.724\times10^{-6}\,(1+30\times0.00396)$$

$$= 1.93\times10^{-6}\ \Omega\text{-cm}$$

$$R_{50} = \frac{\rho_{50}\times l}{A} = \frac{1.93\times10^{-6}\times23.56}{12} = \mathbf{3.79\times10^{-6}\ \Omega}$$

Tutorial Problems No. 1.2

1. It is found that the resistance of a coil of wire increases from 40 ohm at 15°C to 50 ohm at 60°C. Calculate the resistance temperature coefficient at 0°C of the conductor material.

[1/165 per °C] (*Elect. Technology, Indore Univ. May 1977*)

2. A tungsten lamp filament has a temperature of 2,050°C and a resistance of 500 Ω when taking normal working current. Calculate the resistance of the filament when it has a temperature of 25°C. Temperature-coefficient at 0°C is 0.005/°C. **[50 Ω]** (*Elect. Technology, Indore Univ. 1981*)

3. An armature has a resistance of 0.2 Ω at 150°C and the armature Cu loss is to be limited to 600 watts with a temperature rise of 55°C. If α_0 for Cu is 0.0043/°C, what is the maximum current that can be passed through the armature ? **[50.8 A]**

4. A d.c. shunt motor after running for several hours on constant voltage mains of 400 V takes a field current of 1.6 A. If the temperature rise is known to be 40°C, what value of extra circuit resistance is required to adjust the field current to 1.6 A when starting from cold at 20°C ? Temperature coefficient = 0.0043/°C at 20°C. **[36.69 Ω]**

5. In a test to determine the resistance of a single-core cable, an applied voltage of 2.5 V was necessary to produce a current of 2 A in it at 15°C.

(*a*) Calculate the cable resistance at 55°C if the temperature coefficient of resistance of copper at 0°C is 1/235 per °C.

(*b*) if the cable under working conditions carries a current of 10 A at this temperature, calculate the power dissipated in the cable. **[(*a*) 1.45 Ω (*b*) 145 W]**

6. An electric radiator is required to dissipate 1 kW when connected to a 230 V supply. If the coils of the radiator are of wire 0.5 mm in diameter having resistivity of 60 μ Ω-cm, calculate the necessary length of the wire. **[1732 cm]**

7. An electric heating element to dissipate 450 watts on 250 V mains is to be made from nichrome ribbon of width 1 mm and thickness 0.05 mm. Calculate the length of the ribbon required (the resistivity of nichrome is 110×10^{-8} Ω-m). **[631 m]**

8. When burning normally, the temperature of the filament in a 230 V, 150 W gas-filled tungsten lamp is 2,750°C. Assuming a room temperature of 16°C, calculate (*a*) the normal current taken by the lamp (*b*) the current taken at the moment of switching on. Temperature coefficient of tungsten is 0.0047 Ω/Ω/°C at 0°C. **[(*a*) 0.652 A (*b*) 8.45 A]** (*Elect. Engg. Madras Univ. 1977*)

9. An aluminium wire 5 m long and 2 mm diameter is connected in parallel with a wire 3 m long. The total current is 4 A and that in the aluminium wire is 2.5 A. Find the diameter of the copper wire. The respective resistivities of copper and aluminium are 1.7 and 2.6 μ Ω-cm. **[0.97 mm]**

10. The field winding of d.c. motor connected across 230 V supply takes 1.15 A at room temp. of 20°C. After working for some hours the current falls to 0.26 A, the supply voltage remaining constant. Calculate the final working temperature of field winding. Resistance temperature coefficient of copper at 20°C is 1/254.5. **[70.4°C]** (*Elect. Engg. Pune Univ. 1985*)

11. It is required to construct a resistance of 100 Ω having a temperature coefficient of 0.001 per °C. Wires of two materials of suitable cross-sectional area are available. For material *A*, the resistance is 97 Ω per 100 metres and for material *B*, the resistance is 40 Ω per 100 metres. The temperature coefficient of resistance for material *A* is 0.003 per °C and for material *B* is 0.0005 per °C. Determine suitable lengths of wires of materials *A* and *B*. **[A : 19.4 m, B : 200 m]**

12. The resistance of the shunt winding of a d.c machine is measured before and after a run of several hours. The average values are 55 ohms and 63 ohms. Calculate the rise in temperature of the winding. (Temperature coefficient of resistance of copper is 0.00428 ohm per ohm per °C). **[36 °C]** (*London Univ.*)

13. A piece of resistance wire, 15.6 m long and of cross-sectional area 12 mm^2 at a temperature of 0°C, passes a current of 7.9 A when connected to d.c. supply at 240 V. Calculate (*a*) resistivity of the wire (*b*) the current which will flow when the temperature rises to 55°C. The temperature coefficient of the resistance wire is 0.00029 Ω/Ω/°C. **[(*a*) 23.37 μ Ω-m (*b*) 7.78 A]** (*London Univ.*)

14. A coil is connected to a constant d.c. suppy of 100 V. At start, when it was at the room temperature of 25°C, it drew a current of 13 A. After sometime, its temperature was 70°C and the current reduced to 8.5 A. Find the current it will draw when its temperature increases further to 80°C. Also, find the temperature coefficient of resistance of the coil material at 25°C. **[7.9A; 0.01176°C^{-1}]** (*F.Y. Engg. Univ. Nov. 1989*)

15. The resistance of the field coils with copper conductors of a dynamo is 120Ω at 25°C. After working for 6 hours on full load, the resistance of the coil increases to 140 Ω. Calculate the mean temperature rise of the field coil. Take the temperature coefficient of the conductor material as 0.0042 at 0°C. **[43.8°C]** (*Elements of Elec. Engg. Bangalore Univ. 1991*)

1.13. Ohm's Law

This law applies to electric conduction through good conductors and may be stated as follows :

The ratio of potential difference (V) between any two points on a conductor to the current (I) flowing between them, is constant, provided the temperature of the conductor does not change.

In other words, $\frac{V}{I}$ = constant or $\frac{V}{I} = R$

where R is the resistance of the conductor between the two points considered.

Put in another way, it simply means that provided R is kept constant, current is directly proportional to the potential difference across the ends of a conductor. However, this linear relationship between V and I does not apply to all non-metallic conductors. For example, for silicon carbide, the relationship is given by $V = KI^m$ where K and m are constants and m is less than unity. It also does not apply to non-linear devices such as Zener diodes and voltage-regulator (VR) tubes.

Example 1.22. *A coil of copper wire has resistance of 90 Ω at 20°C and is connected to a 230-V supply. By how much must the voltage be increased in order to maintain the current constant if the temperature of the coil rises to 60°C ? Take the temperature coefficient of resistance of copper as 0.00428 from 0°C.*

Solution. As seen from Art. 1.10.

$$\frac{R_{60}}{R_{20}} = \frac{1+60\times 0.00428}{1+20\times 0.00428} \qquad \therefore\ R_{60} = 90\times 1.2568/1.0856 = 104.2\ \Omega$$

Now, current at 20°C = 230/90 = 23/9 A

Since the wire resistance has become 104.2 Ω at 60°C, the new voltage required for keeping the current constant at its previous value = 104.2 × 23/9 = 266.3 V

∴ increase in voltage required = 266.3 − 230 = **36.3 V**

Example 1.23. *Three resistors are connected in series across a 12-V battery. The first resistor has a value of 1 Ω, second has a voltage drop of 4 V and the third has a power dissipation of 12 W. Calculate the value of the circuit current.*

Solution. Let the two unknown resistors be R_2 and R_3 and I the circuit current.

$$\therefore\quad I^2R_3 = 12 \text{ and } IR_2 = 4 \qquad \therefore\ R_3 = \frac{3}{4}R_2^2.\ \text{Also, } I = \frac{4}{R_2}$$

Now, $I(1+R_2+R_3) = 12$

Substituting the values of I and R_3, we get

$$\frac{4}{R_2}\left(1+R_2+\frac{3}{4}R_2^2\right) = 12 \quad \text{or} \quad 3R_2^2 - 8R_2 + 4 = 0$$

$$\therefore\quad R_2 = \frac{8\pm\sqrt{64-48}}{6}. \qquad \therefore\ R_2 = 2\ \Omega \text{ or } \frac{2}{3}\ \Omega$$

$$\therefore\quad R_3 = \frac{3}{4}R_2^2 = \frac{3}{4}\times 2^2 = 3\ \Omega \quad \text{or} \quad \frac{3}{4}\left(\frac{2}{3}\right)^2 = \frac{1}{3}\ \Omega$$

$$\therefore\quad I = \frac{12}{1+2+3} = \mathbf{2A} \quad \text{or} \quad I = \frac{12}{1+(2/3)+(1/3)} = \mathbf{6\ A}$$

1.14. Resistance in Series

When some conductors having resistances R_1, R_2 and R_3 etc. are joined end-on-end as in Fig. 1.12, they are said to be connected in series. It can be proved that the equivalent resistance or total resistance between points A and D is equal to the sum of the three individual resistances. Being a series circuit, it should be remembered that (*i*) current is the same through all the three conductors (*ii*) but voltage drop across each is different due to its different resistance and is given by Ohm's Law and (*iii*) sum of the three voltage drops is equal to the voltage applied across the three conductors. There is a progressive fall in potential as we go from point A to D as shown in Fig. 1.13.

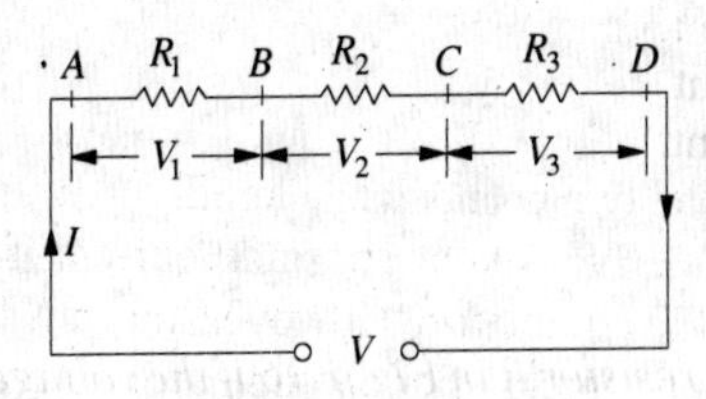

Fig. 1.12

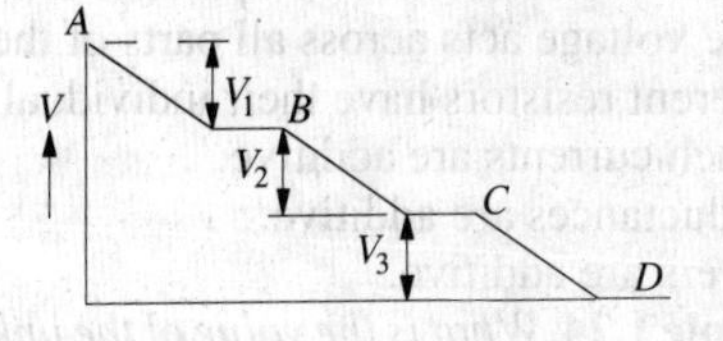

Fig. 1.13

$\therefore$ $V = V_1 + V_2 + V_3 = IR_1 + IR_2 + IR_3$ -Ohm's Law

But $V = IR$

where R is the equivalent resistance of the series combination.

$\therefore$ $IR = IR_1 + IR_2 + IR_3$ or $R = R_1 + R_2 + R_3$

Also $$\frac{1}{G} = \frac{1}{G_1} + \frac{1}{G_2} + \frac{1}{G_3}$$

As seen from above, the main characteristics of a series circuit are :

1. same current flows through all parts of the circuit.
2. different resistors have their individual voltage drops.
3. voltage drops are additive.
4. applied voltage equals the sum of different voltage drops.
5. resistances are additive.
6. powers are additive.

1.15. Voltage Divider Rule

Since in a series circuit, same current flows through each of the given resistors, voltage drop varies directly with its resistance. In Fig. 1.14 is shown a 24-V battery connected across a series combination of three resistors.

Total resistance $R = R_1 + R_2 + R_3 = 12\ \Omega$

According to Voltage Divider Rule, various voltage drops are :

$$V_1 = V \cdot \frac{R_1}{R} = 24 \times \frac{2}{12} = 4\ \text{V}$$

$$V_2 = V \cdot \frac{R_2}{R} = 24 \times \frac{4}{12} = 8\ \text{V}$$

$$V_3 = V \cdot \frac{R_3}{R} = 24 \times \frac{6}{12} = 12\ \text{V}$$

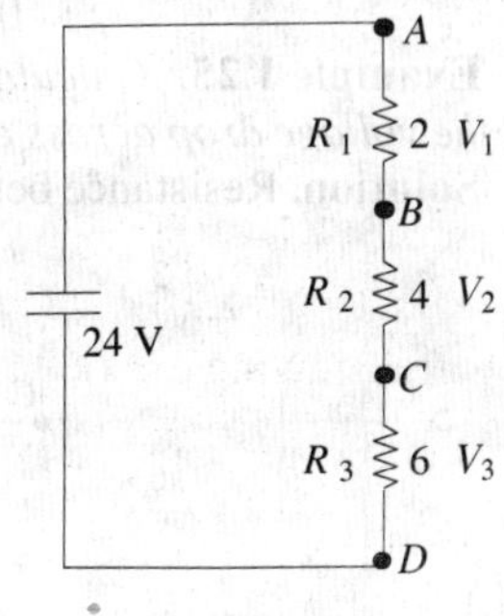

Fig. 1.14

1.16. Resistances in Parallel

Three resistances, as joined in Fig. 1.15 are said to be connected in parallel. In this case (*i*) p.d. across all resistances is the same (*ii*) current in each resistor is different and is given by Ohm's Law and (*iii*) the total current is the sum of the three separate currents.

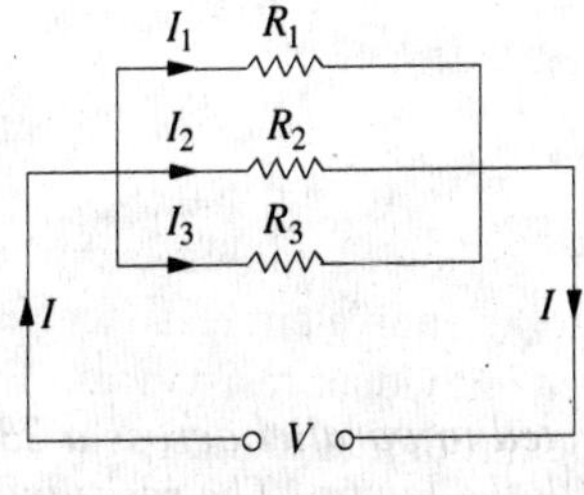

Fig. 1.15

$$I = I_1 + I_2 + I_3 = \frac{V}{R_1} + \frac{V}{R_2} + \frac{V}{R_3}$$

Now, $I = \dfrac{V}{R}$ where V is the applied voltage.

R = equivalent resistance of the parallel combination.

$$\therefore \quad \frac{V}{R} = \frac{V}{R_1} + \frac{V}{R_2} + \frac{V}{R_3} \quad \text{or} \quad \frac{1}{R} = \frac{1}{R_1} + \frac{1}{R_2} + \frac{1}{R_3}$$

Also, $G = G_1 + G_2 + G_3$

The main characteristics of a parallel circuit are :
1. same voltage acts across all parts of the circuit.
2. different resistors have their individual current.
3. branch currents are additive.
4. conductances are additive.
5. powers are additive.

Example 1.24. *What is the value of the unknown resistor R in Fig. 1.16 if the voltage drop across the 500 Ω resistor is 2.5 volts ? All resistances are in ohm.*

(Elect. Technology, Indore Univ. April 1990)

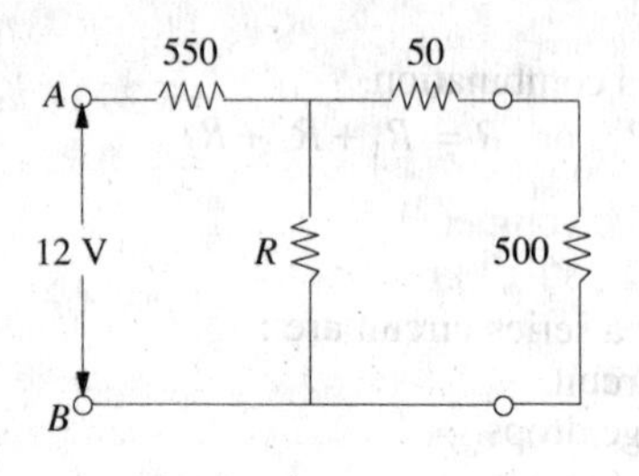

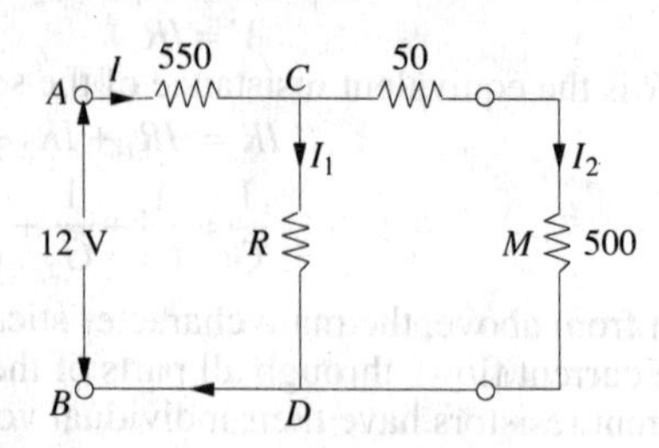

Fig. 1.16

Solution. By direct proportion, drop on 50 Ω resistance

$= 2.5 \times 50/500 = 0.25$ V

Drop across CMD or CD = 2.5 + 0.25 = 2.75 V

Drop across 550 Ω resistance = 12 – 2.75 = 9.25 V

$I = 9.25/550 = 0.0168\,A,\ I_2 = 2.5/500 = 0.005$ A

$I_1 = 0.0168 - 0.005 = 0.0118$ A

$\therefore \quad 0.0118 = 2.75/R;\ \ R = \mathbf{233\ \Omega}$

Example 1.25. *Calculate the effective resistance of the following combination of resistances and the voltage drop across each resistance when a P.D. of 60 V is applied between points A and B.*

Solution. Resistance between *A* and *C* (Fig. 1.17).

$= 6 \parallel 3 = 2\ \Omega$

Resistance of branch *ACD*

$= 18 + 2 = 20\ \Omega$

Now, there are two parallel paths between points *A* and *D* of resistances 20 Ω and 5 Ω. Hence, resistance between *A* and *D*

$= 20 \parallel 5 = 4\ \Omega$

∴ Resistance between *A* and *B*

$= 4 + 8 = 12\ \Omega$

Total circuit current

$= 60/12 = 5$ A

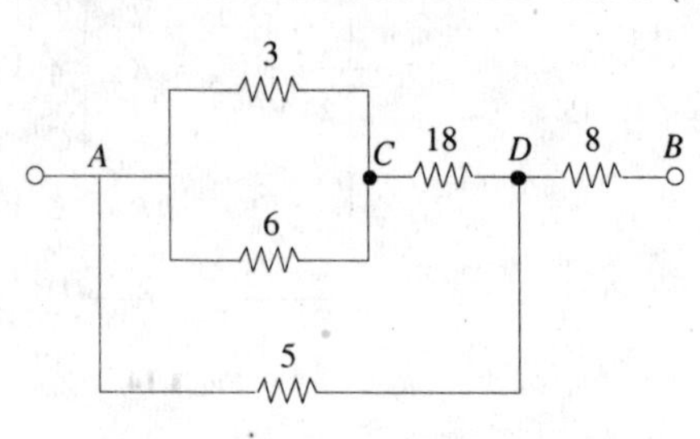

Fig. 1.17

Current through 5 Ω resistance $= 5 \times \dfrac{20}{25} = 4$ A —Art 1.25

Current in branch *ACD* $= 5 \times \dfrac{5}{25} = 1$ A

∴ P.D. across 3 Ω and 6 Ω resistors $= 1 \times 2 = \mathbf{2\ V}$

P.D. across 18 Ω resistor $= 1 \times 18 = \mathbf{18\ V}$

P.D. across 5 Ω resistor $= 4 \times 5 = \mathbf{20\ V}$

P.D. across 8 Ω resistor $= 5 \times 8 = \mathbf{40\ V}$

Example 1.26. *A circuit consists of four 100-W lamps connected in parallel across a 230-V supply. Inadvertently, a voltmeter has been connected in series with the lamps. The resistance of the voltmeter is 1500 Ω and that of the lamps under the conditions stated is six times their value when burning normally. What will be the reading of the voltmeter ?*

Solution. The circuit is shown in Fig. 1.18. The wattage of a lamp is given by :

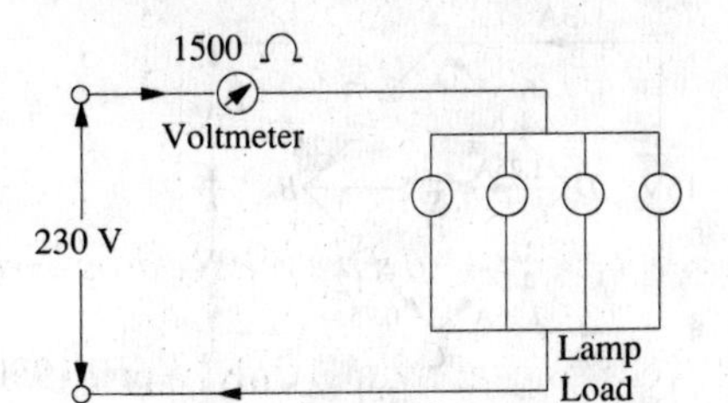

Fig. 1.18

$$W = I^2R = V^2/R$$

$\therefore$ $100 = 230^2/R \quad \therefore R = 529\ \Omega$

Resistance of each lamp under stated condition is $= 6 \times 529 = 3174\ \Omega$

Equivalent resistance of these four lamps connected in parallel $= 3174/4 = 793.5\ \Omega$

This resistance is connected in series with the voltmeter of 1500 Ω resistance.

$\therefore$ total circuit resistance $= 1500 + 793.5 = 2293.5\ \Omega$

$\therefore$ circuit current $= 230/2293.5$ A

According to Ohm's law, voltage drop across the voltmeter

$= 1500 \times 230/2293.5 =$ **150 V (approx)**

Example 1.27. *Determine the value of R and current through it in Fig. 1.19, if current through branch AO is zero.* **(Elect. Engg. & Electronics, Bangalore Univ. 1989)**

Solution. The given circuit can be redrawn as shown Fig. 1.19 (*b*). As seen, it is nothing else but Wheatstone bridge circuit. As is well-known, when current through branch *AO* becomes zero, the bridge is said to be balanced. In that case, products of the resistances of opposite arms of the bridge become equal.

$\therefore$ $4 \times 1.5 = R \times 1 \;;\; R = 6\ \Omega$

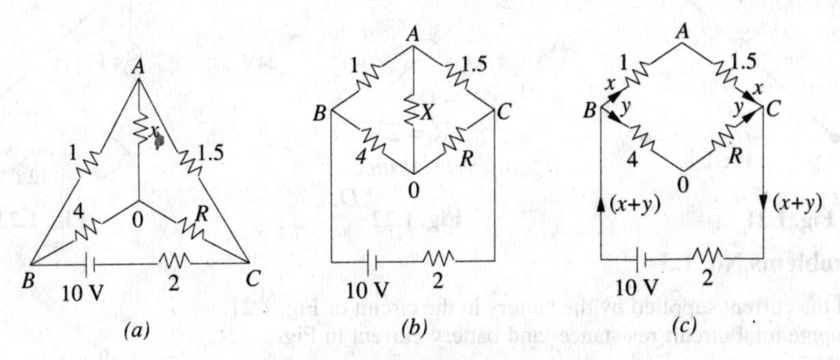

Fig. 1.19

Under condition of balance, it makes no difference if resistance *X* is removed thereby giving us the circuit of Fig. 1.19(*c*). Now, there are two parallel paths between points *B* and *C* of resistances $(1 + 1.5) = 2.5\ \Omega$ and $(4 + 6) = 10\ \Omega$. $R_{BC} = 10 \parallel 2.5 = 2\ \Omega$

Total circuit resistance $= 2 + 2 = 4\ \Omega$. Total circuit current $= 10/4 = 2.5$ A.

This current gets divided into two parts at point *B*. Current through *R* is

$y = 2.5 \times 2.5/12.5 =$ **0.5 A**

Example 1.28. *In the unbalanced bridge circuit of Fig. 1.20(a), find the potential difference that exists across the open switch S. Also, find the current which will flow through the switch when it is closed.*

Solution. With switch open, there are two parallel branches across the 15-V supply. Branch *ABC* has a resistance of $(3 + 12) = 15\ \Omega$ and branch *ADC* has a resistance of $(6 + 4) = 10\ \Omega$. Obviously, each branch has 15 V applied across it.

$$V_B = 12 \times 15/15 = 12\text{ V};\quad V_D = 4 \times 15/(6 + 4) = 6\text{ V}$$

$\therefore$ p.d. across points *B* and $D = V_B - V_D = 12 - 6 = 6$ V

When *S* is closed, the circuit becomes as shown in Fig. 1.20(*b*) where points *B* and *D* become electrically connected together.

$$R_{AB} = 3 \parallel 6 = 2\ \Omega \quad \text{and} \quad R_{BC} = 4 \parallel 12 = 3\ \Omega$$

$$R_{AC} = 2 + 3 = 5\ \Omega\;;\quad I = 15/5 = 3\text{ A}$$

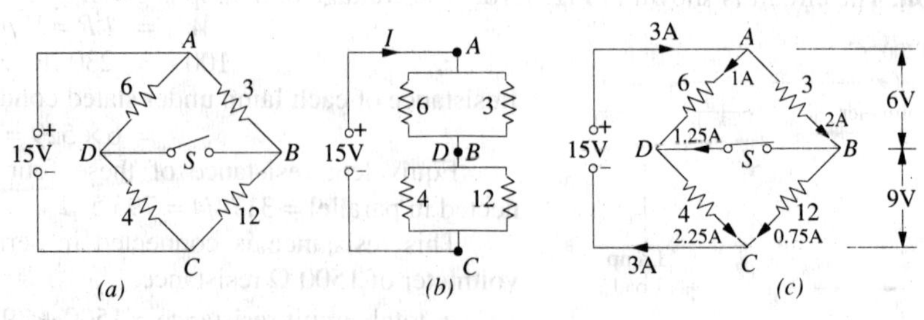

Fig. 1.20

Current through arm $AB = 3 \times 6/9 = 2$A. The voltage drop over arm $AB = 3 \times 2 = 6$ V. Hence, drop over arm $BC = 15 - 6 = 9$ V. Current through $BC = 9/12 = 0.75$ A. It is obvious that at point B, the incoming current is 2A, out of which 0.75A flows along BC, whereas remaining $2 - 0.75 = 1.25$ A passes through the switch.

· As a check, it may be noted that current through $AD = 6/6 = 1$ A. At point D, this current is joined by 1.25 A coming through the switch. Hence, current through $DC = 1.25 + 1 = 2.25$ A. This fact can be further verified by the fact that there is a voltage drop of 9 V across 4 Ω resistor thereby giving a current of $9/4 = 2.25$ A.

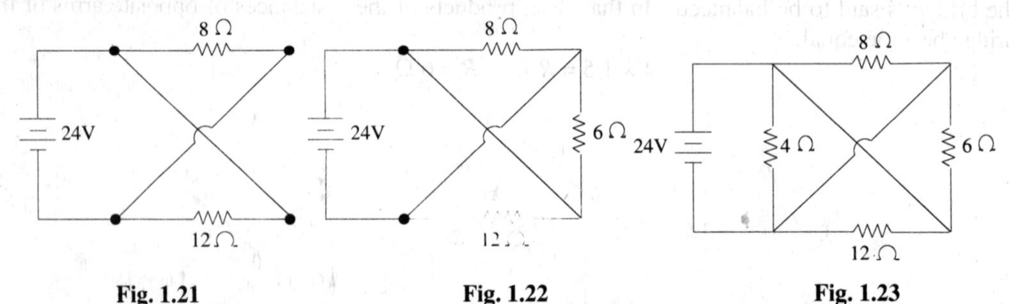

Fig. 1.21 **Fig. 1.22** **Fig. 1.23**

Tutorial Problems No. 1.3

1. Find the current supplied by the battery in the circuit of Fig. 1.21. **[5 A]**

2. Compute total circuit resistance and battery current in Fig. 1.22.**[8/3 Ω, 9A]**

3. Calculate battery current and equivalent resistance of the network shown in Fig. 1.23. **[15A ; 8/5 Ω]**

4. Find the equivalent resistance of the network of Fig. 1.24 between terminals A and B. All resistance values are in ohms. **[6 Ω]**

5. What is the equivalent resistance of the circuit of Fig. 1.25 between terminals A and B ? All resistances are in ohms. **[4 Ω]**

6. Compute the value of battery current I in Fig. 1.26. All resistances are in ohm. **[6 A]**

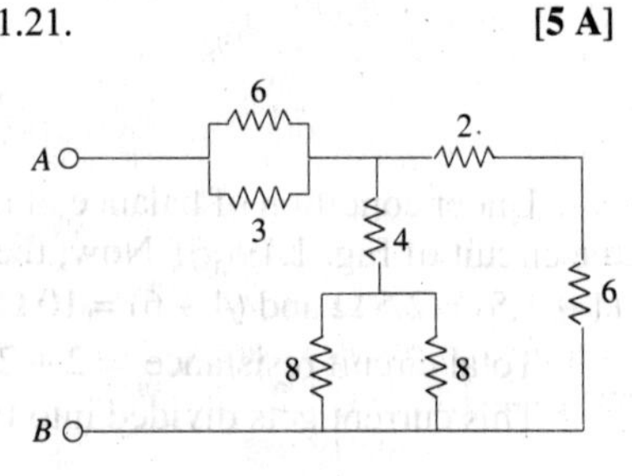

Fig. 1.24

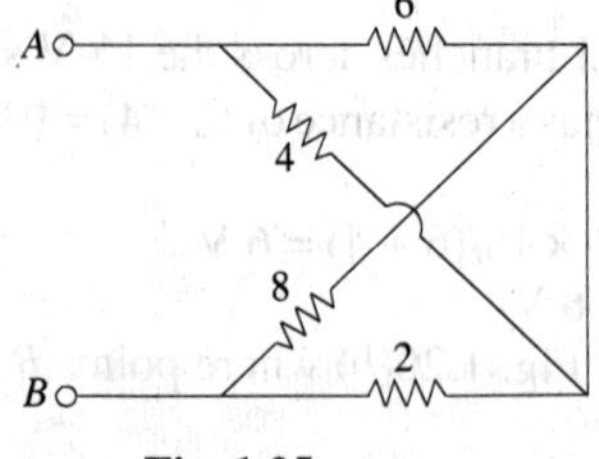

Fig. 1.25

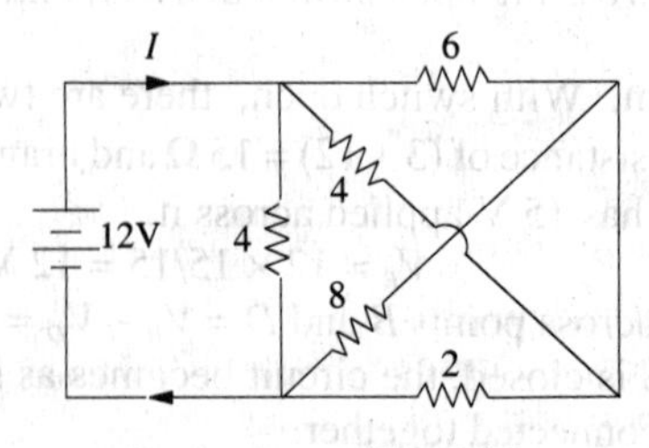

Fig. 1.26

7. Calculate the value of current I supplied by the voltage source in Fig. 1.27. All resistance values are in ohms. (**Hint** : Voltage across each resistor is 6 V) **[1A]**.

8. Compute the equivalent resistance of the circuit of Fig. 1.28 between points (*i*) *ab* (*ii*) *ac* and (*iii*) *bc*. All resistance values are in ohm. **[(*i*) 6 Ω (*ii*) 4.5 Ω (*iii*), 4.5 Ω]**

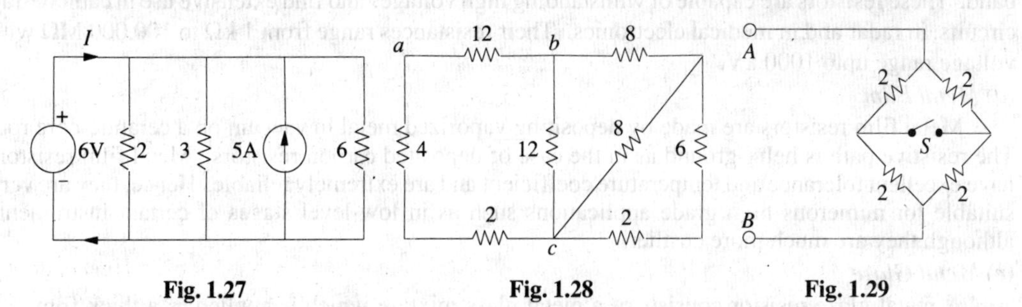

Fig. 1.27 **Fig. 1.28** **Fig. 1.29**

9. In the circuit of Fig. 1.29, find the resistance between terminals *A* and *B* when switch is (*a*) open and (*b*) closed. Why are the two values equal ? **[(*a*) 2 Ω (*b*) 2 Ω]**

10. The total current drawn by a circuit consisting of three resistors connected in parallel is 12A. The voltage drop across the first resistor is 12 V, the value of second resistor is 3 Ω and the power dissipation of the third resistor is 24 W. What are the resistances of the first and third resistors ? **[2 Ω ; 6 Ω]**

11. Three parallel connected resistors when connected across a d.c voltage source dissipate a total power of 72 W. The total current drawn is 6A, the current flowing through the first resistor is 3 A and the second and third resistors have equal value. What are the resistances of the three resistors ? **[4 Ω ; 8 Ω ; 8 Ω]**

12. A bulb rated 110 V, 60 watts is connected with another bulb rated 110-V, 100 W across a 220 V mains. Calculate the resistance which should be joined in parallel with the first bulb so that both the bulbs may take their rated power. **[302.5 Ω]**

13. Two coils connected in parallel across 100 V supply mains take 10 A from the line. The power dissipated in one coil is 600 W. What is the resistance of the other coil ? **[25 Ω]**

14. An electric lamp whose resistance, when in use, is 2 Ω is connected to the terminals of a dry cell whose e.m.f. is 1.5 V. If the current through the lamp is 0.5 A, calculate the internal resistance of the cell and the potential difference between the terminals of the lamp. If two such cells are connected in parallel, find the resistance which must be connected in series with the arrangement to keep the current the same as before. **[1 Ω ; 1V ; 0.5 Ω]** (*Elect. Technology, Indore Univ. 1978*)

1.17. Types of Resistors

(*a*) *Carbon Composition*

It is a combination of carbon particles and a binding resin with different proportions for providing desired resistance. Attached to the ends of the resistive element are metal caps which have axial leads of tinned copper wire for soldering the resistor into a circuit. The resistor is enclosed in a plastic case to prevent the entry of moisture and other harmful elements from outside. Billions of carbon composition resistors are used in the electronic industry every year. They are available in power ratings of 1/8, 1/4, 1/2, 1 and 2 W, in voltage ratings of 250, 350 and 500 V. They have low failure rates when properly used.

Such resistors have a tendency to produce electric noise due to the current passing from one carbon particle to another. This noise appears in the form of a hiss in a loudspeaker connected to a hi-fi system and can overcome very weak signals. That is why carbon composition resistors are used where performance requirements are not demanding and where low cost is the main consideration. Hence, they are extensively used in entertainment electronics although better resistors are used in critical circuits.

(*b*) *Deposited Carbon*

Deposited carbon resistors consist of ceramic rods which have a carbon film deposited on them. They are made by placing a ceramic rod in a methane-filled flask and heating it until, by a gas-cracking process, a carbon film is deposited on them. A helix-grinding process forms the resistive path. As compared to carbon composition resistors, these resistors offer a major improvement in lower current noise and in closer tolerance. These resistors are being replaced by metal film and metal glaze resistors.

(c) High-Voltage Ink Film

These resistors consist of a ceramic base on which a special resistive ink is laid down in a helical band. These resistors are capable of withstanding high voltages and find extensive use in cathode-ray circuits, in radar and in medical electronics. Their resistances range from 1 kΩ to 100,000 MΩ with voltage range upto 1000 kV.

(d) Metal Film

Metal film resistors are made by depositing vaporized metal in vaccum on a ceramic-core rod. The resistive path is helix-ground as in the case of deposited carbon resistors. Metal film resistors have excellent tolerance and temperature coefficient and are extremely reliable. Hence, they are very suitable for numerous high grade applications such as in low-level stages of certain instruments although they are much more costlier.

(e) Metal Glaze

A metal glaze resistor consists of a metal glass mixture which is applied as a thick film to a ceramic substrate and then fired to form a film. The value of resistance depends on the amount of metal in the mixture. With helix-grinding, the resistance can be made to vary from 1 Ω to many megohms.

Another category of metal glaze resistors consists of a tinned oxide film on a glass substrate.

(f) Wire-wound

Wire–wound resistors are different from all other types in the sense that no film or resistive coating is used in their construction. They consist of a ceramic-core wound with a drawn wire having accurately-controlled characteristics. Different wire alloys are used for providing different resistance ranges. These resistors have highest stability and highest power rating.

Because of their bulk, high-power ratings and high cost, they are not suitable for low-cost or high-density, limited-space applications. The completed wire-wound resistor is coated with an insulating material such as baked enamel.

(g) Cermet (Ceramic Metal)

The cermet resistors are made by firing certain metals blended with ceramics on a ceramic substrate. The value of resistance depends on the type of mix and its thickness. These resistors have very accurate resistance values and show high stability even under extreme temperatures. Usually, they are produced as small rectangles having leads for being attached to printed circuit boards (*PCB*).

1.18. Nonlinear Resistors

Those elements whose $V - I$ curves are not straight lines are called nonlinear elements because their resistances are nonlinear resistances. Their $V - I$ characteristics can be represented by an equation of the form $I = kV^n + b$ where n is usually not equal to one and the constant b may or may not be equal to zero.

Examples of nonlinear elements are filaments of incandescent lamps, diodes, thermistors and varistors. A varistor is a special resistor made of carborundum crystals held together by a binder. Fig. 1.30 (*a*) shows how current through a varistor increases rapidly when the applied voltage increases beyond a certain amount (nearly 100 V in the present case).

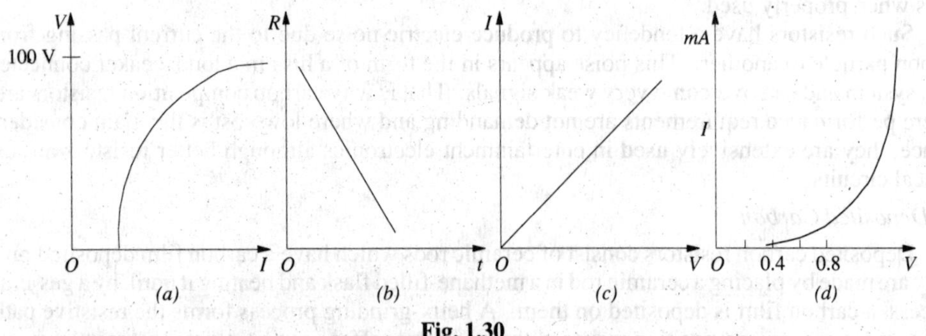

Fig. 1.30

There is a corresponding rapid decrease in resistance when the current increases. Hence, varistors are generally used to provide over-voltage protection in certain circuits.

A thermistor is made of metallic oxides in a suitable binder and has a large negative coefficient of resistance *i.e.* its resistance decreases with increase in temperature as shown in Fig. 1.30(*b*). Fig. 1.30 (*c*) shows how the resistance of an incandescent lamp increases with voltage whereas Fig. 1.30 (*d*) shows the *V-I* characteristics of a typical silicon diode. For a germanium diode, current is related to its voltage by the relation

$$I = I_o(e^{V/0.026} - 1)$$

1.19. Varistor (Nonlinear Resistor)

It is a voltage-dependent metal–oxide material whose resistance decreases sharply with increasing voltage. The relationship between the current flowing through a varistor and the voltage applied across it is given by the relation: $i = ke^n$ where i = instantaneous current, e is the instantaneous voltage and η is a constant whose value depends on the metal oxides used. The value of η for silicon-carbide-based varistors lies between 2 and 6 whereas zinc-oxide-based varistors have a value ranging from 25 to 50.

The zinc-oxide-based varistors are primarily used for protecting solid-state power supplies from low and medium surge voltages in the supply line. Silicon-carbide varistors provide protection against high-voltage surges caused by lightning and by the discharge of electromagnetic energy stored in the magnetic fields of large coils.

1.20. Short and Open Circuits

When two points of a circuit are connected together by a thick metallic wire (Fig. 1.31), they are said to be *short-circuited.* Since 'short' has practically zero resistance, it gives rise to two important facts :

(*i*) no voltage can exist across it because $V = IR = I \times 0 = 0$
(*ii*) current through it (called short-circuit current) is very large (theoretically, infinity)

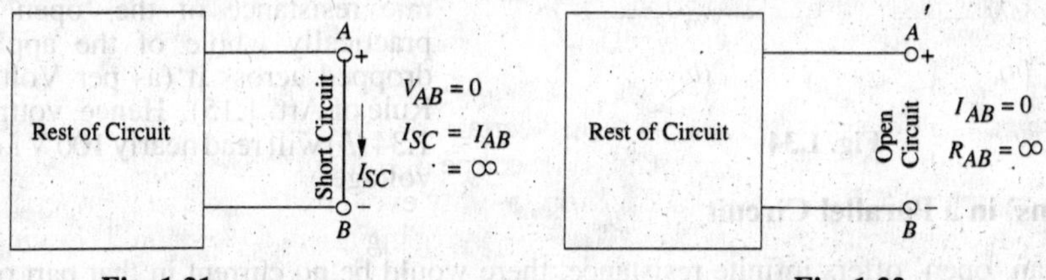

Fig. 1.31 **Fig. 1.32**

Two points are said to be open-circuited when there is no direct connection between them (Fig. 1.32). Oviously, an 'open' represents a break in the continuity of the circuit. Due to this break

(*i*) resistance between the two points is infinite
(*ii*) there is no flow of current between the two points.

1.21. 'Shorts' in a Series Circuit

Since a dead (or solid) short has almost zero resistance, it causes the problem of excessive current which, in turn, causes power dissipation to increase many times and circuit components to burn out.

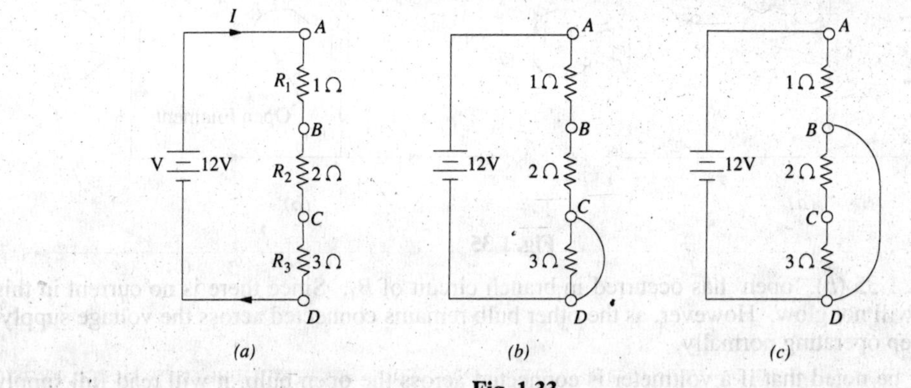

Fig. 1.33

In Fig. 1.33 (*a*) is shown a normal series circuit where

$$V = 12\text{ V}, R = R_1 + R_2 + R_3 = 6\ \Omega$$
$$I = V/R = 12/6 = 2\text{ A}, P = I^2R = 2^2 \times 6 = 24\text{ W}$$

In Fig. 1.33 (*b*), 3-Ω resistor has been shorted out by a reistanceless copper wire so that $R_{CD} = 0$. Now, total circuit resistance $R = 1 + 2 + 0 = 3\ \Omega$. Hence, $I = 12/3 = 4$ A and $P = 4^2 \times 3 = 48$ W.

Fig 1.33 (*c*) shows the situation where both 2 Ω and 3 Ω resistors have been shorted out of the circuit. In this case,

$$R = 1\ \Omega, I = 12/1 = 12\text{ A and } P = 12^2 \times 1 = 144\text{ W}$$

Because of this excessive current (6 times the normal value), connecting wires and other circuit components can become hot enough to ignite and burn out.

1.22. 'Opens' in a Series Circuit

In a normal series circuit like the one shown in Fig. 1.34 (*a*), there exists a current flow and the voltage drops across different resistors are proportional to their resistances. If the circuit becomes 'open' anywhere, following two effects are produced:

(*i*) since 'open' offers infinite resistance, circuit current becomes zero. Consequently, there is no voltage drop across R_1 and R_2.

(*ii*) *whole of the applied voltage (i.e. 100 V in this case) is felt across the 'open' i.e. across terminals A and B* [Fig. 1.34 (*b*)]. The reason for this is that R_1 and R_2 become negligible as compared to the infinite resistance of the 'open' which has practically whole of the applied voltage dropped across it (as per Voltage Divider Rule of Art. 1.15). Hence, voltmeter in Fig. 1.34 (*b*) will read nearly 100 V *i.e.* the supply voltage.

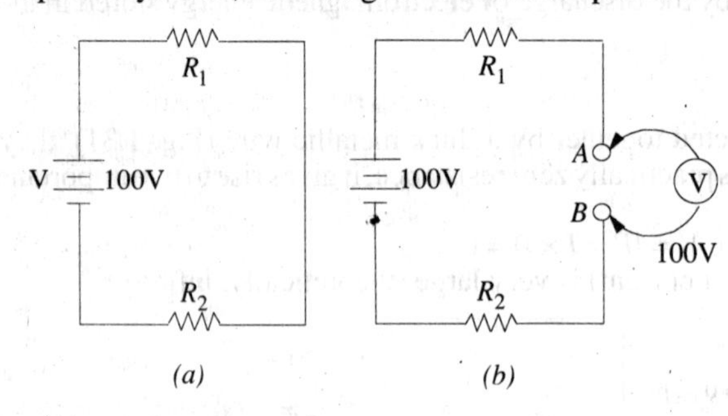

Fig. 1.34

1.23. 'Opens' in a Parallel Circuit

Since an 'open' offers infinite resistance, there would be no current in that part of the circuit where it occurs. In a parallel circuit, an 'open' can occur either in the main line or in any parallel branch.

As shown in Fig. 1.35 (*a*), an open in the main line prevents flow of current *to all branches*. Hence, neither of the two bulbs glows. However, full applied voltage (*i.e.* 220 V in this case) *is available across the open.*

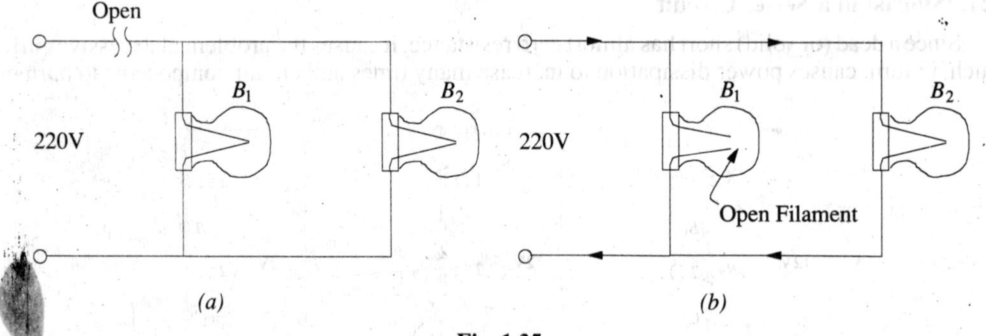

Fig. 1.35

In Fig. 1.35 (*b*), 'open' has occurred in branch circuit of B_1. Since there is no current in this branch, B_1 will not glow. However, as the other bulb remains connected across the voltage supply, it would keep operating normally.

It may be noted that if a voltmeter is connected across the open bulb, it will read full supply voltage of 220 V.

1.24 'Shorts' in Parallel Circuits

Suppose a 'short' is placed across R_3 (Fig. 1.36). It becomes directly connected across the battery and draws almost infinite current because not only its own resistance but that of the connecting wires AC and BD is negligible. Due to this excessive current, the wires may get hot enough to burn out unless the circuit is protected by a fuse.

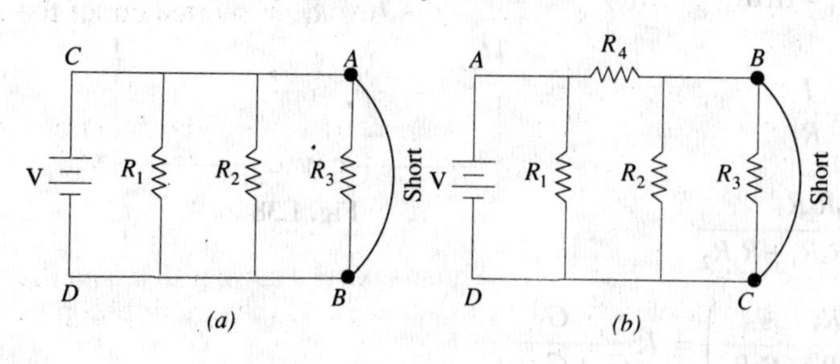

Fig. 1.36

Following points about the circuit of Fig. 1.36 (*a*) are worth noting.

1. not only is R_3 short-circuited but both R_1 and R_2 are also shorted out *i.e. short across one branch means short across all branches.*

2. there is no current in shorted resistors. If these were three bulbs, they will not glow.

3. the shorted components are not damaged. For example, if we had three bulbs in Fig. 1.36 (*a*), they would glow again when circuit is restored to normal conditions by removing the short-circuit.

It may, however, be noted from Fig. 1.36 (*b*) that a short-circuit across R_3 may short out R_2 but not R_1 since it is protected by R_4.

1.25. Division of Current in Parallel Circuits

In Fig. 1.37, two resistances are joined in parallel across a voltage V. The current in each branch, as given by Ohm's law, is

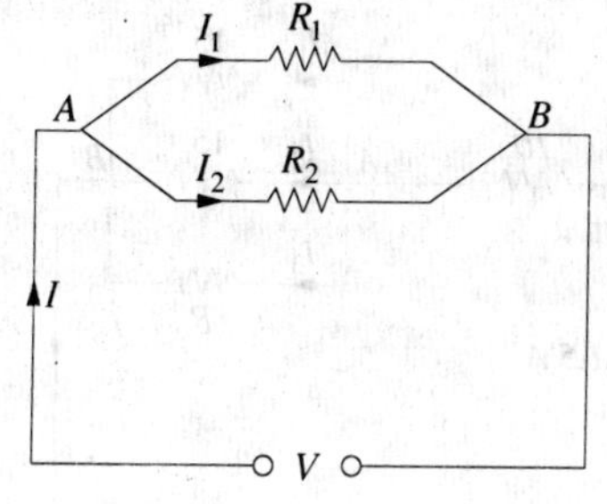

Fig. 1.37

$$I_1 = V/R_1 \text{ and } I_2 = V/R_2$$

$$\therefore \quad \frac{I_1}{I_2} = \frac{R_2}{R_1}$$

As $$\frac{I}{R_1} = G_1 \text{ and } \frac{I}{R_2} = G_2$$

$$\therefore \quad \frac{I_1}{I_2} = \frac{G_1}{G_2}$$

Hence, the division of current in the branches of a parallel circuit is directly proportional to the conductance of the branches or inversely proportional to their resistances. We may also express the branch currents in terms of the total circuit current thus :

$$\text{Now} \quad I_1 + I_2 = I; \quad \therefore \quad I_2 = I - I_1 \quad \therefore \quad \frac{I_1}{I - I_1} = \frac{R_2}{R_1} \quad \text{or} \quad I_1 R_1 = R_2 (I - I_1)$$

$$\therefore \quad I_1 = I \frac{R_2}{R_1 + R_2} = I . \frac{G_1}{G_1 + G_2} \quad \text{and} \quad I_2 = I \frac{R_1}{R_1 + R_2} = I . \frac{G_2}{G_1 + G_2}$$

This Current Divider Rule has direct application in solving electric circuits by Norton's theorem (Art. 2.25).

Take the case of three resistors in parallel connected across a voltage V (Fig. 1.38). Total current is $I = I_1 + I_2 + I_3$. Let the equivalent resistance be R. Then

$$V = IR$$

Also $V = I_1R_1 \quad \therefore \quad IR = I_1R_1$

or $\dfrac{I}{I_1} = \dfrac{R_1}{R}$ or $I_1 = IR/R_1 \quad ...(i)$

Now $\dfrac{I}{R} = \dfrac{I}{R_1} + \dfrac{I}{R_2} + \dfrac{I}{R_3}$

$$R = \frac{R_1R_2R_3}{R_2R_3 + R_3R_1 + R_1R_2}$$

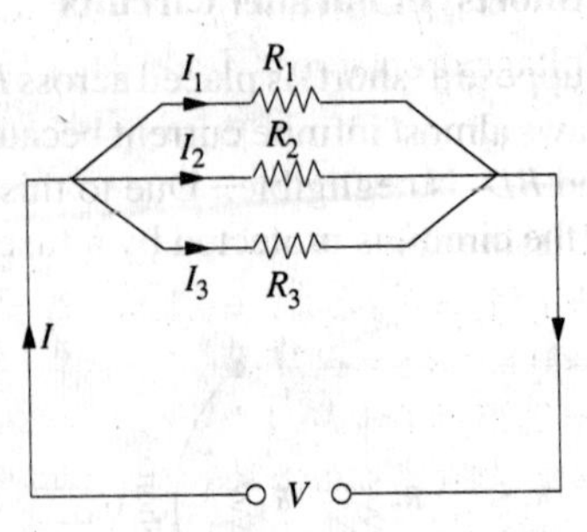

Fig. 1.38

From (i) above, $I_1 = I\left(\dfrac{R_2R_3}{R_1R_2 + R_2R_3 + R_3R_1}\right) = I.\dfrac{G_1}{G_1 + G_2 + G_3}$

Similarly, $I_2 = I\left(\dfrac{R_1R_3}{R_1R_2 + R_2R_3 + R_3R_1}\right) = I.\dfrac{G_2}{G_1 + G_2 + G_3}$

$$I_3 = I\left(\frac{R_1R_2}{R_1R_2 + R_2R_3 + R_3R_1}\right) = I.\frac{G_3}{G_1 + G_2 + G_3}$$

Example 1.29. *A resistance of 10 Ω is connected in series with two resistances each of 15 Ω arranged in parallel. What resistance must be shunted across this parallel combination so that the total current taken shall be 1.5 A with 20 V applied ?*

(Elements of Elect. Engg.-1 ; Bangalore Univ. Jan. 1989)

Solution. The circuit connections are shown in Fig. 1.39.

Drop across 10–Ω resistor = 1.5 × 10 = 15 V

Drop across parallel combination, V_{AB} = 20 – 15 = 5V

Hence, voltage across each parallel resistance is 5V.

I_1 = 5/15 = 1/3A, I_2 = 5/15 = 1/3 A

I_3 = 1.5 – (1/3 + 1/3) = 5/6 A

∴ I_3R = 5 or (5/6) R = 5 or R = **6 Ω**

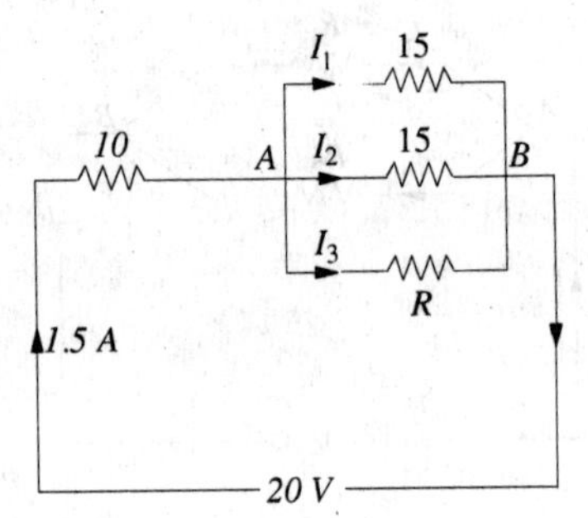

Fig. 1.39

Example 1.30. *If 20 V be applied across AB shown in Fig. 1.40, calculate the total current, the power dissipated in each resistor and the value of the series resistance to halve the total current.*

(Elect. Science-II, Allahabad Univ. 1992)

Solution. As seen from Fig. 1.40. R_{AB} = 370/199 Ω.

Hence, total current = 20 ÷ 370/199 = **10.76 A.**

I_1 = 10.76 × 5(5 + 74/25) = 6.76 A; I_2 = 10.76 – 6.76 = 4 A

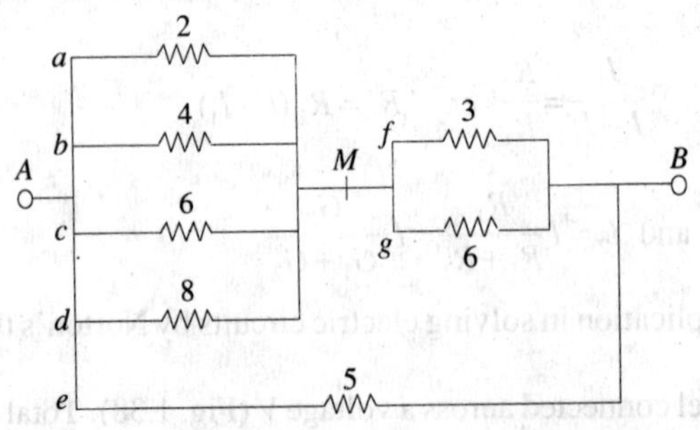

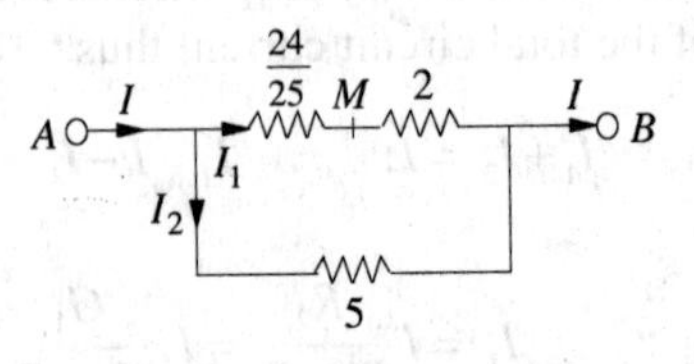

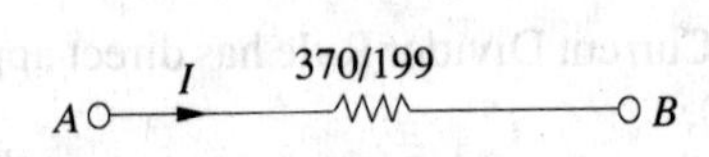

Fig. 1.40

$I_f = 6.76 \times 6/9 = 4.51$ A ; $I_g = 6.76 - 4.51 = 2.25$ A

Voltage drop across A and M, $V_{AM} = 6.76 \times 24/25 = 6.48$ V

$I_a = V_{AM}/2 = 6.48/2 = 3.24$ A; $I_b = 6.48/4 = 1.62$ A; $I_c = 6.48/6 = 1.08$ A

$I_d = 6.48/8 = 0.81$ A, $I_e = 20/5 = 4$A

Power Dissipation

$P_a = I_a^2 R_a = 3.24^2 \times 2 = \mathbf{21\ W}$, $P_b = 1.62^2 \times 4 = \mathbf{10.4\ W}$, $P_c = 1.08^2 \times 6 = \mathbf{7\ W}$

$P_d = 0.81^2 \times 8 = \mathbf{5.25\ W}$, $P_e = 4^2 \times 5 = \mathbf{80\ W}$, $P_f = 4.51^2 \times 3 = \mathbf{61\ W}$

$P_g = 2.25^2 \times 6 = \mathbf{30.4\ W}$

The series resistance required is **370/199 Ω**

Incidentally, total power dissipated $= I^2 R_{AB} = 10.76^2 \times 370/199 = 215.3$ W (as a check).

Example 1.31. *Calculate the values of different currents for the circuit shown in Fig. 1.41. What is the total circuit conductance ? and resistance ?*

Solution. As seen, $I = I_1 + I_2 + I_3$. The current division takes place at point B.

As seen from Art. 1.25

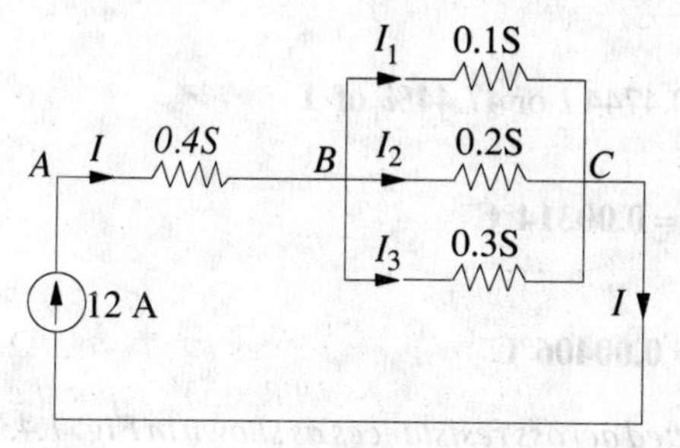

Fig. 1.41

$$I_1 = I \cdot \frac{G_1}{G_1 + G_2 + G_3}$$

$$= 12 \times \frac{0.1}{0.6} = \mathbf{2\ A}$$

$$I_2 = 12 \times 0.2/0.6 = \mathbf{4\ A}$$

$$I_3 = 12 \times 0.3/0.6 = \mathbf{6\ A}$$

$$G_{BC} = 0.1 + 0.2 + 0.3 = 0.6\ \text{S}$$

$$\frac{1}{G_{AC}} = \frac{1}{G_{AB}} + \frac{1}{G_{BC}} = \frac{1}{0.4} + \frac{1}{0.6} = \frac{25}{6}\ \text{S}^{-1} \therefore R_{AC} = 1/G_{AC} = \mathbf{25/6\ \Omega}$$

Example 1.32. *Compute the values of three branch currents for the circuit of Fig. 1.42 (a). What is the potential difference between points A and B ?*

Solution. The two given current sources may be combined together as shown in Fig. 1.42 (*b*). Net current = 25 − 6 = 19 A because the two currents flow in opposite directions.

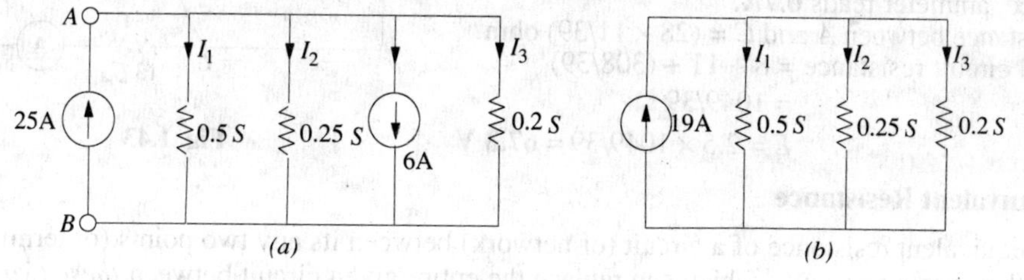

Fig. 1.42

Now, $G = 0.5 + 0.25 + 0.2 = 0.95$ S ; $I_1 = I \dfrac{G_1}{G} = 19 \times \dfrac{0.5}{0.95} = \mathbf{10\ A}$

$$I_2 = I \frac{G_2}{G} = 19 \times \frac{0.25}{0.95} = \mathbf{5\ A}\ ; \ I_3 = I \frac{G_3}{G} = 19 \times \frac{0.2}{0.95} = \mathbf{4\ A}$$

$$V_{AB} = I_1 R_1 = \frac{I_1}{G_1} = \frac{I_2}{G_2} = \frac{I_3}{G_3} \quad \therefore V_{AB} = \frac{10}{0.5} = \mathbf{20\ V}$$

The same voltage acts across the three conductances.

Example 1.33. *Two conductors, one of copper and the other of iron, are connected in parallel and at 20°C carry equal currents. What proportion of current will pass through each if the temperature is raised to 100°C ? Assume α for copper as 0.0042 and for iron as 0.006 per °C at 20°C. Find also the values of temperature coefficients at 100°C.* **(Electrical Engg. Madras Univ. 1987)**

Solution. Since they carry equal currents at 20°C, the two conductors have the same resistance at 20°C *i.e.* R_{20}. As temperature is raised, their resistances increase though unequally.

For Cu, $R_{100} = R_{20}(1 + 80 \times 0.0042) = 1.336\, R_{20}$

For iron $R'_{100} = R_{20}(1 + 80 \times 0.006) = 1.48\, R_{20}$

As seen from Art. 1.25, current through Cu conductor is

$$I_1 = I \times \frac{R'_{100}}{R_{100} + R'_{100}} = I \times \frac{1.48\, R_{20}}{2.816\, R_{20}} = 0.5256\, I \text{ or } \mathbf{52.56\%\ of\ I}$$

Hence, current through Cu conductor is 52.56 per cent of the total current. Obviously, the remaining current *i.e.* 47.44 per cent passes through iron.

Or current through iron conductor is

$$I_2 = I \cdot \frac{R_{100}}{R_{100} + R'_{100}} = I \times \frac{1.336\, R_{20}}{2.816\, R_{20}} = 0.4744\, I \text{ or } \mathbf{47.44\%\ of\ I}$$

For Cu,
$$\alpha_{100} = \frac{1}{(1/0.0042) + 80} = \mathbf{0.00314°C^{-1}}$$

For iron,
$$\alpha_{100} = \frac{1}{(1/0.006) + 80} = \mathbf{0.00406°C^{-1}}$$

Example 1.34. *A battery of unknown e.m.f. is connected across resistances as shown in Fig. 1.43. The voltage drop across the 8-Ω resistor is 20 V. What will be the current reading in the ammeter? What is the e.m.f. of the battery ?* **(Basic Elect Engg. ; Bangladesh Univ., 1990)**

Solution. Current through 8-Ω resistance = 20/8 = 2.5 A

This current is divided into two parts at point *A*; one part going along path *AC* and the other along path *ABC* which has a resistance of 28 Ω.

$$I_2 = 2.5 \times \frac{11}{(11 + 28)} = 0.7 \text{ A}$$

Hence, ammeter reads **0.7A.**

Resistance between *A* and *C* = (28 × 11/39) ohm.

Total circuit resistance = 8 + 11 + (308/39)

= 1049/39 Ω

∴ $E = 2.5 \times 1049/39 = \mathbf{67.3\ V}$

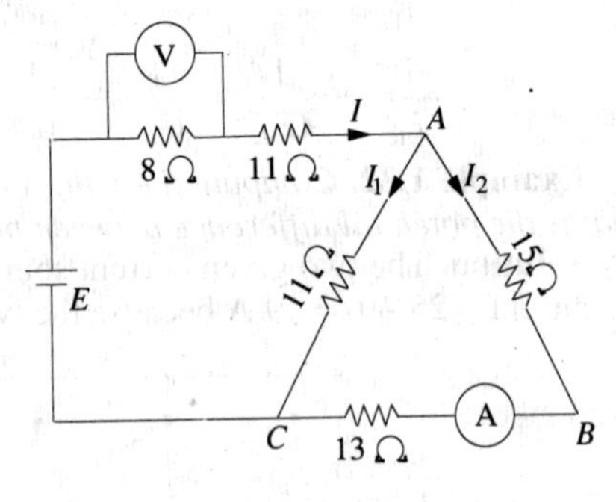

Fig. 1.43

1.26. Equivalent Resistance

The equivalent resistance of a circuit (or network) between its any two points (or terminals) is given by that *single* resistance which can replace the entire given circuit between *these two points.* It should be noted that resistance is always between two *given* points of a circuit and can have different values for different point-pairs as illustrated by Example 1.44. It can usually be found by using series and parallel laws of resistances. Concept of equivalent resistance is essential for understanding network theorems like Thevenin's theorem and Norton's theorem etc. discussed in Chapter 2.

Example 1.35. *Find the equivalent resistance of the circuit given in Fig. 1.44 (a) between the following points (i) A and B (ii) C and D (iii) E and F (iv) A and F and (v) A and C. Numbers represent resistances in ohm.*

Solution. (i) Resistance Between *A* and *B*

In this case, the entire circuit to the right side of *AB* is in parallel with 1 Ω resistance connected directly across points *A* and *B*.

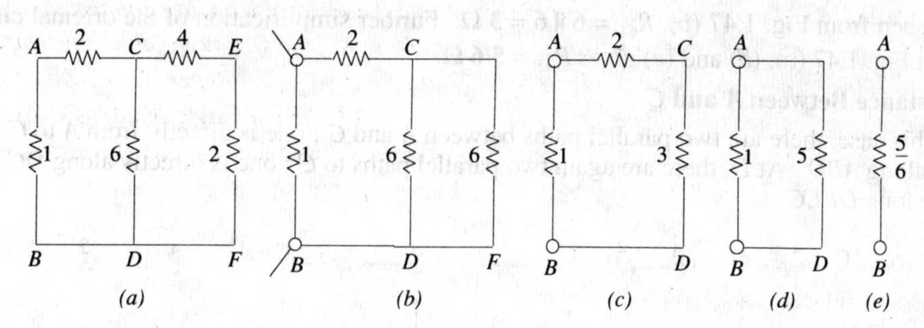

Fig. 1.44

As seen, there are two parallel paths across points C and D ; one having a resistance of 6 Ω and the other of (4 + 2) = 6 Ω. As shown in Fig. 1.44 (*c*), the combined resistance between C and D is = 6 ‖ 6 = 3 Ω. Further simplifications are shown in Fig. 1.44 (*d*) and (*e*). As seen, R_{AD} = **5/6 Ω**.

(ii) Resistance Between *C* and *D*

As seen from Fig. 1.44 (*a*), there are three parallel paths between C and D (*i*) *CD* itself of 6 Ω (*ii*) *CEFD* of (4 + 2) = 6 Ω and (*iii*) *CABD* of (2 + 1) = 3 Ω. It has been shown separately in Fig. 1.45 (*a*). The equivalent resistance R_{CD} = 3 ‖ 6 ‖ 6 = 1.5 Ω as shown in Fig. 1.45 (*b*).

(iii) Resistance Between *E* and *F*

In this case, the circuit to the left side of *EF* is in parallel with the 2 Ω resistance connected directly across points E and F. This circuit consists of a 4 Ω resistance connected in series with a

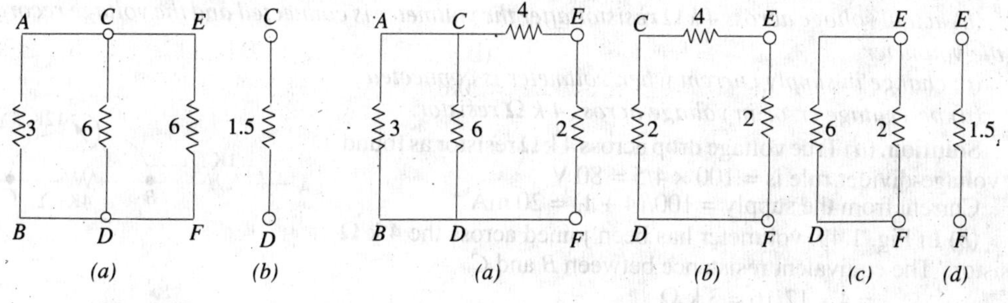

Fig. 1.45 **Fig. 1.46**

parallel circuit of 6 ‖ (2 + 1) = 2 Ω resistance. After various simplifications as shown in Fig. 1.46, R_{EF} = 2 ‖ 6 = **1.5 Ω**.

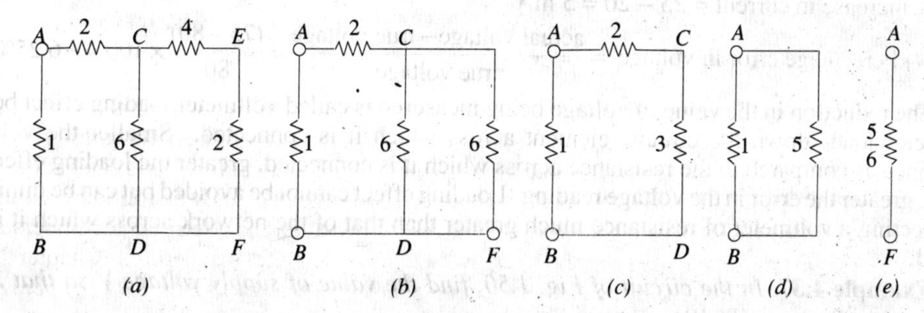

Fig. 1.47

(*iv*) Resistance Between *A* and *F*

As we go from A to F, there are two possible routes to begin with: one along *ABDF* and the other along *AC*. At point C, there are again two alternatives, one along *CDF* and the other along *CEF*.

As seen from Fig. 1.47 (*b*), $R_{CD} = 6 \,||\, 6 = 3\ \Omega$. Further simplification of the original circuit as shown in Fig. 1.47 (*c*), (*d*) and (*e*) gives R_{AF} = **5/6 Ω.**

(*v*) **Resistance Between *A* and *C***

In this case, there are two parallel paths between *A* and *C* ; one is directly from *A* to *C* and the other is along *ABD*. At *D*, there are again two parallel paths to *C* ; one is directly along *DC* and the other is along *DFEC*.

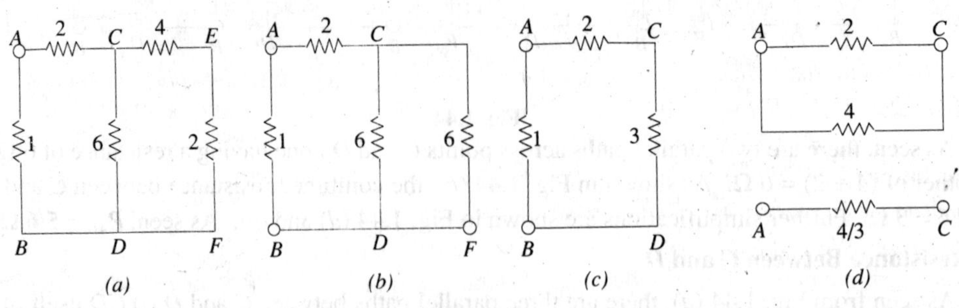

Fig. 1.48

As seen from Fig. 1.48 (*b*), $R_{CD} = 6 \,||\, 6 = 3\ \Omega$. Again, from Fig. 1.48 (*d*), $R_{AC} = 2\,||\,4$ = **4/3 Ω.**

Example 1.36. *Two resistors of values 1 k Ω and 4 k Ω are connected in series across a constant voltage supply of 100 V. A voltmeter having an internal resistance of 12 k Ω is connected across the 4 k Ω resistor. Draw the circuit and calculate*

(*a*) *true voltage across 4 k Ω resistor before the voltmeter was connected.*

(*b*) *actual voltage across 4 k Ω resistor after the voltmeter is connected and the voltage recorded by the voltmeter.*

(*c*) *change in supply current when voltmeter is connected*

(*d*) *percentage error in voltage across 4 k Ω resistor.*

Solution. (*a*) True voltage drop across 4 k Ω resistor as found by voltage-divider rule is = 100 × 4/5 = 80 V

Current from the supply = 100/(4 + 1) = 20 mA

(*b*) In Fig. 1.49, voltmeter has been joined across the 4 k Ω resistor. The equivalent resistance between *B* and *C*

= 4 × 12/16 = 3 k Ω

Drop across *B* and *C* = 100 × 3/(3 + 1) = 75 V.

(*c*) Resistance between *A* and *C* = 3 + 1 = 4 k Ω

New supply current = 100/4 = 25 mA

∴ increase in current = 25 − 20 = 5 mA

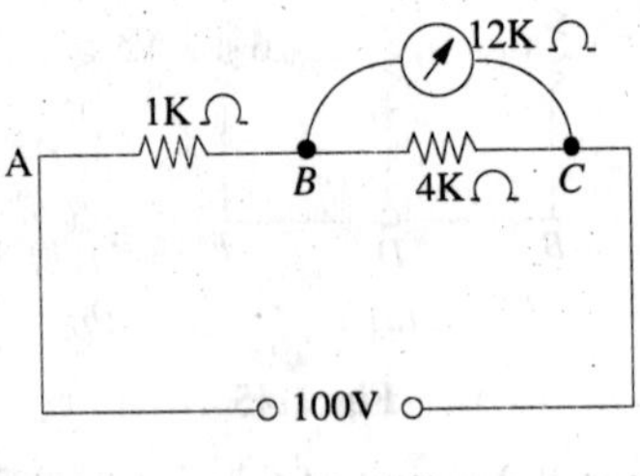

Fig. 1.49

(*d*) Percentage error in voltage $= \dfrac{\text{actual voltage} - \text{true voltage}}{\text{true voltage}} = \dfrac{(75-80)}{80} \times 100 = -6.25\%$

The reduction in the value of voltage being measured is called voltmeter loading effect because voltmeter loads down the circuit element across which it is connected. Smaller the voltmeter resistance as compared to the resistance across which it is connected, greater the loading effect and, hence, greater the error in the voltage reading. Loading effect cannot be avoided but can be minimized by selecting a voltmeter of resistance much greater than that of the network across which it is connected.

Example 1.37. *In the circuit of Fig. 1.50, find the value of supply voltage V so that 20 − Ω resistor can dissipate 180 W.*

Solution. $I_4^2 \times 20 = 180$ W; $I_4 = 3$ A

Since 15 Ω and 20 Ω are in parallel,

$$I_3 \times 15 = 3 \times 20 \therefore I_3 = 4 \text{ A}$$

$$I_2 = I_3 + I_4 = 4 + 3 = 7 \text{ A}$$

Now, resistance of the circuit to the right of point *A* is

$$= 10 + 15 \times 20/35 = 130/7\ \Omega$$

$$\therefore \quad I_1 \times 25 = 7 \times 130/7$$

$$\therefore \quad I_1 = 26/5A = 5.2\text{ A}$$

$$\therefore \quad I = I_1 + I_2 = 5.2 + 7 = 12.2\text{ A}$$

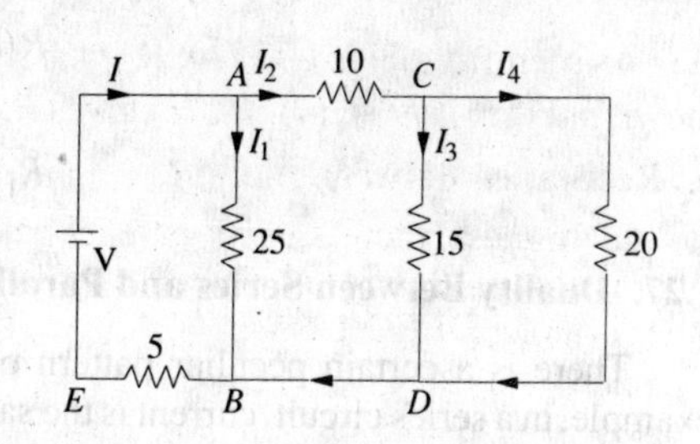

Fig. 1.50

Total circuit resistance

$R_{AE} = 5 + 25 \parallel 130/7 = 955/61\ \Omega$

$\therefore\ V = I.R_{AE} = 12.2 \times 955/61 = \mathbf{191\ V}$

Example 1.38. *For the simple ladder network shown in Fig. 1.51, find the input voltage V_i which produces a current of 0.25 A in the 3-Ω resistor. All resistances are in ohm.*

Solution. We will assume a current of 1 A in the 3-Ω resistor. The voltage necessary to produce 1A bears the same ratio to 1 *A* as V_i does to 0.25 A because of the linearity of the network. It is known as Current Assumption technique.

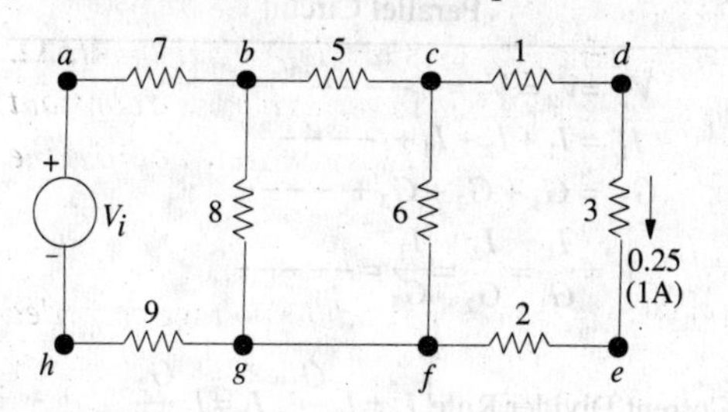

Fig. 1.51

Since $\quad R_{cdef} = R_{cf} = 6\ \Omega$

Hence, $\quad I_{cf} = 1\text{ A}$

and $\quad V_{cf} = V_{cdef} = 1 \times 6 = 6\text{ V.}$

Also, $\quad I_{bc} = 1 + 1 = 2\text{ A}$

$$V_{bg} = V_{bb} + V_{ef} = 2 \times 5 + 6 = 16\text{ V}$$

$$I_{bg} = 16/8 = 2\text{ A}$$

$$I_{ab} = I_{bc} + b_{bg} = 2 + 2 = 4\text{ A}$$

$$V_i = V_{ab} + V_{bg} + V_{gh} = 4 \times 7 + 16 + 4 \times 9 = 80\text{ V}$$

Taking the proportion, we get

$$\frac{80}{1} = \frac{V_i}{0.25} \quad \therefore \quad V_i = 80 \times 0.25 = \mathbf{20\ V}$$

Example 1.39. *In the circuit of Fig. 1.52, find the value of R_1 and R_2 so that $I_2 = I_1/n$ and the input resistance as seen from points A and B is R ohm.*

Solution. As seen, the current through R_2 is $(I_1 - I_2)$. Hence, p.d. across points *C* and *D* is $R_2(I_1 - I_2) = (R_1 + R)\, I_2$ or $R_2 I_1 = (R_1 + R_2 + R)\, I_2$

$$\therefore \quad \frac{I_1}{I_2} = \frac{R_1 + R_2 + R}{R_2} = n \qquad \ldots(i)$$

The input resistance of the circuit as viewed from terminals *A* and *B* is required to be *R*.

$$\therefore \quad R = R_1 + R_2 \parallel (R_1 + R)$$

$$= R_1 + \frac{R_1 + R}{n} \qquad \ldots\text{using Eq. } (i)$$

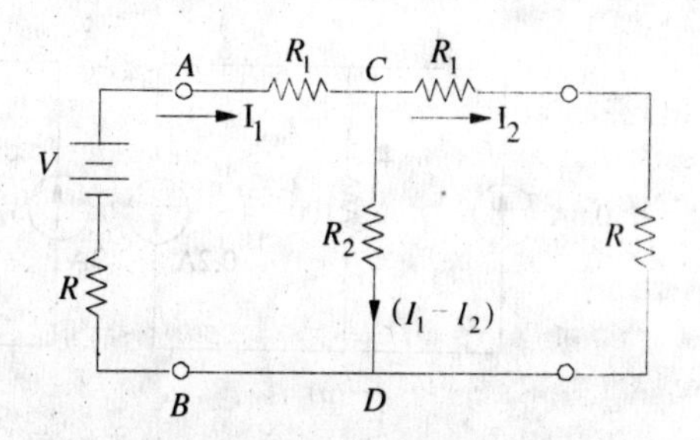

Fig. 1.52

$$\therefore \quad R(n-1) = R_1(n+1)$$

$$\therefore \quad R_1 = \frac{n-1}{n+1}R \text{ and } R_2 = \frac{R_1 + R}{(n-1)} = \frac{2n}{n^2-1}R$$

1.27. Duality Between Series and Parallel Circuits

There is a certain peculiar pattern of relationship between series and parallel circuits. For example, in a series circuit, current is the same whereas in a parallel circuit, voltage is the same. Also, in a series circuit, individual voltages are added and in a parallel circuit, individual currents are added. It is seen that while comparing series and parallel circuits, voltage takes the place of current and current takes the place of voltage. Such a pattern is known as "duality" and the two circuits are said to be duals of each other.

As arranged in Table 1.4 the equations involving voltage, current and resistance in a series circuit have a corresponding dual counterparts in terms of current, voltage and conductance for a parallel circuit.

Table 1.4

Series Circuit	Parallel Circuit
$I_1 = I_2 = I_3 = -----$	$V_1 = V_2 = V_3 = ----$
$V_T = V_1 + V_2 + V_3 + -----$	$I_1 = I_1 + I_2 + I_3 + ----$
$R_T = R_1 + R_2 + R_3 + -----$	$G_T = G_1 + G_2 + G_3 + ----$
$I = \frac{V_1}{R_1} = \frac{V_2}{R_2} = \frac{V_3}{R_3} = -------$	$V = \frac{I_1}{G_1} = \frac{I_2}{G_2} = \frac{I_3}{G_3} = -----$
Voltage Divider Rule $V_1 = V_T\frac{R_1}{R_T}$, $V_2 = V_T\frac{R_2}{R_T}$	Current Divider Rule $I_1 = I_T\frac{G_1}{G_T}$, $I_2 = I_T\frac{G_2}{G_T}$

Tutorial Problems No. 1.4

1. Using the current-divider rule, find the ratio I_L/I_S in the circuit shown in Fig. 1.53. **[0.25]**

2. Find the values of variables indicated in the circuit of Fig. 1.54. All resistances are in ohms. [(a) **40 V** (b) **21** V; **15 V** (c) **–5A; 3A**]

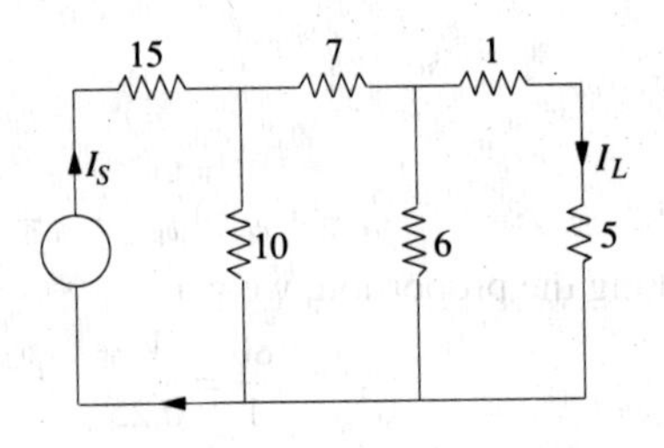

Fig. 1.53

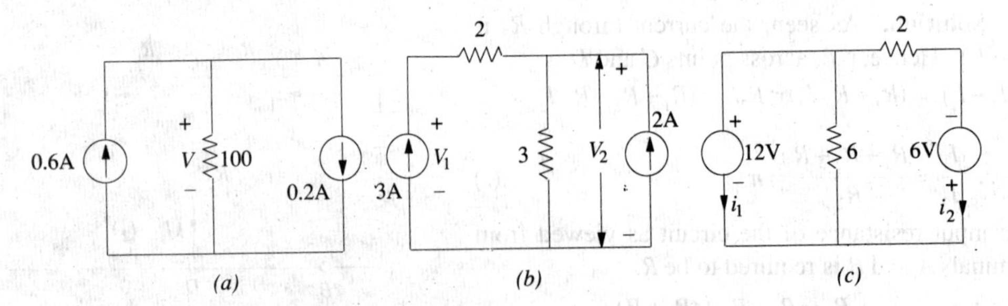

Fig. 1.54

3. An ohmmeter is used for measuring the resistance of a circuit between its two terminals. What would be the reading of such an instrument used for the circuit of Fig. 1.55 at point (a) AB (b) AC and (c) BC ? All resistances are in ohm. [(a) **25 Ω** (b) **24 Ω** (c) **9 Ω**]

4. Find the current and power supplied by the battery to the circuit of Fig. 1.56 (*i*) under normal conditions and (*ii*) when a 'short' occurs across terminals *A* and *B*. All resistances are in kilohm.

[(i) 2mA ; 24 m W; (ii) 3mA ; 36 mW]

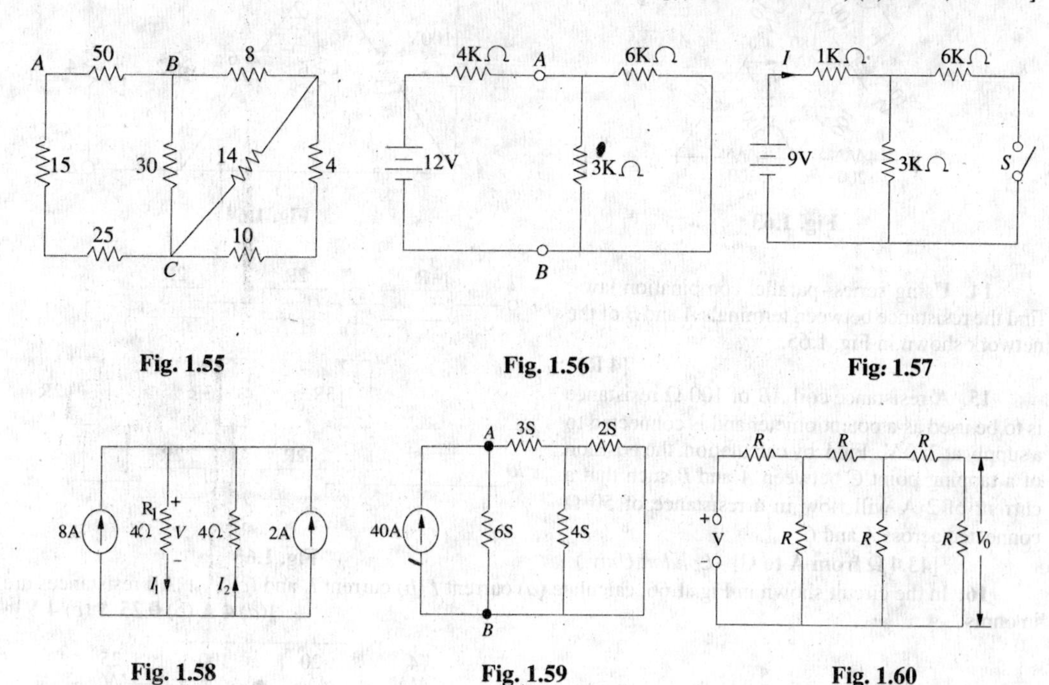

Fig. 1.55 Fig. 1.56 Fig. 1.57

Fig. 1.58 Fig. 1.59 Fig. 1.60

5. Compute the values of battery current *I* and voltage drop across 6 K Ω resistor of Fig. 1.57 when switch *S* is (*a*) closed and (*b*) open. All resistance values are in kilohm. **[(a) 3 mA; 6 V; (b) 2.25 mA; OV]**

6. For the parallel circuit of Fig. 1.58 calculate (*i*) *V* (*ii*) I_1 (*iii*) I_2. **[(*i*) 20 V; (*ii*) 5 A; (*iii*) – 5 A]**

7. Find the voltage across terminals *A* and *B* of the circuit shown in Fig. 1.59. All conductances are in siemens (*S*). **[5 V]**

8. Prove that the output voltage V_0 in the circuit of Fig. 1.60 is *V*/13.

9. A fault has occurred in the circuit of Fig. 1.61. One resistor has burnt out and has become an open. Which is that resistor if current supplied by the battery is 6 A ? All resistances are in ohm. **[4 Ω]**

10. In Fig. 1.62 if resistance between terminals *A* and *B* measures 1000 Ω, which resistor is open-circuted. All conductance values are in milli-siemens (mS). **[0.8 mS]**

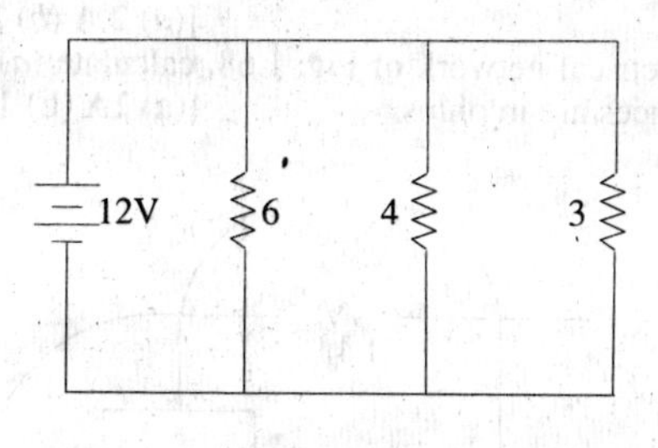

Fig. 1.61

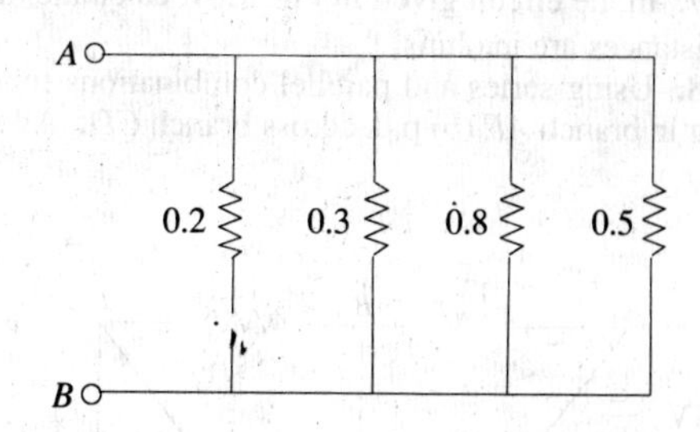

Fig. 1.62

11. In the circuit of Fig. 1.63, find current (*a*) *I* and (*b*) I_1. **[(*a*) 2 A; (*b*) 0.5 A]**

12. Deduce the current *I* in the circuit of Fig. 1.64. All resistances are in ohms. **[25 A]**

13. Two resistors of 100 Ω and 200 Ω are connected in series across a 4-V cell of negligible internal resistance. A voltmeter of 200 Ω resistance is used to measure P.D. across each. What will the voltage be in each case ? **[1 V across 100 Ω; 2 V across 200 Ω]**

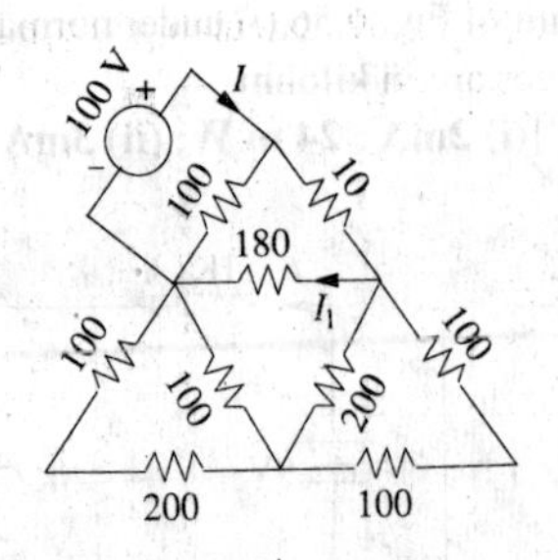

Fig. 1.63

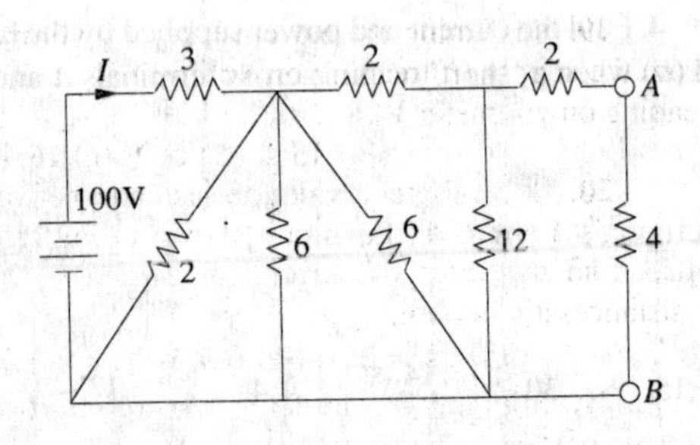

Fig. 1.64

14. Using series–parallel combination laws, find the resistance between terminals *A* and *B* of the network shown in Fig. 1.65.

[4 R]

15. A resistance coil *AB* of 100 Ω resistance is to be used as a potentiometer and is connected to a supply at 230 V. Find, by calculation, the position of a tapping point *C* between *A* and *B* such that a current of 2 A will flow in a resistance of 50 Ω connected across *A* and *C*.

[43.4 Ω from A to C] (*London Univ.*)

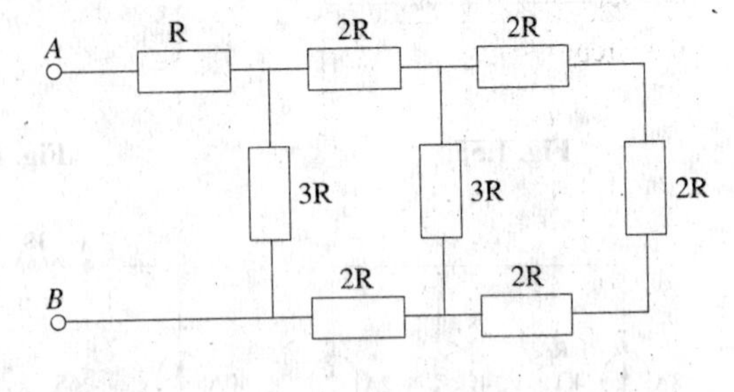

Fig. 1.65

16. In the circuit shown in Fig. 1.66, calculate (*a*) current *I* (*b*) current I_1 and (*c*) V_{AB}. All resistances are in ohms.

[(*a*) 4 A (b) 0.25 A (*c*) 4 V]

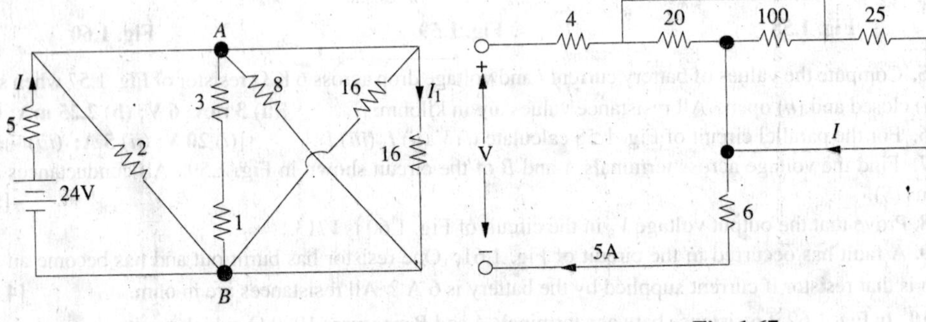

Fig. 1.66

Fig. 1.67

17. In the circuit given in Fig. 1.67, calculate (*a*) current through the 25 Ω resistor (*b*) supply voltage *V*. All resistances are in ohms.

[(*a*) 2 A (*b*) 100 V]

18. Using series and parallel combinations for the electrical network of Fig. 1.68, calculate (*a*) current flowing in branch *AF* (*b*) p.d. across branch *CD*. All resistances are in ohms.

[(a) 2A (b) 1.25 V]

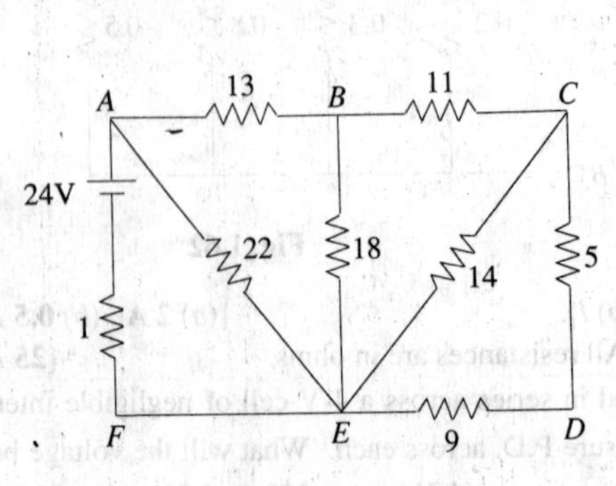

Fig. 1.68

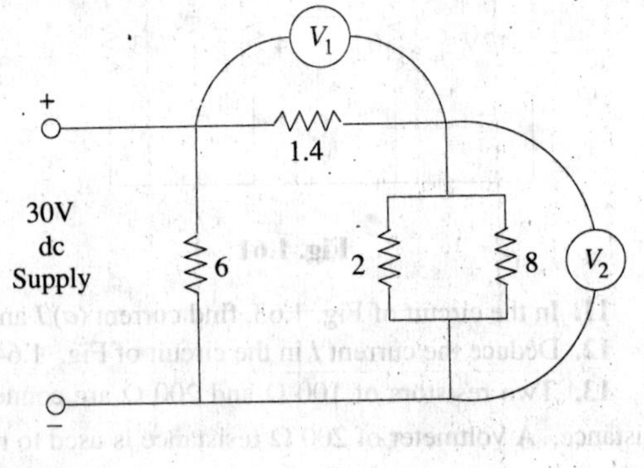

Fig. 1.69

19. Neglecting the current taken by voltmeters V_1 and V_2 in Fig. 1.69, calculate (*a*) total current taken from the supply (*b*) reading on voltmeter V_1 and (*c*) reading on voltmeter V_2.

[(a) 15 A (b) 14 V (c) 16 V]

20. Find the equivalent resistance between terminals *A* and *B* of the circuit shown in Fig. 1.70. Also, find the value of currents I_1, I_2 and I_3. All resistances are in ohm.

[8 Ω ; I_1 = 2 A; I_2 = 0.6 A ; I_3 = 0.4 A]

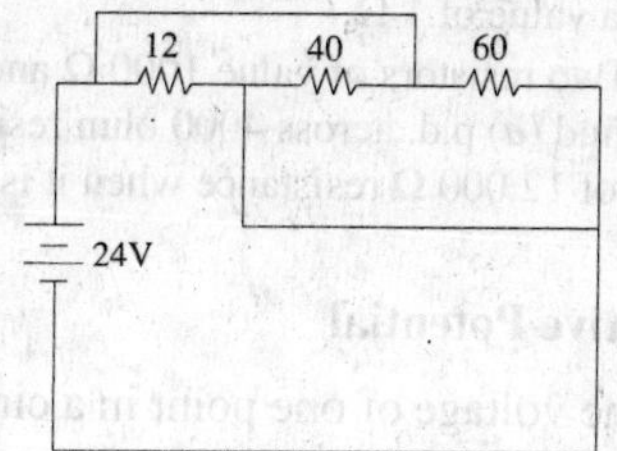

Fig. 1.70.

21. In Fig. 1.71, the 10 Ω resistor dissipates 360 W. What is the voltage drop across the 5 Ω resistor ? **[30 V]**

22. In Fig. 1.72, the power dissipated in the 10 Ω resistor is 250 W. What is the total power dissipated in the circuit ? **[850 W]**

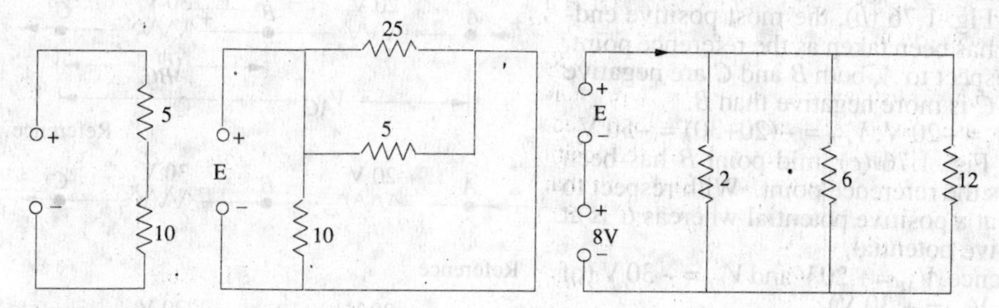

Fig. 1.71 **Fig. 1.72** **Fig. 1.73**

23. What is the value of *E* in the circuit of Fig. 1.73 ? All resistances are in ohms. **[4 V]**

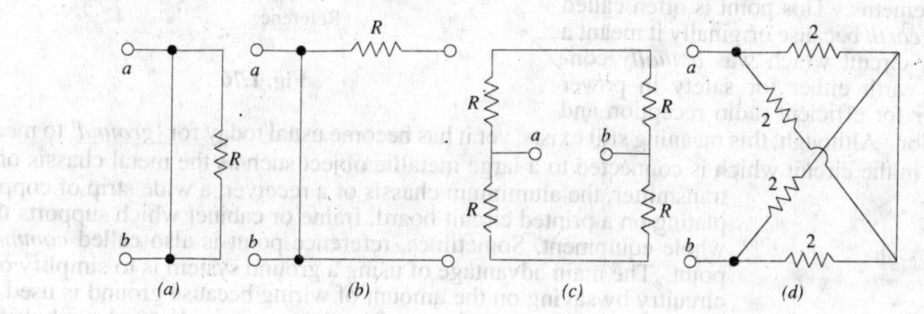

Fig. 1.74

24. Find the equivalent resistance $R_{a\text{-}b}$ at the terminals *a-b* of the networks shown in Fig. 1.74.

[(*a*) 0 (*b*) 0 (*c*) R (*d*) 2 Ω]

25. Find the equivalent resistance between terminals *a* and *b* of the circuit shown in Fig. 1.75 (*a*). Each resistance has a value of 1 Ω. **[5/11 Ω]**

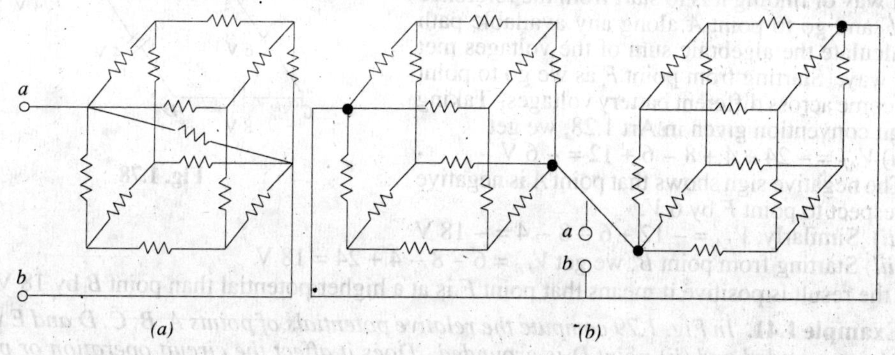

Fig. 1.75

26. Find the equivalent resistance between terminals *a* and *b* of the circuit shown in Fig. 1.75 (*b*). Each resistor has a value of 1 Ω. **[5/12 Ω]**

27. Two resistors of value 1000 Ω and 4000 Ω are connected in series across a constant voltage supply of 150 V. Find (*a*) p.d. across 4000 ohm resistor (*b*) calculate the change in supply current and the reading on a voltmeter of 12,000 Ω resistance when it is connected across the larger resistor. **[(*a*) 120 V (*b*) 7.5 mA; 112.5 V]**

1.28. Relative Potential

It is the voltage of one point in a circuit with respect to that of another point (usually called the reference or common point).

Consider the circuit of Fig. 1.76 (*a*) where the most negative end-point *C* has been taken as the reference. With respect to point *C*, both points *A* and *B* are positive though *A* is more positive than *B*. The voltage of point *B* with respect to that of *C* *i.e.* V_{BC} = +30V.

Similarly, V_{AC} = +(20 + 30) = +50 V.

In Fig. 1.76 (*b*), the most positive end-point *A* has been taken as the reference point. With respect to *A*, both *B* and *C* are negative though *C* is more negative than *B*.

V_{BA} = −20 V, V_{CA} = −(20+30) = −50 V

In Fig. 1.76 (*c*), mid-point *B* has been taken as the reference point. With respect to *B*, *A* is at a positive potential whereas *C* is at a negative potential.

Hence, V_{AB} = + 20 *V* and V_{CB} = − 30 V (of course, V_{BC} = + 30 V)

It may be noted that *any point* in the circuit can be chosen as the reference point to suit our requiremetns. This point is often called *ground* or *earth* because originally it meant a point in a circuit which was *actually* connected to earth either for safety in power systems or for efficient radio reception and transmission. Although, this meaning still exists, yet it has become usual today for '*ground*' to mean any point in the circuit which is connected to a large metallic object such as the metal chassis of a transmitter, the aluminium chassis of a receiver, a wide strip of copper plating on a printed circuit board, frame or cabinet which supports the whole equipment. Sometimes, reference point is also called *common* point. The main advantage of using a ground system is to simplify our circuitry by saving on the amount of wiring because ground is used as the return path for may circuits. The three commonly-used symbols for *ground* are shown in Fig. 1.77.

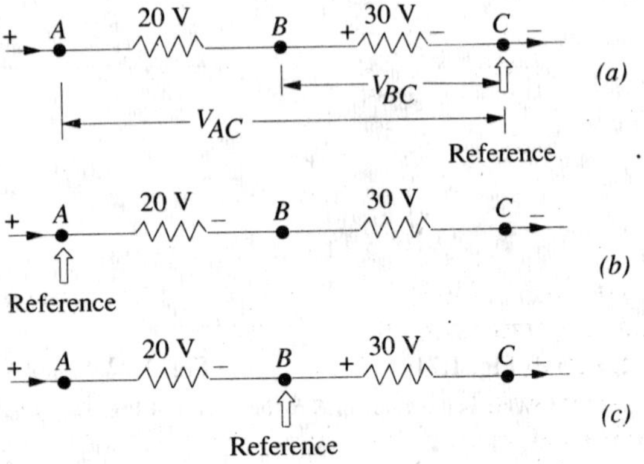

Fig. 1.76

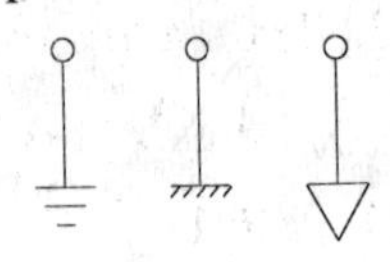

Fig. 1.77

Example 1.40. *In Fig. 1.78, calculate the values of (i)* V_{AF} *(i)* V_{EA} *and (iii)* V_{FB}.

Solution. It should be noted that V_{AF} stands for the potential of point *A* with respect to point *F*. The easiest way of finding it is to start from the reference point *F* and go to point *A* along any available path and calculate the algebraic sum of the voltages met on the way. Starting from point *F* as we go to point A, we come across different battery voltages. Taking the sign convention given in Art 1.28, we get

(*i*) V_{AF} = − 24 + 4 + 8 − 6 + 12 = − 6 V

The negative sign shows that point *A* is negative with respect to point *F* by 6 *V*.

(*ii*) Similarly, V_{EA} = − 12 + 6 − 8 − 4 = − 18 V

(*iii*) Starting from point *B*, we get V_{FB} = 6 − 8 − 4 + 24 = 18 V

Since the result is positive it means that point *F* is at a higher potential than point *B* by 18 V.

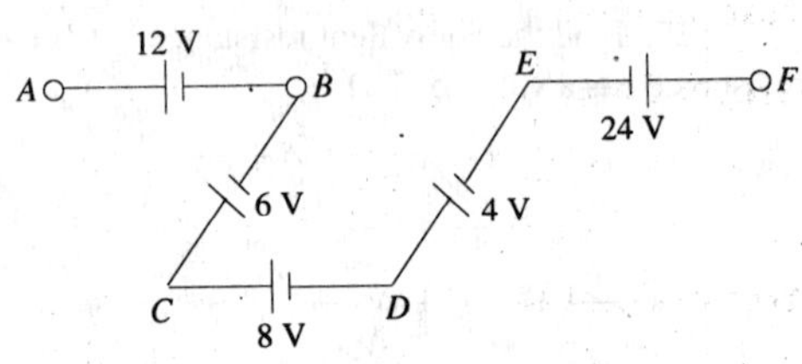

Fig. 1.78

Example 1.41. *In Fig. 1.79 compute the relative potentials of points A, B, C, D and E when (i) point A is grounded and (ii) point D is grounded. Does it affect the circuit operation or potential difference between any pair of points ?*

Solution. As seen, the two batteries have been connected in *series opposition*. Hence, net circuit voltage $= 34 - 10 = 24$ V.

Total circuit resistance $= 6 + 4 + 2 = 12\ \Omega$

Hence, circuit current $= 24/12 = 2$A

Drop across 2 Ω resistor $= 2 \times 2 = 4$ V Drop across 4 Ω resistor $= 2 \times 4 = 8$ V

Drop across 6 Ω resistor $= 2 \times 6 = 12$ V

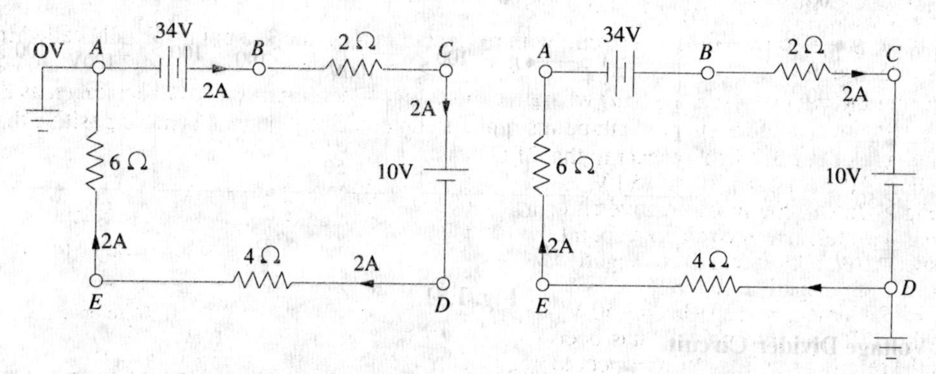

Fig. 1.79 Fig. 1.80

(*i*) Since point *B* is directly connected to the positive terminal of the battery whose negative terminal is earthed, hence, $V_B = +$ **34 V**.

Since there is a fall of 4 V across 2 Ω resistor, $V_C =$ **34 – 4 = 30 V**

As we go from point *C* to *D i.e.* from positive terminal of 10-V battery to its negative terminal, there is a *decrease* in potential of 10 V. Hence, $V_D = 30 - 10 =$ **20 V** *i.e.* point *D* is 20 V above the ground *A*.

Similarly, $V_E = V_D -$ voltage fall across 4 Ω resistor $= 20 - 8 =$ **+ 12 V**.

Also $V_A = V_E -$ fall across 6 Ω resistor = 12 (2 × 6) = **0 V**.

(*ii*) In Fig. 1.80, point *D* has been taken as the ground. Starting from point *D*, as we go to *E* there is a fall of 8 V. Hence, $V_E =$ **– 8 V**. Similarly, $V_A = -(8 + 12) =$ **–20 V**.

As we go from *A* to *B*, there is a sudden increase of 34 V because we are going from negative terminal of the battery to its positive terminal.

$\therefore \qquad V_B = -20 + 34 = +\mathbf{14\ V}$

$V_C = V_B -$ voltage fall across 2Ω resistor = 14–4 = **+10 V**

It should be so because *C* is connected directly to the positive terminal of the 10 V battery.

Choice of a reference point does not in any way affect the operation of a circuit. Moreover, it also does not change the voltage across any resistor or between any *pair* of points (as shown below) because the ground current $i_g = 0$.

Reference Point A

$$V_{CA} = V_C - V_A = 30 - 0 = +30\ V\ ;\ V_{CE} = V_C - V_E = 30 - 12 = +18\text{ V}$$
$$V_{BD} = V_B - V_D = 34 - 20 = +14\text{ V}$$

Reference Point D

$$V_{CA} = V_C - V_A = 10 - (-20) = +30\text{ V};\ V_{CE} = V_C - V_E = 10 - (-8) = +18\text{V}$$
$$V_{BD} = V_B - V_D = 14 - 0 = +\mathbf{14V}$$

Example 1.42. *Find the voltage V in Fig. 1.81* (*a*). *All resistances are in ohms.*

Solution. The given circuit can be simplified to the final form shown in Fig. 1.81 (*d*). As seen, current supplied by the battery is 1 A. At point *A* in Fig. 1.81 (*b*), this current is divided into two equal parts of 0.5 A each.

Obviously, voltage *V* represents the potential of point *B* with respect to the negative terminal of the battery. Point *B* is above the ground by an amount equal to the voltage drop across the series combination of (40 + 50) = 90 Ω.

$\therefore \quad V = 0.5 \times 90 = 45$ V.

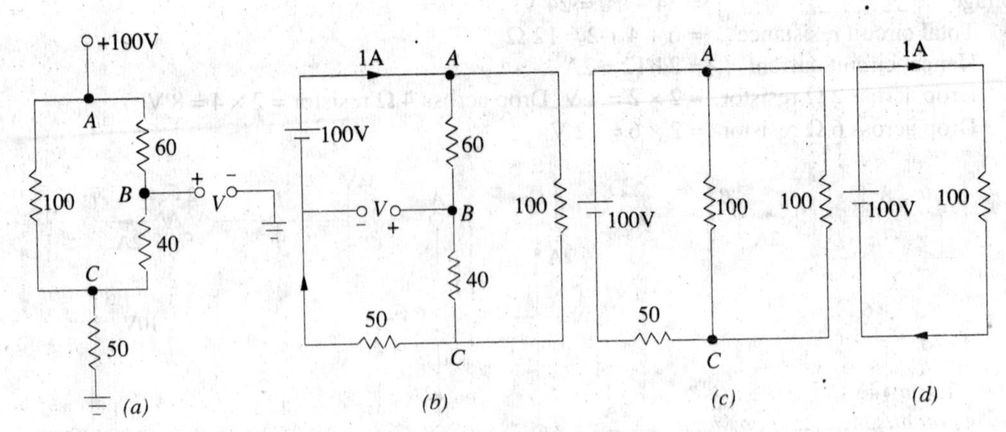

Fig. 1.81

1.29. Voltage Divider Circuit

A voltage divider circuit (also called potential divider) is a series network which is used to feed other networks with a number of different voltages and derived from a single input voltage source.

Fig. 1.82 (*a*) shows a simple voltage divider circuit which provides two output voltages V_1 and V_2. Since no load is connected across the output terminals, it is called an *unloaded* voltage divider.

As seen from Art. 1.15,

$$V_1 = V \frac{R_1}{R_1 + R_2} \text{ and } V_2 = V \cdot \frac{R_2}{R_1 + R_2}$$

The ratio V_2/V is also known as voltage-ratio *transfer function*.

$$\text{As seen, } \quad \frac{V_2}{V} = \frac{R_2}{R_1 + R_2} = \frac{1}{1 + R_1/R_2}$$

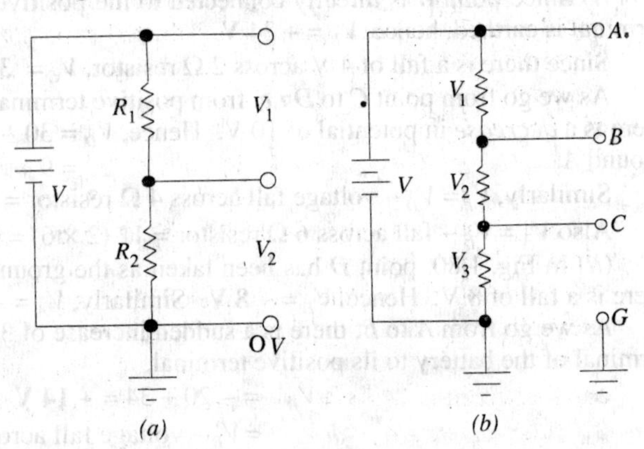

Fig. 1.82

The voltage divider of Fig. 1.82 (*b*) can be used to get six different voltages :

$V_{CG} = V_3$, $V_{BC} = V_2$, $V_{AB} = V_1$, $V_{BG} = (V_2 + V_3)$, $V_{AC} = (V_1 + V_2)$ and $V_{AG} = V$

Example 1.43. *Find the values of different voltages that can be obtained from a 12-V battery with the help of voltage divider circuit of Fig. 1.83.*

Solution. $R = R_1 + R_2 + R_3 = 4 + 3 + 1 = 8\ \Omega$

Drop across $R_1 = 12 \times 4/8 = 6$ V

$\therefore \quad V_B = 12 - 6 = 6$ V above ground

Drop across $R_2 = 12 \times 3/8 = 4.5$ V

$\therefore \quad V_C = V_B - 4.5 = 6 - 4.5 = 1.5$

Drop across $R_3 = 12 \times 1/8 = 1.5$ V

Different available load voltages are :

(*i*) $V_{AB} = V_A - V_B = 12 - 6 =$ **6 V**

(*ii*) $V_{AC} = 12 - 1.5 =$ **10.5 V** (*iii*) $V_{AD} =$ **12 V**

(*iv*) $V_{BC} = 6 - 1.5 =$ **4.5 V** (*v*) $V_{CD} =$ **1.5 V**.

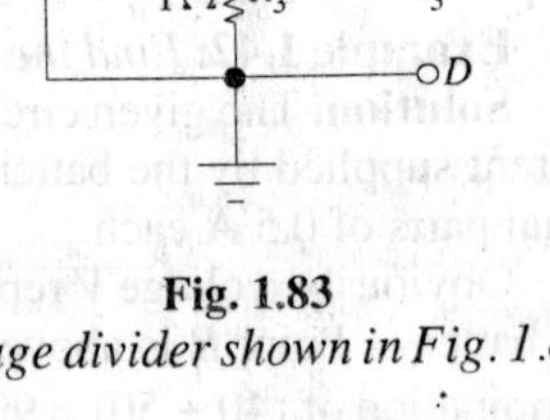

Fig. 1.83

Example 1.44. *What are the output voltages of the unloaded voltage divider shown in Fig. 1.84? What is the direction of current through AB ?*

Solution. It may be remembered that both V_1 and V_2 are with respect to the ground.

Fig. 1.84

$$R = 6 + 4 + 2 = 12\ \Omega$$

$$\therefore \quad V_1 = \text{drop across } R_2$$

$$= 24 \times 4/12 = +\mathbf{8\ V}$$

$$V_2 = \text{drop across } R_3$$

$$= -24 \times 2/12 = -\mathbf{4\ V}$$

It should be noted that point B is at negative potential with respect to the ground.

Current flows from A to B *i.e.* from a point at a higher potentail to a point at a lower potential.

Example 1.45. *Calculate the potentials of point A, B, C and D in Fig. 1.85. What would be the new potential values if connections of 6-V battery are reversed ? All resistances are in ohm.*

Solution. Since the two batteries are connected in additive series, total voltage around the circuit is = 12 + 6 = 18 V. The drops across the three resistors as found by the voltage divider rule as shown in Fig. 1.85 (*a*) which also indicates their proper polarities. The potential of any point in the circuit can be found by starting from the ground point G (assumed to be at 0 V) and going to the point either in clockwise direction or counter-clockwise direction. While going around the circuit, the rise in potential would be taken as positive and the fall in potential as negative. (Art. 2.3). Suppose we start from point G and proceed in the clockwise direction to point A. The only potential met on the way is the battery voltage which is taken as positive because there is a rise of potential since we are going from its negative to positive terminal. Hence, V_A is +12 V.

Similarly,

$$V_B = 12 - 3 = 9\ \text{V};\ V_C = 12 - 3 - 6 = 3\ \text{V}.$$

$$V_D = 12 - 3 - 6 - 9 = -6\ \text{V}.$$

It is also obvious that point D must be at - 6 V because it is directly connected to the negative terminal fo the 6-V battery.

We would also find the potentials of various points by starting from point G and going in the counter–clockwise direction. For example, $V_B = -6 + 9 + 6 = 9$ V as before.

The connections of the 6 – V battery have been reversed in Fig. 1.85 (*b*). Now, the net voltage around the circuit is 12 – 6 = 6 V. The drop over the 1 Ω resistor is = 6 × 1/(1 + 2 + 3) = 1 V; Drop over 2 Ω resistor is = 6 × 2/6 = 2 V. Obviously, V_A = + 12V. V_B = 12 – 1 = 11 V, V_C = 12 – 1 – 2 = 9 V. Similarly, V_D = 12 – 1 – 2 – 3 = + 6V.

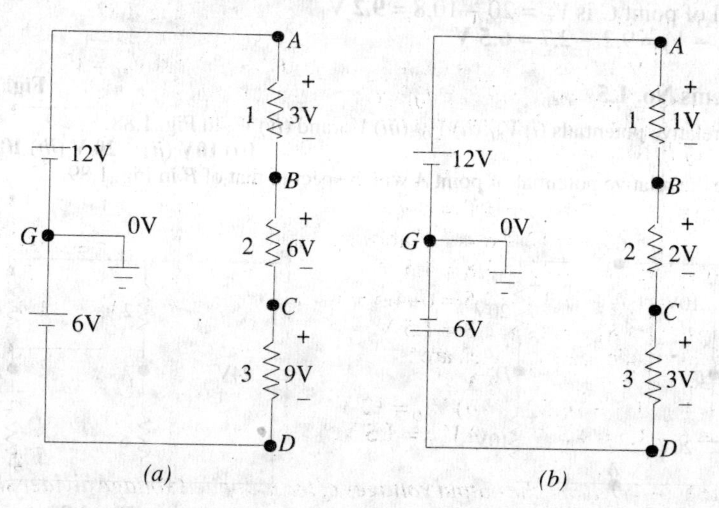

Fig. 1.85

Example 1.46. *Using minimum number of components, design a voltage divider which can deliver 1 W at 100 V, 2 W at –50 V and 1.6 W at – 80 V. The voltage source has an internal resistance of 200 Ω and supplies a current of 100 mA. What is the open-circuit voltage of the voltage source ? All resistances are in ohm.*

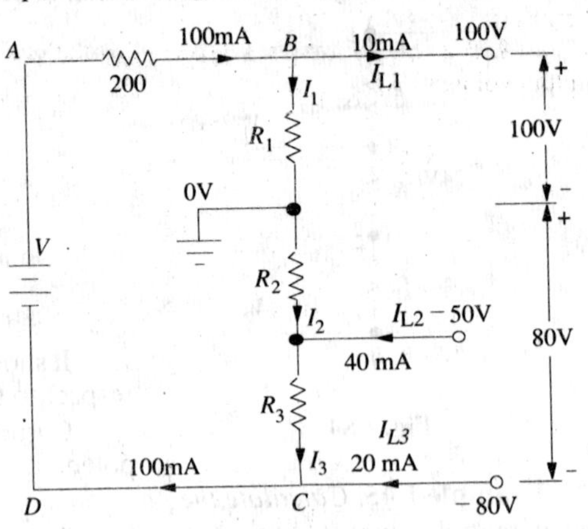

Fig. 1.86

Solution. From the given load conditions, the load currents are as follows :

$I_{L1} = 1/100 = 10$ mA,

$I_{L2} = 2/50 = 40$ mA, $I_{L3} = 1.6/80 = 20$ mA

For economising the number of components, the internal resistance of 200 Ω can be used as the series dropping resistance. The suitable circuit and the ground connection are shown in Fig. 1.86.

Applying Kirchhoff's laws to the closed circuit *ABCDA*, we have

$V - 200 \times 100 \times 10^{-3} - 100 - 80 = 0$ or $V = 200$ V

$I_1 = 100 - 10 = 90$ mA $\therefore R_1 = 100$ V/90 mA = **1.11 k Ω**

$I_3 = 100 - 20 = 80$ mA; voltage drop across $R_3 = -50 - (-80) = 30$ V.

$\therefore R_3 = 30$ V/80 mA = 375 Ω

$I_2 + 40 = 80$; $\therefore I_2 = 40$ mA; $R_2 = 50$ V/40 mA = **1.25 k Ω**

Example 1.47. *Fig. 1.87 shows a transistor with proper voltages established across its base, collector and emitter for proper function. Assume that there is a voltage drop V_{BE} across the base-emitter junction of 0.6 V and collector current I_C is equal to collector current I_E. Calculate (a) V_1 (b) V_2 and V_B (c) V_4 and V_E (d) I_E and I_C (e) V_3 (f) V_C (g) V_{CE}. All resistances are given in kilo-ohm.*

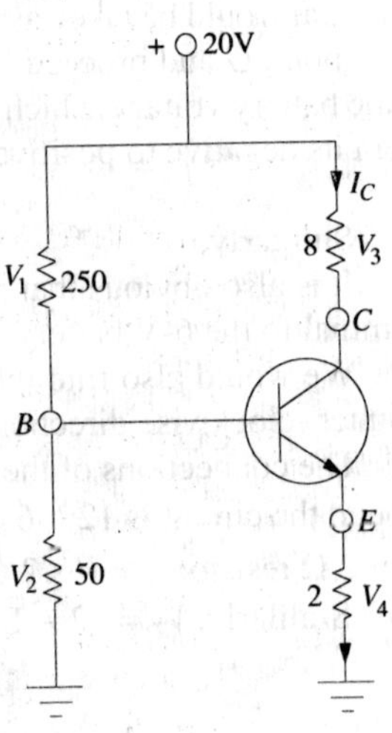

Fig. 1.87

Solution. (*a*) The 250 k and 50 k resistors form a voltage-divider bias network across 20 V supply.

$\therefore V_1 = 20 \times 250/300 =$ **16.7 V**

(*b*) $V_2 = 20 - 16.7 =$ **3.3 V**

The voltage of point V_B with respect to ground is $V_2 =$ **3.3 V.**

(*c*) $V_E = V_V - V_{BE} = 3.3 - 0.6 = 2.7$V. Also, $V_4 = 2.7$ V.

(*d*) $I_E = V_4/2 = 2.7$ V/2 k = **1.35** *mA*. It also equals I_C.

(*e*) V_3 = drop across collector resistor = 1.35 mA × 8 k = 10.8 V

(*f*) Potential of point *C* is $V_C = 20 - 10.8 =$ **9.2 V**

(*g*) $V_{CE} = V_C - V_E = 9.2 - 2.7 =$ **6.5 V**

Tutorial Problems No. 1.5

1. Find the relative potentials (*i*) V_{AD} (*ii*) V_{DC} (*iii*) V_{BD} and (*iv*) V_{AC} in Fig. 1.88. [(*i*) **10V** (*ii*) – **20 V** (*iii*) **10 V** (*iv*) – **30 V**]
2. Calculate the relative potential of point *A* with respect to that of *B* in Fig. 1.89. [2 V]

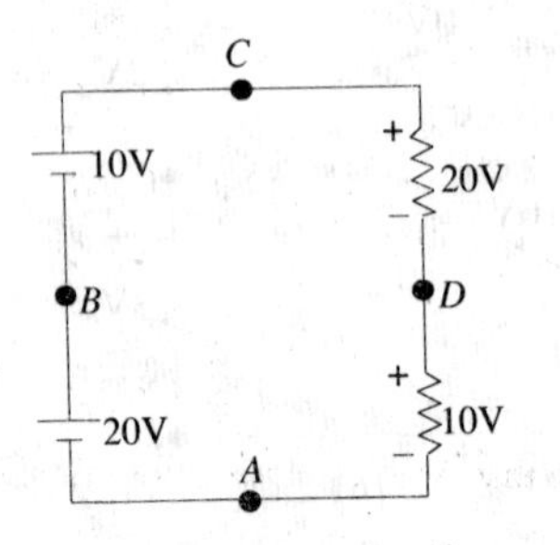

Fig. 1.88

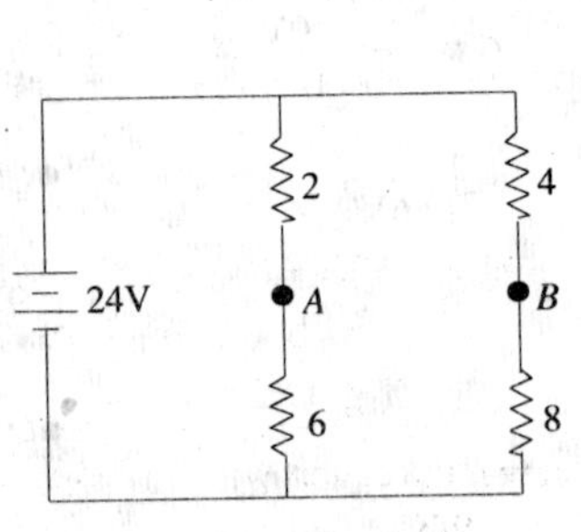

Fig. 1.89

3. Give the magnitude and polarity of the following voltages in the circuit of Fig. 1.90 (*i*) V_1 (*ii*) V_2 (*iii*) V_3 (*iv*) $V_{3\text{-}2}$ (*v*) $V_{1\text{-}2}$ (*vi*) $V_{1\text{-}3}$ **[–75 V, –50V, 125V, 175V, –25V, –200V]**

4. Fig. 1.91 shows the equivalent circuit of a digital-to-analog (D/A) converter. What is the value of the output voltage V_o ?

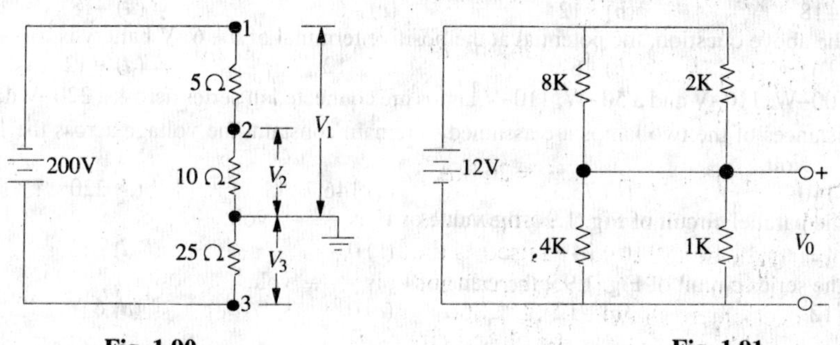

Fig. 1.90 **Fig. 1.91**

Objective Tests–1

1. A good electric conductor is one that
(*a*) has low conductance (*b*) is always made of copper wire
(*c*) produces a minimum voltage drop (*d*) has few free electrons

2. Two wires *A* and *B* have the same cross-section and are made of the same material. $R_A = 600\ \Omega$ and $R_B = 100\ \Omega$. The number of times *A* is longer than *B* is
(*a*) 6 (*b*) 2 (*c*) 4 (*d*) 5

3. A coil has a resistance of 100 Ω at 90°C. At 100°C, its resistance is 101 Ω. The temperature coefficient of the wire at 90°C is
(*a*) 0.01 (*b*) 0.1 (*c*) 0.0001 (*d*) 0.001

4. Which of the following material has nearly zero temperature-coefficient of resistance ?
(*a*) carbon (*b*) porcelain (*c*) copper (*d*) manganin

5. Which of the following material has a negative temperature coefficient of resistance ?
(*a*) brass (*b*) copper (*c*) aluminium (*d*) carbon

6. A cylindrical wire, 1 m in length, has a resistance of 100 Ω. What would be the resistance of a wire made from the same material if both the length and the cross-sectional area are doubled ?
(*a*) 200 Ω (*b*) 400 Ω (*c*) 100 Ω (*d*) 50 Ω

7. Carbon composition resistors are most popular because they
(*a*) cost the least (*b*) are smaller
(*c*) can withstand overloads (*d*) do not produce electric noise

8. A unique feature of a wire-wound resistor is its
(*a*) low power rating (*b*) low cost
(*c*) high stability (*d*) small size

9. If, in Fig. 1.92, resistor R_2 becomes open-circuited, the reading of the voltmeter will become
(*a*) zero (*b*) 150 V (*c*) 50 V (*d*) 200 V

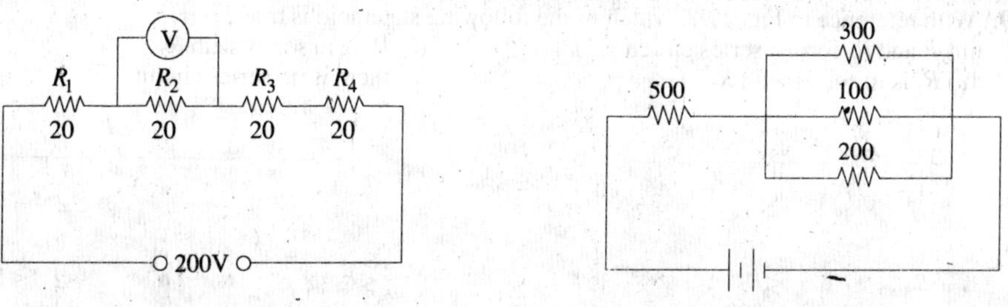

Fig. 1.92 **Fig. 1.93**

10. Whatever the battery voltage in Fig. 1.93, it is certain that smallest current will flow in the resistance of —— ohm.
(*a*) 300 (*b*) 500 (*c*) 200 (*d*) 100

11. Which of the following statement is TRUE both for a series and a parallel d.c circuit ?
(*a*) powers are additive (*b*) voltages are additive
(*c*) currents are additive (*d*) elements have individual currents

12. The positive terminal of a 6-V battery is connected to the negative terminal of a 12-V battery whose positive terminal is grounded. The potential at the negative terminal of the 6-V battery is – volt.
(*a*) +18 (*b*) –12 (*c*) –6 (*d*) –18

13. In the above question, the potential at the positive terminal of the 6–V battery is —— volt.
(*a*) +6 (*b*) –6 (*c*) –12 (*d*) + 12

14. A 100–W, 110–V and a 50–W, 110–V lamps are connected in series across a 220–V dc source. If the resistances of the two lamps are assumed to remain constant, the voltage across the 100–W lamp is —— volt.
(*a*) 110 (*b*) 73.3 (*c*) 146.7 (*d*) 220

15. In the parallel circuit of Fig. 1.94, the value of V_0 is —— volt.
(*a*) 12 (*b*) 24 (*c*) 0 (*d*) –12

16. In the series circuit of Fig. 1.95, the value of V_o is —— volt.
(*a*) 12 (*b*) –12 (*c*) 0 (*d*) 6

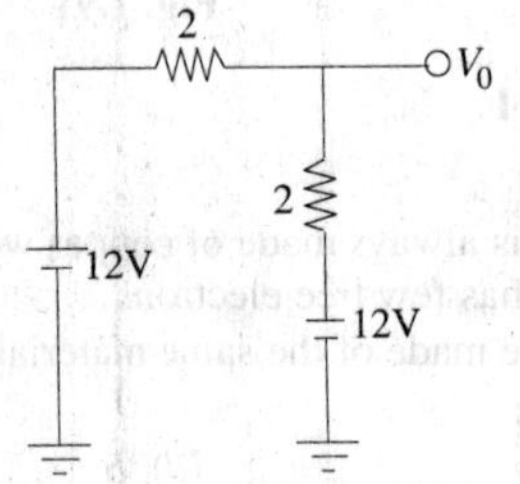

Fig. 1.94 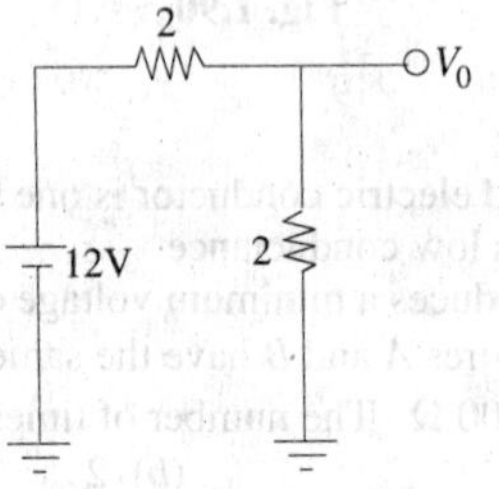 **Fig. 1.95**

17. In Fig. 1.96, there is a drop of 20 *V* on each resistor. The potential of point *A* would be —— volt.
(*a*) + 80 (*b*) –40 (*c*) + 40 (*d*) –80

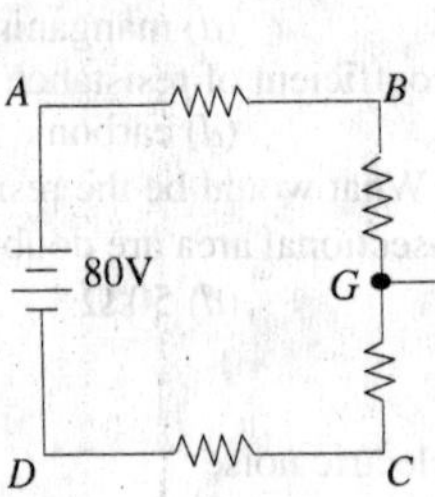

Fig. 1.96 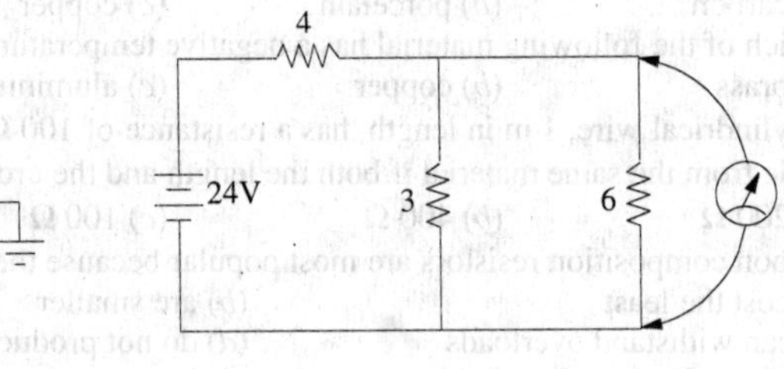 **Fig. 1.97**

18. From the voltmeter reading of Fig. 1.97, it is obvious that
(*a*) 3 Ω resistor is short-circuited (*b*) 6 Ω resistor is short-circuited
(*c*) nothing is wrong with the circuit (*d*) 3 Ω resistor is open-circuited

19. With reference to Fig. 1.98, which of the following statement is true ?
(*a*) *E* and R_1 form a series cirucit (*b*) R_1 is in series with R_3
(*c*) R_1 is in series with R_2 (*d*) there is no series circuit

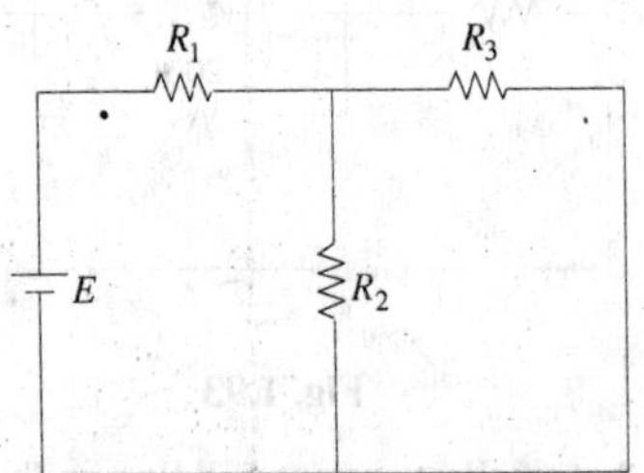

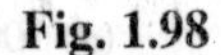
Fig. 1.98

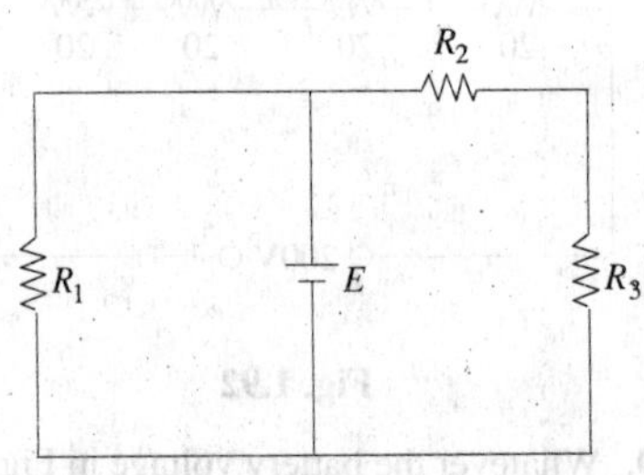

Fig. 1.99

20. Which of the following statements is correct concerning the circuit of Fig. 1.99.
(*a*) R_2 and R_3 form a series path
(*b*) E is in series with R_1
(*c*) R_1 is in parallel with R_3
(*d*) R_1, R_2 and R_3 form a series circuit.

21. What is the equivalent resistance in ohms between points A and B of Fig. 1.100 ? All resistances are in ohms.
(*a*) 12
(*b*) 14.4
(*c*) 22
(*d*) 2

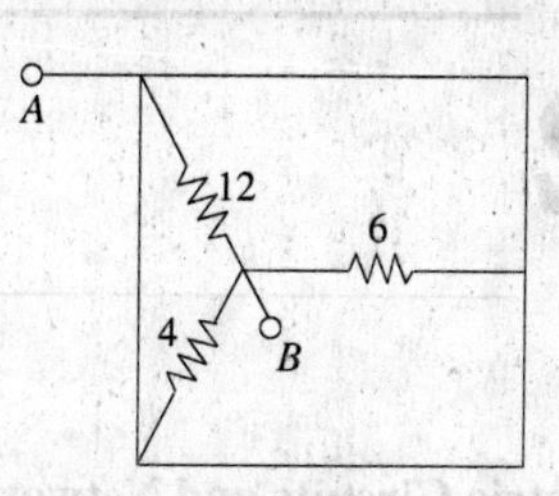

Fig. 1.100

Answers

1. *c*	2. *d*	3. *d*	4. *d*	5. *d*	6. *c*	7. *a*
8. *c*	9. *d*	10. *a*	11. *a*	12. *d*	13. *c*	14. *b*
15. *c*	16. *d*	17. *c*	18. *d*	19. *d*	20. *b*	21. *d*

2 DC NETWORK THEOREMS

2.1. Electric Circuits and Network Theorems

There are certain theorems, which when applied to the solutions of electric networks, either simplify the network itself or render their analytical solution very easy. These theorems can also be applied to an a.c. system, with the only difference that impedances replace the ohmic resistances of d.c. system. Different electric circuits (according to their properties) are defined below :

1. **Circuit**. A circuit is a closed conducting path through which an electric current either flows or is intended to flow.

2. **Parameters**. The various elements of an electric circuit are called its parameters like resistance, inductance and capacitance. These parameters may be *lumped* or *distributed.*

3. **Linear Circuit**. A linear circuit is one whose parameters are constant *i.e.* they do not change with voltage or current.

4. **Non-linear Circuit**. It is that circuit whose parameters change with voltage or current.

5. **Bilateral Circuit**. A bilateral circuit is one whose properties or characteristics are the same in either direction. The usual transmission line is bilateral, because it can be made to perform its function equally well in either direction.

6. **Unilateral Circuit**. It is that circuit whose properties or characteristics change with the direction of its operation. A diode rectifier is a unilateral circuit, because it cannot perform rectification in both directions.

7. **Electric Network**. A combination of various electric elements, connected in any manner whatsoever, is called an electric network.

8. **Passive Network** is one which contains no source of e.m.f. in it.

9. **Active Network** is one which contains one or more than one sources of e.m.f.

10. **Node** is a junction in a circuit where two or more circuit elements are connected together.

11. **Branch** is that part of a network which lies between two junctions.

12. **Loop**. It is a close path in a circuit in which no element or node is encountered more than once.

13. **Mesh**. It is a loop that contains no other loop within it. For example, the circuit of Fig. 2.1 (*a*) has seven branches, six nodes, three loops and two meshes whereas the circuit of Fig. 2.1 (*b*) has four branches, two nodes, six loops and three meshes.

It should be noted that, unless stated otherwise, an electric network would be assumed passive in the following treatment.

We will now discuss the various network theorems which are of great help in solving complicated networks. Incidentally, a network is said to be completely solved or analyzed when all voltages and all currents in its different elements are determined.

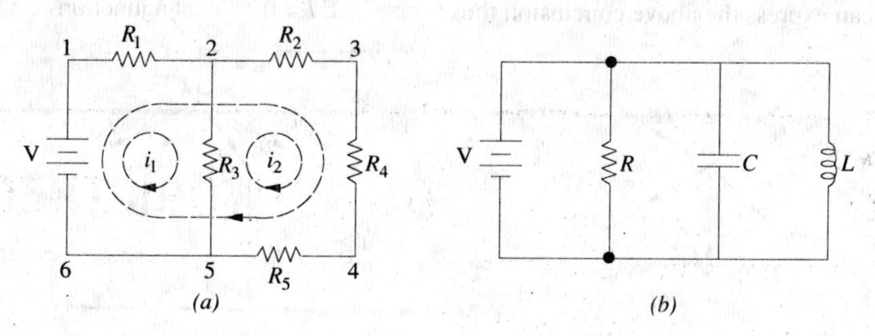

Fig. 2.1

There are two general approaches to network analysis :

(*i*) **Direct Method**

Here, the network is left in its original form while determining its different voltages and currents. Such methods are usually restricted to fairly simple circuits and include Kirchhoff's laws, Loop analysis, Nodal analysis, Superposition theorem, Compensation theorem and Reciprocity theorem etc.

(*ii*) **Network Reduction Method**

Here, the original network is converted into a much simpler equivalent circuit for rapid calculation of different quantities. This method can be applied to simple as well as complicated networks. Examples of this method are : Delta/Star and Star/Delta conversions, Thevenin's theorem and Norton's Theorem etc.

2.2. Kirchhoff's Laws*

These laws are more comprehensive than Ohm's law and are used for solving electrical networks which may not be readily solved by the latter. Kirchhoff's laws, two in number, are particularly useful (*a*) in determining the equivalent resistance of a complicated network of conductors and (*b*) for calculating the currents flowing in the various conductors. The two-laws are :

1. Kirchhoff's Point Law or Current Law (KCL)

It states as follows :

in any electrical network, the algebraic sum of the currents meeting at a point (or junction) is zero.

Put in another way, it simply means that the total current *leaving* a junction is equal to the total current *entering* that junction. It is obviously true because there is no accumulation of charge at the junction of the network.

Consider the case of a few conductors meeting at a point *A* as in Fig. 2.2 (*a*). Some conductors have currents leading to point *A*, whereas some have currents leading away from point *A*. Assuming the incoming currents to be positive and the outgoing currents negative, we have.

$$I_1 + (-I_2) + (-I_3) + (+I_4) + (-I_5) = 0$$

or $\quad I_1 + I_4 - I_2 - I_3 - I_5 = 0 \quad$ or $\quad I_1 + I_4 = I_2 + I_3 + I_5$

or $\quad$ **incoming currents = outgoing currents**

Similarly, in Fig. 2.2 (*b*) for node *A*

$$+I + (-I_1) + (-I_2) + (-I_3) + (-I_4) = 0 \quad \text{or} \quad I = I_1 + I_2 + I_3 + I_4$$

---

*After Gustav Robert Kirchhoff (1824-1887), an outstanding German physicist.

We can express the above conclusion thus ; $\Sigma I = 0$ ...at a junction

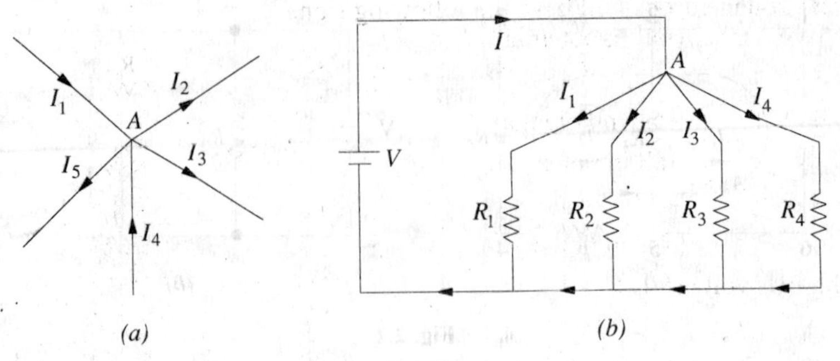

Fig. 2.2

2. Kirchhoff's Mesh Law or Voltage Law (KVL)

It states as follows :

the algebraic sum of the products of currents and resistances in each of the conductors in any closed path (or mesh) in a network plus the algebraic sum of the e.m.fs. in that path is zero.

In other words, $\Sigma\, IR + \Sigma\, e.m.f. = 0$ ...round a mesh

It should be noted that algebraic sum is the sum which takes into account the polarities of the voltage drops.

The basis of this law is this : If we start from a particular junction and go round the mesh till we come back to the starting point, then we must be at the same potential with which we started. Hence, it means that all the sources of e.m.f. met on the way must necessarily be equal to the voltage drops in the resistances, every voltage being given its proper sign, plus or minus.

2.3. Determination of Voltage Sign

In applying Kirchhoff's laws to specific problems, particular attention should be paid to the algebraic signs of voltage drops and e.m.fs., otherwise results will come out to be wrong. Following sign convention is suggested :

(*a*) **Sign of Battery E.M.F.**

A *rise* in voltage should be given a +ve sign and a *fall* in voltage a –ve sign. Keeping this in mind, it is clear that as we go from the –ve terminal of a battery to its +ve terminal (Fig. 2.3), there is a *rise* in potential, hence this voltage should be given a +ve sign. If, on the other hand, we go from +ve terminal to –ve terminal, then there is a *fall* in potential, hence this voltage should be preceded

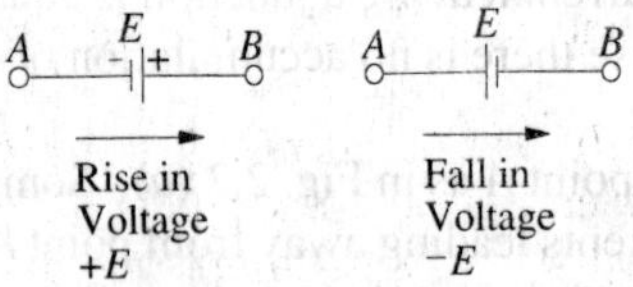

A E + B — Rise in Voltage $+E$ | A E B — Fall in Voltage $-E$

Fig. 2.3

current — A + V – B, R, motion — Fall in Voltage $-V = -IR$ | current — A – V + B, motion — Rise in Voltage $+V = +IR$

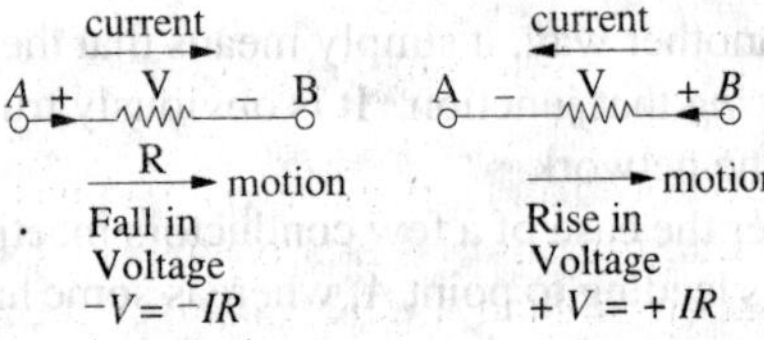

Fig. 2.4

by a –ve sign. *It is important to note that the sign of the battery e.m.f. is independent of the direction of the current through that branch.*

(*b*) **Sign of IR Drop**

Now, take the case of a resistor (Fig. 2.4). If we go through a resistor in the *same* direction as the current, then there is a fall in potential because current flows from a higher to a lower potential. Hence, this voltage fall should be taken –ve. However, if we go in a direction opposite to that of the current, then there is a *rise* in voltage. Hence, this voltage rise should be given a positive sign.

It is clear that the sign of voltage drop across a resistor depends on the direction of current through that resistor but is independent of the polarity of any other source of e.m.f. in the circuit under consideration.

Consider the closed path *ABCDA* in Fig. 2.5. As we travel around the mesh in the clockwise direction, different voltage drops will have the following signs :

I_1R_1 is –ve (fall in potential)

I_2R_2 is –ve (" " ")

I_3R_3 is +ve (rise in potential)

I_4R_4 is –ve (fall in potential)

E_2 is –ve " " ")

E_1 is +ve (rise in potential)

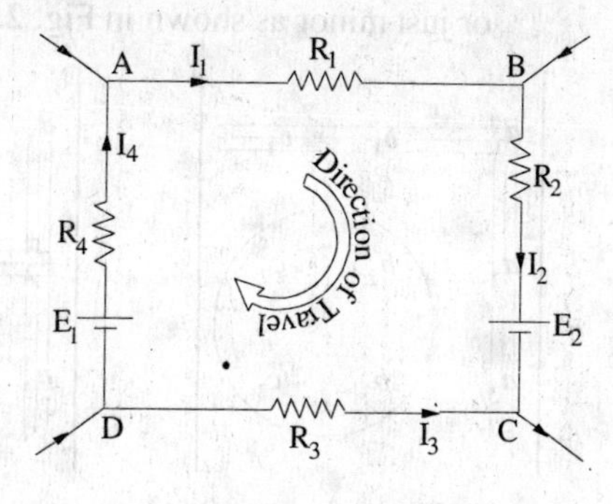

Fig. 2.5

Using Kirchhoff's voltage law, we get

$$-I_1R_1 - I_2R_2 - I_3R_3 - I_4R_4 - E_2 + E_1 = 0$$

or $$I_1R_1 + I_2R_2 - I_3R_3 + I_4R_4 = E_1 - E_2$$

2.4. Assumed Direction of Current

In applying Kirchhoff's laws to electrical networks, the question of assuming proper direction of current usually arises. The direction of current flow may be assumed either clockwise or anti-clockwise. If the assumed direction of current is not the actual direction, then on solving the question, this current will be found to have a minus sign. If the answer is positive, then assumed direction is the same as actual direction (Example 2.10). *However, the important point is that once a particular direction has been assumed, the same should be used throughout the solution of the question.*

Note. It should be noted that Kirchhoff's laws are applicable both to d.c. and a.c. voltages and currents. However, in the case of alternating currents and voltages, any e.m.f. of self-inductance or that existing across a capacitor should be also taken into account (See Example 2.14).

2.5. Solving Simultaneous Equations

Electric circuit analysis with the help of Kirchhoff's laws usually involves solution of two or three simultaneous equations. These equations can be solved by a systematic elimination of the variables but the procedure is often lengthy and laborious and hence more liable to error. Determinants and Cramer's rule provide a simple and straight method for solving network equations through manipulation of their coefficients. Of course, if the number of simultaneous equations happens to be very large, use of a digital computer can make the task easy.

2.6. Determinants

The symbol $\begin{vmatrix} a & b \\ c & d \end{vmatrix}$ is called a determinant of the second order (or 2 × 2 determinant) because it contains two rows (*ab* and *cd*) and two columns (*ac* and *bd*). The numbers *a*, *b*, *c* and *d* are called the elements or constituents of the determinant. Their number in the present case is = $2^2 = 4$.

The evaluation of such a determinant is accomplished by cross-multiplication as illustrated below :

$$\Delta = \begin{vmatrix} a & b \\ c & d \end{vmatrix} = ad - bc$$

The above result for a second order determinant can be remembered as

upper left times lower right **minus** *upper right times lower left*

The symbol $\begin{vmatrix} a_1 & b_1 & c_1 \\ a_2 & b_2 & c_2 \\ a_3 & b_3 & c_3 \end{vmatrix}$ represents a third-order determinant having $3^2 = 9$ elements. It may be evaluated (or expanded) as under :

1. multiply each element of the first row (or alternatively, first column) by the determinant obtained by omitting the row and column in which it occurs. (It is called minor determinant or just minor as shown in Fig. 2.6).

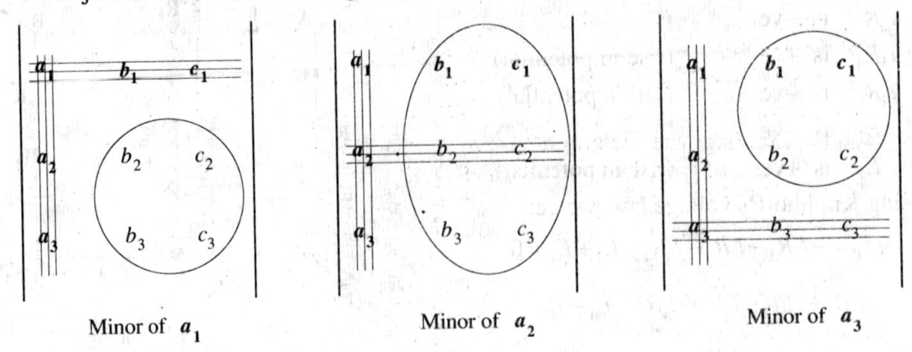

Fig. 2.6

2. prefix + and – sign alternately to the terms so obtained.
3. Add up all these terms together to get the value of the given determinant.

Considering the first column, minors of various elements are as shown in Fig. 2.6.
Expanding in terms of first column, we get

$$\Delta = a_1\begin{vmatrix} b_2 & c_2 \\ b_3 & c_3 \end{vmatrix} - a_2\begin{vmatrix} b_1 & c_1 \\ b_3 & c_3 \end{vmatrix} + a_3\begin{vmatrix} b_1 & c_1 \\ b_2 & c_2 \end{vmatrix}$$

$$= a_1(b_2c_3 - b_3c_2) - a_2(b_1c_3 - b_3c_1) + a_3(b_1c_2 - b_2c_1) \quad \ldots(i)$$

Expanding in terms of the first row, we get

$$\Delta = a_1\begin{vmatrix} b_2 & c_2 \\ b_3 & c_3 \end{vmatrix} - b_1\begin{vmatrix} a_2 & c_2 \\ a_3 & c_3 \end{vmatrix} + c_1\begin{vmatrix} a_2 & b_2 \\ a_3 & b_3 \end{vmatrix}$$

$$= a_1(b_2c_3 - b_3c_2) - b_1(a_2c_3 - a_3c_2) + c_1(a_2b_3 - a_3b_2)$$

which will be found to be the same as above.

Example 2.1. *Evaluate the determinant* $\begin{vmatrix} 7 & -3 & -4 \\ -3 & 6 & -2 \\ -4 & -2 & 11 \end{vmatrix}$

Solution. We will expand with the help of 1st column.

$$\Delta = 7\begin{vmatrix} 6 & -2 \\ -2 & 11 \end{vmatrix} - (-3)\begin{vmatrix} -3 & -4 \\ -2 & 11 \end{vmatrix} + (-4)\begin{vmatrix} -3 & -4 \\ 6 & -2 \end{vmatrix}$$

$$= 7[(6 \times 11) - (-2 \times -2)] + 3[(-3 \times 11) - (-4 \times -2)] - 4[(-3 \times -2) - (-4 \times 6)]$$

$$= 7(66 - 4) + 3(-33 - 8) - 4(6 + 24) = \mathbf{191}$$

2.7. Solving Equations with Two Unknowns

Suppose the two given simultaneous equations are

$$ax + by = c$$
$$dx + ey = f$$

Here, the two unknowns are x and y, a, b, d and e are coefficients of these unknowns whereas c and f are constants. The procedure for solving these equations by the method of determinants is as follows :

1. Write the two equations in the matrix form as $\begin{bmatrix} a & b \\ d & e \end{bmatrix}\begin{bmatrix} x \\ y \end{bmatrix} = \begin{bmatrix} c \\ f \end{bmatrix}$

2. The *common* determinant is given as $\Delta = \begin{bmatrix} a & b \\ d & e \end{bmatrix} = ae - bd$

3. For finding the determinant for x, replace the coefficients of x in the original matrix by the constants so that we get determinant Δ_1 given by $\Delta_1 = \begin{vmatrix} c & b \\ f & e \end{vmatrix}$ $ce - bf)$

4. For finding the determinant for y, replace coefficients of y by the constants so that we get $\Delta_2 = \begin{vmatrix} a & c \\ d & f \end{vmatrix} = (af - cd)$

5. Apply Cramer's rule to get the values of x and y

$$x = \frac{\Delta_1}{\Delta} = \frac{ce - bf}{ae - bd} \quad \text{and} \quad y = \frac{\Delta_2}{\Delta} = \frac{af - cd}{ae - bd}$$

Example 2.2. *Solve the following two simultaneous equations by the method of determinants :*

$$4i_1 - 3i_2 = 1$$

$$3i_1 - 5i_2 = 2$$

Solution. The matrix form of the equations is $\begin{bmatrix} 4 & -3 \\ 3 & -5 \end{bmatrix} \begin{bmatrix} i_1 \\ i_2 \end{bmatrix} = \begin{bmatrix} 1 \\ 2 \end{bmatrix}$

$$\Delta = \begin{vmatrix} 4 & -3 \\ 3 & -5 \end{vmatrix} = (4 \times -5) - (-3 \times 3) = -11$$

$$\Delta_1 = \begin{vmatrix} 1 & -3 \\ 2 & -5 \end{vmatrix} = (1 \times -5) - (-3 \times 2) = 1$$

$$\Delta_2 = \begin{vmatrix} 4 & 1 \\ 3 & 2 \end{vmatrix} = (4 \times 2) - (1 \times 3) = 5$$

$$\therefore \quad i_1 = \frac{\Delta_1}{\Delta} = \frac{1}{-11} = -\frac{1}{11}; \quad i_2 = \frac{\Delta_2}{\Delta} = -\frac{5}{11}$$

2.8. Solving Equations With Three Unknowns

Let the three simultaneous equations be as under :

$$ax + by + cz = d$$
$$ex + fy + gz = h$$
$$jx + ky + lz = m$$

The above equations can be put in the matrix form as under :

$$\begin{bmatrix} a & b & c \\ e & f & g \\ j & k & l \end{bmatrix} \begin{bmatrix} x \\ y \\ z \end{bmatrix} = \begin{bmatrix} d \\ h \\ m \end{bmatrix}$$

The value of common determinant is given by

$$\Delta = \begin{vmatrix} a & b & c \\ e & f & g \\ j & k & l \end{vmatrix} = a(fl - gk) - e(bl - ck) + j(bg - cf)$$

The determinant for x can be found by replacing coefficients of x in the original matrix by the constants.

$$\therefore \quad \Delta_1 = \begin{vmatrix} d & b & c \\ h & f & g \\ m & k & l \end{vmatrix} = d(fl - gk) - h(bl - ck) + m(bg - cf)$$

Similarly, determinant for y is given by replacing coefficients of y with the three constants.

$$\Delta_2 = \begin{vmatrix} a & d & c \\ e & h & g \\ j & m & l \end{vmatrix} = a(hl - mg) - e(dl - mc) + j(dg - hc)$$

In the same way, determinant for z is given by

$$\Delta_3 = \begin{vmatrix} a & b & d \\ e & f & h \\ j & k & m \end{vmatrix} = a(fm - hk) - e(bm - dk) + j(bh - df)$$

As per Cramer's rule $x = \dfrac{\Delta_1}{\Delta}, \quad y = \dfrac{\Delta_2}{\Delta}, \quad z = \dfrac{\Delta_3}{\Delta}$

Example 2.3. *Solve the following three simultaneous equations by the use of determinants and Cramer's rule*

$$i_1 + 3i_2 + 4i_3 = 14$$

$$i_1 + 2i_2 + i_3 = 7$$

$$2i_1 + i_2 + 2i_3 = 2$$

Solution. As explained earlier, the above equations can be written in the form

$$\begin{bmatrix} 1 & 3 & 4 \\ 1 & 2 & 1 \\ 2 & 1 & 2 \end{bmatrix} \begin{bmatrix} i_i \\ i_2 \\ i_3 \end{bmatrix} = \begin{bmatrix} 14 \\ 7 \\ 2 \end{bmatrix}$$

$$\Delta = \begin{bmatrix} 1 & 3 & 4 \\ 1 & 2 & 1 \\ 2 & 1 & 2 \end{bmatrix} = 1(4-1) - 1(6-4) + (3-8) = -9$$

$$\Delta_1 = \begin{bmatrix} 14 & 3 & 4 \\ 7 & 2 & 1 \\ 2 & 1 & 2 \end{bmatrix} = 14(4-1) - 7(6-4) + 2(3-8) = 18$$

$$\Delta_2 = \begin{bmatrix} 1 & 14 & 4 \\ 1 & 7 & 1 \\ 2 & 2 & 2 \end{bmatrix} = 1(14-2) - 1(28-8) + 2(14-28) = -36$$

$$\Delta_3 = \begin{bmatrix} 1 & 3 & 14 \\ 1 & 2 & 7 \\ 2 & 1 & 2 \end{bmatrix} = 1(4-7) - 1(6-14) + 2(21-28) = -9$$

According to Cramer's rule,

$$i_1 = \frac{\Delta_1}{\Delta} = \frac{18}{-9} = -\mathbf{2A}; \; i_2 = \frac{\Delta_2}{\Delta} = \frac{-36}{-9} = \mathbf{4A}\,; \; i_3 = \frac{\Delta_3}{\Delta} = \frac{-9}{-9} = \mathbf{1A}$$

Example 2.4. *What is the voltage V_S across the open switch in the circuit of Fig. 2.7 ?*

Solution. We will apply *KVL* to find V_S. Starting from point *A* in the clockwise direction and using the sign convention given in Art. 2.3, we have

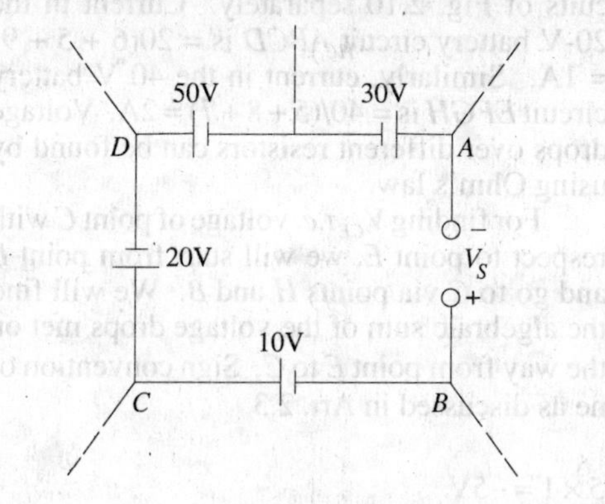

Fig. 2.7

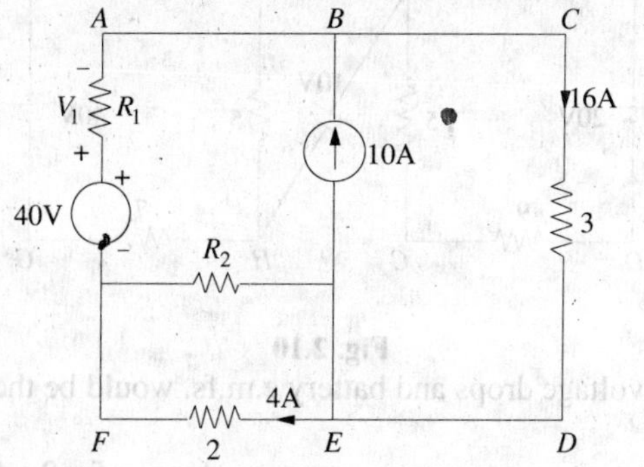

Fig. 2.8

$$+V_S + 10 - 20 - 50 + 30 = 0 \therefore V_S = \mathbf{30\ V}$$

Example 2.5. *Find the unknown voltage V_1 in the circuit of Fig. 2.8.*

Solution. Initially, one may not be clear regarding the solution of this question. One may think of Kirchhoff's laws or mesh analysis etc. But a little thought will show that the question can be solved by the simple application of Kirchhoff's voltage law. Taking the outer closed loop *ABCDEFA* and applying KVL to it, we get

$$-16 \times 3 - 4 \times 2 + 40 - V_1 = 0\ ; \quad \therefore \quad V_1 = \mathbf{-16\ V}$$

The negative sign shows there is a fall in potential.

Example 2.6. *Using Kirchoff's Current Law and Ohm's Law, find the magnitude and polarity of voltage V in Fig. 2.9 (a). Directions of the two current sources are as shown.*

Solution. Let us arbitrarily choose the directions of I_1, I_2 and I_3 and polarity of *V* as shown in Fig. 2.9 (*b*). We will use the sign convention for currents as given in Art. 2.3. Applying KCL to node *A*, we have

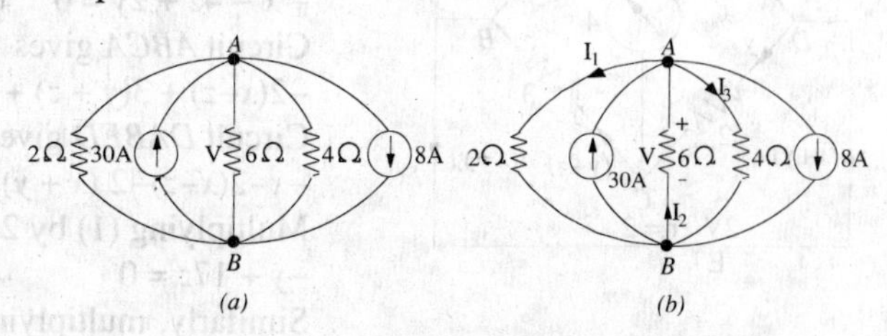

Fig. 2.9

$$-I_1 + 30 + I_2 - I_3 - 8 = 0$$

or $$I_1 - I_2 + I_3 = 22 \qquad \ldots(i)$$

Applying Ohm's law to the three resistive branches in Fig. 2.9(*b*), we have

$$I_1 = \frac{V}{2},\ I_3 = \frac{V}{4},\ I_2 = -\frac{V}{6} \qquad \text{(please note the } - \text{ve sign.)}$$

Substituting these values in (*i*) above, we get

$$\frac{V}{2} - \left(\frac{-V}{6}\right) + \frac{V}{4} = 22 \quad \text{or} \quad V = \mathbf{24\ V}$$

$$\therefore \qquad I_1 = V/2 = 24/2 = 12\text{A},\ I_2 = -24/6 = -4\text{A},\ I_3 = 24/4 = 6\text{A}$$

The negative sign of I_2 indicates that actual direction of its flow is opposite to that shown in Fig. 2.9(*b*). Actually, I_2 flows from *A* to *B* and not from *B* to *A* as shown.

Incidentally, it may be noted that all currents are outgoing except 30A which is an incoming current.

Example 2.7. *For the circuit shown in Fig. 2.10, find V_{CE} and V_{AG}.*

(F.Y. Engg. Pune Univ. May 1988)

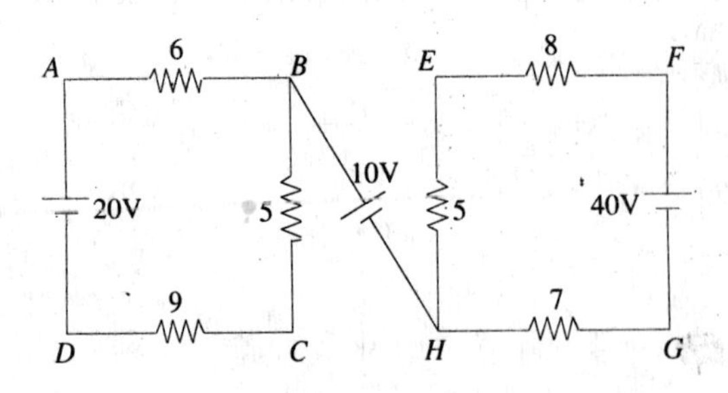

Fig. 2.10

Solution. Consider the two battery circuits of Fig. 2.10 separately. Current in the 20-V battery circuit *ABCD* is = 20(6 + 5 + 9) = 1A. Similarly, current in the 40 V battery circuit *EFGH* is = 40/(5 + 8 + 7) = 2A. Voltage drops over different resistors can be found by using Ohm's law.

For finding V_{CE} *i.e.* voltage of point *C* with respect to point *E*, we will start from point *E* and go to *C* via points *H* and *B*. We will find the algebraic sum of the voltage drops met on the way from point *E* to *C*. Sign convention of the voltage drops and battery e.m.fs. would be the same as discussed in Art. 2.3.

$$\therefore \qquad V_{CE} = -5 \times 2 + 10 - 5 \times 1 = -5\text{V}$$

The negative sign shows that point *C* is negative with respect to point *E*.

$$V_{AG} = 7 \times 2 + 10 + 6 \times 1 = 30\text{V}.$$

The positive sign shows that point *A* is at a positive potential of 30 V with respect to point *G*.

Example 2.8. *Determine the currents in the unbalanced bridge circuit of Fig. 2.11 below. Also, determine the p.d. across BD and the resistance from B to D.*

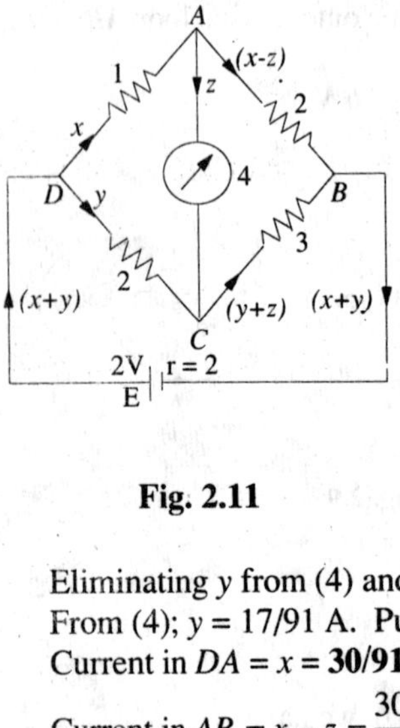

Fig. 2.11

Solution. Assumed current directions are as shown in Fig. 2.11.

Applying Kirchhoff's Second Law to circuit *DACD*, we get

$-x - 4z + 2y = 0$ or $x - 2y + 4z = 0$ ...(1)

Circuit *ABCA* gives

$-2(x-z) + 3(y + z) + 4z = 0$ or $2x - 3y - 9z = 0$ ...(2)

Circuit *DABED* gives

$-x - 2(x-z) - 2(x + y) + 2 = 0$ or $5x + 2y - 2z = 2$ ...(3)

Multiplying (1) by 2 and subtracting (2) from it, we get

$-y + 17z = 0$ ...(4)

Similarly, multiplying (1) by 5 and subtracting (3) from it, we have

$-12y + 22z = -2$ or $-6y + 11z = -1$ ...(5)

Eliminating *y* from (4) and (5), we have $91z = 1$ or $z = 1/91$ A

From (4); $y = 17/91$ A. Putting these values of *y* and *z* in (1), we get $x = 30/91$ A

Current in $DA = x =$ **30/91 A** Current in $DC = y =$ **17/91 A**

$$\text{Current in } AB = x - z = \frac{30}{91} - \frac{1}{91} = \mathbf{\frac{29}{91}\ A}$$

$$\text{Current in } CB = y + z = \frac{17}{91} + \frac{1}{91} = \mathbf{\frac{18}{91}\ A}$$

$$\text{Current in external circuit } = x + y = \frac{30}{91} + \frac{17}{91} = \mathbf{\frac{47}{91}\ A}$$

$$\text{Current in } AC = z = \mathbf{1/91\ A}$$

Internal voltage drop in the cell = $2(x + y) = 2 \times 47/91 =$ **94/91 V**

$$\therefore \text{ P.D. across points } D \text{ and } B = 2 - \frac{94}{91} = \mathbf{\frac{88}{91}\ V}^*$$

P.D. between D and B = drop across DC + drop across CB =* $2 \times 17/91 + 3 \times 18/91 =$ **88/91 V

Equivalent resistance of the bridge between points D and B

$$= \frac{\text{p.d. between points } B \text{ and } D}{\text{current between points } B \text{ and } D} = \frac{88/91}{47/91} = \frac{88}{47} = \mathbf{1.87\ \Omega\ (approx)}$$

Solution By Determinants

The matrix form of the three simultaneous equations (1), (2) and (3) is

$$\begin{bmatrix} 1 & -2 & 4 \\ 2 & -3 & -9 \\ 5 & 2 & -2 \end{bmatrix} \begin{bmatrix} x \\ y \\ z \end{bmatrix} = \begin{bmatrix} 0 \\ 0 \\ 2 \end{bmatrix}$$

$$\Delta_1 = \begin{bmatrix} 1 & -2 & 4 \\ 2 & -3 & -9 \\ 5 & 2 & -2 \end{bmatrix} = 1(6+18) - 2(4-8) + 5(18+12) = 182$$

$$\Delta_1 = \begin{bmatrix} 0 & -2 & 4 \\ 0 & -3 & -9 \\ 2 & 2 & -2 \end{bmatrix} = 0(6+18) - 0(4-8) + 2(18+12) = 60$$

$$\Delta_2 = \begin{bmatrix} 1 & 0 & 4 \\ 2 & 0 & -9 \\ 5 & 2 & -2 \end{bmatrix} = 34, \Delta_3 = \begin{bmatrix} 1 & -2 & 0 \\ 2 & -3 & 0 \\ 5 & 2 & 2 \end{bmatrix} = 2$$

$$\therefore \quad x = \frac{\Delta_1}{\Delta} = \frac{60}{182} = \mathbf{\frac{30}{91}}\ \text{A},\ y = \frac{34}{182} = \mathbf{\frac{17}{91}}\ \text{A},\ z = \frac{2}{182} = \mathbf{\frac{1}{192}}\ \text{A}$$

Example 2.9. *Determine the branch currents in the network of Fig. 2.12 when the value of each branch resistance is one ohm.* **(Elect. Technology. Allahabad Univ. 1992)**

Solution. Let the current directions be as shown in Fig. 2.12.

Apply Kirchhoff's Second law to the closed circuit $ABDA$, we get

$$5 - x - z + y = 0 \text{ or } x - y + z = 5 \qquad \ldots(i)$$

Similarly, circuit $BCDB$ gives

$$-(x-z) + 5 + (y+z) + z = 0$$

$$\text{or } x - y - 3z = 5 \qquad \ldots(ii)$$

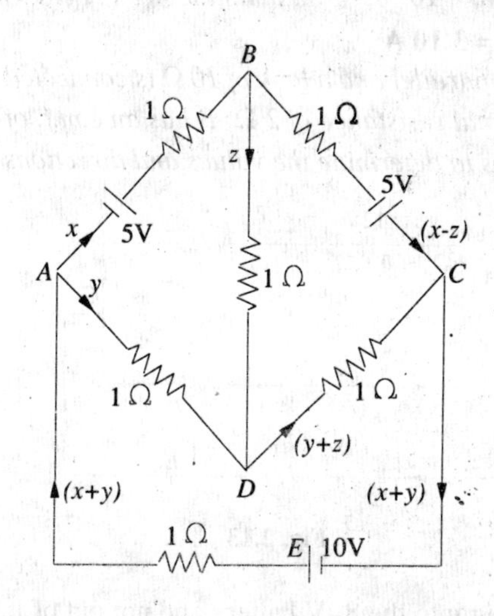

Fig. 2.12

Lastly, from circuit $ADCEA$, we get

$$-y - (y+z) + 10 - (x+y) = 0$$

$$\text{or } x + 3y + z = 10 \qquad \ldots(iii)$$

From Eq. (*i*) and (*ii*), we get, $z = 0$

Substituting $z = 0$ either in Eq. (*i*) or (*ii*) and in Eq. (*iii*), we get

$$x - y = 5 \qquad \ldots(iv)$$

$$x + 3y = 10 \qquad \ldots(v)$$

Subtracting Eq. (*v*) from (*iv*), we get

$-4y = -5$ or $y = 5/4 = \mathbf{1.25\ A}$

Eq. (*iv*) gives $x = 25/4$ A $= 6.25$ A

Current in branch AB = current in branch BC = **6.25 A**

Current in branch BD = 0; current in branch AD = current in branch DC = **1.25 A** ; current in branch CEA = 6.25 + 1.25 = **7.5 A**

Example 2.10. *State and explain Kirchhoff's laws. Determine the current supplied by the battery in the circuit shown in Fig. 2.12A.*

(Elect. Engg.-I, Bombay Univ. 1987).

Solution. Let the current distribution be as shown in the figure. Considering the close circuit *ABCA* and applying Kirchhoff's Second Law, we have

$$-100x - 300z + 500y = 0$$

or $$x - 5y + 3z = 0 \quad ...(i)$$

Similarly, considering the closed loop *BCDB*, we have

$$-300z - 100(y+z) + 500(x-z) = 0$$

or $$5x - y - 9z = 0 \quad ...(ii)$$

Taking the circuit *ABDEA*, we get

$$-100x - 500(x-z) + 100 - 100(x+y) = 0$$

or $$7x + y - 5z = 1 \quad ...(iii)$$

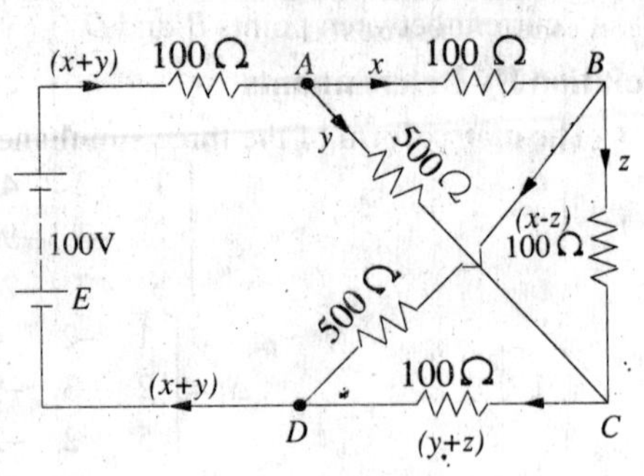

Fig. 2.12A

The value of *x*, *y* and *z* may be found by solving the above three simultaneous equations or by the method of determinants as given below :

Putting the above three equations in the matrix form, we have

$$\begin{bmatrix} 1 & -5 & 3 \\ 5 & -1 & -9 \\ 7 & 1 & -5 \end{bmatrix} \begin{bmatrix} x \\ y \\ z \end{bmatrix} = \begin{bmatrix} 0 \\ 0 \\ 1 \end{bmatrix}$$

$$\Delta = \begin{bmatrix} 1 & -5 & 3 \\ 5 & -1 & -9 \\ 7 & 1 & -5 \end{bmatrix} = 240, \Delta_1 = \begin{bmatrix} 0 & -5 & 3 \\ 0 & -1 & -9 \\ 1 & 1 & -5 \end{bmatrix} = 48$$

$$\Delta_2 = \begin{bmatrix} 1 & 0 & 3 \\ 5 & 0 & -9 \\ 7 & 1 & -5 \end{bmatrix} = 24, \quad \Delta_3 = \begin{bmatrix} 1 & -5 & 0 \\ 5 & -1 & 0 \\ 7 & 1 & 1 \end{bmatrix} = 24$$

$$\therefore \quad x = \frac{48}{240} = \frac{\mathbf{1}}{\mathbf{5}}\mathbf{A}; y = \frac{24}{240} = \frac{\mathbf{1}}{\mathbf{10}}\mathbf{A}; z = \frac{24}{240} = \frac{\mathbf{1}}{\mathbf{10}}\mathbf{A}$$

Current supplied by the battery is $= x + y = 1/5 + 1/10 =$ **3/10 A**

Example 2.11. *Two batteries A and B are connected in parallel and a load of 10 Ω is connected across their terminals. A has an e.m.f. of 12 V and an internal resistance of 2 Ω; B has an e.m.f. of 8 V and an internal resistance of 1 Ω. Use Kirchhoff's laws to determine the values and directions of the currents flowing in each of the batteries and in the external resistance. Also determine the potential difference across the external resistance.*

(F.Y. Engg. Pune Univ. May 1989)

Solution. Applying KVL to the closed circuit *ABCDA* of Fig. 2.13, we get

$-12 + 2x - 1y + 8 = 0$ or $2x - y = 4 \quad ...(i)$

Similarly, from the closed circuit *ADCEA*, we get

$-8 + 1y + 10(x+y) = 0$ or $10x + 11y = 8 \quad ...(ii)$

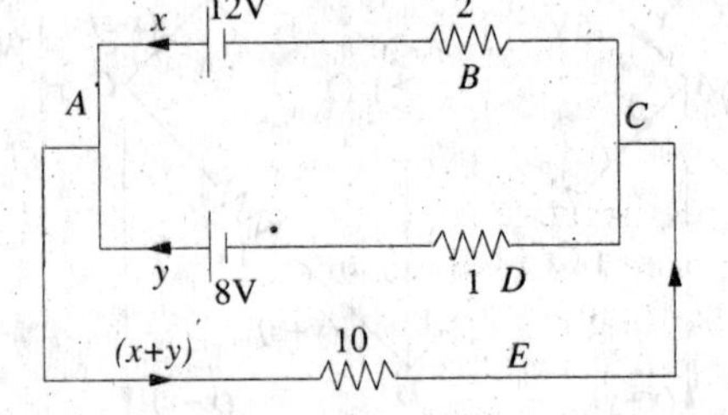

Fig. 2.13

From Eq. (*i*) and (*ii*), we get

$x =$ **1.625 A** and $y = -$ **0.75 A**

The negative sign of *y* shows that the current is flowing into the 8–V battery and not out of it. In other words, it is a charging current and not a discharging current.

Current flowing in the external resistance $= x + y = 1.625 - 0.75 =$ **0.875 A**

P.D. across the external resistance $= 10 \times 0.875 =$ **8.75 V**

Note. To confirm the correctness of the answer, the simple check is to find the value of the external voltage available across point *A* and *C* with the help of the two parallel branches. If the value of the voltage comes out to be the same, then the answer is correct, otherwise it is wrong. For example, $V_{CBA} = -2 \times 1.625 + 12 = 8.75$ V. From the second branch $V_{CDA} = 1 \times 0.75 + 8 = 8.75$ V. Hence, the answer found above is correct.

Example 2.12. *Determine the current x in the 4-Ω resistance of the circuit shown in Fig. 2.13(A).*

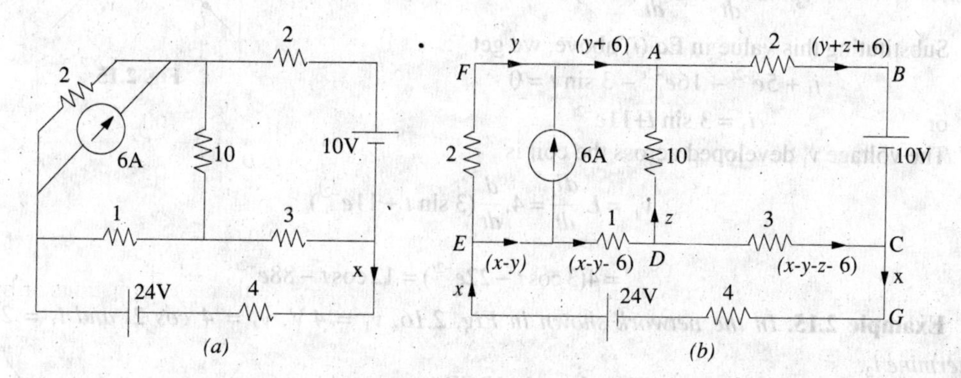

Fig. 2.13A

Solution. The given circuit is redrawn with assumed distribution of currents in Fig. 2.13A(*b*). Applying KVL to different closed loops, we get

Circuit EFADE

$$-2y + 10z + (x - y - 6) = 0 \quad \text{or} \quad x - 3y + 10z = 6 \qquad \ldots(i)$$

Circuit ABCDA

$$2(y + z + 6) - 10 + 3(x - y - z - 6) - 10z = 0 \quad \text{or} \quad 3x - 5y - 15z = 40 \qquad \ldots(ii)$$

Circuit EDCGE

$$-(x - y - 6) - 3(x - y - z - 6) - 4x + 24 = 0 \quad \text{or} \quad 8x - 4y - 3z = 48 \qquad \ldots(iii)$$

From above equations we get $x =$ **4.1 A**

Example 2.13. *Applying Kirchhoff's laws to different loops in Fig. 2.14, find the values of V_1 and V_2.*

Solution. Starting from point *A* and applying Kirchhoff's voltage law to loop No.3, we get

$- V_3 + 5 = 0$ or $V_3 = 5V$

Starting from point *A* and applying Kirchhoff's voltage law to loop No. 1, we get

$10 - 30 - V_1 + 5 = 0$ or $V_1 = -$ **15V**

The negative sign of V_1 denotes that its polarity is opposite to that shown in the figure.

Starting from point *B* in loop No.3, we get

$-(-15) - V_2 + (-15) = 0$ or $V_2 = \mathbf{0}$

Example 2.14. *In the network of Fig. 2.15, the different currents and voltages are as under :*

$i_2 = 5e^{-2t}$, $i_4 = 3 \sin t$ and $v_3 = 4e^{-2t}$

Using KCL, find voltage v_1.

Solution. According to *KCL*, the algebraic sum of the currents meeting at junction *A* is zero *i.e.*

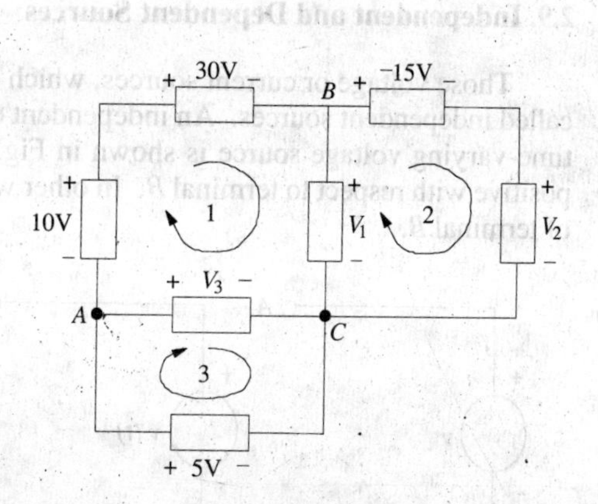

Fig. 2.14

$$i_1 + i_2 + i_3 + (-i_4) = 0$$

$$i_1 + i_2 + i_3 - i_4 = 0 \qquad \ldots(i)$$

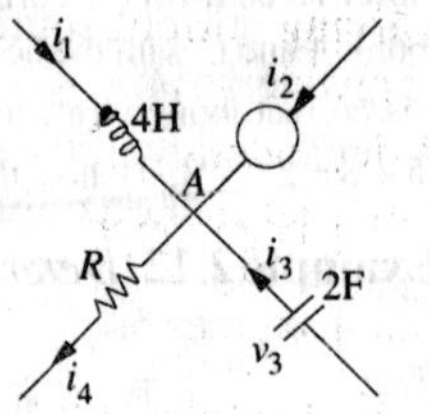

Now, current through a capacitor is given by $i = C\, dv/dt$

$$\therefore \qquad i_3 = C\frac{dv_3}{dt} = 2.\frac{d}{dt}(4e^{-2t}) = -16e^{-2t}$$

Substituting this value in Eq (i) above, we get

$$i_1 + 5e^{-2t} - 16e^{-2t} - 3\sin t = 0$$

or $$i_1 = 3\sin t + 11e^{-2t}$$

Fig. 2.15

The voltage v_1 developed across the coil is

$$v_1 = L\frac{di_i}{dt} = 4.\frac{d}{dt}(3\sin t + 11e^{-2t})$$

$$= 4(3\cos t - 22e^{-2t}) = 12\cos t - 88e^{-2t}$$

Example 2.15. *In the network shown in Fig. 2.16,* $v_1 = 4$ *V,* $v_4 = 4\cos 2t$ *and* $i_3 = 2e^{-t/3}$ *Determine* i_2.

Solution. Applying *KVL* to closed mesh *ABCDA*, we get

$$-v_1 - v_2 + v_3 + v_4 = 0$$

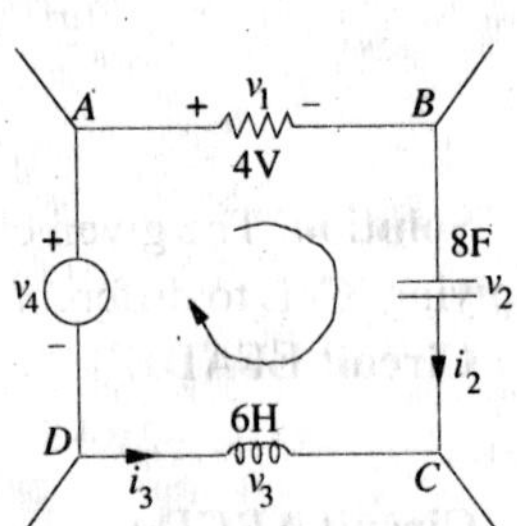

Fig. 2.16

Now, $$v_3 = L\frac{di_3}{dt} = 6.\frac{d}{dt}(2e^{-t/3})$$

$$= -4e^{-t/3}$$

$$\therefore \quad -4 - v_2 - 4e^{-t/3} + 4\cos 2t = 0 \text{ or } v_2 = 4\cos 2t - 4e^{-t/3} - 4$$

Now, $$i_2 = C\frac{dv_2}{dt} = 8\frac{d}{dt}(4\cos 2t - 4e^{-t/3} - 4)$$

$$\therefore \qquad i_2 = 8\left(-8\sin 2t + \frac{4}{3}e^{-t/3}\right) = -64\sin 2t + \frac{32}{3}e^{-t/3}$$

2.9. Independent and Dependent Sources

Those voltage or current sources, which do not depend on any other quantity in the circuit, are called independent sources. An independent d.c. voltage source is shown in Fig. 2.17 (a) whereas a time-varying voltage source is shown in Fig. 2.17(b). The positive sign shows that terminal A is positive with respect to terminal B. In other words, potential of terminal A is v volts higher than that of terminal B.

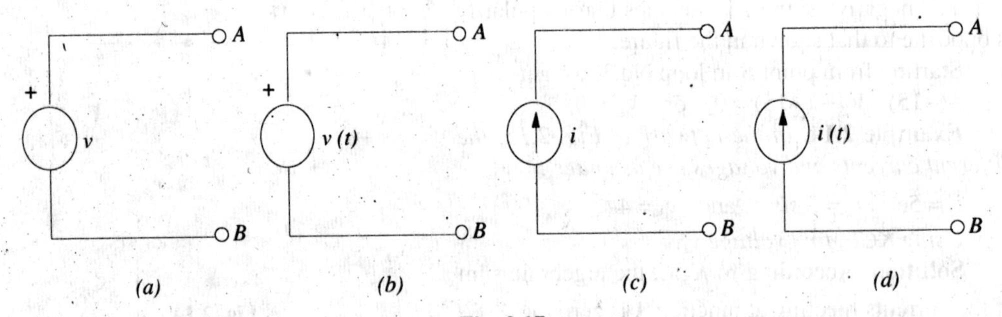

Fig. 2.17

Similarly, Fig. 2.17(*c*) shows an ideal constant current source whereas Fig. 2.17(*d*) depicts a time-varying current source. The arrow shows the direction of flow of the current at any moment under consideration.

A dependent voltage or current source is one which depends on some other quantity in the circuit which may be either a voltage or a current. Such a source is represented by a diamond-shaped symbol as shown in Fig. 2.18 so as not to confuse it with an independent source. There are four possible dependent sources :

1. Voltage-dependent voltage source [Fig. 2.18(*a*)]
2. Current-dependent voltage source [Fig. 2.18(*b*)]
3. Voltage-dependent current source [Fig. 2.18(*c*)]
4. Current-dependent current source [Fig. 2.18(*d*)]

Such sources can also be either constant sources or time-varying sources. Such sources are often met in electronic circuits. As seen above, the voltage or current source is dependent on and is proportional to another current or voltage. The constants of proportionality are written as a, r, g and β. The constants a and β have no units, r has the unit of ohms and g has the unit of siemens.

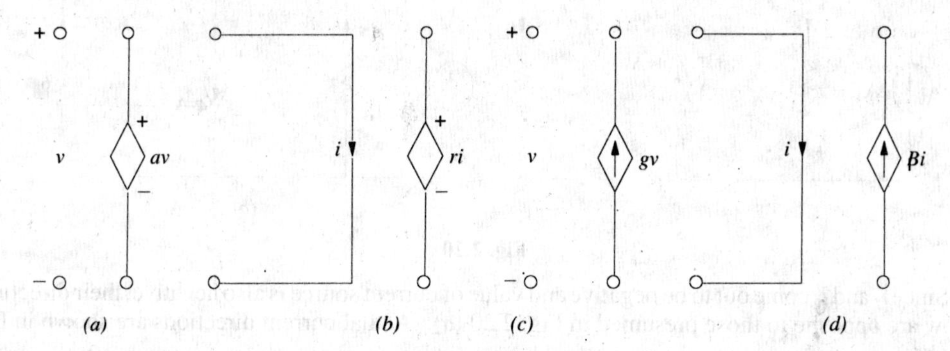

Fig. 2.18

Independent sources actually exist as physical entities such as a battery, a *d.c.* generator and an alternator etc. But dependent sources are parts of *models* that are used to represent electrical properties of electronic devices such as operational amplifiers and transistors etc.

Example 2.15. *Using Kirchhoff's current law, find the values of the currents i_1 and i_2 in the circuit of Fig. 2.19 (a) which contains a current-dependent current source. All resistances are in ohms.*

Solution. Applying *KCL* to node *A*, we get

$$2 - i_1 + 4\,i_1 - i_2 = 0 \quad \text{or} \quad -3i_1 + i_2 = 2$$

By Ohm's law, $i_1 = v/3$ and $i_2 = v/2$

Substituting these values above, we get

$$-3(v/3) + v/2 = 2 \quad \text{or} \quad v = -4\text{ V}$$

$$\therefore \qquad i_1 = -4/3\text{ A and } i_2 = -4/2 = -2\text{ A}$$

The value of the dependent current source is $= 4\,i_1 = 4 \times (-4/3) =$ **−16/3 A.**

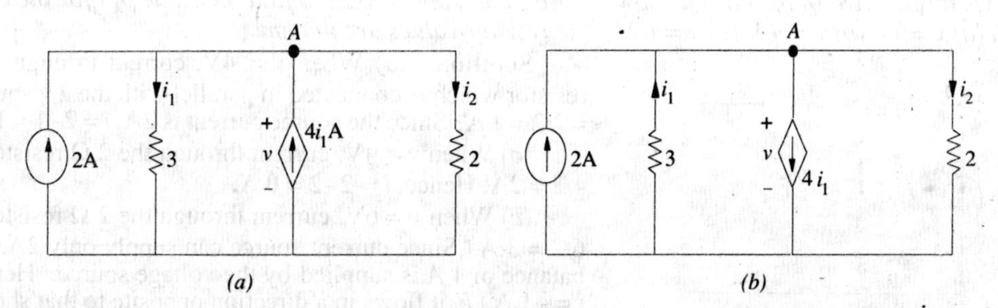

Fig. 2.19

Since i_1 and i_2 come out to be negative, it means that they flow upwards as shown in Fig. 2.19(*b*) and not downwards as presumed. Similarly, the current of the dependent source flows downwards as shown in Fig. 2.19(*b*). It may also be noted that the sum of the upward currents equals that of the downward currents.

Example 2.16. *By applying Kirchhoff's current law, obtain the values of v, i_1 and i_2 in the circuit of Fig. 2.20 (a) which contains a voltage-dependent current source. Resistance values are in ohms.*

Solution. Applying *KCL* to node *A* of the circuit, we get

$$2 - i_1 + 4v - i_2 = 0 \text{ or } i_1 + i_2 - 4v = 2$$

Now, $\quad i_1 = v/3 \quad$ and $\quad i_2 = v/6$

$$\therefore \quad \frac{v}{3} + \frac{v}{6} - 4v = 2 \text{ or } v = \frac{-4}{7}\text{V}$$

$$\therefore \quad i_1 = \frac{-4}{21}\text{A and } i_2 = \frac{-2}{21}\text{A and } 4v = 4 \times \frac{-4}{7} = \mathbf{\frac{-16}{7}V}$$

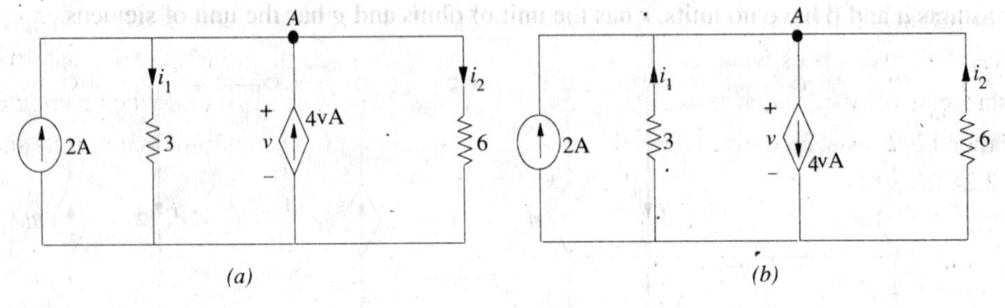

Fig. 2.20

Since i_1 and i_2 come out to be negative and value of current source is also negative, their directions of flow are opposite to those presumed in Fig. 2.20 (*a*). Actual current directions are shown in Fig. 2.20 (*b*).

Example 2.17. *Apply Kirchhoff's voltage law, to find the values of current i and the voltage drops v_1 and v_2 in the circuit of Fig. 2.21 which contains a current-dependent voltage source. What is the voltage of the dependent source? All resistance values are in ohms.*

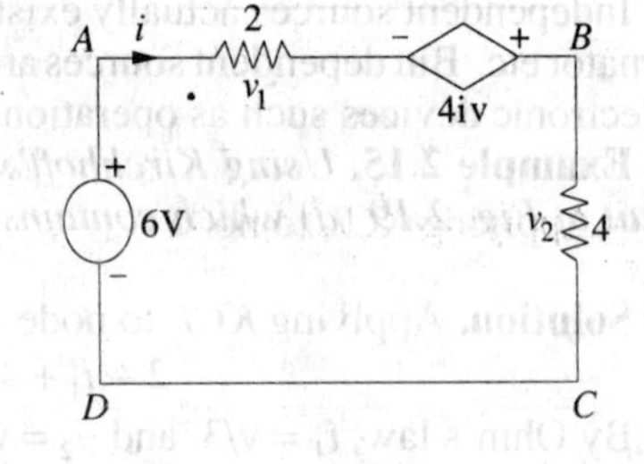

Fig. 2.21

Solution. Applying *KVL* to the circuit of Fig. 2.21 and starting from point *A*, we get

$$-v_1 + 4i - v_2 + 6 = 0 \text{ or } v_1 - 4i + v_2 = 6$$

Now, $\quad v_1 = 2i \quad$ and $v_2 = 4i$

$$\therefore \quad 2i - 4i + 4i = 6 \text{ or } i = 3\text{ A}$$

$$\therefore \quad v_1 = 2 \times 3 = 6\text{V and } v_2 = 4 \times 3 = \mathbf{12\ V}$$

Voltage of the dependent source = $4i = 4 \times 3 =$ **12 V**

Example 2.18. *In the circuit shown in Fig. 2.22, apply KCL to find the value of i for the case when (a) v = 1V (b) v = 4 V (c) v = 6 V. The resistor values are in ohms.*

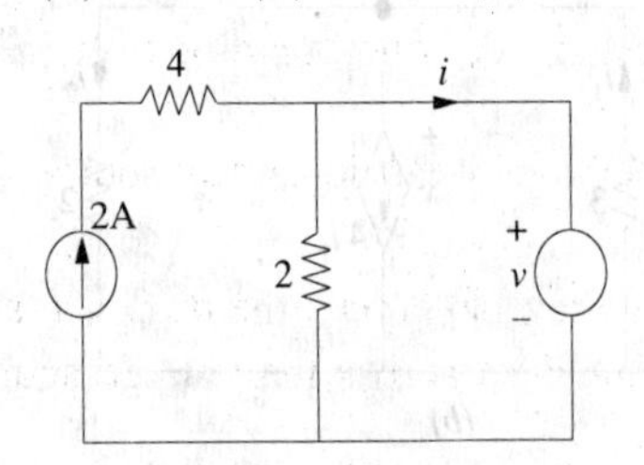

Fig. 2.22

Solution. (*a*) When v = 4V, current through 2 Ω resistor which is connected in parallel with the 2 v source = 2/2 = 1 A. Since the source current is 2A, i = 2−1 = **1 A.**

(*b*) When v = 4V, current through the 2 Ω resistor = 4/2 = **2A.** Hence, i = 2−2 = **0 A.**

(*c*) When v = 6V, current through the 2 Ω resistor = 6/2 = 3 A. Since current source can supply only 2A, the balance of 1 *A* is supplied by the voltage source. Hence, i = −1 A *i.e.* it flows in a direction opposite to that shown in Fig. 2.22.

Example 2.19. *In the circuit of Fig. 2.23, apply KCL to find the value of current i when (a) K = 2 (b) K = 3 and (c) K = 4. Both resistances are in ohms.*

Solution. Since 6 Ω and 3 Ω resistors are connected in parallel across the 24-V battery, $i_1 = 24/6 = 8$ A.

Applying *KCL* to node *A*, we get $i - 4 + 4K - 8 = 0$ or $i = 12 - 4K$.

(*a*) When $K = 2$, $i = 12 - 4 \times 2 =$ **4 A**

(*b*) When $K = 3$, $i = 12 - 4 \times 3 =$ **0 A**

(*c*) When $K = 4$, $i = 12 - 4 \times 4 =$ **–4 A**

It means that current *i* flows in the opposite direction.

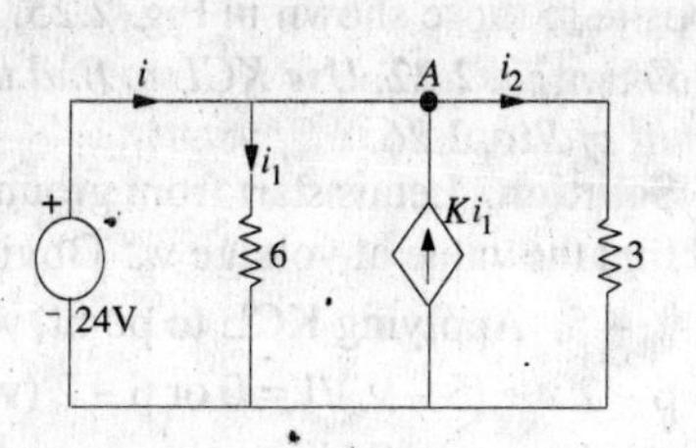

Fig. 2.23

Example 2.20. *Find the current i and also the power and voltage of the dependent source in Fig. 2.24 (a). All resistances are in ohms.*

Solution. The two current sources can be combined into a single source of 8 – 6 = 2 A. The two parallel 4 Ω resistances when combined have a value of 2 Ω which, being in series with the 10 Ω resistance, gives the branch resistance of 10 + 2 = 12 Ω. This 12 Ω resistance when combined with the other 12 Ω resistance gives a combination resistance of 6 Ω. The simplified circuit is shown in Fig. 2.24 (*b*).

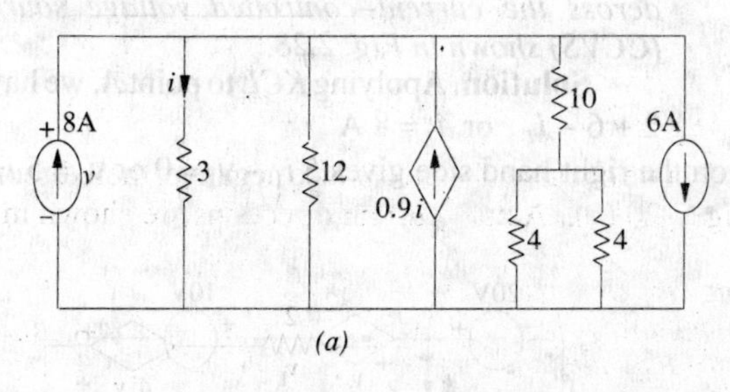

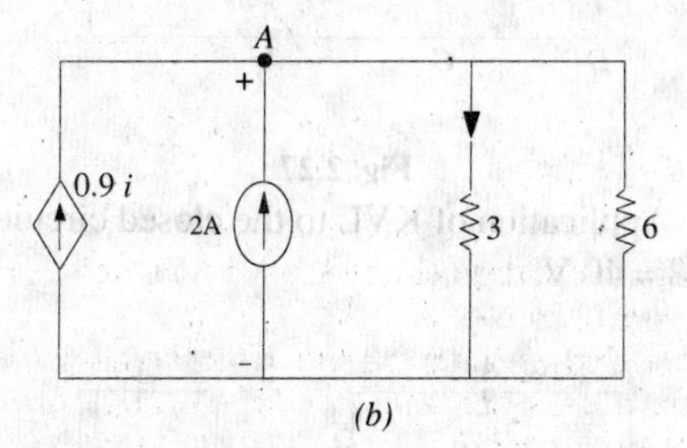

Fig. 2.24

Applying *KCL* to node A, we get

$$0.9i + 2 - i - v/6 = 0 \quad \text{or} \quad 0.6i = 12 - v$$

Also, $v = 3i$ ∴ $i = 10/3$A. Hence, $v =$ **10 V**

The power furnished by the current source = $v \times 0.9\, i = 10 \times 0.9\,(10/3) =$ **30 W**.

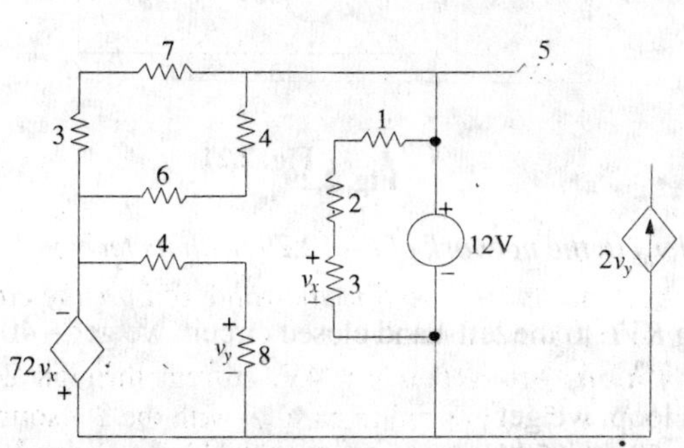

Fig. 2.25

Example 2.21. *By using voltage-divider rule, calculate the voltages v_x and v_y in the network shown in Fig. 2.25.*

Solution. As seen, 12 V drop is over the series combination of 1, 2 and 3 Ω resistors. As per voltage - divider rule, v_x = drop over 3 Ω = 12 × 3/6 = 6 V

The voltage of the dependent source = 12 × 6 = 72 V

The voltage v_y equals the drop across 8 Ω resistor connected across the voltage source of 72 V.

Again using voltage-divider rule, drop over 8 Ω resistor = 72 × 8/12 = 48 *V*.

Hence, $v_y = -48$ V. The negative sign has been given because positive and negative signs of v_y are actually opposite to those shown in Fig. 2.25.

Example 2.22. *Use KCL to find the value of v in the circuit of Fig. 2.26.*

Solution. Let us start from ground and go to point *a* and find the value of voltage v_a. Obviously, $5 + v = v_a$ or $v = v_a - 5$. Applying KCL to point, we get

$6 - 2v + (5 - v_a)/1 = 0$ or $6 - 2(v_a - 5) + (5 - v_a) = 0$

or $\quad v_a = 7$ V

Hence, $v = v_a - 5 = 7 - 5 = 2$ V. Since it turns out to be positive, its sign as indicated in the figure is correct.

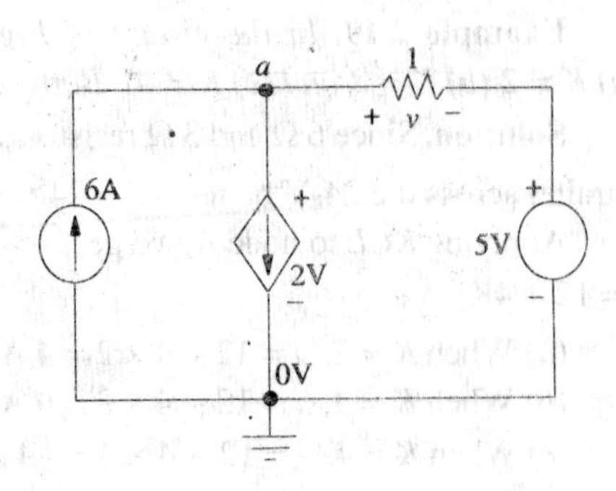

Fig. 2.26

Example 2.23.(a) Basic Electric circuits by Cunninghan. *Find the value of current i_2 supplied by the voltage-controlled current source (VCCS) shown in Fig. 2.27.*

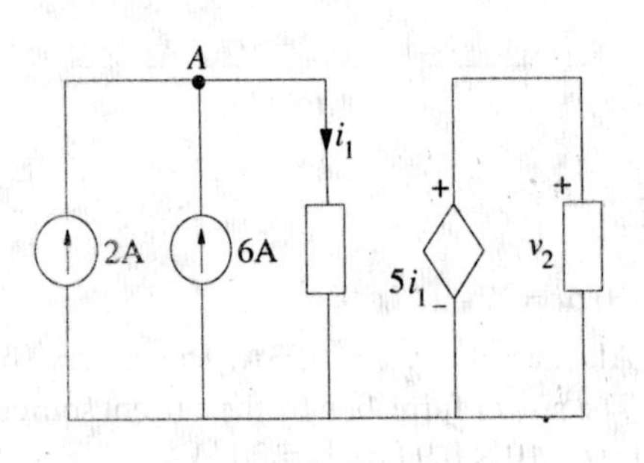

Fig. 2.27

Solution. Applying *KVL* to the closed circuit *ABCD*, we have $-4 + 8 - v_1 = 0 \therefore v_1 = 4$ V

The current supplied by *VCCS* is $10\,v_1 = 10 \times 4 = 40$ A. Since i_2 flows in an opposite direction to this current, hence $i_2 = -$ **40 A.**

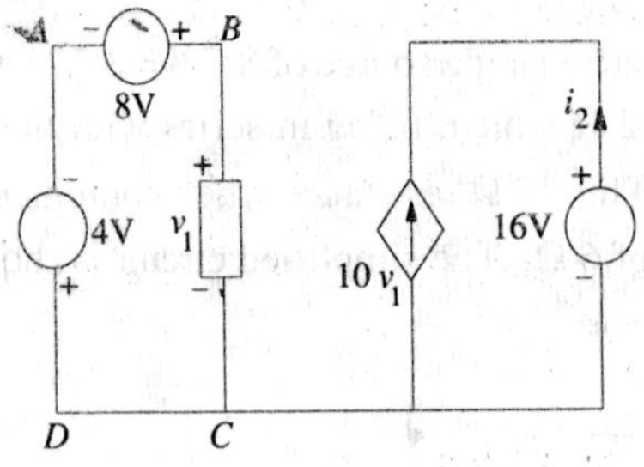

Example 2.23 (b). *Find the voltage drop v_2 across the current–controlled voltage source (CCVS) shown in Fig. 2.28.*

Solution. Applying *KCL* to point *A*, we have

$2 + 6 - i_1$ or $i_1 = 8$ A

Application of KVL to the closed circuit on the right hand side gives $5\,i_1 - v_2 = 0$ or $v_2 = 5\,i_1 = 5 \times 8 = 40$ V.

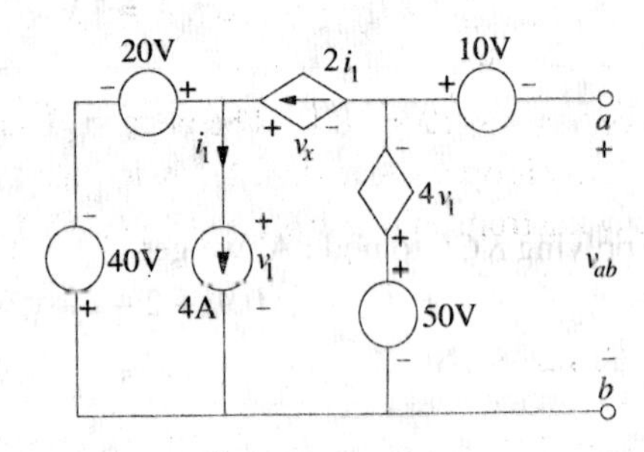

Fig. 2.28

Fig. 2.29

Example 2.24. *Find the values of i_1, v_1, v_x and v_{ab} in the network of Fig. 2.29 with its terminals a and b open.*

Solution. It is obvious that $i_1 = 4$ A. Applying *KVL* to the left-hand closed circuit, we get $-40 + 20 - v_1 = 0$ or $v_1 = -20$ V.

Similarly, applying KVL to the second closed loop, we get

$v_1 - v_x + 4\,v_1 - 50 = 0$ or $v_x = 5\,v_1 - 50 = -5 \times 20 - 50 = -150\ V$

Again applying *KVL* to the right-hand side circuit containing v_{ab}, we get

$50 - 4\,v_1 - 10 - v_{ab} = 0$ or $v_{ab} = 50 - 4(-20) - 10 = 120$ V

Example 2.25(a). *Find the current i in the circuit of Fig. 2.30. All resistances are in ohms.*

Solution. The equivalent resistance of the two parallel paths across point *a* is $3 \,||\, (4 + 2) = 2\ \Omega$. Now, applying KVL to the closed loop, we get $24 - v - 2v - 2i = 0$. Since $v = 2\,i$, we get $24 - 2\,i - 2(2\,i) - 2\,i = 0$ or $i = 3$ A.

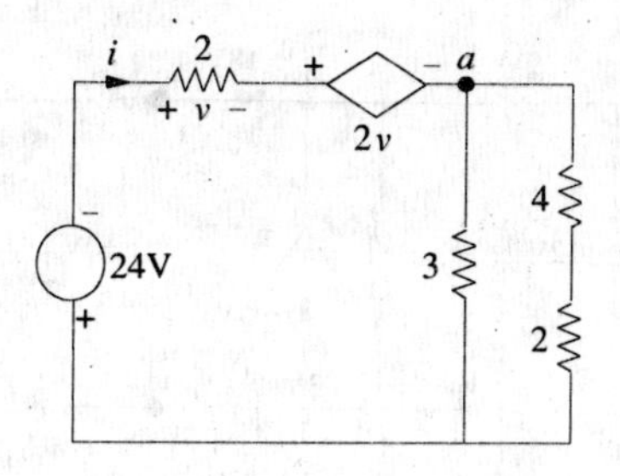

Fig. 2.30

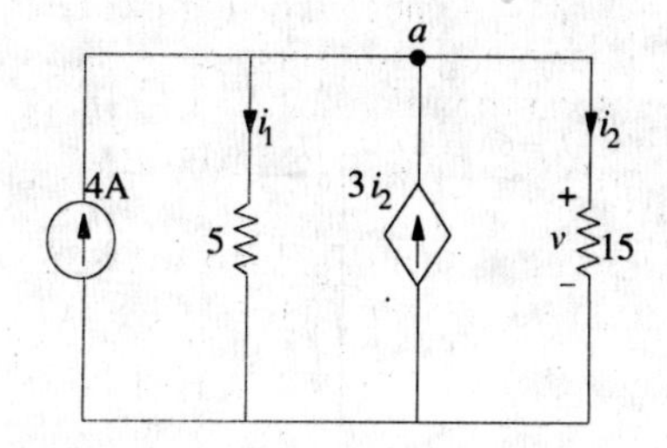

Fig. 2.31

Example 2.25. (b) *Determine the value of current i_2 and voltage drop v across 15 Ω resistor in Fig. 2.31.*

Solution. It will be seen that the dependent current source is related to i_2. Applying KCL to node *a*, we get $4 - i + 3i_2 - i_2 = 0$ or $4 - i_1 + 3i_2 = 0$.

Applying Ohm's law, we get $i_1 = v/5$ and $i_2 = v/15$.

Substituting these values in the above equation, we get $4 - (v/5) + 2(v/15) = 0$ or $v = 60$ V and $i_2 = 4$ A.

Example 2.25(c). *In the circuit of Fig. 2.32, find the values of i and v. All resistances are in ohms.*

Solution. It may be noted that $12 + v = v_a$ or $v = v_a - 12$. Applying KCL to node *a*, we get

$$\frac{0 - v_a}{2} + \frac{v}{4} - \frac{v_a - 12}{2} = 0 \text{ or } v_a = 4 \text{ V}$$

Hence, $v = 4 - 12 = -8$ *V*. The negative sign shows that its polarity is opposite to that shown in Fig. 2.32. The current flowing from point *a* to ground is 4/2 = 2 A. Hence, $i = -2$ A.

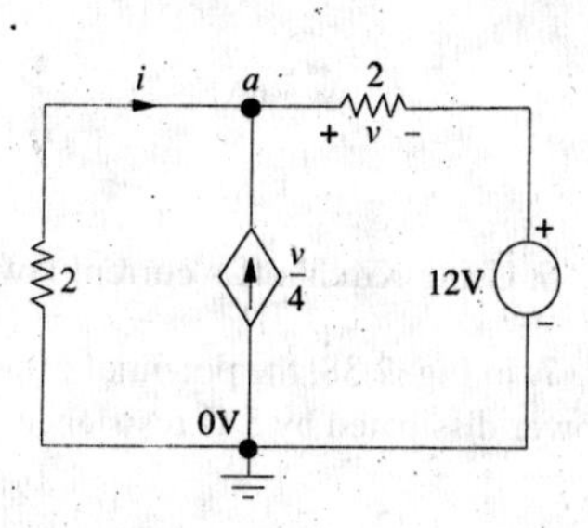

Fig. 2.32

Tutorial Problems No. 2.1

1. Apply KCL to find the value of *I* in Fig. 2.33. **[8A]**

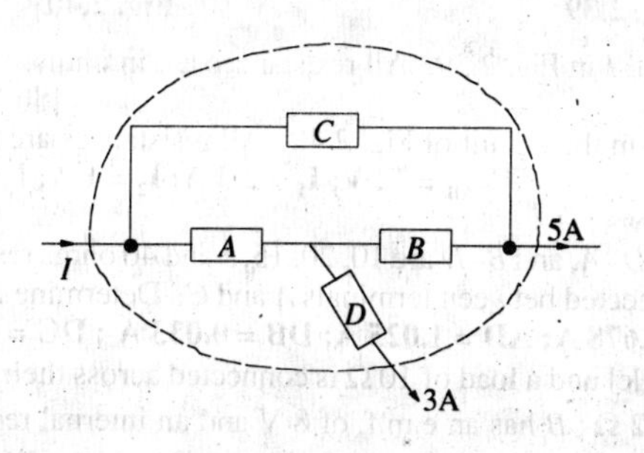

Fig. 2.33

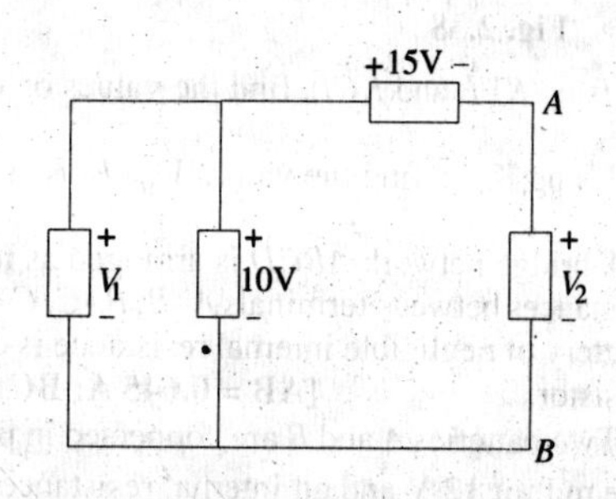

Fig. 2.34

2. Applying Kirchhoff's voltage law, find V_1 and V_2 in Fig. 2.34. **[V_1 = 10 V; V_2 = −5 V]**
3. Find the values of currents I_2 and I_4 in the network of Fig. 2.35. **[I_2 = 4 A ; I_4 = 5 A]**

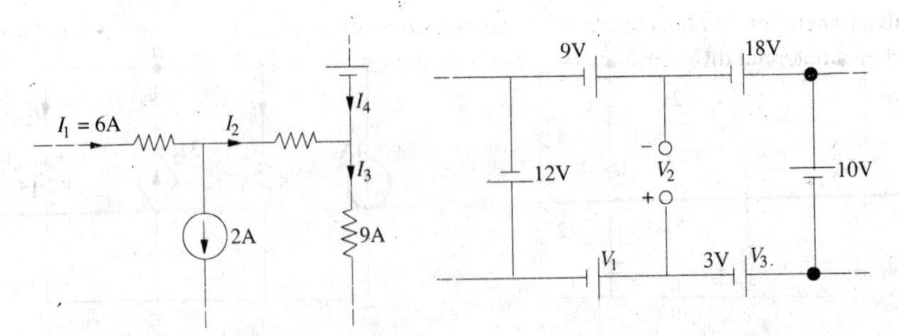

Fig. 2.35 Fig. 2.36

4. Use Kirchhoff's law, to find the values of voltages V_1 and V_2 in the network shown in Fig. 2.36. **[V_1 = 2 V; V_2 = 5 V]**

5. Find the unknown currents in the circuits shown in Fig. 2.37 (*a*). **[I_1 = 2 A; I_2 = 7 A]**

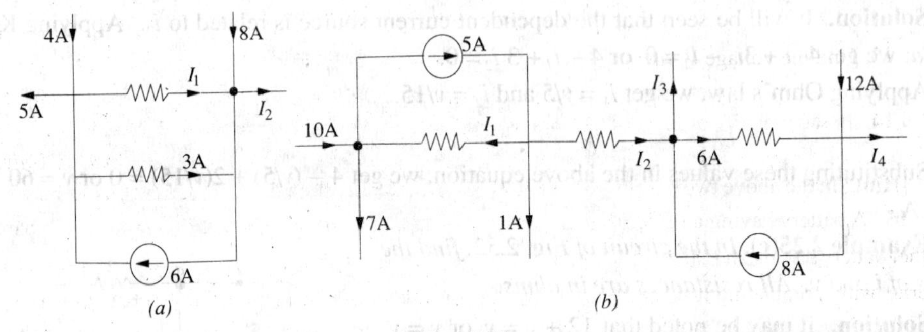

Fig. 2.37

6. Using Kirchhoff's current law, find the values of the unknown currents in Fig.2.37 (*b*). **[I_1 = 2 A ; I_2 = 2 A ; I_3 = 4 A; I_4 = 10 A]**.

7. In Fig. 2.38, the potential of point *A* is –30 V. Using Kirchhoff's voltage law, find (*a*) value of V and (*b*) power dissipated by 5 Ω resistance. All resistances are in ohms. **[100 V; 500 W]**

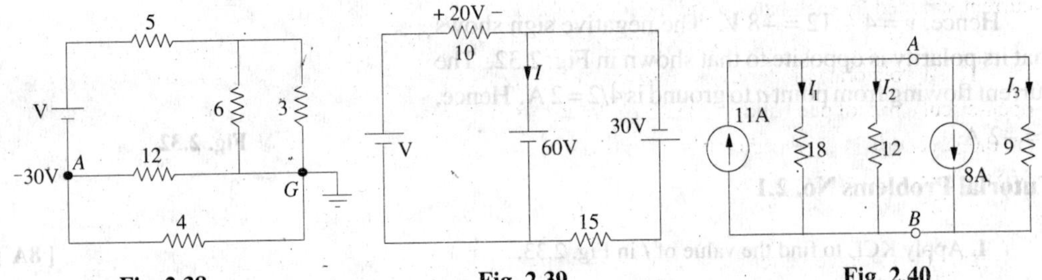

Fig. 2.38 Fig. 2.39 Fig. 2.40

8. Using *KVL* and *KCL*, find the values of *V* and *I* in Fig. 2.39. All resistances are in ohms. **[80 V; –4 A]**

9. Using KCL, find the values V_{AB}, I_1, I_2 and I_3 in the circuit of Fig. 2.40. All resistances are in ohms. **[V_{AB} = 12 V; I_1 = 2/3 A; I_2 = 1 A; I_3 = 4/3 A]**

10. A bridge network *ABCD* is arranged as follows :

Resistances between terminals *A–B*, *B–C*, *C–D*, *D–A*, and *B–D* are 10, 20, 15, 5 and 40 ohms respectively. A 20 V battery of negligible internal resistance is connected between terminals *A* and *C*. Determine the current in each resistor. **[AB = 0.645 A; BC = 0.678 A; AD = 1.025 A; DB = 0.033 A ; DC = 0.992 A]**

11. Two batteries *A* and *B* are connected in parallel and a load of 10 Ω is connected across their terminals. A has an e.m.f. of 12 V and an internal resistance of 2 Ω ; *B* has an e.m.f. of 8 V and an internal resistance of 1 Ω. Use Kirchhoff's laws to determine the values and directions of the currents flowing in each of the batteries and in the external resistance. Also determine the p.d. across the external resistance. **[I_A = 1.625 A (discharge), I_B = 0.75 A (charge) ; 0.875 A; 8.75 V]**

12. The four arms of a Wheatstone bridge have the following resistances ; *AB* = 100, *BC* = 10, *CD* = 4, *DA* = 50 ohms.

A galvanometer of 20 ohms resistance is connected across *BD*. Calculate the current through the galvanometer when a potential difference of 10 volts is maintained across *AC*.

[0.00513 A] [*Elect. Tech. Lond. Univ.*]

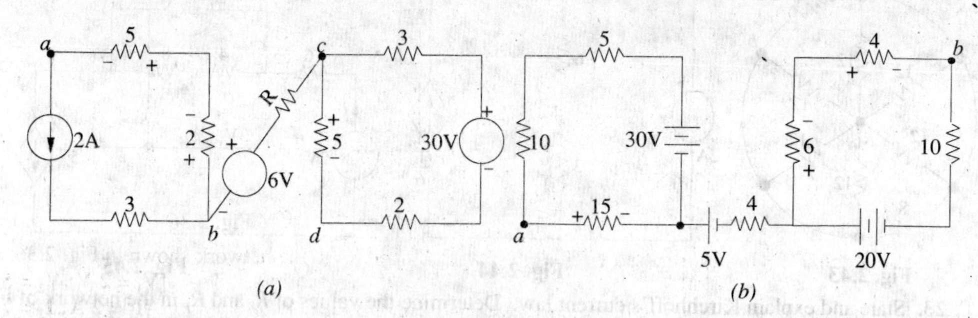

Fig. 2.41

13. Find the voltage V_{da} in the network shown in Fig. 2.41(*a*) if *R* is 10 Ω and (*b*) 20 Ω.

[(*a*) 5 V (*b*) 5 V]

14. In the network of Fig. 2.41(*b*), calculate the voltage between points *a* and *b i.e.* V_{ab}.

[30 V] (*Elect. Engg.-I, Bombay Univ. 1979*)

[**Hint** : In the above two cases, the two closed loops are independent and no current passes between them].

15. A battery having an *E.M.F.* of 110 V and an internal resistance of 0.2 Ω is connected in parallel with another battery having an *E.M.F.* of 100 V and internal resistance 0.25 Ω. The two batteries in parallel are placed in series with a regulating resistance of 5 Ω and connected across 200 V mains. Calculate the magnitude and direction of the current in each battery and the total current taken from the supply mains.

[I_A = 11.96 (discharge); I_B=30.43 A (charge) : 18.47 A]

(*Elect Technology, Sumbhal Univ. May 1978*)

16. Three batteries *P*, *Q* and *R* consisting of 50, 55 and 60 cells in series respectively supply in parallel a common load of 100 A. Each cell has an e.m.f. of 2 V and an internal resistance of 0.005 Ω. Determine the current supplied by each battery and the load voltage.

[1.2 A; 35.4 A : 65.8 A ; 100.3 V] (*Basic Electricity, Bombay Univ. 1980*)

17. Two storage batteries are connected in parallel to supply a load having a resistance of 0.1 Ω. The open-circuit e.m.f. of one battery (*A*) is 12.1 V and that of the other battery (*B*) is 11.8 V. The internal resistances are 0.03 Ω and 0.04 Ω respectively. Calculate (*i*) the current supplied to the lead (*ii*) the current in each battery (*iii*) the terminal voltage of each battery.

[(*i*) 102.2A (*ii*) 62.7 A (A), 39.5 A (B) (*iii*) 10.22 V] (*London Univ.*)

18. Two storage batteries, *A* and *B*, are connected in parallel to supply a load the resistance of which is 1.2 Ω. Calculate (*i*) the current in this lood and (*ii*) the current supplied by each battery if the open-circuit e.m.f. of *A* is 12.5 V and that of *B* is 12.8 *V*, the internal resistance of *A* being 0.05 Ω and that of *B* 0.08 Ω.

[(*i*) 10.25 A (*ii*) 44 (A), 6.25 A (B)] (*London Univ.*)

19. The circuit of Fig. 2.42 contains a voltage – dependent voltage source. Find the current supplied by the battery and power supplied by the voltage source. Both resistances are in ohms. **[8 A; 1920 W]**

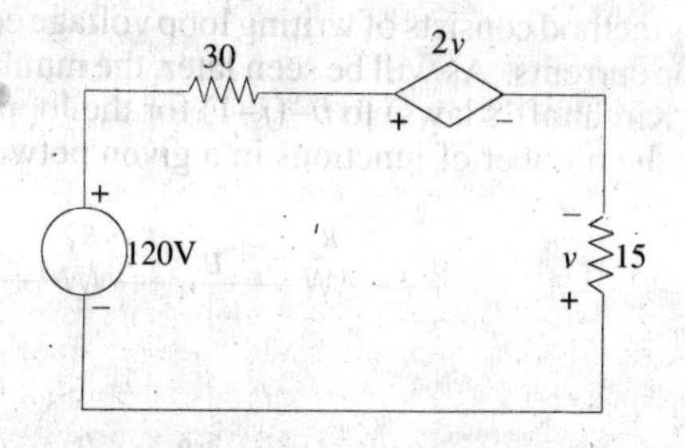

Fig. 2.42

20. Find the equivalent resistance between terminals *a* and *b* of the network shown in Fig. 2.43. [2 Ω].

21. Find the value of the voltage *v* in the network of Fig. 2.44. **[36 V]**

22. Determine the current i for the network shown in Fig. 2.45. **[– 40 A]**

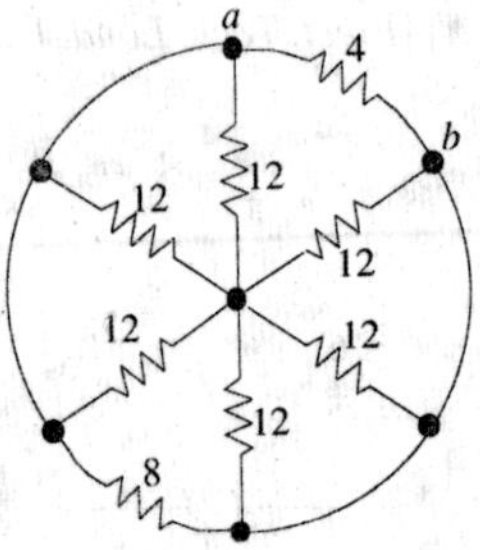

Fig. 2.43 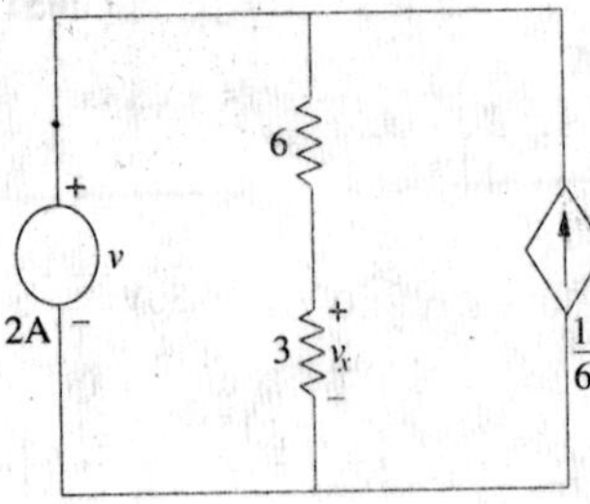 **Fig. 2.44** 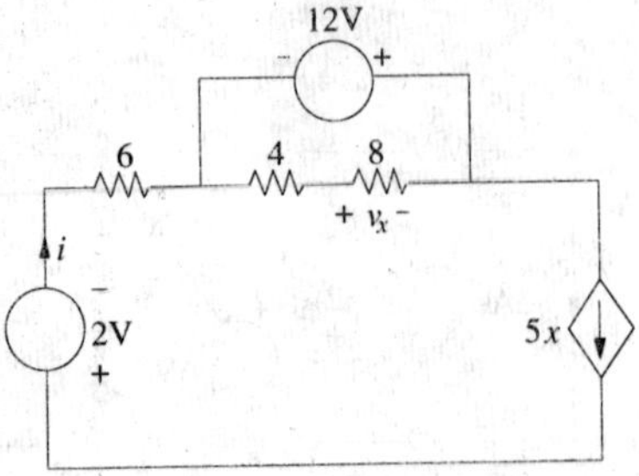 **Fig. 2.45**

23. State and explain Kirchhoff's current law. Determine the values of R_S and R_P in the network of Fig. 2.46 if $V_2 = V_1/2$ and the equivalent resistance of the network between the terminals A and B is 100 Ω.

[R_S= 100/3 Ω . R_P= 400/3 Ω] (*Elect. Engg. I, Bombay Univ./ 1978*)

24. Four resistances each of R ohms and two resistances each of S ohms are connected (as shown in Fig. 2.47) to four terminals AB and CD. A p.d. of V volts is applied across the terminals AB and a resistance of Z ohm is connected across the terminals CD. Find the value of Z in terms of S and R in order that the current at AB may be V/Z.

Find also the relationship that must hold between R and S in order that the p.d. at the points EF be $V/2$.

[$Z = \sqrt{R(R + 2S)}; S = 4R$]

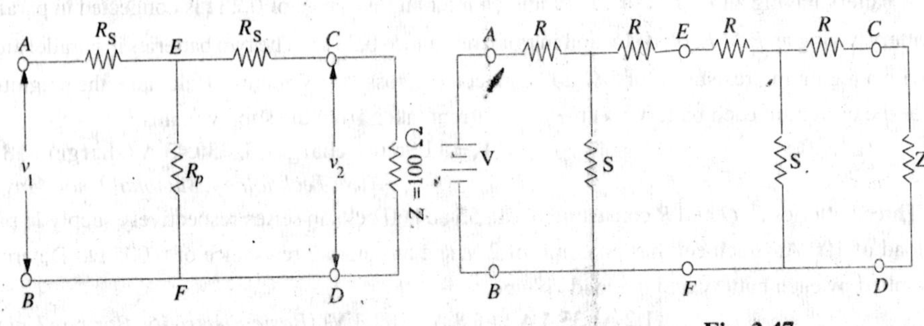

Fig. 2.46 **Fig. 2.47**

2.10. Maxwell's Loop Current Method

This method which is particularly well-suited to coupled circuit solutions employs a system of *loop* or *mesh* currents instead of *branch* currents (as in Kirchhoff's laws). Here, the currents in different meshes are assigned continuous paths so that they do not split at a junction into branch currents. This method eliminates a great deal of tedious work involved in the branch-current method and is best suited when energy sources are voltage sources rather than current sources. Basically, this method consists of writing loop voltage equations by Kirchhoff's voltage law in terms of unknown loop currents. As will be seen later, the number of independent equations to be solved reduces from b by Kirchhoff's laws) to $b-(j-1)$ for the loop current method where b is the number of branches and j is the number of junctions in a given network.

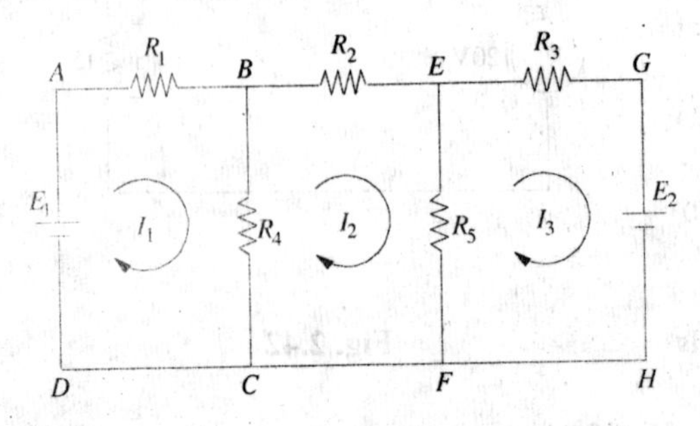

Fig. 2.48

Fig. 2.48 shows two batteries E_1 and E_2 connected in a network consisting of five resistors. Let the loop currents for the three meshes be I_1, I_2 and I_3*. It is obvious that current through R_4 (when considered as a part of the first loop) is $(I_1 - I_2)$ and that through R_5 is $(I_2 - I_3)$. However, when R_4 is considered part of the second loop, current through it is $(I_2 - I_1)$. Similarly, when R_5 is considered part of the third loop, current through it is $(I_3 - I_2)$. Applying Kirchhoff's voltage law to the three loops, we get,

$$E_1 - I_1R_1 - R_4(I_1 - I_2) = 0 \text{ or } I_1(R_1 + R_4) - I_2R_4 - E_1 = 0 \quad \text{...loop 1}$$

Similarly, $$-I_2R_2 - R_5(I_2 - I_3) - R_4(I_2 - I_1) = 0$$

or $$I_1R_4 - I_2(R_2 + R_4 + R_5) + I_3R_5 = 0 \quad \text{...loop 2}$$

Also $$-I_3R_3 - E_2 - R_5(I_3 - I_2) = 0 \text{ or } I_2R_5 - I_3(R_3 + R_5) - E_2 = 0 \quad \text{...loop 3}$$

The above three equations can be solved not only to find loop currents but branch currents as well.

2.11. Mesh Analysis Using Matrix Form

Consider the network of Fig. 2.49, which contains resistances and independent voltage sources and has three meshes. Let the three mesh currents be designated as I_1, I_2 and I_3 and all the three may be assumed to flow in the clockwise direction for obtaining symmetry in mesh equations.

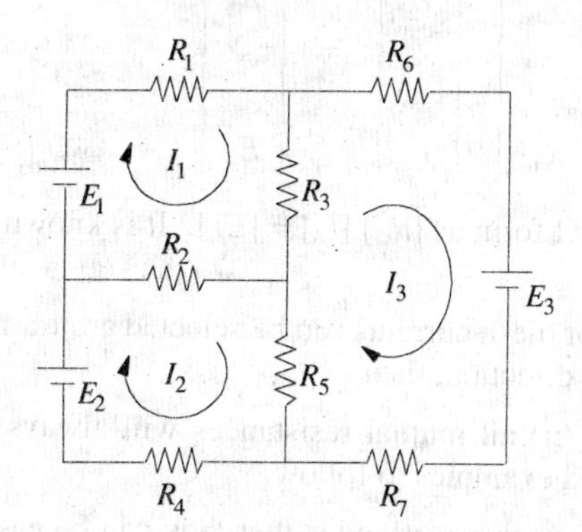

Fig. 2.49

Applying *KVL* to mesh (*i*), we have

$$E_1 - I_1 R_1 - R_3 (I_1 - I_3) - R_2 (I_1 - I_2) = 0$$

or $$(R_1 + R_2 + R_3) I_1 - R_2I_2 - R_3I_3 = E_1 \quad \text{...}(i)$$

Similarly, from mesh (*ii*), we have

$$E_2 - R_2 (I_2 - I_1) - R_5 (I_2 - I_3) - I_2R_4 = 0$$

or $$-R_2I_1 + (R_2 + R_4 + R_5) I_2 - R_5 I_3 = E_2 \quad \text{...}(ii)$$

Applying *KVL* to mesh (*iii*), we have

$$E_3 - I_3 R_7 - R_5 (I_3 - I_2) - R_3 (I_3 - I_1) - I_3R_6 = 0$$

or $$-R_3I_1 - R_5I_2 + (R_3 + R_5 + R_6 + R_7) I_3 = E_3 \text{...}(iii)$$

It should be noted that signs of different items in the above three equations have been so changed as to make the terms containing self resistances positive (please see further).

The matrix equivalent of the above three equations is

$$\begin{bmatrix} +(R_1+R_2+R_3) & -R_2 & -R_3 \\ -R_2 & +(R_2+R_4+R_5) & -R_5 \\ -R_3 & -R_5 & +(R_3+R_5+R_6+R_7) \end{bmatrix} \begin{bmatrix} I_1 \\ I_2 \\ I_3 \end{bmatrix} = \begin{bmatrix} E_1 \\ E_2 \\ E_3 \end{bmatrix}$$

It would be seen that the first item in the first row *i.e.* $(R_1 + R_2 + R_3)$ represents the self resistance of mesh (*i*) which equals the sum of all resistances in mesh (*i*). Similarly, the second item in the first row represents the mutual resistance between meshes (*i*) and (*ii*) *i.e.* the sum of resistances common to mesh (*i*) and (*ii*). Similarly, the third item in the first row represents the mutual–resistance of the mesh (*i*) and mesh (*ii*).

The item E_1, in general, represents the algebraic sum of the voltages of all the voltage sources acting around mesh (*i*). Similar is the case with E_2 and E_3. The sign of the *e.m.f s* is the same as discussed in Art. 2.3 *i.e.* while going along the current, if we pass from negative to the positive terminal of a battery, then its *e.m.f.* is taken positive. If it is the other way around, then battery *e.m.f.* is taken negative.

---

*Although, it is easier to take all loop currents in one direction (usually clockwise), the choice of direction for any loop current is arbitrary and may be chosen independently of the direction for the other loop currents.

In general, let

R_{11} = self-resistance of mesh (*i*)

R_{22} = self-resistance of mesh (*ii*) *i.e.* sum of all resistances in mesh (*ii*)

R_{33} = self-resistance of mesh (*iii*) *i.e.* sum of all resistances in mesh (*iii*)

$R_{12} = R_{21}$ = –[Sum of all the resistances common to meshes (*i*) and (*ii*)]*

$R_{23} = R_{32}$ = – [Sum of all the resistances common to meshes (*ii*) and (*iii*)]*

$R_{31} = R_{13}$ = – [sum of all the resistances common to meshes (*i*) and (*iii*)]*

Using these symbols, the generalized form of the above matrix equivalent can be written as

$$\begin{bmatrix} R_{11} & R_{12} & R_{13} \\ R_{21} & R_{22} & R_{23} \\ R_{31} & R_{32} & R_{33} \end{bmatrix} \begin{bmatrix} I_1 \\ I_2 \\ I_3 \end{bmatrix} = \begin{bmatrix} E_1 \\ E_2 \\ E_3 \end{bmatrix}$$

If there are *m* independent meshes in any linear network, then the mesh equations can be written in the matrix form as under :-

$$\begin{bmatrix} R_{11} & R_{12} & R_{13} & \cdots & R_{1m} \\ R_{21} & R_{22} & R_{23} & \cdots & R_{2m} \\ \cdots & \cdots & \cdots & \cdots & \cdots \\ \cdots & \cdots & \cdots & \cdots & \cdots \\ R_{31} & R_{32} & R_{33} & \cdots & R_{3m} \end{bmatrix} \begin{bmatrix} I_1 \\ I_2 \\ \cdots \\ \cdots \\ I_m \end{bmatrix} = \begin{bmatrix} E_1 \\ E_2 \\ \cdots \\ \cdots \\ E_m \end{bmatrix}$$

The above equation can be written in a more compact form as $[R_m]\,[I_m] = [E_m]$. It is known as Ohm's law in matrix form.

In the end, it may be pointed out that the directions of mesh currents can be selected arbitrarily. If we assume each mesh current to flow in the clockwise direction, then

(*i*) all self-resistances will always be positive and (*ii*) all mutual resistances will always be negative. We will adapt this sign convention in the solved examples to follow.

The main advantage of the generalized form of all mesh equations is that they can be easily remembered because of their symmetry. Moreover, for any given network, these can be written by inspection and then solved by the use of determinants. It eliminates the tedium of deriving simultaneous equations.

Example 2.26. *Write the impedance matrix of the network shown in Fig. 2.50 and find the value of current I_3.* **(Network Analysis A.M.I.E. Sec. B.W. 1980)**

Solution. Different items of the mesh-resistance matrix $[R_m]$ are as under :

$$R_{11} = 1 + 3 + 2 = 6\,\Omega\,;\; R_{22} = 2 + 1 + 4 = 7\,\Omega\,;\; R_{33} = 3 + 2 + 1 = 6\,\Omega\,;$$

$$R_{12} = R_{21} = -2\,\Omega\,;\; R_{23} = R_{32} = -1\,\Omega\,;\; R_{13} = R_{31} = -3\,\Omega\,;$$

$$E_1 = +5\,V : E_2 = 0;\; E_3 = 0.$$

The mesh equations in the matrix form are

$$\begin{bmatrix} R_{11} & R_{12} & R_{13} \\ R_{21} & R_{22} & R_{23} \\ R_{31} & R_{32} & R_{33} \end{bmatrix} \begin{bmatrix} I_1 \\ I_2 \\ I_3 \end{bmatrix} = \begin{bmatrix} E_1 \\ E_2 \\ E_3 \end{bmatrix} \text{ or } \begin{bmatrix} 6 & -2 & -3 \\ -2 & 7 & -1 \\ -3 & -1 & 6 \end{bmatrix} \begin{bmatrix} I_1 \\ I_2 \\ I_3 \end{bmatrix} = \begin{bmatrix} 5 \\ 0 \\ 0 \end{bmatrix}$$

---

*In general, if the two currents through the common resistance flow in the same direction, then the mutual resistance is taken negative. On the other hand, if the two currents flow in the same direction, mutual resistance is taken as positive.

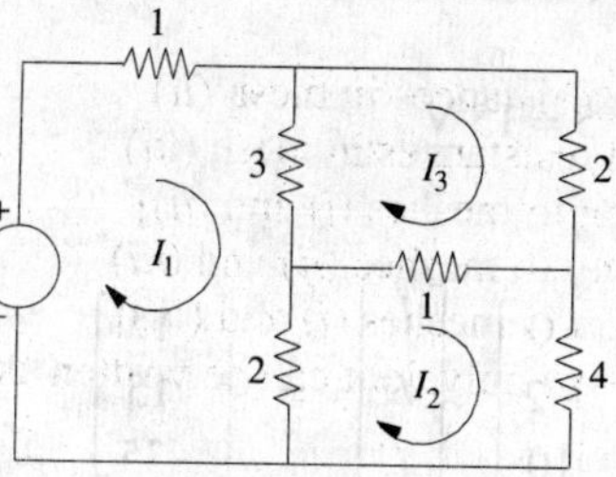

$$\Delta = \begin{bmatrix} 6 & -2 & -3 \\ -2 & 7 & -1 \\ -3 & -1 & 6 \end{bmatrix} = 6(42-1)+2(-12-3)-3(2+21) = 147$$

$$\Delta_3 = \begin{bmatrix} 6 & -2 & 5 \\ -2 & 7 & 0 \\ -3 & -1 & 0 \end{bmatrix} = 6+2(5)-3(-35) = 121$$

$$I_3 = \Delta/\Delta_3 = 147/121 = \mathbf{1.215\ A}$$

Fig. 2.50

Example 2.27. *Determine the current supplied by each battery in the circuit shown in Fig. 2.51.*

(Electrical Engg. Aligarh Univ. 1989)

Solution. Since there are three meshes, let the three loop currents be shown in Fig. 2.51.

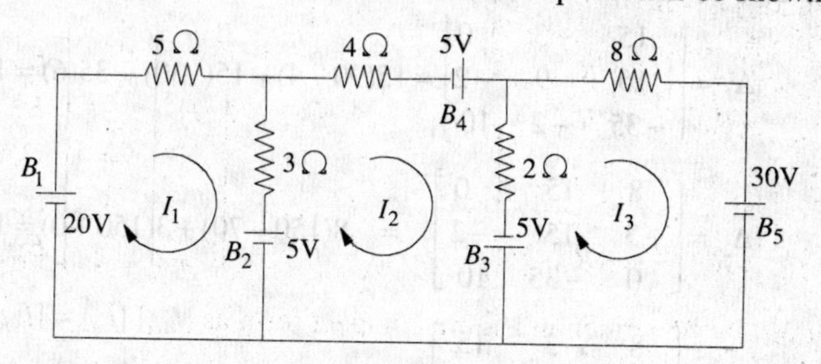

Fig. 2.51

For loop 1 we get

$$20 - 5I_1 - 3(I_1 - I_2) - 5 = 0 \quad \text{or} \quad 8I_1 - 3I_2 = 15 \qquad ...(i)$$

For loop 2, we have

$$-4I_2 + 5 - 2(I_2 - I_3) + 5 + 5 - 3(I_2 - I_1) = 0 \quad \text{or} \quad 3I_1 - 9I_2 + 2I_3 = -15 \qquad ...(ii)$$

Similarly, for loop 3, we get

$$-8I_3 - 30 - 5 - 2(I_3 - I_2) = 0 \quad \text{or} \quad 2I_2 - 10I_3 = 35 \qquad ...(iii)$$

Eliminating I_1 from (*i*) and (*ii*), we get $63I_2 - 16I_3 = 165$ ...(*iv*)

Similarly, for I_2 from (*iii*) and (*iv*), we have $I_2 = 542/299$ A

From (*iv*), $I_3 = -1875/598$ A

Substituting the value of I_2 in (*i*), we get $I_1 = 765/299$ A

Since I_3 turns out to be negative, actual directions of flow of loop currents are as shown in Fig. 2.52.

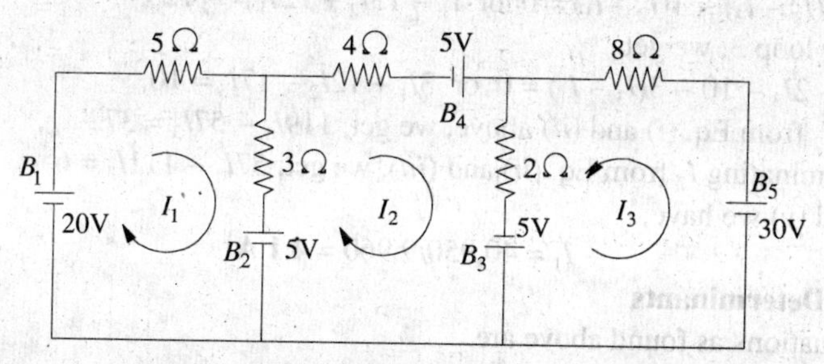

Fig. 2.52

Dicharge current of B_1 = **765/299 A**

Charging current of $B_2 = I_1 - I_2$ = **220/299 A**

Dicharge current of $B_3 = I_2 + I_3$ = **2965/598 A**

Dicharge current of $B_4 = I_2$ = **545/299 A**; Discharge current of B_5 = **1875/598 A**

Solution by Using Mesh Resistance Matrix

The different items of the mesh-resistance matrix $[R_m]$ are as under :

$R_{11} = 5 + 3 = 8\ \Omega$; $R_{22} = 4 + 2 + 3 = 9\ \Omega$; $R_{33} = 8 + 2 = 10\ \Omega$

$R_{12} = R_{21} = -3\ \Omega$; $R_{13} = R_{31} = 0$; $R_{23} = R_{32} = -2\ \Omega$

E_1 = algebraic sum of the voltages around mesh (*i*) = 20 – 5 = 15 V

$E_2 = 5 + 5 + 5 = 15$ V ; $E_3 = -30 - 5 = -35$ V

Hence, the mesh equations in the matrix form are

$$\begin{bmatrix} R_{11} & R_{12} & R_{13} \\ R_{21} & R_{22} & R_{23} \\ R_{31} & R_{32} & R_{33} \end{bmatrix} \begin{bmatrix} I_1 \\ I_2 \\ I_3 \end{bmatrix} = \begin{bmatrix} E_1 \\ E_2 \\ E_3 \end{bmatrix} \text{ or } \begin{bmatrix} 8 & -3 & 0 \\ -3 & 9 & -2 \\ 0 & -2 & 10 \end{bmatrix} \begin{bmatrix} I_1 \\ I_2 \\ I_3 \end{bmatrix} = \begin{bmatrix} 15 \\ 15 \\ -35 \end{bmatrix}$$

$$\Delta = \begin{vmatrix} 8 & -3 & 0 \\ -3 & 9 & -2 \\ 0 & -2 & 10 \end{vmatrix} = 8(90-4)+3(-30) = 598$$

$$\Delta_1 = \begin{vmatrix} 15 & -3 & 0 \\ 15 & 9 & -2 \\ -35 & -2 & 10 \end{vmatrix} = 15(90-4)-15(-30)-35(6) = 1530$$

$$\Delta_2 = \begin{bmatrix} 8 & 15 & 0 \\ -3 & 15 & -2 \\ 0 & -35 & 10 \end{bmatrix} = 8(150-70)+3(150+0) = 1090$$

$$\Delta_3 = \begin{bmatrix} 8 & -3 & 15 \\ -3 & 9 & 15 \\ 0 & -2 & -35 \end{bmatrix} = 8(-315+30)+3(105+30) = -1875$$

$$I_1 = \frac{\Delta_1}{\Delta} = \frac{1530}{598} = \frac{765}{299} \text{ A} ; I_2 = \frac{\Delta_2}{\Delta} = \frac{1090}{598} = \frac{545}{299} \text{ A} ; I_3 = \frac{\Delta_3}{\Delta_1} = \frac{-1875}{598} \text{ A}$$

Example 2.28. *Determine the current in the 4-Ω branch in the circuit shown in Fig. 2.53.*

(Elect. Technology, Nagpur Univ. 1992)

Solution. The three loop currents are as shown in Fig. 2.53 (*b*).

For loop 1, we have

$-1(I_1 - I_2) - 3(I_1 - I_3) - 4I_1 + 24 = 0$ or $8I_1 - I_2 - 3I_3 = 24$ ...(*i*)

For loop 2, we have

$12 - 2I_2 - 12(I_2 - I_3) - 1(I_2 - I_1) = 0$ or $I_1 - 15I_2 + 12I_3 = -12$ ...(*ii*)

Similarly, for loop 3, we get

$-12(I_3 - I_2) - 2I_3 - 10 - 3(I_3 - I_1) = 0$ or $3I_1 + 12I_2 - 17I_3 = 10$ ...(*iii*)

Eliminating I_2 from Eq. (*i*) and (*ii*) above, we get, $119I_1 - 57I_3 = 372$ ...(*iv*)

Similarly, eliminating I_2 from Eq. (*ii*) and (*iii*), we get, $57I_1 - 111I_3 = 6$ ...(*v*)

From (*iv*) and (*v*) we have,

$$I_1 = 40{,}950/9{,}960 = \mathbf{4.1\ A}$$

Solution by Determinants

The three equations as found above are

$$8I_1 - I_2 - 3I_3 = 24$$
$$I_1 - 15I_2 + 12I_3 = -12$$
$$3I_1 + 12I_2 - 17I_3 = 10$$

Their matrix form is
$$\begin{bmatrix} 8 & -1 & -3 \\ 1 & -15 & 12 \\ 3 & 12 & -17 \end{bmatrix} \begin{bmatrix} x \\ y \\ z \end{bmatrix} = \begin{bmatrix} 24 \\ -12 \\ 10 \end{bmatrix}$$

$$\Delta = \begin{bmatrix} 8 & -1 & -3 \\ 1 & -15 & 12 \\ 3 & 12 & -17 \end{bmatrix} = 664, \quad \Delta_1 = \begin{bmatrix} 24 & -1 & -3 \\ -12 & -15 & 12 \\ 10 & 12 & -17 \end{bmatrix} = 2730$$

$\therefore \quad I_1 = \Delta_1/\Delta = 2730/664 = \mathbf{4.1\ A}$

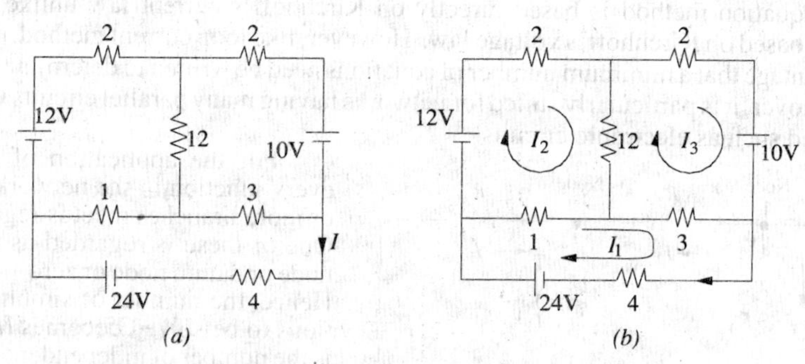

Fig. 2.53

Solution by Using Mesh Resistance Matrix

For the network of Fig. 2.53 (*b*), values of self resistances, mutual resistances and e.m.f's can be written by more inspection of Fig. 2.53.

$R_{11} = 3 + 1 + 4 = 8\ \Omega$; $R_{22} = 2 + 12 + 1 = 15\ \Omega$; $R_{33} = 2 + 3 + 12 = 17\ \Omega$

$R_{12} = R_{21} = -1$; $R_{23} = R_{32} = -12$; $R_{13} = R_{31} = -3$

$E_1 = 24$ V ; $E_2 = 12$ V ; $E_3 = -10$ V

The matrix form of the above three equations can be written by inspection of the given network as under :-

$$\begin{bmatrix} R_{11} & R_{12} & R_{13} \\ R_{21} & R_{22} & R_{23} \\ R_{31} & R_{32} & R_{33} \end{bmatrix} \begin{bmatrix} I_1 \\ I_2 \\ I_3 \end{bmatrix} = \begin{bmatrix} E_1 \\ E_2 \\ E_3 \end{bmatrix} \quad \text{or} \quad \begin{bmatrix} 8 & -1 & -3 \\ -1 & 15 & -12 \\ -3 & -12 & 17 \end{bmatrix} \begin{bmatrix} I_1 \\ I_2 \\ I_3 \end{bmatrix} = \begin{bmatrix} 24 \\ 12 \\ -10 \end{bmatrix}$$

$$\Delta = 8(255 - 144) + 1(-17 - 36) - 3(12 + 45) = 664$$

$$\Delta_1 = \begin{bmatrix} 24 & -1 & -3 \\ 12 & 15 & -12 \\ -10 & -12 & 17 \end{bmatrix} = 24(255 - 144) - 12(-17 - 36) - 10(12 + 45) = 2730$$

$$\therefore \quad I_1 = \frac{\Delta_1}{\Delta} = \frac{2730}{664} = \mathbf{4.1\ A}$$

It is the same answer as found above.

Tutorial Problems No. 2.2

1. Find the ammeter current in Fig. 2.54 by using loop analysis.

[1/7A] (*Network Theory, Indore Univ. 1981*)

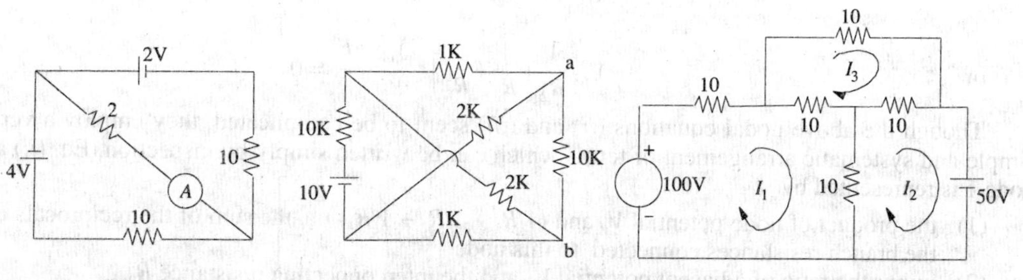

Fig. 2.54 **Fig. 2.55** **Fig. 2.56**

2. Using mesh analysis, determine the voltage across the 10 k Ω resistor at terminals *a-b* of the circuit shown in Fig. 2.55. **(2.65 V)** (*Elect. Technology, Indore Univ. April 1978*)

3. Apply loop current method to find loop currents I_1, I_2 and I_3 in the circuit of Fig. 2.56. **[I_1 = 3.75 A, I_2 = 0, I_3 = 1.25 A]**

2.12. Nodal Analysis With Voltage Sources

The node-equation method is based directly on Kirchhoff's current law unlike loop-current method which is based on Kirchhoff's voltage law. However, like loop current method, nodal method also has the advantage that a minimum number of equations need be written to determine the unknown quantities. Moreover, it is particularly suited for networks having many parallel circuits with common ground connected such as electronic circuits.

For the application of this method, every junction in the network where three or more branches meet is regarded a node. One of these is regarded as the reference node or datum node or zero-potential node. Hence, the number of simultaneous equations to be solved becomes $(n-1)$ where n is the number of independent nodes. These node equations often become simplified if all voltage sources are converted into current sources (Art. 2.12).

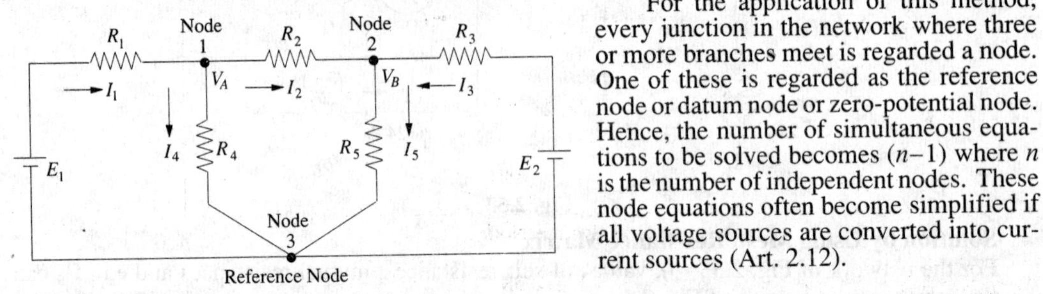

Fig. 2.57

(*i*) **First Case**

Consider the circuit of Fig. 2.57 which has three nodes. One of these *i.e.* node 3 has been taken as the reference node. V_A represents the potential of node 1 with reference to the datum node 3. Similarly, V_B is the potential difference between node 2 and node 3. Let the current directions which have been chosen arbitarily be as shown.

For node 1, the following current equation can be written with the help of *KCL*.

$$I_1 = I_4 + I_2 \qquad \ldots(i)$$

Now $\quad I_1R_1 = E_1 - V_A \quad \therefore \quad I_1 = (E_1 - V_A)/R_1$

Obviously, $\quad I_4 = V_A/R_4$ Also, $I_2R_2 = V_A - V_B \quad (\because \; V_A > V_B)$

$\therefore \quad I_2 = (V_A - V_B)/R_2$

Substituting these values in Eq. (*i*) above, we get,

$$\frac{E_1 - V_A}{R_1} = \frac{V_A}{R_4} + \frac{V_A - V_B}{R_2}$$

Simplifying the above, we have

$$V_A\left(\frac{1}{R_1} + \frac{1}{R_2} + \frac{1}{R_4}\right) - \frac{V_B}{R_2} - \frac{E_1}{R_1} = 0 \qquad \ldots(ii)$$

The current equation for node 2 is $I_5 = I_2 + I_3$

or $$\frac{V_B}{R_5} = \frac{V_A - V_B}{R_2} + \frac{E_2 - V_B}{R_3} \qquad \ldots(iii)$$

or $$V_B\left(\frac{1}{R_2} + \frac{1}{R_3} + \frac{1}{R_5}\right) - \frac{V_A}{R_2} - \frac{E_2}{R_3} = 0 \qquad \ldots(iv)$$

Though the above nodal equations (*ii*) and (*iii*) seem to be complicated, they employ a very simple and systematic arrangement of terms which can be written simply by inspection. Eq. (*ii*) at node 1 is represented by

(1) the product of node potential V_A and $(1/R_1 + 1/R_2 + 1/R_4)$ *i.e.* the sum of the reciprocals of the branch resistances connected to this node.

(2) *minus* the ratio of adjacent potential V_B and the interconnecting resistance R_2.

(3) *minus* ratio of adjacent battery (or generator) voltage E_1 and interconnecting resistance R_1.
(4) all the above set to zero.

Same is the case with Eq. (*iii*) which applies to node 2.

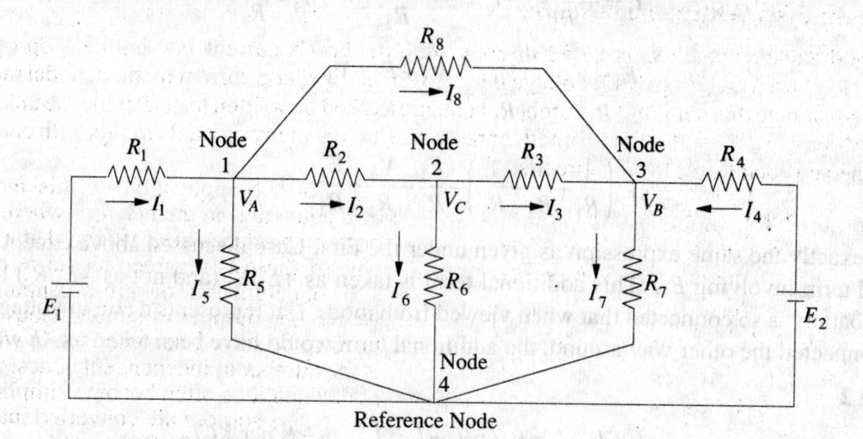

Fig. 2.58

Using conductances instead of resistances, the above two equations may be written as

$$V_A(G_1+G_2+G_4)-V_BG_2-E_1G_1=0 \quad \ldots(iv)$$

$$V_B(G_2+G_3+G_5)-V_AG_2-E_2G_3=0 \quad \ldots(v)$$

To emphasize the procedure given above, consider the circuit of Fig. 2.58.
The three node equations are

$$V_A\left(\frac{1}{R_1}+\frac{1}{R_2}+\frac{1}{R_5}+\frac{1}{R_8}\right)-\frac{V_C}{R_2}-\frac{V_B}{R_8}-\frac{E_1}{R_1}=0 \quad \text{(node 1)}$$

$$V_C\left(\frac{1}{R_2}+\frac{1}{R_3}+\frac{1}{R_6}\right)-\frac{V_A}{R_2}-\frac{V_B}{R_3}=0 \quad \text{(node 2)}$$

$$V_B\left(\frac{1}{R_3}+\frac{1}{R_4}+\frac{1}{R_7}+\frac{1}{R_8}\right)-\frac{V_C}{R_3}-\frac{V_A}{R_8}-\frac{E_4}{R_4}=0 \quad \text{(node 3)}$$

After finding different node voltages, various currents can be calculated by using Ohm's law.

(*ii*) **Second Case**

Now, consider the case when a third battery of e.m.f. E_3 is connected between nodes 1 and 2 as shown in Fig. 2.59.

It must be noted that as we travel from node 1 to node 2, we go from the −ve terminal of E_3 to its +ve terminal. Hence, according to the sign convention given in Art. 2.3, E_3 must be taken as *positive*. However, if we travel from node 2 to node 1, we go from the +ve to the −ve terminal of E_3. Hence, *when viewed from node* 2, E_3 is taken *negative*.

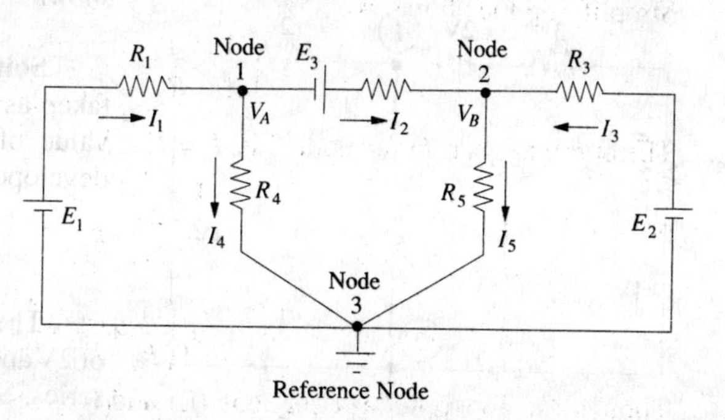

Fig. 2.59

For node 1

$$I_1 - I_4 - I_2 = 0 \text{ or } I_1 = I_4 + I_2 \text{ – as per KCL}$$

Now, $$I_1 = \frac{E_1 - V_A}{R_1};\ I_2 = \frac{V_A + E_3 - V_B}{R_2};\ I_4 = \frac{V_A}{R_4}$$

$$\therefore \quad \frac{E_1 - V_A}{R_1} = \frac{V_A}{R_4} + \frac{V_A + E_3 - V_B}{R_2}$$

or $$V_A\left(\frac{1}{R_1} + \frac{1}{R_2} + \frac{1}{R_3}\right) - \frac{E_1}{R_1} - \frac{V_B}{R_2} + \frac{E_3}{R_2} = 0 \qquad \ldots(i)$$

It is exactly the same expression as given under the First Case discussed above except for the additional term involving E_3. This additional term is taken as $+E_3/R_2$ (and not as $-E_3/R_2$) because this third battery is so connected that when viewed from mode 1, it represents a *rise* in voltage. Had it been connected the other way around, the additional term would have been taken as $-E_3/R_2$.

For node 2

$$I_2 + I_3 - I_5 = 0 \text{ or } I_2 + I_3 = I_5 \qquad \text{–as per } KCL$$

Now, as before, $$I_2 = \frac{V_A + E_3 - V_B}{R_2},\ I_3 = \frac{E_2 - V_B}{R_3},\ I_5 = \frac{V_B}{R_5}$$

$$\therefore \quad \frac{V_A + E_3 - V_B}{R_2} + \frac{E_2 - V_B}{R_3} = \frac{V_B}{R_5}$$

On simplifying, we get $$V_B\left(\frac{1}{R_2} + \frac{1}{R_3} + \frac{1}{R_5}\right) - \frac{E_2}{R_3} - \frac{V_A}{R_2} - \frac{E_3}{R_2} = 0 \qquad \ldots(ii)$$

As seen, the additional term is $-E_3/R_2$ (and not $+E_3/R_2$) because as viewed from *this* node, E_3 represents a *fall* in potential.

It is worth repeating that the additional term in the above Eq. (*i*) and (*ii*) can be either $+E_3/R_2$ or $-E_3/R_2$ depending on whether it represents a rise or fall of potential when viewed from *the node under consideration*.

Example 2.29. *Using Node voltage method, find the current in the 3Ω resistance for the network shown in Fig. 2.60.*

(Elect. Tech. Osmania Univ. Feb. 1992)

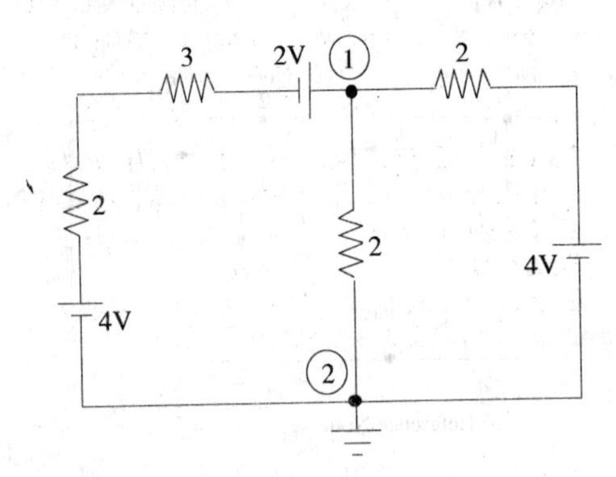

Fig. 2.60

Solution. As shown in the figure node 2 has been taken as the reference node. We will now find the value of node voltage V_1. Using the technique developed in Art. 2.10, we get

$$V_1\left(\frac{1}{5} + \frac{1}{2} + \frac{1}{2}\right) - \frac{4}{2} - \left(\frac{4+2}{5}\right) = 0$$

The reason for adding the two battery voltages of 2V and 4V is because they are connected in additive series. Simplifying the above, we get $V_1 = 8/3$ V. The current flowing through the 3Ω resistance towards node 1 is $= \frac{6 - (8/3)}{(3+2)} = \frac{2}{3}$ A

Alternatively

$$\frac{6-V_1}{5}+\frac{4}{2}-\frac{V_1}{2}=0$$

$$12-2V_1+20-5V_1=0$$

$$7V_1=32$$

Also
$$\frac{6-V_1}{5}+\frac{4-V_1}{2}=\frac{V_1}{2}$$

$$12-2V_1+20-5V_1=5V_1$$

$$12V_1=32\,;\ V_1=8/3$$

Example 2.30. *Frame and solve the node equations of the network of Fig. 2.61. Hence, find the total power consumed by the passive elements of the network.*

(Elect. Circuits Nagpur Univ. 1992)

Solution. The node equation for node 1 is

$$V_1\left(1+1+\frac{1}{0.5}\right)-\frac{V_2}{0.5}-\frac{15}{1}=0$$

or $4V_1-2V_2=15$ ...(*i*)

Similarly, for node 2, we have

$$V_2\left(1+\frac{1}{2}+\frac{1}{0.5}\right)-\frac{V_1}{0.5}-\frac{20}{1}=0$$

or $4V_1-7V_2=-40$ ...(*ii*)

$\therefore$ $V_2=11$ volt and $V_1=37/4$ volt

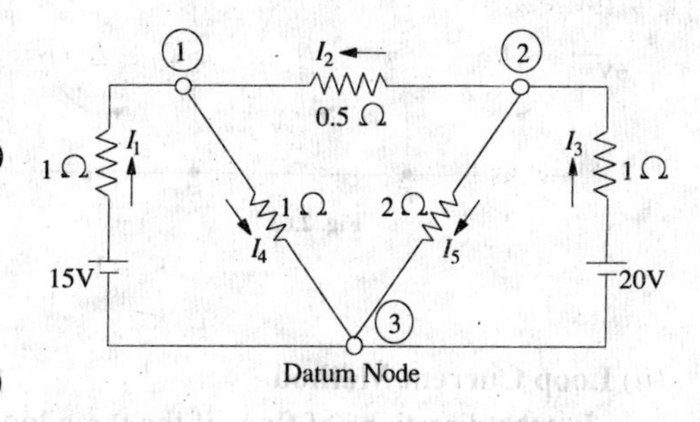

Fig. 2.61

Now,
$$I_1=\frac{15-37/4}{I}=\frac{23}{4}\text{A}=5.75\text{ A}\ ;\ I_2=\frac{11-37/4}{0.5}=3.5\text{ A}$$

$$I_4=5.75+3.5=9.25\text{ A}\ ;\ I_3=\frac{20-11}{1}=9\text{ A};\ I_5=9-3.5=5.5\text{ A}$$

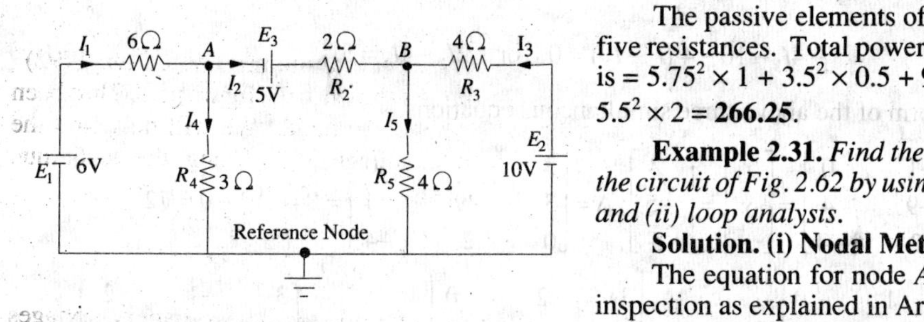

Fig. 2.62

The passive elements of the network are its five resistances. Total power consumed by them is $=5.75^2\times1+3.5^2\times0.5+9^2\times1+9.25^2\times1+5.5^2\times2=$ **266.25.**

Example 2.31. *Find the branch currents in the circuit of Fig. 2.62 by using (i) nodal analysis and (ii) loop analysis.*

Solution. (i) Nodal Method

The equation for node *A* can be written by inspection as explained in Art. 2-12.

$$V_A\left(\frac{1}{R_1}+\frac{1}{R_2}+\frac{1}{R_4}\right)-\frac{E_1}{R_1}-\frac{V_B}{R_2}+\frac{E_3}{R_2}=0$$

Substituting the given data, we get,

$$V_A\left(\frac{1}{6}+\frac{1}{2}+\frac{1}{3}\right)-\frac{6}{6}-\frac{V_B}{2}+\frac{5}{2}=0\text{ or }2V_A-V_B=-3 \qquad ...(i)$$

For node B, the equation becomes

$$V_B\left(\frac{1}{R_2}+\frac{1}{R_3}+\frac{1}{R_5}\right)-\frac{E_2}{R_3}-\frac{V_A}{R_2}-\frac{E_3}{R_2}=0$$

$$\therefore\ V_B\left(\frac{1}{2}+\frac{1}{4}+\frac{1}{4}\right)-\frac{10}{4}-\frac{V_A}{2}-\frac{5}{2}=0 \qquad \therefore\ V_B-\frac{V_A}{2}=5 \qquad \ldots(ii)$$

From Eq. (*i*) and (*ii*), we get,

$$V_A=\frac{4}{3}\text{ V},\ V_B=\frac{17}{3}\text{ V}$$

$$\therefore \qquad I_1=\frac{E_1-V_A}{R_1}=\frac{6-4/3}{6}=\frac{\mathbf{7}}{\mathbf{9}}\ \mathbf{A}$$

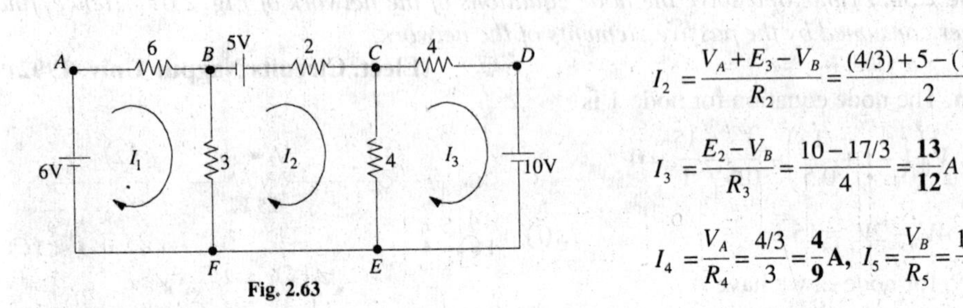

Fig. 2.63

$$I_2=\frac{V_A+E_3-V_B}{R_2}=\frac{(4/3)+5-(17/3)}{2}=\frac{\mathbf{1}}{\mathbf{3}}\ \mathbf{A}$$

$$I_3=\frac{E_2-V_B}{R_3}=\frac{10-17/3}{4}=\frac{\mathbf{13}}{\mathbf{12}}A$$

$$I_4=\frac{V_A}{R_4}=\frac{4/3}{3}=\frac{\mathbf{4}}{\mathbf{9}}\mathbf{A},\ I_5=\frac{V_B}{R_5}=\frac{17/3}{4}=\frac{\mathbf{17}}{\mathbf{12}}\mathbf{A}$$

(*ii*) Loop Current Method

Let the directions of flow of the three loop currents be as shown in Fig. 2.63.

Loop ABFA :

$$-6I_1-3(I_1-I_2)+6=0$$

or $$3I_1-I_2=2 \qquad \ldots(i)$$

Loop BCEFB :

$$+5-2I_2-4(I_2-I_3)-3(I_2-I_1)=0$$

or $$3I_1-9I_2+4I_3=-5 \qquad \ldots(ii)$$

Loop CDEC :

$$-4I_3-10-4(I_3-I_2)=0 \quad \text{or} \quad 2I_2-4I_3=5\ldots \qquad (iii)$$

The matrix form of the above three simultaneous equations is

$$\begin{bmatrix}3 & -1 & 0\\ 3 & -9 & 4\\ 0 & 2 & -4\end{bmatrix}=\begin{bmatrix}x\\ y\\ z\end{bmatrix}=\begin{bmatrix}2\\ -5\\ 5\end{bmatrix};\ \Delta=\begin{bmatrix}3 & -1 & 0\\ 3 & -9 & 4\\ 0 & 2 & -4\end{bmatrix}=84-12-0=72$$

$$\Delta_1=\begin{vmatrix}2 & 1 & 0\\ -5 & -9 & 4\\ 5 & 2 & -4\end{vmatrix}=56;\ \Delta_2=\begin{vmatrix}3 & 2 & 0\\ 3 & -5 & 4\\ 0 & 5 & -4\end{vmatrix}=24;\ \Delta_3=\begin{vmatrix}3 & -1 & 2\\ 3 & -9 & -5\\ 0 & 2 & 5\end{vmatrix}=-78$$

$$\therefore\ I_1=\Delta_1/\Delta=56/72=7/9\text{ A};\ I_2=\Delta_2/\Delta=24/72=1/3\text{ A}$$

$$I_3=\Delta_3/\Delta=-78/72=\mathbf{-13/12\ A}$$

The negative sign of I_3 shows that it is flowing in a direction opposite to that shown in Fig. 2.64

i.e. it flows in the CCW direction. The actual directions are as shown in Fig. 2.64.

The various branch currents are as under :

$$I_{AB} = I_1 = \mathbf{7/9\ A};\ I_{BF} = I_1 - I_2 = \frac{7}{9} - \frac{1}{3} = \frac{\mathbf{4}}{\mathbf{9}}\mathbf{A}$$

$$I_{BC} = I_2 = \frac{\mathbf{1}}{\mathbf{3}}A\ ;\ I_{CE} = I_2 + I_3 = \frac{1}{3} + \frac{13}{12} = \frac{\mathbf{17}}{\mathbf{12\ A}}$$

$$I_{DC} = I_3 = \frac{\mathbf{13}}{\mathbf{12}}\mathbf{A}$$

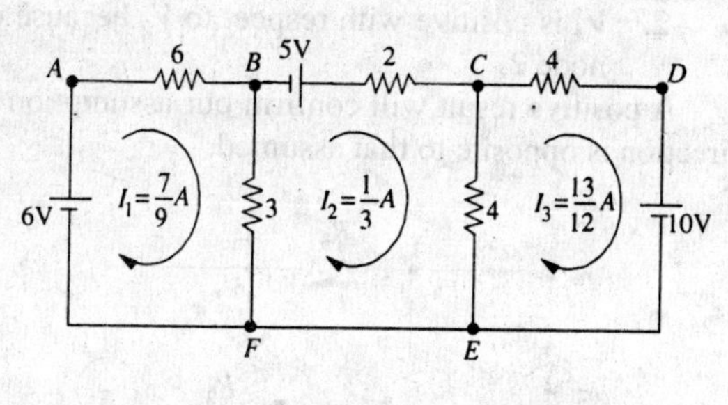

Fig. 2.64

Solution by Using Mesh Resistance Matrix

From inspection of Fig. 2.64, we have

$R_{11} = 9\ ;\ R_{22} = 9\ ;\ R_{33} = 8$

$R_{12} = R_{21} = -3\ \Omega\ ;\ R_{23} = R_{32} = -4\ \Omega;\ R_{13} = R_{31} = 0\ \Omega$

$E_1 = 6\ V;\ E_2 = 5\ V;\ E_3 = -10\ V$

$$\begin{bmatrix} R_{11} & R_{12} & R_{13} \\ R_{21} & R_{22} & R_{23} \\ R_{31} & R_{32} & R_{33} \end{bmatrix} \begin{bmatrix} I_1 \\ I_2 \\ I_3 \end{bmatrix} = \begin{bmatrix} E_1 \\ E_2 \\ E_3 \end{bmatrix} \text{ or } \begin{bmatrix} 9 & -3 & 0 \\ -3 & 9 & -4 \\ 0 & -4 & 8 \end{bmatrix} \begin{bmatrix} I_1 \\ I_2 \\ I_3 \end{bmatrix} = \begin{bmatrix} 6 \\ 5 \\ -10 \end{bmatrix}$$

$$\Delta = \begin{vmatrix} 9 & -3 & 0 \\ -3 & 9 & -4 \\ 0 & -4 & 8 \end{vmatrix} = 9(72-16) + 3(-24) = 432$$

$$\Delta_1 = \begin{vmatrix} 6 & -3 & 0 \\ 5 & 9 & -4 \\ -10 & -4 & 8 \end{vmatrix} = 6(72-16) - 5(-24) - 10(12) = 336$$

$$\Delta_2 = \begin{vmatrix} 9 & 6 & 0 \\ -3 & 5 & -4 \\ 0 & -10 & 8 \end{vmatrix} = 9(40-40) + 3(48) = 144$$

$$\Delta_3 = \begin{vmatrix} 9 & -3 & 6 \\ -3 & 9 & 5 \\ 0 & -4 & -10 \end{vmatrix} = 9(-90+90) + 3(30+24) = -468$$

$I_1 = \Delta_1/\Delta = 336/432 = 7/9\ A$

$I_2 = \Delta_2/\Delta = 144/432 = 1/3\ A$

$I_3 = \Delta_3/\Delta = -468/432 = -13/12\ A$

These are the same values as found above.

2.13. Nodal Analysis with Current Sources

Consider the network of Fig. 2.65 (*a*) which has two current sources and three nodes out of which 1 and 2 are independent ones whereas No. 3 is the reference node.

The given circuit has been redrawn for ease of understanding and is shown in Fig. 2.65 (*b*). The current directions have been taken on the assumptions that

1. both V_1 and V_2 are positive with respect to the reference node. That is why their respective currents flow from nodes 1 and 2 to node 3.

2. V_1 is positive with respect to V_2 because current has been shown flowing from node 1 to node 2.

A positive result will confirm our assumption whereas a negative one will indicate that actual direction is opposite to that assumed.

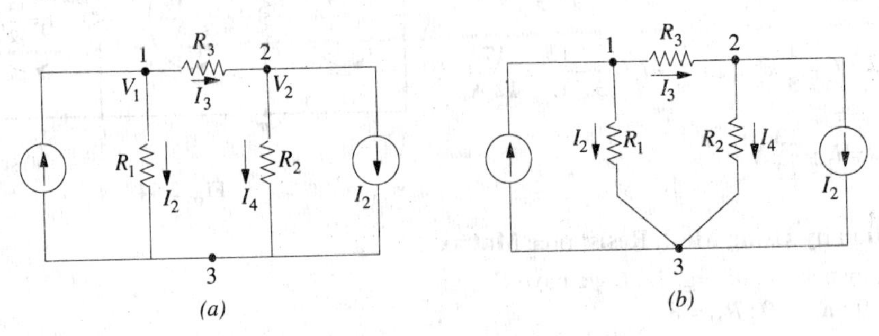

Fig. 2.65

We will now apply KCL to each node and use Ohm's law to express branch currents in terms of node voltages and resistances.

Node 1

$$I_1 - I_2 - I_3 = 0 \quad \text{or} \quad I_1 = I_2 + I_3$$

Now,
$$I_2 = \frac{V_1}{R_1} \quad \text{and} \quad I_3 = \frac{V_1 - V_2}{R_3}$$

$$\therefore \quad I_1 = \frac{V_1}{R_1} + \frac{V_1 - V_2}{R_3} \quad \text{or} \quad V_1\left(\frac{1}{R_1} + \frac{1}{R_3}\right) - \frac{V_2}{R_3} = I_1 \qquad \ldots(i)$$

Node 2

$$I_3 - I_2 - I_4 = 0 \quad \text{or} \quad I_3 = I_2 + I_4$$

Now,
$$I_4 = \frac{V_2}{R_2} \quad \text{and} \quad I_3 = \frac{V_1 - V_2}{-R_3} \quad \text{– as before}$$

$$\therefore \quad \frac{V_1 - V_2}{R_3} = I_2 + \frac{V_2}{R_2} \quad \text{or} \quad V_2\left(\frac{1}{R_2} + \frac{1}{R_3}\right) - \frac{V_1}{R_3} = -I_1 \qquad \ldots(ii)$$

The above two equations can also be written by simple inspection. For example, Eq. (*i*) is represented by

(1) *product* of potential V_1 and $(1/R_1 + 1–R_3)$ *i.e.* sum of the reciprocals of the branch resistances connected to this node.

(2) *minus* the ratio of adjoining potential V_2 and the interconnecting resistance R_3.

(3) all the above equated to the current supplied by the current source connected to this node.

This current is taken *positive* if flowing *into* the node and negative if flowing *out* of it (as per sign convention of Art. 2.3). Same remarks apply to Eq. (*ii*) where I_2 has been taken negative because it flows *away* from node 2.

In terms of branch conductances, the above two equations can be put as

$$V_1(G_1 + G_3) - V_2G_3 = I_1 \quad \text{and} \quad V_2(G_2 + G_3) - V_1G_3 = -I_2$$

Example 2.32. *Use nodal analysis method to find currents in the various resistors of the circuit shown in Fig. 2.66(a).*

Solution. The given circuit is redrawn in Fig. 2.66 (*b*) with its different nodes marked 1, 2, 3 and 4, the last one being taken as the reference or datum node. The different node-voltage equations are as under :

Node 1 $$V_1\left(\frac{1}{2}+\frac{1}{2}+\frac{1}{10}\right)-\frac{V_2}{2}-\frac{V_3}{10}=8$$

or $$11V_1-5V_2-V_3-280=0 \qquad \ldots(i)$$

Node 2 $$V_2\left(\frac{1}{2}+\frac{1}{5}+1\right)-\frac{V_1}{2}-\frac{V_3}{1}=0$$

or $$5V_1-17V_2+10V_3=0 \qquad \ldots(ii)$$

Node 3 $$V_3\left(\frac{1}{4}+1+\frac{1}{10}\right)-\frac{V_2}{1}-\frac{V_1}{10}=-2$$

or $$V_1+10V_2-13.5V_3-20=0 \qquad \ldots(iii)$$

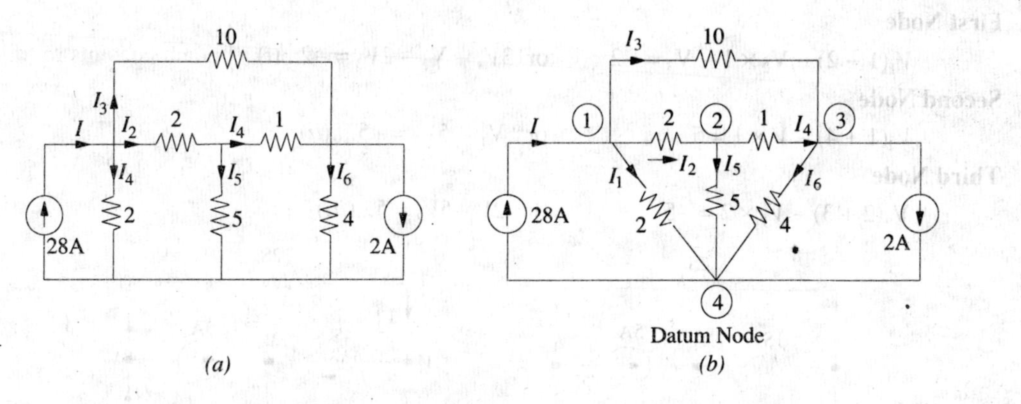

Fig. 2.66

The matrix form of the above three equations is

$$\begin{bmatrix} 11 & -5 & -1 \\ 5 & -17 & 10 \\ 1 & 10 & -13.5 \end{bmatrix}\begin{bmatrix} x \\ y \\ z \end{bmatrix}=\begin{bmatrix} 280 \\ 0 \\ 20 \end{bmatrix}$$

$$\Delta=\begin{vmatrix} 11 & -5 & -1 \\ 5 & -17 & 10 \\ 1 & 10 & -13.5 \end{vmatrix}=1424.5-387.5-67=970$$

$$\Delta_1=\begin{vmatrix} 280 & -5 & -1 \\ 0 & -17 & 10 \\ 20 & 20 & -13.5 \end{vmatrix}=34{,}920, \quad \Delta_2=\begin{vmatrix} 11 & 280 & -1 \\ 5 & 0 & 10 \\ 1 & 20 & -13.5 \end{vmatrix}=19{,}400$$

$$\Delta_3=\begin{vmatrix} 11 & -5 & 280 \\ 5 & -17 & 0 \\ 1 & 10 & 20 \end{vmatrix}=15{,}520$$

$$V_1=\frac{\Delta_1}{\Delta}=\frac{34{,}920}{970}=36\text{ V},\ V_2=\frac{\Delta_2}{\Delta}=\frac{19{,}400}{970}=20\text{ V},\ V_3=\frac{\Delta_3}{\Delta}=\frac{15{,}520}{970}=16\text{ V}$$

It is obvious that all nodes are at a higher potential with respect to the datum node. The various currents shown in Fig. 2.66 (*b*) can now be found easily.

$$I_1 = V_1/2 = 36/2 = \mathbf{18\ A}$$

$$I_2 = (V_1 - V_2)/2 = (36 - 20)/2 = \mathbf{8\ A}$$

$$I_3 = (V_1 - V_3)/10 = (36 - 16)/10 = \mathbf{2\ A}$$

It is seen that total current, as expected, is $18 + 8 + 2 = \mathbf{28\ A}$

$$I_4 = (V_2 - V_3)/1 = (20 - 16)/1 = \mathbf{4\ A}$$

$$I_5 = V_2/5 = 20/5 = \mathbf{4\ A},\ I_6 = V_3/4 = 16/4 = \mathbf{4\ A}$$

Example 2.33. *Using nodal analysis, find the different branch currents in the circuit of Fig. 2.67(a). All branch conductances are in siemens (i.e. mho).*

Solution. Let the various branch currents be as shown in Fig. 2.67 (*b*). Using the procedure detailed in Art. 2.11, we have

First Node

$V_1(1 + 2) - V_2 \times 1 - V_3 = -2$ or $3V_1 - V_2 - 2V_3 = -2$...(*i*)

Second Node

$V_2(1 + 4) - V \times 1 = 5$ or $V_1 - 5V_2 = -5$....(*ii*)

Third Node

$V_3(2 + 3) - V_1 \times 2 = -5$ or $2V_1 - 5V_3 = 5$...(*iii*)

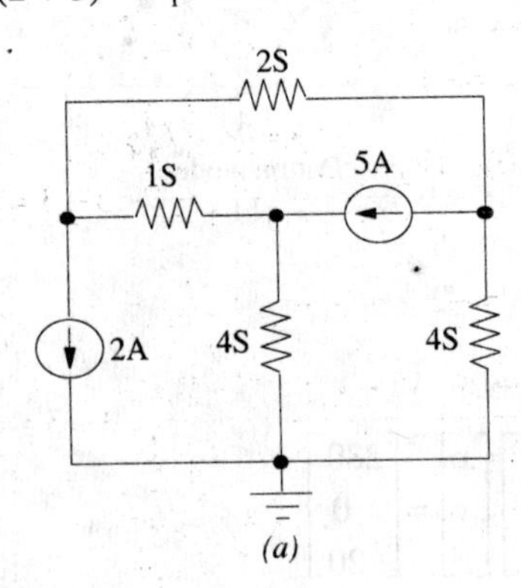

(a)

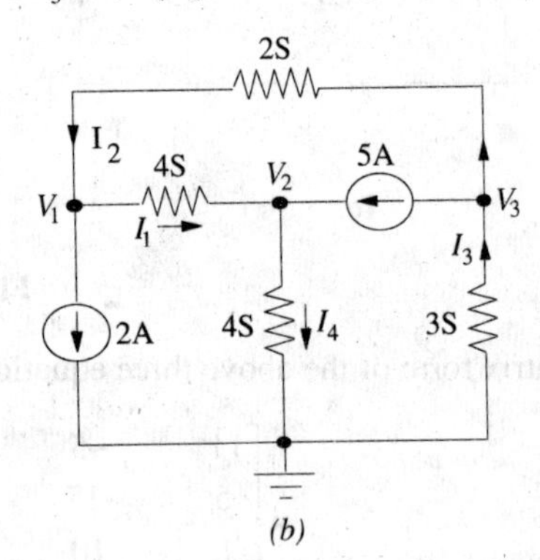

(b)

Fig. 2.67

Solving for the different voltages, we have

$$V_1 = -\frac{3}{2}\text{V},\ V_2 = \frac{7}{10}\text{V and } V_3 = -\frac{8}{5}\text{V}$$

$$I_1 = (V_1 - V_2) \times 1 = (-1.5 - 0.7) \times 1 = \mathbf{-2.2\ A}$$

$$I_2 = (V_3 - V_1) \times 2 = [-1.6 - (-1.5)] \times 2 = \mathbf{-0.2\ A}$$

$$I_4 = V_2 \times 4 = 4 \times (7/10) = \mathbf{2{,}8\ A}$$

$$I_3 = 2 + 2.8 = \mathbf{4.8\ A}$$

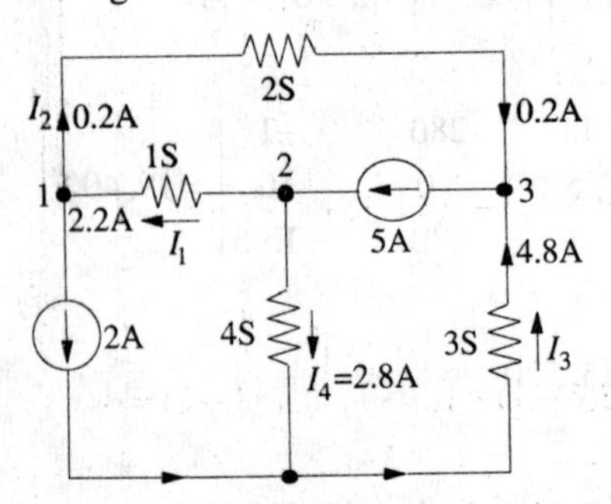

Fig. 2.68

As seen, I_1 and I_2 flow in directions opposite to those originally assumed (Fig. 2.68).

Example 2.34. *Find the current I in Fig. 2.69 (a) by changing the two voltage sources into their equivalent current sources and then using Nodal method. All resistances are in ohms.*

Solution. The two voltage sources have been converted into their equivalent current sources in Fig. 2.69 (*b*). The circuit has been redrawn as shown in Fig. 2.69 (*c*) where node No. 4 has been

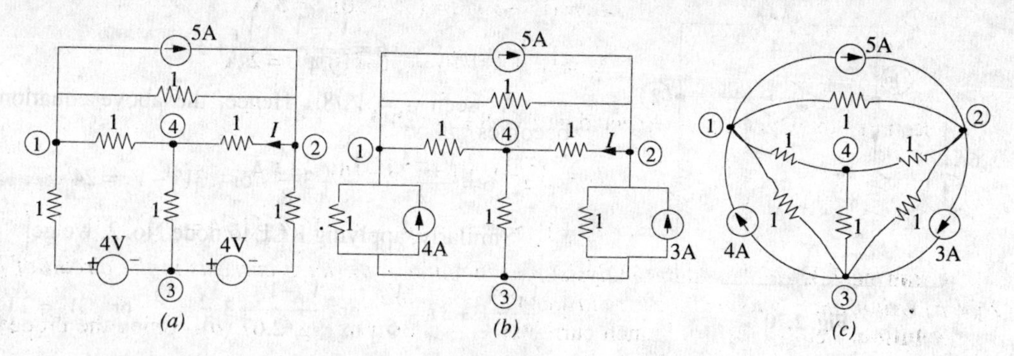

Fig. 2.69

taken as the reference node or common ground for all other nodes. We will apply *KCL* to the three nodes and take currents coming towards the nodes as positive and those going away from them as negative. For example, current going away from node No. 1 is $(V_1 - V_2)/1$ and hence would be taken as negative. Since 4A current is coming towards node No. 1, it would be taken as positive but 5A current would be taken as negative.

Node 1 : $\frac{(V_1-0)}{1}-\frac{(V_1-V_2)}{1}-\frac{(V_1-V_3)}{1}-5+4=0$

or $$3V_1-V_2-V_3=-1 \qquad \ldots(i)$$

Node 2 : $-\frac{(V_2-0)}{1}-\frac{(V_2-V_3)}{1}-\frac{(V_2-V_1)}{1}+5-3=0$

or $$V_1-3V_2+V_3=-2 \qquad \ldots(ii)$$

Node 3 : $-\frac{(V_3-0)}{1}-\frac{V_3-V_1}{1}-\frac{V_3-V_2}{1}-4+3=0$

or $$V_1+V_2-3V_3=1 \qquad \ldots(iii)$$

The matrix form of the above three equations is

$$\begin{bmatrix} 3 & -1 & -1 \\ 1 & -3 & 1 \\ 1 & 1 & -3 \end{bmatrix} = \begin{bmatrix} V_1 \\ V_2 \\ V_3 \end{bmatrix} = \begin{bmatrix} -1 \\ -2 \\ 1 \end{bmatrix}$$

$$\Delta = \begin{vmatrix} 3 & -1 & -1 \\ 1 & -3 & 1 \\ 1 & 1 & -3 \end{vmatrix} = 3(9-1)-1(3+1)+1(-1-3)=16$$

$$\Delta_2 = \begin{vmatrix} 3 & -1 & -1 \\ 1 & -2 & 1 \\ 1 & 1 & -3 \end{vmatrix} = 3(6-1)-1(3+1)+1(-1-2)=8$$

$\therefore$ $$V_2 = \Delta_2/\Delta = 8/16 = 0.5\ \text{V}$$

$\therefore$ $$I = V_2/1 = 0.5\ \text{A}$$

Example 2.35. *Use Nodal analysis to determine the value of current i in the network of Fig. 2.70.*

Solution. We will apply *KCL* to the two nodes 1 and 2. Equating the incoming currents at node 1 to the outgoing currents, we have

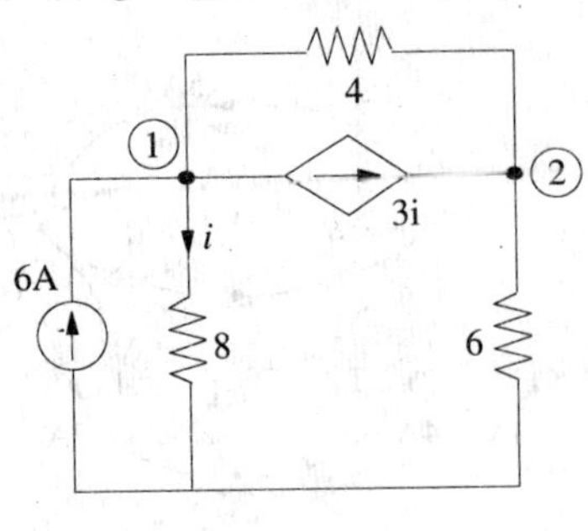

Fig. 2.70

$$6 = \frac{V_1 - V_2}{4} + \frac{V_1}{8} + 3i$$

As seen, $i = V_1/8$. Hence, the above equation becomes

$$6 = \frac{V_1 - V_2}{4} + \frac{V_1}{8} + 3\frac{V_1}{8} \quad \text{or} \quad 3V_1 - V_2 = 24$$

Similarly, applying KCL to node No. 2, we get

$$\frac{V_1 - V_2}{4} + 3i = \frac{V_2}{6} \quad \text{or} \quad \frac{V_1 - V_2}{4} + 3\frac{V_1}{8} = \frac{V_2}{6} \quad \text{or} \quad 3V_1 = 2V_2$$

From the above two equations, we get

$V_1 = 16$ V $\therefore$ $i = 16/8 =$ **2 A.**

Example 2.36. *Using Nodal analysis, find the node voltages V_1 and V_2 in Fig. 2.71.*

Solution. Applying *KCL* to node 1, we get

$$8 - 1 - \frac{V_1}{3} - \frac{(V_1 - V_2)}{6} = 0 \quad \text{or} \quad 3V_1 - V_2 = 42 \quad ...(i)$$

Similarly, applying *KCL* to node 2, we get

$$1 + \frac{(V_1 - V_2)}{6} - \frac{V_2}{15} - \frac{V_2}{10} = 0 \quad \text{or} \quad V_1 - 2V_2 = -6 \quad ...(ii)$$

Solving for V_1 and V_2 from Eqn. (*i*) and (*ii*), we get $V_1 = 18$ V and $V_2 = 12$ V.

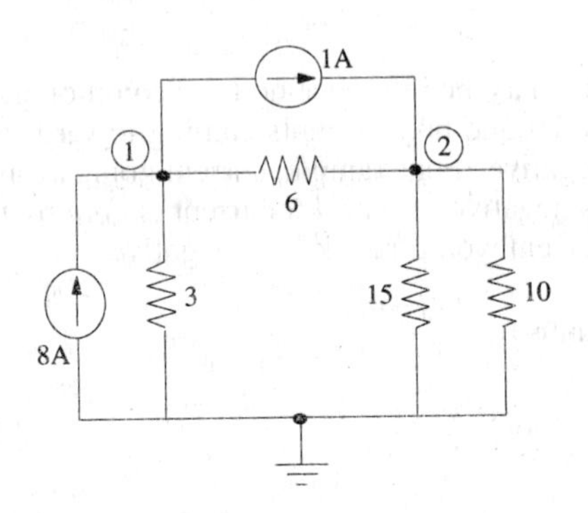

Fig. 2.71

2.14. Source Conversion

A given voltage source with a series resistance can be converted into (or replaced by) an equivalent current source with a parallel resistance. Conversely, a current source with a parallel resistance can be converted into a voltage source with a series resistance. Suppose, we want to convert the voltage source of Fig. 2.72 (*a*) into an equivalent current source. First, we will find the value of current supplied by the source when a 'short' is put across its terminals *A* and *B* as shown in Fig. 2.72 (*b*). This current is $I = V/R$.

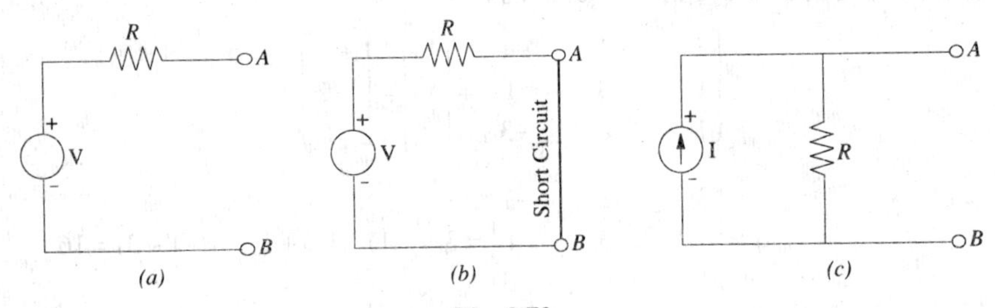

Fig. 2.72

A current source supplying this current *I* and having the same resistance *R* connected in *parallel* with it represents the equivalent source. It is shown in Fig. 2.72 (*c*). Similarly, a current source of *I* and a parallel resistance *R* can be converted into a voltage source of voltage $V = IR$ and a resistance *R* in series with it. It should be kept in mind that a voltage source-series resistance combination is equivalent to (or is replaceable by) a current source-parallel resistance combination if, and only if their

1. respective open-circuit voltages are equal, and
2. respective short-circuit currents are equal.

For example, in Fig. 2.72 (*a*), voltage across terminals *A* and *B* when they are open (*i.e.* open-circuit voltage V_{OC}) is *V* itself because there is no drop across *R*. Short-circuit current across $AB = I = V/R$.

Now, take the circuit of Fig. 2.72 (*c*). The open-circuit voltage across AB = drop across $R = IR = V$. If a short is placed across *AB*, whole of *I* passes through it because *R* is completely shorted out.

Example. 2.37. *Convert the voltage source of Fig. 2.73 (a) into an equivalent current source.*

Solution. As shown in Fig. 2.73 (*b*), current obtained by putting a short across terminals *A* and *B* is = 10/5 = 2A.

Hence, the equivalent current source is as shown in Fig. 2.73 (*c*).

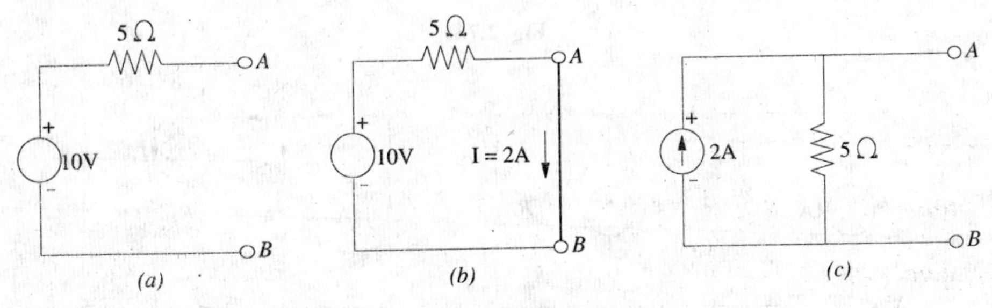

Fig. 2.73

Example 2.38. *Find the equivalent voltage source for the current source in Fig. 2.74 (a).*

Solution. The open-circuit voltage across terminals *A* and *B* in Fig. 2.74 (*a*) is

$$V_{OC} = \text{drop across } R$$
$$= 5 \times 2 = 10 \text{ V}$$

Hence, voltage source has a voltage of 10 V and the same resistance of 2 Ω though connected in series [Fig. 2.74 (*b*)].

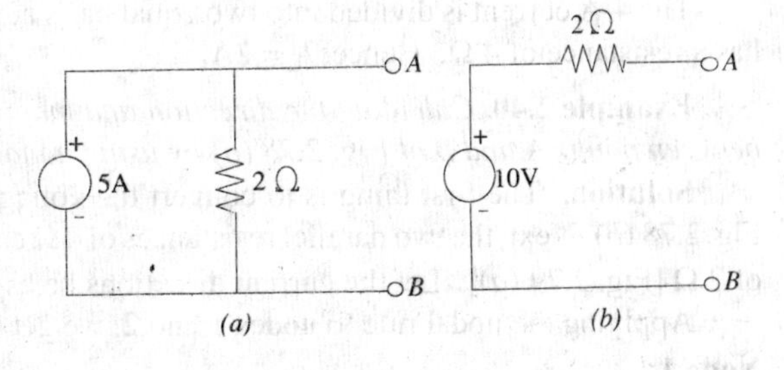

Fig. 2.74

Example 2.39. *Use Source Conversion technique to find the load current I in the circuit of Fig. 2.75 (a).*

Solution. As shown in Fig. 2.75 (*b*). 6-V voltage source with a series resistance of 3 Ω has been converted into an equivalent 2A current source with a 3 Ω resistance in parallel.

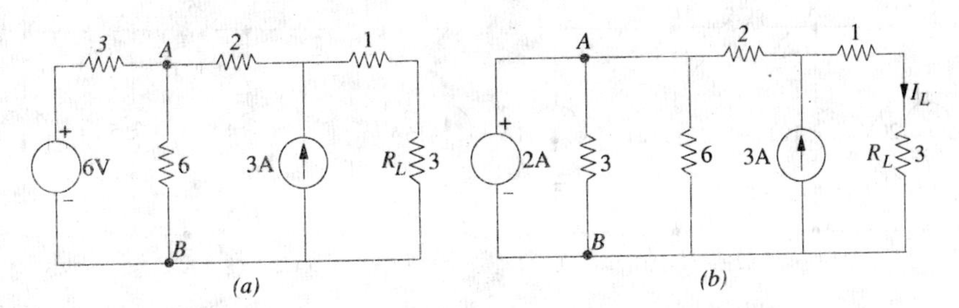

Fig. 2.75

The two parallel resistances of 3 Ω and 6 Ω can be combined into a single resistance of 2 Ω as shown in Fig. 2.76.(a)

The two current sources cannot be combined together because of the 2 Ω resistance present between points *A* and *C*. To remove this hurdle, we convert the 2A current source into the equivalent 4 V voltage source as shown in Fig. 2.76 (*b*). Now, this 4 *V* voltage source with a series resistance of (2 + 2) = 4 Ω can again be converted into the equivalent current source as shown in Fig. 2.77 (*a*). Now, the two current sources can be combined into a single 4-A source as shown in Fig. 2.77 (*b*).

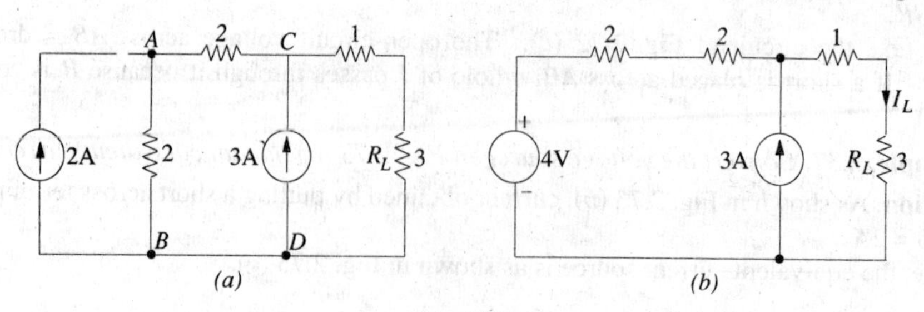

Fig. 2.76

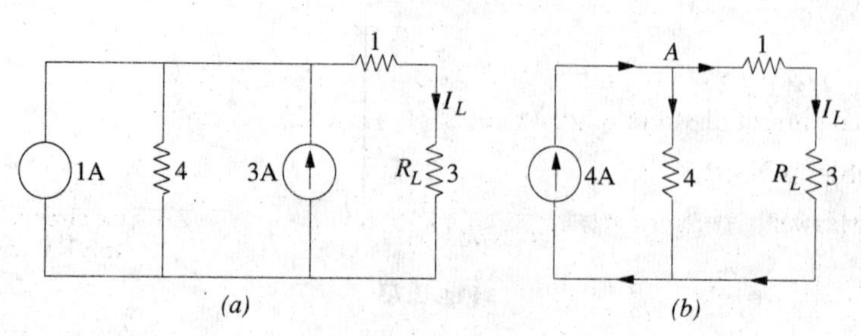

Fig. 2.77

The 4-A current is divided into two equal parts at point *A* because each of the two parallel paths has a resistance of 4 Ω. Hence, I_L = **2A.**

Example 2.40. *Calculate the direction and magnitude of the current through the 5 Ω resistor between points A and B of Fig. 2.78 (a) by using nodal voltage method.*

Solution. The first thing is to convert the voltage source into the current source as shown in Fig. 2.78 (*b*). Next, the two parallel resistances of 4 Ω each can be combined to give a single resistance of 2 Ω [Fig. 2.79 (*a*)]. Let the current directions be as indicated.

Applying the nodal rule to nodes 1 and 2, we get

Node 1

$$V_1\left(\frac{1}{2}+\frac{1}{5}\right)-\frac{V_2}{5} = 5 \text{ or } 7V_1-2V_2=50 \qquad \ldots(i)$$

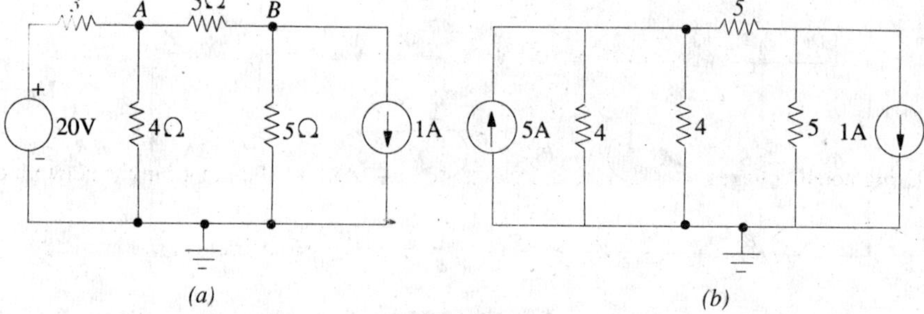

Fig. 2.78

Node 2

$$V_2\left(\frac{1}{5}+\frac{1}{5}\right)-\frac{V_1}{5} = -1 \text{ or } V_1-2V_2=5 \qquad \ldots(ii)$$

Solving for V_1 and V_2, we get $V_1 = \frac{15}{2}$ V and $V_2 = \frac{5}{4}$ V.

$$\therefore \quad I_2 = \frac{V_1 - V_2}{5} = \frac{15/2 - 5/4}{5} = \mathbf{1.25\ A}$$

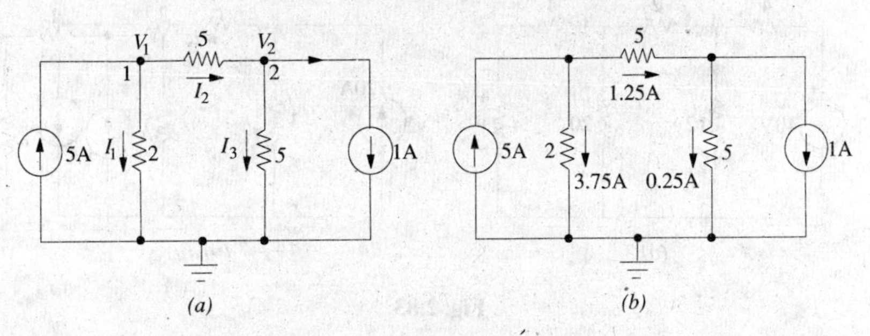

Fig. 2.79

Similarly, $I_1 = V_1/2 = 15/4 = 3.75$ A; $I_3 = V_2/5 = 5/20 = 0.25$ A.

The actual current distribution becomes as shown in Fig. 2.79 (*b*).

Tutorial Problems No. 2.3

1. Using Maxwell's loop current method, calculate the output voltage V_0 for the circuits shown in Fig. 2.80.
[(*a*) **4 V** (*b*) **–150/7 V** (*c*) $\mathbf{V_0 = 0}$ (*d*) $\mathbf{V_0 = 0}$]

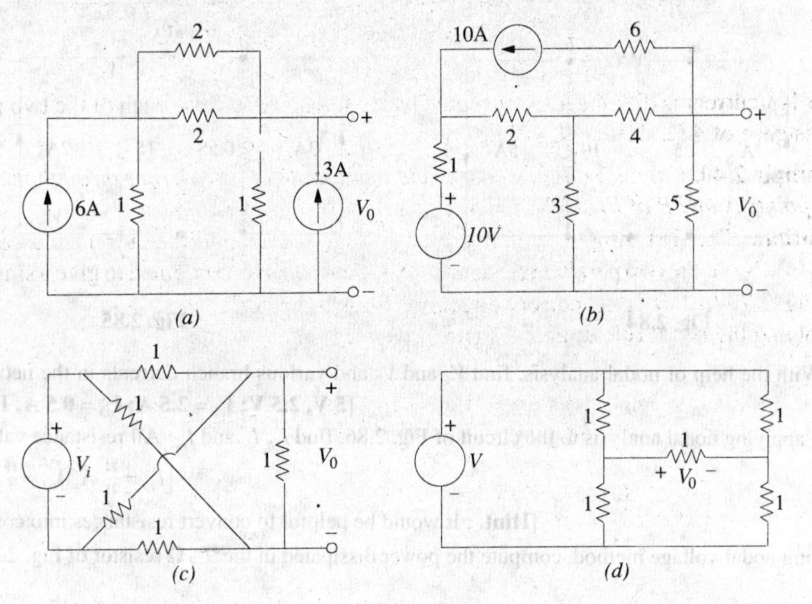

Fig. 2.80

2. Using nodal voltage method, find the magnitude and direction of current I in the network of Fig. 2.81.

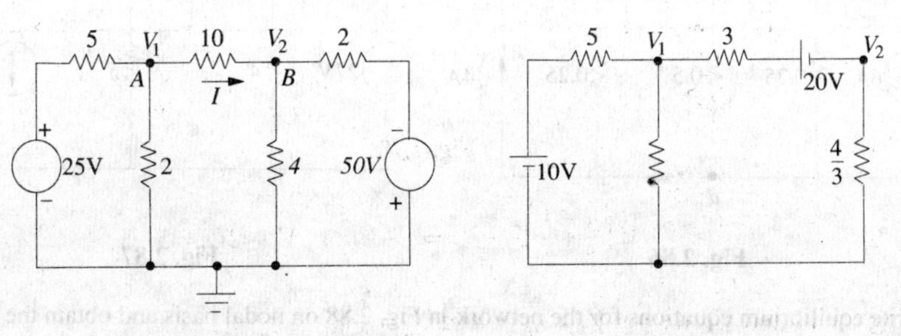

Fig. 2.81 Fig. 2.82

3. By using repeated source transformations, find the value of voltage v in Fig. 2.83(a). **[8 V]**

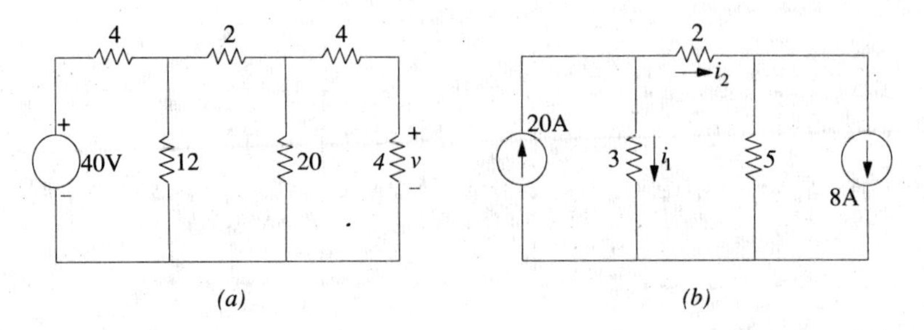

Fig. 2.83

4. Use source transformation technique to find the current flowing through the 2 Ω resistor in Fig. 2.83(b). **[10 A]**

5. With the help of nodal analysis, calculate the values of nodal voltage V_1 and V_2 in the circuit of Fig. 2.82. **[7.1 V; –3.96 V]**

6. Use nodal analysis to find various branch currents in the circuit of Fig. 2.84
[**Hint** : Check by source conversion.] **[I_{ac} = 2A ; I_{ab} = 5 A, I_{bc} = 0]**

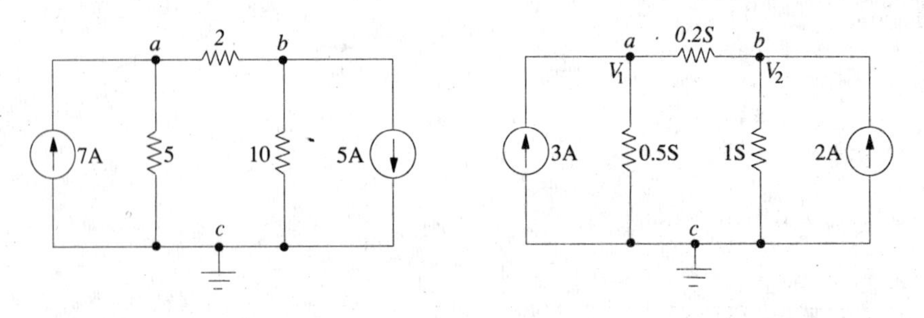

Fig. 2.84 **Fig. 2.85**

7. With the help of nodal analysis, find V_1 and V_2 and various branch currents in the network of Fig. 2.85. **[5 V, 2.5 V; I_{ac} = 2.5 A; I_{ab} = 0.5 A; I_{bc} = 2.5 A]**

8. By applying nodal analysis to the circuit of Fig. 2.86, find I_{ab}, I_{bd} and I_{bc}. All resistance values are in ohms. $\left[I_{ab} = \frac{22}{21}\text{ A}, I_{bd} = \frac{10}{7}\text{ A}, I_{bc} = \frac{-8}{21}\text{ A}\right]$

[Hint. : It would be helpful to convert resistances into conductances.]

9. Using nodal voltage method, compute the power dissipated in the 9– Ω resistor of Fig. 2.87. **[81 W]**

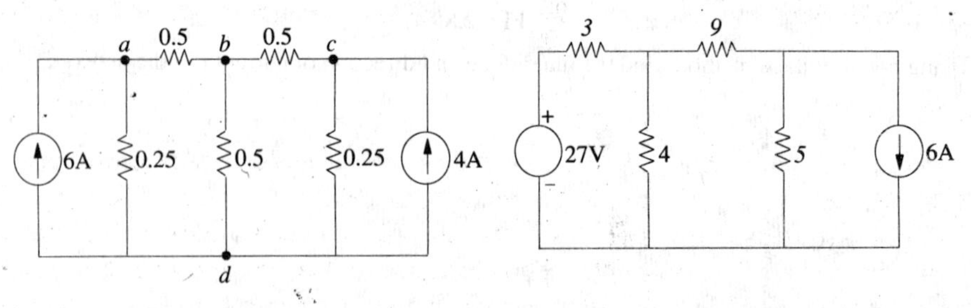

Fig. 2.86 **Fig. 2.87**

10. Write equilibrium equations for the network in Fig. 2.88 on nodal basis and obtain the voltage V_1, V_2 and V_3. All resistors in the network are of 1 Ω. [*Network Theory & Fields, Madras Univ. 1977*]

11. By applying nodal method of network analysis, find current in the 15 Ω resistor of the network shown in Fig. 2.89. **[3.5 A]** [*Elect. Technology-1, Gwalior Univ. 1977*]

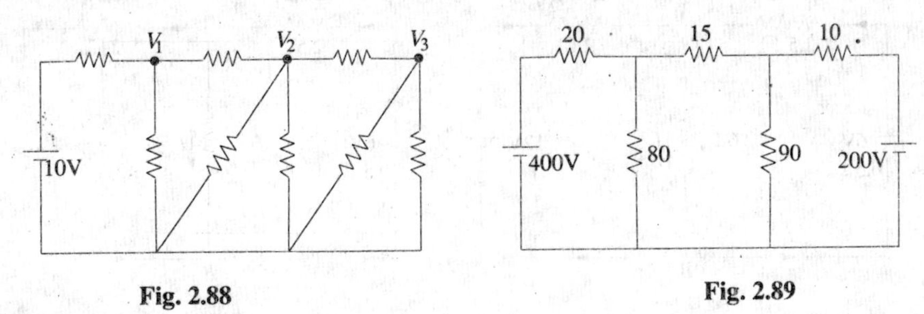

Fig. 2.88 **Fig. 2.89**

2.15. Ideal Constant-Voltage Source

It is that voltage source (or generator) whose output voltage remains absolutely constant whatever the change in load current. Such a voltage source must possess *zero internal resistance so that internal voltage drop in the source is zero.* In that case, output voltage provided by the source would remain constant *irrespective of the amount of current drawn from it.* In practice, none such ideal constant-voltage source can be obtained. However, smaller the internal resistance r of a voltage source, closer it comes to the ideal source described above.

Suppose, a 6–V battery has an internal resistance of 0.005 Ω [Fig. 2.90(*a*)]. When it supplies no current *i.e.* it is on no-load, $V_0 = 6$ V *i.e.* output voltage provided by it at its output terminals A and B is 6 V. If load current increases to 100 A, internal drop = 100 × 0.005 = 0.5 V. Hence, $V_0 = 6 - 0.5 = 5.5$ V.

Obviously an output voltage of 5.5 – 6 V can be considered constant as compared to wide variations in load current from 0 A to 100 A.

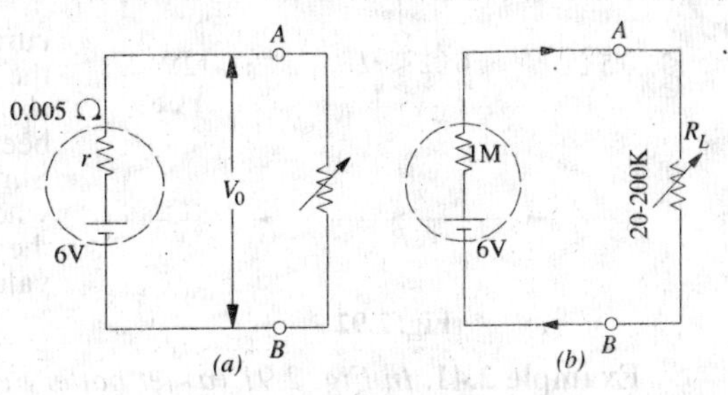

Fig. 2.90

2.16. Ideal Constant–Current Source

It is that voltage source whose internal resistance is infinity. In practice, it is approached by a source which possesses very high resistance as compared to that of the external load resistance. As shown in Fig. 2.90(*b*), let the 6-V battery or voltage source have an internal resistance of 1 M Ω and let the load resistance vary from 20 K to 200 K. The current supplied by the source varies from 6/1.02 = 5.9 μ A to 6/1.2 = 5 μA. As seen, even when load resistance increases 10 times, current decreases by 0.9 μA. Hence, the source can be considered, for all practical purposes, to be a constant-current source.

2.17. Superposition Theorem

According to this theorem, if there are a number of e.m.fs. acting simultaneously in any linear bilateral network, then each e.m.f acts independently of the others *i.e.* as if the other e.m.fs. did not exist. The value of current in any conductor is the algebraic sum of the currents due to each e.m.f. Similarly, voltage across any conductor is the algebraic sum of the voltages which each e.m.f would have produced while acting singly. In other words, current in or voltage across, any conductor of the network is obtained by superimposing the currents and voltages due to each e.m.f. in the network. It is important to keep in mind that this theorem is applicable only to *linear* networks where current is *linearly* related to voltage as per Ohm's law.

Hence, this theorem may be stated as follows :

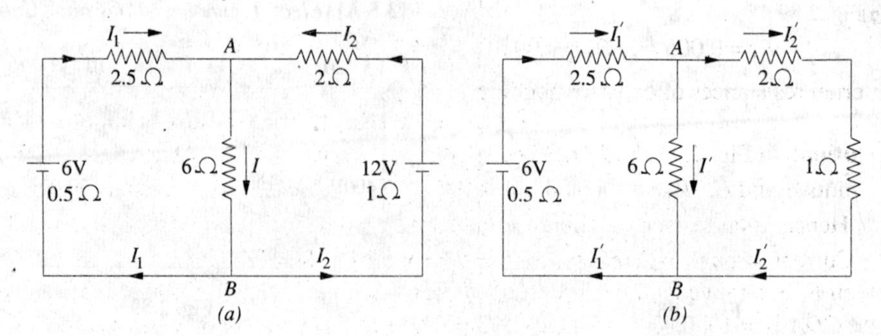

Fig. 2.91

In a network of linear resistances containing more than one generator (or source of e.m.f.), the current which flows at any point is the sum of all the currents which would flow at that point if each generator were considered separately and all the other generators replaced for the time being by resistances equal to their internal resistances.

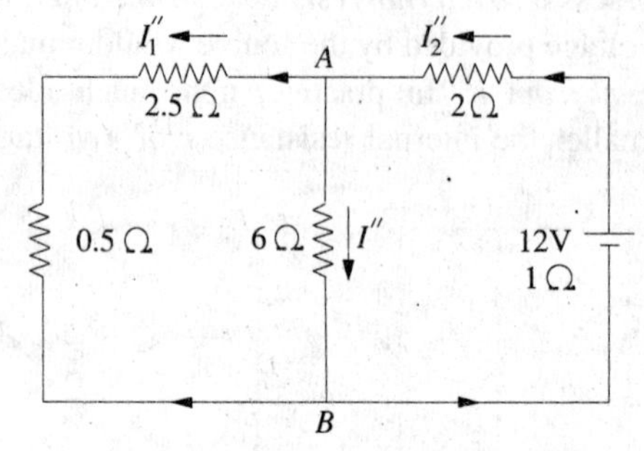

Fig. 2.92

Explanation

In Fig. 2.91 (*a*) I_1, I_2 and I represent the values of currents which are due to the simultaneous action of the two sources of e.m.f. in the network. In Fig. 2.91 (*b*) are shown the current values which would have been obtained if left-hand side battery had acted alone. Similarly, Fig. 2.92 represents conditions obtained when right-hand side battery acts alone. By combining the current values of Fig. 2.91 (*b*) and 2.92 the actual values of Fig. 2.91 (*a*) can be obtained.

Obviously, $I_1 = I_1' - I_1''$, $I_2 = I_2'' - I_2'$, $I = I' + I''$.

Example 2.41. *In Fig. 2.91 (a) let battery e.m.fs. be 6 V and 12 V, their internal resistances 0.5 Ω and 1 Ω. The values of other resistances are as indicated. Find the different currents flowing in the branches and voltage across 6-ohm resistor.*

Solution. In Fig. 2.91 (*b*), 12-volt battery has been removed though its internal resistance of 1 Ω remains. The various currents can be found by applying Ohm's Law.

It is seen that there are two parallel paths between points *A* and *B*, having resistances of 6 Ω and (2+1) = 3 Ω.

∴ equivalent resistance $= 3 \parallel 6 = 2\ \Omega$

Total resistance $= 0.5 + 2.5 + 2 = 5\ \Omega$ ∴ $I_1' = 6/5 = 1.2$ **A**

This current divides at point *A* inversely in the ratio of the resistances of the two parallel paths.

∴ $I' = 1.2 \times (3/9) = 0.4$ A. Similarly, $I_2' = 1.2 \times (6/9) = 0.8$ A

In Fig. 2.92, 6 volt battery has been removed but not its intrenal resistance. The various currents and their directions are as shown.

The equivalent resistance to the left of points *A* and *B* is $= 3 \parallel 6 = 2\ \Omega$

∴ total resistnce $= 1 + 2 + 2 = 5\ \Omega$ ∴ $I_2'' = 12/5 = 2.4$ A

At point *A*, this current is divided into two parts,

$I'' = 2.4 \times 3/9 = 0.8$ A $I_1'' = 2.4 \times 6/9 = 1.6$ A

The actual current values of Fig. 2.91 (*a*) can be obtained by superposition of these two sets of current values.

∴ $I_1 = I_1' - I_1'' = 1.2 - 1.6 = -\mathbf{0.4\ A}$ (it is a charging current)

$I_2 = I_2'' - I_2' = 2.4 - 0.8 = \mathbf{1.6\ A}$

$I = I' + I'' = 0.4 + 0.8 = \mathbf{1.2\ A}$

Voltage drop across 6-ohm resistor $= 6 \times 1.2 = \mathbf{7.2\ V}$

Example 2.42. *By using Superposition Theorem, find the current in resistance R shown in Fig. 2.93 (a)*

$$R_1 = 0.005\ \Omega,\quad R_2 = 0.004\ \Omega,\quad R = 1\ \Omega, E_1 = 2.05\ \text{V},\quad E_2 = 2.15\ \text{V}$$

Internal resistances of cells are negligible.

(Electronic Circuits, Allahabad Univ. 1992)

Solution. In Fig. 2.93 (*b*), E_2 has been removed. Resistances of 1 Ω and 0.04 Ω are in parallel across points *A* and *C*. $R_{AC} = 1 \parallel 0.04 = 1 \times 0.04/1.04 = 0.038\ \Omega$. This resistance is in series with 0.05 Ω. Hence, total resistance offered to battery $E_1 = 0.05 + 0\ 038 = 0.088\ \Omega$. $I = 2.05/0.088 =$ 23.3 A. Current through 1-Ω resistance, $I_1 = 23.3 \times 0.04/1.04 = 0.896$ A from *C* to *A*.

When E_1 is removed, circuit becomes as shown in Fig. 2.93 (*c*). Combied resistance of paths *CBA* and *CDA* is $= 1 \parallel 0.05 = 1 \times 0.05/1.05 = 0.048\ \Omega$. Total resistance offered to E_2 is $= 0.04 + 0.048 = 0.088\ \Omega$. Current $I = 2.15/0.088 = 24.4$ A. Again, $I_2 = 24.4 \times 0.05\ /1.05 = 1.16$ A.

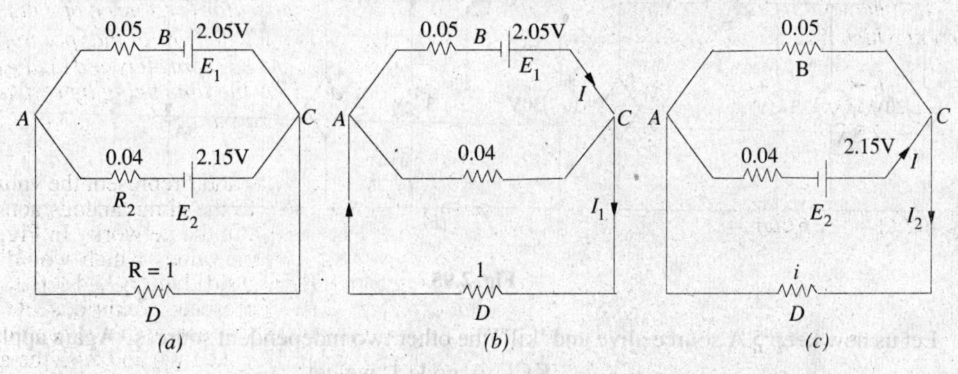

Fig. 2.93

Total current through 1-Ω resistance when both batteries are present

$$= I_1 + I_2 = 0.896 + 1.16 = \mathbf{2.056\ A}.$$

Example 2.43. *Use Superposition theorem to find current I in the circuit shown in Fig. 2.94(a). All resistances are in ohms.* **(Basic Circuit Analysis Osmania Univ. Jan/Feb 1992)**

Solution. In Fig. 2.94 (*b*), the voltage source has been replaced by a short and the 40 A current source by an open. Using the current-divider rule, we get $I_1 = 120 \times 50\ /200 = 30$ A.

In Fig. 2.94 (*c*), only 40 A current source has been considered. Again, using current-divider rule, $I_2 = 40 \times 150/200 = 30$ A.

In Fig. 2.94 (*d*), only voltage source has been considered. Using Ohm's law,

$I_3 = 10/200 = 0.05$ A.

Since I_1 and I_2 cancel out, $I = I_3 = 0.05$ A.

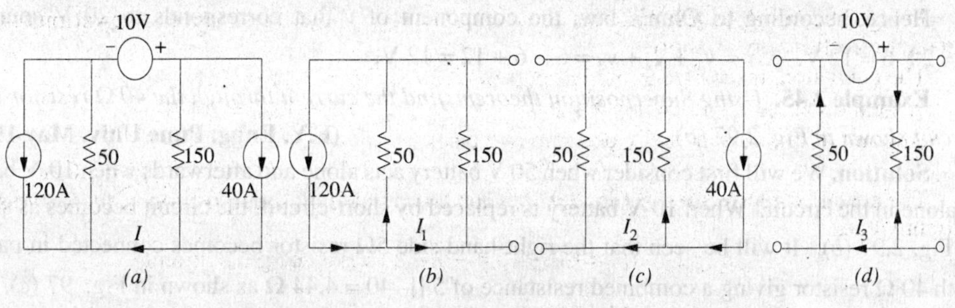

Fig. 2.94

Example 2.44. *Use Superposition theorem to determine the voltage v in the network of Fig. 2.95(a).*

Solution. As seen, there are three independent sources and one dependent source. We will find the value of v produced by each of the three independent sources when acting alone and add the three values to find v. It should be noted that unlike independent sources, a dependent source cannot be set to zero *i.e.* it cannot be 'killed' or deactivated.

Let us find the value of v_1 due to 30 V source only. For this purpose we will replace current source by an open circuit and the 20 V source by a short circuit as shown in Fig. 2.95 (*b*). Applying KCL to node 1, we get

$$\frac{(30-v_1)}{6}-\frac{v_1}{3}+\frac{(v_1/3-v_1)}{2}= \text{ or } v_1=6\text{V}$$

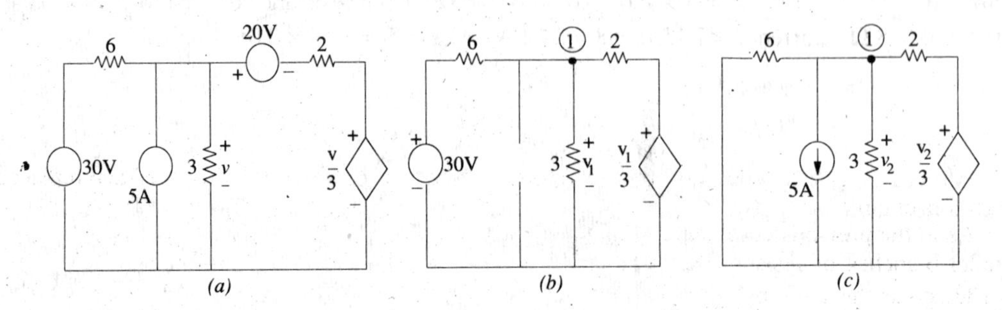

Fig. 2.95

Let us now keep 5 A source alive and 'kill' the other two independent sources. Again applying KCL to node 1, we get

$$\frac{v_2}{6}-5-\frac{v_2}{3}+\frac{(v_2/3-v_2)}{2}=0 \text{ or } v_2=-6\text{ V}$$

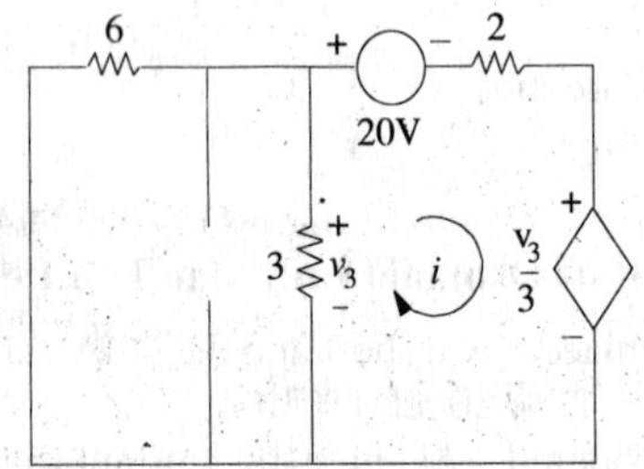

Fig. 2.96

Let us now 'kill' 30 V source and 5 A source and find v_3 due to 20 V source only. The two parallel resistances of 6 Ω and 3 Ω can be combined into a single resistance of 2 Ω. Assuming a circulating current of i and applying KVL to the indicated circuit, we get

$$-2i-20-2i-\frac{1}{3}(-2i)=0 \text{ or } i=6\text{ A}$$

Hence, according to Ohm's law, the component of v that corresponds to 20 V source is $v_3 = 2 \times 6 = 12$ V. $\therefore\ v = v_1 + v_2 + v_3 = 6 - 6 + 12 = 12$ V.

Example 2.45. *Using Superposition theorem, find the current through the 40 Ω resistor of the circuit shown in Fig. 2.97 (a).* **(F.Y. Engg. Pune Univ. May 1990)**

Solution. We will first consider when 50 V battery acts alone and afterwards when 10-V battery is alone in the circuit. When 10-V battery is replaced by short-circuit, the circuit becomes as shown in Fig. 2.97 (*b*). It will be seen that the right-hand side 5Ω resistor becomes connected in parallel with 40 Ω resistor giving a combined resistance of 5 || 40 = 4.44 Ω as shown in Fig. 97 (*c*). This 4.44 Ω resistance is in series with the left-hand side resistor of 5 Ω giving a total resistance of (5 + 4.44) = 9.44 Ω. As seen there are two resistances of 20 Ω and 9.44 Ω connected in parallel. In Fig. 2.97 (*c*), current $I = 50/9.44 = 5.296$ A.

At point A in Fig. 97 (b) there are two resistances of 5 Ω and 40 Ω connected in parallel, hence, current I divides between them as per the current-divider rule. If I_1 is the current flowing through the 40 Ω resistor, then

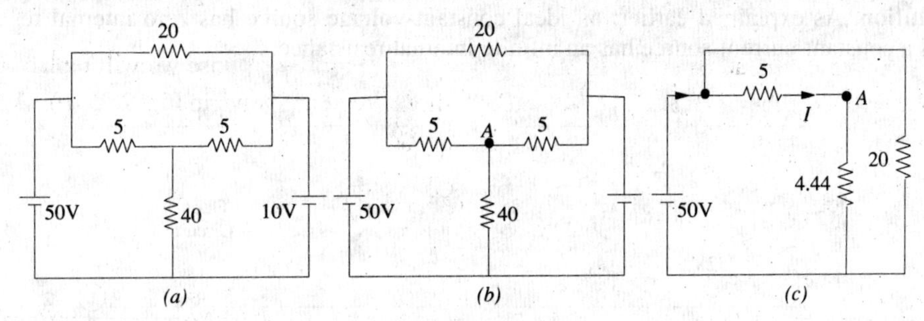

Fig. 2.97

$$I_1 = I \times \frac{5}{5+40} = 5.296 \times \frac{5}{45} = 0.589 \textbf{ A.}$$

In Fig. 2.98 (a), 10 V battery acts alone because 50–V battery has been removed and replaced by a short-circuit.

As in the previous case, there are two parallel branches of resistances 20 Ω and 9.44 Ω across the 10-V battery. Current I through 9.44 Ω branch is I = 10/9 44 = 1.059 A. This current divides at point B between 5 Ω resistor and 40 Ω resistor. Current through 40 Ω resistor I_2 = 1.059 × 5/45 = 0. 118 A.

According to the Superposition theorem, total current through 40 Ω resistance is $= I_1 + I_2 = 0.589 + 0.118 =$ **0.707 A.**

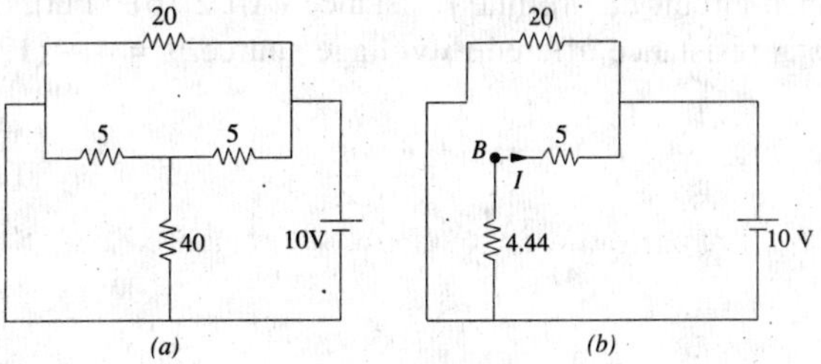

Fig. 2.98

Example 2.46. *Solve for the power delivered to the 10 Ω resistor in the circuit shown in Fig. 2.99 (a) All resistances are in ohms.* **(Elect. Science-I, Allahabad Univ. 1991)**

Solution. The 4-A source and its parallel resistance of 15 Ω can be converted into a voltage source of (15 × 4) = 60 V in series with a 15 Ω resistance as shown in Fig.2.99 (b).

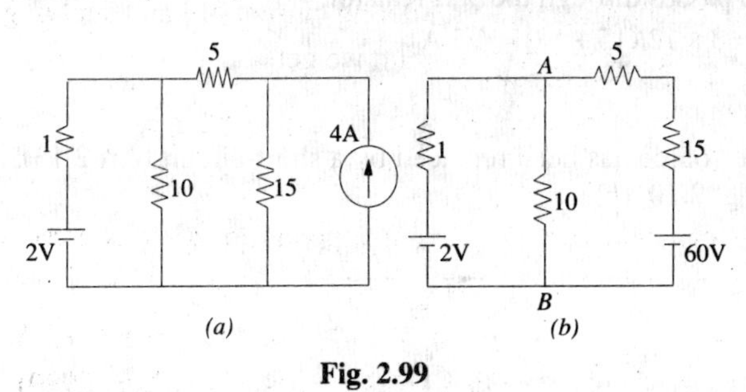

Fig. 2.99

Now, we will use Superposition theorem to find current through the 10 Ω resistance.

When 60 – V Source is Removed

When 60 – V battery is removed the total resistance as seen by 2 V battery is = 1 + 10 || (15 + 5) = 7.67 Ω.
The battery current = 2 /7.67 A = 0.26 A.
At point A, this current is divided into two parts. The current passing through the 10 Ω resistor from A to B is

$$I_1 = 0.26 \times (20/30) = 0.17 \text{ A}$$

When 2-V Battery is Removed

The resistance seen by 60 V battery is = 20+10 || 1 = 20.9 Ω. Hence, battery current = 60/20. 9 = 2.87 A. This current divides at point A. The current flowing through 10 Ω resistor from A to B is

$$I_2 = 2.87 \times 1/(1 + 10) = 0.26 \text{ A}$$

Total current through 10 Ω resistor due to two batteries acting together is $= I_1 + I_2 = 0.43$ A.

Power delivered to the 10 Ω resistor = $0.43^2 \times 10$ = 1.85 W

Example 2.47. *Compute the power dissipated in the 9-Ω resistor of Fig. 2.100 by applying the Superposition principle. The voltage and current sources should be treated as ideal sources. All resistances are in ohms.*

Solution. As explained earlier, an ideal constant-voltage source has zero internal resistance whereas a constant-current source has an infinite internal resistance.

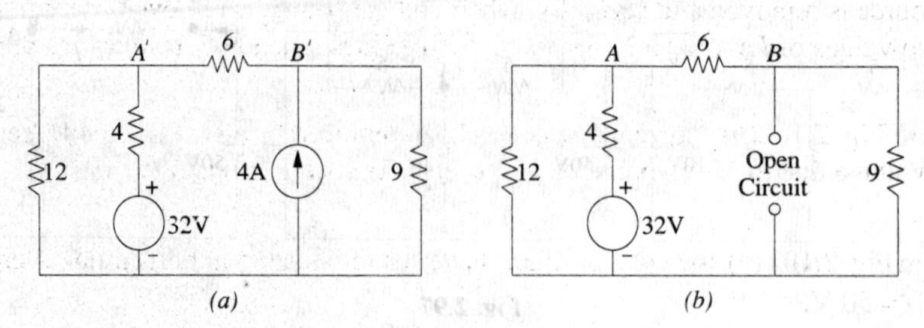

Fig. 2.100 (Contd...)

(*i*) When Voltage Source Acts Alone

This case is shown in Fig. 2.100 (*b*) where constant-current source has been replaced by an open-circuit *i.e.* infinite resistance (Art 2.16). Further circuit simplification leads to the fact that total resistance offered to voltage source is = 4 + (12 || 15) = 32/3 Ω as shown in Fig. 2.100 (*c*).

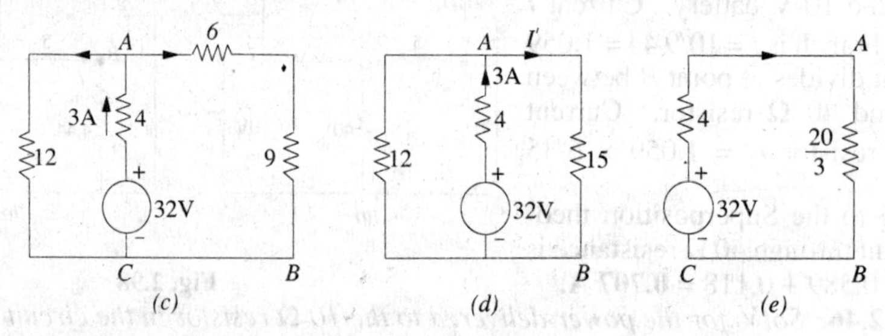

Fig. 2.100

Hence, current = 32 ÷ 32/3 = 3 A. At point *A* in Fig. 2.100 (*d*), this current divides into two parts. The part going along *AB* is the one that also passes through the 9 Ω resistor.

$$\therefore \quad I' = 3 \times 12/(15+12) = 4/3 \text{ A}$$

(*ii*) When Current Source Acts Alone

As shown in Fig. 2.101 (*a*), the voltage source has been replaced by a short-circuit (Art 2.13). Further simplification gives the circuit of Fig. 2.101(*b*).

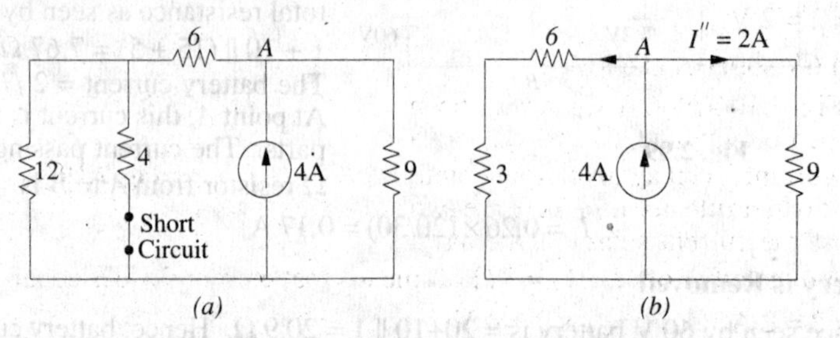

Fig. 2.101

The 4 – A current divides into two equal parts at point *A* in Fig. 2.101(*b*). Hence I = 4/2 = 2A. Since both I' and I'' flow in the *same* direction, total current through 9-Ω resistor is

$$I = I' + I'' = (4/3) + 2 = (10/3) \text{ A}$$

Power dissipated in 9 Ω resistor = $I^2 R = (10/3)^2 \times 9 =$ **100 W**

Example 2.48(a). *With the help of superposition theorem, obtain the value of current I and voltage V_0 in the circuit of Fig. 2.102(a).*

Solution. We will solve this question in three steps. First, we will find the value of I and V_0 when current source is removed and secondly, when voltage source is removed. Thirdly, we would combine the two values of I and V_0 in order to get their values when both sources are present.

First Step

As shown in Fig. 2.102(*b*), current source has been replaced by an open-circuit. Let the values of current and voltage due to 10 V source be I_1 and V_{01}. As seen $I_1 = 0$ and $V_{01} = 10$ V.

Second Step

As shown in Fig. 2.102 (*c*), the voltage source has been replaced by a short circuit. Here $I_2 = -5$ A and $V_{02} = 5 \times 10 = 50$ V.

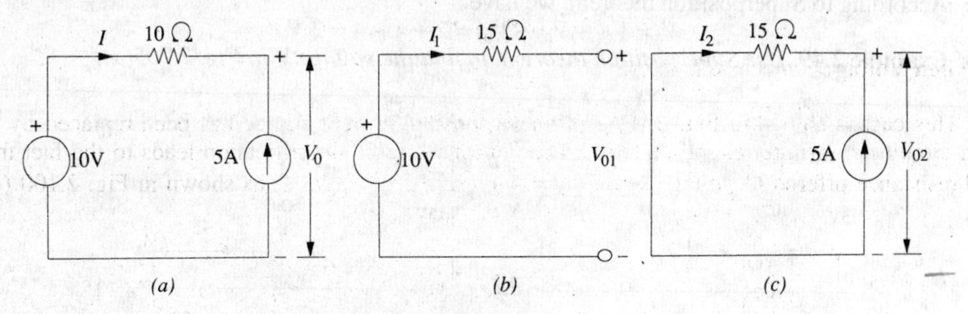

Fig. 2.102

Third Step

By applying superposition theorem, we have

$$I = I_1 + I_2 = 0 + (-5) = -\mathbf{5\ A}$$

$$V_0 = V_{01} + V_{02} = 10 + 50 = \mathbf{60\ V}$$

Example 2.48(b). *Using Superposition theorem, find the value of the output voltage V_0 in the circuit of Fig. 2.103.*

Solution. As usual, we will break down the problem into three parts involving one source each.

(*a*) **When 4 A and 6 V sources are killed***.

As shown in Fig. 2.104 (*a*), 4 A source has been replaced by an open circuit and 6 V source by a short-circuit. Using the current-divider rule, we find current i_1 through the 2Ω resistor = $6 \times 1/(1 + 2 + 3)$ = 1 A $\therefore V_{01} = 1 \times 2 = 2$ V.

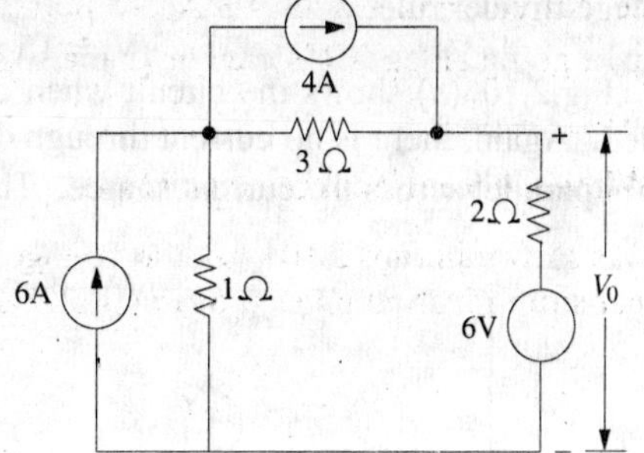

Fig. 2.103

(*b*) **When 6 A and 6V sources are killed**

As shown in Fig. 2.104(*b*), 6 A source has been replaced by an open-circuit and 6*V* source by a short-circuit. The current i_2 can again be found with the help of current-divider rule because there are two parallel paths across the current source. One has a resistance of 3 Ω and the other of (2 + 1) = 3 Ω. It means that current divides equally at point A.

Hence, $i_2 = 4/2 = 2$ A $\therefore V_{02} = 2 \times 2 = 4$ V

(*c*) **When 6 A and 4 A sources are killed**

*The process of setting the voltage source to zero is called *killing* the sources.

As shown in Fig. 2.104(*c*), drop over 2 Ω resistor = 6 × 2/ 6 = 2 V. The potential of point *B* with respect to point *A* is = 6 – 2 = + 4 V. Hence, $V_{03} = -4$ V.

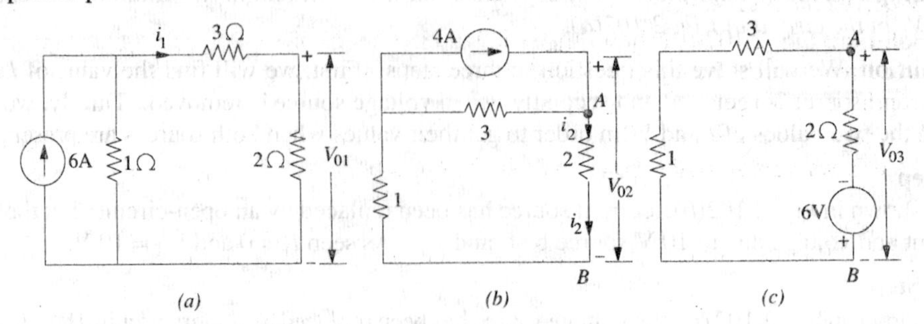

Fig. 2.104

According to Superposition theorem, we have

$$V_0 = V_{01} + V_{02} + V_{03} = 2 + 4 - 4 = \mathbf{2\ V}$$

Example 2.49. *Use Superposition theorem, to find the voltage V in Fig. 2.105(a).*

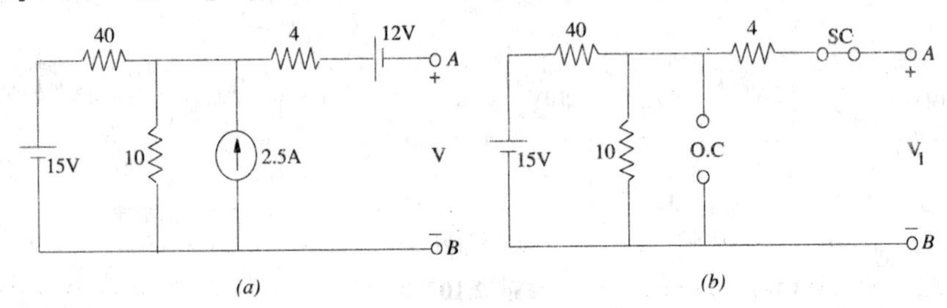

Fig. 2.105

Solution. The given circuit has been redrawn in Fig. 2.105(*b*) with 15 – V battery acting alone while the other two sources have been killed. The 12 – V battery has been replaced by a short-circuit and the current source has been replaced by an open-circuit (O.C) (Art. 2.19). Since the output terminals are open, no current flows through the 4 Ω resistor and hence, there is no voltage drop across it. Obviously V_1 equals the voltage drop over 10 Ω resistor which can be found by using the voltage-divider rule.

$$V_1 = 15 \times 10/ (40 + 50) = 3 \text{ V}$$

Fig.2.106(*a*) shows the circuit when curent source acts alone, while two batteries have been killed. Again, there is no current through 4 Ω resistor. The two resistors of values 10 Ω and 40 Ω are in parallel across the current source. Their combined resistances is 10 || 40 = 8 Ω

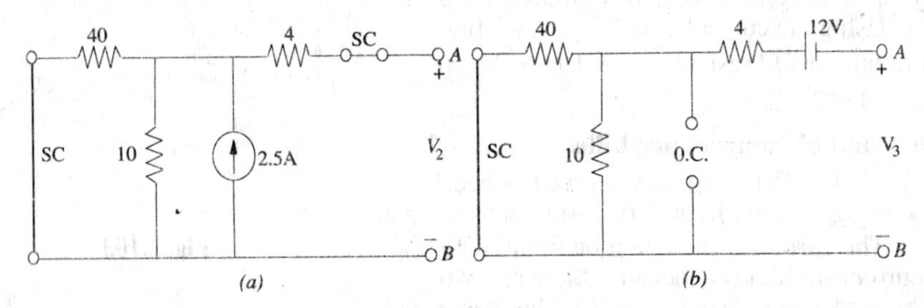

Fig. 2.106

∴ $V_2 = 8 \times 2.5 = 20$ V with point *A* positive.

Fig. 2.106(*b*) shows the case when 12-V battery acts alone. Here, $V_3 = -12$ V*. Minus sign has been taken because negative terminal of the battery is connected to point *A* and the positive terminal to point *B*. As per the Superposition theorem,

*Because Fig. 2.106(*b*) resembles a voltage source with an internal resistance = 4 + 10 || 40 = 12 Ω and which is an open-circuit.

$$V = V_1 + V_2 + V_3 = 3 + 20 - 12 = 11 \text{ V}$$

Example 2.50. *Apply Superposition theorem to the circuit of Fig. 2.107(a) for finding the voltage drop V across the 5 Ω resistor.*

Solution. Fig. 2.107 (*b*) shows the redrawn circuit with the voltage source acting alone while the two current sources have been 'killed' *i.e.* have been replaced by open circuits. Using voltage-divider principle, we get

$V_1 = 60 \times 5/(5 + 2 + 3) = 30$ V. It would be taken as positive, because current through the 5 Ω resistance flows from *A* to *B*, thereby making the upper end of the resistor positive and the lower end negative.

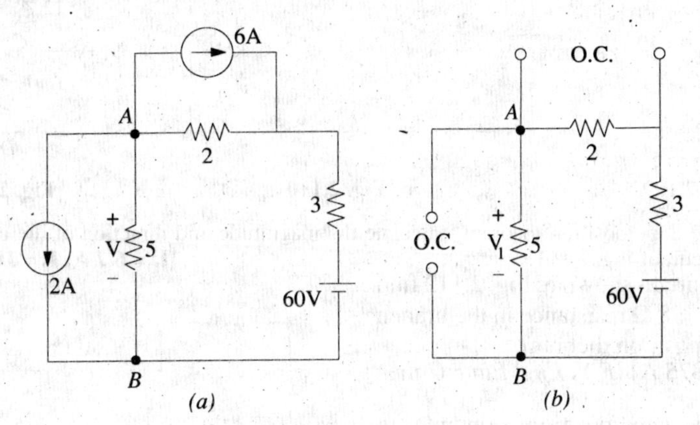

Fig. 2.107

Fig. 2.108(*a*) shows the same circuit with the 6 A source acting alone while the two other sources have been 'killed'. It will be seen that 6 A source has two parallel circuits across it, one having a resistance of 2 Ω and the other (3 + 5) = 8 Ω. Using the current–divider rule, the current through the 5 Ω resistor = 6 × 2/(2 + 3 + 5) = 1.2 A.

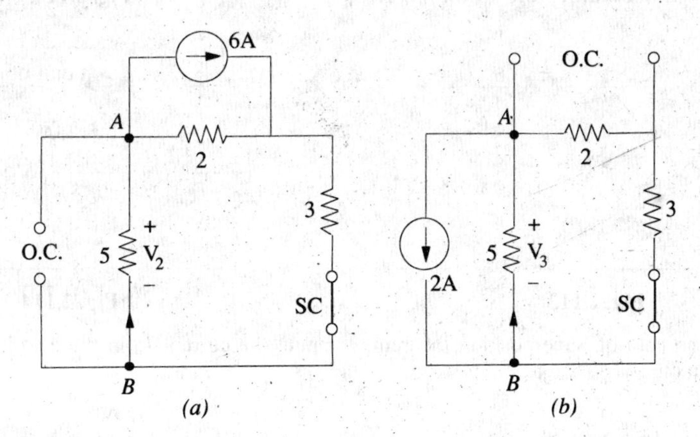

Fig. 2.108

∴ $V_2 = 1.2 \times 5 = 6$ V. It would be taken *negative* because current is flowing from *B* to *A*. *i.e.* point *B* is at a higher potential as compared to point *A*. Hence, $V_2 = -6$ V.

Fig. 2.108(*b*) shows the case when 2-A source acts alone, while the other two sources are dead. As seen, this current divides equally at point *B*, because the two parallel paths have equal resistances of 5 Ω each. Hence, $V_3 = 5 \times 1 = 5$ V. It also would be taken as negative because current flows from *B* to *A*. Hence, $V_3 = -5$ V.

Using Superposition principle, we get

$$V = V_1 + V_2 + V_3 = 30 - 6 - 5 = \mathbf{19\ V}$$

Tutorial Problems No. 2.4.

1. Apply the principle of Superposition to the network shown in Fig. 2.109 to find out the current in the 10 Ω resistance. **[0.464 A]** (*F.Y. Engg. Pune Univ. May 1987*)

2. Find the current through the 3 Ω resistance connected between *C* and *D* Fig. 2.110. **[1 A from C to D]** (*F.Y. Engg. Pune Univ. May 1989*)

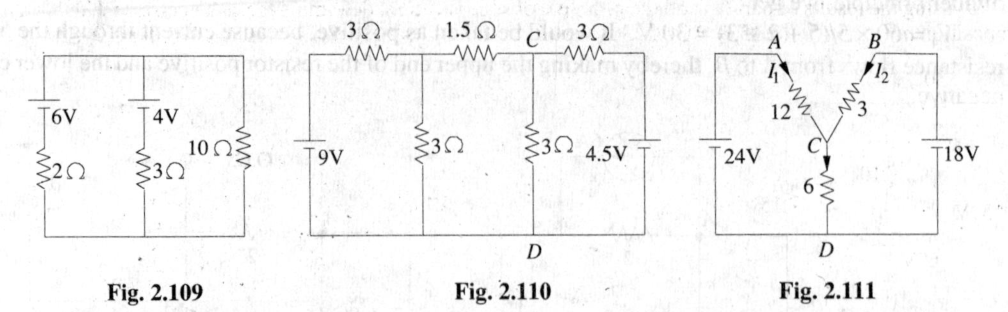

Fig. 2.109 **Fig. 2.110** **Fig. 2.111**

3. Using the Superposition theorem, calculate the magnitude and direction of the current through each resistor in the circuit of Fig. 2.111. **[I_1 = 6/7 A; I_2 = 10/7 A; I_3 = 16/7 A]**

4. For the circuit shown in Fig. 2.112 find the current in *R* = 8 Ω resistance in the branch *AB* using superposition theorem. **[0.875 A]** (*F.Y. Eng. Pune Univ. May 1988*)

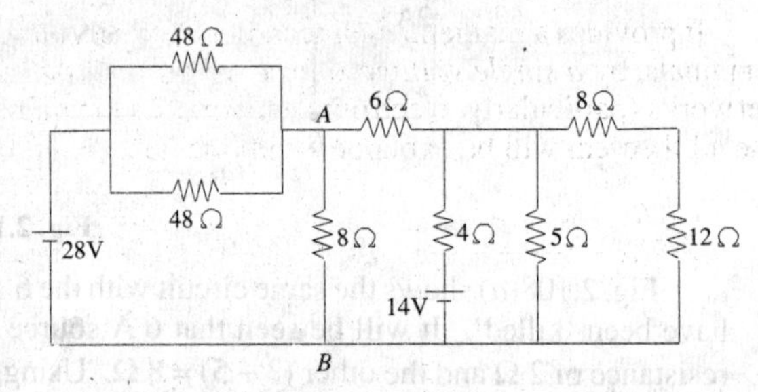

Fig. 2.112

5. Apply superposition principle to compute current in the 2– Ω resistor of Fig. 2.113. All resistors are in ohms. **[I_{ab} = 5 A]**

6. Use Superposition theorem to calculate the voltage drop across the 3 Ω resistor of Fig. 2.114. All resistance values are in ohms. **[18 V]**

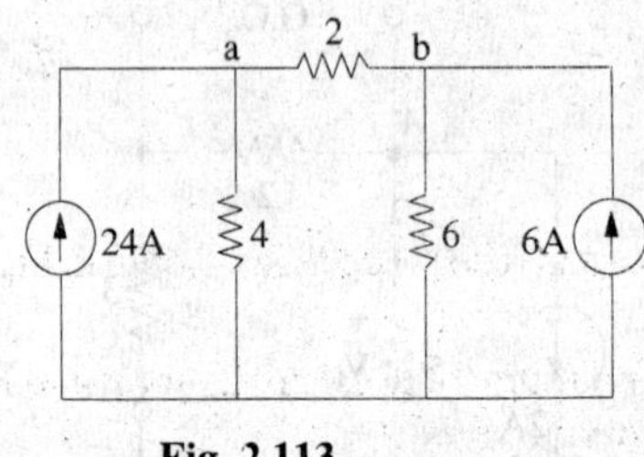

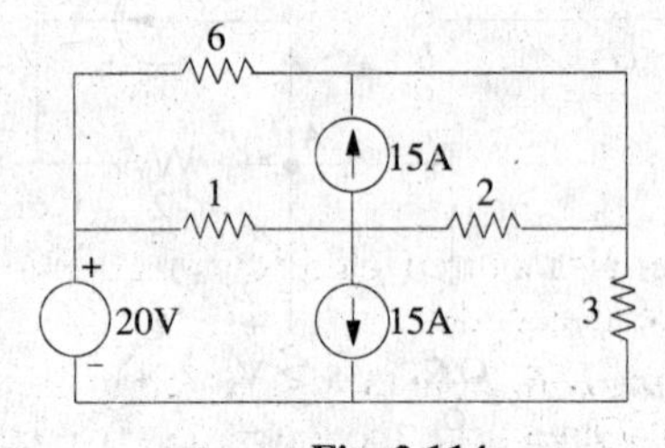

Fig. 2.113 **Fig. 2.114**

7. With the help of Superposition theorem, compute the current I_{ab} in the circuit of Fig. 2.115. All resistances are in ohms. **[I_{ab} = –3 A]**

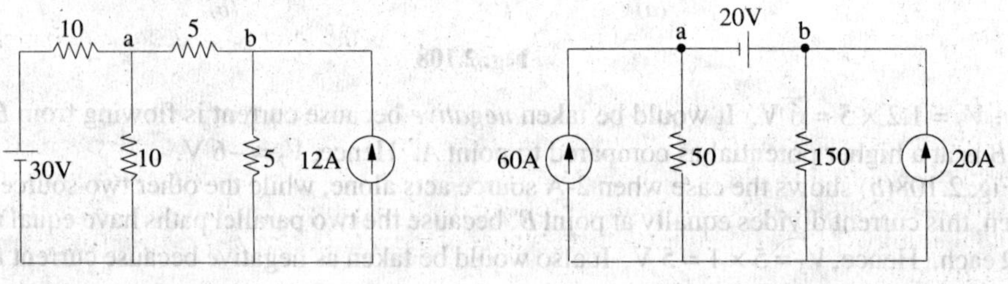

Fig. 2.115 **Fig. 2.116**

8. Use Superposition theorem to find current I_{ab} in the circuit of Fig. 2.116. All resistances are in ohms. **[100 A]**

9. Find current in the 15 Ω resistor of Fig. 2.117 by using Superposition principle. Numbers represent resistances in ohms. **[2.8 A]**

10. Use Superposition principle to find current in the 10–Ω resistor of Fig. 2.118. All resistances are in ohms. **[1 A]**

11. State and explain Superposition theorem. For the circuit of Fig. 2.119.

(*a*) determine currents I_1, I_2 and I_3 when switch *S* is in position *b*.

(*b*) using the results of part (*a*) and the principle of superposition, determine the same currents with switch *S* in position *a*. **[(*a* 15 A, 10 A, 25 A (*b*) 11 A, 16 A, 27 A]**(*Elect. Technology, Vikram Univ. 1978*)

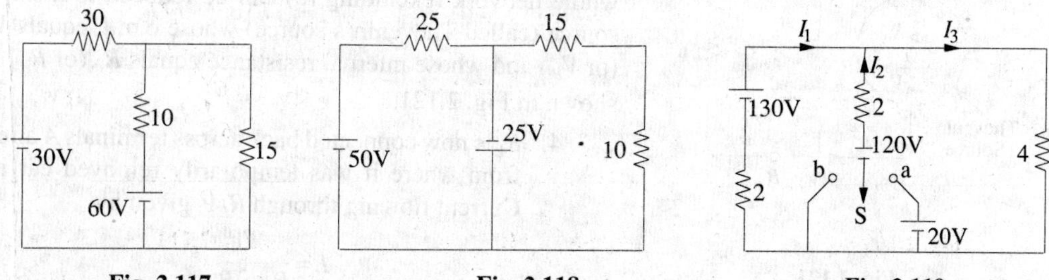

Fig. 2.117 **Fig. 2.118** **Fig. 2.119**

2.18. Thevenin Theorem*

It provides a mathematical technique for replacing a given network, as viewed from two output terminals, by *a single voltage source with a series resistance*. It makes the solution of complicated networks (particularly, electronic networks) quite quick and easy. The application of this extremely useful theorem will be explained with the help of the following simple example.

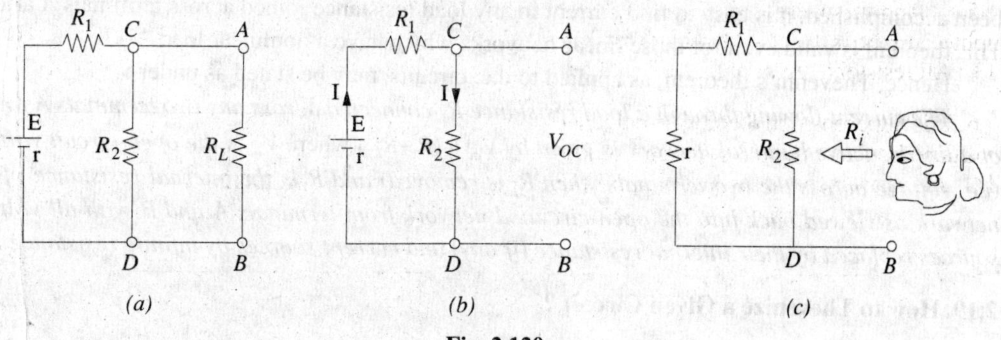

Fig. 2.120

Suppose, it is required to find current flowing through load resistance R_L, as shown in Fig. 120 (*a*). We will proceed as under :

1. Remove R_L from the circuit terminals *A* and *B* and redraw the circuit as shown in Fig. 2.120 (*b*). Obviously, the terminals have become open-circuited.
2. Calculate the open–circuit voltage V_{oc} which appears across terminals *A* and *B* when they are open *i.e.* when R_L is removed.
   As seen, V_{oc} = drop across $R_2 = IR_2$ where *I* is the circuit current when *A* and *B* are open.
   $$I = \frac{E}{R_1 + R_2 + r} \qquad \therefore \quad V_{oc} = IR_2 = \frac{ER_2}{R_1 + R_2 + r} \quad [r \text{ is the internal resistance of battery}]$$
   It is also called 'Thevenin voltage' V_{th}.
3. Now, imagine the battery to be removed from the circuit, leaving its internal resistance *r* behind and redraw the circuit, as shown in Fig. 2.120(*c*). When viewed *inwards* from terminals *A* and *B*, the circuit consists of two parallel paths : one containing R_2 and the other containing $(R_1 + r)$. The equivalent resistance of the network, as viewed from these terminals is given as

*After the French engineer M.L. Thevenin (1857-1926) who while working in Telegraphic Department published a statement of the theorem in 1893.

$$R = R_2 \| (R_1 + r) = \frac{R_2(R_1 + r)}{R_2 + (R_1 + r)}$$

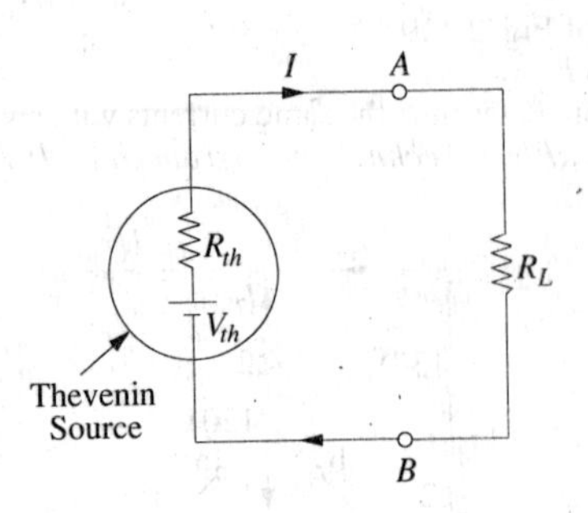

Fig. 2.121

This resistance is also called, *Thevenin resistance R_{th} (though, it is also sometimes written as R_i or R_0).

Consequently, as viewed from terminals A and B, the whole network (excluding R_1) can be reduced to a single source (called Thevenin's source) whose e.m.f. equals V_{oc} (or V_{th}) and whose internal resistance equals R_{th} (or R_i), as shown in Fig. 2.121.

4. R_L is now connected back across terminals A and B from where it was temporarily removed earlier. Current flowing through R_L is given by

$$I = \frac{V_{th}}{R_{th} + R_L}$$

It is clear from above that any network of resistors and voltage sources (and current sources as well) when viewed from any two points A and B in the network, can be replaced by a single voltage source and a single resistance* in series with the voltage source.

After this replacement of the network by a single voltage source with a series resistance has been accomplished, it is easy to find current in any load resistance joined across terminals A and B. This theorem is valid even for those linear networks which have a nonlinear load.

Hence, Thevenin's theorem, as applied to d.c. circuits, may be stated as under :

The current flowing through a load resistance R_L connected across any two terminals A and B of a linear, active bilateral network is given by $V_{oc} \| (R_i + R_L)$ where V_{oc} is the open-circuit voltage (i.e. voltage across the two terminals when R_L is removed) and R_i is the internal resistance of the network as viewed back into the open-circuited network from terminals A and B with all voltage sources replaced by their internal resistance (if any) and current sources by infinite resistance.

2.19. How to Thevenize a Given Circuit ?

(1) Temporarily remove the resistance (called load resistance R_L) whose current is required.

(2) Find the open-circuit voltage V_{oc} which appears across the two terminals from where resistance has been removed. It is also called Thevenin voltage V_{th}.

(3) Compute the resistance of the whole network as looked into from these two terminals after all voltage sources have been removed leaving behind their internal resistances (if any) and current sources have been replaced by open-circuit *i.e.* infinite resistance. It is also called Thevenin resistance R_{th} or R_i.

(4) Replace the entire network by a single Thevenin source, whose voltage is V_{th} or V_{oc} and whose internal resistance is R_{th} or R_i.

(5) Connect R_L back to its terminals from where it was previously removed.

(6) Finally, calculate the current flowing through R_L by using the equation,

$$I = V_{th}/(R_{th} + R_L) \quad \text{or} \quad I = V_{oc}/(R_i + R_L)$$

Example 2.51. *Convert the circuit shown in Fig. 2.122(a), to a single voltage source in series with a single resistor.* **(AMIE Sec. B, Network Analysis Summer 1992)**

*Or impedance in the case of a.c. circuits.

Solution. Obviously, we have to find equivalent Thevenin circuit. For this purpose, we have to calculate (*i*) V_{th} or V_{AB} and (*ii*) R_{th} or R_{AB}.

With terminals *A* and *B* open, the two voltage sources are connected in subtractive series becuase they oppose each other. Net voltage around the circuit is (15 − 10) = 5 V and total resistance is (8 + 4) = 12 Ω. Hence circuit current is = 5/12 A. Drop across 4 Ω resistor = 4 × 5 /12 = 5 /3 V with the polarity as shown in Fig. 2.122 (*a*).

$\therefore \quad V_{AB} = V_{th} = +10 + 5/3 = 35/3\text{V}.$

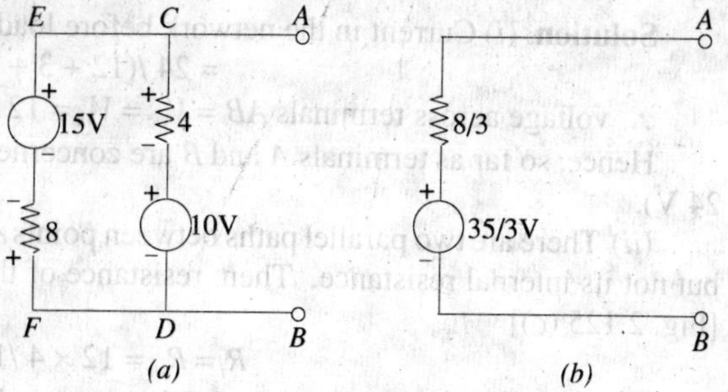

Fig. 2.122

Incidentally, we could also find V_{AB} while going along the parallel route *BFEA*. Drop across 8 Ω resistor = 8 × 5 /12 = 10/3 V. V_{AB} equals the algebraic sum of voltages met on the way from *B* to *A*. Hence, V_{AB} = (−10/3) + 15 = 35 /3 V.

As shown in Fig. 2.122(*b*), the single voltage source has a voltage of 35/3 V.

For finding R_{th}, we will replace the two voltage sources by short-circuits. In that case, $R_{th} = R_{AB}$ = 4 || 8 = 8 /3 Ω.

Example 2.52. *State Thevenin's theorem and give a proof. Apply this theorem to calculate the current through the 4 Ω resistor of the circuit of Fig. 2.123(a).*

(A.M.I.E. Sec. B Network Analysis W. 1989)

Solution. As shown in Fig. 2.125 (*b*), 4 Ω resistance has been removed thereby open-circuiting the terminals *A* and *B*. We will now find V_{AB} and R_{AB} which will give us V_{th} and R_{th} respectively. The potential drop across 5 Ω resistor can be found with the help of voltage-divider rule. Its value is = 15 × 5/ (5 + 10) = 5 V.

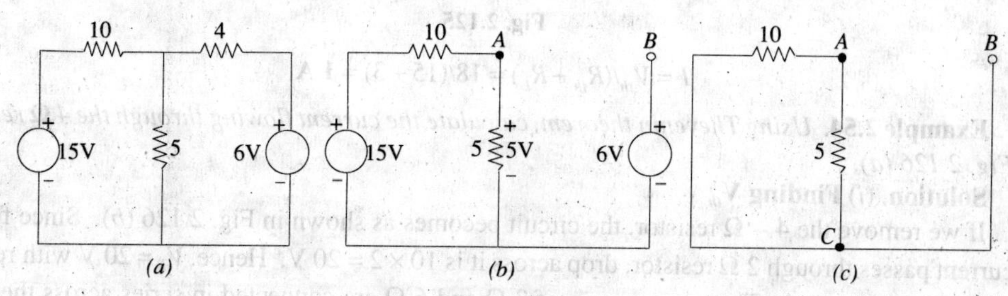

Fig. 2.123

For finding V_{AB}, we will go from point *B* to point *A* in the clockwise direction and find the algebraic sum of the voltages met on the way.

$\therefore V_{AB} = -6 + 5 = -1$ V. It means that point *A* is negative with respect to point *E*, or point *B* is at a higher potential than point *A* by one volt.

In Fig. 2.123 (*c*), the two voltage sources have been short-circuited. The resistance of the network as viewed from points *A* and *B* is the same as viewed from points *A* and *C*.

$$\therefore \qquad R_{AB} = R_{AC} = 5 \parallel 10 = 10/3\ \Omega$$

Thevenin's equivalent source is shown in Fig. 2.124 in which 4 Ω resistor has been joined back across terminals *A* and *B*. Polarity of the voltage source is worth noting.

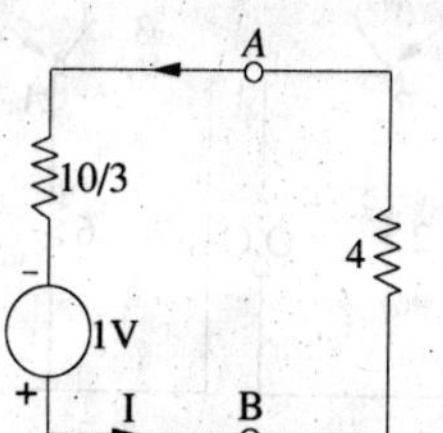

Fig. 2.124

$$\therefore \qquad I = \frac{1}{(10/3)+4} = \frac{3}{22} = 0.136\ \text{A} \qquad \text{from } E \text{ to } A$$

Example 2.53. *With reference to the network of Fig. 2.125 (a), by applying Thevenin's theorem find the following :*

(*i*) *the equivalent e.m.f. of the network when viewed from terminals A and B*

(*ii*) *the equivalent resistance of the network when looked into from terminals A and B* (*iii*) *current in the load resistance* R_L *of 15 Ω.*

(Basic Circuit Analysis, Nagpur Univ. 1993)

Solution. (*i*) Current in the network before load resistance is connected [Fig. 2.125 (*a*)]

$$= 24/(12 + 3 + 1) = 1.5 \text{ A}$$

∴ voltage across terminals $AB = V_{oc} = V_{th} = 12 \times 1.5 = \mathbf{18\ V}$

Hence, so far as terminals *A* and *B* are concerned, the network has an e.m.f. of 18 volt (and not 24 V).

(*ii*) There are two parallel paths between points *A* and *B*. Imagine that battery of 24 V is removed but not its internal resistance. Then, resistance of the circuit as looked into from points *A* and *B* is [Fig. 2.125 (*c*)]

$$R_i = R_{th} = 12 \times 4/12 + 4) = 3\ \Omega$$

(*iii*) When load resistance of 15 Ω is connected across the terminals, the network is reduced to the structure shown in Fig. 2.125 (*d*).

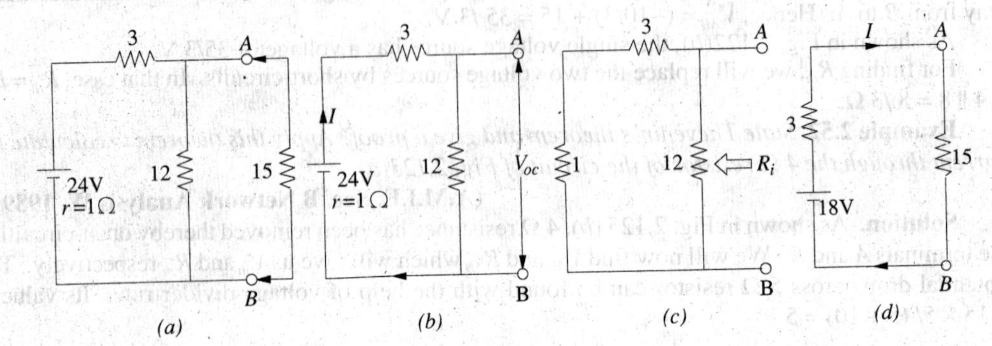

Fig. 2.125

$$I = V_{th}/(R_{th} + R_L) = 18/(15 + 3) = \mathbf{1\ A}$$

Example 2.54. *Using Thevenin theorem, calculate the current flowing through the 4 Ω resistor of Fig. 2.126 (a).*

Solution. (*i*) **Finding V_{th}**

If we remove the 4 – Ω resistor, the circuit becomes as shown in Fig. 2.126 (*b*). Since full 10 A current passes through 2 Ω resistor, drop across it is $10 \times 2 = 20$ V. Hence, $V_B = 20$ V with respect to the common ground. The two resistors of 3 Ω and 6 Ω are connected in series across the 12 V battery. Hence, drop across 6 Ω resistor = 12 × 6/(3 + 6) = 8 **V**.

∴ $V_A = 8$ V with respect to the common ground*

∴ $V_{th} = V_{BA} = V_B - V_A = 20 - 8 = 12$ V –with *B* at a higher potential

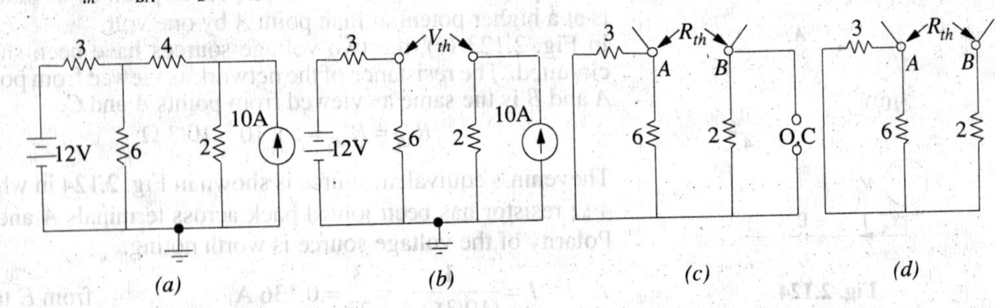

Fig. 2.126

*Also, $V_A = 12$ – drop across 3-Ω resistor = 12–12 × 3/(6 + 3) = 12 – 4 = 8 V (*ii*) **Finding $\mathbf{R_{th}}$**

Now, we will find R_{th} *i.e.* equivalent resistance of the network as looked back into the open-circuited terminals *A* and *B*. For this purpose, we will replace both the voltage and current sources. Since voltage source has no internal resistance, it would be replaced by a short-circuit *i.e.* zero resistance. However, current source would be removed and replaced by an 'open' *i.e.* infinite resistance (Art. 1.18). In that case, the circuit becomes as shown in Fig. 2.126 (*c*). As seen from Fig. 2.126 (*d*), $R_{th} = 6 \parallel 3 + 2 = 4\ \Omega$ Hence, Thevenin's equivalent circuit consists of a voltage source of 12 V and a series resistnce of 4 Ω as shown in Fig. 2.127 (*a*). When 4 Ω resistor is connected across terminals *A* and *B*

$I = 12/(4 + 4) = $ **1.5 A –from B to A**

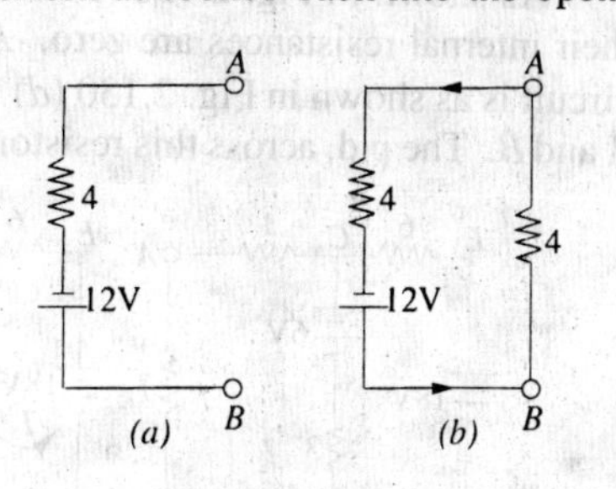

Fig. 2.127

Example 2.55. *For the circuit shown in Fig. 2.128 (a), calculate the current in the 10 ohm resistance. Use Thevenin's theorem only.* **(Elect. Science-I Allahabad Univ. 1992)**

Solution. When the 10 Ω resistance is removed, the circuit becomes as shown in Fig. 2.128 (*b*).

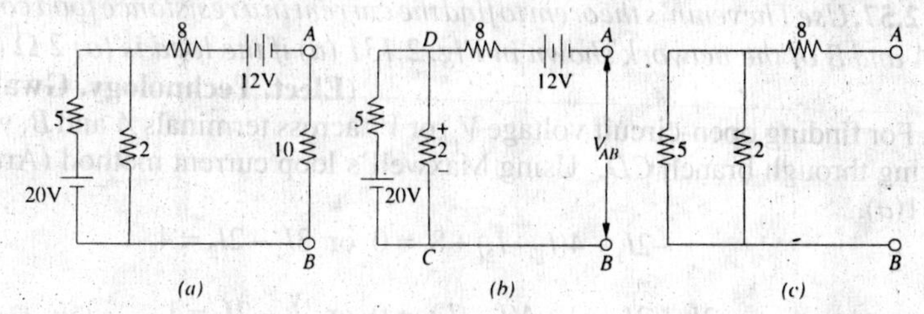

Fig. 2.128

Now, we will find the open-circuit voltage $V_{AB} = V_{th}$. For this purpose, we will go from point *B* to point *A* and find the algebraic sum of the voltages met on the way. It should be noted that with terminals *A* and *B* open, there is no voltage drop on the 8 Ω resistance. However, the two resistances of 5 Ω and 2 Ω are connected in series across the 20-V battery. As per voltage-divider rule, drop on 2 Ω resistance = 20 × 2/(2 + 5) = 5.71 V with the polarity as shown in the figure. As per the sign convention of Art.

$$V_{AB} = V_{th} = +\,5.71 - 12 = -\,6.29 \text{ V}$$

The negative sign shows that point *A* is negative with respect to point *B* or which is the same thing, point *B* is positive with respect to point *A*.

For finding $R_{AB} = R_{th}$, we replace the batteries by short-circuits as shown in Fig. 2.128 (*c*).

$\therefore\ R_{AB} = R_{th} = 8 + 2 \| 5 = 9.43\ \Omega$

Hence, the equivalent Thevenin's source with respect to terminals *A* and *B* is an shown in Fig. 2.129. When 10 Ω resistance is reconnected across *A* and *B*, current through it is $I = 6.24/(9.43 + 10) = $ **0.32 A**.

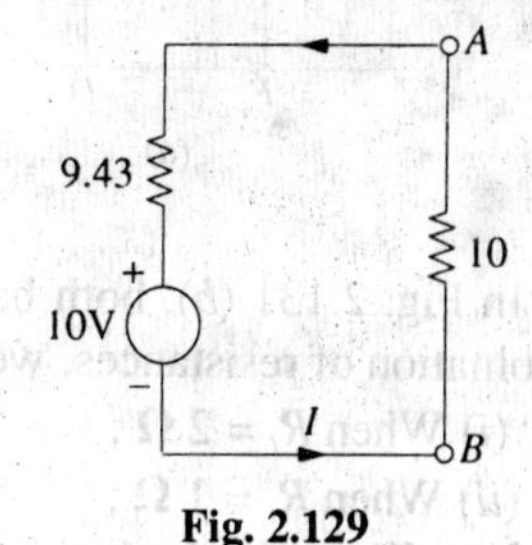

Fig. 2.129

Example 2.56. *Using Thevenin's theorem, calculate the p.d. across terminals A and B in Fig. 2.130 (a).*

Solution. (*i*) **Finding $\mathbf{V_{oc}}$**

First step is to remove 7 Ω resistor thereby open-circuiting terminals *A* and *B* as shown in Fig. 2.130 (*b*). Obviously, there is no current through the 1 Ω resistor and hence no drop across it. Therefore, $V_{AB} = V_{oc} = V_{CD}$. As seen, current *I* flows due to the combined action of the two batteries. Net voltage in the *CDFE* circuit = 18 – 6 = 12 V. Total resistance = 6 + 3 = 9 Ω. Hence, $I = 12/9 = 4/3$ A

$$V_{CD} = 6\ V + \text{drop across } 3\ \Omega \text{ resistor} = 6 + (4/3) \times 3 = 10\ V^{*}$$

$$\therefore \qquad V_{oc} = V_{th} = 10 \text{ V.}$$

*Also, V_{CD} = 18–drop across 6 Ω resistor = **18** – (4/3) × 6 = 10 V (*ii*) **Finding R_i or R_{th}**

As shown in Fig. 2.130 (*c*), the two batteries have been replaced by short-circuits (SC) since their internal resistances are zero. As seen, $R_i = R_{th} = 1 + 3 \parallel 6 = 3\ \Omega$. The Thevenin's equivalent circuit is as shown in Fig. 2.130 (*d*) where the 7 Ω resistance has been reconnected across terminals *A* and *B*. The p.d. across this resistor can be found with the help of Voltage Divider Rule (Art. 1.15).

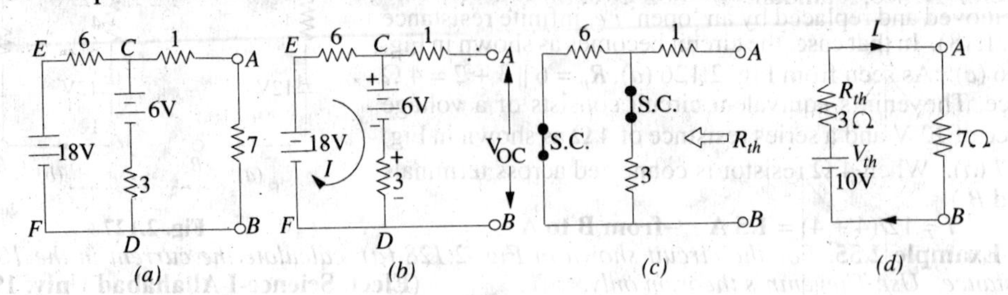

Fig. 2.130

Example 2.57. *Use Thevenin's theorem to find the current in a resistance load connected between the terminals A and B of the network shown in Fig. 2.131 (a) if the load is (a) 2 Ω (b) 1 Ω.*

(Elect. Technology, Gwalior Univ. 1987)

Solution. For finding open-circuit voltage V_{oc} or V_{th} across terminals *A* and *B*, we must first find current I_2 flowing through branch *CD*. Using Maxwell's loop current method (Art. 2.11), we have from Fig. 2.131(*a*).

$$-2I_1 - 4(I_1 - I_2) + 8 = 0 \quad \text{or} \quad 3I_1 - 2I_2 = 4$$

Also $$-2I_2 - 2I_2 - 4 - 4(I_2 - I_1) = 0 \quad \text{or} \quad I_1 - 2I_2 = 1$$

From these two equations, we get $I_2 = 0.25$ A

As we go from point *D* to *C*, voltage rise $= 4 + 2 \times 0.25 = 4.5$V

Hence, $V_{CD} = 4.5$ or $V_{AB} = V_{th} = 4.5$ V. Also, it may be noted that point *A* is positive with respect to point *B*.

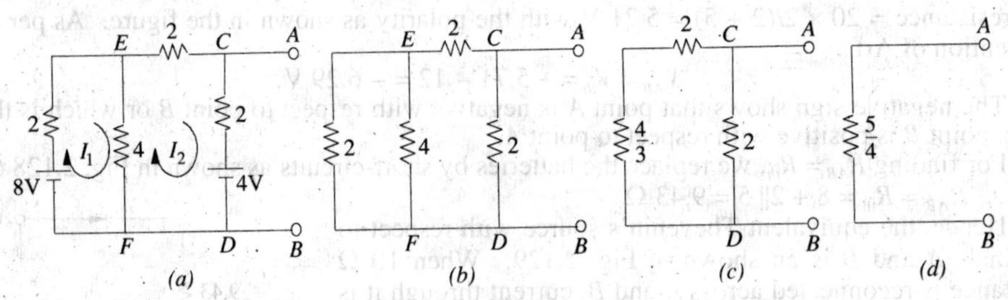

Fig. 2.131

In Fig. 2.131 (*b*), both batteries have been removed. By applying laws of series and parallel combination of resistances, we get $R_i = R_{th} = 5/4\ \Omega = 1.25\ \Omega$

(*i*) When $R_L = 2\ \Omega$; $I = 4.5/(2 + 1.25) = \mathbf{1.38\ A}$

(*ii*) When $R_L = 1\ \Omega$; $I = 4.5(1 + 1.25) = \mathbf{2.0\ A}$

Note. We could also find V_{oc} and R_i by first Thevenizing part of the circuit across terminals *E* and *F* and then across *A* and *B* (Ex. 2.62.).

Example 2.58. *The four arms of a Wheatstone bridge have the following resistances : AB = 100, BC = 10, CD = 4, DA = 50 Ω. A galvanometer of 20 Ω resistance is connected across BD. Use Thevenin's theorem to compute the current through the galvanometer when a p.d. of 10 V is maintained across AC.*

(Elect. Technology, Vikram Univ. Ujjain 1988)

Solution.

(*i*) When galvanometer is removed from Fig. 2.132 (*a*), we get the circuit of Fig. 2.132 (*b*).

(*ii*) Let us next find the open-circuit voltage V_{oc} (also called Thevenin voltage V_{th}) between points *B* and *D*. Remembering that *ABC* (as well as *ADC*) is a potential divider on which a voltage drop of 10 V takes place, we get

Potential of B w.r.t. $C = 10 \times 10/110 = 10/11 = 0.909$ V

Potential of D w.r.t. $C = 10 \times 4/54 = 20/27 = 0.741$ V

∴ p.d. between B and D is V_{oc} or $V_{th} = 0.909–0.741 = 0.168$ V

(*iii*) Now, remove the 10 – V battery retaining its internal resistance which, in this case, happens to be zero. Hence, it amounts to short-circuiting points A and C as shown in Fig. 2.132 (*d*).

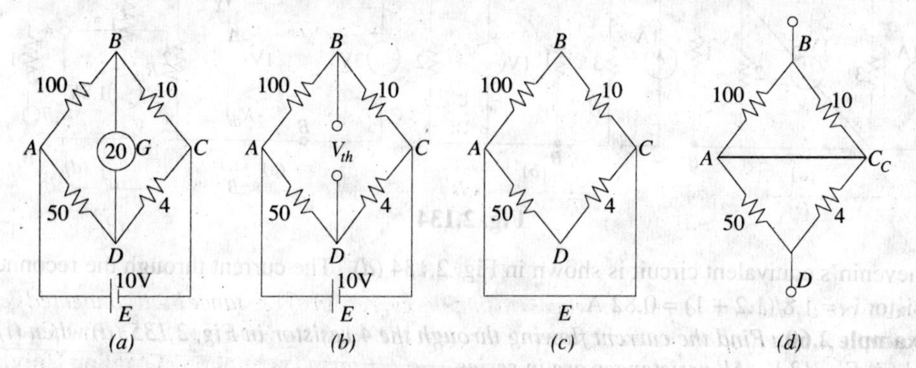

Fig. 2.132

(*iv*) Next, let us find the resistance of the whole network as viewed from points B and D. It may be easily found by noting that electrically speaking, points A and C have become one as shown in Fig. 2.133 (*a*). It is also seen that BA is in parallel with BC and AD is in parallel with CD. Hence, $R_{BD} = 10 \parallel 100 + 50 \parallel 4 = 12.79\ \Omega$.

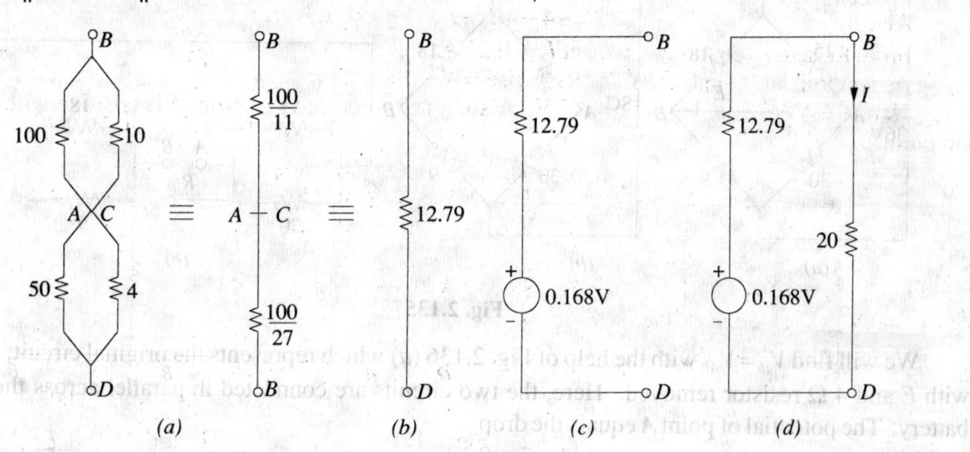

Fig. 2.133

(*v*) Now, so far as points B and D are concerned, the network has a voltage source of 0.168 *V* and internal resistance $R_i = 12.79\ \Omega$. This Thevenin's source is shown in Fig. 2.133 (*c*).

(*vi*) Finally, let us connect the galvanometer (initially removed) to this Thevenin source and calculate the current I flowing through it. As seen from Fig. 2.133 (*d*).

$$I = 0.168/(12.79 + 20) = 0.005 \text{ A} = \mathbf{5\ mA}.$$

Example 2.59. *Determine the current in the 1 Ω resistor across AB of network shown in Fig. 2.134(a) using Thevenin's theorem.* **(Network Analysis, Nagpur Univ. 1993)**

Solution. The given circuit can be redrawn, as shown in Fig. 2.134 (*b*) with the 1 Ω resistor removed from terminals A and B. The current source has been converted into its equivalent voltage source as shown in Fig. 2.134 (*c*). For finding V_{th}, we will find the currents x and y in Fig. 2.134 (*c*). Applying KVL to the first loop, we get

$$3-(3+2)x-1=0 \text{ or } x=0.4 \text{ A}$$

$$\therefore \qquad V_{th} = V_{AB} = 3-3\times 0.4 = 1.8 \text{ V}$$

The value of R_{th} can be found from Fig. 2.134 (*c*) by replacing the two voltage sources by short-circuits. In this case $R_{th} = 2 \parallel 3 = 1.2\ \Omega$.

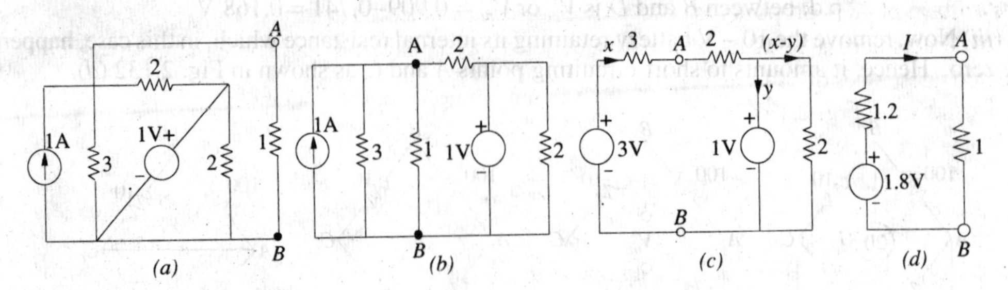

Fig. 2.134

Thevenin's equivalent circuit is shown in Fig. 2.134 (*d*). The current through the reconnected 1 Ω resistor is = 1.8/(1.2 + 1) = 0.82 A.

Example 2.60. *Find the current flowing through the 4 resistor in Fig. 2.135 (a) when (i) E = 2 V and (ii) E = 12 V. All resistances are in series.*

Solution. When we remove E and 4 Ω resistor, the circuit becomes as shown in Fig. 2.135 (*b*). For finding R_{th} *i.e.* the circuit resistance as viewed from terminals A and B, the battery has been short-circuitted, as shown. It is seen from Fig. 2.135 (*c*) that $R_{th} = R_{AB} = 15 \parallel 30 + 18 \parallel 9 = 16\ \Omega$.

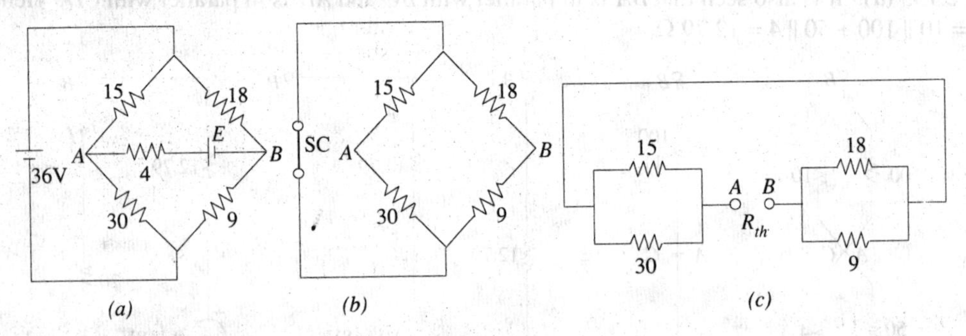

Fig. 2.135

We will find $V_{th} = V_{AB}$ with the help of Fig. 2.136 (*a*) which represents the original circuit, except with E and 4 Ω resistor removed. Here, the two circuits are connected in parallel across the 36 V battery. The potential of point A equals the drop on 30 Ω resistance, whereas potential of point B equals the drop across 9 Ω resistance. Using the voltage, divider rule, we have

$$V_A = 36 \times 30/45 = 24\ \text{V}$$

$$V_B = 36 \times 9/27 = 12\ \text{V}$$

$$\therefore \quad V_{AB} = V_A - V_B = 24 - 12 = \text{V}$$

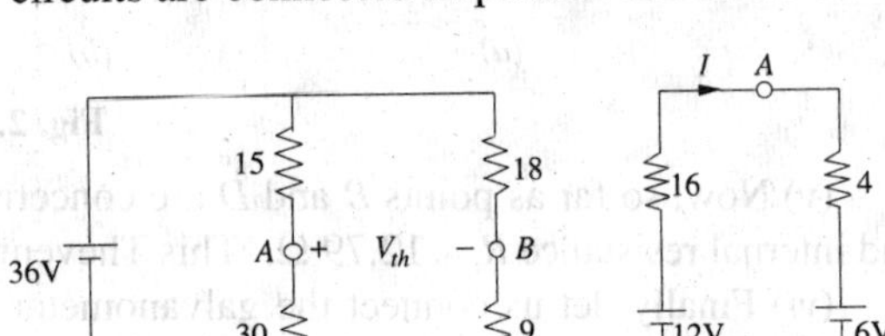

Fig. 2.136

In Fig. 2.136 (*b*), the series combination of E and 4 Ω resistors has been reconnected across terminals A and B of the Thevenin's equivalent circuit.

(*i*) $I = (12 - E)/20 = (12 - 2)/20 = 0.5$ A (*ii*) $I = (12 - 12)/20 = 0$

Example 2.61. *Calculate the value of V_{th} and R_{th} between terminals A and B of the circuit shown in Fig. 2.137 (a). All resistance values are in ohms.*

Solution. Forgetting about the terminal B for the time being, there are two parallel paths between E and F : one consisting of 12 Ω and the other of (4 + 8) = 12 Ω. Hence, R_{EF} = 12 ‖ 12 = 6Ω. The source voltage of 48 V drops across two 6 Ω resistances connected in series. Hence, V_{EF} = 24 V. The same 24 V acts across 12 Ω resistor connected directly between E and F and across two series – connected resistances of 4 Ω and 6 Ω connected across E and F. Drop across 4 Ω resistor = 24×4/(4+8) = 8 V as shown in Fig. 2.137 (*c*).

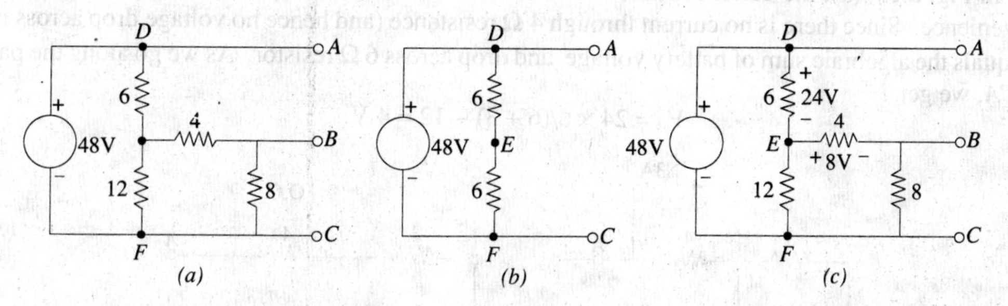

Fig. 2.137

Now, as we go from B to A via point E, there is a rise in voltage of 8 V followed by another rise in voltage of 24 V thereby giving a total voltage drop of 32 V. Hence, V_{th} = 32 V with point A positive.

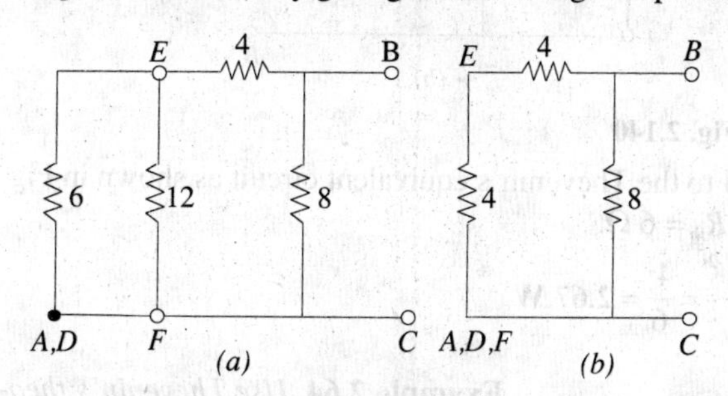

Fig. 2.138

For finding R_{th}, we short-circuit the 48 V source. This short circuiting, in effect, combines the points A, D and F electrically as shown in Fig. 2.138 (*a*). As seen from Fig. 2.138 (*b*),

$$R_{th} = V_{AB} = 8 \parallel (4 + 4) = 4\ \Omega.$$

Example 2.62. *Determine Thevenin's equivalent circuit which may be used to represent the given network (Fig. 2.139) at the terminals AB.*

(Electrical Engineering ; Calcutta Univ. 1987)

Solution. The given circuit of Fig. 2.139 (*a*) would be solved by applying Thevenin's theorem twice, first to the circuit to the left of point C and D and then to the left of points A and B. Using this technique, the network to the left of CD [Fig. 2.139 (*a*)] can be replaced by a source of voltage V_1 and series resistance R_{i1} as shown in Fig. 2.139 (*b*).

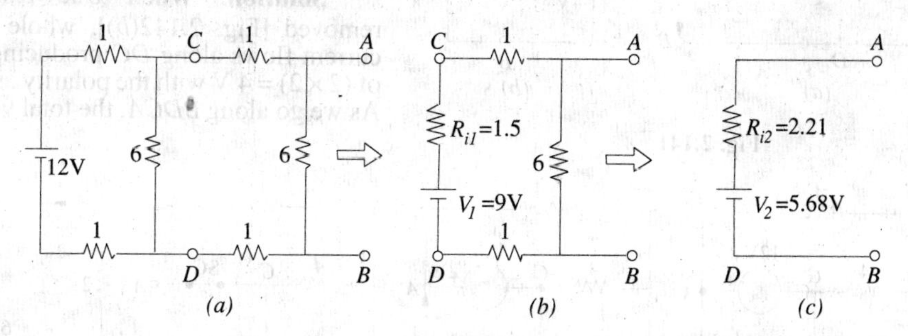

Fig. 2.139

$$V_1 = \frac{12\times 6}{(6+1+1)} = 9 \text{ volts and } R_{i1} = \frac{6\times 2}{(6+2)} = 1.5\ \Omega$$

Similarly, the circuit of Fig. 2.139(*b*) reduced to that shown in Fig. 2.139(*c*)

$$V_2 = \frac{9\times 6}{(6+2+1.5)} = \mathbf{5.68\ V} \text{ and } R_{i2} = \frac{6\times 3.5}{9.5} = \mathbf{2.21\ \Omega}$$

Example 2.63. *Use Thevenin's theorem, to find the value of load resistance R_L in the circuit of Fig. 2.140 (a) which results in the production of maximum power in R_L. Also, find the value of this maximum power. All resistances are in ohms.*

Solution. We will remove the voltage and current sources as well as R_L from terminals A and B in order to find R_{th} as shown in Fig. 2.140 (*b*).

$$R_{th} = 4 + 6 \parallel 3 = 6\ \Omega$$

In Fig. 2.140(*a*), the current source has been converted into the equivalent voltage source for convenience. Since there is no current through 4 Ω resistance (and hence no voltage drop across it), V_{th} equals the algebraic sum of battery voltage and drop across 6 Ω resistor. As we go along the path *BDCA*, we get.

$$V_{th} = 24 \times 6/(6 + 3) - 12 = 4\ \text{V}$$

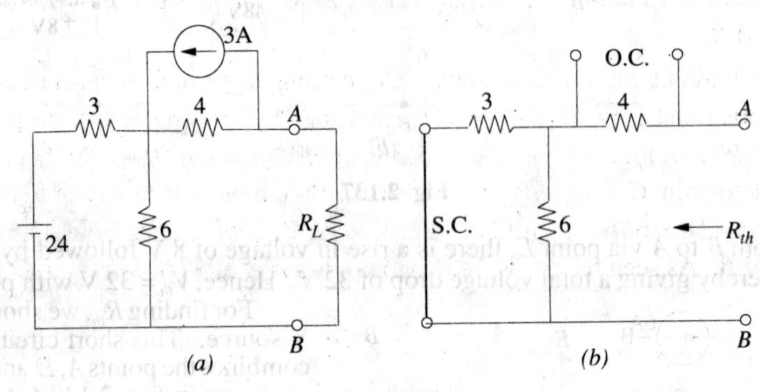

Fig. 2.140

The load resistance has been reconnected to the Thevenin's equivalent circuit as shown in Fig. 2.141(*b*). For maximum power transfer, $R_L = R_{th} = 6\ \Omega$.

Now, $V_L = \frac{1}{2} V_{th} = \frac{1}{2} \times 4 = 2V$; $P_{L\max} = \frac{V_L^2}{R_L} = \frac{4^2}{6} = \mathbf{2.67\ W}$

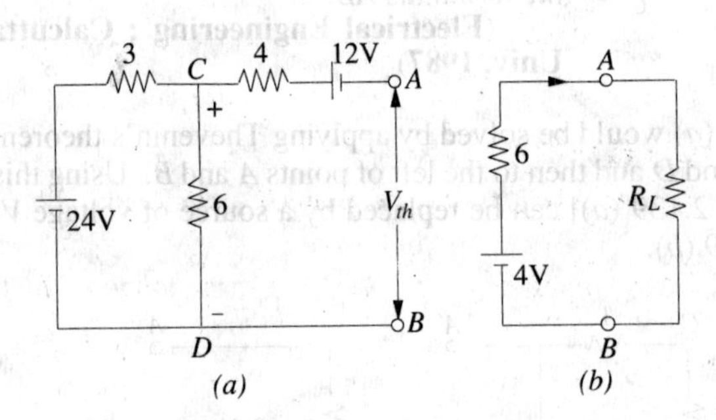

Fig. 2.141

Example 2.64. *Use Thevenin's theorem to find the current flowing through the 6 Ω resistor of the network shown in Fig. 2.142(a). All resistances are in ohms.*

(Network Theory, Nagpur Univ. 1992)

Solution. When 6 Ω resistor is removed [Fig. 2.142(*b*)], whole of 2 A current flows along *DC* producing a drop of (2 × 2) = 4 V with the polarity as shown. As we go along *BDCA*, the total voltage is

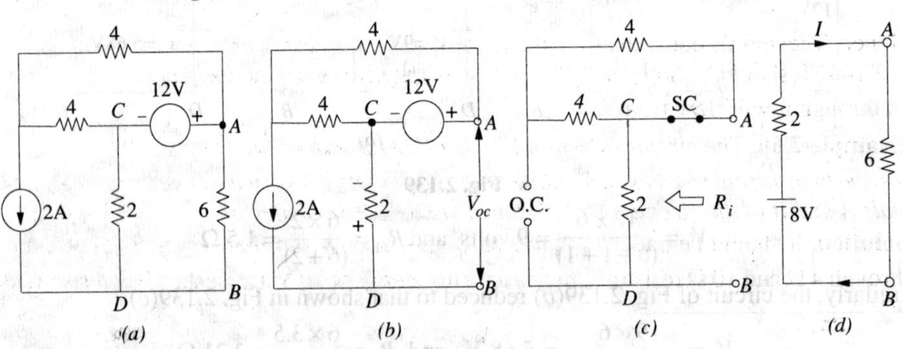

Fig. 2.142

$= -4 + 12 = 8$ V —with A positive w.r.t. B.

Hence, $V_{oc} = V_{th} = 8$ V

For finding R_i or R_{th}, voltage source is replaced by a short-circuit (Art. 2.15) and the current source by an open-circuit as shown in Fig. 2.142 (*c*). The two 4 Ω resistors are in series and are thus equivalent to an 8 Ω resistance. However, this 8 Ω resistor is in parallel with a short of 0 Ω. Hence, their equivalent value is 0 Ω. Now this 0 Ω resistance is in series with the 2 Ω resistor. Hence, $R_i = 2 + 0 = 2$ Ω. The Thevenin's equivalent circuit is shown in Fig. 2.142(*d*).

$\therefore$ $I = 8/(2 + 6) = $ **IA** – from A to B

Example 2.65. *Find Thevenin's equivalent circuit for the network shown in Fig. 2.143(a) for the terminal pair A B.*

Solution. It should be carefully noted that after coming to point D, the 6 A current has only one path to reach its other end C *i.e.* through 4 Ω resistor thereby creating and IR drop of $6 \times 4 = 24$ V with polarity as shown in Fig. 2.143 (*b*). No part of it can go along DE or DF because it would not find any path back to point C. Similarly, current due to 18-V battery is restricted to loop $EDFE$. Drop across 6 Ω resistor $= 18 \times 6/(6 + 3) = 12$ V. For finding V_{AB}, let us start from A and go to B via the *shortest* route $ADFB$. As seen from Fig. 143(*b*), there is a *rise* of 24 V from A to D but a *fall* of 12V

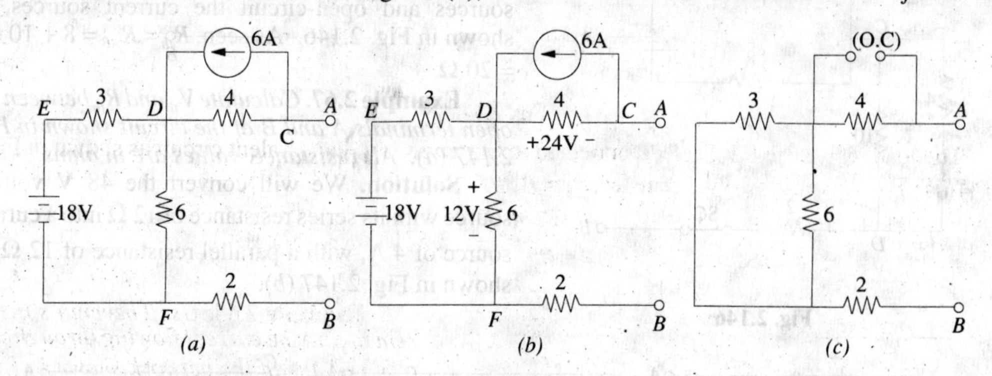

Fig. 2.143

from D to F. Hence, $V_{AB} = 24 - 12 = 12$ V with point A negative w.r.t. point B*. Hence, $V_{th} = V_{AB} = -12$ V (or $V_{BA} = 12$ V).

For finding R_{th}, 18 V battery has been replaced by a short-circuit and 6 A current source by an open-circuit, as shown Fig. 2.143(*c*).

As seen, $R_{th} = 4 + 6 || 3 + 2$

$$= 4 + 2 + 2 = 8\ \Omega$$

Hence, Thevenin's equivalent circuit for terminals A and B is as shown in Fig. 2.144. It should be noted that if a load resistor is connected across AB, current through it will flow from B to A.

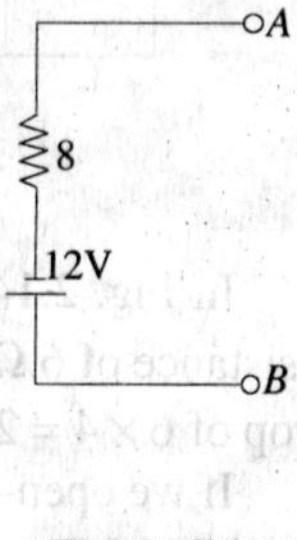

A
8
12V
B

Fig. 2.144

Example 2.66. *The circuit shown in Fig. 2.145(a) contains two voltage sources and two current sources. Calculate (a) V_{th} and (b) R_{th} between the open terminals A and B of the circuit. All resistance vlaues are in ohms.*

Solution. It should be understood that since terminals A and B are open, 2 A current can flow only through 4 Ω and 10 Ω resistors, thus producing a drop of 20 V across the 10 Ω resistor, as shown

*Incidentally, had 6 A current been flowing in the opposite direction, polarity of 24 V drop would have been reversed so that V_{AB} would have equalled $(24 + 12) = 36$ V with A positive w.r.t. point B.

in Fig. 2.145 (*b*). Similarly, 3 A current can flow through its own closed circuit between *A* and *C* thereby producing a drop of 24 V across 8 Ω resistor as shown in Fig. 2.145(*b*). Also, there is no drop across 2 Ω resistor because no current flows through it.

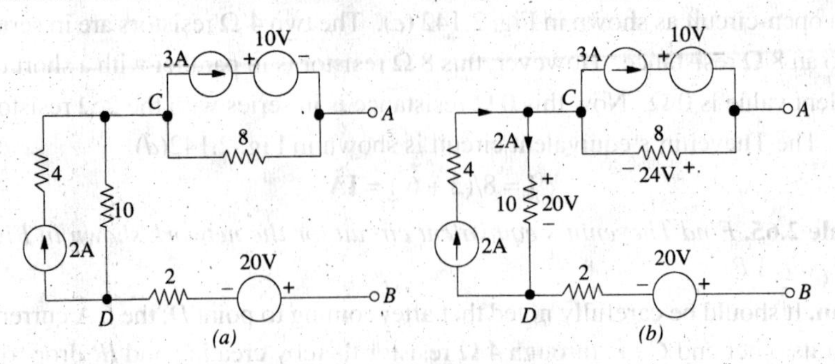

Fig. 2.145

Starting from point *B* and going to point *A* via points *D* and *C*, we get

$V_{th} = -20 + 20 + 24 = 24$ V – with point *A* positive.

For finding R_{th}, we will short-circuit the voltage sources and open-circuit the current sources, as shown in Fig. 2.146. As seen, $R_{th} = R_{AB} = 8 + 10 + 2 = 20\ \Omega$.

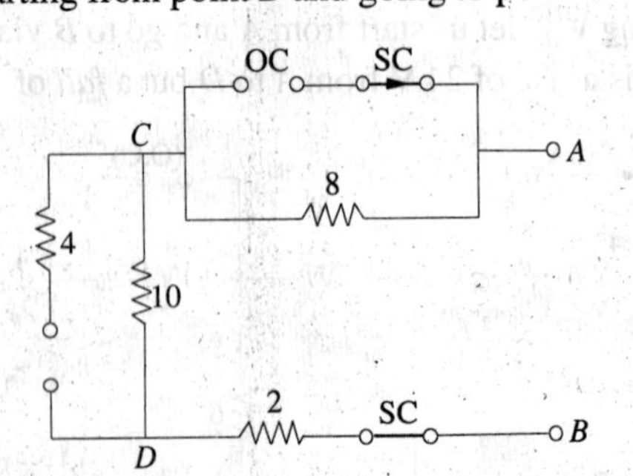

Fig. 2.146

Example 2.67. *Calculate V_{th} and R_{th} between the open terminals A and B of the circuit shown in Fig. 2.147 (a). All resistance values are in ohms.*

Solution. We will convert the 48 V voltage source with its series resistance of 12 Ω into a current source of 4 A, with a parallel resistance of 12 Ω, as shown in Fig. 2.147 (*b*).

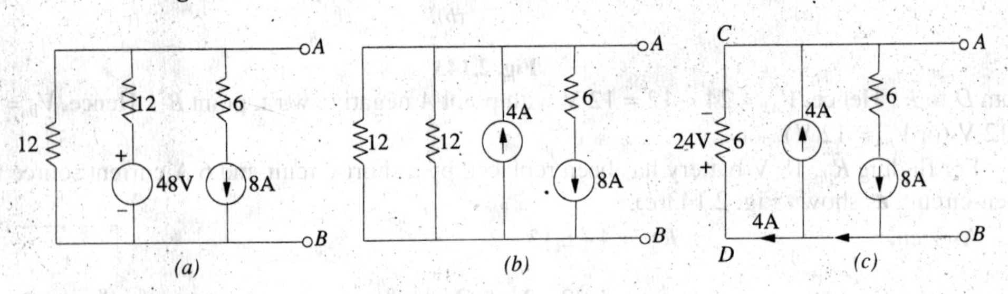

Fig. 2.147

In Fig. 2.147(*c*), the two parallel resistances of 12 Ω each have been combined into a single resistance of 6 Ω. It is obvious that 4 A current flows through the 6 Ω resistor, thereby producing a drop of 6 × 4 = 24 *V*. Hence, $V_{th} = V_{AB} = 24$ V with terminal *A* negative. In others words $V_{th} = -24$V.

If we open-circuit the 8 *A* source and short-circuit the 48-V source in Fig. 2.147 (*a*), $R_{th} = R_{AB} = 12 \parallel 12 = 6\ \Omega$.

Example 2.68. *Calculate the value of V_{th} and R_{th} between the open terminals A and B of the circuit shown in Fig. 2.148 (a). All resistance values are in ohms.*

Solution. It is seen from Fig. 2.148(*a*) that positive end of the 24 V source has been shown connected to point *A*. It is understood that the negative terminal is connected to the ground terminal *G*. Just to make this point clear, the given circuit has been redrawn in Fig. 2.148 (*b*) as well as in Fig. 2.148(*c*).

Let us start from the positive terminal of the battery and go to its negative terminal G via point C. We find that between points C and G, there are two parallel paths : one of 6 Ω resistance and the

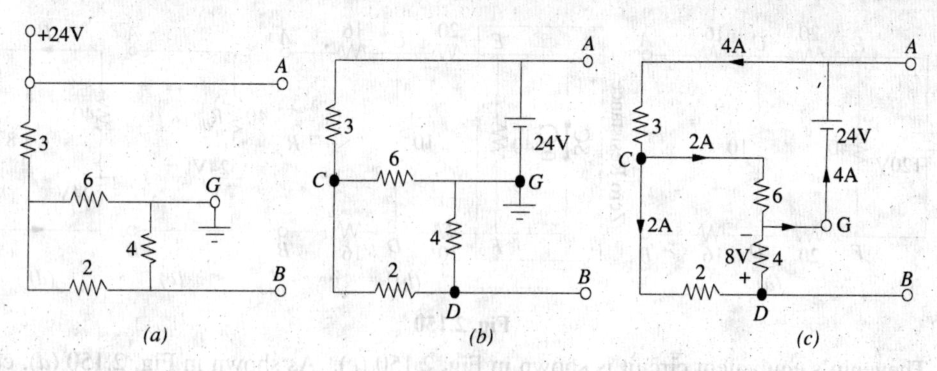

Fig. 2.148

other of (2 + 4) = 6 Ω resistance, giving a combined resistance of 6 || 6 = 3 Ω. Hence, total resistance between positive and negative terminals of the battery = 3 + 3 = 6 Ω. Hence, battery current = 24/6 = 4 A. As shown in Fig. 2.148 (*c*), this current divides equally at point C Let us go from B to A via points D and G and total up the potential difference between the two, $V_{th} = V_{AB} = -8\text{ V} + 24\text{ V} = 16$ V with point A positive.

For finding R_{th}, let us replace the voltage source by a short-circuit, as shown in Fig. 2.149 (*a*). It connects one end each of 6 Ω resistor and 4 Ω resistor directly to point A, as shown in Fig. 2.149(*b*). The resistance of branch $DCG = 2 + 6 \parallel 3 = 4\ \Omega$. Hence $R_{th} = R_{AB} = 4 \parallel 4 = 2\ \Omega$.

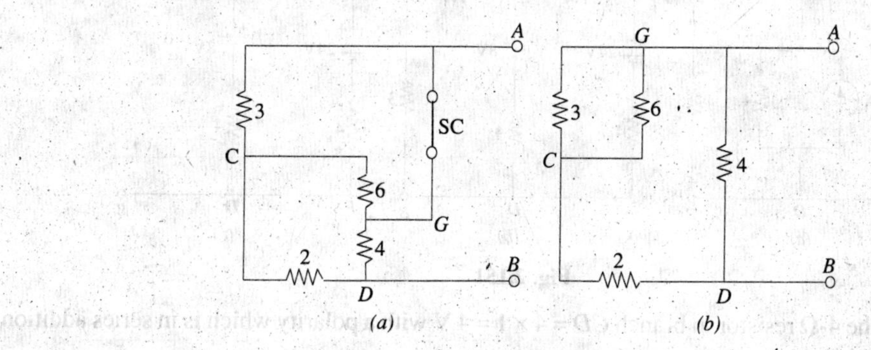

Fig. 2.149

Example 2.69. *Calculate the power which would be dissipated in the 8-Ω resistor connected across terminals A and B of Fig. 2.150(a). All resistance values are in ohms.*

Solution. The open-circuit voltage V_{oc} (also called Thevenin's voltage V_{th}) is that which appears across terminals A and B. This equals the voltage drop across 10 Ω resistor between points C and D. Let us find this voltage. With AB on open-circuit, 120 – V battery voltage acts on the two parallel paths EF and $ECDF$. Hence, current through 10 Ω resistor is

$$I = 120/(20 + 10 + 20) = 2.4\text{ A}$$

Drop across 10 – Ω resistor, $V_{th} = 10 \times 2.4 = 24$ V

Now, let us find Thevenin's resistance R_{th} *i.e.* equivalent resistance of the given circuit when looked into from terminals A and B. For this purpose, 120 V battery is removed. This results in shorting the 40–Ω resistance since internal resistance of the battery is zero as shown in Fig. 2.150 (*b*).

$$\therefore \quad R_i \text{ or } R_{th} = 16 + \frac{10 \times (20+20)}{10+(20+20)} + 16 = 40\ \Omega$$

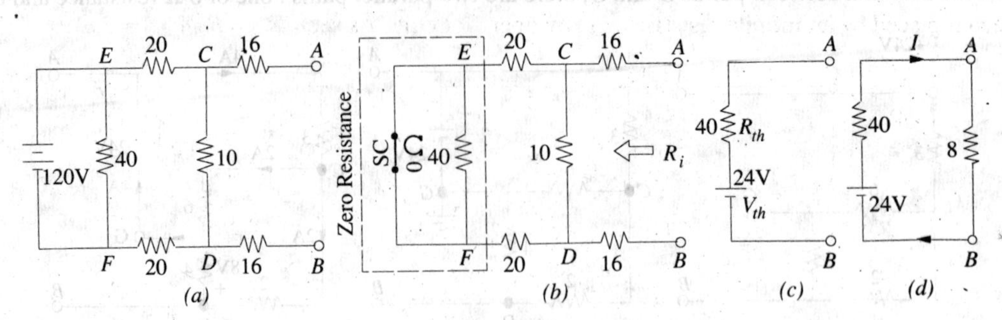

Fig. 2.150

Thevenin's equivalent circuit is shown in Fig. 2.150 (*c*). As shown in Fig. 2.150 (*d*), current through 8-Ω resistor is

$$I = 24/(40+8) = \frac{1}{2}\text{ A} \quad \therefore \quad P = I^2R = \left(\frac{1}{2}\right)^2 \times 8 = \mathbf{2\ W}$$

Example 2.70. *With the help of Thevenin's theorem, calculate the current flowing through the 3-Ω resistor in the network of Fig. 2.151(a). All resistances are in ohms.*

Solution. The current source has been converted into an equivalent voltage source in Fig. 151(*b*).
(*i*) **Finding V_{oc}.** As seen from Fig. 2.151(*c*), $V_{oc} = V_{CD}$. In closed circuit *CDFEC*, net voltage = 24–8 = 16 V and total resistance = 8 + 4 + 4 = 16 Ω. Hence, current = 16/16 = 1 A.

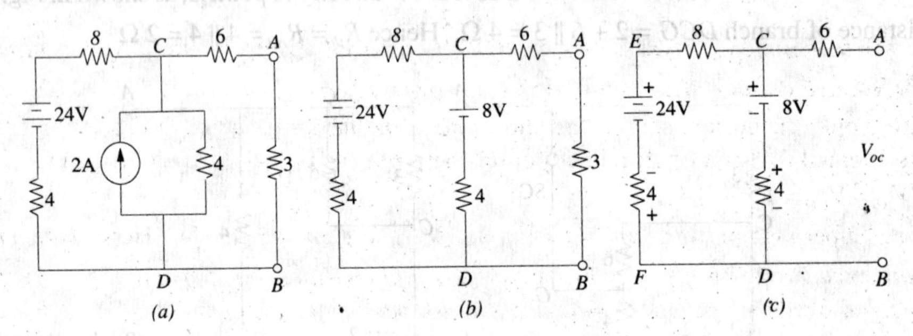

Fig. 2.151

Drop over the 4-Ω resistor in branch $CD = 4 \times 1 = 4$ V with a polarity which is in series addition with 8–V battery.

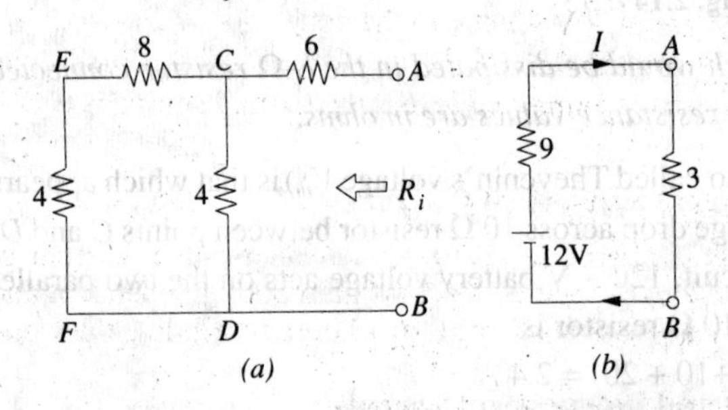

Fig. 2.152

Hence, $V_{oc} = V_{th} = V_{CD} = 8 + 4 = 12$ V

(*ii*) **Finding R_i or R_{th}.** In Fig. 2.152 (*a*), the two batteries have been replaced by short-circuits because they do not have any internal resistance.

As seen, $R_i = 6 + 4 \parallel (8 + 4) = 9\ \Omega$.

The Thevenin's equivalent circuit is as shown in Fig. 2.152(*b*).

$$I = 12/(9+3) = \mathbf{1\ A}$$

Example 2.71. *Using Thevenin and Superposition theorems find complete solution for the network shown in Fig. 2.153 (a).*

Solution. First, we will find R_{th} across open terminals *A* and *B* and then find V_{th} due to the voltage sources only and then due to current source only and then using Superposition theorem, combine the two voltages to get the single V_{th}. After that, we will find the Thevenin equivalent.

In Fig. 2.153 (*b*), the terminals *A* and *E* have been open-circuited by removing the 10 V source and the 1 Ω resistance. Similarly; 24 V source has been replaced by a short and current source has been replaced by an infinite resistance *i.e.* by open - circuit. As seen, $R_{AB} = R_{th} 4 \parallel 4 = 2\ \Omega$.

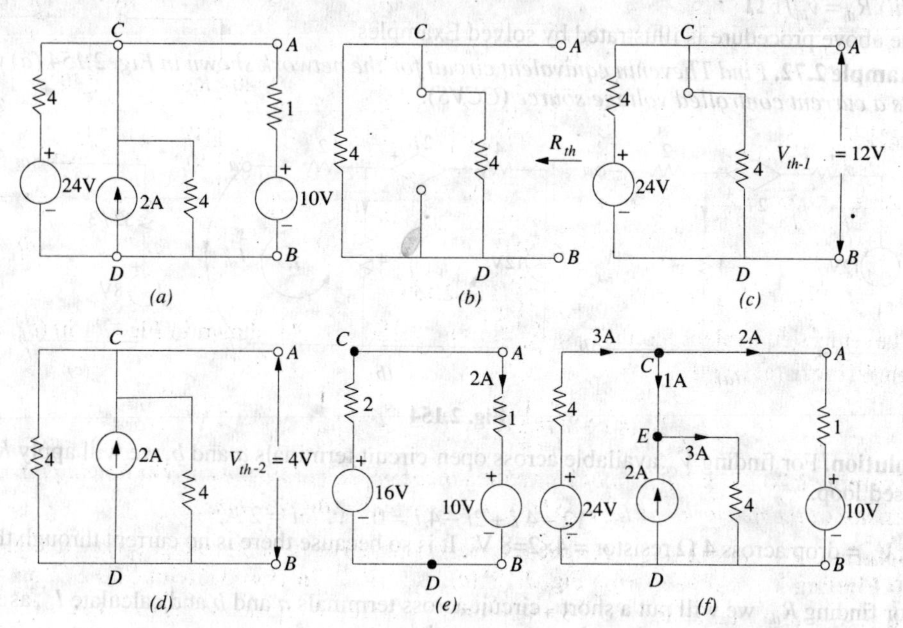

Fig. 2.153

We will now find V_{th-1} across *AB* due to 24 V source only by open-circuiting the current source. Using the voltage-divider rule in Fig. 2.153 (*c*), we get $V_{AB} = V_{CD} = V_{th-1} = 24/2 = 12$ V.

Taking only the current source and short-circuiting the 24 V source in Fig. 2.153 (*d*), we find that there is equal division of current at point *C* between the two 4 Ω parallel resistors. Therefore, $V_{th-2} = V_{AB} = V_{CD} = 1 \times 4 = 4$ V.

Using Superposition theorem, $V_{th} = V_{th-1} + V_{th-2} = 12 + 4 = 16$–V. Hence, the Thevenin's equivalent consists of a 16 *V* source in series with a 2 Ω resistance as shown in Fig. 2.153 (*e*) where the branch removed earlier has been connected back across the terminals *A* and *B*. The net voltage around the circuit is = 16–10 = 6 V and total resistance is = 2 + 1 = 3 Ω. Hence, current in the circuit is = 6/ 3 = 2 A. Also, $V_{AB} = V_{AD} = 16 - (2 \times 2) = 12$ V. Alternatively, V_{AB} equals $(2 \times 1) + 10 = 12$ V.

Since we know that $V_{AB} = V_{CD} = 12$ V , we can find other voltage drops and various circuit currents as shown in Fig. 2.153(*f*). Current delivered by the 24 – V source to the node *C* is = $(24 - V_{CD})/4 = (24-12)/4 = 3$ A. Since current flowing through branch *AB* is 2 A, the balance of 1 A flows along *CE*. As seen, current flowing through the 4 Ω resistor connected across the current source is = (1 + 2) = 3 A.

2.20. General Instructions for Finding Thevenin Equivalent Circuit

So far, we have considered circuits which consisted of resistors and independent current or voltage sources only. However, we often come across circuits which contain both independent and dependent sources or circuits which contain only dependent sources. Procedure for finding the value of V_{th} and R_{th} in such cases is detailed below :

(*a*) When Circuit Contains Both Dependent and Independent Sources

(*i*) The open-circuit voltage v_{oc} is determined as usual with the sources activated or 'alive'.

(*ii*) A short-circuit is applied across the terminals *a* and *b* and the vlaue of short-circuit current i_{sh} is found as usual.

(*iii*) Thevenin resistance $R_{th} = v_{oc}/i_{sh}$. It is the same procedure as adopted for Norton's theorem. Solved examples 2-72, 2-73, 2-74, 2-75 and 2-76 illustrate this procedure.

(b) When Circuit Contains Dependent Sources Only

(i) In this case, $v_{oc} = 0$.

(ii) We connect 1 A source to the terminals a and b and calculate the value of v_{ab}.

(iii) $R_{th} = v_{ab}/1\ \Omega$

The above procedure is illustrated by solved Examples.

Example 2.72. *Find Thevenin equivalent circuit for the network shown in Fig. 2.154 (a) which contains a current controlled voltage source (CCVS).*

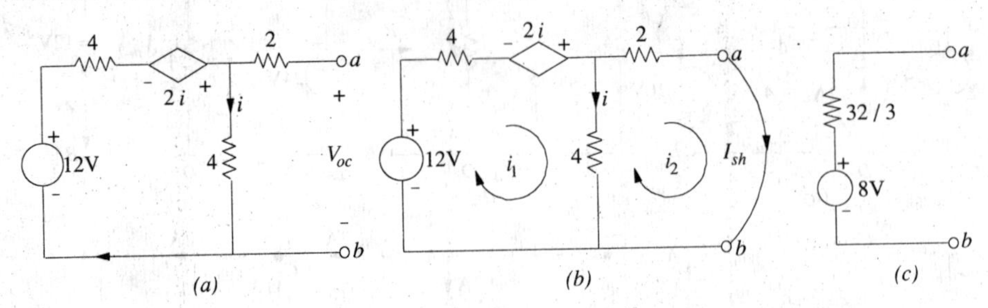

Fig. 2.154

Solution. For finding V_{oc} available across open-circuit terminals a and b, we will apply KVL to the closed loop.

$$12 - 4\,i + 2i - 4\,i = 0 \quad \therefore \quad i = 2\text{ A}.$$

$\therefore$ Hence, V_{oc} = drop across 4 Ω resistor = 4×2=8 V. It is so because there is no current through the 2 Ω resistor.

For finding R_{th}, we will put a short - circuit across terminals a and b and calculate I_{sh}, as shown in Fig. 2.154 (b). Using the two mesh currents, we have

$12 - 4\,i_1 + 2\,i - 4\,(i_1 - i_2) = 0$ and $-8\,i_2 - 4(i_2 - i_1) = 0$. Substituting $i = (i_1 - i_2)$ and Simplifying the above equations, we have

$$12 - 4\,i_1 + 2\,(i_1 - i_2) - 4\,(i_1 - i_2) = 0 \quad \text{or} \quad 3\,i_1 - i_2 = 6 \qquad \text{...(i)}$$

Similarly, from the second equation, we get $i_1 = 3\,i_2$. Hence, $i_2 = 3/4$ and $R_{th} = V_{oc}/I_{sh} = 8/(3/4) = 32/3\ \Omega$. The Thevenin equivalent circuit is as shown in Fig. 2.154 (c).

Example 2.73. *Find the Thevenin equivalent circuit with respect to terminals a and b of the network shown in Fig. 2.155 (a).*

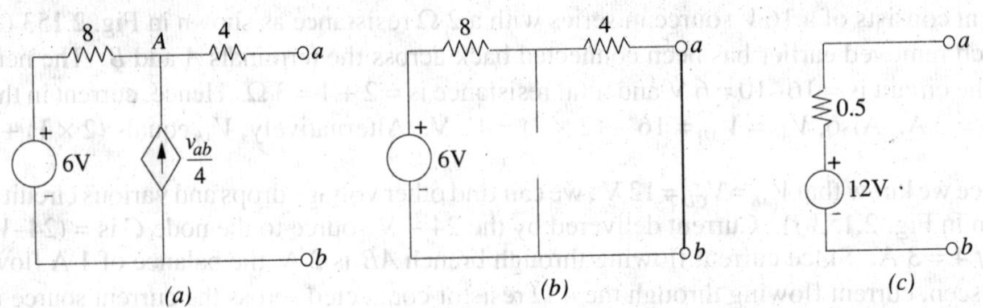

Fig. 2.155

Solution. It will be seen that with terminals a and b open, current through the 8 Ω resistor is $v_{ab}/4$ and potential of point A is the same of that of point a (because there is no current through 4 Ω resistor). Applying KVL to the closed loop of Fig. 2.155 (a), we get

$$6 + (8 \times v_{ab}/4) - v_{ab} = 0 \quad \text{or} \quad v_{ab} = 12\text{ V}$$

It is also the value of the open-circuit voltage v_{oc}.

For finding short-circuit current i_{sh}, we short-circuit the terminals a and b as shown in Fig. 2.155 (b). Since with a and b short-circuited, $v_{ab} = 0$, the dependent current source also becomes zero. Hence, it is replaced by an open-circuit as shown. Going around the closed loop, we get

$$12 - i_{sh}\,(8 + 4) = 0 \quad \text{or} \quad i_{sh} = 6/12 = 0.5\text{ A}$$

Hence, the Thevenin equivalent is as shown in Fig. 2.155 (c).

Example 2.74. *Find the Thevenin equivalent circuit for the network shown in Fig. 2.156 (a) which contains only a dependent source.*

Solution. Since circuit contains no independent source, $i = 0$ when terminals a and b are open. Hence, $v_{oc} = 0$. Moreover, i_{sh} is zero since $v_{oc} = 0$.

Consequently, R_{th} cannot be found from the relation $R_{th} = v_{oc} / i_{sh}$. Hence, as per Art. 2.20, we will connect a 1 A current source to terminals a and b as shown in Fig. 2.156 (b). Then by finding the value of v_{ab}, we will be able to calculate $R_{th} = v_{ab} / 1$.

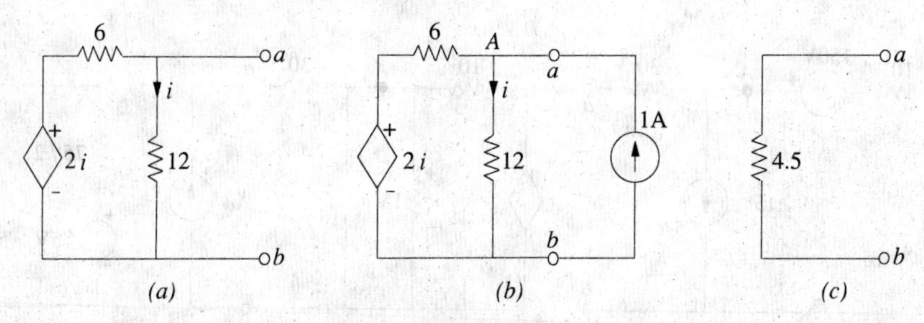

Fig. 2.156

It should be noted that potential of point A is the same as that of point a *i.e.* voltage across 12 Ω resistor is v_{ab}. Applying KCL to point A, we get

$$\frac{2i - v_{ab}}{6} - \frac{v_{ab}}{12} + 1 = 0 \text{ or } 4i - 3v_{ab} = -12$$

Since $i = v_{ab} / 12$, we have $4(v_{ab} / 12) - 3\, v_{ab} = -12$ or $v_{ab} = 4.5$ V $\therefore R_{th} = v_{ab} / 1 = 4.5/1 = \mathbf{4.5\ \Omega}$

The Thevenin equivalent circuit is shown in Fig. 2.156(c).

Example 2.75. *Determine the Thevenin's equivalent circuit as viewed from the open-circuit terminals a and b of the network shown in Fig. 2.157 (a). All resistances are in ohms.*

Solution. It would be seen from Fig. 2.157 (a) that potential of node A equals the open-circuit terminal voltage v_{oc}. Also, $i = (v_s - v_{oc}) / (80 + 20) = (6 - v_{oc}) / 100$.

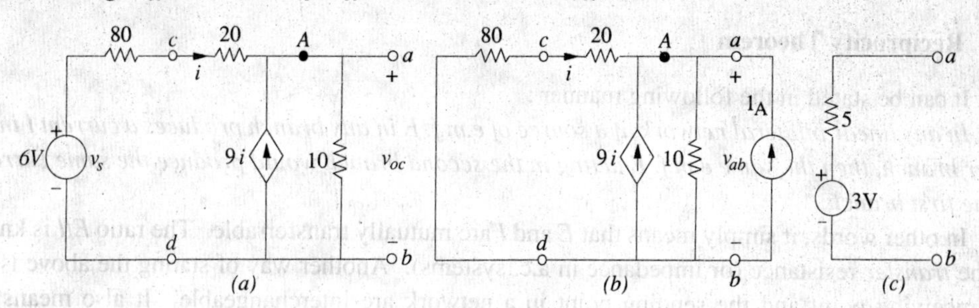

Fig. 2.157

Applying KCL to node A, we get

$$\frac{6 - V_{oc}}{100} + \frac{9 \times (6 - v_{oc})}{100} - \frac{V_{oc}}{10} = \text{or } V_{oc} \;=\; 3\text{V}$$

For finding the Thevenin's resistance with respect to terminals a and b, we would first 'kill' the independent voltage source as shown in Fig. 2.157 (b). However, the dependent current source cannot be 'killed'. Next, we will connect a current source of 1 A at terminals a and b and find the value of v_{ab}. Then, Thevenin's resistance $R_{th} = v_{ab} / 1$. It will be seen that current flowing away from node A *i.e.* from point c to d is $= v_{ab} / 100$. Hence, $i = -\, v_{oc} / 100$. Applying KCL to node A, we get

$$-\frac{v_{ab}}{100} + 9\left(-\frac{v_{ab}}{100}\right) - \frac{v_{ab}}{10} + 1 = 0 \text{ or } v_{ab} = 5\text{ V}$$

$\therefore R_{th} = 5/1 = 5\ \Omega$. Hence, Thevenin's equivalent source is as shown in Fig. 2.157(c).

Example 2.76. *Find the Thevenin's equivalent circuit with respect to terminals a and b of the network shown in Fig. 2.158(a). All resistances are in ohms.*

Solution. It should be noted that with terminals a and b open, potential of node A equals v_{ab}. Moreover, $v = v_{ab}$. Applying KCL to node A, we get

$$-5-\frac{V_{ab}}{15}+\frac{1}{10}\left[\left(\frac{V_{ab}}{3}+150\right)-V_{ab}\right]=0 \text{ or } V_{ab}=75\text{V}$$

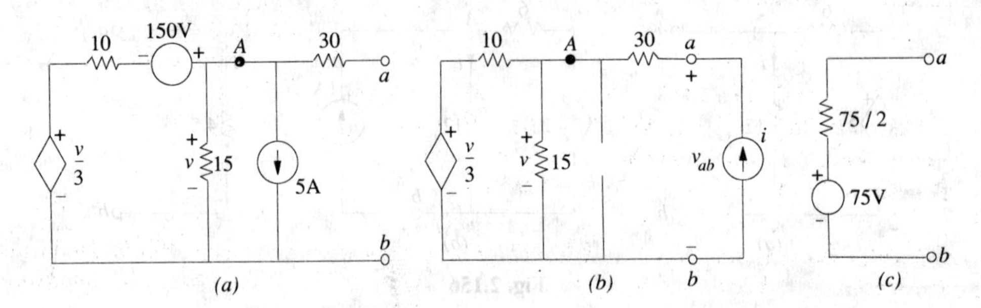

Fig. 2.158

For finding R_{th}, we will connect a current source of i A* across terminals a and b. It should be particularly noted that in this case the potential of node A equals $(v_{ab} - 30\,i)$. Also, $v = (v_{ab} - 30\,i)$ = potential of node A. Applying KCL to node A, we get

$$i=\frac{(v_{ab}-30\,i)}{15}+\frac{1}{10}\left[\left(\frac{v_{ab}-30\,i}{3}\right)-(v_{ab}-30\,i)\right]=0$$

$\therefore$ $4\,v_{ab} = 150\,i$ or $v_{ab}/i = 75/2\ \Omega$. Hence, $R_{th} = v_{ab}/i = 75/2\ \Omega$. The Thevenin's equivalent circuit is shown in Fig. 2.158 (c).

2.21. Reciprocity Theorem

It can be stated in the following manner :

In any linear bilateral network, if a source of e.m.f. E in any branch produces a current I in any other branch, then the same e.m.f. E acting in the second branch would produce the same current I in the first branch.

In other words, it simply means that E and I are mutually transferrable. The ratio E/I is known as the *transfer* resistance (or impedance in a.c. systems). Another way of stating the above is that the receiving point and the sending point in a network are interchangeable. It also means that interchange of an *ideal* voltage source and an *ideal* ammeter in any network will not change the ammeter reading. Same is the case with the interchange of an ideal current source and an *ideal* voltmeter.

Example 2.77. *In the network of Fig. 2.159 (a), find (a) ammeter current when battery is at A and ammeter at B and (b) when battery is at B and ammeter at point A. Values of various resistances are as shown in the diagram. Also, calculate the transfer resistance.*

Solution. (*a*) Equivalent resistance between points C and B in Fig. 2.159(*a*) is

$$= 12 \times 4/16 = 3\ \Omega$$

---

*We could also connect a source of 1 A as done in Ex. 2.74.

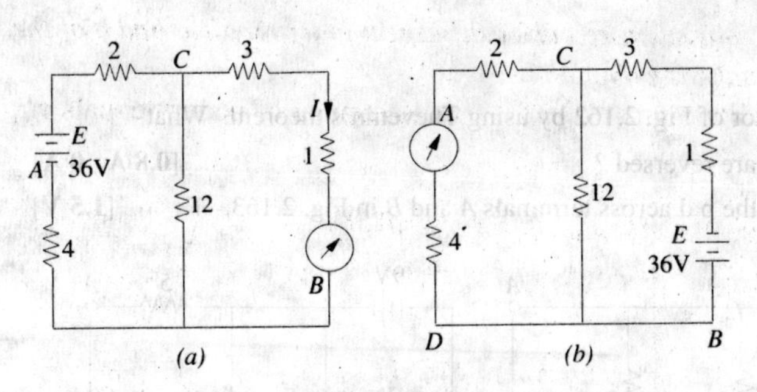

Fig. 2.159

$\therefore$ Total circuit resistance
$= 2 + 3 + 4 = 9\ \Omega$

$\therefore$ Battery current = 36/9 = 4 A

$\therefore$ Ammeter current
$= 4 \times 12/16 =$ **3 A.**

(*b*) Equivalent resistance between points *C* and *D* in Fig. 2.159 (*b*) is
$= 12 \times 6/18 = 4\ \Omega$

Total circuit resistance
$= 4 + 3 + 1 = 8\ \Omega$

Battery current = 36 /8 = 4.5 A

$\therefore$ Ammeter current $= 4.5 \times 12/18 =$ **3 A**

Hence, ammeter current in both cases is the same.

Transfer resistance = 36 /3 = **12 Ω.**

Example 2.78. *Calculate the currents in the various branches of the network shown in Fig. 2.160 and then utilize the principle of Superposition and Reciprocity theorem together to find the value of the current in the 1-volt battery circuit when an e.m.f. of 2 volts is added in branch BD opposing the flow of original current in that branch.*

Solution. Let the currents in the various branches be as shown in the figure. Applying Kirchhoff's second law, we have

For loop *ABDA* ; $-2I_1 - 8I_3 + 6I_2 = 0$ or $I_1 - 3I_2 + 4I_3 = 0$... (*i*)

For loop *BCDB*, $-4(I_1 - I_3) + 5(I_2 + I_3) + 8I_3 = 0$ or $4I_1 - 5I_2 - 17I_3 = 0$... (*ii*)

For loop *ABCEA*, $-2I_1 - 4(I_1 - I_3) - 10(I_1 + I_2) + 1 = 0$ or $16I_1 + 10I_2 - 4I_3 = 1$ ... (*iii*)

Solving for I_1, I_2 and I_3, we get $I_1 =$ **0.0494 A** ; $I_2 =$ **0.0229 A** ; $I_3 =$ **0.0049 A**

$\therefore$ Current in the 1 volt battery circuit is $= I_1 + I_2 =$ **0.0723 A.**

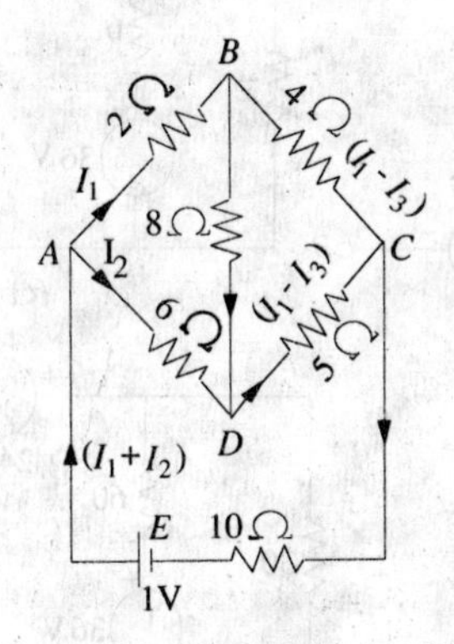

Fig. 2.160

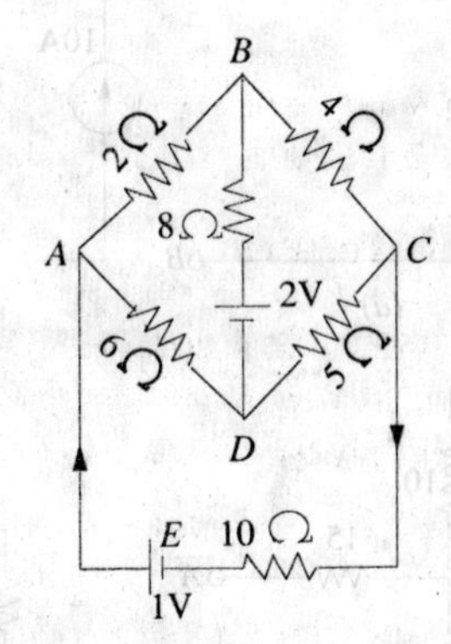

Fig. 2.161

The new circuit having 2 – V battery connected in the branch *BD* is shown in Fig. 2.161. According to the Principle of Superposition, the new current in the 1 – volt battery circuit is due to the superposition of two currents; one due to 1 – volt battery and the other due to the 2 – volt battery when each acts independently.

The current in the external circuit due to 1 – volt battery when 2 – V battery is not there, as found above, is 0.0723 A.

Now, according to Reciprocity theorem; if 1 – volt battery were transferred to the branch *BD* (where it produced a current of 0.0049 A), then it would produce a current of 0.0049 A in the branch *CEA* (where it was before). Hence, a battery of 2 – V would produce a current of $(-2 \times 0.0049) = -0.0098\ A$ (by proportion). The negative sign is used because the 2 – volt battery has been so connected as to oppose the current in branch *BD*.

$\therefore$ new current in branch *CEA* = 0.0723 – 0.0098 = **0.0625 A**

Tutorial Problems No. 2.5

1. Calculate the current in the 8-Ω resistor of Fig. 2.162 by using Thevenin's theorem. What will be its value if connections of 6-V battery are reversed ? **[0.8 A ; 0 A]**

2. Use Thevenin's theorem to calculate the p.d across terminals *A* and *B* in Fig. 2.163. **[1.5 V]**

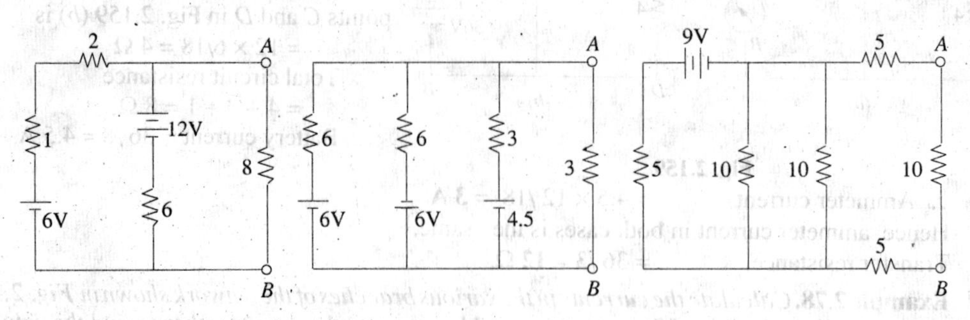

Fig. 2.162 **Fig. 2.163** **Fig. 2.164**

3. Compute the current flowing through the load resistance of 10 Ω connected across terminals *A* and *B* in Fig. 2.164 by using Thevenin's theorem.

4. Find the equivalent Thevenin voltage and equivalent Thevenin resistance respectively as seen from open-circuited terminals *A* and *B* of the circuits shown in Fig. 2.165 All resistances are in ohms.

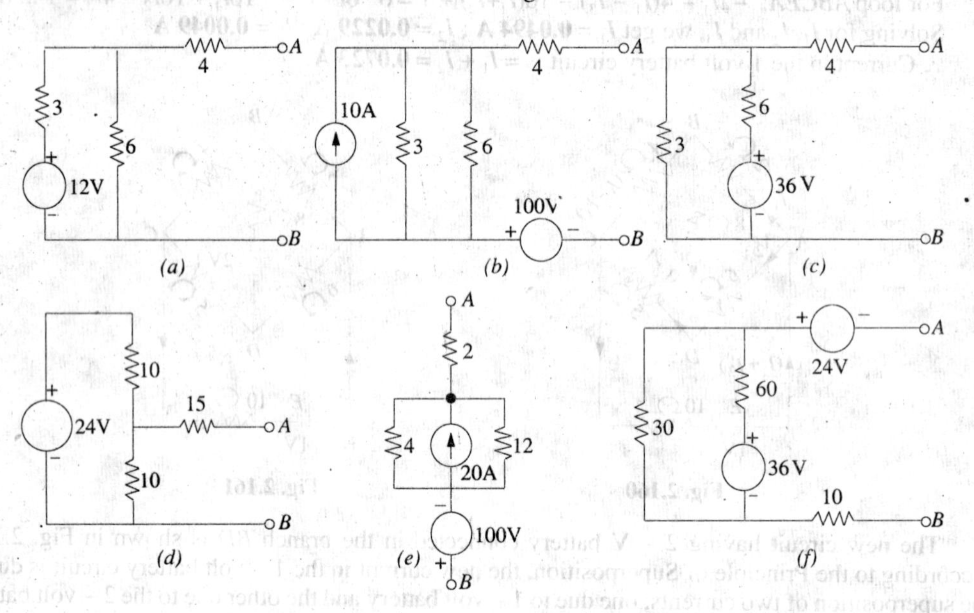

Fig. 2.165

[(*a*) 8 V, 6 Ω ; (*b*) 120 V, 6 Ω ; (*c*) 12 V, 6 Ω ; (*d*) 12 V, 20 Ω ; (*e*) –40 V, 5 Ω ; (*f*) –12 V, 30 Ω]

5. **Find Thevenin's equivalent of the circuits shown in Fig. 2.166 between terminals *A* and *B*.**

[(*a*) $V_{th} = I\dfrac{R_1 R_2}{R_1+R_2} + V\dfrac{R_1}{R_1+R_2}$; $R_{th} = \dfrac{R_1 R_2}{R_1+R_2}$ (*b*) $V_{th} = \dfrac{V_1 R_2 + V_2 R_1}{R_1+R_2}$; $R_{th} = \dfrac{R_1 R_2}{R_1+R_2}$

(*c*) $V_{th} = -IR; R_{th} = R_1$ (*d*) $V_{th} = -V_1 - IR$; $R_{th} = R$ (*e*) Not possible]

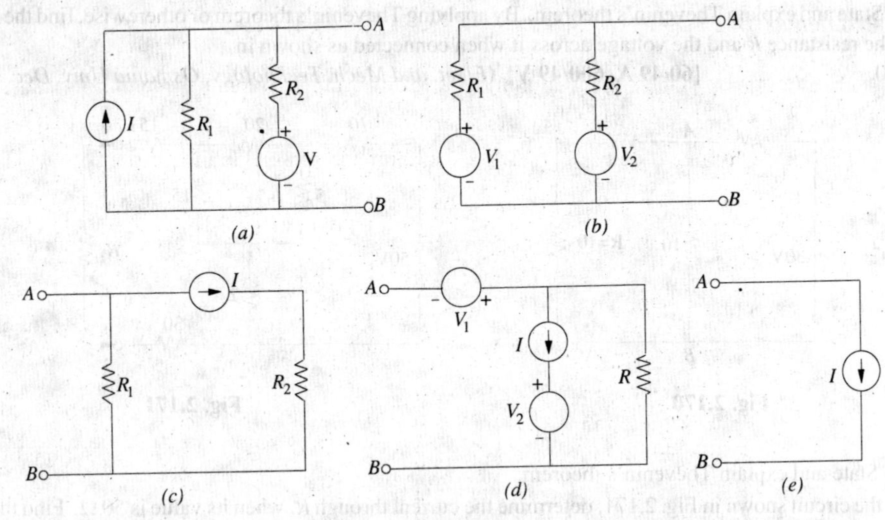

Fig. 2.166

6. The four arms of a Wheatstone bridge have the following resistances in ohms.

AB = 100, BC = 10, CD = 5, DA = 60

A galvenometer of 15 ohm resistance is connected across *BD*. Calculate the current through the galvanometer when a potential difference of 10 V is maintained across *AC*.

[*Elect. Engg. A.M.Ae. S.I. Dec. 1991*] **[4.88 mA]**

7. Find the Thevenin equivalent circuit for the network shown in Fig. 2.167.

[(*a*) 4 V; 8 Ω (*b*) 6 V ; 3 Ω (*c*") OV ; 2/5 Ω]

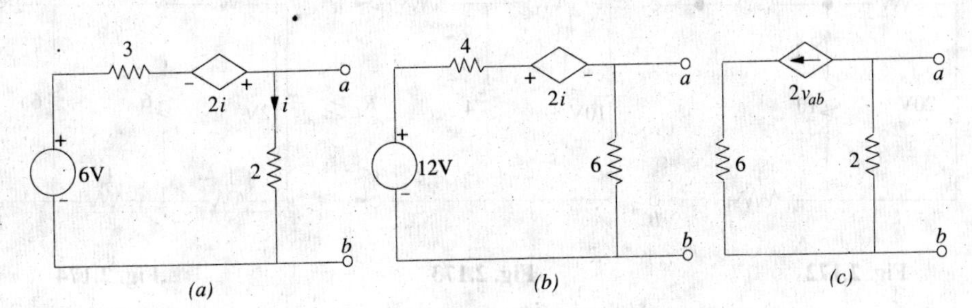

Fig. 2.167

8. Use Thevenin's theorem to find current in the branch *AB* of the network shown in Fig. 2.168.

[1.84 A]

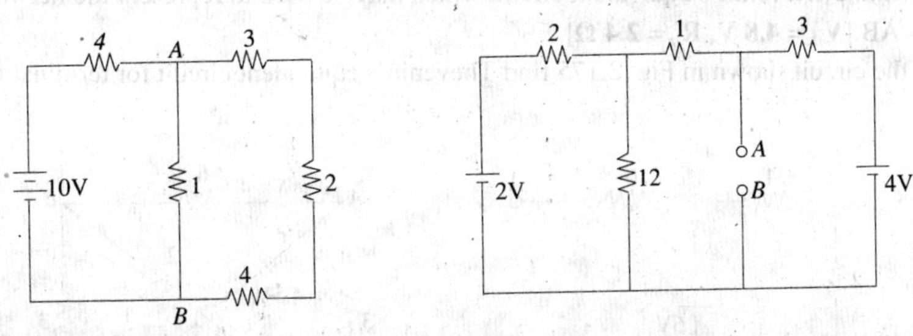

Fig. 2.168 **Fig. 2.169**

9. In the network shown in Fig. 2.169 find the current that would flow if a 2-Ω resistor were connected between points *A* and *B* by using.

(*a*) Thevenin's theorem and (*b*) Superposition theorem. The two batteries have negligible resistance. **[0.82 A]**

10. State and explain Thevenin's theorem. By applying Thevenin's theorem or otherwise, find the current through the resistance R and the voltage across it when connected as shown in Fig. 2.170. **[60/49 A, 600/49 V]** (*Elect. and Mech. Technology, Osmania Univ. Dec. 1978*)

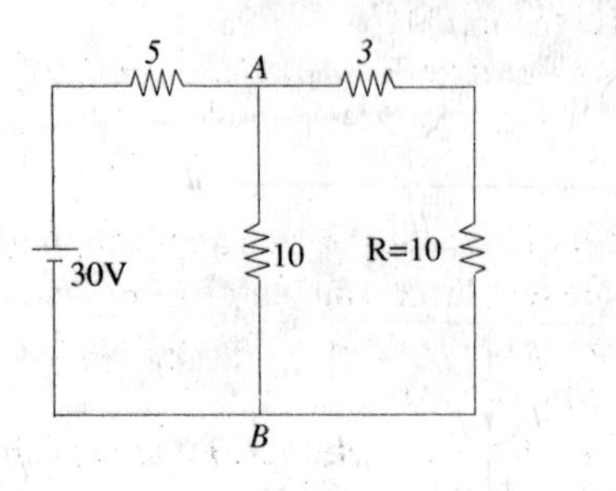

Fig. 2.170

Fig. 2.171

11. State and explain Thevenin's theorem.

For the circuit shown in Fig. 2.171, determine the current through R_L when its value is 50 Ω. Find the value of R_L for which the power drawn from the source is maximum.

(*Elect. Technology I, Gwalior Univ. Nov. 79*)

12. Find the Thevenin's equivalent circuit for terminal pair *AB* for the network shown in Fig. 2.172. **[V_{th} = −16 V and R_{th} = 16 Ω]**

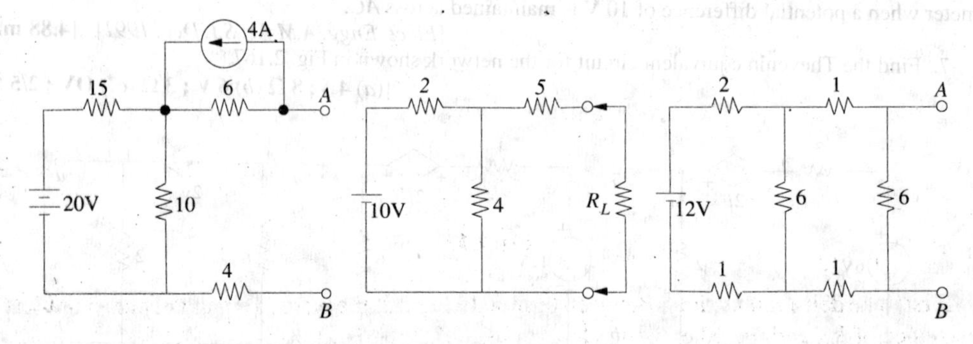

Fig. 2.172 Fig. 2.173 Fig. 2.174

13. For the circuit shown in Fig. 2.173, determine current through R_L when it takes values of 5 and 10 Ω. **[0.588A, 0.408 A]** [*Network Theorem & Fields, Madras Univ. 1980*]

14. Determine Thevenin's equivalent circuit which may be used to represent the network of Fig. 2.174 at the terminals AB.**[V_{th} = 4.8 V, R_{th} = 2.4 Ω]**

15. For the circuit shown in Fig. 2.175 find Thevenin's equivalent circuit for terminal pair *AB*.

[6 V, 6 Ω]

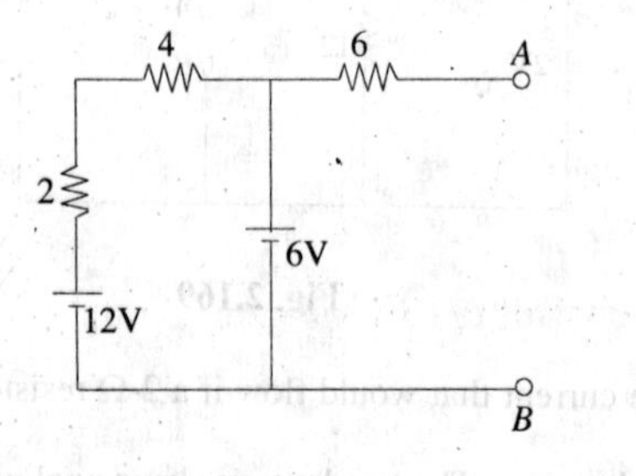

Fig. 2.175

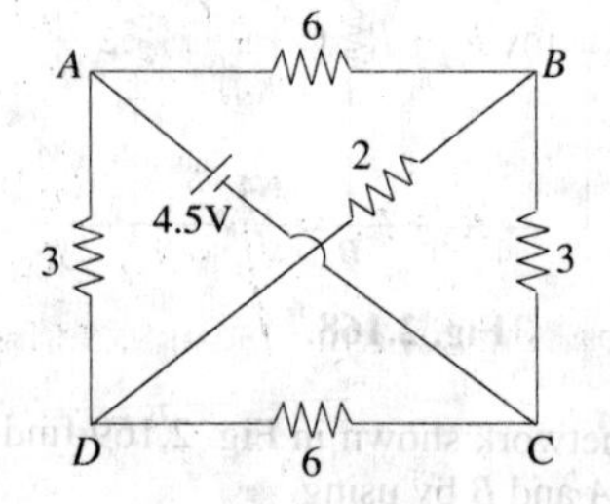

Fig. 2.176

16. *ABCD* is a rectangle whose opposite sides *AB* and *DC* represent resistances of 6 Ω each, while *AD* and *BC* represent 3 Ω each. A battery of e.m.f. 4.5 V and negligible resistance is connected between diagonal points *A* and *C* and a 2 – Ω resistance between *B* and *D*. Find the magnitude and direction of the current in the 2-ohm resistor by using Thevenin's theorem. The positive terminal is connected to *A*. (Fig. 176)[**0.25 A from D to B**] (*Basic Electricity. Bombay Univ. Oct. 1977*)

2.22. Delta/Star* Transformation

In solving networks (having considerable number of branches) by the application of Kirchhoff's Laws, one sometimes experiences great difficulty due to a large number of simultaneous equations that have to be solved. However, such complicated networks can be simplified by successively replacing delta meshes by equivalent star systems and *vice versa*.

Suppose we are given three resistances R_{12}, R_{23} and R_{31} connected in delta fashion between terminals 1, 2 and 3 as in Fig. 2.177(*a*). So far as the respective terminals are concerned, these three given resistances can be replaced by the three resistances R_1, R_2 and R_3 connected in star as shown in Fig. 2.177 (*b*).

These two arrangements will be electrically equivalent if the resistance as measured between any pair of terminals is the same in both the arrangements. Let us find this condition.

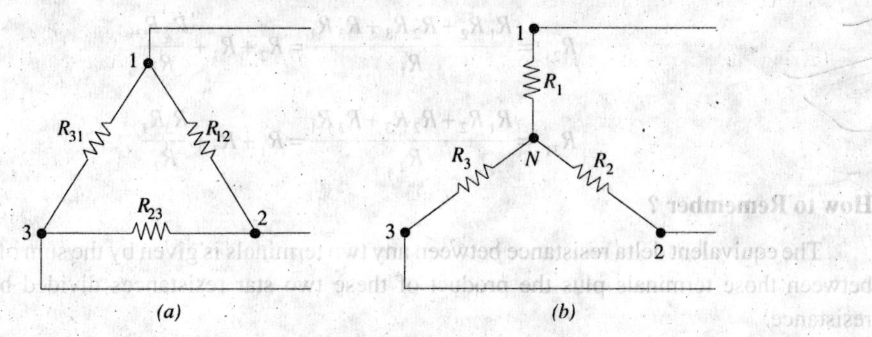

Fig. 2.177

First, take delta connection : Between terminals 1 and 2, there are two parallel paths; one having a resistance of R_{12} and the other having a resistance of $(R_{23} + R_{31})$.

∴ Resistance between terminals 1 and 2 is $= \dfrac{R_{12} \times (R_{23} + R_{31})}{R_{12} + (R_{23} + R_{31})}$

Now, take star connection : The resistance between the same terminals 1 and 2 is $(R_1 + R_2)$. As terminal resistances have to be the same

$$\therefore \qquad R_1 + R_2 = \frac{R_{12} \times (R_{23} + R_{31})}{R_{12} + R_{23} + R_{31}} \qquad \ldots(i)$$

Similarly, for terminals 2 and 3 and terminals 3 and 1, we get

$$R_2 + R_3 = \frac{R_{23}(R_{31} + R_{12})}{R_{12} + R_{23} + R_{31}} \qquad \ldots(ii)$$

$$\text{and} \qquad R_3 + R_1 = \frac{R_{31}(R_{12} + R_{23})}{R_{12} + R_{23} + R_{31}} \qquad \ldots(iii)$$

Now, subtracting (*ii*) from (*i*) and adding the result to (*iii*), we get

*In Electronics, star and delta circuits are generally referred to as *T* and π circuits respectively.

$$R_1 = \frac{R_{12}\,R_{31}}{R_{12}+R_{23}+R_{31}}; \quad R_2 = \frac{R_{23}\,R_{12}}{R_{12}+R_{23}+R_{31}} \quad \text{and} \quad R_3 = \frac{R_{31}\,R_{23}}{R_{12}+R_{23}+R_{31}}$$

How to Remember ?

It is seen from above that each numerator is the product of the two sides of the delta which meet at the point in star. Hence, it should be remembered that : *resistance of each arm of the star is given by the product of the resistances of the two delta sides that meet at its end divided by the sum of the three delta resistances.*

2.23. Star/Delta Transformation

This transformation can be easily done by using equations (*i*), (*ii*), and (*iii*) given above. Multiplying (*i*) and (*ii*), (*ii*) and (*iii*), (*iii*) and (*i*) and adding them together and then simplifying them, we get

$$R_{12} = \frac{R_1 R_2 + R_2 R_3 + R_3 R_1}{R_3} = R_1 + R_2 + \frac{R_1 R_2}{R_3}$$

$$R_{23} = \frac{R_1 R_2 + R_2 R_3 + R_3 R_1}{R_1} = R_2 + R_3 + \frac{R_2 R_3}{R_1}$$

$$R_{31} = \frac{R_1 R_2 + R_2 R_3 + R_3 R_1}{R_2} = R_1 + R_3 + \frac{R_3 R_1}{R_2}$$

How to Remember ?

The equivalent delta resistance between any two terminals is given by the sum of star resistances between those terminals plus the product of these two star resistances divided by the third star resistance.

Example 2.79. *Find the input resistance of the circuit between the points A and B of Fig. 2.178(a).* **(AMIE Sec. B Network Analysis Summer 1992)**

Solution. For finding R_{AB}, we will convert the delta *CDE* of Fig. 2.178 (*a*) into its equivalent star as shown in Fig. 2.178 (*b*).

$R_{CS} = 8 \times 4/18 = 16/9\ \Omega$; $R_{ES} = 8 \times 6/18 = 24/9\ \Omega$; $R_{DS} = 6 \times 4/18 = 12/9\ \Omega$.

The two parallel resistances between *S* and *B* can be reduced to a single resistance of 35/9 Ω.

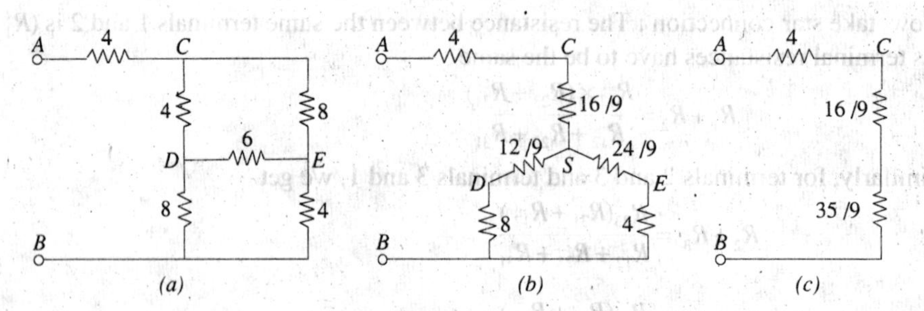

Fig. 2.178.

As seen from Fig. 2.178(*c*), $R_{AB} = 4 + (16/9) + (35/9) = 87/9\ \Omega$.

Example 2.80. *Calculate the equivalent resistance between the terminals A and B in the network shown in Fig. 2.179(a)* **(F.Y. Engg. Pune Univ. May 1987)**

Solution. The given circuit can be redrawn as shown in Fig. 2.179 (*b*). When the delta *BCD* is converted to its equivalent star, the circuit becomes as shown in Fig. 2.179 (*c*).

Each arm of the delta has a resistance of 10 Ω. Hence, each arm of the equivalent star has a resistance = 10 × 10 /30 = 10 /3 Ω. As seen, there are two parallel paths between points A and N, each having a resistance of (10 + 10 /3) = 40 /3 Ω. Their combined resistance is 20 /3 Ω. Hence, R_{AB} = (20 /3) + 10 /3 = 10 Ω.

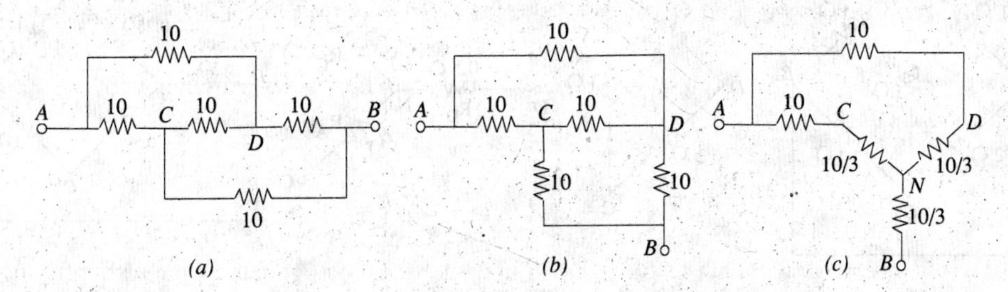

Fig. 2.179

Example 2.81. *Calculate the current flowing through the 10 Ω resistor of Fig. 2.180(b) by using any method.* **(Network Theory, Nagpur Univ. 1993).**

Solution. It will be seen that there are two deltas in the circuit *i.e. ABC* and *DEF*. They have been converted into their equivalent stars as shown in Fig. 2.180 (*b*). Each arm of the delta *ABC* has a resistance of 12 Ω and each arm of the equivalent star has a resistance of 4 Ω. Similarly, each arm of the delta *DEF* has a resistance of 30 Ω and the equivalent star has a resistance of 10 Ω per arm.

The total circuit resistance between A and $F = 4 + 48 \| 24 + 10 = 30$ Ω. Hence, I = 180/30 = 6 A.

Current through 10 Ω resistor as given by current-divider rule = 6 × 48/(48 + 24) = **4 A.**

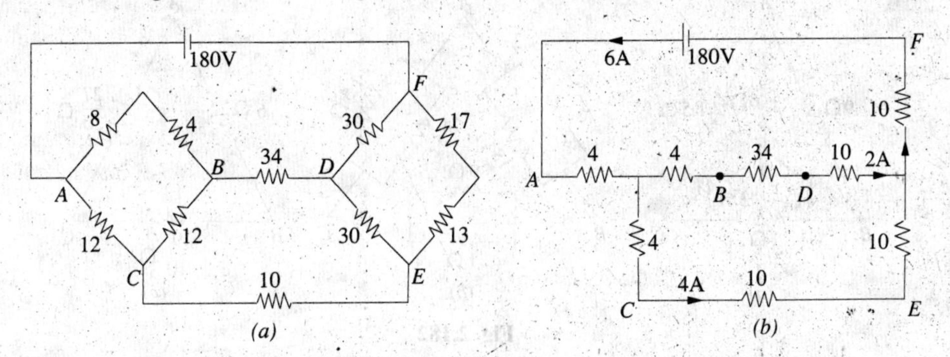

Fig. 2.180

Example 2.82. *A bridge network ABCD has arms AB, BC, CD and DA of resistances 1, 1, 2 and 1 ohm respectively. If the detector AC has a resistnace of I ohm, determine by star/delta transformation, the network resistance as viewed from the battery terminals.*

(Basic Electricity, Bombay Univ. 1980)

Solution. As shown in Fig. 2.181 (*b*), delta *DAC* has been reduced to its equivalent star.

$$R_D = \frac{2\times 1}{2+1+1} = 0.5\ \Omega, \quad R_A = \frac{1}{4} = 0.25\ \Omega, \quad R_C = \frac{2}{4} = 0.5\ \Omega$$

Hence, the original network of Fig. 2.181 (a) is reduced to the one shown in Fig. 2.181(d). As seen, there are two parallel paths between points N and B, one of resistance 1.25 Ω and the other of resistance 1.5 Ω. Their combined resistance is

$$= \frac{1.25 \times 1.5}{1.25 + 1.5} = \frac{15}{22}\ \Omega$$

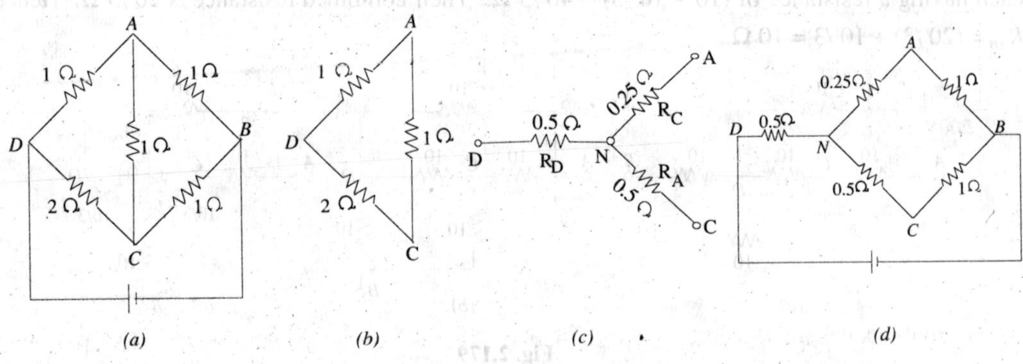

Fig. 2.181

Total resistance of the network between points D and B is

$$= 0.5 + \frac{15}{22} = \frac{\mathbf{13}}{\mathbf{11}}\ \Omega$$

Example 2.83. *A network of resistances is formed as follows :*
$AB = 9\ \Omega$; $BC = 1\ \Omega$; $CA = 1.5\ \Omega$ *forming a delta and* $AD = 6\ \Omega$; $BD = 4\ \Omega$ *and* $CD = 3\ \Omega$ *forming a star. Compute the network resistance measured between (i) A and B (ii) B and C and (iii) C and A.*

(Basic Electricity, Bombay Univ. 1980)

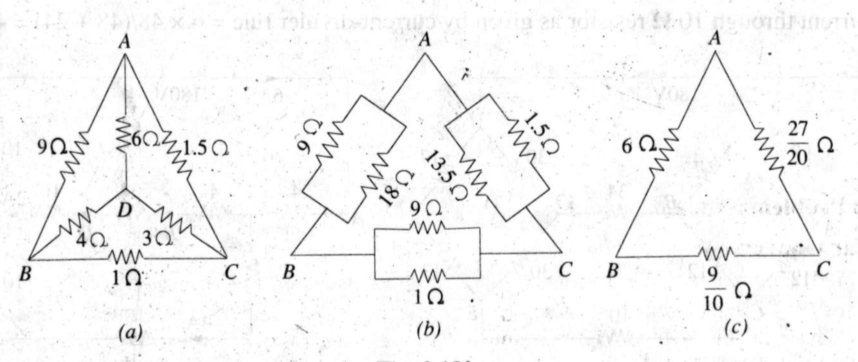

Fig. 2.182

Solution. The star of Fig. 2.182 (*a*) may be converted into the equivalent delta and combined in parallel with the given delta *ABC*. Using the rule given in Art. 2.22, the three equivalent delta resistances of the given star become as shown in Fig. 2.182 (*b*).

When combined together, the final circuit is as shown in Fig. 182 (*c*).

(*i*) As seen, there are two parallel paths across points *A* and *B*.

(*a*) one directly from *A* to *B* having a resistance of 6 Ω and

(*b*) the other *via C* having a total resistance

$$= \left(\frac{27}{20} + \frac{9}{10}\right) = 2.25\ \Omega \quad \therefore \quad R_{AB} = \frac{6 \times 2.25}{(6 + 2.25)} = \frac{\mathbf{18}}{\mathbf{11}}\ \Omega$$

$$(ii)\ R_{BC} = \frac{\frac{9}{10} \times \left(6 + \frac{27}{20}\right)}{\left(\frac{9}{10} + 6 + \frac{27}{10}\right)} = \frac{\mathbf{441}}{\mathbf{550}}\ \Omega \qquad (iii)\ R_{CA} = \frac{\frac{27}{20} \times \left(6 + \frac{9}{10}\right)}{\left(\frac{9}{10} + 6 + \frac{27}{20}\right)} = \frac{\mathbf{621}}{\mathbf{550}}\ \Omega$$

Example 2.84. *State Norton's theorem and find current using Norton's theorem through a load of 8 Ω in the circuit shown in Fig. 2.183 (a)* **(Circuit and Field Theory, A.M.I.E. Sec B, 1993)**

Solution. In Fig. 2.183(*b*), load impedance has been replaced by a short-circuit. $I_{SC} = I_N = 200/2 = 100$ A.

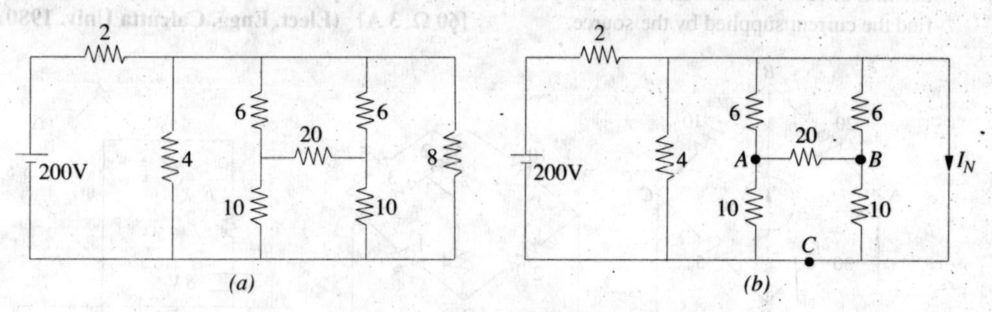

Fig. 2.183

Norton's resistance R_N can be found by looking into the open terminals of Fig. 2.183(*a*). For this purpose Δ *ABC* has been replaced by its equivalent Star. As seen, R_N is equal to 8 /7 Ω.

Hence, Norton's equivalent circuit consists of a 100 A source having a parallel resistance of 8 /7 Ω as shown in Fig. 2.184(*c*). The load current I_L can be found by using the Current Divider rule.

$$I_L = 100 \times \frac{(8/7)}{8 + (8/7)} = 12.5 \text{ A}$$

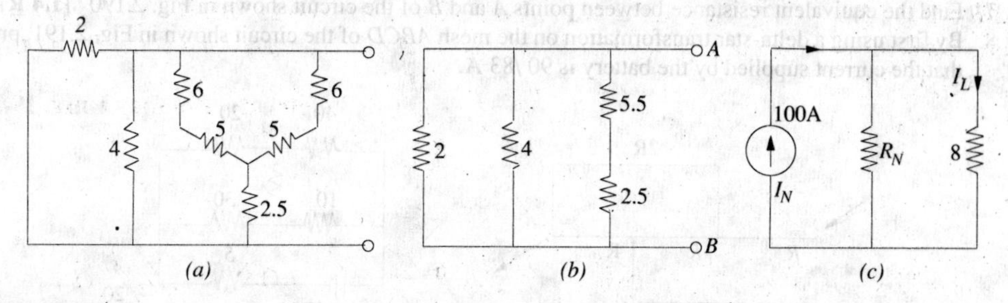

Fig. 2.184

Tutorial Problems No. 2.6

Delta/Star Conversion

1. Find the current in the 17 Ω resistor in the network shown in Fig. 2.185 by using (a) star/delta conversion and (*b*) Thevenin's theorem. The numbers indicate the resistance of each member in ohms. **[10/3 A]**
2. Convert the star circuit of Fig. 2.186 into its equivalent delta circuit. Values shown are in ohms. Derive the formula used. **(Elect. Technology, Indore Univ. 1980)**

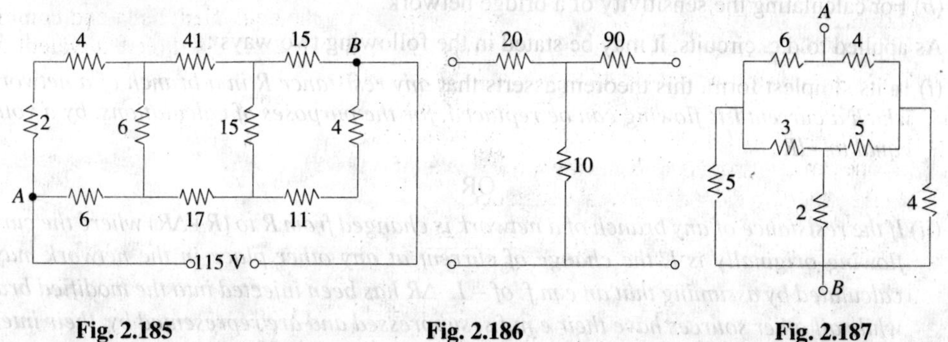

Fig. 2.185 **Fig. 2.186** **Fig. 2.187**

3. Determine the resistance between points *A* and *B* in the network of Fig. 2.187. **[4.23 Ω] (Elect. Technology, Indore Univ. 1977)**

4. Three resistances of 20 Ω each are connected in star. Find the equivalent delta resistance. If a source of e.m.f. of 120 V is connected across any two terminals of the equivalent delta - connected resistances, find the current supplied by the source. **[60 Ω, 3 A]** **(Elect. Engg. Calcutta Univ. 1980)**

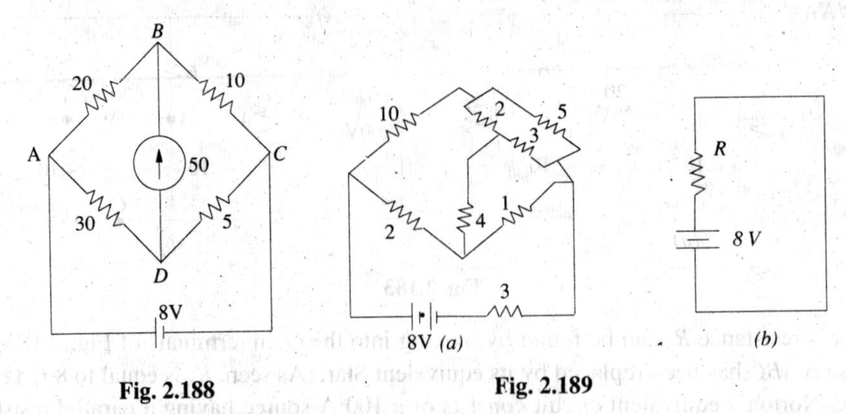

Fig. 2.188 **Fig. 2.189**

5. Using delta/star transformation determine the current through the galvanometer in the Wheatstone bridge of Fig. 2.188. **[0.025 A]**
6. With the aid of the delta star transformation reduce the network given in Fig. 2.189 (*a*) to the equivalent circuit shown at (*b*). **[Ans. R = 5.38 Ω]**
7. Find the equivalent resistance between points *A* and *B* of the circuit shown in Fig. 2.190 **[1.4 R]**
8. By first using a delta-star transformation on the mesh *ABCD* of the circuit shown in Fig. 2.191, prove that the current supplied by the battery is 90 /83 A.

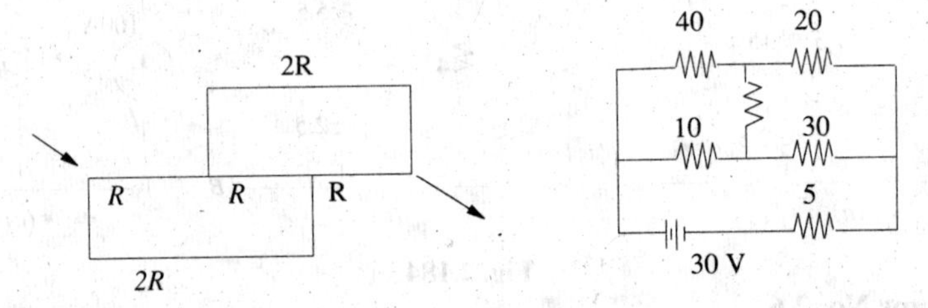

Fig. 2.190 **Fig. 2.191**

2.24. Compensation Theorem

This theorem is particularly useful for the following two purposes :

(*a*) For analysing those networks where the values of the branch elements are varied and for studying the effect of tolerance on such values.

(*b*) For calculating the sensitivity of a bridge network.

As applied to d.c. circuits, it may be stated in the following two ways :

(*i*) In its simplest form, this theorem asserts that *any resistance R in a branch of a network in which a current I is flowing can be replaced, for the purposes of calculations, by a voltage equal to –IR.*

OR

(*ii*) *If the resistance of any branch of a network is changed from R to (R+ΔR) where the current flowing originally is I, the change of current at any other place in the network may be calculated by assuming that an e.m.f. of – I. ΔR has been injected into the modified branch while all other sources have their e.m.f.s. suppressed and are represented by their internal resistances only.*

Example 2.85. *Calculate the values of new currents in the network illustrated in Fig. 2.192 when the resistor R_3 is increased by 30%.*

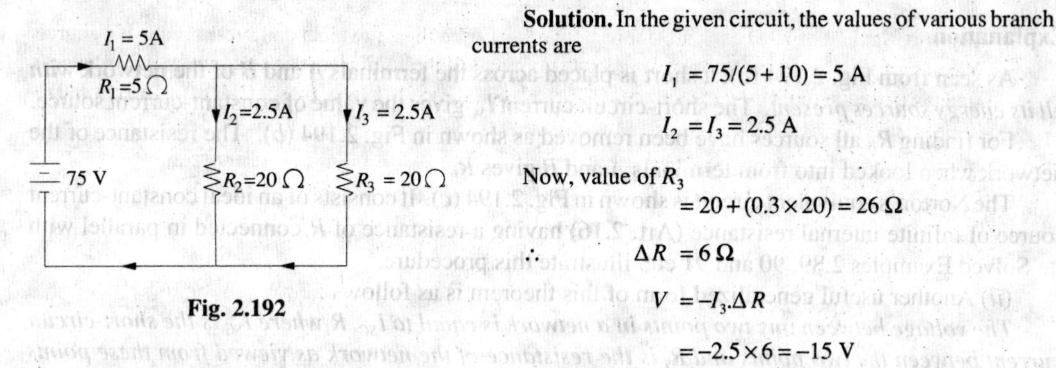

Fig. 2.192

Solution. In the given circuit, the values of various branch currents are

$$I_1 = 75/(5+10) = 5\ \text{A}$$

$$I_2 = I_3 = 2.5\ \text{A}$$

Now, value of R_3

$$= 20 + (0.3 \times 20) = 26\ \Omega$$

$$\therefore \quad \Delta R = 6\ \Omega$$

$$V = -I_3 . \Delta R$$

$$= -2.5 \times 6 = -15\ \text{V}$$

The compensating currents produced by this voltage are as shown in Fig. 2.193 (*a*).

When these currents are added to the original currents in their respective branches the new current distribution becomes as shown in Fig. 2.193 (*b*).

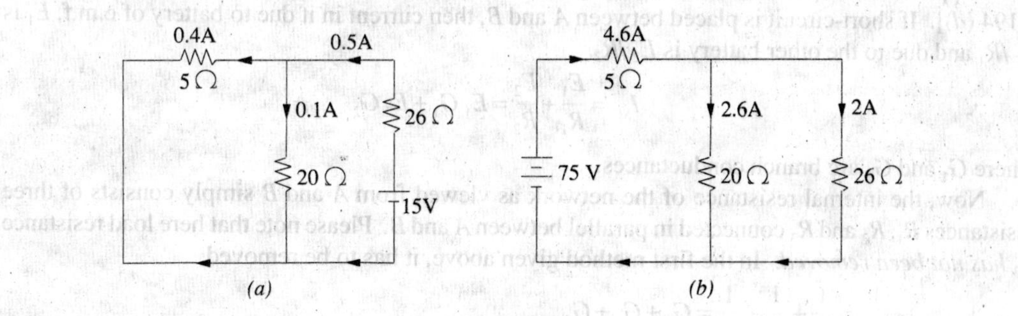

Fig. 2.193

2.25. Norton's Theorem*

This theorem is an alternative to the Thevenin's theorem. In fact, it is the dual of Thevenin's theorem. Whereas Thevenin's theorem reduces a two-terminal active network of linear resistances and generators to an equivalent constant-voltage source and series resistance, Norton's theorem replaces the network by an equivalent constant-current source and a parallel resistance.

This theorem may be stated as follows :–

(i) Any two-terminal active network containing voltage sources and resistances when viewed from its output terminals, is equivalent to a constant-current source and a parallel resistance. The constant current is equal to the current which would flow in a short-circuit placed across the terminals and parallel resistance is the resistance of the network when viewed from these open-circuited terminals after all voltage and current sources have been removed and replaced by their internal resistances.

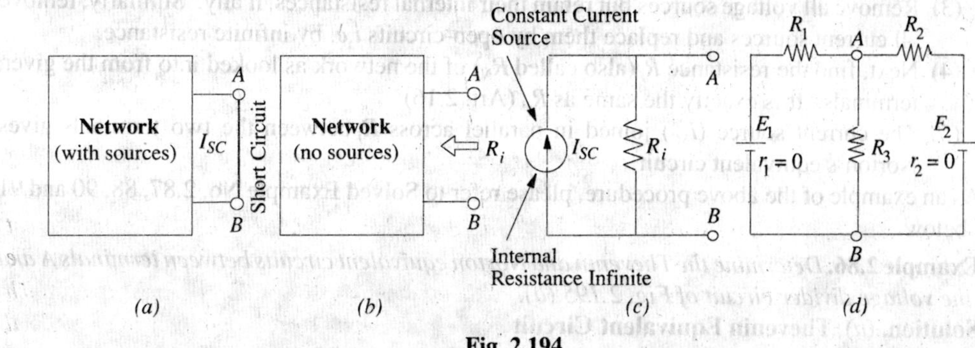

Fig. 2.194

*After E.L. Norton, formerly an engineer at Bell Telephone Laboratory, USA.

Explanation

As seen from Fig. 2.194 (*a*), a short is placed across the terminals *A* and *B* of the network *with all its energy sources present*. The short-circuit current I_{SC} gives the value of constant-current source.

For finding R_i, all sources have been removed as shown in Fig. 2.194 (*b*). The resistance of the network when looked into from terminals *A* and *B* gives R_i.

The Norton's equivalent circuit is shown in Fig. 2.194 (*c*). It consists of an ideal constant-current source of infinite internal resistance (Art. 2.16) having a resistance of R_i connected in parallel with it. Solved Examples 2.89, 90 and 91 etc. illustrate this procedure.

(*ii*) Another useful generalized form of this theorem is as follows :

The voltage between any two points in a network is equal to I_{SC}. R_i where I_{SC} is the short-circuit current between the two points and R_i is the resistance of the network as viewed from these points with all voltage sources being replaced by their internal resistances (if any) and current sources replaced by open-circuits.

Suppose, it is required to find the voltage across resistance R_3 and hence current through it [Fig. 2.194 (*d*)]. If short-circuit is placed between *A* and *B*, then current in it due to battery of e.m.f. E_1 is E_1/R_1 and due to the other battery is E_2/R_2.

$$\therefore \quad I_{SC} = \frac{E_1}{R_1} + \frac{E_2}{R_2} = E_1 G_1 + E_2 G_2$$

where G_1 and G_2 are branch conductances.

Now, the internal resistance of the network as viewed from *A* and *B* simply consists of three resistances R_1, R_2 and R_3 connected in parallel between *A* and *B*. Please note that here load resistance *R_3 has not been removed.* In the first method given above, it has to be removed.

$$\therefore \quad \frac{1}{R_i} = \frac{1}{R_1} + \frac{1}{R_2} + \frac{1}{R_3} = G_1 + G_2 + G_3$$

$$\therefore \quad R_i = \frac{1}{G_1 + G_2 + G_3} \qquad \therefore \quad V_{AB} = I_{SC}.R_i = \frac{E_1 G_1 + E_2 G_2}{G_1 + G_2 + G_3}$$

Current through R_3 is $I_3 = V_{AB}/R_3$.

Solved example No. 2.89 illustrates this approach.

2.26. How To Nortonize a Given Circuit ?

This procedure is based on the first statement of the theorem given above.

(1) Remove the resistance (if any) across the two given terminals and put a short-circuit across them.

(2) Compute the short-circuit current I_{SC}.

(3) Remove all voltage sources but retain their internal resistances, if any. Similarly, remove all current sources and replace them by open-circuits *i.e.* by infinite resistance.

(4) Next, find the resistance R_i (also called R_N) of the network as looked into from the given terminals. It is exactly the same as R_{th} (Art. 2.16)

(5) The current source (I_{SC}) joined in parallel across R_i between the two terminals gives Norton's equivalent circuit.

As an example of the above procedure, please refer to Solved Example No. 2.87, 88, 90 and 91 given below.

Example 2.86. *Determine the Thevenin and Norton equivalent circuits between terminals A and B for the voltage divider circuit of Fig. 2.195 (a).*

Solution. (*a*) Thevenin Equivalent Circuit

Obviosuly, V_{th} = drop across $R_2 = E\dfrac{R_2}{R_1 + R_2}$

• When battery is replaced by a short-circuit.

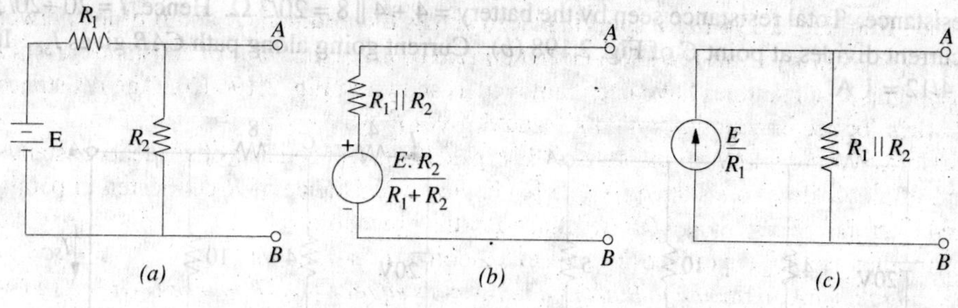

Fig. 2.195

$$R_i = R_1 \| R_2 = R_1 R_2 /(R_1 + R_2)$$

Hence, Thevenin equivalent circuit is as shown in Fig. 2.195 (*b*).

(*b*) **Norton Equivalent Circuit**

A short placed across terminals *A* and *B* will short out R_2 as well. Hence, $I_{sc} = E/R_1$. The Norton equivalent resistance is exactly the same as Thevenin resistance except that it is connected in parallel with the current source as shown in Fig. 2.195 (*c*).

Example 2.87. *Using Norton's theorem, find the constant-current equivalent of the circuit shown in Fig. 2.196 (a).*

Solution. When terminals *A* and *B* are short-circuited as shown in Fig. 2.196 (*b*), total resistance of the circuit, as seen by the battery, consists of a 10-Ω resistance in series with a parallel combination of 10 Ω and 15 Ω resistances.

$$\therefore \quad \text{total resistance} = 10 + \frac{15 \times 10}{15 + 10} = 16\ \Omega$$

$$\therefore \quad \text{battery current} \quad I = 100/16 = 6.25\ \text{A}$$

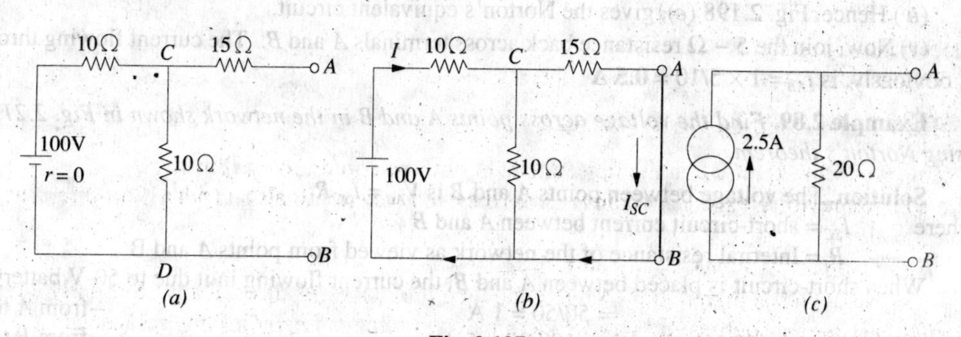

Fig. 2.197

This current is divided into two parts at point *C* of Fig. 2.197 (*b*).

Current through *A B* is $I_{SC} = 6.25 \times 10 /25 = 2.5$ A

Since the battery has no internal resistance, the input resistance of the network when viewed from *A* and *B* consists of a 15 Ω resistance in series with the parallel combination of 10 Ω and 10 Ω. Hence, $R_i = 15 + (10/2) = 20\ \Omega$

Hence, the equivalent constant-current source is as shown in Fig. 2.197 (*c*)

Example 2.88. *Apply Norton's theorem to calculate current flowing through 5 – Ω resistor of Fig. 2.198 (a).*

Solution. (*i*) Remove 5 – Ω resistor and put a short across terminals *A* and *B* as shown in Fig. 2.198 (*b*). As seen, 10 – Ω resistor also becomes short-circuited.

(*ii*) Let us now find I_{SC}. The battery sees a parallel combination of 4 Ω and 8 Ω in series with a 4 Ω resistance. Total resistance seen by the battery = 4 + 4 || 8 = 20/3 Ω. Hence, $I = 20 \div 20/3 = 3$ A. This current divides at point *C* of Fig. 2.198 (*b*). Current going along path *CAB* gives I_{SC}. Its value = 3 × 4/12 = 1 A.

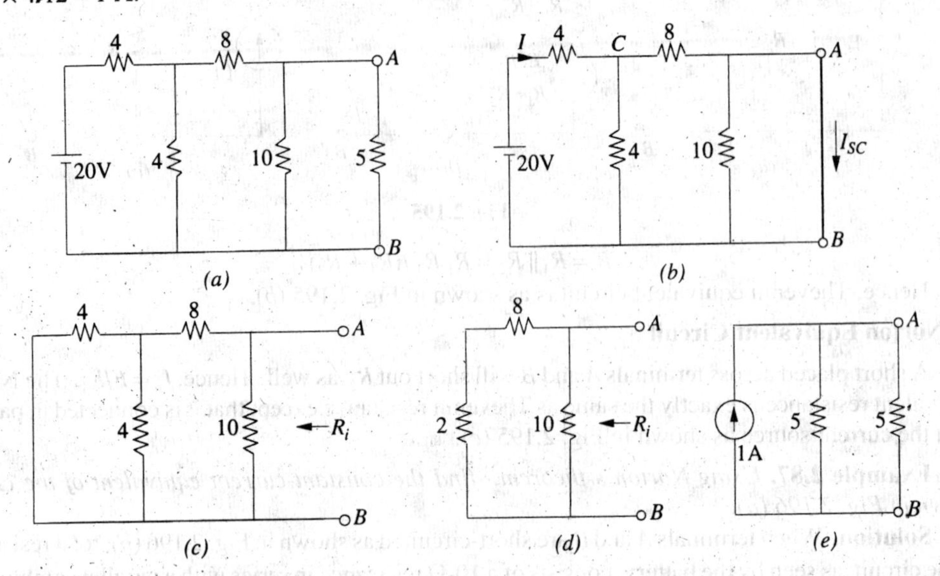

Fig. 2.198

(*iii*) In Fig. 2.198 (*c*), battery has been removed leaving behind its internal resistance which, in this case, is zero.

Resistance of the network looking into the terminals *A* and *B* is

$$R_i = 10 \parallel 10 = 5\ \Omega$$

(*iv*) Hence, Fig. 2.198 (*e*), gives the Norton's equivalent circuit.

(*v*) Now, join the 5 – Ω resistance back across terminals *A* and *B*. The current flowing through it, obviously, is $I_{AB} = 1 \times 5/10 = \mathbf{0.5\ A}$

Example 2.89. *Find the voltage across points A and B in the network shown in Fig. 2.211 by using Norton's theorem.*

Solution. The voltage between points *A* and *B* is $V_{AB} = I_{SC} R_i$

where I_{SC} = short-circuit current between *A* and *B*

R_i = Internal resistance of the network as viewed from points *A* and B

When short-circuit is placed between *A* and *B*, the current flowing in it due to 50-V battery is

= 50/50 = 1 A –from *A* to *B*

Current due to 100 V battery is = 100/20 = 5 A –from *B* to *A*

∴ $I_{SC} = 1 - 5 = -4$ A –from *B* to *A*

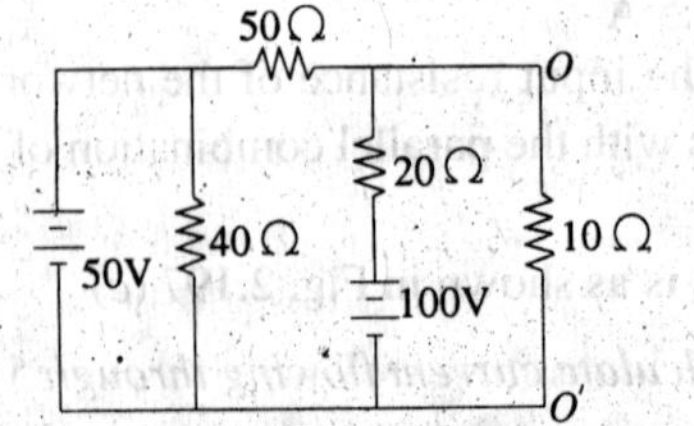

Fig. 2.199

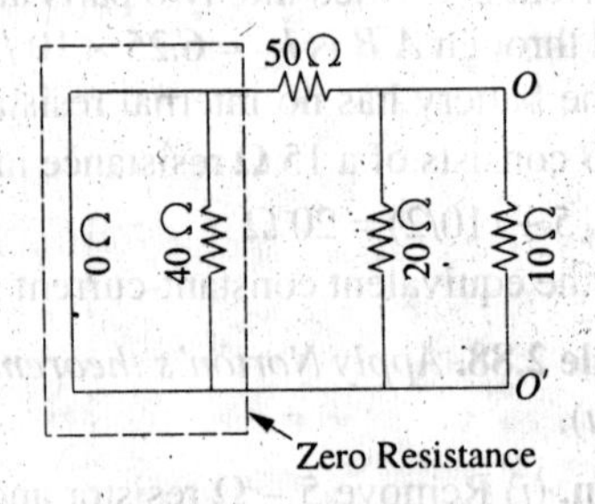

Fig. 2.200

Now, suppose that the two batteries are removed so that the circuit becomes as shown in Fig. 2.200. The resistance of the network as viewed from points *A* and *B* consists of three resistances of 10 Ω, 20 Ω and 50 Ω ohm connected in parallel (as per second stateemnt of Norton's theorem).

$$\therefore \quad \frac{1}{R_i} = \frac{1}{10} + \frac{1}{20} + \frac{1}{50}; \qquad \text{hence } R_i = \frac{100}{17}\ \Omega$$

$$\therefore \quad V_{AB} = -4 \times 100/17 = \mathbf{-23.5\ V}$$

The negative sign merely indicates that point *B* is at a higher potential with respect to the point A.

Example 2.90. *Using Norton's theorem, calculate the current flowing through the 15 Ω load resistor in the circuit of Fig. 2.201 (a). All resistance values are in ohm.*

Solution. (*a*) **Short-Circuit Current I_{SC}**

As shown in Fig. 2.201 (*b*), terminals *A* and *B* have been shorted after removing 15 Ω resistor. We will use Superposition theorem to find I_{SC}.

(*i*) **When Only Current Source is Present**

In this case, 30-V battery is replaced by a short-circuit. The 4 A current divides at point *D* between parallel combination of 4 Ω and 6 Ω. Current through 6 Ω resistor is

$$I_{sc}' = 4 \times 4/(4+6) = 1.6\ \text{A} \qquad \text{–from } B \text{ to } A$$

(*ii*) **When Only Battery is Present**

In this case, current source is replaced by an open-circuit so that no current flows in the branch *CD*. The current supplied by the battery constitutes the short-circuit current

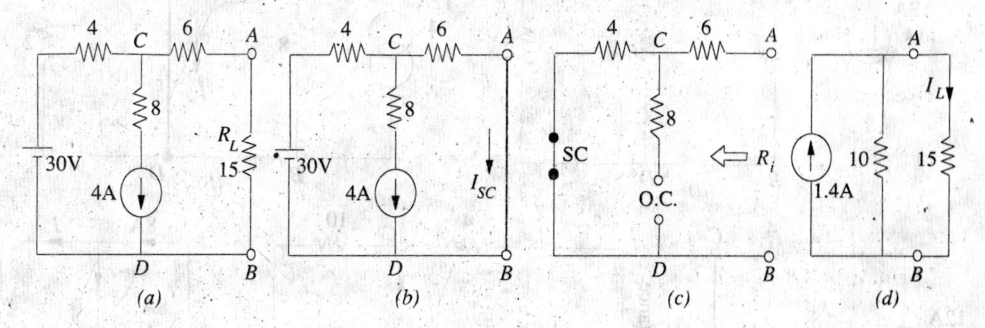

Fig. 2.201

$$\therefore \quad I_{sc}'' = 30/(4+6) = \mathbf{3\ A} \qquad \textbf{–from A to B}$$

$$\therefore \quad I_{sc} = I_{sc}'' - I_{sc}' = 3 - 1.6 = \mathbf{1.4\ A} \qquad \textbf{–from A to B}$$

(*b*) **Norton's Parallel Resistance**

As seen from Fig. 2.201 (*c*), R_i = 4 + 6 = 10 Ω. The 8 Ω resistance does not come into the picture because of an 'open' in the branch *CD*.

Fig. 2.201 (*d*) shows the Norton's equivalent circuit along with the load resistor.

$$I_L = 1.4 \times 10/(10+15) = \mathbf{0.56\ A}$$

Example 2.91. *Using Norton's current-source equivalent circuit of the network shown in Fig. 2.202 (a), find the current that would flow through the resistor R_2 when it takes the values of 12, 24 and 36 Ω respectively.* **(Elect. Circuits, South Gujarat Univ. 1987)**

Solution. In Fig. 2.202 (*b*), terminals *A* and *B* have been short-circuited. Current in the shorted path due to E_1 is = 120/40 = 3 A from *A* to *B*. Current due to E_2 is 180/60 = 3A from *A* to *B*. Hence, I_{SC} = 6 A. With batteries removed, the resistance of the network when viewed from open-circuited terminals is = 40 || 60 = 24 Ω.

(*i*) When $R_L = 12\ \Omega$ $I_L = 6 \times 24(24 + 12) = \mathbf{4\ A}$

(*ii*) When $R_L = 24\ \Omega$ $I_L = 6/2 = \mathbf{3\ A}$

(*iii*) When $R_L = 36\ \Omega$ $I_L = 6 \times 24/(24 + 36) = \mathbf{2.4\ A.}$

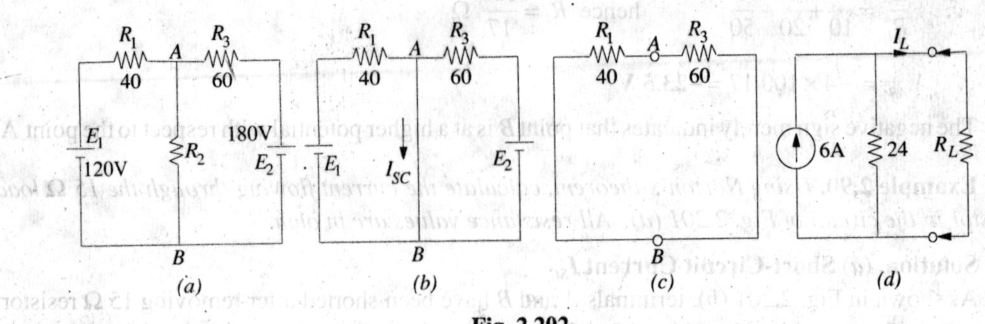

Fig. 2.202

Example 2.92. *Using Norton's theorem, calculate the current in the 6-Ω resistor in the network of Fig. 2.203 (a). All resistances are in ohms.*

Solution. When the branch containing 6 – Ω resistance is short-circuited, the given circuit is reduced to that shown in Fig. 203 (*b*) and finally to Fig. 2.203 (*c*). As seen, the 12 A current divides

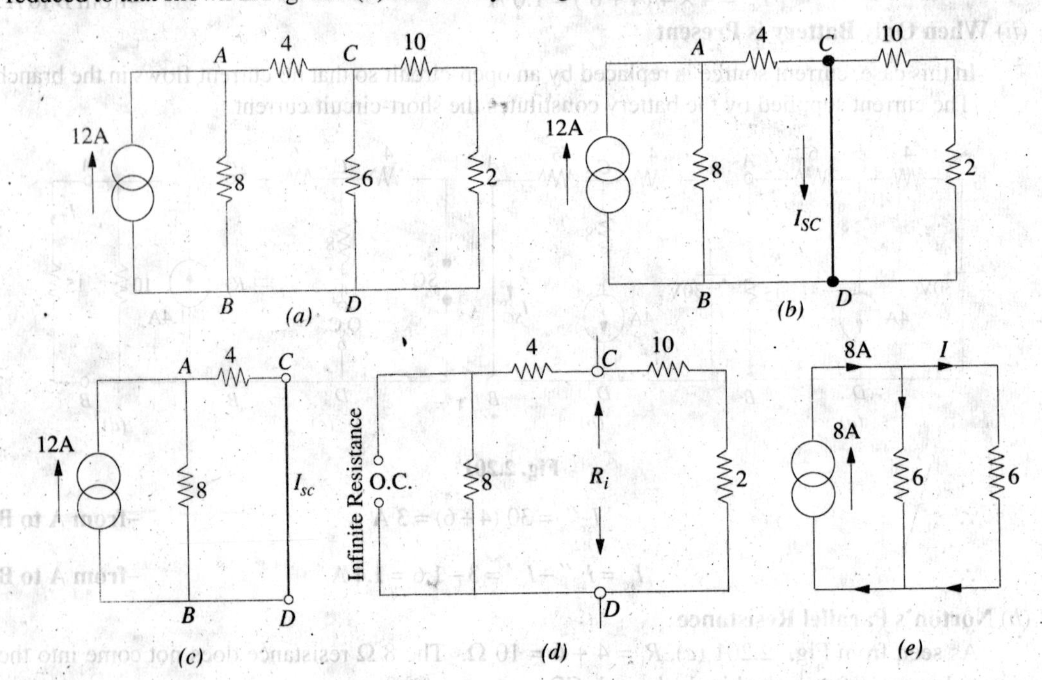

Fig. 2.203

into two unequal parts at point *A*. The current passing through 4 Ω resistor forms the short-circuit current I_{SC}.

$$I_{SC} = 12 \times \frac{8}{8+4} = 8\ \text{A}$$

Resistance R_i between points *C* and *D* when they are open-circuited is

$$R_i = \frac{(4+8)\times(10+2)}{(4+8)+(10+2)} = 6\ \Omega$$

It is so because the constant-current source has *infinite* resistance *i.e.* it behaves like an open circuit as shown in Fig. 2.203 (*d*).

Hence, Norton's equivalent circuit is as shown in Fig. 2.203 (*e*). As seen, the current of 8 A is divided equally between the two equal resistances of 6 Ω each. Hence, current through the required 6 Ω resistor is **4 A.**

Example 2.93. *Using Norton's theorem, find the current which would flow in a 25 – Ω resistor connected between points N and O in Fig. 2.204 (a). All resistance values are in ohms.*

Solution. For ease of understanding, the given circuit may be redrawn as shown in Fig. 2.204 (*b*). Total current in short-circuit across *ON* is equal to the sum of currents driven by different batteries through their respective resistances.

$$I_{SC} = \frac{10}{5} + \frac{20}{10} + \frac{30}{20} = 5.5\text{A}$$

The resistance R_i of the circuit when looked into from points *N* and *O* is

$$\frac{1}{R_i} = \frac{1}{5} + \frac{1}{10} + \frac{1}{20} = \frac{7}{20} \quad \Omega; \; R_i = \frac{20}{7}\,\Omega = 2.86\,\Omega$$

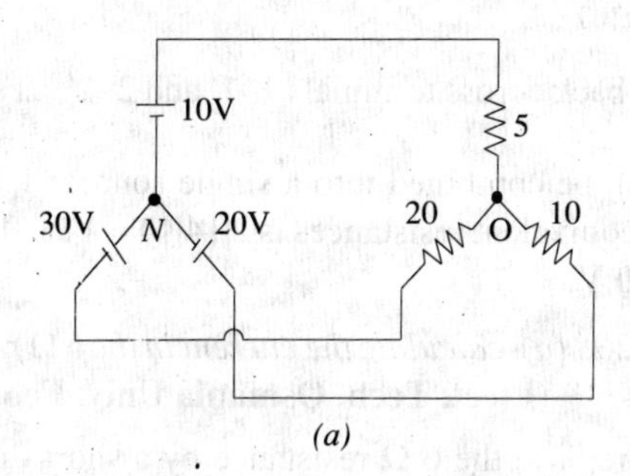

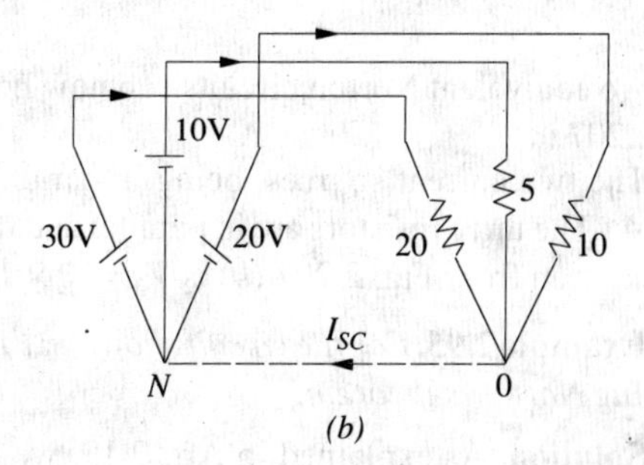

Fig. 2.204

Hence, given circuit reduces to that shown in Fig. 2.205 (*a*)

Open - circuit voltage across *NO* is

$= I_{SC}\,R_i = 5.5 \times 2.86 =$ **15.73 V**

Hence, current through 25- Ω resistor connected across *NO* is [Fig. 2.205 (*b*)]

$I = 15.73/25 =$ **0.56 A**

or $\quad I = 5.5 \times \dfrac{2.86}{2.86 + 25} =$ **0.56 A.**

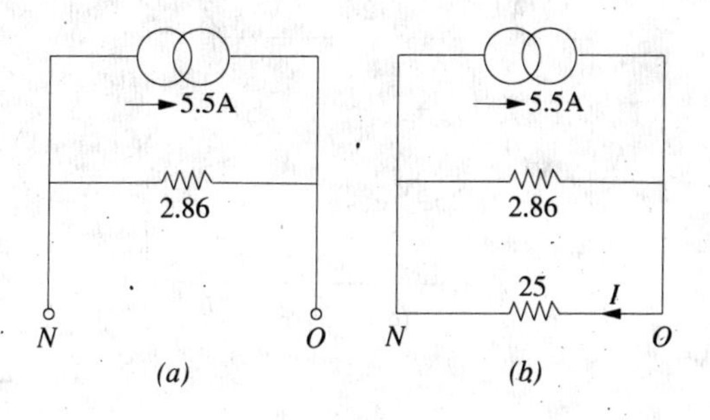

Fig. 2.205

Example 2.94. *With the help of Norton's theorem, find V_o in the circuit shown in Fig. 2.206 (a). All resistances are in ohms.*

Solution. For solving this circuit, we will Nortonise the circuit to the left of the terminals 1–1′ and to the right of terminals 2–2,′ as shown in Fig. 2.206 (*b*) and (*c*) respectively.

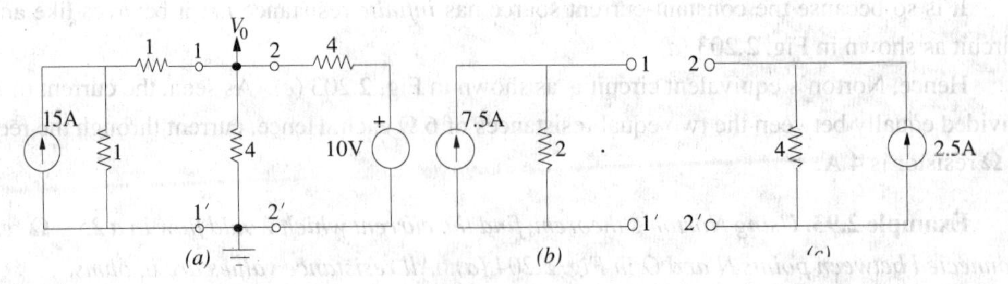

Fig. 2.206

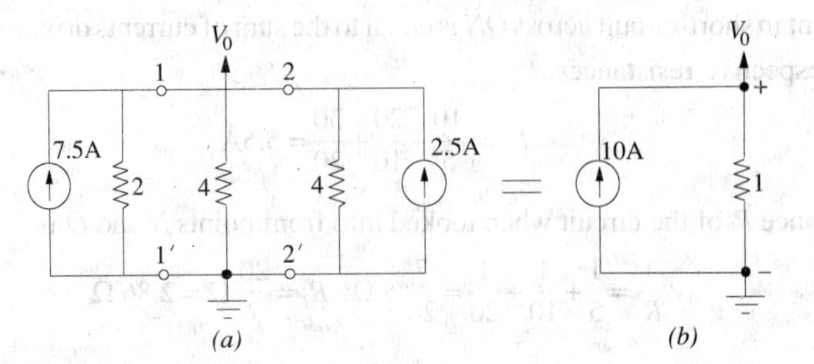

Fig. 2.207

The two equivalent Norton circuits can now be put back across terminals 1–1′ and 2–2′, as shown in Fig. 2.207 (*a*).

The two current sources, being in parallel, can be combined into a single source of 7.5 + 2.5 = 10 A. The three resistors are in parallel and their equivalent resistances is 2||4||4 = 1 Ω. The value of V_0 as seen from Fig. 2.207 (*b*) is $V_0 = 10 \times 1 = \mathbf{10\ V}$.

Example 2.95. *For the circuit shown in Fig. 2.208 (a), calculate the current in the 6 Ω resistance by using Norton's theorem.* **(Elect. Tech. Osmania Univ. Feb. 1992).**

Solution. As explained in Art. 2.19, we will replace the 6 Ω resistance by a short - circuit as shown in Fig. 2.208 (*b*). Now, we have to find the current passing through the short - circuited terminals *A* and *B*. For this purpose we will use the mesh analysis by assuming mesh currents I_1 and I_2.

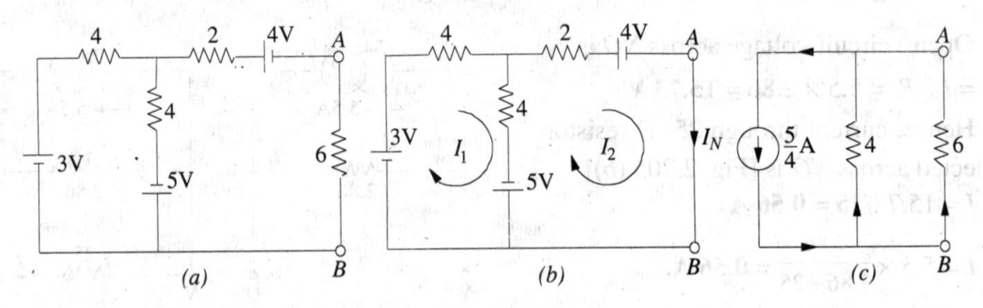

Fig. 2.208

From mesh (*i*), we get

$$3 - 4I_1 - 4(I_1 - I_2) + 5 = 0 \quad \text{or} \quad 2I_1 - I_2 = 2 \qquad \text{...(i)}$$

From mesh (*ii*), we get

$$-2I_2 - 4 - 5 - 4(I_2 - I_1) = 0 \quad \text{or} \quad 4I_1 - 6I_2 = 9 \qquad \text{...(ii)}$$

From (*i*) and (*ii*) above, we get $I_2 = -5/4$

The negative sign shows that the actual direction of flow of I_2 is opposite to that shown in Fig. 2.208 (*b*). Hence, $I_{sh} = I_N = I_2 = -5/4$ A *i.e.* current flows from point *B* to A

After the terminals A and B are open - circuited and the three batteries are replaced by short - circuits (since their internal resistances are zero), the internal resistance of the circuit, as viewed from these terminals' is

$$R_i = R_N = 2 + 4\,||\,4 = 4\ \Omega$$

The Norton's equivalent circuit consists of a constant current source of 5/4 A in parallel with a resistance of 4 Ω as shown in Fig. 2.208 (*c*). When 6 Ω resistance is connected across the equivalent circuit, current through it can be found by the current - divider rule (Art).

$$\text{Current through 6 } \Omega \text{ resistor} = \frac{5}{4} \times \frac{4}{10} = 0.5 \text{ from } B \text{ to } A.$$

2.27. General Instructions For Finding Norton Equivalent Circuit

Procedure for finding Norton equivalent circuit of a given network has already been given in Art. That procedure applies to circuits which contain resistors and independent voltage or current sources. Similar procedures for circuits which contain both dependent and independent sources or only dependent sources are given below :

(a) Circuits Containing Both Dependent and Independent Sources

(*i*) Find the open-circuit voltage v_{oc} with all the sources activated or 'alive'.

(*ii*) Find short-circuit current i_{sh} by short-circuiting the terminals a and b but with all sources activated.

(*iii*) $R_N = v_{oc}/i_{sh}$

(b) Circuits Containing Dependent Sources Only

(*i*) $i_{sh} = 0$.

(*ii*) Connect 1 A source to the terminals a and b and calculate v_{ab}.

(*iii*) $R_N = v_{ab}/1$.

Example 2.96. *Find the Norton equivalent for the transistor amplifier circuit shown in Fig. 2.209 (a). All resistances are in ohms.*

Solution. We have to find the values of i_{sh} and R_N. It should be noted that when terminals a and b are short-circuited, $v_{ab} = 0$. Hence, in that case, we find from the left-hand portion of the circuit that $i = 2/200 = 1/100$ A = 0.01 A. As seen from Fig. 2.209 (*b*), the short-circuit across terminals a and b, short circuits 20 Ω resistance also. Hence, $i_{sh} = -5\,i = -5 \times 0.01 = -0.05$ A.

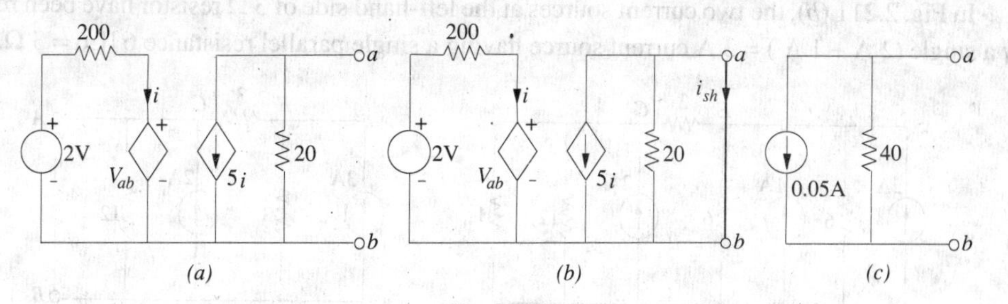

Fig. 2.209

Now, for finding R_N, we need $v_{oc} = v_{ab}$ from the left-hand portion of the Fig. 2.209 (*a*). Applying *KVL* to the closed circuit, we have

$$2 - 200\,i - v_{ab} = 0 \qquad \ldots (i)$$

Now, from the right-hand portion of the circuit, we find v_{ab} = drop over 20 Ω resistance = $-20 \times 5i = -100\,i$. The negative sign is explained by the fact that current flows from point b towards point a. Hence, $i = -v_b/100$. Substituting this value in Eqn. (*i*). above, we get

$$2 - 200\,(-v_b/100) - v_{ab} = 0 \text{ or } v_{ab} = -2 \text{ V}$$

$$\therefore \qquad R_N = v_{ab}/i_{sh} = -2/-0.05 = 40\ \Omega$$

Hence, the Norton equivalent circuit is as shown in Fig. 2.209 (*c*).

Example 2.97. *Using Norton's theorem, compute current through the 1-Ω resistor of Fig. 2.210.*

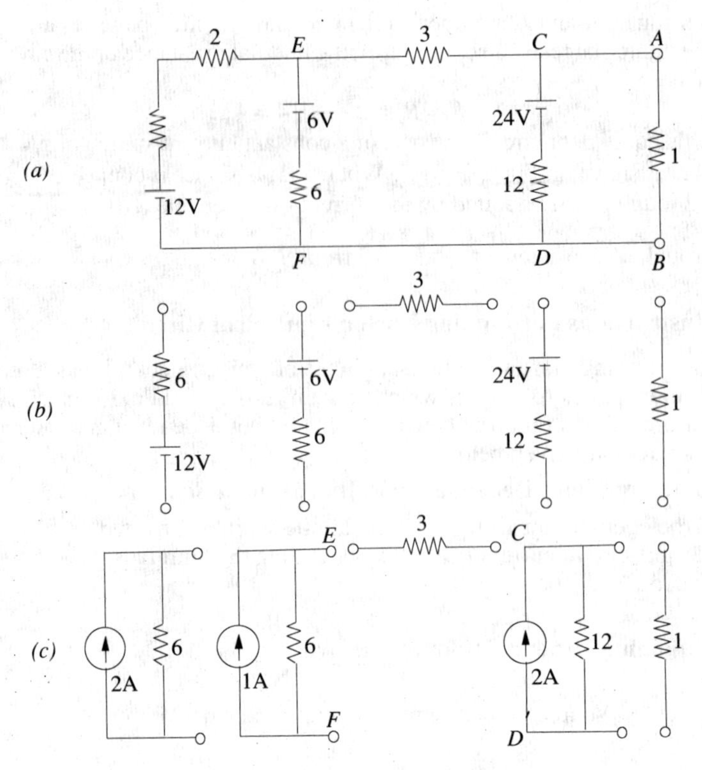

Fig. 2.210

Solution. We will employ source conversion technique to simplify the given circuit. To begin with, we will convert the three voltage sources into their equivalent current sources as shown in Fig. 2.210 (*b*) and (*c*). We can combine together the two current sources on the left of *EF* but cannot combine the 2-A source across *CD* because of the 3-Ω resistance between *C* and *E*.

In Fig. 2.211 (*b*), the two current sources at the left-hand side of 3 Ω resistor have been replaced by a single (2 A + 1 A) = 3 A current source having a single parallel resistance 6 || 6 = 3 Ω.

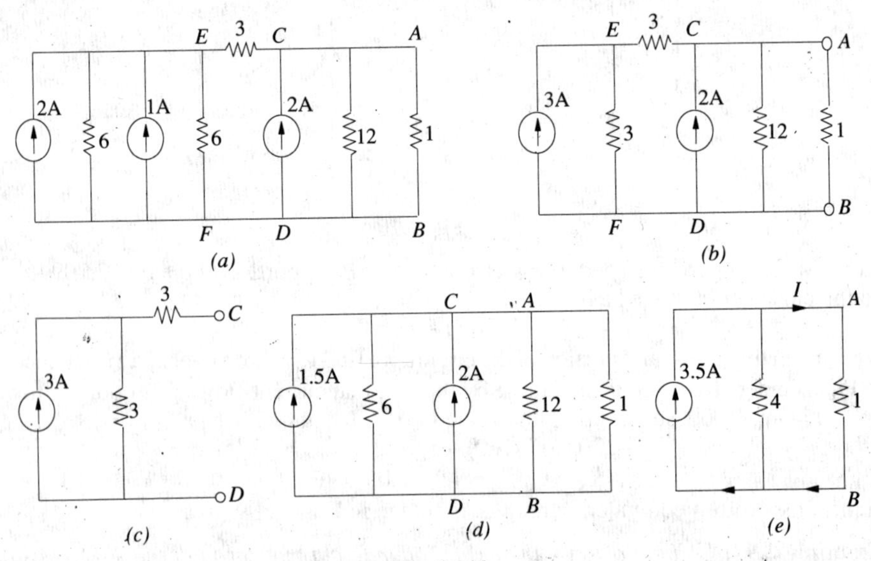

Fig. 2.211

We will now apply Norton's theorem to the circuit on the left-hand side of *CD* [Fig. 2.211 (*c*)] to convert it into a single current source with a single parallel resistor to replace the two 3 Ω resistors. As shown in Fig. 2.211 (*d*), it yields a 1.5 A current source in parallel with a 6 Ω resistor. This current source can now be combined with the one across *CD* as shown in Fig. 2.111 (*e*). The current through the 1-Ω resistor is

$$I = 3.5 \times 4/(4 + 1) = \mathbf{2.8\ A}$$

Example 2.98. *Obtain Thevenin's and Norton's equivalent circuits at AB shown in Fig. 2.212(a).* **(Elect. Network, Analysis Nagpur Univ. 1993)**

Solution. *Thevenin's Equivalent Circuit*

We will find the value of V_{th} by using two methods (*i*) *KVL* and (*ii*) mesh analysis.

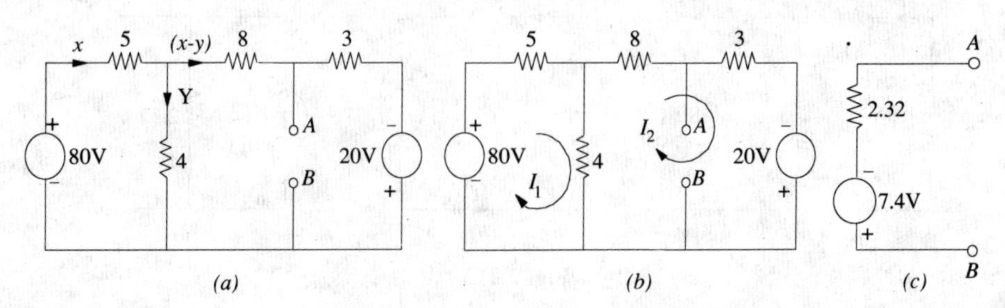

Fig. 2.212

(a) Using KVL

If we apply *KVL* to the first loop of Fig. 2.212 (*a*), we get

$$80 - 5x - 4y = 0 \quad \text{or} \quad 5x + 4y = 80 \qquad \text{... (i)}$$

From the second loop, we have

$$-11(x - y) + 20 + 4y = 0 \quad \text{or} \quad 11x - 15y = 20 \qquad \text{...(ii)}$$

From (*i*) and (*ii*), we get $x = 10.75$ A; $y = 6.56$ A and $(x - y) = 4.2$ A.

Now, $V_{th} = V_{AB}$ *i.e.* voltage of point *A* with respect to point *B*. For finding its value, we start from point *B* and go to point *A* either via 3 Ω resistance or 4 Ω resistance or (5 + 8) = 13 Ω resistance and take the algebraic sum of the voltages met on the way. Taking the first route, we get

$$V_{AB} = -20 + 3(x - y) = -20 + 3 \times 4.2 = -7.4\ \text{V}$$

It shows that point *A* is negative with respect to point *B* or, which is the same thing, point *B* is positive with respect to point *A*.

(b) Mesh Analysis [Fig. 2.212 (*b*)]

Here, $R_{11} = 9$; $R_{22} = 15$; $R_{12} = R_{21} = -4$

$$\therefore \quad \begin{vmatrix} 9 & -4 \\ -4 & 15 \end{vmatrix} \begin{vmatrix} I_1 \\ I_2 \end{vmatrix} = \begin{vmatrix} 80 \\ 20 \end{vmatrix}; \Delta = 135 - 16 = 119$$

$$\Delta_1 = \begin{vmatrix} 80 & -4 \\ 20 & 15 \end{vmatrix} = 1280\ ; \ \Delta_2 = \begin{vmatrix} 9 & 80 \\ -4 & 20 \end{vmatrix} = 500$$

$$I_1 = 1280/119 = 10.75\ \text{A}\ ; \ I_2 = 500/119 = 4.2\ \text{A}$$

Again $V_{AB} = -20 + 12.6 = -7.4$ V

Value of R_{th}

For finding R_{th}, we replace the two voltage sources by short - circuits.

$$\therefore \quad R_{th} = R_{AB} = 3 \,\|\, (8 + 4 \,\|\, 5) = 2.32\ \Omega$$

The Thevenin's equivalent circuit becomes as shown in Fig. 2.212 (*c*). It should be noted that point *B* has been kept positive with respect to point *A* in the figure.

Norton's Equivalent Circuit

For this purpose, we will short-circuit the terminals A and B and find the short-circuit currents produced by the two voltage sources. When viewed from the side of the 80–V source, a short across AB short-circuits everything on the right side of AB. Hence, the circuit becomes as shown in Fig. 2.213 (a). The short-circuit current I_1 can be found with the help of series-parallel circuit technique. The total resistance offered to the 80 – V source is $= 5 + 4 \parallel 8 = 23/3\ \Omega$.

$\therefore\ I = 80 \times 3/23 = 10.43$ A ; $\therefore\ I_1 = 10.43 \times 4/12 = 3.48$ A.

When viewed from the side of the 20 -V source, a short across AB short - circuits everything beyond AB. In that case, the circuit becomes as shown in Fig. 2.213 (b). The short circuit current flowing from B to $A = 20/3 = 6.67$ A.

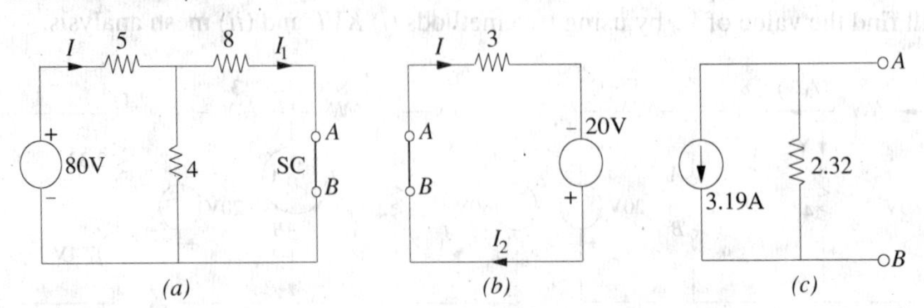

Fig. 2.213

Total short - circuit current = 6.67 – 3.48 = 3.19 A ... from B to A.

$$R_N = R_{th} = 3 \parallel (8 + 4 \parallel 5) = 2.32\ \Omega$$

Hence, the Norton's equivalent circuit becomes as shown in Fig. 2.213 (c).

2.28. Millman's Theorem

This theorem can be stated either in terms of voltage sources or current sources or both.

(a) As Applicable to Voltage Sources

This theorem is a combination of Thevenin's and Norton's theorems. It is used for finding the common voltage across any network which contains a number of parallel voltage sources as shown in Fig. 2.214 (a.) The common voltage V_{AB} which appears across the output terminals A and B is affected by the voltage sources E_1, E_2 and E_3. The value of the voltage is given by

$$V_{AB} = \frac{E_1/R_1 + E_2/R_2 + E_3/R_3}{1/R_1 + 1/R_2 + 1/R_3} = \frac{I_1 + I_2 + I_3}{G_1 + G_2 + G_3} = \frac{\Sigma I}{\Sigma G}$$

This voltage represents the Thevenin's voltage V_{th}. The resistance R_{th} can be found, as usual, by replacing each voltage source by a short circuit. If there is a load resistance R_L across the terminals A and B, then load current I_L is given by

$$I_L = V_{th}/(R_{th} + R_L)$$

If as shown in Fig. 2.214 (b), a branch does not contain any voltage source, the same procedure is used except that the value of the voltage for that branch is equated to zero as illustrated in Example 2.99.

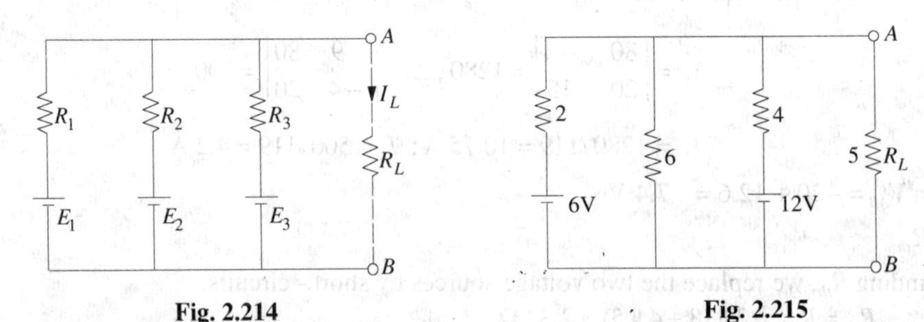

Fig. 2.214 **Fig. 2.215**

Example 2.99. *Use Millman's theorem, to find the common voltage across terminals A and B and the load current in the circuit of Fig. 2.215.*

Solution. As per Millman's theorem,

$$V_{AB} = \frac{6/2 + 0/6 + 12/4}{1/2 + 1/6 + 1/4} = \frac{6}{11/12} = 6.55 \text{ V}$$

$$\therefore \quad V_{th} = 6.55 \text{ V}$$

$$R_{th} = 2 \parallel 6 \parallel 4 = 12/11\ \Omega$$

$$I_L = \frac{V_{th}}{R_{th} + R_L} = \frac{6.55}{(12/11) + 5} = 1.07 \text{ A}$$

(b) As Applicable to Current Sources

This theorem is applicable to a mixture of parallel voltage and current sources that are reduced to a single final equivalent source which is either a constant current or a constant voltage source. This theorem can be stated as follows :

Any number of constant current sources which are directly connected in parallel can be converted into a single current source whose current is the algebraic sum of the individual source currents and whose total internal resistance equals the combined individual source resistances in parallel.

Example 2.100. *Use Millman's theorem, to find the voltage across and current through the load resistor R_L in the circuit of Fig. 2.216 (a).*

Solution. First thing to do is to convert the given voltage sources into equivalent current sources. It should be kept in mind that the two batteries are connected in opposite direction. Using the source conversion technique given in Art. 1.14 we get the circuit of Fig. 2.216 (*b*).

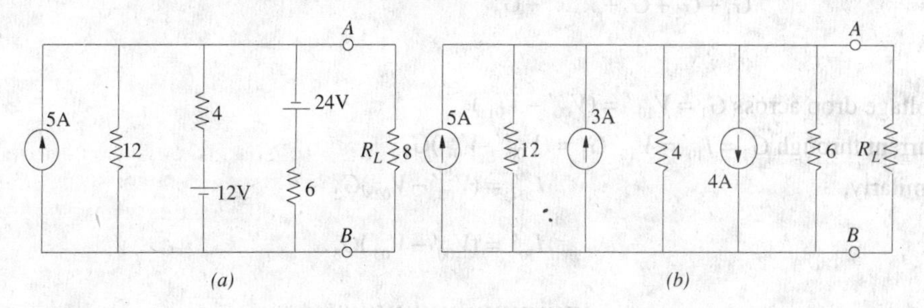

Fig. 2.216

The algebraic sum of the currents = 5 + 3 − 4 = 4 A. The combined resistance is = 12||4||6 = 2 Ω. The simplified circuit is shown in the current - source form in Fig. 2.217 (*a*) or voltage source form in Fig. 2.217 (*b*).

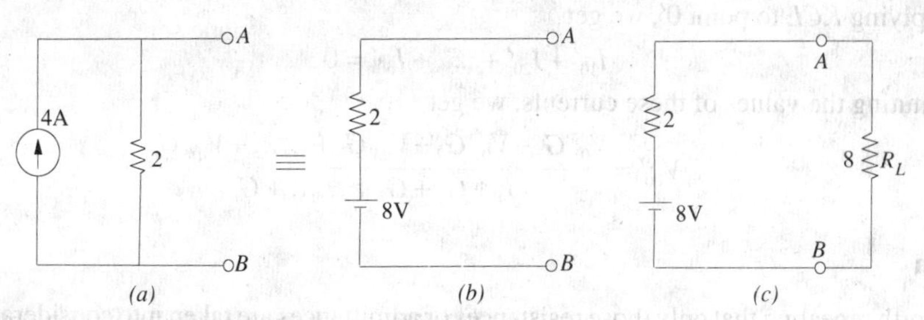

Fig. 2.217

As seen from Fig. 2.217 (*c*).

$$I_L = 8/(2 + 8) = 0.8 \text{ A} \; ; \; V_L = 8 \times 0.8 = 0.64 \text{ V}$$

Alternatively, $\quad V_L = 8 \times 8/(2 + 8) = 6.4 \text{ V}$

Following steps are necessary when using Millman's theorem :

1. convert all voltage sources into their equivalent current sources.

2. calculate the algebraic sum of the individual dual source currents.
3. calculate the total internal resistance by combining the individual source resistance in parallel.
4. if found necessary, convert the final current source into its equivalent voltage source.

As pointed out earlier, this theorem can also be applied to voltage sources which must be initially converted into their constant current equivalents.

2.29. Generalised Form of Millman's Theorem

This theorem is particularly useful for solving many circuits which are frequently encountered in both electronics and power applications.

Consider a number of admittances G_1, G_2, G_3 .... G_n which terminate at common point 0′ (Fig. 2.218). The other ends of the admittances are numbered as 1, 2, 3 .... *n*. Let *O* be any other point in the network. It should be clearly understood that it is not necessary to know anything about the inter-connection between point *O* and the end points 1, 2, 3 .... *n*. However, what is essential to know is the voltage drops from 0 to 1, 0 to 2, ... 0 to *n etc.*

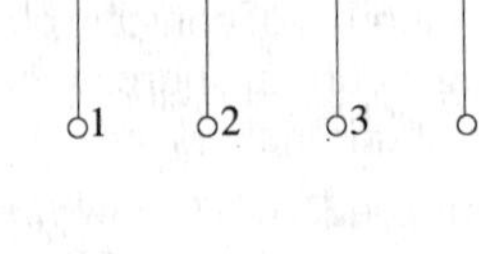

Fig. 2.218

According to this theorem, the voltage drop from 0 to *0′* ($V_{00'}$) is given by

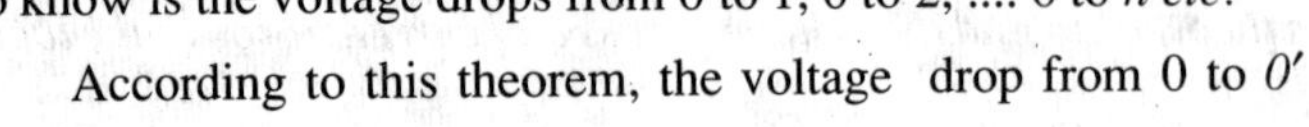

$$V_{oo}' = \frac{V_{01}G_1 + V_{02}G_2 + V_{03}\,G_3 + \ldots. + V_{0n}G_n}{G_1 + G_2 + G_3 + \ldots\ldots + G_n}$$

Proof

Voltage drop across $G_1 = V_{10}' = (V_{00}' - V_{01})$

Current through $G_1 = I_{10}' = V_{10}'\,G_1 = (V_{00}' - V_{01})G_1$

Similarly, $I_{20}' = (V_{00}' - V_{02})G_2$

$$I_{30}' = (V_{00}' - V_{03})G_3$$

....

....

and $I_{n0}' = (V_{00}' - V_{0n})\,G_n$

By applying *KCL* to point 0′, we get

$$I_{10}' + I_{20}' + \ldots.. + I_{n0}' = 0$$

Substituting the values of these currents, we get

$$V_{00}' = \frac{V_{01}\,G_1 + V_{02}\,G_2 + V_{03}\,G_3 + \ldots\ldots + V_{on}\,G_n}{G_1 + G_2 + G_3 + \ldots\ldots + G_n}$$

Precaution

It is worth repeating that only those resistances or admittances are taken into consideration which terminate at the common point. All those admittances are ignored which do not terminate at the common point even though they are connected in the circuit.

Example 2.101. *Use Millman's theorem to calculate the voltage developed across the 40 Ω resistor in the network of Fig. 2.219.*

Solution. Let the two ends of the 40 Ω resistor be marked as 0 and 0′. The end points of the three resistors terminating at the common point 0′ have been marked 1, 2 and 3. As already explained in Art. 2.29, the two resistors of values 10 Ω and 60 Ω will not come into the picture because they are not directly connected to the common point 0′.

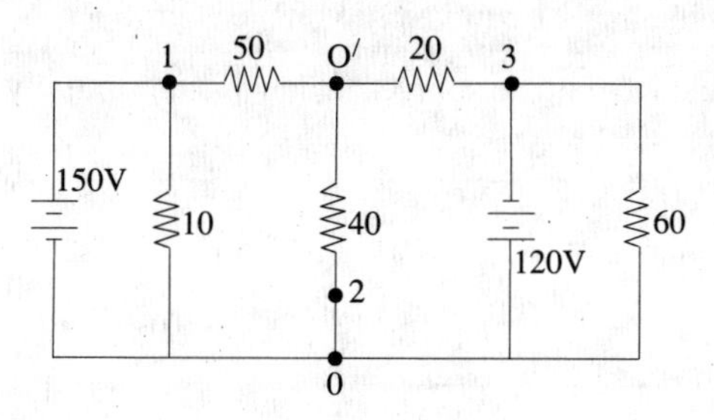

Fig. 2.219

Here, $V_{01} = -150\text{V}$; $V_{02} = 0$; $V_{03} = 120$ V

$G_1 = 1/50$; $G_2 = 1/40$; $G_3 = 1/20$

$$\therefore \quad V_{00}' = \frac{(-150/50) + (0/40) + (120/20)}{(1/50) + (1/40) + (1/20)} = 31.6 \textbf{ V}$$

It shows that point 0 is at a higher potential as compared to point 0′.

Example 2.102. *Calculate the voltage across the 10 Ω resistor in the network of* Fig. 2.220 *by using (a) Millman's theorem (b) any other method.*

Solution. (*a*) As shown in the Fig. 2.220 we are required to calculate voltages V_{00}'. The four resistances are connected to the common terminal 0′.

Let their other ends be marked as 1, 2, 3 and 4 as shown in Fig. 2.220. Now potential of point 0 with respect to point 1 is (Art. 1.25) – 100 V because (see Art 1.25)

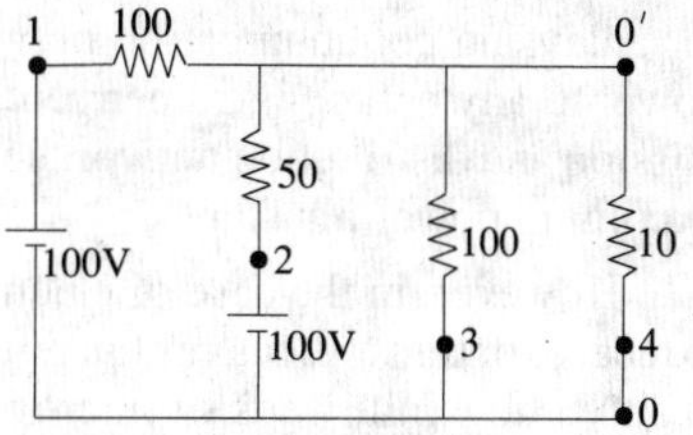

$\therefore V_{01} = -100$ V; $V_{02} = -100$ V; $V_{03} = 0$ V; $V_{04} = 0$ V.

$G_1 = 1/100 = 0.01$ Siemens ; $G_2 = 1/50 = 0.02$ Siemens;

$G_3 = 1/100 = 0.01$ Siemens ; $G_4 = 1/10 = 0.1$ Siemens

Fig. 2.220

$$\therefore \quad V_{00}' = \frac{V_{01}G_1 + V_{02}G_2 + V_{03}G_3 + V_{04}G_4}{G_1 + G_2 + G_3 + G_4}$$

$$= \frac{-100 \times 0.01 + (-100) \times 0.02 + 0. \times 0.01 + 0 \times 0.1}{0.01 + 0.02 + 0.01 + 0.1} = \frac{-3}{0.14} = \textbf{– 21.4 V}$$

Also, $\quad V_{0'0} = -V_{00'} = \textbf{21.4 V}$

(*b*) We could use the source conversion technique (Art. 2.14) to solve this question. As shown in Fig. 2.221 (*a*), the two voltage sources with thier series resistances have been converted into current sources with their parallel reistances. The two current sources have been combined into a single current source of 3 A and the three parallel resistances have been combined into a single resistance of 25 Ω. This current source has been reconverted into a voltage source of 75 V having a series resistance of 25 Ω as shown in Fig. 2.221 (*c*).

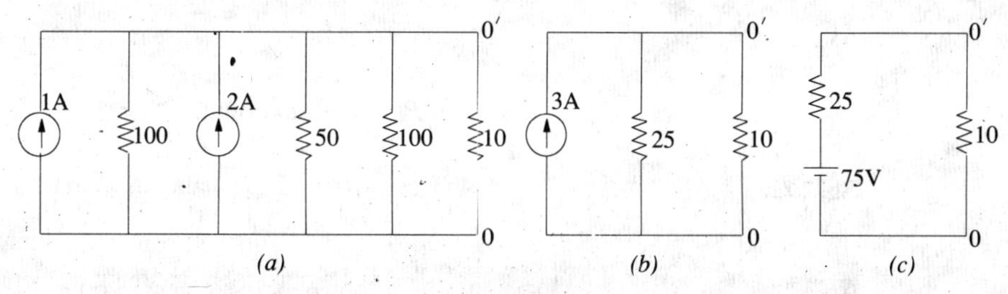

Fig. 2.221

Using the voltage divider formula (Art. 1.15), the voltage drop across 10 Ω resistance is $V_{0'0} = 75 \times 10/(10 + 25) = 21.4$ V.

Example 2.103. *In the network shown in Fig. 2.222, using Millman's theorem, or otherwise find the voltage between A and B.* **(Elect. Engg. Paper-I Indian Engg. Services 1990)**

Solution. The end points of the different admittances which are connected directly to the common point B have been marked as 1, 2 and 3 as shown in the Fig. 2.222. Incidentally, 40 Ω resistance will not be taken into consideration because it is not directly connected to the common point B. Here, $V_{01} = V_{A1} = -50$ V; $V_{02} = V_{A2} = 100$ V; $V_{03} = V_{A3} = 0$ V.

$$\therefore V_{00}' = V_{AB} = \frac{(-50/50)+(100/20)+(0/10)}{(1/50)+(1/20)+(1/10)} = 23.5 \text{ V}$$

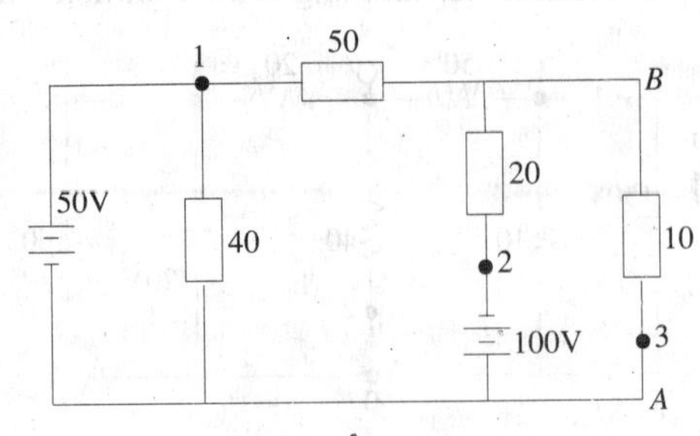

Fig. 2.222

Since the answer comes out to be positive, it means that point A is at a higher potential as compared to point B.

The detail reason for not taking any notice of 40 Ω resistance are given in Art. 2.29.

2.30. Maximum Power Transfer Theorem

Although applicable to all branches of electrical engineering, this theorem is particularly useful for analysing communication networks. The overall efficiency of a network supplying maximum power to any branch is 50 per cent. For this reason, the application of this theorem to power transmission and distribution networks is limited because, in their case, the goal is high efficiency and not maximum power transfer.

However, in the case of electronic and communication networks, very often, the goal is either to receive or transmit maximum power (through at reduced efficiency) specially when power involved is only a few milliwatts or microwatts. Frequently, the problem of maximum power transfer is of crucial significance in the operation of transmission lines and antennas.

As applied to d.c. networks, this theorem may be stated as follows :

A resistive load will abstract maximum power from a network when the load resistance is equal to the resistance of the network as viewed from the output terminals, with all energy sources removed leaving behind their internal resistances.

In Fig. 2.223(*a*), a load resistance of R_L is connected across the terminals A and B of a network which consists of a generator of e.m.f. E and internal resistance R_g and a series resistance R which, in fact, represents the lumped resistance of the connecting wires. Let $R_i = R_g + R$ = internal resistance of the netowrk as viewed from A and B.

According to this theorem, R_L will abstract maximum power from the network when $R_L = R_i$.

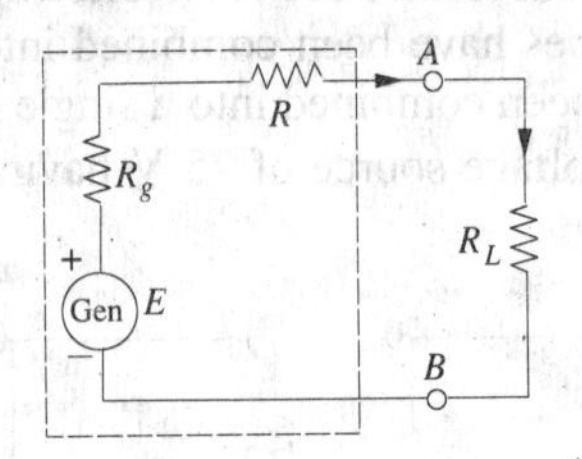

Fig. 2.223

Proof. Circuit current $I = \dfrac{E}{R_L + R_i}$

Power consumed by the load is

$$P_L = I^2 R_L = \frac{E^2 R_L}{(R_L + Ri)^2} \qquad \ldots(i)$$

For P_L to be maximum, $\dfrac{dP_L}{dR_L} = 0$.

Differentiating Eq. (*i*) above, we have

$$\frac{dP_L}{dR_L} = E^2\left[\frac{1}{(R_L+R_i)^2} + R_L\left(\frac{-2}{(R_L+R_i)^3}\right)\right] = E^2\left[\frac{1}{(R_L+R_i)^2} - \frac{2R_L}{(R_L+R_i)^3}\right]$$

$$\therefore \quad 0 = E^2\left[\frac{1}{(R_L+R_i)^2} - \frac{2R_L}{(R_L+R_i)^3}\right] \quad \text{or} \quad 2R_L = R_L + R_i \quad \text{or} \quad R_L = R_i$$

It is worth noting that under these conditions, the voltage across the load is half the open-circuit voltage at the terminals A and B.

$$\therefore \qquad \text{Max. power is } P_{L\,\text{max.}} = \frac{E^2 R_L}{4R_L^2} = \frac{E^2}{4R_L} = \frac{E^2}{4R_i}$$

Let us consider an a.c. source of internal impedance $(R_1 + j X_1)$ supplying power to a load impedance $(R_L + jX_L)$. It can be proved that maximum power transfer will take place when the modules of the load impedance is equal to the modulus of the source impedance i.e. $|Z_L| = |Z_1|$

Where there is a completely free choice about the load, the maximum power transfer is obtained when load impedance is the complex conjugate of the source impedance. For example, if source impedance is $(R_1 + jX_1)$, then maximum transfer power occurs, when load impedance is $(R_1 - jX_1)$. It can be shown that under this condition, the load power is $= E^2/4R_1$.

Example 2.104. *In the network shown in Fig. 2.224 (a), find the value of R_L such that maximum possible power will be transferred to R_L. Find also the value of the maximum power and the power supplied by source under these conditions.* **(Elect. Engg. Paper I Indian Engg. Services 1989)**

Solution. We will remove R_L and find the equivalent Thevenin's source for the circuit to the left of terminals *A* and *B*. As seen from Fig. 2.224 (*b*) V_{th} equals the drop across the vertical resistor of 3 Ω because no current flows through 2 Ω and 1 Ω resistors. Since 15 V drops across two series resistors of 3 Ω each, $V_{th} = 15/2 = 7.5$ V. Thevenin's resistance can be found by replacing 15 V source with a short-circuit. As seen from Fig. 2.224 (*b*), $R_{th} = 2 + (3||3) + 1 = 4.5$ Ω. Maximum power transfer to the load will take place when $R_L = R_{th} = 4.5$ Ω.

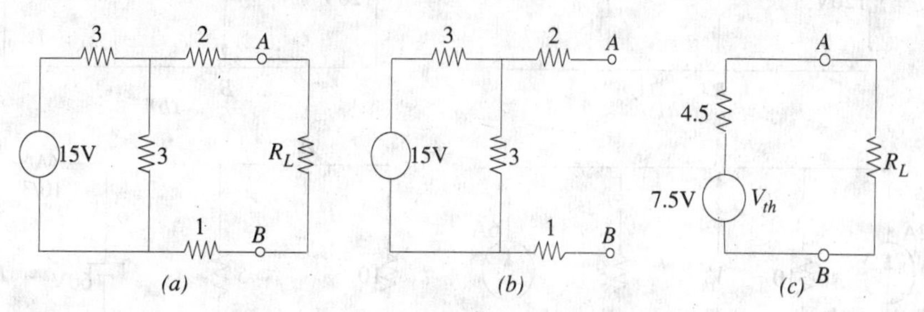

Fig. 2.224

Maximum power drawn by $R_L = V_{th}^2/4 \times R_L = 7.5^2/4 \times 4.5 = 3.125$ W.

Since same power is developed is R_{th}, power supplied by the source $= 2 \times 3.125 = 6.250$ W.

Example 2.105. *In the circuit shown in Fig. 2.225(a) obtain the condition from maximum power transfer to the load R_L. Hence determine the maximum power transferred.*

(Elect. Science-I Allahabad Univ. 1992).

Solution. We will find Thevenin's equivalent circuit to the left of terminals *A* and *B* for which purpose we will convert the battery source into a current source as shown in Fig. 2.225 (*b*). By combining the two current sources, we get the circuit of Fig. 2.225 (*c*), It would be seen that open-circuit voltage V_{AB} equials the drop over 3 Ω resistance because there is no drop on the 5 Ω resistance connected to terminal *A*. Now, there are two parallel paths across the current source each of resistance 5 Ω. Hence, current through 3 Ω resistance equals $1.5/2 = 0.75$ A. Therefore, $V_{AB} = V_{th} = 3 \times 0.75 = 2.25$ V with point *A* positive with respect to point *B*.

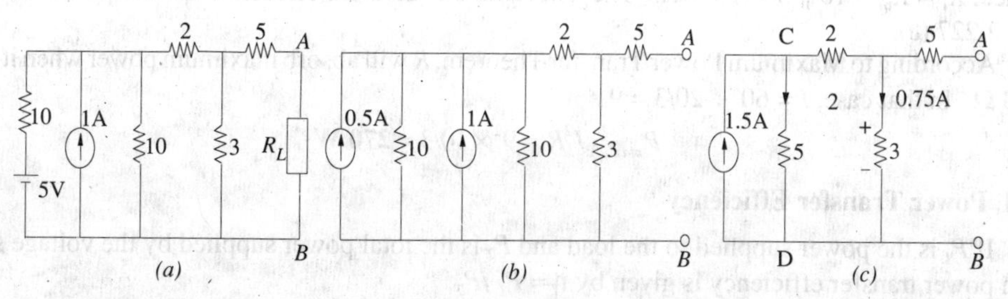

Fig. 2.225

For finding R_{AB}, current source is replaced by an infinite resistance.

$$\therefore \quad R_{AB} = R_{th} = 5 + 3 \parallel (2+5) = 7.1\ \Omega$$

The Thevenin's equivalent circuit alongwith R_L is shown in Fig. 2.226. As per Art. 2.30, the condition for *MPT* is that $R_L = 7.1\ \Omega$.

Maximum power transferred $= V_{th}^2 / 4R_L = 2.25^2/4 \times 7.1$ $= 0.178$ W = 178 mW.

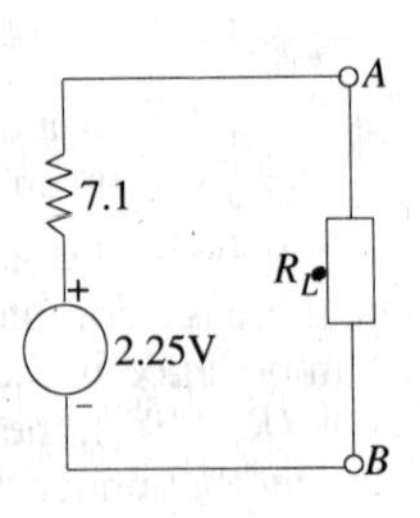

Fig. 2.226

Example 2.106. *Calculate the value of R which will absorb maximum power from the circuit of Fig. 2.227 (a). Also, compute the value of maximum power.*

Solution. For finding power, it is essential to know both *I* and *R*. Hence, it is essential to find an equation relating *I* to *R*.

Let us remove *R* and find Thevenin's voltage V_{th} across *A* and *B* as shown in Fig. 2.227 (*b*).

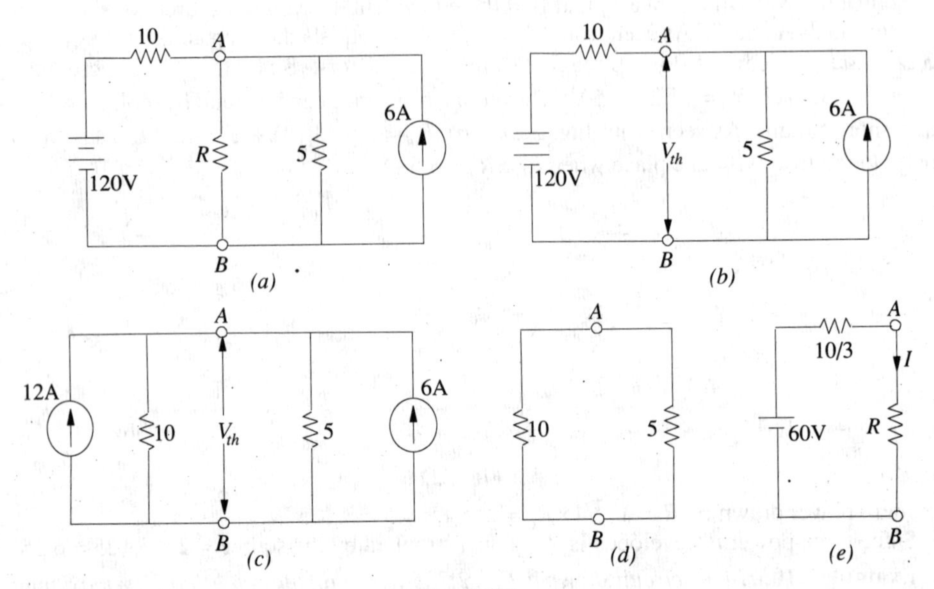

Fig. 2.227

It would be helpful to convert 120 V, 10–Ω source into a constant-current source as shown in Fig. 2.227 (*c*). Applying *KCL* to the circuit, we get

$$\frac{V_{th}}{10} + \frac{V_{th}}{5} = 12 + 6 \quad \text{or} \quad V_{th} = 60 \text{ V}$$

Now, for finding R_i or R_{th}, the two sources are reduced to zero. Voltage of the voltage-source is reduced to zero by short - circuiting it whereas current of the current source is reduced to zero by open-circuitring it. The circuit which results from such source suppression is shown in Fig. 2.227 (*d*). Hence, $R_i = R_{th} = 10 \parallel 5 = 10/3\ \Omega$. The Thevenin's equivalent circuit of the network is shown in Fig. 2.227 (*e*).

According to Maximum Power Transfer Theorem, *R* will absorb maximum power when it equals 10/3 Ω. In that case, $I = 60 \div 20/3 = 9$ A

$$P_{max} = I^2 R = 9^2 \times 10/3 = \mathbf{270\ W}$$

2.31. Power Transfer Efficiency

If P_L is the power supplied to the load and P_T is the total power supplied by the voltage source, then power transfer efficiency is given by $\eta = P_L / P_T$.

Now, the generator or voltage source *E* supplies power to both the load resistance R_L and to the internal resistance $R_i = (R_g + R)$.

$$P_T = P_L + P_i \quad \text{or} \quad E \times I = I^2R_L + I^2R_i$$

$$\therefore \qquad \eta = \frac{P_L}{P_T} = \frac{I^2R_L}{I^2R_L + I^2R_i} = \frac{R_L}{R_L + R_i} = \frac{1}{1 + (R_i/R_L)}$$

The variation of η with R_L is shown in Fig. 2.228 (*a*). The maximum value of η is unity when $R_L = \infty$ and has a value of 0.5 when $R_L = R_i$. It means that under maximum power transfer conditions, the power transfer efficiency is only 50%. As mentioned above, maximum power transfer condition is important in communication applications but in most power systems applications, a 50% efficiency is undesirable because of the wasted energy. Often, a compromise has to be made between the load power and the power transfer efficiency. For example, if we make $R_L = 2\ R_i$, then

$$P_L = 0.222\ E^2/R_i \quad \text{and} \quad \eta = 0.667.$$

It is seen that the load power is only 11% less than its maximum possible value, whereas the power transfer efficiency has improved from 0.5 to 0.667 *i.e.* by 33%.

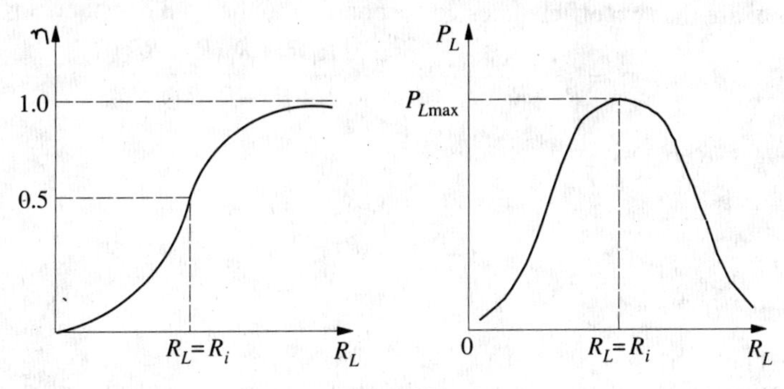

Fig. 2.228

Example 2.107. *A voltage source delivers 4 A when the load connected to it is 5 Ω and 2 A when the load becomes 20 Ω. Calculate.*

(a) maximum power which the source can supply (b) power transfer efficiency of the source with R_L of 20 Ω (c) the power transfer efficiency when the source delivers 60 W.

Solution. We can find the values of E and R_i from the two given load conditions.

(*a*) When $R_L = 5\ \Omega$, $I = 4$ A and $V = IR_L = 4 \times 5 = 20$ V, then $20 = E - 4\ R_i$ ...(*i*)

When $R_L = 20\ \Omega$, $I = 2$ A and $V = IR_L = 2 \times 20 = 40$ V $\therefore$ $40 = E - 2\ R_i$ ...(*ii*)

From (*i*) and (*ii*), we get, $R_i = 10\ \Omega$ and $E = 60$ V

When $R_L = R_i = 10\ \Omega$

$$P_{L\,max} = \frac{E^2}{4R_i} = \frac{60 \times 60}{4 \times 10} = 90 \text{ W}$$

(*b*) When $R_L = 20\ \Omega$, the power transfer efficiency is given by

$$\eta = \frac{R_L}{R_L + R_i} = \frac{20}{30} = 0.667 \quad \text{or} \quad 66.7\%$$

(*c*) For finding the efficiency corresponding to a load power of 60 W, we must first find the value of R_L.

Now, $$P_L = \left(\frac{E}{R_i + R_L}\right)^2 R_L$$

$$\therefore \qquad 60 = \frac{60^2 \times R_L}{(R_L + 10)^2} \quad \text{or} \quad R_L^2 - 40R_L + 100 = 0$$

Hence, $$R_L = 37.32\ \Omega \quad \text{or} \quad 2.68\ \Omega$$

Since there are two values of R_L, there are two efficiencies corresponding to these values.

$$\eta_1 = \frac{37.32}{37.32+10} = 0.789 \text{ or } 78.9\%, \quad \eta_2 = \frac{2.68}{12.68} = 0.211 \text{ or } 21.1\%$$

It will be seen from above, the $\eta_1 + \eta_2 = 1$.

Example 2.108. *Two load resistances R_1 and R_2 dissipate the same power when connected to a voltage source having an internal resistance of R_i. Prove that* (*a*) $R_i^2 = R_1R_2$ and (*b*) $\eta_1 + \eta_2 = 1$.

Solution. (*a*) Since both resistances dissipate the same amount of power, hence

$$P_L = \frac{E^2R_1}{(R_1+R_i)^2} = \frac{E^2R_2}{(R_2+R_i)^2}$$

Cancelling E^2 and cross - multiplying, we get

$$R_1R_2^2 + 2R_1R_2R_i + R_1R_i^2 = R_2R_1^2 + 2R_1R_2R_i + R_2R_i^2.$$

Simplifying the above, we get, $R_i^2 = R_1R_2$

(*b*) If η_1 and η_2 are the two efficiencies corresponding to the load resistances R_1 and R_2, then

$$\eta_1 + \eta_2 = \frac{R_1}{R_1+R_i} + \frac{R_2}{R_2+R_i} = \frac{2R_1R_2 + R_i(R_1+R_2)}{R_1R_2 + R_i^2 + R_i(R_1+R_2)}$$

Substituting $R_i^2 = R_1R_2$, we get

$$\eta_1 + \eta_2 = \frac{2R_i^2 + R_i(R_1+R_2)}{2R_i^2 + R_i(R_1+R_2)} = 1$$

Tutorial Problems No. 2.6

(*a*) Norton Theorem

1. Find the Thevenin and Norton equivalent circuits for the active network shown in Fig. 2.229. All resistances are in ohms. [**Hint** : Use Superposition principle to find contribution of each source.]

[10. V source, series resistor = 5 Ω ; 2A source, parallel resistance = 5 Ω].

2. Obtain the Thevenin and Norton equivalant circuits for the circuit shown in Fig. 2.230. All resistance values are in ohms. **[15-V source, series resistance = 5 Ω ; 3-A source, parallel resistance = 5 Ω]**

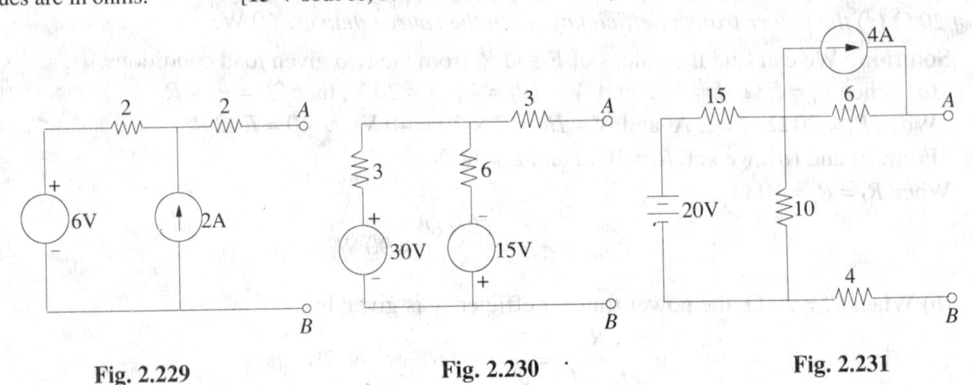

Fig. 2.229 Fig. 2.230 Fig. 2.231

3. Find the Norton equivalent circuit for the active linear network shown in Fi.g 2.231. All resistances are in ohms. [**Hint** : It would be easier to first find Thevenin's equivalent circuit.]

[2 A source ; parallel resistance = 16 Ω]

4. Apply Norton's theorem to find current flowing through the 3.6 Ω resistor of the circuit shown in Fig. 2.232. **[2 A]**

5. Find (*i*) Thevenin and (*ii*) Norton equivalent circuits for the terminals *A* and *B* of the network shown in Fig. 2.233 All resistances are in ohms. Take $E_1 > E_2$.

$$\left[\textbf{(i) } \mathbf{V}_{th} = \frac{\mathbf{E_1R_2 + E_2R_1}}{\mathbf{R_1 + R_2}};\ \mathbf{R}_{th} = \mathbf{R_1 \parallel R_2}\ \textbf{(ii) } \mathbf{I}_{sc} = \frac{\mathbf{E_1}}{\mathbf{R_1}} + \frac{\mathbf{E_2}}{\mathbf{R_2}};\ \mathbf{R}_p = \mathbf{R_1 \parallel R_2}\right]$$

6. Obtain (*i*) Thevenin and (*ii*) Norton equivalent circuit with respect to the terminals *A* and *B* of the network of Fig. 2.234. Numbers represent resistances in ohm.

[(i) V_{th} = 4V ; R_{th} = 14/9 Ω (*ii*) I_{sc} = 2.25A; R_p = 14/9 Ω]

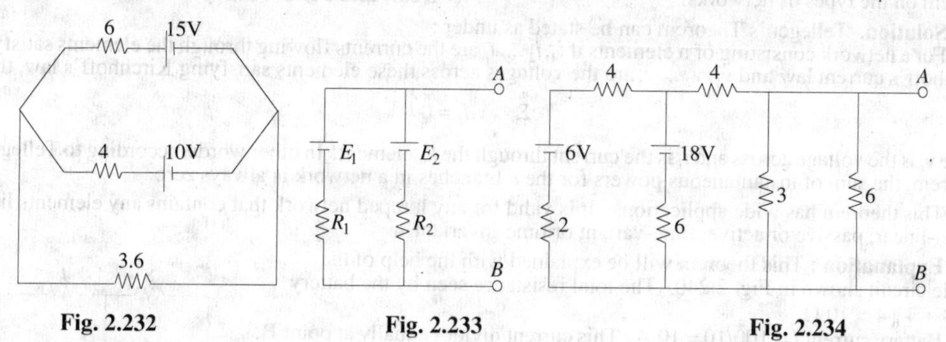

Fig. 2.232 **Fig. 2.233** **Fig. 2.234**

7. The Norton equivalent of the network shown in Fig. 2.235 between terminals *A* and *B* is a parallel resistance of 10 Ω. What is the value of the unknown resistance *R* ? **[60 Ω]**

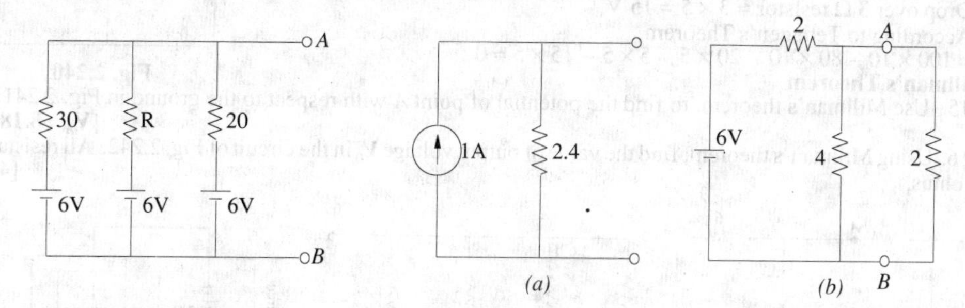

Fig. 2.235 **Fig. 2.236**

8. The circuit shown in Fig. 2.236 (*a*) is the Norton equivalent of the circuit to the left of *AB* in Fig. 2.236 (*b*). What current will flow if a short is placed across *AB* ? **[1 A]**

9. The Norton equivalent circuit of two identical batteries connected in parallel consists of a 2-A source in parallel with a 4 – Ω resistor. Find the value of the resistive load to which a single battery will deliver maximum power. Also calculate the value of this maximum power. **[8 Ω ; 2 W]**

10. The Thevenin equivalent circuit of a certain circuit consists of a 6-V d.c. source in series with a resistance of 3 Ω. The Norton equivalent of another circuit is a 3-A current source in parallel with a resistance of 6 Ω. The two circuits are connected in parallel like polarity to like. For this combination, determine (*i*) Norton equivalent (*ii*) Thevenin equivalent (*iii*) maximum power it can deliver and (*iv*) value of load resistance from maximum power. **[(i) 5 A, 2Ω (ii) 10 V, 2 Ω (iii) 12.5 W (iv) 2 Ω]**

11. For the ladder network shown in Fig. 2.237, find the vlaue of R_L for maximum power transfer. What is the value of this P_{max} **[2 Ω, 9/16 W]**

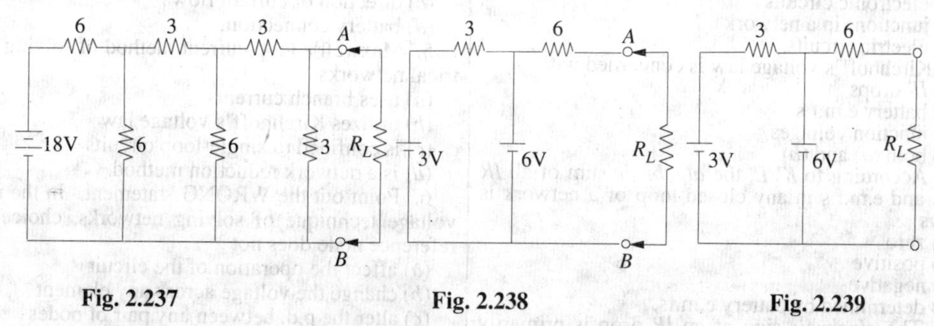

Fig. 2.237 **Fig. 2.238** **Fig. 2.239**

12. Calculate the value of R_L which will draw maximum power from the circuit of Fig. 2.238. Also, find the value of this maximum power. **[6 Ω ; 1.5 W]**

13. Find Norton's equivalent circuit for the network shown in Fig. 2.239. Verify it through its Thevenin's equivalent circuit. **[1 A, Parallel resistance = 6 Ω]**

14. State the Tellegen's theorem and verify it by an illustration. Comment on the applicability of Tellegen's theorem on the types of networks. **(Circuit and Field Theory, A.M.I.E. Sec B, 1993)**

Solution. Tellegen's Theorem can be stated as under :

For a network consisting of n elements if $i_1, i_2 .... i_n$ are the currents flowing through the elements satisfying Kirchhoff's current law and $v_1, v_2 ... v_n$ are the voltages across these elements satisfying Kirchhoff's law, then

$$\sum_{k=1}^{n} v_k i_k = 0$$

where v_k is the voltage across and i_k is the current through the k_{th} element. In other words, according to Tellegen's Theorem, the sum of instantaneous powers for the n branches in a network is always zero.

This theorem has wide applications. It is valid for any lumped network that contains any elements linear or non-linear, passive or active, time-variant or time-invariant.

Explanation : This theorem will be explained with the help of the simple circuit shown in Fig. 2.240. The total resistance seen by the battery is = 8 + 4 || 4 = 10 Ω.

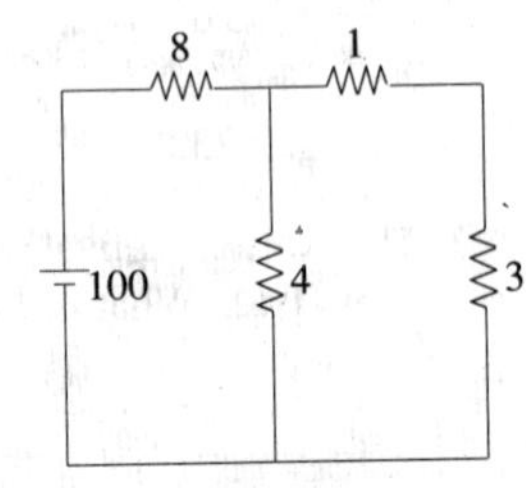

Fig. 2.240

Battery current I = 100/10 = 10 A. This current divides equally at point B,

Drop over 8 Ω resistor = 8 × 10 = 80 V

Drop over 4 Ω resistor = 4 × 5 = 20 V

Drop over 1 Ω resistor = 1 × 5 = 5 V

Drop over 3 Ω resistor = 3 × 5 = 15 V

According to Tellegen's Theorem,

= 100 × 10 – 80 × 10 – 20 × 5 – 5 × 5 – 15 × 5 = 0

(*b*) **Millman's Theorem**

15. Use Millman's theorem, to find the potential of point A with respect to the ground in Fig. 2.241. **[V_A = 8.18 V]**

16. Using Millman's theorem, find the value of output voltage V_0 in the circuit of Fig. 2.242. All resistances are in ohms. **[4 V]**

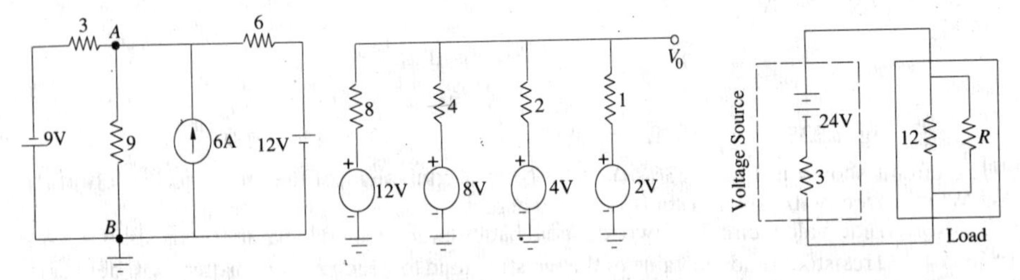

Fig. 2.241 **Fig. 2.242** **Fig. 2.243**

(*c*) **MPT Theorem**

17. In Fig. 2.243 what value of R will allow maximum power transfer to the load ? Also calculate the maximum total load power. All resistances are in ohms. **[4 Ω ; 48 W]**

OBJECTIVE TESTS—2

1. Kirchhoff's current law is applicable to only
(*a*) closed loops in a network
(*b*) electronic circuits
(*c*) junctions in a network
(*d*) electric circuits.

2. Kirchhoff's voltage law is concerned with
(*a*) *IR* drops
(*b*) battery e.m.f.s
(*c*) junction voltages
(*d*) both (*a*) and (*b*)

3. According to *KVL*, the *algebraic* sum of all *IR* drops and e.m.f.s in any closed loop of a network is always
(*a*) zero
(*b*) positive
(*c*) negative
(*d*) determined by battery e.m.fs.

4. The *algebraic* sign of an *IR* drop is primarily dependent upon the
(*a*) amount of current flowing through it
(*b*) value of *R*
(*c*) direction of current flow
(*d*) battery connection.

5. Maxwell's loop current method of solving electrical networks
(*a*) uses branch currents
(*b*) utilizes Kirchhoff's voltage law
(*c*) is confined to single-loop circuits
(*d*) is a network reduction method.

6. Point out the WRONG statement. In the node-voltage technique of solving networks, choice of a reference node does not
(*a*) affect the operation of the circuit
(*b*) change the voltage across any element
(*c*) alter the p.d. between any pair of nodes
(*d*) affect the voltages of various nodes.

7. The nodal analysis is primarily based on the application of
(*a*) *KVL* (*b*) *KCL*
(*c*) Ohm's law
(*d*) both (*b*) and (*c*)
(*e*) both (*a*) and (*b*).

8. Superposition theorem can be applied only to circuits having—elements.
(*a*) non-linear (*b*) passive
(*c*) linear bilateral (*d*) resistive

9. The Superposition theorem is essentially based on the concept of
(*a*) duality (*b*) linearity
(*c*) reciprocity (*d*) non-linearity.

10. While Thevenizing a circuit between two terminals, V_{th} equals
(*a*) short-circuit terminal voltage
(*b*) open-circuit terminal voltage
(*c*) EMF of the battery nearest to the terminals
(*d*) net voltage available in the circuit.

11. Thevenin resistance R_{th} is found
(*a*) between any two 'open' terminals
(*b*) by short-circuiting the given two terminals
(*c*) by removing voltage sources along with their internal resistances
(*d*) between same open terminal as for V_{th}.

12. While calculating R_{th}, constant-current sources in the circuit are
(*a*) replaced by 'opens'
(*b*) replaced by 'shorts'
(*c*) treated in parallel with other voltage sources
(*d*) coverted into equivalent voltage sources.

13. Thevenin resistance of the circuit of Fig. 2.244 across its terminals *A* and *B* is—ohm.
(*a*) 6
(*b*) 3
(*c*) 9
(*d*) 2

Fig. 2.244

14. The load resistance needed to extract maximum power from the circuit of Fig. 2.245 is –ohm.
(*a*) 2 (*b*) 9
(*c*) 6 (*d*) 18

15. The Norton equivalent circuit for the network of Fig. 2.245 between *A* and *B* is–current source with parallel resistance of–
(*a*) 2A, 6 Ω
(*b*) 3A, 2 Ω
(*c*) 2A, 3 Ω
(*d*) 3A, 9 Ω

Fig. 2.245

16. The Norton equivalent of a circuit consists of a 2A current source in parallel with a 4 Ω resistor. Thevenin equivalent of this circuit is a–vot sour… in series with a 4 Ω resistor.
(*a*) 2 (*b*) 0.5
(*c*) 6 (*d*) 8

17. If two identical 3A, 4Ω Norton equivalent circuits are connected in parallel with like polarity to like, the combined Norton equivalent circuit is
(*a*) 6A, 4 Ω (*b*) 6A, 2 Ω
(*c*) 3A, 2 Ω (*d*) 6A, 8 Ω

18. Two 6 V, 2Ω batteries are connected in series siding. This combination can be replaced by a single equivalent current generator of–with a parallel resistance of– ohm.
(*a*) 3A, 4 Ω (*b*) 3A, 2 Ω
(*c*) 3A, Ω (*d*) 6A, 2 Ω

19. Two identical 3-A, 1 Ω batteries are connected in parallel with like polarity to like. The Norton equivalent circuit of this combination is
(*a*) 3A, 0.5 Ω (*b*) 6A, 1 Ω
(*c*) 3A, 1 Ω (*d*) 6A, 0.5 Ω

20. Thevenin equivalent circuit of the network shown in Fig. 2.246 is required. The value of open-circuit voltage across terminals *a* and *b* of this circuit is _____ volt.
(*a*) zero
(*b*) $2i/10$
(*c*) $2i/5$
(*d*) $2i/15$

Fig. 2.246

21. For a linear network containing generators and impedances, the ratio of the voltage to the current produced in other loop is the same as the ratio of voltage and current obtained when the positions of the voltage source and the ammeter measuring the current are interchanged. This network theorem is known as _____ theorem.
(*a*) Millman's (*b*) Norton's
(*c*) Tellegen's (*d*) Reciprocity

(Circuits and Field Theory, A.M.I.E. Sec. B., 1993)

22. A 12 volt source with an internal resistance of 1.2 ohms is connected across a wire-would resistor Maximum power will be dissipated in the resistor whe… its resistance is equal to
(*a*) zero (*b*) 1.2 ohm
(*c*) 12 ohm (*d*) infinity

(Grad. I. E. T. E. Dec. 1985)

ANSWERS

1. *c* 2. *d* 3. *a* 4. *c* 5. *b* 6. *d* 7. *d* 8. *c*
9. *b* 10. *b* 11. *d* 12. *a* 13. *a* 14. *c* 15. *b* 16. *d*
17. *d* 18. *b* 19. *d* 20. *a* 21. *d* 22. *b*

3 WORK, POWER AND ENERGY

3.1. Effect of Electric Current

It is a matter of common experience that a conductor, when carrying current, becomes hot after some time. As explained earlier, an electric current is just a directed flow or drift of electrons through a substance. The moving electrons as they pass *through* molecules or atoms of that substance, collide with other electrons. This electronic collision results in the production of heat. This explains why passage of current is always accompanied by generation of heat.

3.2. Joule's Law of Electric Heating

The amount of work required to maintain a current of I amperes through a resistance of R ohm for t second is

$$\begin{aligned} \text{W.D.} &= I^2Rt \text{ joules} \\ &= VIt \text{ joules} \quad (\because R = V/I) \\ &= Wt \text{ joules} \quad (\because W = VI) \\ &= V^2t/R \text{ joules} \quad (\because I = V/R) \end{aligned}$$

This work is converted into heat and is dissipated away. The amount of heat produced is

$$H = \frac{\text{work done}}{\text{mechanical equivalent of heat}} = \frac{W.D.}{J}$$

where $J = 4,186 \text{ joules/kcal} = 4,200 \text{ joules / kcal (approx)}$

$$\therefore \quad H = I^2Rt/4,200 \text{ kcal} = VIt/4,200 \text{ kcal}$$

$$= Wt/4,200 \text{ kcal} = V^2t/4,200\,R \text{ kcal}$$

3.3. Thermal Efficiency

It is defined as the ratio of the heat actually utilized to the total heat produced electrically. Consider the case of the electric kettle used for boiling water. Out of the total heat produced (*i*) some goes to heat the apparatus itself *i.e.* kettle (*ii*) some is lost by radiation and convection etc. and (*iii*) the rest is utilized for heating the water. Out of these, the heat utilized for useful purpose is that in (*iii*). Hence, thermal efficiency of this electric apparatus is the ratio of the heat utilized for heating the water to the total heat produced.

Hence, the relation between heat produced electrically and heat absorbed usefully becomes

$$\frac{VIt}{J} \times \eta = ms\,(\theta_2 - \theta_1)$$

Example 3.1. *The heater element of an electric kettle has a constant resistance of 100 Ω and the applied voltage is 250 V. Calculate the time taken to raise the temperature of one litre of water from 15°C to 90°C assuming that 85% of the power input to the kettle is usefully employed. If the water equivalent of the kettle is 100 g, find how long will it take to raise a second litre of water through the same temperature range immediately after the first.*

(Electrical Engineering, Calcutta Univ. 1980)

Solution. Mass of water $= 1000$ g $= 1$ kg $\quad$ ($\because$ 1 cm^3 weighs 1 gram)

Heat taken by water $= 1 \times (90 - 15) = 75$ kcal

Heat taken by the kettle $= 0.1 \times (90 - 15) = 7.5$ kcal

Total heat taken $= 75 + 7.5 = 82.5$ kcal

Heat produced electrically $H = I^2Rt/J$ kcal

Now, $I = 250/100 = 2.5$ A, $\quad J = 4{,}200$ J/kcal; $H = 2.5^2 \times 100 \times t/4200$ kcal

Heat actually utilized for heating one litre of water and kettle

$= 0.85 \times 2.5^2 \times 100 \times t/4{,}200$ kcal

$$\therefore \quad \frac{0.85 \times 6.25 \times 100 \times t}{4{,}200} = 82.5 \quad \therefore \quad t = \textbf{10 min 52 second}$$

In the second case, heat would be required only for heating the water because kettle would be already hot.

$$\therefore \quad 75 = \frac{0.85 \times 6.25 \times 100 \times t}{4{,}200} \quad \therefore \quad t = \textbf{9 min 53 second}$$

Example 3.2. *Two heaters A and B are in parallel across supply voltage V. Heater A produces 500 kcal in 20 min. and B produces 1000 kcal in 10 min. The resistance of A is 10 ohm. What is the resistance of B ? If the same heaters are connected in series across the voltage V, how much heat will be produced in kcal in 5 min ?*

(Elect. Science – II, Allahabad Univ. 1992)

Solution. Heat produced $= \dfrac{V^2 t}{JR}$ kcal

For heater A, $$500 = \frac{V^2 \times (20 \times 60)}{10 \times J} \quad \ldots(i)$$

For heater B, $$1000 = \frac{V^2 \times (10 \times 60)}{R \times J} \quad \ldots(ii)$$

From Eq. (*i*) and (*ii*), we get, $\quad R = \mathbf{2.5\ \Omega}$.

When the two heaters are connected in series, let H be the amount of heat produced in kcal. Since combined resistance is $(10 + 2.5) = 12.5\ \Omega$, hence

$$H = \frac{V^2 \times (5 \times 60)}{12.5 \times J} \quad \ldots(iii)$$

Dividing Eq. (*iii*) by Eq. (*i*), we have $H =$ **100 kcal.**

Example 3.3. *An electric kettle needs six minutes to boil 2 kg of water from the initial temperature of 20°C. The cost of electrical energy required for this operation is 12 paise, the rate being 40 paise per kWh. Find the kW-rating and the overall efficiency of the kettle.*

(F.Y. Engg. Pune Univ. Nov. 1989)

Solution. Input energy to the kettle $= \dfrac{12 \text{ paise}}{40 \text{ paise/kWh}} = 0.3$ kWh

Input power $= \dfrac{\text{energy in kWh}}{\text{Time in hours}} = \dfrac{0.3}{(6/60)} = 3$ kW.

Hence, the power rating of the electric kettle is 3 kW.

Energy utilised in heating the water

$= mst = 2 \times 1 \times (100 - 20) = 160$ kcal $= 160 / 860$ kWh $= 0.186$ kWh.

Efficiency = output / input $= 0.186 / 0.3 = 0.62 =$ **62%.**

3.4. S.I. Units

1. Mass. It is the quantity of matter contained in a body.

Unit of mass is kilogram (kg). Other multiples commonly used are :

1 quintal = 100 kg, 1 tonne = 10 quintals = 1000 kg

2. Force. Unit of force is newton (N). Its definition may be obtained from Newton's Second Law of Motion *i.e.* $F = ma$.

If $m = 1$ kg ; $a = 1$ m/s^2, then $F = 1$ newton.

Hence, one newton is that force which can give an acceleration of 1 m/s^2 to a mass of 1 kg.

Gravitational unit of force is kilogram-weight (kg-wt). It may be defined as follows :

It is the force which can impart an acceleration of 9.8 m/s^2 to a mass of 1 kg.

or

It is the force which can impart an acceleration of 1 m/s^2 to a mass of 9.8 kg.

Obviously, 1 kg-wt. = 9.8 N

3. Weight. It is the force with which earth pulls a body downwards. Obviously, its units are the same as for force.

(*a*) Unit of weight is newton (N)

(*b*) Gravitational unit of weight is kg-wt.*

Note. If a body has a mass of *m* kg, then its weight, $W = mg$ newtons = 9.8 newtons.

4. Work. If a force of *F* moves a body through a distance *S* in its direction of application, then

$$\text{Work done} \quad W = F \times S$$

(*a*) Unit of work is joule (J).

If, in the above equation, $F = 1$ N : $S = 1$ m ; then work done = 1 m.N or joule.

Hence, one joule is the work done when a force of 1 N moves a body through a distance of 1 m in the direction of its application.

(*b*) Gravitational unit of work is m-kg.wt or m-kg**.

If $F = 1$ kg-wt; $S = 1$ m; then W.D. = 1 m-kg. wt = 1 m-kg.

Hence, one m-kg is the work done by a force of one kg-wt when applied over a distance of one metre.

Obviously, 1 m-kg = 9.8 m-N or J.

5. Power. It is the rate of doing work. Its units is watt (W) which represents 1 joule per second.

$$1 \text{ W} = 1 \text{ J/s}$$

If a force of *F* newton moves a body with a velocity of ν m/s then

$$\text{power} = F \times \nu \text{ watt}$$

If the velocity ν is in km/s, then

$$\text{power} = F \times \nu \text{ kilowatt}$$

6. Kilowatt-hour (kWh) and kilocalorie (kcal)

$$1 \text{ kWh} = 1000 \times 1 \frac{\text{J}}{\text{s}} \times 3600 \text{ s} = 36 \times 10^5 \text{ J}$$

$$1 \text{ kcal} = 4,186 \text{ J} \quad \therefore \quad 1 \text{ kWh} = 36 \times 10^5/4,186 = 860 \text{ kcal}$$

7. Miscellaneous Units

(*i*) 1 watt hour (Wh) $= 1 \frac{\text{J}}{\text{s}} \times 3600 \text{ s} = 3600 \text{ J}$

(*ii*) 1 horse power (metric) = 75 m-kg/s = 75 × 9.8 = 735.5 J/s or watt

(*iii*) 1 kilowatt (kW) = 1000 W and 1 megawatt (MW) = 10^6 W

3.5. Calculation of Kilo-watt Power of a Hydroelectric Station

Let Q = water discharge rate in cubic metres / second (m^3/s), H = net water head in metre (m).

$g = 9.81$ η ; overall efficiency of the hydroelectric station expressed as a fraction.

Since 1 m^3 of water weighs 1000 kg., discharge rate is 1000 Q kg/s.

When this amount of water falls through a height of *H* meter, then energy or work available per second or available power is

$$= 1000\, QgH \text{ J/s} \quad \text{or} \quad \text{W} = QgH \text{ kW}$$

---

*Often it is referred to as a force of 1 kg, the word '*wt*' being omitted. To avoid confusion with mass of 1 kg, the *force* of 1 kg is written in engineering literature as kgf instead of kg. wt.

**Generally the word '*wt*' is omitted and the unit is simply written as m-kg.

Since the overall station efficiency is η, power actually available is = 9.81 η QH kW.

Example 3.4. *A de-icing equipment fitted to a radio aerial consists of a length of a resistance wire so arranged that when a current is passed through it, parts of the aerial become warm. The resistance wire dissipates 1250 W when 50 V is maintained across its ends. It is connected to a d.c. supply by 100 metres of this copper wire, each conductor of which has a resistance of 0.006 Ω / m. Calculate*

(*a*) *the current in the resistance wire*
(*b*) *the power lost in the copper connecting wire*
(*c*) *the supply voltage required to maintain 50 V across the heater itself.*

Solution. (*a*) Current = wattage/voltage = 1250/50 = **25 A**
(*b*) Resistance of one copper conductor = $0.006 \times 100 = 0.6\ \Omega$
Resistance of both copper conductors = $0.6 \times 2 = 1.2\ \Omega$
Power loss = I^2R watts = $25^2 \times 1.2$ = **750 W**
(*c*) Voltage drop over connecting copper wire = IR volt = $25 \times 1.2 = 30$ V
∴ Supply voltage required = 50 + 30 = **80 V**

Example 3.5. *A factory has a 240-V supply from which the following loads are taken :*
Lighting : Three hundred 150-W, four hundred 100 W and five hundred 60-W lamps
Heating : 100 kW
Motors : A total of 44.76 kW (60 b.h.p.) with an average efficiency of 75 per cent
Misc. : Various loads taking a current of 40 A.

Assuming that the lighting load is on for a period of 4 hours/day, the heating for 10 hours per day and the remainder for 2 hours/day, calculate the weekly consumption of the factory in kWh when working on a 5-day week.

What current is taken when the lighting load only is switched on ?

Solution. The power consumed by each load can be tabulated as given below :

Power consumed

Lighting	300 × 150	=	45,000	=	45 kW
	400 × 100	=	40,000	=	40 kW
	500 × 60	=	30,000	=	30 kW
			Total	=	115 kW

Heating = 100 kW
Motors = 44.76/0.75 = 59.7 kW
Misc. = 240 × 40/1000 = 9.6 kW

Similarly, the energy consumed/day can be tabulated as follows :

Energy consumed / day

Lighting	= 115 kW × 4 hr	= 460 kWh
Heating	= 100 kW × 10 hr	= 1,000 kWh
Motors	= 59.7 kW × 2 hr	= 119.4 kWh
Misc.	= 9.6 kW × 2 hr	= 19.2 kWh

Total daily consumption = 1,598.6 kWh
Weekly consumption = 1,598.6 × 5 = **7,993 kWh**
Current taken by the lighting load alone = 115 × 1000/240 = **479 A**

Example 3.6. *A Diesel-electric generating set supplies an output of 25 kW. The calorific value of the fuel oil used is 12,500 kcal/kg. If the overall efficiency of the unit is 35% (a) calculate the mass of oil required per hour (b) the electric energy generated per tonne of the fuel.*

Solution. Output = 25 kW, Overall η = 0.35, Input = 25/0.35 = 71.4 kW
∴ input per hour = 71.4 kWh = 71.4 × 860 = 61,400 kcal
Since 1 kg of fuel-oil produces 12,500 kcal
(*a*) ∴ mass of oil required = 61,400/12,500 = **4.91 kg**
(*b*) 1 tonne of fuel = 1000 kg

Heat content $= 1000 \times 12{,}500 = 12.5 \times 10^6$ kcal
$= 12.5 \times 10^6/860 = 14{,}530$ kWh

Overall $\eta = 0.35\%$ $\therefore$ energy output $= 14{,}530 \times 0.35 =$ **5,088 kWh**

Example 3.7. *The effective water head for a 100 MW station is 220 metres. The station supplies full load for 12 hours a day. If the overall efficiency of the station is 86.4%, find the volume of water used.*

Solution. Energy supplied in 12 hours $= 100 \times 12 = 1200$ MWh
$= 12 \times 10^5$ kWh $= 12 \times 10^5 \times 36 \times 10^5$ J
$= 43.2 \times 10^{11}$ J

Overall $\eta = 86.4\% = 0.864$ $\therefore$ Energy input $= 43.2 \times 10^{11}/0.864 = 5 \times 10^{12}$ J

Suppose m kg is the mass of water used in 12 hours, then $m \times 9.81 \times 220 = 5 \times 10^{12}$

$\therefore$ $m = 5 \times 10^{12}/9.81 \times 220 = 23.17 \times 10^8$ kg

Volume of water $= 23.17 \times 10^8/10^3 =$ **23.17×10^5 m³**

($\because$ 1 m³ of water weighs 10^3 kg)

Example 3.8. *Calculate the current required by a 1,500 volts d.c. locomotive when drawing 100 tonne load at 45 km.p.h. with a tractive resistance of 5 kg/tonne along (a) level track (b) a gradient of 1 in 50. Assume a motor efficiency of 90 per cent.*

Solution. As shown in Fig. 3.1 (*a*), in this case, force required is equal to the tractive resistance only.

(*a*) Force required at the rate of 5 kg-wt/tonne
$= 100 \times 5$ kg-wt. $= 500 \times 9.81 = 4905$ N

Distance travelled/second $= 45 \times 1000/3600 = 12.5$ m/s

Power output of the locomotive $= 4905 \times 12.5$ J/s or watt $= 61{,}312$ W

$\eta = 0.9$ $\therefore$ Power input $= 61{,}312/0.9 = 68{,}125$ W

$\therefore$ Current drawn $= 68{,}125/1500 =$ **45.41 A**

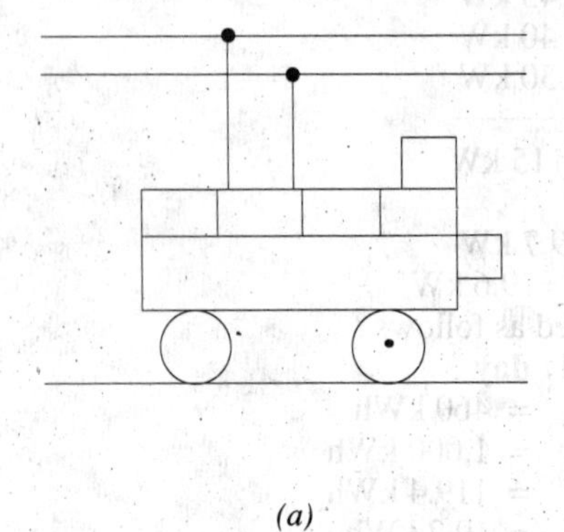
(*a*)

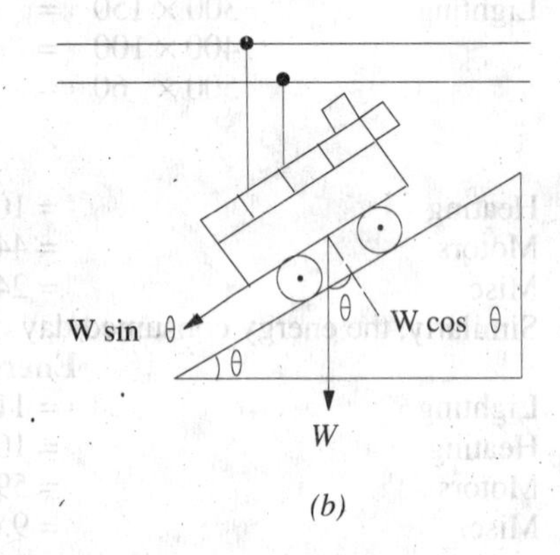

(*b*)

Fig. 3.1

(*b*) When the load is drawn along the gradient [Fig. 3.1 (*b*)], component of the weight acting downwards $= 100 \times 1/50 = 2$ tonne-wt $= 2000$ kg-wt $= 2000 \times 9.81 = 19{,}620$ N

Total force required $= 19{,}620 + 4{,}905 = 24{,}525$ N

Power output $=$ force $\times$ velocity $= 24{,}525 \times 12.5$ watt

Power input $= 24{,}525 \times 12.5/0.9$ W ; Current drawn $= \dfrac{24{,}525 \times 12.5}{0.9 \times 1500} =$ **227 A**

Example 3.9. *A room measures 4 m × 7 m × 5 and the air in it has to be always kept 15°C higher than that of the incoming air. The air inside has to be renewed every 35 minutes. Neglecting radiation loss, calculate the rating of the heater suitable for this purpose. Take specific heat of air as 0.24 and density as 1.27 kg/m³.*

Solution. Volume of air to be changed per second $= 4 \times 7 \times 5/35 \times 60 = 1/15$ m³

Mass of air to be changed/second $= (1/15) \times 1.27$ kg

Heat required/second $=$ mass/second $\times$ sp. heat $\times$ rise in temp.

$= (1.27/15) \times 0.24 \times 15$ kcal/s $= 0.305$ kcal/s

$= 0.305 \times 4186$ J/s $=$ **1277 watt.**

Example 3.10. *A motor is being self-started against a resisting torque of 60 N-m and at each start, the engine is cranked at 75 r.p.m. for 8 seconds. For each start, energy is drawn from a lead-acid battery. If the battery has the capacity of 100 Wh, calculate the number of starts that can be made with such a battery. Assume an overall efficiency of the motor and gears as 25%.*

(Principles of Elect. Engg.-I, Jadavpur Univ. 1987)

Solution. Angular speed $\omega = 2\pi N/60$ rad/s $= 2\pi \times 75/60 = 7.85$ rad/s

Power required for rotating the engine at this angular speed is

$P =$ torque $\times$ angular speed $= \omega T$ watt $= 60 \times 7.85 =$ **471 W**

Energy required per start is $=$ power $\times$ time per start

$= 471 \times 8 = 3{,}768$ watt-s $= 3{,}768$ J

$= 3{,}768/3600 = 1.047$ Wh

Energy drawn from the battery taking into consideration the efficiency of the motor and gearing $= 1.047/0.25 = 4.188$ Wh

No. of starts possible with a fully-charged battery $= 100/4.188 =$ **24** (approx.)

Example 3.11. *Find the amount of electrical energy expended in raising the temperature of 45 litres of water by 75 °C. To what height could a weight of 5 tonnes be raised with the expenditure of the same energy ? Assume efficiencies of the heating equipment and lifting equipment to be 90% and 70% respectively.* **(Elect. Engg. A.M.Ae. S.I. Dec. 1991)**

Solution. Mass of water heated $= 45$ kg; Heat required $= 45 \times 75 = 3{,}375$ kcal

Heat produced electrically $= 3{,}375/0.9 = 3{,}750$ kcal. Now, 1 kcal $= 4{,}186$ J

$\therefore$ electrical energy expended $= 3{,}750 \times 4{,}186$ J

Energy available for lifting the load is $= 0.7 \times 3{,}750 \times 4{,}186$ J

If h metre is the height through which the load of 5 tonnes can be lifted, then potential energy of the load $= mgh$ joules $= 5 \times 1000 \times 9.81\, h$ joules

$\therefore \quad 5000 \times 9.81 \times h = 0.7 \times 3{,}750 \times 4{,}186 \quad \therefore \quad h =$ **224 metres**

Example 3.12. *An hydro-electric station has a turbine of efficiency 86% and a generator of efficiency 92%. The effective head of water is 150 m. Calculate the volume of water used when delivering a load of 40 MW for 6 hours. Water weighs 1000 kg/m^3.*

Solution. Energy output $= 40 \times 6 = 240$ MWh

$= 240 \times 10^3 \times 36 \times 10^5 = 864 \times 10^9$ J

Overall $\eta = 0.86 \times 0.92 \quad \therefore \quad$ Energy input $= \dfrac{864 \times 10^9}{0.86 \times 0.92} = 10.92 \times 10^{11}$ J

Since the head is 150 m and 1m^3 of water weighs 1000 kg, energy contributed by each m^3 of water $= 150 \times 1000$ m-kg (wt) $= 150 \times 1000 \times 9.81$ J $= 147.2 \times 10^4$ J

$\therefore$ Volume of water for the required energy $= \dfrac{10.92 \times 10^{11}}{147.2 \times 10^4} =$ **74.18×10^4 m^3**

Example 3.13. *An hydroelectric generating station is supplied from a reservoir of capacity 6 million m^3 at a head of 170 m.*

(i) What is the available energy in kWh if the hydraulic efficiency be 0.8 and the electrical efficiency 0.9 ?

(ii) Find the fall in reservoir level after a load of 12,000 kW has been supplied for 3 hours, the area of the reservoir is 2.5 km^2.

(iii) If the reservoir is supplied by a river at the rate of 1.2 m^3/s, what does this flow represent in kW and kWh /day ? Assume constant head and efficiency.

Water weighs 1 tonne/m^3. **(Elect. Engineering-I, Osmania Univ. 1987)**

Solution. (*i*) Wt. of water $W = 6 \times 10^6 \times 1000$ kg wt $= 6 \times 10^9 \times 9.81$ N

Water head $= 170$ m

Potential energy stored in this much water

$= Wh = 6 \times 10^9 \times 9.81 \times 170$ J $= 10^{12}$ J

Overall efficiency of the station $= 0.8 \times 0.9 = 0.72$

$\therefore$ energy available $= 0.72 \times 10^{13}$J $= 72 \times 10^{11}/36 \times 10^5$

$= \mathbf{2 \times 10^6}$ **kWh**

(*ii*) Energy supplied $= 12{,}000 \times 3 = 36{,}000$ kWh

Energy drawn from the reservoir after taking into consideration the overall efficiency of the station $= 36{,}000/0.72 = 5 \times 10^4$ kWh

$= 5 \times 10^4 \times 36 \times 10^5 = 18 \times 10^{10}$J

If m kg is the mass of water used in two hours, then, since water head is 170 m

$$mgh = 18 \times 10^{10}$$

or $m \times 9.81 \times 170 = 18 \times 10^{10}$ $\therefore$ $m = 1.08 \times 10^8$ kg

If h metre is the fall in water level, then

$h \times$ area $\times$ density $=$ mass of water

$\therefore$ $h \times (2.5 \times 10^6) \times 1000 = 1.08 \times 10^8$ $\therefore$ $h = 0.0432$ m $=$ **4.32 cm**

(*iii*) Mass of water stored per second $= 1.2 \times 1000 = 1200$ kg

Wt. of water stored per second $= 1200 \times 9.81$ N

Power stored $= 1200 \times 9.81 \times 170$ J/s $=$ **2,000 kW**

Power actually available $= 2{,}000 \times 0.72 = 1440$ kW

Energy delivered / day $= 1440 \times 24 =$ **34,560 kWh**

Example 3.14. *The reservoir for a hydro-electric station is 230 m above the turbine house. The annual replenishment of the reservoir is 45×10^{10} kg. What is the energy available at the generating station bus-bars if the loss of head in the hydraulic system is 30 m and the overall efficiency of the station is 85%. Also, calculate the diameter of the steel pipes needed if a maximum demand of 45 MW is to be supplied using two pipes.* **(Power System, Allahabad Univ. 1991)**

Solution. Actual head available $= 230 - 30 = 200$ m

Energy available at the turbine house $= mgh$

$$= 45 \times 10^{10} \times 9.81 \times 200 = 88.29 \times 10^{13} \text{ J}$$

$$= \frac{88.29 \times 10^{13}}{36 \times 10^5} = \mathbf{24.52 \times 10^7 \text{ kWh}}$$

Overall $\eta = 0.85$

$\therefore$ Energy output $= 24.52 \times 10^7 \times 0.85 = \mathbf{20.84 \times 10^7 \text{ kWh}}$

The kinetic energy of water is just equal to its loss of potential energy.

$$\frac{1}{2}mv^2 = mgh \quad \therefore \quad v = \sqrt{2gh} = \sqrt{2 \times 9.81 \times 200} = 62.65 \text{ m/s}$$

Power available from a mass of m kg when it flows with a velocity of v m/s is

$$P = \frac{1}{2}mv^2 = \frac{1}{2} \times m \times 62.65^2 \text{ J/s} \quad \text{or} \quad \text{W}$$

Equating this to the maximum demand on the station, we get

$$\frac{1}{2}m\ 62.65^2 = 45 \times 10^6 \quad \therefore \quad m = 22{,}930 \text{ kg/s}$$

If A is the total area of the pipes in m^2, then the flow of water is Av m^3/s. Mass of water flowing/second $= Av \times 10^3$ kg $\quad (\because$ 1 m^3 of water $= 1000$ kg)

$$\therefore \quad A \times v \times 10^3 = 22{,}930 \text{ or } A = \frac{22{,}930}{62.65 \times 10^3} = 0.366 \text{ m}^2$$

If 'd' is the diameter of each pipe, then

$$\pi d^2/4 = 0.183 \quad \therefore \quad d = \mathbf{0.4826 \text{ m}}$$

Example 3.15. *A large hydel power station has a head of 324 m and an average flow of 1370 cubic metres / sec. The reservoir is a lake covering an area of 6400 sq. km. Assuming an efficiency of 90% for the turbine and 95% for the generator, calculate*

(*i*) *the available electric power ;*

(*ii*) *the number of days this power could be supplied for a drop in water level by 1 metre.*

(AMIE Sec. B Power System I (E–6) Winter 1991)

Solution. (*i*) Available power = 9.81 η QH kW = $(0.9 \times 0.95) \times 1370 \times 324$ = 379, 524 kW = **379.52 MW**.

(*ii*) If A is the lake area in m^2 and h metre is the fall in water level, the volume of water used is $= A \times h = m^3$. The time required to discharge this water is Ah / Q second.

Now, $A = 6400 \times 10^6$ m^2; h = 1 m; Q = 1370 m^3/s.

$\therefore$ $t = 6400 \times 10^6 \times 1/1370 = 4.67 \times 10^6$ second = **540686 days**

Example 3.16. *The reservoir area of a hydro-electric generating plant is spread over an area of 4 sq km with a storage capacity of 8 million cubic-metres. The net head of water available to the turbine is 70 metres. Assuming an efficiency of 0.87 and 0.93 for water turbine and generator respectively, calculate the electrical energy generated by the plant.*

Estimate the difference in water level if a load of 30 MW is continuously supplied by the generator for 6 hours. **(Power System I–AMIE Sec. B, Summer 1990)**

Solution. Since 1 cubic metre of water weighs 1000 kg., the reservoir capacity = 8×10^6 m^3 = $8 \times 10^6 \times 1000$ kg. = 8×10^9 kg.

Wt. of water, $W = 8 \times 10^9$ kg. Wt. = $8 \times 10^9 \times 9.81 = 78.48 \times 10^9$ N. Net water head = 70 m.

Potential energy stored in this much water = Wh = $78.48 \times 10^9 \times 70 = 549.36 \times 10^{10}$ J

Overall efficiency of the generating plant = $0.87 \times 0.93 = 0.809$

Energy available = $0.809 \times 549.36 \times 10^{10}$ J = 444.4×10^{10} J

= $444.4 \times 10^{10}/36 \times 10^5$ = **12.34×10^5 kWh**

Energy supplied in 6 hours = 30 MW $\times$ 6 h = 180 MWh

= 180,000 kWh

Energy drawn from the reservoir after taking into consideration, the overall efficiency of the station = 180,000/0.809 = 224,500 kWh = $224,500 \times 36 \times 10^5$

= 80.8×10^{10} J

If m kg. is the mass of water used in 6 hours, then since water head is 70 m,

$mgh = 80.8 \times 10^{10}$ or $m \times 9.81 \times 70 = 80.8 \times 10^{10}$ $\quad \therefore \quad m = 1.176 \times 10^9$ kg.

If h is the fall in water level, then $h \times$ area $\times$ density = mass of water

$\therefore \quad h \times (4 \times 10^6) \times 1000 = 1.176 \times 10^9 \quad \therefore \quad h = 0.294$ m = **29.4 cm.**

Example 3.17. *A proposed hydro-electric station has an available head of 30 m, a catchment area of 50×10^6 sq.m, the rainfall for which is 120 cm per annum. If 70% of the total rainfall can be collected, calculate the power that could be generated. Assume the following efficiencies: Penstock 95%, Turbine 80% and Generator 85%.* **(Elect. Engg. AMIETE Sec. A Part II Dec. 1991)**

Solution. Volume of water available = $0.7(50 \times 10^6 \times 1.2) = 4.2 \times 10^7$ m^3

Mass of water available = $4.2 \times 10^7 \times 1000 = 4.2 \times 10^{10}$ kg

This quantity of water is available for a period of one year. Hence, quantity available per second = $4.2 \times 10^{10}/365 \times 24 \times 3600 = 1.33 \times 10^3$.

Available head = 30 m

Potential energy available = $mgh = 1.33 \times 10^3 \times 9.8 \times 30 = 391 \times 10^3$ J

Since this energy is available per second, hence power available is

= 391×10^3 J/s = 391×10^3 W = 391 kW

Overall efficiency = $0.95 \times 0.80 \times 0.85 = 0.646$

The power that could be generated = 391×0.646 = **253 kW.**

Example 3.18. *In a hydro-electric generating station, the mean head (i.e. the difference of height between the mean level of the water in the lake and the generating station) is 400 metres. If the overall efficiency of the generating station is 70%, how many litres of water are required to generate 1 kWh of electrical energy ? Take one litre of water to have a mass of 1 kg.*

(F.Y. Engg. Pune Univ. Nov. 1989)

Solution. Output energy = 1 kWh = 36×10^5 J

Input energy = $36 \times 10^5/0.7 = 5.14 \times 10^6$ J

If m kg. water is required, then

$mgh = 5.14 \times 10^6$ or $m \times 9.81 \times 400 = 5.14 \times 10^6$, $\therefore$ m = **1310 kg.**

Example 3.19. *A 3-tonne electric-motor-operated vehicle is being driven at a speed of 24 km/hr upon an incline of 1 in 20. The tractive resistance may be taken as 20 kg per tonne. Assuming a motor efficiency of 85% and the mechanical efficiency between the motor and road wheels of 80%, calculate*

(*a*) *the output of the motor*

(*b*) *the current taken by motor if it gets power from a 220-V source.*

Calculate also the cost of energy for a run of 48 km, taking energy charge as 40 paise/kWh.

Solution. Different forces acting on the vehicle are shown in Fig. 3.2.

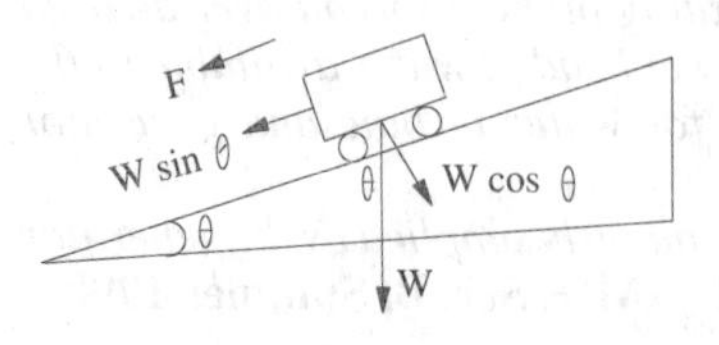

Fig. 3.2

Wt. of the vehicle = 3×10^3 = 3000 kg-wt

Component of the weight of the vehicle acting downwards along the slope = $3000 \times 1/20$ = 150 kg-wt

Tractive resistance = 3×20 = 60 kg-wt

Total downward force = 150 + 60 = 210 kg-wt = 210×9.81 = 2,060 N

Distance travelled/second = 24,000/3600 = 20/3 m/s

Output at road wheels = $2,060 \times 20/3$ watt

Mechanical efficiency = 80% or 0.8

(*a*) Motor output $= \dfrac{2,060 \times 20}{3 \times 0.8}$ = **17,167 W**

(*b*) Motor input = 17,167/0.85 = 20,200 W

Current drawn = 20,200/220 = **91.7 A**

Motor power input = 20,200 W = 20.2 kW

Time for 48 km run = 2 hr.

$\therefore$ Motor energy input = 20.2×2 = 40.4 kW

Cost = Rs. 40.4×0.4 = **Rs. 16 paise 16**

Example 3.20. *Estimate the rating of an induction furnace to melt two tonnes of zinc in one hour if it operates at an efficiency of 70%. Specific heat of zinc is 0.1. Latent heat of fusion of zinc is 26.67 kcal per kg. Melting point is 455°C. Assume the initial temperature to be 25°C.*

(Electric Drives and Utilization Punjab Univ. Jan. 1991)

Solution. Heat required to bring 2000 kg of zinc from 25°C to the melting temperature of 455°C = $2000 \times 0.1 \times (455 - 25)$ = 86,000 kcal.

Heat of fusion or melting = mL = 2000×26.67 = 53,340 kcal.

Total heat reqd. = 86,000 + 53,340 = 139,340 kcal

Furnace input = 139,340/0.7 = 199,057 kcal

Now, 860 kcal = 1 kWh $\therefore$ furnace input = 199.057/860 = 231.5 kWh.

Power rating of furnace = energy input/time = 231.5 kWh/1 h = **231.5 kW.**

Example 3.21. *A pump driven by an electric motor lifts 1.5 m³ of water per minute to a height of 40 m. The pump has an efficiency of 90% and motor has an efficiency of 85%. Determine : (a) the power input to the motor. (b) The current taken from 480 V supply. (c) The electric energy consumed when motor runs at this load for 4 hours. Assume mass of 1 m³ of water to be 1000 kg.*

(Elect. Engg. Pune Univ. 1986)

Solution. (*a*) Weight of the water lifted = 1.5 m³ = 1.5×1000 = 1500 kg.Wt = 1500×9.8 = 14700 N.

Height = 40 m; time taken 1 min. = 60 s

$\therefore$ Motor output power = $14700 \times 40/60$ = 9800 W

Combined pump and motor efficiency = 0.9×0.85

$\therefore$ Motor power input = $9800/0.9 \times 0.85$ = 12810 W = **12.81 kW.**

(*b*) Current drawn by the motor = 12810/480 = **26.7 A**

Electrical energy consumed by the motor = 12.81 kW × 4 h = **51.2 kWh.**

Example 3.22. *An electric lift is required to raise a load of 5 tonne through a height of 30 m. One quarter of electrical energy supplied to the lift is lost in the motor and gearing. Calculate the energy in kWhr supplied. If the time required to raise the load is 27 minutes, find the kW rating of the motor and the current taken by the motor, the supply voltage being 230 V d.c. Assume the efficiency of the motor at 90%.* **(Elect. Engg. A.M.Ae. S.I. June 1991)**

Solution. Work done by the lift = $Wh = mgh = (5 \times 1000) \times 9.8 \times 30 = 1.47 \times 10^6$ J

Since 25% of the electric current input is wasted, the energy supplied to the lift is 75% of the input.

$\therefore$ input energy to the lift = $1.47 \times 10^6/0.75 = 1.96 \times 10^6$ J

Now, 1 kWh = 36×10^5 J

$\therefore$ energy input to the lift = $1.96 \times 10^6/36 \times 10^5 = 0.544$ kWh

Motor energy output = 1.96×10^6 J; $\eta = 0.9$

Motor energy input = $1.96 \times 10^6/0.9 = 2.18 \times 10^6$ J; time taken = $27 \times 60 = 1620$ second

Power rating of the electric motor = work done/time taken

$= 2.18 \times 10^6/1620 = 1.345 \times 10^3$ J/s = 1345 W

Current taken by the motor = 1345/230 = **5.85 A**

Example 3.23. *An electrical lift makes 12 double journeys per hour. A load of 5 tonnes is raised by it through a height 50 m and it returns empty. The lift takes 65 seconds to go up and 48 seconds to return. The weight of the cage is 1/2 tonne and that of the counterweight is 2.5 tonne. The efficiency of the hoist is 80 per cent and that of the motor is 85%. Calculate the hourly consumption in kWh.* **(Elect. Engg. Pune Univ. 1988)**

Solution. The lift is shown in Fig. 3.3.

Weight raised during upward journey

$= 5 + 1/2 - 2.5 = 3$ tonne = 3000 kg-wt

Distance travelled = 50 m

Work done during upward journey

$= 3000 \times 50 = 15 \times 10^4$ m-kg

Weight raised during downward journey

$= 2.5 - 0.5 = 2$ tonne = 2000 kg

Similarly, work done during downward journey

$= 2000 \times 50 = 10 \times 10^4$ m-kg.

Total work done per double journey

$= 15 \times 10^4 + 10 \times 10^4 = 25 \times 10^4$ m-kg

Fig. 3.3

Now, 1, m-kg = 9.8 joules

$\therefore$ Work done per double journey $= 9.8 \times 25 \times 10^4$ J $= 245 \times 10^4$ J

No. of double journeys made per hour = 12

$\therefore$ work done per hour $= 12 \times 245 \times 10^4 = 294 \times 10^5$ J

Energy drawn from supply $= 294 \times 10^5/0.8 \times 0.85 = 432.3 \times 10^5$ J

Now, 1 kWh $= 36 \times 10^5$ J

$\therefore$ Energy consumption per hour $= 432.3 \times 10^5/36 \times 10^5 =$ **12 kWh**

Example 3.24. *An electric hoist makes 10 double journeys per hour. In each journey, a load of 6 tonnes is raised to a height of 60 metres in 90 seconds. The hoist cage weighs 1/2 tonne and has a balance load of 3 tonnes. The efficiency of the hoist is 80% and of the driving motor 88%. Calculate (a) electric energy absorbed per double journey (b) hourly energy consumption in kWh (c) hp (British) rating of the motor required (d) cost of electric energy if hoist works for 4 hours/day for 30 days. Cost per kWh is 50 paise.* **(Elect. Power – 1, Bangalore Univ. 1983)**

Solution. Wt. of cage when fully loaded = $6\frac{1}{2}$ tonne-wt.

Force exerted on upward journey $= 6\frac{1}{2} - 3 = 3\frac{1}{2}$ tonnes-wt.

$= 3\frac{1}{2} \times 1000 = 3{,}500$ kg-wt.

Force exerted on downward journey $= 3 - \frac{1}{2} = 2\frac{1}{2}$ tonnes-wt. = 2500 kg-wt.

Distance moved = 60 m

Work done during upward journey $= 3{,}500 \times 60$ m-kg

Work done during downward journey $= 2{,}500 \times 60$ m-kg

Work done during each double journey $= (3{,}500 + 2{,}500) \times 60 = 36 \times 10^4$ m-kg
$= 36 \times 10^4 \times 9.81 = 354 \times 10^4$ J

Overall η $= 0.80 \times 0.88$

$\therefore$ Energy input per double journey $= 354 \times 10^4/0.8 \times 0.88 = 505 \times 10^4$ J

(*a*) Electric energy absorbed per double journey
$= 505 \times 10^4/36 \times 10^5 =$ **1.402 kWh**

(*b*) Hourly consumption $= 1.402 \times 10 =$ **14.02 kWh**

(*c*) Before calculating the rating of the motor, maximum rate of working should be found. It is seen that maximum rate of working is required in the upward journey.

Work done $= 3{,}500 \times 60 \times 9.81 = 206 \times 10^4$ J

Time taken $= 90$ second

$$\therefore \quad \text{B.H.P. of motor} = \frac{206 \times 10^4}{90 \times 0.8 \times 746} = \mathbf{38.3\ (British\ h.p.)}$$

$$(d)\quad \text{Cost} = 14.02 \times (30 \times 4) \times 50/100 = \mathbf{Rs.\ 841.2}$$

Example 3.25. *A current of 80 A flows for 1 hr. in a resistance across which there is a voltage of 2 V. Determine the velocity with which a weight of 1 tonne must move in order that its kinetic energy shall be equal to the energy dissipated in the resistance.*

(Elect. Engg. A.M.Ae. S.I. June 1989.)

Solution. Energy dissipated in the resistance $= V\,It = 2 \times 80 \times 3600 = 576{,}000$ J

A weight of one tonne represents a mass of one tonne *i.e.* 1000 kg. Its kinetic energy is $= (1/2) \times 1000 \times v^2 = 500\,v^2$

$$\therefore \quad 500\,v^2 = 576{,}000 \quad \therefore \; v = 1152 \text{ m/s.}$$

Tutorial Problems No. 3.1

1. A heater is required to give 900 cal/min on a 100.V d.c. circuit. What length of wire is required for this heater if its resistance is 3 Ω per metre ? **[53 metres]**

2. A coil of resistance 100 Ω is immersed in a vessel containing 500 gram of water at 16°C and is connected to a 220-V electric supply. Calculate the time required to boil away all the water (1 kcal = 4200 joules, latent heat of steam = 536 kcal/kg). **[44 min 50 second]**

3. A resistor, immersed in oil, has 62.5 Ω resistance and is connected to a 500-V d.c. supply. Calculate
(*a*) the current taken
(*b*) the power in watts which expresses the rate of transfer of energy to the oil
(*c*) the kilowatt-hours of energy taken into the oil in 48 minutes. **[8A; 4000 W; 3.2 kWh]**

4. An electric kettle is marked 500-W, 230 V and is found to take 15 minutes to raise 1 kg of water from 15°C to boiling point. Calculate the percentage of energy which is employed in heating the water. **[79 per cent]**

5. An aluminium kettle weighing 2 kg holds 2 litres of water and its heater element consumes a power of 2 kW. If 40 per cent of the heat supplied is wasted, find the time taken to bring the kettle of water to boiling point from an initial temperature of 20°C. (Specific heat of aluminium = 0.2 and Joule's equivalent = 4200 J/kcal.) **[11.2 min]**

6. A small electrically heated drying oven has two independent heating elements each of 1000 Ω in its heating unit. Switching is provided so that the oven temperature can be altered by rearranging the resistor connections. How many different heating positions can be obtained and what is the electrical power drawn in each arrangement from a 200 V battery of negligible resistance ? **[Three, 40, 20 and 80 W]**

7. Ten electric heaters, each taking 200 W were used to dry out on site an electric machine which had been exposed to a water spray. They were used for 60 hours on a 240 V supply at a cost of twenty paise/kWh. Calculate the values of following quantities involved :
(*a*) current (*b*) power in kW (*c*) energy in kWh (*d*) cost of energy.
[(*a*) 8.33A (*b*) 2 kW (*c*) 120 kWh (*d*) Rs.24]

8. An electric furnace smelts 1000 kg of tin per hour. If the furnace takes 50 kW of power from the electric supply, calculate its efficiency, given : the smelting tempt. of tin = 235°C ; latent heat of fusion = 13.31 kcal/kg; initial temperatrure = 15°C ; specific heat = 0.056. Take J = 4200 J/kcal.
[59.8%] (*Electrical Engg. -I, Delhi Univ. 1980*)

9. Find the useful rating of a tin-smelting furnace in order to smelt 50 kg of tin per hour. Given: Smelting temperature of tin = 235°C, Specific heat of tin = 0.055 kcal/kg-K. Latent heat of liquefaction = 13.31 kcal per kg. Take initial temperature of metal as 15°C. **[1.5 kW]** (*F.Y. Engg. Pune Univ. 1990*)

OBJECTIVE TESTS–3

1. In the SI system of units, the unit of force is
(*a*) kg-wt
(*b*) newton
(*c*) joule
(*d*) N-m

2. The basic unit of electric charge is
(*a*) ampere-hour
(*b*) watt-hour
(*c*) coulomb
(*d*) farad

3. The SI unit of energy is
(*a*) joule
(*b*) kWh
(*c*) kcal
(*d*) m-kg

4. Two heating elements, each of 230-V, 3.5 kW rating are first joined in parallel and then in series to heat same amount of water through the same range of temperature. The ratio of the time taken in the two cases would be
(*a*) 1 : 2
(*b*) 2 : 1
(*c*) 1 : 4
(*d*) 4 : 1

5. If a 220 V heater is used on 110 V supply, heat produced by it will be—as much.
(*a*) one-half
(*b*) twice
(*c*) one-fourth
(*d*) four times

6. For a given line voltage, four heating coils will produce maximum heat when connected
(*a*) all in parallel
(*b*) all in series
(*c*) with two parallel pairs in series
(*d*) one pair in parallel with the other two in series

7. The electric energy required to raise the temperature of a given amount of water is 1000 kWh. If heat losses are 25%, the total heating energy required is —— kWh.
(*a*) 1500
(*b*) 1250
(*c*) 1333
(*d*) 1000

8. One kWh of energy equals nearly
(*a*) 1000 W
(*b*) 860 kcal
(*c*) 4186 J
(*d*) 735.5 W

9. One kWh of electric energy equals
(*a*) 3600 J
(*b*) 860 kcal
(*c*) 3600 W
(*d*) 4186 J

10. A force of 10,000 N accelerates a body to a velocity 0.1 km/s. The power developed is ———— kW
(*a*) 1,00,000
(*b*) 36,000
(*c*) 3600
(*d*) 1000

11. A 100 W light bulb burns on an average of 10 hours a day for one week. The weekly consumption of energy will be ———— unit/s.
(*a*) 7
(*b*) 70
(*c*) 0.7
(*d*) 0.07

(Principles of Elect. Engg. Delhi Univ. July, 1984)

12. Two heaters, rated at 1000 W, 250 volts each, are connected in series across a 250 volts 50 Hz A.C. mains. The total power drawn from the supply would be ———— watt.
(*a*) 1000
(*b*) 500
(*c*) 250
(*d*) 2000

(Principles of Elect. Engg. Delhi Univ. July, 1984)

Answers

1. *b* **2.** *c* **3.** *a* **4.** *c* **5.** *c* **6.** *a* **7.** *c* **8.** *b* **9.** *b* **10.** *d* **11.** *a* **12.** *b*

4 ELECTROSTATICS

4.1. Static Electricity

In the preceding chapters, we concerned ourselves exclusively with electric current *i.e.* electricity in motion. Now, we will discuss the behaviour of static electricity and the laws governing it. In fact, electrostatics is that branch of science which deals with the phenomena associated with electricity at rest.

It has been already discussed that generally an atom is electrically neutral *i.e.* in a normal atom the aggregate of positive charge of protons is exactly equal to the aggregate of negative charge of the electrons.

If, somehow, some electrons are removed from the atoms of a body, then it is left with a preponderance of positive charge. It is then said to be positively-charged. If, on the other hand, some electrons are added to it, negative charge out-balances the positive charge and the body is said to be negatively charged.

In brief, we can say that positive electrification of a body results from a deficiency of the electrons whereas negative electrification results from an excess of electrons.

The total deficiency or excess of electrons in a body is known as its charge.

4.2. Absolute and Relative Permittivity of a Medium

While discussing electrostatic phenomenon, a certain property of the medium called its *permittivity* plays an important role. Every medium is supposed to possess two permittivities :

(*i*) absolute permittivity (ε) and (*ii*) relative permittivity (ε_r).

For measuring relative permittivity, vacuum or free space is chosen as the reference medium. It has an absolute permittivity of 8.854×10^{-12} farad/metre. Its symbol is ε_0. Obviously, the relative permittivity of vacuum with reference to itself is unity. Hence, for *free space* or *vacuum*

Absolute permittivity $\varepsilon_0 = 8.854 \times 10^{-12}$ F/m

Relative permittivity, $\varepsilon_r = 1$

Being a ratio of two similar quantities, ε_r has no units.

Now, take any other medium. If its relative permittivity, as compared to vacuum is ε_r, then its absolute permittivity is

$$\varepsilon = \varepsilon_0 \, \varepsilon_r \text{ F/m}$$

If, for example, relative permittivity of mica is 5, then, its absolute permittivity is

$$\varepsilon = \varepsilon_0 \, \varepsilon_r = 8.854 \times 10^{-12} \times 5 = 44.27 \times 10^{-12} \text{ F/m}$$

4.3. Laws of Electrostatics

First Law. Like charges of electricity repel each other, whereas unlike charges attract each other.

Second Law. According to this law, the force exerted between two *point* charges (*i*) is directly proportional to the product of their strengths (*ii*) is inversely proportional to the square of the distance between them.

This law is known as Coulomb's Law and can be expressed mathematically as :

$$F \propto \frac{Q_1 Q_2}{d^2} \quad \text{or} \quad F = k \, \frac{Q_1 Q_2}{d^2}$$

In vector form, the Coulomb's law can be written as

$$\vec{F} = k\frac{Q_1Q_2}{d^2}\hat{d} = k\frac{Q_1Q_2}{d^3}\vec{d}$$

where $\hat{d}$ is the unit vector *i.e.* a vector of unit length in the direction of distance d, *i.e.*, $\hat{d} = \vec{d}/d$ where $\vec{d}$ is the vector notation for d, which is a scalar notation.)

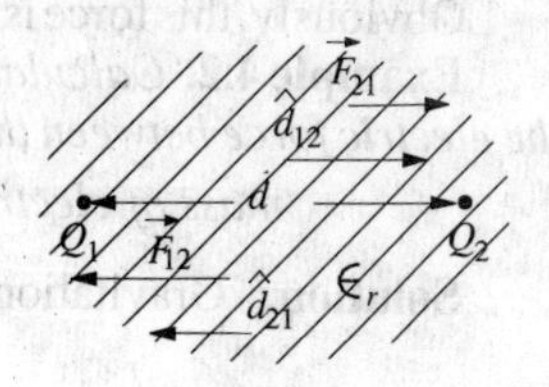

Fig. 4.1

Therefore, explicit forms of this law are :

$$\vec{F}_{21} = k\frac{Q_1Q_2}{d_{12}^2}\hat{d}_{12} = k\frac{Q_1Q_2}{d_{12}^3}\vec{d}_{12}$$

where $\vec{F}_{21}$ is the force on Q_2 due to Q_1 and $\hat{d}_{12}$ is the unit vector in direction from Q_1 to Q_2

and

$$\vec{F}_{12} = k\frac{Q_1Q_2}{d_{21}^2}\hat{d}_{21} = k\frac{Q_1Q_2}{d_{21}^3}\vec{d}_{21}$$

where F_{12} is the force on Q_1 due to Q_2 and $\hat{d}_{21}$ is the unit vector in the direction from Q_2 to Q_1

where k is the constant of proportionality, whose value depends on the system of units employed. In S.I. system, as well as M.K.S.A. system, $k = 1/4\pi\,\varepsilon$. Hence, the above equation becomes.

$$F = \frac{Q_1Q_2}{4\,\pi\,\varepsilon\,d^2} = \frac{Q_1Q_2}{4\,\pi\,\varepsilon_0\,\varepsilon_r\,d^2}$$

If Q_1 and Q_2 are in coulomb, d in metre and ε in farad/metre, then F is in newtons

Now $$\frac{1}{4\,\pi\,\varepsilon_0} = \frac{1}{4\pi\times 8.854\times 10^{-12}} = 8.9878\times 10^9 = 9\times 10^9 \text{(approx.)}$$

Hence, Coloumb's Law can be written as

$$F = 9\times 10^9\frac{Q_1Q_2}{\varepsilon_r\,d^2} \quad \text{–in a medium}$$

$$= 9\times 10^9\frac{Q_1Q_2}{d^2} \quad \text{–in air or vacuum} \ldots(i)$$

If in Eq. (*i*) above $Q_1 = Q_2 = Q$ (say), $d = 1$ metre ; $F = 9\times 10^9$ N
then $Q_2 = 1$ or $Q = \pm 1$ coulomb

Hence, one coulomb of charge may be defined as *that charge (or quantity of electricity) which when placed in air (strictly vacuum) from an equal and similar charge repels it with a force of 9×10^9 N.*

Although coulomb is found to be a unit of convenient size in dealing with electric current, yet, from the standpoint of electrostatics, it is an enormous unit. Hence, its submultiples like micro-coulomb (μ C) and micro-microcoulomb (μ μC) are generally used.

$$1\ \mu\text{C} = 10^{-6}\ \text{C} ;\ 1\ \mu\,\mu\ \text{C} = 10^{-12}\ \text{C}$$

It may be noted here that relative permittivity of air is one, of water 81, of paper between 2 and 3, of glass between 5 and 10 and of mica between 2.5 and 6.

Example 4.1. *Calculate the electrostatic force of repulsion between two α-particles when at a distance of 10^{-13} m from each other. Charge of an α-particle is 3.2×10^{-12} C. If mass of each particle is 6.68×10^{-27} kg, compare this force with the gravitational force between them. Take the gravitational constant as equal to 6.67×10^{-11} N-m²/kg².*

Solution. Here $Q_1 = Q_2 = 3.2\times 10^{-19}$ C, $d = 10^{-13}$ m

$$F = 9\times 10^9\times\frac{3.2\times 10^{-19}\times 3.2\times 10^{-19}}{(10^{-13})^2} = \mathbf{9.2\times 10^{-2}\ N}$$

The force of gravitational attraction between the two particles is given by

$$F = G\frac{m_1m_2}{d^2} = \frac{6.67\times 10^{-11}\times(6.68\times 10^{-27})^2}{(10^{-13})^2} = \mathbf{2.97\times 10^{-37}\ N}$$

Obviously, this force is negligible as compared to the electrostatic force between the two particles.

Example 4.2. *Calculate the distance of separation between two electrons (in vacuum) for which the electric force between them is equal to the gravitational force on one of them at the earth surface.*

$$\text{mass of electron} = 9.1 \times 10^{-31} \text{ kg}, \quad \text{charge of electron} = 1.6 \times 10^{-19} \text{ C}.$$

Solution. Gravitational force on one electron.

$$= mg \text{ newton} = 9.1 \times 10^{-31} \times 9.81 \text{ N}$$

Electrostatic force between the electrons

$$= 9 \times 10^9 \frac{Q^2}{d^2} = \frac{9 \times 10^9 \times (1.6 \times 10^{-19})^2}{d^2} \text{ N}$$

Equating the two forces, we have

$$\frac{9 \times 10^9 \times 2.56 \times 10^{-38}}{d^2} = 9.1 \times 10^{-31} \times 9.81 \quad \therefore \; d = \mathbf{5.08 \text{ m}}$$

Example 4.3. *(a) Three identical point charges, each Q coulombs, are placed at the vertices of an equilateral triangle 10 cm apart. Calculate the force on each charge.*

(b) Two charges Q coulomb each are placed at two opposite corners of a square. What additional charge "q" placed at each of the other two corners will reduce the resultant electric force on each of the charges Q to zero ?

Solution. (*a*) The equilateral triangle with its three charges is shown in Fig. 4.2 (*a*). Consider the charge Q placed at corner B. It is being repelled by charges placed at corners A and C along ABN and CBM respectively. These forces are equal to each other and each is

$$F = 9 \times 10^9 \frac{Q^2}{0.1^2} = 9 \times 10^{11} Q^2 \text{ newton}$$

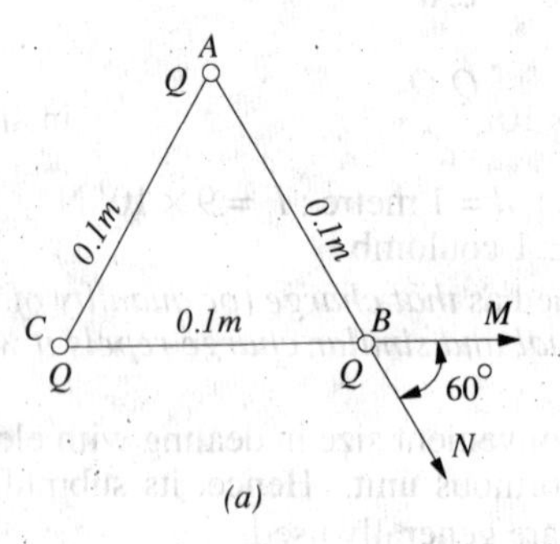

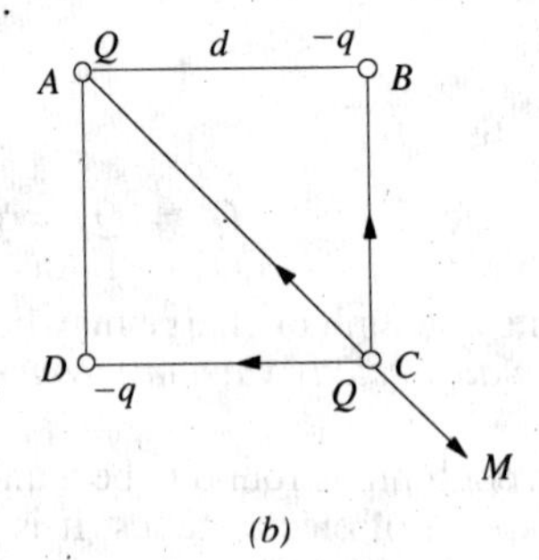

Fig. 4.2

Since the angle between these two equal forces is 60°, their resultant is

$$= 2 \times F \times \cos 60°/2 = \sqrt{3} F = \mathbf{9 \times 10^{11} \times \sqrt{3}\, Q^2 \text{ newton}}$$

The force experienced by other charges is also the same.

(*b*) The various charges are shown in Fig. 4.2 (*b*). The force experienced by the charge Q at point C due to the charge Q at point A acts along ACM and is

$$= 9 \times 10^9 \frac{Q^2}{(\sqrt{2}d)^2} = 4.5 \times 10^9 \, Q^2/d^2 \text{ newton} \quad \text{...(i)}$$

where d is the side of the square in metres.

If the charges q are negative, they will exert attractive forces on the charge Q at point C along CB and CD respectively. Each force is

$$= -9 \times 10^9 \frac{Qq}{d^2} \text{ newton}$$

Since these two forces are at right angles to each other, their resultant is

$$= -\sqrt{2}\times 9\times 10^9\,\frac{qQ}{d^2} \qquad \ldots(ii)$$

If net force on charge Q at point C is to be zero, then (*i*) must equal (*ii*),

$$4.5\times 10^9\,\frac{Q^2}{d^2} = -9\times 10^9\sqrt{2}\,\frac{Qq}{d^2}. \quad \therefore\ q = -\mathbf{Q/2\sqrt{2}\ coulomb}$$

Example 4.4. *Two small identical conducting spheres have charges of 2.0 × 10⁻⁹ C and −0.5 × 10⁻⁹ respectively. When they are placed 4 cm apart, what is the force between them ? If they are brought into contact and then separated by 4 cm, what is the force between them ?*

(Electromagnetic Theory, A.M.I.E. Sec B, 1990)

Solution. $F = 9\times 10^{-9}\,Q_1\,Q_2/d^2 = 9\times 10^{-9}\times 2\times 10^{-9}\times(-0.5\times 10^{-9})/0.04^2 = -56.25\times 10^{-7}$ N
When two identical spheres are brought into contact with each other and then separated, each gets half of the total charge. Hence,

$$Q_1 = Q_2 = [2\times 10^{-9} + (-0.5\times 10^{-9})]/2 = 0.75\times 10^{-9}\ \text{C}$$

When they are separated by 4 cm,

$$F = 9\times 10^{-9}\times(0.75\times 10^{-9})^2/0.04^2 = \mathbf{0.316\times 10^{-5}\ N}$$

4.4. Electric Field

It is found that in the medium around a charge a force acts on a positive or negative charge when placed in that medium. If the charge is sufficiently large, then it may create such a huge stress as to cause the electrical rupture of the medium, followed by the passage of an arc discharge.

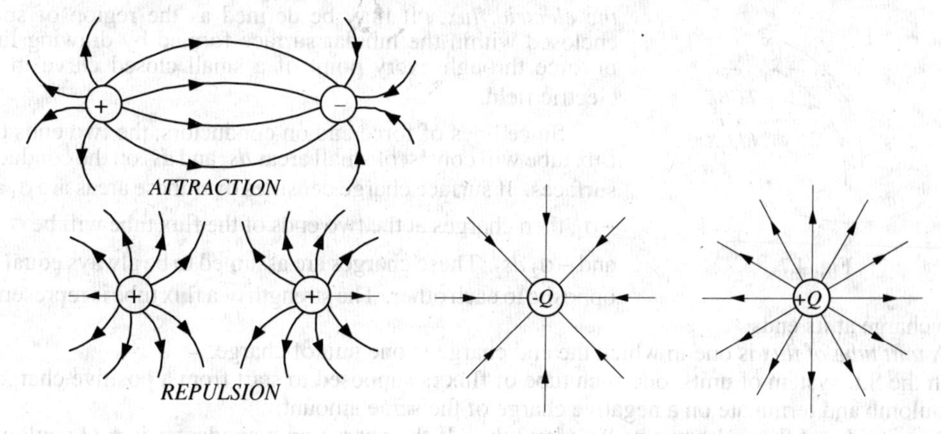

Fig. 4.3 **Fig. 4.4**

The region in which the stress exists or in which electric forces act, is called an electric fie' electrostatic field.

The stress is represented by imaginary lines of force. The direction of the lines of force at any point is the direction along which a unit positive charge placed at that point would move if free so. It was suggested by Faraday that the electric field should be imagined to be divided into *tubes of force* containing a fixed number of *lines* of force. He assumed these tubes to be elastic and having the property of contracting longitudinally and repelling laterally. With the help of these properties, it becomes easy to explain (*i*) why unlike charges attract each other and try to come ne...er to each other and (*ii*) why like charges repel each other (Fig. 4.3).

However, it is more common to use the term lines of force. These lines are supposed to emanate from a positive charge and end on a negative charge (Fig. 4.4). These lines always leave or enter a conducting surface normally.

4.5. Electrostatic Induction

It is found that when an uncharged body is brought near a charged body, it acquires some charge.

This phenomenon of an uncharged body getting charged merely by the nearness of a charged body is known as *induction*. In Fig. 4.5, a positively-charged body A is brought close to a perfectly-insulated uncharged body B. It is found that the end of B nearer to A gets negatively charged whereas further end becomes positively charged. The negative and positive charges of B are known as *induced charges*. The negative charge of B is called 'bound' charge because it must remain on B so long as positive charge of A remains there. However, the positive charge on the farther end of B is called free charge. In Fig. 4.6, the body B has been earthed by a wire. The positive charge flows to earth leaving negative charge behind. If next A is removed, then this negative charge will also go to earth, leaving B uncharged. It is found that :

(*i*) a positive charge induces a negative charge and *vice-versa*.
(*ii*) each of the induced charges is equal to the inducing charge.

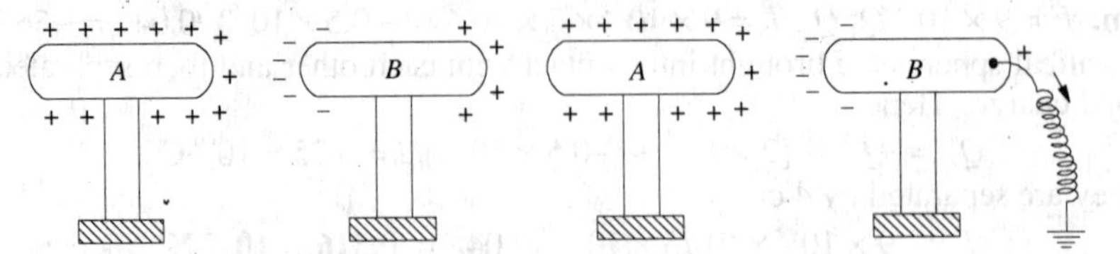

Fig. 4.5 **Fig. 4.6**

4.6. Electric Flux and Faraday Tubes

Consider a small closed curve in an electric field (Fig. 4.7). If we draw lines of force through each point of this closed curve, then we get a tube as shown in the figure. It is called the *tube of the electric flux*. It may be defined as the region of space enclosed within the tubular surface formed by drawing lines of force through every point of a small closed curve in the electric field.

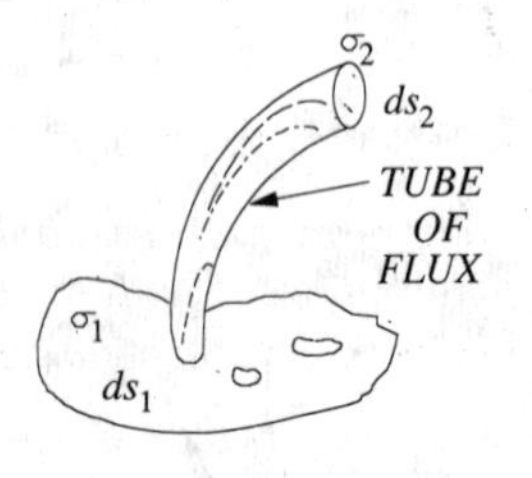

Fig. 4.7

Since lines of force end on conductors, the two ends of a flux tube will consist of small areas ds_1 and ds_2 on the conductor surfaces. If surface charge densities over these areas are σ_1 and $-\sigma_2$, then charges at the two ends of the flux tube will be $\sigma_1\, ds_1$ and $-\sigma_2\, ds_2$. These charges are assumed to be always equal but opposite to each other. The strength of a flux tube is represented by the charge at its ends.

A *unit tube of flux* is one in which the end charge is one unit of charge.

In the S.I. system of units, one such tube of flux is supposed to start from a positive charge of one coulomb and terminate on a negative charge of the same amount.

A *unit tube of flux* is known as Faraday tube. If the charge on a conductor is $\pm Q$ coulombs, then the *number of Faraday tubes starting* or *terminating on it is also* Q.

The number of Faraday tubes of flux passing through a surface in an electric field is called the *electric flux* (or dielectric flux) through that surface. Electric flux is represented by the symbol ψ. Since electric flux is numerically equal to the charge, it is measured in coulombs.

Hence, $\psi = Q$ coulombs

Note. It may also be noted that 'tubes of flux' passing per unit area through a medium are also supposed to measure the 'electric displacement' of that dielectric medium. In that case, they are referred to as *lines of displacement* and are equal to ε times the lines of force (Art. 4.8). It is important to differentiate between the 'tubes of flux' and 'lines of force' and to remember that if Q is the charge, then

$$\text{tubes of flux} = Q \text{ and lines of force} = Q/\varepsilon$$

4.7. Field Strength or Field Intensity or Electric Intensity (E)

Electric intensity at any point within an electric field may be defined in either of the following three ways :

(*a*) It is given by the force experienced by a unit positive charge placed at that point. Its direction is the direction along which the force acts.

Obviously, the unit of E is newton/coulomb (N/C).

For example, if a charge of Q coulombs placed at a particular point P within an electric field experiences a force of F newton, then electric field at that point is given by

$$E = F/Q \text{ N/C}$$

The value of E within the field due to a point charge can be found with help of Coulomb's laws.

Suppose it is required to find the electric field at a point A situated at a distance of d metres from a charge of Q coulombs. Imagine a positive charge of one coulomb placed at that point (Fig. 4.8). The force experienced by this charge is

$$F = \frac{Q \times 1}{4 \pi \varepsilon_0 \varepsilon_r d^2} \text{ N or } \vec{F} = \frac{Q \times 1}{4 \pi \varepsilon_0 \varepsilon_r d_{PA}^2} \hat{d}_{PA}$$

$$\therefore \quad E = \frac{Q}{4 \pi \varepsilon_0 \varepsilon_r d^2_{PA}} \text{ N/C}$$

$$= 9 \times 10^9 \frac{Q}{\varepsilon_r d^2_{PA}} \text{ N/C} \quad \text{in a medium}$$

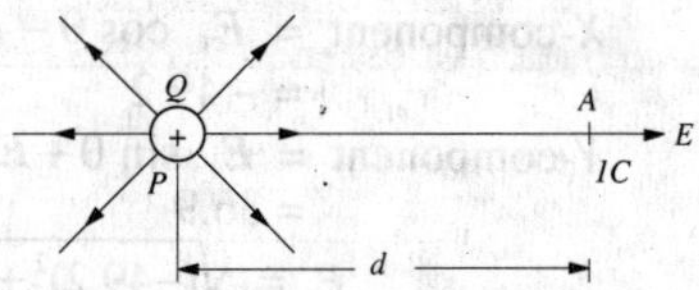

Fig. 4.8

or in vector notation,

$$\vec{E}(d) = 9 \times 10^9 \frac{Q}{\varepsilon_r d^2} \hat{d}$$

where $\vec{E}(d)$ denotes $\vec{E}$ as a function of d

$$= \frac{Q}{4\pi\varepsilon_0 d^2} \text{ N/C}$$

$$= 9 \times 10^9 \frac{Q}{d^2} \text{ N/C} \quad \text{in air}$$

(*b*) Electric intensity at a point may be defined as equal to the *lines of force* passing normally through a unit cross-section at that point. Suppose, there is a charge of Q coulombs. The number of lines of force produced by it is Q/ε. If these lines fall normally on an area of A m^2 surrounding the point, then electric intensity at that point is

$$E = \frac{Q/\varepsilon}{A} = \frac{Q}{\varepsilon A}$$

Now $\quad Q/A = D \quad$ – the flux density over the area

$$\therefore \quad E = \frac{D}{\varepsilon} = \frac{D}{\varepsilon_0 \varepsilon_r} \quad \text{... in a medium}$$

$$= \frac{D}{\varepsilon_0} \quad \text{... in air}$$

The unit of E is volt/metre.

(*c*) Electric intensity at any point in an electric field is *equal to the potential gradient at that point.*

In other words, E is equal to the rate of fall of potential in the direction of the lines of force.

$$\therefore \quad E = \frac{-dV}{dx}$$

Obviously, the unit of E is volt/metre.

It may be noted that E and D are vector quantities having magnitude and direction.

$\therefore$ In vector notation, $\vec{D} = \varepsilon_0 \vec{E}$

Example 4.5. *Point charges in air are located as follows :*
+ 5 × 10^{-8} C at (0, 0) metres, + 4 × 10^{-8} C at (3, 0) metres and −6 × 10^{-8} C at (0, 4) metres. Find electric field intensity at (3, 4) metres.

Solution. Electric intensity at point D (3, 4) due to positive charge at point A is

$$E_1 = 9 \times 10^9\, Q/d^2 = 9 \times 10^9 \times 5 \times 10^{-8}/5^2 = 18 \text{ V/m}$$

As shown in Fig. 4.9, it acts along AD.

Similarly, electric intensity at point D due to positive charge at point B is $E_2 = 9 \times 10^9 \times 4 \times 10^{-8}/4^2 = 22.5$ V/m. It acts along BD.

$E_3 = 9 \times 10^9 \times 6 \times 10^{-8}/3^2 = 60$ V/m. It acts along DC.

The resultant intensity may be found by resolving E_1, E_2 and E_3 into their X-and Y-components. Now, $\tan\theta = 4/3$; $\theta = 53°8'$.

X-component $= E_1 \cos\theta - E_2 = 18 \cos 53°8' - 60 = -49.2$

Y-component $= E_1 \sin\theta + E_2 = 18 \sin 53°8' + 22.5 = 36.9$

$$\therefore \quad E = \sqrt{(-49.2)^2 + 36.9^2} = 61.5 \text{ V/m}.$$

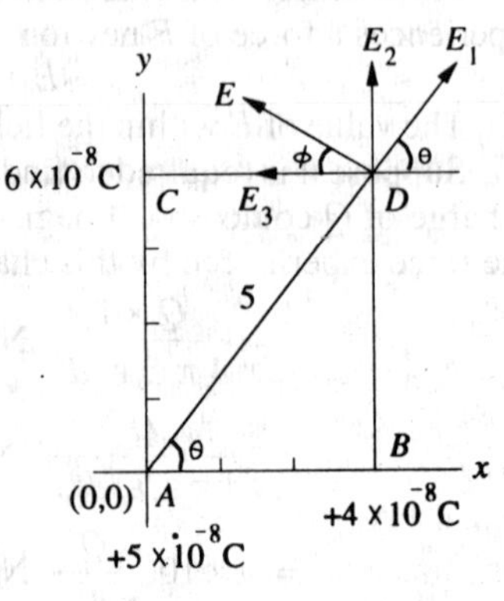

Fig. 4.9

It acts along DE such that $\tan\phi = 36.9/49.2 = 0.75$. Hence $\phi = \mathbf{36.9°}$.

Example 4.6. *An electron has a velocity of 1.5×10^7 m/s at right angles to the uniform electric field between two parallel deflecting plates of a cathode-ray tube. If the plates are 2.5 cm long and spaced 0.9 cm apart and p.d. between the plates is 75 V, calculate how far the electron is deflected sideways during its movement through the electric field. Assume electronic charge to be 1.6×10^{-19} coulomb and electronic mass to be 9.1×10^{-31} kg.*

Solution. The movement of the electron through the electric field is shown in Fig. 4.10. Electric intensity between the plates is $E = dV/dx = 75/0.009 = 8{,}333$ V/m.

Force on the electron is $F = QE = 8{,}333 \times 1.6 \times 10^{-19} = 1.33 \times 10^{-15}$ N.

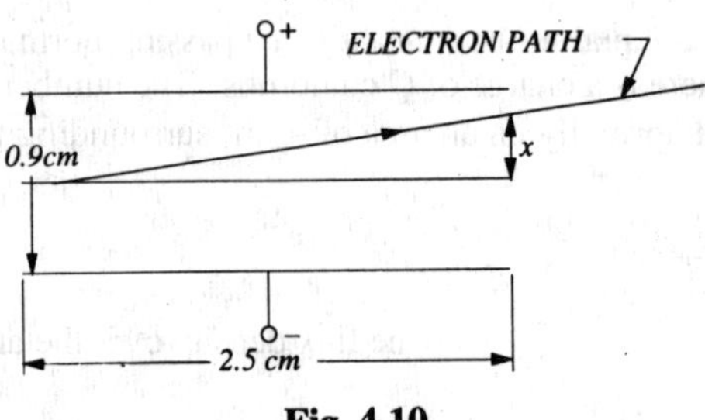

Fig. 4.10

Since the deflection x is small as compared to the length of the plates, time taken by the electron to travel through the electric field is

$$= 0.025/1.5 \times 10^7 = 1.667 \times 10^{-9} \text{ s}$$

Now, force = mass × acceleration

$\therefore$ Transverse acceleration is

$$= \frac{1.33 \times 10^{-15}}{9.1 \times 10^{-31}} = 1.44 \times 10^{15} \text{ m/s}^2$$

Final transverse velocity of the electron = acceleration × time

$$= 1.44 \times 10^{-15} \times 1.667 \times 10^{-9} = 2.4 \times 10^6 \text{ m/s}$$

$\therefore$ sideways or transverse movement of the electron is

$$x = (\text{average velocity}) \times \text{time}$$

$$= \tfrac{1}{2} \times 2.4 \times 10^6 \times 1{,}667 \times 10^{-9} = \mathbf{2\ mm\ (approx.)}^*$$

4.8. Electric Flux Density or Electric Displacement D

It is given by the normal flux per unit area.

If a flux of ψ coulombs passes normally through an area of A m², then flux density is

$$D = \frac{\psi}{A} \text{ C/m}^2$$

---

*The above result could be found by using the general formula

$$x = \frac{1}{2}\left(\frac{e}{m}\right)\left(\frac{V}{d}\right)\left(\frac{l}{v}\right)^2 \text{ metres}$$

where e/m = ratio of the charge and mass of the electron
V = p.d. between plates in volts ; d = separation of the plates in metres
l = length of the plates in metres ; v = axial velocity of the electron in m/s.

It is related to electric field intensity by the relation

$$D = \varepsilon_0 \varepsilon_r E \qquad \text{...in a medium}$$

$$= \varepsilon_0 E \qquad \text{...in free space}$$

In other words, the product of electric intensity E at any point within a dielectric medium and the absolute permittivity ε ($= \varepsilon_0 \varepsilon_r$) at the same point is called the *displacement* at that point.

Like electric intensity E, electric displacement D^* is also a vector quantity (see 4.7) whose direction at every point is the same as that of E but whose magnitude is $\varepsilon_0 \varepsilon_r$ times E. As E is represented by lines of force, similarly D may also be represented by lines called lines of electric *displacement*. The tangent to these lines at any point gives the direction of D at that point and the number of lines per unit area perpendicular to their direction is numerically equal to the electric displacement at that point. Hence, the number of lines of electric displacement per unit area (D) is $\varepsilon_0 \varepsilon_r$ times the number of lines of force per unit area at that point.

It should be noted that whereas the value of E depends on the permittivity of the surrounding medium, that of D is independent of it.

One useful property of D is that its surface integral over any closed surface equals the enclosed charge (Art. 4.9).

Let us find the value of D at a point distant r metres from a point charge of Q coulombs. Imagine a sphere of radius r metres surrounding the charge. Total flux = Q coulombs and it falls normally on a surface area of $4 \pi r^2$ metres. Hence, electric flux density.

$$D = \frac{\psi}{4 \pi r^2} = \frac{Q}{4 \pi r^2} \text{ coulomb/metre}^2 \text{ or } \vec{D} = \frac{Q}{4 \pi r^2} \hat{r} \text{ (in vector notation)}$$

4.9. Gauss** Law

Consider a point charge Q lying at the centre of a sphere of radius r which surrounds it completely [Fig. 4.11(*a*)]. The total number of *tubes of flux* originating from the charge is Q (but number of lines of force is Q/ε_0) and are normal to the surface of the sphere. The electric field E which equals $Q/4 \pi \varepsilon_0 r^2$ is also normal to the surface. As said earlier, total number of lines of force passing perpendicularly through the whole surface of the sphere is

$$= E \times \text{Area} = \frac{Q}{4\pi\varepsilon_0 r^2} \times 4\pi r^2 = \frac{Q}{\varepsilon_0}$$

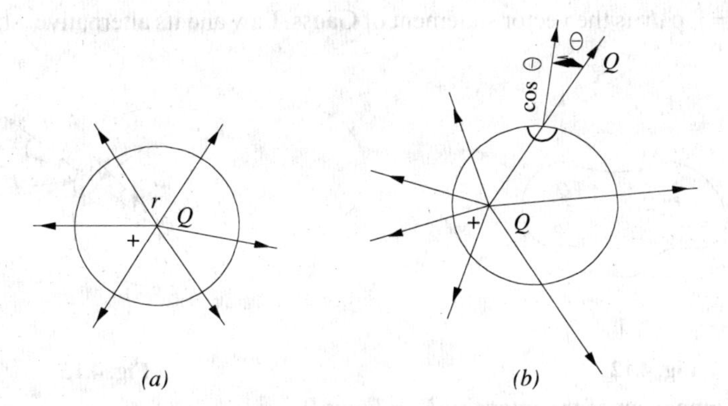

Fig. 4.11

---

*A more general definition of displacement D is that $D = \varepsilon_0 \varepsilon_r E + P$ where P is the polarisation of the dielectric and is equal to the dipole moment per unit volume.

**After the German mathematician and astronomer Karel Freidrich Gauss (1777-1855).

Now, suppose we draw another sphere surrounding the charge [Fig. 4.11 (*b*)] but whose centre does not lie at the charge but elsewhere. In this case also, the number of tubes of flux emanating from the charge is Q and lines of force is Q/ε_0 though they are not normal to the surface. These can, however, be split up into cos θ components and sin θ components. If we add up sin θ components all over the surface, they will be equal to zero. But if we add up cos θ components over the whole surface of the sphere, the normal flux will again come out to be Q (or lines of force will come out to be Q/ε_0). Hence, it shows that irrespective of where the charge Q is placed within a closed surface completely surrounding it, the total normal flux is Q and the total number of lines of force passing out normally is Q/ε_0.

In fact, as shown in Fig. 4.12, if there are placed charges of value $Q_1, Q_2, -Q_3$ inside a closed surface, the total *i.e.* net charge enclosed by the surface is $(Q_1 + Q_2 - Q_3)$ and the total flux is $(Q_1 + Q_2 - Q_3)/\varepsilon_0$ through the closed surface.

This is the meaning of Gauss's law which may be stated thus : the surface integral of the normal component of the electric intensity E over a closed surface is equal to $1/\varepsilon_0$ times the total charge inside it.

Mathematically, $\oint E_n ds = Q/\varepsilon_0$ (where the circle on the integral sign indicates that the surface of integration is a closed surface).

or $$\oint E_n \varepsilon_0 \, ds = Q, \; i.e. \oint D_n \, ds = Q \qquad [\because D_n = \varepsilon_0 E_n]$$

or $$\oint \varepsilon_0 E \cos\theta \, ds = Q, \; i.e. \oint D \cos\theta \, ds = Q$$

or $$\oint \varepsilon_0 E_{ds} \cos\theta = Q, i.e., \oint D \, ds \cos\theta = Q$$

when E & D are not normal to the surface but make an angle θ with the normal (perpendicular) to the surface as shown in Fig. 4.13.

Proof. In Fig. 4.13, let a surface S completely surround a quantity of electricity or charge Q. Consider a small surface area ds subtending a small solid angle $d\omega$ at point charge Q. The field intensity at ds is $E = \dfrac{Q}{4\pi\varepsilon_0 d^2}$ where d is the distance between Q and ds.

In vector rotation, $\oint \varepsilon_0 \vec{E} \cdot \vec{ds} = \text{Q } i.e. \oint \vec{D}\,\vec{ds} = Q = \int_v \rho \, dv$ (where ρ is the volume density of charge in the volume enclosed by closed surface S).

Thus $\oint_s \vec{D}.\vec{ds} = \int_v \rho \, dv$ is the vector statement of Gauss, Law and its alternative* statement is $\nabla \vec{D} = \rho$

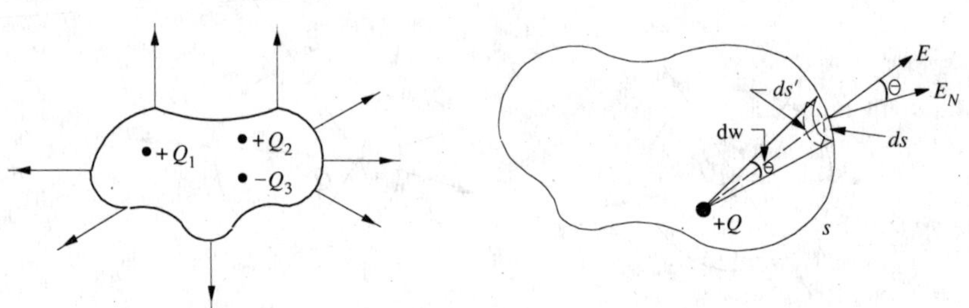

Fig. 4.12 Fig. 4.13

The normal component of the intensity $E_n = E \cos\theta$

∴ No. of lines of force passing normally through the area ds is

*This results from the application of the Divergence theorem, also called the Gauss' Theorem, *viz.*, $\int_v .\nabla \vec{D} d\upsilon = \oint \vec{D}.\vec{ds}$ where vector operator called 'del' is defined as

$$\nabla = \frac{\partial}{\partial x}\hat{x} + \frac{\partial}{\partial y}\hat{y} + \frac{\partial}{\partial z}\hat{z}$$

$$= E_n.ds = E\,ds\cos\theta \;\; = \vec{E}.\vec{ds} \text{ in vector notation}$$

Now $\quad ds\cos\theta = ds' \quad \therefore E.ds' = \dfrac{Q}{4\pi\varepsilon_0 d^2}\cdot ds'$

Now $\quad ds'/d^2 = d\omega$

Hence, the number of lines of force passing normally is $= \dfrac{Q}{4\pi\varepsilon_0}\,d\omega$

Total number of lines of force over the whole surface

$$= \frac{Q}{4\pi\varepsilon_0}\oint_s d\omega = \frac{Q}{4\pi\varepsilon_0}\times 4\pi = \frac{Q}{\varepsilon_0}$$

where sign $\oint$ denotes integration around the whole of the closed surface *i.e.* surface integral.

If the surface passes through a material medium, then the above law can be generalized to include the following :

the surface integral of the normal component of D over a closed surface equals the free charge enclosed by the surface.

As before $D = \dfrac{Q}{4\pi d^2}$. The normal component $D_n = D\cos\theta = \dfrac{Q}{4\pi d^2}\times\cos\theta$

Hence, the normal electric flux from area ds is

$$d\psi = D_n \times ds = \frac{Q}{4\pi d^2}\cdot\cos\theta\cdot ds = \frac{Q}{4\pi d^2}.ds'$$

$$\therefore \quad d\psi = \frac{Q}{4\pi}\left(\frac{ds'}{d^2}\right) = \frac{Q}{4\pi}\,d\omega$$

or $$\psi = \int\frac{Q}{4\pi}\cdot d\omega = \frac{Q}{4\pi}\int d\omega = \frac{Q}{4\pi}\times 4\pi = Q \quad \therefore \psi = Q$$

which proves the statement made above.

Hence, we may state Gauss's law in two slightly different ways.

$$\oint_3 E_n\cdot ds = \oint_3 E\cdot\cos\theta\cdot ds = Q/\varepsilon_0 \text{ or } \varepsilon_0\oint_3 E_n\cdot ds = Q$$

and $$\oint_3 D_n\,.\,ds = \oint_3 D\cos\theta\,.\,ds = Q$$

(vector statement is given above)

4.10. The Equations of Poisson and Laplace

These equations are useful in the solution of many problems concerning electrostatics especially the problem of space charge present in an electronic valve. The two equations can be derived by applying Gauss's theorem. Consider the electric field set up between two charged plates P and Q [Fig. 4.14 (*a*)]. Suppose there is some electric charge present in the space between the two plates. It is, generally, known as the space charge*. Let the space charge density be ρ coulomb/metre^3. It will be assumed that the space charge density varies from one point of space to another but is uniform throughout any thin layer taken parallel to the plates P and Q. If X-axis is taken perpendicular to the plates, then ρ is assumed to depend on the value of x. It will be seen from Fig. 4.14 (*a*) that the value of electric intensity E increases with x because of the space charge.

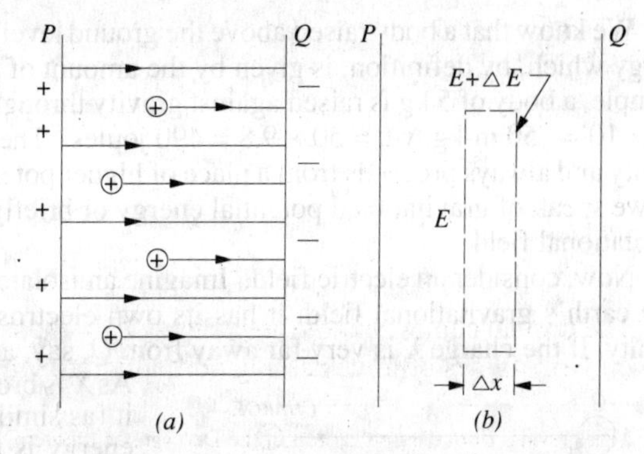

Fig. 4.14

*Such a space charge exists in the space between the cathode and anode of a vacuum tube.

Now, consider a thin volume element of cross-section A and thickness Δx as shown in Fig. 4.14(*b*). The values of electric intensity at the two opposite faces of this element are E and $(E + \Delta E)$. If dE/dx represents the rate of increase of electric intensity with distance, then

$$\Delta E = \frac{\partial E}{\partial x} \times \Delta x \quad \therefore E + \Delta E = E + \frac{\partial E}{\partial x} \times \Delta x$$

The surface integral of electric intensity over the right-hand face of this element is

$$= \left(E + \frac{\partial E}{\partial x} \cdot \Delta x\right) A$$

The surface integral over the left-hand face of the element is $= -E \times A$

The negative sign represents the fact that E is directed inwards over this face.

The surface integral over the entire surface, *i.e.*, the closed surface of the element is $= \left(E + \frac{\partial E}{\partial x} \Delta x\right) A - E \times A = A \cdot \Delta x \cdot \frac{\partial E}{\partial x}$. From symmetry it is evident that along with y & z there is no field.

Now, according to Gauss's theorem (Art. 4.9), the surface integral of electric intensity over a closed surface is equal to $1/\varepsilon_0$ time the charge within that surface.

Volume of the element, $dV = A \times \Delta x$; charge $= \rho A \cdot \Delta x$

$$\therefore \quad A \cdot \Delta x \cdot \frac{\partial E}{\partial x} = \rho A \cdot \Delta x \cdot \frac{1}{\varepsilon_0} \quad \text{or} \quad \frac{\partial E}{\partial x} = \frac{\rho}{\varepsilon_0}$$

$$\text{Now} \quad E = -\frac{\partial V}{\partial x} \quad \therefore \frac{\partial E}{\partial x} = \frac{\partial}{\partial x}\left(-\frac{dV}{dx}\right) = -\frac{\partial^2 V}{\partial x^2} \quad \therefore \frac{\partial^2 V}{\partial x^2} = -\frac{\rho}{\varepsilon_0}$$

It is known as Poisson's equation in one dimension where potential varies with x.

When V varies with x, y and z, then $\frac{\partial^2 V}{\partial x^2} + \frac{\partial^2 V}{\partial y^2} + \frac{\partial^2 V}{\partial z^2} = -\frac{\rho}{\varepsilon_0} = \nabla^2 V$ in vector notation

If, as a special case, where space charge density is zero, then obviously,

$$\partial^2 V / \partial x^2 = 0$$

In general, we have $\frac{\partial^2 V}{\partial x^2} + \frac{\partial^2 V}{\partial y^2} + \frac{\partial^2 V}{\partial z^2} = 0$ or $\nabla^2 V = 0$ in vector notation where ∇^2 is defined (in cartesian co-ordinates) as the operation

$$\nabla^2 = \frac{\partial^2}{\partial x^2} + \frac{\partial^2}{\partial y^2} + \frac{\partial^2}{\partial z^2}$$

It is known as Laplace's equation.

4.11. Electric Potential and Energy

We know that a body raised above the ground level has a certain amount of mechanical potential energy which, by definition, is given by the amount of work done in raising it to that height. If, for example, a body of 5 kg is raised against gravity through 10 m, then the potential energy of the body is $5 \times 10 = 50$ m-kg. wt. $= 50 \times 9.8 = 490$ joules. The body falls because there is attraction due to gravity and always proceeds from a place of higher potential energy to one of lower potential energy. So, we speak of gravitational potential energy or briefly 'potential' at different points in the earth's gravitational field.

Now, consider an electric field. Imagine an isolated positive charge Q placed in air (Fig. 4.15). Like earth's gravitational field, it has its own electrostatic field which theoretically extends upto infinity. If the charge X is very far away from Q, say, at infinity, then force on it is practically zero. As X is brought nearer to Q, a force of repulsion acts on it (as similar charges repel each other), hence work or energy is required to bring it to a point like A in the electric field. Hence, when at point A, charge X has some amount of electric potential energy. Similar other points in the field will also have some potential energy.

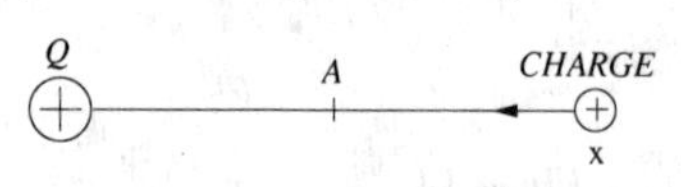

Fig. 4.15

In the gravitational field, usually 'sea level' is chosen as the place of 'zero' potential. In electric field, infinity is chosen as the theoretical place of 'zero' potential although, in practice, earth is chosen as 'zero' potential, because earth is such a large conductor that its potential remains practically constant although it keeps on losing and gaining electric charge every day.

4.12. Potential and Potential Difference

As explained above, the force acting on a charge at infinity is zero, hence 'infinity' is chosen as the theoretical place of zero electric potential. Therefore, potential at any point in an electric field may be defined as

numerically equal to the work done in bringing a positive charge of one coulomb from infinity to that point against the electric field.

The unit of this potential will depend on the unit of charge taken and the work done.

If, in shifting one coulomb from infinity to a certain point in the electric field, the work done is one joule, then potential of that point is one volt.

Obviously, potential is work per unit charge.

$$\therefore \quad 1 \text{ volt} = \frac{1 \text{ joule}}{1 \text{ coulomb}}$$

Similarly, potential difference (p.d.) of one volt exists between two points if one joule of work is done in shifting a charge of one coulomb from one point to the other.

4.13. Potential at a Point

Consider a positive point charge of Q coulombs placed in air. At a point x metres from it, the force on one coulomb positive charge is $Q/4\pi\,\varepsilon_0\,x^2$ (Fig. 4.16). Suppose, this one coulomb charge is moved towards Q through a small distance dx. Then, work done is

$$dW = \frac{Q}{4\,\pi\,\varepsilon_0\,x^2} \times (-\,dx)$$

The negative sign is taken because dx is considered along the negative direction of x.

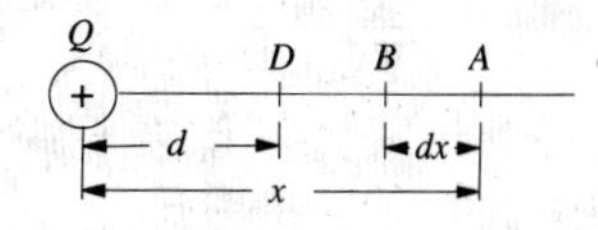

Fig. 4.16

The total work done in bringing this coulomb of positive charge from infinity to any point D which is d metres from Q is given by

$$W = -\int_{x=\infty}^{x=d} Q \cdot \frac{dx}{4\,\pi\,\varepsilon_0\,x^2} = -\frac{Q}{4\,\pi\,\varepsilon_0}\int_{\infty}^{d} \frac{dx}{x^2}$$

$$= -\frac{Q}{4\pi\varepsilon_0}\left| -\frac{1}{x} \right|_{\infty}^{d} = \frac{-Q}{4\,\pi\,\varepsilon_0}\left[-\frac{1}{d} - \left(-\frac{1}{\infty}\right)\right] = \frac{Q}{4\,\pi\,\varepsilon_0\,d} \text{ joules}$$

By definition, this work in joules in numerically equal to the potential of that point in volts.

$$\therefore \quad V = \frac{Q}{4\,\pi\,\varepsilon_0\,d} = 9 \times 10^9\,\frac{Q}{d} \text{ volt} \quad \text{—in air}$$

$$\text{and} \quad V = \frac{Q}{4\,\pi\,\varepsilon_0\,\varepsilon_r\,d} = 9 \times 10^9\,\frac{Q}{\varepsilon_r d} \text{ volt} \quad \text{—in a medium}$$

We find that as d increases, V decreases till it becomes zero at infinity.

4.14. Potential of a Charged Conducting Sphere

The above formula $V = Q/4\pi\,\varepsilon_0\,\varepsilon_r\,d$ applies only to a charge concentrated at a point. The problem of finding potential at a point outside a charged sphere sounds difficult, because the charge on the sphere is distributed over its entire surface and so, is not concentrated at a point. But the problem is easily solved by noting that the lines of force of a charged sphere, like A in Fig. 4.17 spread out normally from its surface. If produced backwards, they meet at the centre of A. Hence for finding the potentials at points outside the sphere, we can imagine the charge on the sphere as concentrated at its centre O. If r is the radius of sphere in metres and Q its charge in coulomb then, potential of its surface is $Q/4\pi\,\varepsilon_0\,r$ volt and electric intensity is $Q/4\pi\,\varepsilon_0\,r^2$ At any other point 'd' metres from the centre of the sphere, the corresponding values are $Q/4\pi\,\varepsilon_0\,d$ and $Q/4\pi\varepsilon_0 d^2$ respectively with $d > r$ as shown in Fig. 4.18 though its starting point is coincident with that of r. The variations of the potential and electric intensity with distance for a charged sphere are shown in Fig. 4.18.

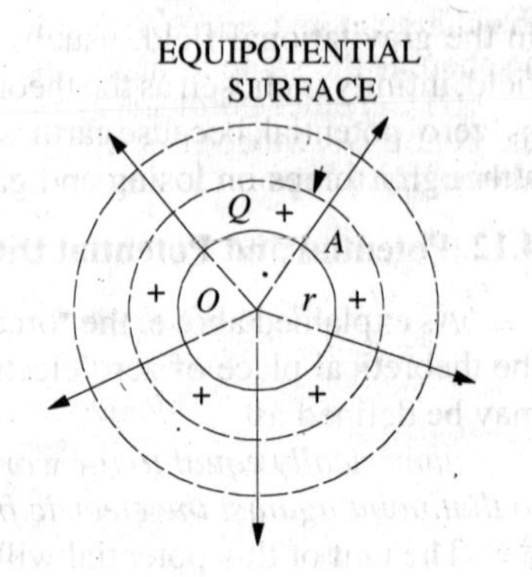

Fig. 4.17

4.15. Equipotential Surfaces

An equipotential surface is a surface in an electric field such that all points on it are at the same potential. For example, different spherical surfaces around a charged sphere are equipotential surfaces. One important property of an equipotential surface is that the direction of the electric field strength and flux density is always at right angles to the surface. Also, electric flux emerges out normal to such a surface. If, it is not so, then there would be some component of E along the surface resulting in potential difference between various points lying on it which is contrary to the definition of an equipotential surface.

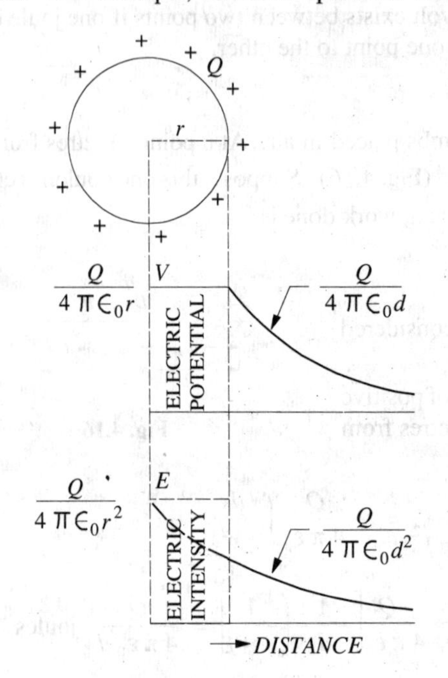

Fig. 4.18

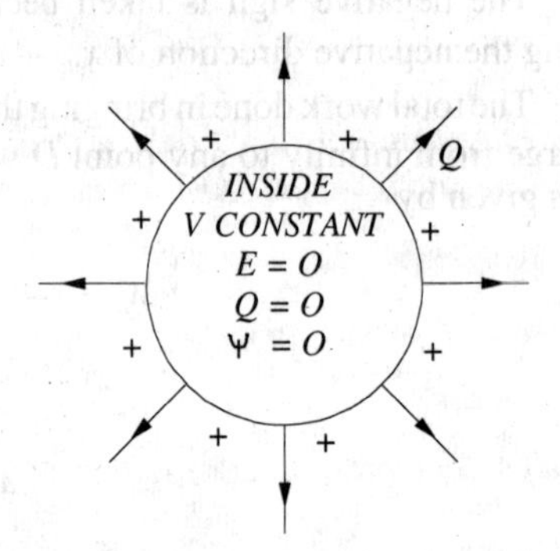

Fig. 4.19

4.16. Potential and Electric Intensity Inside a Conducting Sphere

It has been experimentally found that when charge is given to a conducting body say, a sphere then it resides entirely on its outer surface *i.e.*, within a conducting body whether hollow or solid, the charge is zero. Hence, (*i*) flux is zero (*ii*) field intensity is zero (*iii*) all points within the conductor are at the same potential as at its surface (Fig. 4.19).

Example 4.7. *Three concentric spheres of radii 4, 6 and 8 cm have charges of + 8, − 6 and +4 μμC respectively. What are the potentials and field strengths at points, 2, 5, 7 and 10 cm from the centre.*

Solution. As shown in Fig. 4.20, let the three spheres be marked A, B and C. It should be remembered that (*i*) the field intensity outside a sphere is the same as that obtained by considering the charge at its centre (*ii*) inside the sphere, the field strength is zero (*iii*) potential anywhere inside a sphere is the same as at its surface.

(*i*) Consider point '*a*' at a distance of 2 cm from the centre *O*. Since it is inside all the spheres, field strength at this point is zero.

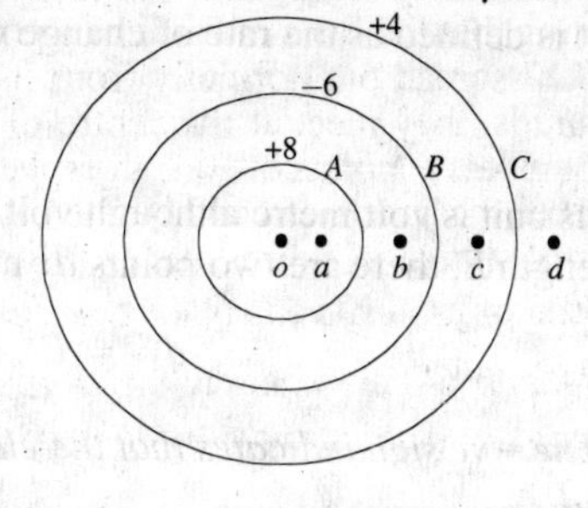

Fig. 4.20

Potential at '*a*'

$$= \Sigma \frac{Q}{4\pi\varepsilon_0 d} = 9\times10^9\,\Sigma\frac{Q}{d}$$

$$= 9\times10^9\left(\frac{8\times10^{-12}}{0.04}-\frac{6\times10^{-12}}{0.06}+\frac{4\times10^{-12}}{0.08}\right) = \mathbf{1.35\ V}$$

(*ii*) Since point '*b*' is outside sphere *A* but inside *B* and *C*.

$$\therefore \quad \text{Electric field} = \frac{Q}{4\pi\varepsilon_0 d^2} = 9\times10^9\frac{Q}{d^2}\ \text{N/C}$$

$$= 9\times10^9\times\frac{8\times10^{-12}}{0.05^2} = \mathbf{28.8\ N/C}$$

$$\text{Potential at '}b\text{'} = 9\times10^9\times\left(\frac{8\times10^{-12}}{0.05}-\frac{6\times10^{-12}}{0.06}+\frac{4\times10^{-12}}{0.08}\right) = \mathbf{0.99\ V}$$

(*iii*) The field strength at point '*c*' distant 7 cm from centre *O*

$$= 9\times10^9\times\left[\frac{8\times10^{-12}}{0.07^2}-\frac{6\times10^{-12}}{0.07^2}\right] = \mathbf{3.67\ N/C}$$

$$\text{Potential at '}c\text{'} = 9\times10^9\times\left[\frac{8\times10^{-12}}{0.07}-\frac{6\times10^{-12}}{0.07}+\frac{4\times10^{-12}}{0.08}\right] = \mathbf{0.71\ V}$$

(*iv*) Field strength at '*d*' distant 10 cm from point *O* is

$$= 9\times10^9\left[\frac{8\times10^{-12}}{0.1^2}-\frac{6\times10^{-12}}{0.1^2}+\frac{4\times10^{-12}}{0.1^2}\right] = \mathbf{5.4\ N/C}$$

$$\text{Potential at '}d\text{' is} = 9\times10^9\left[\frac{8\times10^{-12}}{0.1}-\frac{6\times10^{-12}}{0.1}+\frac{4\times10^{-12}}{0.1}\right] = \mathbf{0.54\ V}$$

Example 4.8. *Two positive point charges of 12×10^{-10} C and 8×10^{-10} C are placed 10 cm apart. Find the work done in bringing the two charges 4 cm closer.*

Solution. Suppose the 12×10^{-10} C charge to be fixed. Now, the potential of a point 10 cm from this charge

$$= 9\times10^9\frac{12\times10^{-10}}{0.1} = 108\ \text{V}$$

The potential of a point distant 6 cm from it

$$= 9\times10^9\times\frac{12\times10^{-10}}{0.06} = 180\ \text{V}$$

$\therefore$ potential difference $= 180 - 108 = 72$ V

Work done $=$ charge × p.d. $= 8\times10^{-10}\times72 = \mathbf{5.76\times10^{-8}\ joule}$

Example 4.9. *A point charge of 10^{-9} C is placed at a point A in free space. Calculate :*
(i) the intensity of electrostatic field on the surface of sphere of radius 5 cm and centre A.
(ii) the difference of potential between two points 20 cm and 10 cm away from the charge at A.

(Elements of Elect. Engg.-I, Bangalore Univ. 1987)

Solution. (*i*) $E = Q/4\pi\varepsilon_0 r^2 = 10^{-9}/4\pi\times8.854\times10^{-12}\times(5\times10^{-2})^2 = \mathbf{3{,}595\ V/m}$

(*ii*) Potential of first point $= Q/4\pi\varepsilon_0 d = 10^{-9}/4\pi\times8.854\times10^{-12}\times0.2 = 45$ V

Potential of second point $= 10^{-9}/4\pi\times8.854\times10^{-12}\times0.1 = 90$ V

$\therefore$ p.d. between two points $= 90 - 45 = \mathbf{45\ V}$

4.17. Potential Gradient

It is defined as the rate of change of potential with distance in the direction of electric force

i.e. $\frac{dV}{dx}$

Its unit is volt/metre although volt/cm is generally used in practice. Suppose in an electric field of strength E, there are two points dx metre apart. The p.d. between them is

$$dV = E\,.\,(-dx) = -E\,.\,dx \quad \therefore E = -\frac{dV}{dx} \qquad \ldots(i)$$

The – ve sign indicates that the electric field is directed outward, while the potential increases inward.

Hence, it means that electric intensity at a point is equal to the negative potential gradient at that point.

4.18. Breakdown Voltage and Dielectric Strength

An insulator or dielectric is a substance within which there are no mobile electrons necessary for electric conduction. However, when the voltage applied to such an insulator exceeds a certain value, then it breaks down and allows a heavy electric current (much larger than the usual leakage current) to flow through it. If the insulator is a solid medium, it gets punctured or cracked.

The disruptive or breakdown voltage of an insulator is the minimum voltage required to break it down.*

Dielectric strength of an insulator or dielectric medium is given by the *maximum potential difference which a unit thickness of the medium can withstand without breaking down.*

In other words, the dielectric strength is given by the potential gradient necessary to cause breakdown of an insulator. Its unit is volt/metre (V/m) although it is usually expressed in kV/mm.

For example, when we say that the dielectric strength of air is 3 kV/mm, then it means that the maximum p.d. which one mm thickness of air can withstand across it without breaking down is 3 kV or 3000 volts. If the p.d. exceeds this value, then air insulation breaks down allowing large electric current to pass through.

Dielectric strength of various insulating materials is very important factor in the design of high-voltage generators, motors and transformers. Its value depends on the thickness of the insulator, temperature, moisture, content, shape and several other factors.

For example, doubling the thickness of insulation does not double the safe working voltage in a machine.**

Note. It is obvious that the electric intensity E, potential gradient and dielectric strength are dimensionally equal.

4.19. Safety Factor of a Dielectric

It is given by the ratio of the dielectric strength of the insulator and the electric field intensity established in it. If we represent the dielectric strength by E_{bd} and the actual field intensity by E, then

$$\text{safety factor, } k = E_{bd}/E$$

For example, for air, $E_{bd} = 3 \times 10^6$ V/m. If we establish a field intensity of 3×10^5 V/m in it, then, $k = 3 \times 10^6 / 3 \times 10^5 = 10$.

4.20. Boundary Conditions

There are discontinuities in electric fields at the boundaries between conductors and dielectrics of different permittivities. The relationships existing between the electric field strengths and flux

---

*Flashover is the disruptive discharge which takes places over the surface of an insulator and occurs when the air surrounding it breaks down. Disruptive conduction is luminous.

**The relation between the breakdown voltage V and the thickness of the dielectric is given approximately by the relation

$$V = At^{2/3}$$

where A is a constant depending on the nature of the medium and also on the thickness t. The above statement is known as Baur's law.

densities at the boundary are called the boundary conditions.

With reference to Fig. 4.21, first boundary condition is that the normal component of flux density is continuous across a surface.

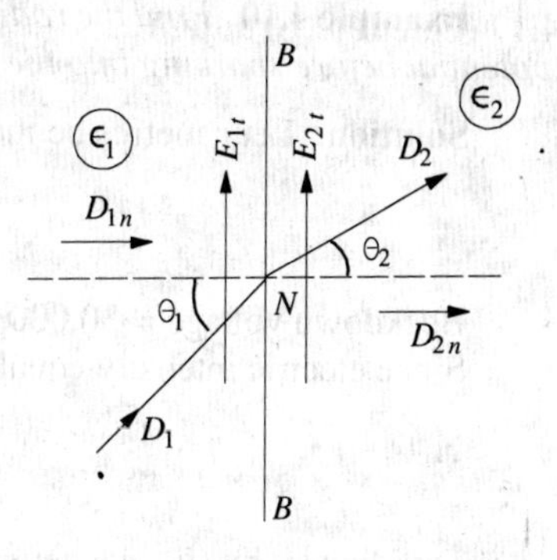

Fig. 4.21

As shown, the electric flux approaches the boundary *BB* at an angle θ_1 and leaves it at θ_2. D_{1n} and D_{2n} are the normal components of D_1 and D_2. According to first boundary condition,

$$D_{1n} = D_{2n} \quad \text{... (i)}$$

The second boundary condition is that the tangential field strength is continuous across the boundary

$$\therefore \quad E_{1t} = E_{2t} \quad \text{... (ii)}$$

In Fig. 4.21, we see that

$$D_{1n} = D_1 \cos\theta_1 \quad \text{and} \quad D_{2n} = D_2 \cos\theta_2$$

Also $\quad E_1 = D_1/\varepsilon_1 \quad$ and $\quad E_{1t} = D_1 \sin\theta_1/\varepsilon_1$

Similarly, $\quad E_2 = D_2/\varepsilon_2 \quad$ and $\quad E_{2t} = D_2 \sin\theta_1/\varepsilon_2$

$$\therefore \quad \frac{D_{1n}}{E_{1t}} = \frac{\varepsilon_1}{\tan\theta_1} \quad \text{and} \quad \frac{D_{2n}}{E_{2t}} = \frac{\varepsilon_2}{\tan\theta_2}$$

Since $\quad D_{1n} = D_{2n} \quad$ and $\quad E_{1t} = E_{2t} \quad \therefore \quad \dfrac{\tan\theta_1}{\tan\theta_2} = \dfrac{\varepsilon_1}{\varepsilon_2}$

This gives the law of electric flux refraction at a boundary.

It is seen that if $\varepsilon_1 > \varepsilon_2$, $\theta_1 > \theta_2$.

TABLE NO. 4.1
Dielectric Constant and Strength

(* indicates average value)

Insulating material	*Dielectric constant or relative permittivity* (ε_r)	*Dielectric Strength in kV/mm*
Air	1.0006	3.2
Asbestos*	2	2
Bakelite	5	15
Epoxy	3.3	20
Glass	5-12	12-100
Marble*	7	2
Mica	4-8	20-200
Micanite	4.5-6	25-35
Mineral Oil	2.2	10
Mylar	3	400
Nylon	4.1	16
Paper	1.8-2.6	18
Paraffin wax	1.7-2.3	30
Polyethylene	2.3	40
Polyurethane	3.6	35
Porcelain	5-6.7	15
PVC	3.7	50
Quartz	4.5-4.7	8
Rubber	2.5-4	12-20
Teflon	2	20
Vacuum	1	infinity
Wood	2.5-7	---

Example 4.10. *Find the radius of an isolated sphere capable of being charged to 1 million volt potential before sparking into the air, given that breakdown voltage of air is 30,000 V/cm.*

Solution. Let r metres be the radius of the sphere, then

$$V = \frac{Q}{4\pi\varepsilon_0 r} = 10^6 \text{ V} \qquad ...(i)$$

Beakdown voltage = 30,000 V/cm = 3×10^6 V/m
Since electric intensity equals breakdown voltage

$$\therefore \qquad E = \frac{Q}{4\pi\varepsilon_0 r^2} = 3\times 10^6 \text{ V/m} \qquad ...(ii)$$

Dividing (*i*) by (*ii*), we get r = 1/3 = **0.33 metre**

Example 4.11. *A parallel plate capacitor having waxed paper as the insulator has a capacitance of 3800 pF, operating voltage of 600 V and safety factor of 2.5. The waxed paper has a relative permittivity of 4.3 and breakdown voltage of 15×10^6 V/m. Find the spacing d between the two plates of the capacitor and the plate area.*

Solution. Breakdown voltage V_{bd} = operating voltage × safety factor = 600 × 2.5 = 1500 V

$V_{bd} = d \times E_{bd}$ or $d = 1500/15 \times 10^6 = 10^{-4}$ m = 0.1 mm

$C = \varepsilon_0 \varepsilon_r A/d$ or $A = Cd/\varepsilon_0\varepsilon_r = 3800 \times 10^{-9} \times 10^{-4}/8.854 \times 10^{-12} \times 4.3 =$ **0.01 m²**

Example 4.12. *Two brass plates arc arranged horizontally, one 2 cm above the other and the lower plate is earthed. The plates are charged to a difference of potential of 6,000 volts. A drop of oil with an electric charge of 1.6×10^{-19} C is in equilibrium between the plates so that it neither rises nor falls. What is the mass of the drop ?*

Solution. The electric intensity is equal to the potential gradient between the plates.

$$g = 6{,}000/2 = 3{,}000 \text{ volt/cm} = 3 \times 10^5 \text{ V/m}$$

$$\therefore \quad E = 3 \times 10^5 \text{ V/m or N/C}$$

$$\therefore \quad \text{force on drop} = E \times Q = 3 \times 10^5 \times 1.6 \times 10^{-19} = 4.8 \times 10^{-14} \text{ N}$$

$$\text{Wt. of drop} = mg \text{ newton}$$

$$\therefore \quad m \times 9.81 = 4.8 \times 10^{-14} \quad \therefore m = \mathbf{4.89 \times 10^{-15} \text{ kg}}$$

Example 4.13. *A parallel-plate capacitor has plates 0.15 mm apart and dielectric with relative permittivity of 3. Find the electric field intensity and the voltage between plates if the surface charge is 5×10^{-4} μC/cm².* **(Electrical Engineering, Calcutta Univ. 1988)**

Solution. The electric intensity between the plates is

$$E = \frac{D}{\varepsilon_0 \varepsilon_r} \text{ volt/metre; Now, } \sigma = 5 \times 10^{-4} \, \mu\text{C/cm}^2 = 5 \times 10^{-6} \text{ C/m}^2$$

Since, charge density equals flux density

$$\therefore \quad E = \frac{D}{\varepsilon_0\varepsilon_r} = \frac{5 \times 10^{-6}}{8.854 \times 10^{-12} \times 3} = 188{,}000 \text{ V/m} = \mathbf{188 \text{ kV/m}}$$

Now potential difference $V = E \times dx = 188{,}000 \times (0.15 \times 10^{-3})$ = **28.2 V**

Example 4.14. *A parallel-plate capacitor consists of two square metal plates 500 mm on a side separated by 10 mm. A slab of Teflon (ε_r = 2.0) 6 mm thick is placed on the lower plate leaving an air gap 4 mm thick between it and the upper plate. If 100 V is applied across the capacitor, find the electric field (E_0) in the air, electric field E_t in Teflon, flux density D_a in air, flux density D_t in Teflon and potential difference V_t across Teflon slab.*

(Circuit and Field Theory, A.M.I.E. Sec B, 1995)

Solution.

$$C = \frac{\varepsilon_0 A}{(d_1/\varepsilon_{r1} + d_2/\varepsilon_{r2})} = \frac{8.854 \times 10^{-12} \times (0.5)^2}{(6 \times 10^{-3}/2) + (4 \times 10^{-3}/1)} = 3.16 \times 10^{-10} \text{ F}$$

$$Q = CV = 3.16 \times 10^{-10} \times 100 = 31.6 \times 10^{-9} \text{ C}$$

$$D = Q/A = 31.6 \times 10^{-9}/(0.5)^2 = 1.265 \times 10^{-7} \text{ C/m}^2$$

The charge or flux density will be the same in both media *i.e.* $D_a = D_t = D$

In air, $E_0 = D/\varepsilon_0 = 1.265 \times 10^{-7}/8.854 \times 10^{-12} = 14{,}280$ V/m

In Teflon, $E_t = D/\varepsilon_0\varepsilon_r = 14{,}280/2 = 7{,}140$ V/m

$V_t = E_t \times d_t = 7{,}140 \times 6 \times 10^{-3} = \mathbf{42.8\ V}$

Example 4.15. *Calculate the dielectric flux in micro-coulombs between two parallel plates each 35 cm square with an air gap of 1.5 mm between them, the p.d. being 3,000 V. A sheet of insulating material 1 mm thick is inserted between the plates, the permittivity of the insulating material being 6. Find out the potential gradient in the insulating material and also in air if the voltage across the plates is raised to 7,500 V.* **(Elect. Engg.-I, Nagpur Univ. 1993)**

Solution. The capacitance of the two parallel plates is

$C = \varepsilon_0\, \varepsilon_r\, A/d$ Now, $\varepsilon_r = 1$ —for air

$A = 35 \times 35 \times 10^{-4} = 1{,}225 \times 10^{-4}$ m² ; $d = 1.5 \times 10^{-2}$ m

$$\therefore \quad C = \frac{8.854 \times 10^{-12} \times 1{,}225 \times 10^{-4}}{1.5 \times 10^{-3}} \text{ F} = 7.22 \times 10^{-10} \text{ F}$$

Charge $Q = CV = 7.22 \times 10^{-10} \times 3{,}000$ coulomb

Dielectric flux $= 7.22 \times 3{,}000 \times 10^{-10}$ C

$= 2.166 \times 10^{-6}$ C $=$ **2.166 μC**

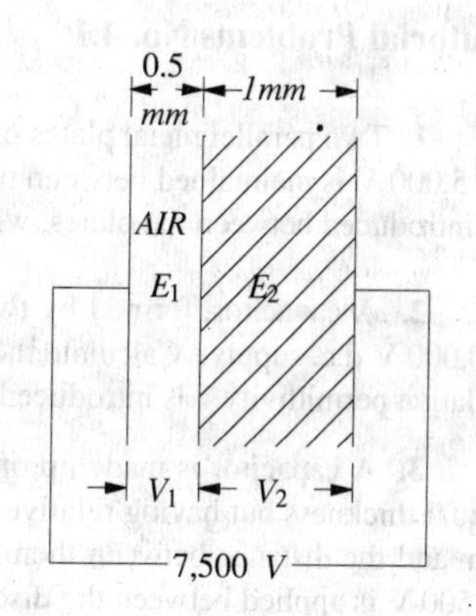

Fig. 4.22

With reference to Fig. 4.23, we have

$V_1 = E_1 x_1 = 0.5 \times 10^{-3}\, E_1$; $V_2 = 10^{-3}\, E_2$

Now $V = V_1 + V_2$

$\therefore \quad 7{,}500 = 0.5 \times 10^{-3}\, E_1 + 10^{-3}\, E_2$

or $\quad E_1 + 2E_2 = 15 \times 10^6 \quad$ ...(*i*)

Also $\quad D = \varepsilon_0\, \varepsilon_{r1}\, E_1 = \varepsilon_0\, \varepsilon_{r2}\, E_2 \quad \therefore E_1 = 6\, E_2 \quad$ ...(*ii*)

From (*i*) and (*ii*), we obtain $E_1 = \mathbf{11.25 \times 10^6\ V/m}$; $E_2 = \mathbf{1.875 \times 10^6\ V/m}$

Example 4.16. *An electric field in a medium with relative permittivity of 7 passes into a medium of relative permittivity 2. If E makes an angle of 60° with the normal to the boundary in the first dielectric, what angle does the field make with the normal in the second dielectric ?* **(Elect. Engg. Nagpur Univ. 1991)**

Solution. As seen from Art. 4.19.

$$\frac{\tan \theta_1}{\tan \theta_2} = \frac{\varepsilon_1}{\varepsilon_2} \quad \text{or} \quad \frac{\tan 60^\circ}{\tan \theta_2} = \frac{7}{2} \qquad \therefore \tan \theta_2 = \sqrt{3} \times 2/7 = 0.495 \text{ or } \theta_2 = \mathbf{26^\circ 20'}$$

Example 4.17. *Two parallel sheets of glass having a uniform air gap between their inner surfaces are sealed around their edges (fig. 4.23). They are immersed in oil having a relative permittivity of 6 and are mounted vertically. The glass has a relative permittivity of 3. Calculate the values of electric field strength in the glass and in the air when that in the oil is 1.2 kV/m. The field enters the glass at 60° to the horizontal.*

Solution. Using the law of electric flux refraction, we get (Fig. 4.23).

$\tan\theta_2/\tan\theta_1 = \varepsilon_2/\varepsilon_1 = \varepsilon_0\,\varepsilon_{r2}/\varepsilon_0\,\varepsilon_{r1} = (\varepsilon_{r2}/\varepsilon_{r1})$

$\therefore \tan\theta_2 = (6/3)\tan 60° = 2 \times 1.732 = 3.464;\ \theta_2 = 73.9°$

Similarly $\tan\theta_3 = (\varepsilon_{r3}/\varepsilon_{r2})\tan\theta_2$

$= (1/6)\tan 73.9° = 0.577;\quad \therefore \theta_3 = 30°$

As shown in Art. 4.20.

$D_{1n} = D_{2n}$ or $D_1\cos\theta_1 = D_2\cos\theta_2$

$\therefore D_2 = D_1 \times \cos\theta_1/\cos\theta_2$ or $\varepsilon_0\,\varepsilon_{r2}\,E_2$

$= \varepsilon_0\,\varepsilon_{r1}\,E_1 \times \cos\theta_1/\cos\theta_2$

$\therefore 6\,E_2 = 3 \times 1.2 \times 10^3 \times \cos 60°/\cos 73.9°$

$\therefore E_2 = 1082$ V/m

Now, $\varepsilon_0\,\varepsilon_{r3}\,E_3\cos\theta_3 = \varepsilon_0\,\varepsilon_{r2}\,E_2\cos\theta_2$

$\therefore E_3 = E_2\,(\varepsilon_{r2}/\varepsilon_{r3}) \times (\cos\theta_2/\cos\theta_3)$

$= 1082(6/1)(\cos 73.9°/\cos 30°) =$ **2079V/m**

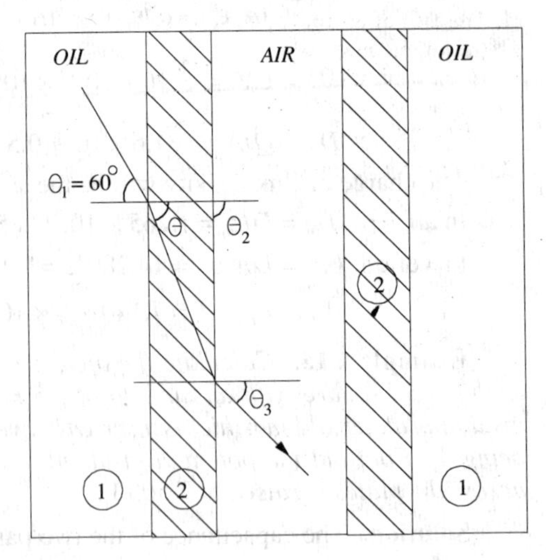

Fig. 4.23

Tutorial Problems No. 4.1

1. Two parallel metal plates of large area are spaced at a distance of 1 cm from each other in air and a p.d. of 5,000 V is maintained between them. If a sheet of glass 0.5 cm thick and having a relative permittivity of 6 is introduced between the plates, what will be the maximum electric stress and where will it occur ? **[8.57 kV/cm ; in air]**

2. A capacitor, formed by two parallel plates of large area, spaced 2 cm apart in air, is connected to a 10,000 V d.c. supply. Calculate the electric stress in the air when a flat sheet of glass of thickness 1.5 cm and relative permittivity 7 is introduced between the plates. **[1.4×10^8 V/m]**

3. A capacitor is made up of two parallel circular metal discs separated by three layers of dielectric of equal thickness but having relative permittivities of 3, 4 and 5 respectively. The diameter of each disc is 25.4 cm and the distance between them is 6 cm. Calculate the potential gradient in each dielectric when a p.d. of 1,500 V is applied between the discs. **[319.2 ; 239.4 ; 191.5 kV/m]**

4. A capacitor, formed by two parallel plates of large area, spaced 2 cm apart in air, is connected to a 10,000 V d.c. supply. Calculate the electric stress in the air when a flat sheet of glass of thickness 0.5 cm and relative permittivity 5 is introduced between the plates. **[0.625×10^4 V/m]**

5. The capacitance of a capacitor formed by two parallel metal plates, each having an effective surface area of 50 cm^2 and separated by a dielectric 1 mm thick, is 0.0001 μ F. The plates are charged to a p.d. of 200 V. Calculate (*a*) the charge stored (*b*) the electric flux density (*c*) the relative permittivity of the dielectric. [(*a*) **0.02 μC** (*b*) **4 μC/m²** (*c*) **2.26**]

6. A capacitor is constructed from two parallel metallic circular plates separated by three layers of dielectric each 0.5 cm thick and having relative permittivity of 4, 6 and 8 respectively. If the metal discs are 15.25 cm in diameter, calculate the potential gradient in each dielectric when the applied voltage is 1,000 volts. **(Elect. Engg.-I Delhi Univ. 1978)**

7. A point electric charge of 8 μC is kept at a distance of 1 metre from another point charge of –4μC in free space. Determine the location of a point along the line joining two charges where in the electric field intensity is zero. **(Elect. Engineering, Kerala Univ. 1981)**

OBJECTIVE TESTS–4

1. The unit of absolute permittivity of a medium is
(*a*) joule/coulomb
(*b*) newton-metre
(*c*) farad/metre
(*d*) farad/coulomb

2. If relative permittivity of mica is 5, its absolute permittivity is
(*a*) $5\ \varepsilon_0$
(*b*) $5/\varepsilon_0$
(*c*) $\varepsilon_0/5$
(*d*) 8.854×10^{-12}

3. Two similar electric charges of 1 C each are placed 1 m apart in air. Force of repulsion between them would be nearly newton.
(*a*) 1
(*b*) 9×10^9
(*c*) 4π
(*d*) 8.854×10^{-12}

4. Electric flux emanating from an electric charge of $+Q$ coulomb is
(*a*) Q/ε_0
(*b*) Q/ε_r
(*c*) $Q/\varepsilon_0\varepsilon_r$
(*d*) Q

5. The unit of electric intensity is
(*a*) joule/coulomb
(*b*) newton/coulomb
(*c*) volt/metre
(*d*) both (*b*) and (*c*)

6. If D is the electric flux density, then value of electric intensity in air is
(*a*) D/ε_0
(*c*) $D/\varepsilon_0\varepsilon_r$
(*c*) dV/dt
(*d*) $Q/\varepsilon A$

7. For any medium, electric flux density D is related to electric intensity E by the equation
(*a*) $D = \varepsilon_0 E$
(*b*) $D = \varepsilon_0 \varepsilon_r E$
(*c*) $D = E/\varepsilon_0 \varepsilon_r$
(*d*) $D = \varepsilon_0 E/\varepsilon_r$

8. Inside a conducting sphere, ----- remains constant
(*a*) electric flux
(*b*) electric intensity
(*c*) charge
(*d*) potential

9. The SI unit of electric intensity is
(*a*) N/m
(*b*) V/m
(*c*) N/C
(*d*) either (*b*) or (*c*)

10. According to Gauss's theorem, the surface integral of the normal component of electric flux density D over a closed surface containing charge Q is
(*a*) Q
(*b*) Q/ε_0
(*c*) $\varepsilon_0 Q$
(*d*) Q^2/ε_0

11. Which of the following is zero inside a charged conducting sphere ?
(*a*) potential
(*b*) electric intensity
(*c*) both (*a*) and (*b*)
(*d*) both (*b*) and (*c*)

12. In practice, earth is chosen as a place of zero electric potential because it
(*a*) is non-conducting
(*b*) is easily available
(*c*) keeps lossing and gaining electric charge every day
(*d*) has almost constant potential.

Answers

1. *c* **2.** *a* **3.** *b* **4.** *d* **5.** *d* **6.** *a* **7.** *b* **8.** *d* **9.** *d* **10.** *a* **11.** *c* **12.** *d*

5 CAPACITANCE

5.1. Capacitor

A capacitor essentially consists of two conducting surfaces separated by a layer of an insulating medium called *dielectric*. The conducting surfaces may be in the form of either circular (or rectangular) plates or be of spherical or cylindrical shape. The purpose of a capacitor is to store electrical energy by electrostatic stress in the dielectric (the word 'condenser' is a misnomer since a capacitor does not 'condense' electricity as such, it merely stores it).

A parallel-plate capacitor is shown in Fig. 5.1. One plate is joined to the positive end of the supply and the other to the negative end or is earthed. It is experimentally found that in the presence of an earthed plate *B*, plate *A* is capable of withholding more charge than when *B* is not there. When such a capacitor is put across a battery, there is a momentary flow of electrons from *A* to *B*. As negatively-charged electrons are withdrawn from *A*, it becomes positive and as these electrons collect on *B*, it becomes negative. Hence, a p.d. is established between plates *A* and *B*. The transient flow of electrons gives rise to charging current. The strength of the charging current is maximum when the two plates are uncharged but it then decreases and finally ceases when p.d. across the plates becomes slowly and slowly equal and opposite to the battery e.m.f.

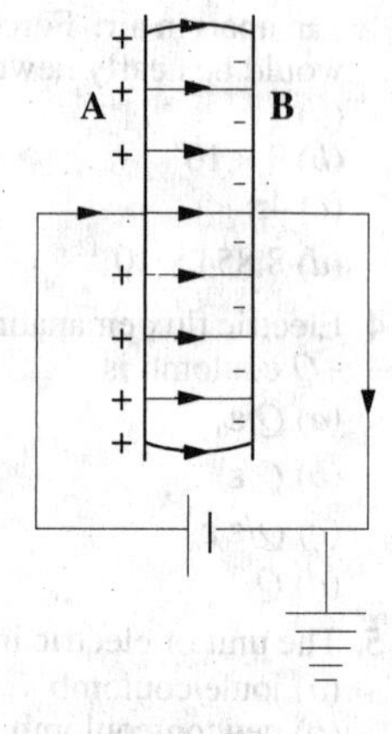

Fig. 5.1

5.2. Capacitance

The property of a capacitor to 'store electricity' may be called its capacitance.

As we may measure the capacity of a tank, not by the total mass or volume of water it can hold, but by the mass in kg of water required to raise its level by one metre, similarly, the capacitance of a capacitor is defined as "*the amount of charge required to create a unit p.d. between its plates.*

Suppose we give Q coulomb of charge to one of the two plates of capacitor and if a p.d. of V volts is established between the two, then its capacitance is

$$C = \frac{Q}{V} = \frac{\text{charge}}{\text{potential difference}}$$

Hence, capacitance is the *charge required per unit potential difference*.

By definition, the unit of capacitance is coulomb/volt which is also called *farad* (in honour of Michael Faraday)

$$\therefore \qquad 1 \text{ farad} = 1 \text{ coulomb/volt}$$

One farad is defined as *the capacitance of a capacitor which requires a charge of one coulomb to establish a p.d. of one volt between its plates.*

One farad is actually too large for practical purposes. Hence, much smaller units like microfarad (μF), nanofarad (nF) and micro-microfarad (μμF) or picofarad (pF) are generally employed.

$$1\ \mu F = 10^{-6}\ F\ ;\ 1\ nF = 10^{-9}\ F\ ;\ 1\ \mu\mu F \text{ or } pF = 10^{-12}\ F$$

Incidentally, capacitance is that property of a capacitor which delays any change of voltage across it.

5.3. Capacitance of an Isolated Sphere

Consider a charged sphere of radius r metres having a charge of Q coulomb placed in a medium of relative permittivity ε_r as shown in Fig. 5.2.

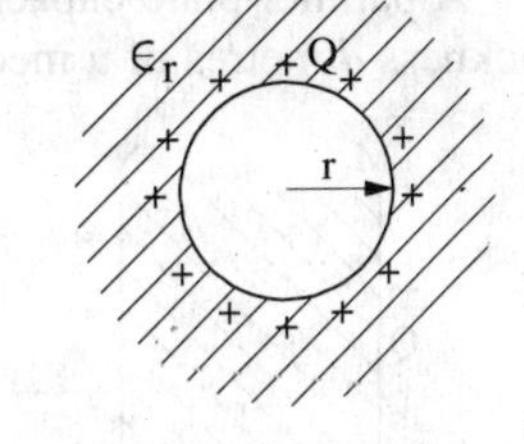

Fig. 5.2

It has been proved in Art 4.13 that the free surface potential V of such a sphere with respect to infinity (in practice, earth) is given by

$$V = \frac{Q}{4\pi\varepsilon_0\varepsilon_r r} \quad \therefore \quad \frac{Q}{V} = 4\pi\varepsilon_0\varepsilon_r r$$

By definition, Q/V = capacitance C

$\therefore \quad C = 4\pi\varepsilon_0\varepsilon_r r$ F — in a medium

$\quad\quad = 4\pi\varepsilon_0 r$ F — in air

Note : It is sometimes felt surprising that an isolated sphere can act as a capacitor because, at first sight, it appears to have one plate only. The question arises as to which is the second surface. But if we remember that the surface potential V is with reference to infinity (actually earth) then it is obvious that the other surface is earth. The capacitance $4\pi\varepsilon_0 r$ exists between the surface of the sphere and earth.

5.4. Spherical Capacitor

(*a*) **When outer sphere is earthed**

Consider a spherical capacitor consisting of two concentric spheres of radii 'a' and 'b' metres as shown in Fig. 5.3. Suppose, the inner sphere is given a charge of $+Q$ coulombs. It will induce a charge of $-Q$ coulombs on the inner surface of the outer sphere and a charge $+Q$ coulombs on its outer surface which will go to earth. If the dielectric medium between the two spheres has a relative permittivity of ε_r, then the free surface potential of the inner sphere due to its own charge is $Q/4\pi\varepsilon_0\varepsilon_r a$ volts. The potential of the inner sphere due to $-Q$ charge on the outer sphere is $-Q/4\pi\varepsilon_0\varepsilon_r b$ (remembering that potential anywhere inside a sphere is the same as at its surface).

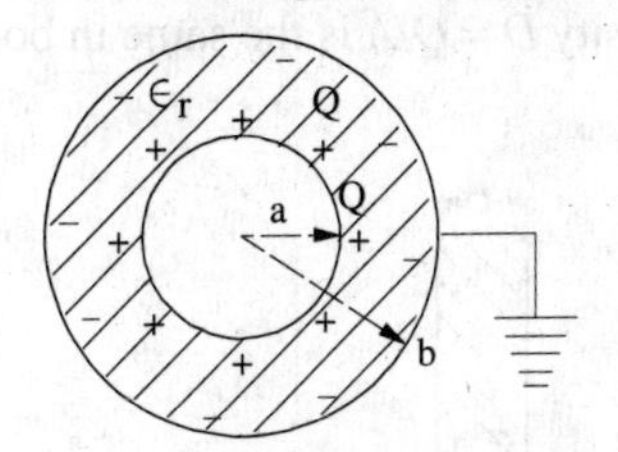

Fig. 5.3

$\therefore$ Total potential difference between two surfaces is

$$V = \frac{Q}{4\pi\varepsilon_0\varepsilon_r a} - \frac{Q}{4\pi\varepsilon_0\varepsilon_r b}$$

$$= \frac{Q}{4\pi\varepsilon_0\varepsilon_r}\left(\frac{1}{a} - \frac{1}{b}\right) = \frac{Q}{4\pi\varepsilon_0\varepsilon_r}\left(\frac{b-a}{ab}\right)$$

$$\frac{Q}{V} = \frac{4\pi\varepsilon_0\varepsilon_r ab}{b-a} \quad \therefore C = 4\pi\varepsilon_0\varepsilon_r \frac{ab}{b-a} \text{ F}$$

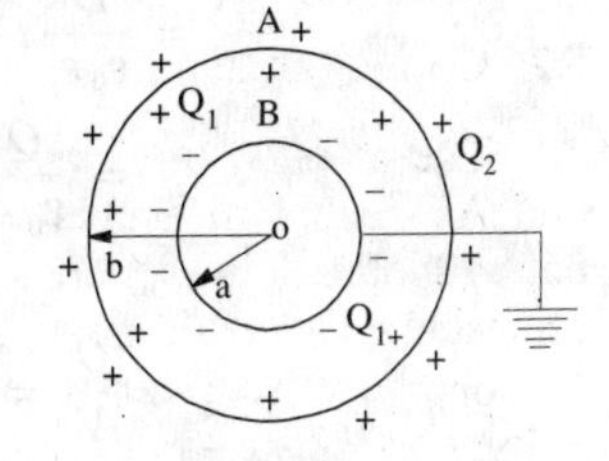

Fig. 5.4

(*b*) **When inner sphere is earthed**

Such a capacitor is shown in Fig. 5.4. If a charge of $+Q$ coulombs is given to the outer sphere A, it will distribute itself over both its inner and outer surfaces. Some charge Q_2 coulomb will remain on the outer surface of A because it is surrounded by earth all around. Also, some charge $+Q_1$ coulombs will shift to its inner side because there is an earthed sphere B inside A.

Obviously, $Q = Q_1 + Q_2$

The inner charge $+Q_1$ coulomb on A induces $-Q_1$ coulomb on B but the other induced charge of $+Q_1$ coulomb goes to earth.

Now, there are two capacitors connected in parallel :

(*i*) One capacitor consists of the inner surface of A and the outer surface of B. Its capacitance, as found earlier,

is $$C_1 = 4\pi\varepsilon_0\varepsilon_r \frac{ab}{b-a}$$

(*ii*) The second capacitor consists of outer surface of B and earth. Its capacitance is $C_2 = 4\pi\varepsilon_0 b$ — if surrounding medium is air. Total capacitance $C = C_1 + C_2$.

5.5. Parallel-plate Capacitor

(*i*) Uniform Dielectric Medium

A parallel-plate capacitor consisting of two plates M and N each of area A m^2 separated by a thickness d metres of a medium of relative permittivity ε_r is shown in Fig. 5.5. If a charge of $+Q$ coulomb is given to plate M, then flux passing through the medium is $\psi = Q$ coulomb. Flux density in the medium is

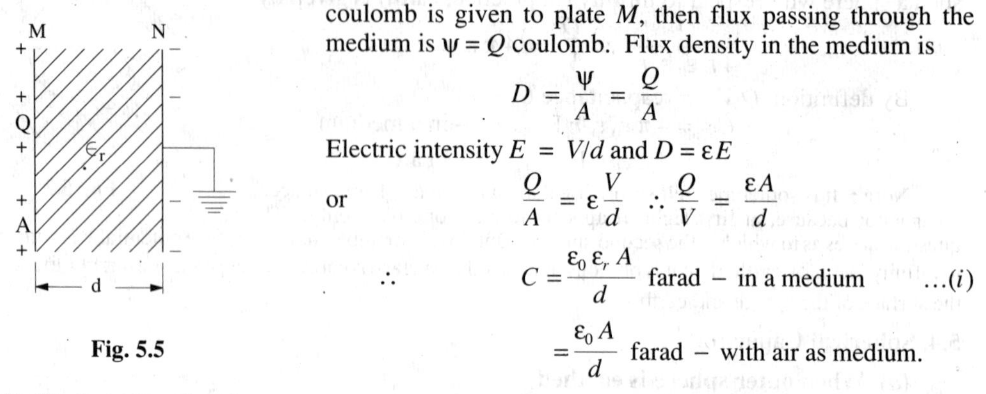

$$D = \frac{\psi}{A} = \frac{Q}{A}$$

Electric intensity $E = V/d$ and $D = \varepsilon E$

$$\text{or} \qquad \frac{Q}{A} = \varepsilon \frac{V}{d} \quad \therefore \quad \frac{Q}{V} = \frac{\varepsilon A}{d}$$

$$\therefore \qquad C = \frac{\varepsilon_0 \varepsilon_r A}{d} \text{ farad} - \text{in a medium} \qquad \ldots(i)$$

$$= \frac{\varepsilon_0 A}{d} \text{ farad} - \text{with air as medium.}$$

Fig. 5.5

(*ii*) Medium Partly Air

As shown in Fig. 5.6, the medium consists partly of air and partly of parallel-sided dielectric slab of thickness t and relative permittivity ε_r. The electric flux density $D = Q/A$ is the same in both media. But electric intensities are different.

$$E_1 = \frac{D}{\varepsilon_0 \varepsilon_r} \qquad \ldots\text{in the medium}$$

$$E_2 = \frac{D}{\varepsilon_0} \qquad \ldots \text{ in air}$$

p.d. between plates, $V = E_1 . t + E_2 (d - t)$

$$= \frac{D}{\varepsilon_0 \varepsilon_r} t + \frac{D}{\varepsilon_0}(d - t) = \frac{D}{\varepsilon_0}\left(\frac{t}{\varepsilon_r} + d - t\right)$$

$$= \frac{Q}{\varepsilon_0 A}\left[d - (t - t/\varepsilon_r)\right]$$

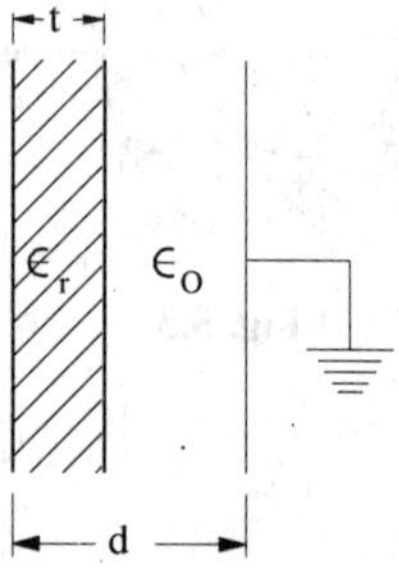

Fig. 5.6

$$\text{or} \qquad \frac{Q}{V} = \frac{\varepsilon_0 A}{\left[d - (t - t/\varepsilon_r)\right]} \quad \text{or} \quad C = \frac{\varepsilon_0 A}{\left[d - (t - t/\varepsilon_r)\right]} \qquad \ldots(ii)$$

If the medium were totally air, then capacitance would have been

$$C = \varepsilon_0 A/d \qquad \ldots(iii)$$

From (*ii*) and (*iii*), it is obvious that when a dielectric slab of thickness t and relative permittivity ε_r is introduced between the plates of an air capacitor, then its capacitance increases because as seen from (*ii*), the denominator decreases. The distance between the plates is effectively reduced by $(t - t/\varepsilon_r)$. To bring the capacitance back to its original value, the capacitor plates will have to be further separated by that much distance in air. Hence, the new separation between the two plates would be $= [d + (t - t/\varepsilon_r)]$

The expression given in (*i*) above can be written as $C = \dfrac{\varepsilon_0 A}{d/\varepsilon_r}$

If the space between the plates is filled with slabs of different thicknesses and relative permittivities, then the above expression can be generalized into $C = \dfrac{\varepsilon_0 A}{\Sigma d/\varepsilon_r}$

The capacitance of the capacitor shown in Fig. 5.7 can be written as

$$C = \frac{\varepsilon_0 A}{\left(\frac{d_1}{\varepsilon_{r1}} + \frac{d_2}{\varepsilon_{r2}} + \frac{d_3}{\varepsilon_{r3}}\right)}$$

(*iii*) **Composite Medium**

The above expression may be derived independently as given under :

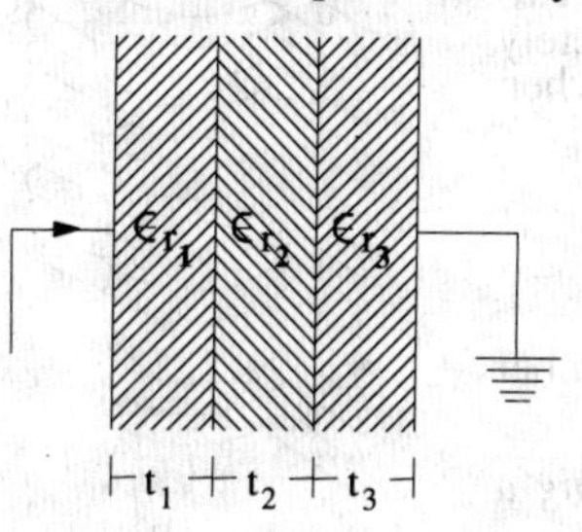

Fig. 5.7

If V is the total potential difference across the capacitor plates and V_1, V_2, V_3, the potential differences across the three dielectric slabs, then

$$V = V_1 + V_2 + V_3 = E_1 t_1 + E_2 t_2 + E_3 t_3$$

$$= \frac{D}{\varepsilon_0\, \varepsilon_{r1}} . t_1 + \frac{D}{\varepsilon_0\, \varepsilon_{r2}} . t_2 + \frac{D}{\varepsilon_0\, \varepsilon_{r3}} . t_3$$

$$= \frac{D}{\varepsilon_0}\left(\frac{t_1}{\varepsilon_{r1}} + \frac{t_2}{\varepsilon_{r2}} + \frac{t_3}{\varepsilon_{r3}}\right) = \frac{Q}{\varepsilon_0 A}\left(\frac{t_1}{\varepsilon_{r1}} + \frac{t_2}{\varepsilon_{r2}} + \frac{t_3}{\varepsilon_{r3}}\right)$$

$$\therefore C = \frac{Q}{V} = \frac{\varepsilon_0 A}{\left(\frac{t_1}{\varepsilon_{r1}} + \frac{t_2}{\varepsilon_{r2}} + \frac{t_3}{\varepsilon_{r3}}\right)}$$

5.6. Special Cases of Parallel-plate Capacitor

Consider the cases illustrated in Fig. 5.8.

(*i*) As shown in Fig. 5.8 (*a*), the dielectric is of thickness d but occupies only a part of the area. This arrangement is equal to two capacitors in parallel. Their capacitances are

$$C_1 = \frac{\varepsilon_0 A_1}{d} \text{ and } C_2 = \frac{\varepsilon_0\, \varepsilon_r A_2}{d}$$

Total capacitance of the parallel-plate capacitor is

$$C = C_1 + C_2 = \frac{\varepsilon_0 A_1}{d} + \frac{\varepsilon_0\, \varepsilon_r A_2}{d}$$

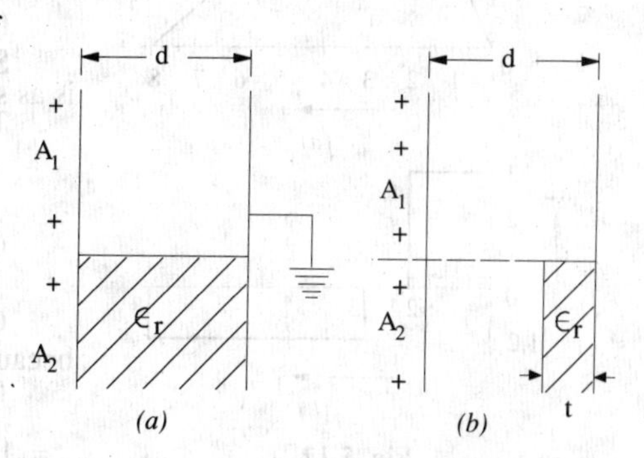

Fig. 5.8

(*ii*) The arrangement shown in Fig. 5.8 (*b*) consists of two capacitors connected in parallel.

(*a*) one capacitor having plate area A_1 and air as dielectric. Its capacitance is $C_1 = \dfrac{\varepsilon_0 A_1}{d}$

(*b*) the other capacitor has dielectric partly air and partly some other medium. Its capacitance is [Art 5.5 (*ii*)]. $C_2 = \dfrac{\varepsilon_0 A_2}{\left[d - (t - t/\varepsilon_r)\right]}$. Total capacitance is $C = C_1 + C_2$

5.7. Multiple and Variable Capacitors

Multiplate capacitors are shown in Fig. 5.9. and Fig. 5.10.

The arrangement of Fig. 5.9. is equivalent to two capacitors joined in parallel. Hence, its capacitance

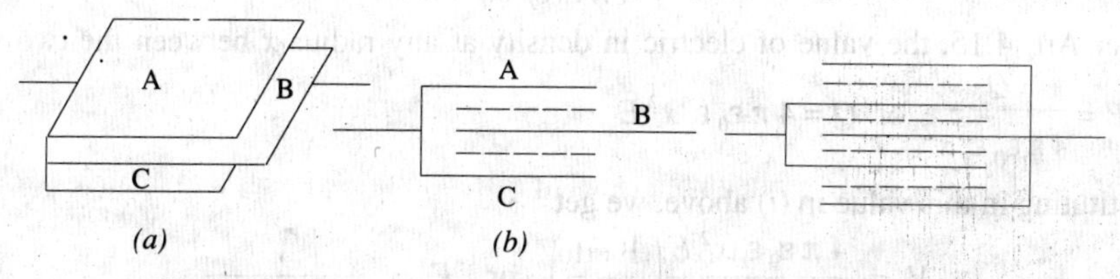

Fig. 5.9 **Fig. 5.10**

is double that of a single capacitor. Similarly, the arrangement of Fig. 5.10 has four times the capacitance of single capacitor.

If one set of plates is fixed and the other is capable of rotation, then capacitance of such a multiplate capacitor can be varied. Such variable-capacitance air capacitors are widely used in radio work (Fig. 5.11). The set of fixed plates F is insulated from the other set R which can be rotated by turning the knob K. The common area between the two sets is varied by rotating K, hence the capacitance between the two is altered. Minimum capacitance is obtained when R is completely rotated out of F and maximum when R is completely rotated in *i.e.* when the two sets of plates completely overlap each other.

The capacitance of such a capacitor is

$$= \frac{(n-1)\cdot \varepsilon_0 \varepsilon_r A}{d}$$

where n is the number of plates which means that $(n-1)$ is the number of capacitors.

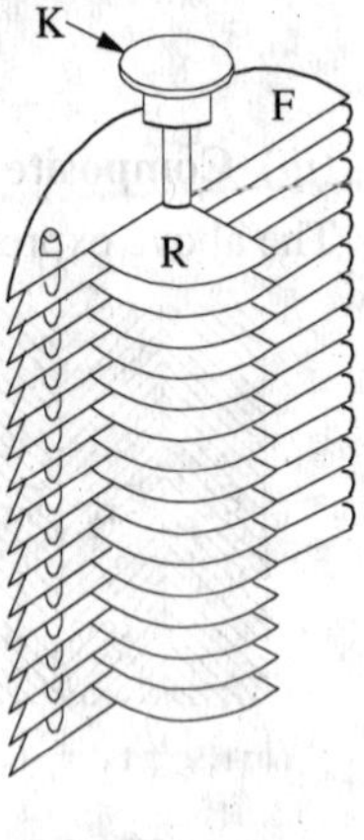

Fig. 5.11

Example 5.1. *The voltage applied across a capacitor having a capacitance of 10 μ F is varied thus :*

The p.d. is increased uniformly from 0 to 600 V in 2 seconds. It is then maintained constant at 600 V for 1 second and subsequently decreased uniformly to zero in five seconds. Plot a graph showing the variation of current during these 8 seconds. Calculate (a) the charge and (b) the energy stored in the capacitor when the terminal voltage is 600.

(Principles of Elect. Engg.-I, Jadavpur Univ. 1987)

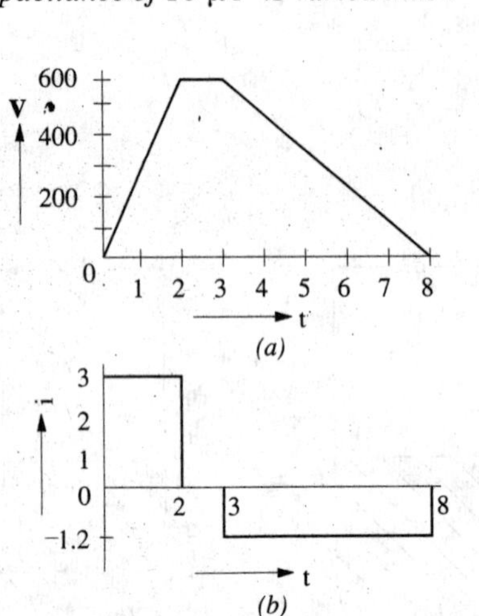

Fig. 5.12

Solution. The variation of voltage across the capacitor is as shown in Fig. 5.12 (*a*).

The charging current is given by

$$i = \frac{dq}{dt} = \frac{d}{dt}(Cv) = C \cdot \frac{dv}{dt}$$

Charging current during the first stage

$$= 10 \times 10^{-6} \times (600/2) = 3 \times 10^{-3}\ \text{A} = 3\ \text{mA}$$

Charging current during the second stage is zero because $dv/dt = 0$ as the voltage remains constant.

Charging current through the third stage

$$= 10 \times 10^{-6} \times \left(\frac{0-600}{5}\right) = -1.2 \times 10^{-3}\ \text{A} = -1.2\ \text{mA}$$

The waveform of the charging current or capacitor current is shown in Fig. 5.12. (*b*).

(*a*) Charge when a steady voltage of 600 V is applied is

$$= 600 \times 10 \times 10^{-6} = \mathbf{6 \times 10^{-3}\ C}$$

(*b*) Energy stored $= \frac{1}{2} C\,V^2 = \frac{1}{2} \times 10^{-5} \times 600^2 = \mathbf{1.8\ J}$

Example 5.2. *A voltage of V is applied to the inner sphere of a spherical capacitor, whereas the outer sphere is earthed. The inner sphere has a radius of a and the outer one of b. If b is fixed and a may be varied, prove that the maximum stress in the dielectric cannot be reduced below a value of 4 V/b.*

Solution. As seen from Art. 5.4,

$$V = \frac{Q}{4\pi\varepsilon_0\varepsilon_r}\left(\frac{1}{a} - \frac{1}{b}\right) \qquad \text{...}(i)$$

As per Art. 4.15, the value of electric in density at any radius x between the two spheres is given by $E = \dfrac{Q}{4\pi\varepsilon_0\varepsilon_r x^2}$ or $Q = 4\pi\varepsilon_0\varepsilon_r x^2 E$

Substituting in this value in (*i*) above, we get

$$V = \frac{4\pi\varepsilon_0\varepsilon_r x^2 E}{4\pi\varepsilon_0\varepsilon_r}\left(\frac{1}{a} - \frac{1}{b}\right) \quad \text{or} \quad E = \frac{V}{(1/a - 1/b)\,x^2}$$

As per Art. 5.9, the maximum value of E occurs at the surface of inner sphere *i.e.* when $x = a$

For E to be maximum or minimum, $dE/da = 0$.

$$\therefore \quad \frac{d}{da}\left(\frac{1}{a}-\frac{1}{b}\right)a^2 = 0 \quad \text{or} \quad \frac{d}{da}(a - a^2/b) = 0$$

or $$1 - 2\,a/b = 0 \quad \text{or} \quad a = b/2$$

Now, $$E = \frac{V}{(1/a - 1/b)\,x^2} \quad \therefore E_{max} = \frac{V}{(1/a - 1/b)a^2} = \frac{V}{(a - a^2/b)}$$

Since, $a = b/2 \; \therefore \; E_{max} = \dfrac{V}{(b/2 - b^2/4b)} = \dfrac{4\,bV}{2\,b^2 - b^2} = \dfrac{4\,bV}{b^2} = \dfrac{\mathbf{4\,V}}{\mathbf{b}}$

Example 5.3. *A.capacitor consists of two similar square aluminium plates, each 10 cm × 10 cm mounted parallel and opposite to each other. What is their capacitance in μμ F when distance between them is 1 cm and the dielectric is air ? If the capacitor is given a charge of 500 μμ C, what will be the difference of potential between plates ? How will this be affected if the space between the plates is filled with wax which has a relative permittivity of 4 ?*

Solution. $$C = \varepsilon_0 A/d \text{ farad}$$

Here $\varepsilon_0 = 8.854 \times 10^{-12}$ F/m ; $A = 10 \times 10 = 100 \text{ cm}^2 = 10^{-2} \text{ m}^2$

$d = 1 \text{ cm} = 10^{-2}$ m

$$\therefore \quad C = \frac{8.854 \times 10^{-12} \times 10^{-2}}{10^{-2}} = 8.854 \times 10^{-12} \text{ F} = \mathbf{8.854\ \mu\mu F}$$

Now $C = \dfrac{Q}{V} \quad \therefore \quad V = \dfrac{Q}{C}$ or $V = \dfrac{500 \times 10^{-12}\ \text{C}}{8.854 \times 10^{-12}\ \text{F}} = \mathbf{56.5\ volts.}$

When wax is introduced, their capacitance is increased four times because

$$C = \varepsilon_0 \varepsilon_r A/d \text{ F} = 4 \times 8.854 = 35.4\ \mu\mu \text{ F}$$

The p.d. will obviously decrease to one fourth value because charge remains constant.

$$\therefore \quad V = 56.5/4 = \mathbf{14.1\ volts.}$$

Example 5.4. *The capacitance of a capacitor formed by two parallel metal plates each 200 cm² in area separated by a dielectric 4 mm thick is 0.0004 microfarads. A p.d. of 20,000 V is applied. Calculate (a) the total charge on the plates (b) the potential gradient in V/m (c) relative permittivity of the dielectric (d) the electric flux density.* **(Elect. Engg. I Osmaina Univ. 1988)**

Solution. $C = 4 \times 10^{-4}\ \mu\text{F}$; $V = 2 \times 10^4$ V

(*a*) ∴ Total charge $Q = CV = 4 \times 10^{-4} \times 2 \times 10^4\ \mu\text{C} = 8\ \mu\text{C} = \mathbf{8 \times 10^{-6}\ C}$

(*b*) Potential gradient $= \dfrac{dV}{dx} = \dfrac{2 \times 10^4}{4 \times 10^{-3}} = \mathbf{5 \times 10^6\ V/m}$

(*c*) $D = Q/A = 8 \times 10^{-6}/200 \times 10^{-4} = \mathbf{4 \times 10^{-4}\ C/m^2}$

(*d*) $E = 5 \times 10^6$ V/m

Since $D = \varepsilon_0 \varepsilon_r E \quad \therefore \varepsilon_r = \dfrac{D}{\varepsilon_0 \times E} = \dfrac{4 \times 10^{-4}}{8.854 \times 10^{-12} \times 5 \times 10^6} = \mathbf{9}$

Example 5.5. *A parallel plate capacitor has 3 dielectrics with relative permittivities of 5.5, 2.2 and 1.5 respectively. The area of each plate is 100 cm² and thickness of each dielectric 1 mm. Calculate the stored charge in the capacitor when a potential difference of 5,000 V is applied across the composite capacitor so formed. Calculate the potential gradient developed in each dielectric of the capacitor.* **(Elect. Engg. A.M.Ae. S.I. June 1990)**

Solution. As seen from Art. 5.5,

$$C = \frac{\varepsilon_0 A}{\left(\frac{d_1}{\varepsilon_{r1}} + \frac{d_2}{\varepsilon_{r2}} + \frac{d_3}{\varepsilon_{r3}}\right)} = \frac{8.854 \times 10^{-12} \times (100 \times 10^{-4})}{\left(\frac{10^{-3}}{5.5} + \frac{10^{-3}}{2.2} + \frac{10^{-3}}{1.5}\right)} = \frac{8.854 \times 10^{-14}}{10^{-3} \times 0.303} = 292 \text{ pF}$$

$$Q = CV = 292 \times 10^{-12} \times 5000 = 146 \times 10^{-8} \text{ coulomb}$$
$$D = Q/A = 146 \times 10^{-8}/(100 \times 10^{-4}) = 146 \times 10^{-6} \text{ C/m}^2$$
$$g_1 = E_1 = D/\varepsilon_0 \varepsilon_{r1} = 146 \times 10^{-6}/8.854 \times 10^{-12} \times 5.5 = 3 \times 10^6 \text{ V/m}$$
$$g_2 = E_2 = D/\varepsilon_0 \varepsilon_{r2} = 7.5 \times 10^6 \text{ V/m}; \; g_3 = D/\varepsilon_0 \varepsilon_{r3} = 11 \times 10^6 \text{ V/m}$$

Example 5.6. *An air capacitor has two parallel plates 10 cm² in area and 0.5 cm apart. When a dielectric slab of area 10 cm² and thickness 0.5 cm was inserted between the plates, one of the plates has to be moved by 0.4 cm to restore the capacitance. What is the dielectric constant of the slab ?* **(Elect. Technology, Hyderabad Univ. 1992)**

Solution. The capacitance in the first case is

$$C_a = \frac{\varepsilon_0 A}{d} = \frac{\varepsilon_0 \times 10 \times 10^{-4}}{0.5 \times 10^{-2}} = \frac{\varepsilon_0}{5}$$

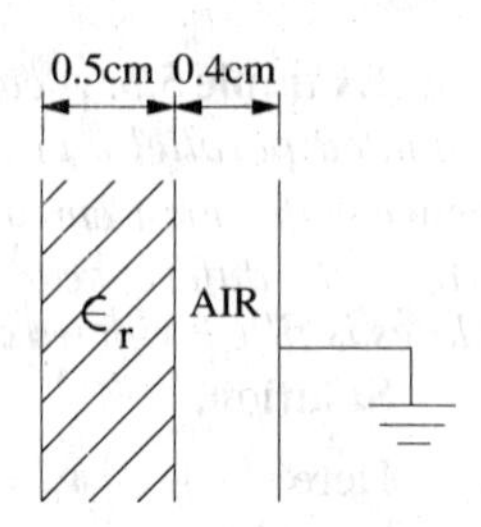

Fig. 5.13

The capacitor, as it becomes in the second case, is shown in Fig. 5.13. The capacitance is

$$C_m = \frac{\varepsilon_0 A}{\Sigma d/\varepsilon_r} = \frac{\varepsilon_0 \times 10^{-3}}{\left(\frac{5 \times 10^{-3}}{\varepsilon_r} + \frac{4 \times 10^{-3}}{1}\right)} = \frac{\varepsilon_0}{\left(\frac{5}{\varepsilon_r} + 4\right)}$$

$$\text{Since, } C_a = C_m \quad \therefore \quad \frac{\varepsilon_0}{5} = \frac{\varepsilon_0}{(5/\varepsilon_r + 4)} \quad \therefore \quad \varepsilon_r = 5$$

Note. We may use the relation derived in Art. 5.5 (*ii*)

Separation $= (t - t/\varepsilon_1)$ $\quad\therefore\quad 0.4 = (0.5 - 0.5/\varepsilon_r)$ or $\varepsilon_r = 5$

Example 5.7. *A parallel plate capacitor of area, A, and plate separation, d, has a voltage, V_0, applied by a battery. The battery is then disconnected and a dielectric slab of permittivity ε_1 and thickness, d_1, ($d_1 < d$) is inserted. (a) Find the new voltage V_1 across the capacitor. (b) Find the capacitance C_0 before and its value C_1 after the slab is introduced. (c) Find the ratio V_1/V_0 and the ratio C_1/C_0 when $d_1 = d/2$ and $\varepsilon_1 = 4\varepsilon_0$.*

(Electromagnetic Fields and Waves AMIETE (New Scheme) June 1990)

Solution. (*b*) $C_0 = \dfrac{\varepsilon_0 A}{d}$; $C_1 = \dfrac{A}{\left(\dfrac{(d - d_1)}{\varepsilon_0} + \dfrac{d_1}{\varepsilon_1}\right)}$

Since $d_1 = d/2$ and $\varepsilon_1 = 4\,\varepsilon_0$ $\therefore$ $C_1 = \dfrac{A}{\left(\dfrac{d}{2\varepsilon_0} + \dfrac{d}{2 \times 4\varepsilon_0}\right)} = \dfrac{8\,\varepsilon_0 A}{5\,d}$

(*a*) Since the capacitor charge remains the same

$$Q = C_0 V_0 = C_1 V_1 \quad \therefore \quad V_1 = V_0 \frac{C_0}{C_1} = V_0 \times \frac{\varepsilon_0 A}{d} \times \frac{5\,d}{8\,\varepsilon_0 A} = \frac{5V_0}{8}$$

(*c*) As seen from above, $V_1 = V_0 5/8$; $C_1 C_0 = \dfrac{8\,\varepsilon_0 A}{5d} \times \dfrac{d}{\varepsilon_0 A} = \dfrac{8}{5}$

Tutorial Problems No. 5.1.

1. Two parallel plate capacitors have plates of an equal area, dielectrics of relative permittivities ε_{r1} and ε_{r2} and plate spacings of d_1 and d_2. Find the ratio of their capacitances if $\varepsilon_{r1}/\varepsilon_{r2} = 2$ and $d_1/d_2 = 0.25$. **[$C_1/C_2 = 8$]**

2. A capacitor is made of two plates with an area of 11 cm² which are separated by a mica sheet 2 mm thick. If for mica $\varepsilon_r = 6$, find its capacitance. If, now, one plate of the capacitor is moved further to give an air gap 0.5 mm wide between the plates and mica, find the change in capacitance. **[29.19 pF, 11.6 pF]**

3. A parallel-plate capacitor is made of two plane circular plates separated by *d* cm of air. When a parallel-faced plane sheet of glass 2 mm thick is placed between the plates, the capacitance of the system is increased by 50% of its initial value. What is the distance between the plates if the dielectric constant of the glass is 6 ? **[0.5×10^{-3} m]**

4. A p.d. of 10 kV is applied to the terminals of a capacitor consisting of two circular plates, each having an area of 100 cm^2 separated by a dielectric 1 mm thick. If the capacitance is 3×10^{-4} μ F, calculate

(*a*) the total electric flux in coulomb (*b*) the electric flux density and

(*c*) the relative permittivity of the dielectric. **[(*a*) 3×10^{-6} C (*b*) 3×10^{-4} μ C/m^2 (*c*) 3.39]**

5. Two slabs of material of dielectric strength 4 and 6 and of thicknesses 2mm and 5 mm respectively are inserted between the plates of a parallel-plate capacitor. Find by how much the distance between the plates should be changed so as to restore the potential of the capacitor to its original value. **[5.67 mm]**

6. The oil dielectric to be used in a parallel-plate capacitor has a relative permittivity of 2.3 and the maximum working potential gradient in the oil is not to exceed 10^6 V/m. Calculate the approximate plate area required for a capacitance of 0.0003 μ F, the maximum working voltage being 10,000 V. **[147×10^{-3}m^2]**

7. A capacitor consists of two metal plates, each 10 cm square placed parallel and 3 mm apart. The space between the plates is occupied by a plate of insulating material 3 mm thick. The capacitor is charged to 300 V.

(*a*) the metal plates are isolated from the 300 V supply and the insulating plate is removed. What is expected to happen to the voltage between the plates ?

(*b*) if the metal plates are moved to a distance of 6 mm apart, what is the further effect on the voltage between them. Assume throughout that the insulation is perfect.

[300 ε_r ; 600 ε_r ; where ε_r is the relative permittivity of the insulating material]

8. A parallel-plate capacitor has an effective plate area of 100 cm^2 (each plate) separated by a dielectric 0.5 mm thick. Its capacitance is 442 μμ F and it is raised to a potential difference of 10 kV. Calculate from first principles

(*a*) potential gradient in the dielectric (*b*) electric flux density in the dielectric

(*c*) the relative permittivity of the dielectric material. **[(*a*) 20 kV/mm (*b*) 442 μC/m^2 (*c*) 2.5]**

9. A parallel-plate capacitor with fixed dimensions has air as dielectric. It is connected to a supply of p.d. V volts and then isolated. The air is then replaced by a dielectric medium of relative permittivity 6. Calculate the change in magnitude of each of the following quantities.

(*a*) the capacitance (*b*) the charge (*c*) the p.d. between the plates

(*d*) the displacement in the dielectric (*e*) the potential gradient in the dielectric.

[(*a*) 6 : 1 increase (*b*) no change (*c*) 6 : 1 decrease (*d*) no change (*e*) 6 : 1 decrease]

5.8. Cylindrical Capacitor

A single-core cable or cylindrical capacitor consisting of two co-axial cylinders of radii *a* and *b* meters, is shown in Fig. 5.14. Let the charge per metre length of the cable on the outer surface of the inner cylinder be + *Q* coulomb and on the inner surface of the outer cylinder be − *Q* coulomb. For all practical purposes, the charge + *Q* coulomb/metre on the surface of the inner cylinder can be supposed to be located along its axis. Let ε_r be the relative permittivity of the medium between the two cylinders. The outer cylinder is earthed.

Now, let us find the value of electric intensity at any point distant *x* metres from the axis of the inner cylinder. As shown in Fig. 5.15, consider an imaginary co-axial cylinder of radius *x* metres and length one metre between the two given cylinders. The electric field between the two cylinders is radial as shown. Total flux coming out radially from the curved surface of this imaginary cylinder is *Q* coulomb. Area of the curved surface $= 2\,\pi\,x \times 1 = 2\,\pi\,x$ m^2.

Hence, the value of electric flux density on the surface of the imaginary cylinder is

$$D = \frac{\text{flux in coulomb}}{\text{area in metre}^2} = \frac{\psi}{A} = \frac{Q}{A}\ \text{C/m}^2 \quad \therefore\ D = \frac{Q}{2\pi x}\ \text{C/m}^2$$

The value of electric intensity is

$$E = \frac{D}{\varepsilon_0\,\varepsilon_r} \quad \text{or} \quad E = \frac{Q}{2\,\pi\,\varepsilon_0\,\varepsilon_r\,x}\ \text{V/m}$$

Now, $dV = -E\,dx$

$$\text{or } V = \int_b^a -E.dx = \int_b^a -\frac{Q\,dx}{2\,\pi\,\varepsilon_0\,\varepsilon_r\,x}$$

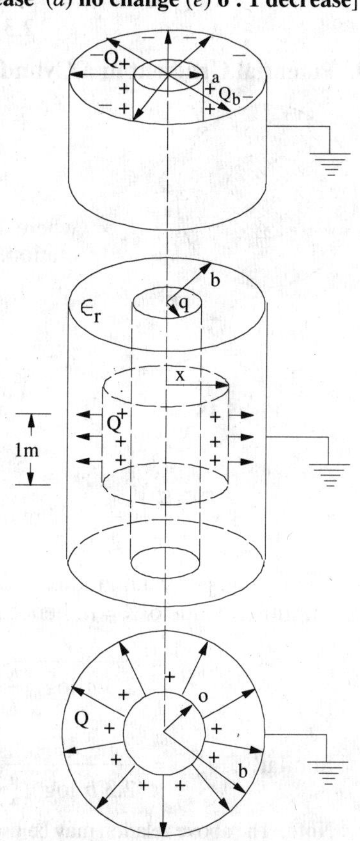

Fig. 5.14

$$= \frac{-Q}{2\pi\varepsilon_0\varepsilon_r}\int_b^a \frac{dx}{x} = \frac{-Q}{2\pi\varepsilon_0\varepsilon_r}\left|\log x\right|_b^a$$

$$= \frac{-Q}{2\pi\varepsilon_0\varepsilon_r}(\log_e a - \log_e b) = \frac{-Q}{2\pi\varepsilon_0\varepsilon_r}\log_e\left(\frac{a}{b}\right)$$

$$= \frac{Q}{2\pi\varepsilon_0\varepsilon_r}\log_e\left(\frac{b}{a}\right)$$

$$\frac{Q}{V} = \frac{2\pi\varepsilon_0\varepsilon_r}{\log_e\left(\frac{b}{a}\right)} \quad \therefore C = \frac{2\pi\varepsilon_0\varepsilon_r}{2.3\log_{10}\left(\frac{b}{a}\right)} \text{ F/m} \quad \left(\because \log_e\left(\frac{b}{a}\right) = 2.3\log_{10}\left(\frac{b}{a}\right)\right)$$

The capacitance of l metre length of this cable is $C = \dfrac{2\pi\varepsilon_0\varepsilon_r l}{2.3\log_{10}\left(\frac{b}{a}\right)}$ F

In case the capacitor has compound dielectric, the relation becomes.

$$C = \frac{2\pi\varepsilon_0 l}{\Sigma\log_e\left(\frac{b}{a}\right)/\varepsilon_r} \text{ F}$$

The capacitance of 1 km length of the cable in μF can be found by putting $l = 1$ km in the above expression.

$$C = \frac{2\pi\times 8.854\times 10^{-12}\times\varepsilon_r\times 1000}{2.3\log_{10}\left(\frac{b}{a}\right)} \text{ F/km} = \frac{0.024\,\varepsilon_r}{\log_{10}\left(\frac{b}{a}\right)}\ \mu\text{F/km}$$

5.9. Potential Gradient in a Cylindrical Capacitor

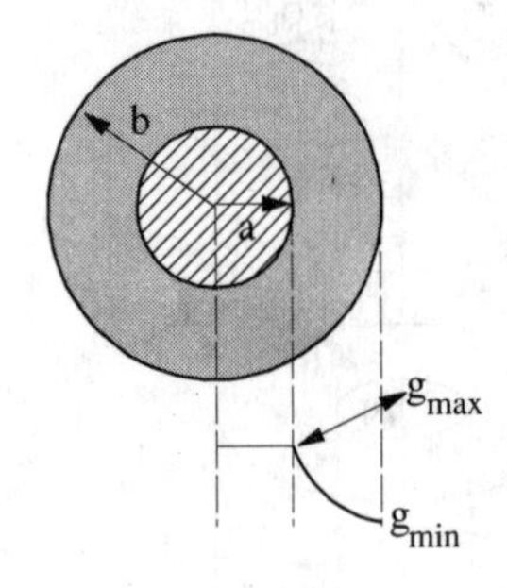

Fig. 5.15

It is seen from Art. 5.8 that in a cable capacitor

$$E = \frac{Q}{2\pi\varepsilon_0\varepsilon_r x} \text{ V/m}$$

where x is the distance from cylinder axis to the point under consideration.

$$\text{Now } E = g \quad \therefore g = \frac{Q}{2\pi\varepsilon_0\varepsilon_r x} \text{ V/m} \qquad \ldots(i)$$

From Art. 5.8, we find that $V = \dfrac{Q}{2\pi\varepsilon_0\varepsilon_r}\log_e\left(\dfrac{b}{a}\right)$ or $Q = \dfrac{2\pi\varepsilon_0\varepsilon_r V}{\log_e\left(\frac{b}{a}\right)}$

Substituting this value of Q in (i) above, we get

$$g = \frac{2\pi\varepsilon_0\varepsilon_r V}{\log_e\left(\frac{b}{a}\right)\times 2\pi\varepsilon_0\varepsilon_r x} \text{ V/m or } g = \frac{V}{x\log_e\left(\frac{b}{a}\right)} \text{ V/m or } g = \frac{V}{2.3\,x\log_{10}\left(\frac{b}{a}\right)} \text{ volt/metre}$$

Obviously, potential gradient varies inversely as x.
Minimum value of $x = a$, hence maximum value of potential gradient is

$$g_{max} = \frac{V}{2.3\,a\log_{10}\left(\frac{b}{a}\right)} \text{ V/m} \qquad \ldots(ii)$$

$$\text{Similarly, } g_{max} = \frac{V}{2.3\,b\log_{10}\left(\frac{b}{a}\right)} \text{ V/m}$$

Note. The above relation may be used to obtain most economical dimensions while designing a cable. As seen, greater the value of permissible maximum stress E_{max}, smaller the cable may be for given value of V. However, E_{max} is dependent on the dielectric strength of the insulating material used.

If V and E_{max} are fixed, then Eq. (*ii*) above may be written as

$$E_{max} = \frac{V}{a \log h_e\left(\frac{b}{a}\right)} \text{ or } a \log h\left(\frac{b}{a}\right) = \frac{V}{E_{max}} \quad \therefore \quad \frac{b}{a} = e^{k/a} \text{ or } b = a\, e^{k/a}$$

For most economical cable $db/da = 0$

$$\therefore \quad \frac{db}{da} = 0 = e^{k/a} + a(-k/a^2)e^{k/a} \text{ or } a = k = V/E_{max} \text{ and } b = ae = 2.718\, a$$

Example 5.8. *A cable is 300 km long and has a conductor of 0.5 cm in diameter with an insulation covering of 0.4 cm thickness. Calculate the capacitance of the cable if relative permittivity of insulation is 4.5.* **(Elec. Engg. A.M.Ae.S.I. June 1987)**

Solution. Capacitance of a cable is $C = \dfrac{0.024\,\varepsilon_r}{\log_{10}\left(\frac{b}{a}\right)}\ \mu$ F/km

Here, $a = 0.5/2 = 0.25$ cm ; $b = 0.25 + 0.4 = 0.65$ cm ; $b/a = 0.65/0.25 = 2.6$; $\log_{10}{}^{2.6} = 0.415$

$$\therefore \quad C = \frac{0.024 \times 4.5}{0.415} = 0.26$$

Total capacitance for 300 km is $= 300 \times 0.26 = 78\ \mu$ F.

Example 5.9. *In a concentric cable capacitor, the diameters of the inner and outer cylinders are 3 and 10 mm respectively. If ε_r for insulation is 3, find its capacitance per metre.*

A p.d. of 600 volts is applied between the two conductors. Calculate the values of the electric force and electric flux density : (a) at the surface of inner conductor (b) at the inner surface of the outer conductor.

Solution. $a = 1.5$ mm; $b = 5$ mm ; $\therefore$ $b/a = 5/1.5 = 10/3$; $\log_{10}\left(\frac{10}{3}\right) = 0.523$

$$C = \frac{2\,\pi\,\varepsilon_0\,\varepsilon_r\, l}{2.3 \log_{10}\left(\frac{b}{a}\right)} = \frac{2\,\pi \times 8.854 \times 10^{-12} \times 3 \times 1}{2.3 \times 0.523} = 138.8 \times 10^{-12} \text{ F} = \mathbf{138.8\ pF}$$

(*a*) $D = Q/2\pi a$

Now $Q = CV = 138.8 \times 10^{-12} \times 600 = 8.33 \times 10^{-9}$ C

$D = 8.33 \times 10^{-8}/2\pi \times 1.5 \times 10^{-3} = \mathbf{8.835\ \mu C/m^2}$

$E = D/\varepsilon_0\,\varepsilon_r = \mathbf{332.6\ V/m}$

(*b*) $D = \dfrac{8.33 \times 10^{-8}}{2\pi \times 5 \times 10^{-3}}$ C/m^2 $= \mathbf{2.65\ \mu\ C/m^2}$; $E = D/\varepsilon_0\varepsilon_r = \mathbf{99.82\ \ V/m.}$

Example 5.10. *The radius of the copper core of a single-core rubber-insulated cable is 2.25 mm. Calculate the radius of the lead sheath which covers the rubber insulation and the cable capacitance per meter. A voltage of 10 kV may be applied between the core and the lead sheath with a safety factor of 3. The rubber insulation has a relative permittivity of 4 and breakdown field strength of 18×10^6 V/m.*

Solution. As shown in Art. 5.9, $g_{max} = \dfrac{V}{2.3\, a \log_{10}\left(\frac{b}{a}\right)}$

Now, $g_{max} = E_{max} = 18 \times 10^6$ V/m ; V = breakdown voltage x
Safety factor = $10^4 \times 3 = 30{,}000$ V

$$\therefore\ 18\times10^6 = \frac{30{,}000}{2.3\times2.25\times10^{-3}\times\log_{10}\left(\frac{b}{a}\right)} \quad \therefore\ \frac{b}{a} = 2.1 \text{ or } b = 2.1\times2.25 = 4.72 \text{ mm}$$

$$C = \frac{2\pi\varepsilon_0\varepsilon_r l}{2.3\log_{10}\left(\frac{b}{a}\right)} = \frac{2\pi\times8.854\times10^{-12}\times4\times1}{2.3\log_{10}(2.1)} = \mathbf{3\times10^{-9}\ F}$$

5.10. Capacitance Between Two Parallel Wires

This case is of practical importance in overhead transmission lines. The simplest system is 2-wire system (either *d.c.* or *a.c.*). In the case of *a.c.* system, if the transmission line is long and voltage high, the charging current drawn by the line due to the capacitance between conductors is appreciable and affects its performance considerably.

With reference to Fig. 5.16, let

d = distance between centres of the wires A and B

r = radius of each wire ($\leq d$)

Q = charge in coulomb/metre of each wire*

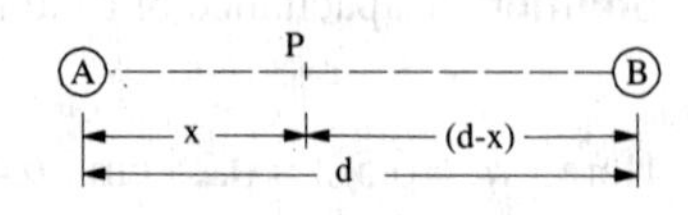

Fig. 5.16

Now, let us consider electric intensity at any point P between conductors A and B.

Electric intensity at P due to charge $+Q$ coulomb/metre on A is

$$= \frac{Q}{2\pi\varepsilon_0\varepsilon_r x}\ \text{V/m} \qquad \text{– towards } B.$$

Electric intensity at P due to charge $-Q$ coulomb/metre on B is

$$= \frac{Q}{2\pi\varepsilon_0\varepsilon_r (d-x)}\ \text{V/m} \qquad \text{– towards } B.$$

Total electric intensity at P

$$E = \frac{Q}{2\pi\varepsilon_0\varepsilon_r}\left(\frac{1}{x}+\frac{1}{d-x}\right)$$

Hence, potential difference between the two wires is

$$V = \int_r^{d-r} E.dx = \frac{Q}{2\pi\varepsilon_0\varepsilon_r}\int_r^{d-r}\left(\frac{1}{x}+\frac{1}{d-x}\right)dx$$

$$V = \frac{Q}{2\pi\varepsilon_0\varepsilon_r}\left|\log_e x - \log_e(d-x)\right|_r^{d-r} = \frac{Q}{\pi\varepsilon_0\varepsilon_r}\log_e\frac{d-r}{r}$$

Now

$$C = Q/V \quad \therefore\ C = \frac{\pi\varepsilon_0\varepsilon_r}{\log_e\frac{(d-r)}{r}} = \frac{\pi\varepsilon_0\varepsilon_r}{2.3\log_{10}\frac{(d-r)}{r}} = \frac{\pi\varepsilon_0\varepsilon_r}{2.3\log_{10}\left(\frac{d}{r}\right)}\ \text{F/m (approx.)}$$

The capacitance for a length of l metres is $C = \dfrac{\pi\varepsilon_0\varepsilon_r l}{2.3\log_{10}\left(\frac{d}{r}\right)}$ F

The capacitance per kilometre is

$$C = \frac{\pi\times8.854\times10^{-12}\times\varepsilon_r\times100\times10^6}{2.3\log_{10}\left(\frac{d}{r}\right)} = \frac{0.0121\varepsilon_r}{\log_{10}\left(\frac{d}{r}\right)} = \mu\ \text{F/km}$$

Example 5.11. *The conductors of a two-wire transmission line (4 km long) are spaced 45 cm between centres. If each conductor has a diameter of 1.5 cm, calculate the capacitance of the line.*

Solution. Formula used

$$C = \frac{\pi\varepsilon_0\varepsilon_r l}{2.3\log_{10}\left(\frac{d}{r}\right)}\ \text{F}$$

---

*If charge on A is $+Q$, then on B will be $-Q$.

Here l = 4000 metres ; r = 1.5/2 cm ; d = 45 cm ; ε_r = 1 – for air $\therefore \frac{d}{r} = \frac{45 \times 2}{1.5} = 60$

$$C = \frac{\pi \times 8.854 \times 10^{-12} \times 4000}{2.3 \log_{10} 60} = \mathbf{0.0272 \times 10^{-6} F}$$

[or $$C = 4 \times \frac{0.0121}{\log_{10} 60} = 0.0272\ \mu F]$$

5.11. Capacitors in Series

With reference to Fig. 5.17, let

C_1, C_2, C_3 = capacitances of three capacitors
V_1, V_2, V_3 = p.ds. across three capacitors.
V = applied voltage across combination
C = combined or equivalent or joint capacitance.

In series combination, charge on all capacitors is the same but p.d. across each is different,

$$\therefore \quad V = V_1 + V_2 + V_3$$

or $$\frac{Q}{C} = \frac{Q}{C_1} + \frac{Q}{C_2} + \frac{Q}{C_3}$$

or $$\frac{1}{C} = \frac{1}{C_1} + \frac{1}{C_2} + \frac{1}{C_3}$$

For a changing applied voltage,

$$\frac{dV}{dt} = \frac{dV_1}{dt} + \frac{dV_2}{dt} + \frac{dV_3}{dt}$$

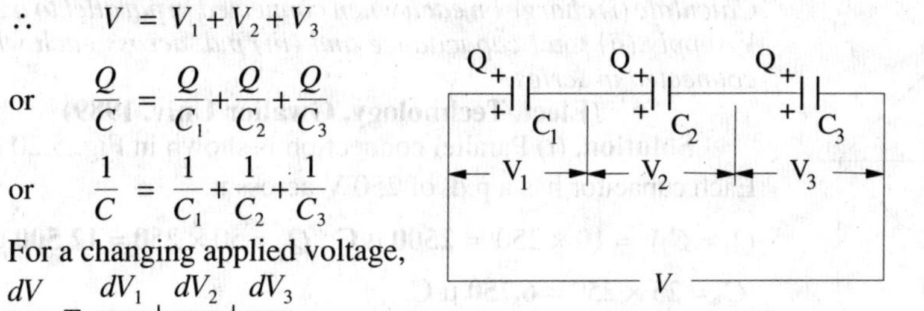

Fig. 5.17

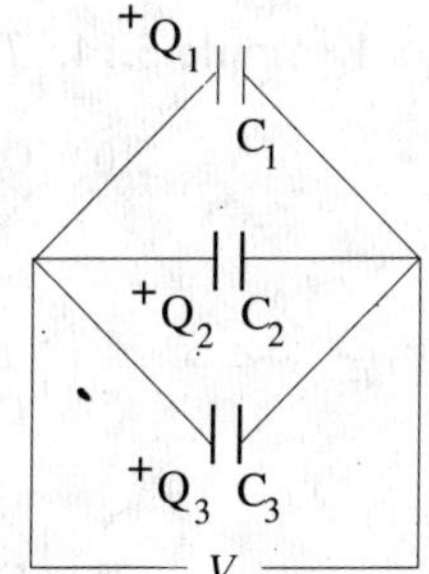

Fig. 5.18

We can also find values of V_1, V_2 and V_3 in terms of V. Now, $Q = C_1 V_1 = C_2 V_2 = C_3 V_3 = CV$

where $$C = \frac{C_1 C_2 C_3}{C_1 C_2 + C_2 C_3 + C_3 C_1} = \frac{C_1 C_2 C_3}{\Sigma C_1 C_2}$$

$$\therefore \quad C_1 V_1 = C\,V \text{ or } V_1 = V \cdot \frac{C}{C_1} = V \frac{C_2 C_3}{\Sigma C_1 C_2}$$

Similarly, $$V_2 = V \cdot \frac{C_1 C_3}{\Sigma C_1 C_2} \text{ and } V_3 = V \cdot \frac{C_1 C_2}{\Sigma C_1 C_2}$$

5.12. Capacitors in Parallel

In this case, p.d. across each is the same but charge on each is different (Fig. 5.18).

$\therefore\ Q = Q_1 + Q_2 + Q_3$ or $CV = C_1 V + C_2 V + C_3 V$ or $C = C_1 + C_2 + C_3$

For such a combination, dV/dt is the same for all capacitors.

Example 5.12. *Find the C_{eq} of the circuit shown in Fig. 5.19. All capacitances are in μ F.*
(Basic Circuit Analysis Osmania Univ. Jan./Feb. 1992)

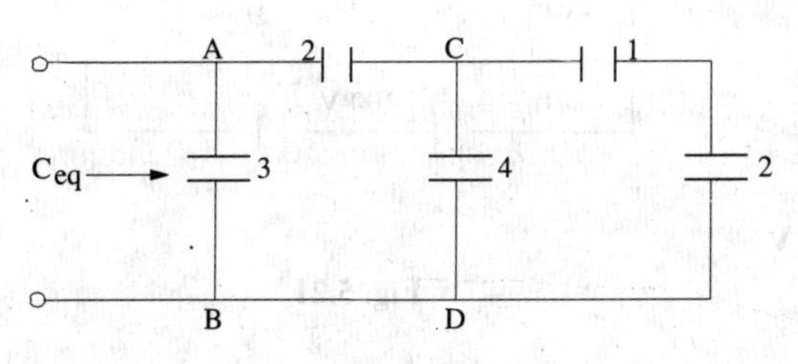

Fig. 5.19

Solution. Capacitance between C and D = 4 + 1 || 2 = 14/3 μ F

Capacitance between A and B *i.e.* C_{eq} = 3 + 2 || 14/3 = 4.4 μ F

Example 5.13. *Two capacitors of capacitance 4 μ F and 2 μ F respectively, are joined in series with a battery of e.m.f. 100 V. The connections are broken and the like terminals of the capacitors are then joined. Find the final charge on each capacitor.*

Solution. When joined in series, let V_1 and V_2 be the voltages across the capacitors. Then as charge across each is the same.

$\therefore \quad 4 \times V_1 = 2V_2 \qquad \therefore \quad V_2 = 2V_1$ Also $V_1 + V_2 = 100$

$\therefore \quad V_1 + 2V_1 = 100 \qquad \therefore \quad V_1 = 100/3$ V and $V_2 = 200/3$ V

$\therefore \quad Q_1 = Q_2 = (200/3) \times 2 = (400/3)\ \mu$ C

$\therefore$ Total charge on both capacitors = 800/3 μ C

When joined in parallel, a redistribution of charge takes place because both capacitors are reduced to a common potential V.

Total charge = 800/3 μ C ; total capacitance = 4 + 2 = 6 μ F

$$\therefore \quad V = \frac{800}{3 \times 6} = \frac{400}{9} \text{ volts}$$

Hence $Q_1 = (400/9) \times 4 = 1600/9 = \mathbf{178\ \mu\ C}$

$Q_2 = (400/9) \times 2 = 800/9 = \mathbf{89\ \mu\ C}$ **(approx.)**

Example 5.14. *Three capacitors A, B, C have capacitances 10, 50 and 25 μ F respectively. Calculate (i) charge on each when connected in parallel to a 250 V supply (ii) total capacitance and (iii) p.d. across each when connected in series.*

(Elect. Technology, Gwalior Univ. 1989)

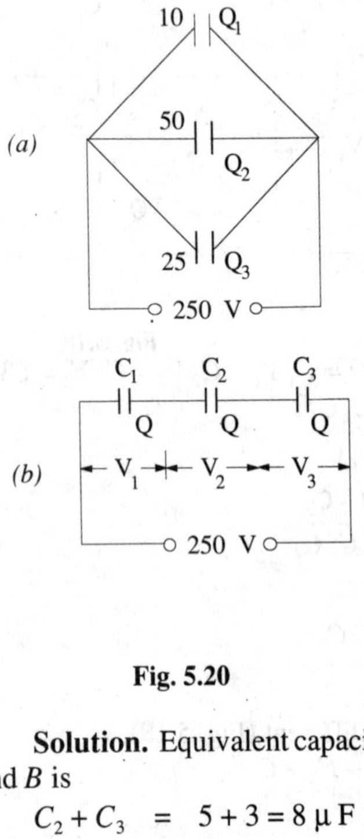

Solution. (*i*) Parallel connection is shown in Fig. 5.20 (*a*). Each capacitor has a p.d. of 250 V across it.

$Q_1 = C_1V = 10 \times 250 = \mathbf{2500\ \mu\ C}$; $Q_2 = 50 \times 250 = \mathbf{12{,}500\ \mu\ C}$

$C_3 = 25 \times 250 = \mathbf{6{,}750\ \mu\ C}$.

(*ii*) $C = C_1 + C_2 + C_3 = 10 + 50 + 25 = \mathbf{85\ \mu\ F}$

(*iii*) Series connection is shown in Fig. 5.20 (*b*). Here, charge on each capacitor is the same and is equal to that on the equivalent single capacitor.

$1/C = 1/C_1 + 1/C_2 + 1/C_3$; $C = 25/4\ \mu$ F

$Q = CV = 25 \times 250/4 = 1562.5\ \mu$ F

$Q = C_1V_1$; $V_1 = 1562.5/10 = \mathbf{156.25\ V}$

$V_2 = 1562.5/25 = \mathbf{62.5\ V}$; $V_3 = 1562.5/50 = \mathbf{31.25\ V}$.

Fig. 5.20

Example 5.15. *Find the charges on capacitors in Fig. 5.21 and the p.d. across them.*

Solution. Equivalent capacitance between points A and B is

$C_2 + C_3 = 5 + 3 = 8\ \mu$ F

Capacitance of the whole combination (Fig. 5.21)

$$C = \frac{8 \times 2}{8 + 2} = 1.6\ \mu\text{F}$$

Charge on the combination is

$Q_1 = CV = 100 \times 1.6 = \mathbf{160\ \mu\ C}$

$$V_1 = \frac{Q_1}{C_1} = \frac{160}{2} = \mathbf{80\ V}; V_2 = 100 - 80 = \mathbf{20\ V}$$

$Q_1 = C_2V_2 = 3 \times 10^{-6} \times 20 = \mathbf{60\ \mu C}$

$Q_3 = C_3V_2 = 5 \times 10^{-6} \times 20 = \mathbf{100\ \mu C}$

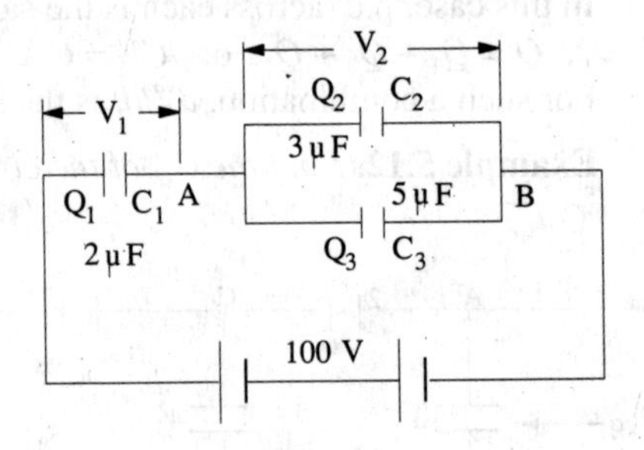

Fig. 5.21

Example 5.16. *Two capacitors A and B are connected in series across a 100 V supply and it is observed that the p.ds. across them are 60 V and 40 V respectively. A capacitor of 2 μF capacitance is now connected in parallel with A and the p.d. across B rises to 90 volts. Calculate the capacitance of A and B in microfarads.*

Solution. Let C_1 and C_2 μF be the capacitances of the two capacitors. Since they are connected in series [Fig. 5.22 (*a*)], the charge across each is the same.

$$\therefore \quad 60\,C_1 = 40\,C_2 \quad \text{or} \quad C_1/C_2 = 2/3 \qquad ...(i)$$

In Fig. 5.22 (*b*) is shown a capacitor of 2 μ F connected across capacitor *A*. Their combined capacitance = $(C_1 + 2)$ μF

$$\therefore \quad (C_1 + 2)\,10 = 90\,C_2 \quad \text{or} \quad (C_1 + 2)/C_2 = 9 \qquad ...(ii)$$

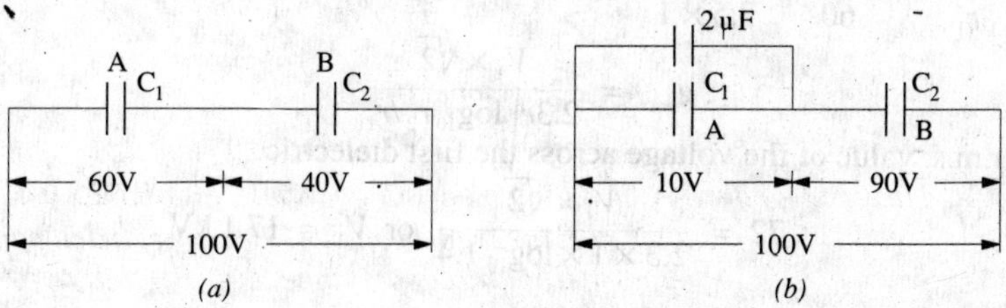

Fig. 5.22

Putting the value of $C_2 = 3C_1/2$ from (*i*) in (*ii*) we get

$$\frac{C_1 + 2}{3C_1/2} = 9 \quad \therefore C_1 + 2 = 13.5\,C_1$$

or
$$C_1 = 2/12.5 = \mathbf{0.16}\ \mu\text{F and}$$
$$C_2 = (3/2) \times 0.16 = \mathbf{0.24\ \mu F}$$

5.13. Cylindrical Capacitor With Compound Dielectric

Such a capacitor is shown in Fig. 5.23.

Let
r_1 = radius of the core
r_2 = radius of inner dielectric ε_{r1}
r_3 = radius of outer dielectric ε_{r2}

Obviously, there are two capacitors joined in series.

$$\text{Now} \quad C_1 = \frac{0.024\,\varepsilon_{r1}}{\log_{10}(r_2/r_1)}\ \mu\text{F/km and } C_2 = \frac{0.024\,\varepsilon_{r2}}{\log_{10}(r_3/r_2)}\ \mu\text{F/km}$$

Total capacitance of the cable is $C = \dfrac{C_1 C_2}{C_1 + C_2}$

Now, for capacitors joined in series, charge is the same.

$$\therefore \quad Q = C_1 V_1 = C_2 V_2$$

$$\text{or} \quad \frac{V_2}{V_1} = \frac{C_1}{C_2} = \frac{\varepsilon_{r1}\log_{10}(r_3/r_2)}{\varepsilon_{r2}\log_{10}(r_2/r_1)}$$

From this relation, V_2 and V_1 can be found,

$$g_{max} \text{ in inner capacitor} = \frac{V_1}{2.3 r_1 \log_{10}(r_2/r_1)} \qquad \text{(Art. 5.9)}$$

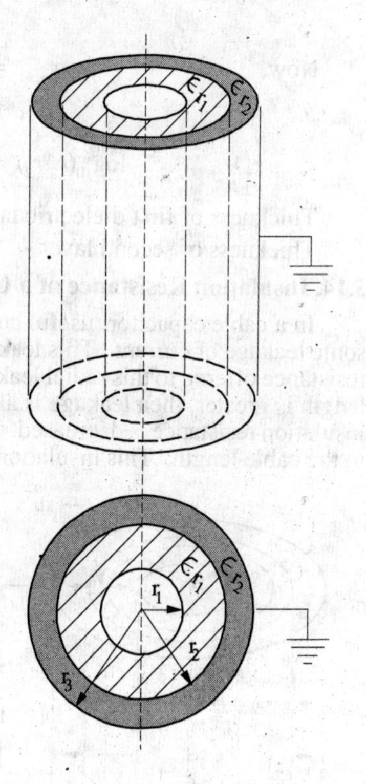

Fig. 5.23

Similarly, g_{max} for outer capacitor $= \dfrac{V_2}{2.3 r_2 \log_{10}(r_3/r_2)}$

$$\therefore \quad \frac{g_{max1}}{g_{max2}} = \frac{V_1}{2.3\,r_1 \log_{10}(r_2/r_1)} \div \frac{V_2}{2.3\,r_2 \log_{10}(r_3/r_2)} = \frac{V_1 r_2}{V_2 r_1} \times \frac{\log_{10}(r_3/r_2)}{\log_{10}(r_2/r_1)} = \frac{C_2 r_2}{C_1 r_1} \times \frac{\log_{10}(r_3/r_2)}{\log_{10}(r_2/r_1)} \quad \left(\because \frac{V_1}{V_2} = \frac{C_2}{C_1}\right)$$

Putting the values of C_1 and C_2, we get

$$\frac{g_{max1}}{g_{max2}} = \frac{0.024\,\varepsilon_{r2}}{\log_{10}(r_3/r_2)} \times \frac{\log_{10}(r_3/r_2)}{0.024\,\varepsilon_{r1}} \times \frac{r_2}{r_1} \times \frac{\log_{10}(r_2/r_1)}{\log_{10}(r_2/r_1)} \quad \therefore \quad \frac{g_{max1}}{g_{max2}} = \frac{\varepsilon_{r2} \cdot r_2}{\varepsilon_{r1} \cdot r_1}$$

Hence, voltage gradient is inversely proportional to the permittivity and the inner radius of the insulating material.

Example 5.17. *A single-core lead-sheathed cable, with a conductor diameter of 2 cm is designed to withstand 66 kV. The dielectric consists of two layers A and B having relative permittivities of 3.5 and 3 respectively. The corresponding maximum permissible electrostatic stresses are 72 and 60 kV/cm. Find the thicknesses of the two layers.* **(Power Systems-I, M.S. Univ. Baroda, 1989)**

Solution. As seen from Art. 5.13.

$$\frac{g_{max1}}{g_{max2}} = \frac{\varepsilon_{r2}.r_2}{\varepsilon_{r1}.r_1} \quad \text{or} \quad \frac{72}{60} = \frac{3 \times r_2}{3.5 \times 1} \quad \text{or} \quad r_2 = 1.4 \text{ cm}$$

Now,
$$g_{max} = \frac{V_1 \times \sqrt{2}}{2.3 r_1 \log_{10} r_2/r_1} \qquad \text{...Art. 5.9}$$

where V_1 is the r.m.s. value of the voltage across the first dielectric.

$$\therefore \quad 72 = \frac{V_1 \times \sqrt{2}}{2.3 \times 1 \times \log_{10} 1.4} \quad \text{or} \quad V_1 = 17.1 \text{ kV}$$

Obviously,
$$V_2 = 66 - 17.1 = 48.9 \text{ kV}$$

Now,
$$g_{max2} = \frac{V_2 \times \sqrt{2}}{2.3 r_2 \log_{10}(r_3/r_2)} \quad \therefore \quad 60 = \frac{48.9}{2.3 \times 1.4 \log_{10}(r_3/r_2)}$$

$$\therefore \quad \log_{10}(r_3/r_2) = 0.2531 = \log_{10}(1.79) \quad \therefore \quad \frac{r_3}{r_2} = 1.79 \text{ or } r_3 = 2.5 \text{ cm}$$

Thickness of first dielectric layer = 1.4 – 1.0 = **0.4 cm.**
Thickness of second layer = 2.5 – 1.4 = **1.1 cm.**

5.14. Insulation Resistance of a Cable Capacitor

In a cable capacitor, useful current flows along the axis of the core but there is always present some leakage of current. This leakage is radial *i.e.* at right angles to the flow of useful current. The resistance offered to this radial leakage of current is called *insulation resistance* of the cable. If cable length is greater, then leakage is also greater. It means that more current will leak. In other words, insulation resistance is decreased. Hence, we find that insulation resistance is inversely proportional to the cable length. This insulation resistance is not to be confused with conductor resistance which is directly proportional to the cable length.

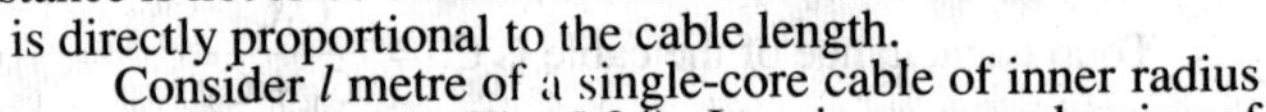

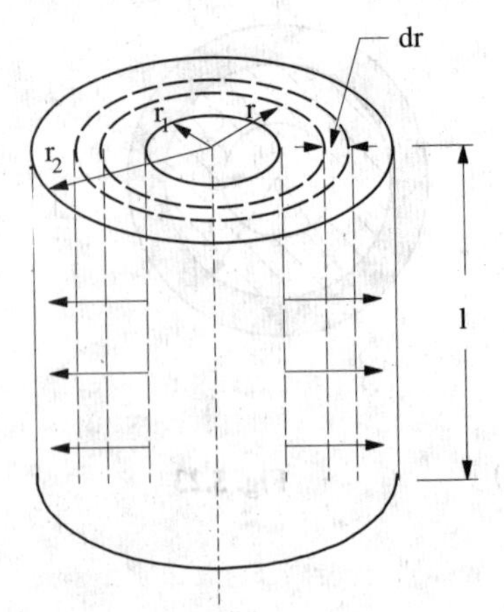

Fig. 5.24

Consider l metre of a single-core cable of inner radius r_1 and outer radius r_2 (Fig. 5.24). Imagine an annular ring of radius 'r' and radial thickness 'dr'.

If resistivity of insulating material is ρ, then resistance of this narrow ring is $dR = \dfrac{\rho\, dr}{2\pi r \times l} = \dfrac{\rho\, dr}{2\pi r l}$

$\therefore$ Insulation resistance of l metre 'length of cable is

$$\int dR = \int_{r_1}^{r_2} \frac{\rho\, dr}{2\pi r l} \quad \text{or} \quad R = \frac{\rho}{2\pi l} \left| \log_e(r) \right|_{r_1}^{r_2}$$

$$R = \frac{\rho}{2\pi l} \log_e(r_2/r_1) = \frac{2.3\rho}{2\pi l} \log_{10}(r_2/r_1)\Omega$$

It should be noted

(i) that R is inversely proportional to the cable length

(ii) that R depends upon the ratio r_2/r_1 and NOT on the thickness of insulator itself.

Example 5.18. *A liquid resistor consists of two concentric metal cylinders of diameters D = 35 cm and d = 20 cm respectively with water of specific resistance ρ = 8000 Ω-cm between them. The length of both cylinders is 60 cm. Calculate the resistance of the liquid resistor.*

(Elect. Engg. Aligarh Univ. 1989)

Solution. $r_1 = 10$ cm ; $r_2 = 17.5$ cm ; $\log_{10}(r_2/r_1) = \log_{10}(1.75) = 0.243$

$$\rho = 8 \times 10^3 \ \Omega\text{-cm} ; l = 60 \text{ cm.}$$

Resistance of the liquid resistor $R = \dfrac{2.3 \times 8 \times 10^3}{2\pi \times 60} \times 0.243 = \mathbf{11.85 \ \Omega}$.

Example 5.19. *Two underground cables having conductor resistances of 0.7 Ω and 0.5 Ω and insulation resistances of 300 M Ω and 600 M Ω respectively are joined (i) in series (ii) in parallel. Find the resultant conductor and insulation resistances.*

(Elect. Enginering, Calcutta Univ. 1987)

Solution. (*i*) The conductor resistances will add like resistances in series. However, the leakage resistances will decrease and would be given by the reciprocal relation.

Total conductor resistance = 0.7 + 0.5 = **1.2 Ω**

If R is the combined leakage resistance, then

$$\frac{1}{R} = \frac{1}{300} + \frac{1}{600} \qquad \therefore \quad \mathbf{R = 200 \ M\Omega}$$

(*ii*) In this case, conductor resistance is = 0.7 × 0.5/(0.7 + 0.5) = **0.3 Ω (approx)**

Insulation resistance = 300 + 600 = **900 MΩ**

Example 5.20. *The insulation resistance of a kilometre of the cable having a conductor diameter of 1.5 cm and an insulation thickness of 1.5 cm is 500 MΩ. What would be the insulation resistance if the thickness of the insulation were increased to 2.5 cm ?*

(Communication Systems, Hyderabad Univ. 1992)

Solution. The insulation resistance of a cable is

$$R = \frac{2.3\rho}{2\pi l} \log_{10}(r_2/r_1)$$

First Case

$r_1 = 1.5/2 = 0.75$ cm ; $r_2 = 0.75 + 1.5 = 2.25$ cm

$$\therefore \ r_2/r_1 = 2.25/0.75 = 3; \ \log_{10}(3) = 0.4771 \ \therefore \ 500 = \frac{2.3\,\rho}{2\,\pi\, l} \times 0.4771 \qquad ...(i)$$

Second Case

$r_1 = 0.75$ cm – as before $r_2 = 0.75 + 2.5 = 3.25$ cm

$$r_2/r_1 = 3.25/0.75 = 4.333 ; \ \log_{10}(4.333) = 0.6368 \ \therefore \ R = \frac{2.3\,\rho}{2\pi l} \times 0.6368 \qquad ...(ii)$$

Dividing Eq. (*ii*) by Eq. (*i*), we get

$$\frac{R}{500} = \frac{0.6368}{0.4771} ; R = 500 \times 0.6368 / 0.4771 = \mathbf{667.4 \ M\,\Omega}$$

5.15. Energy Stored in a Capacitor

Charging of a capacitor always involves some expenditure of energy by the charging agency. This energy is stored up in the electrostatic field set up in the dielectric medium. On discharging the capacitor, the field collapses and the stored energy is released.

To begin with, when the capacitor is uncharged, little work is done in transferring charge from one plate to another. But further instalments of charge have to be carried against the repulsive force due to the charge already collected on the capacitor plates. Let us find the energy spent in charging a capacitor of capacitance C to a voltage V.

Suppose at any stage of charging, the p.d. across the plates is v. By definition, it is equal to the work done in shifting one coulomb of charge from one plate to another. If 'dq' is the charge next transferred, the work done is

$$dW = v.dq$$

Now $\quad q = Cv \quad \therefore\ dq = C.dv \quad \therefore\ dW = Cv.dv$

Total work done in giving V units of potential is

$$W = \int_0^V C\,v.dv = C\left|\frac{v^2}{2}\right|_0^V \quad \therefore\ W = \frac{1}{2}CV^2$$

If C is in farads and V is in volts, then $W = \frac{1}{2}CV^2$ joules $= \frac{1}{2}QV$ joules $= \frac{Q^2}{2C}$ joules

If Q is in coulombs and C is in farads, the energy stored is given in joules.

Note. As seen from above, energy stored in a capacitor is $E = \frac{1}{2}CV^2$

Now, for a capacitor of plate area A m^2 and dielectric of thickness d metre, energy per unit volume of dielectric medium.

$$= \frac{1}{2}\frac{CV^2}{Ad} = \frac{1}{2}\frac{\varepsilon A}{d}\cdot\frac{V^2}{Ad} = \frac{1}{2}\varepsilon\left(\frac{V}{d}\right)^2 = \frac{1}{2}\varepsilon E^2 = \frac{1}{2}DE = D^2/2\varepsilon \text{ joules/m}^3\text{*}$$

It will be noted that the formula $\frac{1}{2}DE$ is similar to the expression $\frac{1}{2}$ stress × strain which is used for calculating the mechanical energy stored per unit volume of a body subjected to elastic stress.

Example 5.21. *Since a capacitor can store charge just like a lead-acid battery, it can be used at least theoretically as an electrostatic battery. Calculate the capacitance of 12-V electrostatic battery which the same capacity as a 40 Ah, 12 V lead-acid battery.*

Solution. Capacity of the lead-acid battery = 40 Ah = 40 × 36 As = 144000 Coulomb

Energy stored in the battery = QV = 144000 × 12 = 1.728 × 10^6 J

Energy stored in an electrostatic battery $= \frac{1}{2}CV^2$

$\therefore\ \frac{1}{2}\times C\times 12^2 = 1.728\times 10^6 \quad \therefore\ C = 2.4\times 10^4$ F = 24 kF

Example 5.22. *A capacitor-type stored-energy welder is to deliver the same heat to a single weld as a conventional welder that draws 20 kVA at 0.8 pf for 0.0625 second/weld. If C = 2000 μF, find the voltage to which it is charged.* **(Power Electronics, A.M.I.E. Sec B, 1993)**

Solution. The energy supplied per weld in a conventional welder is

$$W = VA\times\cos\phi\times\text{time} = 20{,}000\times 0.8\times 0.0625 = 1000 \text{ J}$$

Now, energy stored in a capacitor is $(1/2)CV^2$

$$\therefore \qquad W = \frac{1}{2}CV^2 \text{ or } V = \sqrt{\frac{2W}{C}} = \sqrt{\frac{2\times 1000}{2000\times 10^{-6}}} = \mathbf{1000\ V}$$

Example 5.23. *A parallel-plate capacitor is charged to 50 μC at 150 V. It is then connected to another capacitor of capacitance 4 times the capacitance of the first capacitor. Find the loss of energy.* **(Elect. Engg. Aligarh Univ. 1989)**

Solution. $C_1 = 50/150 = 1/3\ \mu F\ ;\ C_2 = 4\times 1/3 = 4/3\ \mu F$

Before Joining

$$E_1 = \frac{1}{2}C_1V_1^2 = \frac{1}{2}\times\left(\frac{1}{3}\right)\times 10^{-6}\times 150^2 = 37.5\times 10^{-4}\text{ J}\ ;\ E_2 = 0$$

$$\text{Total energy} = 37.5\times 10^{-4}\text{ J}$$

After Joining

When the two capacitors are connected in parallel, the charge of 50 μ C gets redistributed and the two capacitors come to a common potential V.

*It is similar to the expression for the energy stored per unit volume of a magenetic field.

$$V = \frac{\text{total charge}}{\text{total capacitance}} = \frac{50\ \mu C}{[(1/3) + (4/3)]\ \mu F} = 30\ V$$

$$E_1 = \frac{1}{2} \times (1/3) \times 10^{-6} \times 30^2 = 1.5 \times 10^{-4}\ J$$

$$E_2 = \frac{1}{2} \times (4/3) \times 10^{-6} \times 30^2 = 6.0 \times 10^{-4}\ J$$

Total energy $= 7.5 \times 10^{-4}$ J ; Loss of energy $= (37.5 - 7.5) \times 10^{-4} = \mathbf{3 \times 10^{-2}\ J}$

This energy is wasted away as heat in the conductor connecting the two capacitors.

Example 5.24. *An air-capacitor of capacitance 0.005 μF is connected to a direct voltage of 500 V, is disconnected and then immersed in oil with a relative permittivity of 2.5. Find the energy stored in the capacitor before and after immersion.* **(Elect. Technology ; London Univ.)**

Solution. Energy before immersion is

$$E_1 = \frac{1}{2} CV^2 = \frac{1}{2} \times 0.005 \times 10^{-6} \times 500^2 = \mathbf{625 \times 10^{-6}\ J}$$

When immersed in oil, its capacitance is increased 2.5 times. Since charge is constant, voltage must become 2.5 times. Hence, new capacitance is 2.5 × 0.005 = 0.0125 μF and new voltage is 500/2.5 = 200 V.

$$E_2 = \frac{1}{2} \times 0.0125 \times 10^{-6} \times (200)^2 = \mathbf{250 \times 10^{-6}\ J}$$

Example 5.25. *A parallel-plate air capacitor is charged to 100 V. Its plate separation is 2 mm and the area of each of its plates is 120 cm².*

Calculate and account for the increase or decrease of stored energy when plate separation is reduced to 1 mm

(a) at constant voltage (b) at constant charge.

Solution. Capacitance in the first case

$$C_1 = \frac{\varepsilon_0 A}{d} = \frac{8.854 \times 10^{-12} \times 120 \times 10^{-4}}{2 \times 10^{-3}} = 53.1 \times 10^{-12}\ F$$

Capacitance in the second case *i.e.* with reduced spacing

$$C_2 = \frac{8.854 \times 10^{-12} \times 120 \times 10^{-4}}{1 \times 10^{-3}} = 106.2 \times 10^{-12}\ F$$

(*a*) **When Voltage is Constant**

Change in stored energy $dE = \frac{1}{2} C_2 V^2 - \frac{1}{2} C_1 V^2$

$$= \frac{1}{2} \times 100^2 \times (106.2 - 53.1) \times 10^{-12} = \mathbf{26.55 \times 10^{-8}\ J}$$

This represents an increase in the energy of the capacitor. This extra work has been done by the external supply source because charge has to be given to the capacitor when its capacitance increases, voltage remaining constant.

(*b*) **When Charge Remains Constant**

Energy in the first case $E_1 = \frac{1}{2}\frac{Q^2}{C_1}$; Energy in the second case, $E_2 = \frac{1}{2}\frac{Q^2}{C_2}$

$\therefore$ change in energy is $dE = \frac{1}{2} Q^2 \left(\frac{1}{53.1} - \frac{1}{106.2}\right) \times 10^{12}\ J$

$$= \frac{1}{2}(C_1V_1)^2\left(\frac{1}{53.1} - \frac{1}{106.2}\right) \times 10^{12}\,\text{J}$$

$$= \frac{1}{2} \times (53.1 \times 10^{-12})^2 \times 10^4 \times 0.0094 \times 10^{12}$$

$$= \mathbf{13.3 \times 10^{-8}\ joules}$$

Hence, there is a decrease in the stored energy. The reason is that charge remaining constant, when the capacitance is increased, then voltage must fall with a consequent decrease in stored energy $(E = \frac{1}{2}QV)$

Example 5.26. *A point charge of 100 μC is embedded in an extensive mass of bakelite which has a relative permittivity of 5. Calculate the total energy contained in the electric field outside a radial distance of (i) 100 m (ii) 10 m (iii) 1 m and (iv) 1 cm.*

Solution. As per the Coulomb's law, the electric field intensity at any distance x from the point charge is given by $E = Q/4\pi\varepsilon x^2$. Let us draw a spherical shell of radius x as shown in Fig. Another spherical shell of radius $(x + dx)$ has also been drawn. A differential volume of the space enclosed between the two shells is $dv = 4\pi x^2 dx$. As per Art. 5.15, the energy stored per unit volume of the electric field is (1/2) DE. Hence, differential energy contained in the small volume is

$$dW = \frac{1}{2}DE\ dv = \frac{1}{2}\varepsilon E^2\,dv = \frac{1}{2}\varepsilon\left(\frac{Q}{4\pi\varepsilon x^2}\right)^2 4\pi x^2\,dx = \frac{Q^2}{8\pi\varepsilon}\cdot\frac{dx}{x^2}$$

Total energy of the electric field extending from $x = R$ to $x = \infty$ is

$$W = \frac{Q^2}{8\pi\varepsilon}\int_R^{\infty} x^{-2}\,dx = \frac{Q^2}{8\pi\varepsilon R} = \frac{Q^2}{8\pi\varepsilon_0\varepsilon_r R}$$

(*i*) The energy contained in the electric field lying outside a radius of $R = 100$ m is

$$W = \frac{(100 \times 10^{-6})^2}{8\pi \times 8.854 \times 10^{-12} \times 5 \times 100} = \mathbf{0.09\ J}$$

(*ii*) For $R = 10$ m, $W = 10 \times 0.09 = \mathbf{0.9\ J}$
(*iii*) For $R = 1$ m, $W = 100 \times 0.09 = \mathbf{9\ J}$
(*iv*) For $R = 1$ cm, $W = 10{,}000 \times 0.09 = \mathbf{900\ J}$

Example 5.27. *Calculate the change in the stored energy of a parallel-plate capacitor if a dielectric slab of relative permittivity 5 is introduced between its two plates.*

Solution. Let A be the plate area, d the plate separation, E the electric field intensity and D the electric flux density of the capacitor. As per Art. 5.15, energy stored per unit volume of the field is $= (1/2)\,DE$. Since the space volume is $d \times A$, hence,

$$W_1 = \frac{1}{2}D_1E_1 \times dA = \frac{1}{2}\varepsilon_0 E_1^2 \times dA = \frac{1}{2}\varepsilon_0\,dA\left(\frac{V_1}{d}\right)^2$$

When the dielectric slab is introduced,

$$W_2 = \frac{1}{2}D_2E_2 \times dA = \frac{1}{2}\varepsilon E_2^2 \times dA = \frac{1}{2}\varepsilon_0\varepsilon_r\,dA\left(\frac{V_2}{d}\right)^2$$

$$= \frac{1}{2}\varepsilon_0\varepsilon_r\,dA\left(\frac{V_1}{\varepsilon_r d}\right)^2 = \frac{1}{2}\varepsilon_0\,dA\left(\frac{V_1}{d}\right)^2\frac{1}{\varepsilon_r} \quad \therefore \quad W_2 = \frac{W_1}{\varepsilon_r}$$

It is seen that the stored energy is reduced by a factor of ε_r. Hence, change in energy is

$$dW = W_1 - W_2 = W_1\left(1 - \frac{1}{\varepsilon_r}\right) = W_1\left(1 - \frac{1}{5}\right) = W_1 \times \frac{4}{5} \quad \therefore \quad \frac{dW}{W_1} = \mathbf{0.8}$$

5.16. Force of Attraction Between Oppositely-charged Plates

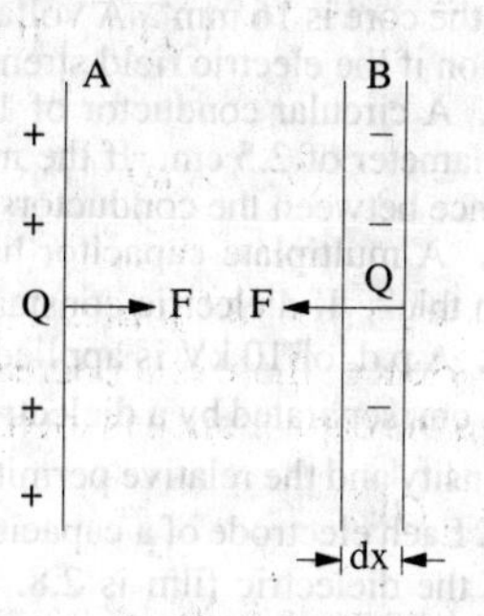

Fig. 5.25

In Fig. 5.25 are shown two parallel conducting plates A and B carrying constant charges of $+Q$ and $-Q$ coulombs respectively. Let the force of attraction between the two be F newtons. If one of the plates is pulled apart by distance dx, then work done is

$$= F \times dx \text{ joules} \qquad \ldots(i)$$

Since the plate charges remain constant, no electrical energy comes into the arrangement during the movement dx.

$\therefore$ work done = change in stored energy

Initial stored energy $= \dfrac{1}{2}\dfrac{Q_2}{C}$ joules

If capacitance becomes $(C - dC)$ due to the movement dx, then

$$\text{Final stored energy} = \frac{1}{2}\frac{Q^2}{(C-dC)} = \frac{1}{2}\cdot\frac{Q^2}{C}\cdot\frac{1}{\left(1-\dfrac{dC}{C}\right)} = \frac{1}{2}\frac{Q^2}{C}\left(1+\frac{dC}{C}\right) \text{ if } dC \ll C$$

$$\therefore \text{ change in stored energy} = \frac{1}{2}\frac{Q^2}{C}\left(1+\frac{dC}{C}\right) - \frac{1}{2}\frac{Q^2}{C} = \frac{1}{2}\frac{Q^2}{C^2}.dC \qquad \ldots(ii)$$

Equating Eq. (i) and (ii), we have $F\,.\,dx = \dfrac{1}{2}\dfrac{Q^2}{C^2}\,.\,dC$

$$F = \frac{1}{2}\frac{Q^2}{C^2}\cdot\frac{dC}{dx} = \frac{1}{2}V^2\frac{dC}{dx} \qquad (\because V = Q/C)$$

Now $$C = \frac{\varepsilon A}{x} \quad \therefore \quad \frac{dC}{dx} = -\frac{\varepsilon A}{x^2}$$

$$\therefore \quad F = -\frac{1}{2}V^2\cdot\frac{\varepsilon A}{x^2} = -\frac{1}{2}\varepsilon A\left(\frac{V}{x}\right)^2 \text{ newtons} = -\frac{1}{2}\varepsilon A E^2 \text{ newtons}$$

This represents the force between the plates of a parallel-plate capacitor charged to a p.d. of V volts. The negative sign shows that it is a force of attraction.

Example 5.28. *A parallel-plate capacitor is made of plates 1 m square and has a separation of 1 mm. The space between the plates is filled with dielectric of $\varepsilon_r = 25.0$. If 1 kV plotential difference is applied to the plates, find the force squeezing the plates together.*

(Electromagnetic Theory, A.M.I.E. Sec B, 1993)

Solution. As seen from Art. 5.16, $F = -(1/2)\,\varepsilon_0\,\varepsilon_r\,AE^2$ newton

Now, $E = V/d = 1000/1 \times 10^{-3} = 10^{-6}$ V/m

$$\therefore F = -\frac{1}{2}\varepsilon_0\,\varepsilon_r\,AE^2 = -\frac{1}{2}\times 8.854\times 10^{-12}\times 25\times 1\times(10^6)^2 = -1.1\times 10^{-4}\,\text{N}$$

Tutorial Problems No. 5.2

1. Find the capacitance per unit length of a cylindrical capacitor of which the two conductors have radii 2.5 and 4.5 cm and the dielectric consists of two layers whose cylinder of contact is 3.5 cm in radius, the inner layer having a dielectric constant of 4 and the outer one of 6. **[440 pF/m]**

2. A parallel-plate capacitor, having plates 100 cm^2 area, has three dielectrics 1 mm each and of permittivities 3, 4 and 6. If a peak voltage of 2,000 V is applied to the plates, calculate :

(a) potential gradient across each dielectric

(b) energy stored in each dielectric.

[8.89 kV/cm; 6.67 kV/cm ; 4.44 kV/cm ; 1047, 786, 524 × 10^{-7} joule]

3. The core and lead-sheath of a single-core cable are separated by a rubber covering. The cross-sectional area of the core is 16 mm^2. A voltage of 10 kV is applied to the cable. What must be the thickness of the rubber insulation if the electric field strength in it is not to exceed 6×10^6 V/m ? **[2.5 mm (approx)]**

4. A circular conductor of 1 cm diameter is surrounded by a concentric conducting cylinder having an inner diameter of 2.5 cm. If the maximum electric stress in the dielectric is 40 kV/cm, calculate the potential difference between the conductors and also the minimum value of the electric stress. **[18.4 kV ; 16 kV/cm]**

5. A multiplate capacitor has parallel plates each of area 12 cm^2 and each separated by a mica sheet 0.2 mm thick. If dielectric constant for mica is 5, calculate the capacitance. **[265.6 μμF]**

6. A p.d. of 10 kV is applied to the terminals of a capacitor of two circular plates each having an area of 100 sq. cm. separated by a dielectric 1 mm thick. If the capacitance is 3×10^{-4} microfarad, calculate the electric flux density and the relative permittivity of the dielectric. **[D = 3×10^{-4} C/m^2, ε_r=3.39]** (*City & Guilds, London*)

7. Each electrode of a capacitor of the electrolytic type has an area of 0.02 sq. metre. The relative permittivity of the dielectric film is 2.8. If the capacitor has a capacitance of 10 μF, estimate the thickness of the dielectric film. **[4.95×10^{-8} m]** (*I.E.E. London*)

5.17. Current-Voltage Relationships in a Capacitor

The charge on a capacitor is given by the expression $Q = CV$. By differentiating this relation, we get

$$i = \frac{dQ}{dt} = \frac{d}{dt}(CV) = C\frac{dV}{dt}$$

Following important facts can be deduced from the above relations :

(*i*) since $Q = CV$, it means that the voltage across a capacitor is proportional to *charge*, not the *current*.

(*ii*) a capacitor has the ability to store charge and hence to provide a sort of memory.

(*iii*) a capacitor can have a voltage across it even when there is *no current flowing*.

(*iv*) from $i = C\,dV/dt$, it is clear that current in the capacitor is present only when voltage on it changes with time. If $dV/dt = 0$ *i.e.* when its voltage is constant or for d.c. voltage, $i = 0$. Hence, the capacitor behaves like an *open circuit*.

(*v*) from $i = C\,dV/dt$, we have $dV/dt = i/C$. It shows that for a given value of (charge or discharge) current i, rate of change in voltage is inversely proportional to capacitance. Larger the value of C, slower the rate of change in capacitive voltage. Also, capacitor voltage *cannot change instantaneously*.

(*vi*) the above equation can be put as $dv = \frac{i}{C} \cdot dt$

Integrating the above, we get $\int dv = \frac{1}{C}\int i \cdot dt$ or $v = \frac{1}{C}\int_0^t i\,dt$

Example 5.29. *The voltage across a 5 μF capacitor changes uniformly from 10 to 70 V in 5 ms. Calculate (i) change in capacitor charge (ii) charging current.*

Solution. $Q = CV \quad \therefore \; dQ = C \cdot dV$ and $i = C\,dV/dt$

(*i*) $dV = 70 - 10 = 60$ V, $\therefore \; dQ = 5 \times 60 =$ **300 μC.**

(*ii*) $i = C \cdot dV/dt = 5 \times 60/5 =$ **60 mA**

Example 5.30. *An uncharged capacitor of 0.01 F is charged first by a current of 2 mA for 30 seconds and then by a current of 4 mA for 30 seconds. Find the final voltage in it.*

Solution. Since the capacitor is initially uncharged, we will use the principle of Superposition.

$$V_1 = \frac{1}{0.01}\int_0^{30} 2 \times 10^{-3} \cdot dt = 100 \times 2 \times 10^{-3} \times 30 = 6 \text{ V}$$

$$V_2 = \frac{1}{0.01}\int_0^{30} 4 \times 10^{-3} \cdot dt = 100 \times 4 \times 10^{-3} \times 30 = 12 \text{ V} ; \; \therefore \; V = V_1 + V_2 = 6 + 12 = \mathbf{18\ V}$$

Example 5.31. *The voltage across two series-connected 10 μF and 5 μF capacitors changes uniformly from 30 to 150 V in 1 ms. Calculate the rate of change of voltage for (i) each capacitor and (ii) combination.*

Solution. For series combination

$$V_1 = V\frac{C_2}{C_1 + C_2} = \frac{V}{3} \text{ and } V_2 = V \cdot \frac{C_1}{C_1 + C_2} = \frac{2V}{3}$$

When V = 30 V $V_1 = V/3 = 30/3 = 10\text{ V}$; $V_2 = 2V/3 = 2 \times 30/3 = 20\text{ V}$

When V = 150 V $V_1 = 150/3 = 50\text{ V}$ and $V_2 = 2 \times 150/3 = 100\text{ V}$

(*i*) $\therefore \quad \frac{dV_1}{dt} = \frac{(50-10)\text{V}}{1\text{ ms}} = \mathbf{40\ kV/s}\ ; \quad \frac{dV_2}{dt} = \frac{(100-20)\text{ V}}{1\text{ ms}} = \mathbf{80\ kV/s}$

(*ii*) $\quad \frac{dV}{dt} = \frac{(150-30)}{1\text{ ms}} = \mathbf{120\ kV/s}$

It is seen that $dV/dt = dV_1/dt + dV_2/dt$.

5.18. Charging of a Capacitor

In Fig. 5.26. (*a*) is shown an arrangement by which a capacitor *C* may be charged through a high resistance *R* from a battery of *V* volts. The voltage across *C* can be measured by a suitable voltmeter. When switch *S* is connected to terminal (*a*), *C* is charged but when it is connected to *b*, *C* is short-circuited through *R* and is thus discharged. As shown in Fig. 5.26. (*b*), switch *S* is shifted to *a* for charging the capacitor from the battery. The voltage across *C* does not rise to *V* instantaneously but builds up slowly *i.e.* exponentially and not linearly. Charging current i_c is maximum at the start *i.e.* when *C* is uncharged, then it decreases exponentially and finally ceases when p.d. across capacitor plates becomes equal and opposite to the battery voltage *V*. At any instant during charging, let

v_c = p.d. across *C*; i_c = charging current

q = charge on capacitor plates

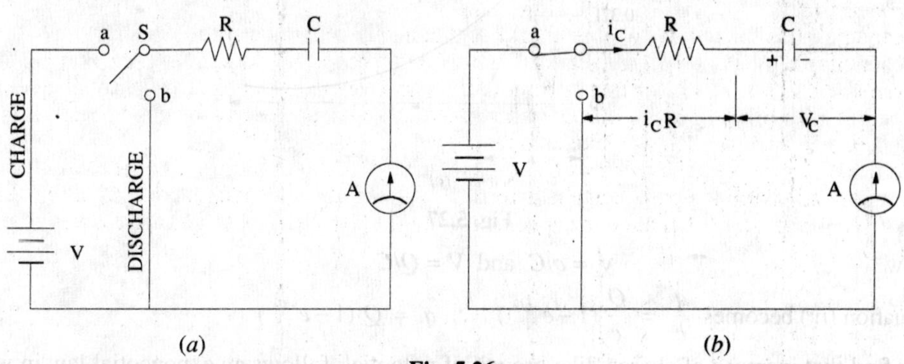

Fig. 5.26

The applied voltage *V* is always equal to the sum of :

(*i*) resistive drop ($i_c R$) and (*ii*) voltage across capacitor (v_c)

$$\therefore \quad V = i_c R + v_c \quad \text{...(i)}$$

$$\text{Now} \quad i_c = \frac{dq}{dt} = \frac{d}{dt}(Cv_c) = C\frac{dv_c}{dt} \quad \therefore \quad V = v_c + CR\frac{dv_c}{dt} \quad \text{...(ii)}$$

or
$$-\frac{dv_c}{V - v_c} = -\frac{dt}{CR}$$

Integrating both sides, we get $\int \frac{-d\,v_c}{V - v_c} = -\frac{1}{CR}\int dt;\ \therefore \quad \log_e(V - v_c) = -\frac{t}{CR} + K \quad \text{...(iii)}$

where *K* is the constant of integration whose value can be found from initial known conditions. We know that at the start of charging when $t = 0$, $v_c = 0$.

Substituting these values in (*iii*), we get $\log_e V = K$

Hence, Eq. (*iii*) becomes $\log_e(V - v_c) = \frac{-t}{CR} + \log_e V$

or $\quad \log_e \frac{V - v_c}{V} = \frac{-t}{CR} = -\frac{t}{\lambda}$ where $\lambda = CR$ = time constant

$$\therefore \quad \frac{V - v_c}{V} = e^{-t/\lambda} \text{ or } v_c = V(1 - e^{-t/\lambda}) \quad \text{...(iv)}$$

This gives variation with time of voltage across the capacitor plates and is shown in Fig. 5.27. (*a*)

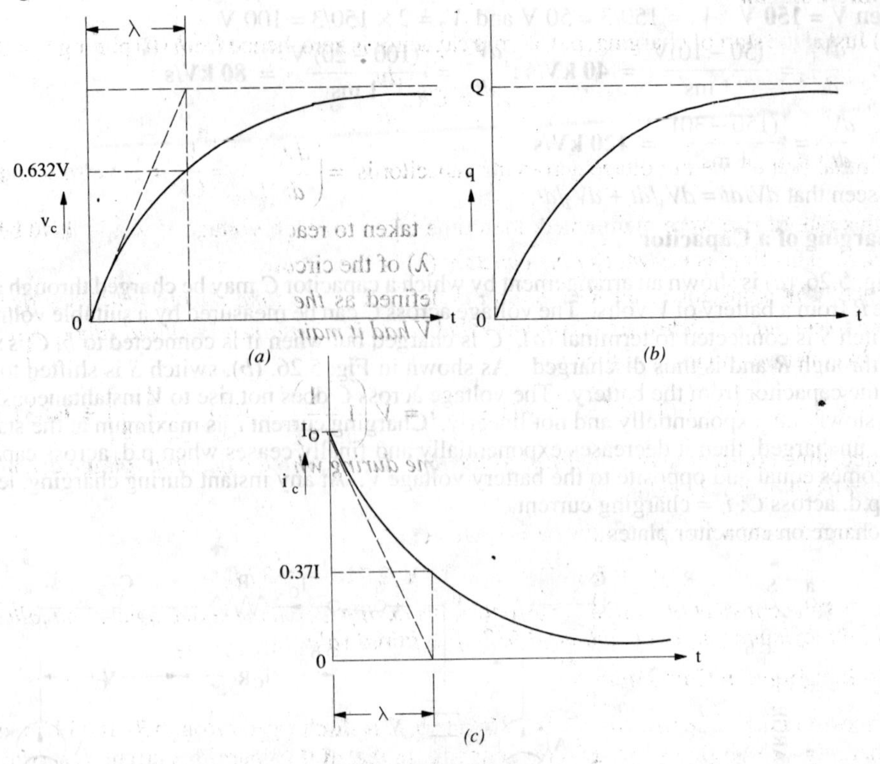

Fig. 5.27

Now $\quad v_c = q/C$ and $V = Q/C$

Equation (*iv*) becomes $\quad \dfrac{q}{c} = \dfrac{Q}{c}(1-e^{-t/\lambda}) \quad \therefore \; q = Q\,(1-e^{-t/\lambda}) \qquad \ldots(v)$

We find that increase of charge, like growth of potential, follows an exponential law in which the steady value is reached after infinite time (Fig. 5.27 b). Now, $i_c = dq/dt$.

Differentiating both sides of Eq. (*v*), we get

$$\frac{dq}{dt} = i_c = Q\,\frac{d}{dt}\,(1-e^{-t/\lambda}) = Q\left(+\frac{1}{\lambda}\,e^{-t/\lambda}\right)$$

$$= \frac{Q}{\lambda}\,e^{-t/\lambda} = \frac{CV}{CR}\,e^{-t/\lambda} \qquad (\because Q = CV \text{ and } \lambda = CR)$$

$$\therefore \qquad i_c = \frac{V}{R}\,.\,e^{-t/\lambda} \text{ or } i_c = I_0\,e^{-t/\lambda} \qquad \ldots(vi)$$

where I_0 = maximum current = V/R

Exponentially rising curves for v_c and q are shown in Fig. 5.27 (*a*) and (*b*) respectively. Fig. 5.27 (*c*) shows the curve for exponentially decreasing charging current. It should be particularly noted that i_c decreases in magnitude only but its direction of flow remains the same *i.e.* positive.

As charging continues, charging current decreases according to equation (*vi*) as shown in Fig. 5.27 (*c*). It becomes zero when $t = \infty$ (though it is almost zero in about 5 time constants). Under steady-state conditions, the circuit appears only as a capacitor which means it acts as an open-circuit. Similarly, it can be proved that v_R decreases from its initial maximum value of V to zero exponentially as given by the relation $v_R = V\,e^{-t/\lambda}$.

5.19. Time Constant

(*a*) Just at the start of charging, p.d. across capacitor is zero, hence from (*ii*) putting $v_c = 0$, we get

$$V = CR\frac{dv_c}{dt}$$

$\therefore$ initial rate of rise of voltage across the capacitor is* $= \left(\frac{dv_c}{dt}\right)_{t=0} = \frac{V}{CR} = \frac{V}{\lambda}$ volt/second

If this rate of rise were maintained, then time taken to reach voltage V would have been $V \div V/CR = CR$. This time is known as *time constant* (λ) of the circuit.

Hence, time constant of an R–C circuit is defined as *the time during which voltage across capacitor would have reached its maximum value V had it maintained its initial rate of rise.*

(*b*) In equation (*iv*) if $t = \lambda$, then

$$v_c = V(1-e^{-t/\lambda}) = V(1-e^{-\lambda/\lambda}) = V(1-e^{-1}) = V\left(1-\frac{1}{e}\right) = V\left(1-\frac{1}{2.718}\right) = \mathbf{0.632\ V}$$

Hence, time constant may be defined as *the time during which capacitor voltage actually rises to 0.632 of its final steady value.*

(*c*) From equation (*vi*), by putting $t = \lambda$, we get

$$i_c = I_0 e^{-\lambda/\lambda} = I_0 e^{-1} = I_0/2.718 \cong 0.37\, I_0$$

Hence, time constant of a circuit is also the *time during which the charging current falls to 0.37 of its initial maximum value (or falls by 0.632 of its initial value).*

5.20. Discharging of a Capacitor

As shown in Fig. 5.28 (*a*), when *S* is shifted to *b*, *C* is discharged through *R*. It will be seen that the discharging current flows in a direction opposite to that of the charging current as shown in Fig. 5.28 (*b*). Hence, if the direction of the charging current is taken positive, then that of the discharging current will be taken as negative. To begin with, the discharge current is maximum but then decreases exponentially till it ceases when capacitor is fully discharged.

Since battery is cut out of the circuit, therefore, by putting $V = 0$ in equation (*ii*) of Art. 5.18, we get

$$0 = CR\frac{dv_c}{dt} + v_c \quad \text{or} \quad v_c = -CR\frac{dv_c}{dt} \qquad \left(\because\ i_c = C\frac{dv_c}{dt}\right)$$

$$\therefore \quad \frac{dv_c}{v_c} = -\frac{dt}{CR} \quad \text{or} \quad \int\frac{dv_c}{v_c} = -\frac{1}{CR}\int dt \quad \therefore\ \log_e v_c = -\frac{t}{CR} + k$$

At the start of discharge, when $t = 0$, $v_c = V$ $\therefore$ $\log_e V = 0 + K$; or $\log_e V = K$

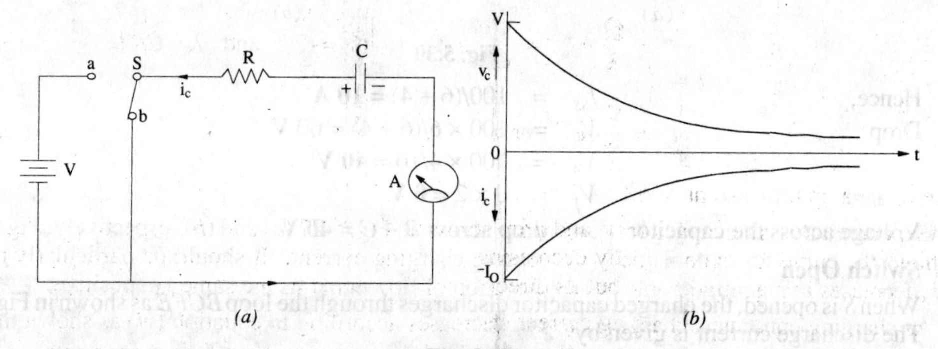

Fig. 5.28

*It can also be found by differentiating Eq. (*iv*) with respect to time and then putting $t = 0$.

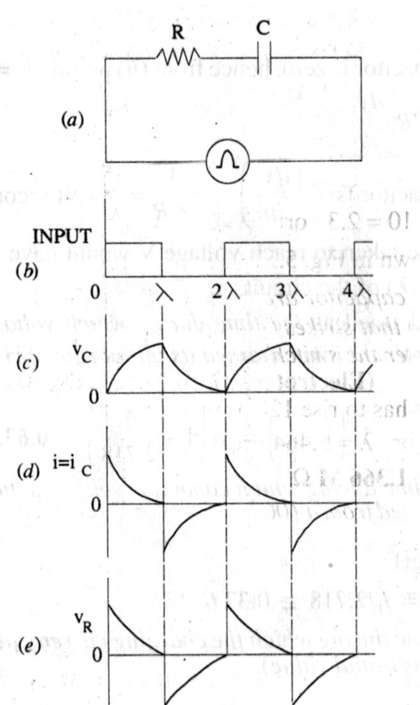

Fig. 5.29

Putting this value above, we get

$$\log_e v_c = -\frac{t}{\lambda} + \log_e V \text{ or } \log_e v_c/V = -t/\lambda$$

$$\text{or } \frac{v_c}{V} = e^{-t/\lambda} \text{ or } v_c = Ve^{-t/\lambda}$$

Similarly, $q = Qe^{-t/\lambda}$ and $i_c = -I_0 e^{-t/\lambda}$

It can be proved that

$$v_R = -V e^{-t/\lambda}$$

The fall of capacitor potential and its discharging current are shown in Fig. 5.28(b).

One practical application of the above charging and discharging of a capacitor is found in digital control circuits where a square-wave input is applied across an *R-C* circuit as shown in Fig. 5.29 (*a*). The different waveforms of the current and voltages are shown in Fig. 5.29 (*b*), (*c*), (*d*), (*e*). The sharp voltage pulses of V_R are used for control circuits.

Example 5.32. *Calculate the current in and voltage drop across each element of the circuit shown in Fig. 5.30 (a) after switch S has been closed long enough for steady-state conditions to prevail.*

Also, calculate voltage drop across the capacitor and the discharge current at the instant when S is opened.

Solution. Under steady-state conditions, the capacitor becomes fully charged and draws no current. In fact, it acts like an open circuit with the result that no current flows through the 1-Ω resistor. The steady state current I_{SS} flows through loop *ABCD* only:

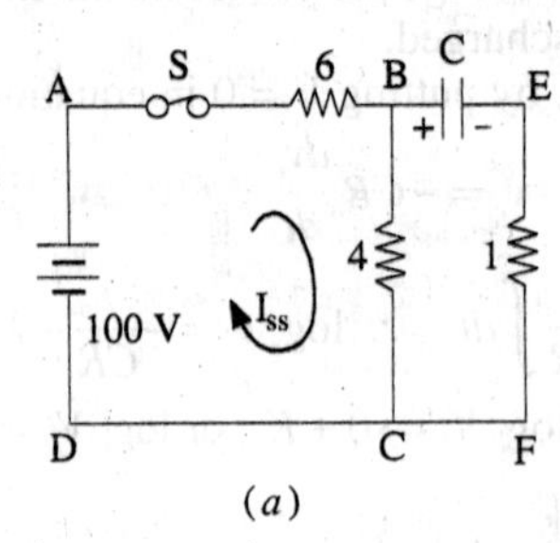

(*a*)

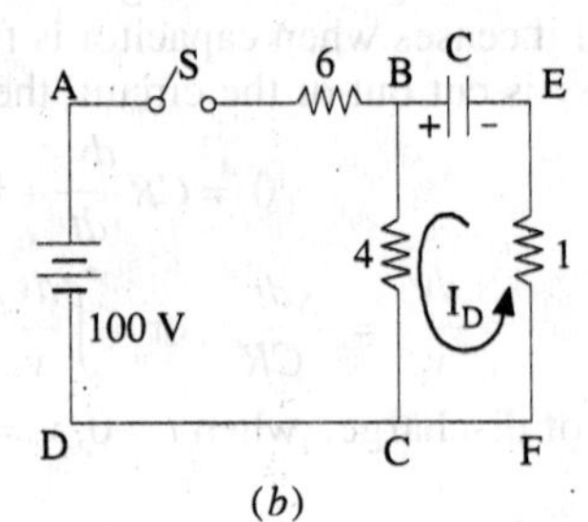

(*b*)

Fig. 5.30

Hence, $I_{SS} = 100/(6+4) = \mathbf{10\ A}$

Drop $V_6 = 100 \times 6/(6+4) = 60$ V

$V_4 = 100 \times 4/10 = \mathbf{40\ V}$

$V_1 = 0 \times 2 = \mathbf{0\ V}$

Voltage across the capacitor = drop across $B - C$ = **40 V**

Switch Open

When *S* is opened, the charged capacitor discharges through the loop *BCFE* as shown in Fig. 5.30 (*b*). The discharge current is given by

$$I_D = 40/(4+1) = \mathbf{8\ A}$$

As seen, it flows in a direction opposite to that of I_{SS}.

Example 5.33. *(a) A capacitor is charged through a large non-reactive resistance by a battery of constant voltage V. Derive an expression for the instantaneous charge on the capacitor.*

(b) For the above arrangement, if the capacitor has a capacitance of 10 μF and the resistance is 1 M Ω, calculate the time taken for the capacitor to receive 90% of its final charge. Also, draw the charge/time curve.

Solution. (*a*) For this part, please refer to Art. 5.18.

(*b*) $\lambda = CR = 10 \times 10^{-6} \times 1 \times 10^{6} = 10$ s ; $q = 0.9\ Q$

Now, $q = Q\,(1 - e^{-t/\lambda})$ $\therefore$ $0.9\,Q = Q\,(1 - e^{-t/10})$ or $e^{t/10} = 10$

$\therefore$ $0.1\, t \log_e e = \log_e 10$ or $0.1\, t = 2.3 \log_{10} 10 = 2.3$ or $t =$ **23 s**

The charge/time curve is similar to that shown in Fig. 5.27 (*b*).

Example 5.34. *A resistance R and a 4 μF capacitor are connected in series across a 200 V. d.c. supply. Across the capacitor is a neon lamp that strikes (glows) at 120 V. Calculate the value of R to make the lamp strike (glow) 5 seconds after the switch has been closed.*

(Electrotechnics-I M.S. Univ. Baroda 1988)

Solution. Obviously, the capacitor voltage has to rise 120 V in 5 seconds.

$\therefore$ $120 = 200\,(1 - e^{-5/\lambda})$ or $e^{5/\lambda} = 2.5$ or $\lambda = 5.464$ second.

Now, $\lambda = CR$ $\therefore$ $R = 5.464/4 \times 10^{-6} =$ **1.366 M Ω**

Example 5.35. *A capacitor of 0.1 μF is charged from a 100-V battery through a series resistance of 1,000 ohms. Find*

(a) the time for the capacitor to receive 63.2% of its final charge.
(b) the charge received in this time *(c) the final rate of charging*
(d) the rate of charging when the charge is 63.2% of the final charge.

(Elect. Engineering, Bombay Univ. 1985)

Solution. (*a*) As seen from Art. 5.18 (*b*), 63.2% of charge is received in a time equal to the time constant of the circuit.

Time required $= \lambda = CR = 0.1 \times 10^{-6} \times 1000 = 0.1 \times 10^{-3} =$ **10^{-4} second**

(*b*) Final charge, $Q = CV = 0.1 \times 100 = 10$ μC

Charge received during this time is $= 0.632 \times 10 =$ **6.32 μC**

(*c*) The rate of charging at any time is given by Eq. (*ii*) of Art. 5.18.

$$\frac{dv}{dt} = \frac{V - v}{CR}$$

Initially $v = 0$, Hence $\dfrac{dv}{dt} = \dfrac{V}{CR} = \dfrac{100}{0.1 \times 10^{-6} \times 10^{3}} =$ **10^{6} V/s**

(*d*) Here $v = 0.632\, V = 0.632 \times 100 = 63.2$ volts

$$\therefore \quad \frac{dv}{dt} = \frac{100 - 63.2}{10^{-4}} = \textbf{368 kV/s}$$

Example 5.36. *A series combination having R = 2 M Ω and C = 0.01 μF is connected across a d.c. voltage source of 50 V. Determine*

(a) capacitor voltage after 0.02 s, 0.04 s, 0.06 s and 1 hour
(b) charging current after 0,02 s, 0.04 s, 0.06 s and 0.1 s.

Solution. $\lambda = CR = 2 \times 10^{6} \times 0.01 \times 10^{-6} = 0.02$ second

$I_m = V/R = 50/2 \times 10^{6} = 25$ μA.

While solving this question, it should be remembered that (*i*) in each time constant, v_e increases further by 63.2% of its *balance* value and (*ii*) in each constant, i_c decreases to 37% its previous value.

(*a*) (*i*) **t = 0.02 s**

Since, initially at $t = 0$, $v_c = 0$ V and $V_e = 50$ V, hence, in one time constant

$v_c = 0.632\,(50 - 0) =$ **31.6 V**

(*ii*) **t = 0.04 s**

This time equals two time-constants.

$\therefore$ $v_c = 31.6 + 0.632(50 - 31.6) =$ **43.2 V**

(*iii*) **t = 0.06 s**

This time equals three time-constants.

$\therefore \qquad v_c = 43.2 + 0.632(50 - 43.2) = \mathbf{47.5\ V}$

Since in one hour, steady-state conditions would be established, v_c would have achieved its maximum possible value of **50 V**.

(*b*) (*i*) **t = 0.02 s,** $\quad i_c = 0.37 \times 25 = \mathbf{9.25\ \mu A}$

(*ii*) **t = 0.4 s,** $\quad i_c = 0.37 \times 9.25 = \mathbf{3.4\ \mu A}$

(*iii*) **t = 0.06 s,** $\quad i_c = 0.37 \times 3.4 = \mathbf{1.26\ \mu A}$

(*iv*) **t = 0.1 s,** This time equals 5 time constants. In this time, current falls almost to **zero** value.

5.21. Transient Relations During Capacitor Charging Cycle

Whenever a circuit goes from one steady-state condition to another steady-state condition, it passes through a transient state which is of short duration. The first steady-state condition is called the *initial condition* and the second steady-state condition is called the *final condition*. In fact, transient condition lies in between the initial and final conditions. For example, when switch *S* in Fig. 5.31(*a*) is not connected either to *a* or *b*, the *RC* circuit is in its initial steady state with no current and hence no voltage drops. When *S* is shifted to point *a*, current starts flowing through *R* and hence, transient voltages are developed across *R* and *C* till they achieve their final steady values. The period during which current and voltage changes take place is called *transient condition.*

The moment switch *S* is shifted to point '*a*' as shown in Fig. 5.31 (*b*), a charging current i_c is set up which starts charging *C* that is initially uncharged. At the beginning of the transient state, i_c is maximum because there is no potential across *C* to oppose the applied voltage V. It has maximum value = $V/R = I_0$. It produces maximum voltage drop across $R = i_c R = I_0 R$. Also, initially, $v_c = 0$, but as time passes, i_c decreases gradually so does v_R but v_C increases exponentially till it reaches the final steady value of *V*. Although V is constant, v_R and v_C are variable. However, at any time $V = v_R + v_C = i_C R + v_C$.

At the beginning of the transient state, $i_C = I_0$, $v_C = 0$ but v_R = V. At the end of the transient state, $i_C = 0$ hence, $v_R = 0$ but v_C = V.

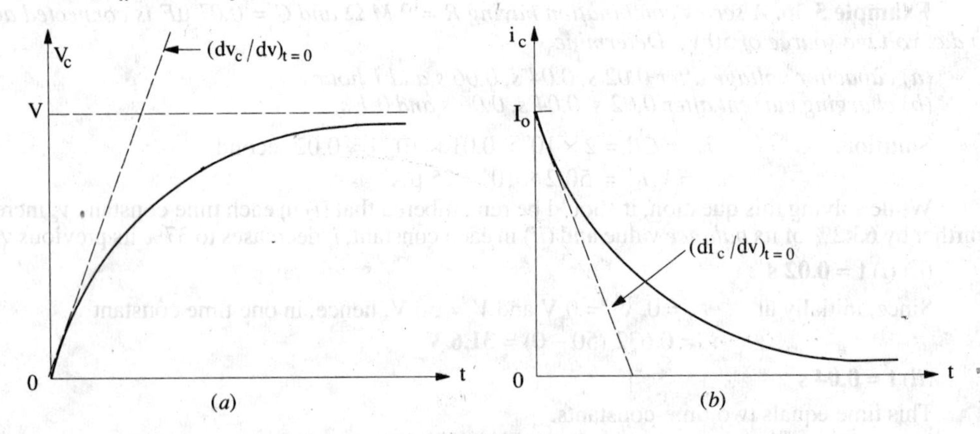

Fig. 5.31

The initial rates of change of v_C, v_R and i_C are given by

$$\left(\frac{dv_c}{dt}\right)_{t=0} = \frac{V}{\lambda} \text{ volt/second,}$$

$$\left(\frac{dv_R}{dt}\right)_{t=0} = -\frac{I_0 R}{\lambda} = -\frac{V}{\lambda} \text{ volt/second}$$

$$\left(\frac{di_c}{dt}\right)_{t=0} = -\frac{I_0}{\lambda} \text{ where } I_0 = \frac{V}{R}$$

These are the initial rates of change. However, their rate of change at any time during the charging transient are given as under :

$$\frac{dv_c}{dt} = \frac{V}{\lambda} e^{-t/\lambda} ; \frac{di_C}{dt} = -\frac{I_0}{\lambda} e^{-t/\lambda} ; \frac{dv_R}{dt} = -\frac{V}{\lambda} e^{-t/\lambda}$$

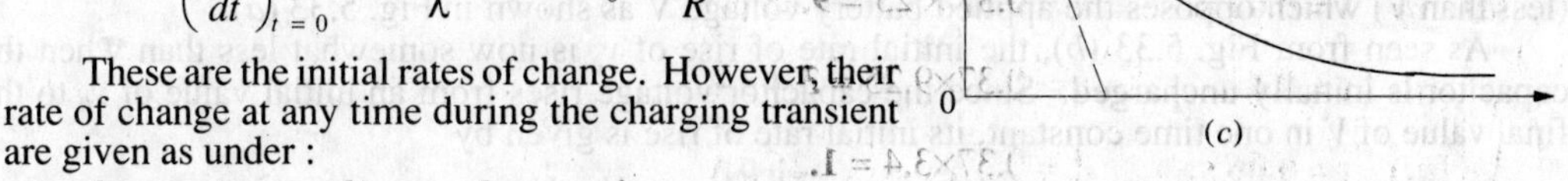

Fig. 5.31

It is shown in Fig. 5.31 (*c*).

It should be clearly understood that a negative rate of change means a decreasing rate of change. It does not mean that the concerned quantity has reversed its direction.

5.22. Transient Relations During Capacitor Discharging Cycle

As shown in Fig. 5.28 (*b*), switch *S* has been shifted to *b*. Hence, the capacitor undergoes the discharge cycle. Just before the transient state starts, $i_c = 0$, $v_R = 0$ and $v_c = $ V. The moment transient state begins, i_c has maximum value and decreases exponentially to zero at the end of the transient state. So does v_c. However, during discharge, all rates of change have polarity opposite to that during charge. For example, dv_c/dt has a positive rate of change during charging and negative rate of change during discharging.

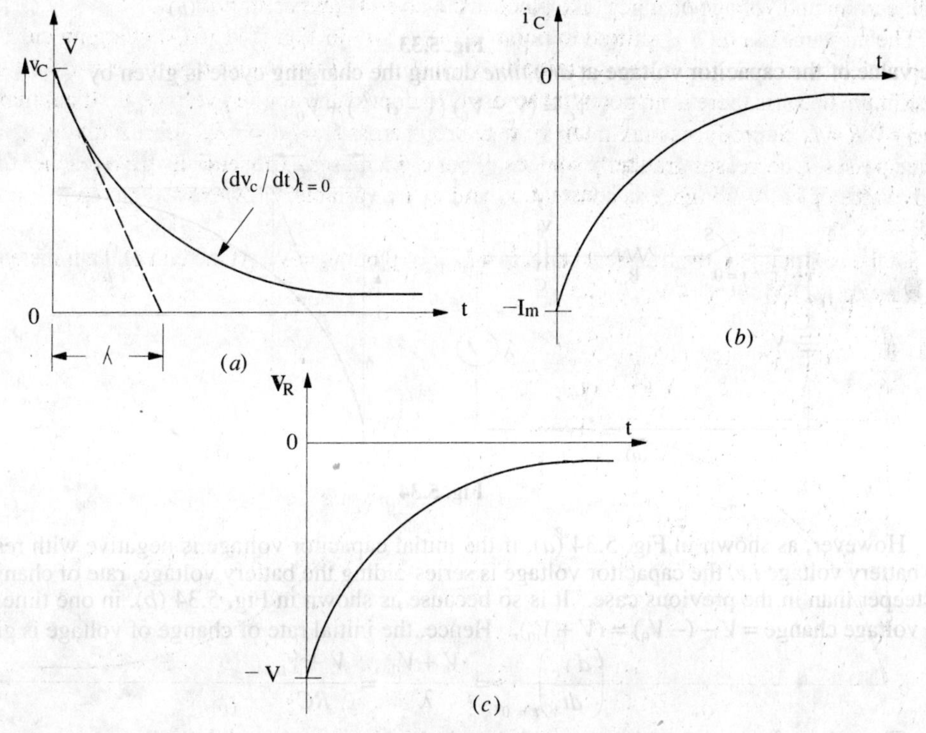

Fig. 5.32

Also, it should be noted that during discharge, v_c maintains its original polarity whereas i_c reverses its direction of flow. Consequently, during capacitor discharge, v_R also reverses its direction.

The various rates of change at *any time* during the discharge transients are as given in Art.

$$\frac{dv_c}{dt} = -\frac{V}{\lambda}e^{-t/\lambda};\quad \frac{di_c}{dt} = \frac{I_0}{\lambda}e^{-t/\lambda};\quad \frac{dv_R}{dt} = \frac{V}{\lambda}e^{-t/\lambda}$$

These are represented by the curves of Fig. 5.32.

5.23. Charging and Discharging of a Capacitor with Initial Charge

In Art. 5.18, we considered the case when the capacitor was initially uncharged and hence, had no voltage across it. Let us now consider the case, when the capacitor has an initial potential of V_0 (less than V) which opposes the applied battery voltage V as shown in Fig. 5.33 (*a*).

As seen from Fig. 5.33 (*b*), the initial rate of rise of v_c is now somewhat less than when the capacitor is initially uncharged. Since the capacitor voltage rises from an initial value of v_0 to the final value of V in one time constant, its initial rate of rise is given by

$$\left(\frac{dv_C}{dt}\right)_{t=0} = \frac{V - V_0}{\lambda} = \frac{V - V_0}{RC}$$

(*a*) (*b*)

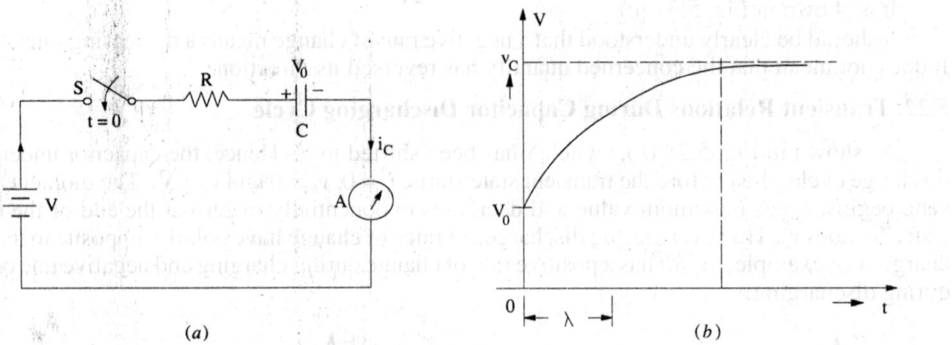

Fig. 5.33

The value of the capacitor voltage at *any time* during the charging cycle is given by

$$v_C = (V - V_0)(1 - e^{-t/\lambda}) + V_0$$

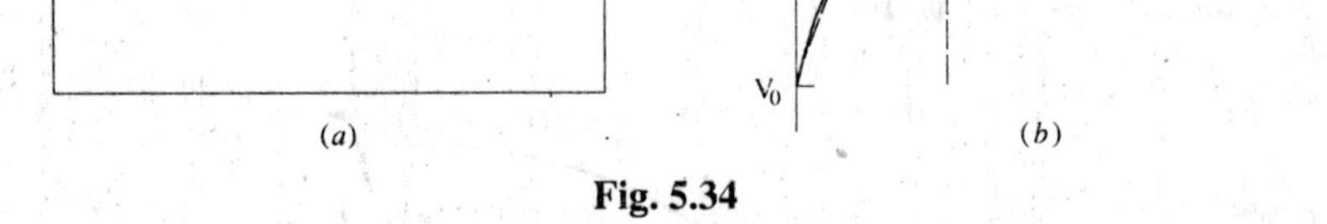

(*a*) (*b*)

Fig. 5.34

However, as shown in Fig. 5.34 (*a*), if the initial capacitor voltage is negative with respect to the battery voltage *i.e.* the capacitor voltage is series-aiding the battery voltage, rate of change of v_C is steeper than in the previous case. It is so because as shown in Fig. 5.34 (*b*), in one time period, the voltage change = $V - (-V_0) = (V + V_0)$. Hence, the initial rate of change of voltage is given by

$$\left(\frac{dv_c}{dt}\right)_{t=0} = \frac{V + V_0}{\lambda} = \frac{V + V_0}{RC}$$

The value of capacitor voltage at *any time* during the charging cycle is given by

$$v_C = (V + V_0)(1 - e^{-t/\lambda}) - V_0$$

The time required for the capacitor voltage to attain any value of v_C during the charging cycle is given by

$$t = \lambda \ ln \left(\frac{V - V_0}{V - v_C}\right) = RC \ ln \left(\frac{V - V_0}{V - v_C}\right) \quad \text{...when } V_0 \text{ is positive}$$

$$t = \lambda \ ln \left(\frac{V + V_0}{V - v_C}\right) = RC \ ln \left(\frac{V + V_0}{V - v_C}\right) \quad \text{...when } V_0 \text{ is negative}$$

Example 5.37. *In Fig. 5.35, the capacitor is initially uncharged and the switch S is then closed. Find the values of I, I_1, I_2 and the voltage at the point A at the start and finish of the transient state.*

Solution. At the moment of closing the switch *i.e.* at the start of the transient state, the capacitor acts as a short-circuit. Hence, there is only a resistance of 2 Ω in the circuit because 1 Ω resistance is shorted out thereby grounding point *A*. Hence, $I_1 = \mathbf{0}$; $I = I_2 = 12/2 = \mathbf{6A}$. Obviously, $V_A = \mathbf{0\ V}$.

At the end of the transient state, the capacitor acts as an open-circuit. Hence,

$I_2 = 0$ and $I = I_1 = 12/(2 + 1) = \mathbf{4\ A}$. $V_A = \mathbf{6\ V}$.

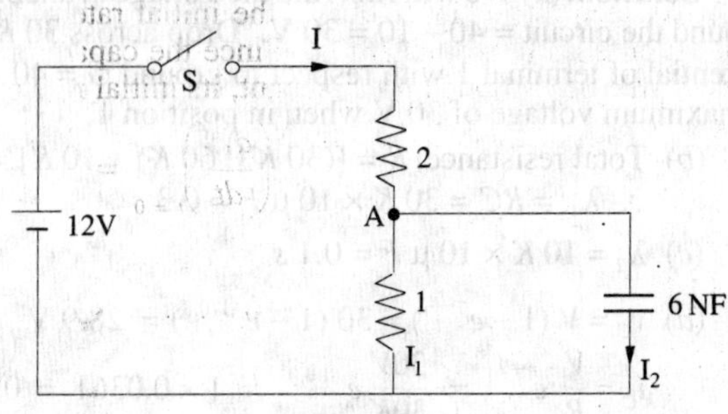

Fig. 5.35

Example 5.38. *Calculate the values of i_2, i_3, v_2, v_3, v_a, v_c and v_L of the network shown in Fig. 5.36 at the following times :*

(i) At time, t = 0 + immediately after the switch S is closed ;

(ii) At time, t → ∞ i.e. in the steady state. **(Network Analysis AMIE Sec. B Winter 1990)**

Solution. (*i*) In this case the coil acts as an open circuit, hence $i_2 = 0$; $v_2 = 0$ and $v_L = 20$ V.

Since a capacitor acts as a short circuit $i_3 = 20/(5 + 4) = 9 = 20/9$ A. Hence, $v_3 = (20/9) \times 4 = 80/9$ V and $v_C = 0$.

(*ii*) Under steady state conditions, capacitor acts as an open circuit and coil as a short circuit. Hence, $i_2 = 20/(5 + 7) = 20/12 = 5/3$ A; $v_2 = 7 \times 5/3 = 35/3$ V; $v_L = 0$. Also $i_3 = 0$, $v_3 = 0$ but $v_C = \mathbf{20\ V}$.

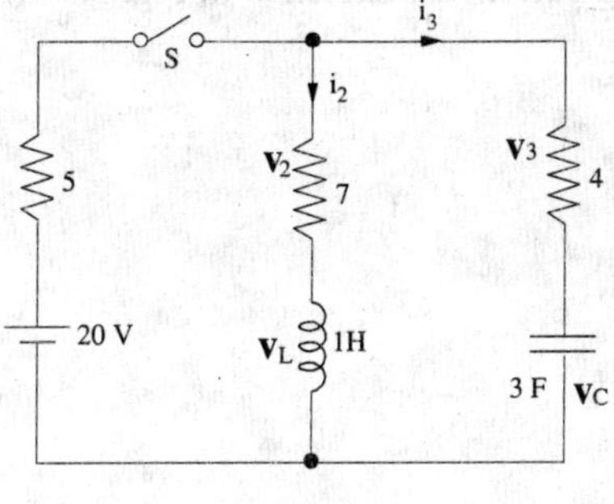

Fig. 5.36

Example 5.39. If in the *RC* circuit of Fig. 5.36; $R = 2$ M Ω, $C = 5$ μ F and $V = 100$ V, calculate

(a) initial rate of change of capacitor voltage

(b) initial rate of change of capacitor current

(c) initial rate of change of voltage across the 2 M Ω resistor

(d) all of the above at t = 80 s.

Solution. (*a*) $\left(\frac{dv_C}{dt}\right)_{t=0} = \frac{V}{\lambda} = \frac{100}{2 \times 10^6 \times 5 \times 10^{-6}} = \frac{100}{10} = 10 \text{ V/s}$

(*b*) $\left(\frac{di_C}{dt}\right)_{t=0} = -\frac{I_0}{\lambda} = -\frac{V/R}{\lambda} = -\frac{100/2 \times 10^6}{10} = -5 \ \mu\text{A/s}$

(*c*) $\left(\frac{dv_R}{dt}\right)_{t=0} = -\frac{V}{\lambda} = -\frac{100}{10} = -10 \text{ V/s}$

(*d*) All the above rates of change would be zero because the transient disappears after about $5\lambda = 5 \times 10 = 50$ s.

Example 5.40. *In Fig. 5.37 (a), the capacitor C is fully discharged, since the switch is in position 2.*

At time t = 0, the switch is shifted to position 1 for 2 seconds. It is then returned to position 2 where it remains indefinitely. Calculate

(a) the maximum voltage to which the capacitor is charged when in position 1.
(b) charging time constant λ_1 in position 1.
(c) discharging time constant λ_2 in position 2.
(d) v_C and i_C at the end of 1 second in position 1.
(e) v_C and i_C at the instant the switch is shifted to position 2 at t = 1 second.
(f) v_C and i_C after a lapse of 1 second when in position 2.
(g) sketch the waveforms for v_C and i_C for the first 2 seconds of the above switching sequence.

Solution. (*a*) We will first find the voltage available at terminal 1. As seen the net battery voltage around the circuit = 40 – 10 = 30 V. Drop across 30 *K* resistor = 30 × 30/(30 + 60) = 10 V. Hence, potential of terminal 1 with respect to ground *G* = 40 – 10 = 30 V. Hence, capacitor will charge to a maximum voltage of 30 V when in position 1.

(*b*) Total resistance, $R = [(30\,K \,||\, 60\,K) + 10\,K] = 30\,K$

$\therefore\ \lambda_1 = RC = 30\,K \times 10\,\mu F = 0.3\,s$

(*c*) $\lambda_2 = 10\,K \times 10\,\mu F = 0.1\,s$

(*d*) $v_C = V(1 - e^{-t/\lambda_1}) = 30\,(1 - e^{-1/0.3}) = 28.9$ V

$$i_C = \frac{V}{R}\,e^{-t/\lambda_1} = \frac{30V}{30K}\,e^{-1/0.03} = 1 \times 0.0361 = 0.036 \text{ mA}$$

(*e*) $v_C = 28.9$ V at $t = 1^+$ s at position 2 but $i_C = -28.9\text{ V}/10\,K = -2.89$ mA at $t = 1^+$s in position 2.

(*f*) $v_C = 28.9\,e^{-t/\lambda_2} = 28.9\,e^{-1/0.1} = 0.0013\text{ V} \cong 0\text{ V}.$

$i_C = -2.89\,e^{-t/\lambda_2} = -2.89\,e^{-1/0.1} = 0.00013\text{ mA} \cong 0.$

The waveforms of the capacitor voltage and charging current are sketched in Fig. 5.37 (*b*).

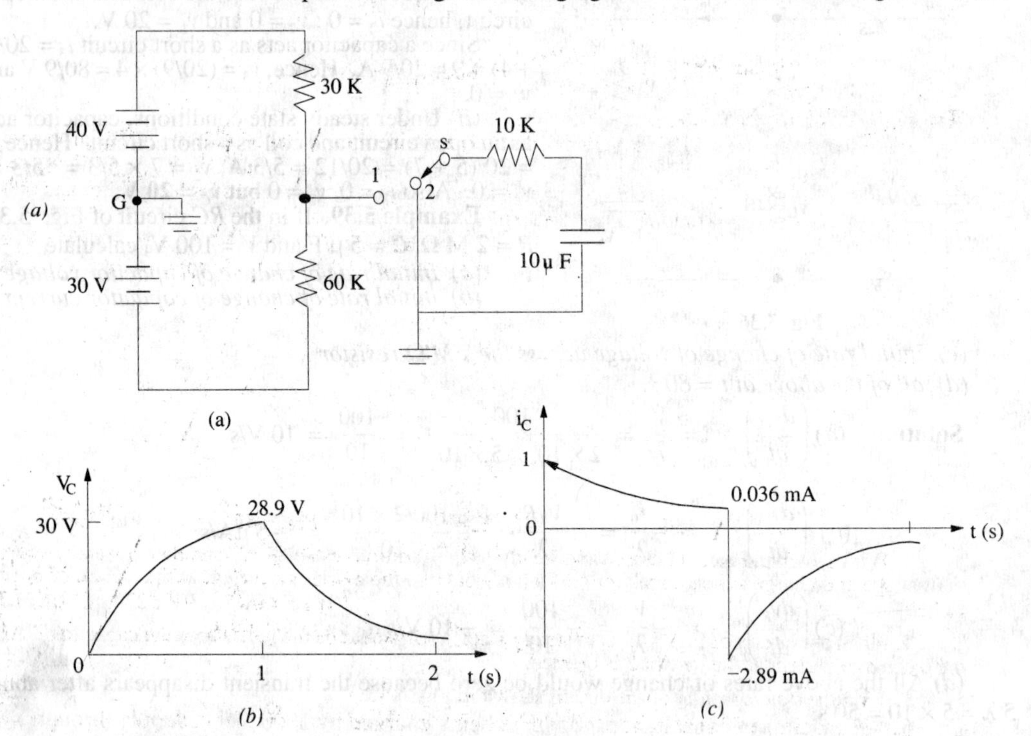

Fig. 5.37

Example 5.41. *In the RC circuit of Fig. 5.38, R = 2 MΩ and C = 5 μF, the capacitor is charged to an initial potential of 50 V. When the switch is closed at t = 0^+, calculate*
(a) initial rate of change of capacitor voltage and
(b) capacitor voltage after a lapse of 5 times the time constant i.e. 5 λ.
If the polarity of capacitor voltage is reversed, calculate
(c) the values of the above quantities and
(d) time for v_c to reach – 10 V, 0 V and 95 V.

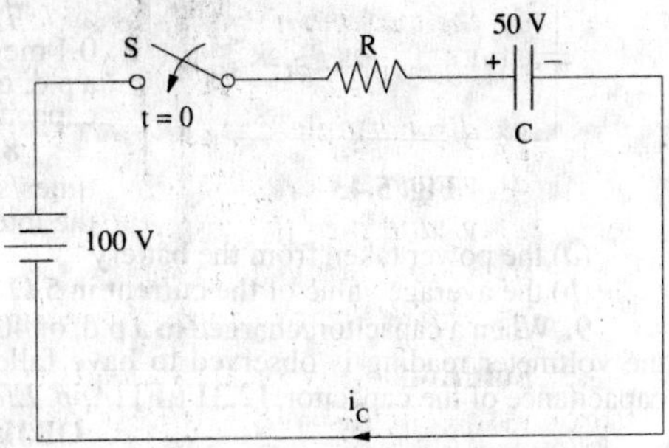

Fig. 5.38

Solution. (a) $\left(\dfrac{d\,v_C}{dt}\right)_{t=0} = \dfrac{V - V_0}{\lambda}$

$$= \frac{V - V_0}{RC} = \frac{100 - 50}{10} = \mathbf{5\ V/s}$$

(b) $v_C = (V - V_0)(1 - e^{-t/\lambda}) + V_0 = (100 - 50)(1 - e^{-5\lambda/\lambda}) + 50 = 49.7 + 50 = \mathbf{99.7\ V}$

(c) When $V_0 = -50$ V, $\left(\dfrac{dv_C}{dt}\right)_{t=0} = \dfrac{V - (-V_0)}{\lambda} = \dfrac{V + V_0}{\lambda} = \dfrac{150}{10} = \mathbf{15\ V/s}$

$$v_C = (V - V_0)(1 - e^{-t/\lambda}) + V_0 = [100 - (-50)](1 - e^{-5}) + (-50)$$

$$= 150(1 - e^{-5}) - 50 = \mathbf{99\ V.}$$

(d) $t = \lambda \ln\left(\dfrac{V - V_0}{V - v_C}\right) = 10\,ln\left[\dfrac{100 - (-50)}{100 - (-10)}\right] = 10\,ln\left(\dfrac{150}{110}\right) = \mathbf{3.1\ s}$

$$t = 10\,ln\left[\frac{100 - (-50)}{100 - (0)}\right] = 10\,ln\left(\frac{150}{100}\right) = \mathbf{4.055\ s}$$

$$t = 10\,ln\left[\frac{100 - (-50)}{100 - 95}\right] = 10\,ln\left(\frac{150}{5}\right) = \mathbf{34\ s}$$

Tutorial Problems No. 5.3

1. For the circuit shown in Fig. 5.39 calculate (*i*) equivalent capacitance and (*ii*) voltage drop across each capacitor. All capacitance values are in μ F. **[(*i*) 6μF (*ii*) V_{AB} = 60 V, V_{BC} = 40 V]**

2. In the circuit of Fig. 5.40 find (*i*) equivalent capacitance (*ii*) drop across each capacitor and (*iii*) charge on each capacitor. All capacitance values are in μF.
[(*i*) 1.82 μV (*ii*) V_1 = 50 V ; V_2 = V_3 = 20 V ; V_4 = 40 V
[(*iii*) Q_1 = 200 μC ; Q_2 = 160 μC ; Q_3 = 40 μC ; Q_4 = 200 μC]

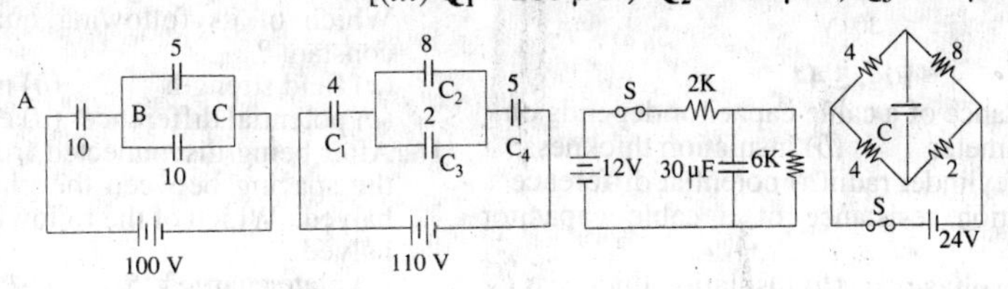

Fig. 5.39 Fig. 5.40 Fig. 5.41 Fig. 5.42

3. With switch in Fig. 5.41 closed and steady-state conditions established, calculate (*i*) steady-state current (*ii*) voltage and charge across capacitor (*iii*) what would be the discharge current at the instant of opening the switch ? **[(*i*) 1.5 mA (*ii*) 9V ; 270 μC (*iii*) 1.5 mA]**

4. When the circuit of Fig. 5.42 is in steady state, what would be the p.d. across the capacitor ? Also, find the discharge current at the instant *S* is opened. **[8V ; 1.8A]**

5. Find the time constant of the circuit shown in Fig. 5.43. **[200 μS]**

6. A capacitor of capacitance 0.01 μF is being charged by 1000 V d.c. supply through a resistor of 0.1 megaohm. Determine the voltage to which the capacitor has been charged when the charging

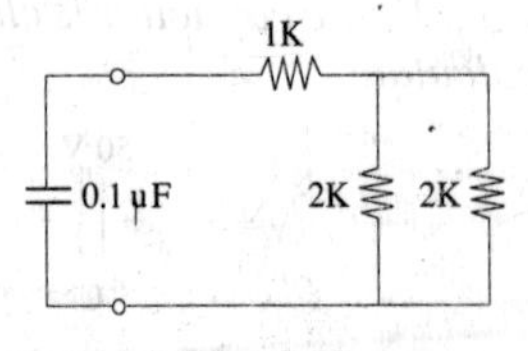

Fig. 5.43

current has decreased to 90% of its initial value. Find also the time taken for the current to decrease to 90% of its initial value. **[100 V, 0.1056 ms]**

7. An 8 μF capacitor is being charged by a 400 V supply through 0.1 mega-ohm resistor. How long will it take the capacitor to develop a p.d. of 300 V ? Also what fraction of the final energy is stored in the capacitor ? **[1.11 Second, 56.3% of full energy]**

8. A 10 μF capacitor is charged from a 200 V battery 250 times/second and completely discharged through a 5 Ω resistor during the interval between charges. Determine

(*a*) the power taken from the battery
(*b*) the average value of the current in 5 Ω resistor. **[(*a*) 50 W (*b*) 0.5 A]**

9. When a capacitor, charged to a p.d. of 400 V, is connected to a voltmeter having a resistance of 25 M Ω, the voltmeter reading is observed to have fallen to 50 V at the end of an interval of 2 minutes. Find the capacitance of the capacitor. **[2.31 μF]** (*App. Elect. London Univ.*)

OBJECTIVE TESTS–5

1. A capacitor consists of two
(*a*) insulation separated by a dielectric
(*b*) conductors separated by an insulator
(*c*) ceramic plates and one mica disc
(*d*) silver-coated insulators

2. The capacitance of a capacitor is NOT influenced by
(*a*) plate thickness (*b*) plate area
(*c*) plate separation (*d*) nature of the dielectric

3. A capacitor that stores a charge of 0.5 C at 10 volts has a capacitance of ----------- farad.
(*a*) 5 (*b*) 20 (*c*) 10 (*d*) 0.05

4. If a dielectric slab of thickness 5 mm and $\varepsilon_r = 6$ is inserted between the plates of an air capacitor with plate separation of 8 mm, its capacitance is
(*a*) decreased (*b*) almost doubled
(*c*) almost halved (*d*) unaffected

5. In a cable capacitor, voltage gradient is maximum at the surface of the
(*a*) sheath (*b*) conductor
(*c*) insulator (*d*) earth

6. In Fig. 5.44 voltage across C_1 will be-----volt.
(*a*) 100 (*b*) 200 (*c*) 150 (*d*) 300

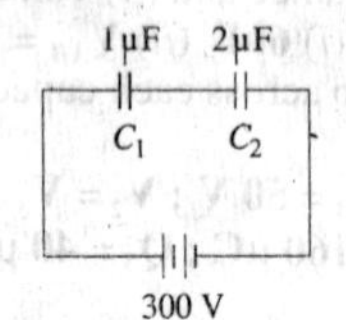

Fig. 5.44

7. The capacitance of a cable capacitor depends on
(*a*) core diameter (*b*) insulation thickness
(*c*) ratio of cylinder radii (*d*) potential difference

8. The insulation resistance of a cable capacitor depends on
(*a*) applied voltage (*b*) insulation thickness
(*c*) core diameter (*d*) ratio of inner and outer radii

9. The time constant of an *R-C* circuit is defined as the time during which capacitor charging current becomes ----percent of its ----value.
(*a*) 37, final (*b*) 63, final
(*c*) 63, initial (*d*) 37, initial

10. The period during which current and voltage changes take place in a circuit is called ----------- condition.
(*a*) varying (*b*) permanent
(*c*) transient (*d*) steady

11. In an *R-C* circuit connected across a *d.c.* voltage source, which of the following is zero at the beginning of the transient state ?
(*a*) drop across *R* (*b*) charging current
(*c*) capacitor voltage (*d*) none of the above

12. When an *R-C* circuit is suddenly connected across a *d.c.* voltage source, the initial rate of change of capacitor current is
(*a*) $-I_0/\lambda$ (*b*) I_0/λ (*c*) V/R (*d*) $-V/\lambda$

13. Which of the following quantity maintains the same polarity during charging and discharging of a capacitor ?
(*a*) capacitor voltage (*b*) capacitor current
(*c*) resistive drop (*d*) none of the above

14. In a cable capacitor with compound dielectric, voltage gradient is inversely proportional to
(*a*) permittivity (*b*) radius of insulating material
(*c*) cable length (*d*) both (*a*) and (*b*)

15. While a capacitor is still connected to a power source, the spacing between its plates is halved. Which of its following quantity would remain constant ?
(*a*) field strength (*b*) plate charge
(*c*) potential difference (*d*) electric flux density

16. After being disconnected from the power source, the spacing between the plates of a capacitor is halved. Which of the following quantity would be halved ?
(*a*) plate charge (*b*) field strength
(*c*) electric flux density (*d*) potential difference

ANSWERS

1. *b* **2.** *a* **3.** *d* **4.** *b* **5.** *b* **6.** *b* **7.** *c* **8.** *d* **9.** *d* **10.** *c* **11.** *c* **12.** *a* **13.** *a* **14.** *d* **15.** *c* **16.** *d*

6 MAGNETISM AND ELECTROMAGNETISM

6.1. Absolute and Relative Permeabilities of a Medium

The phenomena of magnetism and electromagnetism are dependent upon a certain property of the medium called its permeability. Every medium is supposed to possess two permeabilities :

(*i*) absolute permeability (μ) and (*ii*) relative permeability (μ_r).

For measuring relative permeability, vacuum or free space is chosen as the reference medium. It is allotted an absolute permeability of $\mu_0 = 4\pi \times 10^{-7}$ henry/metre. Obviously, relative permeability of vacuum with reference to itself is unity. Hence, for free space,

absolute permeability $\quad \mu_0 = 4\pi \times 10^{-7}$H/m

relative permeability $\quad \mu_r = 1.$

Now, take any medium other than vacuum. If its relative permeability, as compared to vacuum is μ_r, then its absolute permeability is $\mu = \mu_0\, \mu_r$ H/m.

6.2. Laws of Magnetic Force

Coulomb was the first to determine experimentally the quantitative expression for the magnetic force between two *isolated* point poles. It may be noted here that, in view of the fact that magnetic poles always exist in pairs, it is impossible, in practice, to get an isolated pole. The concept of an isolated pole is purely theoretical. However, poles of a thin but long magnet may be assumed to be point poles for all practical purposes (Fig. 6.1). By using a torsion balance, he found that the force between two magnetic poles placed in a medium is

(*i*) directly proportional to their pole strengths

(*ii*) inversely proportional to the square of the distance between them and

(*iii*) inversely proportional to the absolute permeability of the surrounding medium.

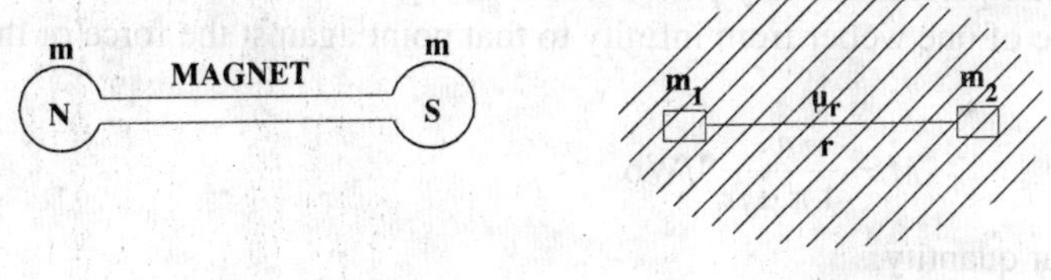

Fig. 6.1 Fig. 6.2

For example, if m_1 and m_2 represent the magnetic strengths of the two poles (its unit as yet being undefined), r the distance between them (Fig. 6.2) and μ the absolute permeability of the surrounding medium, then the force F is given by

$$F \propto \frac{m_1 m_2}{\mu\, r^2} \quad \text{or} \quad F = k\frac{m_1 m_2}{\mu\, r^2} \quad \text{or} \quad \vec{F} = \frac{k\, m_1 m_2}{\mu\, r^2}\hat{r} \quad \text{in vector from}$$

where $\hat{r}$ is a unit vector to indicate direction of r.

$$\text{or } \vec{F} = k\frac{m_1 m_2}{r^3}\,\vec{r} \text{ where } \vec{F} \,\&\, \vec{r} \text{ are vectors}$$

In the S.I. system of units, the value of the constant k is = $1/4\pi$.

$$\therefore \quad F = \frac{m_1 m_2}{4\pi\mu r^2} \text{N} \quad \text{or} \quad F = \frac{m_1 m_2}{4\pi\mu_0\mu_r r^2} \text{N} \qquad \text{–in a medium}$$

In vector form, $\vec{F} = \frac{m_1 m_2}{4\pi\mu r^3}\vec{r} = \frac{m_1 m_2}{4\pi\mu_0 r^2}$ N –in air

If, in the above equation,

$$m_1 = m_2 = m \text{ (say)} ;\ r = 1 \text{ metre} ;\ F = \frac{1}{4\pi\mu_0} \text{N.}$$

Then $m^2 = 1$ or $m = \pm 1$ weber*

Hence, a unit magnetic pole may be defined as *that pole which when placed in vacuum at a distance of one metre* from a similar and equal pole repels it with a force of $1/4\pi\mu_0$ *newtons.***

6.3. Magnetic Field Strength (H)

Magnetic Field strength at any point within a magnetic field is numerically equal to the force experienced by a *N*-pole of one weber placed at that point. Hence, unit of *H* is N/Wb.

Suppose, it is required to find the field intensity at a point *A* distant *r* metres from a pole of *m* webers. Imagine a similar pole of one weber placed at point *A*. The force experienced by this pole is

$$F = \frac{m \times 1}{4\pi\mu_0 r^2} \text{N} \qquad \therefore \quad H = \frac{m}{4\pi\mu_0 r^2} \text{N/Wb (or A/m)}^{***} \text{ or oersted}$$

Also, if a pole of *m* Wb is placed in a uniform field of strength *H* N/Wb, then force experienced by the pole is = *mH* newtons.

It should be noted that field strength is a vector quantity having both magnitude and direction

$$\therefore \quad \vec{H} = \frac{m}{4\pi\mu_0 r^2}\hat{r} = \frac{m}{4\times\mu_0 r^3}\vec{r}$$

It would be helpful to remember that following terms are sometimes interchangeably used with field intensity : Magnetising force, strength of field, magnetic intensity and intensity of magnetic field.

6.4. Magnetic Potential

The magnetic potential at any point within a magnetic field is measured by the work done in shifting a *N*-pole of one weber from infinity to that point against the force of the magnetic field. It is given by

$$M = \frac{m}{4\pi\mu_0 r} \text{J/Wb} \qquad \text{...(Art. 4.13)}$$

It is a scalar quantity.

6.5. Flux per Unit Pole

A unit *N*-pole is supposed to radiate out a flux of one weber. Its symbol is Φ. Therefore, the flux coming out of a *N*-pole of *m* weber is given by

$$\Phi = m \text{ Wb}$$

---

*To commemorate the memory of German physicist Wilhelm Edward Weber (1804-1891).

**A unit magnetic pole is also defined as that magnetic pole which when placed at a distance of one metre from a very long straight conductor carrying a current of one ampere experiences a force of $1/2\pi$ netwons (Art. 6.18).

***It should be noted that N/Wb is the same thing as ampere/metre (A/m) or just A/m cause 'turn' has no units.

6.6. Flux Density (B)

It is given by the flux passing per unit area through a plane at right angles to the flux. It is usually designated by the capital letter B and is measured in weber/metre^2. It is a Vector Quantity.

It Φ Wb is the total magnetic flux passing normally through an area of A m^2, then

$$B = \Phi/A \text{ Wb/m}^2 \text{ or tesla (T)}$$

Note. Let us find an expression for the flux density at a point distant r metres from a unit N-pole (*i.e.* a pole of strength 1 Wb.) Imagine a sphere of radius r metres drawn round the unit pole. The flux of 1 Wb radiated out by the unit pole falls normally on a surface of $4\pi r^2$.m^2. Hence

$$B = \frac{\Phi}{A} = \frac{1}{4\pi r^2} \text{ Wb/m}^2$$

6.7. Absolute Permeability (μ) and Relative Permeability (μ_r)

In Fig. 6.3 is shown a bar of a magnetic material, say, iron placed in a uniform field of strength H N/Wb. Suppose, a flux density of B Wb/m^2 is developed in the rod.

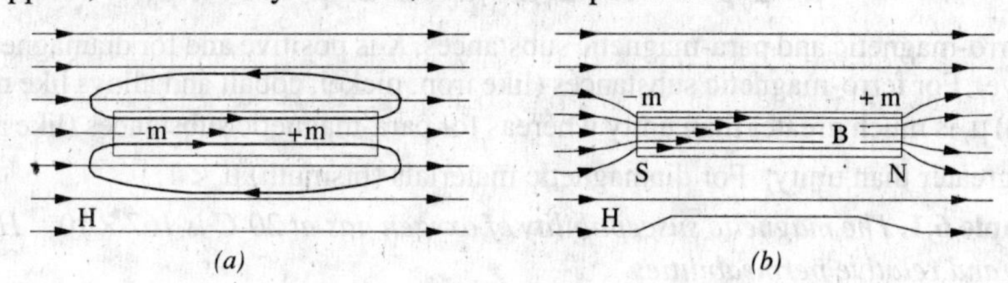

Fig. 6.3

Then, the absolute permeability of the material of the rod is defined as

$$\mu = B/H \text{ henry/metre or } B = \mu H = \mu_0 \mu_r H \text{ Wb/m}^2 \quad \ldots(i)$$

When H is established in air (or vacuum), then corresponding flux density developed in air is

$$B_0 = \mu_0 H$$

Now, when iron rod is placed in the field, it gets magnetised by induction. If induced pole strength in the rod is m Wb, then a flux of m Wb emanates from its N-pole, re-enters its S-pole and continues from S to N-pole within the magnet. If A is the face or pole area of the magentised iron bar, the *induction* flux density in the rod is

$$B_i = m/A \text{ Wb/m}^2$$

Hence, total flux density in the iron rod consists of two parts [Fig. 6.3 (*b*)].

(*i*) B_0 –flux density in air even when rod is not present

(*ii*) B_i–induction flux density in the rod

$$\therefore \quad B = B_0 + B_i = \mu_0 H + m/A$$

Eq. (*i*) above may be written as $B = \mu_r.\mu_0 H = \mu_r B_0$

$$\therefore \quad \mu_r = \frac{B}{B_0} = \frac{B\,(\text{material})}{B_0(\text{vacuum})} \quad \ldots\text{for same } H$$

Hence, relative permeability of a material is equal to *the ratio of the flux density produced in that material to the flux density produced in vacuum by the same magnetising force.*

6.8. Intensity of Magnetisation (I)

It may be defined as the induced pole strength developed per unit area of the bar. Also, it is the magnetic moment developed per unit volume of the bar.

Let m = pole strength induced in the bar in Wb

A = face or pole area of the bar in m^2

Then $I = m/A$ Wb/m^2

Hence, it is seen that intensity of magnetisation of a substance may be defined as *the flux density produced in it due to its own induced magnetism.*

If l is the magnetic length of the bar, then the product $(m \times l)$ is known as its magnetic moment M

$$\therefore \quad I = \frac{m}{A} = \frac{m \times l}{A \times l} = \frac{M}{V} = \text{magnetic moment/volume}$$

6.9. Susceptibility (K)

Susceptibility is defined as *the ratio of intensity of magnetisation I to the magnetising force H.*

$$\therefore \quad K = I/H \text{ henry/metre.}$$

6.10. Relation Between B, H, I and K

It is obvious from the above discussion in Art. 6.7 that flux density B in a material is given by

$$B = B_0 + m/A = B_0 + I \qquad \therefore \quad B = \mu_0 H + I$$

Now absolute permeability is $\mu = \frac{B}{H} = \frac{\mu_0 H + I}{H} = \mu_0 + \frac{I}{H} \quad \therefore \quad \mu = \mu_0 + K$

Also $\quad \mu = \mu_0 \mu_r \quad \therefore \quad \mu_0 \mu_r = \mu_0 + K$ or $\mu_r = 1 + K/\mu_0$

For ferro-magnetic and para-magnetic substances, K is positive and for diamagnetic substances, it is negative. For ferro-magnetic substances (like iron, nickel, cobalt and alloys like nickel-iron and cobalt-iron) μ_r is much greater than unity whereas for para-magnetic substances (like aluminium), μ_r is slightly greater than unity. For diamagnetic materials (bismuth) $\mu_r < 1$.

Example 6.1. *The magnetic susceptibility of oxygen gas at 20°C is 167×10^{-11} H/m. Calculate its absoute and relative permeabilities.*

Solution. $\mu_r = 1 + \frac{K}{\mu_0} = 1 + \frac{167 \times 10^{-11}}{4\pi \times 10^{-7}} = \mathbf{1.00133}$

Now, absolute permeability

$$\mu = \mu_0 \mu_r = 4\pi \times 10^{-7} \times 1.00133 = \mathbf{12.59 \times 10^{-7}} \; H/m$$

6.11. Boundary Conditions

The case of boundary conditions between two materials of different permeabilities is similar to that discussed in Art. 4.19.

As before, the two boundary conditions are

(*i*) the normal component of flux density is continuous across a boundary.

$$\therefore \quad B_{1n} = B_{2n} \qquad \ldots(i)$$

(*ii*) the tangential component of H is continuous across boundary $\quad H_{1t} = H_{2t}$

As proved in Art. 4.19, in a similar way, it can be shown that

$$\frac{\tan \theta_1}{\tan \theta_2} = \frac{\mu_1}{\mu_2}$$

This is called the law of magnetic flux refraction.

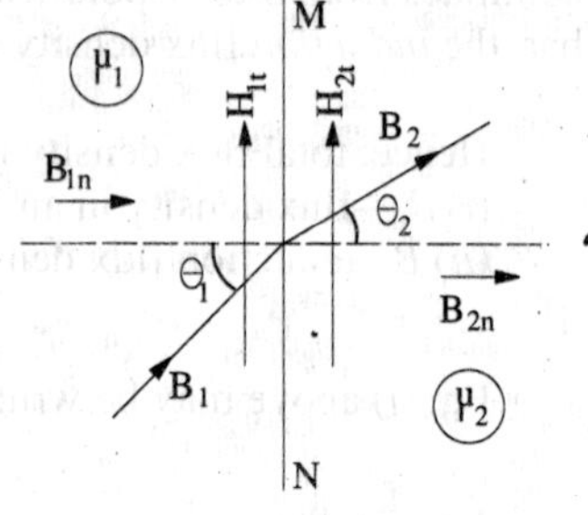

Fig. 6.4

6.12. Weber and Ewing's Molecular Theory

This theory was first advanced by Weber in 1852 and was, later on, further developed by Ewing in 1890. The basic assumption of this theory is *that molecules of all substances are inherently magnets in themselves, each having a N and S pole.* In an unmagnetised state, it is supposed that these small molecular magnets lie in all sorts of haphazard manner forming more or less closed loops (Fig. 6.5). According to the laws of attraction and repulsion, these closed magnetic circuits are satisfied internally, hence there is no resultant external magnetism exhibited by the iron bar. But when such an iron bar is placed in a magnetic field or under the influence of a magnetising force, then these molecular magnets start turning round their axes and orientate themselves more or

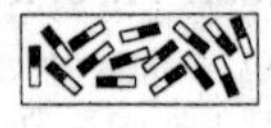
Fig. 6.5

less along straight lines parallel to the direction of the magnetising force. This linear arrangement of the molecular magnets results in *N* polarity at one end of the bar and *S* polarity at the other (Fig. 6.6). As the small magnets turn more nearly in the direction of the magnetising force, it requires more and more of this force to produce a given turning moment, thus accounting for the magnetic saturation. On this theory, the hysteresis loss is supposed to be due to molecular friction of these turning magnets.

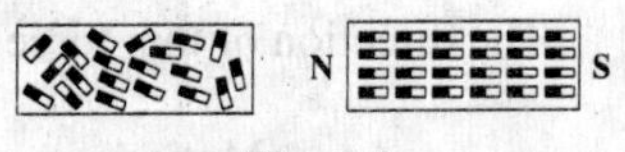

Fig. 6.6

Because of the limited knowledge of molecular structure available at the time of Weber, it was not possible to explain firstly, as to why the molecules themselves are magnets and secondly, why it is impossible to magnetise certain substances like wood etc. The first objection was explained by Ampere who maintained that orbital movement of the electrons round the atom of a molecule constituted a flow of current which, due to its associated magnetic effect, made the molecule a magnet.

Later on, it became difficult to explain the phenomenon of diamagnetism (shown by materials like water, quartz, silver and copper etc.), erratic behaviour of ferromagnetic (intensely magnetisable) substances like iron, steel, cobalt, nickel and some of their alloys etc. and the paramagnetic (weakly magnetisable) substances like oxygen and aluminium etc. Moreover, it was asked : if molecules of all substances are magnets, then why does not wood or air etc. become magnetised ?

All this has been explained satisfactorily by the atom-domain theory which has superseded the molecular theory. It is beyond the scope of this book to go into the details of this theory. The interested reader is advised to refer to some standard book on magnetism. However, it may just be mentioned that this theory takes into account not only the planetary motion of an electron but its rotation about its own axis as well. This latter rotation is called 'electron spin'. The gyroscopic behaviour of an electron gives rise to a magnetic moment which may be either positive or negative. A substance is ferromagnetic or diamagnetic accordingly as there is an excess of unbalanced positive spins or negative spins. Substances like wood or air are non-magnetisable because in their case, the positive and negative electron spins are equal, hence they cancel each other out.

6.13. Curie Point

As a magnetic material is heated, its molecules vibrate more violently. As a consequence, individual molecular magnets get out of alignment as the temperature is increased, thereby reducing the magnetic strength of the magnetised substance. Fig. 6.7 shows the approximate decrease of magnetic strength with rise in temperature. Obviously, it is possible to partially or even completely destroy the magnetic properties of a material by heating. The temperature at which the vibrations of the molecular magnets become so random and out of alignment as to reduce the magnetic strength to zero is called Curie point. More accurately, it is that critical temperature above which a ferromagnetic material becomes paramagnetic.

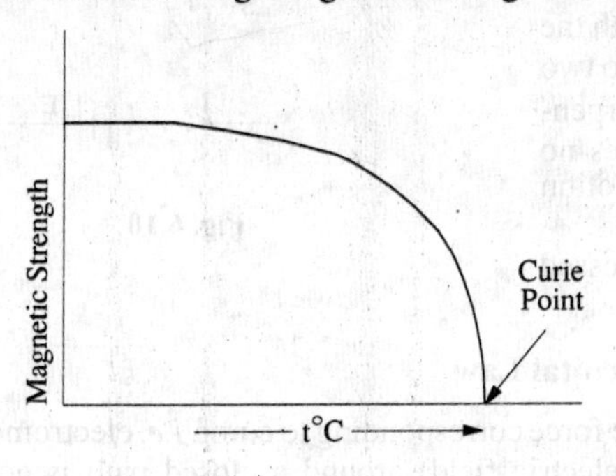

Fig. 6.7

ELECTROMAGNETISM

6.14. Force on a Current-carrying Conductor Lying in a Magnetic Field

It is found that whenever a current-carrying conductor is placed in magnetic field, it experiences a force which acts in a direction perpendicular both to the direction of the current and the field. In Fig. 6.8 is shown a conductor *XY* lying at right angles to the uniform horizontal field of flux density *B* Wb/m^2 produced by two solenoids *A* and *B*. If *l* is the length of the conductor lying within this field and *I* ampere the current carried by it, then the magnitude of the force experienced by it is

$$F = BIl = \mu_0\mu_r HIl \text{ newton}$$

Using vector notation $\vec{F} = I\,\vec{l} \times \vec{B}$ & $F = IlB \sin\theta$ where θ is the angle between $\vec{l}$ & $\vec{B}$ which is 90° in the present case

$$\text{or } F = Il\,B \sin 90° = Il\,\text{B newtons} \qquad (\because \sin 90° = 1)$$

The direction of this force may be easily found by Fleming's left-hand rule.

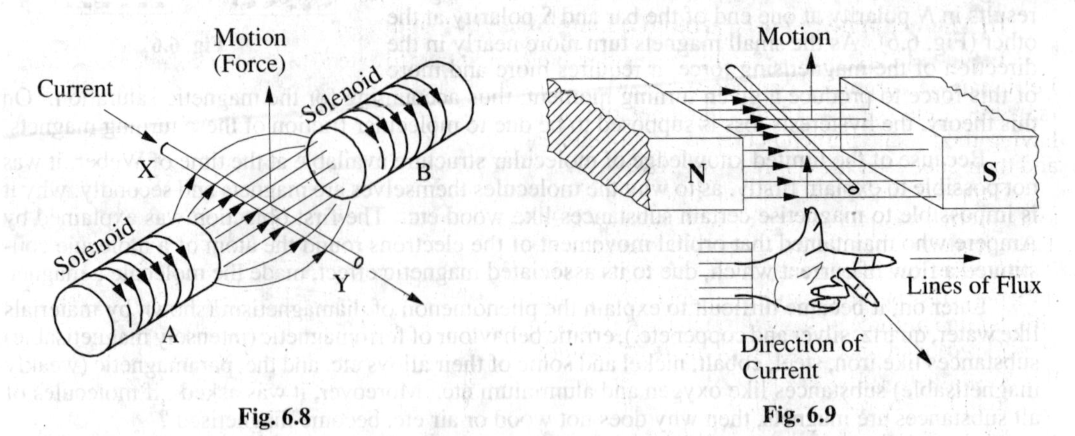

Fig. 6.8 **Fig. 6.9**

Hold out your left hand with forefinger, second finger and thumb at right angles to one another. If the forefinger represents the direction of the field and the second finger that of the current, then thumb gives the direction of the motion. It is illustrated in Fig. 6.9.

Fig. 6.10 shows another method of finding the direction of force acting on a current carrying conductor. It is known as Flat Left Hand rule. The force acts in the direction of the thumb obviously, the direction of motor of the conductor is the same as that of the force.

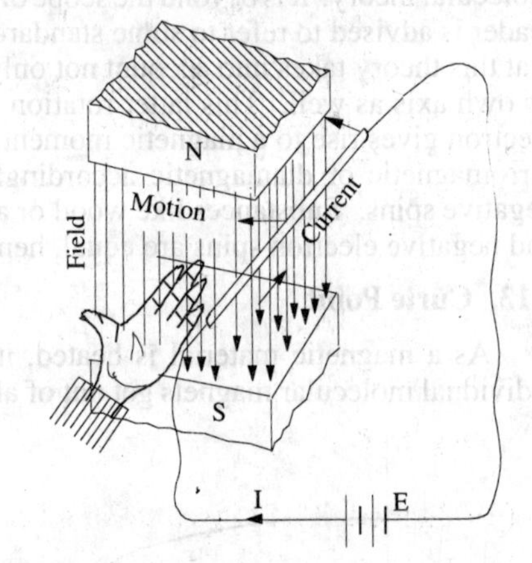

Fig. 6.10

It should be noted that no force is exerted on a conductor when it lies parallel to the magnetic field. In general, if the conductor lies at an angle θ with the direction of the field, then B can be resolved into two components, $B \cos \theta$ parallel to and $B \sin \theta$ perpendicular to the conductor. The former produces no effect whereas the latter is responsible for the motion observed. In that case,

$F = BIl \sin \theta$ newton, which has been expressed as cross product of vector above.*

6.15. Ampere's Work Law or Ampere's Circuital Law

The law states that m.m.f.** (magnetomotive force corresponding to e.m.f. *i.e.* electromotive force of electric field) around a closed path is equal to the current enclosed by the path. Mathematically, $\oint \vec{H}.\vec{ds} =$ I amperes where $\vec{H}$ is the vector representing magnetic field strength in dot product with vector $\vec{ds}$ of the enclosing path S around current I ampere and that is why line integral ($\oint$) of dot product $\vec{H} \cdot \vec{ds}$ is taken.

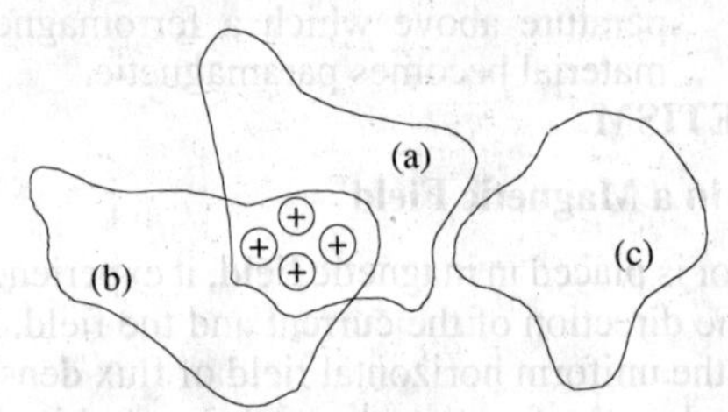

Fig. 6.11

Work law is very comprehensive and is applicable to all magnetic fields whatever the shape of enclosing path *e.g.* (*a*) & (*b*) in Fig. 6.11. Since path *c* does not enclose the conductor, the m.m.f. around it is zero.

*It is simpler to find direction of Force (Motion) through cross product of given vectors $I \vec{l}$ & $\vec{B}$.

**M.M.F. is not a force, but is the work done.

The above Work Law is used for obtaining the value of the magnetomotive force around simple idealized circuits like (*i*) a long straight current-carrying conductor and (*ii*) a long solenoid.

(i) Magnetomotive Force around a Long Straight Conductor

In Fig. 6.12 is shown a straight conductor which is assumed to extend to infinity in either direction. Let it carry a current of I amperes upwards. The magnetic field consists of circular lines of force having their plane perpendicular to the conductor and their centres at the centre of the conductor.

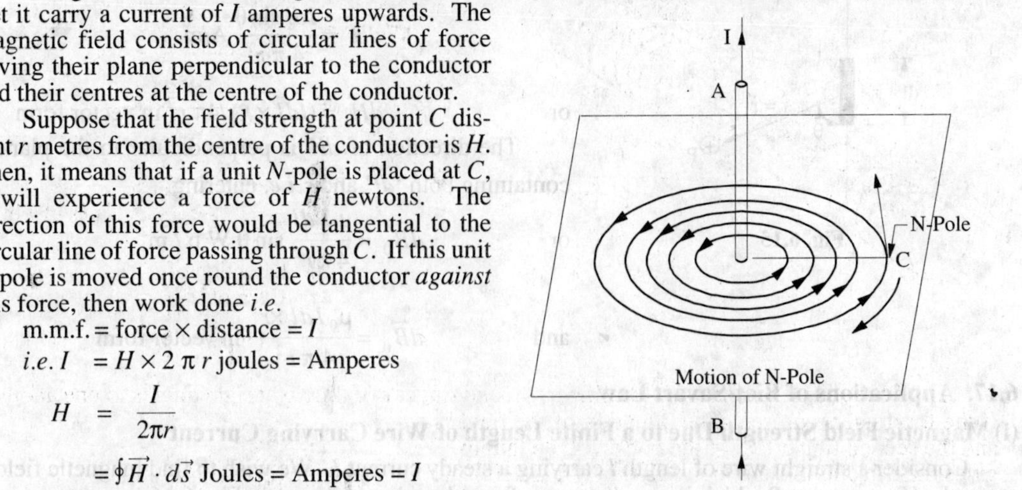

Fig. 6.12

Suppose that the field strength at point C distant r metres from the centre of the conductor is H. Then, it means that if a unit N-pole is placed at C, it will experience a force of H newtons. The direction of this force would be tangential to the circular line of force passing through C. If this unit N-pole is moved once round the conductor *against* this force, then work done *i.e.*

m.m.f. = force × distance = I

i.e. $I = H \times 2\pi r$ joules = Amperes

or $H = \dfrac{I}{2\pi r}$

$= \oint \vec{H} \cdot d\vec{s}$ Joules = Amperes = I

Obviously, if there are N conductors (as shown in Fig. 6.13), then

$$H = \frac{NI}{2\pi r} \text{ A/m or Oersted}$$

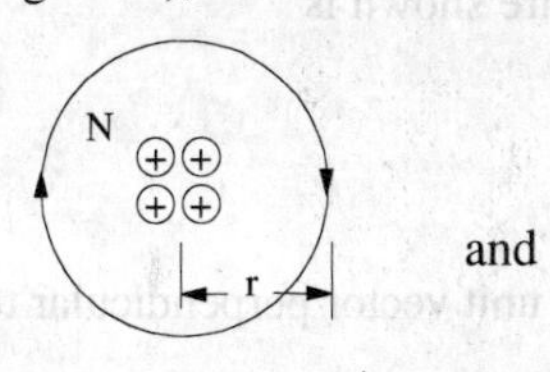

Fig. 6.13

and $B = \mu_0 \dfrac{NI}{2\pi r}$ Wb/m² tesla ...in air

$= \dfrac{\mu_0 \mu_r NI}{2\pi r}$ Wb/m² tesla ...in a medium

(ii) Magnetic Field Strength of a Long Solenoid

Let the Magnetic Field Strength along the axis of the solenoid be H. Let us assume that

(*i*) the value of H remains constant throughout the length l of the solenoid and

(*ii*) the volume of H outside the solenoid is negligible.

Suppose, a unit N-pole is placed at point A outside the solenoid and is taken once round the completed path (shown dotted in Fig. 6.14) in a direction opposite to that of H. Remembering that the force of H newtons acts on the N-pole only over the length l (it being negligible elsewhere), the work done in one round is

$= H \times l$ joules = Amperes

The 'ampere-turns' linked with this path are NI where N is the number of turns of the solenoid and I the current in amperes passing through it. According to Work Law

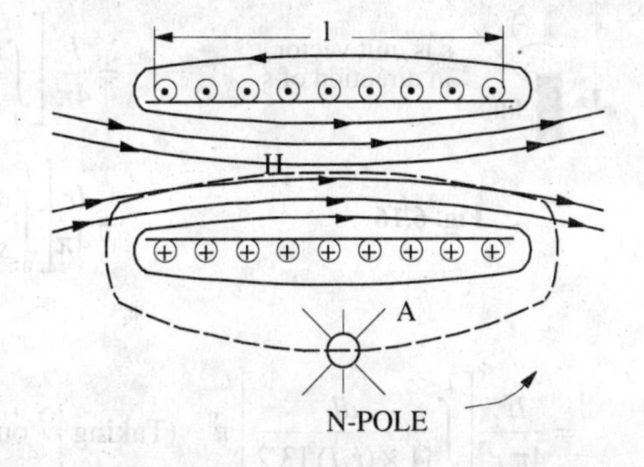

Fig. 6.14

$H \times l = NI$ or $H = \dfrac{NI}{l}$ A/m or Oersted

Also $B = \dfrac{\mu_0 NI}{l}$ Wb/m² or tesla...in air

$= \dfrac{\mu_0 \mu_r NI}{l}$ Wb/m² or tesla...in a medium

6.16. Biot-Savart Law*

The expression for the magnetic field strength dH produced at point P by a vanishingly small length dl of a conductor carrying a current of I amperes (Fig. 6.15) is given by

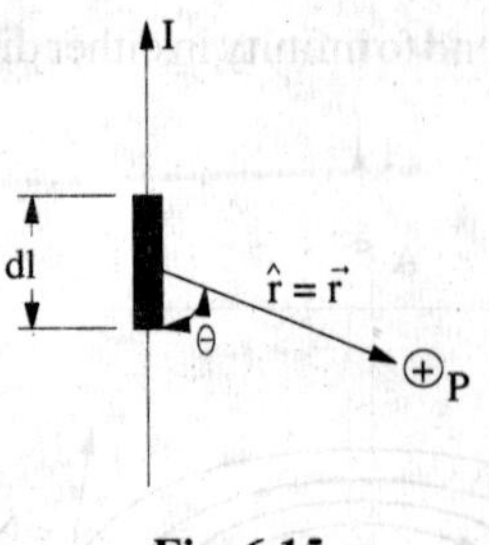

Fig. 6.15

$$dH = \frac{I\,dl\sin\theta}{4\pi r^2}\ \text{A/m}$$

or $\quad d\vec{H} = (I d\vec{l} \times \hat{r})/4\pi r^2$ in vector form

The direction of $d\vec{H}$ is perpendicular to the plane containing both '$d\vec{l}$' and $\vec{r}$ *i.e.* entering.

or $\quad dB_0 = \frac{\mu_0 I\,dl}{4\pi r^2}\sin\theta\ \text{W b / m}^2$

and $\quad d\vec{B}_0 = \frac{\mu_0 \vec{I} d\vec{l} \times \hat{r}}{4\pi r^2}$ in vector form

6.17. Applications of Biot-Savart Law

(i) Magnetic Field Strength Due to a Finite Length of Wire Carrying Current

Consider a straight wire of length l carrying a steady current I. We wish to find magnetic field strength (H) at a point P which is at a distance r from the wire as shown in Fig. 6.16.

The magnetic field strength $\vec{dH}$ due to a small element dl of the wire shown is

$$d\vec{H} = \frac{I\,d\vec{l}\times\hat{s}}{4\pi s^2}\ \text{(By Biot-Savart Law)}$$

or $\quad d\vec{H} = \frac{I\,dl\sin\theta}{4\pi\times s^2}\hat{u}$ (where $\hat{u}$ is unit vector perpendicular to plane containing $d\vec{l}$ and $\hat{s}$ and into the plane.)

or $\quad d\vec{H} = \frac{I\,dl\cos\phi}{4\pi s^2}\hat{u}$ ...[$\because$ θ & φ are complementary angles]

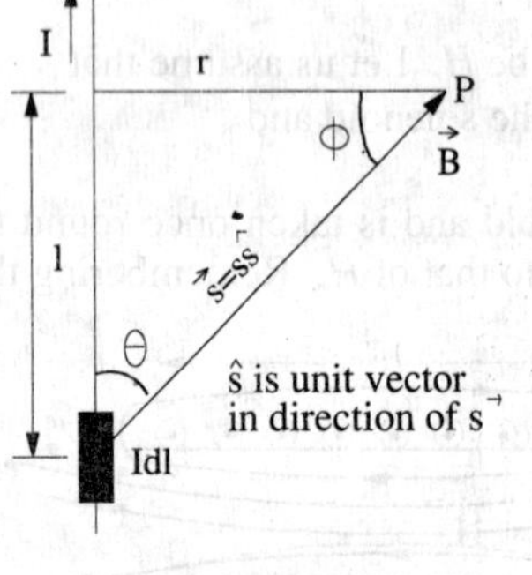

Fig. 6.16

The magnetic field strength due to entire length l:

$$\vec{H} = \frac{I}{4\pi}\left[\int_0^l \frac{\cos\phi\,dl}{s^2}\right]\hat{u}$$

$$= \frac{I}{4\pi}\left[\int_0^l \frac{r/s}{s^2}dl\right]\hat{u} \qquad \left(\because \cos\phi = \frac{r}{s}\ \text{in Fig. 6.16}\right)$$

$$= \frac{Ir}{4\pi}\left[\int_0^l \frac{dl}{s^3}\right]\hat{u} = \frac{Ir}{4\pi}\left[\int_0^l \frac{dl}{(r^2+l^2)^{3/2}}\right]\hat{u}$$

($\because$ r is constant) ; $s = \sqrt{r^2+l^2}$ in Fig. 6.16

$$= \frac{Ir}{4\pi r^3}\left[\int_0^l \frac{dl}{[1\times(r/l)^2]3/2}\right]\hat{u}\quad \text{(Taking } r^3 \text{ out from denominator)}$$

*After the French mathematician and physicist Jean Baptiste Biot (1774–1862) and Felix Savart (1791-1841) a well-known French physicist.

To evaluate the integral most simply, make the following substitution

$$\frac{l}{r} = \tan\phi \text{ in Fig. 6.16}$$

$\therefore$ $l = r\tan\phi$ $\therefore$ $dl = r\sec^2\phi\, d\phi$ & $1 + (r/l)^2 = 1 + \tan^2\phi = \sec^2\phi$ and limits get transformed *i.e.* become 0 to ϕ.

$$\vec{H} = \frac{Ir}{4\pi r^3}\left[\int_0^{\phi} \frac{r\sec^2\phi}{\sec^3\phi} d\phi\right]\hat{u} = \frac{Ir^2}{4\pi r^3}\left[\int_0^{\phi} \cos\phi\, d\phi\right]\hat{u} = \frac{I}{4\pi r}\Big[\sin\phi\Big]_0^{\phi}\hat{u}$$

$$= \frac{I}{4\pi r}\sin\phi\,\hat{u}$$

N.B. For wire of infinite length extending it at both ends *i.e.* $-\infty$ to $+\infty$ the limits of integration would be $-\frac{\pi}{2}$ to $+\frac{\pi}{2}$, giving $\vec{H} = \frac{I}{4\pi r} \times 2\,\hat{u}$ or $\vec{H} = \frac{1}{2\pi r}\hat{u}$.

(ii) Magnetic Field Strength along the Axis of a Square Coil

This is similar to (*i*) above except that there are four conductors each of length, say, $2a$ metres and carrying a current of I amperes as shown in Fig. 6.17. The Magnetic Field Strengths at the axial point P due to the opposite sides ab and cd are H_{ab} and H_{cd} directed at right angles to the planes containing P and ab and P and cd respectively. Now, H_{ab} and H_{cd} are numerically equal, hence their components at right angles to the axis of the coil will cancel out, but the axial components will add together. Similarly, the other two sides da and bc will also give a resultant axial component only.

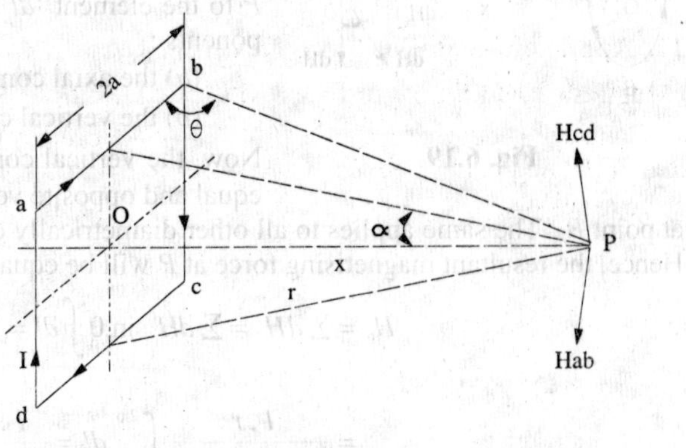

Fig. 6.17

As seen from Eq. (*ii*) above,

$$H_{ab} = \frac{I}{4\pi r}[\cos\theta - \cos(180° - \theta)] = \frac{I.2\cos\theta}{4\pi r} = \frac{I\cos\theta}{2\pi r}$$

Now $\quad r = \sqrt{a^2+x^2} \quad \therefore \quad H_{ab} = \dfrac{I.\cos\theta}{2\pi\sqrt{a^2+x^2}}$

Its axial component is $H_{ab'} = H_{ab}.\sin\alpha = \dfrac{I\cos\theta}{2\pi\sqrt{a^2+x^2}}.\sin\alpha$

All the four sides of the rectangular coil will contribute an equal amount to the resultant magnetic field at P. Hence, resultant magnetising force at P is

$$H = 4 \times \frac{I\cos\theta}{2\pi\sqrt{a^2+x^2}}.\sin\alpha,$$

Now $\quad \cos\theta = \dfrac{a}{\sqrt{(2a^2+x^2)}}$ and $\sin\alpha = \dfrac{a}{\sqrt{a^2+x^2}}$

$$\therefore \quad H = \frac{2a^2.I}{\pi(a^2+x^2).\sqrt{(x^2+2a^2)}} \text{AT/m}$$

In case, value of H is required at the centre O of the coil, then putting $x = 0$ in the above expression, we get

$$H = \frac{2a^2 \cdot I}{\pi a^2 \cdot \sqrt{2}\cdot a} = \frac{\sqrt{2}\cdot I}{\pi a} \text{ AT/m.}$$

Note. The last result can be found directly as under. As seen from Fig. 6.18, the field at point O due to any side is, as given by Eq (*i*)

$$= \frac{I}{4\pi a}\int_{\pi/4}^{-\pi/4} \sin\theta . d\theta = \frac{I}{4\pi a} \left| -\cos\theta \right|_{45°}^{-45°} = \frac{I}{4\pi a}.2\cos 45° = \frac{I}{4\pi a}.\frac{2}{\sqrt{2}}$$

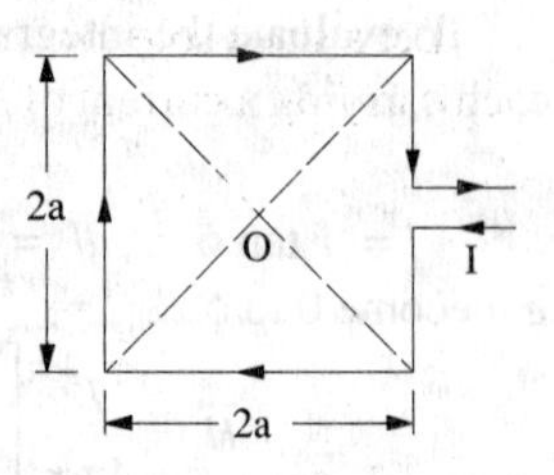

Fig. 6.18

Resultant magnetising force due to all sides is

$$H = 4\times\frac{1}{4\pi\, a}.\frac{2}{\sqrt{2}} = \frac{\sqrt{2}I}{\pi\, a} \text{ AT/m} \qquad \text{...as found above}$$

(iii) Magnetising Force on the Axis of a Circular Coil

In Fig. 6.19 is shown a circular one-turn coil carrying a current of I amperes. The magnetising force at the axial point P due to a small element 'dl' as given by Laplace's Law is

$$\left|\vec{dH}\right| = \frac{I\,dl}{4\,\pi\,(r^2+x^2)}$$

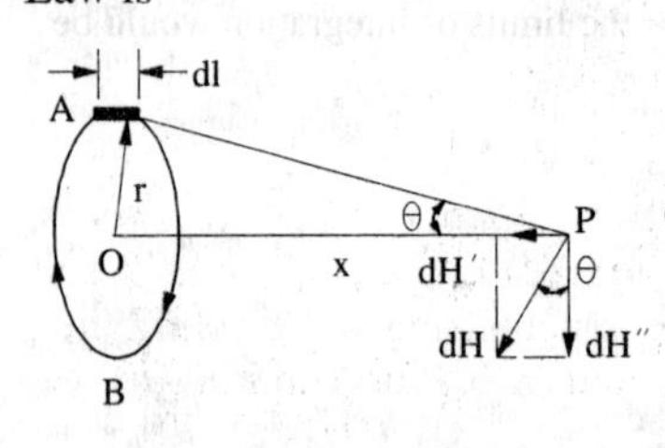

Fig. 6.19

The direction of dH is at right angles to the line AP joining point P to the element 'dl'. Now, dH can be resolved into two components :

(*a*) the axial component $dH' = dH \sin\theta$

(*b*) the vertical component $dH'' = dH \cos\theta$

Now, the vertical component $dH \cos\theta$ will be cancelled by an equal and opposite vertical component of dH due to element 'dl' at point B. The same applies to all other diametrically opposite pairs of dl's taken around the coil. Hence, the resultant magnetising force at P will be equal to the sum of all the axial components.

$$\therefore \quad H = \Sigma\, dH' = \Sigma\, dH \sin\theta \int dl = \Sigma \frac{I\,.\,dl\,.\,r}{4\pi\,(r^2+x^2)^{3/2}}\int dl \quad \left(\because \quad \sin\theta = \frac{r}{\sqrt{r^2+x^2}}\right)$$

$$= \frac{I\,.\,r}{4\pi\,(r^2+x^2)^{3/2}}\int_0^{2\pi r} dl = \frac{I\,.\,r\,.2\pi\,r}{4\pi\,(r^2+x^2)^{3/2}} = \frac{Ir^2}{2(r^2+x^2)^{3/2}}$$

$$= \frac{I}{2\,r}.\frac{r^3}{(r^2+x^2)^{3/2}} \qquad \therefore H = \frac{I\sin^3\theta}{2r} \text{ AT/m}$$

or $$H = \frac{NI}{2r}\sin^3\theta \text{ AT/m} \qquad \text{–for an } N\text{- turn coil} \qquad ...(iii)$$

In case the value of H is required at the centre O of the coil, then putting $\theta = 90°$ and $\sin\theta = 1$ in the above expression, we get

$$H = \frac{I}{2\,r} \text{ – for single-turn coil} \quad \text{or} \quad H = \frac{NI}{2\,r} \text{ –for } N\text{-turn coil}$$

Note. The magnetising force H at the centre of a circular coil can be directly found as follows :

With reference to the coil shown in the Fig. 6.20, the magnetising force dH produced at O due to the small element dl (as given by Laplace's law) is

$$dH = \frac{I.dl\sin\theta}{4\pi\,r^2} = \frac{I.dl}{4\pi\,r^2} \qquad (\because \sin\theta = 90° = 1)$$

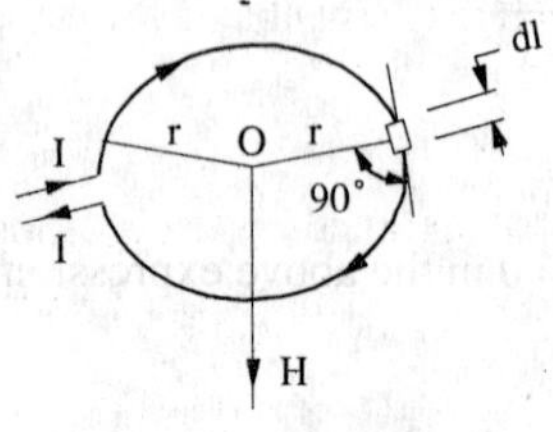

Fig. 6.20

$$\therefore \qquad \Sigma\,dH = \Sigma\frac{I.dl}{4\pi\,r^2} = \frac{I}{4\pi\,r^2}\Sigma\,dl \text{ or } H = \frac{I.2\pi\,r}{4\pi\,r^2} = \frac{I}{2r}$$

$$\therefore \quad H = \frac{I}{2r}\text{AT/m – for 1-turn coil;} = \frac{NI}{2r}\text{ AT/m – for } N\text{-turn coil.}$$

(iv) Magnetising Force on the Axis of a Short Solenoid

Let a short solenoid having a length of l and radius of turns r be uniformly wound with N turns each carrying a current of I as shown in Fig. 6.21. The winding density *i.e.* number of turns per unit

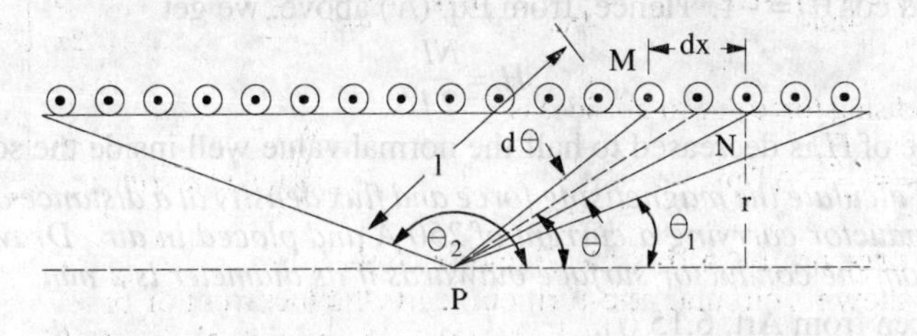

Fig. 6.21

length of the solenoid is N/l. Hence, in a small element of length dx, there will be $N.dx/l$ turns. Obviously, a very short element of length of the solenoid can be regarded as a concentrated coil of very short axial length and having $N.dx/l$ turns. Let dH be the magnetising force contributed by the element dx at any axial point P. Then, substituting dH for H and $N.dx/l$ for N in Eq. (*iii*), we get

$$dH = \frac{N.dx}{l} \cdot \frac{I}{2r} \cdot \sin^3 \theta$$

Now $\qquad dx. \sin \theta = r \,.\, d\theta / \sin \theta^* \quad \therefore \quad dx = r.\, d\theta / \sin^2 \theta$

Substituting this value of dx in the above equation, we get

$$dH = \frac{NI}{2\,l} \sin \theta . d\theta$$

Total value of the magnetising force at P due to the whole length of the solenoid may be found by integrating the above expression between proper limits.

$$\therefore \quad H = \frac{NI}{2\,l} \int_{\theta_1}^{\theta_2} \sin \theta . d\theta = \frac{NI}{2l} \Big| -\cos \theta \Big|_{\theta_1}^{\theta_2}$$

$$= \frac{NI}{2\,l} (\cos \theta_1 - \cos \theta_2) \qquad \ldots(iv)$$

The above expression may be used to find the value of H at any point of the axis, either inside or outside the solenoid.

(*i*) At mid-point, $\theta_2 = (\pi - \theta_1)$, hence $\cos \theta_2 = -\cos \theta_1$

$$\therefore \qquad H = \frac{2NI}{2\,l} \cos \theta_1 = \frac{NI}{l} \cos \theta_1$$

Obviously, when the solenoid is very long, $\cos \theta_1$ becomes nearly unity. In that case,

$$H = \frac{NI}{l} \text{ AT/m} \qquad \text{–Art. 6.15 } (ii)$$

(*ii*) At any point on the axis inside a very *long* solenoid but not too close to either end, $\theta_1 \cong 0$ and $\theta_2 \cong \pi$ so that $\cos \theta_1 \cong 1$ and $\cos \theta_2 \cong -1$. Then, putting these values in Eq. (*iv*) above, we have

$$H \cong \frac{NI}{2\,l} \times 2 = \frac{NI}{l}$$

It proves that inside a very long solenoid, H is practically constant at all axial points except those lying too close to either end of the solenoid.

---

*Because $l \sin \theta = r \;\therefore\; l = r/\sin \theta$. Now, $MN = l.d\theta = r d\theta / \sin \theta$. Also, $MN = dx. \sin \theta$, hence $dx = r\, d\theta / \sin^2 \theta$.

(*iii*) Towards either end of the solenoid, H decreases and exactly at the ends, $\theta_1 = \pi/2$ and $\theta_2 \cong \pi$, so that $\cos\theta_1 = 0$ and $\cos\theta_2 = -1$. Hence, from Eq. (*iv*) above, we get

$$H = \frac{NI}{2l}$$

In other words, value of H is decreased to half the normal value well inside the solenoid.

Example 6.2. *Calculate the magnetising force and flux density at a distance of 5 cm from a long straight circular conductor carrying a current of 250 A and placed in air. Draw a curve showing the variation of B from the conductor surface outwards if its diameter is 2 mm.*

Solution. As seen from Art. 6.15 (*i*),

$$H = \frac{I}{2\pi r} = \frac{250}{2\pi \times 0.05} = \mathbf{795.6\ AT/m}$$

$$B = \mu_0 H = 4\pi \times 10^{-7} \times 795.6 = \mathbf{10^{-3}\ WB/m^2}$$

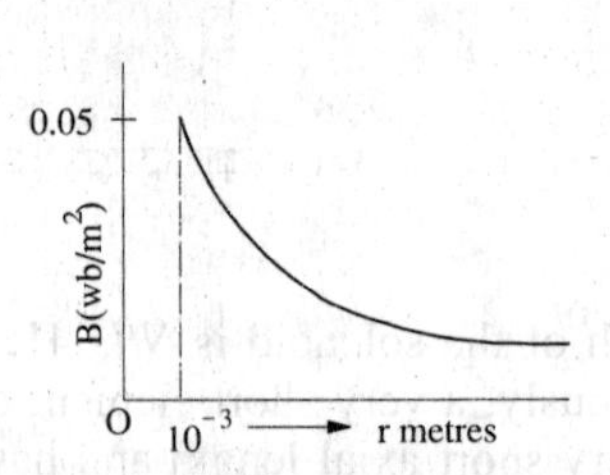

Fig. 6.22

In general, $B = \dfrac{\mu_0 I}{2\pi r}$

Now, at the conductor surface, $r = 1\text{ mm} = 10^{-3}$ m

$$\therefore \quad B = \frac{4\pi \times 10^{-7} \times 250}{2\pi \times 10^{-3}} = \mathbf{0.05\ Wb/m^2}$$

The variation of B outside the conductor is shown in Fig. 6.22.

Example 6.3. *A wire 2.5 m long is bent (i) into a square and (ii) into a circle. If the current flowing through the wire is 100 A, find the magnetising force at the centre of the square and the centre of the circle.* **(Elect. Measurements; Nagpur Univ. 1992)**

Solution. (*i*) Each side of the square is $2a = 2.5/4 = 0.625$ m

Value of H at the centre of the square is [Art. 6.17(*ii*)]

$$= \frac{\sqrt{2}I}{\pi a} = \frac{\sqrt{2} \times 100}{\pi \times 0.3125} = \mathbf{144\ AT/m}$$

(*ii*) $2\pi r = 2.5$; $r = 0.398$ m

Value of H at the centre is $= I/2r = 100/2 \times 0.398 = \mathbf{125.6\ AT/m}$

Example 6.4. *A current of 15 A is passing along a straight wire. Calculate the force on a unit magnetic pole placed 0.15 metre from the wire. If the wire is bent to form into a loop, calculate the diameter of the loop so as to produce the same force at the centre of the coil upon a unit magnetic pole when carrying a current of 15 A.* **(Elect. Engg. Calcutta Univ. 1987)**.

Solution. By the force on a unit magnetic pole is meant the magnetising force H.

For a straight conductor [Art. 6.15 (*i*)], $H = I/2\pi r = 15/2\pi \times 0.15 = 50/\pi$ AT/m

Now, the magnetising force at the centre of a loop of wire is [Art. 6.17 (*iii*)]

$$= I/2r = I/D = 15/D \text{ AT/m}$$

Since the two magnetising forces are equal

$$\therefore \quad 50/\pi = 15/D;\ D = 15\pi/50 = 0.9426\text{ m} = \mathbf{94.26\ cm}.$$

Example 6.5. *A single-turn circular coil of 50 m. diameter carries a direct current of 28×10^4 A. Assuming Laplace's expression for the magnetising force due to a current element, determine the magnetising force at a point on the axis of the coil and 100 m. from the coil. The relative permeability of the space surrounding the coil is unity.*

Solution. As seen from Art 6.17 (*iii*), $H = \dfrac{I}{2r} \cdot \sin^3\theta$ AT/m

Here $$\sin\theta = \frac{r}{\sqrt{r^2 + x^2}} = \frac{25}{\sqrt{25^2 + 100^2}} = 0.2425$$

$$\sin^3\theta = (0.2425)^3 = 0.01426 \quad \therefore H = \frac{28 \times 10^4}{2 \times 25} \times 0.01426 = \mathbf{79.8\ AT/m}$$

6.18. Force Between Two Parallel Conductors

(i) Currents in the same direction. In Fig. 6.23 are shown two parallel conductors P and Q carrying currents I_1 and I_2 amperes in the same direction *i.e.* upwards. The field strength in the space between the two conductors is decreased due to the two fields there being in opposition to each other. Hence, the resultant field is as shown in the figure. Obviously, the two conductors are attracted towards each other.

(ii) Currents in opposite directions. If, as shown in Fig. 6.24, the parallel conductors carry currents in opposite directions, then field strength is increased in the space between the two conductors due to the two fields being in the same direction there. Because of the lateral repulsion of the lines of force, the two conductors experience a mutual force of repulsion as shown separately in Fig. 6.24 (*b*).

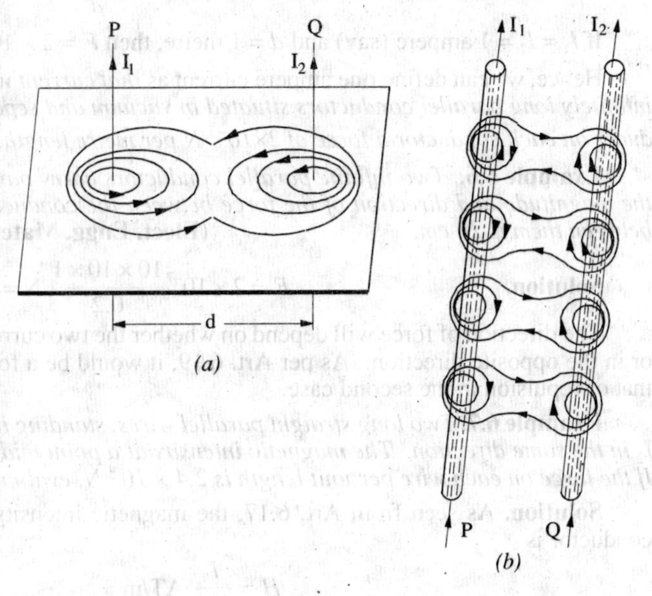

Fig. 6.23

6.19. Magnitude of Mutual Force

It is obvious that each of the two parallel conductors lies in the magnetic field of the other conductor. For example, conductor P lies in the magnetic field of Q and Q lies in the field of P. If 'd' metres is the distance between them, then flux density at Q due to P is [Art. 6.15 (*i*)]

$$B = \frac{\mu_0 I_1}{2\pi d} \text{Wb/m}^2$$

If l is the length of conductor Q lying in this flux density, then force (either of attraction or repulsion) as given in Art. 6.14 is

$$F = BI_2\, l \text{ newton} \quad \text{or} \quad F = \frac{\mu_0 I_1 I_2\, l}{2\pi d} \text{ N}$$

Obviously, conductor P will experience an equal force in the opposite direction.

The above facts are known as Laws of Parallel Currents and may be stated as follows :

(*i*) Two parallel conductors attract each other if currents through them flow in the *same* direction and repel each other if the currents through them flow in the *opposite* directions.

(*ii*) The force between two such parallel conductors is proportional to the product of the current strengths and to the length of the conductors considered and varies inversely as the distance between them.

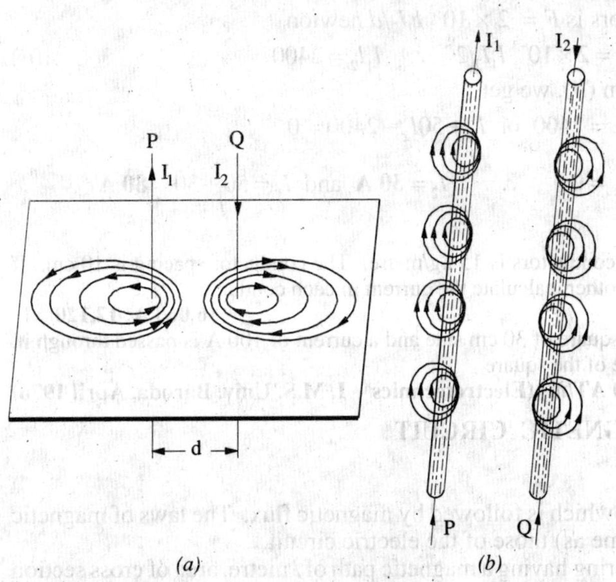

Fig. 6.24

6.20. Definition of Ampere

It has been proved in Art. 6.19 above that the force between two infinitely long parallel current-carrying conductors is given by the expression

$$F = \frac{\mu_0 I_1 I_2 l}{2\pi d} \text{ N} \quad \text{or} \quad F = \frac{4\pi \times 10^{-7} I_1 I_2 l}{2\pi d} = 2 \times 10^{-7} \frac{I_1 I_2 l}{d} \text{ N}$$

The force per metre run of the conductors is

$$F = 2 \times 10^{-7} \frac{I_1 I_2}{d} \text{ N/m}$$

If $I_1 = I_2 = 1$ ampere (say) and $d = 1$ metre, then $F = 2 \times 10^{-7}$ N

Hence, we can define one ampere current as *that current which when flowing in each of the two infinitely long parallel conductors situated in vacuum and separated 1 metre between centres, produces on each conductor a force of* 2×10^{-7} *N per metre length.*

Example 6.6. *Two infinite parallel conductors carry parallel currents of 10 amp. each. Find the magnitude and direction of the force between the conductors per meter length if the distance between them is 20 cm.* **(Elect. Engg. Materials - II Punjab Univ. May 1990)**

Solution. $F = 2 \times 10^{-7} \frac{10 \times 10 \times 1}{0.2} \text{ N} = \mathbf{10^{-4}\ N}$

The direction of force will depend on whether the two currents are flowing in the same direction or in the opposite direction. As per Art. 6.19, it would be a force of attraction in the first case and that of repulsion in the second case.

Example 6.7. *Two long straight parallel wires, standing in air 2 m apart, carry currents* I_1 *and* I_2 *in the same direction. The magnetic intensity at a point midway between the wires is 7.95 AT/m. If the force on each wire per unit length is* 2.4×10^{-4} *N, evaluate* I_1 *and* I_2.

Solution. As seen from Art. 6.17, the magnetic intensity of a long straight current-carrying conductor is

$$H = \frac{I}{2\pi r} \text{ AT/m}$$

Also, it is seen from Fig. 6.23 that when the two currents flow in the same direction, net field strength midway between the two conductors is the difference of the two field strengths.

Now, $H_1 = I_1/2\pi$ and $H_2 = I_2/2\pi$ because $r = 2/1 = 1$ metre

$$\therefore \quad \frac{I_1}{2\pi} - \frac{I_2}{2\pi} = 7.95 \quad \therefore \quad I_1 - I_2 = 50 \qquad \ldots(i)$$

Force per unit length of the conductors is $F = 2 \times 10^{-7} I_1 I_2 / d$ newton

$$\therefore \quad 2.4 \times 10^{-4} = 2 \times 10^{-7} I_1 I_2 / 2 \quad \therefore \quad I_1 I_2 = 2400 \qquad \ldots(ii)$$

Substituting the value of I_1 from (*i*) in (*ii*), we get

$$(50 + I_2) I_2 = 2400 \quad \text{or} \quad I_2^2 + 50 I_2 - 2400 = 0$$

$$\text{or} \quad (I_2 + 80)(I_2 - 30) = 0 \quad \therefore \quad I_2 = \mathbf{30\ A} \text{ and } I_1 = 50 + 30 = \mathbf{80\ A}$$

Tutorial Problems No : 6.1

1. The force between two long parallel conductors is 15 kg/metre. The conductor spacing is 10 cm. If one conductor carries twice the current of the other, calculate the current in each conductor.

[6,060 A; 12,120 A]

2. A wire is bent into a plane to form a square of 30 cm side and a current of 100 A is passed through it. Calculate the field strength set up at the centre of the square.

[300 AT/m] (Electrotechnics – I, M.S. Univ. Baroda, April 1976)

MAGNETIC CIRCUIT

6.21. Magnetic Circuit

It may be defined as the route or path which is followed by magnetic flux. The laws of magnetic circuit are quite similar to (but not the same as) those of the electric circuit.

Consider a solenoid or a toroidal iron ring having a magnetic path of l metre, area of cross section A m^2 and a coil of N turns carrying I amperes wound anywhere on it as in Fig. 6.25.

Then, as seen from Art. 6.15, field strength inside the solenoid is

$$H = \frac{NI}{l} \text{ AT/m}$$

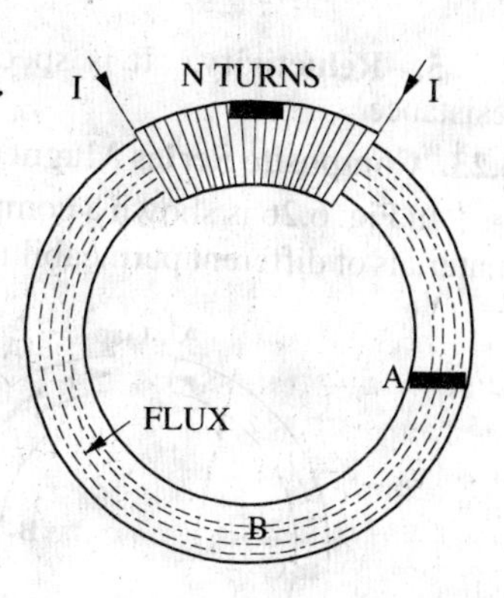

Fig. 6.25

Now $\quad B = \mu_0 \mu_r H = \frac{\mu_0 \mu_r NI}{l} \text{Wb/m}^2$

Total flux produced $\Phi = B \times A = \frac{\mu_0 \mu_r ANI}{l}$ Wb

$$\therefore \quad \Phi = \frac{NI}{l/\mu_0 \mu_r A} \text{ Wb}$$

The numerator '*NI*' which produces magnetisation in the magnetic circuit is known as magnetomotive force (m.m.f.). Obviously, its unit is ampere-turn (AT)*. It is analogous to e.m.f. in an electric circuit.

The denominator $\frac{l}{\mu_0 \mu_r A}$ is called the *reluctance* of the circuit and is analogous to resistance in electric circuits.

$$\therefore \quad \text{flux} = \frac{\text{m.m.f.}}{\text{reluctance}} \quad \text{or} \quad \Phi = \frac{F}{S}$$

Sometimes, the above equation is called the "Ohm's Law of Magnetic Circuit" because it resembles a similar expression in electric circuits *i.e.*

$$\text{current} = \frac{\text{e.m.f.}}{\text{resistance}} \quad \text{or} \quad I = \frac{V}{R}$$

6.22. Definitions Concerning Magnetic Circuit

1. Magnetomotive force (m.m.f.). It drives or tends to drive flux through a magnetic circuit and corresponds to electromotive force (e.m.f.) in an electric circuit.

M.M.F. is equal to the work done in joules in carrying a unit magnetic pole once through the entire magnetic circuit. It is measured in ampere-turns.

In fact, as p.d. between any two points is measured by the work done in carrying a unit charge from one point to another, similarly, m.m.f. between two points is measured by the work done in joules in carrying a unit magnetic pole from one point to another.

2. Ampere-turns (AT). It is the unit of magnetometre force (m.m.f.) and is given by the product of number of turns of a magnetic circuit and the current in amperes in those turns.

3. Reluctance. It is the name given to that property of a material which opposes the creation of magnetic flux in it. It, in fact, measures the opposition offered to the passage of magnetic flux through a material and is analogous to resistance in an electric circuit even in form. Its unit is **AT/Wb.****

$$\text{reluctance} = \frac{l}{\mu A} = \frac{l}{\mu_0 \mu_r A}; \text{ resistance} = \rho \frac{l}{A} = \frac{l}{\sigma A}$$

In other words, the reluctance of a magnetic circuit is the number of amp-turns required per weber of magnetic flux in the circuit. Since 1 AT/Wb = 1/henry, the unit of reluctance is "reciprocal henry."

4. Permeance. It is reciprocal of reluctance and implies the case or readiness with which magnetic flux is developed. It is analogous to conductance in electric circuits. It is measured in terms of Wb/AT or henry.

---

*Strictly speaking, it should be only 'ampere' because turns have no unit.

**From the relation $\Phi = \frac{\text{m.m.f.}}{\text{reluctance}}$, it is obvious that reluctance = m.m.f./Φ. Since m.m.f. is in ampre-turns and flux in webers, unit of reluctance is ampre-turn/weber (AT/Wb) or A/Wb.

5. Reluctivity. It is specific reluctance and corresponds to resistivity which is 'specific resistance'.

6.23. Composite Series Magnetic Circuit

In Fig. 6.26 is shown a composite series magnetic circuit consisting of three different magnetic materials of different permeabilities and lengths and one air gap ($\mu_r = 1$). Each path will have its own reluctance. The total reluctance is the sum of individual reluctances as they are joined in series.

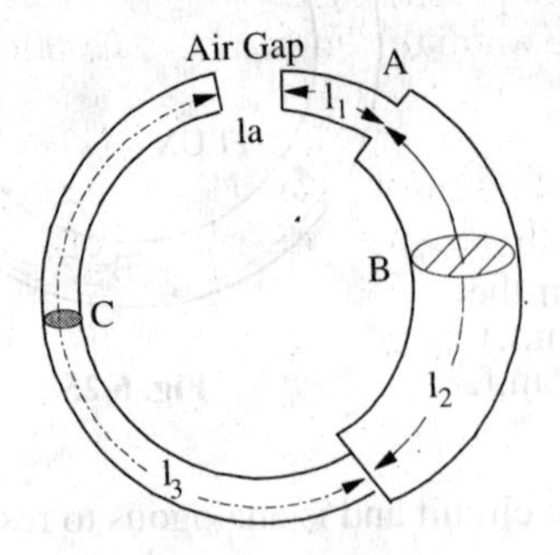

Fig. 6.26

$$\therefore \quad \text{total reluctance} = \Sigma \frac{l}{\mu_0 \mu_r A}$$

$$= \frac{l_1}{\mu_0 \mu_{r1} A_1} + \frac{l_2}{\mu_0 \mu_{r_2} A_2} + \frac{l_3}{\mu_0 \mu_{r3} A_3} + \frac{l_a}{\mu_0 A_g}$$

$$\therefore \quad \text{flux } \Phi = \frac{\text{m.m.f.}}{\frac{l}{\mu_0 \mu_r A}}$$

6.24 How to Find Ampere-turns ?

It has been shown in Art. 6.15 that

$H = NI/l$ AT/m or $NI = H \times l$ $\therefore$ ampere-turns AT $= H \times l$

Hence, following procedure should be adopted for calculating the total ampere turns of a composite magnetic path.

(*i*) Find H for each portion of the composite circuit. For air, $H = B/\mu_0$, otherwise $H = B/\mu_0\mu_r$.

(*ii*) Find ampere-turns for each path separately by using the relation AT $= H \times l$.

(*iii*) Add up these ampere-turns to get the total ampere-turns for the entire circuit.

6.25. Comparison Between Magnetic and Electric Circuits

SIMILARITIES

Magnetic Circuit	Electric Circuit
Fig. 6.27	Fig. 6.28
1. Flux $= \dfrac{\text{m.m.f.}}{\text{reluctance}}$	Current $= \dfrac{\text{e.m.f.}}{\text{resistance}}$
2. M.M.F. (ampere-turns)	E.M.F. (volts)
3. Flux Φ (webers)	Current I (amperes)
4. Flux density B (Wb/m^2)	Current density (A/m^2)
5. Reluctance $S = \dfrac{l}{\mu A}\left(= \dfrac{l}{\mu_0 \mu_r A}\right)$	resistance $R = \rho \dfrac{l}{A} = \dfrac{l}{\rho A}$
6. Permeance (= 1/reluctance)	Conductance (= 1/resistance)
7. Relucitivity	Resistivity
8. Permeability (= 1/reluctivity)	Conductivity (= 1/resistivity)
9. Total m.m.f. $= \Phi S_1 + \Phi S_2 + \Phi S_3 + \ldots$	9. Total e.m.f. $= IR_1 + IR_2 + IR_3 + \ldots$

DIFFERENCES

1. Strictly speaking, flux does not actually 'flow' in the sense in which an electric current flows.
2. If temperature is kept constant, then resistance of an electric circuit is constant and is

independent of the current strength (or current density). On the other hand, the reluctance of a magnetic circuit does depend on flux (and hence flux density) established in it. It is so because μ (which equals the slope of *B/H* curve) is not constant even for a given material as it depends on the flux density *B*. Value of μ is large for low values of *B* and *vice versa*. Hence, reluctance is small ($S = l/\mu A$) for small values of *B* and large for large values of *B*.

3. Flow of current in an electric circuit involves continuous expenditure of energy but in a magnetic circuit, energy is needed only for creating the flux initially but not for maintaining it.

6.26. Parallel Magnetic Circuits

Fig. 6.29 (*a*) shows a parallel magneitc circuit consisting of two parallel magnetic paths *ACB* and *ADB* acted upon by the same *m.m.f.* Each magnetic path has an average length of $2\,(l_1 + l_2)$.

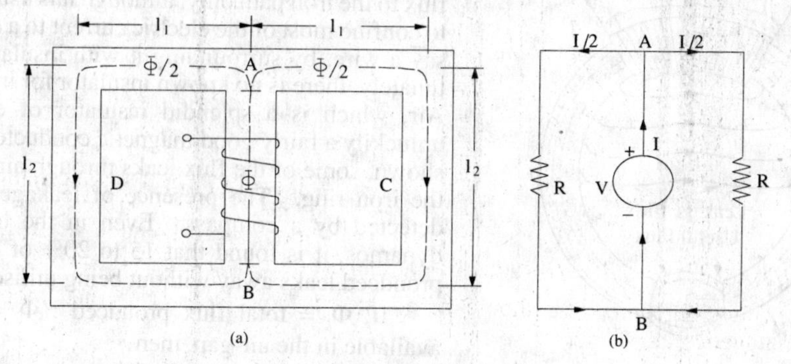

Fig. 6.29

The flux produced by the coil wound on the central core is divided equally at point *A* between the two outer parallel paths. The reluctance offered by the two parallel paths is = half the reluctance of each path.

Fig. 6.29 (*b*) shows the equivalent electrical circuit where resistance offered to the voltage source is $= R\|R = R/2$.

It should be noted that reluctance offered by the central core *AB* has been neglected in the above treatment.

6.27. Series-Parallel Magnetic Circuits

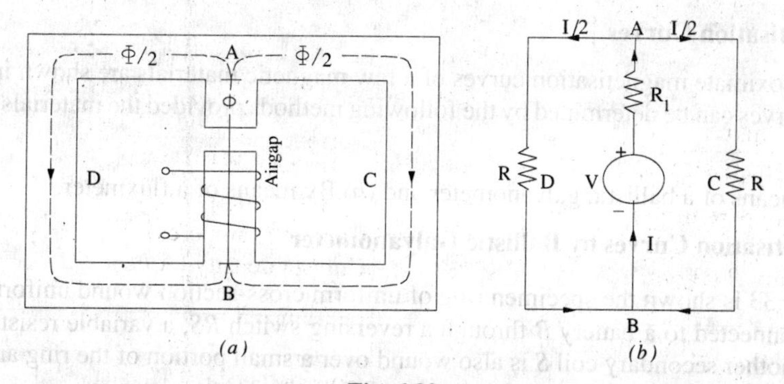

Fig. 6.30

Such a circuit is shown in Fig. 6.30 (*a*). It shows two parallel magnetic circuits *ACB* and *ACD* connected across the common magnetic path *AB* which contains an air-gap of length l_g. As usual, the flux Φ in the common core is divided equally at point *A* between the two parallel paths which have equal reluctance. The reluctance of the path *AB* consists of (*i*) air gap reluctance and (*ii*) the reluctance of the central core which is comparatively negligible. Hence, the reluctance of the central core *AB* equals only the air-gap reluctance across which are connected two equal parallel reluctances. Hence, the *m.m.f.* required for this circuit would be the sum of (*i*) that required for the air-gap and (*ii*) that required for either of the two paths (not both) as illustrated in Ex. 6.19, 6.20 and 6.21.

The equivalent electrical circuit is shown in Fig. 6.30 (*b*) where the total resistance offered to the voltage source is $= R_1 + R \,||\, R = R_1 + R/2$.

6.28. Leakage Flux and Hopkinson's Leakage Coefficient

Leakage flux is the flux which follows a path not intended for it. In Fig. 6.31 is shown an iron ring wound with a coil and having an air-gap. The flux in the air-gap is known as the useful flux because it is only this flux which can be utilized for various useful purposes.

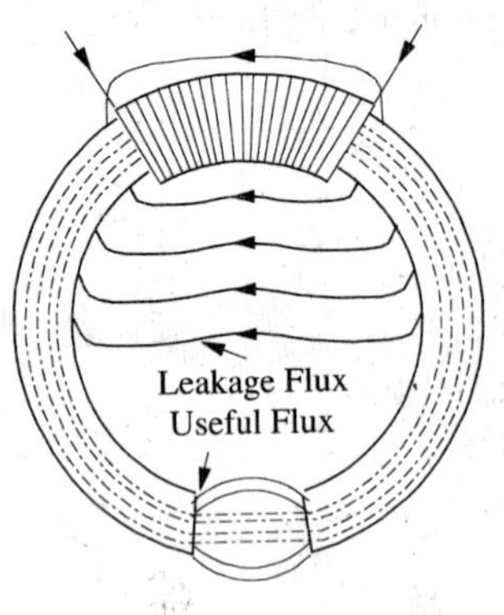

Fig. 6.31

It is found that it is impossible to confine all the flux to the iron path only, although it is usually possible to confine most of the electric current to a definite path, say a wire, by surrounding it with insulation. Unfortunately, there is no known insulator for magnetic flux. Air, which is a splendid insulator of electricity, is unluckily a fairly good magnetic conductor. Hence, as shown, some of the flux leaks through air surrounding the iron ring. The presence of leakage flux can be detected by a compass. Even in the best designed dynamos, it is found that 15 to 20% of the total flux proudced leaks away without being utilised usefully.

If, Φ_t = total flux proudced ; Φ = useful flux available in the air-gap, then

$$\text{leakage coefficient } \lambda = \frac{\text{total flux}}{\text{useful flux}} \quad \text{or} \quad \lambda = \frac{\Phi_t}{\Phi}$$

In electric machines like motors and generators, magnetic leakage is undesirable, because, although it does not lower their power efficiency, yet it leads to their increased weight and cost of manufacture. Magnetic leakage can be minimised by placing the exciting coils or windings as close as possible to the air-gap or to the points in the magnetic circuit where flux is to be utilized for useful purposes.

It is also seen from Fig. 6.31 that there is fringing or spreading of lines of flux at the edges of the air-gap. This fringing increases the effective area of the air-gap.

The value of λ for modern electric machines varies between 1.1 and 1.25.

6.29. Magnetisation Curves

The approximate magnetisation curves of a few magnetic materials are shown in Fig. 6.32

These curves can be determined by the following methods provided the materials are in the form of a ring :

(*a*) By means of a ballistic galvanometer and (*b*) By means of a fluxmeter.

6.30. Magnetisation Curves by Ballistic Galvanometer

In Fig. 6.33 is shown the specimen ring of uniform cross-section wound uniformly with a coil *P* which is connected to a battery *B* through a reversing switch *RS*, a variable resistance R_1 and an ammeter. Another secondary coil *S* is also wound over a small portion of the ring and is connected through a resistance *R* to a ballistic galvanometer *BG*.

The current through the primary *P* can be adjusted with the help of R_1. Suppose the primary current is *I*. When the primary current is reversed by menas of *RS*, then flux is reversed through *S*, hence an induced e.m.f is produced in it which sends a current through *BG*. This current is of very short duration. The first deflection or 'throw' of the *BG* is proportional to the quantity of electricity or charge passing through it so long as the time taken for this charge to flow is short as compared with the time of one oscillation.

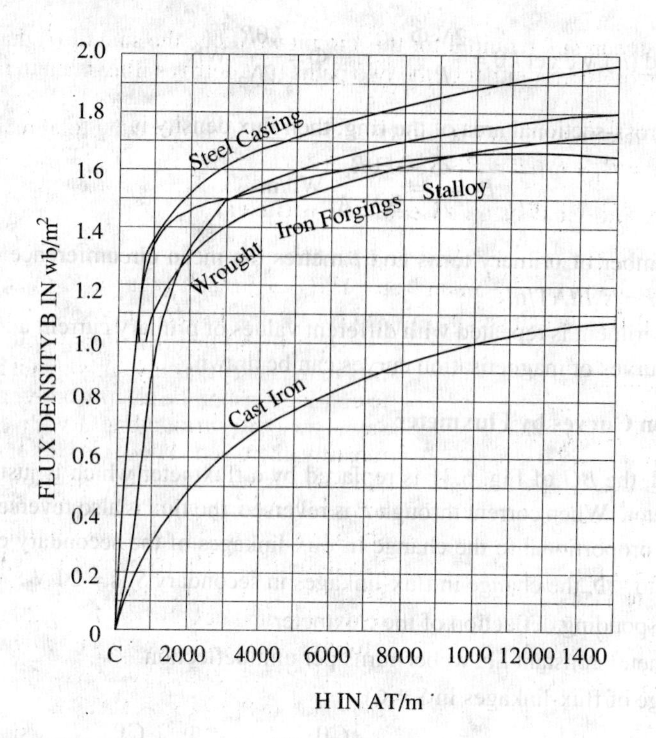

Fig. 6.32

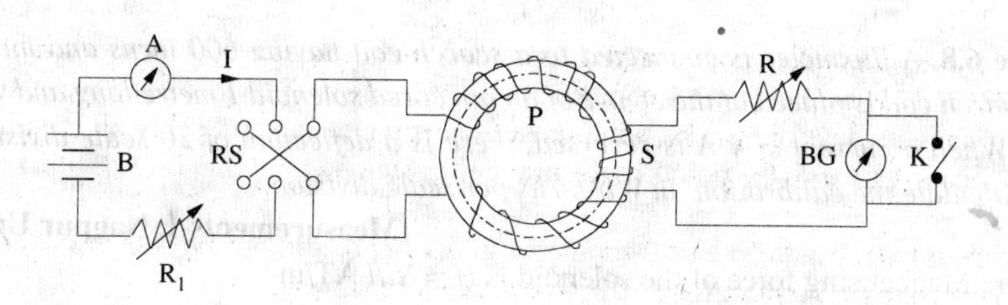

Fig. 6.33

If θ = first deflection or 'throw' of the galvanometer when primary current I is reversed.

k = ballistic constant of the galvanometer *i.e.* charge per unit deflection.

then, charge passing through BG is = $k\theta$ coloumbs ...(*i*)

Let Φ = flux in Wb produced by primary current of I amperes ; t = time of reversal of flux ; then

rate of change of flux $= \dfrac{2\Phi}{t}$ Wb/s

If N_2 is the number of turns in secondary coil S, then average e.m.f. induced in it is

$$= N_2 \cdot \frac{2\Phi}{t} \text{ volt.}$$

Secondary current or current through $BG = \dfrac{2N_2\Phi}{R_s t}$ amperes

where R_s is the total resistance of the secondary circuit.

Charge flowing through BG = average current $\times$ time $= \dfrac{2\,N_2\,\Phi}{R_s\,t} \times t = \dfrac{2\,N_2\,\Phi}{R_s}$ coulomb ...(*ii*)

Equation (*i*) and (*ii*), we get $k\theta = \dfrac{2N_2\Phi}{R_s}$ $\therefore$ $\Phi = \dfrac{k\theta R_s}{2N_2}$ Wb

If A m^2 is the cross-sectional area of the ring, then flux density is

$$B = \frac{\Phi}{A} = \frac{k\theta R_s}{2\,N_2\,A}\ \text{Wb/m}^2$$

If N_1 is the number of primary turns and l metres the mean circumference of the ring, then, magnetising force $H = N_1I/l$ AT/m

The above experiment is repeated with different values of primary current and from the data so obtained, the B/H curves or magnetisation curves can be drawn.

6.31 Magnetisation Curves by Fluxmeter

In this method, the *BG* of Fig. 6.31 is replaced by a fluxmeter which is just a special type of ballistic galvanometer. When current through *P* is reversed, the flux is also reversed. The deflection of the fluxmeter is proportional to the change in flux-linkages of the secondary coil. If the flux is reversed from $+\Phi$ to $-\Phi$, the change in flux-linkages in secondary *S* is $= 2\Phi N_2$.

If θ = corresponding deflection of the fluxmeter

C = fluxmeter constant *i.e.* weber-turns per unit deflection.

then, change of flux-linkages in $S = C\theta$

$\therefore$ $2\Phi\,N_2 = C\theta$ or $\Phi = \dfrac{C\theta}{2N_2}$ Wb ; $B = \dfrac{\Phi}{A} = \dfrac{C\theta}{2N_2A}$ Wb/m^2

Example 6.8. *A fluxmeter is connected to a search-coil having 600 turns and mean area of 4 cm^2. The search coil is placed at the centre of an air-cored solenoid 1 metre long and wound with 1000 turns. When a current of 4 A is reversed, there is a deflection of 20 scale divisions on the fluxmeter. Calculate the calibration in Wb-turns per scale division.*

(Measurements-I, Nagpur Univ. 1991)

Solution. Magnetising force of the solenoid is $H = NI/l$ AT/m

$B = \mu_0 H = \mu_0 NI/l = 4\pi \times 10^{-7} \times 1000 \times 4/1 = 16\pi \times 10^{-4}$ Wb/m^2

Flux linked with the search coil is $\Phi = BA = 64\pi \times 10^{-8}$ Wb

Total change of flux-linkages on reversal

$= 2 \times 64\pi \times 10^{-8} \times 600$ Wb-turns —Art. 6.29

$= 7.68\pi \times 10^{-4}$ Wb - turns

Fluxmeter constant C is given by $= \dfrac{\text{Change in flux-linkages}}{\text{deflection produced}}$

$= 7.68\pi \times 10^{-4}/20 =$ $\mathbf{1.206 \times 10^{-4}}$ **Wb-turns/division**

Example 6.9. *A ballistic galvanometer, connected to a search coil for measuring flux density in a core, gives a throw of 100 scale divisions on reversal of flux. The galvanometer coil has a resistance of 180 ohm. The galvanometer constant is 100 μC per scale division. The search coil has an area of 50 cm^2, wound with 1000 turns having a resistance of 20 ohm. Calculate the flux density in the core.*

(Elect. Instru & Measu. Nagpur Univ. 1992)

Solution. As seen from Art. 6.28.

$$k\theta = 2N_2\Phi/R_s \text{ or } \Phi = k\theta R_s/2N_2 \text{ Wb}$$

$\therefore$ $$BA = k\theta R_s/2N_2 \text{ or } B = k\theta R_s/2N_2A$$

Here $$k = 100\,\mu\text{C/division} = 100\times 10^{-6} = 10^{-4}\text{ C / division}$$

$$\theta = 100;\ A = 50\text{ cm}^2 = 5\times 10^{-3}\text{ m}^2$$

$$R_s = 180 + 20 = 200\ \Omega$$

$\therefore$ $$B = 10^{-4}\times 100\times 200/2\times 1000\times 5\times 10^{-3} = \mathbf{0.2\ Wb/m^2}$$

Example 6.10. *A ring sample of iron, fitted with a primary and a secondary winding is to be tested by the method of reversals to obtain its B/H curve. Give a diagram of connections and explain briefly how the test could be carried out.*

In such a test, the primary winding of 400 turns carries a current of 1.8 A. On reversal, a change of 8×10^{-3} Wb-turns is recorded in the secondary winding of 10 turns. The ring is made up of 50 laminations, each 0.5 mm thick with outer and inner diameters of 25 and 23 cm respectively. Assuming uniform flux distribution, determine the values of B, H and the permeability.

Solution. Here, change of flux-linkages $= 2\,\Phi N_2 = 8\times 10^{-3}$ Wb-turns

$\therefore$ $$2\,\Phi\times 10 = 8\times 10^{-3} \text{ or } \Phi = 4\times 10^{-4}\text{ Wb and } A = 2.5\times 10^{-4}\text{m}^2$$

$\therefore$ $$B = \frac{4\times 10^{-4}}{2.5\times 10^{-4}} = 1.6\text{ Wb/m}^2;\ H = \frac{NI}{l} = \frac{400\times 1.8}{0.24\,\pi} = 955\text{ AT/m}$$

Now $$\mu_0\,\mu_r = \frac{B}{H};\ \mu_r = \frac{B}{\mu_0 H} = \frac{1.6}{4\pi\times 10^{-7}\times 955} = \mathbf{1333}$$

Example 6.11. *An iron ring of 3.5 cm² cross-sectional area with a mean length of 100 cm is wound with a magnetising winding of 100 turns. A secondary coil of 200 turns of wire is connected to a ballistic galvanometer having a constant of 1 micro-coulomb per scale division, the total resistance of the secondary circuit being 2000 Ω. On reversing a current of 10 A in the magnetising coil, the galvanometer gave a throw of 100 scale divisions. Calculate the flux density in the specimen and the value of the permeability at this flux density.* **(Elect. Measure, A.M.I.E Sec. B. 1992)**

Solution. Reference may please be made to Art. 6.28.

Here $$N_1 = 100;\ N_2 = 200 : A = 3.5\times 10^{-4}\text{ m}^2;\ l = 100\text{ cm} = 1\text{ m}$$

$$k = 10^{-6}\text{C/division},\ \theta = 100\text{ divisions};\ R_s = 2000\ \Omega;\ I = 10\text{ A}$$

$$B = \frac{k\theta R_s}{2\,N_2\,A} = \frac{10^{-6}\times 100\times 2000}{2\times 200\times 3.5\times 10^{-4}} = 1.43\text{ Wb/m}^2$$

Magnetising force $H = N_1\,I/l = 100\times 10/1 = 1000$ AT/m

$$\mu = \frac{B}{H} = \frac{1.43}{1000} = \mathbf{1.43\times 10^{-3} H/m}$$

Note. The relative permeability is given by $\mu_r = \mu/\mu_0 = 1.43\times 10^{-3}/4\pi\times 10^{-7} = \mathbf{1137}$.

Example 6.12. *An iron ring has a mean diameter of 0.1 m and a cross-section of $33.5\times 10^{-6}m^2$. It is wound with a magnetising winding of 320 turns and the secondary winding of 220 turns. On reversing a current of 10A in the magnetising winding, a ballistic galvanometer gives a throw of 272 scale divisions, while a Hilbert Magnetic standard with 10 turns and a flux of 2.5×10^{-4} gives a reading of 102 scale divisions, other conditions remaining the same. Find the relative permeability of the specimen.* **(Elect. Measu. A.M.I.E Sec B, 1991)**

Solution. Length of the magnetic path $l = \pi D = 0.1\,\pi$ m

Magnetising Force, $H = NI/l = 320\times 10/0.1\,\pi = 10{,}186$ AT/m

Flux density $B = \mu_0 \mu_r \mathbf{H} = 4\pi \times 10^{-7} \times \mu_r \times 10{,}186 = 0.0128\,\mu_r$ .... (i)

Now, from Hilbert's Magnetic standard, we have

$2.5 \times 10^{-4} \times 10 = K \times 102, K = 2.45 \times 10^{-5}$

On reversing a current of 10 A in the magnetising winding, total change in Weber-turns is

$2\Phi N_s = 2.45 \times 10^{-5} \times 272$ or $2 \times 220 \times \Phi = 2.45 \times 10^{-5} \times 272$ or $\Phi = 1.51 \times 10^{-5}$ Wb

$\therefore\ B = \Phi/A = 1.51 \times 10^{-5}/33.5 \times 10^{-6} = 0.45$ Wb/m^2

Substituting this value in Eq. (*i*), we have $0.0128\,\mu_r = 0.45$, $\therefore\ \mu_r =$ **35.1**

Example 6.13. *A laminated soft iron ring of relative permeability 1000 has a mean circumference of 800 mm and a cross-sectional area 500 mm². A radial air-gap of 1 mm width is cut in the ring which is wound with 1000 turns. Calculate the current required to produce an air-gap flux of 0.5 mWb if leakage factor is 1.2 and stacking factor 0.9. Neglect fringing.*

Solution. Total AT reqd. $= \Phi_g S_g + \Phi_i S_i = \dfrac{\Phi_g l_g}{\mu_0 A_g} + \dfrac{\Phi_i l_i}{\mu_0 \mu_r A_i B}$

Now, air-gap flux $\Phi_g = 0.5$ mWb $= 0.5 \times 10^{-3}$ Wb, $l_g = 1$ mm $= 1 \times 10^{-3}$ m ; $A_g = 500$ mm^2 $= 500 \times 10^{-6}$ m^2

Flux in the iron ring, $\Phi_i = 1.2 \times 0.5 \times 10^{-3}$ Wb

Net cross-sectional area $= A_i \times$ stacking factor $= 500 \times 10^{-6} \times 0.9$ m^2

$$\therefore \quad \text{total AT reqd.} = \frac{0.5 \times 10^{-3} \times 1 \times 10^{-3}}{4\pi \times 10^{-7} \times 500 \times 10^{-6}} + \frac{1.2 \times 0.5 \times 10^{-3} \times 800 \times 10^{-3}}{4\pi \times 10^{-7} \times 1000 \times (0.9 \times 500 \times 10^{-6})} = 1644$$

$\therefore \quad I = 1644/1000 =$ **1.64 A**

Example 6.14. *A ring has a mean diameter of 21 cm and a cross-sectional area of 10 cm². The ring is made up of semicircular sections of cast iron and cast steel, with each joint having a reluctance equal to an air-gap of 0.2 mm. Find the ampere-turns required to produce a flux of 8 × 10⁻⁴ Wb. The relative permeabilities of cast steel and cast iron are 800 and 166 respectively. Neglect fringing and leakage effects.* **(Elect. Circuits, South Gujarat Univ. 1987)**

Solution. $\Phi = 8 \times 10^{-4}$ Wb; $A = 10$ cm$^2 = 10^{-3}$ m^2; $B = 8 \times 10^{-4}/10^{-3} = 0.8$ Wb/m^2

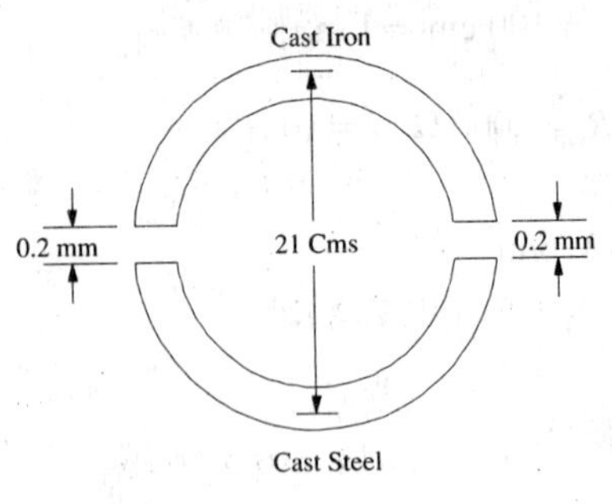

Fig. 6.34

Air gap

$H = B/\mu_0 = 0.8/4\pi \times 10^{-7} = 6.366 \times 10^5$ AT/m

Total air-gap length $= 2 \times 0.2 = 0.4$ mm $= 4 \times 10^{-4}$ m

$\therefore$ AT required $= H \times l = 6.366 \times 10^5 \times 4 \times 10^{-4} = 255$

Cast Steel Path (Fig. 6.34)

$H = B/\mu_0 \mu_r = 0.8/4\pi \times 10^{-7} \times 800 = 796$ AT/m

path $= \pi D/2 = 21\,\pi/2 = 33$ cm $= 0.33$ m

AT required $= H \times l = 796 \times 0.33 = 263$

Cast Iron Path

$H = 0.8/\pi \times 10^{-7} \times 166 = 3{,}835$ AT/m; path $= 0.33$ m

AT required $= 3{,}835 \times 0.33 = 1265$

Total AT required $= 255 + 263 + 1265 =$ **1783.**

Example 6.15. *A mild steel ring of 30 cm mean circumference has a cross-sectional area of 6 cm² and has a winding of 500 turns on it. The ring is cut through at a point so as to provide an air-gap of 1 mm in the magnetic circuit. It is found that a current of 4 A in the winding, produces a flux density of 1 T in the air-gap. Find (i) the relative permeability of the mild steel and (ii) inductance of the winding.* **(F.E. Engg. Pune Univ. Nov. 1988)**

Solution. (*a*) **Steel ring**

$$H = B/\mu_o\mu_r = 1/4\pi\times10^{-7}\times\mu_r \text{ AT/m} = 0.7957\times10^7/\mu_r \text{ AT/m}$$

$$\text{m.m.f.} = H\times l = (0.7957\times10^7/\mu_r)\times29.9\times10^{-2} = .2379\times10^6/\mu_r \text{ AT}$$

(b) **Air-gap**

$$H = B/\mu_0 = 1/4\pi\times10^{-7} = 0.7957\times10^6 \text{ AT/m}$$

$$\text{m.m.f. reqd.} = H\times1 = 0.7957\times10^6\times(1\times10^{-3}) = 795.7 \text{ AT}$$

Total m.m.f $= (0.2379\times10^6/\mu_r) + 795.7$

Total mmf available $= NI = 500\times4 = 2000$ AT

(i) $\therefore$ $2000 = (0.2379\times10^6/\mu_r) + 795.7$ $\therefore$ $\mu_r =$ **197.5**

(ii) Inductance of the winding $= \dfrac{N\Phi}{I} = \dfrac{NBA}{I} = \dfrac{500\times1\times6\times10^{-4}}{4} =$ **0.075 H**

Example 6.16. *An iron ring has a X-section of 3 cm² and a mean diameter of 25 cm. An air-gap of 0.4 mm has been cut across the section of the ring. The ring is wound with a coil of 200 turns through which a current of 2 A is passed. If the total magnetic flux is 0.24 mWb, find the relative permeability of iron, assuming no magnetic leakage.* **(Elect. Engg. A.M.Ae. S.I, June 1992)**

Solution. $\Phi = 0.24$ mWb; $A = 3 \text{ cm}^2 = 3\times10^{-4} \text{ m}^2$;

$B = \Phi/A = 0.24\times10^{-3}/3\times10^{-4} = 0.8 \text{ Wb/m}^2$

AT for iron ring $= H\times l = (B/\mu_o\mu_r)\times l = (0.8/4\pi\times10^{-7}\times\mu_r)\times0.25 = 1.59\times10^{-5}/\mu_r$

AT for air-gap $= H\times l = (B/\mu_o)\times l = (0.8/4\pi\times10^{-7})\times0.4\times10^{-3} = 255$

Total AT reqd. $= (1.59\times10^5/\mu_r) + 255$; total AT provided $= 200\times2 = 400$

$\therefore$ $(1.59\times10^5/\mu_r) + 255 = 400$ or $\mu_r =$ **1096.**

Example 6.17. *A rectangular iron core is shown in Fig. 6.35. It has a mean length of magnetic path of 100 cm, cross-section of (2 cm × 2 cm), relative permeability of 1400 and an air-gap of 5 mm cut in the core. The three coils carried by the core have number of turns $N_a = 335$, $N_b = 600$, and $N_c = 600$; and the, respective currents are 1.6 A, 4 A and 3A. The directions of the currents are as shown. Find the flux in the air-gap.*

(F.Y. Engg. Pune Univ. Nov. 1987)

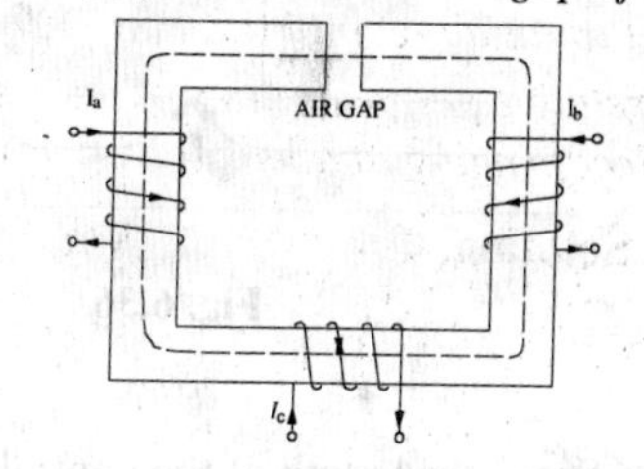

Fig. 6.35

Solution. By applying the Right-Hand Thumb rule, it is found that fluxes produced by the current I_a and I_b are directed in the clockwise direction through the iron core whereas that produced by current I_c is directed in the anticlockwise direction through the core.

$\therefore$ total mmf $= N_a I_a + N_b I_b - N_c I_c = 335\times1.6 + 600\times4 - 600\times3 = 1136$ AT

$$\text{Reluctance of the air-gap} = \frac{1}{\mu_0 A} = \frac{5\times10^3}{4\pi\times10^{-7}\times4\times10^{-4}} = 9.946\times10^6 \text{AT/Wb}$$

$$\text{Reluctance of the iron path} = \frac{l}{\mu_0\mu_r A} = \frac{100-(0.5)\times10^{-2}}{4\pi\times10^{-7}\times1400\times4\times10^{-4}} = 1.414\times10^6 \text{AT/Wb}$$

Total reluctance $= (9.946 + 1.414)\times10^6 = 11.36\times10^6$ AT/Wb

The flux in the air-gap is the same as in the iron core.

$$\text{Air-gap flux} = \frac{\text{m.m.f.}}{\text{reluctance}} = \frac{1136}{11.36\times10^6} = 100\times10^{-6} \text{ Wb} = \mathbf{100\,\mu Wb}$$

Example 6.18. *A series magnetic circuit comprises of three sections (i) length of 80 mm with cross-sectional area 60 mm², (ii) length of 70 mm with cross-sectional area 80 mm² and (iii) and air-gap of length 0.5 mm with cross-sectional area of 60 mm². Sections (i) and (ii) are of a material having magnetic characteristics given by the following table :*

H (AT/m)	*100*	*210*	*340*	*500*	*800*	*1500*
B (Tesla)	*0.2*	*0.4*	*0.6*	*0.8*	*1.0*	*1.2*

Determine the current necessary in a coil of 4000 turns wound on section (ii) to produce a flux density of 0.7 Tesla in the air-gap. Neglect magnetic leakage. **(F.E. Pune Univ. May 1990)**

Solution. Section (*i*) It has the same cross-sectional area as the air-gap. Hence, it has the same flux density *i.e.* 0.7 Tesla as in the air-gap. The value of the magnetising force H corresponding to this flux density of 0.7 T as read from the *B-H* plot is 415 AT/m

m.m.f. reqd. = $H \times 1 = 415 \times (80 \times 10^{-3}) = 33.2$ AT

Section (*ii*) Since its cross-sectional area is different from that of the air-gap, its flux density would also be different even though, being a series circuit, its flux would be the same.

Air-gap flux = $B \times L = 0.7 \times (60 \times 10^{-6}) = 42 \times 10^{-6}$ Wb.

Flux density in *this* section = $42 \times 10^{-6}/80 \times 10^{-6} = 0.525$ T

The corresponding value of the *H* from the given graph is 285 AT/m

m.m.f. reqd. for this section = $285 \times (70 \times 10^{-3}) = 19.95$ AT.

Air-gap

$H = B/\mu_0 = 0.7/4\pi \times 10^{-7} = 0.557 \times 10^6$ AT/m

$\therefore$ m.m.f. reqd. $= 0.557 \times 10^{-6} \times (0.5 \times 10^{-3}) = 278.5$ AT

Total m.m.f. reqd. = $33.2 + 19.95 + 278.5 = 331.6$

$\therefore NI = 331.6$ or $I = 331.6/4000 =$ **0.083 A**

Example 6.19. *A magnetic circuit made of mild steel is arranged as shown in Fig. 6.36. The central limb is wound with 500 turns and has a cross-sectional area of 800 mm². Each of the outer limbs has a cross-sectional area of 500 mm². The air-gap has a length of 1 mm. Calculate the current required to set up a flux of 1.3 mWb in the central limb assuming no magnetic leakage and fringing. Mild steel required 3800 AT/m to produce flux density of 1.625 T and 850 AT/m to produce flux density of 1.3 T.* **(F. Y. Engg. Pune Univ. May 1987)**

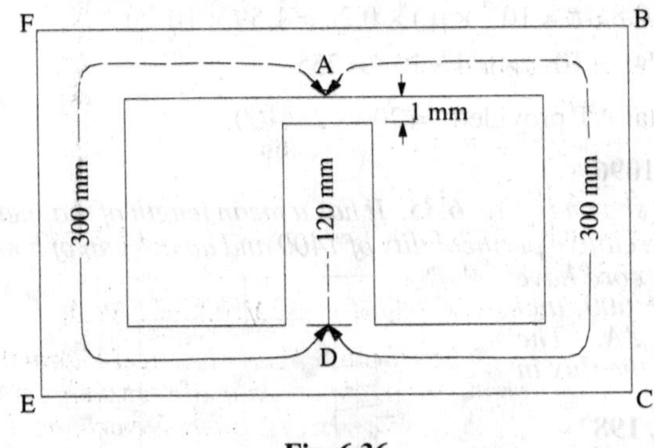

Fig. 6.36

Solution. Flux in the central limb is = 1.3 mWb = 1.3×10^{-3} Wb

Cross section A = 800 mm² = 800×10^{-6} m²

$\therefore B = \Phi/A = 1.3 \times 10^{-6}/800 \times 10^{-6} = 1.625$ T

Corresponding value of *H* for this flux density is given as 3800 AT/m.

Since the length of the central limb is 120 mm, m.m.f. required is = $H \times l = 3800 \times (120 \times 10^{-3}) = 456$ AT/m.

Air-gap

Flux density in the air-gap is the same as that in the central limb.

$H = B/\mu_0 = 1.625/4\pi \times 10^{-7} = 0.1293 \times 10^{-7}$ AT/m

Length of the air-gap = 1 mm = 10^{-3} m

m.m.f. reqd. for the air-gap = $H \times 1 = 0.1293 \times 10^7 \times 10^{-3} = 1293$ AT.

The flux of the central limb divides equally at point *A* in figure along two parallel path *ABCD* and *AFED*. We may consider either path, say, *ABCD* and calculate the m.m.f. required for it. The same m.m.f. will also send the flux through the other parallel path *AFED*.

Flux through *ABCD* = $1.3 \times 10^{-3}/2 = 0.65 \times 10^{-3}$ Wb

Flux density $B = 0.65 \times 10^{-3}/500 \times 10^{-6} = 1.3$ T

The corresponding value of *H* for this value of *B* is given as 850 AT/m.

$\therefore$ *m.m.f.* reqd. for path ABCD = $H \times 1 = 850 \times (300 \times 10^{-3}) = 255$ AT

As said above, this m.m.f. will also send the flux in the parallel path AFED.

Total m.m.f. reqd. = 456 + 1293 + 255 = 2004 AT

Since the number of turns is 500, I = 2004/500 = **4A.**

Example 6.20. *A cast steel d.c. electromagnet shown in Fig. 6.37 has a coil of 1000 turns on its central limb. Determine the current that the coil should carry to produce a flux of 2.5 mWb in the air-gap. Neglect leakage. Dimensions are given in cm. The magnetisation curve for cast steel is as under :*

Flux density (Wb/m²) :	*0.2*	*0.5*	*0.7*	*1.0*	*1.2*
Amp-turns/metre :	*300*	*540*	*650*	*900*	*1150*

(Electrotechnics-I, ; M.S. Univ. Baroda 1988)

Solution. Two points should be noted

(*i*) there are two (equal) parallel paths *ACDE* and *AGE* across the central path *AE*

(*ii*) flux density in either parallel path is half of that in the central path because flux divides into two equal parts at point *A*.

Total m.m.f. required for the whole electromagnet is equal to the sum of the following three m.m.fs.

(*i*) that required for path *EF*

(*ii*) that required for air-gap

and (*iii*) that required for either of the two parallel paths; say, path $ACDE_2$

Fig. 6.37

Flux density in the central limb and air gap is

$$= 2.5 \times 10^{-3}/(5 \times 5) \times 10^{-4} = 1 \text{ Wb/m}^2$$

Corresponding value of H as found from the given data is 900 AT/m.

$\therefore$AT for central limb $= 900 \times 0.3 = 270$

H in air-gap $= B/\mu_0 = 1/4\pi \times 10^{-7} = 79.56 \times 10^4$ AT/m

AT required $= 79.56 \times 10^4 \times 10^{-3} = 795.6$

Flux density in path *ACDE* is 0.5 Wb/m² for which corresponding value of H is 540 AT/m.

$\therefore$ AT for path *ACDE* $= 540 \times 0.6 = 324$

Total AT required = 270 + 795.6 + 324 = 1390; Current required = 1390/1000 = **1.39 A**

Example 6.21. *A cast steel magnetic structure made of a bar of section 8 cm × 2 cm is shown in Fig. 6.35. Determine the current that the 500 turn-magnetising coil on the left limb should carry so that a flux of 2 mWb is produced in the right limb. Take μ_r = 600 and neglect leakage.*

(Elect. Technology Allahabad Univ. 1993)

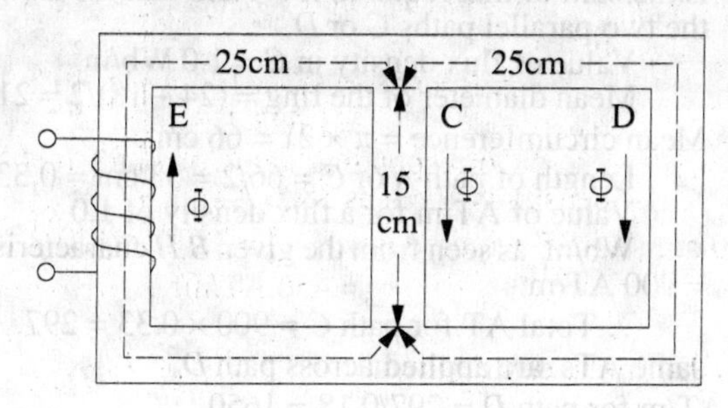

Fig. 6.38

Solution. Since paths *C* and *D* are in parallel with each other w.r.t. path *E* (Fig. 6.38), the m.m.f. across the two is the same.

$$\Phi_1 S_1 = \Phi_2 S_2$$

$$\therefore \qquad \Phi_1 \times \frac{15}{\mu A} = 2 \times \frac{25}{\mu A}$$

$$\therefore \qquad \Phi_1 = 10/3 \text{ mWb}$$

$$\therefore \qquad \Phi = \Phi_1 + \Phi_2 = 16/3 \text{ mWb}$$

Total AT required for the whole circuit is equal to the sum of

(*i*) that required for path *E* and (*ii*) that required for either of the two paths *C* or *D*.

Flux density in path $E = \dfrac{16 \times 10^{-3}}{3 \times 4 \times 10^{-4}} = \dfrac{40}{3}$ Wb/m^2

AT reqd. $= \dfrac{40 \times 0.25}{3 \times 4\pi \times 10^{-7} \times 600} = 4{,}420$

Flux density in path $D = \dfrac{2 \times 10^{-3}}{4 \times 10^{-4}} = 5$ Wb/m^2

AT reqd. $= \dfrac{5}{4\pi \times 10^{-7} \times 600} \times 0.25 = 1658$

Total AT $= 4{,}420 + 1{,}658 = 6{,}078$;

Current needed $= 6078/500 =$ **12.16 A**

Example 6.22. *A ring of cast steel has an external diameter of 24 cm and a square cross-section of 3 cm side. Inside and across the ring, an ordinary steel bar 18 cm × 3 cm × 0.4 cm is fitted with negligible gap. Calculate the number of ampere-turns required to be applied to one half of the ring to produce a flux density of 1.0 weber per metre² in the other half. Neglect leakage. The B-H characteristics are as below :*

For Cast Steel			
B in Wb/m²	*1.0*	*1.1*	*1.2*
Amp-turn/m	*900*	*1020*	*1220*

For Ordinary Plate			
B in Wb/m²	*1.2*	*1.4*	*1.45*
Amp-turn/m	*590*	*1200*	*1650*

(Elect. Technology, Indore Univ. 1985)

Solution. The magnetic circuit is shown in Fig. 6.39.

The m.m.f. (or AT) produced on the half A acts across the parallel magnetic circuits C and D. First, total AT across C is calculated and since these amp-turns are also applied across D, the flux density B in D can be estimated. Next, flux density in A is calculated and therefore the AT required for this flux density. In fact, the total AT (or m.m.f.) required is the sum of that required for A and that for either of the two parallel paths C or D.

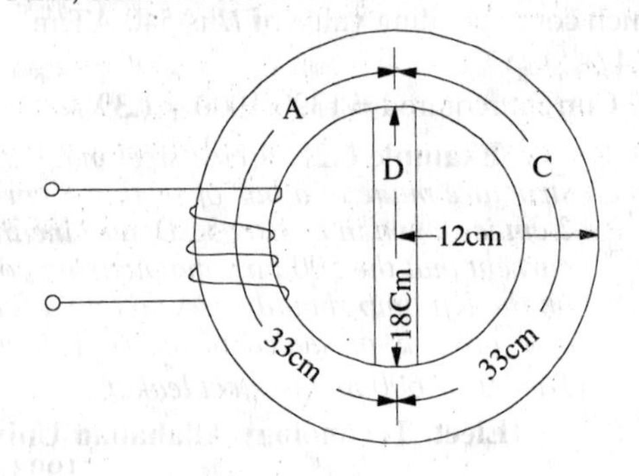

Fig. 6.39

Value of flux density in $C = 1.0$ Wb/m^2

Mean diameter of the ring $= (24 + 18)/2 = 21$ cm

Mean circumference $= \pi \times 21 = 66$ cm

Length of path A or $C = 66/2 = 33$ cm $= 0.33$ m

Value of AT/m for a flux density of 1.0 Wb/m^2 as seen from the given $B.H$ characteristics $= 900$ AT/m

$\therefore$ Total AT for path $C = 900 \times 0.33 = 297$. The same ATs. are applied across path D

Length of path $D = 18$ cm $= 0.18$ m $\therefore$ AT/m for path $D = 297/0.18 = 1650$

Value of B corresponding to this AT/m from given table is $= 1.45$ Wb/m^2

Flux through $C = B \times A = 1.0 \times 9 \times 10^{-4} = 9 \times 10^{-4}$ Wb

Flux through $D = 1.45 \times (3 \times 0.4 \times 10^{-4}) = 1.74 \times 10^{-4}$ Wb

$\therefore$ Total flux through $A = 9 \times 10^{-4} + 1.74 \times 10^{-4} = 10.74 \times 10^{-4}$ Wb.

Flux density through $A = 10.74 \times 10^{-4}/9 \times 10^{-4} = 1.193$ Wb/m^2

No. of AT/m reqd. to produce this flux density as read from the given table = 1200 (approx.)

$\therefore$ Amp-turns required for limb $A = 1200 \times 0.33 = 396$

Total AT required = 396 + 297 = **693.**

Example 6.23. *Show how the ampere-turns per pole required to produce a given flux in a d.c. generator are calculated.*

Find the amp-turns per pole required to produce a flux of 40 mWb per pole in a machine with a smooth core armature and having the following dimensions :

Length of air gap = 5 mm — *Area of air-gap = 500 cm²*

Length of pole = 12 cm — *Sectional area of pole core = 325 cm²*

Relative permeability of pole core = 1,500

Length of magnetic path in yoke between pole = 65 cm

Cross-sectional area of yoke = 450 cm² ; Relative permeability of yoke = 1,200

Leakage coefficient = 1.2

The ampere-turns for the armature core may be neglected.

Solution. Air-gap $\Phi = 40$ mWb $= 4 \times 10^{-2}$ Wb ; $A = 500 \times 10^{-4} = 5 \times 10^{-2}$ m²

$\therefore B = 4 \times 10^{-2}/5 \times 10^{-2} = 0.8$ Wb/m²; $H = B/\mu_0 = 0.8/4\pi \times 10^{-7} = 63.63 \times 10^4$ AT/m

Air-gap length $= 5 \times 10^{-3}$ m; AT reqd. $= 63.63 \times 10^4 \times 5 \times 10^{-3} = 3181.5$

Pole Core

$$\Phi = 1.2 \times 4 \times 10^{-2} = 4.8 \times 10^{-2} \text{ Wb}; \quad A = 325 \times 10^{-4} \text{m}^2$$

$$B = 4.8 \times 10^{-2}/325 \times 10^{-4} = 1.477 \text{ Wb/m}^2$$

$$H = B/\mu_0 \mu_r = 1.477/4\pi \times 10^{-7} \times 1{,}500 = 783 \text{ AT/m}$$

Pole length $= 0.12$ m ; AT reqd. $= 783 \times 0.12 = 94$

Yoke Path

flux = half the pole flux $= 0.5 \times 4 \times 10^{-2} = 2 \times 10^{-2}$ Wb

$$A = 450 \text{ cm}^2 = 45 \times 10^{-3} \text{m}^2; \quad B = 2 \times 10^{-2}/45 \times 10^{-3} = 4/9 \text{ Wb/m}^2$$

$$H = \frac{4/9}{4\pi \times 10^{-7} \times 1{,}200} = 294.5 \text{ AT/m} \quad \text{Yoke length} = 0.65 \text{ m}$$

At reqd. $= 294.5 \times 0.65 = 191.4$, Total AT/Pole $= 3181.5 + 94 + 191.4 =$ **3,467**

Example 6.24. *A shunt field coil is required to develop 1,500 AT with an applied voltage of 60 V. The rectangular coil is having a mean length of turn of 50 cm. Calculate the wire size. Resistivity of copper may be assumed to be 2 μΩ-cm at the operating temperature of the coil. Estimate also the number of turns if the coil is to be worked at a current density of 3 A/mm².*

(Basis Elect. Machines Nagpur Univ. 1992)

Solution. $NI = 1{,}500$ (given) or $N.\dfrac{V}{R} = N.\dfrac{60}{R} = 1{,}500$

$$\therefore \quad R = \frac{N}{25} \text{ ohm} \qquad \text{Also } R = \rho.\frac{l}{A} = \frac{2 \times 10^{-6} \times 50\,N}{A}$$

$$\therefore \quad \frac{N}{25} = \frac{10^{-4} N}{A} \qquad \text{or } A = 25 \times 10^{-4} \text{cm}^2 \text{ or } A = 0.25 \text{ mm}^2$$

$$\therefore \quad \frac{\pi D^2}{4} = 0.25 \qquad \text{or} \quad D = \mathbf{0.568 \text{ mm}}$$

Current in the coil $= 3 \times 0.25 = 0.75$ A

Now, $NI = 1{,}500$; $\therefore N = 1{,}500/0.75 =$ **2,000**

Tutorial Problems No. 6.2

1. An iron specimen in the form of a closed ring has a 350-turn magnetizing winding through which is passed a current of 4A. The mean length of the magnetic path is 75 cm and its cross-sectional area is 1.5 cm².

Wound closely over the specimen is a secondary winding of 50 turns. This is connected to a ballistic galvanometer in series with the secondary coil of 9-mH mutual inductance and a limiting resistor. When the magnetising current is suddenly reversed, the galvanometer deflection is equal to that produced by the reversal of a current of 1.2 A in the primary coil of the mutual inductance. Calculate the B and H values for the iron under these conditions, deriving any formula used. **[1.44 Wb/m^2 ; 1865 AT/m] (London Univ.)**

2. A moving-coil ballistic galvanometer of 150 Ω gives a throw of 75 divisions when the flux through a search coil, to which it is connected, is reversed.

Find the flux density in which the reversal of the coil takes place, given that the galvanometer constant is 110 μ*C* per scale division and the search coil has 1400 turns, a mean area of 50 cm^2 and a resistance of 20 Ω. **[0.1 Wb/m^2] (Elect. Meas. & Measuring Inst. Gujarat Univ. Oct. 1979)**

3. A fluxmeter is connected to a search coil having 500 turns and mean area of 5 cm^2. The search coil is placed at the centre of a solenoid one metre long wound with 800 turns. When a current of 5 A is reversed, there is a deflection of 25 scale divisions on the fluxmeter. Calculate the fluxmeter constant. **[10^{-4} Wb-turn/division] (Elect. Means. & Measuring Inst., M.S. Univ. Baroda, 1977)**

4. An iron ring of mean length 50 cms has an air gap of 1 mm and a winding of 200 turns. If the permeability of iron is 300 when a current of 1 A flows through the coil, find the flux density. **[94.2 mWb/m^2] (Elect. Engg. A.M.Ae. S.I. June 1989)**

5. An iron ring of mean length 100 cm with an air gap of 2 mm has a winding of 500 turns. The relative permeability of iron is 600. When a current of 3 A flows in the winding, determine the flux density. Neglect fringing. **[0.523 Wb/m^2] (Elect. Engg. & Electronic Bangalore Univ. 1990)**

6. A coil is wound uniformly with 300 turns over a steel ring of relative permeability 900, having a mean circumference of 40 mm and cross-sectional area of 50 mm^2. If a current of 25 amps is passed through the coil, find (*i*) m.m.f. (*ii*) reluctance of the ring and (*iii*) flux. **[(i) 7500 AT (ii) 0.7 × 10^6 AT/Wb (iii) 10.7 mWb] (Elect. Engg. & Electronics Bangalore Univ. 1983)**

7. A specimen ring of transformer stampings has a mean circumference of 40 cm and is wound with a coil of 1,000 turns. When the currents through the coil are 0.25 A, 1 A and 4 A the flux densities in the stampings are 1.08, 1.36 and 1.64 Wb/m^2 respectively. Calculate the relative permeability for each current and explain the differences in the values obtained. **[1,375, 434, 131]**

8. A magnetic circuit consists of an iron ring of mean circumference 80 cm with crossectional area 12 cm^2 throughout. A current of 2A in the magnetising coil of 200 turns produces a total flux of 1.2 mWb in the iron. Calculate :

(a) the flux density in the iron **(b)** the absolute and relative permeabilities of iron
(c) the reluctance of the circuit **[1 Wb/m^2 ; 0.002, 1,590 ; 3.33 × 10^5 AT/Wb]**

9. A coil of 500 turns and resistance 20 Ω is wound uniformly on an iron ring of mean circumference 50 cm and cross-sectional area 4 cm^2. It is connected to a 24-V d.c. supply. Under these conditions, the relative permeability of iron is 800. Calculate the values of :

(a) the magnetomotive force of the coil **(b)** the magnetizing force
(c) the total flux in the iron **(d)** the reluctance of the ring
[(a) 600 AT (b) 1,200 AT/m (c) 0.483 mWb (d) 1.24 × 10^6 AT/Wb)

10. A series magnetic circuit has an iron path of length 50 cm and an air-gap of length 1 mm. The cross-sectional area of the iron is 6 cm^2 and the exciting coil has 400 turns. Determine the current required to produce a flux of 0.9 mWb in the circuit. The following points are taken from the magnetisation characteristic:

Flux density (Wb/m^2) :	1.2	1.35	1.45	1.55
Magnetizing force (AT/m) :	500	1,000	2,000	4,500

[6.35 A]

11. An iron-ring of mean length 30 cm is made of three pieces of cast iron, each has the same length but their respective diameters are 4, 3 and 2.5 cm. An air-gap of length 0.5 mm is cut in the 2.5 cm piece. If a coil of 1,000 turns is wound on the ring, find the value of the current it has to carry to produce a flux density of 0.5 Wb/m^2 in the air gap. B/H characteristic of cast-iron may be drawn from the following :

B (Wb/m^2) :	0.1	0.2	0.3	0.4	0.5	0.6
(AT/m) :	280	620	990	1,400	2,000	2,800

[0.58 A]

Permeability of free space = $4\pi \times 10^{-7}$ H/m. Neglect leakage and fringing.

12. The length of the magnetic circuit of a relay is 25 cm and the cross-sectional area is 6.25 cm^2. The length of the air-gap in the operated position of the relay is 0.2 mm. Calculate the magnetomotive force required to produce a flux of 1.25 mWb in the air gap. The relative permeability of magnetic material at this flux density is 200. Calculate also the reluctance of the magnetic circuit when the relay is in the unoperated position, the air-gap then being 8 mm long (assume μ_r remains constant). **[2307 AT, 1.18 × 10^7 AT/Wb]**

13. For the magnetic circuit shown in Fig. 6.40, all dimensions are in cm and all the air-gaps are 0.5 mm wide. Net thickness of the core is 3.75 cm throughout. The turns are arranged on the centre limb as shown.

Calculate the m.m.f. required to produce a flux of 1.7 m Wb in the centre limb. Neglect the leakage and fringing. The magnetisation data for the material is as follows :

H (AT/m) :	400	440	500	600	800	
B (Wb/m^2) :	0.8	0.9	1.0	1.1	1.2	**[1,052 AT]**

14. In the magnetic circuit shown in Fig. 6.41 a coil of 500 turns is wound on the centre limb. The magnetic paths A to B by way of the outer limbs have a mean length of 100 cm each and an effective cross-sectional area of 2.5 cm^2. The centre limb is 25 cm long and 5 cm^2 cross-sectional area. The air-gap is 0.8 cm long. A current of 9.2 A through the coil is found to produce a flux of 0.3 m Wb.

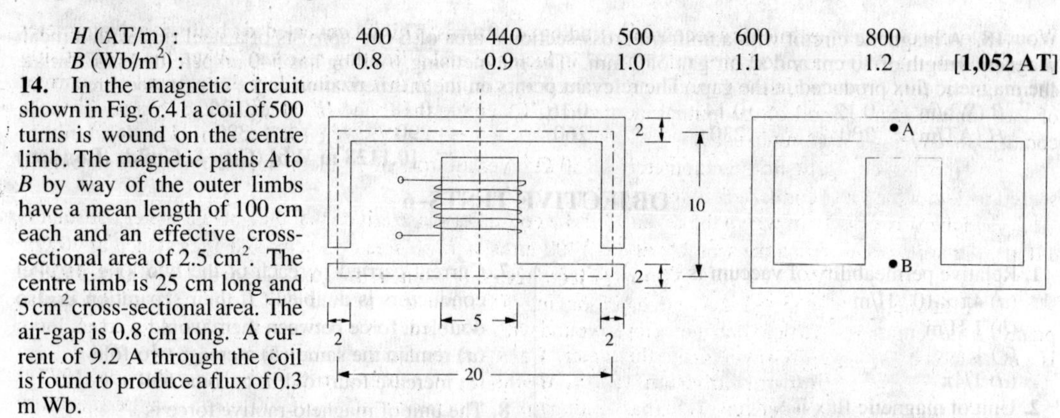

Fig. 6.40 **Fig. 6.41**

15. The magnetic circuit of a choke is shown in Fig. 6.42. It is designed so that the flux in the central core is 0.003 Wb. The cross-section is square and a coil of 500 turns is wound on the central core. Calculate the exciting current. Neglect leakage and assume the flux to be uniformly distributed along the mean path shown dotted. Dimensions are in cm.

The characteristics of magnetic circuit are as given below :

B (Wb/m^2) :	0.38	0.67	1.07	1.2	1.26
H (AT/m) :	100	200	600	1000	1400

(Elect. Technology I. Gwalior Univ. Nov. 1976)

16. A 680-turn coil is wound on the central limb of the cast steel sheet frame as shown in Fig. 6.43 where dimensions are in cm. A total flux of 1.6 mWb is required to be in the gap. Find the current required in the magnetising coil. Assume gap density is uniform and all lines pass straight across the gap. Following data is given :

H (AT/m) :	300	500	700	900	1100
B (Wb/m^2) :	0.2	0.45	0.775	1.0	1.13

(Elect. Technology; Indore Univ. Jan. 1975)

17. In the magnetic circuit of Fig. 6.44, the core is composed of annealed sheet steel for which a stacking factor of 0.9 should be assumed. The core is 5 cm thick. When $\Phi_A = 0.002$ Wb, $\Phi_B = 0.0008$ Wb and $\Phi_C = 0.0012$ Wb. How many amperes must each coil carry and in what direction ? Use of the following magnetisation curves can be made for solving the problem.

B (Wb/m^2) :	0.2	0.4	0.6	0.8	1.0	1.4	1.6	1.8
H (AT/m^2) :	50	100	130	200	320	1200	3800	10,000

(Elect. Technology, Vikram Univ. 1975)

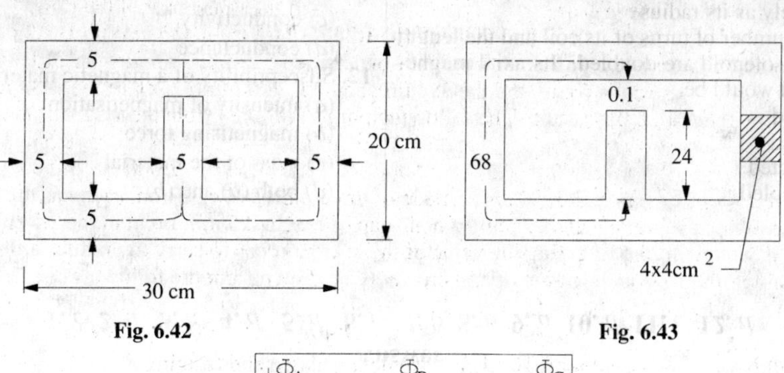

Fig. 6.42 **Fig. 6.43**

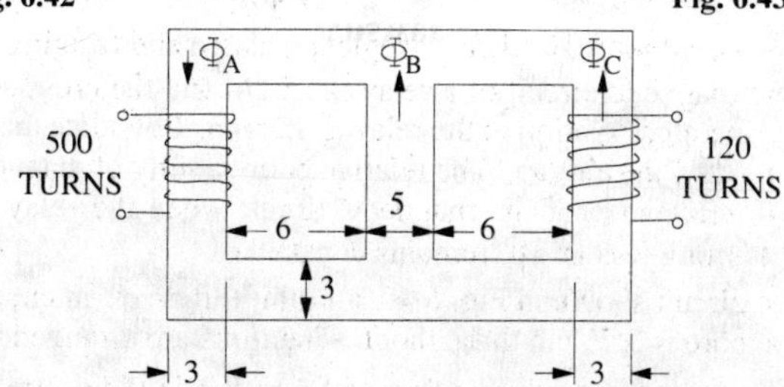

Fig. 6.44

18. A magnetic circuit with a uniform cross-sectional area of 6 cm^2 consists of a steel ring with a mean magnetic length of 80 cm and an air gap of 2 mm. The magnetising winding has 540 ampere-turns. Estimate the magnetic flux produced in the gap. The relevant points on the magnetization curve of cast steel are :

B (Wb/m^2) :	0.12	0.14	0.16	0.18	0.20
H (AT/m) :	200	230	260	290	320

[0.1128 m Wb] (City & Guilds, London)

OBJECTIVE TESTS–6

1. Relative permeability of vaccum is
(*a*) $4\pi \times 10^{-7}$ H/m
(*b*) 1 H/m
(*c*) 1
(*d*) $1/4\pi$

2. Unit of magnetic flux is
(*a*) weber
(*b*) ampere-turn
(*c*) tesla
(*d*) coulomb

3. The magnetising force (H) and magnetic flux density (B) are connected by the relation.
(*a*) $B = \mu H$
(*b*) $B = H/\mu_0\mu_r$
(*c*) $B = \mu_0 H/\mu_r$
(*d*) $B = \mu_r H/\mu_0$

4. The force experience by a current-carrying conductor lying *parallel* to a magnetic field is
(*a*) BIl
(*b*) $BIl \sin\theta$
(*c*) HIl
(*d*) zero

5. Point out the WRONG statement.
The magnetising force at the centre of a circular coil varies.
(*a*) directly as the number of its turns
(*b*) directly as the current
(*c*) directly as its radius
(*d*) inversely as its radius

6. Both the number of turns of its coil and the length of a short solenoid are doubled. Its axial magnetising field would be
(*a*) doubled
(*b*) halved
(*c*) unaffected
(*d*) quadrupled

7. Current carried by each of the two long parallel conductors is doubled. If their separation is also doubled, force between them would
(*a*) remain the same (*b*) increase two-fold
(*c*) increase four-fold(*d*) become half

8. The unit of magneto-motive force is
(*a*) weber
(*b*) ampere/metre
(*c*) henry
(*d*) ampere-turn/weber

9. Permeability in a magnetic circuit corresponds to—— in an electric circuit.
(*a*) conductivity
(*b*) resistivity
(*c*) conductance
(*d*) resistance

10. Point out the WRONG statement.
Magnetic leakage is undesirable in electric machines because it
(*a*) leads to their increased weight
(*b*) increases their cost of manufacture
(*c*) produces fringing
(*d*) lowers their power efficiency.

11. Permeability in a magnetic circuit corresponds to—— in an electric circuit.
(*a*) reluctivity
(*b*) resistivity
(*c*) conductivity
(*d*) conductance

12. Susceptibility of a magnetic material depends on
(*a*) intensity of magnetisation
(*b*) magnetising force
(*c*) mass of the material
(*d*) both (*a*) and (*b*)

Answer

1. *c* **2.** *a* **3.** *a* **4.** *d* **5.** *d* **6.** *c* **7.** *b* **8.** *b* **9.** *a* **10.** *d* **11.** *c* **12.** *d*

7 ELECTROMAGNETIC INDUCTION

7.1. Relation Between Magnetism and Electricity

It is well known that whenever an electric current flows through a conductor, a magnetic field is immediately brought into existence in the space surrounding the conductor. It can be said that when electrons are in motion, they produce a magnetic field. The converse of this is also true *i.e.* when a magnetic field embracing a conductor moves *relative* to the conductor, it produces a flow of electrons in the conductor. This phenomenon whereby an e.m.f. and hence current (*i.e.* flow of electrons) is induced in any conductor which is cut across or is cut by a magnetic flux is known as *electromagnetic induction.* The historical background of this phenomenon is this :

After the discovery (by Oersted) that electric current produces a magnetic field, scientists began to search for the converse phenomenon from about 1821 onwards. The problem they put to themselves was how to 'convert' magnetism into electricity. It is recorded that Michael Faraday* was in the habit of walking about with magnets in his pockets so as to constantly remind him of the problem. After nine years of continuous research and experimentation, he succeeded in producing electricity by 'converting magnetism'. In 1831, he formulated basic laws underlying the phenomenon of electromagnetic induction (known after his name), upon which is based the operation of most of the commercial apparatus like motors, generators and transformers etc.

7.2. Production of Induced E.M.F and Current

In Fig. 7.1 is shown an insulated coil whose terminals are connected to a sensitive galvanometer *G*. It is placed close to a stationary bar magnet initially at position *AB* (shown dotted). As seen, some flux from the *N*-pole of the magnet is linked with or threads through the coil but, as yet, there is no deflection of the galvanometer. Now, suppose that the magnet is *suddenly* brought closer to the coil in position *CD* (see figure). Then, it is found that there is a jerk or a sudden but a momentary

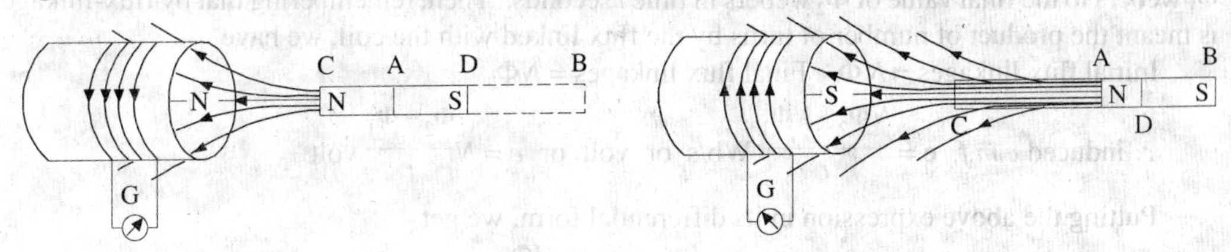

Fig. 7.1 Fig. 7.2

deflection in the galvanometer and that this *lasts so long as the magnet is in motion relative to the coil, not otherwise*. The deflection is reduced to zero when the magnet becomes again stationary at its new position *CD*. It should be noted that due to the approach of the magnet, flux linked with the coil is increased.

Next, the magnet is *suddenly* withdrawn away from the coil as in Fig. 7.2. It is found that again

*Michael Faraday (1791–1867), an English physicist and chemist.

there is a *momentary* deflection in the galvanometer and it persists so long as the magnet is in motion, not when it becomes stationary. It is important to note that this deflection is in a direction opposite to that of Fig. 7.1. Obviously, due to the withdrawal of the magnet, flux linked with the coil is decreased.

The deflection of the galvanometer indicates the production of e.m.f. in the coil. The only cause of the production can be the sudden approach or withdrawal of the magnet from the coil. It is found that the actual cause of this e.m.f. is the change of flux linking with the coil. This e.m.f. exists so long as the change in flux exists. Stationary flux, however strong, will never induce any e.m.f. in a stationary conductor. In fact, the same results can be obtained by keeping the bar magnet stationary and moving the coil suddenly away or towards the magnet.

The direction of current set up by the induced e.m.f. is as shown in the two figures given above.

The production of this electromagnetically-induced e.m.f. is further illustrated by considering a conductor *AB* lying within a magnetic field and connected to a galvanometer as shown in Fig. 7.3. It is found that whenever this conductor is moved up or down, *a momentary* deflection is produced in the galvanometer. It means that some transient e.m.f. is induced in *AB*. The magnitude of this induced e.m.f. (and hence the amount of deflection in the galvanometer) *depends on the quickness of the movement of AB*.

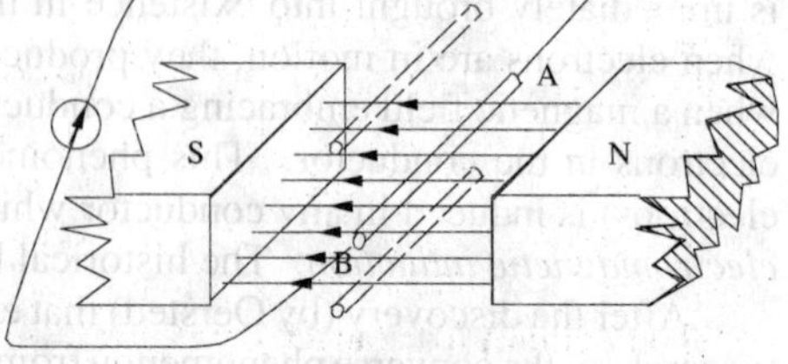

Fig. 7.3

From this experiment we conclude that whenever a conductor cuts or *shears* the magnetic flux, an e.m.f. is always induced in it.

It is also found that if the conductor is moved parallel to the direction of the flux so that it does not cut it, then no e.m.f. is induced in it.

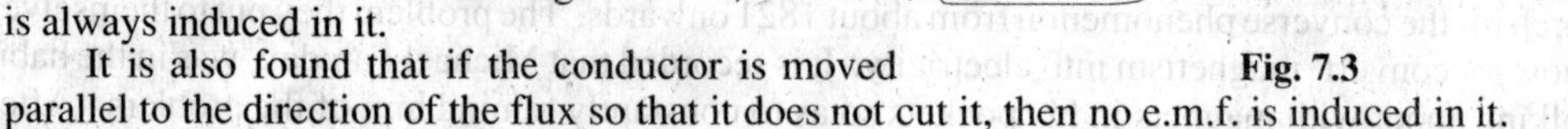

7.3. Faraday's Laws of Electromagnetic Induction

Faraday summed up the above facts into two laws known as Faraday's Laws of Electromagnetic Induction.

First Law. It states :

Whenever the magnetic flux linked with a circuit changes, an e.m.f. is always induced in it.

or

Whenever a conductor cuts magnetic flux, an e.m.f. is induced in that conductor.

Second Law. It states :

The magnitude of the induced e.m.f. is equal to the rate of change of *flux-linkages*.

Explanation. Suppose a coil has *N* turns and flux through it changes from an initial value of Φ_1 webers to the final value of Φ_2 webers in time *t* seconds. Then, remembering that by flux-linkages is meant the product of number of turns by the flux linked with the coil, we have

Initial flux linkages = $N\Phi_1$ Final flux linkages = $N\Phi_2$

$$\therefore \text{ induced } e.m.f. \; e = \frac{N\Phi_2 - N\Phi_1}{t} \text{ Wb/s or volt or } e = N\frac{\Phi_2 - \Phi_1}{t} \text{ volt}$$

Putting the above expression in its differential form, we get

$$e = \frac{d}{dt}(N\Phi) = N\frac{d\Phi}{dt} \text{ volt}$$

Usually, a minus sign is given to the right-hand side expression to signify the fact that the induced e.m.f. sets up current in such a direction that magnetic effect produced by it opposes the very cause producing it (Art. 7.5).

$$\therefore \qquad e = -N\frac{d\Phi}{dt} \text{ volt}$$

Example 7.1. *The field coils of a 6-pole d.c. generator each having 500 turns, are connected in series. When the field is excited, there is a magnetic flux of 0.02 Wb/pole. If the field circuit is opened in 0.02 second and residual magnetism is 0.002 Wb/pole, calculate the average voltage which is induced across the field terminals. In which direction is this voltage directed relative to the direction of the current.*

Solution. Total number of turns, $N = 6 \times 500 = 3000$
Total initial flux $= 6 \times 0.02 = 0.12$ Wb
Total residual flux $= 6 \times 0.002 = 0.012$ Wb
Change in flux, $d\Phi = 0.12 - 0.012 = 0.108$ Wb
Time of opening the circuit, $dt = 0.02$ second

$$\therefore \text{ induced e.m.f.} = N\frac{d\Phi}{dt} \text{ volt} = 3000 \times \frac{0.108}{0.02} = \mathbf{16{,}200\ V}$$

The direction of this induced e.m.f. is the same as the initial direction of the exciting current.

Example 7.2. *A coil of resistance 100 Ω is placed in a magnetic field of 1 mWb. The coil has 100 turns and a galvanometer of 400 Ω resistance is connected in series with it. Find the average e.m.f. and the current if the coil is moved in 1/10th second from the given field to a field of 0.2 mWb.*

Solution. Induced e.m.f. $= N.\dfrac{d\Phi}{dt}$ volt

Here $d\Phi = 1 - 0.2 = 0.8$ m Wb $= 0.8 \times 10^{-3}$ Wb

$dt = 1/10 = 0.1$ second ; $N = 100$

$e = 100 \times 0.8 \times 10^{-3}/0.1 = \mathbf{0.8\ V}$

Total circuit resistance $= 100 + 400 = 500\ \Omega$

$\therefore$ current induced $= 0.8/500 = 1.6 \times 10^{-3}$ A $= \mathbf{1.6\ mA}$

Example 7.3. *The time variation of the flux linked with a coil of 500 turns during a complete cycle is as follows :*

$\Phi = 0.04\,(1 - 4\,t/T)$ *Weber* $\qquad 0 < t < T/2$

$\Phi = 0.04\,(4t/T - 3)$ *Weber* $\qquad T/2 < t < T$

where T represents time period and equals 0.04 second. Sketch the waveforms of the flux and induced emf and also determine the maximum value of the induced emf.

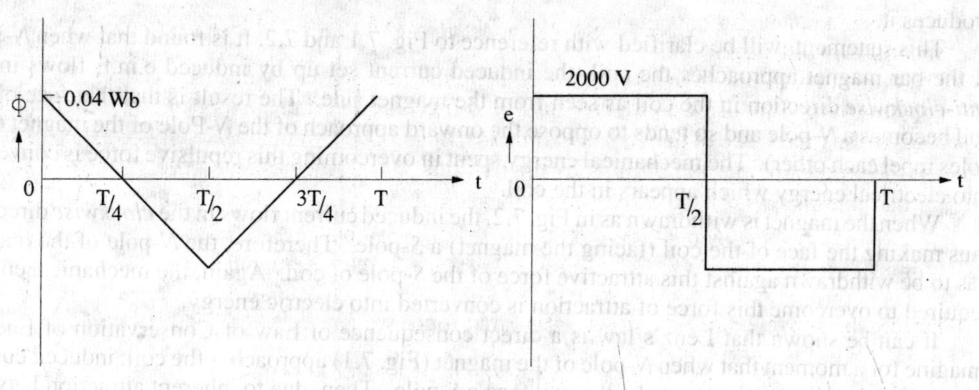

Fig. 7.4

Solution. The variation of flux is linear as seen from the following table.

t (second) :	0	$T/4$	$T/2$	$3T/4$	T
Φ(weber) :	0.04	0	−0.04	0	0.04

The induced e.m.f. is given by $e = -Nd\,\Phi/dt$
From $t = 0$ to $t = T/2$, $d\Phi/dt = -0.04 \times 4/T = -4$ Wb/s $\therefore e = -500(-4) = 2000$ V
From $t = T/2$ to $t = T$ $d\Phi/dt = 0.04 \times 4/T = 4$ Wb/s $\therefore e = -500 \times 4 = -2000$ V.
The waveforms are selected in Fig. 7.4.

7.4. Direction of Induced E.M.F. and Currents

There exists a definite relation between the direction of the induced current, the direction of the flux and the direction of motion of the conductor. The direction of the induced current may be found easily by applying either Fleming's Right-hand Rule or Flat-hand rule or Lenz's Law. Fleming's rule (Fig. 7.5) is used where induced e.m.f. is due to flux-cutting (*i.e.* dynamically induced e.m.f.) and Lenz's when it is due to change by flux-linkages (*i.e.* statically induced e.m.f).

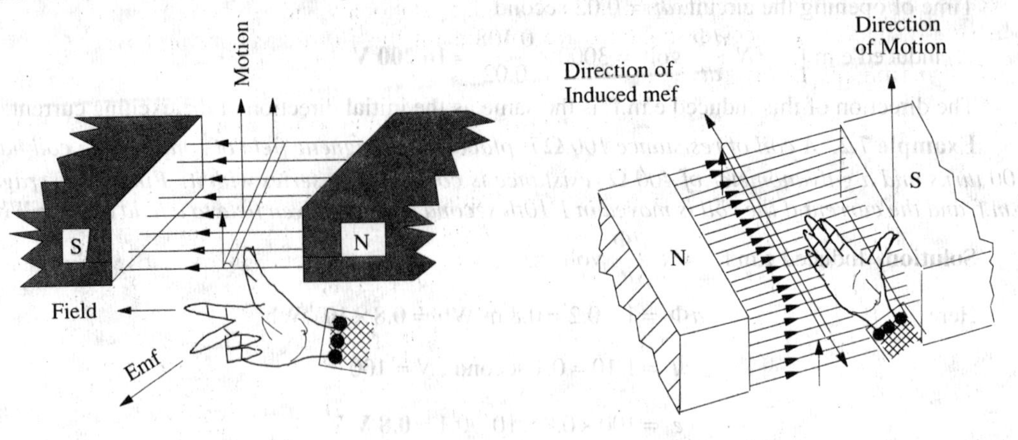

Fig. 7.5 **Fig. 7.6**

Fig. 7.6 shows another way of finding the direction of the induced e.m.f. It is known as Right Flat-hand rule. Here, the front side of the hand is held perpendicular to the incident flux with the thumb pointing in the direction of the motion of the conductor. The direction of the fingers give the direction of the induced e.m.f. and current.

7.5. Lenz's Law

The direction of the induced current may also be found by this law which was formulated by Lenz* in 1835. This law states, in effect, that electromagnetically induced current always flows in such direction that the action of the magnetic field set up by it tends to oppose the very cause which produces it.

This statement will be clarified with reference to Fig. 7.1 and 7.2. It is found that when *N*-pole of the bar magnet approaches the coil, the induced current set up by induced e.m.f. flows in the *anti-clockwise* direction in the coil as seen from the magnet side. The result is that that face of the coil becomes a *N*-pole and so tends to oppose the onward approach of the *N*-Pole of the magnet (like poles repel each other). The mechanical energy spent in overcoming this repulsive force is converted into electrical energy which appears in the coil.

When the magnet is withdrawn as in Fig. 7.2, the induced current flows in the *clockwise* direction thus making the face of the coil (facing the magnet) a *S*-pole. Therefore, the *N*-pole of the magnet has to be withdrawn against this attractive force of the *S*-pole of coil. Again, the mechanical energy required to overcome this force of attraction is converted into electric energy.

It can be shown that Lenz's law is a direct consequence of Law of Conservation of Energy. Imagine for a moment that when *N*-pole of the magnet (Fig. 7.1) approaches the coil, induced current flows in such a direction as to make the coil face a *S*-pole. Then, due to inherent attraction between unlike poles, the magnet would be automatically pulled towards the coil without the expenditure of any mechanical energy. It means that we would be able to create electric energy out of nothing, which is denied by the inviolable Law of Conservation of Energy. In fact, to maintain the sanctity of this law, it is imperative for the induced current to flow in such a direction that the magnetic effect produced by it tends to oppose the very cause which produces it. In the present case, it is the relative motion of the magnet with respect to the coil which is the cause of the production of the induced current. Hence, the induced current always flows in such a direction as to tend to oppose this relative motion *i.e.* the approach or withdrawal of the magnet.

---

*After the Russian born geologist and physicist Heinrich Friedrich Emil Lenz (1808-1865).

7.6. Induced E.M.F.

Induced e.m.f. can be either *(i)* **dynamically induced** or *(ii)* **statically induced**. In the first case, usually the field is stationary and conductors cut across it (as in d.c. generators). But in the second case, usually the conductor or the coil remains stationary and flux linked with it is changed by simply increasing or decreasing the current producing this flux (as in transformers).

7.7. Dynamically Induced E.M.F.

In Fig. 7.7, a conductor *A* is shown in cross-section, lying within a uniform magnetic field of flux density *B* Wb/m^3. The arrow attached to *A* shows its direction of motion. Consider the conditions shown in Fig. 7.7(*a*) when *A* cuts across at right angles to the flux. Suppose '*l*' is its length lying within the field and let it move a distance *dx* in time *dt*. Then area swept by it is = *ldx*. Hence, flux cut = $l.dx \times B$ webers.

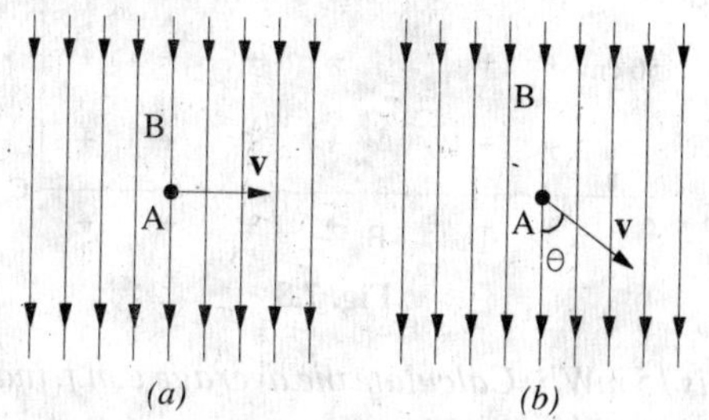

Fig. 7.7

Change in flux = *Bldx* weber

Time taken = *dt* second

Hence, according to Faraday's Laws (Art. 7.3) the e.m.f. induced in it (known as dynamically induced e.m.f.) is

$$= \text{rate of change of flux linkages} = \frac{Bldx}{dt} = Bl\frac{dx}{dt} = Blv \text{ volt where } \frac{dx}{dt} = \text{velocity}$$

If the conductor *A* moves at an angle θ with the direction of flux [Fig. 7.7 (*b*)] then the induced e.m.f. is $e = Bl\upsilon \sin\theta$ volts $= \vec{l\upsilon} \times \vec{B}$ (*i.e.* as cross product vector $\vec{\upsilon}$ and $\vec{B}$).

The direction of the induced e.m.f. is given by Fleming's Right-hand rule (Art. 7.5) or Flat-hand rule and most easily by vector cross product given above.

It should be noted that generators work on the production of dynamically induced e.m.f. in the conductors housed in a revolving armature lying within a strong magnetic field.

Example 7.4. *A conductor of length 1 metre moves at right angles to a uniform magnetic field of flux density 1.5 Wb/m² with a velocity of 50 metre/second. Calculate the e.m.f. induced in it. Find also the value of induced e.m.f. when the conductor moves at an angle of 30° to the direction of the field.*

Solution. Here $B = 1.5\ Wb/m^2$ $l = 1$ m $\upsilon = 50$ m/s ; $e = ?$

Now $e = Bl\upsilon = 1.5 \times 1 \times 50 =$ **75V.**

In the second case $\theta = 30°$ ∴ $\sin 30° = 0.5$ ∴ $e = 75 \times 0.5 =$ **37.5 V**

Example 7.5. *A square coil of 10 cm side and with 100 turns is rotated at a uniform speed of 500 rpm about an axis at right angle to a uniform field of 0.5 Wb/m². Calculate the instantaneous value of induced e.m.f. when the plane of the coil is (i) at right angle to the plane of the field. (ii) in the plane of the field. (iii) at 45° with the field direction.* **(Elect. Engg. A.M.Ae. S.I. Dec. 1991)**

Solution. As seen from Art. 12.2, e.m.f. induced in the coil would be zero when its plane is at right angles to the plane of the field, even though it will have maximum flux linked with it. However, the coil will have maximum e.m.f. induced in it when its plane lies parallel to the plane of the field even though it will have minimum flux linked with it. In general, the value of the induced e.m.f. is given by $e = \omega N \Phi_m \sin\theta = E_m \sin\theta$ where θ is the angle between the axis of zero *e.m.f.* and the plane of the coil.

Here, $f = 500/60 = 25/3$ r.p.s ; $N = 100$; $B = 0.5\ Wb/m^2$; $A = (10 \times 10) \times 10^{-4} = 10^{-2}\ m^2$.

∴ $E_m = 2\pi f NBA = 2\pi (25/3) \times 100 \times 0.5 \times 10^{-2} = 26.2$ V (*i*) since $\theta = 0$; $\sin\theta = 0$; therefore, $e = 0$. (*ii*) Here, $\theta = 90°$; $e = E_m \sin 90° = 26.2 \times 1 = 26.2$ V (*iii*) $\sin 45° = 1/\sqrt{2}$; $e = 26.2 \times 1/\sqrt{2}$ **= 18.5 V**

Example 7.6. *A conducting rod AB (Fig. 7.8) makes contact with metal rails AD and BC which are 50 cm apart in a uniform magnetic field of B = 1.0 Wb/m² perpendicular to the plane ABCD. The total resistance (assumed constant) of the circuit ABCD is 0.4 Ω.*

(a) What is the direction and magnitude of the e.m.f. induced in the rod when it is moved to the left with a velocity of 8 m/s ?

(b) *What force is required to keep the rod in motion ?*

(c) *Compare the rate at which mechanical work is done by the force F with the rate of development of electric power in the circuit.*

Solution. (a) Since AB moves to the left, direction of the induced current, as found by applying Fleming's Right-hand rule is from A to B. Magnitude of the induced e.m.f. is given by

$e = \beta l \upsilon$ volt $= 1 \times 0.5 \times 8 =$ **4 volt**

(b) Current through $AB = 4/0.4 = 10$ A

Force on AB *i.e.* $F = BIl = 1 \times 10 \times 0.5 =$ **5 N**

The direction of this force, as found by applying Fleming's left-hand rule, is to the right.

(c) Rate of doing mechanical work

$= F \times \upsilon = 5 \times 8 =$ **40 J/s or W**

Electric power produced

$= e\,i = 4 \times 10 = 40$ W

From the above, it is obvious that the mechanical work done in moving the conductor against force F is converted into electric energy.

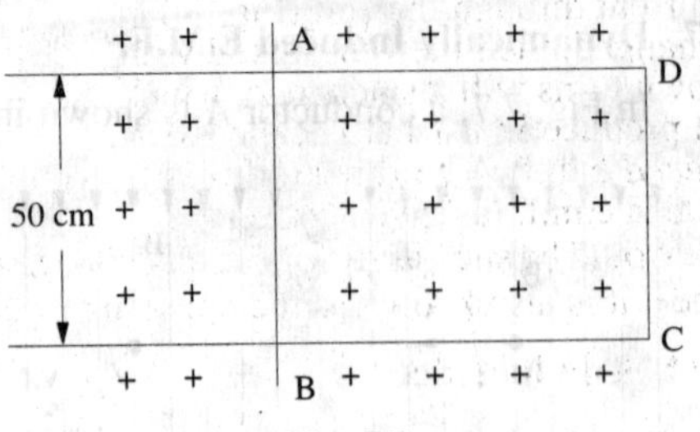

Fig. 7.8

Example 7.7. *In a 4-pole dynamo, the flux/pole is 15 mWb. Calculate the average e.m.f. induced in one of the armature conductors, if armature is driven at 600 r.p.m.*

Solution. It should be noted that each time the conductor passes under a pole (whether N or S) it cuts a flux of 15 mWb. Hence, the flux cut in one revolution is $15 \times 4 = 60$ mWb. Since conductor is rotating at $600/60 = 10$ r.p.s, time taken for one revolution is $1/10 = 0.1$ second.

$\therefore$ average e.m.f. generated $= N\dfrac{d\Phi}{dt}$ volt

$N = 1$; $d\Phi = 60$ mWb $= 6 \times 10^{-2}$ Wb; $dt = 0.1$ second

$\therefore\ e = 1 \times 6 \times 10^{-2}/0.1 =$ **0.6 V**

Tutorial Problems No. 7.1

1. A conductor of active length 30 cm carries a current of 100 A and lies at right angles to a magnetic field of strength 0.4 Wb/m^2. Calculate the force in newtons exerted on it. If the force causes the conductor to move at a velocity of 10 m/s, calculate (a) the e.m.f. induced in it and (b) the power in watts developed by it. **[12 N; 1.2 V, 120 W]**

2. A straight horizontal wire carries a steady current of 150 A and is situated in a uniform magnetic field of 0.6 Wb/m^2 acting vertically downwards. Determine the magnitude of the force in kg/metre length of conductor and the direction in which it works. **[9.175 kg/m horizontally]**

3. A conductor, 10 cm in length, moves with a uniform velocity of 2 m/s at right angles to itself and to a uniform magnetic field having a flux density of 1 Wb/m^2. Calculate the induced e.m.f. between the ends of the conductor. **[0.2 V]**

7.8. Statically Induced E.M.F.

It can be further sub-divided into (a) *mutually induced e.m.f.* and (b) *self-induced e.m.f.*

(a) **Mutually-induced e.m.f.** Consider two coils A and B lying close to each other (Fig. 7.9).

Coil A is joined to a battery, a switch and a variable resistance R whereas coil B is connected to a sensitive voltmeter V. When current through A is established by closing the switch, its magnetic field is set up which partly links with or threads through the coil B. As current through A is changed, the flux linked with B is also changed. Hence, mutually induce e.m.f. is produced in B whose magnitude is given by Faraday's Laws (Art. 7.3) and direction by Lenz's Law (Art. 7.5).

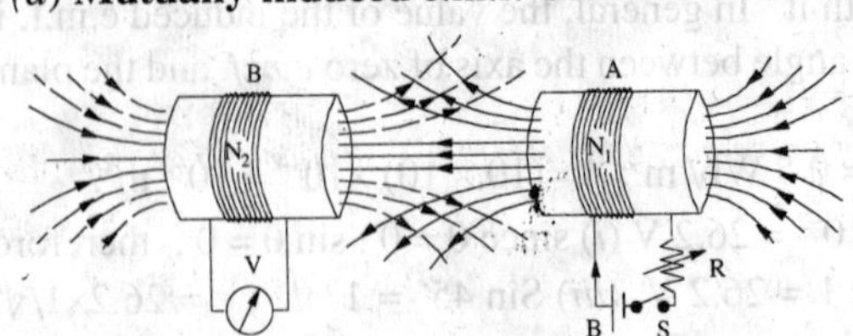

Fig. 7.9

If, now, battery is connected to B and the voltmeter across A (Fig. 7.10), then the situation is reversed and now a change of current in B will produce mutually-induced e.m.f. in A.

It is obvious that in the examples considered above, there is no movement of any conductor, the flux variations being brought about by variations in current strength only. Such an **e.m.f.** induced in one coil by the influence of the other coil is called (statically but) mutually induced e.m.f.

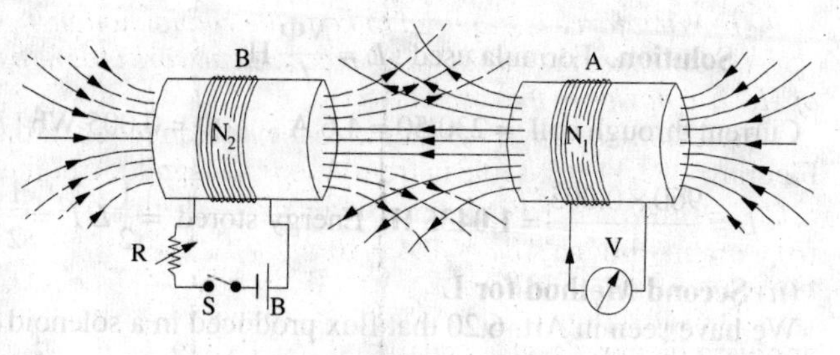

Fig. 7.10

(*b*) **Self-induced e.m.f.** This is the e.m.f. induced in a coil due *to the change of its own flux linked with it.* If current through the coil (Fig. 7.11) is changed, then the flux linked with its own turns will also change, which will produce in it what is called *self-induced* e.m.f. The direction of this induced e.m.f. (as given by Lenz's law) would be such as to oppose any change of flux which is, in fact, the very cause of its production. Hence, it is also known as the opposing or counter e.m.f. of self-induction.

7.9. Self-inductance

Imagine a coil of wire similar to the one shown in Fig. 7.11 connected to a battery through a rheostat. It is found that whenever an effort is made to increase current (and hence flux) through it, it is always opposed by the instantaneous production of counter e.m.f. of self-induction. Energy required to overcome this opposition is supplied by the battery. As will be fully explained later on, this energy is stored in the additional flux produced.

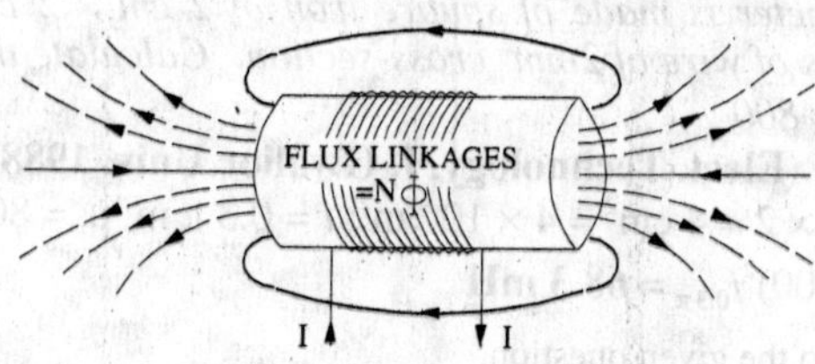

Fig. 7.11

If, now, an effort is made to decrease the current (and hence the flux), then again it is delayed due to the production of self-induced e.m.f., this time in the opposite direction. This property of the coil due to which it opposes any increase or decrease of current or flux through it, is known as *self-inductance*. It is quantitatively measured in terms of coefficient of self induction *L*. This property is analogous to inertia in a material body. We know by experience that initially it is difficult to set a heavy body into motion, but once in motion, it is equally difficult to stop it. Similarly, in a coil having large self-induction, it is initially difficult to establish a current through it, but once established, it is equally difficult to withdraw it. Hence, self-induction is sometimes analogously called *electrical inertia or electromagnetic inertia.*

7.10. Coefficient of Self-induction (L)

It may be defined in any one of the three ways given below :

(*i*) **First Method for L**

The coefficient of self-induction of a coil is defined as

the weber-turns per ampere in the coil.

By 'weber-turns' is meant the product of flux in webers and the number of turns with which the flux is linked. In other words, it is the flux-linkages of the coil.

Consider a solenoid having N turns and carrying a current of I amperes. If the flux produced is Φ webers, the weber-turns are $N\Phi$. Hence, weber-turns per ampere are $N\Phi/I$.

By definition, $L = \dfrac{N\Phi}{I}$. The unit of self-induction is henry*

If in the above relation,

$N\Phi = 1$ Wb-turn, $I = 1$ ampere, then $L = 1$ henry (H)

Hence, a coil is said to have a self-inductance of one henry if a current of 1 ampere when flowing through it produces flux-linkages of 1 Wb-turn in it.

Therefore, the above relation becomes $L = \dfrac{N\Phi}{I}$ henry

Example 7.8. *The field winding of a d.c. electromagnet is wound with 960 turns and has resistance of 50 Ω. When the exciting voltage is 230 V, the magnetic flux linking the coil is 0.005 Wb. Calculate the self-inductance of the coil and the energy stored in the magnetic field.*

*After the American scientist Joseph Henry (1797-1878), à comtemporary of Faraday.

Solution. Formula used : $L = \frac{N\Phi}{I}$ H

Current through coil $= 230/50 = 4.6$ A $\Phi = 0.005$ Wb ; $N = 960$

$L = \frac{960 \times 0.005}{4.6} =$ **1.0435 H.** Energy stored $= \frac{1}{2} L I^2 = \frac{1}{2} \times 1.0435 \times 4.6^2 =$ **11.04 J**

(ii) **Second Method for L**

We have seen in Art. 6.20 that flux produced in a solenoid is

$$\Phi = \frac{NI}{l/\mu_0\mu_r A} \quad \therefore \quad \frac{\Phi}{I} = \frac{N}{l/\mu_0\mu_r A} \quad \text{Now } L = N\frac{\Phi}{I} = N.\frac{N}{l/\mu_0\mu_r A} \text{ H} \quad \therefore L = \frac{N^2}{l/\mu_0\mu_r A} = \frac{N^2}{S} \text{ H}$$

or $$L = \frac{\mu_0\mu_r A N^2}{l} \text{ H}$$

It gives the value of self-induction in terms of the dimensions of the solenoid.*

Example 7.9. *An iron ring 30 cm mean diameter is made of square iron of 2 cm × 2 cm cross-section and is uniformly wound with 400 turns of wire of 2mm² cross-section. Calculate the value of the self-inductance of the coil. Assume $\mu_r = 800$*

(Elect. Technology. I, Gwalior Univ. 1988)

Solution. $L = \mu_0\, \mu_r\, AN^2/l$ Here $N = 400$; $A = 2 \times 2 = 4$ cm$^2 = 4 \times 10^{-4}$ m^2; $l = 0.3\,\pi$ m; $\mu_r = 800$

$\therefore$ $L = 4\pi \times 10^{-7} \times 800 \times 4 \times 10^{-4} \times (400)^2/_{0.3\,\pi} =$ **68.3 mH**

Note : The cross-section of the wire is not relevant to the given question.

(iii) **Third Method for L**

It will be seen from Art. 7.10 *(i)* above that $L = \frac{N\Phi}{I}$ $\therefore$ $N\Phi = L\,I$ or $-N\Phi = -L\,I$

Differentiating both sides, we get $-\frac{d}{dt}(N\Phi) = -L.\frac{dI}{dt}$ (assuming L to be constant);

$$-N.\frac{d\Phi}{dt} = -L.\frac{dI}{dt}$$

As seen from Art. 7.3, $-N.\frac{d\Phi}{dt}$ = self-induced e.m.f. $\therefore e_L = -L\frac{dI}{dt}$

If $\frac{dI}{dt} = 1$ ampere/second and $e_L = 1$ volt, then $L = 1$ H

Hence, a coil has a self-inductance of one henry if one volt is induced in it when current through it changes at the rate of one ampere/second.

Example 7.10. *If a coil of 150 turns is linked with a flux of 0.01 Wb when carrying current of 10 A, calculate the inductance of the coil. If this current is uniformly reversed in 0.01 second, calculate the induced electromotive force.*

Solution. $L = N\Phi/I = 150 \times 0.01/10 =$ **0.15 H**

Now, $e_L = L\, dI/dt$; $dI = -10 - (-10) = 20$ A

$\therefore$ $e_L = 0.15 \times 20/0.01 =$ **300 V**

Example 7.11. *An iron rod, 2 cm in diameter and 20 cm long is bent into a closed ring and is wound with 3000 turns of wire. It is found that when a current of 0.5 A is passed through this coil, the flux density in the coil is 0.5 Wb/m². Assuming that all the flux is linked with every turn of the coil, what is (a) the B/H ratio for the iron (b) the inductance of the coil ? What voltage would be developed across the coil if the current through the coil is interrupted and the flux in the iron falls to 10% of its former value in 0.001 second ?* **(Principles of Elect. Engg. Jadavpur Univ. 1986)**

*In practice, the inductance of a short solenoid is given by $L = K$..

Solution. $H = NI/l = 3000 \times 0.5/0.2 = 7500$ AT/m $B = 0.5$ Wb/m^2

(*a*) Now, $\frac{B}{H} = \frac{0.5}{7500} = \mathbf{6.67 \times 10^{-5} H/m}$ Also $\mu_r = B/\mu_o H = 6.67 \times 10^{-5}/4\pi \times 10^{-7} = 53$

(*b*) $L = \frac{N\Phi}{I} = \frac{3000 \times \pi \times (0.02)^2 \times 0.5}{4 \times 0.5} = \mathbf{0.94\ H}$

$e_L = N\frac{d\Phi}{dt}$ volt ; $d\Phi = 90\%$ of original flux $= \frac{0.9 \times \pi \times (0.02)^2 \times 0.5}{4} = 0.45\,\pi \times 10^{-4}$ Wb

$dt = 0.001$ second $\therefore$ $e_L = 3000 \times 0.45\pi \times 10^{-4}/0.001 = \mathbf{424\ V}$

Example 7.12. *A circuit has 1000 turns enclosing a magnetic circuit 20 cm² in section. With 4 A, the flux density is 1.0 Wb/m² and with 9 A, it is 1.4 Wb/m². Find the mean value of the inductance between these current limits and the induced e.m.f. if the current falls from 9 A to 4 A in 0.05 seconds.*

(Elect. Engineering–1, Delhi Univ. 1987)

Solution. $L = N\frac{d\Phi}{dI} = N\frac{d}{dI}(BA) = NA\frac{dB}{dI}$ henry $= 1000 \times 20 \times 10^{-4}(1.4 - 1)/(9 - 4) = \mathbf{0.16\ H}$

Now, $e_L = L.dI/dt$; $dI = (9 - 4) = 5$A, $dt = 0.05$ s $\therefore e_L = 0.16 \times 5/0.05 = \mathbf{16\ V}$

Example 7.13. *A direct current of one ampere is passed through a coil of 5000 turns and produces a flux of 0.1 mWb. Assuming that whole of this flux threads all the turns, what is the inductance of the coil ? What would be the voltage developed across the coil if the current were interrupted in* 10^{-3} *second ? What would be the maximum voltage developed across the coil if a capacitor of 10 μF were connected across the switch breaking the d.c. supply ?*

Solution. $L = N\Phi/I = 5000 \times 10^{-4}$ **0.5 H** ; Induced e.m.f. $= L.\frac{dI}{dt} = \frac{0.5 \times 1}{10^{-3}} = \mathbf{500\ V}$

The energy stored in the coil is $= \frac{1}{2}L\,I^2 = \frac{1}{2} \times 0.5 \times I^2 = 0.25$ J

When the capacitor is connected, then the voltage developed would be equal to the p.d. developed across the capacitor plates due to the energy stored in the coil. If V is the value of the voltage, then

$\frac{1}{2}CV^2 = \frac{1}{2}LI^2$; $\frac{1}{2} \times 10 \times 10^{-6}\,V^2 = 0.25$ or $V = \mathbf{224\ volt}$

Example 7.14. (*a*) *A coil of 1000 turns is wound on a torroidal magnetic core having a reluctance of* 10^4 *AT/Wb. When the coil current is 5 A and is increasing at the rate of 200 A/s, determine.*

(i) energy stored in the magnetic circuit (ii) voltage applied across the coil

Assume coil resistance as zero.

(b) How are your answers affected if the coil resistance is 2 Ω.

(Elect. Technology, Hyderabad Univ. 1991)

Solution. (*a*) $L = N^2/S = 1000^2/10^6 = 1$ H

(*i*) Energy stored $= \frac{1}{2}LI^2 = \frac{1}{2} \times 1 \times 5^2 = \mathbf{12.5\ J}$

(*ii*) Voltage applied across coil

= self-induced e.m.f. in the coil $= L.dI/dt = 1 \times 200 = \mathbf{200\ V}$

(*b*) Though there would be additional energy loss of $5^2 \times 2 = 50$ W over the coil resistance, energy stored in the coil would remain the same. However, voltage across the coil would increase by an amount $= 5 \times 2 = 10$ V *i.e.* now its value would be **210 V.**

7.11. Mutual Inductance

In Art. 7.8 (Fig. 7.9) we have seen that any change of current in coil A is always accompanied by the production of mutually-induced e.m.f. in coil B. Mutual inductance may, therefore, be defined as the ability of one coil (or circuit) to produce an e.m.f. in a nearby coil by induction when the current in the first coil changes. This action being reciprocal, the second coil can also induce an e.m.f. in the first when current in the second coil changes. This ability of reciprocal induction is measured in terms of the coefficient of mutual induction M.

7.12. Coefficient of Mutual Inductance (M)

It can also be defined in three ways as given below :

(*i*) **First Method for M**

Let there be two magnetically-coupled coils having N_1 and N_2 turns respectively (Fig. 7.9). Coefficient of mutual inductance between the two coils is defined as

the weber-turns in one coil due to one ampere current in the other.

Let a current of I_1 ampere when flowing in the first coil produce a flux Φ_1 webers in it. *It is supposed that whole of this flux links with the turns of the second coil.** Then, flux-linkages *i.e.* weber-turns in the *second* coil for unit current in the *first* coil are $N_2\Phi_1/I_1$. Hence, by definition

$$M = \frac{N_2\Phi_1}{I_1}$$

If weber-turns in *second* coil due to one ampere current in the *first* coil *i.e.* $N_2\Phi_1/I_1 = 1$ then, as seen from above, M = 1 H.

Hence, two coils are said to have a mutual inductance of 1 henry if one ampere current when flowing in one coil produces flux-linkages of one Wb-turn in the other.

Example 7.15. *Two identical coils X and Y of 1,000 turns each lie in parallel planes such that 80% of flux produced by one coil links with the other. If a current of 5 A flowing in X produces a flux of 0.5 mWb in it, find the mutual inductance between X and Y.*

(Elect. Engg. A.M.Ae. S.I. 1989)

Solution. Formula used $M = \frac{N_2\Phi_1}{I_1}$ H ; Flux produced in $X = 0.5$ mWb $= 0.5 \times 10^{-3}$ Wb

Flux linked with $Y = 0.5 \times 10^{-3} \times 0.8 = 0.4 \times 10^{-3}$ Wb ; $M = \frac{1000 \times 0.4 \times 10^{-3}}{5} = \mathbf{0.08\ H}$

Example 7.16. *A long single-layer solenoid has an effective diameter of 10 cm and is wound with 2500 AT/meter. There is a small concentrated coil having its plane lying in the centre cross-sectional plane of the solenoid. Calculate the mutual inductance between the two coils in each case if the concentrated coil has 120 turns on an effective diameter of (a) 8 cm and (b) 12 cm.*

(Elect. Science - II. Allahabad Univ. 1992)

Solution. The two cases (*a*) and (*b*) are shown in Fig. 7.12 (*a*) and (*b*) respectively.

(*a*) Let I_1 be the current flowing through the solenoid. Then

$$B = \mu_0 H = \mu_0 N I_1/l = 2500\ \mu_0 I_1 \text{ Wb/m}^2 \qquad ...l = 1 \text{ m}$$

Area of search coil $A_1 = \frac{\pi}{4} \times 8^2 \times 10^{-4} = 16\pi \times 10^{-4}$ m^2

Flux linked with search coil is

$$\Phi = BA_1 = 2500\ \mu_0 I_1 \times 16\pi \times 10^{-4} = 15.79\ I_1 \times 10^{-6} \text{ Wb}$$

$$\therefore \quad M = \frac{N_2\Phi}{I_1} = \frac{120 \times 15.39 I_1 \times 10^{-6}}{I_1} = \mathbf{1.895 \times 10^{-3}\ H}$$

---

*If whole of this flux does not link with turns of the second coil, then only that part of the flux which is actually linked is taken instead. (Ex. 7.13 and 7.17). In general, $M = N_2\Phi_2/I_1$.

(*b*) Since the field strength outside the solenoid is negligible, the effective area of the search coil, in this case, equals the area of the long solenoid.

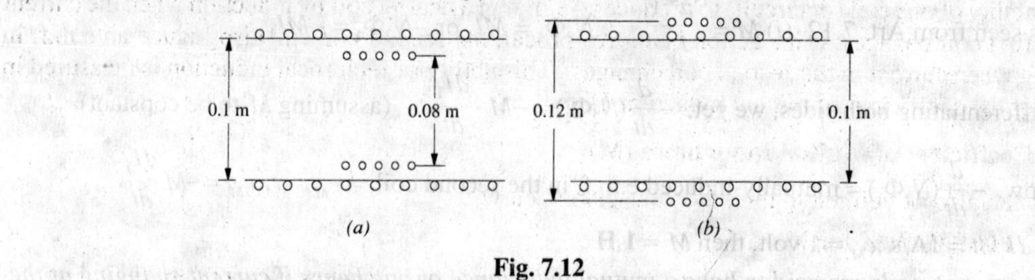

Fig. 7.12

$$A_2 = \frac{\pi}{4} \times 10^2 \times 10^{-4} = \frac{\pi}{4} 10^{-2} \text{m}^2;$$

$$\Phi = BA_2 = 2500\, \mu_0 I_1 \times \frac{\pi}{4} \times 10^{-2} = 24.68\, I_1 \times 10^{-6} \text{ Wb}$$

$$M = \frac{120 \times 24.68\, I_1 \times 10^{-6}}{I_1} = \mathbf{2.962 \times 10^{-3}\ H}$$

Example 7.17. *A flux of 0.5 mWb is produced by a coil of 900 turns wound on a ring with a current of 3 A in it. Calculate (i) the inductance of the coil (ii) the e.m.f. induced in the coil when a current of 5 A is switched off, assuming the current to fall to zero in 1 milli second and (iii) the mutual inductance between the coils, if a second coil of 600 turns is uniformly wound over the first coil.*

(F.E. Pune Univ. May 1987)

Solution. (*i*) Inductance of the first coil $= \frac{N\Phi}{I} = \frac{900 \times 0.5 \times 10^{-3}}{3} = \mathbf{0.15\ H}$

(*ii*) e.m.f. induced $e_1 = L\frac{di}{dt} = 0.15 \times \frac{(5-0)}{1 \times 10^{-3}} = \mathbf{750\ V}$

(*iii*) $M \frac{N_2 \Phi_1}{I_1} = \frac{600 \times 0.5 \times 10^{-3}}{3} = \mathbf{0.1\ H}$

(*ii*) **Second Method for M**

We will now deduce an expression for coefficient of mutual inductance in terms of the dimensions of the two coils.

$$\text{Flux in the first coil } \Phi_1 = \frac{N_1 I_1}{l/\mu_0 \mu_r A} \text{ Wb}; \quad \text{Flux/ampere} = \frac{\Phi_1}{I_1} = \frac{N_1}{l/\mu_0 \mu_r A}$$

Assuming that whole of this flux (it usually is some percentage of it) is linked with the other coil having N_2 turns, the weber-turns in it due to the flux/ampere in the first coil is

$$M = \frac{N_2 \Phi_1}{I_1} = \frac{N_2 . N_1}{l/\mu_0\, \mu_r\, A} \quad \therefore \quad M = \frac{\mu_0 \mu_r\, A N_1\, N_2}{l} \text{ H}$$

Also $$M = \frac{N_1 N_2}{l/\mu_0 \mu_r A} = \frac{N_1 N_2}{\text{reluctance}} = \frac{N_1 N_2}{S} \text{ H}$$

Example 7.18. *If a coil of 150 turns is linked with a flux of 0.01 Wb when carrying a current of 10 A; calculate the inductance of the coil. If this current is uniformly reversed in 0.1 second, calculate the induced e.m.f. If a second coil of 100 turns is uniformly wound over the first coil, find the mutual inductance between the coils.* **(F.E. Pune Univ. May 1989)**

Solution. $L_1 = N_1\Phi_1/I_1 = 150 \times 0.01/10 = \mathbf{0.15\ H}$

$e = L \times di/dt = 0.15 \times [10-(-10)]\ /0.1 = 1 = \mathbf{30\ V}$

$M = N_2\Phi/I_1 = 100 \times 0.01/10 = \mathbf{0.1\ H}$

(iii) **Third Method for M**

As seen from Art. 7.12 *(i)* $M = \dfrac{N_2\Phi_1}{I_1}$ $\quad\therefore N_2\Phi_1 = MI_1$ or $-N_2\Phi_1 = -MI_1$

Differentiating both sides, we get: $-\dfrac{d}{dt}(N_2\Phi_1) = -M \cdot \dfrac{dI_1}{dt}$ (assuming M to be constant)

Now, $-\dfrac{d}{dt}(N_2\Phi_1) =$ mutually-induced e.m.f. in the second coil $= e_M$ $\quad\therefore e_M = -M\dfrac{dI_1}{dt}$

If $dI_1/dt = 1$ A/s ; $e_M = 1$ volt, then $M =$ **1 H**

Hence, two coils are said to have a mutual inductance of one henry if current changing at the rate of 1 ampere/second in one coil induces an e.m.f. of one volt in the other.

Example 7.19. *Two coils having 30 and 600 turns respectively are wound side-by-side on a closed iron circuit of area of cross-section 100 sq. cm. and mean length 200 cm. Estimate the mutual inductance between the coils if the relative permeability of the iron is 2000. If a current of zero ampere grows to 20 A in a time of 0.02 second in the first coil, find the e.m.f. induced in the second coil.* **(Elect. Engg.I, JNT Univ. Warangal 1985)**

Solution. Formula used : $M = \dfrac{N_1N_2}{l/\mu_0\,\mu_r\,A}$ **H**, $N_1 = 30$; $N_2 = 600$; $A = 100 \times 10^{-4} = 10^{-2}$ m^2, $l = 2$ m

$\therefore \quad M = \mu_0\,\mu_r\,A\,N_1N_2/l = 4\pi \times 10^{-7} \times 2000 \times 10^{-2} \times 30 \times 600/2 =$ **0.226 H**

$dI_1 = 20 - 0 = 20$ A ; $dt = 0.02$ s ; $e_M = MdI_1/dt = 0.226 \times 20/0.2 =$ **226 V**

Example 7.20. *Two coils A and B each having 1200 turns are placed near each other. When coil B is open-circuited and coil A carries a current of 5 A, the flux produced by coil A is 0.2 Wb and 30% of this flux links with all the turns of coil B. Determine the voltage induced in coil B on open-circuit when the current in the coil A is changing at the rate of 2 A/s.*

Soluton. Coefficient of mutual induction between the two coils is $M = N_2\Phi_2/I_1$

Flux linked with coil B is 30 per cent of 0.2 Wb *i.e.* 0.06 Wb

$\therefore \quad M = 1200 \times 0.06/5 = 14.4$ H

Mutually-induced e.m.f. in coil B is $e_M = MdI_1/dt = 14.4 \times 2 =$ **28.8 V**

Example 7.21. *Two coils are wound side by side on a paper-tube former. An e.m.f. of 0.25 V is induced in coil A when the flux linking it changes at the rate of 10^3 Wb/s. A current of 2 A in coil B causes a flux of 10^{-5} Wb to link coil A. What is the mutual inductance between the coils ?* **(Elect. Engg-1, Bombay Univ. 1985)**

Solution. Induced e.m.f. in coil A is $e = N_1\dfrac{d\Phi}{dt}$ where N_1 is the number of turns of coil A.

$\therefore \quad 0.25 = N_1 \times 10^{-3} \quad \therefore N_1 = 250$

Now, flux linkages in coil A due to 2 A current in coil $B = 250 \times 10^{-5}$

$\therefore \quad M = \dfrac{\text{flux linkages in coil } A}{\text{current in coil } B} = 250 \times 10^{-5}/2 =$ **1.25 mH**

7.13. Coefficient of Coupling

Consider two magnetically-coupled coils A and B having N_1 and N_2 turns respectively. Their individual coefficients of self-induction are,

$$L_1 = \frac{N_1^{\,2}}{l/\mu_0\,\mu_r\,A} \quad \text{and} \quad L_2 = \frac{N_2^{\,2}}{l/\mu_0\,\mu_r\,A}$$

The flux Φ_1 produced in A due to a current I_1 ampere is $\Phi_1 = \dfrac{N_1I_1}{l/\mu_0\,\mu_r\,A}$

Suppose a fraction k_1 of this flux *i.e.* $k_1\,\Phi_1$ is linked with coil B.

Then $\quad M = \dfrac{k_1\Phi_1 \times N_2}{I_1}$ $\quad$ where $k_1 \le 1$.

Substituting the value of Φ_1, we have, $M = k_1 \times \dfrac{N_1 N_2}{l/\mu_0 \mu_r A}$ ...(*i*)

Similarly, the flux Φ_2 produced in B due to I_2 ampere in it is $\Phi_2 = \dfrac{N_2 I_2}{l/\mu_0 \mu_r A}$

Suppose a fraction k_2 of this flux *i.e.* $k_2\Phi_2$ is linked with A.

Then $M = \dfrac{k_2\Phi_2 \times N_1}{I_2} = k_2 \dfrac{N_1 N_2}{l/\mu_0 \mu_r A}$ ...(*ii*)

Multiplying Eq. (*i*) and (*ii*), we get,

$$M^2 = k_1 k_2 \frac{N_1^{\,2}}{l/\mu_0 \mu_r A} \times \frac{N_2^{\,2}}{l/\mu_0 \mu_r A} \quad \text{or} \quad M^2 = k_1\, k_2\, L_1\, L_2$$

Putting $\sqrt{k_1 k_2} = k$, we have $M = k\sqrt{L_1 L_2}$ or $k = \dfrac{M}{\sqrt{L_1 L_2}}$

The constant k is called the *coefficient of coupling* and may be defined as the ratio *of mutual inductance actually present between the two coils to the maximum possible value.* If the flux due to one coil completely links with the other, then value of k is unity. If the flux of one coil does not at all link with the other, then $k = 0$. In the first case, when $k = 1$, coils are said to be tightly coupled and when $k = 0$, the coils are magnetically isolated from each other.

Example 7.22. *Two identical 750 turn coils A and B lie in parallel planes. A current changing at the rate of 1500 A/s in A induces an e.m.f. of 11.25 V in B. Calculate the mutual inductance of the arrangement. If the self-inductance of each coil is 15 mH, calculate the flux produced in coil A per ampere and the percentage of this flux which links the turns of B.*

Solution. Now, $e_M = M\, dI_1/dt$ ...Art. 7.12

$$M = \frac{e_M}{dI_1/dt} = \frac{11.25}{1500} = 7.5 \times 10^{-3}\ \text{H} = \mathbf{7.5\ mH}$$

Now $L_1 = \dfrac{N_1 \Phi_1}{I_1}$ $\therefore$ $\dfrac{\Phi_1}{I_1} = \dfrac{L_1}{N_1} = \dfrac{15 \times 10^{-3}}{750} = \mathbf{2 \times 10^{-5}\ Wb/A}$ ...Art. 7.10

Now $k = \dfrac{M}{\sqrt{L_1 L_2}} = \dfrac{7.5 \times 10^{-3}}{\sqrt{L^2}} = \dfrac{7.5 \times 10^{-3}}{15 \times 10^{-3}} = \mathbf{0.5 = 50\%}$ $(\because L_1 = L_2 = L)$ ...Art. 7.13

Example 7.23. *Two coils, A of 12,500 turns and B of 16,000 turns, lie in parallel planes so that 60% of flux produced in A links coil B. It is found that a current of 5A in A produces a flux of 0.6 mWb while the same current in B produces 0.8 mWb. Determine (i) mutual inductance and (ii) coupling coefficient.*

Solution. (*i*) Flux/ampere in A $= 0.6/5 = 0.12$ mWb

Flux linked with B $= 0.12 \times 0.6 = 0.072$ mWb

$\therefore$ $M = 0.072 \times 10^{-3} \times 16{,}000 = \mathbf{1.15\ H}$

Now $L_1 = \dfrac{12,500 \times 0.6}{5} = 150 \times 10^{-3}\ \text{H}$; $L_2 = \dfrac{16,000 \times 0.8}{5} = 256 \times 10^{-3}\ \text{H}$

(*ii*) $k = M/\sqrt{L_1 L_2} = 1.15/\sqrt{1.5 \times 2.56} = \mathbf{0.586}$

Note. We could find k in another way also. Value of $k_1 = 0.6$, that of k_2 could also be found, then $k = \sqrt{k_1 k_2}$.

Example 7.24. *Two magnetically-coupled coils have a mutual inductance of 32 mH. What is the average e.m.f. induced in one, if the current through the other changes from 3 to 15 mA in 0.004 second? Given that one coil has twice the number of turns in the other, calculate the inductance of each coil. Neglect leakage.*

Solution. $M = 32 \times 10^{-3}$ H ; $dI_1 = 15 - 3 = 12$ mA $= 12 \times 10^{-3}$ A ; $dt = 0.004$ second

Average e.m.f. induced $= M\dfrac{dI_1}{dt} = \dfrac{32\times10^{-3}\times12\times10^{-3}}{0.004} = \mathbf{96\times10^{-3}\ V}$

Now $\qquad L_1 = \mu_0 N^2A/l = kN^2 \quad \text{where } k = \mu_0 A/l \qquad$ (taking $\mu_r = 1$)

$$L_2 = \frac{(2N)^2\mu_0 A}{2\,l} = 2kN^2\,;\ \frac{L_2}{L_1} = \frac{2kN^2}{kN^2} = 2 \quad \therefore \quad L_2 = 2L_1$$

Now $M = \sqrt{L_1L_2} = \sqrt{2L_1\times L_1} = 32, L_1 = 32/\sqrt{2} = \mathbf{16\sqrt{2}\,mH}\,;\ L_2 = 2\times16\sqrt{2} = \mathbf{32\sqrt{2}\,mH.}$

Example 7.25. *Two coils, A and B, have self inductances of 120 μ H and 300 μ H respectively. A current of 1 A through coil A produces flux linkages of 100 μWb turns in coil B. Calculate (i) the mutual inductance between the coils (ii) the coupling coefficient and (iii) the average e.m.f. induced in coil B if a current of 1 A in coil A is reversed at a uniform rate in 0.1 sec.*

(F.E. Pune Univ. Nov. 1989)

Solution. (*i*) $M = \dfrac{\text{flux-linkages of coil } B}{\text{current in coil } A} = \dfrac{100\times10^{-6}}{1} = \mathbf{100\ \mu\ H}$

(*ii*) $\qquad M = k\sqrt{L_1L_2} \quad \therefore k = \dfrac{M}{\sqrt{L_1\,L_2}} = \dfrac{100\times10^{-6}}{\sqrt{120\times10^{-6}\times300\times10^{-6}}} = \mathbf{0.527}$

(*iii*) $\qquad e_2 = M\times di/dt = (100\times10^{-6})\times2/0.1 = 0.002\text{V}$ or $\mathbf{2\ mV.}$

7.14. Inductances in Series

(*i*) Let the two coils be so joined in series that their fluxes (or m.m.fs) are additive *i.e.* in the same direction (Fig. 7.13).

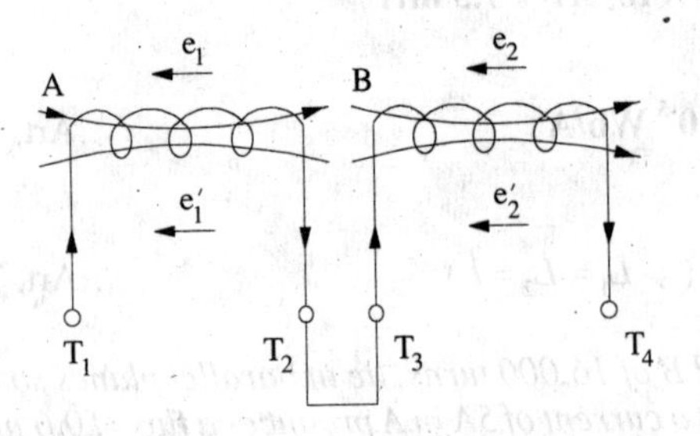

Fig. 7.13

Let M = coefficient of mutual inductance

L_1 = coefficient of self-inductance of 1st coil

L_2 = coefficient of self-inductance of 2nd coil

Then, self-induced e.m.f in A is $= e_1 = -L_1.\dfrac{di}{dt}$

Mutually-induced e.m.f. in A due to change of current in B is $= e_1' = -M.\dfrac{di}{dt}$

Self-induced e.m.f. in B is $= e_2 = -L_2.\dfrac{di}{dt}$

Mutually-induced e.m.f. in B due to change of current in A is $= e_2' = -M.\dfrac{di}{dt}$

(All have −ve sign, because both self and mutually-induced e.m.fs. are in opposition to the applied e.m.f.). Total induced e.m.f. in the combination $= -\dfrac{di}{dt}(L_1 + L_2 + 2M)$ ...(*i*)

If L is the equivalent inductance then total induced e.m.f. in that single coil would have been

$$= -L\frac{di}{dt} \qquad ...(ii)$$

Equating (*i*) and (*ii*) above, we have $L = L_1 + L_2 + 2M$

(*ii*) When the coils are so joined that their fluxes are in opposite directions (Fig. 7.14).

As before $e_1 = -L_1 \frac{di}{dt}$

$e_1' = +M.\frac{di}{dt}$ (mark this direction)

$e_2 = -L_2\frac{di}{dt}$ and $e_2' = +M.\frac{di}{dt}$

Total induced e.m.f. $= -\frac{di}{dt}(L_1 + L_2 - 2M)$

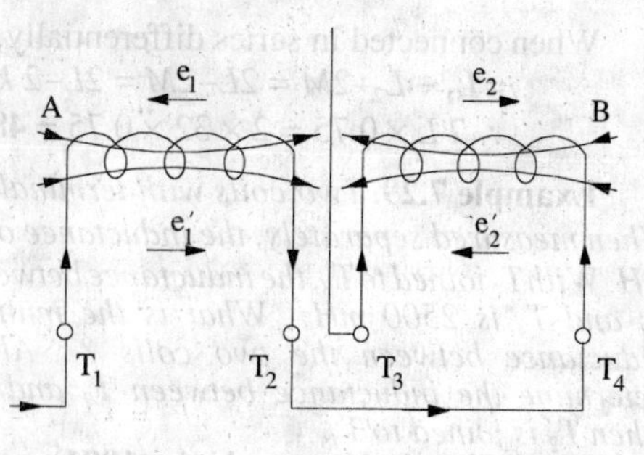

Fig 7.14

∴ Equivalent inductance

$L = L_1 + L_2 - 2M$

In general, we have : $L = L_1 + L_2 + 2M$ ... if m.m.fs are additive
and $L = L_1 + L_2 - 2M$ ... if m.m.fs. are subtractive

Example 7.26. *Two coils with a coefficient of coupling of 0.5 between them, are connected in series so as to magnetise (a) in the same direction (b) in the opposite direction. The corresponding values of total inductances are for (a) 1.9 H and for (b) 0.7 H. Find the self-inductances of the two coils and the mutual inductance between them.*

Solution. (*a*) $L = L_1 + L_2 + 2M$ or $1.9 = L_1 + L_2 + 2M$ ...(*i*)

(*b*) Here $L = L_1 + L_2 - 2M$ or $0.7 = L_1 + L_2 - 2M$ ...(*ii*)

Substracting (*ii*) from (*i*), we get

$1.2 = 4\,M$ ∴ $M = 0.3$ H

Putting this value in (*i*) above, we get $L_1 + L_2 = 1.3$ H ...(*iii*)

We know that, in general, $M = k\sqrt{L_1L_2}$

∴ $\sqrt{L_1L_2} = \frac{M}{k} = \frac{0.3}{0.5} = 0.6$ ∴ $L_1L_2 = \mathbf{0.36}$

From (*iii*), we get $(L_1 + L_2)^2 - 4L_1L_2 = (L_1 - L_2)^2$

∴ $(L_1 - L_2)^2 = 0.25$ or $L_1 - L_2 = 0.5$ ...(*iv*)

From (*iii*) and (*iv*), we get $L_1 = \mathbf{0.9\ H}$ and $L_2 = \mathbf{0.4\ H}$

Example 7.27. *The combined inductance of two coils connected in series is 0.6 H or 0.1 H depending on the relative directions of the currents in the coils. If one of the coils when isolated has a self-inductance of 0.2 H, calculate (a) mutual inductance and (b) coupling coefficient.*

(Elect. Technology, Univ. of Indore, 1987)

Solution. (*i*) $L = L_1 + L_2 + 2M$ or $0.6 = L_1 + L_2 + 2M$ ...(*i*)

and $0.1 = L_1 + L_2 - 2M$ ...(*ii*)

(*a*) From (*i*) and (*ii*) we get, $M = \mathbf{0.125\ H}$

Let $L_1 = 0.2$ H, then substituting this value in (*i*) above, we get $L_2 = 0.15$ H

(*b*) Coupling coefficient $k = M/\sqrt{L_1L_2} = 0.125/\sqrt{0.2 \times 0.15} = \mathbf{0.72}$

Example 7.28. *Two similar coils have a coupling coefficient of 0.25. When they are connected in series cumulatively, the total inductance is 80 mH. Calculate the self inductance of each coil. Also calculate the total inductance when the coils are connected in series differentially.*

(F.E. Pune Univ. 1988)

Solution. If each coil has an inductance of L henry, then $L_1 = L_2 = L$; $M = k\sqrt{L_1L_2} = k\sqrt{L \times L} = kL$.

When connected in series cumulatively, the total inductance of the coils is

$= L_1 + L_2 + 2M = 2L + 2M = 2L + 2kL = 2L\,(1 + 0.25) = 2.5L$

∴ $2.5\,L = 80$ or $L = \mathbf{32\ mH}$

When connected in series differentially, the total inductance of the coils is

$= L_1 + L_2 - 2M = 2L - 2M = 2L - 2\,kL = 2L\,(1-k) = 2L\,(1-0.25)$

$\therefore\ 2\,L \times 0.75 = 2 \times 32 \times 0.75 =$ **48 mH.**

Example 7.29. *Two coils with terminals T_1, T_2 and T_3, T_4 respectively are placed side by side. When measured separately, the inductance of the first coil is 1200 mH and that of the second is 800 mH. With T_2 joined to T_3, the inductance between T_1 and T_4 is 2500 mH. What is the mutual inductance between the two coils ? Also, determine the inductance between T_1 and T_3 when T_2 is joined to T_4.*

(Electrical Circuit, Nagpur Univ. 1991)

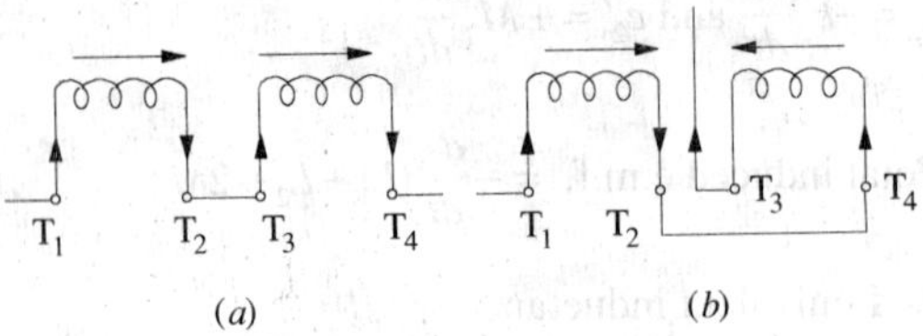

T1 T2 T3 T4 T1 T2 T3 T4

(a) (b)

Fig. 7.15

Solution. $L_1 = 1200$ mH, $L_2 = 800$ mH

Fig. 7.15 (*a*) shows additive series.

$\therefore\quad L = L_1 + L_2 + 2M$

or $\quad 2500 = 1200 + 800 + 2M$; $M =$ **250 mH**

Fig. 7.15 (*b*) shows the case of subtractive or opposing series.

Here, $\quad L = L_1 + L_2 - 2M = 1200 + 800 - 2 \times 250 =$ **1500 mH**

Example 7.30. *The total inductance of two coils, A and B, when connected in series, is 0.5 H or 0.2 H, depending on the relative directions of the current in the coils. Coil A, when isolated from coil B, has a self-inductance of 0.2 H. Calculate*

(a) the mutual inductance between the two coils (b) the self-inductance of coil B

(c) the coupling factor between the coils.

(d) the two possible values of the induced e.m.f in coil A when the current is decreasing at 1000 A per second in the series circuit.

(Elect. Technology, Hyderabad Univ. 1992)

Solution. (*a*) Combined inductance is given by $L = L_1 + L_2 \pm 2M$

$\therefore\quad 0.5 = L_1 + L_2 + 2M$ ...(*i*), $0.2 = L_1 + L_2 - 2M$ ...(*ii*)

Subtracting (*ii*) from (*i*), we have $4M = 0.3$ or $M =$ **0.075 H**

(*b*) Adding (*i*) and (*ii*) we have $\quad 0.7 = 2 \times 0.2 + 2L_2$; $L_2 =$ **0.15 H**

(*c*) Coupling factor or coefficient is $k = M/\sqrt{L_1 L_2} = 0.075/\sqrt{0.2 \times 0.15} = 0.433$ or **43.3%**

(*d*) $\quad e_1 = L_1 \dfrac{di}{dt} \pm M \dfrac{di}{dt}$

$\therefore\quad e_1 = (0.2 + 0.075) \times 1000 =$ **275 V** $\quad$ ...'cumulative connection'

$\quad = (0.2 - 0.075) \times 1000 =$ **125 V** $\quad$ ...'differential connection'

7.15. Inductances in Parallel

In Fig. 7.16, two inductances of values L_1 and L_2 henry are connected in parallel. Let the coefficient of mutual inductance between the two be M. Let i be the main supply current and i_1 and i_2 be the branch currents

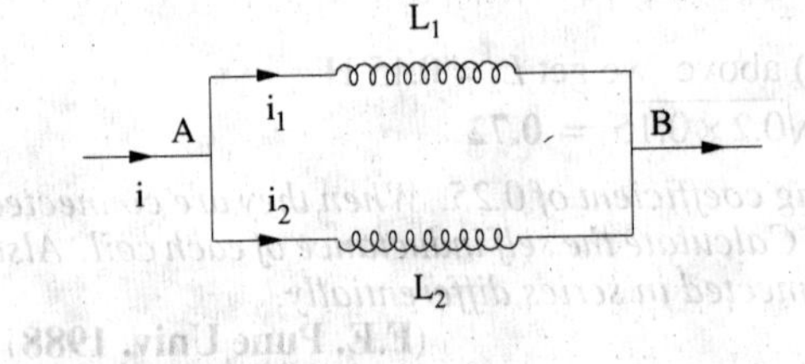

Fig. 7.16

Obviously, $\quad i = i_1 + i_2$

$$\therefore\quad \frac{di}{dt} = \frac{di_1}{dt} + \frac{di_2}{dt} \qquad \ldots(i)$$

In each coil, both self and mutually induced e.m.fs. are produced. Since the coils are in parallel, these e.m.fs. are equal. For a case when self-induced e.m.f. assists the mutually-induced e.m.f., we get

$$e = L_1 \frac{di_1}{dt} + M \frac{di_2}{dt} = L_2 \frac{di_2}{dt} + M \frac{di_1}{dt} \qquad \therefore\ L_1 \frac{di_1}{dt} + M \frac{di_2}{dt} = L_2 \frac{di_2}{dt} + M \frac{di_1}{dt}$$

or $\frac{di_1}{dt}(L_1-M)=\frac{di_2}{dt}(L_2-M)\;\therefore\;\frac{di_1}{dt}=\left(\frac{L_2-M}{L_1-M}\right)\frac{di_2}{dt}$ ...(ii)

Hence, (i) above becomes $\frac{di}{dt}=\left[\left(\frac{L_2-M}{L_1-M}\right)+1\right]\frac{di_2}{dt}$ ...(iii)

If L is the equivalent inductance, then $e=L.\frac{di}{dt}$ = induced e.m.f. in the parallel combination

= induced e.m.f. in any one coil $=L_1.\frac{di_1}{dt}+M\frac{di_2}{dt}$

$$\therefore\quad \frac{di}{dt}=\frac{1}{L}\left(L_1\frac{di_1}{dt}+M\frac{di_2}{dt}\right)\quad ...(iv)$$

Substituting the value of $d\,i_1/dt$ from (ii) in (iv), we get $\frac{di}{dt}=\frac{1}{L}\left[L_1\left(\frac{L_2-M}{L_1-M}\right)+M\right]\frac{di_2}{dt}$ ...(v)

Hence, equating (iii) to (iv), we have $\frac{L_2-M}{L_1-M}+1=\frac{1}{L}\left[L_1\left(\frac{L_2-M}{L_1-M}\right)+M\right]$ or $\frac{L_1+L_2-2M}{L_1-M}=\frac{1}{L}\left(\frac{L_1L_2-M^2}{L_1-M}\right)$

$$\therefore\quad L=\frac{L_1L_2-M^2}{L_1+L_2-2M}$$ when mutual field assists the separate fields.

Similarly $L=\frac{L_1L_2-M^2}{L_1+L_2+2M}$ when the two fields oppose each other.

Example 7.31. *Two coils of inductances 4 and 6 henry are connected in parallel. If their mutual inductance is 3 henry, calculate the equivalent inductance of the combination if (i) mutual inductance assists the self-inductance (ii) mutual inductance opposes the self-inductance.*

Solution. (i) $L=\frac{L_1L_2-M^2}{L_1+L_2-2M}=\frac{4\times6-3^2}{4+6-2\times3}=\frac{15}{4}=$ **3.75 H**

(ii) $L=\frac{L_1L_2-M^2}{L_1+L_2+2M}=\frac{24-9}{16}=\frac{15}{16}=$ **0.94 H (approx.)**

Tutorial Problems No. 7.2

1. Two coils are wound close together on the same paxolin tube. Current is passed through the first coil and is varied at a uniform rate of 500 mA per second, inducing an e.m.f. of 0.1 V in the second coil. The second coil has 100 turns. Calculate the number of turns in the first coil if its inductance is 0.4 H. **[200 turns]**

2. Two coils having 50 and 500 turns respectively are wound side by side on a closed iron circuit of section 50 cm^2 and mean length 120 cm. Estimate the mutual inductance between the coils if the permeability of iron is 1000. Also, find the self-inductance of each coil. If the current in one coil grows steadily from zero to 5A in 0.01 second, find the e.m.f. induced in the other coil.

[M = 0.131 H, L_1 = 0.0131, H, L_2 = 1.31 H, E = 65.4 V]

3. An iron-cored choke is designed to have an inductance of 20 H when operating at a flux density of 1 Wb/m^2, the corresponding relative permeability of iron core is 4000. Determine the number of turns in the winding, given that the magnetic flux path has a mean length of 22 cm in the iron core and of 1 mm in air-gap that its cross-section is 10 cm^2. Neglect leakage and fringing. **[4100]**

4. A non-magnetic ring having a mean diameter of 30 cm and a cross-sectional area of 4 cm^2 is uniformly wound with two coils *A* and *B*, one over the other. *A* has 90 turns and *B* has 240 turns. Calculate from first principles the mutual inductance between the coils.

Also, calculate the e.m.f. induced in *B* when a current of 6 A in *A* is reversed in 0.02 second.

[11.52 μ H, 6.912 mV]

5. Two coils *A* and *B*, of 600 and 100 turns respectively are wound uniformly around a wooden ring having a mean circumference of 30 cm. The cross-sectional area of the ring is 4 cm^2. Calculate (*a*) the mutual inductance of the coils and (*b*) the e.m.f. induced in coil *B* when a current of 2 *A* in coil *A* is reversed in 0.01 second.

[(*a*) 100.5 μ H(*b*) 40.2 mV]

6. A coil consists of 1,000 turns of wire uniformly wound on a non-magnetic ring of mean diameter 40 cm and cross-sectional area 20 cm^2.

Calculate (*a*) the inductance of the coil (*b*) the energy stored in the magnetic field when the coil is carrying a current of 15 A (*c*) the e.m.f. induced in the coil if this current is completely interrupted in 0.01 second. **[(*a*) 2 mH (*b*) 0.225 joule (*c*) 3V]**

7. A coil of 50 turns having a mean diameter of 3 cm is placed co-axially at the centre of a solenoid 60 cm long, wound with 2,500 turns and carrying a current of 2 A. Determine mutual inductance of the arrangement. **[0.185 mH]**

8. A coil having a resistance of 2 Ω and an inductance of 0.5 H has a current passed through it which varies in the following manner; (*a*) a uniform change from zero to 50 A in 1 second (*b*) constant at 50 A for 1 second (*c*) a uniform change from 50 A to zero in 2 seconds. Plot the current graph to a time base. Tabulate the p.d. applied to the coil during each of the above periods and plot the graph of p.d. to a time base. **[(*a*) 25 to 125 V (*b*) 100 V (*c*) 87.5 V to – 12.5 V]**

9. A primary coil having an inductance of 100 μH is connected in series with a secondary coil of 240 μH and the total inductance of the combination is measured as 146 μH. Determine the coefficient of coupling. **[62.6%]** (*Circuit Theory, Jadavpur Univ. 1987*)

OBJECTIVE TESTS – 7

1. With the switch *S* open in Fig. 7.17 as the magnet is moved to and fro
(*a*) current reverses through the galvanometer
(*b*) energy is needed to move the magnet toward or away from the coil
(*c*) magnet is repelled as it approaches the coil
(*d*) galvanometer needle does not move.

2. According to Faraday's Laws of Electromagnetic Induction, an e.m.f. is induced in a conductor whenever it
(*a*) lies in a magnetic field
(*b*) cuts magnetic flux
(*c*) moves parallel to the direction of the magnetic field
(*d*) lies perpendicular to the magnetic flux

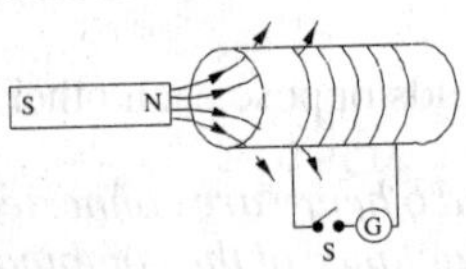

Fig. 7.17

3. The magnitude of the induced e.m.f. in a conductor depends on the
(*a*) amount of flux cut
(*b*) amount of flux-linkages
(*c*) rate of change of flux-linkages
(*d*) flux density of the magnetic field

4. The direction of induced e.m.f. can be found with the help of
(*a*) Lenz's law
(*b*) Fleming's right-hand rule
(*c*) Kirchhoff's voltage law
(*d*) Laplace's law

5. If a current of 5A flowing in a coil of inductance 0.1 H is reversed in 10 ms, e.m.f. induced in it is —— volt.
(*a*) 100 (*b*) 50
(*c*) 1 (*d*) 10,000

6. Higher the self-inductance of a coil,
(*a*) lower the e.m.f. induced in it
(*b*) longer the delay in establishing steady current through it
(*c*) greater the flux produced by it
(*d*) lesser its weber-turns

7. Mutual inductance between two magnetically-coupled coils depends on
(*a*) the number of their turns
(*b*) permeability of the core
(*c*) cross-sectional area of their common core
(*d*) all of the above.

8. Both the number of turns and the core length of an inductive coil are doubled. Its self-inductance will be
(*a*) doubled (*b*) quadrupled
(*c*) halved (*d*) unaffected

9. Two coils having self-inductances of 0.6 H and 0.4 H and a mutual inductance of 0.2 H are connected in series. What is their combined self-inductance ?
(*a*) 1.4 H (*b*) 0.6 H
(*c*) 1.2 H (*d*) either (*a*) or (*b*)

10. Two similar coils have a coupling coefficient of 0.25 and a mutual inductance of 0.9 H. The self-inductance of each coil is —— henry.
(*a*) 0.4 (*b*) 0.6
(*c*) 0.2 (*d*) 0.36

Answers

1. *d* **2.** *b* **3.** *c* **4.** *a* **5.** *a* **6.** *b* **7.** *d* **8.** *a* **9.** *d* **10.** *b*

8 MAGNETIC HYSTERESIS

8.1. Magnetic Hysteresis

It may be defined as the lagging of magnetisation or induction flux density (B) behind the magnetising force (H). Alternatively, it may be defined as that quality of a magnetic substance, due to which energy is dissipated in it, on the reversal of its magnetism.

Let us take an unmagnetised bar of iron AB and magnetise it by placing it within the field of a solenoid (Fig. 8.1). The field H ($= NI/l$) produced by the solenoid is called the magnetising force. The value of H can be increased or decreased by increasing or decreasing current through the coil. Let H be increased in steps from zero up to a certain maximum value and the corresponding values of flux density (B) be noted. If we plot the relation between H and B, a curve like OA, as shown in Fig. 8.2, is obtained. The material becomes magnetically saturated for $H = OM$ and has at that time

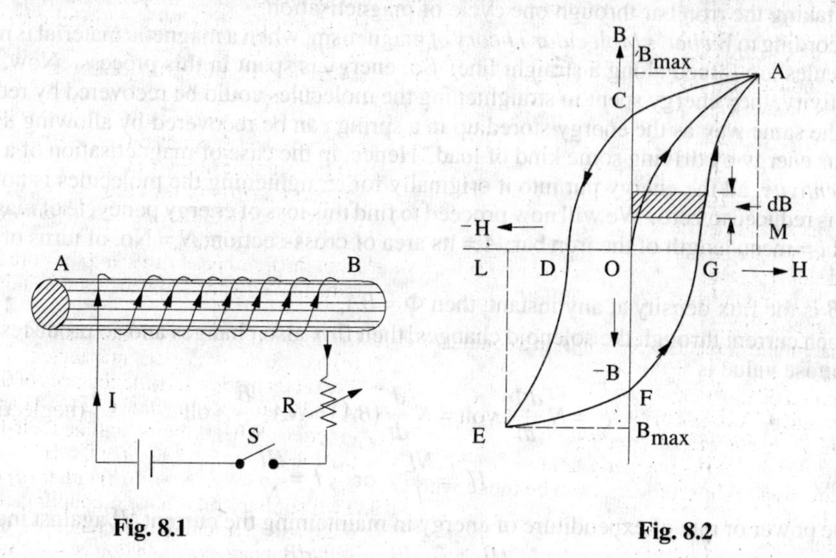

Fig. 8.1

Fig. 8.2

a maximum flux density of B_{max} established through it.

If H is now decreased gradually (by decreasing solenoid current), flux density B will not decrease along AO, as might be expected, but will decrease less rapidly along AC. When H is zero, B is not but has a definite value $B_r = OC$. It means that on removing the magnetising force H, the iron bar is not completely demagnetised. This value of $B(= OC)$ measures the *retentivity or remanence* of the material and is called the remanent or residual flux density B_r.

To demagnetise the iron bar, we have to apply the magnetising force in the reverse direction. When H is reversed (by reversing current through the solenoid), then B is reduced to zero at point D where $H = OD$. This value of H required to wipe off residual magnetism is known as *coercive force* (H_c) and is a measure of the *coercivity* of the material *i.e.* its 'tenacity' with which it holds on to its magnetism.

If, after the magnetisation has been reduced to zero, value of H is further increased in the 'negative' *i.e.* reversed direction, the iron bar again reaches a state of magnetic saturation, represented by point L. By taking H back from its value corresponding to negative saturation, (= OL) to its value for positive saturation (= OM), a similar curve $EFGA$ is obtained. If we again start from G, the same curve $GACDEFG$ is obtained once again.*

It is seen that B always lags behind H. The two never attain zero value simultaneously. This lagging of B behind H is given the name 'hystereis' which literally means '*to lag behind*' The closed loop $ACDEFGA$ which is obtained when iron bar is taken through one complete cycle of magnetisation is known as 'hypothesis loop'.

By one cycle of magnetisation of a magnetic material is meant its being carried through one reversal of magnetisation, as shown in Fig. 8.3.

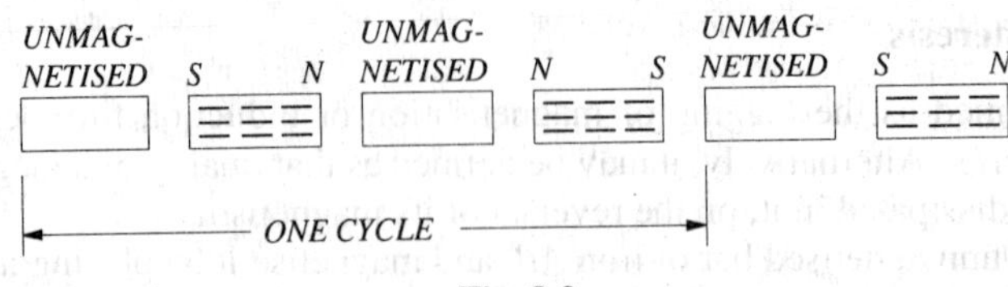

Fig. 8.3

8.2. Area of Hysteresis Loop

Just as the area of an indicator diagram measures the energy made available in a machine, when taken through one cycle of operation, so also the area of the hysteresis loop represents the net energy spent in taking the iron bar through one cycle of magnetisation.

According to *Weber's Molecular Theory of* magnetism, when a magnetic material is magnetised, its molecules are forced along a straight line. So, energy is spent in this process. Now, if iron had no retentivity, then energy spent in straightening the molecules could be recovered by reducing H to zero in the same way as the energy stored up in a spring can be recovered by allowing the spring to release its energy by driving some kind of load. Hence, in the case of magnetisation of a material of *high retentivity*, all the energy put into it originally for straightening the molecules is not recovered when H is reduced to zero. We will now proceed to find this loss of energy per cycle of magnetisation.

Let l = mean length of the iron bar; A = its area of cross-section; N = No. of turns of wire of the solenoid.

If B is the flux density at any instant, then $\Phi = BA$.

When current through the solenoid changes, then flux also changes and so produces an induced e.m.f. whose value is

$$e = N\frac{d\Phi}{dt} \text{ volt} = N\frac{d}{dt}(BA) = NA\frac{dB}{dt} \text{ volt} \qquad \text{(neglecting-ve sign)}$$

Now $$H = \frac{NI}{l} \quad \text{or} \quad I = \frac{Hl}{N}$$

The power or rate of expenditure of energy in maintaining the current 'I' against induced e.m.f. 'e' is

$$= eI \text{ watt} = \frac{Hl}{N} \times NA\frac{dB}{dt} = AlH\frac{dB}{dt} \text{ watt}$$

Energy spent in time 'dt' $= Al.H\frac{dB}{dt} \times dt = Al.H.dB$ joule

Total net work done for one cycle of magnetisation is $W = al \oint H\, dB$ joule

where $\oint$ stands for integration over the whole cycle. Now, '$H\,dB$' represents the shaded area in Fig. 8.2. Hence, $\oint HdB$ = area of the loop *i.e.* the area between the B/H curve and the B-axis.

*In fact, when H is varied a number of times between fixed positive and negative maxima, the size of the loop becomes smaller and smaller till the material is cyclically magnetised. A material is said to be cyclically magnetised when for each increasing (or decreasing) value of H, B has the same value in successive cycles.

$\therefore$ work done/cycle = $Al \times$ (area of the loop) joule. Now Al = volume of the material

$\therefore$ net work done/cycle/m^3 = (loop area) joule, or W_h = (Area of B/H loop) joule m^3/cycle

Precaution

Scale factors of B and H should be taken into consideration while calculating the actual loop area. For example, if the scales are, 1 cm = x AT/m –for H and 1 cm = y Wb/m^2 –for B then

$$W_h = xy \text{ (area of } B/H \text{ loop) joule/m}^3\text{/cycle}$$

In the above expression, loop area has to be in cm^2.

As seen from above, hysteresis loop measures the energy dissipated due to hysteresis which appears in the form of heat and so raises the temperature of that portion of the magnetic circuit which is subjected to magnetic reversals. The shape of the hysteresis loop depends on the nature of the magnetic material (Fig. 8.4).

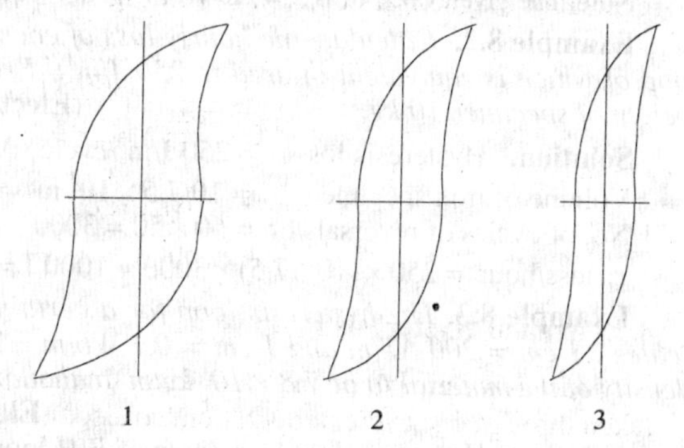

Fig. 8.4

Loop 1 is for hard steel. Due to its high retentivity and collectivity, it is well suited for making permanent magnets. But due to large hysteresis loss (as shown by large loop area) it is not suitable for rapid reversals of magnetisation. Certain alloys of aluminium, nickel and steel called Alnico alloys have been found extremely suitable for making permanent magnets.

Loop 2 is for wrought iron and cast steel. It shows that these materials have high permeability and fairly good coercivity, hence making them suitable for cores of electromagnets.

Loop 3 is for alloyed sheet steel and it shows high permeability and low hysteresis loss. Hence, such materials are most suited for making armature and transformer cores which are subjected to rapid reversals of magnetisation.

8.3. Properties and Applications of Ferromagnetic Materials

Ferromagnetic materials having low retentivities are widely used in power and communication apparatus. Since silicon iron has high permeability and saturation flux density, it is extensively used in the magnetic circuits of electrical machines and heavy current apparatus where a high flux density is desirable in order to limit the cross-sectional area and, therefore, the weight and cost. Thin silicon-iron laminations (clamped together but insulated from each other by varnish, paper or their own surface scale) are used in the construction of transformer and armature cores where it is essential to minimize hysteresis and eddy-current losses.

In field systems (where flux remains constant), a little residual magnetism is desirable. For such systems, high permeability and high saturation flux density are the only important requirements which are adequately met by fabricated rolled steel or cast or forged steel.

Frequencies used in line communication extend up to 10 MHz whereas those used in radio vary from about 100 kHz to 10 GHz. Hence, such material which have high permeability and low losses are very desirable. For these applications, nickel-iron alloys containing up to 80 per cent of nickel and a small percentage of molybdenum or copper, cold rolled and annealed are very suitable.

8.4. Permanent Magnet Materials

Permanent magnets find wide application in electrical measuring instruments, magnetos, magnetic chucks and moving-coil loudspeakers etc. In permanent magnets, high retentivity as well as high coercivity are most desirable in order to resist demagnetisation. In fact, the product B_rH_c is the best criterion for the merit of a permanent magnet. The material commonly used for such purposes are carbon-free iron-nickel-aluminium copper-cobalt alloys which are made anisotropic by heating to a very high temperature and then cooling in a strong magnetic field. This alloy possesses $B_r H_c$ value of about 40,000 J/m^3 as compared with 2,500 J/m^3 for chromium-steel.

Example 8.1. *The hysteresis loop of a sample of sheet steel subjected to a maximum flux density of 1.3 Wb/m^2 has an area of 93 cm^2, the scales being 1 cm = 0.1 Wb/m^2 and 1 cm = 50 AT/m. Calculate the hysteresis loss in watts when 1500 cm^3 of the same material is subjected to an alternating flux density of 1.3 Wb/m^2 peak value of a frequency of 65 Hz.*

(Electromechanics, Allahabad Univ, 1992)

Solution. Loss $= xy$ (area of B/H loop) J/m^3/cycle
$= 0.1 \times 50 \times 93 = 465$ J/m^3/cycle

Volume $= 1500$ cm^3 $= 15 \times 10^{-4}$ m^3; No. of reversals/second = 65

$\therefore$ $W_h = 465 \times 15 \times 10^{-4} \times 65$ J/s $=$ **45.3 W**

Note. The given value of $B_{max} = 1.3$ Wb/m^2 is not required for solution.

Example 8.2. *Calculate the hourly loss of energy in kWh in a specimen of iron, the hystersis loop of which is equivalent in area to 250 J/m^3. Frequency 50 Hz ; specific gravity of iron 7.5 ; weight of specimen 10 kg.*

(Electrical Engg. Materials, Nagpur Univ. 1991)

Solution. Hysteresis loss $= 250$ J/m^3/cycle, Mass of iron = 10 kg

Volume of iron specimen $= 10/7.5 \times 10^3$ m^3 $= 10^{-2}/7.5$ m^3

No. of cycles of reversals/hr $= 60 \times 50 = 3000$

$\therefore$ loss/hour $= 250 \times (10^{-2}/7.5) \times 3000 = 1000$ J $= 1000/36 \times 10^5 =$ **27.8 × 10^{-5} kWh**

Example 8.3. *The hysteresis loop for a certain magnetic material is drawn to the following scales : 1 cm = 200 AT/m and 1 cm = 0.1 Wb/m^2. The area of the loop is 48 cm^2. Assuming the density of the material to be 7.8 × 10^3 kg/m^3, calculate the hysteresis loss in watt/kg at 50 Hz.*

(Elect. Circuits & Fields, Gujarat Univ. 1985)

Solution. Hysteresis loss $= xy$ (area of B/H loop) J/m^3/cycle

Now, 1 cm = 200 AT/m ; 1 cm = 0.1 Wb/m^2

$\therefore$ $x = 200$, $y = 0.1$, area of loop = 48 cm^2

$\therefore$ loss $= 200 \times 0.1 \times 48 = 960$ J/m^3/cycles, Density $= 7.8 \times 10^2$ kg/m^2

Volume of 1 kg of material = mass/density ' $1/7.8 \times 10^3$ m^3

$\therefore$ loss $= 960 \times 1/7.8 \times 10^3$ J/cycle No. of reversals/second = 50

$\therefore$ loss $= 960 \times 50 \times 10^{-3}/7.8 = 6.15$ J/s or watt

$\therefore$ hysteresis loss= **6.15 watt/kg.**

Example 8.4. *Determine the hysteresis loss in an iron core weighing 50 kg having a density of 7.8 × 10 kg/m^3 when the area of the hysteresis loop is 150 cm^2, frequency is 50 Hz and scales on X and Y axes are : 1 cm = 30 AT/cm and 1 cm = 0.2 Wb/m^2 respectively.*

(Elements of Elect. Engg-1, Bangalore Univ. 1987)

Solution. Hysteresis loss $= xy$ (area of B/H loop) J/m^3/cycle

1 cm = 30 AT/cm = 3000 AT/m ; 1 cm = 0.2 Wb/m^2

$x = 3000$, $y = 0.2$, $A = 150$ cm^2

$\therefore$ loss $= 3000 \times 0.2 \times 150 = 90{,}000$ J/m^3/cycle

Volume of 50 kg of iron $= m/\rho = 50/7.8 \times 10^{-3} = 6.4 \times 10^{-3}$ m^3

$\therefore$ loss $= 90{,}000 \times 6.4 \times 10^{-3} \times 50 = 28{,}800$ J/s or watts = **28.8 kW.**

Example 8.5. *In a transformer core of volume 0.16 m^3, the total iron loss was found to be 2,170 W at 50 Hz. The hysteresis loop of the core material, taken to the same maximum flux density, had an area of 9.0 cm^2 when drawn to scales of 1 cm = 0.1 Wb/m^2 and 1 cm = 250 AT/m. Calculate the total iron loss in the transformer core if it is energised to the same maximum flux density but at a frequency of 60 Hz.*

Solution. $W_h = xy \times$ (area of hysteresis loop) where x and y are the scale factors.

$W_h = 9 \times 0.1 \times 250 = 225$ J/m^3/cycle

At 50 Hz

Hysteresis loss $= 225 \times 0.16 \times 50 = 1{,}800$ W; Eddy-current loss $= 2{,}170 - 1{,}800 = 370$ W

At 60 Hz

Hysteresis loss= $1800 \times 60/50 = 2{,}160$ W ; Eddy-current loss $= 370 \times (60/50)^2 = 533$ W

Total iron loss $= 2{,}160 + 533 =$ **2,693 W**

Tutorial Problems No. 8.1

1. The area of a hysteresis loop of a material is 30 cm^2. The scales of the co-ordinates are : 1 cm = 0.4

Wb/m^2 and 1 cm = 400 AT/m. Determine the hysteresis power loss if 1.2×10^{-3} m^3 of the material is subjected to alternating flux density at 50 Hz. **[288 W]** (*Elect. Engg., Aligarh Univ 1980*)

2. Calculate the loss of energy caused by hysteresis in one hour in 50 kg of iron when subjected to cyclic magnetic changes. The frequency is 25 Hz, the area of the hysteresis loop represents 240 joules/m^3 and the density of iron is 7800 kg/m^3. **[138,240]** (*Principles of Elect. Engg. 1, Jadvapur Univ. 1979*)

3. The hysteresis loop of a specimen weighing 12 kg is equivalent to 300 joules/m^3. Find the loss of energy per hour at 50 Hz. Density of iron is 7500 kg/m^3. **[86,400]** (*Electrotechnics – I, Gauhati Univ. 1981*)

4. The area of the hysteresis loop for a steel specimen is 3.84 cm^2. If the ordinates are to the scales : 1 cm = 400 AT/m and 1cm = 0.5 Wb/m^2, determine the power loss due to hysteresis in 1,200 cm^3 of the steel if it is magnetised from a supply having a frequency of 50 Hz. **[46.08 W]**

5. The armature of a 4-pole d.c. motor has a volume of 0.012 m^3. In a test on the steel iron used in the armature carried out to the same value of maximum flux density as exists in the armature, the area of the hysteresis loop obtained represented a loss of 200 J/m^3. Determine the hysteresis loss in watts when the armature rotates at a speed of 900 r.p.m. **[72 W]**

6. In a magnetisation test on a sample of iron, the following values were obtained.

H (AT/m)	1,900	2,000	3,000	4,000	4,500	3,000	1,000	0	−1,000	−1,900
B(Wb/m^2)	0	0.2	0.58	0.7	0.73	0.72	0.63	0.54	0.38	0

Draw the hysteresis loop and find the loss in watts if the volume of iron is 0.1 m^3 and frequency is 50 Hz. **[22 kW]**

8.5. Steinmetz Hysteresis Law

It was experimentally found by Steinmetz that hysteresis loss per m^3 per cycle of magnetisation of a magnetic material depends on (*i*) the maximum flux density established in it *i.e.* B_{max} and (*ii*) the magnetic quality of the material.

$$\therefore \quad \text{Hysteresis loss } W_h \propto B_{max}^{1.6} \text{ joule/m}^3\text{/cycle} = \eta B_{max}^{1.6} \text{ joule/m}^3\text{/cycle}$$

where η is a constant depending on the nature of the magnetic material and is known as **Steinmetz hysteresis coefficient.** The index 1.6 is empirical and holds good if the value of B_{max} lies between 0.1 and 1.2 Wb/m^2. If B_{max} is either lesser than 0.1 Wb/m^2 or greater than 1.2 Wb/m^2, the index is greater than 1.6.

$$\therefore \quad W_h = \eta B_{max}^{1.6} f V \text{ J/s or watt}$$

where f is frequency of reversals of magnetisation and V is the volume of the magnetic material.

The armatures of electric motors and generators and transformer cores etc. which are subjected to rapid reversals of magnetisation should, obviously, be made of substances having low hysteresis coefficient in order to reduce the hysteresis loss.

Example 8.6. *A cylinder of iron of volume 8×10^{-3} m^3 revolves for 20 minutes at a speed of 3,000 r.p.m in a two-pole field of flux density 0.8 Wb/m^2. If the hysteresis coefficient of iron is 753.6 joule/m^3, specific heat of iron is 0.11, the loss due to eddy current is equal to that due to hysteresis and 25% of the heat produced is lost by radiation, find the temperature rise of iron. Take density of iron as 7.8×10^3 kg/m^3.* **(Elect. Engineering-I, Osmania Univ. 1987)**

Solution. An armature revolving in a multipolar field undergoes one magnetic reversal after passing under a pair of poles. In other words, number of magnetic reversals is the same as the number of *pair* of poles. If P is the number of poles, the magnetic reversals in one revolution are $P/2$. If speed of armature rotation is N r.p.m, then number of revolutions /second = N /60.

No. of reversals/second = reversals in one revolution × No. of revolutions/second

$$= \frac{P}{2} \times \frac{N}{60} = \frac{PN}{120} \text{ reversals/second}$$

Here $\quad N = 3,000$ r.p.m ; $P = 2 \therefore f = \dfrac{3,000 \times 2}{120} = 50$ reversals/second

According to Steinmetz's hysteresis law, $W_h = \eta B_{max}^{1.6} f V$ watt

Note that f here stands for magnetic reversals/second and not for mechanical frequency of armature rotation.

$$\therefore \quad W_h = 753.6 \times (0.8)^{1.6} \times 50 \times 8 \times 10^{-3} = 211 \text{ J/s}$$

Loss in 20 minutes $= 211 \times 1,200 = 253.2 \times 10^3$ J

Eddy current loss $= 253.2 \times 10^3$ J; Total loss $= 506.4 \times 10^3$ J

Heat produced $= 506.4 \times 10^3/4200 = 120.57$ kcal ; Heat utilized $= 120.57 \times 0.75 = 90.43$ kcal

Heat absorbed by iron $= (8 \times 10^{-3} \times 7.8 \times 10^3) \times 0.11\, t$ kcal

$\therefore \quad (8 \times 10^{-3} \times 7.8 \times 10^3) \times 0.11 \times t = 90.43 \qquad \therefore \qquad t = \mathbf{13.17°C}$

Example 8.7. *The area of the hysteresis loop obtained with a certain specimen of iron was 9.3 cm². The coordinates were such that 1 cm = 1,000 AT/m and 1 cm = 0.2 Wb/m². Calculate (a) the hysteresis loss per m³ per cycle and (b) the hysteresis loss per m³ at a frequency of 50 Hz if the maximum flux density were 1.5 Wb/m² (c) calculate the hysteresis loss per m³ for a maximum flux density of 1.2 Wb/m² and a frequency of 30 Hz, assuming the loss to be proportional to* $B_{max}^{1.8}$.

(Elect. Technology, Allahabad Univ. 1991)

Solution. $W_h = xy \times$ (area of B/H loop)

(*a*) $= 1,000 \times 0.2 \times 9.3 = \mathbf{1860\ J/m^3/cycle}$

(*b*) $W_h = 1,860 \times 50$ J/s/m³ $= \mathbf{93,000\ W/m^3}$

(c) $W_h = \eta B_{max}^{1.8} f\, VW$ For a given specimen, $W_h \propto B_{max}^{1.8} f$

In (*b*) above, $93,000 \propto 1.5^{1.8} \times 50$ and $W_h \propto 1.2^{1.8} \times 30$

$$\therefore \quad \frac{W_h}{93,000} = \left(\frac{1.2}{1.5}\right)^{1.8} \times \frac{30}{50}; \qquad W_h = 93,000 \times 0.669 \times 0.6 = \mathbf{37.360\ W}$$

Example 8.8. *Calculate the loss of energy caused by hysteresis in one hour in 50 kg of iron if the peak density reached is 1.3 Wb/m² and the frequency is 25 Hz. Assume Steinmetz coefficient as 628 J/m³ and density of iron as 7.8 × 10³ kg/m³. What will be the area of B/H curve of this specimen if 1 cm = 12.5 AT/m and 1 cm = 0.1 Wb/m².*

(Elect. Engg. ; Madras Univ. 1987)

Solution. $W_h = \eta B_{max}^{1.6} f V$ watt ; volume $V = \dfrac{50}{7.8 \times 10^3} = 6.41 \times 10^{-3}$ m³

$\therefore \quad W_h = 628 \times 1.3^{1.6} \times 25 \times 6.41 \times 10^{-3} = 152$ J/s

Loss in one hour $= 153 \times 3,600 = 551,300$ J

As per Steinmetz law, hysteresis loss $= \eta B_{max}^{1.6}$ J/m³/cycle

Also, hysteresis loss $= xy$ (area of B/H loop)

Equating the two, we get

$628 \times 1.3^{1.6} = 12.5 \times 0.1 \times$ loop area

$\therefore$ loop area $= 628 \times 1.3^{1.6}/1.25 = \mathbf{764.3\ cm^2}$

Tutorial Problems No. 8.2

1. In a certain transformer, the hysteresis loss is 300 *W* when the maximum flux density is 0.9 Wb/m² and the frequency 50 *Hz*. What would be the hysteresis loss if the maximum flux density were increased to 1.1 Wb/m² and the frequency reduced to 40 Hz. Assume the hysteresis loss over this range to be proportional to $B_{max}^{1.7}$ **[337 W]**

2. It a transformer, the hysteresis loss is 160 *W* when the value of B_{max} = 1.1 Wb/m² and when supply frequency is 60 Hz. What would be the loss when the value of B_{max} is reduced to 0.9 Wb/m² and the supply frequency is reduced to 50 Hz. **[97W]** (*Elect. Engg. II, Bangalore Univ. Jan. 1980*)

8.6. Energy Stored in a Magnetic Field

For establishing a magnetic field, energy must be spent, though no energy is required to *maintain* it. Take the example of the exciting coils of an electromagnet. The energy supplied to it is spent in two ways (*i*) part of it goes to meet I^2R loss and is lost once for all (*ii*) part of it goes to create flux and is stored in the magnetic field as potential energy and is similar to the potential energy of a raised weight. When a weight W is raised through a height of h, the potential energy stored in it is Wh. Work is done in *raising* this weight but once raised to a certain height, no further expenditure of energy is required to *maintain* it at that position. This mechanical potential energy can be recovered, so can be the electrical energy stored in a magnetic field.

When current through an inductive coil is gradually changed from zero to maximum value I, then every change of it is opposed by the self-induced e.m.f. produced due to this change. Energy is needed to overcome this opposition. This energy is stored in the magnetic field of the coil and is, later on, recovered when that field collapses. The value of this stored energy may be found in the following two ways :-

(*i*) **First Method.** Let, at any instant,

i = instantaneous value of current ; e = induced e.m.f. at that instant = $L.di/dt$

Then, work done in time dt in overcoming this opposition is

$$dW = ei\ dt = L.\frac{di}{dt} \times i \times dt = Li\ di$$

Total work done in establishing the maximum steady current of I is

$$\int_0^W dW = \int_0^I L.i.di = LI^2 \text{ or } W = \frac{1}{2}LI^2$$

This work is stored as the energy of the magnetic field. $\therefore E = \frac{1}{2}LI^2$ joules

(*ii*) **Second Method**

If current grows uniformly from zero value to its maximum steady value I, then average current is $I/2$. If L is the inductance of the circuit, then self-induced e.m.f. is $e = LI/t$ where 't' is the time for the current change from zero to I.

$\therefore$ Average power absorbed = induced e.m.f. × average current

$$= L\frac{I}{t} \times \frac{1}{2}I = \frac{1}{2}\frac{LI^2}{t}$$

Total energy absorbed = power × time $= \frac{1}{2}\frac{LI^2}{t} \times t = \frac{1}{2}LI^2$

$\therefore$ energy stored $E = \frac{1}{2}LI^2$ joule

It may be noted that in the case of series-aiding coils, energy stored is

$$E = \frac{1}{2}(L_1 + L_2 + 2M)I^2 = \frac{1}{2}L_1I^2 + \frac{1}{2}L_2I^2 + MI^2$$

Similarly, for series-opposing coils, $E = \frac{1}{2}L_1I^2 + \frac{1}{2}L_2I^2 - MI^2$

Example 8.9. *Reluctance of a magnetic circuit is known to be 10^5 AT/Wb and excitation coil has 200 turns. Current in the coil is changing uniformly at 200 A/s. Calculate (a) inductance of the coil (b) voltage induced across the coil and (c) energy stored in the coil when instantaneous current at t = 1 second is 1 A. Neglect resistance of the coil.* **(Elect. Technology, Univ. of Indore, 1987)**

Solution. (*a*) $L = N^2/S = 200^2/10^5 =$ **0.4 H**

(*b*) $e_L = L\ dI/dt = 0.4 \times 200 =$ **80 V**

(*c*) $E = \frac{1}{2}LI^2 = 0.5 \times 0.4 \times 1^2 =$ **0.2 J**

Example 8.10. *An iron ring of 20 cm mean diameter having a cross-section of 100 cm² is wound with 400 turns of wire. Calculate the exciting current required to establish a flux density of 1 Wb/m² if the relative permeability of iron is 1000. What is the value of energy stored ?*

(Elect. Engg-I, Nagpur Univ. 1992)

Solution. $B = \mu_0 \mu_r NI/l$ Wb/m²

$\therefore$ $1 = 4\pi \times 10^{-7} \times 1000 \times 400.I/0.2\pi$ or $I =$ **1.25 A**

Now, $L = \mu_0 \mu_r AN^2/l = 4\pi \times 10^{-7} \times 10^3 \times (100 \times 10^{-4}) \times (400)^2/0.2\pi = 3.2$ H

$E = \frac{1}{2}LI^2 = \frac{1}{2} \times 3.2 \times 1.25^2 =$ **5 J**

8.7. Rate of Change of Stored Energy

As seen from Art. 8.6, $E = \frac{1}{2} L I^2$. The rate of change of energy can be found by differentiating the above equation

$$\frac{dE}{dt} = \frac{1}{2}\left[L.2.I.\frac{dI}{dt} + I^2\frac{dL}{dt}\right] = LI.\frac{dI}{dt} + \frac{1}{2}I^2\frac{dL}{dt}$$

Example 8.11. *A relay (Fig. 8.5) has a coil of 1000 turns and an air-gap of area 10 cm² and length 1.0 mm. Calculate the rate of change of stored energy in the air-gap of the relay when*

(i) armature is stationary at 1.0 mm from the core and current is 10 mA but is increasing at the rate of 25 A/s

(ii) current is constant at 20 mA but inductance is changing at the rate of 100 H/s.

Solution. $L = \frac{\mu_0 N^2 A}{l_g}$

$$= \frac{4\pi \times 10^{-7} \times (10^3)^2 \times 10 \times 10^{-4}}{1 \times 10^{-3}} = 1.26 \text{ H}$$

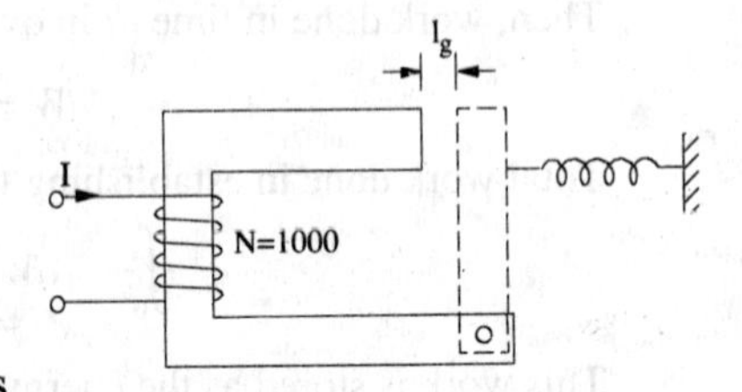

Fig. 8.5

(*i*) Here, dI/dt = 25 A/s, dL/dt = 0 because armature is stationary.

$$\therefore \quad \frac{dE}{dt} = L\,I\frac{dI}{dt} = 1.26 \times 10 \times 10^{-3} \times 25 = \mathbf{0.315\ W}$$

(*ii*) Here, dL/dt = 100 H/s; dI/dt = 0 because current is constant.

$$\therefore \quad \frac{dE}{dt} = \frac{1}{2}I^2\frac{dL}{dt} = \frac{1}{2}(20 \times 10^{-3})^2 \times 100 = \mathbf{0.02\ W}$$

8.8. Energy Stored Per Unit Volume

It has already been shown that the energy stored in a magnetic field of length l metre and of cross-section A m² is $E = \frac{1}{2} L\, I^2$ joule or $E = \frac{1}{2} \times \frac{\mu_0 \mu_r A N^2}{l} . I^2$ joule

Now $H = \frac{NI}{l}$ $\quad\therefore\quad E = \left(\frac{NI}{l}\right)^2 \times \frac{1}{2}\mu_0 \mu_r Al = \frac{1}{2}\mu_0 \mu_r H^2 \times Al$ joule

Now, Al = volume of the magneitc field in m³

$\therefore$ energy stored/m³ $= \frac{1}{2}\mu_0 \mu_r H^2 = \frac{1}{2} BH$ joule $\quad (\because \quad \mu_0 \mu_r H = B)$

$= \frac{B^2}{2\,\mu_0\,\mu_r}$ joule — in a medium

or $= \frac{B^2}{2\,\mu_0}$ joule —in air

8.9. Lifting Power of a Magnet

In Fig. 8.6 let, P = pulling force in newtons between two poles & A = pole area in m²

If one of the poles (say, upper one) is pulled apart against this attractive force through a distance of dx metres, then work done = $P \times dx$ joule ...(*i*)

This work goes to provide energy for the additional volume of the magnetic field so created.

Additional volume of the magnetic field created is

$$= A \times dx \text{ m}^3$$

Rate of energy requirement is $= \dfrac{B^2}{2\mu_0}$ joule/m^3

$\therefore$ energy required for the new volume $\dfrac{B^2}{2\mu_0} \times A\,dx$ ...(*ii*)

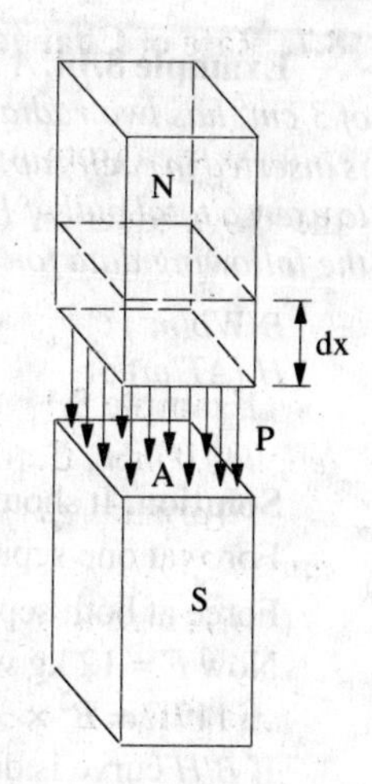

Fig. 8.6

Equating (*i*) and (*ii*), we get,

$$P.dx = \frac{B^2 \times A.dx}{2\mu_0}$$

$$\therefore \quad P = \frac{B^2 A}{2\mu_0} \text{ N} = 4,00,000 \frac{B^2}{A} \text{ N}$$

$$\text{or} \quad P = \frac{B^2}{2\mu_0} \text{ N/m}^2 = 4,00,000\, B^2 \text{ N/m}^2$$

$$\text{Also} \quad P = \frac{B^2 A}{9.81 \times 2\mu_0} = \frac{B^2 A}{19.62\, \mu_0} \text{ kg-wt}$$

Example 8.12. *A horse-shoe magnet is formed out of a bar of wrought iron 45.7 cm long, having a cross-section of 6.45 cm². Exciting coils of 500 turns are placed on each limb and connected in series. Find the exciting current necessary for the magnet to lift a load of 68 kg assuming that the load has negligible reluctance and makes close contact with the magnet. Relative permeability of iron = 700.*

(Elect. Engg. A.M.Ae. S.I, June, 1992)

Solution. Horse-shoe magnet is shown in Fig. 8.7.

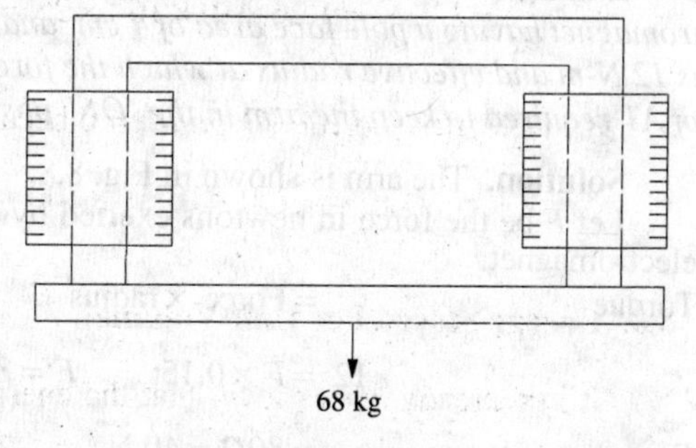

Fig. 8.7

Force of attraction of each pole = 68/2 = 34 kg = 34 × 9.81 = 333.5 N

$$A = 6.45 \text{ cm}^2 = 6.45 \times 10^{-4} \text{m}^2$$

$$\text{Since } F = \frac{B^2 A}{2\mu_0} \text{N}$$

$$\therefore \quad 333.5 = \frac{B^2 \times 6.45 \times 10^{-4}}{2 \times 4\pi \times 10^{-7}} \quad \therefore \quad B = \sqrt{1.3} = 1.14 \text{ Wb/m}^2$$

$$\text{and } H = B/\mu_0\,\mu_r = 1.14/4\pi \times 10^{-7} \times 700 = 1296 \text{ AT/m}$$

Length of the path = 45.7 cm = 0.457 m

$\therefore$ AT required = 1296 × 0.457 = 592.6

No. of turns = 500 × 2 = 1000 $\therefore$ current required = 592.6/1000 = **0.593A**

Example 8.13. *The pole face area of an electromagnet is 0.5 m²/pole. It has to lift an iron ingot weighing 1000 kg. If the pole faces are parallel to the surface of the ingot at a distance of 1 millimetre, determine the coil m.m.f. required. Assume permeability of iron to be infinity and the permeability of free space is $4\pi \times 10^{-7}$ H/m.* **(Elect. Technology, Univ. of Indore, 1985)**

Solution. Since iron has a permeability of infinity, it offers zero reluctance to the magnetic flux.

Force at two poles $= 2 \times B^2 A/2\mu_0 = B^2 A/\mu_0$

$\therefore \quad B^2 \times 0.5/4\pi \times 10^{-7} = 1000 \times 9.8, B = 0.157 \text{ Wb/m}^2$

$\therefore H = 0.157/4\pi \times 10^{-7} = 125 \times 10^3 \text{ AT/m}, l = 2 \times 1 = 2 \text{ mm} = 2 \times 10^{-3}\text{m}$

$\therefore$ AT required $= 125 \times 10^3 \times 2 \times 10^{-3} =$ **250.**

Example 8.14. *A soft iron ring having a mean circumference of 40 cm and cross-sectional area of 3 cm² has two radial saw cuts made at diametrically opposite points. A brass plate 0.5 mm thick is inserted in each gap. The ring is wound with 800 turns. Calculate the magnetising current required to exert a total pull of 12 kg between two halves. Neglect any magnetic leakage and fringing. Assume the following data for soft iron :*

$B(Wb/m^2)$:	0.76	1.13	1.31	1.41	1.5
H (AT/m) :	50	100	150	200	250

(Elect. Engineering-I, Delhi Univ. 1984)

Solution. It should be noted that brass is a non-magnetic material.

Force at one separation = $B^2A/2\mu_0$ newton.

Force at both separations = B^2A/μ_0 newton.

Now F = 12 kg wt = 12 × 9.81 = 117.7 N

$\therefore$ $117.7 = B^2 \times 3 \times 10^{-4}/4\pi \times 10^{-7}$; $B = 0.7$ Wb/m²

If B/H curve is drawn, it will be found that for B = 0.7 Wb/m², value of H = 45 AT/m.

Now, length of iron path = 40 cm = 0.4 m. AT required for iron path = 45 × 0.4 = 18

Value of H in the non-magnetic brass plates = $B/\mu_0 = 0.7/4\pi \times 10^{-7}$ = 557,042 AT/m

Total thickness of brass plates = 0.5 × 2 = 1 mm

AT required = $557{,}042 \times 1 \times 10^{-3}$ = 557, Total AT needed = 18 + 557 = 575

$\therefore$ magnetising current required = 557/800 = **0.72 A**

Example 8.15. *The arm of a d.c. shunt motor starter is held in the 'ON' position by an electromagnet having a pole face area of 4 cm² and air gap of 0.6 mm. The torque exerted by the spring is 12 N-m and effective radius at which the force is exerted is 15 cm. What is the minimum number of AT required to keep the arm in the 'ON' position ?*

Solution. The arm is shown in Fig. 8.8.

Let F be the force in newtons exerted by the two poles of the electromagnet.

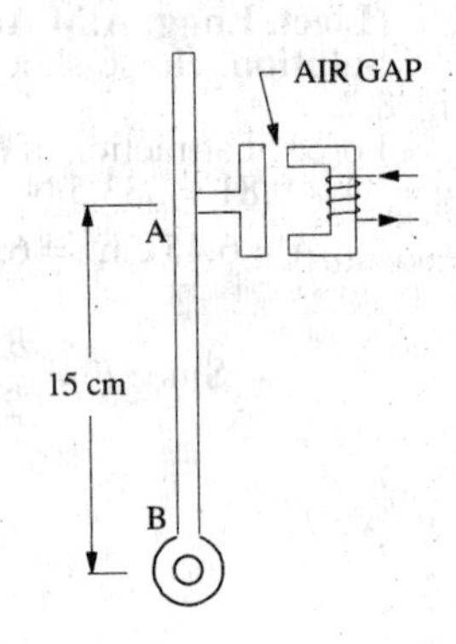

Fig. 8.8

Torque = Force × radius

$\therefore$ $12 = F \times 0.15$; $F = 80$ N

Force per pole = 80/2 = 40 N

Now $F = \dfrac{B^2A}{2\mu_0}$ N $\quad \therefore 40 = \dfrac{B^2 \times 4 \times 10^{-4}}{2 \times 4\pi \times 10^{-7}}$

$\therefore$ $B = 0.5$ Wb/m² $\therefore H = 0.5/\pi \times 10^{-7}$ AT/m

Total air-gap $= 2 \times 0.6 \times 10^{-3} = 1.2 \times 10^{-3}$ m

$\therefore$ AT reqd. $= Hl = \dfrac{0.5 \times 1.2 \times 10^{-3}}{4\pi \times 10^{-7}} = \mathbf{477}$

Example 8.16. *The following particulars are taken from the magnetic circuit of a relay ; Mean length of iron circuit = 20 cm ; length of air gap = 2 mm, number of turns on core = 8000, current through coil = 50 mA, relative permeability of iron = 500. Neglecting leakage, what is the flux density in the air-gap ? If the area of the core is 0.5 cm², what is the pull exerted on the armature ?*

Solution. Flux $\Phi = \dfrac{NI}{\Sigma l/\mu_0 \mu_r A}$.

Now, m.m.f $= NI = 8000 \times 50 \times 10^{-3} = 400$ AT

$$\text{Total circuit reluctance } = \frac{l}{\Sigma \mu_0 \mu_r A} \text{ AT/Wb or } H^{-1}$$

$$= \frac{0.2}{500 \times 4\pi \times 10^{-7} \times 0.5 \times 10^{-4}} + \frac{.2 \times 10^{-3}}{4\pi \times 10^{-7} \times 0.5 \times 10^{-4}}$$

$$= \frac{12 \times 10^7}{\pi} \text{ AT/Wb}$$

$$\Phi = \frac{400}{12 \times 10^7/\phi} \text{ Wb; Flux density } B = \frac{\Phi}{A} = \frac{400\pi}{12 \times 10^7 \times 0.5 \times 10^{-4}} = \mathbf{0.21\ Wb/m^2}$$

$$\text{The pull on the armature } = \frac{B^2 A}{2\mu_0} \text{ N} = \frac{0.21^2 \times 0.5 \times 10^{-4}}{2 \times 4\pi \times 10^{-7}} = \mathbf{0.87\ N}$$

Tutorial Problems No. 8.2

1. An air-cored solenoid has a length of 50 cm and a diameter of 2 cm. Calculate its inductance if it has 1,000 turns and also find the energy stored in it if the current rises from zero to 5A.
[0.7 mH; 8.7 mJ] (*Elect. Engg. & Electronics Bangalore Univ. 1988*)

2. An air-cored solenoid 1 m in length and 10 cm in diameter has 5000 turns. Calculate (*i*) the self inductance (*ii*) the energy stored in the magnetic field when a current of 2 A flows in the solenoid.
[(*i*) **0.2468 H** (*ii*) **0.4936 J**] (*F.E. Pune Univ. Nov. 1986*)

3. Determine the force required to separate two magnetic surfaces with contact area of 100 cm^2 if the magnetic flux density across the surface is 0.1 Wb/m^2. Derive formula used, if any.
[39.8 N] (*Elect. Engg. A.M.Ae. S.I. June 1990*)

4. In a telephone receiver, the size of each of the two poles is 1.2 cm × 0.2 cm and the flux between each pole and the diaphragm is 3×10^{-6} Wb; with what force is the diaphragm attracted to the poles ?
[0.125 N] (*Elect. Engg. A.M.Ae. S.I. June 1991*)

5. A lifting magnet is required to raise a load of 1,000 kg with a factor of safety of 1.5. If the flux density across the pole faces is 0.8 Wb/m^2, calculate the area of each pole. **[577 cm^2]**

6. Magnetic material having a surface of 100 cm^2 are in contact with each other. They are in a magnetic circuit of flux 0.01 Wb uniformly distributed across the surface. Calculate the force required to detach the two surfaces. **[3,978 N]** (*Elect. Engg. Kerala Univ. 1976*)

7. A steel ring having a mean diameter of 35 cm and a cross-sectional area of 2.4 cm^2 is broken by a parallel-sided air-gap of length 1.2 cm. Short pole pieces of negligible reluctance extend the effective cross-sectional area of the air-gap to 12 cm^2. Taking the relative permeability of steel as 700 and neglecting leakage, determine (*a*) the current necessary in 300 turns of wire wound on the ring to produce a flux density in the air-gap of 0.25 Wb/m^2 (*b*) the tractive force between the poles. [(*a*) **13.16 A** (*b*) **29.9 N**]

8. A cast iron ring having a mean circumference of 40 cm and a cross-sectional area of 3 cm^2 has two radial saw-cuts at diametrically opposite points. A brass plate is inserted in each gap (thickness 0.5 mm). If the ring is wound with 800 turns, calculate the magnetising current to exert a total pull of 3 kg between the two halves. Neglect any magnetic leakage and fringing and assume the magnetic data for the cast iron to be :

B (Wb/m^2)	:	0.2	0.3	0.4	0.5
H (AT)	:	850	1150	1500	2000

[1.04 A]

9. A magnetic circuit in the form of an inverted U has an air-gap between each pole and the armature of 0.05 cm. The cross-section of the magnetic circuit is 5 cm^2. Neglecting magnetic leakage and fringing, calculate the necessary exciting ampere-turns in order that the armature may exert a pull of 15 kg. The ampere-turns for the iron portion of the magnetic circuit may be taken as 20 percent of those required for the double air-gap.

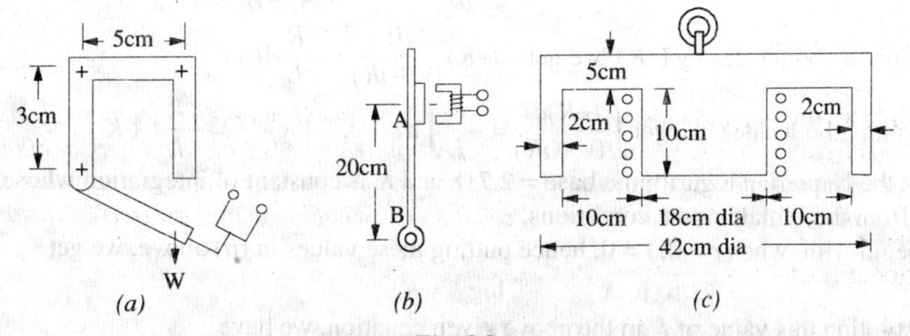

Fig. 8.9

10. In Fig. 8.9 (*a*) is shown the overload trip for a shunt motor starter. The force required to lift the armature is equivalent to a weight of $W = 0.8165$ kg positioned as shown. The air-gaps in the magnetic circuit are equivalent to a single gap of 0.5 cm. The corss-sectional area of the circuit is 1.5 cm^2 throughout and the magnetisation curve is as follows :

H (AT/m) :	1000	2000	3000	4000
B (Wb/m^2) :	0.3	0.5	0.62	0.68

Calculate the number of turns required if the trip is to operate when 80 A passes through the coil.

[21 turns]

11. The armature of a d.c. motor starter is held in the 'ON' position by means of an electromagnet [Fig. 8.9 (*b*)]. A spiral spring exerts a mean counter torque of 8 N-m on the armature in this position after making allowance for the weight of the starter arm. The length between the centre of the armature and the pivot on the starter arm is 20 cm and cross-sectional area of each pole face of the electromagnet 3.5 cm^2.

Find the minimum number of AT required on the electromagnet to keep the arm in the 'ON' position when the air-gap between the armature and the electromagnet is 0.5 mm. (Neglect the AT needed for the iron of the electromagnet). **[301 AT]**

12. A cylindrical lifting magnet of the form shown in Fig. 8.9 (*c*) has a winding of 200 turns which carries a current of 5 A. Calculate the maximum lifting force which could be exerted by the magnet on a flat iron sheet 5 cm thick. Why would this value not be realized in practice ? The relative permeability of the iron can be taken as 500. **[698 N]**

8.10. Rise of Current in an Inductive Circuit

In Fig. 8.10 is shown a resistance of R in series with a coil of self-inductance L henry, the two being put across a battery of V volt. The R-L combination becomes connected to battery when swich S is connected to terminal '*a*' and is short-circuited when S is connected to '*b*'. The inductive coil is assumed to be resistanceless, its actual small resistance being included in R.

When S is connected to '*a*' the R-L combination is suddenly put across the voltage of V volt. Let us take the instant of closing S as the starting zero time. It is found that current does not reach its maximum value instantaneously but takes some finite time. It is easily explained by recalling that the coil possesses electrical inertia *i.e.* self-inductance and hence, due to the production of the counter e.m.f. of self-inductance, delays the instantaneous full establishment of current through it.

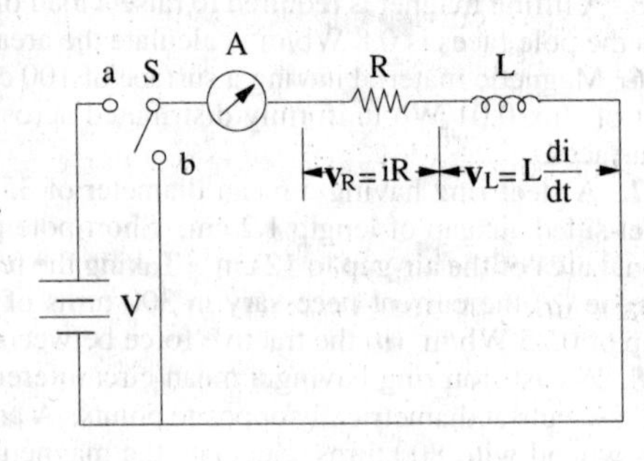

Fig. 8.10

We will now investigate the growth of current i through such an inductive circuit.

The applied voltage V must, at any instant, supply not only the ohmic drop iR over the resistance R but must also overcome the e.m.f. of self-inductance *i.e.* L *di*/*dt*.

$$\therefore \quad V = \upsilon_R + \upsilon_L = iR + \frac{di}{dt}$$

or

$$(V - iR) = L\frac{di}{dt} \quad \therefore \quad \frac{di}{V - iR} = \frac{dt}{L} \quad \ldots(i)$$

Multiplying both sides by $(-R)$. we get $\quad (-R)\dfrac{di}{(V - iR)} = -\dfrac{R}{L}dt$

Integrating both sides, we get $\displaystyle\int \frac{(-R)di}{(V - iR)} = -\frac{R}{L}\int dt \quad \therefore \quad \log_e^{V-iR} = -\frac{R}{L}t + K \quad \ldots(ii)$

where e is the Napierian logarithmic base = 2.718 and K is constant of integration whose value can be found from the initial known conditions.

To begin with, when $t = 0$, $i = 0$, hence putting these values in (*ii*) above, we get

$$\log_e^{V} = K$$

Substituting this value of K in the above given equation, we have

$$\log_e^{V-iR} = \frac{R}{L}t + \log_e^V \quad \text{or} \quad \log_e^{V-iR} - \log_e^V = -\frac{R}{L}t$$

or
$$\log_e \frac{V-iR}{V} = -\frac{R}{L}t = -\frac{t}{\lambda} \quad \text{where } L/R = \lambda \text{ 'time constant'}$$

$$\therefore \quad \frac{V-iR}{V} = e^{-t/\lambda} \quad \text{or} \quad i = \frac{V}{R}(1 - e^{-t/\lambda})$$

Now, V/R represents the maximum steady value of current I_m that would eventually be established through the R-L circuit.

$$\therefore \quad i = I_m(1 - e^{-t/\lambda}) \qquad \ldots(iii)$$

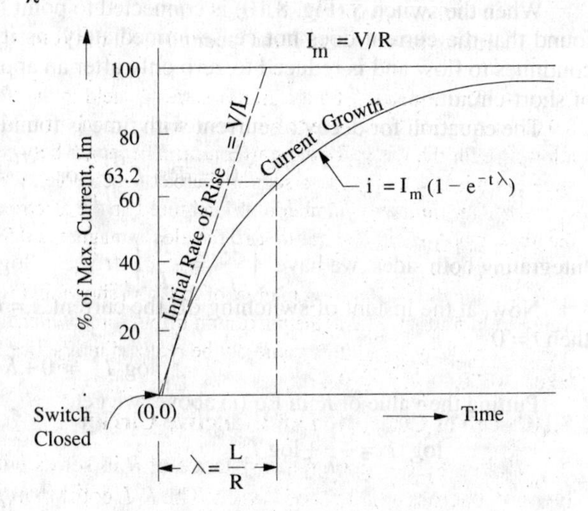

Fig. 8.11

This is an exponential equation whose graph is shown in Fig. 8.11. It is seen from it that current rise is rapid at first and then decreases until at $t = \infty$, it becomes zero. Theoretically, current does not reach its maximum steady value I_m until infinite time. However, in practice, it reaches this value in a relatively short time of about 5λ.

The rate of rise of current di/dt at any stage can be found by differentiating Eq. (*iii*) above w.r.t. time. However, the *initial* rate of rise of current can be obtained by putting $t = 0$ and $i = 0$ in (*i*)* above.

$$\therefore \quad V = 0 \times R + L\frac{di}{dt} \quad \text{or} \quad \frac{di}{dt} = \frac{V}{L}$$

The constant $\lambda = L/R$ is known as the *time-constant* of the circuit. It can be variously defined as :

(*i*) It is the *time* during which current would have reached its maximum value of $I_m (= V/R)$ had it maintained its initial rate of rise.

$$\text{Time taken} = \frac{I_m}{\text{initial rate of rise}} = \frac{V/R}{V/L} = \frac{L}{R}$$

But actually the current takes more time because its rate of rise decreases gradually. In actual practice, in a time equal to the time constant, it merely reaches 0.632 of its maximum value as shown below :

Putting $t = L/R = \lambda$ in Eq. (*iii*) above, we get

$$i = I_m(1 - e^{-\lambda/\lambda}) = I_m\left(1 - \frac{1}{e}\right) = I_m\left(1 - \frac{1}{2.718}\right) = 0.632\, I_m$$

(*ii*) Hence, the time-constant λ of an R–L circuit may also be defined as the time during which the current *actually* rises to 0.632 of its maximum steady value (Fig. 8.11).

---

*Initial value of di/dt can also be found by differentiating Eq. (*iii*) and putting $t = 0$ in it. In fact, the three quantities V, L, R give the following various combinations :

$V/R = I_m$ – the maximum final steady current.
V/L = initial rate of rise of current.
L/R = time constant of the circuit.

The first rule of switching is that the current flowing through an inductance cannot change instantaneously. The second rule of switching is that the voltage across a capacitor cannot change instantaneously.

This delayed rise of current in an inductive circuit is utilized in providing time lag in the operation of electric relays and trip coils etc.

8.11. Decay of Current in an Inductive Circuit

When the switch S (Fig. 8.10) is connected to point 'b', the R–L circuit is short-circuited. It is found that the current does not cease immediately, as it would do in a non-inductive circuit, but continues to flow and is reduced to zero only after an appreciable time has elapsed since the instant of short-circuit.

The equation for decay of current with time is found by putting $V = 0$ in Eq. (i) of Art. 8.10

$$0 = iR + L\frac{di}{dt} \quad \text{or} \quad \frac{di}{i} = -\frac{R}{L}dt$$

Integrating both sides, we have $\int \frac{di}{i} = -\frac{R}{L}\int dt \quad \therefore \quad \log i = -\frac{R}{L}t + K$ ...(i)

Now, at the instant of switching off the current, $i = I_m$ and if time is counted from this instant, . then $t = 0$

$$\therefore \quad \log_e I_m = 0 + K$$

Putting the value of K in Eq (i) above, we get,

$$\log_e i = -\frac{t}{\lambda} + \log_e I_m$$

$$\therefore \quad \log_e i / I_m = -\frac{t}{\lambda}$$

$$\therefore \quad \frac{i}{I_m} = e^{-t/\lambda}$$

$$\text{or} \quad i = I_m e^{-t/\lambda} \quad ...(ii)$$

It is a decaying exponential function and is plotted in Fig. 8.12. It can be shown again that theoretically, current should take infinite time to reach zero value although, in actual practice, it does so in a relatively short time of about 5λ.

Again, putting $t = \lambda$ in Eq. (ii) above, we get

$$i = \frac{I_m}{e} = \frac{I_m}{2.178} = 0.37\, I_m.$$

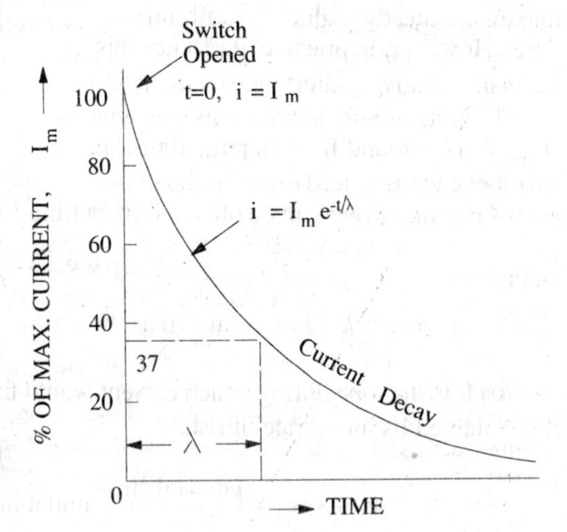

Fig. 8.12

Hence, time constant (λ) of an R–L circuit may also be defined as the time during which current falls to 0.37 or 37% of its maximum steady value while decaying (Fig. 8.12).

Example 8.17. *A coil having an effective resistance of 20 Ω and an inductance of 5 H. is suddenly connected across a 50-V dc supply. What is the rate at which energy is stored in the field of the coil when current is (a) 0.5 A (b) 1.0 A and (c) steady ? Also find the induced EMF in the coil under the above conditions.*

Solution. (a) Power input = 50 × 0.5 = 25 W

Power wasted as heat = $i^2R = 0.5^2 \times 25 = 6.25$ W. Hence, rate of energy storage in the coil field is = 25 – 6.25 = 18.75 W or J/s. (b) Power input = 50 × 1 = 50 W

Power lost as heat = $1^2 \times 25 = 25$ W. ∴ Rate of energy storage in field = 50 – 25 = 25 W or J/s.

(c) Steady value of current = 50/25 = 2 A. Power input = 50 × 2 = 100 W

Power lost as heat = $2^2 \times 25 = 100$ W

Rate of energy storage in field = 100 – 100 = 0; Now, $V = iR + e_L \therefore e_L = V - iR$

(a) $e_L = 50 - 0.5 \times 25 = 37.5$ V (b) $e_L = 50 - 1 \times 25 = 25$ V (c) $e_L = 50 - 2 \times 25 = 0$ V

Example 8.18. *A coil having a resistance of 10 Ω and an inductance of 4 H is switched across a 20-W dc source. Calculating (a) time required by the current to reach 50% of its final steady value and (b) value of the current after 0.5 second.*

Solution. The rise of current through an inductive circuit is given by the equation $i = I\,(1 - e^{-t/\lambda})$. It may be written as

$$e^{-t/\lambda} = \frac{I-i}{I} \quad \text{or} \quad \frac{1}{e^{t/h}} = \frac{I-i}{I} \quad \text{or} \quad e^{e/h} = \frac{I}{I-i}$$

Taking logs of both sides, we have

$$\frac{t}{\lambda}.\log^{e} = \log^{I/(I-i)} = l_n \frac{I}{(I-i)}$$

$$\therefore \qquad \frac{Rt}{L} = R_n \frac{I}{(I-i)} \quad \text{or} \quad t = \frac{L}{R} l_n \frac{I}{I-i}$$

(*a*) Now, $I = V/R = 20/10 = 2$ A

$$\therefore \qquad t = \frac{4}{10} l_n \frac{2}{(2-1)} = \frac{4}{10} \times 0.693 = 0.277 \text{ s.}$$

(*b*) $\lambda = L/R = 4/10 = 0.4$ s and $t = 0.5$ s

$$\therefore \qquad i = 2\,(1 - e^{-t/0.4})$$

Example 8.19. *With reference to the circuit shown in Fig. 8-13, calculate :*

(i) the current taken from the d.c. supply at the instant of closing the switch

(ii) the rate of increase of current in the coil at the instant of switching on

(iii) the supply and coil currents after the switch has been closed for a long time

(iv) the maximum energy stored in the coil

(v) the e.m.f. induced in the coil when the switch is opened.

Solution. (*i*) When switch *S* is closed (Fig. 8.13), the supply d.c. voltage of 120 *V* is applied across both arms. The current in R_2 will immediately become 120/30 = 4*A*. However, due to high inductance of the second arm, there would be no instantaneous flow of current in it. Hence current taken from the supply at the instant of switching on will be **4A.**

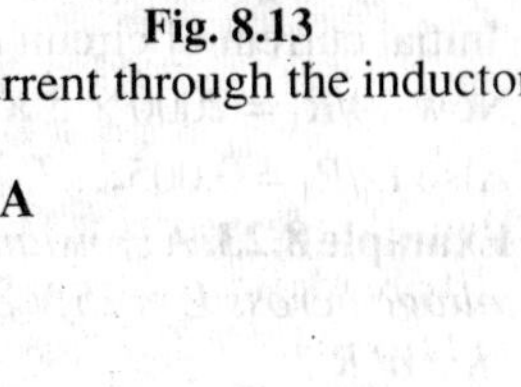

(*ii*) Since at the *instant* of switching on, there is no current through the inductor arm, no potential drop will develop across R_1. The whole of the supply voltage will be applied across the inductor. If *di*/*dt* is the rate of increase of current through the inductor at the instant of switching on, the back e.m.f. produced in it is *L. di*/*dt*. This e.m.f. is equal and opposite to the applied voltage.

Fig. 8.13

$120 = L\,di/dt$ or $di/dt = 120/2$ = **60 A/s**

(*iii*) When switch has been closed for a sufficiently long time, current through the inductor arm reaches a steady value = $120/R_1 = 120/15$ = **8A**

Current through $R_2 = 120/30 = 4$A ; Supply current = 8 + 4 = **12A**

(*iv*) Maximum energy stored in the inductor arm

$$= \frac{1}{2} LI^2 = \frac{1}{2} \times 2 \times 8^2 = \mathbf{64\ J}$$

(*v*) When switch is opened, current through the inductor arm cannot change immediately because of high self-inductance of the inductor. Hence, inductance current remains at 8 A. But the current through R_2 can change immediately. After the switch is opened, the inductor current path lies through R_1 and R_2. Hence, e.m.f. induced in the inductor at the instant of switching off is = 8 × (30 + 15) = **360 V.**

Example 8.20. *A coil has a time constant of 1 second and an inductance of 8 H. If the coil is connected to a 100 V d.c. source, determine :*

(i) the rate of rise of current at the instant of switching (ii) the steady value of the current and (iii) the time taken by the current to reach 60% of the steady value of the current.

(Electrotechnics-I, M.S. Univ. Baroda 1985)

Solution. $\lambda = L/R$; $R = L/\lambda = 8/1 = 8$ ohm (*i*) Initial $di/dt = V/L = 100/8 =$ **12.5 A/s** (*ii*) $I_M = V/R = 100/8 =$ **12.5 A**

(*iii*) Here, $i = 60\%$ of $12.5 = 7.5$ A

Now, $i = I_m(1 - e^{-t/\lambda})$ $\therefore$ $7.5 = 12.5(1 - e^{-t/1})$; $t =$ **0.915 second**

Example 8.21. *A d.c voltage of 80 V is applied to a circuit containing a resistance of 80 Ω in series with an inductance of 20 H. Calculate the growth of current at the instant (i) of completing the circuit (ii) when the current is 0.5 A and (iii) when the current is 1 A.*

(Circuit Theory, Jadavpur Univ. 1986)

Solution. The voltage equation for an *R-L* circuit is

$$V = iR + L\frac{di}{dt} \text{ or } L\frac{di}{dt} = V - iR \text{ or } \frac{di}{dt} = \frac{1}{L}(V - iR)$$

(*i*) when $i = 0$; $\dfrac{di}{dt} = \dfrac{1}{L}(V - 0 \times R) = \dfrac{V}{L} = \dfrac{80}{20} =$ **4 A/s**

(*ii*) when $i = 0.5\text{A}$; $\dfrac{di}{dt} = \dfrac{80 - 0.5 \times 80}{20} =$ **2 A/s**

(*iii*) when $i = 1\text{A}$; $\dfrac{di}{dt} = \dfrac{80 - 80 \times 1}{20} =$ **0.**

In other words, the current has become steady at 1 ampere.

Example 8.22. *The two circuits of Fig. 8.14 have the same time constant of 0.005 second. With the same d.c. voltage applied to the two circuits, it is found that the steady state current of circuit (a) is 2000 times the initial current of circuit (b). Find R_1, L_1 and C.*

(Elect. Engg.-I, Bombay Univ. 1985)

Solution. The time constant of circuit 8.14(*a*) is $\lambda = L_1/R_1$ second, and that of circuit 8.14(*b*) is $\lambda = CR_2$ second.

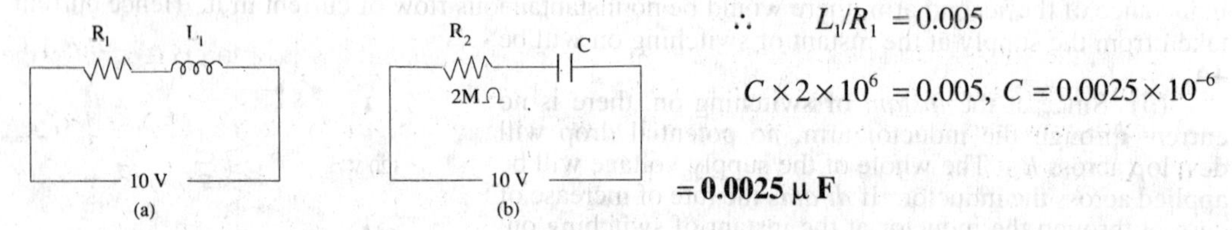

$\therefore$ $L_1/R_1 = 0.005$

$C \times 2 \times 10^6 = 0.005$, $C = 0.0025 \times 10^{-6}$

= 0.0025 μ F

Fig. 8.14

Steady-state current of circuit 8.14 (*a*) is $= V/R_1 = 10/R_1$ amperes.

Initial current of circuit 8-14 (*b*) $= V/R_2 = 10/2 \times 10^6 = 5 \times 10^{-6}$ A*

Now $10/R_1 = 2000 \times 5 \times 10^{-6}$ $\therefore$ $R_1 = 1000\Omega$.

Also $L_1/R_1 = 0.005$ $\therefore$ $L_1 = 1000 \times 0.005 =$ **5 H**

Example 8.23. *A constant voltage is applied to a series R–L circuit at t = 0 by closing a switch. The voltage across L is 25 V at t = 0 and drops to 5 V at t = 0.025 second. If L = 2H, what must be the value of R ?* **(Elect. Engg. I Bombay Univ. 1987)**

Solution. At $t = 0$, $i = 0$, hence there is no *iR* drop and the applied voltage must equal the back e.m.f. in the coil. Hence, the voltage across *L* at $t = 0$ represents the applied voltage.

At $t = 0.025$ second, voltage across *L* is 5 V, hence voltage across

$R = 25 - 5 = 20$ V $\therefore$ $iR = 20\text{V}$ – at $t = 0.025$ second. Now $i = I_m(I - e^{-t/\lambda})$

Here $I_m = 25/R$ ampere, $t = 0.025$ second $\therefore$ $i = \dfrac{25}{R}(1 - e^{-0.025/\lambda})$

*Because just at the time of starting the current, there is no potential drop across *C* so that all the applied voltage is dropped across R_2. Hence, the initial charging current $= V/R_2$.

$R \times \frac{25}{R}(I - e^{-0.025/\lambda}) = 20$ or $e^{0.025/\lambda} = 5$ ∴ $0.025/\lambda = 2.3 \log_{10}5 = 1.6077$

∴ $\lambda = 0.025/1.6077$ Now $\lambda = L/R = 2/R$ ∴ $2/R = 0.025/1.6077$ ∴ $R =$ **128.56 Ω**

Example 8.24. *A circuit of resistance R ohms and inductance L henries has a direct voltage of 230 V applied to it. 0.3 second after switching on, the current in the circuit was found to be 5A. After the current had reached its final steady value, the circuit was suddenly short-circuited. The current was again found to be 5A at 0.3 second after short-circuiting the coil. Find the value of R and L.*

(Basic Electricity, Bombay Univ. 1984)

Solution. For growth ; $5 = I_m(1 - e^{-0.3/\lambda})$

For decay; $5 = I_m e^{-0.3/\lambda}$ ...(*i*)

Equating the two, we get, $I_m e^{-0.3/\lambda} = (1 - e^{-0.3/\lambda})I_m$

or $2e^{-0.3/\lambda} = 1$ ∴ $e^{-0.3/\lambda} = 0.5$ or $\lambda = 0.4328$

Putting this value in (*i*), we get,

$5 = I_m e^{0.3/0.4328}$ or $I_m = 5e^{+0.3/0.4328} = 5 \times 2 = 10$A.

Now $I_m = V/R$ ∴ $10 = 230/R$ or $R = 230/10 = 23$ Ω(approx.)

As $\lambda = L/R = 0.4328$; $L = 0.4328 \times 23 =$ **9.95 H**

Example 8.25. *A relay has a coil resistance of 20 Ω and an inductance of 0.5 H. It is energized by a direct voltage pulse which rises from 0–10 V instantaneously, remains constant for 0.25 second and then falls instantaneously to zero. If the relay contacts close when the current is 200 mA (increasing) and open when it is 100 mA (decreasing), find the total time during which the contacts are closed.*

Solution. The time constant of the relay coil is

$\lambda = L/R = 0.5/20 = 0.025$ second

Now, the voltage pulse remains constant at 10 V for 0.25 second which is long enough for the relay coil current to reach its steady value of $V/R = 10/20 = 0.5$ A

Let us now find the value of time required by the relay coil current to reach a value of 200 mA = 0.2 A. Now $i = I_m(1 - e^{-t/\lambda})$ ∴ $0.2 = 0.5(1 - e^{-t/0.025})$ ∴ $e^{40t} = 5/3$

∴ $t = 0.01276$ second

Hence, relay contacts close at $t = 0.01276$ second and will remain closed till current falls to 100 mA. Let us find the time required by the current to fall from 0.5 A to 0.1 A

At the end of the voltage pulse, the relay current decays according to the relation

$i = I_m e^{e-t/\lambda}$ ∴ $0.1 = 0.5\, e^{-t/0.025}$ ∴ $e^{40t} = 5$

∴ $t = 0.04025$ second after the end of the voltage pulse.

Hence, the time for which contacts remain closed is

$= (0.25 - 0.01276) + 0.04025$ second = **277.5 milli-second (approx)**

8.12. Details of Transient Current Rise in an R–L Circuit

As shown in Fig. 8.15 (*a*), when switch *S* is shifted to position *a*, the *R–L* circuit is suddenly energised by V. Since a coil opposes any change in current, the initial value of current is zero at t = 0 and but then it rises exponentially, although its rate of rise keeps decreasing. After some time, it reaches a maximum value of I_m when it becomes constant *i.e.* its rate of rise becomes zero. Hence, just at the start of the transient state, $i = 0$, $V_R = 0$ and $V_L = V$ with its polarity opposite to that of battery voltage as shown in Fig. 8.15 (*a*). Both *i* and V_R rise exponentially during the transient state, as shown in Fig. 8.15 (*b*) and (*c*) respectively. However, V_L decreases exponentially to zero from its initial maximum value of $V = I_m R$. It does not become negative during the transient rise of current through the circuit.

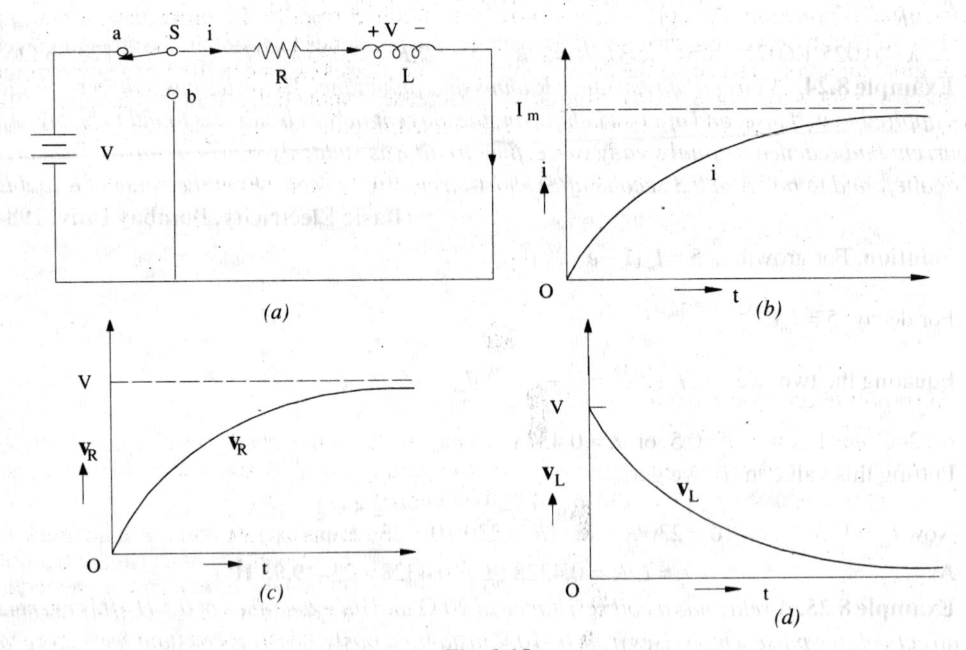

Fig. 8.15

Hence, during the transient rise of current, the following equations hold good :

$$i = \frac{V}{R}(1 - e^{-t/\lambda}) = I_m(1 - e^{-t/\lambda});\ v_R = iR = V(1 - e^{-t/\lambda}) = I_m R(1 - e^{-t/\lambda});\ v_L = V e^{-t/\lambda}$$

If S remains at 'a' long enough, i reaches a steady value of I_m and V_R equals $I_m R$ but since $di/dt = 0$, $v_L = 0$.

8:13. Details of Transient Current Decay in an R–L Circuit

Now, let us consider the conditions during the transient decay of current when S is shifted to point 'b'. Just at the start of the decay condition, the following values exist in the circuit.

$i = I_m = V/R$, $v_R = I_m R = V$ and since initial di/dt is maximum, $v_L = -V = -I_m R$.

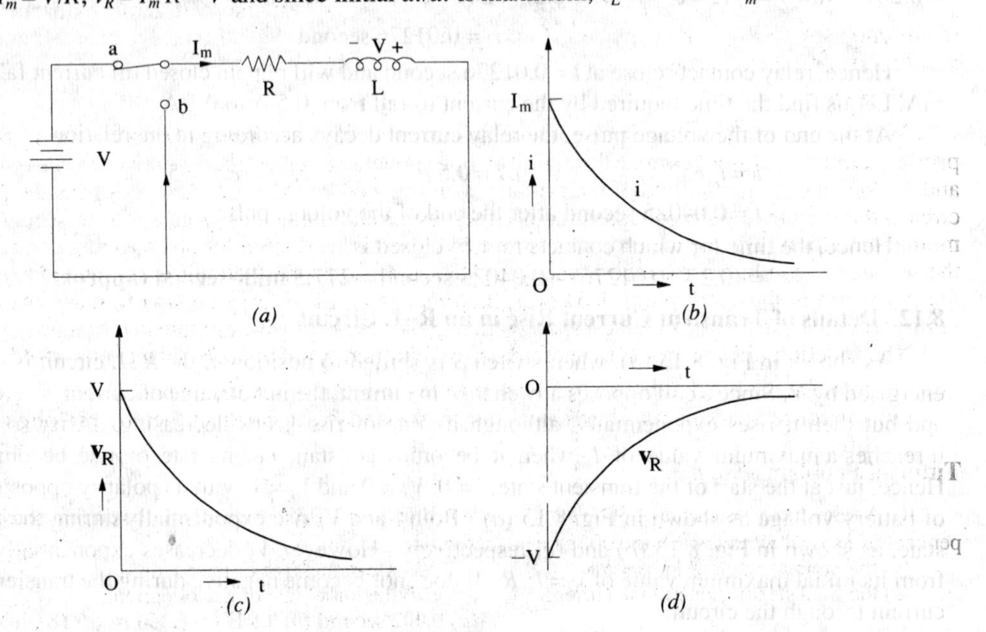

Fig. 8.16

The change in the polarity of voltage across the coil in Fig. 8.16 (*a*) is worth noting. Due to its property of self-induction, the coil will not allow the circuit current to die immediately, but only gradually. In fact, by reversing the sign of its voltage, the coil tends to maintain the flow of current in the original direction. Hence, as the decay continues, *i* decreases exponentially from its maximum value to zero, as shown in Fig. 8.16 (*b*). Similarly, v_R decreases exponetially from its maximum value to zero, as shown in Fig. 8.16 (*c*). However, v_L is reversed in polarity and decreases exponentially from its initial value of $-V$ to zero as shown in Fig. 8.16 (*d*).

During the transient decay of current and voltage, the following relations hold good :

$$i = i_L = I_m e^{-t/\lambda} = \frac{V}{R} e^{-t/\lambda}$$

$$v_R = V e^{-t/\lambda} = I_m R e^{-t/\lambda}$$

$$v_L = -V e^{-t/\lambda} = -I_m R e^{-t/\lambda}$$

8.14. Automobile Ignition System

Practical application of mutual induction is found in the single-spark petrol-engine ignition system extensively employed in automobiles and air-engines. Fig. 8.17 shows the circuit diagram of such a system as applied to a 4-cylinder automobile engine.

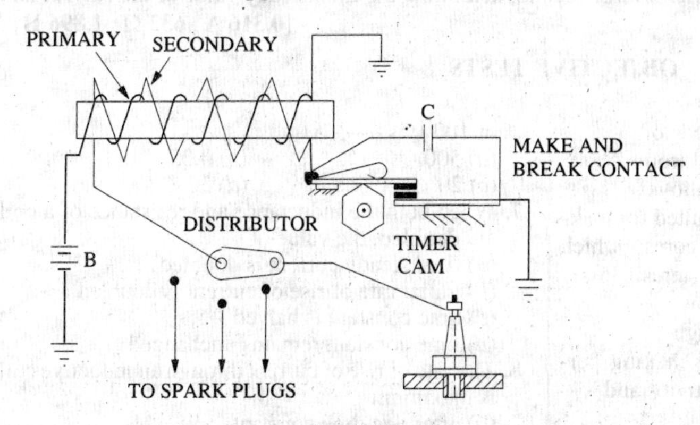

Fig. 8.17

It has a spark coil (or induction coil) which consists of a primary winding (of a few turns) and a secondary winding (of a large number of turns) wound on a common iron core (for increasing mutual induction). The primary circuit (containing battery *B*) includes a 'make and break contact' actuated by a timer cam. The secondary circuit includes the rotating blade of the distributor and the spark gap in the spark plug as shown in Fig. 8.17. The timer cam and the distributor are mounted on the same shaft and are geared to rotate at exactly half the speed of the engine shaft. It means that in the case of automobile engines (which are four-cycle engines) each cylinder is fired only once for every two revolutions of the engine shaft.

Working

When timer cam rotates, it alternately closes and opens the primary circuit. During the time primary circuit is closed, current through it rises exponentially after the manner shown in Fig. 8.11 and so does the magnetic field of the primary winding. When the cam suddenly opens the primary circuit, the magnetic field collapses rapidly thereby producing a very large e.m.f. in secondary by mutual induction. During the time this large e.m.f. exists, the distributor blade rotates and connects the secondary winding across the proper plug and so the secondary circuit is completed except for the spark gap in the spark plug. However, the induced e.m.f. is large enough to make the current jump across the gap thus producing a spark which ignites the explosive mixture in the engine cylinder.

The function of capacitor *C* connected across the 'make and break' contact is two-fold :

(*i*) to make the break rapid so that large e.m.f is induced in secondary and

(*ii*) to reduce sparking and burning at the 'make-and-break' contact thereby prolonging their life.

Tutorial Problems No. 8.3

1. A relay has a resistance of 300 Ω and is switched on to a 110 V d.c. supply. If the current reaches 63.2 per cent of its final steady value in 0.002 second, determine

(*a*) the time-constant of the circuit (*b*) the inductance of the circuit

(*c*) the final steady value of the current (*d*) the initial rate of rise of current.

[(*a*) **0.002 second** (*b*) **0.6 H** (*c*) **0.366 A** (*d*) **183 A/second**]

2. A coil with a self-inductance of 2.4 H and resistance 12 Ω is suddenly switched across a 120-V d.c. supply of negligible internal resistance. Determine the time constant of the coil, the instantaneous value of the current after 0.1 second, the final steady value of the current and the time taken for the current to reach 5 A.
[(*a*) 0.2 second ; 3.94 A ; 10 A ; 0.139 second]

3. A circuit whose resistance is 20 Ω and inductance 10 H has a steady voltage of 100 V suddenly applied to it. For the instant 0.5 second after the voltage is applied, determine (*a*) the total power input to the circuit (*b*) the power dissipated in the resistance. Explain the reason for the difference between (*a*) and (*b*).
[(*a*) 316 W (*b*) 200 W]

4. A lighting circuit is operated by a relay of which the coil has a resistance of 5 Ω and an inductance of 0.5 H. The relay coil is supplied from a 6-V d.c. source through a push-button switch. The relay operates when the current in the relay coil attains a value of 500 mA. Find the time interval between the pressing of the push-button and the closing of lighting circuit. **[53.8 ms]**

5. The field winding of a separately-excited d.c. generator has an inductance of 60 H and a resistance of 30 Ω. A discharge resistance of 50 Ω is permanently connected in parallel with the winding which is excited from a 200-V supply. Find the vlaue of decay current 0.6 second after the supply has been switched off.
[3.0 A]

6. The field winding of a dynamo may be taken to have a constant inductance of 120 H and an effective resistance of 30 Ω. When it is carrying a current of 5A, the supply is interrupted and a resistance of 50 Ω is connected across the winding. How long will it take for the current to fall to 1-0 ? **[2.415 s]**

7. A 200-V d.c. supply is suddenly switched to a relay coil which has a time constant of 3 milli-second. If the current in the coil reaches 0.2 A after 3 milli-second, determine the final steady value of the current and the resistance of the coil. **[0.316 A ; 632 Ω ; 1.896 H]**

OBJECTIVE TESTS–8

1. Permanent magnets are normally made of
(*a*) aluminium (*b*) wrought iron
(*c*) cast iron (*d*) alnico alloys

2. Those magnetic materials are best suited for making armature and transformer cores which have —— permeability and —— hysteresis loss.
(*a*) low, high (*b*) high, low
(*c*) low, low (*d*) high, high

3. Those materials are well-suited for making permanent magnets which have–retentivity and—— coercivity.
(*a*) high, high (*b*) high, low
(*c*) low, low (*d*) low, high

4. In a magnetic material, hysteresis loss takes place primarily due to
(*a*) flux density lagging behind magnetising force
(*b*) molecular friction
(*c*) its high retentivity
(*d*) rapid reversals of its magnetisation

5. Energy stored by a coil is doubled when its current is increased by —— per cent.
(*a*) 100 (*b*) 41.4
(*c*) 50 (*d*) 25

6. The initial rate of rise of current through a coil of $L = 5H$ when suddenly connected to a.d.c. supply of 100 V is —— A/s.
(*a*) 500 (*b*) 0.05
(*c*) 20 (*d*) 50

7. When both the inductance and resistance of a coil are doubled, the value of
(*a*) final steady current is doubled
(*b*) initial rate of rise of current is doubled
(*c*) time constant is halved
(*d*) time constant remains unchanged

8. The rate of rise of current through an inductive coil is maximum
(*a*) after one time constant
(*b*) at the start of current flow
(*c*) near the final maximum value of current
(*d*) at 63.2% of its maximum steady value

9. The lifting power of an electromagnetic depends on
(*a*) its pole area
(*b*) magnetic flux density
(*c*) its shape
(*d*) both (a) and (b)

10. During the transient rise of current through an $R - L$ circuit, which of the following has maximum initial value ?
(*a*) circuit current
(*b*) resistive drop
(*c*) coil energy
(*d*) coil voltage

Answers

1. *d* 2. *b* 3. *a* 4. *c* 5. *b* 6. *c* 7. *d* 8. *b* 9. *d* 10. *d*

9 ELECTROCHEMICAL POWER SOURCES

9.1. Faraday's Laws of Electrolysis

From his experiments, Faraday deduced two fundamental laws which govern the phenomenon of electrolysis. These are :

(i) First Law. The mass of ions liberated at an electrode is directly proportional to the quantity of electricity *i.e.* charge which passes through the electrolyte.

(ii) Second Law. The masses of ions of different substances liberated by the same quantity of electricity are proportional to their chemical equivalent weights.

Explanation of the First Law

If m = mass of the ions liberated, Q = quantity of electricity = $I \times t$ where I is the current and t is the time, then according to the first law $m \propto Q$ or $m = Z\,Q$ or $m = Z\,I\,t$

where Z is a constant and is known as the electrochemical equivalent (*E.C.E.*) of the substance.

If $Q = 1$ coulomb *i.e.* $I = 1$ ampere and $t = 1$ second, then $m = Z$.

Hence, *E.C.E.* of a substance is *equal to the mass of its ions liberated by the passage of one ampere current for one second through its electrolytic solution or by the passage of a charge of one coulomb.*

In fact, the constant Z is composite and it depends on the valency and atomic weight of the substance concerned. Its value is given by $Z = \left(\frac{1}{F}.\frac{a}{v}\right)$ where a is the atomic weight, v the valency and F is Faraday's constant. It is so because m is proportional to atomic weight, since each ion carries a definite charge. Obviously, the charge carried by an ion is proportional to its valency. Now, consider the molecules of sulphuric acid and copper sulphate. The sulphion SO_4^{--} in the acid molecule is combined with two positive hydrogen ions, whereas in $CuSO_4$ molecule, it is combined only with one positive (bivalent) Cu^{++} ion. It is seen that a copper ion being bivalent carries twice the charge of a hydrogen ion which is univalent (monovalent). It means that in order to transfer a given quantity of electricity, only one-half as many bivalent copper ions as univalent hydrogen ions will be required. In other words, greater is the valency of an ion, smaller is the number of ions needed to carry a given quantity of electricity or charge which means that the mass of an ion liberated is inversely proportional to its valency.

$$\therefore \qquad m = \left(\frac{1}{F}.\frac{a}{v}\right)It = \left(\frac{1}{F}.\frac{a}{v}\right)Q = \frac{E}{F}.Q$$

where E is the chemical equivalent weight (= a/v).

The constant F is known as Faraday's constant. The value of Faraday's constant can be found thus. It is found that one coulomb liberates 0.001118 gram of silver. Moreover, silver is univalent and its atomic weight is 107.88. Hence, substituting these values above, we find that

$$0.001118 = \frac{1}{F}.107.88 \times 1$$

$$\therefore \qquad F = 107.88 \,\|\, 0.00118 = 96,500 \text{ coulomb } = 96,500/3600 = 26.8 \text{ Ah}$$

Faraday's constant is defined as *the charge required to liberate one gram-equivalent of any substance.*

For all substances, $\dfrac{\text{chemical equivalent }(E)}{\text{electrochemical equivalent }(Z)}$ = Faraday's constant (F) = 96,500 coulomb or $F = E/Z$

Explanation of the Second Law

Suppose an electric current is passed for the same time through acidulated water, solution of $CuSO_4$ and $AgNO_3$, then for every 1.0078 (or 1.008) gram of hydrogen evolved, 107.88 gram of silver and 31.54 gram of Cu are liberated. The values 107.88 and 31.54 represent the equivalent weights* of silver and copper respectively *i.e.* their atomic weights (as referred to hydrogen) divided by their respective valencies.

Example 9.1. *Calculate the time taken to deposit a coating of nickel 0.05 cm thick on a metal surface by means of a current of 8 A per cm² of surface. Nickel is a divalent metal of atomic weight 59 and of density 9 gram/cm³. Silver has an atomic weight of 108 and an E.C.E. of 1.118 mg/C.*

Solution. Wt. of nickel to be deposited per cm² of surface = $1 \times 0.05 \times 9 = 0.45$ g

Now $$\frac{E.C.E.\text{ of Ni}}{E.C.E.\text{ of Ag}} = \frac{\text{chemical equivalent of } Ni}{\text{chemical equivalent of Ag}}$$

$$\therefore \quad E.C.E.\text{ of } Ni = 1.118 \times 10^{-3} \times \frac{(59/2)}{108} = 0.0003053 \text{ g/C}$$

(chemical equivalent = atomic wt./valency)

Now $m = ZIt$ $\therefore$ $0.45 = 0.0003053 \times 8 \times t$; $t = 184$ second = **3 min. 4 second**

Example 9.2. *If 18.258 gm of nickel are deposited by 100 amp flowing for 10 minutes, how much copper would be deposited by 50 amp for 6 minutes ? Atomic weight of nickel = 58.6 and that of copper 63.18. Valency of both is 2.* **(Electric Power AMIE Sec. Summer 1991)**

Solution. From Faraday's first law, we get $m = ZIt = m\left(\frac{1}{F} \cdot \frac{a}{v}\right)$ It.

If m_1 is the mass of nickel deposited and m_2 that of copper, then

$$m_1 = 18.258 = \left(\frac{1}{F} \cdot \frac{58.6}{2}\right) \times 100(10 \times 60), \left(\frac{1}{F} \cdot \frac{63.18}{2}\right) \times 50 \times (6 \times 60)$$

$$\therefore \quad \frac{m_2}{18.258} = \frac{31.59}{29.3} \times \frac{18,000}{60,000} \quad \therefore m_2 = \mathbf{5.905} \text{ gm}$$

Example 9.3 *The cylindrical surface of a shaft of diameter 12 cm and length 24 cm is to be repaired by electrodeposition of 0.1 cm thick nickel on it. Calculate the time taken if the current used is 100 A. The following data may be used :*

Specific gravity of nickel = 8.9 ; Atomic weight of nickel = 58.7 (divalent); E.C.E of silver = 1.2 mg/C ; Atomic weight of silver = 107.9. **(Elect. Engg. A.M.Ae. S.I. June, 1991)**

Solution. Curved surface of the salt $= \pi D \times l = \pi \times 12 \times 24$ cm²

Thickness of nickel layer = 0.1 cm

Volume of nickel to be deposited $= 12\pi \times 24 \times 0.1 = 90.5$ cm³

Mass of nickel deposited $= 90.5 \times 8.9 = 805.4$ g

---

*The electro-chemical equivalents and chemical equivalents of different substances are inter-related thus :

$$\frac{E.C.E.\text{ of } A}{E.C.E.\text{ of } B} = \frac{\text{chemical equivalent of } A}{\text{chemical equivalent of } B}$$

Further, if m_1 and m_2 are masses of ions deposited at or liberated from an electrode, E_1 and E_2 their chemical equivalents and Z_1 and Z_2 their electrochemical equivalent weights, then

$$m_1/m_2 = E_1/E_2 = Z_1/Z_2$$

$$\text{Chemical equivalent of } Ni = \frac{\text{atomic weight}}{\text{valency}} = \frac{58.7}{2} = 29.35$$

Now $$\frac{E.C.E. \text{ of Ni}}{E.C.E. \text{ of Ag}} = \frac{\text{chemical equivalent of } Ni}{\text{chemical equivalent of Ag}}$$

$\therefore$ $$\frac{E.C.E. \text{ of Ni}}{1.12} = \frac{29.35}{107.9}$$

$\therefore$ $$E.C.E. \text{ of Ni} = 1.12 \times 29.35/107.9 = 0.305 \text{ mg/C}$$

Now $$m = ZIt$$

$\therefore$ $805.4 = 0.305 \times 10^{-3} \times 100 \times t$ $\therefore$ $t = 26,406$ second or **7 hr, 20 min 7 s**

Example 9.4. *The worn-out part of a circular shaft 15 cm in diameter and 30 cm long is to be repaired by depositing on it 0.15 cm of Nickel by an electro-depositing process. Estimate the quantity of electricity required and the time if the current density is to be 25 mA/cm². The current efficiency of the process may be taken as 95 per cent. Take E.C.E. for nickel as 0.3043 mg/coulomb and the density of nickel as 8.9 g/cm³.* **(Elect. Power-I, Bangalore Univ. 1987)**

Solution. Curved surface area of shaft $= \pi D \times l = \pi \times 15 \times 30 = 1414$ cm².
Thickness of nickel layer $= 0.15$ cm
Volume of nickel to be deposited $= 1414 \times 0.15 = 212$ cm³
Mass of nickel to be deposited $= 212 \times 8.9 = 1887$ gram
Now, $m = ZQ$; $Q = m/Z$ $= 1887/0.3043 \times 10^{-3} =$ $\mathbf{62 \times 10^5}$ **C**
Now, current density $= 25 \times 10^{-3}$ A/cm²; $A = 1414$ cm²
$I = 25 \times 10^{-3} \times 1414 = 35.35$ A
Since $Q = It$ $\therefore$ $t = 62 \times 10^5/35.35 = 1.7 \times 10^5$ s = **47.2 hr.**

Example 9.5. *A refining plant employs 1000 cells for copper refining. A current of 5000 A is used and the voltage per cell is 0.25 volt. If the plant works for 100 hours/week, determine the annual output of refined copper and the energy consumption in kWh per tonne. The electrochemical equivalent of copper is 1.1844 kg/1000 Ah.*

(Electric Drives and Utilization, Punjab Univ. Jan. 1991)

Solution. Total cell voltage = 0.25 × 1000 = 250 V; I = 5000 A; plant working time = 100 hour/week = 100 × 52 = 5200 hour/year ; Z = 1.1844 kg/1000 Ah; 1 Ah = 1 × 60 × 60 = 3600 C; $\therefore$ $Z = 1.1844$ kg/1000 × 3600 = 0.329×10^{-6} *kg/C*.

According to Faraday's Law of Electrolysis, the amount of refined copper produced per year is $m = Z\,I\,t = 0.329 \times 10^{-6} \times 5000 \times (5200 \times 3600) = 3079$ kg = **3.079 tonne**
Hence, annual output of refined copper = 3.079 tonne
Energy consumed per year = 250 × 5000 × 5200/1000 = 6500 kWh
This is the energy consumed for refining 3.079 tonne of copper
$\therefore$ Energy consumed per tonne = 6500/3.079 = **2110 kWh/tonne.**

Example 9.6. *A sheet of iron having a total surface area of 0.36 m² is to be electroplated with copper to a thickness of 0.0254 mm. What quantity of electricity will be required ? The iron will be made the cathode and immersed, together with an anode of pure copper, in a solution of copper sulphate.*

[Assume the mass density of copper = 8.96×10^3 kg m^{-3}; E.C.E. of copper —— 32.9×10^{-8} kg C^{-1} Current density = 300 Am^{-2}]

(AMIE Sec. B Utilisation of Electric Power Summer 1992)

Solution. Area over which copper is to be deposited = 0.36 m²
Thickness of the deposited copper = 0.0254×10^{-3} m
Volume of deposited copper = $0.36 \times 0.0254 \times 10^{-3} = 9.144 \times 10^{-6}$ m³
Mass of copper deposited = volume × density
$= 9.144 \times 10^{-6} \times 8.96 \times 10^3 = 0.0819$ kg
Now, $m = Z\,Q$ $\therefore$ $Q = m/Z$ $= 0.0819/32.9 \times 10^{-8} =$ **248936 C**

Tutorial Problems No. 9.1

1. A steady current was passed for 10 minutes through an ammeter in series with a silver voltameter and 3.489 grams of silver were deposited. The reading of the ammeter was 5A. Calculate the percentage error. Electrochemical equivalent of silver = 1.1183 mg/C. **[3.85%]**(*City and Guilds, London*)

2. Calculate the ampere-hours required to deposit a coating of silver 0.05 mm thick on a sphere of 5 cm radius. Assume electrochemical equivalent of silver = 0.001118 and density of silver to be 10.5 g cm^3.
[*Utilization of Elect. Power, A.M.I.E. Summer, 1979*]

9.2. Polarisation or Back E.M.F.

Let us consider the case of two platinum electrodes dipped in dilute sulphuric acid solution. When a small potential difference is applied across the electrodes, no current is found to flow. When, however, the applied voltage is increased, a time comes when a temporary flow of current takes place. The H^+ ions move towards the cathode and O^- ions move towards the anode and are absorbed there. These absorbed ions have a tendency to go back into the electrolytic solution, thereby leaving them as oppositely-charged electrodes. This tendency produces an e.m.f. that is in opposition to the applied voltage which is consequently reduced.

This opposing e.m.f. which is produced in an electrolyte due to the absorption of gaseous ions by the electrolyte from the two electrodes is known as the back e.m.f. of electrolysis or polarisation.

The value of this back e.m.f. is different for different electrolytes. The minimum voltage required to decompose an electrolyte is called the *decomposition* voltage for that electrolyte.

9.3. Value of Back E.M.F.

For producing electrolysis, it is necessary that the applied voltage must be greater than the back e.m.f. of electrolysis for that electrolyte. The value of this back e.m.f. of electrolysis can be found thus :

Let us, for example, find the decomposition voltage of water. We will assume that the energy required to separate water into its constituents (*i.e.* oxygen and hydrogen) is equal to the energy liberated when hydrogen and oxygen combine to form water. Let H kcal be the amount of heat energy absorbed when 9 kg of water are decomposed into 1 kg of hydrogen and 8 kg of oxygen. If the electro-chemical equivalent of hydrogen is Z kg/coulomb, then the passage of q coulomb liberates Zq kg of hydrogen. Now, H is the heat energy required to release 1 kg of hydrogen, hence for releasing Zq kg of hydrogen, heat energy required is HZq kcal or $JHZq$ joules. If E is the decomposition voltage, then energy spent in circulating q coulomb of charge is Eq joule. Equating the two amounts of energies, we have

$$Eq = JHZq \quad \text{or} \quad E = JHZ$$

where J is 4200 joule/kcal.

The e.m.f. of a cell can be calculated by determining the two electrode potentials. The electrode potential is calculated on the assumption that the electrical energy comes entirely from the heat of the reactions of the constituents. Let us take a zinc electrode. Suppose it is given that 1 kg of zinc when dissolved liberates 540 kcal of heat and that the electrochemical equivalent of zinc is 0.338×10^{-6} kg/coulomb. As calculated above,

$$E = JHZ = 4200 \times 540 \times 0.338 \times 10^{-6} = 0.76 \text{ volt}$$

The electrode potentials are usually referred to in terms of the potential of a standard hydrogen electrode *i.e.* an electrode of hydrogen gas at normal atmospheric pressure and in contact with a normal acid solution. In Table No. 9.1 are given the electrode potentials of various elements as referred to the standard hydrogen electrode. The elements are assumed to be in normal solution and at atmospheric pressure.

In the case of Daniel cell having copper and zinc electrodes, the copper electrode potential with respect to hydrogen ion is + 0.345 V and that of the zinc electrode is −0.758 V. Hence, the cell e.m.f. is = 0.345−(−0.758) = 1.103 volt. The e.m.f. of other primary cells can be found in a similar way.

Table No. 9.1

Electrode	*Potential (volt)*	*Electrode*	*Potential (volt)*
Cadmium	−0.398	Mercury	+0.799
Copper	+0.345	Nickel	−0.231
Hydrogen	0	Potassium	−2.922
Iron	−0.441	Silver	+0.80
Lead	−0.122	Zinc	−0.758

Example 9.7. *Calculate the weight of zinc and MnO_2 required to produce 1 ampere-hour in a Leclanche cell.*

Atomic weights : Mn, 55 ; 0, 16; Zn, 65. E.C.E. of hydrogen = 1.04×10^{-8} kg/C.

Solution. 1 ampere-hour = 3600 A-s = 3600 C

Wt. of hydrogen liberated = $Zq = 1.04 \times 10^{-8} \times 3600 = 37.44 \times 10^{-6}$ kg

Now, the chemical reactions in the cell are

$$Zn + 2NH_4Cl = ZnCl_2 + 2NH_3 + H_2$$

It is seen that 1 *atom* of zinc is used up in liberating *two atoms* of hydrogen. In other words, to produce 2 kg of hydrogen, 65 kg of zinc will have to go into chemical combination.

$\therefore$ zinc required to produce 37.44×10^{-6} kg of hydrogen = $37.44 \times 10^{-6} \times 65/2$

$$= 1.217 \times 10^{-3} \text{ kg}$$

The hydrogen liberated combines with manganese dioxide as under :

$$2MnO_2 + H_2 = H_2O + Mn_2O_3$$

Atomic weight of $\qquad MnO_2 = 2(55 + 16 \times 2) = 174$

It is seen that 174 kg of MnO_2 combine with 2 kg of hydrogen, hence Wt. of MnO_2 needed to combine with 37.44×10^{-6} kg of hydrogen = $37.44 \times 10^{-6} \times 174/2 = 3.258 \times 10^{-3}$ kg.

Hence, for 1 ampre-hour, **1.217×10^{-3} kg** of zinc and **3.258×10^{-3} kg** of MnO_2 are needed.

9.4. Primary and Secondary Batteries

An electric battery consists of a number of electrochemical cells, connected either in series or parallel. A cell, which is the basic unit of a battery, may be defined as a power generating device, which is capable of converting stored chemical energy into electrical energy. If the stored energy is inherently present in the chemical substances, it is called a primary cell or a non-rechargeable cell. Accordingly, the battery made of these cells is called primary battery. The examples of primary cells are Leclanche cell, zinc-chlorine cell, alkaline-manganese cell and metal air cells etc.

If, on the other hand, energy is induced in the chemical substances by applying an external source, it is called a secondary cell or rechargeable cell. A battery made out of these cells is called a secondary battery or storage battery or rechargeable battery. Examples of secondary cells are lead-acid cell, nickel-cadmium cell, nickel-iron cell, nickel-zinc cell, nickel-hydrogen cell, silver-zinc cell and high temperature cells like lithium-chlorine cell, lithium-sulphur cell, sodium-sulphur cell etc.

9.5. Classification of Secondary Batteries based on their Use

Various types of secondary batteries can be grouped into the following categories as per their use :

1. *Automotive Batteries* or *SLI Batteries* or *Portable Batteries.*

These are used for starting, lighting and ignition (*SLI*) in internal-combustion-engined vehicles. Examples are; lead-acid batteries, nickel-cadmium batteries etc.

2. *Vehicle Traction Batteries* or *Motive Power Batteries* or *Industrial Batteries.*

These are used as a motive power source for a wide variety of vehicles. Lead-acid batteries, nickel-iron batteries, silver-zinc batteries have been used for this purpose. A number of advance batteries including high-temperature batteries are under development for electric vehicle (*EV*) use. These high-temperature batteries like sodium-sulphur and lithium-iron sulphide have energy densities in the range of 100–120 Wh/kg.

3. *Stationary Batteries.*

These fall into two groups (*a*) standby power system which is used intermittently and (*b*) load-levelling system which stores energy when demand is low and, later on, uses it to meet peak demand.

9.6. Classification of Lead Storage Batteries

Lead storage batteries may be classified according to the service which they provide.

1. *SLI Batteries*

The primary purpose of these batteries is to supply power for engine starting, lighting and ignition (*SLI*) of vehicles propelled by IC engines such as automobiles, buses, lorries and other heavy road vehicles and motor cycles etc. Usually, these batteries provide 12 V and consist of six series–connected lead-acid cells with capacity of the order of 100 Ah. Their present-day energy density is about 45 Wh/kg and 75 Wh/dm^3.

These days 'maintenance–free' (*MF*) *SLI* batteries have been designed, which do not require the addition of water throughout their normal service life of 2–5 years. *MF* versions of the *SLI* batteries are constructed of such material that no gassing occurs during charging. In *MF* batteries, the electrolyte is either absorbed within the microporous separators and the plates or is immobilized with suitable gelling agents.

These days the *SLI* batteries are charged from an alternator (*AC* generator) and not from dynamo (*DC* generator). The alternating current produced by the alternator is converted into direct current by a full-wave bridge rectifier, which uses semi-conductor diodes. In this arrangement, no cutout is needed and the transistorised voltage controller regulates the alternator output to suit the electrical load and the state of charge of the battery. The battery is charged under constant-voltage condition.

2. *Vehicle Traction Batteries*

The recent universal concern over the levels of toxic gases (particularly in urban areas) emitted by the *IC* engines has revived interest in electric traction. There has been great development in the use of battery-powered vehicles, primarily industrial trucks and commercial road vehicles of various types like 'milk floats' (*i.e.* bottled-milk delivery trucks), fork lift trucks, mining, airport tractors, aircraft service vehicles, electric cars and, more recently, in robotics and guided vehicles.

Traction batteries are of higher quality than *SLI* batteries. They provide constant output voltage, high volumetric capacity, good resistance to vibration and a long service life. They can withstand prolonged and deep discharges followed by deep recharges usually on a daily basis. The voltage of traction batteries varies from 12 V to 240 V and they have a cycle life of 1000–1500 cycles.

A number of advanced batteries are under development for *EV* use (*i*) room temperature batteries like zinc-nickel oxide battery (75 Wh/kg) and zinc–chlorine hydrate battery (80 Wh/kg) and (*ii*) high-temperature batteries like sodium-sulphur battery (120 Wh/kg) and lithium-iron sulphide battery (100 wh/kg).

3. *Stationary Batteries*

Their use falls into two groups :

(*a*) as standby power system and (*b*) as load-levelling system.

In the standby applications, the battery is used to power essential equipment or to provide alarms or emergency lighting, in case of break-down in the main power supply. Standby applications have increased in recent years with increasing demand for uninterruptable power systems (*UPS*) and a tremendous growth in new telecommunication networks. The *UPS* provides 'clean' a.c supply free of sags or surges in the line voltage, frequency variations, spikes and transients to modern computer and electronic equipment. Banks of sealed lead-acid (*SLA*) standby batteries have been recently used in telecommunication systems and for *UPS* applications.

Recently, advanced lead-acid batteries have been used for load-levelling purposes in the electric generating plants. A 100 *M* Wh lead-acid battery load-levelling system could occupy a building two and a half storey high and an area of about 250,000 m^2.

9.7. Parts of a Lead-acid Battery

A battery consists of a number of cells and each cell of the battery –consists of (*a*) positive and negative plates (*b*) separators and (*c*) electrolyte, all contained in one of the many compartments of the battery container.* Different parts of a lead-acid battery are as under :

---

*The most common form of lead-acid cell used for marine applications is the tubular cell which consists of 'armoured' tubular positive plate and standard flat negative plate.

(*i*) *Plates.* A plate consists of a lattice type of grid of cast antimonial lead alloy which is covered with *active* material (Art. 9.8). The grid not only serves as a support for the fragile active material but also conducts electric current. Grids for the positive and negative plates are often of the same design although negative plate grids are made somewhat lighter. As discussed in Art. 9.10, positive plates are usually Plante plates whereas negative plates are generally of Faure or pasted type.

(*ii*) *Separators.* These are thin sheets of a porous material placed between the positive and negative plates for preventing contact between them and thus avoiding internal short-circuiting of the battery. A separator must, however, be sufficiently porous to allow diffusion or circulation of electrolyte between the plates. These are made of especially-treated cedar wood, glass wool mat, microporous rubber (mipor), microporous plastics (plastipore, miplast) and perforated p.v.c. as shown in Fig. 9.1. In addition to good porosity, a separator must possess high electrical resistance and mechanical strength.

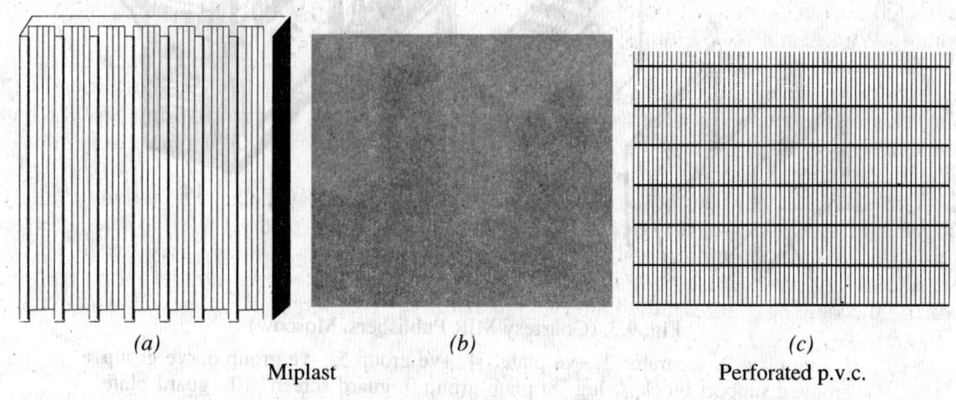

(*a*) (*b*) Miplast (*c*) Perforated p.v.c.

Fig. 9.1

(*iii*) *Electrolyte.* It is dilute sulphuric acid which fills the cell compartment to immerse the plates completely.

(*iv*) *Container.* It may be made of vulcanised rubber or moulded hard rubber (ebonite), moulded plastic, ceramics, glass or celluloid. The vulcanised rubber containers are used for car service, while glass containers are superior for lighting plants and wireless sets. Celluloid containers are mostly used for portable wireless set batteries. A single monoblock type container with 6 compartments generally used for starting batteries is shown in Fig. 9.2. Full details of a Russian 12-CAM-28 lead-acid battery parts are shown in Fig. 9.3. Details of some of these parts are as follows :

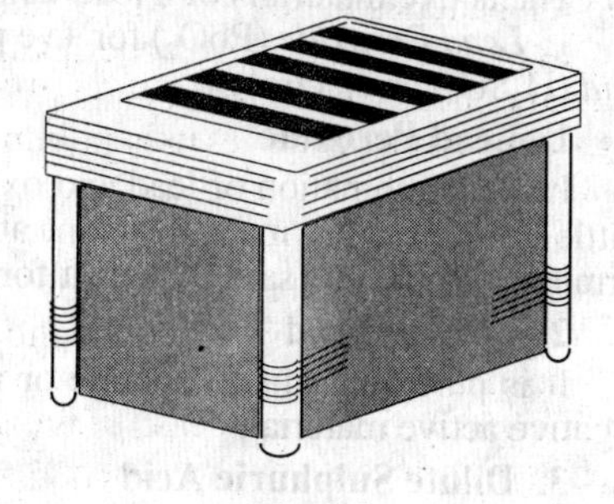

Fig. 9.2

(*a*) *Bottom Grooved Support Blocks.* These are raised ribs, either fitted in the bottom of the container or made with the container itself. Their function is to support the plates and hold them in position and at the same time protect them from short-circuits that would otherwise occur as a result of fall of the active material from the plates onto the bottom of the container.

(*b*) *Connecting Bar.* It is the lead alloy link which joins the cells together in series connecting the positive pillar of one cell to the negative pillar of the next one.

(*c*) *Terminal Post* or *Pillar.* It is the upward extension from each connecting bar which passes through the cell cover for cable connections to the outside circuits. For easy indentification, the negative terminal post is smaller in diameter than the positive terminal post.

(*d*) *Vent Plugs* or *Filler Caps.* These are made of polystyrene or rubber and are usually screwed in the cover. Their function is to prevent escape of electrolyte but allow the free exit of the gas. These can be easily removed for topping up or taking hydrometer readings.

(*e*) *External Connecting Straps.* These are the antimonial lead alloy flat bars which connect the positive terminal post of one cell to the negative of the next across the top of the cover. These are of very solid construction especially in starting batteries because they have to carry very heavy currents.

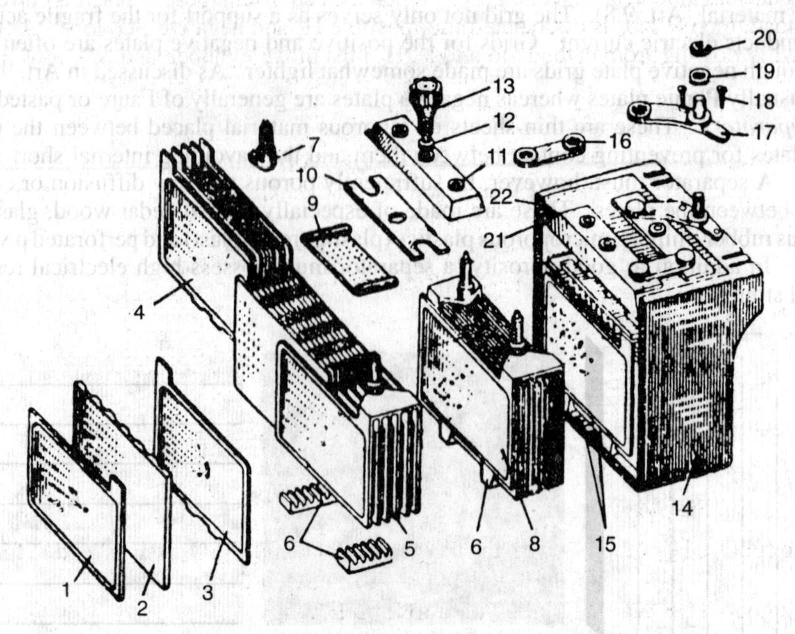

Fig. 9.3. (Courtesy MIR Publishers. Moscow)

1. –ve plate 2. separator 3. +ve plate. 4. +ve group 5. –ve group 6.–ve group grooved support block 7. lug 8. plate group 9. guard screen 10. guard plate 11. cell cover 12. plug washer 13. vent plug 14. monoblock jar 15. supporting prisms of + ve group 16. inter-cell connector 17. terminal lug 18. screw 19. washer 20. nut 21. rubber packing 22. sealing compound.

9.8. Active Materials of a Lead-acid Cell

Those substances of the cell which take active part in chemical combination and hence absorb or produce electricity during charging or discharging, are known as *active materials* of the cell.

The active materials of a lead-acid cell are :

1. *Lead Peroxide* (PbO_2) for +ve plate 2. *Sponge Lead* (Pb) for –ve plate 3. *Dilute Sulphuric Acid* (H_2SO_4) as electrolyte.

1. Lead Peroxide

It is a combination of lead and oxygen, is dark chocolate brown in colour and is quite hard but brittle substance. It is made up of one atom of lead (Pb) and two atoms of oxygen (O_2) and its chemical formula is PbO_2. As said earlier, it forms the positive active material.

2. Sponge Lead

It is pure lead in soft sponge or porous condition. Its chemical formula is Pb and forms the negative active material.

3. Dilute Sulphuric Acid

It is approximately 3 parts water and one part sulphuric acid. The chemical formula of the acid is H_2SO_4. The positive and negative plates are immersed in this solution which is known as electrolyte. It is this medium through which the current produces chemical changes.

Hence, the lead-acid cell depends for its action on the presence of two plates covered with PbO_2 and Pb in a solution of dilute H_2SO_4 of specific gravity 1.21 or nearabout.

Lead in the form of PbO_2 or sponge Pb has very little mechanical strength, hence it is supported by plates of pure lead. Those plates covered with or otherwise supporting PbO_2 are known as +ve plates and those supporting sponge lead are called –ve plates. The +ve and –ve plates are arranged alternately and are connected to two common +ve and –ve terminals. These plates are assembled in a suitable jar or container to make a complete cell as discussed in Art. 9.4 above.

9.9. Chemical Changes

(*i*) DISCHARGING (Fig. 9.4)

When the cell is fully charged, its positive plate or anode is PbO_2 (dark chocolate brown) and the negative plate or cathode is Pb (slate grey). When the cell discharges *i.e.* it sends current through the external load, then H_2SO_4 is dissociated into positive H_2 and negative SO_4 ions. As the current within the cell is flowing from cathode to anode, H_2 ions move to anode and SO_4 ions move to the cathode.

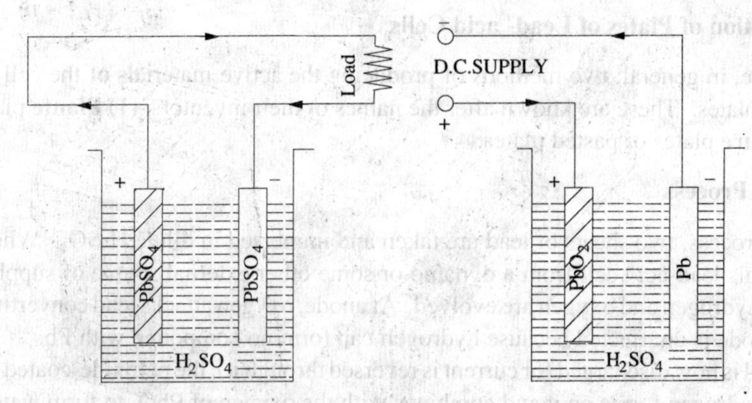

Fig. 9.4 **Fig. 9.5**

At anode (PbO_2), H_2 combines with the oxygen of PbO_2 and H_2SO_4 attacks lead to form $PbSO_4$.

$$PbO_2 + H_2 + H_2SO_4 \longrightarrow PbSO_4 + 2H_2O$$

At the cathode (Pb), SO_4 combines with it to form $PbSO_4$

$$Pb + SO_4 \longrightarrow PbSO_4$$

It will be noted that during discharging :

(*i*) *Both anode and cathode become $PbSO_4$ which is somewhat whitish in colour.*

(*ii*) *Due to formation of water, specific gravity of the acid decreases.*

(*iii*) *Voltage of the cell decreases.* (*iv*) *The cell gives out energy.*

(*ii*) **CHARGING (Fig. 9.5)**

When the cell is recharged, the H_2 ions move to cathode and SO_4 ions go to anode and the following changes take place :

At Cathode $PbSO_4 + H_2 \longrightarrow Pb + H_2SO_4$

At Anode $PbSO_4 + SO_4 + 2H_2O \longrightarrow PbO_2 + 2H_2SO_4$

Hence, the anode and cathode again become PbO_2 and Pb respectively.

It will be noticed that during charging :

(*i*) *The anode becomes dark chocolate brown in colour (PbO_2) and cathode becomes grey metallic lead (Pb)*

(*ii*) *Due to consumption of water, specific gravity of H_2SO_4 is increased.*

(*iii*) *There is a rise in voltage.* (*iv*) *Energy is absorbed by the cell.*

The charging and discharging of the cell can be represented by a single reversible equation given below :

$$\underset{\text{Pos. Plate}}{PbO_2} + 2H_2SO_4 + \underset{\text{Neg. Plate}}{Pb} \underset{\text{Charge}}{\overset{\text{Discharge}}{\rightleftharpoons}} \underset{\text{Pos. Plate}}{PbSO_4} + 2H_2O + \underset{\text{Neg. Plate}}{PbSO_4}$$

For discharge, the equation should be read from left to right and for charge from right to left.

Example 9.8. *Estimate the necessary weight of active material in the positive and negative plates of a lead-acid secondary cell per ampere-hour output (atomic weight of lead 207, valency 2, E.C.E. of hydrogen 0.0104×10^{-6} kg/C)*

Solution. Wt. of hydrogen evolved per ampere-hour $= 0.0104 \times 10^{-6} \times 3{,}600$

$= 37.44 \times 10^{-6}$ kg

During discharge, reaction at cathode is $Pb + H_2SO_4 = PbSO_4 + H_2$

As seen, 207 kg of lead react chemically to liberate 2 kg of hydrogen.

Hence, weight of Pb needed per ampere-hour $= 37.44 \times 10^{-6} \times 207/2 = 3.876 \times 10^{-3}$ kg

At anode the reaction is: $PbO_2 + H_2 \rightarrow PbO + H_2O$

Atomic weight of $PbO_2 = (207 + 32) = 239$

$\therefore$ Wt. of PbO_2 going into combination per ampere-hour = $37.44\times10^{-6} \times 239/2 =$ **4.474×10^{-3} kg**.

Therefore, quantity of active material required per ampere-hour is : lead **3.876×10^{-3} kg** and lead peroxide **4.474×10^{-3} kg.**

9.10. Formation of Plates of Lead- acid Cells

There are, in general, two methods of producing the active materials of the cell and attaching them to lead plates. These are known after the names of their inventors. (1) **Plante** plates or formed plates (2) **Faure** plates or pasted plates.

9.11. Plante Process

In this process, two sheets of lead are taken and immersed in dilute H_2SO_4. When a current is passed into this lead-acid cell from a dynamo or some other external source of supply, then due to electrolysis, hydrogen and oxygen are evolved. At anode, oxygen attacks lead converting it into PbO_2 whereas cathode is unaffected because hydrogen can form no compound with Pb.

If the cell is now discharged (or current is reversed through it), the peroxide-coated plate becomes cathode, so hydrogen forms on it and combines with the oxygen of PbO_2 to form water thus :

$$PbO_2 + 2H_2 \longrightarrow Pb + 2H_2O$$

At the same time, oxygen goes to anode (the plate previously unattacked) which is lead and reacts to form PbO_2. Hence, the anode becomes covered with a thin film of PbO_2.

By continuous reversal of the current or by charging and discharging the above electrolytic cell, the thin film of PbO_2 will become thicker and thicker and the polarity of the cell will take increasingly longer time to reverse. Two lead plates after being subjected to hundreds of reversals will acquire a skin of PbO_2 thick enough to possess sufficiently high capacity. This process of making positive plates is known as *formation*. The negative plates are also made by the same process. They are turned from positive to negative plates by reversing the current through them until whole PbO_2 is converted into sponge lead. Although Plante positives are very commonly used for stationary work, Plante negatives have been completely replaced by the Faure or pasted type plates as discussed in Art. 9.13. However, owing to the length of time required and enormous expenditure of electrical energy, this process is commercially impracticable. The process of formation can be accelerated by forming agents such as acetic, nitric or hydrochloric acid or their salts but still this method is expensive and slow and plates are heavy.

9.12. Structure of Plante Plates

It is seen that since active material on a Plante plate consists of a thin layer of PbO_2 formed on and from the surface of the lead plate, it must be made of large superficial area in order to get an appreciable volume of it. An ordinary lead plate subjected to the forming process as discussed above will have very small capacity. Its superficial area and hence its capacity, can be increased by grooving or laminating. Fig. 9.6 shows a Plante positive plate which consists of a pure lead grid with finely laminated surfaces. The construction of these plates consists of a large number of thin vertical laminations which are strengthened at intervals by horizontal binding ribs. This results in an increase of the superficial area 10 to 12 times that possessed by a plain lead sheet of the same overall dimensions. The above design makes possible the expansion of the plate structure to accommodate the increase in mass and the value of the active material (PbO_2) which takes place when the cell goes through a series of chemical changes during each cycle of charge or discharge. The expansion of the plate structure takes place downwards where there is room left for such purpose. Usually, a Plante positive plate expands by about 10% or so of its length during the course of its useful life.

Another type of Plante positive plate is the 'rosette' plate which consists of a perforated cast grid or framework of lead alloy with 5 to 12 per cent of antimony holding rosettes or spirals of corrugated pure lead tape. The rosettes (Fig. 9.7) provide the active material of the positive plate and, during formation, they expand in the holes of the grid which are countersunk on both sides of the grid. The advantages of such plates are that the lead-antimony grid is itself unaffected by the chemical action and the complete plate is exceptionally strong.

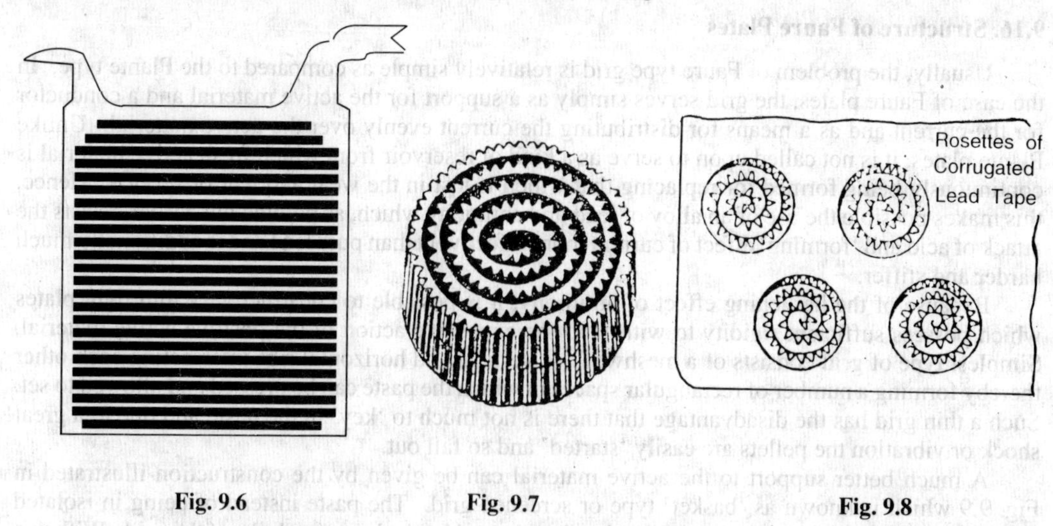

Fig. 9.6 **Fig. 9.7** **Fig. 9.8**

Other things being equal, the life of a Plante plate is in direct proportion to the weight of lead metal in it, because as the original layer of PbO_2 slowly crumbles away during the routine charging and discharging of the cell, fresh active material is formed out of the underlying lead metal. Hence, the capacity of such a plate lasts as long as the plate itself. In this respect, Plante plate is superior to the Faure or pasted plate.

9.13. Faure Process

In the making of Faure plates, the active material is mechanically applied instead of being electrochemically produced out of lead plate itself as in Plante process. The active material which is in the form of red lead (Pb_3O_4) or litharge PbO or the mixture of the two in various proportions, is pressed into the interstices of a thin lead grid or lattice work of intersecting ribs which also serves as conductor of current. The plates after being thus pasted are allowed to dry and harden, are then assembled in weak solution of H_2SO_4 of specific gravity 1.1 to 1.2 and are formed by passing an electric current between them. If plates are meant to be positive, they are connected up as anodes, if negative, then as cathodes. The oxygen evolved at the anode converts the lead oxide (Pb_3O_4) into peroxide (PbO_2) and at cathode the hydrogen reduces PbO to sponge lead by abstracting the oxygen.

9.14. Positive Pasted Plates

Formation of positive plates involves converting lead oxide into PbO_2. A high lead oxide like Pb_3O_4 is used for economy both in current and time, although in practice, a mixture of Pb_3O_4 and PbO is taken–the latter being added to assist in the setting or cementation of the plate.

9.15. Negative Pasted Plates

Faure process is much better adopted for making a negative rather than a positive plate. The negative material *i.e.* sponge lead is quite tough instead of being hard and brittle like PbO_2 and, moreover, it undergoes a comparatively negligible change in volume during the charging and discharging of the cell. Hence, it has no tendency to disintegrate or shed out of the grid although it does tend to lose its porosity and become dense and so lose capacity. Hence, in the manufacture of the pasted negatives, a small percentage of certain substances like powdered pumic or graphite or magnesium sulphate or barium sulphate is added to increase the porosity of the material. If properly handled, a paste made with H_2SO_4, glycerine and PbO (or mixture of PbO and Pb_3O_4) results in a very good negative, because glycerine is carbonised during formation and so helps in keeping the paste porous.

Faure plates are in more general use because they are cheaper and have a high (capacity/weight) ratio than Plante plates. Because of the lightness and high capacity/weight ratio, such plates are used practically for all kinds of portable service like electric vehicles, train lighting, car-lighting and starting etc. But their life is shorter as compared to Plante plates.

9.16. Structure of Faure Plates

Usually, the problem of Faure type grid is relatively simple as compared to the Plante type. In the case of Faure plates, the grid serves simply as a support for the active material and a conductor for the current and as a means for distributing the current evenly over the active material. Unlike Plante plates, it is not called upon to serve as a kind of reservoir from which fresh active material is continuously being formed for replacing that which is lost in the wear and tear of service. Hence, this makes possible the use of an alloy of lead and antimony which, as pointed out earlier, resists the attack of acid and 'forming' effect of current more effectively than pure lead and is additionally much harder and stiffer.

Because of the hardening effect of antimony, it is possible to construct very thin light plates which possess sufficient rigidity to withstand the expensive action of the positive active material. Simplest type of grid consists of a meshwork of vertical and horizontal ribs intersecting each other thereby forming a number of rectangular spaces in which the paste can be pressed and allowed to set. Such a thin grid has the disadvantage that there is not much to 'key' in the paste and due to a great shock or vibration the pellets are easily 'started' and so fall out.

A much better support to the active material can be given by the construction illustrated in Fig. 9.9 which is known as 'basket' type or screened grid. The paste instead of being in isolated pellets forms a continuous sheet contained and supported by the horizontal ribs of the grid. With this arrangement the material can be very effectively keyed in.

Another type of grid structure used in pasted plates is shown in Fig. 9.10.

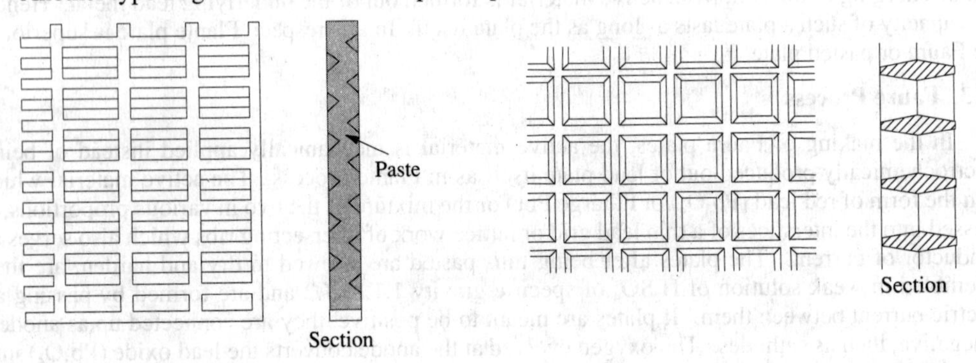

Fig. 9.9

Fig. 9.10

9.17. Comparison : Plante and Faure Plates

1. Plante plates have a longer life and can withstand rapid discharging (as in traction work) better than Faure's.
2. They are less liable to disintegration when in use than Faure's plates.
3. They are heavier and more expensive than Faure plates.
4. Plante plates have less capacity-to-weight ratio, values being 12 to 21 Ah per kg of plate, the corresponding values for Faure plate being 65 to 90 Ah/kg.

9.18. Internal Resistance and Capacity of a Cell

The secondary cell possesses internal resistance due to which some voltage is lost in the form of potential drop across it when current is flowing. Hence, the internal resistance of the cell has to be kept to the minimum.

One obvious way to lessen internal resistance is to increase the size of the plates. However, there is a limit to this because the cell will become too big to handle. Hence, in practice, it is usual to multiply the number of plates inside the cell and to join all the negative plates together and all the positives ones together as shown in Fig. 9.11.

The effect is equivalent to joining many cells in parallel. At the same time, the length of the electrolyte between the electrodes is decreased with a consequent reduction in the internal resistance.

The 'capacity' of a cell is given by the product of current in amperes and the time in hours during which the cell can supply current until its e.m.f. falls to 1.8 volt. It is expressed in ampere-hour (Ah).

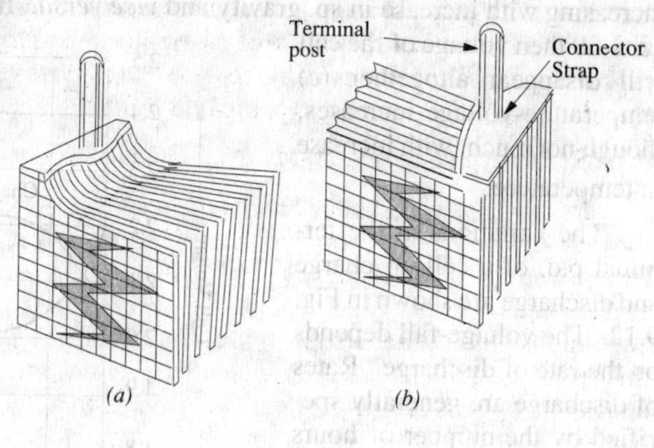

Fig. 9.11

The interlacing of plates not only decreases the internal resistance but additionally increases the capacity of the cell also. There is always one more negative plate than the positive plates *i.e.* there is a negative plate at both ends. This gives not only more mechanical strength but also assures that both sides of a positive plate are used.

Since in this arrangement, the plates are quite close to each other, something must be done to make sure that a positive plate does not touch the negative plate otherwise an internal short-circuit will take place. The separation between the two plates is achieved by using separators which, in the case of small cells, are made of treated cedar wood, glass, wool mat, microporous rubber and microporous plastic and in the case of large stationary cells, they are in the form of glass rods.

9.19. Two Efficiencies of the Cell

The efficiency of a cell can be considered in two ways :

1. *The quantity or ampere-hour (Ah) efficiency*
2. *The energy or watt-hour (Wh) efficiency*

The Ah efficiency does not take into account the varying voltages of charge and discharge. The Wh efficiency does so and is always less than Ah efficiency because average p.d. during discharging is less than that during charging. Usually, during discharge the e.m.f. falls from about 2.1 V to 1.8 V whereas during charge it rises from 1.8 volt to about 2.6 V.

$$\text{Ah efficiency} = \frac{\text{amp-hour discharge}}{\text{amp-hour charge}}$$

The Ah efficiency of a lead-acid cell is normally between 90 to 95%, meaning that about 100 Ah must be put back into the cell for every 90–95 Ah taken out of it. Because of gassing which takes place during the charge, the Ah available for delivery from the battery decreases. It also decreases (*i*) due to self-discharge of the plates caused due to local reactions and (*ii*) due to leakage of current because of faulty insulation between the cells of the battery.

The Wh efficiency varies between 72–80%.

If Ah efficiency is given, Wh efficiency can be found from the following relation :

$$\text{Wh efficiency} = \text{Ah efficiency} \times \frac{\text{average volts on discharge}}{\text{average volts on charge}}$$

From the above, it is clear that anything that increases the charge volts or reduces the discharge volts will decrease Wh efficiency. Because high charge and discharge rates will do this, hence it is advisable to avoid these.

9.20. Electrical Characteristics of the Lead-acid Cell

The three important features of an accumulator, of interest to an engineer, are (1) voltage (2) capacity and (3) efficiency.

1. Voltage

The open-circuit voltage of a fully-charged cell is approximately 2.2 volt. This value is not fixed but depends on (*a*) length of time since it was last charged (*b*) specific gravity-voltage increasing with increase in sp. gravity and *vice versa*. If sp. gravity comes near to density of water *i.e.* 1.00 then voltage of the cell will disappear altogether (*c*) temperature-voltage increases, though not much, with increase in temperature.

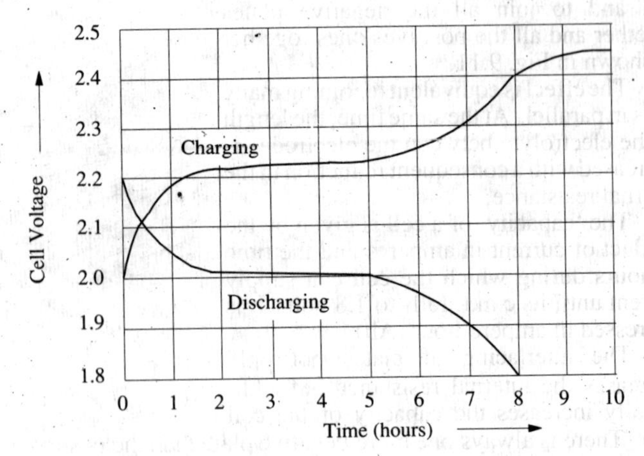

Fig. 9.12

The variations in the terminal p.d. of a cell on charge and discharge are shown in Fig. 9.12. The voltage-fall depends on the rate of discharge. Rates of discharge are generally specified by the number of hours during which the cell will sustain the rate in question before falling to 1.8 V. The voltage falls rapidly in the beginning (rate of fall depending on the rate of discharge), then very slowly up to 1.85 and again suddenly to 1.8 V. The voltage should not be allowed to fall to lower than 1.8 V, otherwise hard insoluble lead sulphate is formed on the plate which increases the internal resistance of the cell.

The general form of the voltage-time curves corresponding to 1-, 3-, 5- and 10-hour rates of discharge, is shown in Fig. 9.13 *i.e.* corresponding to the steady currents which would discharge the cell in the above-mentioned times (in hours). It will be seen that both the terminal voltage and the rate at which the voltage falls, depend on the rate of discharge. The more rapid fall in voltage at higher rates of discharge is due to the rapid increase in the internal resistance of the cell.

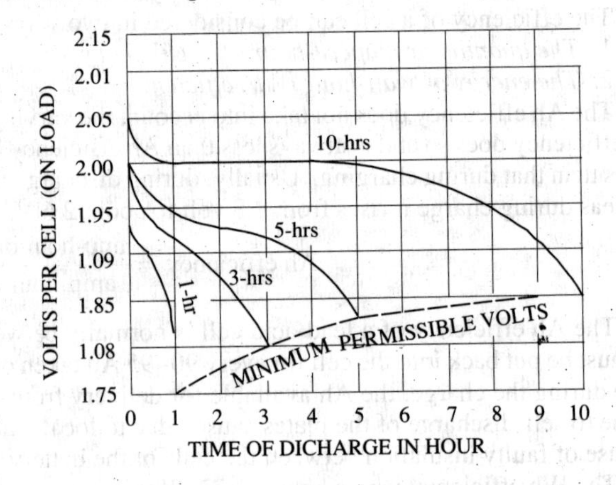

Fig. 9.13

During charging, the p.d. increases (Fig. 9.12). The curve is similar to the discharge curve reversed but is everywhere higher due to the increased density of H_2SO_4 in the pores of the positive plate.

2. Capacity

It is measured in amp-hours (Ah). The capacity is always given at a specified rate of discharge (10-hour discharge in U.K., 8-hour discharge in U.S.A.). However, motor-cycle battery capacity is based on a 20-hour rate (at 30°C). The capacity depends upon the following :

(*a*) *Rate of discharge*. The capacity of a cell, as measured in Ah, depends on the discharge rate. It decreases with increased rate of discharge. Rapid rate of discharge means greater fall in p.d. of the cell due to internal resistance of the cell. Moreover, with rapid discharge the weakening of the acid in the pores of the plates is also greater. Hence, the chemical change produced at the plates by

1 ampere for 10 hours is not the same as produced by 2 A for 5 hours or 4 A for 2.5 hours. It is found that a cell having a 100 Ah capacity at 10 hour discharge rate, has its capacity reduced to 82.5 Ah at 5-hour rate and 50 Ah at 1-hour rate.

The variation of capacity with discharge rate is shown in Fig. 9.14.

(*b*) *Temperature*. At high temperature,

(*i*) chemical reactions within the cell take place more vigorously.

(*ii*) the resistance of the acid is decreased and

(*iii*) there is a better diffusion of the electrolyte.

Hence, high temperature increases the capacity of the lead-acid cell. Apparently, it is better to operate the battery at a high temperature. However, at high temperatures:

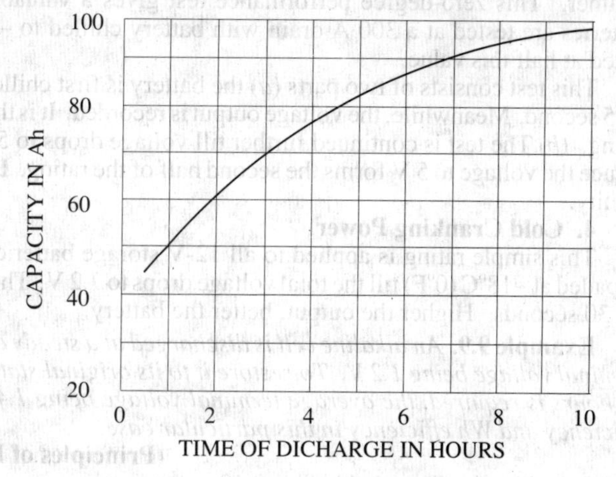

Fig. 9.14

(*a*) the acid attacks the antimony-lead alloy grid, terminal posts and wooden separators.

(*b*) the paste is rapidly changed into lead sulphate. Sulphation is always accompanied by expansion of paste particularly at the positive plates and results in buckling and cracking of the grid.

Hence, it is not advisable to work batteries above 40°C.

As temperature is lowered, the speed of chemical reactions is decreased. Moreover, cell resistance also increases. Consequently, the capacity of the cell decreases with decreases in temperature till at freezing point the capacity is reduced to zero even though the battery otherwise be fully charged.

(*c*) *Density of electrolyte*. As the density of electrolyte affects the internal resistance and the vigour of chemical reaction, it has an important effect on the capacity. Capacity increases with the density.

(*d*) *Quantity of active material*. Since production of electricity depends on chemical action taking place within the cells, it is obvious that the capacity of the battery must depend directly upon the kind and amount of the active material employed. Consider the following calculations :

The gram-equivalent of lead is 103.6 gram and Faraday's constant is 96,500 coulombs which is = 96,500/3600 = 26.8 Ah. Hence, during the delivery of one Ah by the cell, the quantity of lead expended to form lead sulphate at the negative plate is 103.6/26.8 = 3.86 gram.

Similarly, it can be calculated that, at the same time, 4.46 gram of PbO_2 would be converted into lead sulphate at the positive plate while 3.66 gram of acid would be expended to form 0.672 gram of water. It is obvious that for obtaining a cell of a greater capacity, it is necessary to provide the plates with larger amounts of active material.

3. Efficiency

It has already been discussed in Art. 9.19

9.21. Battery Ratings

Following standards have been adopted, both by industry and government organisations to get a fair picture of battery quality :

1. Ampere-hour Capacity

It is a function of the total plate area *i.e.* size of the individual plate multiplied by the number of plates. For measuring this capacity, the battery is discharged continuously for 20 hours and its current output supplied to a standard load is measured. Suppose that a battery delivers 4 A current for 20 hours. Hence, its rating is 80 Ah which is stamped on the battery case.

2. Reserve Capacity

It is one of the newly-developed rating standards and is more realistic because it provides a double-check on the Ah figures. This capacity is given by the number of minutes a battery will tolerate a 25 A drain without dropping below 10.5 V. Higher this rating, better the battery.

3. Zero Cranking Power

It was the first cold weather rating and is applicable in relation to crafts which ply in freezing weather. This zero-degree performance test gives a valuable insight into battery quality. Large batteries are tested at a 300 A drain with battery chilled to – 18°C (0°F) whereas smaller sizes are tested at half this value.

This test consists of two parts (*a*) the battery is first chilled to – 18°C (0°F) and the load applied for 5 second. Meanwhile, the voltage output is recorded. It is the first part of the zero–cranking-power rating. (*b*) The test is continued further till voltage drops to 5 V. The number of minutes it takes to reduce the voltage to 5 V forms the second half of the rating. Higher both the digits, better the battery quality.

4. Cold Cranking Power

This simple rating is applied to all 12-V storage batteries regardless of their size. The battery is loaded at –18°C (0°F) till the total voltage drops to 7.2 V. The output current in amperes is measured for 30 seconds. Higher the output, better the battery.

Example 9.9. *An alkaline cell is discharged at a steady current of 4 A for 12 hours, the average terminal voltage being 1.2 V. To restore it to its original state of charge, a steady current at 3 A for 20 hours is required, the average terminal voltage being 1.44 V. Calculate the ampere-hour (Ah) efficiency and Wh efficiency in this particular case.*

(Principles of Elect. Engg.-I, Jadavpur Univ. 1987)

Solution. As discussed in Art. 9.19

$$\text{Ah efficiency} = \frac{\text{Ah of discharge}}{\text{Ah of charge}} = \frac{12 \times 4}{20 \times 3} = 0.8 \text{ or } \mathbf{80\%}$$

$$\text{Wh efficiency} = \text{Ah effi.} \times \frac{\text{Av. volts on discharge}}{\text{Av. volts on charge}} = \frac{0.8 \times 1.2}{1.44} = \mathbf{0.667} \text{ or } \mathbf{66.7\%}$$

Example 9.10. *A discharged battery is charged at 8 A for 2 hours after which it is discharged through a resistor of R Ω. If discharge period is 6 hours and the terminal voltage remains fixed at 12 V, find the value of R assuming the Ah efficiency of the battery as 80%.*

Solution. Input amp-hours $= 8 \times 2 = 16$

Efficiency $= 0.8 \quad \therefore$ Ah output $16 \times 0.8 = 12.8$

Discharge current $= 12.8/6\text{A} \quad \therefore \quad R = \dfrac{12}{12.8/6} = \dfrac{6 \times 12}{12.8} = \mathbf{5.6\ \Omega}$

9.22. Indications of a Fully-charged Cell

The indications of a fully-charged cell are :

(*i*) *gassing* (*ii*) *voltage* (*iii*) *specific gravity and* (*iv*) *colour of the plates.*

(i) Gassing

When the cell is fully charged, it freely gives off hydrogen at cathode and oxygen at the anode, the process being known as "Gassing". Gassing at *both* plates indicates that the current is no longer doing any useful work and hence should be stopped. Moreover, when the cell is fully charged, the electrolyte assumes a milky appearance.

(ii) Voltage

The voltage ceases to rise when the cell becomes fully-charged. The value of the voltage of a fully-charged cell is a variable quantity being affected by the rate of charging, the temperature and specific gravity of the electrolyte etc. The approximate value of the e.m.f. is 2.1 V or so.

(iii) Specific Gravity of the Electrolyte

A third indication of the state of charge of a battery is given by the specific gravity of the electrolyte. We have seen from the chemical equations of Art. 9.9, that during discharging, the density of electrolyte decreases due to the production of water, whereas it increases during charging due to the absorption of water. The value of density when the cell is fully charged is 1.21 and 1.18 when discharged up to 1.8 V. Specific gravity can be measured with a suitable hydrometer which consists of a float, a chamber for the electrolyte and a squeeze bulb.

(iv) Colour

The colour of plates, on full charge, is deep chocolate brown for positive plate and clear slate gray for negative plate and the cell looks quite brisk and alive.

9.23. Applications of Lead-acid Batteries

Storage batteries are these days used for a great variety and range of purposes, some of which are summarised below :

1. In Central Stations for supplying the whole load during light load periods, also to assist the generating plant during peak load periods, for providing reserve emergency supply during periods of plant breakdown and finally, to store energy at times when load is light for use at times when load is at its peak value.

2. In private generating plants both for industrial and domestic use, for much the same purpose as in Central Stations.

3. In sub-stations, they assist in maintaining the declared voltage by meeting a part of the demand and so reducing the load on and the voltage drop in, the feeder during peak-load periods.

4. As a power source for industrial and mining battery locomotives and for road vehicles like cars and trucks.

5. As a power source for submarines when submerged.

6. Marine aplications include emergency or stand-by duties in case of failure of ship's electric supply, normal operations where batteries are subjected to regular cycles of charge and discharge and for supplying low-voltage current to bells, telephones, indicators and warning systems etc.

7. For petrol motor-car starting and ignition etc.

8. As a low voltage supply for operating purposes in many different ways such as high-tension switchgear, automatic telephone exchange and repeater stations, broadcasting stations and for wireless receiving sets.

9.24. Voltage Regulators

As explained in Art. 9.20, the voltage of a battery varies over a considerable range while under discharge. Hence, it is necessary to find some means to control the battery voltage upto the end so as to confine variations within reasonable limits–these limits being supplied by the battery.

The voltage control systems may be hand-operated or automatic. The simplest form of hand-operated control consists of a rheostat having a sufficient number of steps so that assistance can be inserted in the circuit when battery is fully charged and gradually cut out as the discharge continues, as shown in Fig. 9.15.

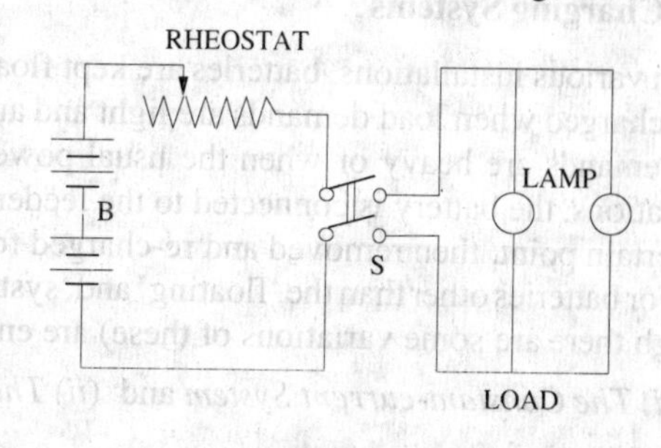

Fig. 9.15

The above system can be designed for automatic operation as shown in Fig. 9.16. A rise in voltage results in the release of pressure on the carbon block rheostat, thereby increasing its resistance whereas a fall in voltage results in increasing the pressure on the block thereby decreasing its resistance. By this automatic variation of control resistance, variations in battery voltage are automatically controlled.

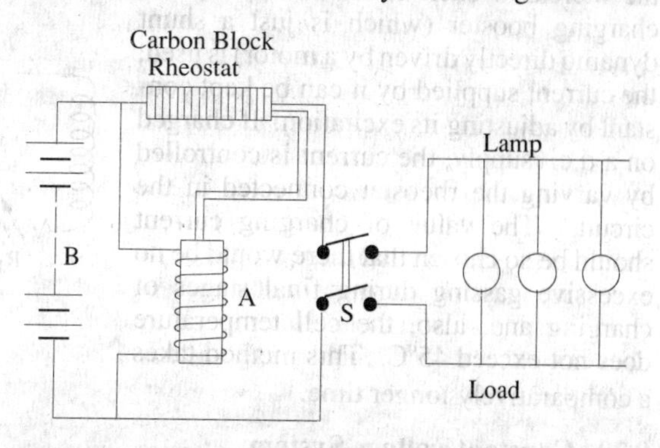

Fig. 9.16

9.25. End-cell Control System

The use of rheostat for controlling the battery voltage is objectionable especially in large-capacity installations where the I^2R loss would be considerable. Hence other more economical systems have been developed and put into use. One such system is the end-cell control system. It consists of suitable regulator switches which cut one or more of a selected number of cells out of the circuit when the battery is fully charged and into the circuit again as the discharge continues. To make the process of cutting cells in and out at the battery circuit simple, the group of cells selected for this control is situated at one end of the battery whereform it derives the name *end cell*. By moving the

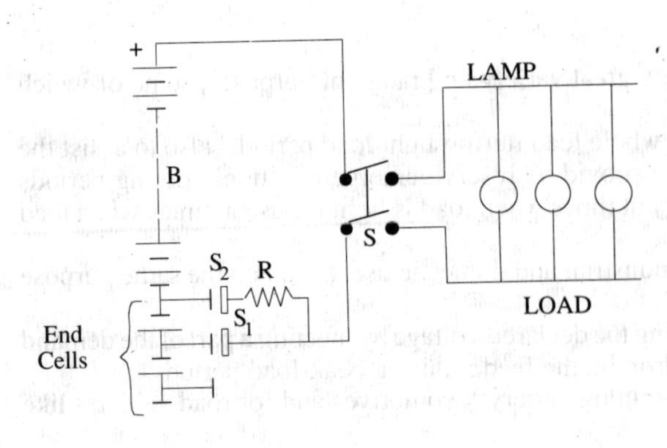

Fig. 9.17.

contact arm of the switch to the left or right, cells are cut in or out of the discharge circuit and so the voltage is varied accordingly.

For making the end-cell switch operate without opening the circuit or short-circuiting the cells during its passage from one cell to another, an auxiliary constraint S_2 is employed. S_2 prevents the circuit from being open entirely but has sufficient resistance R between it and the main contact arm S_1 to prevent any objectionably large current to flow on short-circuit. The above mechanism usually incorporates devices for preventing the stoppage of the switch in the short-circuit position.

9.26. Number of End-cells

For maintaining a supply voltage of V volts from a battery of lead-acid cells when the latter are approaching their discharge voltage of 1.83 (depending on the discharge rate), the number of cells required is $V/1.83$. When the battery is fully charged with each cell having an e.m.f. of 2.1 V, then the number of cells required is $V/2.1$. Hence, the number of end-cells required is $(V/1.83 - V/2.1)$. These are connected to a regulating switch which adds them in series with the battery one or two at a time, as the discharge proceeds.

9.27. Charging Systems

In various installations, batteries are kept floating on the line and are so connected that they are being charged when load demands are light and automatically discharged during peak periods when load demands are heavy or when the usual power supply fails or is disconnected. In some other installations, the battery is connected to the feeder circuit as and when desired, allowed to discharge to a certain point, then removed and re-charged for further requirements.

For batteries other than the 'floating' and 'system-governed' type, following two general methods (though there are some variations of these) are employed.

(*i*) *The Constant-current System* and (*ii*) *The Constant-voltage System.*

9.28. Constant-current System

In this method, the charging current is kept constant by varying the supply voltage to overcome the increased back e.m.f. of cells. If a charging booster (which is just a shunt dynamo directly driven by a motor) is used, the current supplied by it can be kept constant by adjusting its excitation. If charged on a d.c. supply, the current is controlled by varying the rheostat connected in the circuit. The value of charging current should be so chosen that there would be no excessive gassing during final stages of charging and, also, the cell temperature does not exceed 45°C. This method takes a comparatively longer time.

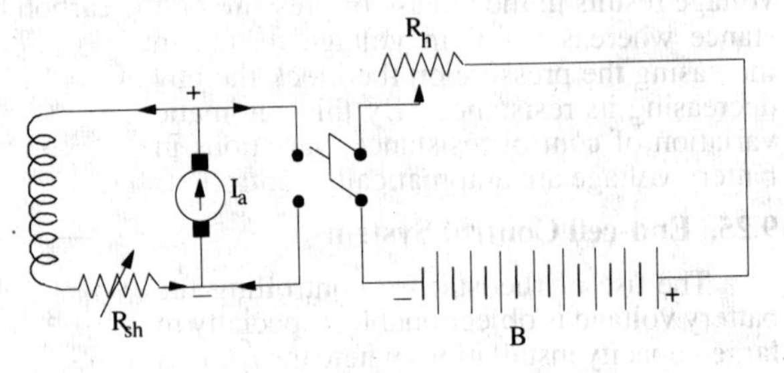

Fig. 9.18

9.29. Constant-voltage System

In this method, the voltage is kept constant but it results in very large charging curient in the beginning when the back e.m.f. of the cells is low and a small current when their back e.m.f. increases on being charged.

With this method, time of charging is almost reduced to half. It increases the capacity, by approximately 20% but reduces the efficiency by 10% or so.

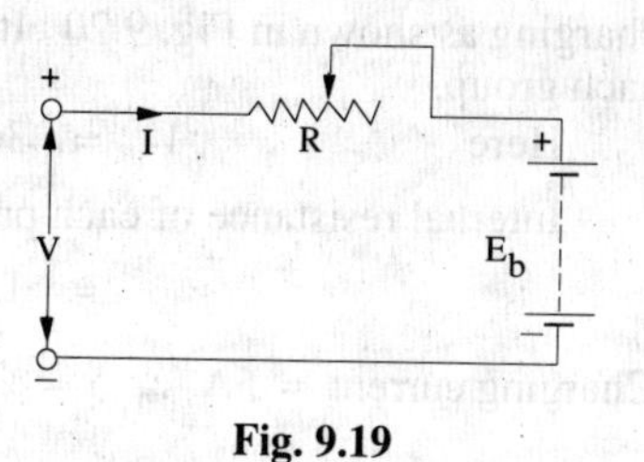

Fig. 9.19

Calculations

When a secondary cell or a battery of such cells is being charged, then the emf of the cells acts in opposition to the applied voltage. If V is the supply voltage which sends a charging current of I against the back e.m.f. E_b, then input is VI but the power spent in overcoming the opposition (Fig. 9.19) is E_bI. This power E_bI is converted into the chemical energy which is stored in the cell. The charging current can be found from the following equation :

$$I = \frac{V - E_b}{R}$$

where R = total circuit resistance including internal resistance of the battery

I = charging current

By varying R, the charging current can be kept constant throughout.

Example 9.11. *A battery of accumulators of e.m.f. 50 volt and internal resistance 2 Ω is charged on 100 volt direct mains. What series resistance will be required to give a charging current of 2 A? If the price of energy is 50 paise per kWh, what will it cost to charge the battery for 8 hours and what percentage of energy supplied will be used in the form of heat?*

Solution. Applied voltage = 100 V; Back e.m.f. of the battery = 50 V

Net charging voltage = 100 − 50 = 50 V

Let R be the required resistance, then 2 = $50/(R + 2)$; $R = 46/2 = 23$ Ω

Input for eight hours = $100 \times 2 \times 8 = 1600$ Wh = 1.6 kWh

Cost = 50×1.6 = **80 paise** ; Power wasted on total resistance = $25 \times 2^2 = 100$ W

Total input = $100 \times 2 = 200$ W : Percentage waste = $100 \times 100/200$ = **50%**

Example 9.12. *A 6-cell, 12-V battery is to be charged at a constant rate of 10 A from a 24-V d.c. supply. If the e.m.f. of each cell at the beginning and end of the charge is 1.9 V and 2.4 V, what should be the value of maximum resistance to be connected in series with the battery. Resistance of the battery is negligible.*

Solution. Beginning of Charging

Total back e.m.f. of battery = $6 \times 1.9 = 11.4$ volt

Net driving voltage = $24 - 11.4 = 12.6$ V ; $R_{max} = 12.6/10 =$ **1.26 Ω**

End of Charging

Back e.m.f. of battery = $6 \times 2.4 = 14.4$ volt

Net driving voltage = $24 - 14.4 = 9.6$ V ; $R_{min} = 9.6/10 =$ **0.96 Ω**

Example 9.13. *Thirty accumulators have to be charged from their initial voltage of 1.8 V using a direct current supply of 36 volt. Each cell has an internal resistance of 0.02 Ω and can be charged at 5 amperes. Sketch a circuit by which this can be done, calculating the value of any resistance or resistances used. What will be the current taken from the mains towards the end of the charging period when the voltage has risen to 2.1 volt per cell ?*

Solution. Since the supply voltage (36 V) is less than the back e.m.f. of the 30 cell battery (54 V), hence the cells are divided into two equal groups and placed in parallel across the supply for charging as shown in Fig. 9.20. It would be economical to use a separate resistance R in series with each group.

Here $V = 36\text{ V}, E_b = 15 \times 1.8 = 27\text{ V}$

Internal resistance of each parallel group

$= 15 \times 0.02 = 0.3\ \Omega$

Charging current = 5A $\therefore\ 5 = \dfrac{36-27}{R+0.3}$

$R = \mathbf{1.5\ \Omega}$

Now, when the voltage per cell becomes 2.1 V, then back e.m.f. of each parallel group = $15 \times 2.1 = 31.5$ V

$\therefore$ Charging current $= \dfrac{36-31.5}{1.5+0.3} = \mathbf{2.5\ A}$

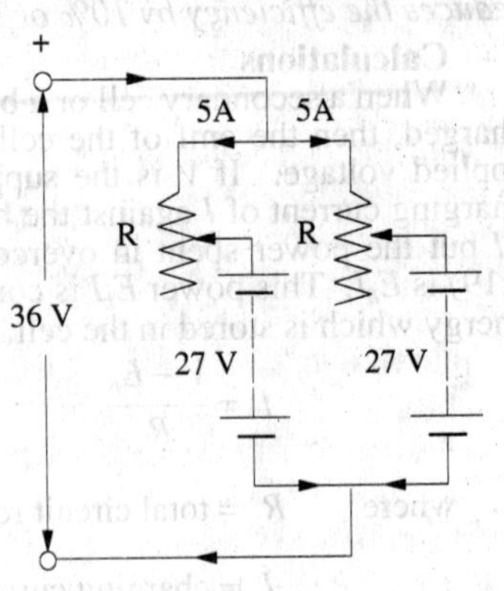

Fig. 9.20

Example 9.14. *A battery of 60 cells is charged from a supply of 250 V. Each cell has an e.m.f. of 2 volts at the start of charge and 2.5 V at the end. If internal resistance of each cell is 0.1 Ω and if there is an external resistance of 19 Ω in the circuit, calculate (a) the initial charging current (b) the final charging current and (c) the additional resistance which must be added to give a finishing charge of 2 A rate.*

Solution. (*a*) Supply voltage $V = 250$ V

Back e.m.f. of the battery E_b at start = 60 × 2 = 120 V and at the end = 60 × 2.5 = 150 V

Internal resistance of the battery = 60 × 0.1 = 6 Ω

Total circuit resistance = 19 + 6 = 25 Ω

(*a*) Net charging voltage at start= 250 − 120 = 130 V

$\therefore$ Initial charging current = 130 / 25 = **5.2 A**

(*b*) Final charging current = 100 / 25 = **4 A**

(*c*) Let R be the external resistance, then

$2 = \dfrac{100}{R+6} \quad \therefore \quad R = 88/2 = 44\ \Omega$

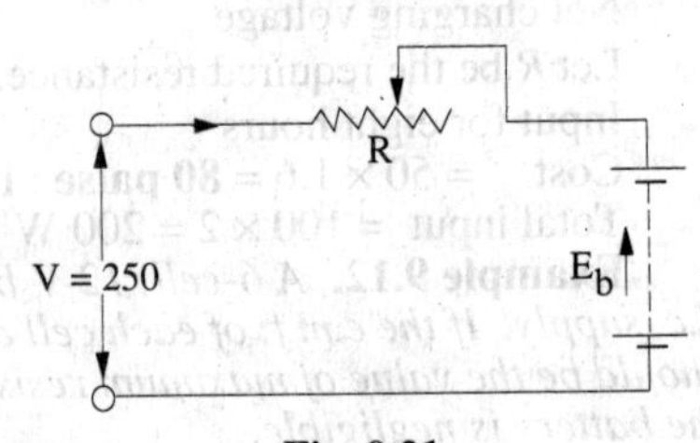

Fig. 9.21

$\therefore$ Additional resistance required = 44 − 19 = **25 Ω.**

Example 9.15. *Two hundred and twenty lamps of 100 W each are to be run on a battery supply at 110 V. The cells of the battery when fully charged have an e.m.f. of 2.1 V each and when discharged 1.83 V each. If the internal resistance per cell is 0.00015 Ω (i) find the number of cells in the battery and (ii) the number of end cells. Take the resistance of the connecting wires as 0.005 Ω.*

Solution. Current drawn by lamps = 220 × 100/110 = 200 A

Voltage drop on the resistance of the connecting wires = 0.005 × 200 = 1.0 V

Battery supply voltage = 110 + 1 = 111 V

Terminal voltage/cell when fully charged and supplying the load

= 2.1 −(200 × 0.00015) = 2.08 V

Terminal voltage/cell when discharged = 1.83 −(200 × 0.00015) = 1.8 V

(*i*) No. of cells in the battery

= 111/2.08 = 53.4 say, **54**

(*ii*) No. of cells required when discharged = 111/1.8 = 62

Hence, number of end cells

= 62−54 = **8**

The connections are shown in Fig. 9.22

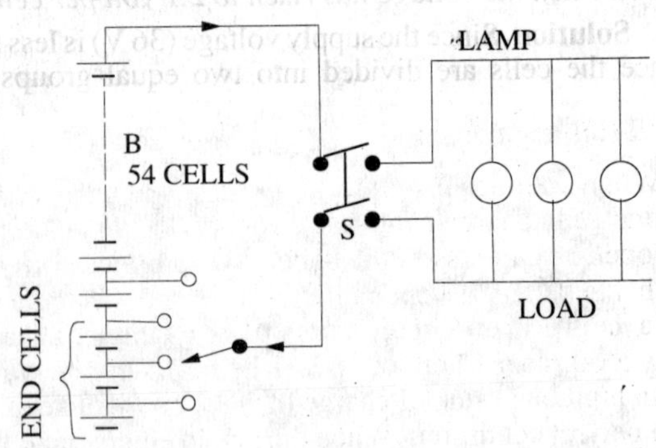

Fig. 9.22

Example 9.16. *A storage battery consists of 55 series-connected cells each of internal resistance 0.001 Ω and e.m.f. 2.1 V. Each cell consists of 21 plates, ten positive and eleven negative, each plate measuring 20 × 25 cm. If full-load current per cell is 0.01 A per cm² of positive plate surface, find (i) full-load terminal voltage of the battery and (ii) power wasted in the battery if the connectors have a total resistance of 0.025 Ω.*

Solution. Since both sides of a positive plate are utilized, the area of both sides will be taken into consideration.

Total area (both sides) of ten positive plates = $2 \times 20 \times 25 \times 10 = 10{,}000\ \text{cm}^2$

Full load current $= 10{,}000 \times 0.01 = 100$ A

Voltage drop in battery and across connectors $= 100\,[(55 \times 0.001) + 0.025] = 8$ V

Battery e.m.f. $= 55 \times 2.1 = 115.5$ V

(*i*) Battery terminal voltage on full-load $= 115.5 - 8 =$ **107.5 V**

(*ii*) Total resistance $= (55 \times 0.001) + 0.025 = 0.08\ \Omega$; Power loss $= 100^2 \times 0.08 =$ **800 W.**

Example 9.17. *A charging booster (shunt generator) is to charge a storage battery of 100 cells each of internal resistance 0.001 Ω. Terminal p.d. of each cell at completion of charge is 2.55 V. Calculate the e.m.f. which the booster must generate to give a charging current of 20 A at the end of charge. The armature and shunt field resistances of the generator are 0.2 and 258 Ω respectively and the resistance of the cable connectors is 0.05 Ω.*

Solution. Terminal p.d. per cell = 2.55 volt

The charging voltage across the battery must be capable of overcoming the back e.m.f and also to supply the voltage drop across the internal resistance of the battery.

Back e.m.f. $= 100 \times 2.55 = 255$ V

Voltage drop on internal resistance

$= 100 \times 0.001 \times 20 = 2$ V

∴ P.D. across points *A* and *B* $= 255 + 2 = 257$ V

P.D. across terminals C and D of the generator

$= 257 + (20 \times 0.05) = 258$ V

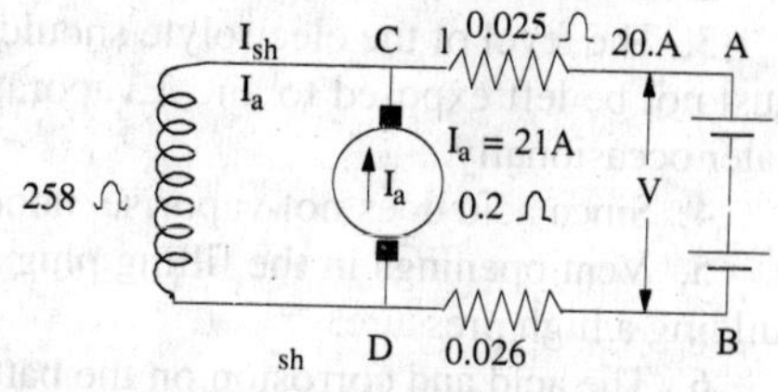

Fig. 9.23

$$\therefore \quad I_{sh} = 258/258 = 1\ \text{A};\ I_a = 20 + 1 = 21\ \text{A}$$

$$\therefore \quad I_a R_a = 21 \times 0.2 = 4.2\ \text{V}$$

$$\therefore \quad \text{Generated e.m.f.} = 258 + 4.2 = \mathbf{262.2\ V}$$

Tutorial Problems No. 9.2

1. A 60-cell storage battery having a capacity of 360 Ah takes 8 hours when charged by a dc generator at a voltage of 220 V. Calculate the charging current and the range of the rheostat required to ensure a constant charging current. The emf of each cell is 1.8 V at the beginning of charging and 2.7 V at the end of charging. Ignore the internal resistance of the cell. **[45 A; 2.45 to 1.29 Ω]**

2. A storage battery consists of 55 series connected cells each of internal resistance 0.001 Ω and e.m.f. 2.1 V. Each cell consists of 21 plates, ten positive and eleven negative, each plate measuring 20 × 25 cm. If full-load current per cell is 0.01 A per cm² of positive plate surface, find (*i*) full-load terminal voltage of the battery and (*ii*) power wasted in the battery if the connectors have a total resistance of 0.025 Ω. [(*i*) **107.5 V** (*ii*) **800 W**]

9.30. Trickle Charging

When a storage battery is kept entirely as an emergency reserve, it is very essential that it should be found fully charged and ready for service when an emergency arises. Due to leakage action and other open-circuit losses, the battery deteriorates even when idle or on open-circuit. Hence, to keep it fresh, the battery is kept on a trickle charge. The rate of trickle charge is small and is just sufficient to balance the open-circuit losses. For example, a standby battery for station bus-bars capable of giving 2000 A for 1 hour or 400 Ah at the 10-hr rate, will be having a normal charging rate of 555 A, but a continuous 'trickle' charge of 1 A or so will keep the cells fully charged (without any gassing) and in perfect condition. When during an emergency, the battery gets discharged, it is re-charged at its normal charging rate and then is kept on a continuous trickle charge.

9.31. Sulphation – Causes and Cure

If a cell is left incompletely charged or is not fully charged periodically, then the lead sulphate formed during discharge, is not converted back into PbO_2 and Pb. Some of the unreduced $PbSO_4$ which is left, gets deposited on the plates which are then said to be sulphated. $PbSO_4$ is in the form of minute crystals which gradually increase in size if not reduced by thoroughly charging the cells. It increases the internal resistance of the cell thereby reducing its efficiency and capacity. Sulphation also sets in if the battery is overcharged or left discharged for a long time.

Sulphated cells can be cured by giving them successive overcharges, for which purpose they are cut out of the battery during discharge, so that they can get two charges with no intervening discharge. The other method, in which sulphated cells need not be cut out of the battery, is to continue charging them with a '*milking booster*' even after the battery as a whole has been charged. A milking booster is a motor-driven low-voltage dynamo which can be connected directly across the terminals of the sulphated cells.

9.32. Maintenance of Lead-acid Cells

The following important points should be kept in mind for keeping the battery in good condition :

1. Discharging should not be prolonged after the minimum value of the voltage for the particular rate of discharge is reached.
2. It should not be left in discharged condition for long.
3. The level of the electrolyte should always be 10 to 15 mm above the top of the plates which must not be left exposed to air. Evaporation of electrolyte should be made up by adding distilled water occasionally.
4. Since acid does not vaporise, none should be added.
5. Vent openings in the filling plug should be kept open to prevent gases formed within from building a high pressure.
6. The acid and corrosion on the battery top should be washed off with a cloth moistened with baking soda or ammonia and water.
7. The battery terminals and metal supports should be cleaned down to bare metal and covered with vaseline or petroleum jelly.

9.33. Mains Operated Battery Chargers

A battery charger is an electrical device that is used for putting energy into a battery. The battery charger changes the a.c. from the power line into d.c. suitable for charger. However, d.c. generator and alternators are also used as charging sources for secondary batteries.

In general, a mains-operated battery charger consists of the following elements :

1. A step-down transformer for reducing the high a.c mains voltage to a low a.c. voltage.
2. A half-wave or full-wave rectifier for converting alternating current into direct current.
3. A charge-current limiting element for preventing the flow of excessive charging current into the battery under charge.
4. A device for preventing the reversal of current *i.e.* discharging of the battery through the charging source when the source voltage happens to fall below the battery voltage.

In addition to the above, a battery charger may also have circuitry to monitor the battery voltage and automatically adjust the charging current. It may also terminate the charging process when the battery becomes fully charged. However, in many cases, the charging process is not totally terminated but only the charging rate is reduced so as to keep the battery on trickle charging. These requirements have been illustrated in Fig. 9.24.

Most of the modern battery chargers are fully protected against the following eventualities:

(*a*) They are able to operate into a short-circuit.
(*b*) They are not damaged by a reverse-connected battery.
(*c*) They can operate into a totally flat battery.
(*d*) They can be regulated both for current and voltage.

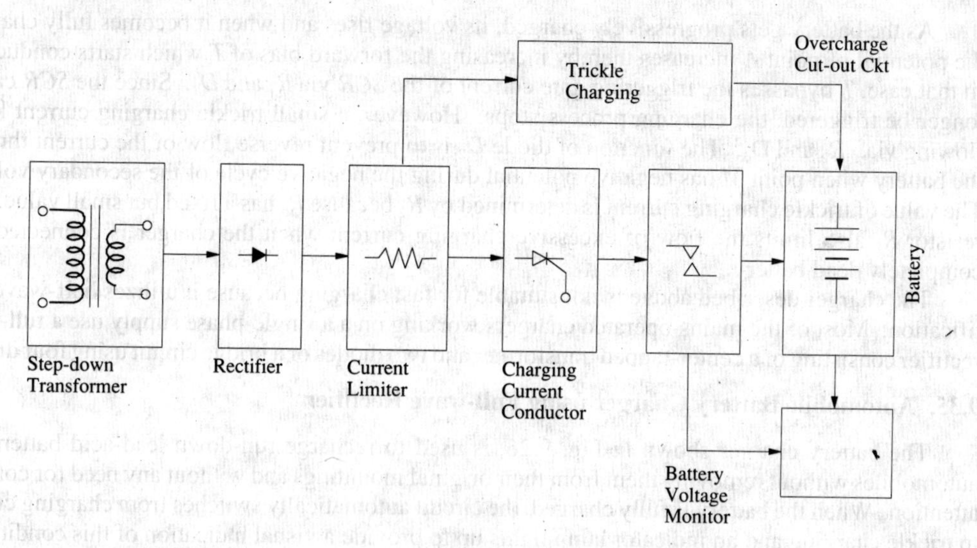

Fig. 9.24

9.34. Car Battery Charger Using SCR

Fig. 9.25 shows the circuitry of a very simple lead-acid battery charger which has been provided protection from load short-circuit and from reverse battery polarity. The *SCR* is used as a half-wave rectifier as well as a switching element to terminate the high-current charging process when battery gets fully-charged.

Working

The *SCR* acts as a half-wave rectifier during only the positive half-cycles of the secondary voltage when point *M* in Fig. 9.25 is at a positive potential. The *SCR* does not conduct during the negative half-cycles of the secondary voltage when point *M* achieves negative potential. When *M* is at positive potential, the *SCR* is triggered into conduction because of the small gate current I_g passing via R_1 and diode D_1. In this way, the charging current *I* after passing through R_5 enters the battery which is being charged.

In the initial state, when the battery voltage is low, the potential of point *A* is also low (remember that R_3, R_4 and preset resistor R_6 are connected across the battery via R_5) which means that the forward bias on the base of transistor *T* is not sufficient to make it conduct and thereby stop the conduction of *SCR*. Hence, *SCR* keeps conducting and consequently, keeps charging the battery through the current-limiting resistor R_5.

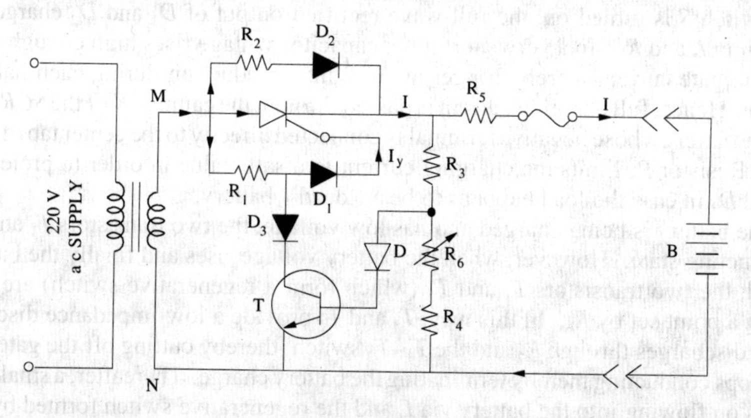

Fig. 9.25

As the battery gets progressively charged, its voltage rises and when it becomes fully charged, the potential of point A increases thereby increasing the forward bias of T which starts conducting. In that case, T bypasses the triggering gate current of the SCR via R_1 and D_3. Since the SCR can no longer be triggered, the charging process stops. However, a small trickle charging current keeps flowing via. R_2 and D_2. The function of diode D_2 is to prevent reverse flow of the current through the battery when point M has negative potential during the negative cycle of the secondary voltage. The value of trickle charging current is determined by R_2 because R_5 has a fixed but small value. The resistor R_5 also limits the flow of excessive charging current when the charger is connected to a completely dead battery.

The charger described above is not suitable for fast charging because it utilizes half-wave rectification. Most of the mains-operated chargers working on a a single-phase supply use a full-wave rectifier consisting of a center-tapped transformer and two diodes or a bridge circuit using four diodes.

9.35. Automobile Battery Charger using Full-wave Rectifier

The battery charger shown in Fig. 9.26, is used to recharge run-down lead-acid batteries in automobiles without removing them from their original mountings and without any need for constant attention. When the battery is fully charged, the circuit automatically switches from charging current to trickle charging and an indicator lamp lights up to provide a visual indication of this condition.

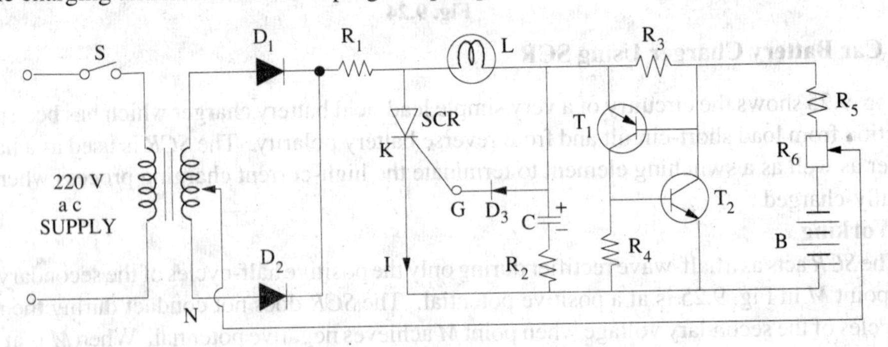

Fig. 9.26

As shown in Fig. 9.26, diodes D_1, and D_2 form a full-wave rectifier to provide pulsating direct current for charging the battery. The battery is charged through the SCR which is also used as switch to terminate the charging process when the battery becomes fully charged. The two transistors T_1 and T_2 together form an electronic switch that has two stable states *i.e.* the *ON* state in which T_1 and T_2 conduct and the OFF state in which T_1 and T_2 do not conduct. The ON-OFF state of this switch is decided by the battery voltage and setting of the "current adjust" potentiometer R_6.

Working

When switch S is turned on, the full-wave rectified output of D_1 and D_2 charges capacitor C through R_1, lamp L and R_2. In a very short time, capacitor voltage rises high enough to make diode D_3 conduct the gate current thereby triggering SCR into conduction during each half-cycle of the output voltage. Hence, full charging current is passed through the cathode K of the SCR to the positive terminal of the battery whose negative terminal is connected directly to the center tap of the step-down transformer. Resistor R_1 limits the charging current to a safe value in order to protect the rectifier diodes D_1 and D_2 in case the load happens to be a ''dead'' battery.

When the battery is being charged and has low voltage, the two transistors T_1 and T_2 remain in the non-conducting state. However, when the battery voltage rises and finally the battery becomes fully-charged, the two transistors T_1 and T_2 (which form a regenerative switch) are triggered into conduction at a point set by R_6. In this way, T_1 and T_2 provide a low-impedance discharge path for C. Hence, C discharges through R_2 and the T_1–T_2 switch, thereby cutting off the gate current of the SCR which stops conducting thereby terminating the battery charge. Thereafter, a small trickle charge current keeps on flowing into the battery via L and the regenerative switch formed by T_1 and T_2. A glowing lamp L indicates that the battery is under trickle charging.

Fig. 9.27 shows the same circuit as shown in Fig. 9.26 except that the two-diode full-wave rectifier has been replaced by a full-wave bridge rectifier using four diodes.

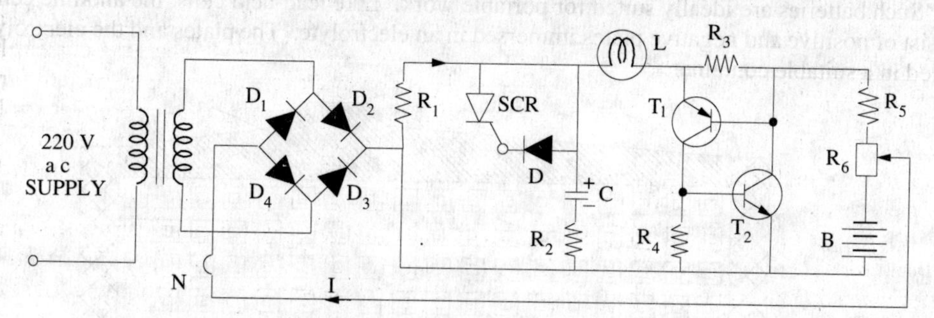

Fig. 9.27

9.36. Static Uninterruptable Power Systems (UPS)

The function of a *UPS* is to ensure absolute continuity of power to the computerised control systems thereby protecting critical equipment from electrical supply failure. A *UPS* makes it possible to provide a 'clean' reliable supply of alternating current free of sags or surges in the line voltage, frequency variation, spikes and transients. *UPS* systems achieve this by rectifying the standard mains supply, using the direct current to charge the standby battery and to produce 'clean' alternating current by passing through an inverter and filter system.

Components of a *UPS* System

The essential components of a *UPS* system as shown in Fig. 9.28 are as under :

1. A rectifier and thyristor-controlled battery charger which converts the *AC* input into regulated *DC* output and keeps the standby battery fully charged.
2. A standby battery which provides *DC* input power to inverter during voltage drops or on failure of the normal mains *AC* supply.
3. An inverter which converts *DC* to clean *AC* thus providing precisely regulated output voltage and frequency to the load as shown.

Working

As shown in Fig. 9.28 the main flow of energy is from the rectifier to the inverter with the standby battery kept on 'float'. If the supply voltage falls below a certain level or fails completely, the battery output to the inverter maintains a clean *a.c* supply. When the mains power supply is restored, the main energy flow again starts from the rectifier to the inverter but, in addition, the rectifier recharges the battery. When the standby battery gets fully charged, the charging current is automatically throttled back due to steep rise in the back emf of the battery. An automatic/manual bypass switch is used to connect the load either directly to the mains a.c supply or to the inverter a.c. supply.

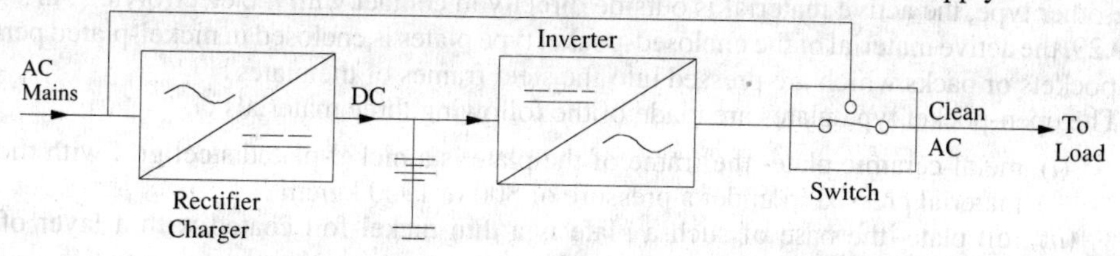

Fig. 9.28

Depending on the application, the voltage of the *UPS* standby batteries may be anywhere between 12 V and 400 V. Typical values are 24 V, 48 V, 110 V and 220 V with currents ranging from a few amperes to 2000 A. Fig. 9.29 shows Everon 4-kVA on-line *UPS* system which works on 170 V–270 V a.c. input and provides an a.c. output voltage of 230 V at 50 Hz frequency with a voltage stability of ± 2% and frequency stability of ± 1%. It has zero changeover time and has audio beeper which indicates mains fail and battery discharge. It provides 100% protection against line noise, spikes, surges and radio frequency interference. It is manufactured by Everon Electro Systems Pvt. Ltd., New Delhi.

9.37. Alkaline Batteries

Such batteries are ideally suited for portable work. Like lead-acid cells, the alkaline cells also consist of positive and negative plates immersed in an electrolyte. The plates and the electrolyte are placed in a suitable container.

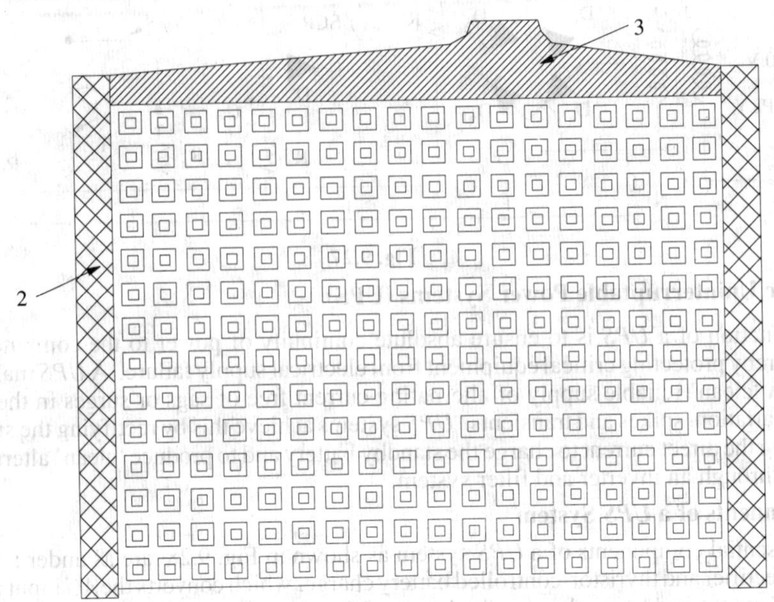

Fig. 9.29

The two types of alkaline batteries which are in general use are :

(*i*) nickel-iron type or Edison type.

(*ii*) nickel-cadmium type or Jungnor type which is commercially known as NIFE battery.

Another alkaline battery which differs from the above only in the mechanical details of its plates is known as Alkum battery which uses nickel hydroxide and graphite in the positive plates and a powdered alloy of iron and chromium in the negative plates.

Silver-zinc type of alkaline batteries are also made whose active material for the positive plate is silver oxide (Ag_2O) and for negative plate is zinc oxide and zinc powder. The electrodes or plates of the alkaline cells are designed to be either of the enclosed-pocket type or open-pocket type. In the case of enclosed-pocket type plates, the active material is inside perforated metal envelopes whereas in the other type, the active material is outside directly in contact with the electrolyte. As shown in Fig. 9.29, the active material of the enclosed-pocket type plates is enclosed in nickel-plated perforated steel pockets or packs which are pressed into the steel frames of the plates.

The open-pocket type plates are made of the following three materials :

(*i*) metal-ceramic plate–the frame of the plate is a nickel-plated steel grid with the active material pressed in under a pressure of 800 to 1900 kg/cm^2.

(*ii*) foil plate–the base of such a plate is a thin nickel foil coated with a layer of nickel suspension deposited by a spray technique.

(*iii*) pressed plates–the base member of these plates is a nickel-plated pressed steel grid. The active material is pressed into them at a pressure of about 400 kg/cm^2.

9.38. Nickel-iron or Edison Batteries

There is revived interest in the nickel-iron battery because it seems to be one of the few systems which may be developed into a high-energy density battery for electric vehicles. Since long the two main designs for this battery have been the tubular positive type and the flat pocket plate type although cells with sintered type negative are also being manufactured.

The active materials in a nickel-iron cell are :

(*i*) Nickel hydroxide $Ni(OH)_4$ or apple green nickel peroxide NiO_2 for the positive plate. About 17 per cent of graphite is added to increase conductivity. It also contains an activating additive barium hydroxide which is about 2 per cent of the active material. This additive increases the service life of the plates.

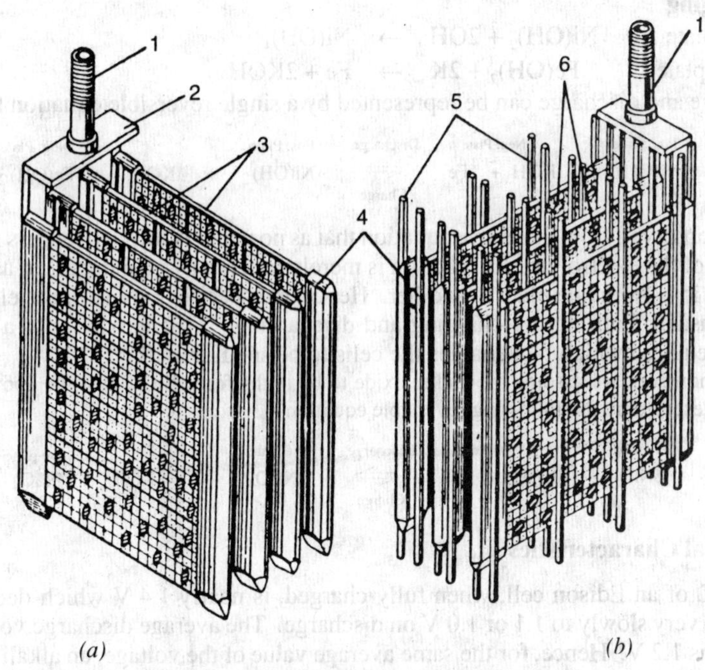

Fig. 9.30 Plate groups of an alkaline cell

(*a*) +ve group (*b*) –ve group. 1–terminal post. 2–connecting strap. 3.–plates. 4–plates side members. 5–ebonite spacer sticks. 6–pockets.

(*ii*) powdered iron and its oxides for the negative plate. Small quantities of nickel sulphate and ferrous sulphide are added to improve the performance of the cell.

(*iii*) the electrolyte is 21 per cent solution of caustic potash KOH (potassium hydrate) to which is added a small quantity of lithium hydrate LiOH for increasing the capacity of the cell.

As shown in Fig. 9.30, plates of the same polarity with their pockets filled, are assembled into cell groups for which purpose they are welded to a common strap having a threaded post.

The number of negative plates is one more than the positive plates. The extreme negative plates are electrically connected to the container. Ebonite separating sticks are placed between the positive and negative plates to prevent any short-circuiting.

The steel containers of the batteries are press-formed from steel and the joints are welded. The body and the cover are nickel-plated and have a dull finish. However, it should be kept in mind that since these containers are electrically alive, no loose wires should touch them owing to the danger of severe sparking from short circuits.

9.39. Chemical Changes

The exact nature of the chemical changes taking place in such a cell is not clearly understood because the exact formula for the nickel oxide is not yet well established but the action of the cell can be understood by assuming the peroxide NiO_2 or its hydrated form $Ni(OH)_4$.

First, let us assume that at positive plate, nickel oxide is in its hydrated form $Ni(OH)_4$. During discharge, electrolyte KOH splits up into positive K ions and negative OH ions. The K ions go to anode and reduce $Ni(OH)_4$ to $Ni(OH)_2$. The OH ions travel towards the cathode and oxidise iron. During charging, just the opposite reactions take place *i.e.* K ions go to cathode and OH ions go to anode. The chemical reactions can be written thus :

$$KOH \rightarrow K + OH$$

During discharge

Positive plate : $Ni(OH)_4 + 2K \rightarrow Ni(OH)_2 + 2KOH$

Negative plate : $Fe + 2OH \rightarrow Fe(OH)_2$

During Charging

Positive plate : $Ni(OH)_2 + 2OH \rightarrow Ni(OH)_4$

Negative plate : $Fe(OH)_2 + 2K \rightarrow Fe + 2KOH$

The charge and discharge can be represented by a single reversible equation thus :

$$\underset{\text{Pos. Plate}}{Ni(OH)_4} + KOH + \underset{\text{Neg. Plate}}{Fe} \underset{\text{Charge}}{\overset{\text{Discharge}}{\rightleftharpoons}} \underset{\text{Pos. Plate}}{Ni(OH)_2} + KOH + \underset{\text{Neg. Plate}}{Fe(OH)_2}$$

It will be observed from the above equation that as no water is formed, there is no overall change in the strength of the electrolyte. Its function is merely to serve as a conductor or as a vehicle for the transfer of OH ions from one plate to another. Hence, the specific gravity of the electrolyte remains practically constant, both during charging and discharging. That is why only a small amount of electrolyte is required which fact enables the cells to be small in bulk.

Note. If, however, we assume the nickel oxide to be in the form NiO_2, then the above reactions can be represented by the following reversible equation :

$$\underset{\text{+ve plate}}{6NiO_2} + 8KOH + \underset{\text{-ve plate}}{3Fe} \underset{\text{Charge}}{\overset{\text{Discharge}}{\rightleftharpoons}} \underset{\text{+ve plate}}{2Ni_3O_4} + 8KOH + \underset{\text{-ve plate}}{Fe_3O_4}$$

9.40. Electrical Characteristics

The e.m.f. of an Edison cell, when fully charged, is nearly 1.4 V which decreases rapidly to 1.3 V and then very slowly to 1.1 or 1.0 V on discharge. The average discharge voltage for a 5-hour discharge rate is 1.2 V. Hence, for the same average value of the voltage, an alkali accumulator will consist of 1.6 to 1.7 times as many cells as in a lead-acid battery. Internal resistance of an alkali cell is nearly five times that of the lead-acid cell, hence there is a relatively greater difference between its terminal voltage when charging and discharging.

The average charging voltage for an alkali cell is about 1.7 V. The general shapes of the charge and discharge curves for such cells are, however, similar to those for lead-acid cells. The rated capacity of nickel accumulators usually refers to either 5-hour or 8-hour discharge rate unless stated otherwise.

The plates of such cells have greater mechanical strength because of all-steel construction. They are comparatively lighter because (*i*) their plates are lighter and (*ii*) they require less quantity of electrolyte. They can withstand heavy charge and discharge currents and do not deteriorate even if left discharged for long periods.

Due to its relatively higher internal resistance, the efficiencies of an Edison cell are power than those of the lead acid cell. On the average, its Ah efficiency is about 80% and Wh efficiency 60 or 50%. It has an average density of 50 Wh/kg.

With increase in temperature, e.m.f. is increased slightly but capacity increases by an appreciable amount. With decrease in temperature, the capacity decreases becoming practically zero at 4°C even though the cell is fully charged. This is a serious drawback in the case of electrically-driven vehicles in cold weather and precautions have to be taken to heat up the battery before starting, though, in practice, the I^2R loss in the internal resistance of the battery is sufficient to keep the battery cells warm when running.

The principal disadvantage of the Edison battery or nickel-iron battery is its high initial cost (which will probably be sufficiently reduced when patents expire). At present, an Edision battery costs approximately twice as much as a lead-acid battery designed for similar service. But since the alkaline battery outlasts an indeterminate number of lead-acid batteries, it is cheaper in the end.

Because of their lightness, compact construction, increased mechanical strength, ability to withstand rapid charging and discharging without injury and freedom from corrosive liquids and fumes, alkaline batteries are ideally suited for traction work such as propulsion of electric factory trucks, mine locomotives, miner's lamps, lighting and starting of public service vehicles and other services involving rough usage etc.

9.41. Nickel-Cadmium Batteries

The active materials in a nickel-cadmium cell (Fig. 9.31) are :

(*i*) $Ni(OH)_4$ for the positive plate exactly as in the nickel-iron cell.

(*ii*) a mixture of cadmium or cadmium oxide and iron mass to which is added about 3 per cent of solar oil for stabilizing the electrode capacity. The use of cadmium results in reduced internal resistance of the cell.

(*iii*) the electrolyte is the same as in the nickel-iron cell.

The cell grouping and plate arrangement is identical with nickel-iron batteries except that the number of positive plates is more than the negative plates. Such batteries are more suitable than nickel-iron batteries for floating duties in conjunction with a charging dynamo because, in their case, the difference between charging and discharging e.m.f.s is not as great as in nickel-iron batteries.

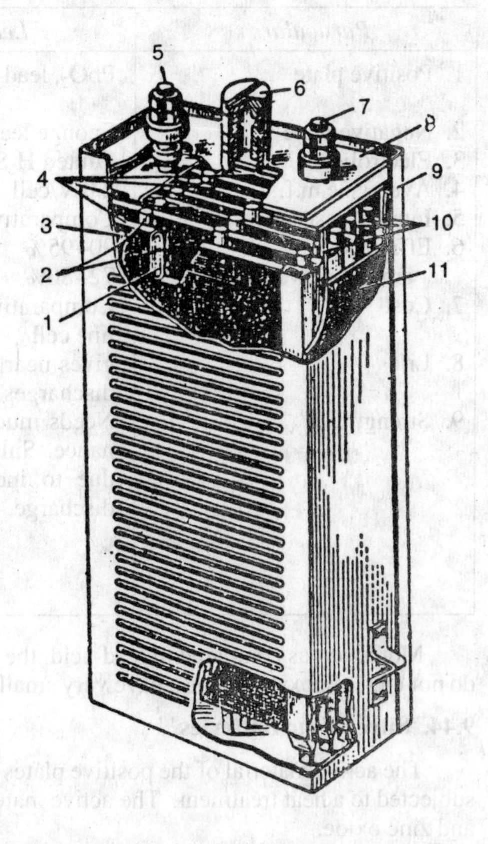

Fig. 9.31. Parts of a Nickel-cadium alkaline cell. 1–active material 2–ebonite spacer sticks 3–pocket element 4–positive plates 5–positive terminal post 6–vent plug 7–negative terminal post 8–cover 9–container 10–negative plates 11–ebonite plate.

Nickel-cadmium sintered plate batteries were first manufactured by Germans for military air-crafts and rockets. Presently, they are available in a variety of designs and sizes and have energy density going upto 55 Wh/kg. Their capacity is less affected by high discharge rates and low operating temperature than any other rechargeable batteries. Since such batteries have very low open-circuit losses, they are well-suited for plea-sure yatches and launches which may be laid up for long periods. They are also used in commercial airliners, military aeroplanes and helicopters for starting main engines or auxiliary turbines and for emergency power supply.

9.42. Chemical Changes

The chemical changes are more or less similar to those taking place in nickel-iron cell. As before, the electrolyte is split up into positive K ions and negative OH ions. The chemical reactions at the two plates are as under :

During discharge

Positive plate : $Ni(OH)_4 + 2K = Ni(OH)_2 + 2KOH$

Negative plate : $Cd + 2\,OH = Cd(OH)_2$

During Charging

Positive plate : $Ni(OH)_2 + 2OH = Ni(OH)_4$

Negative plate : $Cd(OH)_2 + 2K = Cd + 2KOH$

The above reaction can be represented by the following reversible equation :

$$Ni(OH)_4 + KOH + Cd \rightleftharpoons Ni(OH)_2 + KOH + Cd(OH)_2$$

9.43. Comparison : Lead-acid and Edison Cells

The relative strong and weak points of the cells have been summarised below :

Particulars	*Lead-acid cell*	*Edison cell*
1. Positive plate	PbO_2, lead peroxide	Nickel hydroxide $Ni(OH)_4$ or NiO_2
2. Negative plate	Sponge lead	Iron
3. Electrolyte	diluted H_2SO_4	KOH
4. Average e.m.f.	2.0 V/cell	1.2 V/cell
5. Internal resistance	Comparatively low	Comparatively higher
6. Efficiency :	90–95%	nearly 80%
amp-hour watt-hour	72–80%	about 60%
7. Cost	Comparatively less than alkaline cell	almost twice that of Pb-acid cell Easy maintenance
8. Life	gives nearly 1250 charges and discharges	five years at least
9. Strength	Needs much care and maintenance. Sulphation occurs often due to incomplete charge or discharge.	robust, mechanically strong, can withstand vibration, is light, unlimited rates of charge and discharge. Can be left discharged, free from corrosive liquids and fumes.

Moreover, as compared to lead-acid, the alkaline cells operate much better at low temperature, do not emit obnoxious fumes, have very small self-discharge and their plates do not buckle or smell.

9.44. Silver-Zinc Batteries

The active material of the positive plates is silver oxide which is pressed into the plate and then subjected to a heat treatment. The active material of the negative plates is a mixture of zinc powder and zinc oxide.

The chemical changes taking place within the cell can be represented by the following single equation :

$$\underset{+\text{ ve plate}}{Ag_2O} + \underset{-\text{ ve plate}}{Zn} \underset{\text{charge}}{\overset{\text{discharge}}{\rightleftharpoons}} \underset{+\text{ ve plate}}{2\,Ag} + \underset{-\text{ ve plate}}{ZnO}$$

A silver-zinc cell has a specific capacity (*i.e.* capacity per unit weight) 4 to 5 times greater than that of other type of cells. Their ground applications are mainly military *i.e.* communications equipment, portable radar sets and night-vision equipment. Moreover, comparatively speaking, their efficiency is high and self-discharge is small. Silver-zinc batteries can withstand much heavier discharge currents than are permissible for other types and can operate over a temperature range of –20°C to +60°C. Hence, they are used in heavy - weight torpedoes and for submarine propulsion. It has energy density of 150 Wh/kg. Its life time in wet condition is 1-2 years and the dry storage life is upto 5 years. However, the only disadvantage of siver-zinc battery or cell is its higher cost.

9.45. High Temperature Batteries

It is a new group of power source which requires operating temperatures above the ambient. They possess the advantages of high specific energy and power coupled with low cost. They are particularly suitable for vehicle traction and load levelling purposes in the electric supply industry. We will briefly describe the following cell from which high-temperature batteries are made.

1. *Lithium/Chlorine Cell*

It has an emf of 3.5 V, a theoretical specific energy of 2200 Wh/kg at 614°C and operating temperature of 650°C.

2. *Lithium/Sulphur Cell*

It has an emf of 2.25 V, specific energy of 2625 Wh/kg and an operating temperature of 365°C.

3. *Lithium-Aluminium/Iron-Sulphide Cells*

The emf of these cells is 1.3 V and a theoretical specific energy of 450 Wh/kg.

4. *Sodium/Sulphur Cell*

It utilises liquid sodium as negative electrode and sulphur as positive electrode and employs polycrystalline beta alumina as solid electrolyte. It was conceived in the 1960s by J.T. Kummer and N. Weber. The cell reaction can be written as $2Na + 3S = Na_2S_3$. The announcement of sodium/sulphur battery based on beta alumina was made by Ford Motor company of USA in 1966. The open-circuit voltage of the cell is 2.1 V and it has a specific energy of 750 Wh/kg with an operating temperature of 350°C. The two unique features of this cell are (1) a Faradaic efficiency of 100% and an ampere-hour capacity which is invarient with discharge rate and (2) high shelf-life (which is critical for certain space applications).

9.46. Secondary Hybrid Cells

A hybrid cell may be defined as a galvanic electrochemical generator in which one of the active reagents is in the gaseous state i.e. the oxygen of the air. Such cells take advantage of both battery and fuel cell technology. Examples of such cells are:

1. *Metal-air cells such as iron-oxygen and zinc-oxygen cells.*

The Zn/O_2 cell has an open-circuit voltage of 1.65 V and a theoretical energy density of 1090 Wh/kg. The Fe/O_2 cell has an OCV of 1.27 V and energy density of 970 Wh/kg.

2. *Metal-halogen cells such as zinc-chlorine and zinc-bromine cells.*

The zinc-chlorine cell has an OCV of 2.12 V at 25°C and a theoretical energy density of 100 Wh/kg. Such batteries are being developed for EV and load levelling applications. The zinc-bromine cell has an OCV of 1.83 V at 25°C and energy density of 400 Wh/kg.

3. *Metal-hydrogen cells such as nickel-hydrogen cell.*

Such cells have an OCV of 1.4 V and a specific energy of about 65 Wh/kg. Nickel-hydrogen batteries have captured large share of the space battery market in recent years and are rapidly replacing Nickel/cadmium batteries as the energy storage system of choice. They are acceptable for geo-synchronous orbit applications where not many cycles are required over the life of the system (1000 cycles, 10 years).

The impetus for research and development of metal-air cells has arisen from possible EV applications where energy density is a critical parameter. An interesting application suggested for a secondary zinc-oxygen battery is for energy storage on-board space craft where the cell could be installed inside one of the oxygen tanks thereby eliminating need for gas supply pipes and valves etc. These cells could be recharged using solar converters.

Some of the likely future developments for nickel-hydrogen batteries are (1) increase in cycle life for low earth orbit applications upto 40,000 cycles (7 years) (2) increase in the specific energy upto 100 W/kg for geosynchronous orbit applications and (3) development of a bipolar nickel-hydrogen battery for high pulse power applications.

9.47. Fuel Cells

As discussed earlier, a secondary battery produces electric current by oxidation-reduction chemical reaction. Similar chemical reactions take place in fuel cells but there is a basic difference between the two. Whereas in secondary batteries, the chemical energy is stored in the positive and negative electrodes, in fuel cells the oxidant and the fuel are stored outside the cells and must be fed to the electrodes continuously during the time the fuel cell supplies electric current. This gives an advantage to the fuel cells over the storage battery because fuels can be quickly replenished which is similar to filling up the petrol tank of a car. Moreover, storage batteries when fully discharged take several hours to be recharged.

9.48. Hydrogen-Oxygen Fuel Cells

The first fuel battery was designed by F.T. Bacon in 1959. The construction of a simple fuel cell is shown in Fig. 9.32. The electrodes are made from sintered nickel plaques having a coarse pore surface and a fine pore surface, the two surfaces being for gas and electrolyte respectively. The electrolyte used in KOH is of about 85 per cent concentration. The water vapour formed as a by-product of the reaction is removed by condensation from the stream of hydrogen passing over the back of the fuel.

The two electrodes of the fuel cell are fed with a continuous stream of hydrogen and oxygen (or air) as shown. The oxygen and hydrogen ions react with the potassium hydroxide electrolyte at the surface of the electrodes and produce water. The overall cell reaction is

$2H_2 + O_2 = 2H_2O$

The basic reaction taking place in the cells are shown in more details in the Fig. 9.32.

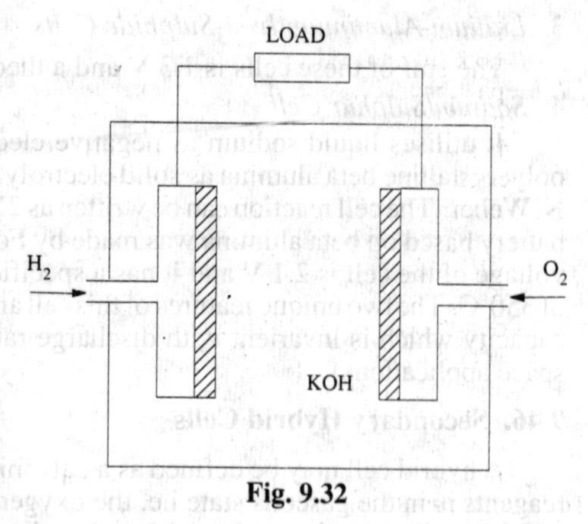

Fig. 9.32

Fuel cell batteries have been used in the manned Apollo space mission for on-board power supply and also for power supply in unmanned satellites and space probes. These batteries have also been used for tractors, fork-lift trucks and golf carts etc. Research is being carried out to run these batteries with natural gas and alcohol. Fuel cell systems are particularly useful where electrical energy is required for long periods. Such applications include (1) road and rail traction (2) industrial trucks (3) naval craft and submarine (4) navigational aids and radio repeater stations etc.

9.49. Batteries for Aircraft

The on-board power requirements in aircraft have undergone many changes during the last three or four decades. The jet engines of the aircraft which require starting current of about 1000 A, impose a heavy burden on the batteries. However, these days this load is provided by small turbo-generator sets and since batteries are needed only to start them, the power required is much less. These batteries possess good high-rate capabilities in order to supply emergency power for upto 1 h in the event of the generator failure. However, their main service is as a standby power for miscellaneous on-board equipment. Usually, batteries having 12 cells (of a nominal voltage of 24 V) with capacities of 18 and 34 Ah at the 10 h rate are used. In order to reduce weight, only light-weight high-impact polystyrene containers and covers are used and the cells are fitted with non-spill vent-plugs to ensure complete unspillability in any aircraft position during aerobatics. Similarly, special plastic manifolds are moulded into the covers to provide outlet for gases evolved during cycling.

9.50. Batteries for Submarines

These batteries are the largest units in the traction service. In older types of submarines, the lead storage battery was the sole means of propulsion when the submarine was fully submerged and, additionally, supplied the 'hotel load' power for lights, instruments and other electric equipment. When the introduction of the snorkel breathing tube made it possible to use diesel engines for propulsion, battery was kept in reserve for emergency use only. Even modern nuclear-powered submarines use storage batteries for this purpose. These lead-acid batteries may be flat, pasted plate or tubular positive plate type with 5 h capacities ranging from 10,000 to 12,000 Ah. One critical requirement for this service is that the rate of evolution of hydrogen gas on open-circuit should not exceed the specified low limit.

Double plate separation with the help of felted glass fibre mats and microporous separators is used in order to ensure durability, high performance and low standing losses.

OBJECTIVE TESTS–9

1. Active materials of a lead-acid cell are :
(*a*) lead peroxide (*b*) sponge lead
(*c*) dilute sulphuric acid (*d*) all of the above

2. During the charging of a lead-acid cell :
(*a*) its cathode becomes dark chocolate brown in colour
(*b*) its voltage increases (*c*) it gives out energy
(*d*) specific gravity of H_2SO_4 is decreased

3. The ratio of Ah efficiency to Wh efficiency of a lead-acid cell is :
(*a*) always less than one (*b*) just one
(*c*) always greater than one (*d*) either (*a*) or (*b*)

4. The capacity of a cell is measured in :
(*a*) watt-hours (*b*) watts
(*c*) amperes (*d*) ampere-hours

5. The capacity of a lead-acid cell does NOT depend on its :
(*a*) rate of charge (*b*) rate of discharge
(*c*) temperature (*d*) quantity of active material

6. As compared to constant-current system, the constant-voltage system of charging a lead-acid cell has the advantage of :
(*a*) avoiding excessive gassing
(*b*) reducing time of charging

(*c*) increasing cell capacity
(*d*) both (*b*) and (*c*).

7. Sulphation in a lead-acid battery occurs due to :
(*a*) trickle charging (*b*) incomplete charging
(*c*) heavy discharging (*d*) fast charging

8. The active materials of a nickel-iron battery are.
(*a*) nickel hydroxide
(*b*) powdered iron and its oxides
(*c*) 21% solution of caustic potash
(*d*) all of the above.

9. During the charging and discharging of a nickel-iron cell :
(*a*) its e.m.f. remains constant
(*b*) water is neither formed nor absorbed
(*c*) corrosive fumes are produced
(*d*) nickel hydroxide remains unsplit

10. As compared to a lead-acid cell, the efficiency of a nickel-iron cell is less due to its :
(*a*) lower e.m.f.
(*b*) smaller quantity of electrolyte used
(*c*) higher internal resistance
(*d*) compactness

11. Trickle charging of a storage battery helps to :
(*a*) prevent sulphation
(*b*) keep it fresh and fully charged
(*c*) maintain proper electrolyte level
(*d*) increase its reserve capacity.

12. A dead storage battery can be revived by :
(*a*) a dose of H_2SO_4
(*b*) adding so-called battery restorer
(*c*) adding distilled water
(*d*) none of the above

13. The sediment which accumulates at the bottom of a lead-acid battery consists largely of :
(*a*) lead-peroxide (*b*) lead-sulphate
(*c*) antimony-lead alloy (*d*) graphite.

14. The reduction of battery capacity at high rates of discharge is primarily due to :
(*a*) increase in its internal resistance
(*b*) decrease in its terminal voltage
(*c*) rapid formation of $PbSO_4$ on the plates
(*d*) non-diffusion of acid to the inside active materials

15. Floating battery systems are widely used for :
(*a*) power stations (*b*) emergency lighting
(*c*) telephone exchange installation
(*d*) all of the above

16. Any charge given to the battery when taken off the vehicle is called :
(*a*) bench charge (*b*) step charge
(*c*) float charge (*d*) trickle charge

Answers

1. *d* **2.** *b* **3.** *c* **4.** *d* **5.** *a* **6.** *d* **7.** *b* **8.** *d* **9.** *b* **10.** *c* **11.** *b* **12.** *d* **13.** *c* **14.** *c* **15.** *d* **16.** *a*

10 ELECTRICAL INSTRUMENTS AND MEASUREMENTS

10.1. Absolute and Secondary Instruments

The various electrical instruments may, in a very broad sense, be divided into (*i*) *absolute* instruments and (*ii*) *secondary* instruments. Absolute instruments *are those which give the value of the quantity to be measured, in terms of the constants of the instrument and their deflection only.* No previous calibration or comparison is necessary in their case. The example of such an instrument is tangent galvanometer, which gives the value of current, in terms of the tangent of deflection produced by the current, the radius and number of turns of wire used and the horizontal component of earth's field.

Secondary instruments are those, in which the value of electrical quantity to be measured can be determined from the deflection of the instruments, only when they have been pre-calibrated by comparison with an absolute instrument. Without calibration, the deflection of such instruments is meaningless.

It is the secondary instruments, which are most generally used in everyday work; the use of the absolute instruments being merely confined within laboratories, as standardizing instruments.

10.2. Electrical Principles of Operation

All electrical measuirng instruments depend for their action on one of the many physical effects of an electric current or potential and are generally classified according to which of these effects is utilized in their operation. The effects generally utilized are :

1. Magnetic effect –for ammeters and voltmeters usually.
2. Electrodynamic effect –for ammeters, voltmeters and wattmeters.
3. Electromagnetic effect –for ammeters, voltmeters, wattmeters and watthour meters.
4. Thermal effect –for ammeters and voltmeters.
5. Chemical effect –for d.c. ampere-hour meters.
6. Electrostatic effect –for voltmeters only.

Another way to classify secondary instruments is to divide them into (*i*) *indicating instruments (ii) recording instruments and (iii) integrating instruments.*

Indicating instruments are those which indicate the instantaneous value of the electrical quantity being measured *at the time* at which it is being measured. Their indications are given by pointers moving over calibrated dials. Ordinary ammeters, voltmeters and wattmeters belong to this class.

Recording instruments are those, which, instead of indicating by means of a pointer and a scale the instantaneous value of an electrical quantity, give a *continuous record* of the variations of such a quantity over a selected period of time. The moving system of the instrument carries an inked pen which rests lightly on a chart or graph, that is moved at a uniform and low speed, in a direction perpendicular to that of the deflection of the pen. The path traced out by the pen presents a continuous record of the variations in the deflection of the instrument.

Integrating instruments are those which measure and register by a set of dials and pointers either the *total* quantity of electricity (in amp-hours) or the *total* amount of electrical energy (in watt-hours or kWh) supplied to a circuit in a given time. This summation gives the product of time and the electrical quantity but gives no direct indication as to the *rate* at which the quantity or energy is being supplied because their registrations are independent of this rate provided the current flowing through the instrument is sufficient to operate it.

Ampere-hour and watt-hour meters fall in this class.

10.3. Essentials of Indicating Instruments

As defined above, indicating instruments are those which indicate the value of the quantity that is being measured at the time at which it is measured. Such instruments consist essentially of a pointer which moves over a calibrated scale and which is attached to a moving system pivoted in jewelled bearings. The moving system is subjected to the following three torques :

1. A deflecting (or operating) torque 2. A controlling (or restoring) torque 3. A damping torque.

10.4. Deflecting Torque

The deflecting or operating torque (T_d) is produced by utilizing one or other effects mentioned in Art. 10.2 *i.e.* magnetic, electrostatic, electrodynamic, thermal or inductive etc. The actual method of torque production depends on the type of instrument and will be discussed in the succeeding paragraphs. This deflecting torque causes the moving system (and hence the pointer attached to it) to move from its '*zero*' position *i.e.* its position when the instrument is disconnected from the supply.

10.5. Controlling Torque

The deflection of the moving system would be indefinite if there were no controlling or restoring torque. This torque opposes the deflecting torque and increases with the deflection of the moving system. The pointer is brought to rest at a position where the two opposing torques are equal. The deflecting torque ensures that currents of different magnitudes shall produce deflections of the moving system in proportion to their size. Without such a torque, the pointer would swing over to the maximum deflected position irrespective of the magnitude of the current to be measured. Moreover, in the absence of a restoring torque, the pointer once deflected, would not return to its zero position on removing the current. The controlling or restoring or balancing torque in indicating instruments is obtained either by a spring or by gravity as described below :

(a) Spring Control

A hair-spring, usually of phosphor-bronze, is attached to the moving system of the instrument as shown in Fig. 10.1 (*a*).

With the deflection of the pointer, the spring is twisted in the opposite direction. This twist in the spring produces restoring torque which is directly proportional to the angle of deflection of the moving system. The pointer comes to a position of rest (or equilibrium) when the deflecting torque (T_d) and controlling torque (T_c) are equal. For example, in permanent-magnet, moving-coil type of instruments, the deflecting torque is proportional to the current passing through them.

$\therefore \quad T_d \propto I$

and for spring control $T_c \propto \theta$

As $\quad T_c = T_d$

$\therefore \quad \theta \propto I$

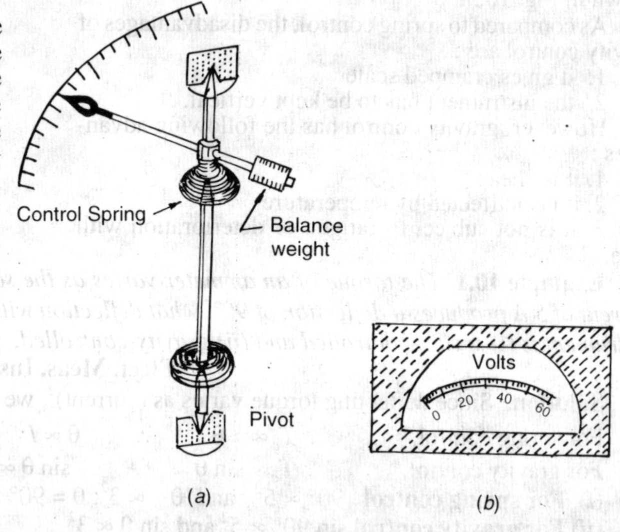

Fig. 10.1

Since deflection θ is directly proportional to current I, the spring-controlled instruments have a uniform or equally-spaced scales over the whole of thier range as shown in Fig. 10-1 (*b*).

To ensure that controlling torque is proportional to the angle of deflection, the spring should have a fairly large number of turns so that angular deformation per unit length, on full-scale deflection, is small. Moreover, the stress in the spring should be restricted to such a value that it does not produce a permanent set in it.

Springs are made of such materials which

(*i*) are non-magnetic (*ii*) are not subject to much fatigue

(*iii*) have low specific resistance–especially in cases where they are used for leading current in or out of the instrument

(*iv*) have low temperature-resistance coefficient.

The exact expression for controlling torque is $T_c = C\theta$ where C is *spring constant.* Its value is given by $C = \frac{Ebt^3}{L}$ N-m/rad. The angle θ is in radians.

(b) Gravity Control

Gravity control is obtained by attaching a small adjustable weight to some part of the moving system such that the two exert torques in the opposite directions. The usual arrangements is shown in Fig. 10.2 (*a*).

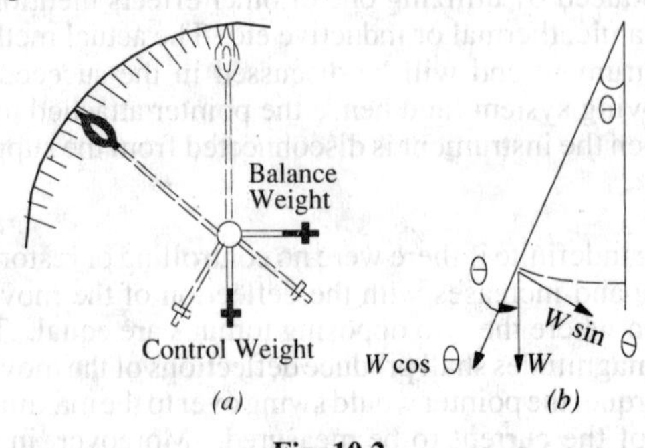

Fig. 10.2

It is seen from Fig. 10.2 (*b*) that the controlling or restoring torque is proportional to the sine of the angle of deflection *i.e.*

$$T_c \propto \sin\theta$$

The degree of control is adjusted by screwing the weight up or down the carrying system

If $\quad T_d \propto I$

then for position of rest

$$T_d = T_c$$

or $\quad I \propto \sin\theta$ (*not* θ)

It will be seen from Fig. 10.2 (*b*) that as θ approaches 90°, the distance AB increases by a relatively small amount for a given change in the angle than when θ is just increasing from its zero value. Hence, gravity-controlled instruments have scales which are not uniform but are cramped or crowded at their lower ends as shown in Fig. 10.3.

As compared to spring control, the disadvantages of gravity control are :

1. it gives cramped scale
2. the instrument has to be kept vertical.

However, gravity control has the following advantages :

1. it is cheap
2. it is unaffected by temperature
3. it is not subject to fatigue or deterioration with time.

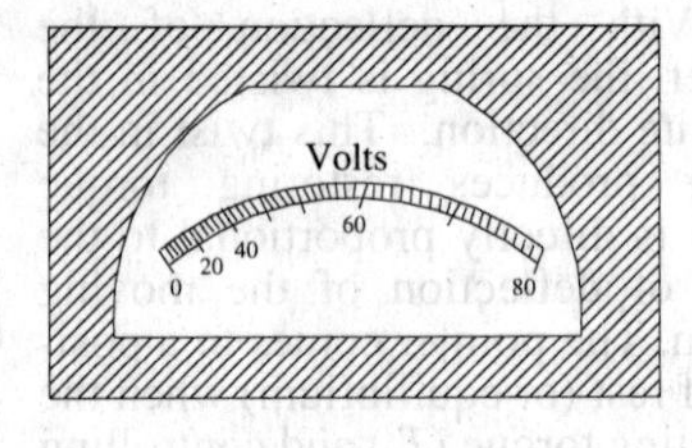

Fig. 10.3

Example 10.1 *The torque of an ammeter varies as the square of the current through it. If a current of 5 A produces a deflection of 90°, what deflection will occur for a current of 3 A when the instrument is* (*i*) *spring-controlled and* (*ii*) *gravity-controlled.*

(Elect. Meas. Inst and Meas. Jadavpur Univ. 1981)

Solution. Since deflecting torque varies as (current)², we have $T_d \propto I^2$

For spring control, $\quad T_c \propto \theta \quad \therefore \quad \theta \propto I^2$

For gravity control, $\quad T_c \propto \sin\theta \quad \therefore \quad \sin\theta \propto I^2$

(*i*) **For spring control** $90° \propto 5^2$ and $\theta \propto 3^2$; $\theta = 90° \times 3^2/5^2 =$ **32.4°**

(*ii*) **For gravity control** $\sin 90° \propto 5^2$ and $\sin\theta \propto 3^2$

$$\sin\theta = 9/25 = 0.36\ ;\ \theta = \sin^{-1}(0.36) = \mathbf{21.1°}.$$

10.6. Damping Torque

A damping force is one which acts on the moving system of the instrument *only when it is moving* and always opposes its motion. Such stabilizing or damping force is necessary to bring the pointer to rest *quickly*, otherwise due to inertia of the moving system, the pointer will oscillate about its final deflected position for quite some time before coming to rest in the steady position. The degree of damping should be adjusted to a value which is sufficient to enable the pointer to rise quickly to its deflected position without overshooting. In that case, the instrument is said to be *dead-beat*. Any increase of damping above this limit *i.e.* overdamping will make the instruments slow and lethargic. In Fig. 10.4 is shown the effect of damping on the variation of position with time of the moving system of an instrument.

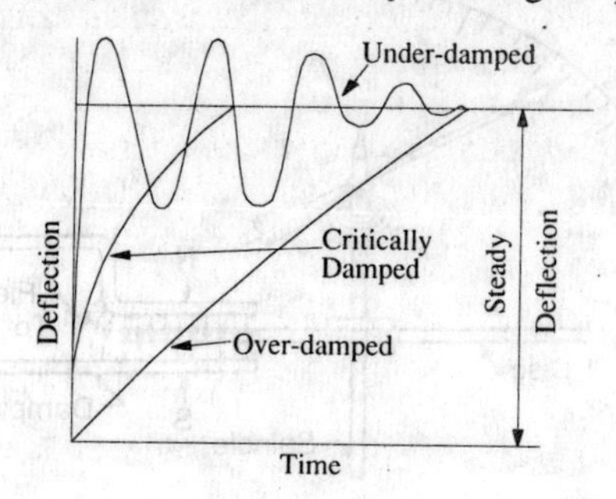

Fig. 10.4

The damping force can be produced by (*i*) *air frictions* (*ii*) *eddy currents* and (*iii*) *fluid friction* (ued occasionally).

Two methods of air-friction damping are shown in Fig. 10.5 (*a*) and 10.5 (*b*). In Fig. 10.5 (*a*), the light aluminium piston attached to the moving system of the instrument is arranged to travel with

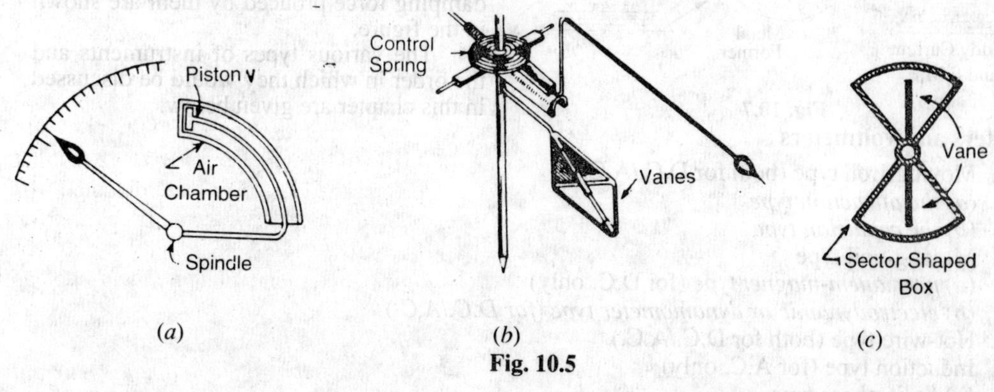

Fig. 10.5

a very small clearance in a fixed air chamber closed at one end. The cross-section of the chamber is either circular or rectangular. Damping of the oscillation is affected by the compression and suction actions of the piston on the air enclosed in the chamber. Such a system of damping is not much favoured these days, those shown in Fig. 10.5 (*b*) and (*c*) being preferred. In the latter method, one or two light aluminium vanes are mounted on the spindle of the moving system which move in a closed sector-shaped box as shown.

Fluid-friction is similar in action to the air friction. Due to greater viscosity of oil, the damping is more effective. However, oil damping is not much used because of several disadvantages such as objectionable creeping of oil, the necessity of using the instrument always in the vertical position and its obvious unsuitability for use in portable instruments.

The eddy-current form of damping is the most efficient of the three. The two forms of such a damping are shown in Fig. 10.6 and 10.7. In Fig. 10.6 (*a*) is shown a thin disc of a conducting but *non-magnetic* material like copper or aluminium mounted on the spindle which carries the moving system and the pointer of the instrument. The disc is so positioned that its edges, when in rotation, cut the magnetic flux between the poles of a permanent magnet. Hence, eddy currents are produced in the disc which flow and so produce a damping force in such a direction as to oppose the very cause producing them (Lenz's Law Art. 7.5). Since the cause producing them is the rotation of the disc, these eddy currents retard the motion of the disc and the moving system as a whole.

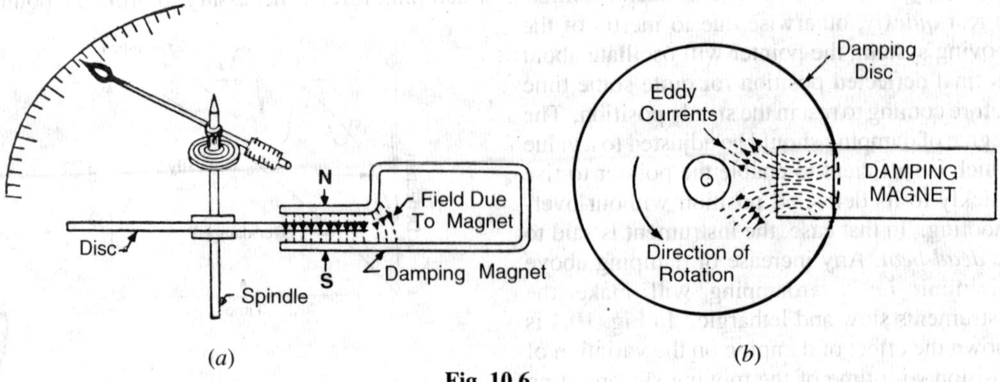

Fig. 10.6

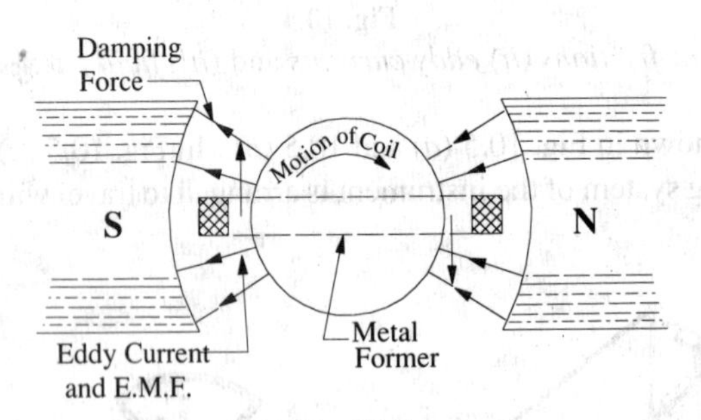

Fig. 10.7

In Fig. 10.7 is shown the second type of eddy-current damping generally employed in permanent-magnet moving-coil instruments. The coil is wound on a thin light aluminium former in which eddy currents are produced when the coil moves in the field of the permanent magnet. The directions of the induced currents and of the damping force prouced by them are shown in the figure.

The various types of instruments and the order in which they would be discussed in this chapter are given below.

Ammeters and voltmeters

1. Moving-iron type (both for D.C./A.C.)
   (*a*) *the attraction type*
   (*b*) *the repulsion type*
2. Moving-coil type
   (*a*) *permanent-magnet type* (for D.C. only)
   (*b*) *electrodynamic or dynamometer type (for D.C./A.C.)*
3. Hot-wire type (both for D.C./A.C.)
4. Induction type (for A.C. only)
   (*a*) *Split-phase type*
   (*b*) *Shaded-pole type*
5. Electrostatic type–for voltmeters only (for D.C./A.C).

Wattmeters

6. *Dynamometer type* (both for D.C./A.C.),
7. *Induction type* (for A.C. only)
8. *Electrostatic type* (for D.C. only)

Energy Meters

9. *Electrolytic type (for* D.C. only)
10. **Motor Meters**
    (*i*) *Mercury Motor Meter*. For d.c. work only. Can be used as amp-hour or watt-hour meter.
    (*ii*) *Commutator Motor Meter*. Used on D.C./A.C. Can be used as Ah or Wh meter.
    (*iii*) *Induction type*. For A.C only.
11. Clock meters (as Wh-meters).

10.7. Moving-iron Ammeters and Voltmeters

There are two basic forms of these instruments *i.e.* the *attraction* type and the *repulsion* type. The operation of the attraction type depends on the attraction of a single piece of soft iron into a magnetic field and that of repulsion type depends on the repulsion of two adjacent pieces of iron magnetised by the same magnetic field. For both types of these instruments, the necessary magnetic field is produced by the ampere-turns of a current-carrying coil. In case the instrument is to be used as an ammeter, the coil has comparatively fewer turns of thick wire so that the ammeter has low resistance because it is connected in series with the circuit. In case it is to be used as a voltmeter, the coil has high impedance so as to draw as small a current as possible since it is connected in parallel with the circuit. As the current through the coil is small, it has large number of turns in order to produce sufficient ampere-turns.

10.8. Attraction Type M.I. Instruments

The basic working principle of an attraction-type moving-iron instrument is illustrated in Fig. 10.8. It is well-known that if a piece of an unmagnetised soft iron is brought up near either of the two ends of a current-carrying coil, it would be attracted into the coil in the same way as it would be attracted by the pole of a bar magnet. Hence, if we pivot an oval-shaped disc of soft iron on a spindle between bearings near the coil (Fig. 10.8), the iron disc will swing into the coil when the latter has an electric current passing through it. As the field strength would be strongest at the centre of the coil, the oval-shaped iron disc is pivoted in such a way that the greatest bulk of iron moves into the centre of the coil. If a pointer is fixed to the spindle carrying the disc, then the passage of current through the coil will cause the pointer to deflect. The amount of deflection produced would be greater when the current producing the magnetic field is greater. Another point worth noting is that *whatever the direction of current through the coil, the iron disc would always be magnetised in such a way that it is pulled inwards.* Hence, such instruments can be used both for direct as well as alternating currents.

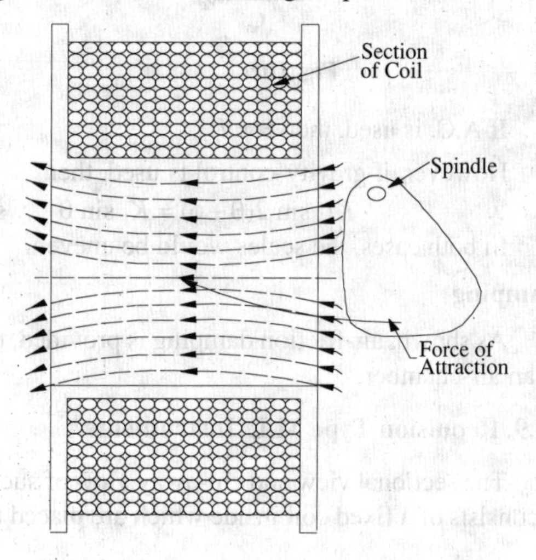

Fig. 10.8

A sectional view of the actual instrument is shown in Fig. 10. 9. When the current to be measured is passed through the coil or solenoid, a magnetic field is produced, which attracts the eccentrically-mounted disc inwards, thereby deflecting the pointer, which moves over a calibrated scale.

Deflecting Torque

Let the axis of the iron disc, when in zero position, subtend an angle of ϕ with a direction perpendicular to the direction of the field H produced by the coil. Let the deflection produced be θ corresponding to a current I through the coil. The magnetisation of iron disc is proportional to the component of H acting along the axis of the disc *i.e.* proportional to $H \cos [90-(\phi + \theta)]$ or $H \sin (\theta + \phi)$. The force F pulling the disc inwards is proportional to MH or $H^2 \sin (\theta + \phi)$. If the permeability of iron is assumed constant, then, $H \propto I$. Hence, $F \propto I^2 \sin (\theta+\phi)$. If this force acted at a distance of l from the pivot of the rotating disc, then deflecting torque $T_d = Fl \cos (\theta + \phi)$. Putting the value of F, we get

$$T_d \propto I^2 \sin (\theta + \phi) \times l \cos (\theta + \phi) \propto I^2 \sin 2(\theta + \phi) = KI^2 \sin 2(\theta + \phi)$$

...since l is constant

If spring-control is used, then controlling torque $T_c = K' \theta$

In the steady position of deflection, $T_d = T_c$

$\therefore\ KI^2 \sin 2(\theta + \phi) = K'\theta$; Hence $\theta \propto I^2$

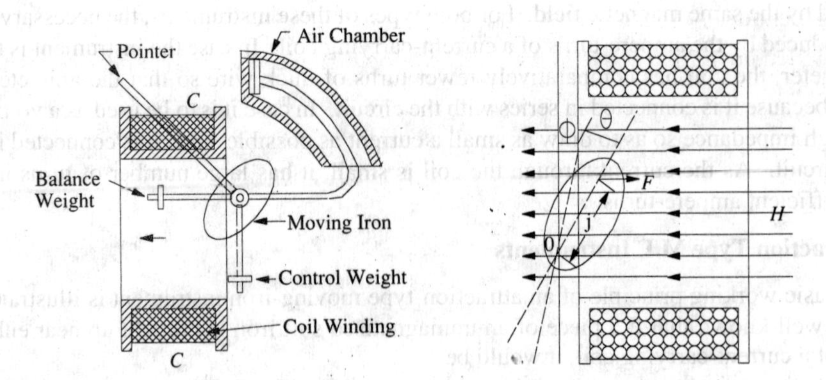

Fig. 10.9 **Fig. 10.10**

If A.C. is used, then $\theta \propto I^2_{r.m.s.}$

However, if gravity-control is used, then $T_c = K_1 \sin\theta$

$\therefore \quad KI^2 \sin 2(\theta + \phi) = K_1 \sin\theta \quad \therefore \sin\theta \propto I^2 \sin 2(\theta + \phi)$

In both cases, the scales would be uneven.

Damping

As shown, air-friction damping is provided, the actual arrangement being a light piston moving in an air-chamber.

10.9. Repulsion Type M.I. Instruments

The sectional view and cut-away view of such an instrument are shown in Fig. 10.11 and 10.12. It consists of a fixed coil inside which are placed two soft-iron rods (or bars) *A* and *B* parallel to one

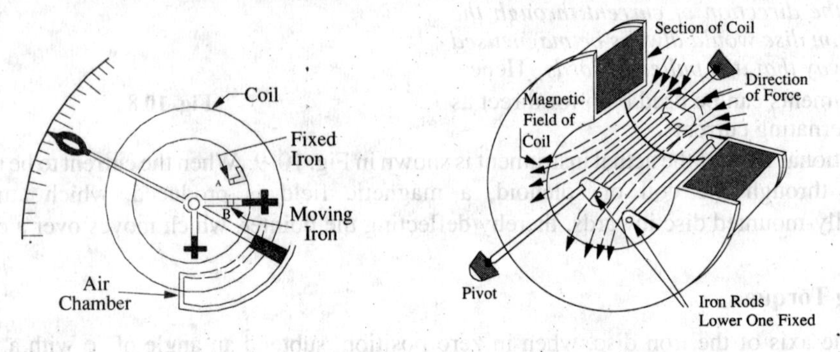

Fig. 10.11 **Fig. 10.12**

another and along the axis of the coil. One of them *i.e.* *A* is fixed and the other *B* which is movable carries a pointer that moves over a calibrated scale. When the current to be measured is passed through the fixed coil, it sets up its own magnetic field which magnetises the two rods similarly *i.e.* the adjacent points on the lengths of the rods will have the same magnetic polarity. Hence, they repel each other with the result that the pointer is deflected against the controlling torque of a spring or gravity. The force of repulsion is approximately proportional to the square of the current passing through the coil. Moreover, whatever may be the direction of the current through the coil, the two rods will be magnetised similarly and hence will repel each other.

In order to achieve uniformity of scale, two tongue-shaped strips of iron are used instead of two rods. As shown in Fig. 10-13(*a*), the fixed iron consists of a tongue-shaped sheet iron bent into a cylindrical form, the moving iron also consists of another sheet of iron and is so mounted as to move parallel to the fixed iron and towards its narrower end [Fig. 10.13(*b*)].

Deflecting Torque

The deflecting torque is due to the repulsive force between the two similarly magnetised iron rods or sheets.

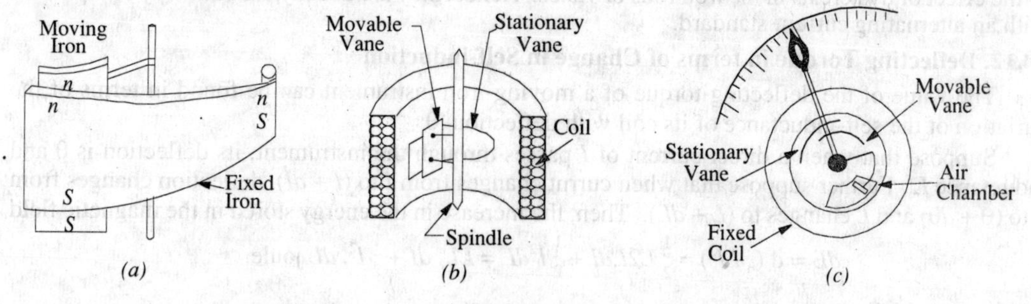

Fig. 10.13

Instantaneous torque ∝ repulsive force ∝ $m_1 m_2$ ....product of pole strengths

Since pole strengths are proportional to the magnetising force H of the coil,

∴ instantaneous torque ∝ H^2

Since H itself is proportional to current (assuming *constant* permeability) passing through the coil, ∴ instantaneous torque ∝ I^2

Hence, the deflecting torque, which is proportional to the mean torque is, in effect, proportional to the mean value of I^2. Therefore, when used on a.c. circuits, the instrument reads the r.m.s value of current.

Scales of such instruments are uneven if rods are used and uniform if suitably-shaped pieces of iron sheet are used.

The instrument is either gravity-controlled or as in modern makes, is spring-controlled.

Damping is pneumatic, eddy current damping cannot be employed because the presence of a permanent magnet required for such a purpose would affect the deflection and hence, the reading of the instrument.

Since the polarity of both iron rods reverses simultaneously, the instrument can be used both for a.c. and d.c. circuits *i.e.* instrument belongs to the unpolarised class.

10.10. Sources of Error

There are two types of possible errors in such instruments, firstly, those which occur both in a.c. and d.c. work and secondly, those which occur in a.c. work alone.

(a) Errors with both d.c. and a.c. work

(*i*) *Error due to hysteresis.* Because of hysteresis in the iron parts of the moving system, readings are higher for descending values but lower for ascending values.

The hysteresis error is almost completely eliminated by using Mumetal or Perm-alloy, which have negligible hysteresis loss.

(*ii*) *Error due to stray fields.* Unless shielded effectively from the effects of stray external fields, it will give wrong readings. Magnetic shielding of the working parts is obtained by using a covering case of cast-rion.

(b) Errors with a.c. work only

Changes of frequency produce (*i*) change in the impedance of the coil and (*ii*) change in the magnitude of the eddy currents. The increase in impedance of the coil with increase in the frequency of the alternating current is of importance in voltmeters (Ex. 10.2). For frequencies higher than the one used for calibration, the instrument gives lower values. However, this error can be removed by connecting a capacitor of suitable value in parallel with the swamp resistance R of the instrument. It can be shown that the impedance of the whole circuit of the instrument becomes independent of frequency if $C = L/R^2$ where C is the capacitance of the capacitor.

10.11. Advantages and Disadvantages

Such instruments are cheap and robust, give a reliable service and can be used both on a.c. and d.c. circuits, although they cannot be calibrated with a high degree of precision with d.c. on account of the effect of hysteresis in the iron rods or vanes. Hence, they are usually calibrated by comparison with an alternating current standard.

10.12. Deflecting Torque in terms of Change in Self-induction

The value of the deflecting torque of a moving-iron instrument can be found in terms of the variation of the self-inductance of its coil with deflection θ.

Suppose that when a direct current of I passes through the instrument, its deflection is θ and inductance L. Further suppose that when currnt changes from I to $(I + dI)$, deflection changes from θ to $(\theta + d\theta)$ and L changes to $(L + dL)$. Then, the increase in the energy stored in the magnetic field is

$$dE = d\left(\frac{1}{2}LI^2\right) = \frac{1}{2}L2I.dI + \frac{1}{2}I^2dL = LI\,.\,dI + \frac{1}{2}I^2.\,dL \text{ joule.}$$

If $T\frac{1}{2}$ N-m is the controlling torque for deflection θ, then extra energy stored in the control system is $T \times d\theta$ joules. Hence, the total increase in the stored energy of the system is

$$LI.dI + \frac{1}{2}I^2.dL + T \times d\,\theta \qquad \ldots(i)$$

The e.m.f. induced in the coil of the instrumetn is $e = N.\frac{d\Phi}{dt}$ volt

where $d\phi$ = change in flux linked with the coil due to change in the position of the disc or the bars

dt = time taken for the above change ; N = No of turns in the coil

Now $L = N\Phi/I \quad \therefore \quad \Phi = LI/N \quad \therefore \quad \frac{d\Phi}{dt} = \frac{1}{N}\cdot\frac{d}{dt}(LI)$

Induced e.m.f. $e = N.\frac{1}{N}\frac{d}{dt}(LI) = \frac{d}{dt}(LI)$

The energy drawn from the supply to overcome this back e.m.f is

$$= e.Idt = \frac{d}{dt}(LI).Idt = I.d(LI) = I(L.dI + I.dL) = LI.dI + I^2.dL \qquad \ldots(ii)$$

Equating (*i*) and (*ii*) above, we get $LI.dI + \frac{1}{2}I^2\,dL + T.d\theta = LI.dI + I^2.dL \quad \therefore \quad T = \frac{1}{2}I^2\frac{dL}{d\theta}$ N-m

where $dL/d\theta$ is henry/radian and I in amperes.

10.13. Extension of Range by Shunts and Multipliers

(*i*) **As Ammeter.** The range of the moving-iron instrument, when used as an ammeter, can be extended by using a suitable shunt across its terminals. So far as the operation with direct currents is concerned, there is no trouble, but with alternating current, the division of current between the instrument and shunt changes with the change in the applied frequency. For a.c. work, both the inductance and resistance of the instrument and shunt have to be taken into account.

Obviously, $\frac{\text{current through instrument, } i}{\text{current through shunt, } I_s} = \frac{R_s + j\omega L_S}{R + j\omega L} = \frac{\mathbf{Z}_s}{\mathbf{Z}}$

where R, L = resistance and inductance of the instrument

R_s, L_s = resistance and inductance of the shunt

It can be shown that the above ratio *i.e.* the division of current between the instrument and shunt would be independent of frequency if the time-constants of the instrument coil and shunt are the same *i.e.* if $L/R = L_S/R_S$. The multiplying power (N) of the shunt is given by

$$N = \frac{I}{i} = 1 + \frac{R}{R_S}$$

where I = line current ; i = full-scale deflection current of the instrument.

(*ii*) **As Voltmeter**. The range of this instrument, when used as a voltmeter, can be extended or multiplied by using a high non-inductive resistance R connected in series with it, as shown in

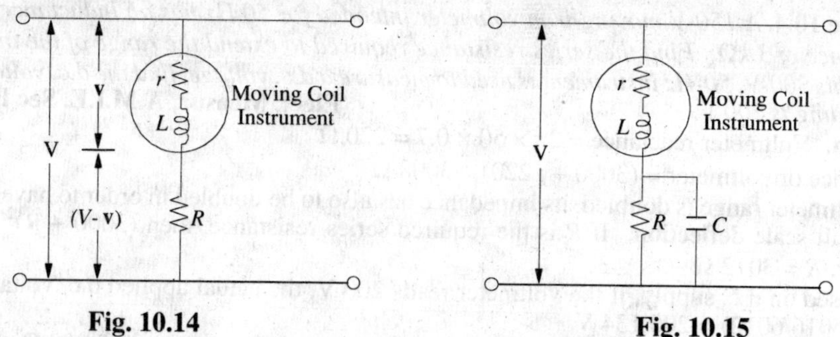

Fig. 10.14 **Fig. 10.15**

Fig. 10.14. This series resistance is known as 'multiplier' when used on d.c. circuits. Suppose, the range of the instrument is to be extended from v to V. Then obviously, the excess voltage of $(V-v)$ is to be dropped across R. If i is the full-scale deflection current of the instrument, then

$$iR = V - v; \quad R = \frac{V-v}{i} = \frac{V-ir}{i} = \frac{V}{i} - r$$

$$\text{Voltage magnification} = V/v. \qquad \text{Since } iR = V - v; \therefore \frac{iR}{v} = \frac{V}{v} - 1$$

or

$$\frac{iR}{ir} = \frac{V}{v} - 1 \qquad \therefore \frac{V}{v} = \left(1 + \frac{R}{r}\right)$$

Hence, greater the value of R, greater is the extension in the voltage range of the instrument.

For d.c. work, the principal requirement of R is that its value should remain constant *i.e.* it should have low temperature-coefficient. But for a.c. work it is essential that total impedance of the voltmeter and the series resistance R should remain as nearly constant as possible at different frequencies. That is why R is made as non-inductive as possible in order to keep the inductance of the whole circuit to the minimum. The frequency error introduced by the inductance of the instrument coil can be compensated by shunting R by a capacitor C as shown in Fig. 10.15. In case $r \ll R$, the impedance of the voltmeter circuit will remain practically constant (for frequenceis upto 1000 Hz) provided.

$$C = \frac{L}{(1+\sqrt{2})R^2} = 0.41\frac{L}{R^2}$$

Example 10.2. *A 250-volt moving-iron voltmeter takes a current of 0.05 A when connected to a 250-volt d.c. supply. The coil has an inductance of 1 henry. Determine the reading on the meter when connected to a 250-volt, 100-Hz a.c. supply.* **(Elect. Engg., Kerala Univ. 1987)**

Solution. When used on d.c. supply, the instrument offers ohmic resistance only. Hence, resistance of the instrument = 250/0.05 = 5000 Ω.

When used on a.c. supply, the instrument offers *impedance* instead of ohmic resistance.

$$\text{impedance at 100 Hz} = \sqrt{5000^2 + (2\pi \times 100 \times 1)^2} = 5039.3\ \Omega$$

$$\therefore \text{ instrument reading} = 250 \times 5000/5039.3 = \mathbf{248\ V}$$

Example 10.3. *A spring-controlled moving-iron voltmeter reads correctly on 250-V d.c. Calculate the scale reading when 250-V a.c. is applied at 50 Hz. The instrument coil has a resistance of 500 Ω and an inductance of 1 H and the series (non-reactive) resistance is 2000 Ω.*

(Elect. Instru. & Measur. Nagpur Univ. 1992)

Solution. Total circuit resistance of the voltmeter is

$$= (r + R) = 500 + 2{,}000 = 2{,}500\ \Omega$$

Since the voltmeter reads correctly on direct current supply, its full-scale deflection current is = 250/2500 = 0.1 A.

When used on a.c. supply, instrument offers an impedance

$$Z = \sqrt{2500^2 + (2\pi \times 50 \times 1)^2} = 2{,}520\ \Omega \qquad \therefore \quad I = 0.099\ \text{A}$$

$$\therefore \text{ Volmeter reading on a.c. supply} = 250 \times 0.099/1 = \mathbf{248\ V}^*$$

Note. Since swamp resistance $R = 2{,}000\ \Omega$, capacitor required for compensating the frequency error is

$$C = 0.41\ L/R^2 = 0.41 \times 1/2000^2 = 0.1\ \mu F.$$

Example 10.4. *A 150-V moving-iron voltmeter intended for 50 Hz has an inductance of 0.7 H and a resistance of 3 kΩ. Find the series resistance required to extend the range of the instrument to 300 V. If this 300-V, 50-Hz instrument is used to measure a d.c. voltage, find the d.c. voltage when the scale reading is 200 V.* **(Elect. Measur, A.M.I.E. Sec B, 1991)**

Solution. Voltmeter reactance $= 2\pi \times 50 \times 0.7 = 220\ \Omega$

Impedance of voltmeter $= (3000 + j\,220) = 3008\ \Omega$

When the voltmeter range is doubled, its impedance has also to be doubled in order to have the same current for full-scale deflection. If R is the required series resistance, then $(3000 + R)^2 + 220^2 = (2 \times 3008)^2 \therefore R = 3012\ \Omega$

When used on d.c. supply, if the voltmeter reads 200 V, the actual applied d.c. voltage would be $= 200 \times (6016/6012) = 200.134$ V

Example 10.5. *The coil of a moving-iron voltmeter has a resistance of 5,000 Ω at 15°C at which temperature it reads correctly when connected to a supply of 200 V. If the coil is wound with wire whose temperature coefficient at 15°C is 0.004, find the percentage error in the reading when the temperature is 50°C.*

In the above instrument, the coil is replaced by one of 2,000 Ω but having the same number of turns and the full 5,000 Ω resistance is obtained by connecting in series a 3,000 Ω resistor of negligible temperature-coefficient. If this instrument reads correctly at 15°C, what will be its percentage error at 50°C.

Solution. Current at 15°C $= 200/5{,}000 = 0.04$ A

Resistance at 50°C is $R_{50} = R_{15}(I + \alpha_{15} \times 35)$

$\therefore R_{50} = 5{,}000(1 + 35 \times 0.004) = 5{,}700\ \Omega$

$\therefore$ current at 50° C $= 200/5{,}700$

$\therefore$ reading at 50°C $= \dfrac{200 \times (200/5{,}700)}{0.04} = 175.4$ V

or $= 200 \times 5000/5700 = 175.4$ V

$\therefore$ % error $= \dfrac{175.4 - 200}{200} \times 100 = \mathbf{-12.3\%}$

In the second case, swamp resistance is 3,000 Ω whereas the resistance of the instrument is only 2,000 Ω

Instrument resistance at 50°C $= 2{,}000(1 + 35 \times 0.004) = 2{,}280\ \Omega$

$\therefore$ total resistance at 50°C $= 3{,}00 + 2{,}280 = 5{,}280\ \Omega$

$\therefore$ current at 50°C $= 200/5{,}280$ A

$\therefore$ instrument reading $= 200 \times \dfrac{200/5{,}280}{0.04} = 189.3$ V

$\therefore$ percentage error $= \dfrac{189.3 - 200}{200} \times 100 = \mathbf{-5.4\ \%}$

Example 10.6. *The change of inductance for a moving-iron ammeter is 2μH/degree. The control spring constant is 5×10^{-7} N-m/degree.*

The maximum deflection of the pointer is 100°, what is the current corresponding to maximum deflection? **(Measurement & Instrumentation Nagpur Univ. 1993)**

Solution. As seen from Art. 10.12 the deflecting torque is given by

$$T_d = \frac{1}{2} I^2 \frac{dL}{d\theta} \text{ N-m}$$

Control spring constant $= 5 \times 10^{-7}$ N-m/degree

Deflection torque for 100° deflection $= 5 \times 10^{-7} \times 100 .81 = 5 \times 10^{-5}$ N-m; $dL/d\theta = 2\mu$H/degree $= 2 \times 10^{-6}$ H/degree.

$\therefore 5 \times 10^{-5} = \frac{1}{2} \times 2 \times 10^{-6} \quad \therefore$ **I = 50 A.**

---

*or reading $= 250 \times 2500/2520 =$ **248 V.**

Example 10.7. *The inductance of attraction type instrument is given by* $L = (10 + 5\theta - \theta^2)\ \mu H$ *where* θ *is the deflection in radian from zero position. The spring constant is* 12×10^{-6} *N-m/rad. Find out the deflection for a current of 5 A.*

(Elect. and Electronics Measurements and Measuring Instruments Nagpur Univ. 1993.)

Solution. $L = (10 + 5\theta - \theta^2) \times 10^{-6}$ H

$$\therefore \frac{dL}{d\theta} = (0 + 5 - 2 \times \theta) \times 10^{-6} = (5 - 2\theta) \times 10^{-6} \text{ H/rad}$$

Let the deflection be θ radians for a current of 5A, then deflecting torque,

$$T_d = 12 \times 10^{-6} \times \theta \text{ N-m}$$

Also, $$T_d = \frac{1}{2} I^2 \frac{dL}{d\theta}$$ ...Art.

Equating the two torques, we get

$$12 \times 10^{-6} \times \theta = \frac{1}{2} \times 5^2 \times (5 - 2\theta) \times 10^{-6} \quad \therefore \theta = \mathbf{1.689 \text{ radian}}$$

Tutorial Problems No. 10-1

1. Derive an expression for the torque of a moving-iron ammeter. The inductance of a certain moving-iron ammeter is $(8 + 4\theta - \frac{1}{2}\theta^2)\ \mu H$ where θ is the deflection in radians from the zero position. The control-spring torque is 12×10^{-6} N-m/rad. Calaculate the scale position in radians for a current of 3A.

[1.09 rad] *(I.E.E. London)*

2. An a.c. voltmeter with a maximum scale reading of 50-V has a resistance of 500 Ω and an inductance of 0.09 henry, The magnetising coil is wound with 50 turns of copper wire and the remainder of the circuit is a non-inductive resistance in series with it. What additional apparatus is needed to make this instrument read correctly on both d.c. and a.c. circuits of frequency 60 ? **[0.44 μF in parallel with series resistance]**

3. A 10-V moving-iron ammeter has a full-scale deflection of 40 mA on d.c. circuit. It reads 0.8% low on 50 Hz a.c. Hence, calculate the inductance of the ammeter. **[115.5 mH]**

4. It is proposed to use a non-inductive shunt to increase the range of a 10-A moving iron ammeter to 100 A. The resistance of the instrument, including the leads to the shunt, is 0.06 Ω and the inductance is 15 μH at full scale. If the combination is correct on a.d.c circuit, find the error at full scale on a 50 Hz a.c. circuit.

[3.5%] *(London Univ.)*

10.14. Moving-coil Instruments

There are two types of such instruments (*i*) *permanent-magnet type* which can be used for d.c. work only and (*ii*) the *dynamometer type* which can be used both for a.c. and d.c. work.

10.15. Permanent Magnet Type Instruments

The operation of a permanent-magnet moving-coil type instrument is based upon the principle that when a current-carrying conductor is placed in a magnetic field, it is acted upon by a force which tends to move it to one side and out of the field.

Construction

As its name indicates, the instrument consists of a permanent magnet and a rectangular coil of many turns wound on a light aluminium or copper former inside which is an iron core as shown in

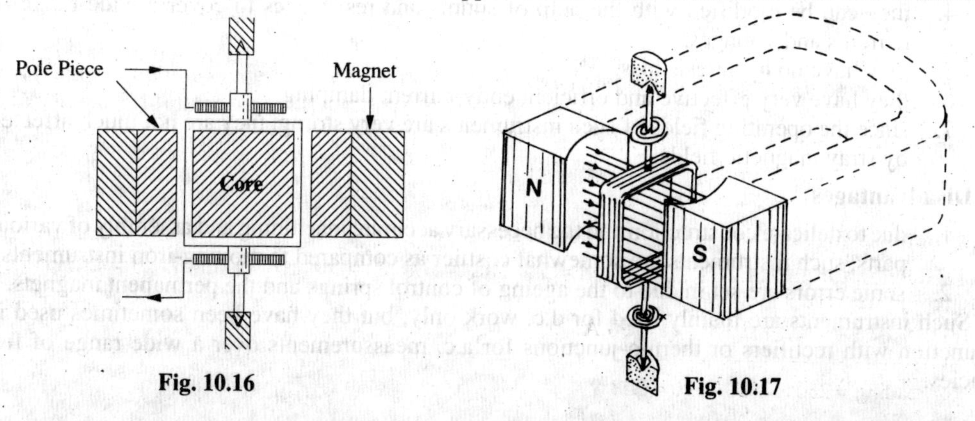

Fig. 10.16 Fig. 10.17

Fig. 10.16. The powerful U-shaped permanent magnet is made of Alnico and has soft-iron end-pole pieces which are bored out cylindrically. Between the magnetic poles is fixed a soft iron cylinder whose function is (*i*) to make the field radial and uniform and (*ii*) to decrease the reluctance of the air path between the poles and hence increase the magnetic flux. Surrounding the core is a rectangular coil of many turns wound on a light aluminium frame which is supported by delicate bearings and to which is attached a light pointer. The aluminium frame not only provides support for the coil but also provides damping by eddy currents induced in it. The sides of the coil are free to move in the two air-gaps between the poles and core as shown in Fig. 10.16 and Fig. 10.17. Control of the coil movement is affected by two phosphor-bronze hair springs, one above and one below, which additionally serve the purpose of leading the current in and out of the coil. The two springs are spiralled in opposite directions in order to neutralize the effects of temperature changes.

Deflecting Torque

When current is passed through the coil, force acts upon its both sides which produce a deflecting torque as shown in Fig. 10.18. Let

B= flux density in Wb/m^2
l = length or depth of the coil in metre
b = breadth of coil in metre
N= number of turns in the coil

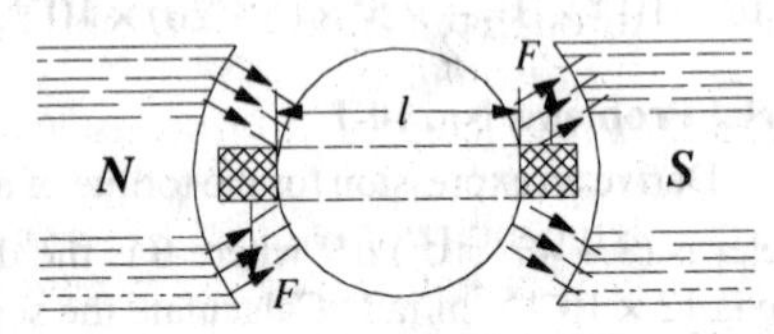

Fig. 10.18

If I ampere is the current passing through the coil, then the magnitude of the force experienced by each of its sides is = BIl newton

For N turns, the force on each side of the coil is = $NBIl$ newton

$\therefore$ deflecting torque T_d = force × perpendicular distance

$$= NBIl \times b = NBI(l \times b) = NBIA \text{ N-m}$$

where A is the face area of the coil.

It is seen that if B is constant, then T_d is proportional to the current passing through the coil *i.e.* $T_d \propto I$.

Such instruments are invariably spring-controlled so that $T_c \propto$ deflection θ.

Since at the final deflected position, $T_d = T_c \quad \therefore \quad \theta \propto I$

Hence, such instruments have uniform scales. Damping is electromagnetic *i.e.* by eddy currents induced in the metal frame over which the coil is wound. Since the frame moves in an intense magnetic field, the induced eddy currents are large and damping is very effective.

10.16. Advantages and Disadvantages

The permanent-magnet moving-coil (PMMC) type instruments have the following advantages and disadvantages :

Advantages

1. They have low power consumption.
2. their scales are uniform and can be designed to extend over an arc of 270° or so.
3. they possess high (torque/weight) ratio.
4. they can be modified with the help of shunts and resistances to cover a wide range of currents and voltages.
5. they have no hysteresis loss.
6. they have very effective and efficient eddy-current damping.
7. since the operating fields of such instruments are very storng, they are not much affected by stray magnetic fields.

Disadvantages

1. due to delicate construction and the necessary accurate machining and assembly of various parts, such instruments are somewhat costlier as compared to moving-iron instruments.
2. some errors are set in due to the ageing of control springs and the permanent magnets.

Such instruments are mainly used for d.c. work only, but they have been sometimes used in conjunction with rectifiers or thermo-junctions for a.c. meassurements over a wide range of frequencies.

Permanent-magnet moving-coil instruments can be used as ammeters (with the help of a low resistance shunt) or as voltmeters (with the help of a high series resistance).

The principle of permanent-magnet moving-coil type instruments has been utilized in the construction of the following :

1. For a.c. galvanometer which can be used for detecting extremely small d.c. currents. A galvanometer may be used either as an ammeter (with the help of a low resistance) or as a voltmeter (with the help of a high series resistance). Such a galvanometer (of pivoted type) is shown in Fig. 10.19.
2. By eliminating the control springs, the instrument can be used for measuring the quantity of electricity passing through the coil. This method is used for *fluxmeters*.
3. If the control springs of such an instrument are purposely made of large moment of inertia, then it can be used as ballistic galvanometer.

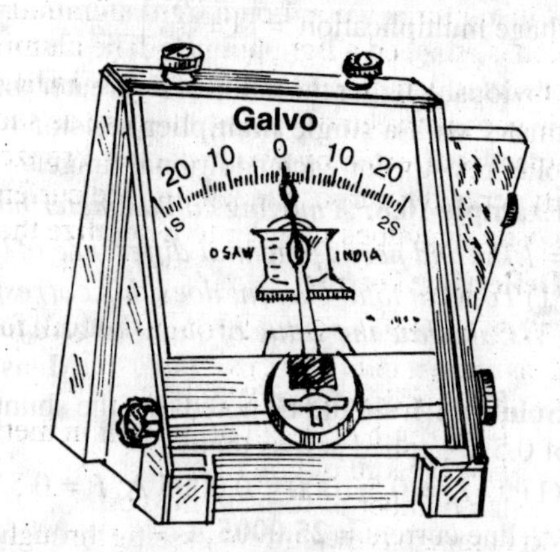

Fig. 10.19

10.17. Extension of Range

(*i*) As Ammeter

When such an instrument is used as an ammeter, its range can be extended with the help of a low-resistance shunt as shown in Fig. 10.20 (*a*). This shunt provides a bypath for extra current because it is connected across (*i.e.* in parallel with) the instrument. These shunted instruments can be made to record currents many times greater than their normal full-scale deflection currents. The ratio of maximum current (with shunt) to the full-scale deflection current (without shunt) is known as the 'multiplying power' or 'multiplying factor' of the shunt.

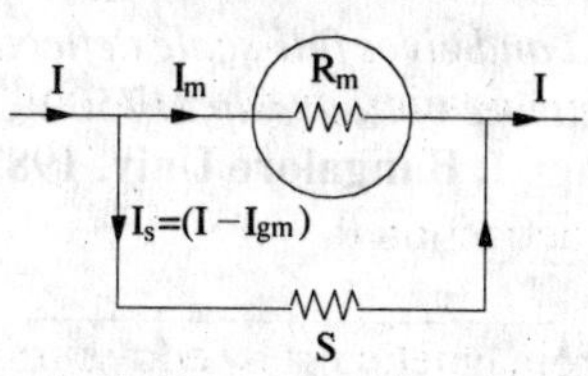

Fig. 10.20(*a*)

Let R_m = instrument resistance
S = shunt resistance
I_m = full-scale deflection current of the instrument
I = line current to be measured

As seen from Fig. 10.20 (*a*), the voltage across the instrument coil and the shunt is the same since both are joined in parallel.

$$\therefore \quad I_m \times R_m = S\,I_m = S(I - I_m) \qquad \therefore \quad S = \frac{I_m R_m}{(I - I_m)}; \text{ Also } \frac{I}{I_m} = \left(1 + \frac{R_m}{S}\right)$$

$$\therefore \text{ multiplying power } = \left(1 + \frac{R_m}{S}\right)$$

Obviously, lower the value of shunt resistance, greater its multiplying power.

(*ii*) As voltmeter

The range of this instrument when used as a voltmeter can be increased by using a high resistance in series with it [Fig. 10.20 (*b*)].

Let I_m = full-scale deflection current
R_m = galvanometer resistance
v = $R_m I_m$ = full-scale p.d. across it
V = voltage to be measured
R = series resistance required

Then it is seen that the voltage drop across R is = $V - v$

$$\therefore \quad R = \frac{V - v}{I_m} \quad \text{or} \quad R \,.\, I_m = V - v$$

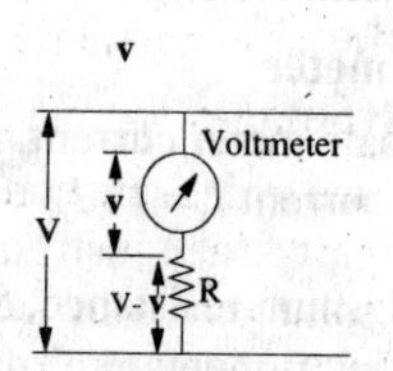

Fig. 10.20(*b*)

Dividing both sides by *v*, we get

$$\frac{RI_m}{v} = \frac{V}{v} - 1 \quad \text{or} \quad \frac{R.I_m}{I_m R_m} = \frac{V}{v} - 1 \quad \therefore \quad \frac{V}{v} = \left(1 + \frac{R}{R_m}\right)$$

$\therefore$ voltage multiplication $= \left(1 + \frac{R}{R_m}\right)$

Obviously, larger the value of R, greater the voltage multiplication or range. Fig. 10.20 (b) shows a voltmeter with a single multiplier resistor for one range. A multi-range voltmeter requires one multiplier resistor for each additional range.

Example 10.8. *A moving coil ammeter has a fixed shunt of 0.02 Ω with a coil circuit resistance of R = 1 kΩ and needs potential difference of 0.5 V across it for full-scale deflection.*

(1) *To what total current does this correspond ?*

(2) *Calculate the value of shunt to give full scale deflection when the total current is 10 A and 75 A.* **(Measurement & Instrumentation Nagpur Univ. 1993)**

Solution. It should be noted that the shunt and the meter coil are in parallel and have a common p.d. of 0.5 V applied across them.

(1) $\therefore$ $I_m = 0.5/1000 = 0.0005$ A; $I_s = 0.5/0.02 = 25$ A

$\therefore$ line current = **25.0005 A**

(2) When total current is 10 A, $I_s = (10-0.0005) = 9.9995$ A

$$\therefore \quad S = \frac{I_m R_m}{I_s} = \frac{0.0005 \times 1000}{9.9995} = 0.05\ \Omega$$

When total current is 75 A, $I_s = (75-0.0005) = 74.9995$ A

$\therefore S = 0.0005 \times 1000/74.9995 =$ **0.00667 Ω**

Example 10.9. *A moving-coil instrument has a resistance of 10 Ω and gives full-scale deflection when carrying a current of 50 mA. Show how it can be adopted to measure voltages up to 750 V and currents upto 1000 A.* **(Elements of Elect. Engg.I, Bangalore Univ. 1987)**

Solution. (*a*) **As Ammeter**. As discussed above, current range of the meter can be extended by using a shunt across it [Fig. 10.21 (*a*)].

Obviously,

$10 \times 0.05 = S \times 99.95$

$\therefore$ $S =$ **0.005 Ω**

(*b*) **As Voltmeter**. In this case, the range can be extended by using a high resistance R in series with it.

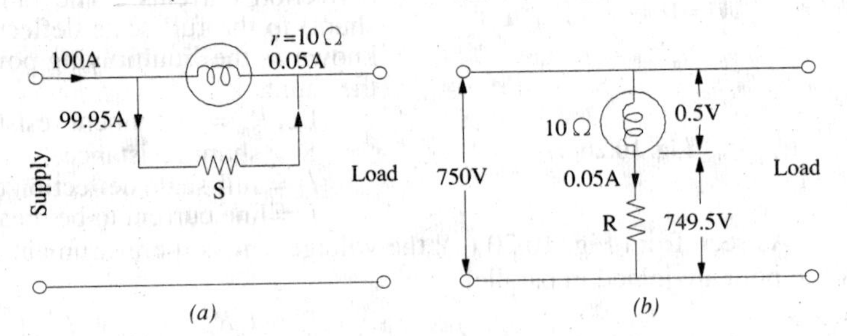

Fig. 10.21

[Fig. 10.21 (*b*)]. Obviously, R must drop a voltage of (750–0.5) = 749.5 V while carrying 0.05 A.

$\therefore$ $0.05\ R = 749.5$ or $R =$ **14.990 Ω**

Example 10.10. *How will you use a P.M.M.C. instrument which gives full scale deflection at 50 mV p.d. and 10 mA current as*

(1) *Ammeter 0–10 A range*

(2) *Voltmeter 0–250 V range* **(Elect. Instruments & Measurements Nagpur Univ. 1993)**

Solution. Resistance of the instrument $R_m = 50$ mV/10 mA = 5 Ω

(i) As Ammeter

full-scale meter current, $I_m = 10$ mA = 0.01 A

shunt current $I_s = I - I_m = 10 - 0.01 = 9.99$ A

$$\text{Reqd. shunt resistance, } S = \frac{I_m.R_m}{(I - I_m)} = \frac{0.01 \times 5}{9.99} = 0.0005\ \Omega$$

(ii) As Voltmeter

Full–scale deflection voltage, $v = 50\ mV = 0.05\ V$; $V = 250$ V

Reqd. series resistance, $R = \frac{V - v}{I_m} = \frac{250 - 0.05}{0.01} = 24,995\ \Omega$

Example 10.11 *A current galvanometer has the following parameters :-*
$B = 10 \times 10^{-3}$ Wb/m^2 ; N = 200 turns, l = 16 mm;
d = 16 mm; k = 12×10^{-9} Nm/radian.
Calculate the deflection of the galvanometer when a current of 1μA flows through it.

(Elect. Measurement Nagpur Univ. 1993)

Solution. Deflecting torque $T_d = NBIA$ N–m $= 200 \times (10 \times 10^{-3}) \times (1 \times 10^{-6}) \times (16 \times 10^{-3} \times 16 \times 10^{-3})$ N-m $= 512 \times 10^{-12}$ N–m

Controlling torque T_c = controlling spring constant × deflection = $12 \times 10^{-9} \times \theta$ N–m

Equating the deflecting and controlling torques, we have $12 \times 110^{-9} \times \theta = 512 \times 10^{-12}$

$\therefore\ \theta = 0.0427$ radian = **2.45°**

Example 10.12. *The coil of a moving coil permanent magnet voltmeter is 40 mm long and 30 mm wide and has 100 turns on it. The control spring exerts a torque of 120×10^{-6} N-m when the deflection is 100 divisions on full scale. If the flux density of the magnetic field in the air gap is 0.5 Wb/m^2, estimate the resistance that must be put in series with the coil to give one volt per division. The resistance of the voltmeter coil may be neglected.*

(Elect. Measur. AMIE Sec. B Summer 1991)

Solution. Let I be the current for full-scale deflection. Deflection torque T_d = NBIA
$= 100 \times 0.5 \times I \times (1200 \times 10^{-6}) = 0.06\ I$ N-m

Controlling torque $T_c = 120 \times 10^{-6}$ N-m

In the equilibrium position, the two torques are equal *i.e.* $T_d = T_c$.

$\therefore\ 0.06\ I = 120 \times 10^{-6} \quad \therefore I = 2 \times 10^{-3}$ A.

Since the instrument is meant to read 1 volt per division, its full-scale reading is 100 V.

Total resistance = $100/2 \times 10^{-3} = 50,000\ \Omega$

Since voltmeter coil resistance is negligible, it represents the additional required resistance.

Example 10.13. *Show that the torque produced in a permanent-magnet moving-coil instrument is proportional to the area of the moving coil.*

A moving-coil voltmeter gives full-scale deflection with a current of 5 mA. The coil has 100 turns, effective depth of 3 cm and width of 2.5 cm. The controlling torque of the spring is 0.5 cm for full-scale deflection. Estimate the flux density in the gap.

(Elect. Meas, Marathwada Univ. 1985)

Solution. The full-scale deflecting torque is $T_d = NBIA$ N–m
where I is the full-scale deflection current ; I = 5 mA = 0.005 A

$T_d = 100 \times B \times 0.005 \times (3 \times 2.5 \times 10^{-4}) = 3.75 \times 10^{-4}\ B$ N-m

The controlling torque is

T_c = 0.5 g-cm = 0.5 g. wt.cm = $0.5 \times 10^{-3} \times 10^{-2}$ kg wt-m
$= 0.5 \times 10^{-5} \times 9.8 = 4.9 \times 10^{-5}$ N-m

For equilibrium, the two torques are equal and opposite.

$\therefore\ 4.9 \times 10^{-5} = 3.75 \times 10^{-4}\ B \qquad \therefore \qquad B =$ **0.13 Wb/m^2**

Example 10.14. *A moving-coil milliammeter has a resistance of 5 Ω and a full-scale deflection of 20 mA. Determine the resistance of a shunt to be used so that the instrument could measure currents upto 500 mA at 20°C. What is the percentage error in the instrument operating at a temperature of 40°C ? Temperature co-efficient of copper = 0.0039 per °C.*

(Measu. & Instrumentation, Allahabad Univ. 1991)

Solution. Let R_{20} be the shunt resistance at 20°C. When the temperature is 20°C, line current is 500 mA and shunt current is = (500–20) = 480 mA.

$\therefore\ 5 \times 20 = R_{20} \times 480, \qquad R_{20} = 1/4.8\ \Omega$

If R_{40} is the shunt resistance at 40°C, then

$$R_{40} = R_{20}(1+20\alpha) = \frac{1}{4.8}(1+0.0039\times 20) = \frac{1.078}{4.8}\ \Omega$$

Shunt current at 40°C is $= \frac{5\times 20}{1.078/4.8} = 445$ mA

Line current = 445 + 20 = 465 mA

Although, line current would be only 465 mA, the instrument will indicate 500 mA.

$\therefore$ error = 35/500 = 0.07 or 7%

Example 10.15. *A moving-coil millivoltmeter has a resistance of 20Ω and full-scale deflection of 120° is reached when a potential difference of 100 mV is applied across its terminals. The moving coil has the effective dimensions of 3.1 cm × 2.6 cm and is wound with 120 turns. The flux density in the gap is 0.15 Wb/m². Determine the control constant of the spring and suitable diameter of copper wire for coil winding if 55% of total instrument resistance is due to coil winding. ρ for copper = 1.73 × 10⁻⁶ Ω cm.* **(Elect. Inst. and Meas. M.S. Univ. Baroda, 1985)**

Solution. Full-scale deflection current is = 100/20 = 5 mA

Deflecting torque for full-scale deflection of 120° is

$T_d = NBIA = 120\times 0.15\times(5\times 10^{-3})\times(3.1\times 2.6\times 10^{-4}) = 72.5\times 10^{-6}$ N-m

Control constant is defined as the deflecting torque per radian (or degree) of deflection of moving coil. Since this deflecting torque is for 120° deflection.

Control constant = $72.5\times 10^{-6}/120 =$ **6.04×10^{-7} N-m/degree**

Now, resistance of copper wire = 55% of 20 Ω = 11 Ω

Total length of copper wire = 120 × 2 (3.1 + 2.6) = 1368 cm

Now $R = \rho l/A \quad \therefore \quad A = 1.73\times 10^{-6}\times 1368/11 = 215.2\times 10^{-6}\ \text{cm}^2$

$\therefore \quad \pi d^2/4 = 215.2\times 10^{-6}$

$\therefore \quad d = \sqrt{215.\times 4\times 10^{-6}/\pi} = 16.55\times 10^{-3}\ \text{cm} =$ **0.1655 mm**

10.18. Voltmeter Sensitivity

It is defined in terms of resistance per volt (Ω/V). Suppose a meter movement of 1 kΩ internal resistance has a full-scale deflection current of 50 μ A. Obviously, full-scale voltage drop of the meter movement is = 50 μ A × 1000 Ω = 50 mV. When used as a voltmeter, its sensitivity would be $1000/50\times 10^{-3}$ = 20 kΩ/V. It should be clearly understood that a sensitivity of 20 kΩ/V means that the total resistance of the circuit in which the above movement is used should be 20 kΩ for a full-scale deflection of 1 V.

10.19. Multi–range Voltmeter

It is a voltmeter which measures a number of voltage ranges with the help of different series resistances. The resistance required for each range can be easily calculated provided we remember one basic fact that the sensitivity of a meter movement is always the same regardless of the range selected. Moreover, the full-scale deflection current is the same in every range. For any range, the total circuit resistance is found by multiplying the sensitivity by the full-scale voltage for that range.

For example, in the case of the above-mentioned 50 μA, 1 kΩ meter movement, total resistance required for 1*V* full-scale deflection is 20 kΩ. It means that an additional series resistance of 19 kΩ is required for the purpose as shown in Fig. 10.22(*a*).

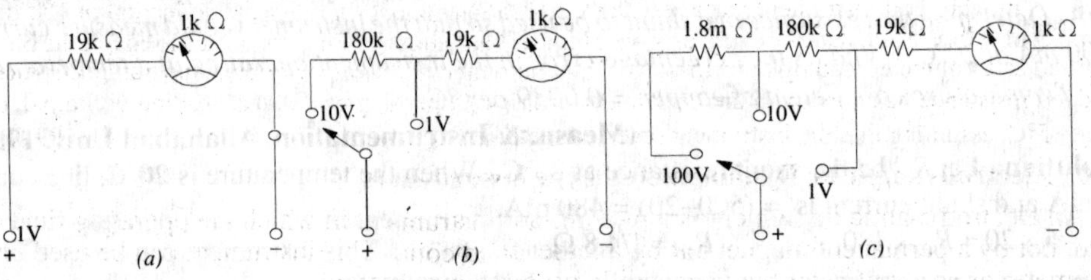

Fig. 10.22

For 10-V range, total circuit resistance must be (20 kΩ/V) (10 V) = 200 kΩ. Since total resistance for 1 V range is 20 kΩ, the series resistance R for 10-V range = 200–20 = 180 kΩ as shown in Fig. 10.22 (*b*).

For the range of 100 V, total resistance required is (20 kΩ/V) (100 V) = 2 MΩ. The additional resistance required can be found by subtracting the existing two-range resistance from the total resistance of 2 MΩ. Its value is

= 2 M Ω–180 kΩ – 19 kΩ – 1 kΩ = 1.8 MΩ

It is shown in Fig. 10.22 (*c*).

Example 10.16. *A basic d'Arsonval movement with internal resistance R_m = 100 Ω and full scale deflection current I_f = 1 mA is to be converted into a multirange d.c. voltmeter with voltage ranges of 0–10 V, 0–50 V, 0–250 V and 0–500 V. Draw the necessary circuit arrangement and find the values of suitable multipliers.* **(Instrumentation AMIE Sec. B Winter 1991)**

Solution. Full-scale voltage drop = (1 mA) (100 Ω) = 100 mV. Hence, sensitivity of this movement is $100/100 \times 10^{-3}$ = 1 kΩ/V.

(i) 0–10 V range

Total resistance required = (1 kΩ/V) (10 V) = 10 kΩ. Since meter resistance is 1 kΩ additional series resistance required for this range R_1 = 10 – 1 = **9 kΩ**

(ii) 0–50 V range

R_T = (1 kΩ/V) (50 V) = 50 kΩ; R_2 = 50–9–1 = **40 kΩ**

(iii) 0–250 V range

R_T = (1 kΩ/V) (250 V) = 250 kΩ; R_3 = 250–50 = **200 kΩ**

(iv) 0–500 V range

R_T = (1 kΩ/V) (500 V) = 500 kΩ ; R_4 = 500–250 = **250 kΩ**

The circuit arrangement is similar to the one shown in Fig. 10.22.

Tutorial problems No. 10.2

1. The flux density in the gap of a 1-mA (full scale) moving '-coil ammeter is 0.1 Wb/m². The rectangular moving-coil is 8 mm wide by 1 cm deep and is wound with 50 turns. Calculate the full-scale torque which must be provided by the springs. [**4×10^{-7} N-m**] (*App. Elect. London Univ.*)

2. A moving-coil instrument has 100 turns of wire with a resistance of 10 Ω,an active length in the gap of 3 cm and width of 2 cm. A p.d of 45 mV produces full-scale deflection. The control spring exerts a torque of 490.5×10^{-7} N-m at full-scale deflection. Calculate the flux density in the gap. [**0.1817 Wb/m²**] (*I.E.E. London*)

3. A moving-coil instrument, which gives full-scale deflection with 0.015 A has a copper coil having a resistance of 1.5 Ω at 15°C and a temperature coefficient of 1/234.5 at 0°C in series with a swamp resistance of 3.5 Ω having a negligible temperature coefficient.

Determine (*a*) the resistance of shunt required for a full-scale deflection of 20 A and (*b*) the resistance required for a full-scale deflection of 250 *V*.

If the instruent reads correctly at 15°C, determine the percentage error in each case when the temperature is 25°C. [(*a*) **0.00376 Ω ; 1.3%** (*b*) **16,662 Ω, negligible**] (*App. Elect. London Univ.*)

4. A direct current ammeter and leads have a total resistance of 1.5 Ω. The instrument gives a full-scale deflection for a current of 50 mA. Calculate the resistance of the shunts necessary to give full-scale ranges of 2-5, 5.0 and 25.0 amperes. [**0.0306 ; 0.01515 ; 0.00301 Ω**] (*I.E.E. London*)

5. The following data refer to a moving-coil voltmeter : resistance = 10,000 Ω, dimensions of coil = 3 cm×3 cm ; number of turns on coil = 100, flux density in air-gap = 0.08 Wb/m², stiffness of springs = 3×10^{-6} N-m per degree. Find the deflection produced by 110 V. [**48.2°**] (*London Univ.*)

6. A moving-coil instrument has a resistance of 1.0 Ω and gives a full-scale deflection of 150 divisions with a.p.d. of 0.15 V. Calculate the extra resistance required and show how it is connected to enable the instrument to be used as a voltmeter reading upto 15 volts. If the moving coil has a negligible temperature coefficient but the added resistance has a temperature coefficient of 0.004 Ω per degree *C*, what reading will a p.d. of 10 V give at 15°C, assuming that the instrument reads correctly at 0°C. [**99 Ω, 9.45**]

10.20. Electrodynamic or Dynamometer Type Instruments

An electrodynamic instrument is a moving-coil instrument in which the operating field is produced, not by a permanent magnet but by another fixed coil. This instrument can be used either as an ammeter or as a voltmeter but is generally used as a wattmeter.

As shown in Fig. 10.23, the fixed coil is usually arranged in two equal sections F and F placed close together and parallel to each other. The two fixed coils are air-cored to avoid hysteresis effects when used on a.c. circuits. This has the effect of making the magnetic field in which moves the moving coil M, more uniform. The moving coil is spring-controlled and has a pointer attached to it as shown.

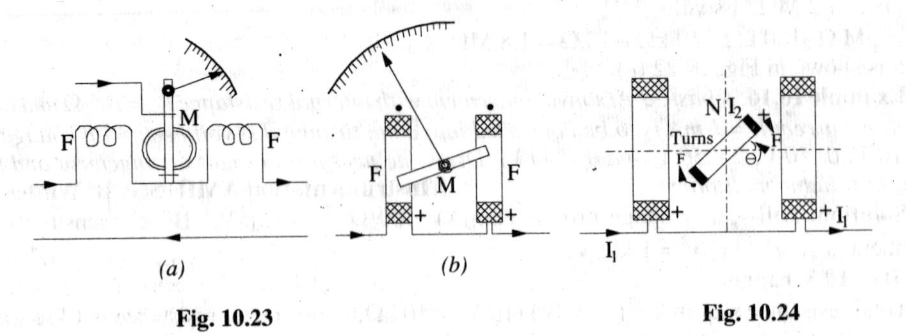

Fig. 10.23

Fig. 10.24

Deflecting Torque*

The production of the deflecting torque can be understood from Fig. 10.24. Let the current passing through the fixed coil be I_1 and that through the moving coil be I_2. Since there is no iron, the field strength and hence the flux density is proportional to I_1.

$\therefore \quad B = KI_1 \quad$ where K is a constant

Let us assume for simplicity that the moving coil is rectangular (it can be circular also) and of dimensions $l \times b$. Then, force on each side of the coil having N turns is (NBI_2l) newton.

The turning moment or deflecting torque on the coil is given by

$$T_d = NBI_2lb = NKI_1I_2lb \text{ N-m}$$

Now, putting $NKlb = K_1$, we have $T_d = K_1I_1I_2$ where K_1 is another constant.

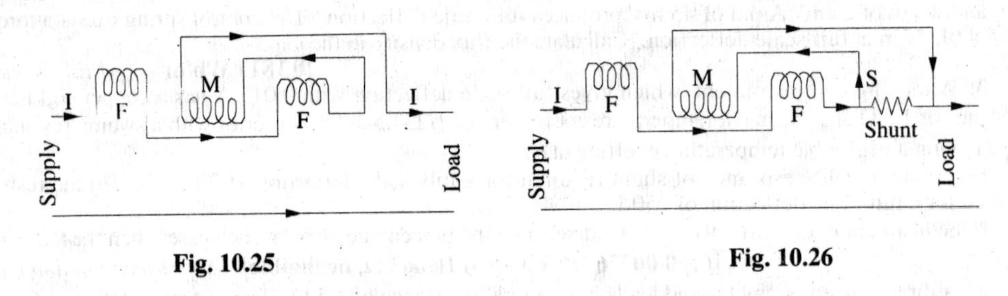

Fig. 10.25

Fig. 10.26

*As shown in Art. 10.12, the value of torque of a moving-coil instrument is

$$T_d = \frac{1}{2} I^2 \, dL/d\theta \; N-m$$

The equivalent inductance of the fixed and moving coils of the electrodynamic instrument is

$$L = L_1 + L_2 + 2M$$

where M is the mutual inductance between the two coils and L_1 and L_2 are their individual self-inductances.

Since L_1 and L_2 are fixed and only M varies,

$$\therefore \quad dL/d\theta = 2dM/d\theta \quad \therefore \quad T_d = \frac{1}{2} I^2 \times 2.dM/d\theta = I^2.dM/d\theta$$

If the currents in the fixed and moving coils are different, say I_1 and I_2, then

$$T_d = I_1 \cdot I_2.dM/d\theta \text{ N-m}$$

It shows that the deflecting torque is proportional to the product of the currents flowing in the fixed coils and the moving coil. Since the instrument is spring-controlled, the restoring or control torque is proportional to the angular deflection θ.

i.e. $T_c \propto \theta = K_2\theta \quad \therefore \quad K_1 I_1 I_2 = K_2\theta$ or $\theta \propto I_1 I_2$

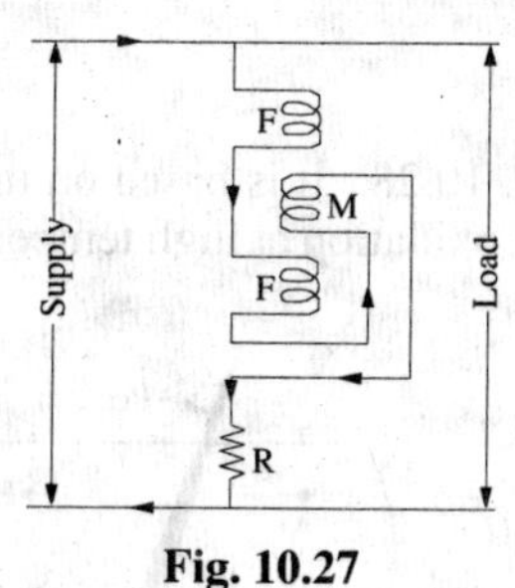

Fig. 10.27

When the instrument is used as an ammeter, the same current passes through both the fixed and the moving coils as shown in Fig. 10.25.

In that case $I_1 = I_2 = I$, hence $\theta \propto I^2$ or $I \propto \sqrt{\theta}$. The connections of Fig. 10.24 are used when small currents are to be measured. In the case of heavy currents, a shunt S is used to limit current through the moving coil as shown in Fig. 10.26.

When used as a voltmeter, the fixed and moving coils are joined in series along with a high resistance and connected as shown in Fig. 10.27. Here, again $I_1 = I_2 = I$, where $I = \frac{V}{R}$ in d.c. circuits and $I = V/Z$ in a.c. circuits.

$\therefore \quad \theta \propto V \times V$ or $\theta \propto V^2$ or $V \propto \sqrt{\theta}$

Hence, it is found that whether the instrument is used as an ammeter or voltmeter, its scale is uneven throughout the whole of its range and is particularly cramped or crowded near the zero.

Damping is pneumatic, since owing to weak operating field, eddy current damping is inadmissible. Such instruments can be used for both a.c. and d.c. measurements. But it is more expensive and inferior to a moving-coil instrument for d.c. measurements.

As mentioned earlier, the most important application of electrodynamic principle is the wattmeter and is discussed in detail in Art. 10.34.

Errors

Since the coils are air-cored, the operating field produced is small. For producing an appreciable deflecting torque, a large number of turns is necessary for the moving coil. The magnitude of the current is also limited because two control springs are used both for leading in and for leading out the current. Both these factors lead to a heavy moving system resulting in frictional losses which are somewhat larger than in other types and so frictional errors tend to be relatively higher. The current in the field coils is limited for the fear of heating the coils which results in the increase of their resistance. A good amount of screening is necessary to avoid the influence of stray fields.

Advantages and Disadvantages

1. Such instruments are free from hysteresis and eddy-current errors.
2. Since (torque/weight) ratio is small, such instruments have low sensitivity.

Example 10.17. *The mutual inductance of a 25-A electrodynamic ammeter changes uniformly at a rate of 0.0035 μH/degree. The torsion constant of the controlling spring is 10^{-6} N-m per degree. Determine the angular deflection for full-scale.*

(Elect. Measurements, Poona Univ. 1985)

Solution. By torsion constant is meant the deflecting torque per degree of deflection. If full-scale deflectin is θ degree, then deflecting torque on full-scale is $10^{-6} \times \theta$ N-m.

Now, $T_d = I^2 dM/d\theta$ Also, $I = 25$ A

$$dM/d\theta = 0.0035 \times 10^{-6} \text{H/degree} = 0.0035 \times 10^{-6} \times 180/\pi \text{ H/radian}$$

$$10^{-6} \times \theta = 25^2 \times 0.0035 \times 10^{-6} \times 180/\pi \quad \therefore \quad \theta = \mathbf{125.4°}$$

Example 10.18. *The spring constant of a 10-A dynamometer wattmeter is 10.5×10^{-6} N-m per radian. The variation of inductance with angular position of moving system is practically linear over the operating range, the rate of change being 0.078 mH per radian. If the full-scale deflection of the instrument is 83 degrees, calculate the current required in the voltage coil at full scale on d.c. circuit.*

(Elect. Inst. & Meas. Nagpur Univ. 1991)

Solution. As seen from foot-note of Art. 10.20, $T_d = I_1 I_2\, dM/d\theta$ N-m

Spring constant = 10.5×10^{-6} N/m/rad = $10.5 \times 10^{-6} \times \pi/180$ N-m/degree

T_d = spring constant × deflection = $(10.5 \times 10^{-6} \times \pi/180) \times 83 = 15.2 \times 10^{-6}$ N-m

$\therefore\ 15.2 \times 10^{-6} = 10 \times I_2 \times 0.078$; I_2 = **19.5 μA.**

10.21. Hot-wire Instruments

The working parts of the instrument are shown in Fig. 10.28. It is based on the heating effect of current. It consists of platinum-iridium (it can withstand oxidation at high temperatures) wire AB stretched between a fixed end B and a tension-adjusting screw at A. When current is passed through AB, it expands according to I^2R formula. This sag in AB produces a slack in phosphor-bronze wire CD attached to the centre of AB. This slack in CD is taken up by the silk fibre which after passing round the pulley is attached to a spring S. As the silk thread is pulled by S, the pulley moves, thereby deflecting the pointer. It would be noted that even a small sag in AB is magnified (Art. 10.22) many times and is conveyed to the pointer. Expansion of AB is magnified by CD which is further magnified by the silk thread.

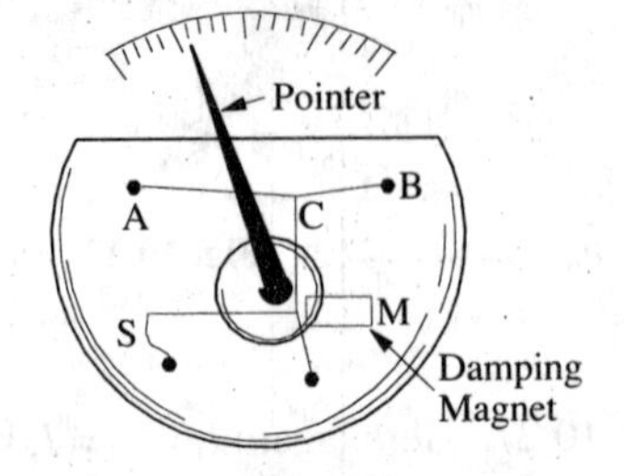

Fig. 10.28

It will be seen that the deflection of the pointer is proportional to the extension of AB which is itself proportional to I^2. Hence, deflection is $\propto I^2$. If spring control is used, then $T_c \propto \theta$.

Hence $\theta \propto I^2$

So, these instruments have a 'square law' type scale. They read the r.m.s. value of current and their readings are independent of its form and frequency.

Damping

A thin light aluminium disc is attached to the pulley such that its edge moves between the poles of a permanent magnet M. Eddy currents produced in this disc give the necessary damping.

These instruments are primarily meant for being used as ammeters but can be adopted as voltmeters by connecting a high resistance in series with them. These instruments are suited both for a.c. and d.c. work.

Advantages of Hot-wire Instruments :

1. As their deflection depends on the r.m.s. value of the alternating current, they can be used on direct current also.
2. Their readings are independent of waveform and frequency.
3. They are unaffected by stray fields.

Disadvantages

1. They are sluggish owing to the time taken by the wire to heat up.
2. They have a high power consumption as compared to moving-coil instruments. Current consumption is 200 mA at full load.
3. Their zero position needs frequent adjustment. 4 They are fragile.

10.22. Magnification of the Expansion

As shown in Fig. 10.29 (a), let L be the length of the wire AB and dL its expansion after steady temperature is reached. The sag S produced in the wire as seen from Fig. 10.29 (a) is given by

$$S^2 = \left(\frac{L + dL}{2}\right)^2 - \left(\frac{L}{2}\right)^2 = \frac{2L.dL + (dL)^2}{4}$$

Neglecting $(dL)^2$, we have $S = \sqrt{L.dL/2}$

Magnification produced is $= \dfrac{S}{dL} = \dfrac{\sqrt{L.dL/2}}{dL} = \sqrt{\dfrac{L}{2.dL}}$

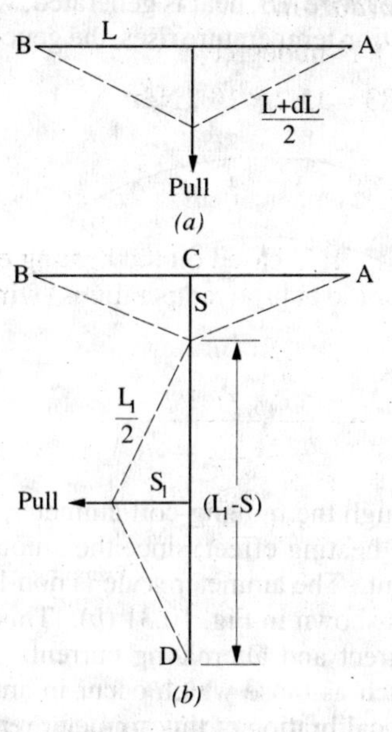

Fig. 10.29

As shown in Fig. 10.29 (*b*), in the case of double-sag instruments, this sag is picked up by wire *CD* which is under the constant pull of the spring. Let L_1 be the length of wire *CD* and let it be pulled at its centre, so as to take up the slack produced by the sag *S* of the wire *AB*.

$$S_1^2 = \left(\frac{L_1}{2}\right)^2 - \left(\frac{L_1 - S}{2}\right)^2 = \frac{2L_1S - S^2}{4}$$

Neglecting S^2 as compared to $2L_1S$, we have $S_1 = \sqrt{L_1S/2}$

Substituting the value of *S*, we get $S_1 = \sqrt{\frac{L_1}{2}\sqrt{\frac{L.dL}{2}}}$

$$= \sqrt{\frac{L_1}{2\sqrt{2}} \times 4\sqrt{L.dL}}$$

Example 10.19. *The working wire of a single-sag hot-wire instrument is 15 cm long and is made up of platinum-silver with a coefficient of linear expansion of 16 × 10⁻⁶. The temperature rise of the wire is 85°C and the sag is taken up at the centre. Find the magnification*

(i) with no initial sag and

(ii) with an initial sag of 1 mm. **[Elect. Meas and Meas. Inst., Calcutta Univ. 1987]**

Solution. (*i*) Length of the wire at room temperature = 15 cm

Length when heated through 85°C is $= 15\,(1 + 16 \times 10^{-6} \times 85) = 15.02$ cm

Increase in length, $dL = 15.02 - 15 = 0.02$ cm

Magnification $= \sqrt{\frac{L}{2.dL}} = \sqrt{\frac{15}{0.004}} = \mathbf{19.36}$

(*ii*) When there is an initial sag of 1 mm, the wire is in the position *ACB* (Fig. 10.30). With rise in temperature, the new position becomes *ADB*. From the right-angled Δ *ADE*,

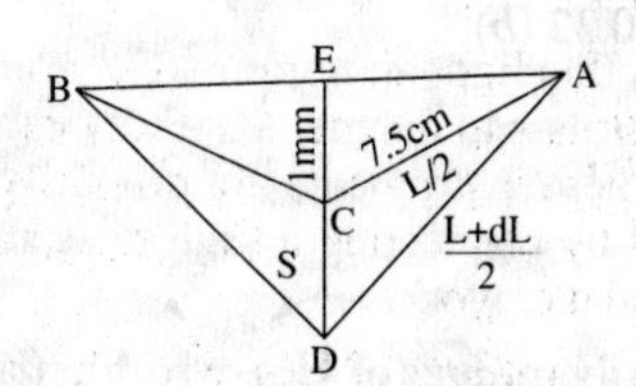

Fig. 10.30

we have $(S + 0.1)^2 = \left(\frac{L + dL}{2}\right)^2 - AE^2$

Now $AE^2 = AC^2 - EC^2 = \left(\frac{L}{2}\right)^2 - 0.1^2 = \frac{L^2}{4} - 0.1^2$

$$\therefore \quad (S + 0.1)^2 = \frac{L^2}{4} + \frac{(dL)^2}{4} + \frac{L.dL}{2} - \left[\frac{L^2}{4} - 0.1^2\right] = \frac{L^2}{4} + \frac{(dL)^2}{4} + \frac{L.dL}{2} - \frac{L^2}{4} + 0.01$$

$$= \frac{L.dL}{2} + 0.01 \qquad \text{...neglecting } (dL)^2/4$$

$$= \frac{15 \times 0.02}{2} + 0.01 = 0.16 \quad \therefore \quad (S + 0.1) = 0.4 \quad \therefore \quad S = 0.3 \text{ cm}$$

Magnification $= S/dL = 0.3/0.02 = \mathbf{15}$

10.23. Thermocouple Ammeter

The working principle of this ammeter is based on the Seebeck effect, which was discovered in 1821. A thermocouple, made of two dissimilar metals (usually bismuth and antimony) is used in the construction of this ammeter. The hot junction of the thermocouple is welded to a heater wire *AB*, both of which are kept in vacuum as shown in Fig. 10.31 (*a*). The cold junction of the thermocouple is connected to a moving-coil ammeter.

When the current to be measured is passed through the heater wire *AB*, heat is generated, which raises the temperature of the thermocouple junction *J*. As the junction temperature rises, the generated

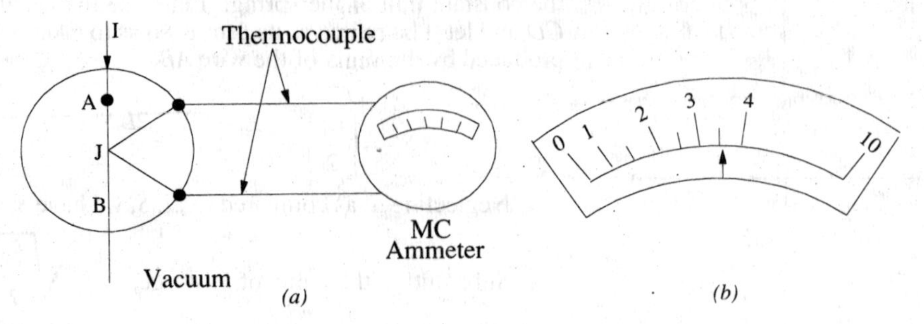

Fig. 10.31

thermoelectric EMF increases and drives a greater current through the moving-coil ammeter. The amount of deflection on the *MC* ammeter scale depends on the heating effect, since the amount of heat produced is directly proportional to the square of the current. The ammeter scale is non-linear so that, it is cramped at the low end and open at the high end as shown in Fig. 10.31 (*b*). This type of "current-squared" ammeter is suitable for reading both direct and alternating currents. It is particularly suitable for measuring radio-frequency currents such as those which occur in antenna systems of broadcast transmiters. Once calibrated properly, the calibration of this ammeter remains accurate from dc upto very high frequency currents.

10.24 Megger

It is a portable instrument used for testing the insulation resistance of a circuit and for measuring resistances of the order of megohms which are connected across the outside terminals XY in Fig. 10.32 (*b*).

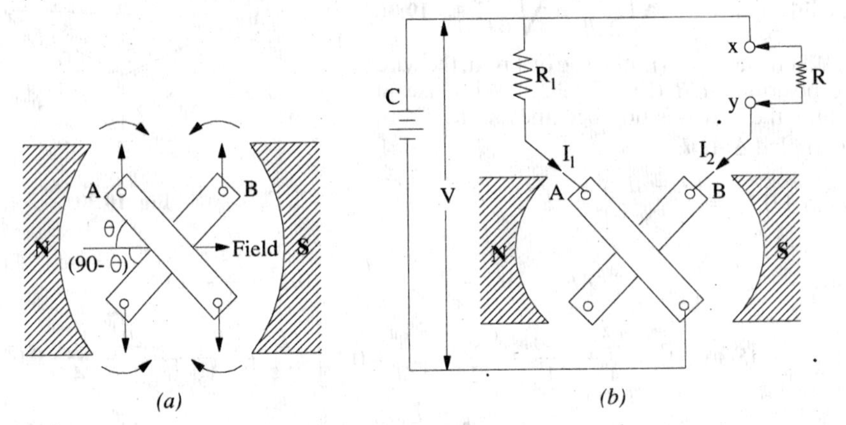

Fig. 10.32

1. Working Principle

The working principle of a 'cross-coil' type megger may be understood from Fig. 10.32 (*a*) which shows two coils *A* and *B* mounted rigidly at right angles to each other on a common axis and free to rotate in a magnetic field. When currents are passed through them, the two coils are acted upon by torques which are in opposite directions. The torque of coil *A* is proportional to $I_1 \cos \theta$ and that of *B* is proportional to $I_2 \cos (90 - \theta)$ or $I_2 \sin \theta$. The two coils come to a position of equilibrium where the two torques are equal and opposite *i.e.* where

$$I_1 \cos \theta = I_2 \sin \theta \quad \text{or} \quad \tan \theta = I_1/I_2$$

In practice, however, by modifying the shape of pole faces and the angle between the two coils, the ratio I_1/I_2 is made proportional to θ instead of tan θ in order to achieve a linear scale.

Suppose the two coils are connected across a common source of voltage *i.e.* battery *C*, as shown in Fig. 10.32 (*b*). Coil *A*, which is connected directly across *V*, is called the voltage (or control) coil. Its current $I_1 = V/R_1$. The coil *B* called current or deflecting coil, carries the current $I_2 = V/R$, where *R* is the external resistance to be measured. This resistance may vary from infinity (for good insulation or open circuit) to zero (for poor insulation or a short-circuit). The two coils are free to rotate in the field of a permanent magnet. The deflection θ of the instrument is proportional to I_1/I_2 which is equal to R/R_1. If R_1 is fixed, then the scale can be calibrated to read *R* directly (in practice, a current-limiting resistance is connected in the circuit of coil *B* but the presence of this resistance can be allowed for in scaling). The value of *V* is immaterial so long as it remains constant and is large enough to give suitable currents with the high resistance to be measured.

2. Construction

The essential parts of a megger are shown in Fig. 10.33. Instead of battery *C* of Fig. 10.32 (*b*), there is a hand-driven d.c. generator. The crank turns the generator armature through a clutch mechanism which is designed to slip at a pre-determined speed. In this way, the generator speed and voltage are kept constant and at their correct values when testing.

The generator voltage is applied across the voltage coil *A* through a fixed resistance R_1 and across deflecting coil *B* through a current-limiting resistance *R′* and the external resistance is connected across the testing terminals *XY*. The two coils, in fact, constitute a moving-coil voltmeter and an ammeter combined into one instrument.

(*i*) Suppose the terminals *XY* are open-circuited. Now, when crank is operated, the generator voltage so produced is applied across coil *A* ande current I_1 flows through it but no current flows through coil *B*. The torque so produced rotates the moving element of the megger until the scale points to 'infinity', thus indicating that the resistance of the external circuit is too large for the instrument to measure.

(*ii*) When the testing terminals *XY* are closed through a low resistance or are short-circuited, then a large current (limited only by *R′*) passes through the deflecting coil *B*. The deflecting torque produced by coil *B* overcomes the small opposing torque of coil *A* and rotates the moving element until the needle points to 'zero', thus showing that the external resistance is too small for the instrument to measure.

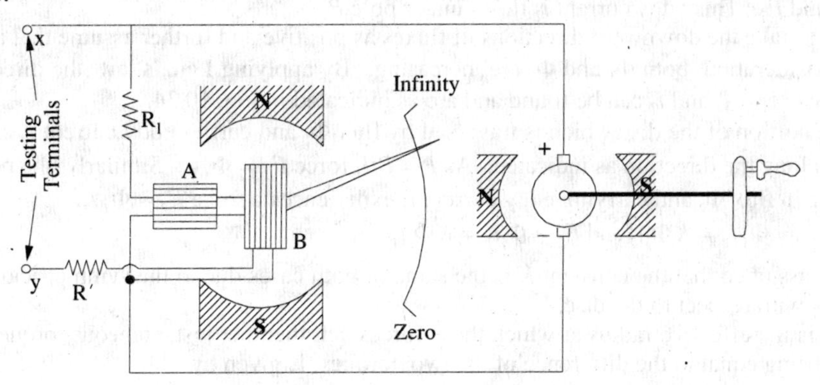

Fig. 10.33

Although, a megger can meassure all resistances lying between zero and infinity, essentially it is a high-resistance measuring device. Usually, zero is the first mark and 10 kΩ is the second mark on its scale, so one can appreciate that it is impossible to accurately measure small resistances with the help of a megger.

The instrument described above is simple to operate, portable, very robust and independent of the external supplies.

10.25. Induction type Voltmeters and Ammeters

Induction type instruments are used only for a.c. measurements and can be used either as ammeter, voltmeter or wattmeter. However, the induction principle finds its widest application as a watt-hour or energy meter. In such instruments, the deflecting torque is produced due to the reaction between the flux of an a.c. magnet and the eddy currents induced by this flux. Before discussing the two types of most commonly-used induction instruments, we will first discuss the underlying principle of thier operation.

Principle

The operation of all induction instruments depends on the production of torque due to the reaction between a flux Φ_1 (whose magnitude depends on the current or voltage to be measured) and eddy currents induced in a metal disc or drum by another flux Φ_2 (whose magnitude also depends on the current or voltage to be measured). Since the magnitude of eddy currents also depends on the flux producing them, the *instantaneous* value of torque is proportional to the square of current or voltage under measurement and the value of *mean* torque is proportional to the mean square value of this current or voltage.

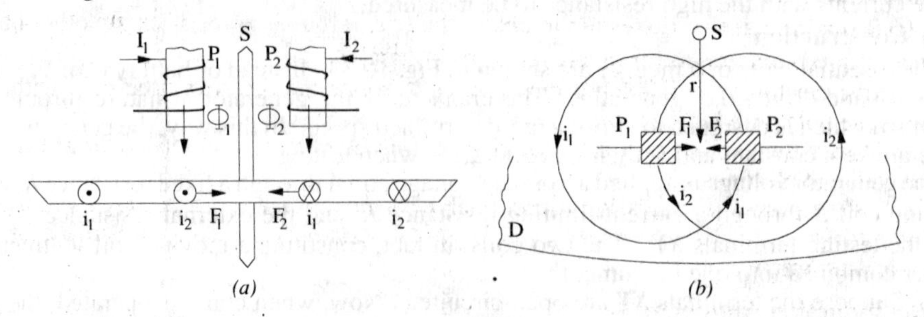

Fig. 10.34

Consider a thin aluminium or Cu disc D free to rotate about an axis passing through its centre as shown in Fig. 10.34. Two a.c. magnetic poles P_1 and P_2 produce alternating fluxes Φ_1 and Φ_2 respectively which cut this disc. Consider any annular portion of the disc around P_1 with centre on the axis of P_1. This portion will be linked by flux Φ_1 and so an alternating e.m.f. e_1 be induced in it. This e.m.f will circulate an eddy current i_1 which, as shown in Fig. 10.34, will pass under P_2. Similarly, Φ_2 will induce an e.m.f. e_2 which will further induce an eddy current i_2 in an annular portion of the disc around P_2. This eddy current i_2 flows under pole P_1.

Let us take the downward directions of fluxes as positive and further assume that at the instant under consideration, both Φ_1 and Φ_2 are increasing. By applying Lenz's law, the directions of the induced currents i_1 and i_2 can be found and are as indicated in Fig. 10.34.

The portion of the disc which is traversed by flux Φ_1 and carries eddy current i_2 experiences a force F_1 along the direction as indicated. As $F = Bil$, force $F_1 \propto \Phi_1\, i_2$. Similarly, the portion of the disc lying in flux Φ_2 and carrying eddy current i_1 experiences a force $F_2 \propto \Phi_2\, i_1$.

$\therefore \quad F_1 \propto \Phi_1 i_2 = K\Phi_1 i_2$ and $F_2 \propto \Phi_2 i_1 = K\Phi_2 i_1$

It is assumed that the constant K is the same in both cases due to the symmetrical positions of P_1 and P_2 with respect to the disc.

If r is the effective radius at which these forces act, then net instantaneous torque T acting on the disc being equal to the difference of the two torques, is given by

$$T = r(K\Phi_1 i_2 - K\Phi_2 i_1) = K_1(\Phi_1 i_2 - \Phi_2 i_1) \qquad \ldots(i)$$

Let the alternating flux Φ_1 be given by $\Phi_1 = \Phi_{1m} \sin \omega t$. The flux Φ_2 which is assumed to lag Φ_1 by an angle α radian is given by $\Phi_2 = \Phi_{2m} \sin(\omega t - \alpha)$

Induced e.m.f.* $\qquad e_1 = \dfrac{d\Phi_1}{dt} = \dfrac{d}{dt}(\Phi_{1m} \sin \omega t) = \omega\, \Phi_{1m} \cos \omega t$

Assuming the eddy current path to be purely resistive and of value R^*, the value of eddy current is

$$i_1 = \frac{e_1}{R} = \frac{\omega\Phi_{1m}}{R} \cos \omega t \text{ Similarly } e_2 = \omega\, \Phi_{2m}(wt - \alpha) \text{ and } i_2 = \frac{\omega\, \Phi_{2m}}{R} \cos(\omega t - \alpha)^{**}$$

---

*If it has a reactance of X, then impedance Z should be taken, whose value is given by $Z = \sqrt{R^2 + X^2}$.

**It being assumed that both paths have the same resistance.

Substituting these values of i_1 and i_2 in Eq. (*i*) above, we get

$$T = \frac{K_1\omega}{R}[\Phi_{1m}\sin\omega t.\Phi_{2m}\cos(\omega t-\alpha)-\Phi_{2m}\sin(\omega t-\alpha)\Phi_{1m}\cos\omega t]$$

$$= \frac{K_1\omega}{R}.\Phi_{1m}\,\Phi_{2m}[\sin\omega t.\cos(\omega t-\alpha)-\cos\omega t.\sin(\omega t-\alpha)]$$

$$= \frac{K_1\omega}{R}.\Phi_{1m}\,\Phi_{2m}\sin\alpha = K_2\omega\,\Phi_{1m}\Phi_{2m}\sin\alpha \text{ (putting } K_1/R = K_2)$$

. It is obvious that

(*i*) if $\alpha = 0$ *i.e.* if two fluxes are in phase, then *net* torque is zero. If on the other hand, $\alpha = 90°$, the net torque is maximum for given values of Φ_{1m} and Φ_{2m}.

(*ii*) the net torque is in such a direction as to rotate the disc from the pole with leading flux towards the pole with lagging flux.

(*iii*) since the expression for torque does not involve '*t*', it is independent of time *i.e.* it has a steady value at all times.

(*iv*) the torque *T* is inversely proportional to *R*–the resistance of the eddy current path. Hence, for large torques, the disc material should have low resistivity. Usually, it is made of Cu or, more often, of aluminium.

10.26. Induction Ammeters

It has been shown in Art. 10.22 above that the net torque acting on the disc is

$$T = K_2\,\omega\Phi_{1m}\,\Phi_{2m}\,\sin\alpha$$

Obviously, if both fluxes are produced by the same alternating current (of maximum value I_m) to be measured, then

$$T = K_3\omega\,I_m^2\,\sin\alpha$$

Hence, for a given frequency ω and angle α, the torque is proportional to the square of the current. If the disc has spring control, it will take up a steady deflected position where controlling torque becomes equal to the deflecting torque. By attaching a suitable pointer to the disc, the apparatus can be used as an ammeter.

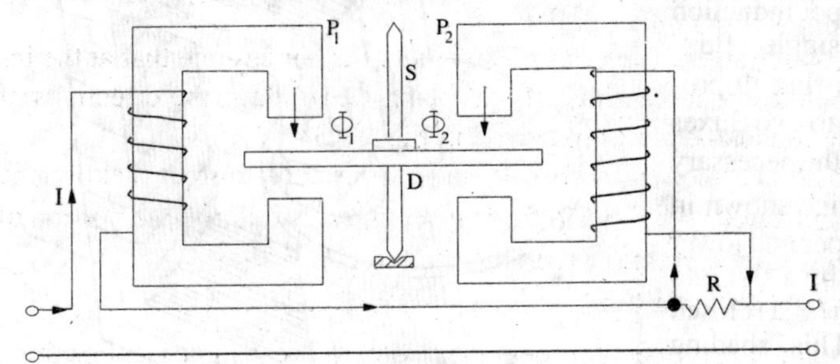

Fig. 10.35

There are three different possible arrangements by which the operational requirements of induction ammeters can be met as discussed below.

(i) Disc Instrument with Split-phase Winding

In this arrangement, the windings on the two laminated a.c. magnets P_1 and P_2 are connected in series (Fig. 10.35). But, the winding of P_2 is shunted by a resistance *R* with the result that the current in this winding lags with respect to the total line current. In this way, the necessary phase angle α is produced between two fluxes Φ_1 and Φ_2 produced by P_1 and P_2 respectively. This angle is of the order of 60°. If the hysteresis effects etc. are neglected, then each flux would be proportional to the current to be measured *i.e.* line current *I*

$$T_d \propto \Phi_{1m}\Phi_{2m}\sin\alpha$$

or $$T_d \propto I^2 \text{ where } I \text{ is the r.m.s. value.}$$

If spring control is used, then $T_c \propto \theta$

In the final deflected position, $T_c = T_d \quad \therefore \quad \theta \propto I^2$

Eddy current damping is employed in this instrument. When the disc rotates, it cuts the flux in the air-gap of the magnet and has eddy currents induced in it which provide efficient damping.

(ii) Cylindrical type with Split-phase Winding

The operating principle of this instrument is the same as that of the above instrument except that instead of a rotating disc, it employs a hollow aluminium drum as shown in Fig. 10.36. The poles

P_1 produce the alternating flux Φ_1 which produces eddy current i_1 in those portions of the drum that lie under poles P_2. Similarly, flux Φ_2 due to poles P_2 produces eddy current i_2 in those parts of the drum that lie under poles P_1. The force F_1 which is $\propto \Phi_1 i_2$ and F_2 which is $\propto \Phi_2 i_1$ are tangential to the surface of the drum and the resulting torque tends to rotate the drum about its own axis. Again, the winding of P_2 is shunted by resistance R which helps to introduce the necessary phase difference α between Φ_1 and Φ_2.

The spiral control springs (not shown in the figure) prevent any continuous rotation of the drum and ultimately bring it to rest at a position where the deflecting torque becomes equal to the controlling torque of the springs. The drum has a pointer attached to it and is itself carried by a spindle whose two ends fit in jewelled bearings. There is a cylindrical laminated core inside the hollow drum whose function is to strengthen the flux cutting the drum. The poles are laminated and magnetic circuits are completed by the yoke Y and the core.

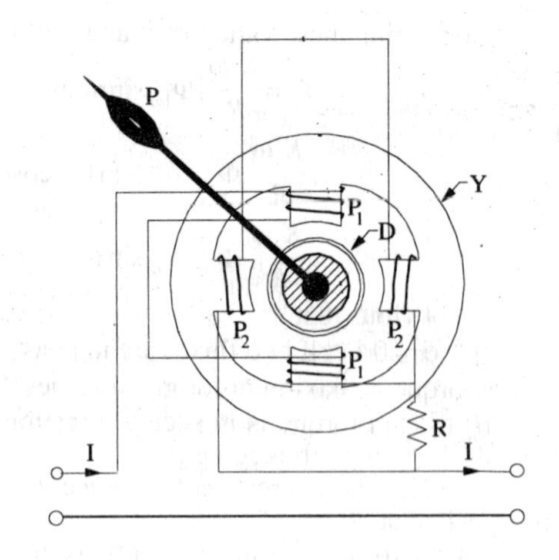

Fig. 10.36

Damping is by eddy currents induced in a separate aluminium disc (not shown in the figure) carried by the spindle when it moves in the air-gap flux of a horse-shoe magnet (also not shown in the figure).

(iii) Shaded-pole Induction Ammeter

In the shaded-pole disc type induction ammeter (Fig. 10.37) only single flux-producing winding is used. The flux Φ produced by this winding is split up into two fluxes Φ_1 and Φ_2 which are made to have the necessary phase difference of α by the device shown in Fig. 10.37. The portions of the upper and lower poles near the disc D are divided by a slot into two halves one of which carries a closed 'shading' winding or ring. This shading winding or ring acts as a short-circuited secondary and the main winding as a primary. The current induced in the ring by transformer action retards the phase of flux Φ_2 with respect to that of Φ_1 by about 50° or so. The two fluxes Φ_1 and Φ_2 passing through the unshaded and shaded parts respectively, react with eddy currents i_2 and i_1 respectively and so produce the net driving torque whose value is

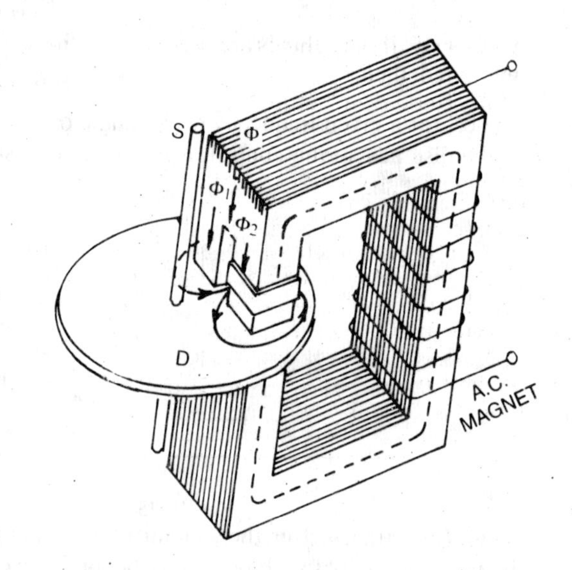

Fig. 10.37

$$T_d \propto \Phi_{1m}\Phi_{2m} \sin \alpha$$

Assuming that both Φ_1 and Φ_2 are proportional to the current I, we have

$$T_d \propto I^2$$

This torque is balanced by the controlling torque provided by the spiral springs.

The actual shaded-pole type induction instruments is shown in Fig. 10.38. It consists of a suitably-shaped aluminium or copper disc mounted on a spindle which is supported by jewelled bearings. The spindle carries a pointer and has a control spring attached to it. The edge or periphery of the disc moves in the air-gap of a laminated a.c. electromagnet which is energised either by the current to be measured (as ammeter) or by a current proportional to the voltage to be measured

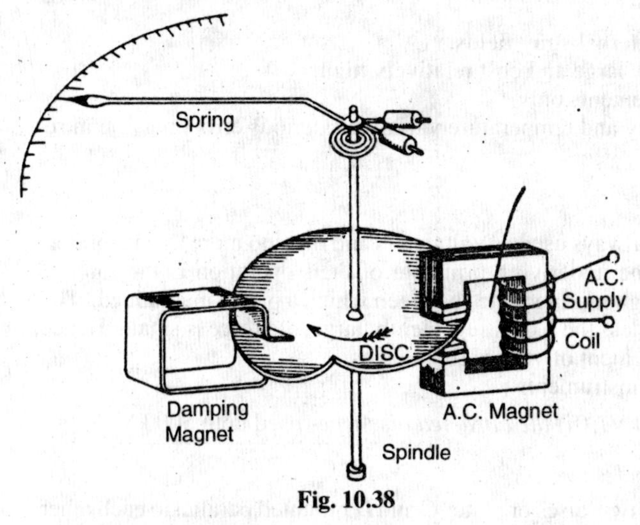

Fig. 10.38

(as a voltmeter). Damping is by eddy currents induced by a permanent magnet embracing another portion of the *same* disc. As seen, the disc serves both for damping as well as operating purposes. The main flux is split into two component fluxes by shading one-half of each pole. These two fluxes have a phase difference of 40° to 50° between them and they induce two eddy currents in the disc. Each eddy current has a component in phase with the *other* flux, so that two torques are produced which are oppositely-directed. The resultant torque is equal to the difference between the two. This torque *deflects* the disc–continuous rotation being prevented by the control spring and the deflection produced is proportional to the square of the current or voltage being measured.

As seen, for a given frequency, $T_d \propto I^2 = KI^2$

For spring control $T_c \propto \theta$ or $T_c = K_c\theta$

For steady deflection, we have $T_c = T_d$ or $\theta \propto I^2$

Hence, such instruments have uneven scales *i.e.* scales which are cramped at their lower ends. A more even scale can, however, be obtained by using a cam-shaped disc as shown in Fig. 10.38.

10.27. Induction Voltmeter

Its construction is similar to that of an induction ammeter except for the difference that its winding is wound with a large number of turns of fine wire. Since it is connected across the lines and carries very small current (5–10 mA), the number of turns of its wire has to be large in order to produce an adequate amount of m.m.f. Split phase windings are obtained by connecting a high resistance R in series with the winding of one magnet and an inductive coil in series with the winding of the other magnet as shown in Fig. 10.39.

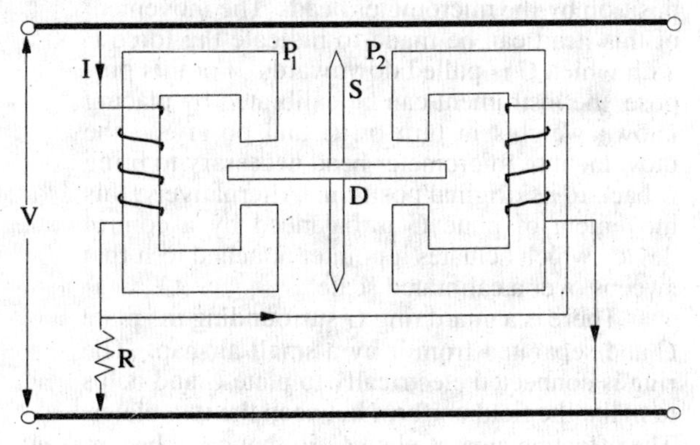

Fig. 10.39

10.28. Errors in Induction Instruments

There are two types of errors (*i*) *frequency* error and (*ii*) *temperature* error.

1. Since deflecting torque depends on frequency, hence unless the alternating current to be measured has the same frequency with which the instrument was calibrated, there will be large error in its readings. Frequency errors can be compensated for by the use of a non-inductive shunt in the case of ammeters. In voltmeters, such errors are not large and, to a great extent, are self-compensating.
2. Serious errors may occur due to the variation of temperature because the resistances of eddy current paths depend on the temperature. Such errors can, however, be compensated for by shunting in the case of ammeters and by a combination of shunt and swamping resistance in the case of voltmeters.

10.29. Advantages and Disadvantages

1. A full-scale deflection of over 200° can be obtained with such instruments. Hence, they have long open scales.
2. Damping is very efficient.

3. They are not much affected by external stray fields.
4. Their power consumption is fairly large and cost relatively high.
5. They can be used for a.c. measurements only.
6. Unless compensated for frequency and temperature variations, serious errors may be introduced.

10.30. Electrostatic Voltmeters

Electrostatic instruments are almost always used as voltmeters and that too more as a laboratory rather than as industrial instruments. The underlying principle of their operation is the force of attraction between electric charges on neighbouring plates between which a p.d. is maintained. This force gives rise to a deflecting torque. Unless the p.d. is sufficiently large, the force is small. Hence, such instruments are used for the measurement of very high voltages.

There are two general types of such instruments :

(*i*) *the quadrant type*–used upto 20 kV. (*ii*) *the attracted disc type*–used upto 500 kV.

10.31. Attracted-disc Type Voltmeter

As shown in Fig. 10.40, it consists of two-discs or plates C and D mounted parallel to each other. Plate D is fixed and is earthed while C is suspended by a coach spring, the support for which carries a micrometer head for adjustment. Plate C is connected to the positive end of the supply voltage. When a p.d. (whether direct or alternating) is applied between the two plates, then C is attracted towards D but may be returned to its original position by the micrometer head. The movement of this head can be made to indicate the force F with which C is pulled downwards. For this purpose, the instrument can be calibrated by placing known weights in turn on C and observing the movement of micrometer head necessary to bring C back to its original position. Alternatively, this movement of plate C is balanced by a control device which actuates a pointer attached to it that sweeps over a calibrated scale.

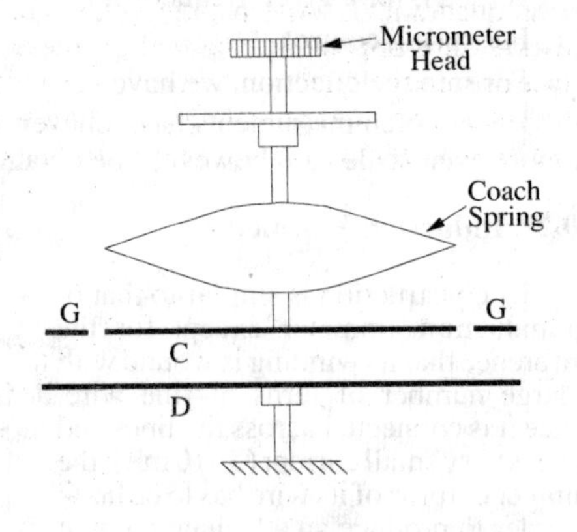

Fig. 10.40

There is a guard ring G surrounding the plate C and separated from it by a small air-gap. The ring is connected electrically to plate C and helps to make the field uniform between the two plates. The effective area of plate C, in that case, becomes equal to its actual area plus half the area of the air-gap.

Theory

In Fig. 10.41 are shown two parallel plates separated by a distance of x metres. Suppose the lower plate is fixed and carries a charge of $-Q$ coulomb whereas the upper plate is movable and carries a charge of $+Q$ coulomb. Let the mutual force of attraction between the two plates be F newtons. Suppose the upper plate is moved apart by a distance dx. Then mechanical work done during this movement is $F \times dx$ joule. Since charge on the plate is constant, no electrical energy can move into the system from outside. This work is done at the cost of the energy stored in the parallel-plate capacitor formed by the two plates.

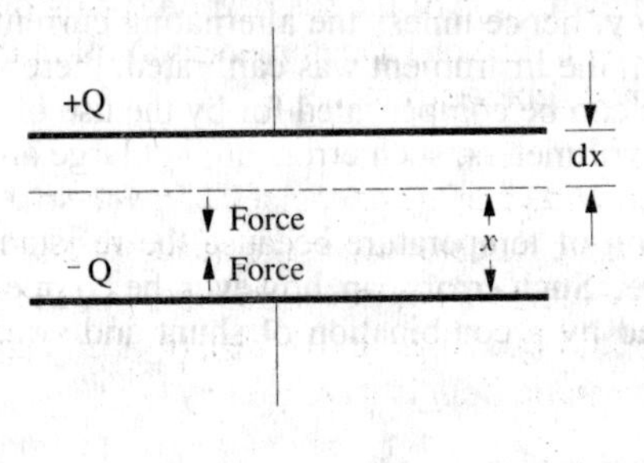

Fig. 10.41

Before movement, let the capacitance of the capacitor be C farad. Then,

$$\text{Initial energy stored} = \frac{1}{2} \cdot \frac{Q^2}{C}$$

If the capacitance changes to $(C + dC)$ because of the movement of plate, then

$$\text{Final energy stored} = \frac{1}{2}\frac{Q^2}{(C+dC)} = \frac{1}{2}\frac{Q^2}{C}\cdot\frac{1}{\left(1+\frac{dC}{C}\right)} = \frac{1}{2}\frac{Q^2}{C}\left(1+\frac{dC}{C}\right)^{-1} = \frac{1}{2}\frac{Q^2}{C}\cdot\left(1-\frac{dC}{C}\right) \text{ if } dC \leq C$$

$$\text{Change in stored energy} = \frac{1}{2}\frac{Q^2}{C} - \frac{1}{2}\frac{Q^2}{C}\left(1-\frac{dC}{C}\right) = \frac{1}{2}\frac{Q^2}{C}\cdot\frac{dC}{C}$$

$$\therefore \quad F \times dx = -\frac{1}{2}\frac{Q^2}{C}\cdot\frac{dC}{C} \quad \text{or} \quad F = \frac{1}{2}\frac{Q^2}{C^2}\cdot\frac{dC}{dx} = \frac{1}{2}V^2\frac{dC}{dx}$$

$$\text{Now,} \quad C = \frac{\varepsilon_0 A}{x} \quad \therefore \quad \frac{dC}{dx} = -\frac{\varepsilon_0 A}{x^2} \quad \therefore \quad F = -\frac{1}{2}V^2\frac{\varepsilon_0 A}{x^2}\text{ N}$$

Hence, we find that force is directly proportional to the square of the voltage to be measured. The negative sign merely shows that it is a force of attraction.

10.32. Quadrant Type Voltmeters

The working principle and basic construction of such instruments can be understood from Fig. 10.42. A light aluminium vane C is mounted on a spindle S and is situated partially within a hollow metal quadrant B. Alternatively, the vane be suspended in the quadrant. When the vane and the quadrant are oppositely charged by the voltage under measurement, the vane is further attracted inwards into the quadrant thereby causing the spindle and hence the pointer to rotate. The amount of rotation and hence the deflecting torque is found proportional to V^2. The deflecting torque in the case of arrangement shown in Fig. 10.42 is very small unless V is extremely large.

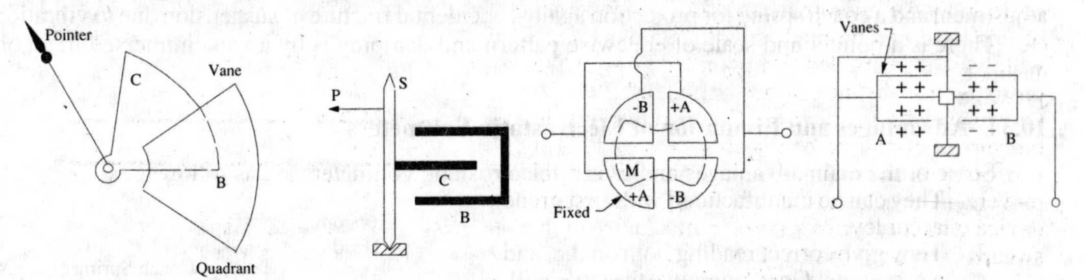

Fig. 10.42 Fig. 10.43

The force on the vane may be increased by using a larger number of quadrants and a double-ended vane. In Fig. 10.43 are shown four fixed metallic double quadrants arranged so as to form a circular box with short air-gaps between the quadrants in which is suspended or pivoted an aluminium vane. Opposite quandrants AA and BB are joined together and each pair is connected to one terminal of the a.c. or d.c. supply and at the same time, one pair is connected to the moving vane M. Under these conditions [Fig. 10.43] the moving vane is repelled by quandrants AA and attracted by quadratns BB. Hence, a deflecting torque is produced which is proportional to (p.d.)^2. Therefore, such voltmeters have an uneven scale. Controlling torque is produced by torsion of the suspension spring or by the spring (used in pivoted type voltmeters). Damping is by a disc or vane immersed in oíl in the case of suspended type or by air friction in the case of pivoted type instruments.

Theory

With reference to Fig. 10.42, suppose the quadrant and vane are connected across a source of V volts and let the resulting deflection be θ. If C is the capacitance between the quadrant and vane in the deflected position, then the charge on the instrument will be CV coulomb. Suppose that the voltage is changed from V to $(V + dV)$, then as a result, let θ, C and Q change to $(\theta + d\theta)$, $(C + dC)$ and $(Q + dQ)$ respectively. Then, the energy stored in the electrostatic field is increased by

$$dE = d\left(\frac{1}{2}CV^2\right) = \frac{1}{2}V^2.dC + C.V\,dV \text{ joule}$$

If T is the value of controlling torque corresponding to a deflection of θ, then the additional energy stored in the control will be $T \times d\theta$ joule.

Total increase in stored energy $= T \times d\theta + \frac{1}{2} V^2 dC + CVdV$ joule

It is seen that during this change, the source supplies a charge dQ at potential V. Hence, the value of energy supplied is

$$= V \times dQ = V \times d(CV) = V^2 \times dC + CV \, . \, dV$$

Since the energy supplied by the source must be equal to the extra energy stored in the field and the control

$$\therefore \qquad T \times d\theta + \frac{1}{2} V^2 dC + CV.dV = V^2.dC + CVdV$$

$$\text{or} \qquad T \times d\theta = \frac{1}{2} V^2.dc \qquad \therefore T = \frac{1}{2} V^2 \frac{dC}{d\theta} \text{ N-m}$$

The torque is found to be proportional to the square of the voltage to be measured whether that voltage is alternating or direct. However, on alternating circuits the scale will read r.m.s. values.

10.33. Kelvin's Multicellular Voltmeter

As shown in Fig. 10.44, it is essentially a quadrant type instrument, as described above, but with the difference that instead of four quadrants and one vane, it has a large number of fixed quandrants and vanes mounted on the same spindle. In this way, the deflecting torque for a given voltage is increased many times. Such voltmeters can be used to measure voltages as low as 30 V. As said above, this reduction in the minimum limit of voltage is due to the increase in the operating force in proportion to the number of component units. Such an instrument has a torsion head for zero adjustment and a coach spring for protection against accidental fracture of suspension due to vibration etc. There is a pointer and scale of edgewise pattern and damping is by a vane immersed in an oil dashpot.

10.34. Advantages and Limitation of Electrostatic Voltmeters

Some of the main advantages and uses of electrostatic voltmeters are as follows :

1. They can be manufactured with first grade accuracy.
2. They give correct reading both on d.c. and a.c. circuits. On a.c. circuits, the scale will, however, read r.m.s. values whatever the wave-form.
3. Since no iron is used in their construction, such instruments are free from hysteresis and eddy current losses and temperature errors.
4. They do not draw any continuous current on d.c. circuits and that drawn on a.c. circuits (due to the capacitance of the instrument) is extremely small. Hence, such voltmeters do not cause any disturbance to the circuits to which they are connected.
5. Their power loss is negligibly small.
6. They are unaffected by stray magnetic fields although they have to be guarded against any stray electrostatic field.

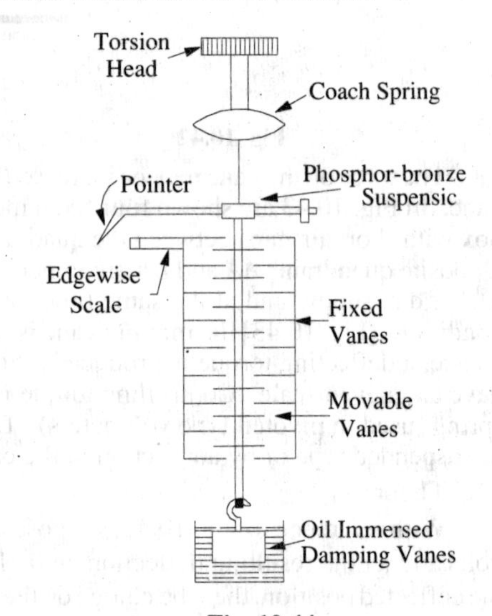

Fig. 10.44

7. They can be used up to 1000 kHz without any serious loss of accuracy.

However, their main limitations are :

1. Low-voltage voltmeters (like Kelvin's Multicellular voltmeter) are liable to friction errors.
2. Since torque is proportional to the square of the voltage, their scales are not uniform although some uniformity can be obtained by suitably shaping the quadrants of the voltmeters.
3. They are expensive and cannot be made robust.

10.35. Range Extension of Electrostic Voltmeters

The range of such voltmeters can be extended by the use of multipliers which are in the form of a resistance potential divider or capacitance potential divider. The former method can be used both for direct and alternating voltages whereas the latter method is useful only for alternating voltages.

(i) Resistance Potential Divider

This divider consists of a high non-inductive resistance across a small portion which is attached the electrostatic voltmeter as shown in Fig. 10.45. Let R be the resistance of the whole of the potential divider across which is applied the voltage V under measurement. Suppose v is the maximum value of the voltage which the voltmeter can measure without the multiplier. If r is the resistance of the portion of the divider across which voltmeter is connected, then the multiplying factor is given by

$$\frac{V}{v} = \frac{R}{r}$$

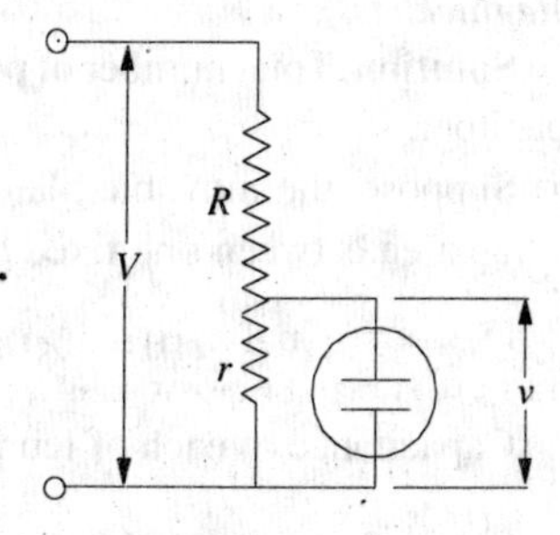

Fig. 10.45

The above expression is true for d.c. circuits but for a.c. circuits, the capacitance of the voltmeter (which is in parallel with r) has to be taken into account, Since this capacitance is variable, it is advisable to calibrate the voltmeter along with its multiplier.

(ii) Capacitance Potential Divider

In this method, the voltmeter may be connected in series with a single capacitor C and put across the voltage V which is to be measured [Fig. 10.46 (*a*)] or a number of capacitors may be joined in series to form the potential divider and the voltmeter may be connected across one of the capacitors as shown in Fig. 10.46 (*b*).

Consider the connection shown in Fig. 10.46 (*a*). It is seen that the multiplying factor is given by

$$\frac{V}{v} = \frac{\text{reactance of total circuit}}{\text{reactance of voltmeter}}$$

Now, capacitance of the total circuit is $\dfrac{CC_v}{C + C_v}$

and its reactance is

$$= \frac{1}{\omega \times \text{capacitance}} = \frac{C + C_v}{\omega C\, C_v}$$

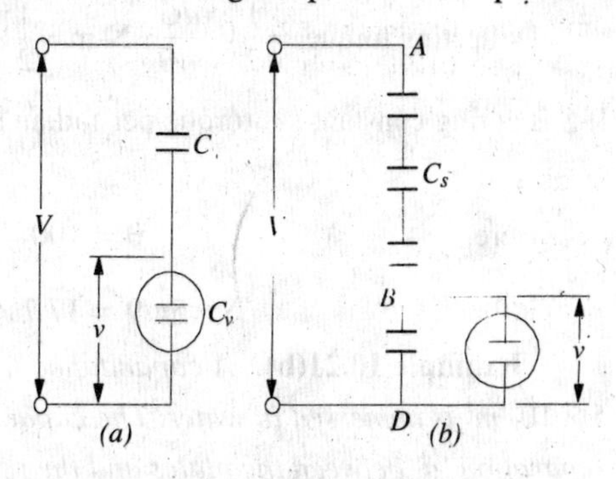

Fig. 10.46

Reactance of the voltmeter $= \dfrac{1}{\omega C_v}$

$$\therefore \quad \frac{V}{v} = \frac{(C + C_v)/\omega C C_v}{1/\omega C_v} = \frac{C + C_v}{C} \quad \therefore \quad \text{Multiplying factor} = \frac{C + C_v}{C} = 1 + \frac{C_v}{C}^{*}.$$

Example 10.20. *The reading '100' of a 120-V electrostatic voltmeter is to represent 10,000 volts when its range is extended by the use of a capacitor in series. If the capacitance of the voltmeter at the above reading is 70μμF, find the capacitance of the capacitor multiplier required.*

Solution. Multiplying factor $= \dfrac{V}{v} = 1 + \dfrac{C_v}{C}$

Here, $V = 10,000$ volt $v = 100$ volt; C_v = capacitance of the voltmeter $= 70\mu\mu F$

C = capacitance of the multiplier $\quad \therefore \quad \dfrac{10,000}{100} = 1 + \dfrac{70}{C}$

or $70/C = 99 \quad \therefore C = 70/99\ \mu\mu F = 0.707\ \mu\mu$**F (approx)**

*It is helpful to compare it with a similar expression in Art. 10.17 for permanent magnet moving-coil instruments.

Example 10.21(a). *An electrostatic voltmeter is constructed with 6 parallel, semicircular fixed plates equi-spaced at 4 mm intervals and 5 interleaved semi-circular movable plates that move in planes midway between the fixed plates, in air. The movement of the movable plates is about an axis through the centre of the circles of the plate system, perpendicular to the planes of the plates. The instrument is spring-controlled. If the radius of the movable plates is 4 cm, calculate the spring constant if 10 kV corresponds to a full-scale deflection of 100°. Neglect fringing, edge effects and plate thickness.* **(Elect. Measurements, Bombay Univ. 1985)**

Solution. Total number of plates (both fixed and movable) is 11, hence there are 10 parallel-plate capacitors.

Suppose, the movable plates are rotated into the fixed plates by an angle of θ radian. Then, overlap area between one fixed and one movable semi-circular plate is

$$A = \frac{1}{2}r^2\theta = \frac{1}{2}\times 0.04^2 \times \theta = 8\times 10^{-4}\,\theta\ \text{m}^2;\ d = 4/2 = 2\ \text{mm} = 2\times 10^{-3}\ \text{m}$$

Capacitance of each of ten parallel-plate capacitors is

$$C = \frac{\varepsilon_0 A}{d} = \frac{8.854\times 10^{-12}\times 8\times 10^{-4}\,\theta}{2\times 10^{-3}} = 3.54\times 10^{-12}\,\theta\,F$$

Total capacitance $C = 10\times 3.54\times 10^{-12}\,\theta = 35.4\times 10^{12}\,\theta F$ $\therefore$ $dC/d\theta = 35.4\times 10^{-12}$ farad/radian

Deflecting torque $= \frac{1}{2}V^2\frac{dC}{d\theta}$ N-m $= \frac{1}{2}\times (10{,}000)^2\times 35.4\times 10^{-12} = 17.7\times 10^{-4}$ N-m

If S is spring constant *i.e.* torque per radian and θ is the plate deflection, then control torque is

$$T_c = S\theta$$

Here, $\theta = 100° = 100\times\pi/180 = 5\pi/9$ radian

$\therefore$ $S\times 5\pi/9 = 17.7\times 10^{-4}$ $\therefore$ $S = \mathbf{10.1\times 10^{-4}}$ **N-m/rad.**

Example 10.21(b). *A capacitance transducer of two parallel plates of overlapping area of $5\times 10^{-4} m^2$ is immersed in water. The capacitance 'C' has been found to be 9.50 pF. Calculate the separation 'd' between the plates and the sensitivity, $S = \partial C/\partial d$, of this transducer, given : ε_r water = 81; ε_0 = 8.854 pF/m.* **(Elect. Measur. A.M.I.E. Sec. B, 1992)**

Solution. Since $C = \varepsilon_0\varepsilon_r A/d$, $d = 3\,\varepsilon_0\varepsilon_r A/C$.

Substituting the given values we get, $d = 37.7\times 10^{-3}$ m

Sensitivity $\frac{\partial C}{\partial d} = \frac{\partial}{d}\left(\frac{\varepsilon_0\varepsilon_r A}{d}\right) = -\frac{8.854\times 10^{-12}\times 81\times 5\times 10^{-4}}{(37.7\times 10^{-3})^2} = -0.025\times 10^{-8}$ F/m

10.36. Wattmeters

We will discuss the two main types of wattmeters in general use, that is, (*i*) the dynamometer or electrodynamic type and (*ii*) the induction type.

10.37. Dynamometer Wattmeter

The basic principle of dynamometer instrument has already been explained in detail in Art. 10.20. The connections of a dynamometer type wattmeter are shown in Fig. 10.47. The fixed circular coil which carries the main circuit current I_1 is wound in two halves positioned parallel to each other. The distance between the two halves can be adjusted to give a uniform magnetic field. The moving coil which is pivoted centrally carries a current I_2 which is proportional to the voltage V. Current I_2 is led into the moving coil by two springs which also supply the necessary controlling torque. The equivalent diagrammatic view is shown in Fig. 10.48.

Deflecting Torque

Since coils are air-cored, the flux density produced is directly proportional to the current I_1.

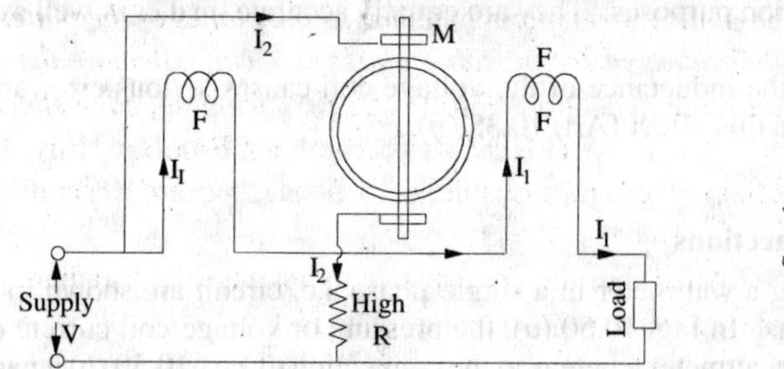

Fig. 10.47

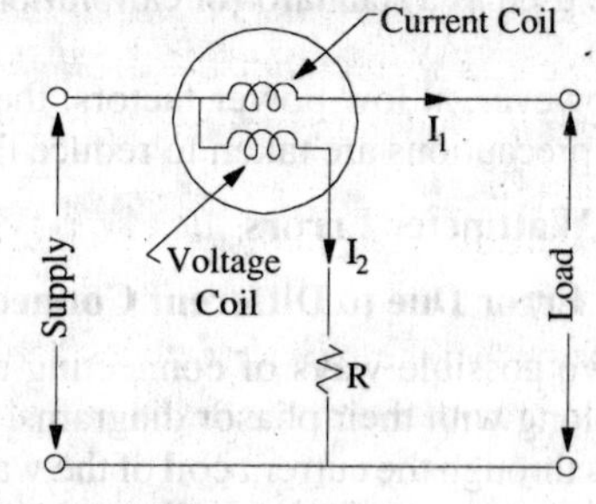

Fig. 10.48

$\therefore \quad B \propto I_1$ or $B = K_1 I_1$; $\qquad$ current $I_2 \propto V$ or $I_2 = K_2 V$

Now $T_d \propto BI_2 \propto I_1 V$ $\qquad \therefore \quad T_d = KV.I_1 = K \times$ power

In d.c. circuits, power is given by the product of voltage and curent in amperes, hence torque is directly proportional to the power.

Let us see how this instrument indicates true power on a.c. circuits.

For a.c. supply, the value of instantaneous torque is given by $T_{inst} \propto vi = K\, vi$

where $\quad v$ = instantaneous value of voltage across the moving coil

$\quad i$ = instantaneous value of current through the *fixed* coils.

However, owing to the large inertia of the moving system, the instrument indicates the mean or average power.

$\therefore$ Mean deflecting torque $T_m \propto$ average value of vi

Let $v = V_{max} \sin\theta$ and $i = I_{max} \sin(\theta - \phi)$ $\quad \therefore \quad T_m \propto \frac{1}{2\pi}\int_0^{2\pi} V_{max} \sin\theta \times I_{max} \sin(\theta - \phi) d\theta$

$$\propto \frac{V_{max} I_{max}}{2\pi}\int_0^{2\pi} \sin\theta \sin(\theta-\phi)\, d\theta \propto \frac{V_{max} I_{max}}{2\pi}\int_0^{2\pi} \frac{\cos\phi - \cos(2\theta-\phi)}{2} d\theta$$

$$\propto \frac{V_{max} I_{max}}{4\pi}\left[\theta\cos\phi - \frac{\sin(2\theta-\phi)}{2}\right]_0^{2\pi} \propto \frac{V_{max}}{\sqrt{2}} \cdot \frac{I_{max}}{\sqrt{2}} \cdot \cos\phi \propto VI\cos\phi$$

where V and I are the r.m.s. values. $\quad \therefore \quad T_m \propto VI\cos\phi \propto$ true power.

Hence, we find that in the case of a.c. supply also, the deflection is proportional to the true power in the circuit.

Scales of dynamometer wattmeters are more or less uniform because the deflection is proportional to the average power and for spring control, controlling torque is proportional to the deflection. Hence $\theta \propto$ power. Damping is pneumatic with the help of a piston moving in an air chamber as shown in Fig. 10.49.

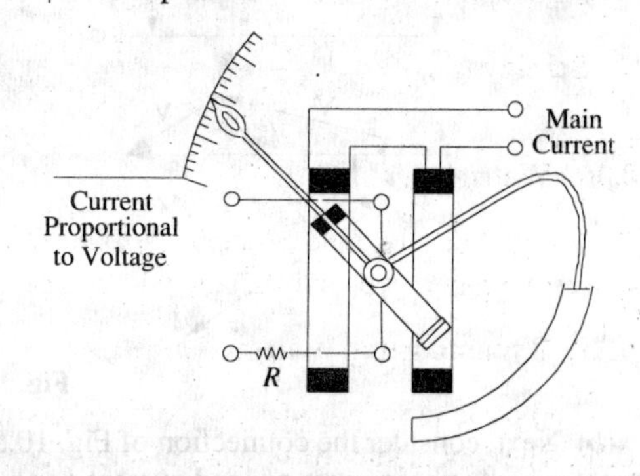

Fig. 10.49

Errors

The inductance of the moving or voltage coil is liable to cause error but the high non-inductive resistance connected in series with the coil swamps, to a great extent, the phasing effect of the voltage-coil inductance.

Another possible error in the indicated power may be due to (*i*) some voltage drop in the circuit or (*ii*) the current taken by the voltage coil. In standard wattmeters, this defect is overcome by having an additional compensating winding which is connected in series with the voltage coil but is so placed that it produces a field in opposite direction to that of the fixed or current coils.

Advantages and Disadvantages

By careful design, such instruments can be built to give a very high degree of accuracy. Hence, they are used as a standard for calibration purposes. They are equally accurate on d.c. as well as a.c. circuits.

However, at low power factors, the inductance of the voltage coil causes serious error unless special precautions are taken to reduce this effect [Art. 10.38 (*ii*)].

10.38. Wattmeter Errors

(i) Error Due to Different Connections

Two possible ways of connecting a wattmeter in a single-phase a.c. circuit are shown in Fig. 10.50 along with their phasor diagrams. In Fig. 10.50 (*a*), the pressure or voltage-coil current does not pass through the current coil of the wattmeter whereas in the connection of Fig. 10.50 (*b*) it passes. A wattmeter is supposed to indicate the power consumed by the load but its actual reading is slightly higher due to power losses in the instrument circuits. The amount of error introduced depends on the connection.

(*a*) Consider the connection of Fig. 10.50 (*a*). If cos ϕ is the power factor of the load, then power in the load is $VI\cos\theta$.

Now, voltage across the pressure-coil of the wattmeter is V_1 which is the phasor sum of the load voltage V and p.d. across current-coil of the instrument *i.e.* V' (= Ir where r is the resistance of the current coil).

Hence, power reading as indicated by the wattmeter is $= V_1 I\cos\theta$

where θ = phase difference between V_1 and I as shown in the phasor diagram of Fig. 10.50(*a*).

As seen from the phasor diagram, $V_1\cos\theta = (V\cos\phi + V')$

$\therefore$ wattmeter reading $= V_1\cos\theta \,.\, I = (V\cos\phi + V')\,I$

$= VI\cos\phi + V'I = VI\cos\phi + I^2 r$ = power in load + power in current coil.

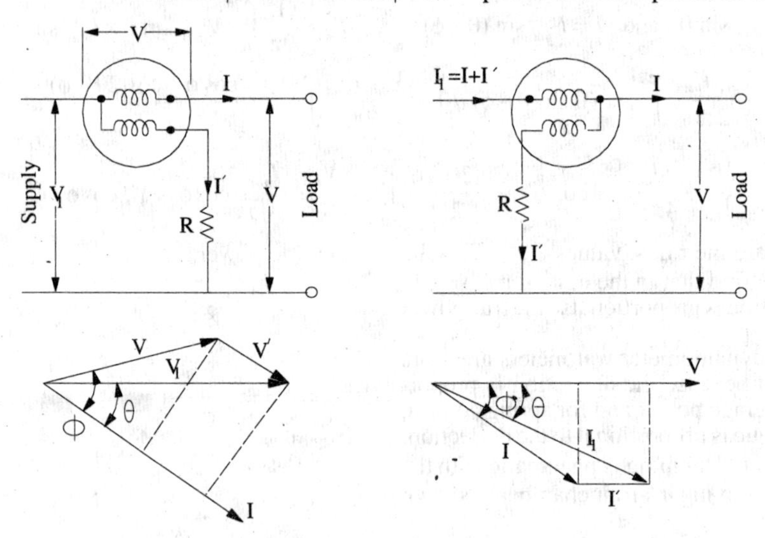

Fig. 10.50

(*b*) Next, consider the connection of Fig. 10.50 (*b*). The current through the current-coil of the wattmeter is the phasor sum of load current I and voltage-coil current $I' = V/R$. The power reading indicated by the wattmeter is $= VI_1\cos\theta$.

As seen from the phasor diagram of Fig. 10.50 (*b*), $I_1\cos\theta = (I\cos\phi + I')$

$\therefore$ wattmeter reading $= V(I\cos\phi + I') = VI\cos\phi + VI' = VI\cos\phi + V^2/R$

= power in load + power in pressure-coil circuit

(ii) Error Due to Voltage-coil Inductance

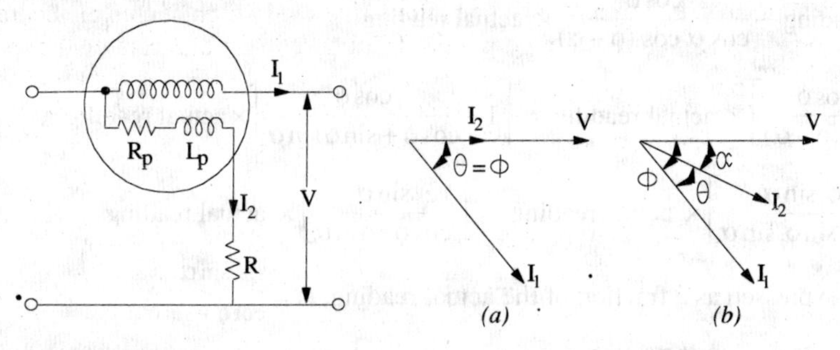

Fig. 10.51 **Fig. 10.52**

While developing the theory of electrodynamic instruments, it was assumed that pressure-coil does not possess any inductance (and hence reactance) so that current drawn by it was = V/R. The wattmeter reading is proportional to the mean deflecting torque, which is itself proportional to $I_1 I_2 \cos\theta$, where θ is the angle between two currents (Fig. 10.52).

In case the inductance of the voltage-coil in neglected,

$$I_2 = V/(R + R_p) = V/R \text{ approximately}$$

and $\theta = \phi$ as shown in the phasor diagram of 10-52(*a*)

$$\therefore \text{ wattmeter reading } \propto \frac{I_1 V}{R}\cos\phi \qquad \ldots(i)$$

In case, inductance of the voltage coil is taken into consideration, then

$$I_2 = \frac{V}{\sqrt{(R_p + R)^2 + X_L^2}} = \frac{V}{\sqrt{R^2 + X_L^2}} = \frac{V}{Z_p}$$

It lags behing V by an angle α [Fig. 10.52 (*b*)] such that

$$\tan\alpha = X_L/(R_p + R) = X_L/R \text{ (approx.)} = \omega L_p/R$$

$$\therefore \text{ wattmeter reading } \propto \frac{I_1 V}{Z_p}\cos\theta \propto \frac{I_1 V}{Z_p}\cos(\phi - \alpha)$$

Now $$\cos\alpha = \frac{R_p + R}{Z_p} = \frac{R}{Z_p} \quad \therefore \quad Z_p = \frac{R}{\cos\alpha}$$

$$\therefore \text{ wattmeter reading in this case is } \propto I_1\frac{V}{R}\cos\alpha\cos(\phi - \alpha) \qquad \ldots(ii)$$

Eq. (*i*) above, gives wattmeter reading when inductance of the voltage coil is neglected and Eq. (*ii*) gives the reading when it is taken into account.

The correction factor which is given by the ratio of the true reading (W_1) and the actual or indicated reading (W_a) of the wattmeter is

$$\frac{W_t}{W_a} = \frac{\frac{VI_1}{R_1}\cos\phi}{\frac{VI_1}{R}\cos\alpha\cos(\phi - \alpha)} = \frac{\cos\phi}{\cos\alpha\cos(\phi - \alpha)}$$

Since, in practice, α is very small, $\cos\alpha = 1$. Hence the correction factor becomes = $\frac{\cos\phi}{\cos(\phi}$

$$\therefore \text{ True reading} = \frac{\cos\phi}{\cos\alpha\cos(\phi - \alpha)} \times \text{actual reading} \cong \frac{\cos\phi}{\cos(\phi - \alpha)} \times \text{actual reading}$$

The error in terms of the actual wattmeter reading can be found as follows :

Actual reading–ture reading

$$= \text{actual reading} - \frac{\cos\phi}{\cos\alpha\cos(\phi-\alpha)} \times \text{actual reading}$$

$$= \left[1 - \frac{\cos\phi}{\cos(\phi-\alpha)}\right] \times \text{actual reading} = \left[1 - \frac{\cos\phi}{\cos\phi + \sin\phi\sin\alpha}\right] \times \text{actual reading}$$

$$= \left[\frac{\sin\phi.\sin\alpha}{\cos\phi + \sin\phi.\sin\alpha}\right] \times \text{actual reading} = \left[\frac{\sin\alpha}{\cos\phi + \sin\alpha}\right] \times \text{actual reading}$$

The error, expressed as a fraction of the actual reading, is $= \dfrac{\sin\alpha}{\cot\phi + \sin\alpha}$

$$\text{Percentage error} = \frac{\sin\alpha}{\cot\phi + \sin\alpha} \times 100$$

(iii) Error Due to Capacitance in Voltage-coil Circuit

There is always present a small amount of capacitance in the voltage-coil circuit, particularly in the series resistor. Its effect is to reduce angle α and thus reduce error due to the inductance of the voltage coil circuit. In fact, in some wattmeters, a small capacitor is purposely connected in parallel with the series resistor for obtaining practically non-inductive voltage-coil circuit. Obviously, over-compensation will make resultant reactance capacitive thus making α negative in the above expressions.

(iv) Error Due to Stray Fields

Since operating field of such an instrument is small, it is very liable to stray field errors. Hence, it should be kept as far away as possible from stray fields. However, errors due to stray fields are, in general, negligible in a properly-constructed instrument.

(v) Error Due to Eddy Currents

The eddy current produced in the solid metallic parts of the instrument by the alternating field of the current coil changes the magnitude and strength of this operating field thus producing an error in the reading of the wattmeter. This error is not easily calculable although it can be serious if care is not taken to remove away solid masses of metal from the proximity of the current coil.

Example 10.22. *A dynamometer type wattmeter with its voltage coil connected across the load side of the instrument reads 250 W. If the load voltage be 200 V, what power is being taken by load? The voltage coil branch has a resistance of 2,000 Ω.*

(Elect. Engineering, Madras Univ. 1985)

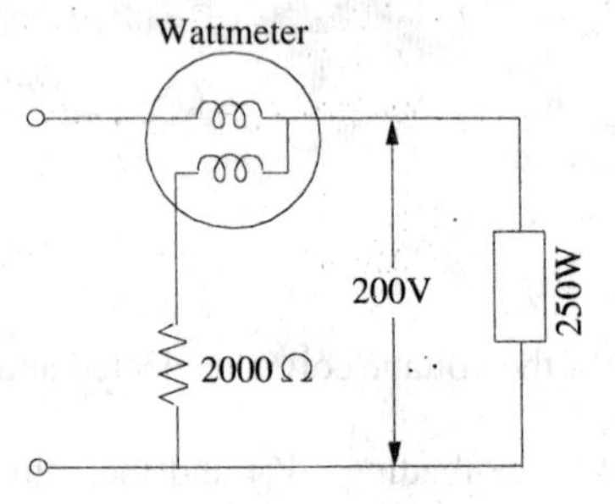

Fig. 10.53

Solution. Since voltage coil is connected across the load side of the wattmeter (Fig. 10.53), the power consumed by it is also included in the meter reading.

Power consumed by voltage coil is

$$= V^2/R = 200^2/2,000 = 20 \text{ W}$$

$\therefore$ Power being taken by load $= 250 - 20$

$= \mathbf{230\ W}$

Example 10.23. *A 250-V, 10-A dynamometer type wattmeter has resistance of current and potential coils of 0.5 and 12,500 ohms respectively. Find the percentage error due to each of the two methods of connection when unity p.f. loads at 250 volts are of (a) 4A (b) 12A.*

Neglect the error due to the inductance of pressure coil.

(Elect. Measurements, Pune. Univ. 1985)

Solution. (*a*) **When I = 4 A**

(*i*) Consider the type of connection shown in Fig. 10.50 (*a*)

Power loss in current coil of wattmeter $= I^2r = 4^2 \times 0.5 = \mathbf{8\ W}$

Load power $= 250 \times 4 \times 1 = 1000$ W ; Wattmeter reading = 1008 W

$\therefore$ percentage error $= (8/1008) \times 100 = \mathbf{0.794\%}$

(*ii*) Power loss in pressure coil resistance = $V^2/R = 250^2/12{,}500 = 5$ W

$\therefore$ percentage error = $5 \times 100/1005$ = **0.497%**

(*b*) **When I = 12A**

(*i*) Power loss in current coil = $12^2 \times 0.5 = 72$ W

Load power = $250 \times 12 \times 1 = 3000$ W : wattmeter reading = 3072 W

$\therefore$ percentage error = $72 \times 100/3072$ = **2.34%**

(*ii*) Power loss in the resistance of pressure coil is = $250^2/12{,}500 = 5$ W

$\therefore$ percentage error = $5 \times 100/3005$ = **0.166%**

Example 10.24. *An electrodynamic wattmeter has a voltage circuit of resistance of 8000 Ω and inductance of 63.6 mH which is connected directly across a load carrying a current of 8A at a 50-Hz voltage of 240-V and p.f. of 0.1 logging. Estimate the percentage error in the wattmeter reading caused by the loading and inductance of the voltage circuit.*

(Elect & Electronic Measu. & Measu. Instru. Nagpur, Univ. 1992)

Solution. The circuit connections are shown in Fig. 10.54.

Load power = $240 \times 8 \times 0.1 = 192$ W

$\cos \phi = 0.1, \phi = \cos^{-1}(0.1) = 84°16'$

Power loss in voltage coil circuit is = V^2/R

$= 240^2/8000 = 7.2$ W

Neglecting the inductance of the voltage coil, the wattmeter reading would be

$= 192 + 7.2 =$ **199.2 W**

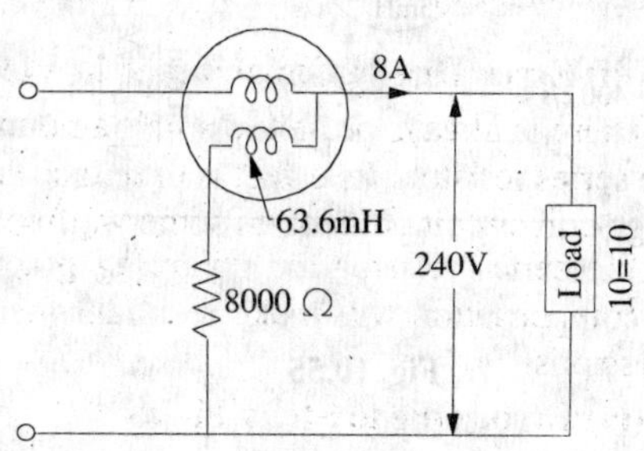

Fig. 10.54

Now, $X_p = 2\pi \times 50 \times 63.3 \times 10^{-3} = 20\Omega$

$\alpha = \tan^{-1}(20/8000) = \tan^{-1}(0.0025) = 0°\,9'$

Error factor due to inductance of the voltage coil = $\dfrac{\cos(\phi - \alpha)}{\cos\phi} = \dfrac{\cos 84°7'}{\cos 84°16'} = 1.026$

Wattmeter reading $= 1.026 \times 199.2 = 204.4$ W

Percentage error $= \left(\dfrac{204.4 - 199.2}{199.2}\right) \times 100 = \mathbf{2.6\%}$

Example 10.25. *The inductive reactance of the pressure-coil circuit of a dynamometer wattmeter is 0.4% of its resistance at normal frequency and the capacitance is negligible.*

Calculate the percentage error and correction factor due to the reactance for loads at (i) 0.707 p.f. lagging and (ii) 0.5 p.f. lagging.

(Elect. Measurements, Bombay Univ. 1987)

Solution. It is given that $X_p/R = 0.4\% = 0.004$

$\tan \alpha = X_p/R = 0.004 \quad \therefore \quad \alpha = 0°\,14'$ and $\sin \alpha = 0.004$

(*i*) **When p.f. = 0.707** (*i.e.* $\phi = 45°$)

Correction factor $= \dfrac{\cos\phi}{\cos(\phi - \alpha)} = \dfrac{\cos 45°}{\cos 44°46'} = \mathbf{0.996}$

Percentage error $= \dfrac{\sin\alpha}{\cot\phi + \sin\alpha} \times 100 = \dfrac{\sin 0°14'}{\cot 45° + \sin 0°14'} \times 100$

$= \dfrac{0.004 \times 100}{1 + 0.004} = \dfrac{0.4}{1.004} =$ **0.4 (approx.)**

(*ii*) **When p.f.** = **0.5** (*i.e.* $\phi = 60°$)

Correction factor $= \dfrac{\cos 60°}{\cos 59°46'} = \mathbf{0.993}$

Percentage error $= \dfrac{\sin 0°\,14'}{\cos 60° + \sin 0°14'} \times 100 = \dfrac{0.004 \times 100}{0.577 + 0.004} = \dfrac{0.4}{0.581} = \mathbf{0.7}$

Example 10.26. *The current coil of wattmeter is connected in series with an ammeter and an inductive load. A voltmeter and the voltage circuit of the wattmeter are connected across a 400-Hz supply. The ammeter reading is 4.5 A and voltmeter and wattmeter readings are respectively 240 V and 29 W. The inductance of the voltage circuit is 5 mH and its resistance is 4 kΩ. If the voltage drops across the ammeter and current coil are negligible, what is the percentage error in wattmeter reading ?*

Solution. The reactance of the voltage-coil circuit is $X_p = 2\pi \times 400 \times 5 \times 10^{-3} = 4\,\pi$ ohm

$\tan \alpha = X_p/R = 4\pi/4000 = 0.003142$

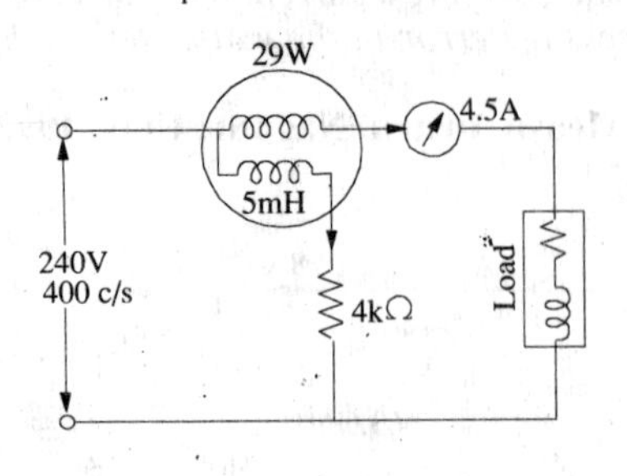

Fig. 10.55

$$\therefore\ \alpha = 0.003142 \text{ radian } (\therefore \text{ angle is very small}) = 0.18° \text{ or } 0°11'$$

$$\text{Now, true reading} = \frac{\cos\phi}{\cos\alpha.\cos(\phi-\alpha)} \times \text{actual reading}$$

$$\text{or} \quad VI\cos\phi = \frac{\cos\phi}{\cos\alpha.\cos(\phi-\alpha)} \times \text{actual reading}$$

$$\text{or} \quad VI = \frac{\text{actual reading}}{\cos(\phi-\alpha)}$$

(taking $\cos\alpha = 1$)

$$\therefore \quad \cos(\phi-\alpha) = 29/240 \times 4.5 = 0.02685$$

$$\therefore \quad \phi - \alpha = 88°28' \text{ or } \phi = 88°39'$$

$$\therefore \text{ Percentage error} = \frac{\sin\alpha}{\cot\phi + \sin\alpha} \times 100 = \frac{\sin 11'}{\cot 88°39' + \sin 11'} \times 100$$

$$= \frac{0.0032}{0.0235 + 0.0032} \times 100 = \mathbf{12\%}$$

10.39. Induction Wattmeters

Principle of induction wattmeters is the same as that of induction ammeters and voltmeters. They can be used on a.c. supply only in contrast with dynamometer wattmeters, which can be used both on d.c. and a.c. supply. Induction wattmeters are useful only when the frequency and supply voltage are constant.

Since, both a current and a pressure element are required in such instruments, it is not essential to use the shaded-pole principle. Instead of this, two separate a.c. magnets are used, which produce two fluxes, which have the required phase difference.

Construction

The wattmeter has two laminated electromagnets, one of which is excited by the current in the main circuit–exciting winding being joined in series with the circuit, hence it is also callled a *series* magnet. The other is excited by current which is proportional to the voltage of the circuit. Its exciting coil is joined in parallel with the circuit, hence this magnet is sometimes referred to as *shunt* magnet.

A thin aluminium disc is so mounted that it cuts the fluxes of both magnets. Hence, two eddy currents are produced in the disc. The deflection torque is produced due to the interaction of these eddy currents and the inducing fluxes., Two or three copper rings are fitted on the central limb of the shunt magnet and can be so adjusted as to make the resultant flux in the shunt magnet lag behind the applied voltage by 90°.

Two most common forms of the electromagnets are shown in Fig. 10.56 and 10.57. It is seen that in both cases, one magnet is placed above and the other below the disc. The magnets are so positioned and shaped that their fluxes are cut by the disc.

In Fig. 10.56, the two pressure coils are joined in series and are so wound that both send the flux through the central limb in the same direction. The series magnet carries two coils joined in series and so wound that they magnetise their respective cores in the same direction. Correct phase displacement between the shunt and series magnet fluxes can be obtained by adjusting the position of the copper shading bands as shown.

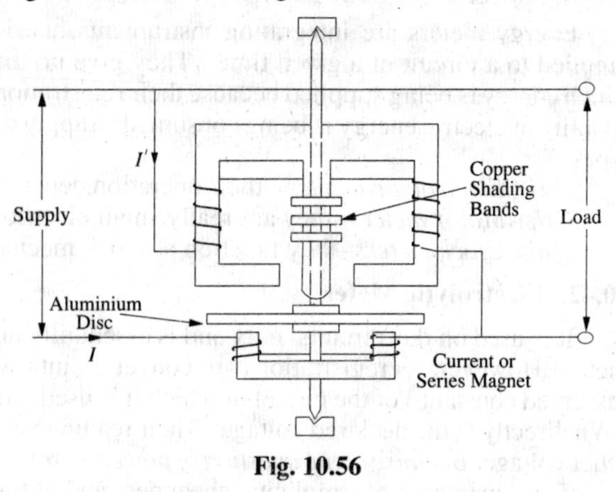

Fig. 10.56

In the type of instrument shown in Fig. 10.57, there is only one pressure winding and one current winding. The two projecting poles of the shunt magnet are surrounded by a copper shading band whose position can be adjusted for correcting the phase of the flux of this magnet with respect to the voltage.

Both types of induction wattmeters shown above, are spring-controlled, the spring being fitted to the spindle of the moving system which also carries the pointer. The scale is uniformly even and extends over 300°.

Currents upto 100 A can be handled by such wattmeters directly but for currents greater than this value, they are used in conjunction with current transformers. The pressure coil is purposely made as much inductive as possible in order that the flux through it should lag behind the voltage by 90°.

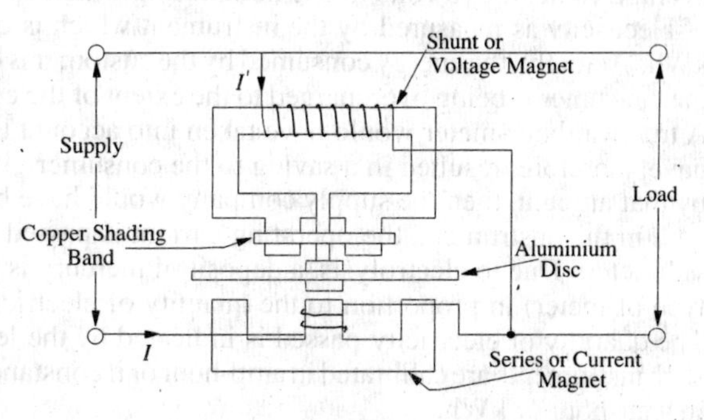

Fig. 10.57

Theory

The winding of one magnet carries line current I so that $\Phi_1 \propto I$ and is in phase with I (Fig. 10.58). The other coil *i.e.* pressure or voltage coil is made highly inductive having an inductance of L and negligible resistance. This is connected across the supply voltage V. The current in the pressure coil is, therefore, equal to $V/\omega L$. Hence, $\Phi_2 \propto V/\omega L$ and lags behind the voltage by 90°. Let the load current I lag behind V by ϕ *i.e.* let the load power factor angle be ϕ. As shown in Fig. 10.56, the phase angle between Φ_1 and Φ_2 is $\alpha = (90 - \phi)$.

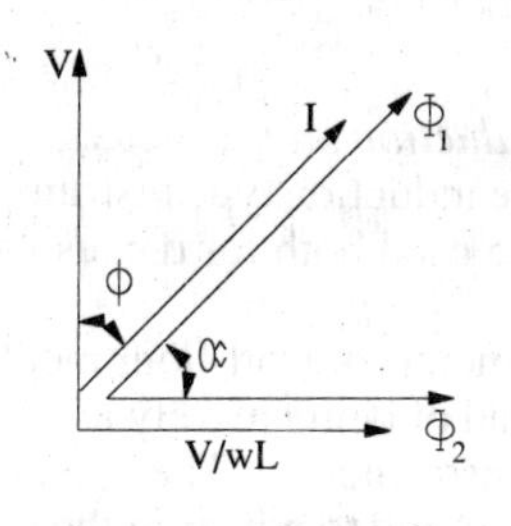

Fig. 10.58

The value of the torque acting on the disc is given by

$$T = K\omega\, \Phi_{1m}\, \Phi_{2m} \sin \alpha \qquad \text{– Art. 10.25}$$

or
$$T \propto \omega . I . \frac{V}{\omega L} \cdot \sin(90 - \phi) \propto VI \cos \phi \propto \text{power}$$

Hence, the torque is proportional to the power in the load circuit, For spring control, the controlling torque $T_c \propto \theta$. $\therefore\ \theta \propto$ power. Hence, the scale is even.

10.40. Advantages and Limitations of Induction Wattmeters

These wattmeters possess the advantage of fairly long scales (extending over 300°), are free from the effects of stray fields and have good damping. They are practically free from frequency errors. However, they are subject to (sometimes) serious temperature errors because the main effect of temperature is on the resistance of the eddy current paths.

10.41. Energy Meters

Energy meters are integrating instruments, used to measure the quantity of electric energy supplied to a circuit in a given time. They give no direct indication of power *i.e.* as to the rate at which energy is being supplied because their registrations are independent of the rate at which a given quantity of electric energy is being consumed. Supply or energy meters are generally of the following types :

(*i*) *Electrolytic meters*– their operation depends on electrolytic action.
(*ii*) *Motor meters*–they are really small electric motors.
(*iii*) *Clock meters*–they function as clock mechanisms.

10.42. Electrolytic Meter

It is used on d.c. circuits* only and is essentailly an ampere-hour meter and not a true watt-hour meter. However, its registrations are converted into watt-hour by multiplying them by the voltage (assumed constant) of the circuit in which it is used. Such instruments are usually calibrated to read kWh directly at the declared voltage. Their readings would obviously be incorrect when used on any other voltage. *Because of the question of power factor, such instrument cannot be used on a.c. circuits.*

The advantages of simplicity, cheapness and of low power consumption of ampere-hour meters are, to a large extent, discounted by the fact that variations in supply voltage are not taken into account by them. As an example suppose that the voltage of a supply whose nominal value is 220 V, has an average value of 216 volts in one hour during which a consumer draws a current of 100 A. Quantity of electricity as measured by the instrument which is calibrated on 220 V, is $220 \times 100/1000 = 22$ kWh. Actually, the energy consumed by the customer is only $216 \times 100/1000 = 21.6$ kWh. Obviously, the consumer is being overcharged to the extent of the cost of $22 - 21.6 = 0.4$ kWh of energy per hour. A true watthour-meter would have taken into account the decrease in the supply voltage and would have, therefore, resulted in a saving to the consumer. If the supply voltage would have been higher by that amount, then the supply company would have been the loser (Ex. 10.27).

In this instrument, the operating current is passed through a suitable electrolyte contained in a voltmeter. Due to electrolysis, a deposit of mercury is given or a gas is liberated (depending on the type of meter) in proportion to the quantity of electricity passed (Faraday's Laws of Electrolysis). The quantity of electricity passed is indicated by the level of mercury in a graduated tube. Hence, such instruments are calibrated in amp-hour or if constancy of supply voltage is assumed, are calibrated in watt-hour or kWh.

Such instruments are cheap, simple and are accurate even at very small loads. They are not affected by stray magnetic fields and due to the absence of any moving parts are free from friction errors.

10.43. Motor Meters

Most commonly-used instruments of this type are :

(*i*) *Mercury motor* meters (*ii*) *Commutator motor* meters and (*iii*) *Induction motor* meters.

Of these, mercury motor meter is normally used on d.c. circuits whereas the induction type instrument is used only on a.c. circuits. However, the commutator type meter can be used both for d.c. as well as a.c. work.

Instruments used for d.c. work can be either in the form of amp-hour meters or watt-hour meters. In both cases, the moving system is allowed to revolve continuously instead of being merely allowed to deflect or rotate through a fraction of a revolution as in indicating instruments. The speed of rotation is directly proportional to the current in the case of amp-hour meter and to power in the case of watt-hour meter. Hence, the number of revolutions made in a given time is proportional, in the case of an amp-hour meter, to the quantity of electricity ($Q = 1 \times t$) and in the case of Wh meter, to the quantity of energy supplied to the circuit. The number of revolutions made are registered by a counting mechanism consisting of a train of gear wheels and dials.

The control of speed of the rotating system is brought about by a permanent magnet (known as braking magnet) which is so placed as to set up eddy currents in some part of the rotating system. These eddy currents produce a retarding torque which is proportional to their magnitude–their

*Recently such instruments have been marketed for measurement of kilovoltampere-hours on a.c. supply, using a small rectifier unit, which consists of a current transformer and full-wave copper oxide rectifier.

magnitude itself depending on the speed of rotation of the rotataing system. The rotating system attains a *steady speed* when the braking torque exactly balances the driving torque which is produced either by the current or power in the circuit.

The essential parts of motor meters are :

1. An operating system which produces an operating torque proportional to the current or power in the circuit and which causes the rotation of the rotating system.
2. A retarding or braking device, usually a permanent magnet, which produces a braking torque in proportional to the speed of rotation. Steady speed of rotation is achieved when braking torque becomes equal to the operating torque.
3. A registering mechanism for the revolutions of the rotating system. Usually, it consists of a train of wheels driven by the spindle of the rotating system. A worm which is cut on the spindle engages a pinion and so drives a wheel-train.

10.44. Errors in Motor Meters

The two main errors in such instruments are : (*i*) friction error and (*ii*) braking error. Friction error is of much more importance in their case than the corresponding error in indicating instruments because (*a*) it operates continuously and (*b*) it affects the speed of the rotor. The braking action in such meters corresponds to damping in indicating instruments. The braking torque directly affects the speed for a given driving torque and also the number of revolutions made in a given time.

Friction torque can be compensated for by providing a small constant driving torque which is applied to the moving system independent of the load.

As said earlier, steady speed of such instruments is reached when driving torque is equal to the braking torque. The braking torque is proportional to the flux of the braking magnet and the eddy current induced in the moving system due to its rotation in the field of the braking magnet

$$\therefore \qquad T_B \propto \Phi_i$$

where Φ is the flux of the braking magnet and i the induced current. Now $i = e/R$ where e is the induced e.m.f. and R the resistance of the eddy current path. Also $e \propto \Phi\, n$ where n is the speed of the moving part of the instrument.

$$\therefore \qquad T_B \propto \phi \times \frac{\Phi_n}{R} \propto \frac{\Phi^2 n}{R}$$

The torque $T_B{}'$ at the steady speed of N is given by $T_B{}' \propto \Phi^2\, N/R$

Now $$T_B{}' = T_D \qquad \text{- the driving torque}$$

$$\therefore \qquad T_D \propto \Phi^2 N/R \quad \text{or} \quad N \propto T_D R/\Phi^2$$

Hence, for a given driving torque, the steady speed is directly proportional to the resistance of the eddy current path and inversely to the square of the flux.

Obviously, it is very important that the strength of the field of the brake magnet should be constant throughout the time the meter is in service. This constancy of field strength can be assured by careful design and treatment during the manufacture of the brake magnet. Variations in temperature will affect the braking torque since the resistance of the eddy current path will change. This error is difficult to fully compensate for.

10.45. Quantity or Ampere-hour Meters

The use of such meters is mostly confined to d.c. circuits. Their operation depends on the production of two torques (*i*) a driving torque which is proportional to the current I in the circuit and (*ii*) a braking torque which is proportional to the speed n of the spindle. This speed attains a steady value N when these two torques become numerically equal. In that case, speed becomes propotional to current *i.e.* $N \propto I$. Over a certain period of time, the total number of revolutions $\int N dt$ will be proportional to the quantity of electricity $\int I dt$ passing through the meter. A worm cut in the spindle at its top engages gear wheels of the recording mechanism which has suitably marked dials reading directly in ampere hours. Since electric supply charges are based on watt-hours rather than ampere-hours, the dials of ampere-hour meters are frequently marked in corresponding watt-hours at the normal supply voltage. Hence, their indications of watt-hours are correct only when the supply voltage remains constant, otherwise reading will be wrong.

10.46. Ampere-hour Mercury Motor Meter

It is one of the best and most popular forms of mercury Ah meter used for d.c. work.

Construction

It consists of a thin Cu disc D mounted at the base of a spindle, working in jewelled cup bearings and revolving between a pair of *permanent* magnets M_1 and M_2. One of the two magnets *i.e.* M_2 is used for driving purposes whereas M_1 is used for braking. In between the poles of M_1 and M_2 is a hollow circular box B in which rotates the Cu disc and the rest of the space is filled up with mercury which exerts considerable upward thrust on the disc, thereby reducing the pressure on the bearings. The spindle is so weighted that it just sinks in the mercury bath. A worm cut in the spindle at its top engages the gear wheels of the recording mechanism as shown in Fig. 10.59.

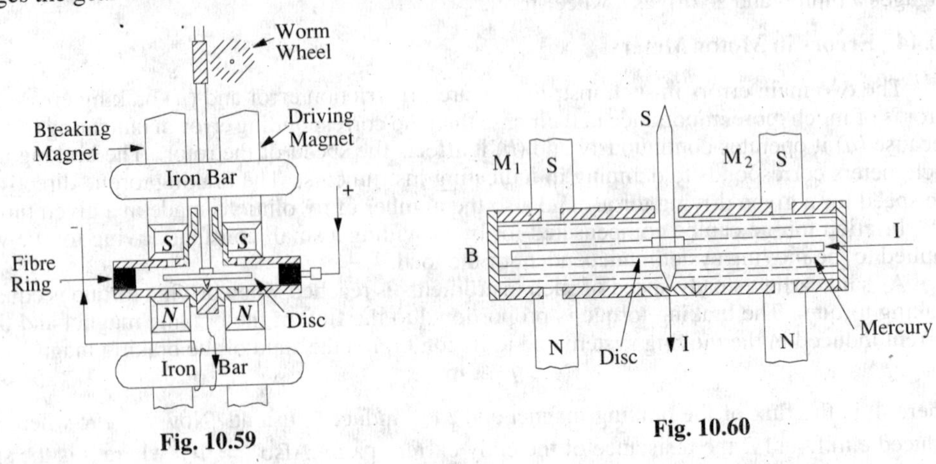

Fig. 10.59

Fig. 10.60

Principle of Action

Its principle of action can be understood from Fig. 10.61 which shows a separate line drawing of the motor element.

The current to be measured is led into the disc through the mercury at a point at its circumference on the right-hand side. As shown by arrows, it flows radially to the centre of the disc where it passes out to the external circuit through the spindle and its bearings. It is worth noting *that current flow takes place only under the right-hand side magnet* M_2 and not under the left-hand side magnet M_1. The field of M_2 will, therefore, exert a force on the right– side portion of the disc which carries the current (motor action). The direction of the force, as found by Fleming's Left-hand rule, is as shown by the arrow. The magnitude of the force depends on the flux density and current ($\because$ $F = BIl$). The driving or motoring torque T_d so produced is given by the product of the force and the distance from the spindle at which this force acts. When the disc rotates under the influence of this torque, it cuts through the field of left-hand side magnet M_1 and hence eddy currents are produced in it which result in the production of braking torque. The magnitude of the retarding or braking torque is proportional to the speed of rotation of the disc.

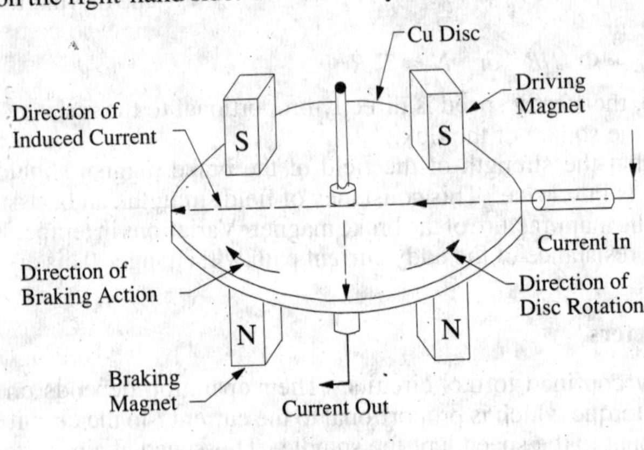

Fig. 10.61

Theory

Driving torque $T_d \propto$ force on the disc $\propto B\,I$

If the flux density of M_2 remains constant, then $T_d \propto I$

The braking torque T_B is proportional to the flux Φ of braking magnet M_1 and the eddy current i induced in the disc due to its rotation in the field of M_1.

$\therefore \quad T_B \propto \Phi_i$

Now $i = e/R$ where e is the induced e.m.f. and R the resistance of eddy current path.

Also $e \propto \Phi n$ - where n is the speed of the disc $\therefore \; T_B \propto \Phi \times \dfrac{\Phi n}{R} \propto \dfrac{\Phi^2 n}{R}$

The speed of the disc will attain a steady value N when the driving and braking torques become equal. In that case, $T_B \propto \Phi^2/N/R$

If Φ and R are constant, then $I \propto N$

The total number of revolutions in any given time t *i.e.* $\int_0^t N.\,dt$ will become proportional to $\int_0^t I.\,dt$ *i.e.* to the total quantity of electricity passed through the meter.

10.47. Friction Compensation

There are two types of frictions in this ampere-hour meter.

(*i*) *Bearing Friction.* The effect of this friction is normally negligible because the disc and spindle float in mercury. Due to upward thrust, the pressure on bearings is considerably reduced which results in freedom from wear as well as a great reduction in the bearing friction.

(*ii*) *Mercury Friction.* Since the disc revolves in mercury, there is friction between mercury and the disc, which gives rise to a torque, approximately proportional to the square of the speed of rotation. Hence, this friction causes the meter to run slow on heavy loads. It can be compensated for in the following two ways :

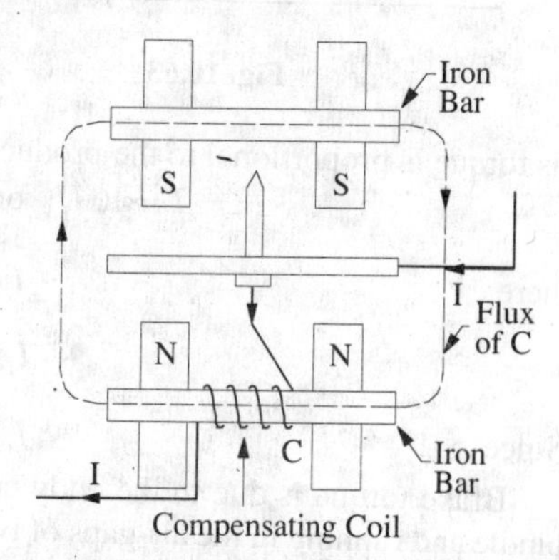

Fig.10.62

(*a*) a coil of few turns is wound on one of the poles of the driving magnet M_2 and the meter current is passed through it in a suitable direction so as to increase the strength of M_2. The additional driving torque so produced can be made just sufficient to compensate for the mercury friction.

(*b*) in the other method, two iron bars are placed across the two permanent magnets, one above and one below the mercury chamber as shown in Fig. 10.62. The lower bar carries a small compensating coil through which is passed the load current. The local magnetic field set up by this coil strengthens the field of driving magnet M_2 and weakens that of the braking magnet M_1, thereby compensating for mercury friction.

10.48. Mercury Meter Modified as Watt-hour Meter

If the permanent magnet M_2 of the amp-hour meter, used for producing the driving torque, is replaced by a wound electromagnet connected across the supply, the result is a watt-hour meter. The exciting current of this electromagnet is proportional to the voltage of the supply. The driving torque is exerted on the aluminium disc immersed in the mercury chamber below which is placed this electromagnet. The aluminium disc has radial slots cut in it for ensuring the radial flow of current through it-the current being led into and out of this disc through mercury contacts situated at diametrically opposite points. These radial slots, moreover, prevent the same disc being used for braking purposes. Braking is by a separate aluminium disc mounted on the same spindle and revolving in the air-gap of a separate braking magnet.

10.49. Commutator Motor Meters

These meters may be either ampere-hour or true watt-hour meters. In Fig. 10.63 is shown the principle of a common type of watt-hour meter known as Elihu-Thomson meter. It is based on the dynamometer principle (Art. 10.20) and is essentially an ironless motor with a wound armatu·· having a commutator.

Construction

There are two fixed coils C_1 and C_2 each consisting of a few turns of heavy copper strip and joined in series with each other and with the supply circuit so that they carry the main current in the circuit (a shunt is used if the current is too heavy). The field produced by them is proportional to the current to be measured. In this field rotates an armature carrying a number of coils which are connected to the segments of a small commutator. The armature coils are wound on a former made of non-magnetic material and are connected through the brushes and in series with a large resistance across the supply lines. The commutator is made of silver and the brushes are silver tipped in order to reduce friction. Obviously, the current passing through the armature is proportional to the supply voltage.

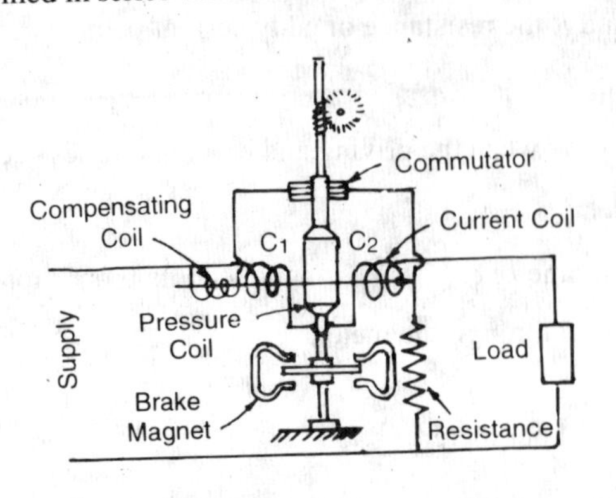

Fig. 10.63

The operating torque is produced due to the reaction of the field produced by the fixed coils and the armature coils. The magnitude of this torque is proportional to the product of the two currents *i.e.*

$$T_d \propto \phi I_1 \quad \text{or} \quad T_d \propto I_1 \times I \qquad (\because \quad \Phi \propto I)$$

where I = main circuit current

I_1 = current in armature coils.

Since $I_1 \propto V \quad \therefore \quad T_d \propto VI$ - power

Brake torque is due to the eddy currents induced in an aluminium disc mounted on the *same* spindle and running in the air-gaps of two permanent magnets. As shown in Art. 10.44, this braking torque is proportional to the speed of the disc if the flux of the braking magnet and the resistance of the eddy current paths are assumed constant.

When steady speed of rotation is reached, then

$$T_B = T_d \qquad \therefore \qquad N \propto VI \propto \text{power } W$$

Hence, steady number of revolutions in a given time is proportional to $\int Wt$ = the energy in the circuit.

The friction effect is compensated for by means of a small compensating coil placed coaxially with the two current coils and connected in series with the armature such that it strengthens the field of current coil. But its position is so adjusted that with zero line current the armature just fails to rotate. Such meters are now employed mainly for switchboard use, house service meters being invariably of the mercury ampere-hour type.

10.50. Induction Type Single-phase Watthour Meter

Induction type meters are, by far, the most common form of a.c. meters met with in every day domestic and industrial installations. These meters measure electric energy in kilo-watthours. The principle of these meters is practically the same as that of the induction wattmeters. Constructionally, the two are similar except that the control spring and the pointer of the watt-meter are replaced, in the case of watthour meter, by a brake magnet and by a spindle of the meter.

The brake magnet induces eddy currents in the disc which revolves continuously instead of rotating through only a fraction of a revolution as in the case of wattmeters.

Construction

The meter consists of two a.c. electromagnets as shown in Fig. 10.64 (*a*), one of which *i.e.* M_1 is excited by the line current and is known as *series* magnet. The alternating flux Φ_1 produced by it is proportional to and in phase with the line current (provided effects of hysteresis and iron saturation are neglected). The winding of the other magnet M_2, called *shunt* magnet, is connected across the supply line and carries current proportional to the supply voltage V. The flux Φ_2 produced by it is proportional to supply voltage V and lags behind it by 90°. This phase displacement of exact 90° is achieved by adjustment of the copper shading band C (also known as power factor compensator) on the shunt magnet M_2. Major portion of Φ_2 crosses the narrow gap between the centre and side limbs of M_2 but a small amount, which is the useful flux, passes through the disc D. The two fluxes Φ_1 and Φ_2 induce e.m.fs. in the disc which further produce the circulatory eddy currents. The reaction between these fluxes and eddy currents produces the driving torque on the disc in a manner similar to that explained in Art. 10.39. The braking torque is produced by a pair of magnets [Fig. 10.64(*b*)] which are mounted diametrically opposite to the magnets M_1 and M_2. This arrangement minimizes the interaction between the fluxes of M_1 and M_2. This arrangement minimizes the interaction between the fluxes of M_1 and M_2 and that of the braking magnet. When the peripheral portion of the rotating disc passes through the air-gap of the braking magnet, the eddy currents are induced in it which give rise to the necessary torque. The braking torque $T_B \propto \Phi^2 N/R$ where Φ is the flux of braking magnet, N the speed of the rotating disc and R the resistance of the eddy current path. If Φ and R are constant, then $T_B \propto N$.

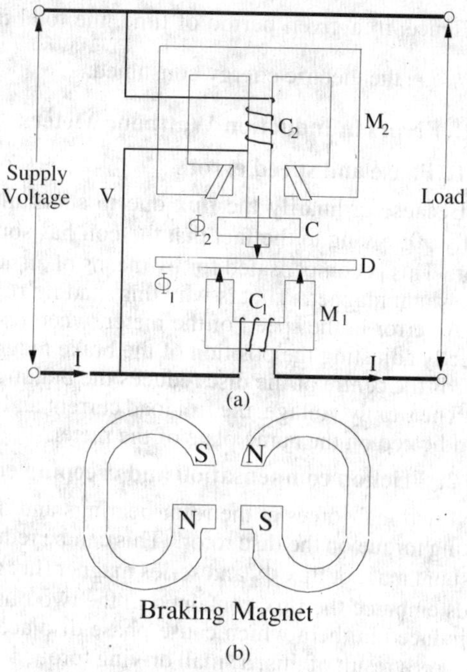

(a)

Braking Magnet

(b)

Fig. 10.64

The register mechanism is either of pointer type of cyclometer type. In the former type, the pinion on the rotor shaft drives, with the help of a suitable train of reduction gears, a series of five or six pointers rotating on dials marked with ten equal divisions. The gearing between different pointers is such that each pointer advances by 1/10th of a revolution for a complete revolution of the adjacent pointer on the main rotor disc in the train of gearing as shown in Fig. 10.65.

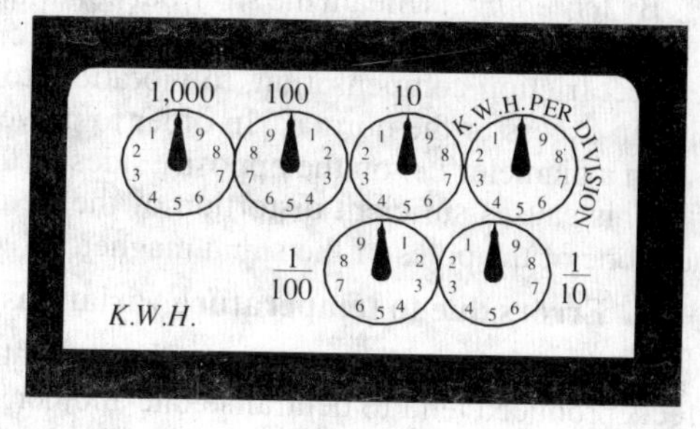

Fig. 10.65

Theory

As shown in Art. 10.39 and with reference to Fig. 10.58, the driving torque is given by $T_d \propto \omega\Phi_{1m}\,\Phi_{2m}$ where sin $\alpha\,\Phi_{1m}$ and Φ_{2m} are the maximum fluxes produced by magnets M_1 and M_2 and α the angle between these fluxes. Assuming that fluxes are proportional to the currents, we have

Current through the winding of $M_1 = I$ - the line current

Current through the winding of $M_2 = V/\omega L$

$$\alpha = 90 - \phi \quad \text{where } \phi \text{ is the load p.f. angle}$$

$$T_d \propto \omega \cdot \frac{V}{\omega L} \cdot I\cos(90 - \phi) \propto VI\,\cos\phi \propto \text{power}$$

Also, $T_B \propto N$

The disc achieves a steady speed N when the two torques are equal *i.e.* when

$$T_d = T_B \qquad \therefore\ N \propto \text{power } W$$

Hence, in a given period of time, the total number of revolutions $\int_0^t N\,dt$ is proportional to $\int_0^t W.dt$ *i.e.* the electric energy consumed.

10.51. Errors in Induction Watthour Meters

1. Phase and speed errors

Because ordinarily the flux due to shunt magnet does not lag behind the supply voltage by exactly 90° owing to the fact that the coil has some resistance, the torque is not zero at zero power factor. This is compensated for by means of an adjustable shading ring placed over the central limb of the shunt magnet. That is why this shading ring is known as *power factor compensator.*

An error in the speed of the meter, when tested on a non-inductive load, can be eliminated by correctly adjusting the position of the brake magnet. Movement of the poles of the braking magnet towards the centre of the disc reduces the braking torque and *vice-versa.*

The supply voltage, the full load current and the correct number of revolutions per kilowatthour are indicated on the name plate of the meter.

2. Friction compensation and creeping error

Frictional forces at the rotor bearings and in the register mechanism give rise to an unwanted braking torque on the disc rotor. This can be reduced to an unimportant level by making the ratio of the shunt magnet flux Φ_2 and series magnet flux Φ_1 large with the help of two shading bands. These bands embrace the flux contained in the two outer limbs of the shunt magnet and so eddy currents are induced in them which cause phase displacement between the enclosed flux and the main-gap flux. As a result of this, a small driving torque is exerted on the disc rotor solely by the pressure coil and independent of the main driving torque. The amount of this corrective torque is adjusted by the variation of the position of the two bands, so as to exactly compensate for firctional torque in the instrument. Correctness of friction compensation is achieved when the rotor does not run on no-load *with only the supply voltage connected.*

By '*creeping*' is meant the slow but continuous rotation of the rotor when only the pressure coils are excited but with no current flowing in the circuit. It may be caused due to various factors like incorrect friction compensation, to vibration, to stray magnetic fields or due to the voltage supply being in excess of the normal. In order to prevent creeping on no-load, two holes are drilled in the disc on a diameter *i.e.* on the opposite sides of the spindle.

This causes suficient distortion of the field to prevent rotation when one of the holes comes under one of the poles of the shunt magnet.

3. Errors due to temperature variations

The errors due to temperature variations of the instruments are usually small, because the various effects produced tend to neutralise one another.

Example 10.27. *An ampere-hour meter, calibrated at 210 V, is used on 230 V circuit and indicates a consumption of 730 units in a certain period. What is the actual energy supplied ?*

If this period is reckoned as 200 hours, what is the average value of the current ?

(Elect. Technology, Utkal Univ. 1987)

Solution. As explained in Art. 10.42, ampere-hour meters are calibrated to read directly in kWh at the *declared voltage*. Obviously, their readings would be incorrect when used on any other voltage.

Reading on 210 volt $= 730$ kWh

Reading on 230 volt $= 730 \times 230/210 =$ **800 kWh (approx.)**

Average current $= 800{,}000/230 \times 200 =$ **17.4 A**

Example 10.28. *In a test run of 30 min. duration with a constant current of 5 A, mercury-motor amp-hour meter was found to register 0.51 kWh. If the meter is to be used in a 200-V circuit, find its error and state whether it is running fast or slow. How can the instrument be adjusted to read correctly ?*

(Elect. Meas Inst. and Meas., Jadavpur Univ. 1985)

Solution. Ah passed in 30 minutes $= 5 \times 1/2 = 2.5$

Assumed voltage $= 0.51 \times 1000/2.5 = 204$ V

When used on 200-V supply, it would obviously show higher values because actual voltage is less than the assumed voltage. It would be fast by $4 \times 100/200 =$ **2%**.

Example 10.29. *An amp-hour meter is calibrated to read kWh on a 220-V supply. In one part of the gear train from the rotor to the first counting dial, there is a pinion driving a 75-tooth wheel. Calculate the number of teeth on a wheel which is required to replace 75-tooth wheel, in order to render the meter suitable for operation on 250-V supply.*

Solution. An amp-hour meter, which is calibrated on 220-V supply would run fast when operated on 250-V supply in the ratio 250/220 or 25/22. Hence, to neutralize the effect of increased voltage, the number of teeth in the wheel should be reduced by the same ratio.

∴ Teeth on the new wheel = $75 \times 22/25 =$ **66**

Example 10.30. *A meter, whose constant is 600 revolutions per kWh, makes five revolutions in 20 seconds. Calculate the load in kW.*

(Elect. Meas. and Meas. Inst. Gujarat Univ. 1989)

Solution. Time taken to make 600 revolutions is $600 \times 20/5 = 2{,}400$ second

During this time, the load consumes 1 kWh of energy. If W is load in kW, then

$$W \times (2400/60 \times 60) = 1 \quad \text{or} \quad W = \mathbf{1.5\ kW}$$

Example 10.31. *A current of 6 A flows for 20 minutes through a 220-V ampere hour meter. If during a test the initial and final readings on the meter are 3.53 and 4.00 kWh respectively, calculate the meter error as a percentage of the meter reading.*

If during the test, the spindle makes 480 revolutions, calculate the testing constant in coulomb/rev and rev/kWh.

Solution. Energy actually consumed $= 6 \times \left(\dfrac{20}{60}\right) \times \dfrac{220}{1000} = 0.44$ kWh

Energy as registered by meter $= 4.00 - 3.53 = 0.47$ kWh

Error $= 0.47 - 0.44 = 0.03$ kWh ; % error $= 0.03 \times 100/0.47 = 6.38\%$

No. of coulombs passed through in 20 minutes $= 6 \times 20 \times 60 = 7{,}200$ coulomb

Testing constant = 7,200/480 = **15 C/rev.**

Since for 480 revolutions, only 0.44 kWh are consumed, hence testing constant = 480/0.44 = **1091 rev/kWh.**

Example 10.32. *A 230-V, single-phase domestic energy meter has a constant load of 4A passing through it for 6 hours at unity power factor. If the meter disc makes 2208 revolutions during this period, what is the meter constant in rev. per kWh ? Calculate the power factor of the load, if the No. of rev. made by the meter are 1472 when operating at 230 V and 5 V for 4 hours.*

(Elect. Measur. A.M.I.E. Sec B, 1991)

Solution. Energy consumption in 6 hr $= 230 \times 4 \times 1 \times 6 = 5520$ W = 5.52 kW

Meter constant = 2208/5.52 = 400 rev/kWh.

Now, 1472 revolutions represent energy consumption of 1472/400 = 3.68 kWh

$\therefore VI \cos\phi \times \text{hours} = 3.68 \times 10^{-3}$ or $230 \times 5 \times \cos\phi \times 4 = 3680$, $\therefore \cos\phi =$ **0.8**

Example 10.33. *A 230 V, single-phase domestic energymeter has a constant load of 4 A passing through it for 6 hours at unity power factor. If the motor disc makes 2208 revolutions during this period, what is the meter constant in rev per kWh ? Calculate the power factor of the load if the No. of rev made by the meter are 1472 when operating at 230 V and 5 A for 4 hours.*

(Elect. Measuring. AMIE Sec. Winter 1991)

Solution. Energy supplied at unity p.f. $= 230 \times 4 \times 6 \times 1/1000 = 5.52$

∴ meter constant = 2208/5.52 = 400 rev/kWh

Energy consumed during 1472 meter revolutions = 1472/400 = **3.68 kWh.**

$\therefore 230 \times 5 \times 4 \times \cos\phi/1000 = 3.68 \quad \therefore \cos\phi =$ **0.8.**

Example 10.34. *The testing constant of a supply meter of the amp-hour type is given as 60 coulomb/revolution. It is found that with a steady current of 50 A, the spindle makes 153 revolutions in 3 minutes. Calculate the factor by which dial indications of the meter must be multiplied to give the consumption.* **(City and Guilds, London)**

Solution. Coulombs supplied in 3 min. $= 50 \times 3 \times 60$

At the rate of 60 C/rev., the correct number of revolutions should have been

$$= 50 \times 3 \times 60/60 = 150$$

Registered No. of revolutions. $= 153$

Obviously, the meter is fast. The registered readings should be multiplied by 150/153 = **0.9804** for correction.

Example 10.35. *A single phase kWhr meter makes 500 revolutions per kWh. It is found on testing as making 40 revolutions in 58.1 seconds at 5 kW full load. Find out the percentage error.* **(Elect. Measurements & Measuring Instruments Nagpur Univ. 1993)**

Solution. The number of revolutions the meter will make in one hour on testing = 40 × 3600/58.1 = 2478.5

These revolutions correspond to an energy of 5 × 1 = 5 kWh.

∴ No. of revolutions/kWh = 2478.5/5 = 495.7

Percentage error = (500 – 495.7) × 100/500 = **0.86%**

Example 10.36. *An energy meter is designed to make 100 revolutions of the disc for one unit of energy. Calculate the number of revolutions made by it when connected to a load carrying 40 A at 230-V and 0.4 power factor for an hour. If it actually makes 360 revolutions, find the percentage error.* **(Elect. Engg - I Nagpur Univ. 1993)**

Solution. Energy consumed in one hour = 230 × 40 × 0.4 × 1/1000 = 3.68 kWh

No. of revolutions the meter should make if it is correct = 3.68 × 100 = 368

No. of revolutions actually made = 360

∴ percentage error = (368–360) × 100/368 = **2.17%**

Example 10.37. *The constant of a 25-ampere, 220-V meter is 500 rev/kWh. During a test at full load of 4,400 watt, the disc makes 50 revolutions in 83 seconds. Calculate the meter error.*

Solution. In one hour, at full-load the meter should make (4400 × 1) × 500/1000 = 2200 revolutions. This corresponds to a speed of 2200/60 = 36.7 r.p.m.

Correct time for 50 rev. = (50 × 60)/36.7 = 81.7 s

Hence, meter is slow by 83 – 81.7 = 1.3 s

∴ Percentage error = 1.3 × 100/81.7 = **1.59%**

Example 10.38. *A 16-A amp-hour meter with a dial marked in kWh, has an error of +2.5% when used on 250-V circuit. Find the percentage error in the registration of the meter if it is connected for an hour in series with a load taking 3.2 kW at 200 V.*

Solution. In one hour, the reading given by the meter is = 16 × 250/1000 = 4 kWh

Correct reading = 4 + 2.5% of 4 = 4.1 kWh

Meter current on a 3.2 kW load at 200 V = 3200/200 = 16 A

Since on the given load, meter current is the same as the normal current of the meter, hence in one hour it would give a *corrected* reading of 4.1 kWh.

But actual load is 3.2 kWh. ∴ Error = 4.1–3.2 = 0.9 kWh

% error = 0.9×100/3.2 = **28.12%**

Example 10.39. *The disc of an energy meter makes 600 revolutions per unit of energy. When a 1000 watt load is connected, the disc rotates at 10-2 r.p.s. If the load is on for 12 hours, how many units are recorded as error ?* **(Measurs Instru. Allahabad Univ, 1992)**

Solution. Since load power is one kW, energy actually consumed is

= 1 × 12 = 12 kWh

Total number of revolutions made by the disc during that period of 12 hours

= 10.2 × 60 × 12 = 7,344

since 600 revolutions record one kWh, energy recorded by the meter is

= 7,344/600 = 12,24 kWh

Hence, **0.24** unit is recorded extra.

Example 10.40. *A. d.c. ampere-hour meter is rated at 5-A, 250-V. The declared constant is 5 A-s/rev. Express this constant in rev/kWh. Also calculate the full-load speed of the meter.* **(Elect. Meas Inst. and Meas., Jadavpur Univ. 1987)**

Solution. Meter constant = 5 A-s/rev

Now, 1 kWh = 10^3 Wh = 10^3 × 3600 volt-second

= 36 × 10^5 volt × amp × second

Hence, on a 250-V circuit, this corresponds to 36 × 10^5/250 = 14,400 A-s

Since for every 5 A-s, there is one revolution, the number of revolutions in one kWh is

= 14,400/5 = 2,880 revolutions

∴ Meter constant = **2,880 rev/kWh**

Since full-load meter current is 5 A and its constant is 5 A-s/rev, it is obvious that it makes one revolution every second.

$\therefore$ full-load speed = **60 r.p.m**

Example 10.41. *The declared constant of a 5-A, 200-V amp-hour meter is 5 coulomb per revolution. Express the constant in rev/kWh and calculate the full-load speed of the meter.*

In a test run at half load, the meter disc completed 60 revolutions in 119.5 seconds. Calculate the meter error.

Solution. Meter constant = 5 C/rev. or 5 A-s/rev.

1 k Wh = 1000 Wh = 1000 × 3600 watt-second
= 1000 × 3600 volt × amp × second

Hence, on a 200-V circuit, this corresponds to
= 100 × 3600/200 = 18,000 A-s

Since for every 5 A-s, there is one revolution, hence number of revolutions in one kWh = 18,000/5 = 3600 revolutions. $\therefore$ Meter constant = **3600 rev/kWh**

Since full-load meter current is 5 A and its constant 5 A-s/rev, it is obvious that it makes one revolution every one second.

$\therefore$ its full-load speed = **60 r.p.m.**

At half-load

Quantity passed in 60 revolutions = 119.5 × 2.5 A-s or 298.75 C

Correct No. of revolutions = 298.75/5 = 59.75

Obviously the meter is running fast because instead of making 59.75 revolutions, it is making 60 revolutions.

Error = 60 − 59.75 = 0.25 $\therefore$ % error = 0.25×100/59.75 = **0.418%**

Example 10.42. *A single-phase energy meter of the induction type is rated 230-V; 10-A, 50-Hz and has a meter constant of 600 rev/kWh when correctly adjusted. If 'quadrature' adjustment is slightly disturbed so that the lag is 85°, calculate the percentage error at full-load 0.8 p.f. lag*

(Measu & Instru. Nagpur Univ 1991)

Solution. As seen from Art. 10.50, the driving torque T_d depends, among other factors, on sin α where α is the angle between the two alternating fluxes. $\therefore$ $T_d \propto \sin\alpha$

If the voltage flux-lagging adjustment is disturbed so that the phase angle between the voltage flux and the voltage is less than 90° (instead of being exactly 90°) the error is introduced.

Now $\cos\phi = 0.8, \quad \phi = \cos^{-1}(0.8) = 36°52'$

$\therefore \quad \alpha = 85° - 36°52' = 48°8'$ whereas it should be $= 90 - 36°52' = 53°8'$

$$\therefore \text{ error} = \frac{\sin 53°8' - \sin 48°8'}{\sin 53°8'} \times 100 = 7\%$$

10.52. Ballistic Galvanometer

It is used principally for measuring small electric charges such as those obtained in magnetic flux measurements. Constructionally, it is similar to a moving-coil galvanometer except that (*i*) it has extremely small electromagnetic damping and (*ii*) has long period of undamped oscillations (several seconds). These conditions are necessary if the galvanometer is to measure electric charge. In fact, the moment of inertia of the coil is made so large that whole of the charge passes through the galvanometer before its coil has had time to move sufficiently. In that case, the first swing of the coil is proportional to the charge passing through the galvanometer. After this swing has been observed, the oscillating coil may be rapidly brought to rest by using eddy-current damping.

As explained above, the coil moves after the charge to be measured has passed through it. Obviously, during the movement of the coil, there is no current flowing through it. Hence, the equation of its motion is

$$J\frac{d^2\theta}{dt^2} + D\frac{d\theta}{dt} + C\theta = 0$$

where J is the moment of inertia, D is damping constant and C is restoring constant.

Since damping is extremely small, the approximate solution of the above equation is

$$\theta = Ue^{-(D/2J)t}\sin(\omega_0 t + \phi)$$

At the start of motion, when $t = 0$, $\theta = 0$, hence $\phi = 0$ $\therefore$ $\theta = Ue^{-(D/2J)t}\sin\omega_0 t$ ...(i)

During the passage of charge, at any instant, there will be a deflecting torque of Gi acting on the coil. If t is the time taken by the whole charge to pass through, the torque impulse due to this charge is

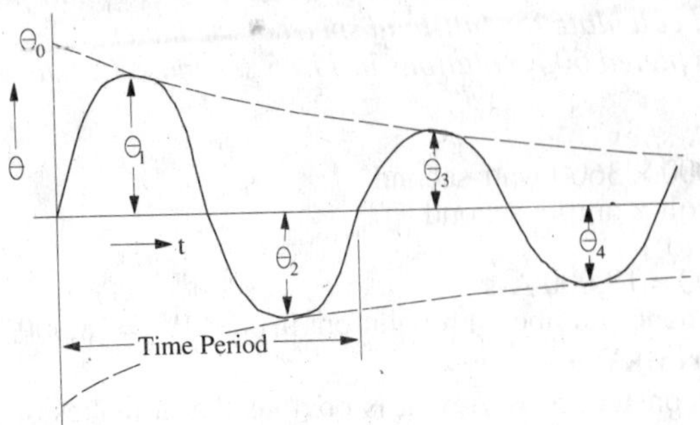

Fig. 10.66.

$\int_0^t Gi\,dt$. Now $\int_0^t i\,dt = Q$

Hence, torque impulse $= GQ$. This must be equal to the change of angular momentum produced *i.e.* $J\alpha$ where α is the angular velocity of the coil at the end of the impulse period.

$\therefore$ $GQ = J\alpha$

or $\alpha = GQ/J$

Differentiating Eq. (i) above, we get

$$\frac{d\theta}{dt} = U\left(e^{-(D/2J)t}\omega_0\cos\omega_0 t - \frac{D}{2J}e^{-(D/2J)t}\sin\omega_0 t\right)$$

Since duration of the passage of charge is very small, at the end of the passage, $t \simeq 0$, so that from above, $d\theta/dt = U\omega_0$.

However, at this time, $\frac{d\theta}{dt} = \alpha = \frac{GQ}{J}$ $\therefore$ $\frac{GQ}{J} = U\,\omega_0$ or $Q = \frac{J\omega_0}{G}U$

Now, U being the amplitude which the oscillations would have if the damping were zero, it may be called undamped swing θ_0.

$$\therefore \quad Q = \frac{J\omega_0}{G}.\theta_0 \quad \text{or} \quad Q \propto \theta_0 \qquad ...(ii)$$

However, in practice, due to the presence of small amount of damping, the successive oscillations diminish exponentially (Fig. 10.66). Even the first swing θ_1 is much less than θ_0. Hence, it becomes necessary to obtain the value of θ_0 from the observed value of first maximum swing θ_1.

As seen from Fig. 10.66, the successive peak values θ_1, θ_2, θ_3 etc. are π radian apart or π/ω_0 second apart. The ratio of the amplitudes of any two successive peaks is

$$= \frac{\theta_0 e^{-(D/2J)t}\sin\omega_0 t}{\theta_0\, e^{-(D/2J)t+(\pi/\omega_0)}\sin(\omega_0 t+\pi)} = \frac{e^{-(D/2J)t}\sin\omega_0 t}{e^{-(D/2J)t}\,e^{-(D/2J)(\pi/\omega_0)}(-\sin\omega_0 t)} = -e^{(D/2J)(\pi/\omega_0)}$$

$$\therefore \quad \frac{\theta_1}{\theta_2} = \frac{\theta_2}{\theta_3} = \frac{\theta_3}{\theta_4} = \text{.........} = e^{(D/2J)(\pi/\omega_0)}$$

Let $e^{-(D/2J)(\pi/\omega_0)} = \Delta^2$ where Δ is called the *damping factor*.*

The time period of oscillation $T_0 = 2\pi/\omega_0$. If damping is very small $\theta_0 = \theta_1$, $t = T_0/4 = \pi/2\omega_0$ as a very close approximation.

Hence, from Eq. (i) above, putting $t = \pi/2\omega_0$, we have

$$\theta = \theta_0^{-(D/2J)(\pi/2\omega_0)}\sin\pi/2 = \theta_0\Delta^{-1} \quad \therefore \quad \theta_0 = \Delta\,\theta_1 \qquad ...(iii)$$

---

* Now, $\frac{\theta_1}{\theta_2} = \frac{\theta_2}{\theta_3} = \frac{\theta_{n-1}}{\theta_n} = \Delta^2$ $\therefore$ $\frac{\theta_1}{\theta_2}\times\frac{\theta_2}{\theta_3}\times\frac{\theta_3}{\theta_4}\times\ldots\times\frac{\theta_{n-1}}{\theta_n} = (\Delta^2)^{n-1}$ $\therefore$ $\frac{\theta_1}{\theta_n} = (\Delta^2)^{n-1}$ or $\Delta = \left(\frac{\theta_1}{\theta_n}\right)^{1/2(n-1)}$

Hence, Δ may be obtained by observing the first and nth swing.

or undamped swing = damping factor × 1st swing

Suppose, a steady current of I_S flowing through the galvanometer produces a steady deflection θ_S, then $C\theta_s = GI_S$ or $G = C\theta_S/I_S$

Since damping is small, $\omega_0 = \sqrt{C/J}$ $\therefore$ $C = J\omega_0^2 = J \times \left(\frac{2\pi}{T_0}\right)^2 = 4\pi^2\frac{J^2}{T_0^2}$ $\therefore$ $G = \frac{4\pi^2 J\theta s}{T_0^2 I_S}$

Substituting this value of G in Eq. (*ii*), we get

$$Q = \frac{J\omega_0\theta_0}{4\pi^2 J\theta_S / T_0^2 I_S^2} \text{ or } Q = \frac{T_0}{2\pi}\cdot\frac{I_s}{\theta_s}.\theta_0 = \frac{T_0}{2\pi}\cdot\frac{I_S}{\theta_S}.\Delta.\theta_1 \qquad \text{...(iv)}$$

Alternatively, let quantity (D/2J)(π / ω_0) be called the logarithmic decrement λ. Since, $\Delta^2 = e^\lambda$, we have

$$\Delta = e^{\lambda/2} = 1 + (\lambda/2) + \frac{(\lambda/2)^2}{2!} + \ldots \cong \left(1 + \frac{\lambda}{2}\right) \text{ when } \lambda \text{ is small}^*$$

Hence, from Eq. (iv) above, we have $Q = \frac{T_0}{2\pi}, \frac{I_S}{\theta_S}\left(1+\frac{\lambda}{2}\right)\theta_1$ ...(*v*)

In general, Eq. (*iv*) may be put as $Q = K\theta_1$

Example 10.43. *A ballistic galvanometer has a free period of 10 seconds and gives a steady deflection of 200 divisions with a steady current of 0.1 mA. A charge of 121 μC is instantaneously discharged through the galvanometer giving rise to a first maximum deflection of 100 divisions. Calculate the 'decrement' of the resulting oscillations.*

(Electrical Measurements, Bombay Univ. 1987)

Solution. From Eq. (*v*) of Art 10.52, we have $Q = \frac{T_0}{2\pi}\cdot\frac{I_S}{\theta_S}\left(1+\frac{\lambda}{2}\right)\theta_1$

Here, $Q = 121\mu C = 121 \times 10^{-6} C$; $T_0 = 10s$; $I_s = 0.1$ mA $= 10^{-4}$ A; $\theta_s = 200$; $\theta_1 = 100$

$$\therefore \quad 121 \times 10^{-6} = \frac{10}{2\pi} \times \frac{10^{-4}}{200}\left(1 + \frac{\lambda}{2}\right) \times 100; \quad \therefore \lambda = \mathbf{1.04}$$

10.53. Vibration Galvanometer

Such galvanometers are widely used as null-point detectors in a.c. bridges.

Construction

As shown in Fig. 10.67 (*a*), it consists of a moving coil suspended between the poles of a strong permanent magnet. The natural frequency of oscillation of the coil is very high, this being achieved by the use of a large value of control constant and a moving system of very small inertia. The suspension (which provides control) is either a phosphor-bronze strip or is a bifilar suspension in which case the two suspension wires carry the coil and a small piece of mirror (or in some cases the two suspension wires themselves from the coil).

---

*Since $\Delta^2 = e^\lambda$, taking logs, we have $2 \log_e \Delta = \lambda \log_e e = \lambda$

$$\therefore \quad \frac{\lambda}{2} = \log_e\Delta = \log_e\left(\frac{\theta_1}{\theta_n}\right)^{1/2(n-1)} = \frac{1}{2(n-)}\log_e \theta_1/\theta_n \quad \therefore, \quad \lambda = \frac{1}{(n-1)}\log_e\frac{\theta_1}{\theta_n}$$

As seen, W is the suspension, C is the moving coil and M the mirror on which is cast a beam of light. From mirror M, this beam is deflected on to a scale. When alternating current is passed through

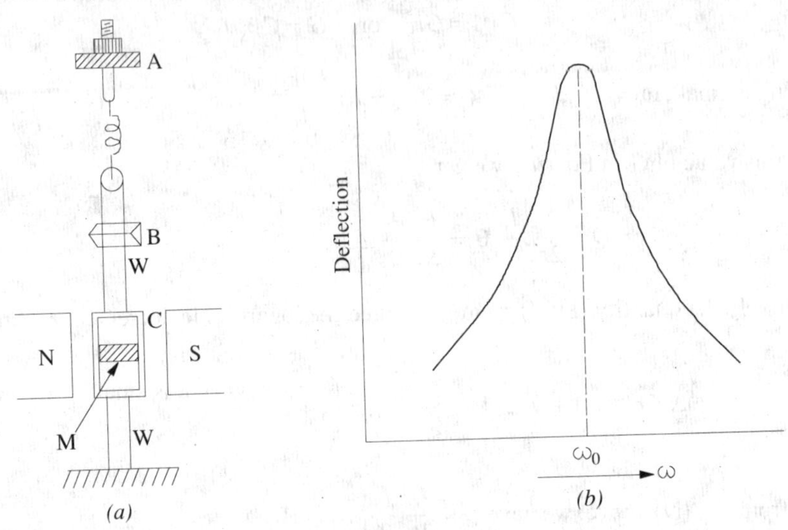

Fig. 10.67

C, an alternating torque is applied to it so that the reflected spot of light on the scale is drawn out in the form of a band of light. The length of this band of light is maximum if the natural frequency of oscillation of C coincides with the supply frequency due to resonance. The tuning of C may be done in the following two ways :

(*i*) by changing the length of suspension W. This is achieved by raising or lowering bridge piece B against which the bifilar loop presses.

(*ii*) by adjusting tension in the suspension. This is achieved by turning the knurled knob A.

By making the damping very small, the resonance curve of the galvanometer can be made sharply-peaked [Fig. 10.67 (*b*)]. In that case, the instrument discriminates sharply against frequencies other than its own natural frequency. In oher words, its deflection becomes very small even when the frequency of the applied current differs by a very small amount from its resonance frequency.

Theory

If the equation of the current passing through the galvanometer is $i = I_m \sin \omega t$, then the equation of motion of the coil is :

$$J\frac{d^2\theta}{dt^2} + D\frac{d\theta}{dt} + C\theta = Gi = GI_m \sin \omega t \qquad \ldots(i)$$

where J, D and C have the usual meaning and G is the deflection constant.

The complementary function of the solution represents the transient motion, which in the case of vibration galvanometers, is of no practical importance. The particular integral is of the form

$$\theta = A \sin (\omega t - \phi)$$

where A and ϕ are constant.

Now, $d\theta/dt = \omega A \cos (\omega t - \phi)$ and $d^2\theta/dt^2 = -\omega^2 A \sin (\omega t - \phi)$. Substituting these values in Eq. (*i*) above, we get

$$-\omega^2 JA \sin (\omega t - \phi) + \omega DA \cos(\omega t - \phi) + CA \sin(\omega t - \phi) = GI_m \sin \omega t$$

It must be true for all values of i

When $\omega t = \phi$ $\qquad DA\omega = G\, I_m \sin \phi \qquad \ldots(ii)$

When $(\omega t - \phi) = \pi/2$ $\qquad -\omega^2 JA + CA = G\, I_m \cos \phi \qquad \ldots(iii)$

Since the phase angle ϕ of oscillations is of no practical significance, it may be eliminated by squaring and adding Eq. (*ii*) and (*iii*).

$$\therefore \quad \omega^2D^2A^2 + A^2(C - \omega^2J)^2 = G^2{I_m}^2 \quad \text{or} \quad A = \frac{G\,I_m}{\sqrt{[D^2\omega^2 + (C - \omega^2J)^2]}} \quad \ldots(iv)$$

This represents the amplitude A of the resulting oscillation for a sinusoidally alternating current of peak value I_m flowing through the moving coil of the galvanometer.

10.54. Vibrating-reed Frequency Meter

1. Working Principle

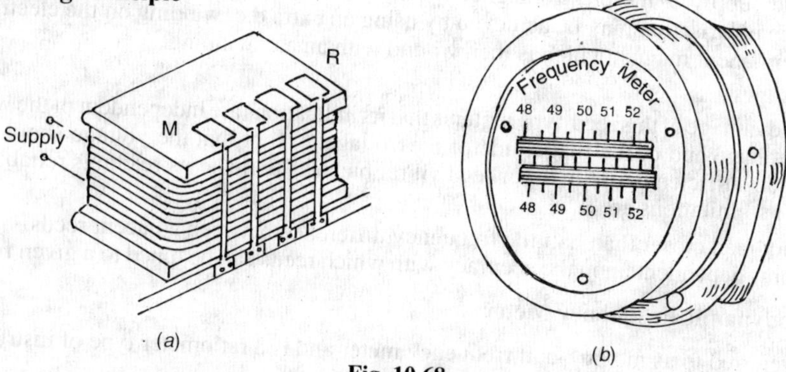

Fig. 10.68

The meter depends for its indication on the mechanical resonance of thin flat steel reeds arranged alongside and, close to, an electromagnet as shown in Fig. 10.68.

2. Construction

The electromagnet has a laminated armature and its winding, in series with a resistance, is connected across a.c. supply whose frequency is required. In that respect, the external connection of this meter is the same as that of a voltmeter.

The metallic reeds (about 4 mm wide and 0.5 mm thick) are arranged in a row and are mounted side by side on a common and slightly flexible base which also carries the armature of the electromagnet. The upper free ends of the reeds are bent over at right angles so as to serve as flags or targets and enamelled white for better visibility. The successive reeds are not exactly similar, their natural frequencies of vibration differing by $\frac{1}{2}$ cycle. The reeds are arranged in ascending order of natural frequency.

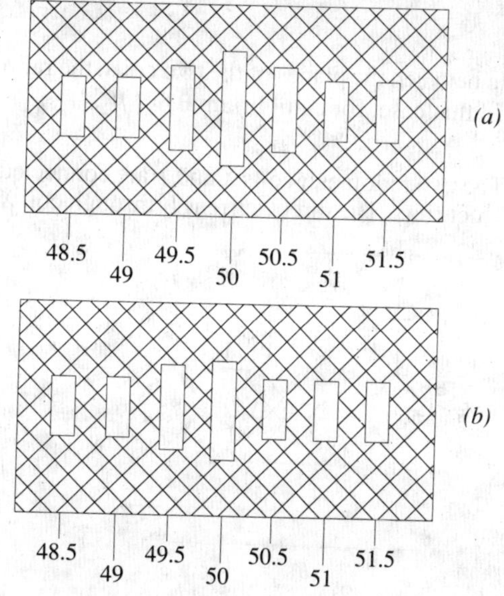

Fig. 10.69

3. Working

When the electromagnet is connected across the supply whose frequency is to be measured, its magnetism alternates with the same frequency. Hence the electromagnet exerts attracting force on each reed once every half cycle. All reeds tend to vibrate but only that whose natural frequency is exactly double the supply frequency vibrates with maximum amplitude due to mechanical resonance [Fig. 10.69 (*a*)]. The supply frequency is read directly by noting the scale mark opposite the white painted flag which is vibrating the most ($f = 50$ *Hz*). The vibrations of other reeds would be so small as to be almost unobservable. For a frequency exactly midway between the natural frequencies of the two reeds ($f = 49.75$ *Hz*), both will vibrate with amplitudes which are equal but much less than when the supply frequency exactly coincides with that of one of the reeds.

4. Range

Such meters have a small range usually from 47 to 53 Hz or from 57 to 63 Hz etc.

The frequency range of a given set of reeds may be doubled by polarising the electromagnet as explained below. As seen from above description, each reed is attracted twice per cycle of the supply *i.e.* once every half-cycle and the reed whose natural frequency is twice that of the current is the one which responds most. Suppose the electromagnet carries an additional winding carrying direct current whose steady flux is equal in magnitude to the alternating flux of the a.c. winding. The resultant flux would be zero in one half-cycle and double in the other half-cycle when the two fluxes reinforce each other so that the reeds would receive one impulse per cycle. Obviously, a reed will indicate the frequency of the supply if the electromagnet is polarised and half the supply frequency if it is unpolarised. The polarisation may be achieved by using an extra d.c. winding on the electromagnet or by using a permanent magnet which is then wound with an a.c. winding.

5. Advantages

One great advantage of this reed-type meter is that its indications are independent of the waveform of the applied voltage and of the magnitude of the voltage, except that the voltage should be high enough to provide sufficient amplitude for reed vibration so as to make its readings reliable.

However, its limitations are :

(*a*) it cannot read closer than half the frequency difference between adjacent reeds.

(*b*) its error is dependent upon the accuracy with which reeds can be tuned to a given frequency.

10.55. Electrodynamic Frequency Meter

It is also referred to as moving-coil frequency meter and is a ratiometer type of instrument.

1. Working Principle

The working principle may be understood from Fig. 10.70 which shows two moving coils rigidly fixed together with their planes at right angles to each other and mounted on the shaft or spindle situated in the field of a permanent magnet. There is no mechanical control torque acting on the two coils. If G_1 and G_2 are displacement constants of the two coils and I_1 and I_2 are the two currents, then their respective torques are $T_1 = G_1 I_1 \cos\theta$ and $T_2 = G_2 I_2 \sin\theta$. These torques act in the opposite directions.

Obviously, T_1 decreases with θ where as T_2 increases but an equilibrium position is possible for same angle θ for which

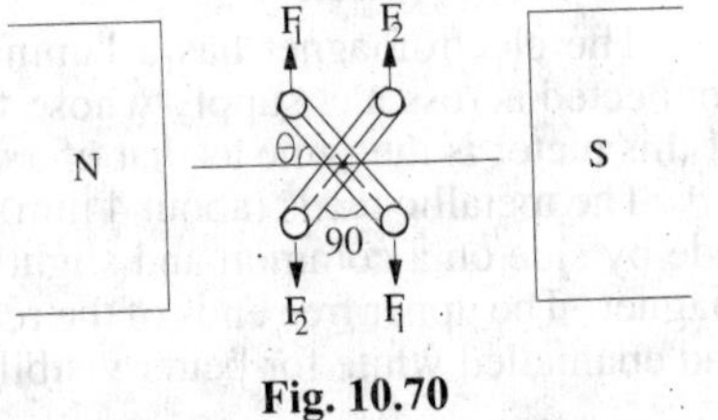

Fig. 10.70

$$G_1 I_1 \cos\theta = G_2 I_2 \sin\theta \quad \text{or} \quad \tan\theta = \frac{G_1}{G_2} \cdot \frac{I_1}{I_2}$$

By modifying the shape of pole faces and the angle between the planes of the two coils, the ratio I_1/I_2 is made proportional to angle θ instead of $\tan\theta$. In that case, for equilibrium $\theta \propto I_1/I_2$.

2. Construction

The circuit connections are shown in Fig. 10.71. The two ratiometer coils X and Y are connected across the supply lines through their respective bridge rectifiers. The direct current I_1 through coil X

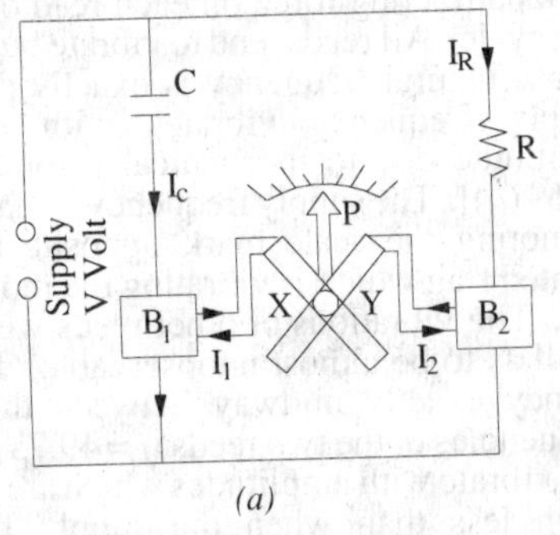

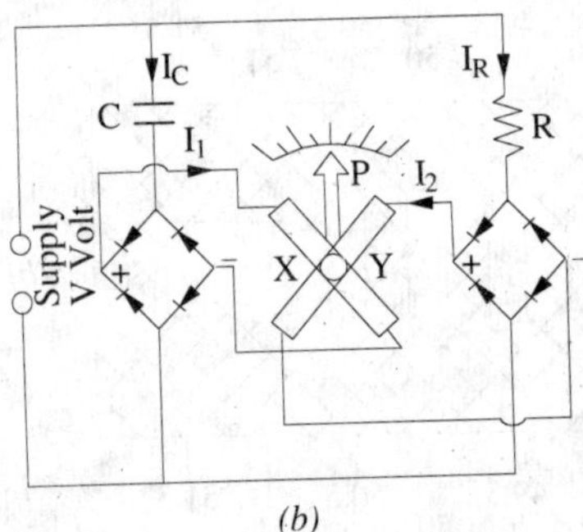

Fig. 10.71

represents the R.M.S. value of capacitor current I_C as rectified by B_1. Similarly, direct current I_2 flowing through Y is the rectified current I_R passing through series resistance R.

3. Working

When the meter is connected across supply lines, rectified currents I_1 and I_2 pass through coils X and Y and they come to rest at an angular position where their torques are equal but opposite. This angular position is dependent on the supply frequency which is read by a pointer attached to the coil.

As proved above, $\theta \propto I_1/I_2$

Assuming sinusoidal waveform, mean values of I_1 and I_2 are proportional to the R.M.S. values of I_C and I_R respectively.

$$\therefore \quad \theta \propto \frac{I_1}{I_2} \propto \frac{I_C}{I_R} \qquad \text{Also } I_C \propto V_m \omega C \text{ and } I_R \propto V_m / R$$

where V_m is the maximum value of the supply voltage whose equation is assumed as $v = V_m \sin \omega t$.

$$\therefore \quad \theta \propto \frac{V_m \omega C}{V_m/R} \propto \omega CR \propto \omega \qquad \therefore \theta \propto f \qquad (\because \quad \omega = 2\pi f)$$

Obviously, such meters have linear frequency scales. Moreover, since their readings are independent of voltage, they can be used over a fairly wide range of voltage although at too low voltages, the distortions introduced by rectifier prevent an accurate indication of frequency.

It will be seen that the range of frequency covered by the meter depends on the values of R and C and these may be chosen to get ranges of 40–60 Hz, 1200–2000 Hz or 8000–12,000 Hz.

10.56. Moving-iron Frequency Meter

1. Working Principle

The action of this meter depends on the variation in current drawn by two parallel circuits–one inductive and the other non-inductive–when the frequency changes.

2. Construction

The construction and internal connections are shown in Fig. 10.72. The two coils A and B are so fixed that their magnetic axes are perpendicular to each other. At their centres is pivoted a long and thin soft-iron needle which aligns itself along the resultant magnetic field of the two coils. There is no control device used in the instrument.

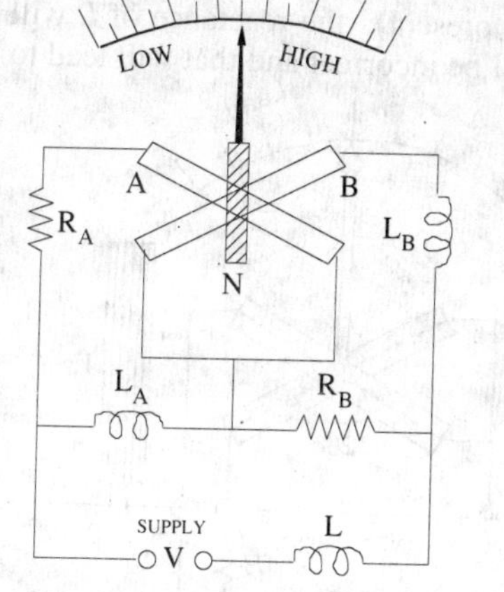

Fig. 10.72

It will be noted that the various circuit elements constitute a Wheatstone bridge which becomes balanced at the supply frequency. Coil A has a resistance R_A in series with it and a reactance L_A in parallel. Similarly, R_B is in series with coil B and L_B is in parallel. The series inductance L helps to suppress higher harmonics in the current waveform and hence, tends to minimize the waveform errors in the indications of the instrument.

3. Working

On connecting the instrument across the supply, currents pass through coils A and B and produce opposing torques. When supply frequency is high, current through coil A is more whereas that through coil B is less due to the increase in the reactance offered by L_B. Hence, magnetic field of coil A is stronger than that of coil B. Consequently, the iron needle lies more nearly to the magnetic axis of coil A than to that of B. For low frequencies, coil B draws more current than coil A and, hence, the needle lies more nearly parallel to the magnetic axis of B than to that of coil A. The variations of frequency are followed by the needle as explained above.

The instrument can be designed to cover a broad or narrow range of frequencies determined by the parameters of the circuit.

10.57. Electrodynamic Power Factor Meter

1. Working Principle

The instrument is based on the dynamometer principle with spring control removed.

2. Construction

As shown in Fig. 10.73 and 10.74, the instrument has a stationary coil which is divided into two sections F_1 and F_2. Being connected in series with the supply line, it carries the load current. Obviously, the uniform field produced by F_1 and F_2 is proportional to the line current. In this field are situated two moving coils C_1 and C_2 rigidly attached to each other and mounted on the same shaft or spindle. The two moving coils are 'voltage' coils but C_1 has a series resistance R whereas C_2 has a series inductance L. The values of R and L as well as turns on C_1 and C_2 are so adjusted that the ampere-turns of C_1 and C_2 are exactly equal. However, I_1 is in phase with the supply voltage V whereas I_2 lags behind V by nearly 90°. As mentioned earlier, there is no control torque acting on C_1 and C_2–the currents being led into them by fine ligaments which exert no control torque.

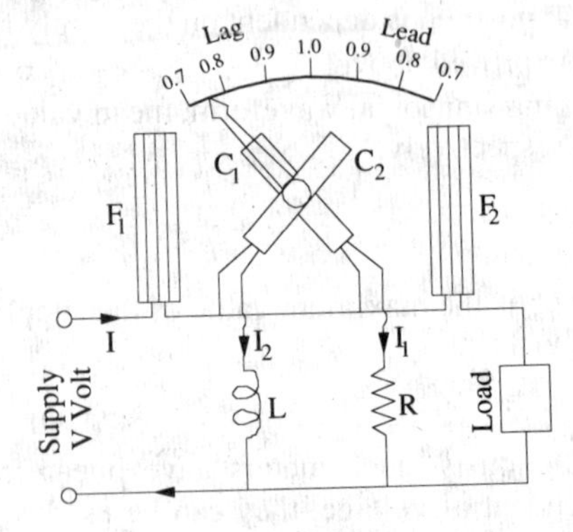

Fig. 10.73

3. Working

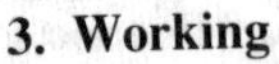

Consider the case when load power factor is unity *i.e.* I is in phase with V. Then, I_1 is in phase with I whereas I_2 lags behind by 90°. Consequently, a torque will act on C_1 which will set its plane perpendicular to the common magnetic axis of coils F_1 and F_2 *i.e.* corresponding to the pointer position of unity p.f. However, there will be no torque acting on coil C_2.

Now, consider the case when load power factor is zero *i.e.* I lags behind V by 90° (like current I_2). In that case, I_2 will be in phase with I whereas I_1 will be 90° out of phase. As a result, there will be no torque on C_1 but that acting on C_2 will bring its plane perpendicular to the common magnetic axis of F_1 and F_2. For intermediate values of power factor, the deflection of the pointer corresponds to the load power factor angle ϕ or to cos ϕ, if the instrument has been calibrated to read the power factor directly.

For reliable readings, the instrument has to be calibrated at the frequency of the supply on which it is to be used. At any other frequency (or when harmonics are present), the reactance of L will change so that the magnitude and phase of current through C_2 will be incorrect and that will lead to serious errors in the instrument readings.

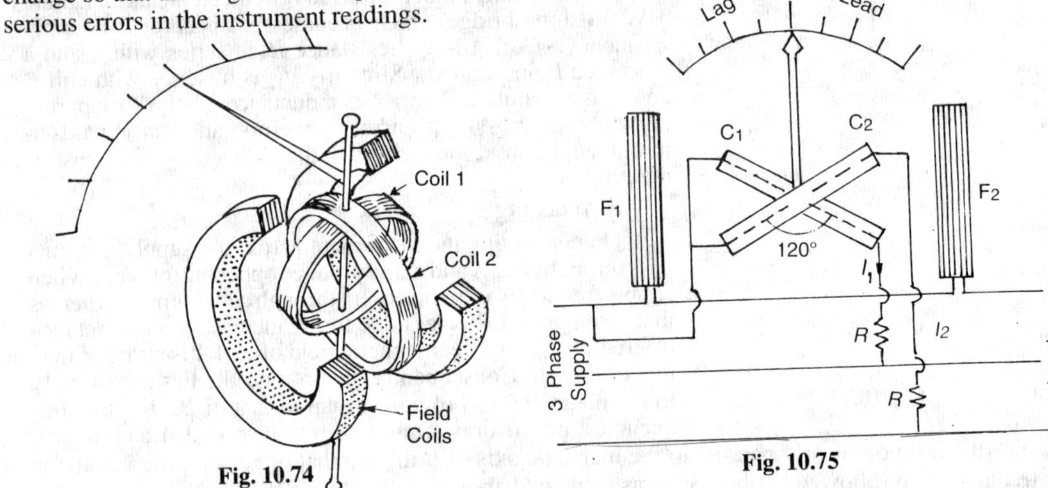

Fig. 10.74

Fig. 10.75

For use on balanced 3-phase load, the instrument is modified, so as to have C_1 and C_2 at 120° to each other, instead of 90°, as in 1-phase supply. As shown in Fig. 10.75, C_1 and C_2 are connected across two different phases of the supply circuit, the stationary coils F_1 and F_2 being connected in series with the third phase (so that it carries the line current). Since there is no need of phase splitting between the currents of C_1 and C_2, I_1 and I_2 are not determined by the phase-splitting circuit and consequently, the instrument is not affected by variations in frequency or waveform.

10.58. Moving-iron Power Factor Meter

1. Construction

One type of power factor meter suitable for 3-phase balanced circuits is shown in Fig. 10.76. It consists of three fixed coils R, Y and B with axes mutually at 120° and intersecting on the centre line of the instrument. These coils are connected respectively in R, Y and B lines of the 3-phase supply through current transformers. When so energised, the three coils produce a synchronously rotating flux.

There is a fixed coil B at the centre of three fixed coils and is connected in series with a high resistance across one of the pair of lines, say, across R and Y lines as shown. Coil B is threaded by the instrument spindle which carries an iron cylinder C (Fig. 10.76 (*b*)] to which are fixed sector-shaped iron vanes V_1 and V_2. The same spindle also carries damping vanes and pointer (not shown in the figure) but *there are no control springs.* The moving system is shown separately in Fig. 10.76 (*b*).

2. Working

The alternating flux produced by coil B interacts with the fluxes produced by the three current coils and causes the moving system to take up a position determined by the power factor angle of the load. However, the instrument is calibrated to read the power factor cos ϕ directly instead of ϕ. In other words, the angular deflection ϕ of the iron vanes from the line $M\,N$ in Fig. 10.76 (*a*) is equal to the phase angle ϕ.

Because of the rotating field produced by coils R, Y and B, there is a slight induction-motor action which tends to continuously turn the moving iron in the direction of the rotating flux. Hence. it becomes essential to so design the moving iron as to make this torque negligibly small *i.e.* by using high-resistance metal for the moving iron in order to reduce eddy currents in it.

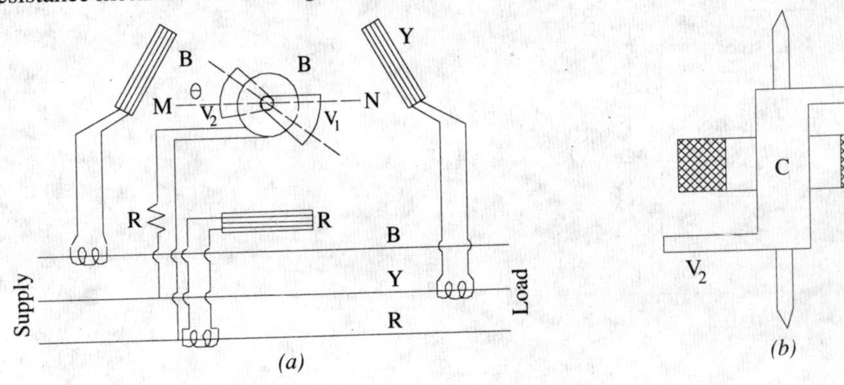

Fig. 10.76

3. Merits and Demerits

Moving-iron p.f. meters are more commonly used as compared to the electrodynamic type because

(*i*) they are robust and comparatively cheap (*ii*) they have scales upto 360° and

(*ii*) in their case, all coils being fixed, there are no electrical connections to the moving parts.

On the other hand, they are not as accurate as the electrodynamic type of instruments and, moreover, suffer from errors introduced by the hysteresis and eddy-current losses in the iron parts–these losses varying with load and frequency.

10.59. Nalder-Lipman Moving-iron Power Factor Meter

1. Construction

The moving system of this instrument (Fig. 10.77) consists of three iron elements similar to the one shown in Fig. 10.76 (*b*). They are all mounted on a common shaft, one above the other, and are separated from one another by non-magnetic distance pieces D_1 and D_2. The three pairs of sectors are displaced in space by 120° relative to each other. Each iron vane is magnetised by one of the three voltage coils B_1, B_2 and B_3 which are connected (in series with a high resistance R) in star across the supply lines. The whole system is free to move in the space between two parallel halves F_1 and F_2 of a single current coil connected in one line of the supply. The common spindle also carries the damping vanes (not shown) and the pointer P.

2. Working

The angular position of the moving system is determined by the phase angle ϕ between the line current and the respective phase voltage. In other words, deflection θ is equal to ϕ, although, in practice, the instrument is calibrated to read the power factor directly.

3. Advantages

(*i*) Since no *rotating* magnetic field is produced, there is no tendency for the moving system to be dragged around continuously in one direction.

(*ii*) This instrument is not much affected by the type of variations of frequency, voltage and waveform as might be expected in an ordinary supply.

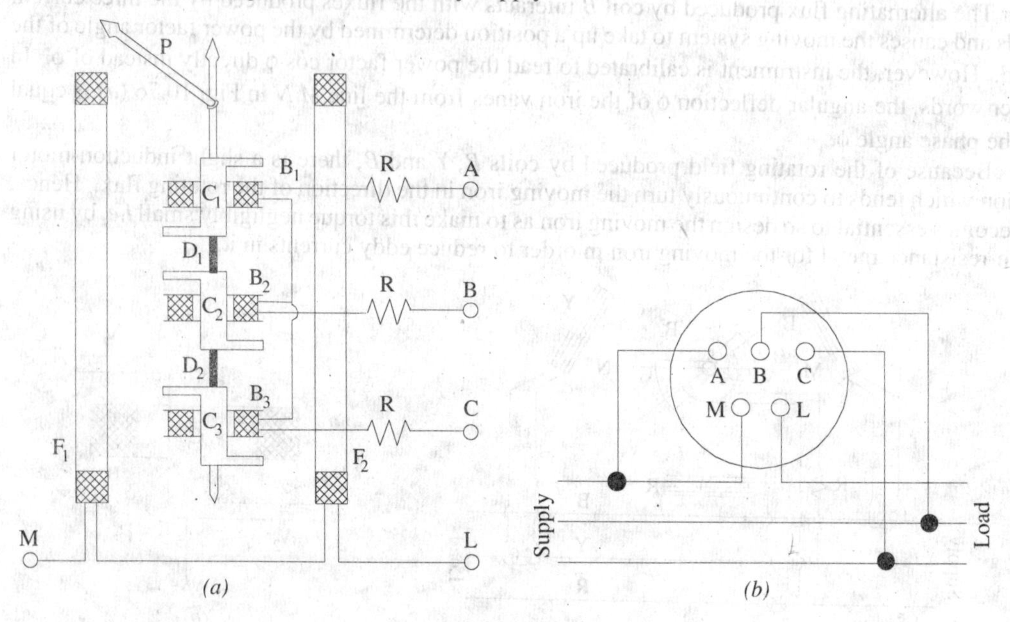

Fig. 10.77

10.60. D.C. Potentiometer

A potentiometer is used for measuring and comparing the e.m.fs. of different cells and for calibrating and standardizing voltmeters, ammeters etc. In its simplest form, it consists of a German silver or manganin wire usually one meter long and stretched between two terminals as shown in Fig. 10.78.

This wire is connected in series with a suitable rheostat and battery B which sends a steady current through the resistance wire AC. As the wire is of uniform cross-section throughout, the fall in potential across it is uniform and the drop between any two points is proportional to the distance between them. As seen, the battery voltage is spread over the rheostat and the resistance wire AC. As we go along AC, there is a progressive fall of potential. If ρ is the resistance/cm of this wire, L its length, then for a current of I amperes, the fall of potential over the whole length of the wire is ρLI volts.

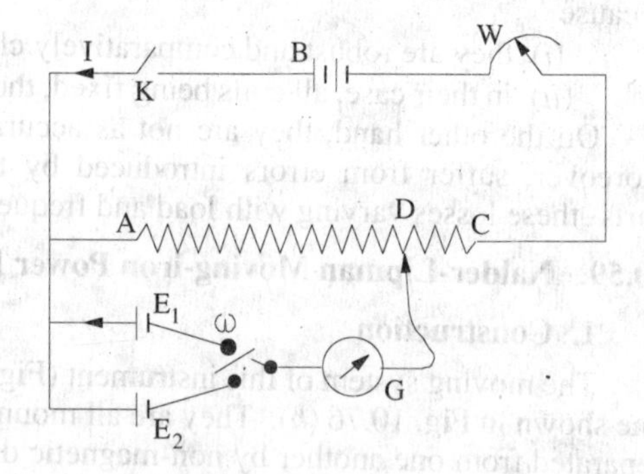

Fig. 10.78

The two cells whose e.m.fs are to be compared are joined as shown in Fig. 10.78, always remembering that *positive terminals of the cells and the battery must be joined together*. The cells can be joined with the galvanometer in turn through a two-way key. The other end of the galvanometer is connected to a movable contact on AC. By this movable contact, a point like D is found when there is no current in and hence no deflection of G. Then, it means that the e.m.f. of the cell just balances the potential fall on AD due to the battery current passing through it.

Suppose that the balance or null point for first cell of e.m.f. E_1 occurs at a length L_1 as measured from point A. Then $E_1 = \rho L_1 I$

Similarly, if the balance point is at L_2 for the other cell, then $E_2 = \rho L_2 I$.

Dividing one equation by the other, we have $\dfrac{E_1}{E_2} = \dfrac{\rho L_1 I}{\rho L_2 I} = \dfrac{L_1}{L_2}$

If one of the cells is a standard cell, the e.m.f. of the other cell can be found.

10.61. Direct-reading Potentiometer

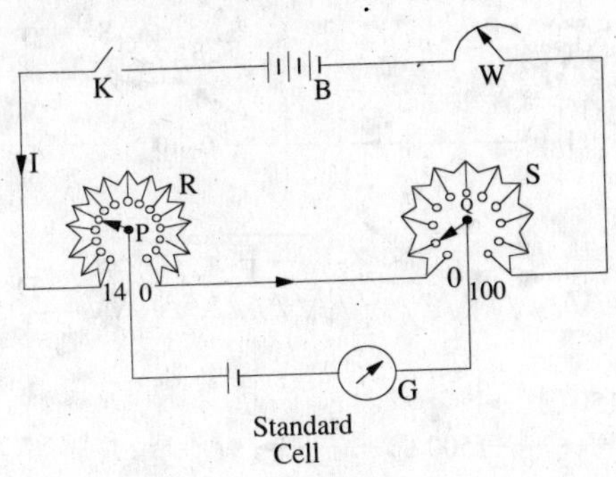

Fig. 10.79

The simple potentiometer described above is used for educational purposes only. But in its commercial form, it is so calibrated that the readings of the potentiometer give the voltage directly, thereby eliminating tedious arithmetical calculations and so saving appreciable time.

Such a direct-reading potentiometer is shown in Fig. 10.79. The resistance R consists of 14 equal resistances joined in series, the resistance of each unit being equal to that of the whole slide were S (which is divided into 100 equal parts). The battery current is controlled by slide wire resistance W.

10.62. Standardizing the Potentiometer

A standard cell *i.e.* Weston cadmium cell of e.m.f. 1.0183 V is connected to sliding contacts P and Q through a sensitive galvanometer G. First, P is put on stud No. 10 and Q on 18.3 division on S and then W is adjusted for zero deflection on G. In that case, potential difference between P and Q is equal to cell voltage *i.e.* 1.0183 V so that potential drop on each resistance of R is 1/10 = 0.1 V and every division of S represents 0.1/100 = 0.001 V. After standardizing this way, the *position of W is not to be changed in any case* otherwise the whole adjustment would go wrong. After this, the instrument becomes direct reading. Suppose in a subsequent experiment, for balance, P is moved to stud No. 7 and Q to 84 division, then voltage would be = $(7 \times 0.1) + (84 \times 0.001) = 0.784$ V.

It should be noted that since most potentiometers have fourteen steps on R, it is usually not possible to measure p.ds. exceeding 1.5 V. For measuring higher voltages, it is necessary to use a volt box.

10.63. Calibration of Ammeters

The ammeter to be calibrated is connected in series with a variable resistance and a standard resistance F, say, of 0.1 Ω across a battery B_1 of ample current capacity as shown in Fig. 10.80 Obviously, the resistance of F should be such that with maximum current flowing through the ammeter A, the potential drop across F should not exceed 1.5 V. Some convenient current, say, 6 amperes (as indicated by A) is passed through the circuit by adjusting the rheostat RH.

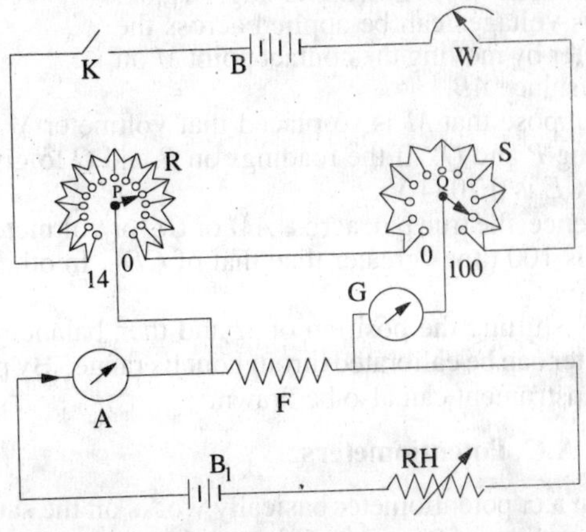

Fig. 10.80

The potential drop across F is applied

between P and Q as shown. Next, the sliding contacts P and Q are adjusted for zero deflection on G. Suppose P reads 5 and Q reads 86.7, Then it means that p.d. across F is 0.5867 V and since F is of 0.1 Ω, hence *true* value of current through F is 0.5867/0.1 = 5.867 amperes. Hence, the ammeter reads high by (6 – 5.867) = 0.133 A. The test is repeated for various values of current over the entire range of the ammeter.

10.64. Calibration of Voltmeters

As pointed out in Art. 10.62, a voltage higher than 1.5 cannot be measured by the potentiometer directly, the limit being set by the standard cell and the type of the potentiometer (since it has only 14 resistances on R as in Fig. 10.79). However, with the help of a volt-box which is nothing else but a voltage reducer, measurements of voltages up to 150 V or 300 V can be made, the upper limit of voltage depending on the design of the volt-box.

The diagram of connections for calibration of voltmeters is shown in Fig. 10.81. By calibration is meant the determination of the extent of error in the reading of the volt-meter throughout its range. A high value resistor AB is connected across the supply terminals of high voltage battery B_1 so that it acts as a voltage divider. The volt-box consists of a high resistance CD with tappings at accurately determined points like E and F etc. The resistance CD is usually 15,000 to 30,000 Ω. The two tappings E and F are such that the resistances of portions CE and CF are 1/100th and I/10th the resistance of CD. Obviously, whatever the potential drop across CD, the corresponding potential drop across CE is 1/100th and that across CF, 1/10th of that across CD.

If supply voltage is 150 V, then p.d. across AB is also 150 V and if M coincides with B, then p.d. across CD is also 150 V, so across CF is 15 volts and across CE is 1.5 V. The p.d. across CE can be balanced over the potentiometer as shown in Fig. 10.81. Various voltages can be applied across the voltmeter by moving the contact point M on the resistance AB.

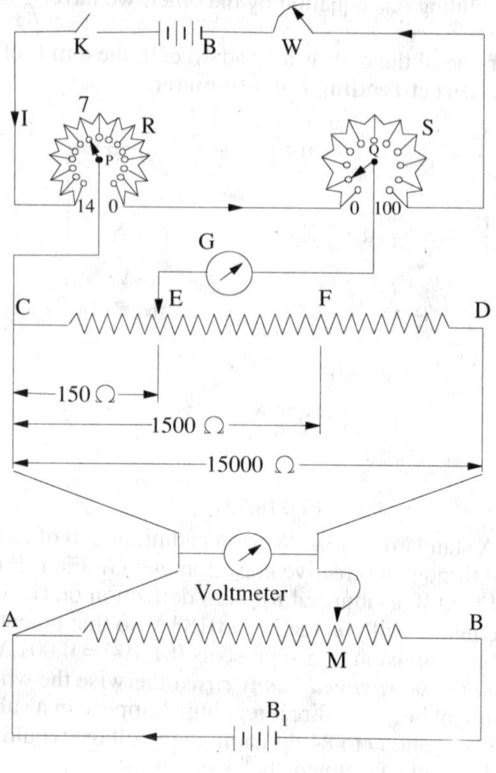

Fig. 10.81

Suppose that M is so placed that voltmeter V reads 70 V and p.d. across CE is balanced by adjusting P and Q. If the readings on P and Q to give balance are 7 and 8.4 respectively, then p.d. across CE is 0.7084 V.

Hence, the true p.d. across AM or CD or voltmeter is 0.7048 × 100 = 70.84 V (because resistance of CD is 100 times greater than that of CE). In other words, the reading of the voltmeter is low by 0.84 V.

By shifting the position of M and then balancing the p.d. across CE on the potentiometer, the voltmeter can be calibrated throughout its range. By plotting the errors on a graph, a calibration curve of the instrument can also be drawn.

10.65. A.C. Potentiometers

An a.c. potentiometer basically works on the same principle as a d.c. potentiometer. However, there is one very important difference between the two. In a d.c. potentiometer, only the *magnitudes* of the unknown e.m.f. and slide-wire voltage drop are made equal for obtaining balance. But in an a.c. potentiometer, not only the magnitudes but *phases* as well have to be equal for obtaining balance. Moreover, to avoid frequency and waveform errors, the a.c. supply for slide-wire must be taken from same source as the voltage or current to be measured.

A.C. potentiometers are of two general types differing in the manner in which the value of the unknown voltage is presented by the instrument dials or scales. The two types are :

(*i*) Polar potentiometers in which the unknown voltage is measured in polar form *i.e.* in terms of magnitude and relative phase.

(*ii*) Co-ordinate potentiometers which measure the rectangular co-ordinates of the voltage under test.

The two procedures are illustrated in Fig. 10.82. In Fig. 10.82 (*a*), vector *OQ* denotes the test voltage whose magnitude and phase are to be imitated. In polar potentiometer, the length *r* of the vector *OP* can be varied with the help of a sliding contact on the slide-wire while its phase ϕ is varied independently with the help of a phase-shifter. Drysdale potentiometer is of this type.

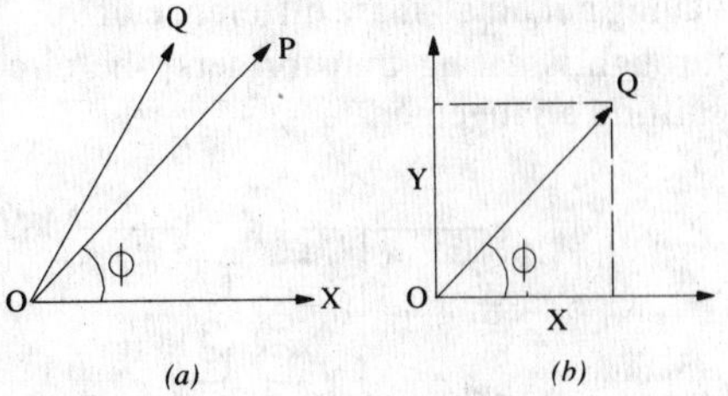

In co-ordinate type potentiometers, the unknown voltage vector *OQ* is copied by the adjustment of 'in phase' and 'quadrature' components *X* and *Y*. Their values are read from two scales of the potentiometer. The magnitude of the required vector is = $\sqrt{X^2 + Y^2}$ and its phase is given by $\phi = \tan^{-1}(X/Y)$. Examples of this type are (*i*) Gall potentiometer and (*ii*) Campbell-Larsen potentiometer.

(*a*) (*b*)

Fig. 10.82

10.66. Drysdale Polar Potentiometer

As shown in Fig. 10.83 for a.c. measurements, the slide-wire *MN* is supplied from a phase shifting circuit so arranged that magnitude of the voltage supplied by it remains constant while its phase

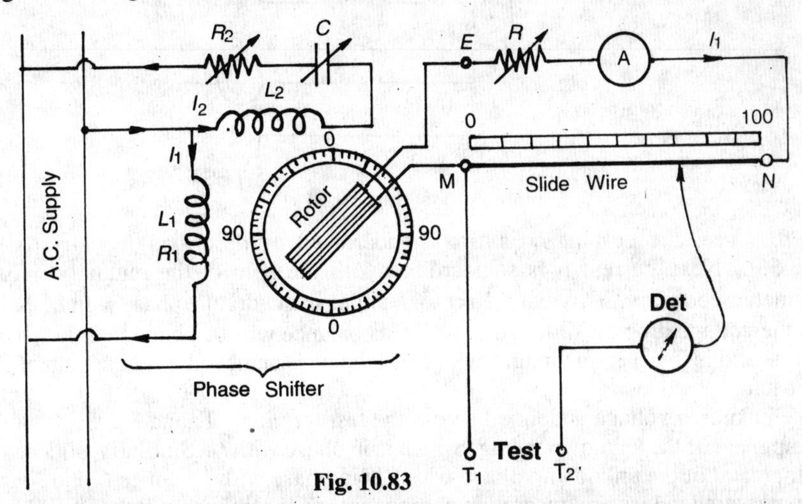

Fig. 10.83

can be varied through 360°. Consequently, slide-wire current *I* can be maintained constant in magnitude but varied in phase. The phase-shifting circuit consists of :

(*i*) Two stator coils supplied from the same source in parallel. Their currents I_1 and I_2 are made to differ by 90° by using well-known phase-splitting technique.

(*ii*) The two windings produce a rotating flux which induces a secondary e.m.f. in the rotor windings which is of constant magnitude but the phase of which can be varied by rotating the rotor in any position either manually or otherwise. The phase of the rotor e.m.f. is read from the circular graduated dial provided for the purpose.

The ammeter *A* in the slide-wire circuit is of electrodynamic or thermal type. Before using it for a.c. measurements, the potentiometer is first calibrated by using d.c. supply for slidewire and a standard cell for test terminals T_1 and T_2.

The unknown alternating voltage to be measured is applied across test terminals T_1 and T_2 and balance is effected by the alternate adjustment of the slide-wire contact and the position of the phase-shifting rotor. The slide-wire reading represents the magnitude of the test voltage and phase-shifter reading gives its phase with reference to an arbitrary reference vector.

10.67. Gall Co-ordinate Potentiometer

This potentiometer uses two slide-wires *CD* and *MN* with their currents I_1 and I_2 (Fig. 10.84) having a mutual phase difference of 90°. The two currents are obtained from the single phase supply through isolating transformers, the circuit for 'quadrature' slidewire *MN* incorporating a phase-shifting arrangement.

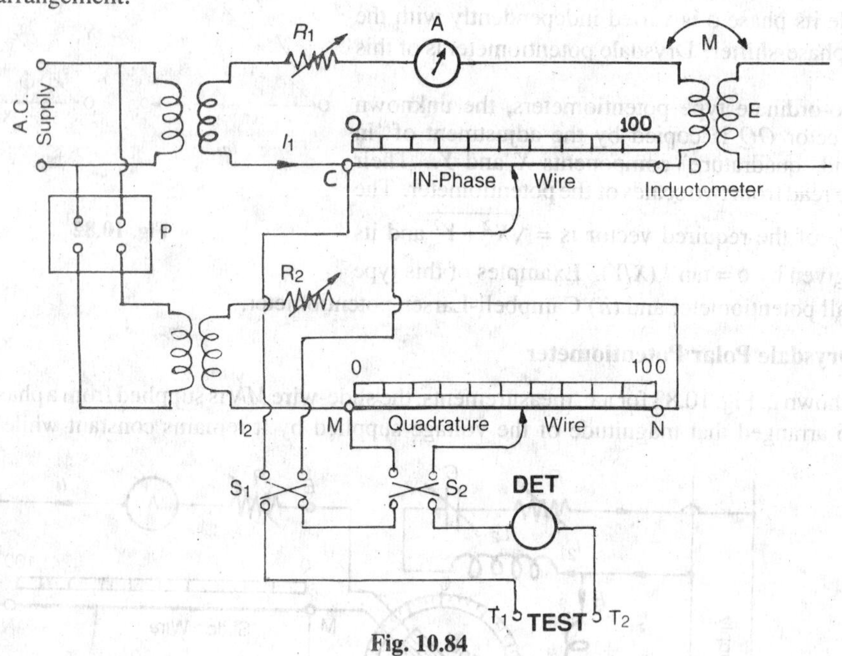

Fig. 10.84

Before use, the current I_1 is first standardised as described for Drysdale potentiometer (Art. 10.66). Next, current I_2 is standardised with the help of the mutually induced e.m.f. *E* in inductometer secondary. This e.m.f. $E = \omega MI_1$ and is in quadrature phase with I_1. Now, *E* is balanced against the voltage drop on slide-wire *MN*. This balance will be obtained only when I_2 is of correct magnitude and is in exact quadrature with I_1. Balance is achieved with the help of the phase-shifter and rheostat R_2.

The unknown voltage is applied across the test terminals T_1 and T_2. Slide-wire *MN* measures that component of the unknown voltage which is in phase with I_2. Similarly, slide-wire *CD* measures that component of the unknown voltage which is in phase with I_1. Since I_1 and I_2 are in quadrature, the two measured values are quadrature components of the unknown voltage. If V_1 and V_2 are these values, then

$$V = \sqrt{V_1^2 + V_2^2} \quad \text{and} \quad \phi = \tan^{-1}(V_2/V_1) \quad \text{—with respect to } I_1$$

Reversing switches S_1 and S_2 are used for measuring both positive and negative in-phase and quadrature components of the unknown voltage.

10.68. Instrument Transformers

In d.c. circuits when large currents are to be measured, it is usual to use low-range ammeters with suitable shunts. For measuring high voltages, low-range voltmeters are used with high resistances connected in series with them. But it is neither convenient nor practical to use this method with alternating current and voltage instruments. For this purpose, specially constructed accurate-ratio instrument transformers are employed in conjunction with standard low-range a.c. instruments. Their purpose is to reduce the line current or supply voltage to a value small enough to be easily measured with meters of moderate size and capacity. In other words, they are used for extending the range of a.c. ammeters and voltmeters. Instrument transformers are of two types :

(*i*) current transformers (*CT*) –for measuring large alternating currents.
(*ii*) potential transformers (*VT*) –for measuring high alternating voltages.

Advantages of using instrument transformers for range extension of a.c. meters are as follows : (1) the instrument is insulated from the line voltage, hence it can be grounded. (2) the cost of the instrument (or meter) together with the instrument transformer is less than that of the instrument alone if it were to be insulated for high voltages. (3) it is possible to achieve standardisation of instruments and meters at secondary ratings of 100-120 volts and 5 or 1 amperes (4) if necessary, several instruments can be operated from a single transformer and 5. power consumed in the measuring circuits is low.

In using instrument transformers for current (or voltage) measurements, we must know the ratio of primary current (or voltage) to the secondary current (or voltage). These ratios give us the multiplying factor for finding the primary values from the instrument readings on the secondary side.

However, for energy or power measurements, it is essential to know not only the transformation ratio but also the phase angle between the primary and secondary currents (or voltages) because it necessitates further correction to the meter reading.

For range extension on a.c. circuits, instrument transformers are more desirable than shunts (for current) and multipliers (for voltage measurements) for the following reasons :

1. time constant of the shunt must closely match the time constant of the instrument. Hence, a different shunt is needed for each instrument.
2. range extension is limited by the current-caying capacity of the shunt *i.e.* upto a few hundred amperes at the most.
3. if current is at high voltage, instrument insulation becomes a very difficult problem.
4. use of multipliers above 1000 becomes almost impracticable.
5. insulation of multipliers against leakage current and reduction of their distributed capacitance becomes not only more difficult but expensive above a few thousand volts.

10.69. Ratio and Phase-angle Errors

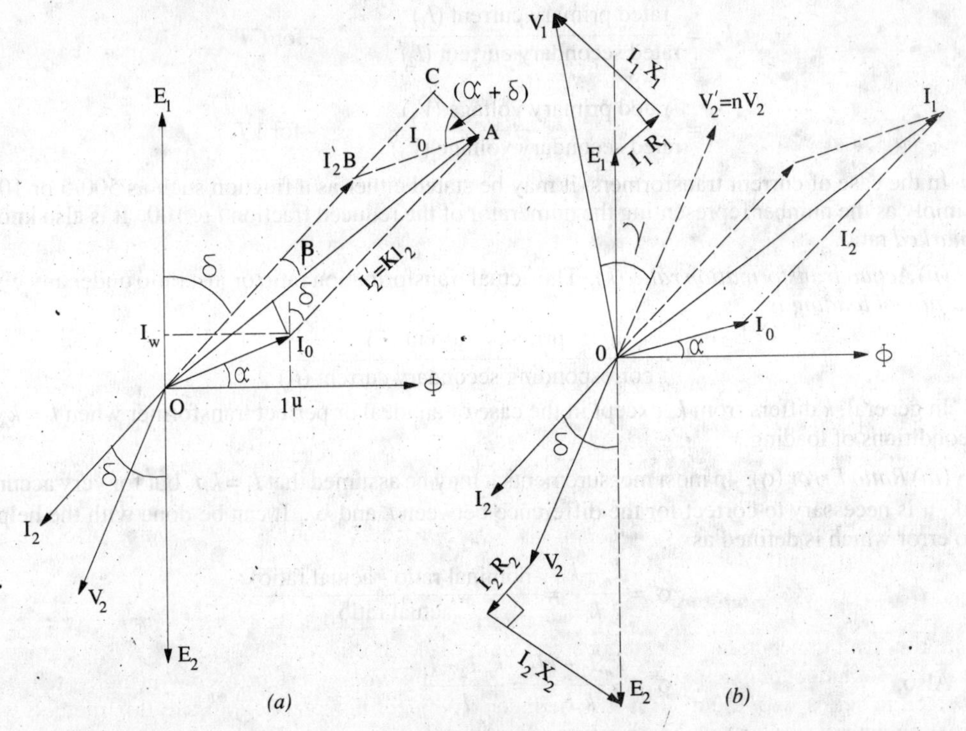

Fig. 10.85

For satisfactory and accurate performance, it is necessary that the ratio of transformation of the instrument transformer should be constant within close limits. However, in practice, it is found that neither current transformation rtio I_1/I_2 (in the case of current transformers) nor voltage transformation ratio V_1/V_2 (in the case of potential transformers) remains constant. The transformation ratio is found to depend on the exciting current as well as the current and the power factor of the secondary circuit. This fact leads to an error called ratio error of the transformer which depends on the working component of primary.

It is seen from Fig. 10.85 (*a*) that the phase angle betwen the primary and secondary currents is not exactly 180° but slightly less than this value. This difference angle β may be found by reversing vector I_2. The angular displacement between I_1 and I_2 reversed is called the phase angle error of the current transformer. This angle is reckoned positive if the reversed secondary current *leads* the primary current. However, on very low power factors, the phase angle may be negative. Similarly, there is an angle of γ between the primary voltage V_1 and secondary voltage reversed-this angle represents the phase angle error of a voltage transformer. In either case, the phase error depends on the magnetising component $I\mu$ of the primary current. It may be noted that ratio error is primarily due to the reason that the *terminal* voltage transformation ratio of a transformer is not exactly equal to its turn ratio. The divergence between the two depends on the resistance and reactance of the transformer windings as well as upon the value of the exciting current of the transformer. Accuracy of voltage ratio is of utmost importance in a voltage transformer althougl phase angle error does not matter if it is to be merely connected *to a voltmeter*. Phase-angle error becomes important only when voltage transformer supplies the voltage coil of a wattmeter *i.e.* in power measurement. In that case, phase angle error causes the wattmeter to indicate on a wrong power factor.

In the case of current transformers, constancy of current ratio is of paramount importance. Again, phase angle error is of no significance if the current transformer is merely feeding an ammeter but it assumes importance when feeding the current coil of a wattmeter. While discussing errors, it is worthwhile to define the following terms :

(*i*) *Nominal transformation ratio* (k_n). It is the ratio of the *rated* primary to the *rated* secondary current (or voltage).

$$k_n = \frac{\text{rated primary current } (I_1)}{\text{rated secondary current } (I_2)} \quad \text{– for } CT$$

$$= \frac{\text{rated primary voltage } (V_1)}{\text{rated secondary voltage } (V_2)} \quad \text{–for } VT$$

In the case of current transformers, it may be stated either as a fraction such as 500/5 or 100/1 or simply as the number representing the numerator of the reduced fraction *i.e.* 100. It is also known as *marked* ratio.

(*ii*) *Actual transformation ratio (k).* The actual transformation ratio or just ratio under any given *condition of loading is*

$$k = \frac{\text{primary current } (I_1)}{\text{corresponding secondary current } (I_2)}$$

In general, k differs from k_n except in the case of an ideal or perfect transformer when $k = k_n$ for all conditions of loading.

(*iii*) *Ratio Error* (σ). In most measurements it may be assumed that $I_1 = k_n I_2$ but for very accurate work, it is necessary to correct for the difference between k and k_m. It can be done with the help of ratio error which is defined as

$$\sigma = \frac{k_n - k}{k} = \frac{\text{nominal ratio - actual ratio}}{\text{actual ratio}}$$

Also, $$\sigma = \frac{k_n.I_2 - kI_2}{k.I_2} = \frac{k_n.I_2 - I_1}{I_1}$$

Accordingly, ratio error may be defined as the difference *between the primary current read (assuming the nominal ratio) and the true primary current divided by the true primary current.*

(*iv*) *Ratio Correction Factor (R.C.F.* It is given by

$$R.C.F. = \frac{\text{actual ratio}}{\text{nominal ratio}} = \frac{k}{k_n}$$

10.70. Current Transformer

A current transformer takes the place of shunt in d.c. measurements and enables heavy alternating current to be measured with the help of a standard 5-A range a.c. ammeter.

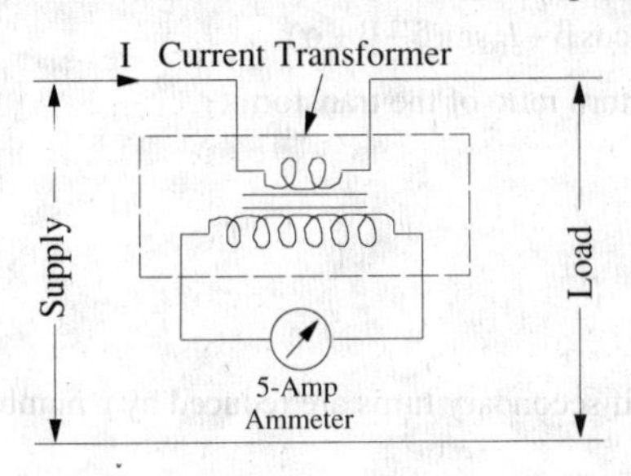

Fig. 10.86

As shown in Fig. 10.86, the current - or series-transformer has a primary winding of one or more turns of thick wire connected in series with the line carrying the current to be measured. The secondary consists of a large number of turns of fine wire and feeds a standard 5-A ammeter (Fig. 10.86) or the current coil of a watt-meter or watthour-meter (Fig. 10.87).

For example, a 1,000/5A current transformer with a single-turn primary will have 200 secondary turns, Obviously, it steps down the current in the 200 : 1 ratio whereas it steps up the voltage drop across the single-turn primary (an extremely small quantity) in the ratio 1 : 200. Hence, if we know the current ratio of the transformer and the reading of the a.c. ammeter, the line current can be calculated.

It is worth noting that ammeter resistance being extremely low, a current transformer operates with its secondary under nearly short-circuit conditions. Should it be necessary to remove the ammeter or the current coils of the wattmeter or a relay, the secondary winding must, first of all, be short-circuited *before* the instrument is disconnected.

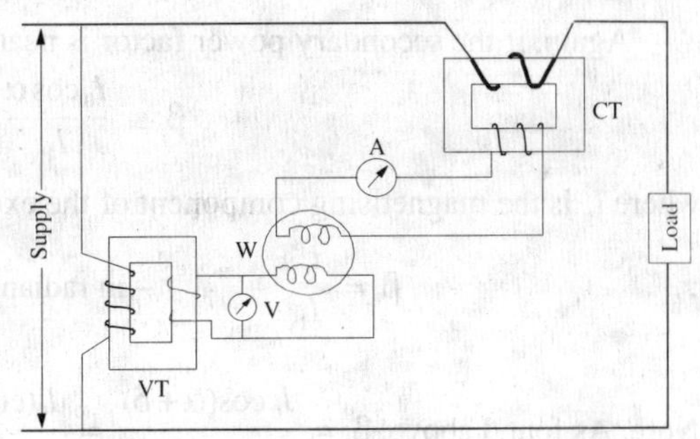

Fig. 10.87

If it is not done, then due to the absence of counter ampere-turns of the secondary, the unopposed primary m.m.f will set up an abnormally high flux in the core which will produce excessive core loss with subsequent heating of and damage to the transformer insulation and a high voltage across the secondary terminals. This is not the case with the ordinary constant-potential transformers because their primary current is determined by the load on their secondary whereas in a current transformer, primary current is determined entirely by the load on the system and not by the load on its own secondary. Hence, the secondary of a current transformer should *never be left open under any circumstances.*

10.71. Theory of Current Transformer

Fig. 10.85 (*b*) represents the general phase diagram for a current transformer. Current I_0 has been exaggerated for clarity.

(*a*) **Ratio Error**. For obtaining an expression for the ratio error, it will be assumed that the turn ratio n (= secondary turns, N_2/primary turns N_1) is made equal to the nominal current ratio *i.e.* $n = k_n$. In other words, it will be assumed that $I_1/I_2 = n$ although actually $n = I_1 / I_2'$. As seen from Art. 10.63.

$$\sigma = \frac{nI_2 - I_1}{I_1} = \frac{I_2' - I_1}{I_1} = \frac{OB - OA}{OA} \qquad [\because\ n = k_n]$$

$$\cong \frac{OB - OC}{OA} \qquad (\because\ \beta \text{ is very small angle})$$

$$= -\frac{BC}{OA} = -\frac{AB\sin(\alpha+\delta)}{OA} = -\frac{I_0\sin(\alpha+\delta)}{I_1} = -\frac{I_0\sin(\alpha+\delta)}{nI_2}$$

For most instrument transformers, the power factor of the secondary burden is nearly unity so that δ is very small. Hence, very approximately.

$$\sigma = -\frac{I_0 \sin \alpha}{I_1} = -\frac{I_\omega}{I_1}$$

where I_ω is the iron-loss or working or wattful component fo the exciting current I_0

Note. The transformation ratio R may be found from Fig. 10.85 (*a*) as under :

$I_1 = OA = OB + BC = nI_2 \cos \beta + I_0 \cos[90 - (\delta + \beta + \alpha)] = nI_2 \cos \beta + I_0 \sin(\delta + \beta + \alpha)$

Now β ≪ $(\alpha + \delta)$hence $I_1 = nI_2 + I_0 \sin(\alpha + \delta)$ where n is the turn ratio of the transformer.

$$\therefore \quad \text{ratio } R = \frac{I_1}{I_2} = \frac{nI_2 + I_0 \sin(\alpha + \delta)}{I_2} \quad \text{or} \quad R = n + \frac{I_0 \sin(\alpha + \delta)}{I_2} \qquad \ldots(i)$$

If δ is negligibly small, then $R = n + \dfrac{I_0 \sin \alpha}{I_2} = n + \dfrac{I_\omega}{I_2}$

It is obvious from (*i*) above that ratio error can be eliminated if secondary turns are reduced by a number

$$= I_0 \sin(\alpha + \delta)/I_2$$

(*b*) **Phase angle** (β)

Again from Fig. 10.85 (*a*), we find that

$$\beta \cong \sin \beta = \frac{AC}{OA} = \frac{AB \cos(\alpha + \delta)}{OA} = \frac{I_0 \cos(\alpha + \delta)}{I_1} = \frac{I_0 \cos(\alpha + \delta)}{nI_2}$$

Again, if the secondary power factor is nearly unity, then δ is very small, hence

$$\beta \cong \frac{I_0 \cos \alpha}{I_1} = \frac{I_\mu}{I_1} \quad \text{or} \quad \frac{I_\mu}{nI_2}$$

where I_μ is the magnetising component of the exciting current I_0.

$$\therefore \quad \beta = \frac{I_\mu}{I_1} - \text{in radian} \; ; \; = \frac{180}{\pi} \times \frac{I_\mu}{I_1} - \text{in degrees}$$

Note. As found above, $\beta = \dfrac{I_0 \cos(\alpha + \delta)}{I_1} = \dfrac{I_0(\cos \alpha \cos \delta - \sin \alpha \sin \delta)}{I_1}$

$$= \frac{I_\mu \cos \delta - I_\omega \sin \delta}{I_1} = \frac{I_\mu \cos \delta - I_\omega \sin \delta}{nI_2} \text{ radian}$$

$$\therefore \quad \beta = \frac{180}{\pi} \times \frac{1_\mu \cos \delta - I_\omega \sin \delta}{nI_2} \text{ degrees.}$$

Dependence of ratio error on working component of I_0 and that of phase angle on the magnetising component is obvious. If R is to come closer to k and β is to become negligibly small, then I_μ and I_ω and hence I_0 should be very small.

10.72. Clip-on Type Current Transformer

It has a laminated core which is so arranged that it can be opened out at a hinged section by merely pressing a trigger-like projection (Fig. 10.88). When the core is thus opened, it permits the admission of very heavy current-carrying bus-bars or feeders whereupon the trigger is released and the core is tightly closed by a spring. The current-carrying conductor or feeder acts as a single-turn primary whereas the secondary is connected across the standard ammeter conveniently mounted in the handle itself.

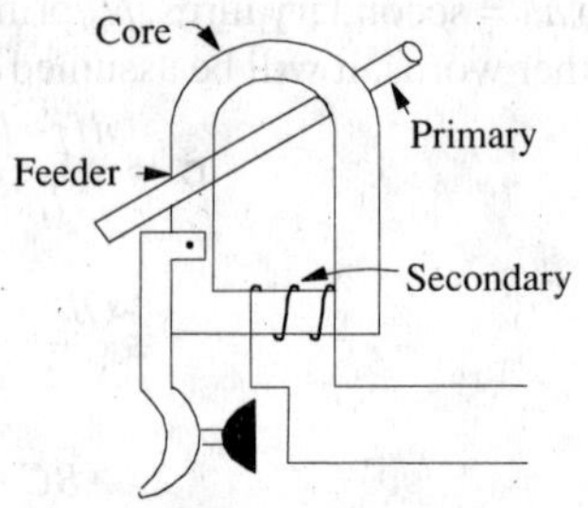

Fig. 10.88

10.73. Potential Transformers

These transformers are extremely accurate-ratio stepdown transformers and are used in conjunction with standard low-range voltmeters (100–120 V) whose deflection when divided by transformation ratio, gives the true voltage on the primary or high-voltage side. In general, they are of the shell type and do not differ much from the ordinary two-winding transformers except that their power rating is extremely small. Since their secondary windings are required to operate instruments or relays or pilot lights, their ratings are usually of 40 to 100 W. For safety, the secondary is completely insulated from the high-voltage primary and is, in addition, grounded for affording protection to the operator. Fig. 10.89 shows the connections of such a transformer.

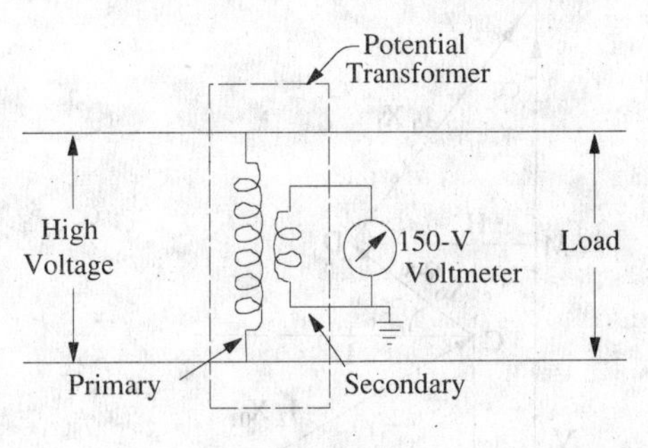

Fig. 10.89

10.74. Ratio and Phase-angle Errors

In the case of a potential transformer, we are interested in the ratio of the primary to the secondary terminal voltage and in the phase angle γ between the primary and reversed secondary terminal voltage V'_2.

The general theory of voltage transformer is the same as for the power transformers except that, as the current in the secondary burden is very small, the total primary current I_1 is not much greater than I_0.

In the phasor diagram of Fig. 10.90, vectors *AB*, *BC*, *CD* and *DE* represent small voltage drops due to resistances and reactances of the transformer winding (they have been exaggerated for the sake of clarity). Since the drops as well as the phase angle γ are small, the top portion of diagram 10.90 (*a*) can be drawn with negligible loss of accuracy as in Fig. 10.90 (*b*) where V'_2 vector has been drawn parallel to the vector for V_1.

In these diagrams, V_2' is the secondary terminal voltage as referrerd to primary assuming transformation without voltage drops. All actual voltage drops have been referred to the primary. Vector *AB* represents total resistive drop as referred to primary *i.e.* $I_2' R_{01}$. Similarly, *BC* represents total reactive drop as referred to primary *i.e.* $I_2' X_{01}$.

In a voltage transformer, the relatively large no-load current produces appreciable resistive drops which have been represented by vectors *CD* and *DE* respectively. Their values are $I_0 R_1$ and $I_0 X_1$ respectively.

(*a*) **Ratio Error**

In the following theory, *n* would be taken to represent the ratio of *primary turns to secondary turns* (Art. 10.69). Further, it would be assumed, as before, that *n* equals the nominal transformation ratio *i.e.* $n = k_n$.

In other words, it would be assumed that $V_1/V_2 = n$, although, actually, $V_1/V_2' = n$.

Then
$$\sigma = \frac{k_n - k}{k} = \frac{k_n.V_2 - kV_2}{kV_2} = \frac{V_2' - V_1}{V_1} = -\frac{EN}{OE} \quad \text{...Fig.10.90}(a)$$

$$= -\frac{AG + FC + LD + EM}{OE} \quad \text{...Fig.10.90}(b)$$

$$= -\frac{I_2'R_{02}\cos\delta + I_2'X_{02}\sin\delta + I_0R_1\sin\alpha + I_0X_1\cos\alpha}{V_1}$$

$$= -\frac{I_2'R_{02}\cos\delta + I_2'X_{02}\sin\delta + I_\mu R_1 + I_\mu X_1}{V_1}$$

where I_ω and I_μ are the iron-loss and magnetising components of the no-load primary current I_0.

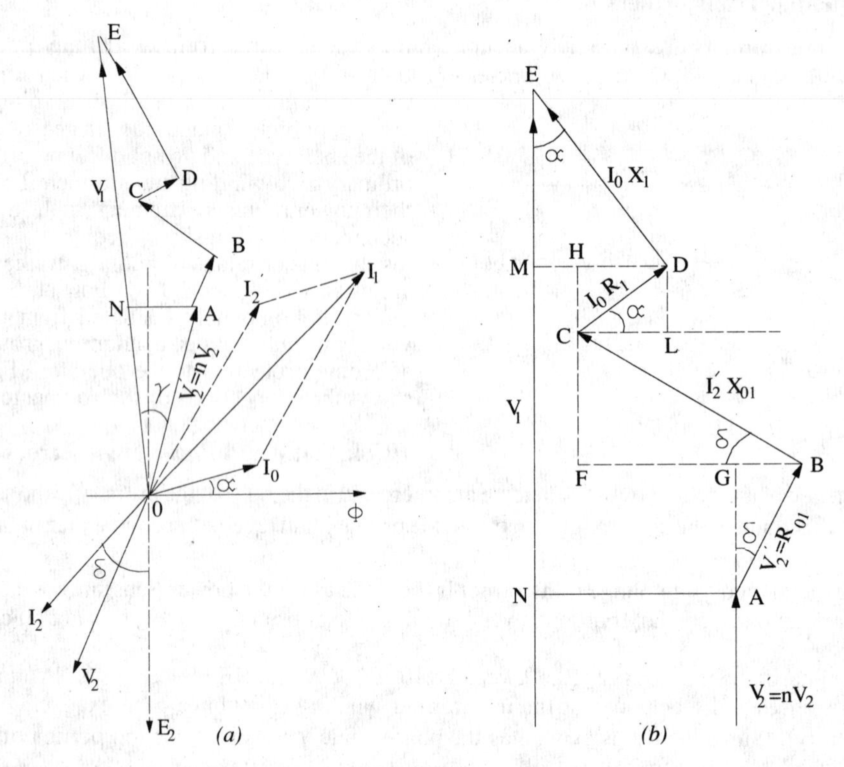

Fig. 10.90

(b) Phase Angle (γ)

To a very close approximation, value of γ is given by $\gamma = AN/OA$ –in radian

Now, $OA \cong OE$ provided ratio error is neglected. In that case,

$$\gamma = \frac{AN}{OE} = -\frac{GF + HM}{OE} \qquad \text{...Fig. 10.90}(b)$$

$$= -\frac{(BE - BG +) + (DM - DH)}{OE}$$

$$= -\frac{I_2'X_{01}\cos\delta - I_2'R_{01}\sin\delta + I_0X_1\sin\alpha - I_0R_1\cos\alpha}{V_1} = -\frac{I_2'X_{01}\cos\delta - I_2'R_{01}\sin\delta + I_\omega X_1 - I_\mu R_1}{V_1}$$

The negative sign has been given because reversed secondary voltage *i.e.* V_2' lags behind V_1.

Example 10.44. *A current transformer with 5 primary turns and a nominal ratio of 1000/5 is operating with a total secondary impedance of 0.4 + j 0.3 Ω. At rated load, the iron loss and magnetising components of no-load primary current are 1.5 A and 6 A respectively. Calculate the ratio error and phase angle at rated primary current if the secondary has (a) 1000 turns and (b) 990 turns.*

Solution. Phasor diagram of Fig. 10.87 may please be referred to.

(*a*) $\tan\delta = 0.3/0.4$ or $\delta = \tan^{-1}(0.3/04) = 36°52'$

$$\alpha = \tan^{-1}(1.5/6) = 14°2' \qquad \therefore \qquad (\alpha + \beta) = 50°54'$$

$$I_0 = \sqrt{I_\omega^2 + I_\mu^2} = \sqrt{1.5^2 + 6^2} = 6.186 \text{ A}$$

$$BC = I_0 \sin(\alpha + \beta) = 6.186 \times \sin 50°54' = 4.8 \text{ A}$$

Since β is small, $OC \cong OA = 1000$ A $\therefore I_2' = OC - BC = 1000 - 4.8 = 995.2$ A

$$\therefore \quad \sigma = \frac{k_n I_2 - I_1}{I_1} = \frac{nI_2 - I_1}{I_1} \text{ because in this case } n (= N_2 / N_1 = 1000/5)$$

is equal to nominal ratio k_n (= 1000/5 A).

$$\text{or} \quad \sigma = \frac{I_2' - I_1}{I_1} = \frac{995.2 - 1000}{1000} = -0.0048 \text{ or } -\mathbf{0.48\%}$$

$$\beta = \tan^{-1}(AC/OC)$$

$$\text{Now } AC = I_0 \cos(\alpha + \beta) = 6.186 \cos 50°54' = 3.9 \text{ A}$$

$$\beta = \tan^{-1}(AC/OC) = \tan^{-1}(3.9/1000) = \mathbf{0°13'}$$

(*b*) In this case, $I_2 = 995.2 \times \frac{5}{990}$; $k_n I_2 = \frac{1000}{5} \times 995.2 \times \frac{5}{990} = 1005.3$ A

$$\therefore \quad \sigma = \frac{k_n I_2 - I_1}{I_1} = \frac{1005.3 - 1000}{1000} = +0.0053 \text{ or } \mathbf{0.53\%}$$

The value of phase angle β would not be significantly different from the value obtained in (*a*) above.

Example 10.45. *A relay current-transformer has a bar primary and 200 secondary turns. The secondary burden is an ammeter of resistance 1.2 Ω and reactance of 0.5 Ω and the secondary winding has a resistance of 0.2 Ω and reactance of 0.3 Ω. The core requires the equivalent of 100 AT for magnetisation and 50 AT for core losses.*

(*i*) *Find the primary current and the ratio error when the secondary ammeter indicates 5.0 A.*

(*ii*) *By how many turns should the secondary winding be reduced to eliminate the ratio error for this condition ?* **(Electrical Measurements, Bombay Univ. 1985)**

Solution. Total secondary impedance is

$$\mathbf{Z}_2 = 1.4 + j0.8 = 1.612 \angle 29°45' \quad \therefore \quad \delta = 29°45'$$

$$\mathbf{I}_0 = 100 + j50 = 111.8 \angle 26°34' \quad \therefore \quad \alpha = 26°34'$$

Turn ratio, $n = 200/I = 200$; Transformation ratio $R = n + \frac{I_0 \sin(\alpha + \delta)}{I_2}$

$$\therefore \quad R = 200 + \frac{111.8 \sin 56°19'}{5} = 218.6$$

(*i*) Primary current $= 5 \times 218.6 = \mathbf{1093\ A}$

$$\text{Ratio error } \sigma = -\frac{I_0 \sin(\alpha + \delta)}{nI_2} = -\frac{111.8 \times 0.8321}{200 \times 5} = -\mathbf{0.093 \text{ or } -9.3\%}$$

(*ii*) No. of secondary turns to be reduced $= I_0 \sin(\alpha + \delta)/I_2 = 93/5 =$ **19 (approx.)**

Example 10.46. *A current transformer has 3 primary turns and 300 secondary turns. The total impedance of the secondary is (0.583 + j 0.25) ohm. The secondary current is 5 A. The ampere-turns required to supply excitation and iron losses are respectively 10 and 5 per volt induced in the secondary.*

Determine the primary current and phase angle of the transformer.

(Elect Meas ; M.S. Univ, Baroda, 1984)

Solution. $\mathbf{Z}_2 = 0.583 + j\,0.25 = 0.6343 \angle 23°\ 10'$ $\therefore E_2 = I_2 Z_2 = 5 \times 0.6343$ $= \mathbf{3.17\ V}$

Now, there are 10 megnetising *AT* per *secondary* volt induced in secondary.

$\therefore$ total magnetising $AT = 3.17 \times 10 = 31.7$; Similarly, iron-loss $AT = 3.17 \times 5 = 15.85$

Remembering that there are 3 primary turns, the magnetising and iron-loss components of primary current are as under :

Magnetising current, $I_\mu = 31.7/3 = 10.6$ A; iron-loss current $I_\omega = 15.85/3 = 5.28$ A

$$I_0 = \sqrt{10.6^2 + 5.28^2} = 11.84 \text{ A}$$

Now,
$$R = n + \frac{I_0 \sin(\propto + \delta)}{I_2}$$

Here,
$$n = 300/3 = 100; \quad \alpha = \tan^{-1}(I_\omega / I_\mu) = \tan^{-1}(5.28/10.6)$$
$$= \tan^{-1}(0.498) = 26°30'$$

$$\delta = \text{secondary load angle } = 23°10' \qquad \text{–found earlier}$$

$\therefore$
$$R = 100 + \frac{11.86}{5}(\sin 49°40') = 100 + 1.81 = 101.81$$

$$I_1 = R \times I_2 = 101.81 \times 5 = \mathbf{509.05\ A}$$

$$\beta = \frac{180}{\pi} \times \frac{I_\mu \cos\delta - I_\omega \sin\delta}{nI_2}$$

$$= \frac{180}{\pi} \times \frac{10.6 \cos 23°\,10' - 5.28 \sin 23°10'}{100 \times 5} = \mathbf{0.88°}$$

Example 10.47. *A current transformer with a bar primary has 300 turns in its secondary winding. The resistance and reactance of the secondary circuit are 1.5 Ω and 1.0 Ω respectively including the transformer winding. With 5 A flowing in the secondary circuit the magnetising ampre-turns required are 100 and the iron loss is 1.2 W. Determine the ratio error at this condition.*

(Elect. Measure, A.M.I.E. Sec. B, 1992)

Solution. Turn ratio $n = 300/1 = 300$

Secondary impedance is $Z_2 = 1.5 + j\,1.0 = 1.8 \angle 33°\,42'$

Secondary induced e.m.f. $E_2 = I_2 Z_2 = 5 \times 1.8 = 9$ V

$E_1 = E_2 / n = 9/300 = 9/300 = 0.03$ V

Let us now find the magnetising and working components of primary no-load current I_0

Magnetising $AT = 100$. Since there is one primary turn, $\therefore \quad I_\mu = 100/1 = 100$A

Now, $\quad E_1 I_\omega = 1.2 \qquad \therefore \qquad I_\omega = 1.2 / 0.03 = 40$ A

$$\mathbf{I}_0 = 100 + j40 = 107.7 \quad \angle 21°48'; \ \sigma = -\frac{I_0 \sin(\alpha + \delta)}{nI_2}$$

Now
$$\alpha = 21°48' \text{ and } \delta = 33°42'$$

$\therefore$
$$\sigma = -\frac{107.7 \sin 55°30'}{300 \times 5} = \mathbf{-\,0.0592\ ro\ -\ 5.92\%}$$

Phase angle
$$\beta = \frac{I_0 \cos(\alpha + \delta)}{nI_2} = \frac{107.7 \times \cos 55°30'}{1500}$$

$$= \frac{180}{\pi} \times \frac{107.7 \times 0.5664}{1500} = \mathbf{2°20'}$$

Tutorial Problems No. 10.4

1. A current transformer with 5 primary turns has a secondary burden consisting of a resistance of 0.16 Ω and an inductive reactance of 0.12 Ω. When primary current is 200 A, the magnetising current is 1.5 A and the iron-loss component is 0.4 A. Determine the number of secondary turns needed to make the current ratio 100/1 and also the phase angle under these conditions. **[407 : 0.275°]**

2. A current transformer having a 1-turn primary is rated at 500/5 A, 50-Hz, with an output of 1.5 VA. At rated load with the non-inductive burden, the in-phase and quadrature components (referred to the flux) of the exciting ampere-turns are 8 and 10 respectively. The number of turns in the secondary is 98 and the resistance and leakage reactance of the secondary winding are 0.35 Ω and 0.3 Ω respectively. Calculate the current ratio and the phase angle error. **[501.95/5; 0.533°]** (*Elect. Inst. and Meas, M.S. Univ. Baroda, 1979*)

3. A ring-core current transformer with a nominal ratio of 500/5 and a bar primary has a secondary resistance of 0.5 Ω and negligible secondary reactance. The resultant of the magnetising and iron-loss components of the primary current associated with a full-load secondary current of 5 A in a burden of 1.0 Ω (non-inductive) is 3 A at a power factor of 0.4. Calculate the true ratio and the phase-angle error of the transformer on full-load. Calculate also the total flux in the core, assuming that frequency is 50 Hz. **[501.2/5 ; 0.314° ; 337 μWb]**

4. A current transformer has a single-turn primary and a 200-turn secondary winding. The secondary supplies a current of 5 A to a non-inductive burden of 1 Ω resistance, the requisite flux is set up in the core by 80 AT. The frequency is 50 Hz and the net cross-section of the core is 10 cm^2. Calculate the ratio and phase angle and the flux density in the core.
[200.64 ; 4° 35′ ; 0.079 Wb/m²] [(*Electrical Measurements, Osmania Univ. 1978*)

5. A potential transformer, ratio 1000/100-V, has the following constants :
primary resistance = 94.5 Ω ; secondary resistance = 0.86 Ω
primary reactance = 66.2 Ω ; equivalent reactance = 66.2 Ω
magnetising current = 0.02 A at 0.4 p.f.
Calculate (*i*) the phase angle at no-load between primary and secondary voltages (*ii*) the load in VA at u.p.f. at which the phase angle would be zero. [(*i*) **0°4′** (*ii*) **18.1 VA]**

OBJECTIVE TESTS—10

1. The kWh meter can be classified as a/an –instrument :
(*a*) deflecting
(*b*) digital
(*c*) recording
(*d*) indicating.

2. The moving system of an indicating type of electrical instrument is subjected to :
(*a*) a deflecting torque
(*b*) a controlling torque
(*c*) a damping torque
(*d*) all of the above.

3. The damping force acts on the moving system of an indicating instrument only when it is :
(*a*) moving
(*b*) stationary
(*c*) near its full deflection
(*d*) just starting to move.

4. The most efficient form of damping employed in electrical instruments is :
(*a*) air friction
(*b*) fluid friction
(*c*) eddy currents
(*d*) none of the above.

5. Moving-iron instruments can be used for measuring :
(*a*) direct currents and voltages
(*b*) alternating currents and voltages
(*c*) radio frequency currents
(*d*) both (*a*) and (*b*).

6. Permanent-magnet moving-coil ammeters have uniform scales because :
(*a*) of eddy current damping
(*b*) they are spring-controlled
(*c*) their deflecting troque varies directly as current
(*d*) both (*b*) and (*c*).

7. The meter that is suitable for *only* direct current measurements is :
(*a*) moving-iron type
(*b*) permanent-magnet type
(*c*) electrodynamic type
(*d*) hot-wire type.

8. A moving coil voltmeter measures——
(*a*) only a.c. voltages
(*b*) only d.c. voltages
(*c*) both a.c. and d.c. voltages.
(Principles of Elect. Engg. Delhi Univ. June 1985)

9. The reading of the voltmeter in Fig. 10.91 would be nearest to—volt :
(*a*) 80
(*b*) 120
(*c*) 200
(*d*) 0

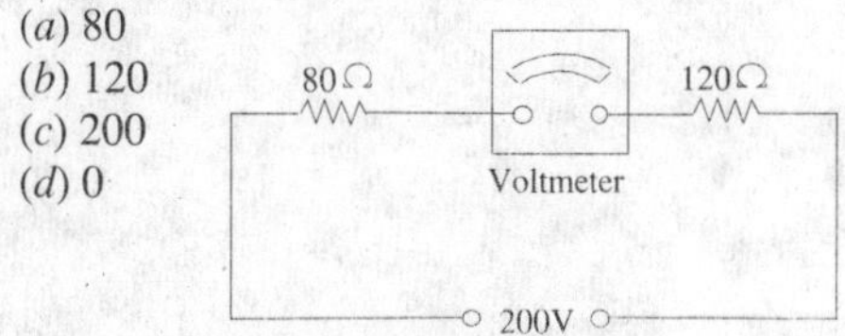

Fig. 10.91

10. The hot wire ammeter :

(*a*) is used only for d.c. circuits
(*b*) is a high precision instrument
(*c*) is used only for a.c. circuits
(*d*) reads equally well on d.c. and/or a.c. circuits.

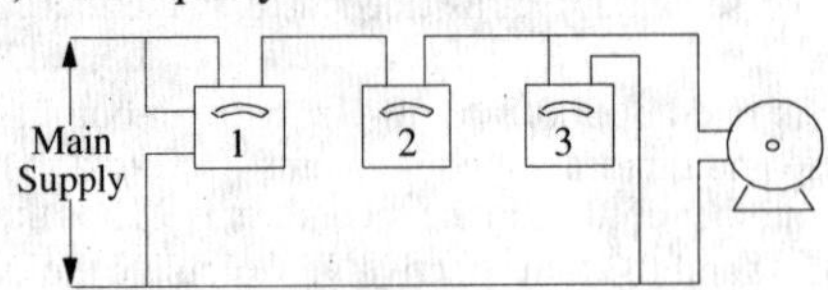

Fig. 10.92

11. In Fig. 10.92, meter No. 1 is a/an :
(*a*) frequency meter
(*b*) voltmeter
(*c*) wattmeter
(*d*) ammeter.

12. Which of the following ammeter will be used to measure alternating currents *only* ?
(*a*) electrodynamic type
(*b*) permanent-magnet type
(*c*) induction-type
(*d*) moving-iron.

13. Damping torque in an indicating instrument is always ——
(*a*) opposite to deflection torque
(*b*) in the same direction as the controlling torque
(*c*) opposite to the direction of motion of moving system
(*d*) opposite to the controlling torque.
(Principles of Elect. Engg. Delhi Univ. July 1984)

14. Mark the WRONG statement. In induction type kWh meters :
(*a*) there is no control spring
(*b*) there is a brake magnet
(*c*) the disc revolves continuosuly
(*d*) the disc stops when braking torque equals deflecting torque.

15. Induction instruments have found widest application as :
(*a*) voltmeter
(*b*) ammeter
(*c*) frequency meter
(*d*) watthour meter.

16. Which of the following instruments has its reading independent of the waveform and frequency of the a.c. supply ?
(*a*) moving-iron
(*b*) hot-wire
(*c*) induction
(*d*) electrostatic.

17. Which of the following instruments is equally accurate on d.c. as well as ac circuits ?
(*a*) dynamometer wattmeter
(*b*) moving-iron ammeter
(*c*) PMMC voltmeter
(*d*) induction wattmeter.

18. Induction watthour meters are free from—errors :
(*a*) phase
(*b*) creeping
(*c*) temperature
(*d*) frequency.

19. The main purpose of using instrument transformers in a.c measurements is to :
(*a*) reduce the possibility of shock
(*b*) extend the range of ac instruments
(*c*) provide high transformation ratio
(*d*) eliminate instrument corrections.

20. A current transformer has a single-turn primary and a 200-turn secondary and is used to measure a.c current with the help of a standard 5-A a.c. ammeter. This arrangement can measure a line current of upto —— ampere.
(*a*) 1000 (*b*) 5000
(*c*) 40 (*d*) 200

Answers

1. *c* **2.** *d* **3.** *a* **4.** *c* **5.** *d* **6.** *d* **7.** *b* **8.** *b* **9.** *c* **10.** *d* **11.** *c* **12.** *c*
13. *c* **14.** *d* **15.** *d* **16.** *b* **17.** *a* **18.** *d.* **19.** *b* **20.** *a*

11 A.C. FUNDAMENTALS

11.1. Generation of Alternating Voltages and Currents

Alternating voltage may be generated by rotating a coil in a magnetic field, as shown in Fig. 11.1 (*a*) or by rotating a magnetic field within a stationary coil, as shown in Fig. 11.1 (*b*).

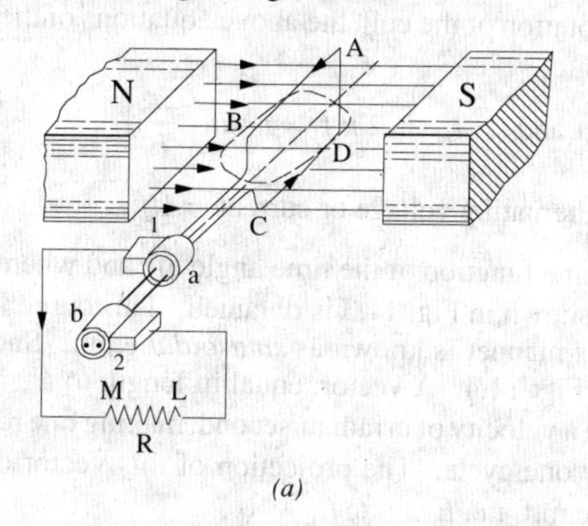

(*a*)

(*b*)

Fig. 11.1

The value of the voltage generated depends, in each case, upon the number of turns in the coil, strength of the field and the speed at which the coil or magnetic field rotates. Alternating voltage may be generated in either of the two ways shown above, but rotating-field method is the one which is mostly used in practice.

11.2. Equations of the Alternating Voltages and Currents

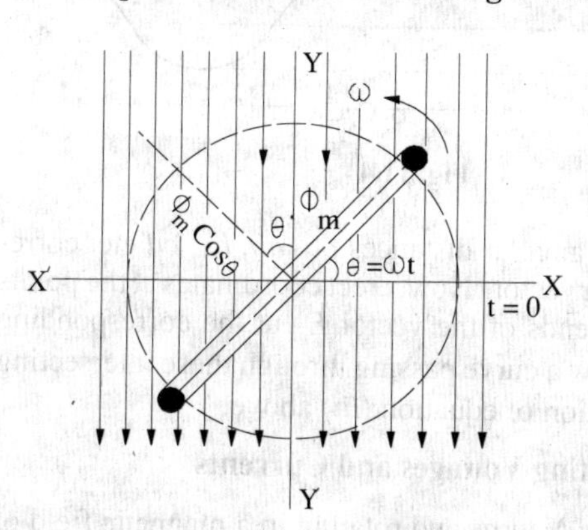

Fig. 11.2

Consider a rectangular coil, having N turns and rotating in a uniform magnetic field, with an angular velocity of ω radian/second, as shown in Fig. 11.2. Let time be measured from the X-axis. Maximum flux Φ_m is linked with the coil, when its plane coincides with the X-axis. In time t seconds, this coil rotates through an angle $\theta = \omega t$. In this deflected position, the component of the flux which is perpendicular to the plane of the coil, is $\Phi = \Phi_m \cos \omega t$. Hence, *flux linkages* of the coil at any time are $N\Phi = N\Phi_m \cos \omega t$.

According to Faraday's Laws of Electromagnetic Induction, the e.m.f. induced in the coil is given by the rate of change of flux-linkages of the coil. Hence, the value of the induced e.m.f. at this instant (*i.e.* when $\theta = \omega t$) or the instantaneous value of the induced e.m.f. is

$$e = -\frac{d}{dt}(N\Phi) \text{ volt} = -N.\frac{d}{dt}(\Phi_m \cos \omega t) \text{ volt} = -N\Phi_m\, \omega\, (-\sin \omega t) \text{ volt}$$

$$= \omega N\Phi_m \sin \omega t \text{ volt} = \omega N\Phi_m \sin \theta \text{ volt} \quad \ldots(i)$$

When the coil has turned through 90° *i.e.* when $\theta = 90°$, then $\sin \theta = 1$, hence e has maximum value, say E_m. Therefore, from Eq. (*i*) we get

$$E_m = \omega N \Phi_m = \omega NB_m A = 2 \pi f NB_m A \text{ volt} \quad \ldots(ii)$$

where B_m = maximum flux density in Wb/ m^2; A = area of the coil in m^2

f = frequency of rotation of the coil in rev/second

Substituting this value of E_m in Eq. (*i*), we get $e = E_m \sin \theta = E_m \sin \omega t$ ...(*iii*)

Similarly, the equation of induced alternating current is $i = I_m \sin \omega t$ ...(*iv*)

provided the coil circuit has been closed through a resistive load.

Since $\omega = 2\pi f$, where f is the frequency of rotation of the coil, the above equations of the voltage and current can be written as

$$e = E_m \sin 2 \pi f t = E_m \sin \left(\frac{2\pi}{T}\right) t \text{ and } i = I_m \sin 2 \pi ft = I_m \sin \left(\frac{2\pi}{T}\right) t$$

where T = time-period of the alternating voltage or current $= 1/f$

It is seen that the induced e.m.f. varies as sine function of the time angle ωt and when e.m.f. is plotted against time, a curve similar to the one shown in Fig. 11.3 is obtained. This curve is known as sine curve and the e.m.f. which varies in this manner is known as *sinusoidal* e.m.f. Such a sine curve can be conveniently drawn, as shown in Fig. 11.4. A vector, equal in length to E_m, is drawn. It rotates in the counter-clockwise direction with a velocity of ω radian/second, making one revolution while the generated e.m.f. makes two loops or one cycle. The projection of this vector on Y-axis gives the instantaneous value e of the induced e.m.f. *i.e.* $E_m \sin \omega t$.

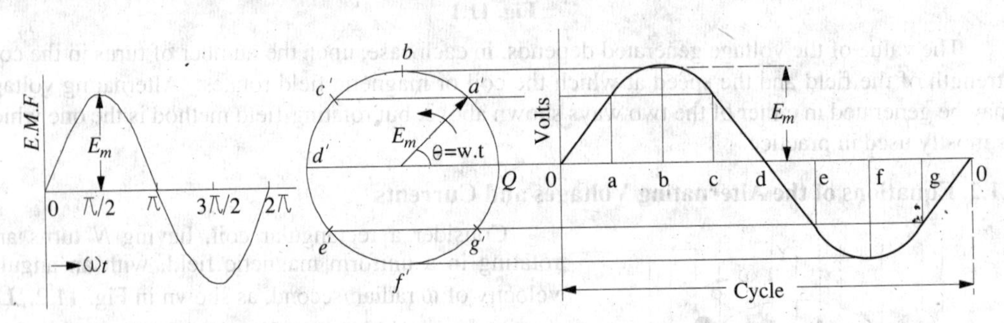

Fig. 11.3 **Fig. 11.4**

To construct the curve, lay off along X-axis equal angular distances oa, ab, bc, cd etc. corresponding to suitable angular displacements of the rotating vector. Now, erect coordinates at the points a, b, c and d etc. (Fig. 11.4) and then project the free ends of the vector E_m at the corresponding positions a', b', c', etc to meet these ordinates. Next draw a curve passing through these intersecting points. The curve so obtained is the graphic representation of equation (*iii*) above.

11.3. Alternate Method for the Equations of Alternating Voltages and Currents

In Fig. 11.5 is shown a rectangular coil AC having N turns and rotating in a magnetic field of flux density B Wb/m^2. Let the length of each of its sides A and C be l meters and their peripheral velocity v metre/second. Let angle be measured from the horizontal position *i.e.* from the X-axis. When in horizontal position, the two sides A and C move parallel to the lines of the magnetic flux. Hence, no flux is cut and so no e.m.f. is generated in the coil.

When the coil has turned through angle θ, its velocity can be resolved into two mutually perpendicular components (*i*) $v \cos\theta$ component-parallel to the direction of the magnetic flux and (*ii*) $v \sin\theta$ component–perpendicular to the direction of the magnetic flux. The e.m.f. is generated due entirely to the perpendicular component *i.e.* $v \sin\theta$.

Hence, the e.m.f. generated in one side of the coil which contains N conductors, as seen from Art. 7.7, is given by, $e = N \times Bl\, v \sin\theta$.

Total e.m.f. generated in both sides of the coil is

$$e = 2BNl\, v \sin\theta \text{ volt} \qquad ...(i)$$

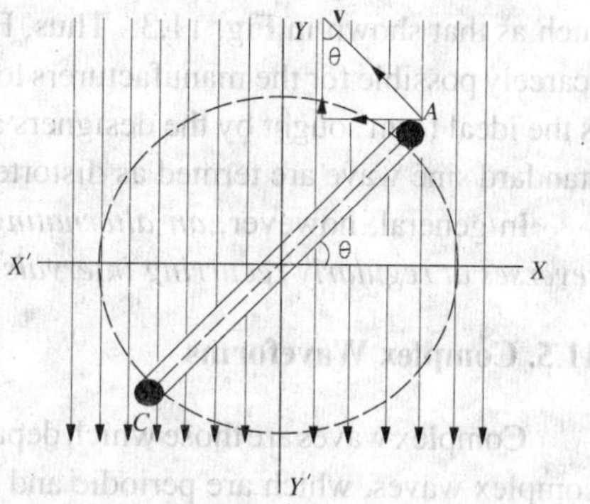

Now, e has maximum value of E_m (say) when $\theta = 90°$. Hence, from Eq. (*i*) above, we get,

Fig. 11.5

$E_m = 2\,B\,N\,l\,v$ volt. Therefore Eq. (*i*) can be rewritten as $e = E_m \sin\theta$ ...as before

If b = width of the coil in meters ; f = frequency of rotation of coil in Hz, then $v = \pi bf$

$\therefore \qquad E_m = 2\,B\,N\,l \times \pi\, b\, f = 2\,p\,i\,f N\,B\,A$ volts ...as before.

Example 11.1. *A square coil of 10 cm side and 100 turns is rotated at a uniform speed of 1000 revolutions per minute, about an axis at right angles to a uniform magnetic field of 0.5 Wb/m². Calculate the instantaneous value of the induced electromotive force, when the plane of the coil is (i) at right angles to the field (ii) in the plane of the field.*

(Electromagnetic Theory, A.M.I.E. Sec B, 1992)

Solution. Let the magnetic field lie in the vertical plane and the coil in the horizontal plane. Also, let the angle θ be measured from X-axis.

Maximum value of the induced emf, $E_m = 2\,\pi f \;\; NB_m A$ volt.

Instantaneous value of the induced e.m.f. $e = E_m \sin\theta$

Now $f = 100/60 = (50/3)$ rps, $N = 100$ $B_m = 0.5$ Wb/m², $A = 10^{-2}$m²

(*i*) In this case, $\theta = 0°$

$\therefore \qquad e = 0$ (*ii*) Here $\theta = 90°$, $\therefore\ e = E_m \sin 90° = E_m$

Substituting the given values, we get

$$e = 2\pi \times (50/3) \times 100 \times 0.5 \times 10^{-2} = \mathbf{52.3\ V}$$

11.4. Simple Waveforms

The shape of the curve obtained by plotting the instantaneous values of voltage or current as the ordinate against time as a abscissa is called its waveform or wave-shape.

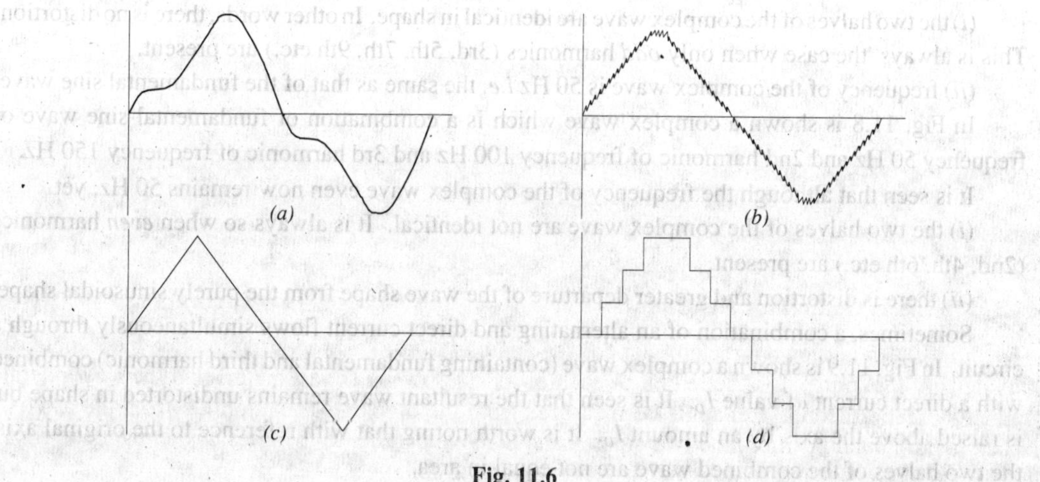

Fig. 11.6

An alternating voltage or current may not always take the form of a systematical or smooth wave such as that shown in Fig. 11.3. Thus, Fig, 11.6 also represents alternating waves. But while it is scarcely possible for the manufacturers to produce sine-wave generators or alternators, yet sine wave is the ideal form sought by the designers and is the accepted standard. The waves deviating from the standard sine wave are termed as distorted waves.

In general, however, *an alternating current or voltage is one the circuit direction of which reverses at regularly recurring intervals.*

11.5. Complex Waveforms

Complex waves are those which depart from the ideal sinusoidal form of Fig. 11.4. All alternating complex waves, which are periodic and have equal positive and negative half cycles can be shown to be made up of a number of pure sine waves, having different frequencies but all these frequencies are integral multiples of that of the lowest alternating wave, called the *fundamental* (or first harmonic). These waves of higher frequencies are called *harmonics*. If the fundamental frequency is 50 Hz, then the frequency of the second *harmonic* is 100 Hz and of the third is 150 Hz and so on. The complex wave may be composed of the fundamental wave (or first harmonic) and any number of other harmonics.

In Fig. 11.7 is shown a complex wave which is made up of a fundamental sine wave of frequency of 50 Hz and third harmonic of frequency of 150 Hz. It is seen that

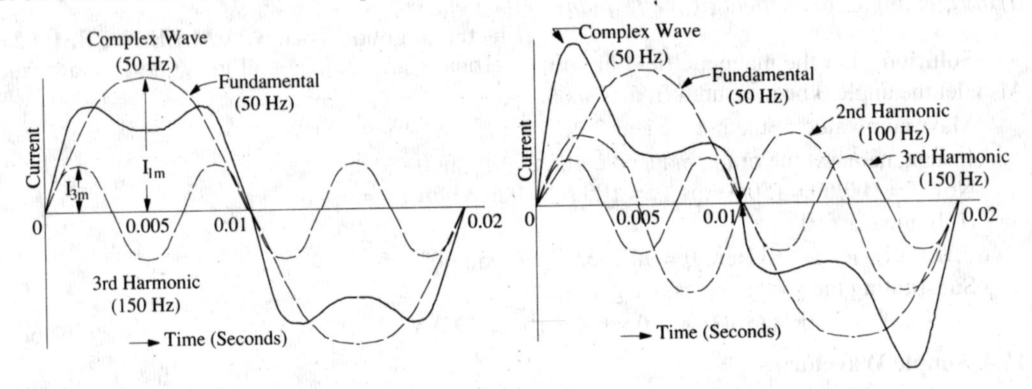

Fig. 11.7

(*i*) the two halves of the complex wave are identical in shape. In other words, there is no distortion. This is always the case when only *odd* harmonics (3rd, 5th, 7th, 9th etc.) are present.

(*ii*) frequency of the complex wave is 50 Hz *i.e.* the same as that of the fundamental sine wave.

In Fig. 11.8 is shown a complex wave which is a combination of fundamental sine wave of frequency 50 Hz and 2nd harmonic of frequency 100 Hz and 3rd harmonic of frequency 150 Hz.

It is seen that although the frequency of the complex wave even now remains 50 Hz, yet

(*i*) the two halves of the complex wave are not identical. It is always so when *even* harmonics (2nd, 4th, 6th etc.) are present.

(*ii*) there is distortion and greater departure of the wave shape from the purely sinusoidal shape.

Sometimes, a combination of an alternating and direct current flows simultaneously through a circuit. In Fig. 11.9 is shown a complex wave (containing fundamental and third harmonic) combined with a direct current of value I_D. It is seen that the resultant wave remains undistorted in shape but is raised above the axis by an amount I_D. It is worth noting that with reference to the original axis, the two halves of the combined wave are not equal in area.

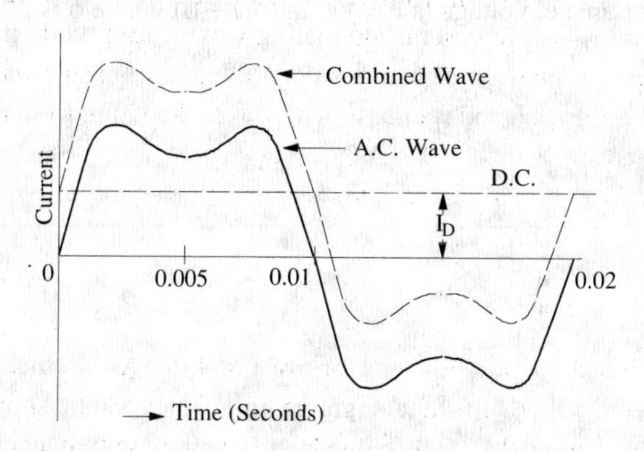

Fig. 11.9

11.6. Cycle

One complete set of positive and negative values of alternating quantity is known as cycle. Hence, each diagram of Fig. 11.6 represents one complete cycle.

A cycle may also be sometimes specified in terms of angular measure. In that case, one complete cycle is said to spread over 360° or 2π radians.

11.7. Time Period

The time taken by an alternating quantity to complete one cycle is called its time period *T*. For example, a 50-Hz alternating current has a time period of 1/50 second.

11.8. Frequency

The number of cycles / second is called the frequency of the alternating quantity. Its unit is hertz (Hz).

In the simple 2-pole alternator of Fig. 24.1 (*b*), one cycle of alternating current is generated in one revolution of the rotating field. However, if there were 4 poles, then two cycles would have been produced in each revolution. In fact, the frequency of the alternating voltage produced is a function of the speed and the number of poles of the generator. The relation connecting the above three quantities is given as

$f = PN/120$ where N = revolutions in r.p.m. and P = number of poles

For example, an alternator having 20 poles and running at 300 r.p.m. will generate alternating voltage and current whose frequency is $20 \times 300/120 = 50$ hertz (Hz).

It may be noted that the frequency is given by the reciprocal of the time period of the alternating quantity.

$$\therefore \qquad f = 1/T \quad \text{or} \quad T = 1/f$$

11.9. Amplitude

The maximum value, positive or negative, of an alternating quantity is known as its amplitude.

11.10. Different Forms of E.M.F. Equation

The standard form of an alternating voltage, as already given in Art. 11.2, is

$$e = E_m \sin \theta = E_m \sin \omega t = E_m \sin 2\pi f t = E_m \sin \frac{2\pi}{T} t$$

By closely looking at the above equations, we find that

(*i*) the maximum value or peak value or amplitude of an *alternating voltage is given by the coefficient of the sine of the time angle.*

(*ii*) *the frequency f is given by the coefficient of time divided by* 2π.

For example, if the equation of an alternating voltage is given by $e = 50 \sin 314t$ then its maximum value is 50 V and its frequency is $f = 314/2\pi = 50$ Hz.

Similarly, if the equation is of the form $e = I_m \sqrt{(R^2 + 4\omega^2 L^2)} \sin 2\omega t$, then its maximum value is $E_m = I_m \sqrt{(R^2 + 4\omega^2 L^2)}$ and the frequency is $2\omega / 2\pi$ or ω/π Hz.

Example 11.2. *The maximum values of the alternating voltage and current are 400 V and 20 A respectively in a circuit connected to 50 Hz supply and these quantities are sinusoidal. The instantaneous values of the voltage and current are 283 V and 10 A respectively at t = 0 both increasing positively.*

(*i*) *Write down the expression for voltage and current at time t,*

(*ii*) *Determine the power consumed in the circuit.* **(Elect. Engg. Pune Univ. 1985)**

Solution. (*i*) In general, the expression for an a.c. voltage is $v = V_m \sin(\omega t + \phi)$ where ϕ is the phase difference with respect to the point where $t = 0$.

Now, $v = 283$ V; $V_m = 400$ V. Substituting $t = 0$ in the above equation, we get

$283 = 400 (\sin \omega \times 0 + \phi) \therefore \sin \phi = 400/283 = 0.707; \therefore \phi = 45°$ or $\pi/4$ radian.

Hence, general expression for voltage is

$v = 400 (\sin 2\pi \times 50 \times t + \pi/4)$

$= 400 \sin (100\, \pi\, t + \pi/4)$

Similarly, at $t = 0$, $10 = 20 \sin (\omega \times 0 + \phi)$

$\therefore \sin \phi = 0.5 \therefore \phi = 30°$ or $\pi/6$ radian

Hence, the general expression for the current is

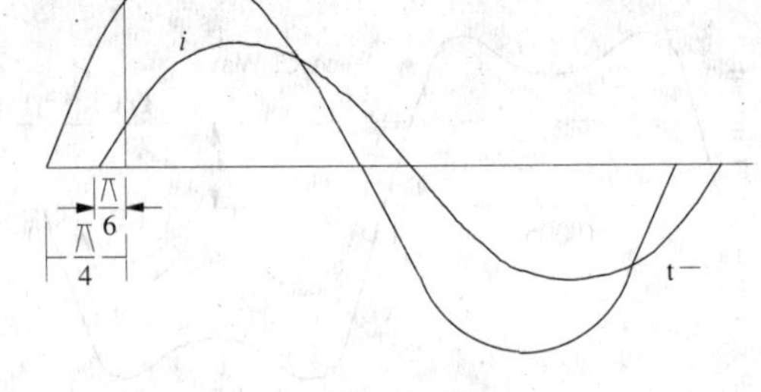

Fig. 11.10

$i = 20 (\sin 100\, \pi\, t + 30°) = 20 \sin (100\, \pi\, t + \pi/6)$

(*ii*) $P = VI \cos \theta$ where V and I are rms values and θ is the phase difference between the voltage and current.

Now, $V = V_m/\sqrt{2} = 400/\sqrt{2}$; $I = 20/\sqrt{2}$; $\theta = 45° - 30° = 15°$(See Fig. 11.10)

$\therefore \quad P = (400/\sqrt{2}) \times (20/\sqrt{2}) \times \cos 15° = \mathbf{3864\ W}$

Example 11.3. *An alternating current of frequency 60 Hz has a maximum value of 120 A. Write down the equation for its instantaneous value. Reckoning time from the instant the current is zero and is becoming positive, find (a) the instantaneous value after 1/360 second and (b) the time taken to reach 96 A for the first time.*

Solution. The instantaneous current equation is

$i = 120 \sin 2\pi ft = 120 \sin 120\, \pi\, t$

Now when $t = 1/360$ second, then

(*a*) $i = 120 \sin (120 \times \pi \times 1/360)$ ...angle in radians

$= 120 \sin (120 \times 180 \times 1/360)$ ...angle in degrees

$= 120 \sin 60° = \mathbf{103.9\ A}$

(*b*) $96 = 120 \times \sin 2 \times 180 \times 60 \times t$ ...angle in degrees

or $\sin(360 \times 60 \times t) = 96/120 = 0.8 \quad \therefore \quad 360 \times 60 \times t = \sin^{-1} 0.8 = 53°$(approx).

$\therefore \quad t = \theta/2\pi f = 53/360 \times 60 = \mathbf{0.00245\ second.}$

11.11. Phase

By phase of an alternating current is meant the fraction of the time period of that alternating current which has elapsed since the current last passed through the zero position of reference. For example, the phase of current at point *A* is *T*/4 second, where *T* is time period or expressed in terms of angle, it is $\pi/2$ radians (Fig. 11.11). Similarly, the phase of the rotating coil at the instant shown in Fig. 11.1 is ωt which is, therefore, called its phase angle.

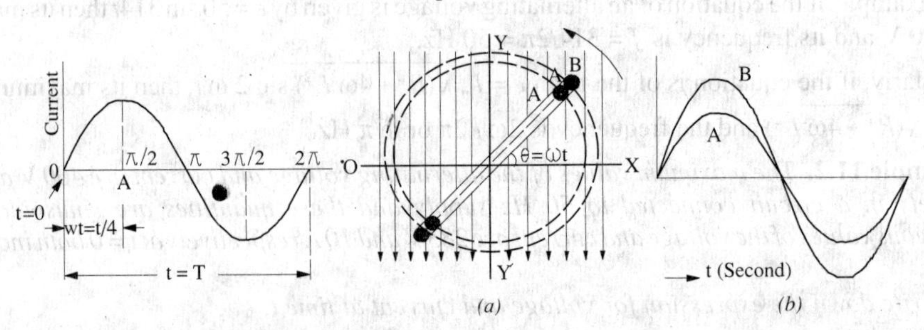

Fig. 11.11 Fig. 11.12

In electrical engineering, we are, however, more concerned with relative phases or phase differences between different alternating quantities, rather than with their absolute phases. Consider two single-turn coils of different sizes [Fig. 11.12 (*a*)] arranged radially in the same plane and rotating with the same angular velocity in a common magnetic field of uniform intensity. The e.m.fs. induced in both coils will be of the same frequency and of sinusoidal shape, although the values of instantaneous e.m.fs. induced would be different. However, the two alternating e.m.fs. would reach their maximum and zero values at the same time as shown in Fig. 11.12 (*b*). Such alternating voltages (or currents) are said to be in phase with each other. The two voltages will have the equations

$$e_1 = E_{m1} \sin \omega t \quad \text{and} \quad e_2 = E_{m2} \sin \omega t$$

11.12. Phase Difference

Now, consider three similar single-turn coils displaced from each other by angles α and β and

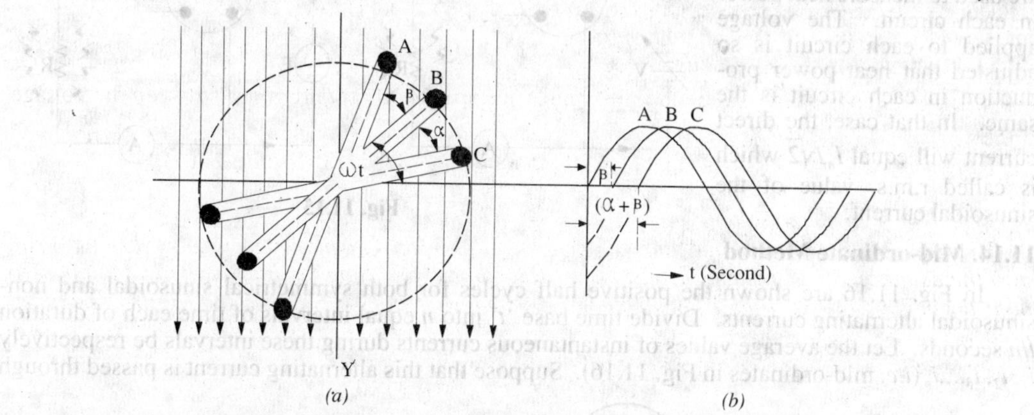

Fig. 11.13

rotating in a uniform magnetic field with the same angular velocity [Fig. 11.13 (*a*)].

In this case, the value of induced e.m.fs. in the three coils are the same, but there is one important difference. The e.m.fs. in these coils do not reach their maximum or zero values simultaneously but one after another. The three sinusoidal waves are shown in Fig. 11.13 (*b*). It is seen that curves *B* and *C* are displaced from curve *A* by angles β and (α + β) respectively. Hence, it means that phase difference between *A* and *B* is β and between *B* and *C* is α but between *A* and *C* is (α + β). The statement, however, does not give indication as to which e.m.f. reaches its maximum value first. This deficiency is supplied by using the terms '*lag*' or '*lead*.'

A leading alternating quantity is one which reaches its maximum (or zero) value earlier as compared to the other quantity.

Similarly, a lagging alternating quantity is one which reaches its maximum or zero value later than the other quantity. For example, in Fig. 11.13 (*b*), *B* lags behind *A* by β and *C* lags behind *A* by (α+β) because they reach their maximum values later.

The three equations for the instantaneous induced e.m.fs. are (Fig. 11.14)

$e_A = E_m \sin \omega t$ ...reference quantity

$e_B = E_m \sin (\omega t - \beta)$

$e_C = E_m \sin [\omega t - (\alpha + \beta)]$.

In Fig. 11.14, quantity *B* leads *A* by an angle ϕ. Hence, their equations are

$e_A = E_m \sin \omega t$ ...reference quantity

$e_B = E_m \sin (\omega t + \phi)$

A plus (+) sign when used in connection with phase difference denotes 'lead' whereas a minus (–) sigh denotes 'lag'.

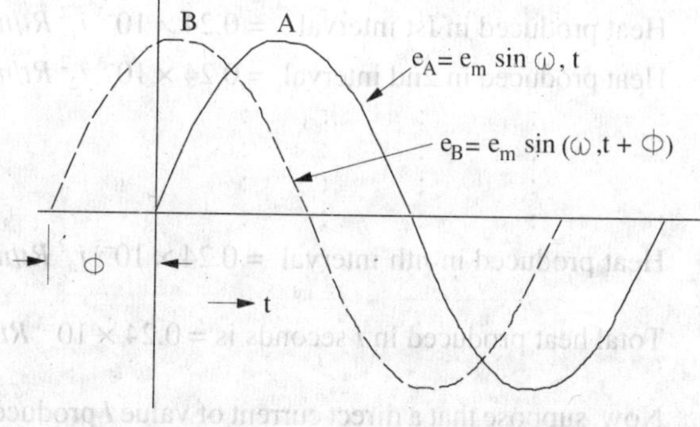

Fig. 11.14

11.13. Root-Mean-Square (R.M.S) Value

The r.m.s. value of an alternating current *is given by that steady (d.c.) current which when flowing through a given circuit for a given time produces the same* **heat** *as produced by the alternating current when flowing through the same circuit for the same time.*

It is also known as the *effective* or *virtual* value of the alternating current, the former term being used more extensively. For computing the r.m.s value of symmetrical sinusoidal alternating currents, either mid-ordinate method or analytical method may be used, although for symmetrical but non-sinusoidal waves, the mid-ordinate method would be found more convenient.

A simple experimental arrangement for measuring the equivalent d.c. value of a sinusoidal current is shown in Fig. 11.15. The two circuits have identical resistances but one is connected to battery and the other to a sinusoidal generator. Wattmeters are used to measure heat power in each circuit. The voltage applied to each circuit is so adjusted that heat power production in each circuit is the same. In that case, the direct current will equal $I_m/\sqrt{2}$ which is called r.m.s. value of the sinusoidal current.

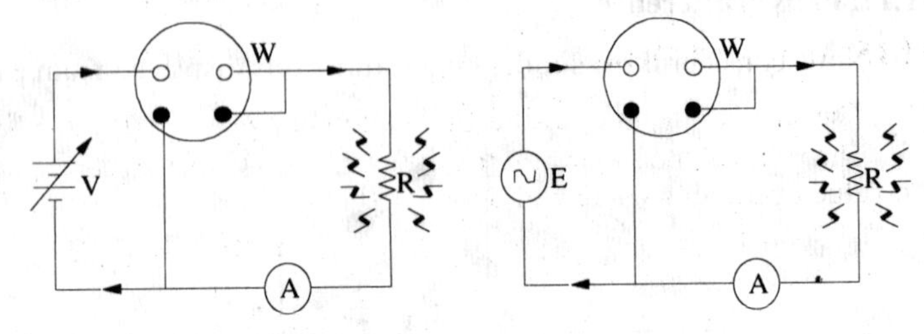

Fig. 11.15

11.14. Mid-ordinate Method

In Fig. 11.16 are shown the positive half cycles for both symmetrical sinusoidal and non-sinusoidal alternating currents. Divide time base 't' into n equal intervals of time each of duration t/n seconds. Let the average values of instantaneous currents during these intervals be respectively $i_1, i_2, i_3....i_n$ (*i.e.* mid-ordinates in Fig. 11.16). Suppose that this alternating current is passed through

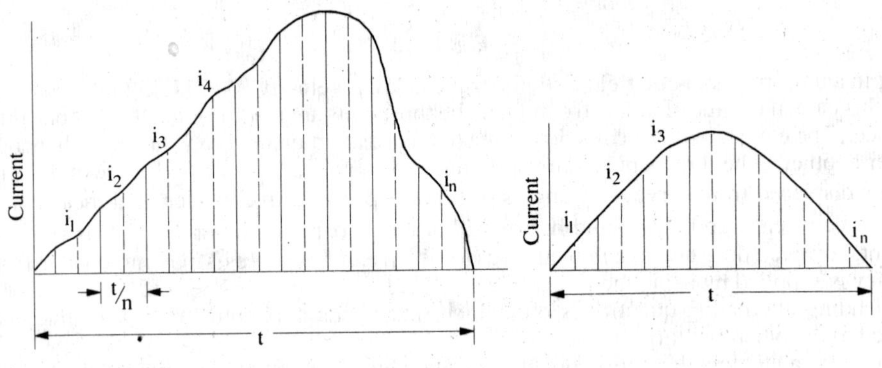

Fig. 11.16

a circuit of resistance R ohms. Then,

Heat produced in Ist interval $= 0.24 \times 10^{-3}\, i_1^2\, Rt/n$ kcal ($\because$ $1/J = 1/4200 = 0.24 \times 10^{-3}$)

Heat produced in 2nd interval $= 0.24 \times 10^{-3}\, i_2^2\, Rt/n$ kcal

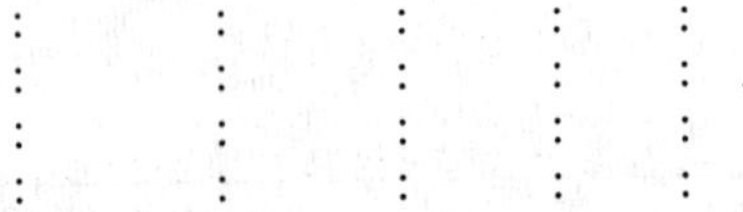

Heat produced in nth interval $= 0.24 \times 10^{-3}\, i_n^2\, Rt/n$ kcal

$$\text{Total heat produced in } t \text{ seconds is} = 0.24 \times 10^{-3}\, Rt\left(\frac{i_1^2 + i_2^2 + \ldots + i_n^2}{n}\right) \text{kcal}$$

Now, suppose that a direct current of value I produces the same heat through the same resistance during the same time t. Heat produced by it is $= 0.24 \times 10^{-3}\, I^2Rt$ kcal. By definition, the two amounts of heat produced should be equal.

$$\therefore \quad 0.24\times10^{-3}I^2Rt = 0.24\times10^{-3}Rt\left(\frac{i_1^2+i_2^2+\ldots+i_n^2}{n}\right)$$

$$\therefore \quad I^2 = \frac{i_1^2+i_2^2+\ldots+i_n^2}{n} \quad \therefore \quad I = \sqrt{\left(\frac{i_1^2+i_2^2+\ldots+i_n^2}{n}\right)}$$

= square root of the mean of the squares of the instantaneous currents

Similarly, the r.m.s. value of alternating voltage is given by the expression

$$V = \sqrt{\left(\frac{v_1^2+v_2^2+\ldots+v_n^2}{n}\right)}$$

11.15. Analytical Method

The standard form of a sinusoidal alternating current is $i = I_m \sin \omega t = I_m \sin \theta$.

The mean of the squares of the instantaneous values of current over one complete cycle is (even the value over half a cycle will do).

$$= \int_0^{2\pi} \frac{i^2 d\theta}{(2\pi - 0)}$$

The square root of this value is $= \sqrt{\left(\int_0^{2\pi} \frac{i^2 d\theta}{2\pi}\right)}$

Hence, the r.m.s. value of the alternating current is

$$I = \sqrt{\left(\int_0^{2\pi} \frac{i^2 d\theta}{2\pi}\right)} = \sqrt{\left(\frac{I_m^2}{2\pi}\int_0^{2\pi} \sin^2\theta \, d\theta\right)} \quad (\text{put } i = I_m \sin\theta)$$

Now, $\cos 2\theta = 1 - 2\sin^2\theta \quad \therefore \quad \sin^2\theta = \frac{1-\cos 2\theta}{2}$

$$\therefore \quad I = \sqrt{\left(\frac{I_m^2}{4\pi}\int_0^{2\pi}(1-\cos 2\theta)\, d\theta\right)} = \sqrt{\left(\frac{I_m^2}{4\pi}\left|\theta - \frac{\sin 2\theta}{2}\right|_0^{2\pi}\right)}$$

$$= \sqrt{\left(\frac{I_m^2}{4\pi}\times 2\pi\right)} = \sqrt{\left(\frac{I_m^2}{2}\right)} \quad \therefore \quad I = \frac{I_m}{\sqrt{2}} = 0.707\, I_m$$

Hence, we find that for a symmetrical sinusoidal current

r.m.s. value of current = 0.707 × max. value of current

The r.m.s. value of an alternating current is of considerable importance in practice, because the ammeters and voltmeters record the r.m.s. value of alternating current and voltage respectively. In electrical engineering work, *unless indicated otherwise, the values of the given current and voltage are always the r.m.s. values.*

It should be noted that the average heating effect produced during one cycle is

$$= I^2R = (I_m/\sqrt{2})^2 R = \frac{1}{2} I_m^{\,2} R$$

11.16. R.M.S. Value of a Complex Wave

In their case also, either the mid-ordinate method (when equation of the wave is not known) or analytical method (when equation of the wave is known) may be used. Suppose a current having the equation $i = 12 \sin \omega t + 6 \sin (3\omega t - \pi/6) + 4 \sin (5\omega t + \pi/3)$ flows through a resistor of R ohm. Then, in the time period T second of the wave, the effect due to each component is as follows :

Fundamental .... $(12/\sqrt{2})^2\, RT$ watt

3rd harmonic .... $(6/\sqrt{2})^2\, RT$ watt

5th harmonic .... $(4/\sqrt{2})^2\, RT$ watt

∴

$\therefore$ Total heating effect $= RT[(12/\sqrt{2})^2 + 6/\sqrt{2})^2 + (4/\sqrt{2})^2]$

If I is the r.m.s. value of the complex wave, then equivalent heating effect is I^2RT

$$\therefore \quad I^2RT = RT[(12/\sqrt{2})^2 + (6/\sqrt{2})^2 + (4/\sqrt{2})^2]$$

$$\therefore \quad I = \sqrt{[(12/\sqrt{2})^2 + (6/\sqrt{2})^2 + (4/\sqrt{2})^2]} = 9.74 \text{ A}$$

Had there been a direct current of (say) 5 amperes flowing in the circuit also*, then the r.m.s. value would have been

$$= \sqrt{(12 + \sqrt{2})^2 + (6/\sqrt{2})^2 + (4/\sqrt{2})^2 + 5^2]} = 10.93 \text{ A}$$

Hence, for complex waves the rule is as follows : *The r.m.s. value of a complex current wave is equal to the square root of the sum of the squares of the r.m.s. values of its individual components.*

11.17. Average Value

The average value I_a of an alternating current is expressed *by that steady current which transfers across any circuit the same* **charge as** *is transferred by that alternating current during the same time.*

In the case of a symmetrical alternating current (*i.e.* one whose two half-cycles are exactly similar, whether sinusoidal or non-sinusoidal), the average value over a complete cycle is zero. Hence, in their case, the average value is obtained by adding or integrating the instantaneous values of current over one half-cycle only. *But in the case of an unsymmetrical alternating current (like half-wave rectified current) the average value must always be taken over the whole cycle.*

(i) Mid-ordinate Method

With reference to Fig. 11.16, $I_{av} = \dfrac{i_1 + i_2 + \dots + i_n}{n}$

This method may be used both for sinusoidal and non-sinusoidal waves, although it is specially convenient for the latter.

(ii) Analytical Method

The standard equation of an alternating current is, $i = I_m \sin\theta$

$$I_{av} = \int_0^{\pi} \frac{i\,d\theta}{(\pi - 0)} = \frac{I_m}{\pi}\int_0^{\pi} \sin\theta\, d\theta \qquad \text{(putting value of } i\text{)}$$

$$= \frac{I_m}{\pi}\left|-\cos\theta\right|_0^{\pi} = \frac{I_m}{\pi}\left|+1-(-1)\right| = \frac{2I_m}{\pi} = \frac{I_m}{\pi/2} = \frac{\text{twice the maximum current}}{\pi}$$

$\therefore I_{av} = I_m/\frac{1}{2}\pi = 0.637\, I_m$ $\therefore$ **average value of current = 0.637 × maximum value**

Note. R.M.S. value is always greater than average value except in the case of a rectangular wave when both are equal.

11.18. Form Factor

It is defined as the ratio, $K_f = \dfrac{\text{r.m.s. value}}{\text{average value}} = \dfrac{0.707\, I_m}{0.637\, I_m} = 1.1$ (for sinusoidal alternating currents only)

In the case of sinusoidal alternating voltage also, $K_f = \dfrac{0.707\, E_m}{0.637\, E_m} = 1.11$

---

*The equation of the complex wave, in that case, would be,
$i = 5 + 12 \sin .\omega t + 6 \sin (3\omega t - \pi/6) + 4 \sin (5\omega t + \pi/3)$.

As is clear, the knowledge of form factor will enable the r.m.s. value to be found from the arithmetic mean value and *vice versa.*

11.19. Crest or Peak or Amplitude Factor

It is defined as the ratio $K_a = \dfrac{\text{maximum value}}{\text{r.m.s. value}} = \dfrac{I_m}{I_m/\sqrt{2}} = \sqrt{2} = 1.414$ (for sinusoidal a.c. only)

For sinusoidal alternating voltage also, $K_a = \dfrac{E_m}{E_m/\sqrt{2}} = 1.414$

Knowledge of this factor is of importance in dielectric insulation testing, because the dielectric stress to which the insulation is subjected, is proportional to the maximum or peak value of the applied voltage. The knowledge is also necessary when measuring iron losses, because the iron loss depends on the value of maximum flux.

Example 11.4. *An alternating current varying sinusoidally with a frequency of 50 Hz has an RMS value of 20 A. Write down the equation for the instantaneous value and find this value (a) 0.0025 second (b) 0.0125 second after passing through a positive maximum value. At what time, measured from a positive maximum value, will the instantaneous current be 14.14 A ?*

(Elect. Science-I Allahabad Univ. 1992)

Solution. $I_m = 20\sqrt{2} = 28.2$ A, $\omega = 2\pi \times 50 = 100\,\pi$ rad/s.

The equation of the sinusoidal current wave with reference to point O (Fig. 11.17) as zero time point is

$$i = \mathbf{28.2 \sin 100\,\pi t} \textbf{ ampere}$$

Since time values are given from point A where voltage has positive and maximum value, the equation may itself be referred to point A. In the case, the equation becomes :

$$i = 28.2 \cos 100\,\pi t$$

(i) When t = 0.0025 second

$i = 28.2 \cos 100\pi \times 0.0025$ ...angle in radian

$= 28.2 \cos 100 \times 180 \times 0.0025$ ...angle in degrees

$= 28.2 \cos 45° = 20\text{A}$ ...point B

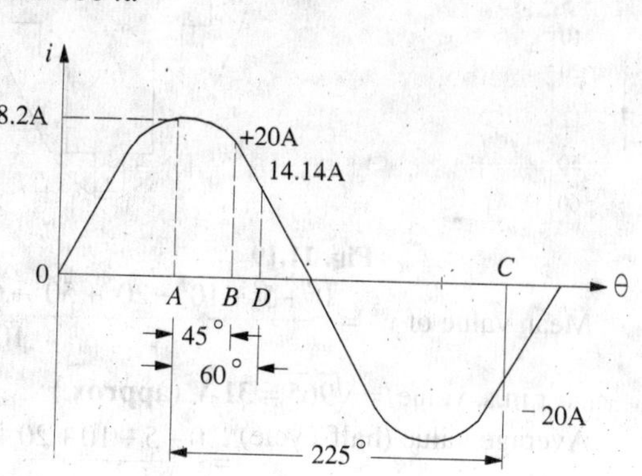

Fig. 11.17

(ii) When t = 0.0125 second

$i = 28.2 \cos 100 \times 180 \times 0.0125$

$= 28.2 \cos 225° = 28.2 \times (-1/\sqrt{2})$

$= -20$ A ...point C

(iii) Here i = 14.14 A

$\therefore \quad 14.14 = 28.2 \cos 100 \times 180\, t \qquad \therefore \quad \cos 100 \times 180\, t = \dfrac{1}{2}$

or $\quad 100 \times 180\, t = \cos^{-1}(0.5) = 60°, t = \mathbf{1/300}$ **second** ...point D

Example 11.5. *An alternating current of frequency 50 Hz has a maximum value of 100 A. Calculate (a) its value 1/600 second after the instant the current is zero and its value decreasing thereafter/wards (b) how many seconds after the instant the current is zero (increasing thereafter wards) will the current attain the value of 86.6 A ?* **(Elect. Technology. Allahabad Univ. 1991)**

Solution. The equation of the alternating current (assumed sinusoidal) with respect to the origin O (Fig. 11.18) is

$i = 100 \sin 2\pi \times 50t = 100 \sin 100\,\pi t$

(*a*) It should be noted that, in this case, time is being measured from point *A* (where current is zero and decreasing thereafter) and not from point *O*.

If the above equation is to be utilized, then, this time must be referred to point *O*. For this purpose, half time-period *i.e.* 1/100 second has to be added to 1/600 second. The given time as referred to point *O* becomes

$$= \frac{1}{100} + \frac{1}{600} = \frac{7}{600} \text{ second}$$

$$\therefore\ i = 100 \sin 100 \times 180 \times 7/600 = 100 \sin 210°$$

$$= 100 \times -1/2 = \mathbf{-50\ A} \qquad \text{...point } B$$

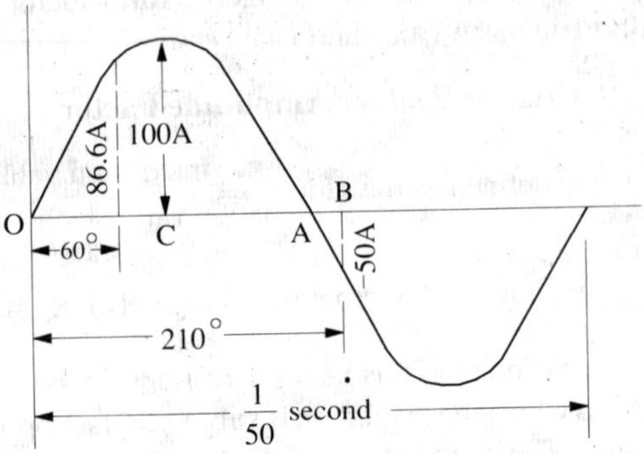

Fig. 11.18

(*b*) In this case, the reference point is *O*.

$\therefore\ 86.6 = 100 \sin 100{\times}180\,t$ or $\sin 18{,}000\,t = 0.866$

or $18{,}000\,t = \sin^{-1}(0.866) = 60°$ $\therefore\ t = 60/18{,}000 =$ **1/300 second**

Example 11.6. *Calculate the r.m.s. value, the form factor and peak factor of a periodic voltage having the following values for equal time intervals changing suddenly from one value to the next : 0, 5, 10, 20, 50, 60, 50, 20, 10, 5, 0, –5, –10 V etc. What would be the r.m.s value of sine wave having the same peak value ?*

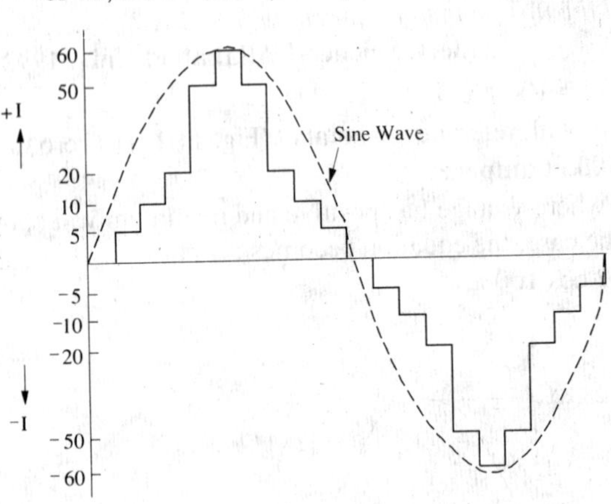

Fig. 11.19

Solution. The waveform of the alternating voltage is shown in Fig. 11.19. Obviously, it is not sinusoidal but it is symmetrical. Hence, though r.m.s. value may be full one cycle, the average value has necessarily to be considered for half-cycle only, otherwise the symmetrical negative and positive half-cycles will cancel each other out.

$$\text{Mean value of } v^2 = \frac{0^2 + 5^2 + 10^2 + 20^2 + 50^2 + 60^2 + 50^2 + 20^2 + 10^2 + 5^2}{10} = \mathbf{965\ V}$$

$$\therefore \text{ r.m.s. value } = \sqrt{965} = \mathbf{31\ V\ (approx.)}$$

$$\text{Average value (half-cycle)} = \frac{0 + 5 + 10 + 20 + 50 + 60 + 50 + 20 + 10 + 5}{10} = \mathbf{23\ V}$$

$$\text{Form factor } = \frac{\text{r.m.s. value}}{\text{average value}} = \frac{31}{23} = \mathbf{1.35}. \text{ Peak factor } = 60/31 = \mathbf{2\ (approx.)}$$

R.M.S. value of a sine wave of the same peak value $= 0.707 \times 60 =$ **42.2 V**.

Alternative Solution

If '*t*' be the regular time interval, then area of the half-cycle is

$= (5t + 10t + 20t + 50t)\ 2 + 60t = 230t$, Base $= 10t$ $\therefore$ Mean value $= 230t/10t = 23$ V.

Area when ordinates are squared $= (25t + 100t + 400t + 2500t)\ 2 + 3600t = 9650t$, Base $= 10t$

$\therefore$ Mean height of the squared curve $= 9650t/10t = 965$

$\therefore$ r.m.s. value $= \sqrt{965} = 31$ V

Further solution is as before.

Example 11.7. *Calculate the reading which will be given by a hot-wire voltmeter if it is connected across the terminals of a generator whose voltage waveform is represented by*

$$v = 200 \sin \omega t + 100 \sin 3\omega t + 50 \sin 5\,\omega t$$

Solution. Since hot-wire voltmeter reads only r.m.s value, we will have to find the r.m.s. value of the given voltage. Considering one complete cycle,

R.M.S. value $V = \sqrt{\frac{1}{2\pi}\int_0^{2\pi} v^2 d\theta}$ where $\theta = \omega t$

or $V^2 = \frac{2}{2\pi}\int_0^{2\pi} (200\sin\theta + 100\sin 3\theta + 50\sin 5\theta)^2\, d\theta$

$$= \frac{1}{2\pi}\int_0^{2\pi} (200^2\sin^2\theta + 100^2\sin^2 3\theta + 50^2\sin^2 5\theta + 2\times 200.100\sin\theta.\sin 3\theta$$

$$+ 2\times 100.50.\sin 3\theta.\sin 5\theta + 2\times 50.200.\sin 5\theta.\sin\theta)d\theta$$

$$= \frac{1}{2\pi}\left(\frac{200^2}{2} + \frac{100^2}{2} + \frac{50^2}{2}\right)2\pi = 26,250 \quad \therefore \quad V = \sqrt{26,250} = \mathbf{162\ V}$$

Alternative Solution

The r.m.s. value of individual components are $(200/\sqrt{2})$, $(100/\sqrt{2})$ and $(50/\sqrt{2})$. Hence, as stated in Art. 11.16,

$$V = \sqrt{V_1^2 + V_2^2 + V_3^2} = \sqrt{(200/\sqrt{2})^2 + (100/\sqrt{2})^2 + (50/\sqrt{2})^2} = \mathbf{162\ V}$$

11.20. R.M.S. Value of H.W. Rectified Alternating Current

Half-wave (H.W.) rectified alternating current is one whose one half-cycle has been suppressed *i.e.* one which flows for half the time during one cycle. It is shown in Fig. 11.20 where suppressed half-cycle is shown dotted.

As said earlier, for finding r.m.s. value of such an alternating current, summation would be carried over the period for which current *actually* flows *i.e.* from 0 to π, though it would be averaged for the whole cycle *i.e.* from 0 to 2π.

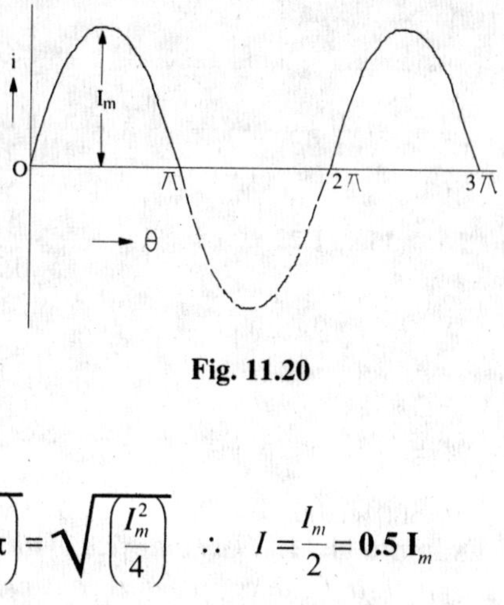

$\therefore$ R.M.S. current

$$I = \sqrt{\left(\int_0^{\pi}\frac{i^2 d\theta}{2\pi}\right)} = \sqrt{\left(\frac{I_m^2}{2\pi}\int_0^{\pi}\sin^2\theta\, d\theta\right)}$$

$$= \sqrt{\frac{I_m^2}{4\pi}\int_0^{\pi}(1-\cos 2\theta)d\theta}$$

$$= \sqrt{\left(\frac{I_m^2}{4\pi}\left|\theta - \frac{\sin 2\theta}{2}\right|_0^{\pi}\right)} = \sqrt{\left(\frac{I_m^2}{4\pi}\times\pi\right)} = \sqrt{\left(\frac{I_m^2}{4}\right)} \quad \therefore \quad I = \frac{I_m}{2} = \mathbf{0.5\ I_m}$$

Fig. 11.20

11.21. Average Value of H.W. Rectified Alternating Current

For the same reasons as given in Art. 11.20, integration would be carried over from $0 - \pi$

$$\therefore \quad I_{av} = \int_0^{\pi}\frac{i\,d\theta}{2\pi} = \frac{I_m}{2\pi}\int_0^{\pi}\sin\theta\, d\theta \qquad (\because\ i = I_m\sin\theta)$$

$$= \frac{I_m}{2\pi}\left|-\cos\theta\right|_0^{\pi} = \frac{I_m}{2\pi}\times 2 = \frac{I_m}{\pi}$$

11.22. Form Factor of H.W. Rectified Alternating Current

$$\text{Form factor} = \frac{\text{r.m.s. value}}{\text{average value}} = \frac{I_m/2}{I_m/\pi} = \frac{\pi}{2} = 1.57$$

Example 11.8. *An alternating voltage e = 200 sin 314t is applied to a device which offers an ohmic resistance of 20 Ω to the flow of current in one direction, while preventing the flow of current in opposite direction. Calculate RMS value, average value and form factor for the current over one cycle.* **(Elect. Engg. Nagpur Univ. 1992)**

Solution. Comparing the given voltage equation with the standard form of alternating voltage equation, we find that $V_m = 200$ V, $R = 20\ \Omega$, $I_m = 200/20 = 10$ A. For such a half-wave rectified current, RMS value $= I_m/2 = 10/2 =$ **5A.**

Average current $= I_m/\pi = 10/\pi =$ **3.18 A** ; Form factor = 5/3.18 = **1.57**

Example 11.9. *Compute the average and effective values of the square voltage wave shown in Fig. 11.21.*

Solution. As seen, for $0 < t < 0.1$ *i.e.* for the time interval 0 to 0.1 second, $v = 20$ V. Similarly, for $0.1 < t < 0.3$, $v = 0$. Also time-period of the voltage wave is 0.3 second.

$$\therefore \quad V_{av} = \frac{1}{T}\int_0^T v\,dt = \frac{1}{0.3}\int_0^{0.1} 20\,dt$$

$$= \frac{1}{0.3}(20 \times 0.1) = \mathbf{6.67\ V}$$

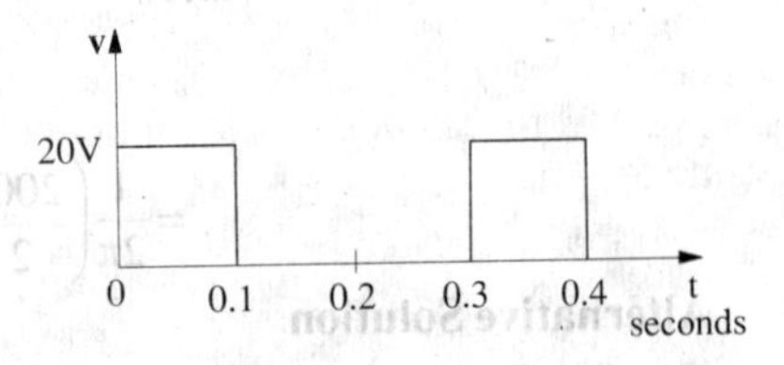

Fig. 11.21

$$V^2 = \frac{1}{T}\int_0^T v^2 dt = \frac{1}{0.3}\int_0^{0.1} 20^2 dt = \frac{1}{0.3}(400 \times 0.1) = 133.3;\ V = \mathbf{11.5\ V}$$

Example 11.10. *Calculate the RMS value of the function shown in Fig. 11.22 if it is given that for $0 < t < 0.1$, $y = 10(1-e^{-100t})$ and for $0.1 < t < 0.2$, $y = 10\,e^{-50(t-0.1)}$*

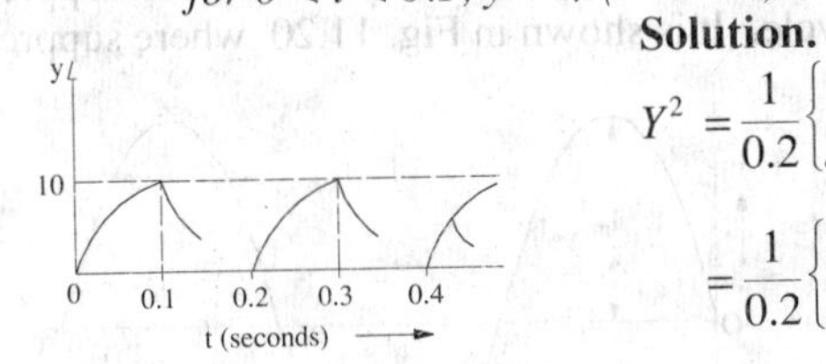

Fig. 11.22

Solution.

$$Y^2 = \frac{1}{0.2}\left\{\int_0^{0.1} y^2 dt + \int_{0.1}^{0.2} y^2 dt\right\}$$

$$= \frac{1}{0.2}\left\{\int_0^{0.1} 10^2(1-e^{-100t})^2 dt + \int_{0.1}^{0.2}(10e^{-50(t-0.1)})^2 dt\right\}$$

$$= \frac{1}{0.2}\left\{\int_0^{0.1} 100\,(1+e^{-200t} - 2e^{-100t})dt + \int_{0.1}^{0.2} 100\,e^{-100(t-0.1)}dt\right\}$$

$$= 500\left\{\left[t - 0.005e^{-200t} + 0.02e^{-100t}\right]_0^{0.1} + \left[-0.01e^{-100(t-0.1)}\right]_{0.1}^{0.2}\right\}$$

$$= 500\left\{\left[(0.1 - 0.005e^{-20} + 0.02e^{-10}) - (0 - 0.005 + 0.02)\right] + \left[(-0.01e^{-10}) - (-0.01)\right]\right\}$$

$$= 500 \times 0.095 = 47.5 \qquad \therefore \quad Y = \sqrt{47.5} = \mathbf{6.9}$$

Example 11.11. *The half cycle of an alternating signal is as follows ; It increases uniformly from zero at 0° to F_m at α°, remains constant from α° to (180 – α)°, decreases uniformly from F_m at (180-α)° to zero at 180°. Calculate the average and effective values of the signal.* **(Elect. Science-I, Allahabad Univ. 1992)**

Solution. For finding the average value, we would find the total area of the trapezium and divide it by π (Fig. 11.23).

Area = $2 \times \Delta\,OAE$ + rectangle $ABDE = 2 \times (1/2) \times F_m\alpha + (\pi - 2\alpha)F_m = (\pi - \alpha)F_m$

$\therefore$ average value $= (\pi - \alpha)F_m/\pi$

RMS Value From similar triangles, we get $\dfrac{y}{\theta} = \dfrac{F_m}{\alpha}$ or $y^2 = \dfrac{Fm^2}{\alpha^2}\theta^2$

This gives the equation of the signal over the two triangles *OAE* and *DBC*. The signal remains constant over the angle α to (π – α) *i.e.* over an angular distance of (π – α) – α = (π – 2α)

Sum of the squares $= \frac{2F_m^2}{\alpha^2}\int_0^{\alpha} \theta^2\, d\theta + F_m^2(\pi - 2\alpha) = F_m^2(\pi - 4\alpha/3).$

The mean value of the squares is $= \frac{1}{\pi}F_m^2\left(\pi - \frac{4\alpha}{3}\right) = F_m^2\left(1 - \frac{4\alpha}{3\pi}\right)$

r.m.s value $= F_m\sqrt{\left(1 - \frac{4\alpha}{3\pi}\right)}$

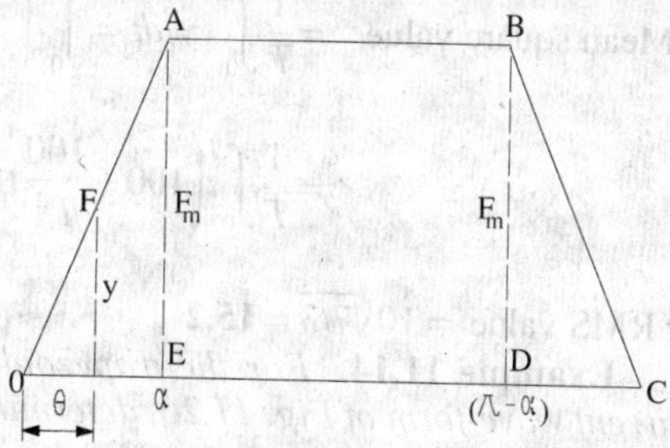

Fig. 11.23

Example 11.12. *Find the average and r.m.s. values of the a.c. voltage whose waveform is given in Fig. 11.24 (a).*

Solution. It is seen [(Fig. 11.24(a)] that the time period of the waveform is 5s. For finding the average value of the waveform, we will calculate the net area of the waveform over one period and then find its average value for one cycle.

$A_1 = 20 \times 1 = 20$ V– s, $A_2 = -5 \times 2 = -10$ V–s

Net area over the full cycle $= A_1 + A_2 = 20 - 10 = 10$ V–s.

Average value = 10 V–s/5s = 2 V.

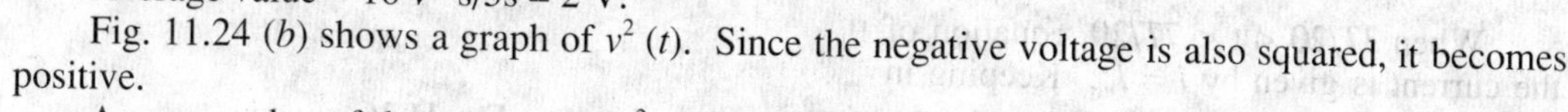

Fig. 11.24 (*b*) shows a graph of $v^2(t)$. Since the negative voltage is also squared, it becomes positive.

Average value of the area = 400 $V^2 \times 1$ s + 25 $V^2 \times 2$ s = 450 V^2–s. The average value of the sum of the square = 450 V^2/–s/5 s = 90V^2 rms value $= \sqrt{90\ V^2} = 9.49$ V.

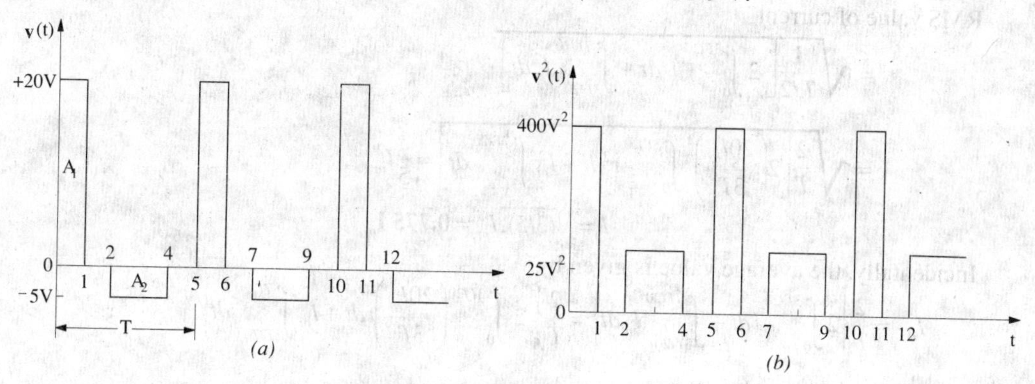

Fig. 11.24

Example 11.13. *What is the significance of the r.m.s. and average values of a wave ? Determine the r.m.s. and average value of the waveform shown in Fig. 11.25.*

(Elect. Technology, Indore Univ., 1984)

Solution. The slope of the curve *AB* is *BC*/*AC* = 20/*T*. Next, consider the function *y* at any time *t*. It is seen that *DE*/*AE* = *BC*/*AC* = 10/*T*

$$\text{or } (y - 10)/t = 10/T$$

$$\text{or } y = 10 + (10/T)t$$

This gives us the equation for the function for one cycle.

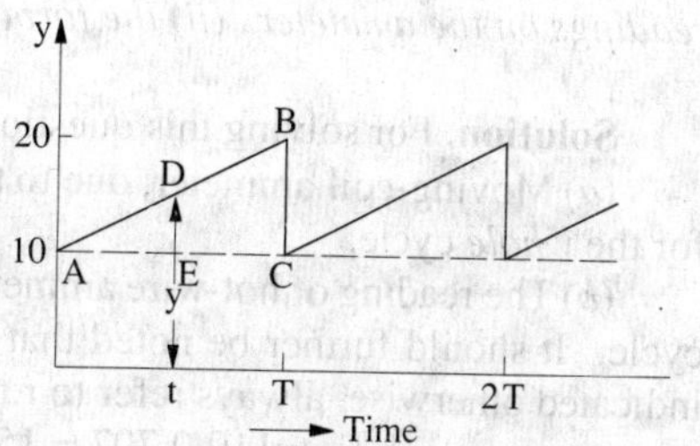

Fig. 11.25

$$Y_{av} = \frac{1}{T}\int_0^T y\, dt = \frac{1}{T}\int_0^T \left(10 + \frac{10}{T}t\right) dt$$

$$= \frac{1}{T}\int_0^T \left[10\,.\,dt + \frac{10}{T}\cdot t\cdot dt\right] = \frac{1}{T}\left|10t + \frac{5t^2}{T}\right|_0^T = \mathbf{15}$$

$$\text{Mean square value} = \frac{1}{T}\int_0^T y^2 dt = \int_0^T\left(10+\frac{10}{T}t\right)^2 dt$$

$$= \frac{1}{T}\int_0^T\left(100+\frac{100}{T^2}t^2+\frac{200}{T}.t\right)dt = \frac{1}{T}\left|100t+\frac{100t^3}{3T^2}+\frac{100t^2}{T}\right|_0^T = \frac{700}{3}$$

or RMS value $= 10\sqrt{7/3} = \mathbf{15.2}$

Example 11.14. *For the trapezoidal current wave-form of Fig. 11.26, determine the effective value.*

(Elect. Technology, Vikram Univ. Ujjain 1985)

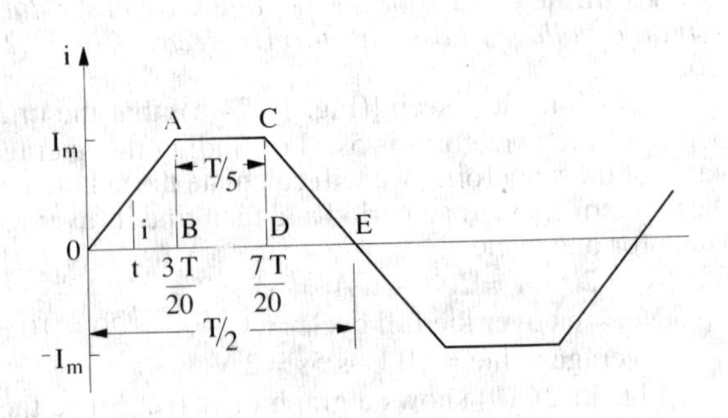

Fig. 11.26

Solution. For $0 < t < 3T/20$, equation of the current can be found from the relation

$$\frac{i}{t} = \frac{I_m}{3T/20} \quad \text{or} \quad i = \frac{20I_m}{3T}.t$$

When $3T/20 < t < 7T/20$, equation of the current is given by $i = I_m$. Keeping in mind the fact that ΔOAB is identical with ΔCDE,

RMS value of current

$$= \sqrt{\frac{1}{T/2}\left[2\int_0^{3T/20} i^2 dt + \int_{3T/20}^{7T/20} I_m^2 dt\right]}$$

$$= \sqrt{\frac{2}{T}\left[2\left(\frac{20I_m}{3T}\right)^2\int_0^{3T/20} t^2 dt + I_m^2\int_{3T/20}^{7T/20} dt\right]} = \frac{3}{5}I_m^2$$

$$\therefore \quad I = \sqrt{(3/5)}.I_m = \mathbf{0.775\ I_m}$$

Incidentally, the average value is given by

$$I_{ac} = \frac{2}{T}\left\{2\int_0^{3T/20} i\,dt + \int_{3T/20.}^{7T/20} I_m dt\right\} = \frac{2}{T}\left\{2\int_0^{3T/20}\left(\frac{20I_m}{3T}\right)t\,dt + I_m\int_{3T/20}^{7T/20} dt\right\}$$

$$= \frac{2}{T}\left\{2\left(\frac{20I_m}{3T}\right)\left|\frac{t^2}{2}\right|_0^{3T/20} + I_m\left|t\right|_{3T/20}^{7T/20}\right\} = \frac{\mathbf{7}}{\mathbf{10}}\cdot\mathbf{I_m}$$

Example 11.15. *A sinusoidal alternating voltage of 110 V is applied across a moving-coil ammeter, a hot-wire ammeter and a half-wave rectifier, all connected in series. The rectifier offers a resistance of 25 Ω in one direction and infinite resistance in opposite direction. Calculate (i) the readings on the ammeters (ii) the form factor and peak factor of the current wave.*

(Elect. Engg. -I Nagpur Univ. 1992)

Solution. For solving this question, it should be noted that

(*a*) Moving-coil ammeter, due to the inertia of its moving system, registers the average current for the *whole* cycle.

(*b*) The reading of hot-wire ammeter is proportional to the average heating effect over the *whole* cycle. It should further be noted that in a.c. circuits, the given voltage and current values, unless indicated otherwise, always refer to r.m.s. values.

$$E_m = 110/0.707 = 155.5\text{ V (approx.)} ; I_m/2 = 155.5/25 = 6.22\text{ A}$$

Average value of current for positive half cycle = 0.637 × 6.22 = 3.96 A

Value of current in the negative half cycle is zero. But, as said earlier, due to inertia of the coil, M.C. ammeter reads the average value for the *whole* cycle.

(*i*) M.C. ammeter reading = 3.96/2 = **1.98 A**

Let R be the resistance of hot-wire ammeter. Average heating effect over the positive half cycle is $\frac{1}{2} I_m^2 \cdot R$ watts. But as there is no generation of heat in the negative half cycle, the average heating effect over the whole cycle is $\frac{1}{4} I_m^2 R$ watt.

Let I be the d.c. current which produces the same heating effect, then

$$I^2 R = \frac{1}{4} I_m^2 R \quad \therefore \quad I = I_m\ 2 = 6.22/2 = 3.11\ A.$$

Hence, hot-wire ammeter will read **3.11 A**

(*ii*) Form factor $= \dfrac{\text{r.m.s value}}{\text{average value}} = \dfrac{3.11}{1.98} = \mathbf{1.57}$; Peak factor $= \dfrac{\text{max. value}}{\text{r.m.s. value}} = \dfrac{6.22}{3.11} = 2$

Example 11.16. *Find the average and effective values of the saw-tooth waveform shown in Fig. 11.27.* **(Elect. Technology, Sumbhal Univ. 1985)**

Solution. Since voltage increases linearly,

$V_{av} = (50 + 0)/2 = \mathbf{25\ V}$

For the interval $0 < t < 2$, slope is $50/2 = 25$. Hence, instantaneous voltage is given by $v = 25t = 25t$ volt.

The r.m.s. or effective value of the voltage is

$$V^2 = \frac{1}{T}\int_0^T v^2\, dt = \frac{1}{2}\int_0^2 625t^2 dt = \frac{625}{2}\int_0^2 t^2 dt$$

$$= \frac{625}{2}\left|\frac{t^3}{3}\right|_0^2 = 834 \quad \therefore \quad V = \sqrt{834} = \mathbf{28.9\ V}.$$

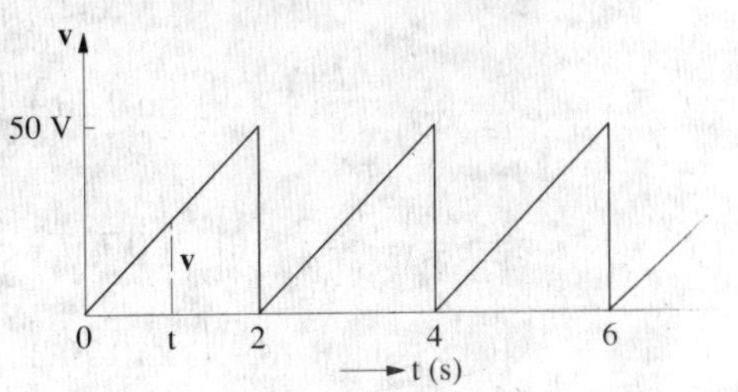

Fig. 11.27

Example 11.17. *A half-wave rectifier which prevents current flowing in one direction is connected in series with an a.c. ammeter and a permanent-magnet moving-coil ammeter. The supply is sinusoidal. The reading on the a.c. ammeter is 10A. Find the reading given by the other ammeter. What should be the readings on the ammeters, if the other half-wave were rectified instead of being cut off ?*

Solution. It should be noted that an a.c. ammeter reads r.m.s. value whereas the d.c. ammeter reads the average value of the rectified current.

As shown in Art. 11.20 from H.W. rectified alternating current, $I = I_m / 2$ and $I_{av} = I_m / \pi$

As a.c. ammeter reads 10 A, hence r.m.s. value of the current is 10 A.

$\therefore\ 10 = I_m/2$ or $I_m = 20$ A

$\therefore\ I_{av} = 20/\pi = \mathbf{6.365\ A}$ –reading of d.c. ammeter.

The full-wave rectified current wave is shown in Fig. 11.28. In this case, mean value of i^2 over a complete cycle is given as

$$= 2\int_0^{\pi} \frac{i^2 d\theta}{2\pi - 0} = \frac{1}{\pi}\int_0^{\pi} I_m^2 \sin^2\theta\, d\theta$$

$$= \frac{I_m^2}{2\pi}\int_0^{\pi}(1 - \cos 2\theta)d\theta = \frac{I_m^2}{2\pi}\left|\theta - \frac{\sin 2\theta}{2}\right|_0^{\pi} = \frac{I_m^2}{2}$$

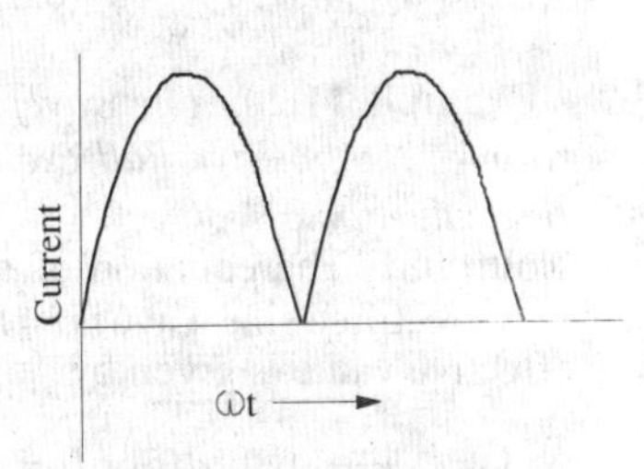

Fig. 11.28

$\therefore\quad I = I_m/\sqrt{2} = 20/\sqrt{2} = 14.14$ A $\therefore$ a.c. ammeter will read **14.14 A**

Now, mean value of i over a complete cycle

$$= \frac{2\int_0^{\pi} I_m \sin\theta\, d\theta}{2\pi} = \frac{I_m}{\pi}\int_0^{\pi}\sin\theta d\theta = \frac{I_m}{\pi}\left|-\cos\theta\right|_0^{\pi} = \frac{2I_m}{\pi} = \frac{2\times 20}{\pi} = \mathbf{12.73\ A}$$

This value, as might have been expected, is twice the value obtained in the previous case.

$\therefore$ d.c. ammeter will read **12.73 A.**

Example 11.18. *A full-wave rectified sinusodial voltage is clipped at $1/\sqrt{2}$ of its maximum value. Calculate the average and RMS values of such a voltage.*

Solution. As seen from Fig. 11.29, the rectified voltage has a period of π and is represented by the following equations during the different intervals.

$$0 < \theta < \pi/4 \; ; v = V_m \sin\theta$$

$$\pi/4 < \theta < 3\pi/4 \; ; v = V_m/\sqrt{2} = 0.707\, V_m$$

$$3\pi/4 < \theta < \pi \; ; \quad v = V_m \sin\theta$$

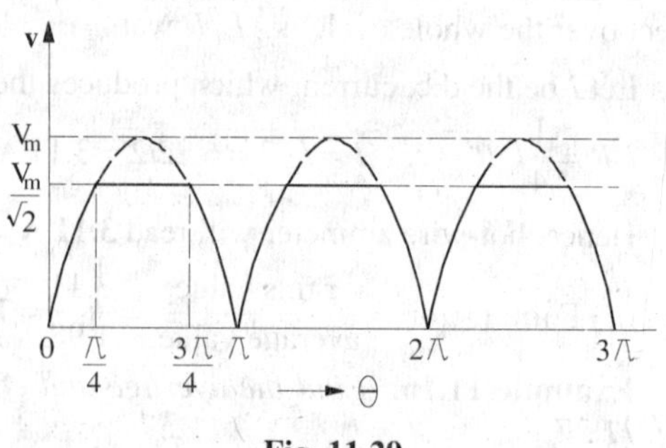

Fig. 11.29

$$\therefore \quad Y_{av} = \frac{1}{\pi}\left\{\int_0^{\pi/4} v\,d\theta + \int_{\pi/4}^{3\pi/4} v\,d\theta + \int_{3\pi/4}^{\pi} v\,d\theta\right\}$$

$$= \frac{I}{\pi}\left\{\int_0^{\pi/4} V_m \sin\theta\,d\theta + \int_{\pi/4}^{3\pi/4} 0.707 V_m\,d\theta + \int_{3\pi/4}^{\pi} V_m \sin\theta\,d\theta\right\}$$

$$= \frac{V_m}{\pi}\left\{\left|-\cos\theta\right|_0^{\pi/4} + 0.707\left|\theta\right|_{\pi/4}^{3\pi/4} + \left|-\cos\theta\right|_{3\pi/4}^{\pi}\right\} = \frac{V_m}{\pi}(0.293 + 1.111 + 0.293) = \mathbf{0.54\ V_m}$$

$$V^2 = \frac{I}{\pi}\left\{\int_0^{\pi/4} V_m^2 \sin^2\theta\,d\theta + \int_{\pi/4}^{3\pi/4} (0.707 V_m)^2 d\theta + \int_{3\pi/4}^{\pi} V_m^2 \sin^2\theta\,d\theta\right\} = 0.341\, V_m^2 \;\therefore\; V = \mathbf{0.584\ V_m}$$

Example 11.19. *A delayed full-wave rectified sinusoidal current has an average value equal to half its maximum value. Find the delay angle θ.* **(Basic Circuit Analysis, Nagpur 1992)**

Solution. The current waveform is shown in Fig. 11.30.

$$I_{av} = \frac{1}{\pi}\int_0^{\pi} I_m \sin\theta\,d\theta = \frac{I_m}{\pi}(-\cos\pi + \cos\theta)$$

Now, $I_{av} = I_m/2$

$$\therefore \quad \frac{I_m}{\pi}(-\cos\pi + \cos\theta) = \frac{I_m}{2}$$

$$\therefore \; \cos\theta = 0.57, \theta = \cos^{-1}(0.57) = \mathbf{55.25°}$$

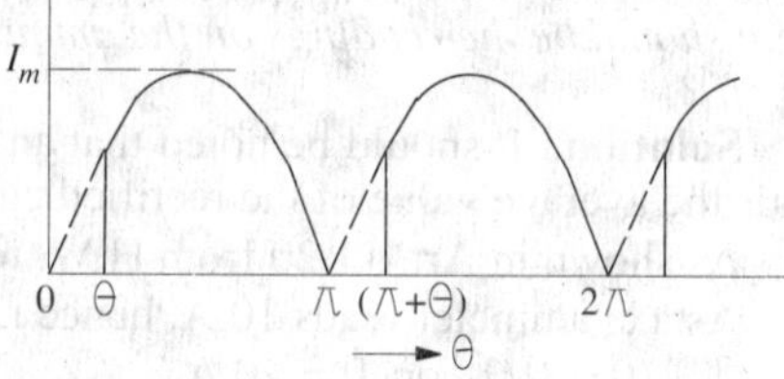

Fig. 11.30

Example 11.20. *The waveform of an output current is as shown in Fig. 11.31. It consists of a portion of the positive half cycle of a sine wave between the angle θ and 180°. Determine the effective value for θ = 30°.* **(Elect. Technology, Vikram Univ. 1984)**

Solution. The equation of the given delayed half-wave rectified sine wave is $i = I_m \sin\omega t = I_m \sin\theta$. The effective value is given by

$$I = \sqrt{\frac{1}{2\pi}\int_{\pi/6}^{\pi} i^2 d\theta} \text{ or } I^2 = \frac{1}{2\pi}\int_{\pi/6}^{\pi} I_m^2 \sin^2\theta . d\theta$$

$$= \frac{I_m^2}{4\pi}\int_{\pi/6}^{\pi}(1 - \cos 2\theta)d\theta = \frac{I_m^2}{4\pi}\left|\left(\theta - \frac{\sin 2\theta}{2}\right)\right|_{\pi/6}^{\pi}$$

$$= 0.242\, I_m^2$$

$$\text{or } I = \sqrt{0.242\, I_m^2} = \mathbf{0.492\ I_m}$$

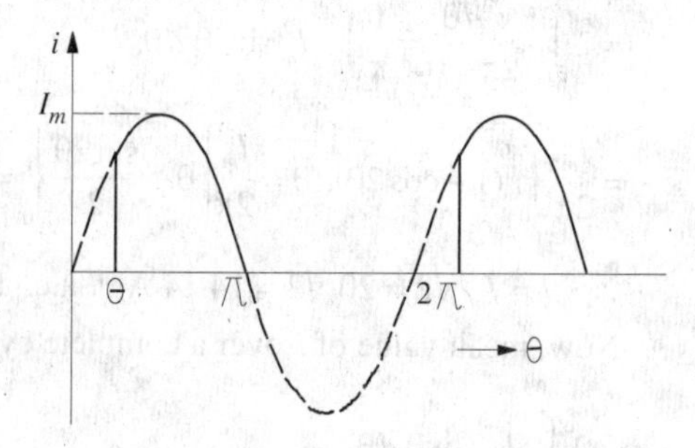

Fig. 11.31

Example 11.21. *Calculate the "form factor" and "peak factor" of the sine wave shown in Fig. 11.32.* **(Elect. Technology-I, Gwalior Univ. 1988)**

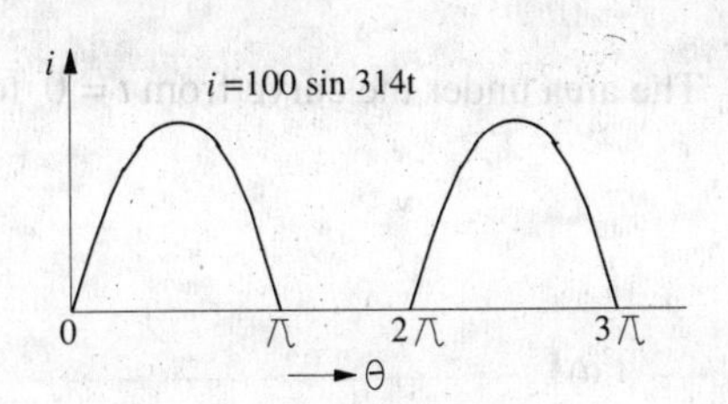

Fig. 11.32

Solution. For $0 < \theta < \pi$, $i = 100 \sin \theta$ and for $\pi < \theta < 2\pi$, $i = 0$. The period is 2π.

$$\therefore \quad I_{av} = \frac{I}{2\pi}\left\{\int_0^{\pi} i\,d\theta + \int_{\pi}^{2\pi} 0\,d\theta\right\}$$

$$= \frac{1}{2\pi}\left\{100 \int_0^{\pi} \sin\theta\,d\theta\right\} = 31.8 \text{ A}$$

$$I^2 = \frac{1}{2\pi}\int_0^{\pi} i^2 d\theta = \frac{100^2}{2\pi}\int_0^{\pi} \sin^2\theta\,d\theta = \frac{100^2}{4} = 2500\;;\; I = 50 \text{ A}$$

$\therefore$ form factor = 50/31.8 = **1.57** ; peak factor = 100/50 = 2.

Example 11.22. *Find the average and effective values of voltage for sinusodial waveform shown in Fig. 11.33.*

(Elect. Science-I Allahabad Univ.1991)

Solution. Although, the given waveform would be integrated from $\pi/4$ to π, it would be averaged over the whole cycle because it is unsymmetrical. The equation of the given sinusoidal waveform is $v = 100 \sin \theta$.

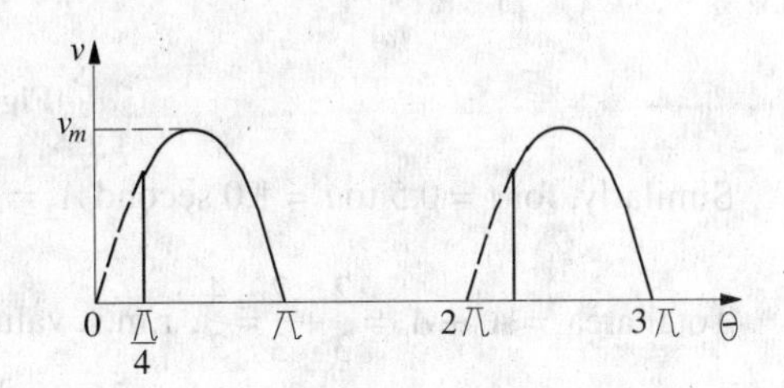

Fig. 11.33

$$\therefore \quad V_{av} = \frac{1}{2\pi}\int_{\pi/4}^{\pi} 100 \sin\theta\,d\theta = \frac{100}{2\pi}\left|-\cos\theta\right|_{\pi/4}^{\pi} = 27.2 \text{ V}$$

$$V^2 = \frac{1}{2\pi}\int_{\pi/4}^{\pi} 100^2 \sin^2\theta\,d\theta = \frac{100^2}{4\pi}(1-\cos 2\theta)d\theta = \frac{100^2}{4\pi}\left|\theta - \frac{\sin 2\theta}{2}\right|_{\pi/4}^{\pi} = \frac{100^2}{4\pi}\left(\pi - \frac{\pi}{4} + \frac{1}{2}\right);$$

$$\therefore \quad v = 47.7 \text{ V}$$

Example 11.23. *Find the r.m.s and average values of the saw tooth waveform shown in Fig. 11.34 (a).*

Solution. The required values can be found by using either graphical method or analytical method.

Graphical Method

The average value can be found by averaging the function from $t = 0$ to $t = 1$ in parts as given below :

Average value of $f(t) = \frac{1}{T}\int_0^T f(t)\,dt = \frac{1}{T} \times$ (net area over one cycle)

Now, area of a right-angled triangle = (1/2) × (base) × (altitude).
Hence, area of the triangle during $t = 0$ to $t = 0.5$ second is

$$A_1 = \frac{1}{2} \times (\Delta t) \times (-2) = \frac{1}{2} \times \frac{1}{2} \times -2 = -\frac{1}{2}$$

Similarly, area of the triangle from $t = 0.5$ to $t = 1$ second is

$$A_2 = \frac{1}{2} \times (\Delta t) \times (+2) = \frac{1}{2} \times \frac{1}{2} \times 2 = -\frac{1}{2}$$

Net area from $t = 0$ to $t = 1.0$ second is $A_1 + A_2 = -\frac{1}{2} + \frac{1}{2} = 0$

Hence, average value of $f(t)$ over one cycle is zero.

For finding the r.m.s. value, we will first square the ordinates of the given function and draw a new plot for $f^2(t)$ as shown in Fig. 11.34(*b*). It would be seen that the squared ordinates form a parabola.

Area under a parabolic curve = $\frac{1}{3}\times$ base $\times$ altitude. The area under the curve from $t = 0$ to $t =$ 0.5 second is; $A_1 = \frac{1}{3}(\Delta t)\times 2^2 = \frac{1}{3}\times\frac{1}{2}\times 4 = \frac{2}{3}$

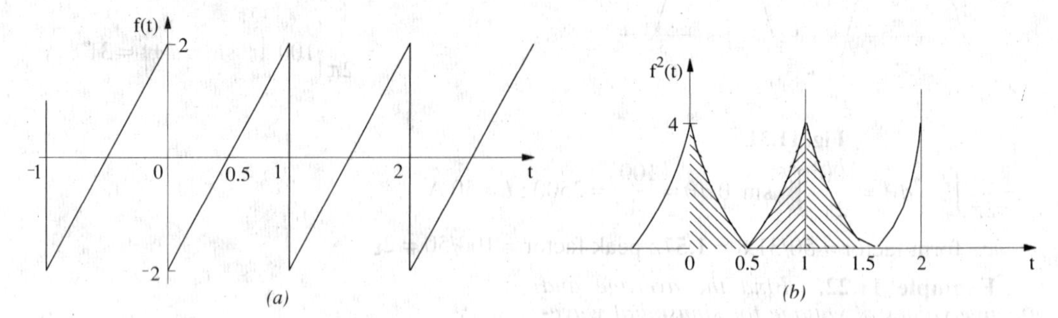

Fig. 11.34

Similarly, for $t = 0.5$ to $t = 1.0$ second $A_2 = \frac{1}{3}(\Delta t)\times 4 = \frac{1}{3}\times\frac{1}{2}\times 4 = \frac{2}{3}$

Total area $= A_1 + A_2 = \frac{2}{3}+\frac{2}{3}=\frac{4}{3}$, r.m.s. value $=\sqrt{\frac{1}{T}\int_0^T f^2(t)dt} = \sqrt{\text{average of } f^2(t)}$

$\therefore$ r.m.s. value $= \sqrt{4/3} = \mathbf{1.15}$

Analytical Method

The equation of the straight line from $t = 0$ to $t = 1$ in Fig. 11.34 (*a*) is

$$f(t) = 4t - 2;\ f^2(t) = 16t^2 - 16t + 4$$

$$\text{Average value } = \frac{1}{T}\int_0^T (4t-2)dt = \frac{1}{T}\left|\frac{4t^2}{2} - 2t\right|_0^T = 0$$

$$\text{rms value } = \sqrt{\frac{1}{T}\int_0^T (16t^2 - 16t + 4)dt} = \sqrt{\frac{1}{T}\left|\frac{16t^3}{3} - \frac{16t^2}{2} + 4t\right|_0^T} = \mathbf{1.15}$$

Example 11.24. *A circuit offers a resistance of 20 Ω in one direction and 100 Ω in the reverse direction. A sinusoidal voltage of maximum value 200 V is applied to the above circuit in series with*

(a) a moving-iron ammeter *(b) a moving-coil ammeter*

(c) a moving-coil instrument with a full-wave rectifier(d) a hot-wire ammeter.

Calculate the reading of each instrument.

Solution. (*a*) The deflecting torque of an *M/I* instrument is proportional to (current)^2. Hence, its reading will be proportional to the average value of i^2 over the whole cycle. Therefore, the reading of such an instrument is :

$$= \sqrt{\left[\frac{1}{2\pi}\left(\int_0^\pi 10^2\sin^2\theta d\theta + \int_\pi^{2\pi} 2^2\sin^2\theta d\theta\right)\right]}$$

$$= \sqrt{\left[\frac{1}{2\pi}\left(\frac{100}{2}\left|\theta - \frac{\sin 2\theta}{2}\right|_0^\pi + \frac{4}{2}\left|\theta - \frac{\sin 2\theta}{2}\right|_\pi^{2\pi}\right)\right]} = \sqrt{26} = \mathbf{5.1\ A}$$

(*b*) An M/C ammeter reads the average current over the whole cycle.

Average current over positive half-cycle is = 10×0.637 = 6.37 A

Average current over positive half-cycle is = -2×0.637 = -1.27 A

$\therefore$ average value over the whole cycle is = (6.37 – 1.27)/2 = **2.55 A**

(*c*) In this case, due to the full-wave rectifier, the current passing through the operating coil of the instrument would flow in the positive direction during both the positive and negative half cycles.

$\therefore$ reading = (6.37 + 1.27)/2 = **3.82 A**

(*d*) Average heating effect over the positive half-cycle is $= \frac{1}{2} I^2_{m1} R$

Average heating effect over the negative half-cycle is $= \frac{1}{2} I m_2^2 R$

where $I_{m1} = 200/20 = 10$ A ; $I_{m2} = 200 / 100 = 2$ A

Average heating effect over the whole cycle is $= (\frac{1}{2} \times 10^2 R + \frac{1}{2} \times 2^2 \times R)/2 = 26 R$

If I is the direct current which produces the same heating effect, then

$$I^2 R = 26 R \qquad \therefore I = \sqrt{26} = \mathbf{5.1\ A}$$

Example 11.25. *A moving-coil ammeter, a hot-wire ammeter and a resistance of 100 Ω are connected in series with a rectifying device across a sinusoidal alternating supply of 200 V. If the device has a resistance of 100 Ω to the current in one direction and 500 Ω to current in opposite direction, calculate the readings of the two ammeters.*

(Elect. Theory and Meas. Madras University, 1985)

Solution. R.M.S. current in one direction is = 200 / (100 + 100) = 1 A
Average current in the first *i.e.* positive half cycle is = 1 / 1.11 = 0.9 A
Similarly, r.m.s. value in the negative half-cycle is = –200 / (100 + 500) = – 1 / 3 A
Average value = (–1/3)/1.11 = –0.3 A
Average value over the whole cycle is = (0.9 – 0.3)/2 = 0.3 A
Hence, M/C ammeter reads **0.3 A**
Average heating effect during the +ve half cycle $= I^2_{rms} \times R = I^2 \times R = R$
Similarly, average heating effect during the –ve half-cycle is $=(-1/3)^2 \times R = R/9$
Here, R is the resistance of the hot-wire ammeter.

Average heating effect over the whole cycle is $= \frac{1}{2}\left(R + \frac{R}{9}\right) = \frac{5R}{9}$

If I is the direct current which produces the same heating effect, then

$$I^2 R = 5R/9 \qquad \therefore \qquad I = \sqrt{5}/3 = 0.745 \text{ A}$$

Hence, hot-wire ammeter indicates **0.745 A**

Example 11.26. *A resultant current wave is made up of two components : a 5A d.c. component and a 50-Hz a.c. component, which is of sinusoidal waveform and which has a maximum value of 5 A.*

(*i*) *Draw a sketch of the resultant wave.*

(*ii*) *Write an analytical expression for the current wave, reckoning t = 0 at a point where the a.c. component is at zero value and where di/dt is positive.*

(*iii*) *What is the average value of the resultant current over a cycle ?*

(*iv*) *What is the effective or r.m.s. value of the resultant current ?*

Solution. (*i*) The two current components and resultant current wave have been shown in Fig. 11.35.

(*ii*) Obviously, the instantaneous value of the resultant current is given by

$$i = (5 + 5 \sin \omega t) = (5 + 5 \sin \theta)$$

(*iii*) Over one complete cycle, the average value of the alternating current is zero. Hence, the average value of the resultant current is equal to the value of d.c. component *i.e.* **5A**

(*iv*) Mean value of i^2 over complete cycle is

$$= \frac{1}{2\pi}\int_0^{2\pi} i^2 d\theta = \frac{1}{2\pi}\int_0^{2\pi} (5 + 5\sin\theta)^2 d\theta$$

$$= \frac{1}{2\pi}\int_0^{2\pi} (25 + 50\sin\theta + 25\sin^2\theta)\, d\theta$$

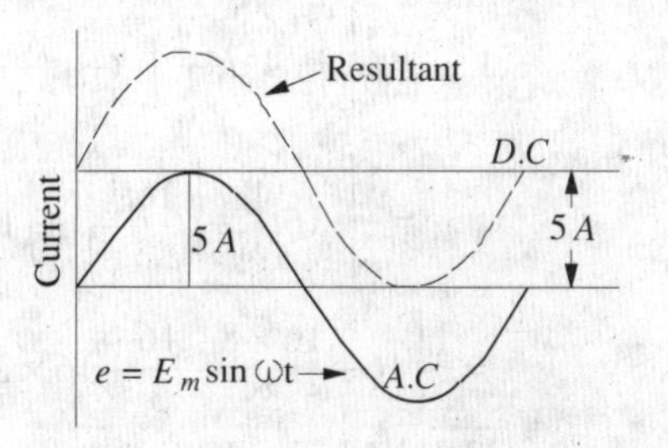

Fig. 11.35

$$= \frac{1}{2\pi}\int_0^{2\pi}\left[25 + 50\sin\theta + 25\left(\frac{1-\cos 2\theta}{2}\right)\right]d\theta = \frac{1}{2\pi}\int_0^{2\pi}(37.5 + 50\sin\theta - 12.5\cos 2\theta)d\theta$$

$$= \frac{1}{2\pi}\left|37.5\,\theta - 50\cos\theta - \frac{12.5}{2}\sin 2\theta\right|_0^{2\pi} = \frac{75\pi}{2\pi} = 37.5\text{ A} \quad \therefore \text{ R.M.S. value } I = \sqrt{37.5} = \mathbf{6.12\ A}$$

Note. In general, let the combined current be given by $i = A + B\sqrt{2}\sin\omega t = A + B\sqrt{2}\sin\theta$ where A represents the value of direct current and B the r.m.s. value of alternating current.

The r.m.s. value of combined current is given by

$$I_{rms} = \sqrt{\left(\frac{1}{2\pi}\int_0^{2\pi} i^2 d\theta\right)} \quad \text{or} \quad I_{r.m.s.}^2 = \frac{1}{2\pi}\int_0^{2\pi} i^2 d\theta = \frac{1}{2\pi}\int_0^{2\pi}(A + B\sqrt{2}\sin\theta)^2 d\theta$$

$$= \frac{1}{2\pi}\int_0^{2\pi}(A^2 + 2B^2\sin^2\theta + 2\sqrt{2}AB\sin\theta)d\theta = \frac{1}{2\pi}\int_0^{2\pi}(A^2 + B^2 - B^2\cos 2\theta + 2\sqrt{2}AB\sin\theta)d\theta$$

$$= \frac{1}{2\pi}\left[A^2\theta + B^2\theta - \frac{B^2\sin 2\theta}{2} - 2\sqrt{2}AB\cos\right]_0^{2\pi} = \frac{1}{2\pi}[2\pi A^2 + 2\pi B^2 - 2\sqrt{2}2AB + 2\sqrt{2}AB]$$

$$= A^2 + B^2 \qquad \therefore \qquad I_{rms} = \sqrt{(A^2 + B^2)}$$

The above example could be easily solved by putting A = 5 and B = 5 $\sqrt{2}$ (because B_{max} = 5)

$\therefore \quad I_{rms} = \sqrt{5^2 + 5/\sqrt{2})^2]} = \mathbf{6.12\ A}$

Example 11.27. *Determine the r.m.s. value of a semi-circular current wave which has a maximum value of a.*

Solution. The equation of a semi-circular wave (shown in Fig. 11.36) is

$$x^2 + y^2 = a^2 \text{ or } y^2 = a^2 - x^2$$

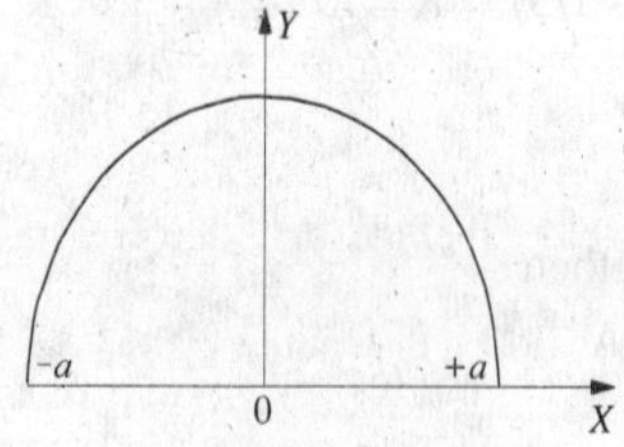

$$\therefore I_{rms} = \sqrt{\frac{1}{2a}\int_{-a}^{+a} y^2 dx} \text{ or } I_{rms}^2 = \frac{1}{2a}\int_{-a}^{+a}(a^2 - x^2)dx$$

$$= \frac{1}{2a}\int_{-a}^{+a}(a^2 dx - x^2 dx) = \frac{1}{2a}\left|a^2 x - \frac{x^3}{3}\right|_{-a}^{+a} = \frac{1}{2a}\left(a^3 - \frac{a^3}{3} + a^3 - \frac{a^3}{3}\right) = \frac{2a^2}{3}$$

Fig. 11.36

$$\therefore \qquad I_{rms} = \sqrt{2a^2/3} = \mathbf{0.816\ a}$$

Example 11.28. *Calculate the r.m.s. and average value of the voltage wave shown in Fig. 11.37.* **(Elect. Measurements, Bangalore Univ. 1985)**

Solution. In such cases, it is difficult to develop a single equation. Hence, it is usual to consider two equations, one applicable from 0 to 1 and an other form 1 to 2 milliseconds.

For t lying between 0 and 1 ms, $v_1 = 4$, For t lying between 1 and 2 ms, $v_2 = -4t + 4$

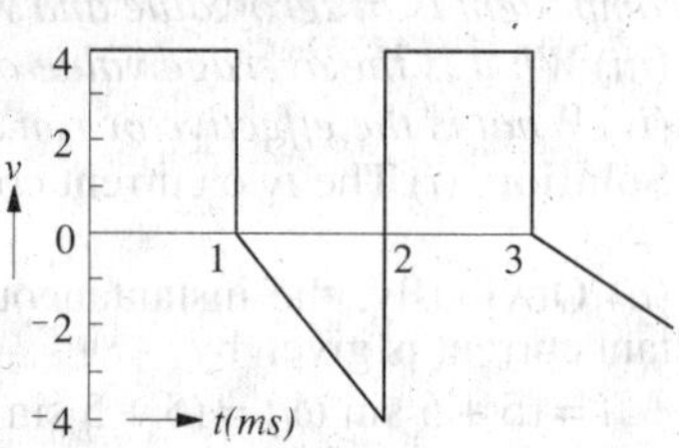

$$\therefore \quad v_{rms} = \sqrt{\frac{1}{2}\left(\int_0^1 v_1^2 dt + \int_1^2 v_2^2\,dt\right)}$$

$$V_{rms}^2 = \frac{1}{2}\left[\int_0^1 4^2 dt + \int_1^2(-4t + 4)^2 dt\right]$$

$$= \frac{1}{2}\left[\left|16t\right|_0^1 + \left|\frac{16t^3}{3}\right|_1^2 + \left|16t\right|_1^2 - \left|\frac{32t^2}{2}\right|_1^2\right]$$

$$= \frac{1}{2}\left[16 + \frac{16\times 8}{3} - \frac{16}{3} + 16\times 2 - 16\times 1 - \frac{32\times 4}{2} + \frac{32\times 1}{2}\right] = \frac{32}{3}$$

Fig. 11.37

$\therefore \quad V_{rms} = \sqrt{32/3} = \mathbf{3.265\ volt}$

$$V_{av} = \frac{1}{2}\left[\int_0^1 v_1 dt + \int_1^2 v_2 dt\right] = \frac{1}{2}\left[\int_0^1 4dt + \int_1^2(-4t + 4)dt\right] = \frac{1}{2}[|4t|_0^1 + | -\frac{4t^2}{2} + 4t|_1^2] = \mathbf{1\ volt}$$

Tutorial Problems No. 11.1

1. Calculate the maximum value of the e.m.f. generated in a coil which is rotating at 50 rev/s in a uniform magnetic field of 0.8 Wb/m^2. The coil is wound on a square former having sides 5 cm in length and is wound with 300 turns. **[188.5 V]**

2. (*a*) What is the peak value of a sinusoidal alternating current of 4.78 r.m.s. amperes ?
(*b*) What is the r.m.s. value of a rectangular voltage wave with an amplitude of 9.87 V ?
(*c*) What is the average value of a sinusoidal alternating current of 31 A maximum value ?
(*d*) An alternating current has a periodic time of 0.03 second. What is its frequency ?
(*e*) An alternating current is represented by i = 70.7 sin 520 t. Determine (*i*) the frequency (*ii*) the current 0.0015 second after passing through zero, increasing positively.
[6.76 A; 9.87 V; 19.75 A ; 33.3 Hz ; 82.8 Hz ; 49.7 A]

3. A sinusoidal alternating voltage has an r.m.s. value of 200 V and a frequency of 50 Hz. It crosses the zero axis in a positive direction when t = 0. Determine (*i*) the time when voltage first reaches the instantaneous value of 200 V and (*ii*) the time when voltage after passing through its maximum positive value reaches the value of 141.4 V. **[(*i*) (0.0025 second (*ii*) 1/300 second]**

4. Find the form factor and peak factor of the triangular wave shown in Fig. 11.38. **[1.155; 1.732]**

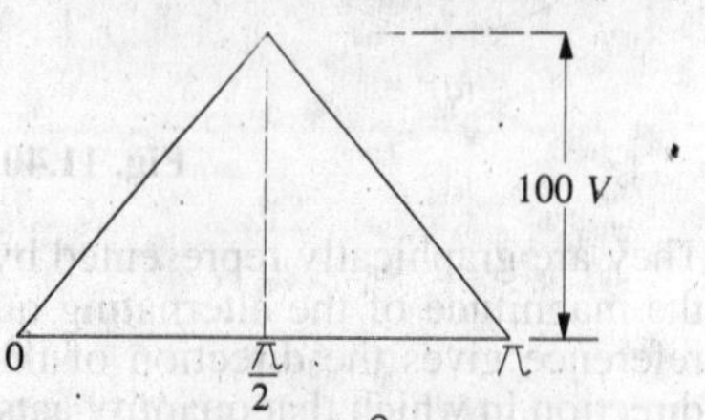

5. An alternating voltage of 200 sin 471 t is applied to a h.w. rectifier which is in series with a resistance of 40 Ω. If the resistance of the rectifier is infinite in one direction and zero in the other, find the r.m.s. value of the current drawn from the supply source. **[2.5 A]**

6. A sinusoidally varying alternating current has an average value of 127.4 A. When its value is zero, then its rate of change is 62,800 A/s. Find an analytical expression for the sine wave. **[i = 200 sin 100πt]**

Fig. 11.38

7. A resistor carries two alternating currents having the same frequency and phase and having the same value of maximum current *i.e.* 10 A. One is sinusoidal and the other is rectangular in waveform. Find the r.m.s. value of the resultant current. **[12.24 A]**

8. A copper-oxide rectifier and a non-inductive resistance of 20 Ω are connected in series across a sinusoidal a.c. supply of 230 V (r.ms.). The resistance of the rectifier is 2.5 Ω in forward direction and 3,000 Ω in the reverse direction. Calculate the r.m.s. and average values of the current.
[r.m.s. value = 5.1 A, average value = 3.22 A)]

9. Find the average and effective values for the waveshape shown in Fig. 11.39 if the curves are parts of a sine wave.
[27.2V, 47.7V] (*Elect. Technology, Indore Univ. 1979*)

10. Find the effective value of the resultant current in a wire which carries simultaneously a direct current of 10 A and a sinusoidal alternating current with a peak value of 15 A.
[14.58A] (*Elect. Technology, Vikram Univ. Ujjain 1978*)

11. Determine the r.m.s. value of the voltage defined by $e = 5 + 5 \sin (314t + \pi/6)$
[6.12 V] (*Elect. Technology, Indore Univ. July, 1979*)

12. Find the r.m.s value of the resultant current in a wire which carries simultaneously a direct current of 10 A and a sinusoidal alternating current with a peak value of 10 A.
[12.25 A] (*Elect. Technology-I ; Delhi Univ. 1981*)

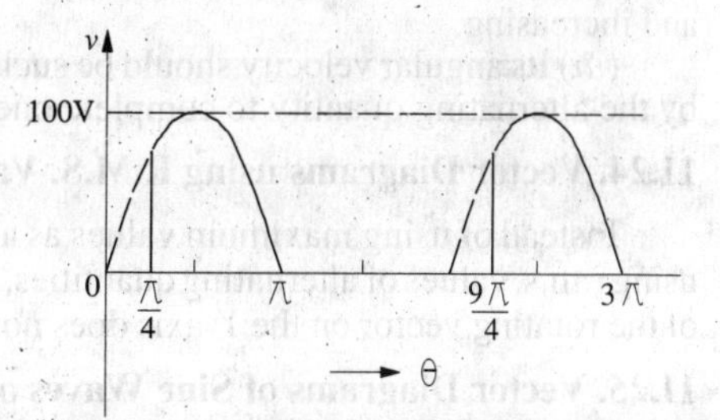

Fig. 11.39

13. An alternating voltage given by e = 150 sin 100πt is applied to a circuit which offers a resistance of 50 ohms to the current in one direction and completely prevents the flow of current in the opposite direction. Find the r.m.s. and average values of this current and its form factor.
([1.5 A, 0.95 A, 1.57] (*Elect. Technology, Indore Univ. Nov. 1978*)

14. Find the relative heating effects of three current waves of equal maximum value, one rectangular, the second semi-circular and the third sinusoidal in waveform. [1 : 2/3 : 1/2] (*Sheffield Univ. U.K*).

15. Calculate the average and root mean-square value, the form factor and peak factor of a periodic current wave having the following values for equal time intervals over half-cycle, changing suddenly from one value to the next :

0, 40, 60, 80, 100, 80, 60, 40, 0 (*A.M.I.E. June 1992*)

16. A sinusoidal alternating voltage of amplitude 100 V is applied across a circuit containing a rectifying device which entirely prevents current flowing in one direction and offers a resistance of 10 ohm to the flow of current in the other direction. A hot wire ammeter is used for measuring the current. Find the reading of instrument. (*Elect. Technology, Punjab Univ. May 1988*)

11.23. Representation of Alternating Quantities

It has already been pointed out that an attempt is made to obtain alternating voltages and currents having sine waveform. In any case, a.c. computations are based on the assumption of sinusoidal voltages and currents. It is, however, cumbersome to continuously handle the instantaneous values in the form of equations of waves like $e = E_m \sin \omega t$ etc. A conventional method is to employ vector method of representing these sine waves. These vectors may then be manipulated instead of the sine functions to achieve the desired result. *In fact, vectors are a short-hand for the representation of alternating voltages and currents and their use greatly simplifies the problems in a.c. work.*

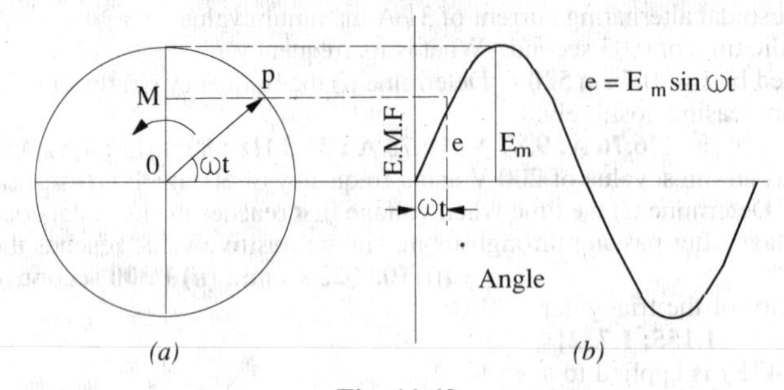

Fig. 11.40

A vector is a physical quantity which has magnitude as well as direction. Such vector quantities are completely known when particulars of their magnitude, direction and the sense in which they act, are given. They are graphically represented by straight lines called vectors. The length of the line represents the magnitude of the alternating quantity, the inclination of the line with respect to some axis of reference gives the direction of that quantity and an arrow-head placed at one end indicates the direction in which that quantity acts.

The alternating voltages and currents are represented by such vectors rotating counter-clockwise with the same frequency as that of the alternating quantity. In Fig. 11.40 (*a*), *OP* is such a vector which represents the maximum value of the alternating current and its angle with *X* axis gives its phase. Let the alternating current be represented by the equation $e = E_m \sin \omega t$. It will be seen that the projection of *OP* and *Y*-axis at any instant gives the instantaneous value of that alternating current.

$$\therefore \quad OM = OP \sin \omega t \quad \text{or} \quad e = OP \sin \omega t = E_m \sin \omega t$$

It should be noted that a line like *OP* can be made to represent an alternating voltage or current if it satisfies the following conditions :

(*i*) Its length should be equal to the peak or maximum value of the sinusoidal alternating current to a suitable scale.

(*ii*) It should be in the horizontal position at the same instant as the alternating quantity is zero and increasing.

(*iii*) Its angular velocity should be such that it completes one revolution in the same time as taken by the alternating quantity to complete one cycle.

11.24. Vector Diagrams using R.M.S. Values

Instead of using maximum values as above, it is very common practice to draw vector diagrams using r.m.s. values of alternating quantities. But it should be understood that in that case, the projection of the rotating vector on the *Y*-axis does not give the instantaneous value of that alternating quantity.

11.25. Vector Diagrams of Sine Waves of Same Frequency

Two or more sine waves of the same frequency can be shown on the same vector diagram because the various vectors representing different waves all rotate counter-clockwise at the same frequency and maintain a fixed position relative to each other. This is illustrated in Fig. 11.41 where a voltage *e* and current *i* of the same frequency are shown. The current wave is supposed to pass upward through zero at the instant when $t = 0$ while at the same time the voltage wave has already advanced an angle α from its zero value. Hence, their equations can be written as

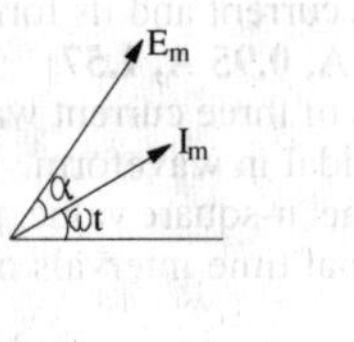

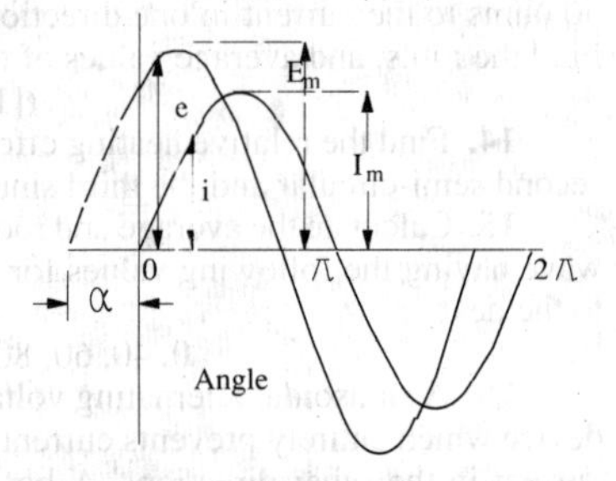

Fig. 11.41

$$i = I_m \sin \omega t$$

and $$e = E_m \sin (\omega t + \alpha)$$

Sine wave of different frequencies cannot be represented on the same vector diagram in a still picture because due to difference in speed of different vectors, the phase angles between them will be continuously changing.

11.26. Addition of Two Alternating Quantities

In Fig. 11.42 (*a*) are shown two rotating vectors representing the maximum values of two sinusoidal voltage waves represented by $e_1 = E_{m1} \sin \omega t$ and $e_2 = E_{m2} \sin (\omega t - \phi)$. It is seen that the sum of the two sine waves of the same frequency is another sine wave of the same frequency but of a different maximum value and phase. The value of the instantaneous resultant voltage e_r, at any instant is obtained by algebraically adding the projections of the two vectors on the *Y*-axis. If these projections are e_1 and e_2, then, $e_r = e_1 + e_2$ at that time. The resultant curve is drawn in this way by adding the ordinates. It is found that the resultant wave is a sine wave of the same frequency as the component waves but lagging behind E_{m1} by an angle α. The vector diagram of Fig. 11.42 (*a*) can be very easily drawn. Lay off E_{m2} lagging ϕ^0 behind E_{m1} and then complete the parallelogram to get E_r.

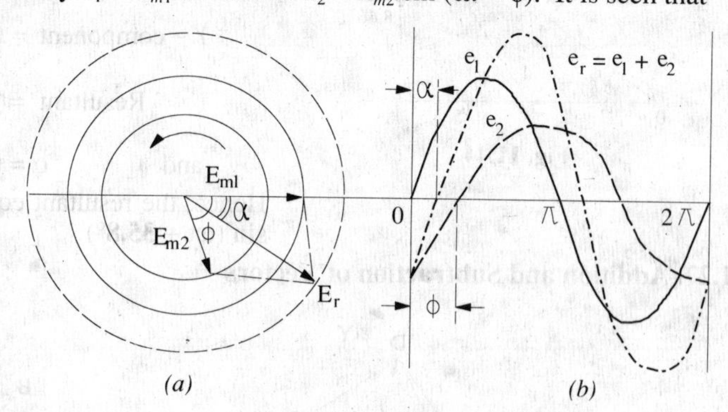

Fig. 11.42

Example 11.29. *Add the following currents as waves and as vectors.*

$$i_1 = 7 \sin \omega t \quad \text{and} \quad i_2 = 10 \sin (\omega t + \pi/3)$$

Solution. As Waves

$$i_r = i_1 + i_2 = 7 \sin \omega t + 10 \sin (\omega t + 60°)$$

$$= 7 \sin \omega t + 10 \sin \omega t \cos 60° + 10 \cos \omega t \sin 60°$$

$$= 12 \sin \omega t + 8.66 \cos \omega t$$

Dividing both sides by $\sqrt{(12^2 + 8.66^2)} = 14.8$, we get

$$i_r = 14.8\left(\frac{12}{14.8} \sin \omega t + \frac{8.66}{14.8} \cos \omega t\right)$$

$$= 14.8(\cos \alpha \sin \omega t + \sin \alpha \cos \omega t)$$

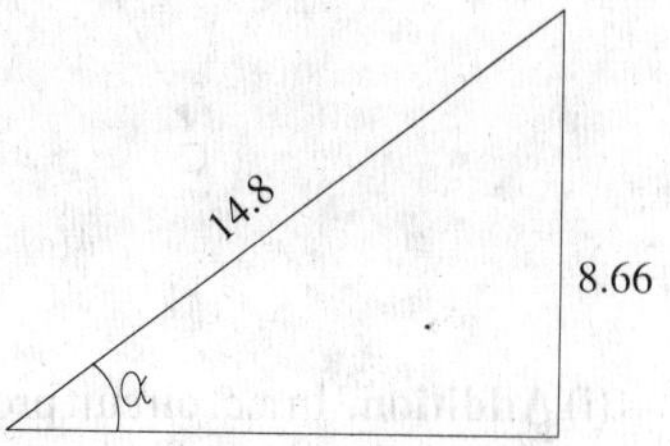

where $\cos \alpha = 12/14.8$ and $\sin \alpha = 8.66/14.8$ –as shown in Fig. 11.43

14.8

8.66

α

Fig. 11.43

$\therefore \quad i_r = 14.8 \sin (\omega t + \alpha)$

where $\tan \alpha = 8.66/12$ or $\alpha = \tan^{-1}(8.66/12) = 35.8°$

$\therefore \quad i_r = 14.8 \sin(\omega t + 35.8°)$

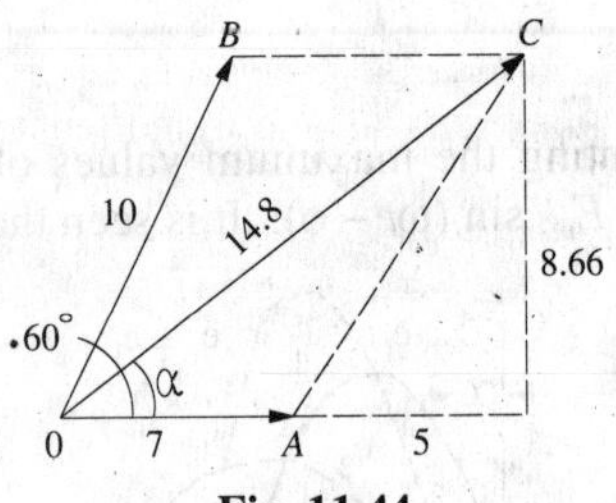

Fig. 11.44

As Vectors

Vector diagram is shown in Fig. 11.44. Resolving the vectors into their horizontal and vertical components, we have

$$X-\text{component} = 7 + 10\cos 60° = 12$$

$$Y-\text{component} = 0 + 10\sin 60° = 8.66$$

$$\text{Resultant} = \sqrt{(12^2 + 8.66^2)} = 14.8 \text{ A}$$

and $$\alpha = \tan^{-1}(8.66/12) = 35.8°$$

Hence, the resultant equation can be written as $\mathbf{i_r = 14.8 \sin(\omega t + 35.8°)}$

11.27. Addition and Subtraction of Vectors

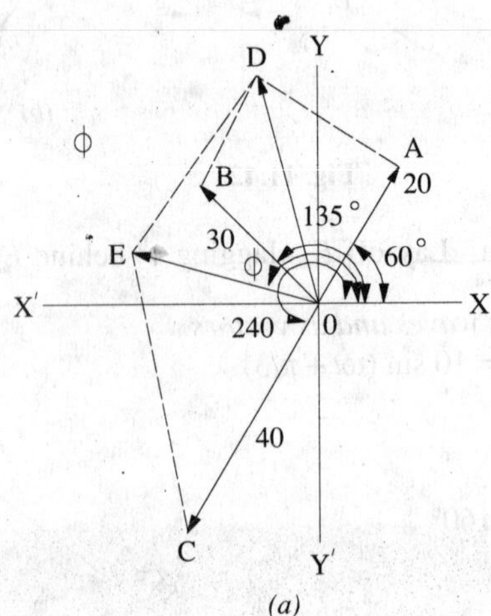

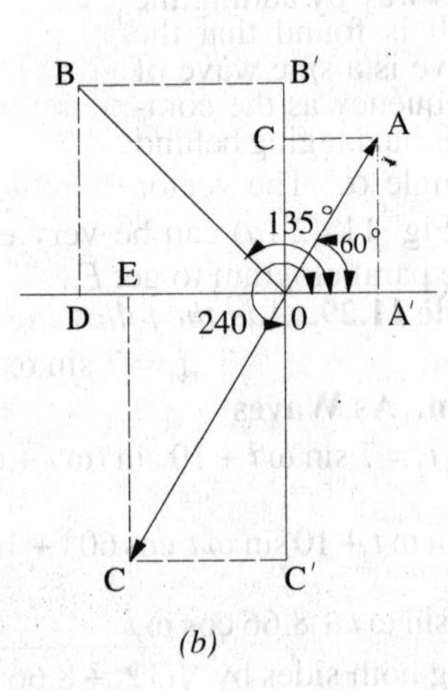

Fig. 11.45

(i) Addition. In a.c. circuit problems we may be concerned with a number of alternating voltages or currents of the same frequency but of different phases and it may be required to obtain the resultant voltage or current. As explained earlier (Art. 11.23) if the quantities are sinusoidal, they may be represented by a number of rotating vectors having a common axis of rotation and displaced from one another by fixed angles which are equal to the phase differences between the respective alternating quantities. The instantaneous value of the resultant voltage is given by the algebraic sum of the projections of the different vectors on Y-axis. The maximum value (or r.m.s. value if the vectors represent that value) is obtained by compounding the several vectors by using the parallelogram and polygon laws of vector addition.

However, another easier method is to resolve the various vectors into their X-and Y-components and then to add them up us as shown in Example 11.30 and 31.

Suppose we are given the following three alternating e.m.fs. and it is required to find the equation of the resultant e.m.f.

$$e_1 = 20\sin(\omega t + \pi/3)$$

$$e_2 = 30\sin(\omega t + 3\pi/4)$$

$$e_3 = 40\sin(\omega t + 4\pi/3)$$

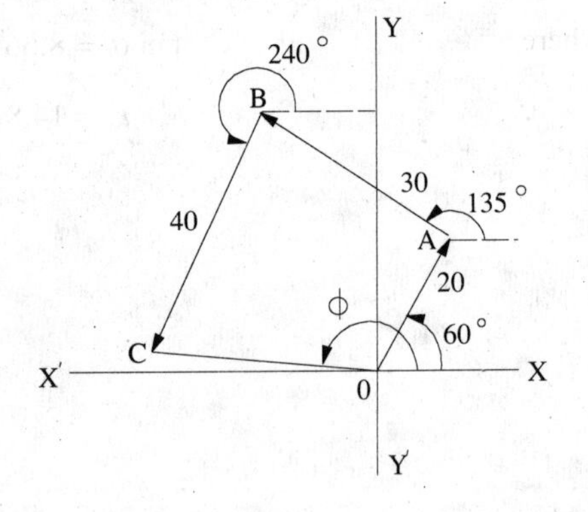

Fig. 11.46

Then the vector diagram can be drawn as explained before and solved in any of the following three ways :

(*i*) By compounding according to parallelogram law as in Fig. 11.45 (*a*).

(*ii*) By resolving the various vectors into their *X*- and *Y*-components as in Fig. 11.45 (*b*).

(*iii*) By laying off various vectors end-on end at their proper phase angles and then measuring the closing vector as shown in Fig. 11.46.

Knowing the magnitude of the resultant vector and its inclination ϕ with *X* axis, the equation of the resultant e.m.f. can be written as $e = E_m \sin(\omega t + \phi)$.

Example 11.30. *Represent the following quantities by vectors :-*
5 sin (2π ft - 1); 3 cos (2π ft + 1) ; 2 sin (2πft + 2.5) and 4 sin (2π ft–1)

Add the vectors and express the result in the form : A sin (2π ft ± Φ)

Solution. It should be noted that all quantities have the same frequency *f*, hence they can be represented vertically on the same vector diagram and added as outlined in Art. 11.27. But before doing this, it would be helpful to express all the quantities as sine functions. Therefore, the second expression 3 cos (2π *ft* + 1) can be written as

$$= 3 \sin\left(2\pi ft + 1 + \frac{\pi}{2}\right) = 3 \sin(2\pi ft + 1 + 1.57) = 3 \sin(2\pi ft + 2.57)$$

The maximum value of each quantity, its phase with respect to the quantity of reference *i.e.* *X* sin 2π*ft*, its horizontal and vertical components are given in the table below :-

Quantity	*Max. value*	*Phase*		*Horizontal component*	*Vertical component*
		radians	*angles*		
(*i*) 5 sin (2π*ft*–1)	5	–1	–57.3°	5 × cos(–57.3°) = 2.7	5 sin (–57.3°) = –4.21
(*ii*) 3 sin (2π*ft*+2.57)	3	+2.57	147.2°	3 × cos 147.2°=–2.52	3 sin 147.2° = 1.63
(*iii*) 2 sin (2π*ft* + 2.5)	2	+2.5	143.2°	2 cos 143.2° =–1.6	2 sin 143.2° = 1.2
(*iv*) 4 sin (2πft–1)	4	–1	–57.3°	4 cos (–57.3°) = 2.16	4 sin (–57.3°) =–3.07
Total				0.74	–4.75

The vector diagram is shown in Fig. 11.47 in which *OA*, *OB*, *OC* and *OD* represent quantities (*i*), (*ii*), (*iii*) and (*iv*) given in the table.

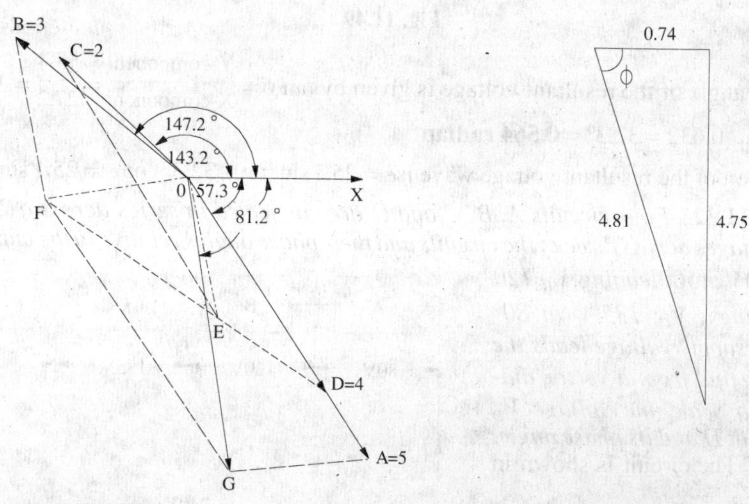

Fig. 11.47 Fig. 11.48

Their resultant is given by *OG* and the net horizontal and vertical components are shown in Fig. 11.48. Resultant $= \sqrt{[0.74^2 + (-4.75)^2]} = 4.81$ and $\tan\theta = -4.75 / 0.74$

$$\therefore \quad \theta = \tan^{-1}(-4.75/0.74) = -81.2° = -1.43 \text{ radians}$$

The equation of the resultant quantity is **4.81 sin (2π ft – 1.43)**

Example 11.31. *Three voltages represented by*

$$e_1 = 20 \sin \omega t \; ; \; e_2 = 30 \sin (\omega t - \pi / 4) \text{ and } e_3 = 40 \cos (\omega t + \pi / 6)$$

act together in a circuit. Find an expression for the resultant voltage. Represent them by appropriate vectors. **(Electro-technics Madras Univ. 1981)** (*Elect. Circuit Nagpur Univ. 1991*)

Solution. First, let us draw the three vectors representing the maximum values of the given alternating voltages.

$e_1 = 20 \sin \omega t$ – here phase angle with X-axis is zero, hence the vector will be drawn parallel to the X-axis

$e_2 = 30 \sin (\omega t - \pi/4)$ –its vector will be below OX by 45°

$e_3 = 40 \cos (\omega t + \pi/6) = 40 \sin (90° + \omega t + \pi/6)$*

$= 40 \sin (\omega t + 120°)$ – its vector will be at 120° with OX in CCW direction.

These vectors are shown in Fig. 11.49 (*a*). Resolving them into X-and Y-components, we get

X-component $= 20 + 30 \cos 45° - 40 \cos 60° = 21.2$ V

Y-component $= 40 \sin 60° - 30 \sin 45° = 13.4$ V

As seen from Fig. 11.49 (*b*), the maximum value of the resultant voltage is

$$OD = \sqrt{21.2^2 + 13.4^2} = 25.1 \text{ V}$$

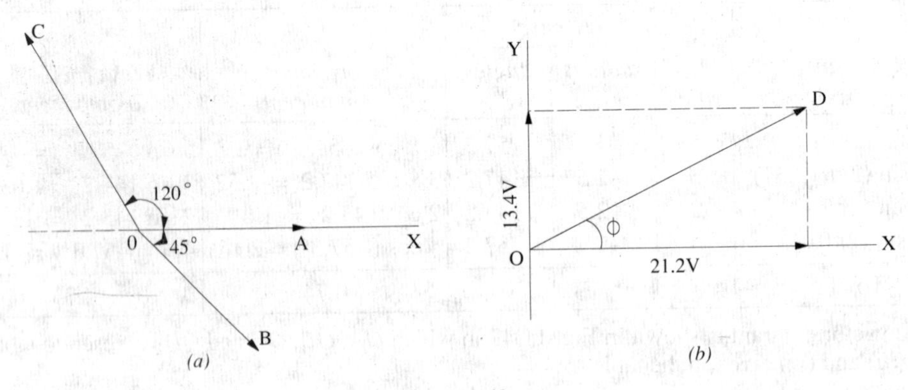

Fig. 11.49

The phase angle of the resultant voltage is given by $\tan \phi = \dfrac{\text{Y-component}}{\text{X-component}} = \dfrac{13.4}{27.2} = \mathbf{0.632}$

$\therefore \; \phi = \tan^{-1} 0.632 = 32.3° = \mathbf{0.564 \text{ radian}}$

The equation of the resultant voltage wave is $e = 25.1 \sin (\omega t + 32.3°)$ or $e = 25.1 \sin (\omega t + 0.564)$

Example 11.32. *Four circuits A, B, C and D are connected in series across a 240 V, 50-Hz supply. The voltages across three of the circuits and their phase angles relative to the current through them are, V_A, 80 V at 50° leading, V_B, 120 V at 65° lagging ; V_C, 135 V at 80° leading. If the supply voltage leads the current by 15°, find from a vector diagram drawn to scale the voltage V_D across the circuit D and its phase angle.*

Solution. The circuit is shown in Fig. 11.50.

(*a*) The vector diagram is shown in Fig. 11.50. The current vector OM is drawn horizontally and is taken as reference vector. Taking a scale of 1 cm

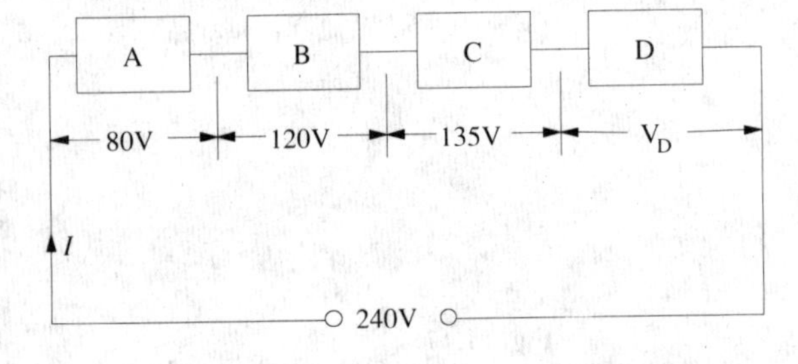

Fig. 11.50

*$\cos \theta = \sin (90° + \theta)$

= 20 V, vector OA is drawn 4 cm in length and leading OM by an angle of 50°. Vector OB represents 120 V and is drawn lagging behind OM by 65°. Their vector sum, as found by Parallelogram Law of Vectors, is given by vector OG.

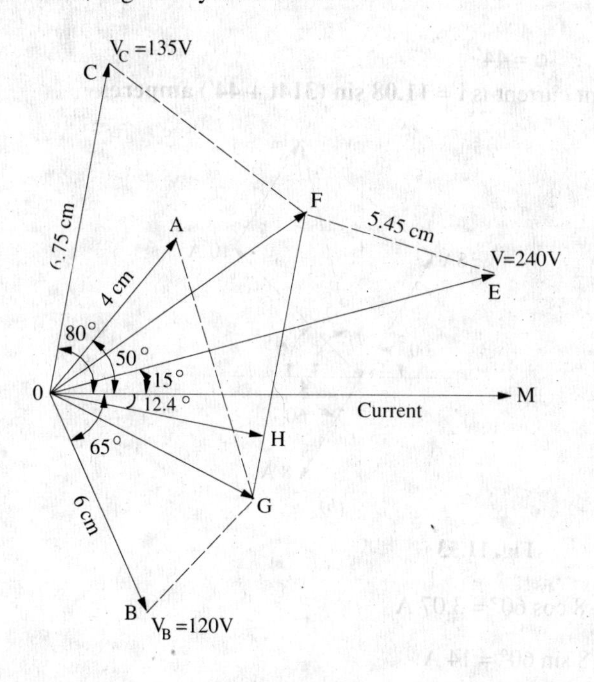

Fig. 11.51

Next, vecor OC is drawn ahead of OM by 80° representing 135 V. Vector OF represents the vector sum of OG and OC. Vector OE represents the applied voltage of 240 V and is drawn 15° ahead of current vector OM. The vector difference of OE and OF gives the required voltage V_D. It is equal to FE. It measures 5.45 cm which means that it represents 20 × 5.45 = 109 V. This vector is transferred to position OH by drawing OH parallel to FE. It is seen that OH lags behind the current vector OM by 12.4°.

Hence, V_D' = **109** volts lagging behind the current by **12.4°**.

(ii) Subtraction of Vectors

If difference of two vectors is required, then one of the vectors is reversed and this reversed vector is then compounded with the other vector as usual.

Suppose it is required to subtract vector OB from vector OA. Then OB is reversed as shown in Fig. 11.52 (*a*) and compounded with OA according to parallelogram law. The vector difference (**A–B**) is given by vector OC.

Similarly, the vector OC in Fig. 12.52 (*b*) represents (**B–A**) *i.e.* the subtraction of **OA** from **OB**.

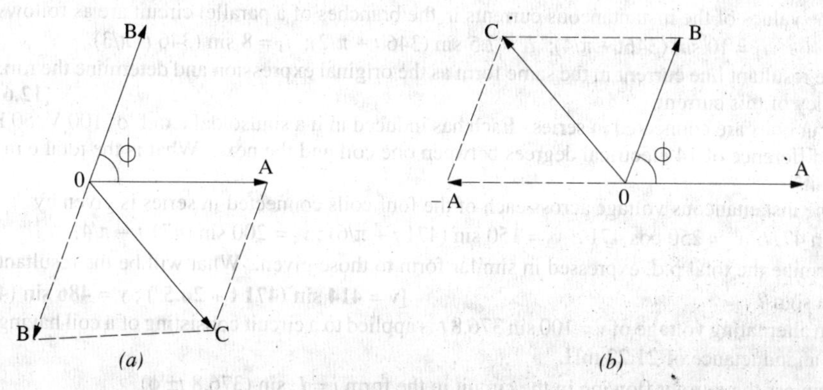

Fig. 11.52

Example 12.33. *Two currents i_1 and i_2 are given by the expressions*

$$i_1 = 10 \sin (314\, t + \pi/4) \text{ amperes and } i_2 = 8 \sin (314\, t - \pi/3) \text{ amperes}$$

Find (a) $i_1 + i_2$ and (b) $i_1 - i_2$. Express the answer in the form $i = I_m \sin (314\, t \pm \phi)$

Solution. (*a*) The current vectors representing maximum values of the two currents are shown in Fig. 11.53 (*a*). Resolving the currents into their *X*-and *Y*-components, we get

X – component $= 10 \cos 45° + 8 \cos 60° = 10/\sqrt{2} + 8/2 = 11.07$ A

Y – component $= 10 \sin 45° - 8 \sin 60° = 0.14$ A

$\therefore I_m = \sqrt{11.07^2 + 0.14^2} = 11.08$ A

$\tan \phi = (0.14/11.07) = 0.01265 \quad \therefore \quad \phi = 44'$

Hence, the equation for the resultant current is **i = 11.08 sin (314t + 44′) amperes**

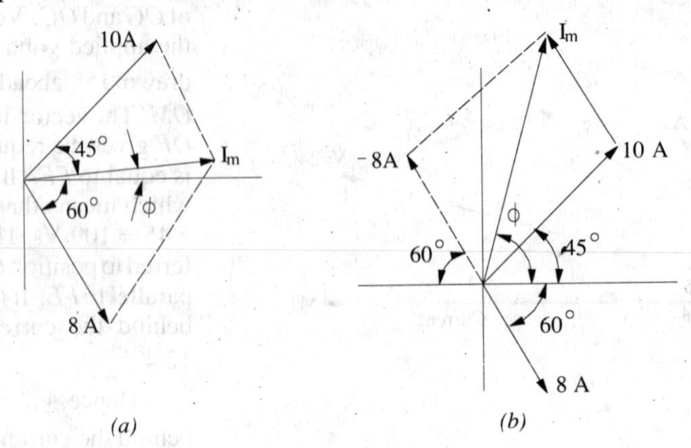

Fig. 11.53

(*b*) X – component $= 10 \cos 45° - 8 \cos 60° = 3.07$ A

Y – component $= 10 \sin 45° + 8 \sin 60° = 14$ A

$\therefore \quad I_m = \sqrt{3.07^2 + 14^2} = 14.33$ A ...Fig. 11.53 (*b*)

$\phi = \tan^{-1}(14/3.07) = 77° \, 38'$

Hence, the equation of the resultant current is

i = 14.33 sin (314 + 77° 38′) amperes

Tutorial Problems No. 11.2

1. The values of the instantaneous currents in the branches of a parallel circuit are as follows :
$i_1 = 5 \sin 346\, t$; $i_2 = 10 \sin (346t + \pi/4)$; $i_3 = 7.5 \sin (346\, t + \pi/2)$; $i_4 = 8 \sin (346\, t - \pi/3)$
Express the resultant line current in the same form as the original expression and determine the r.m.s. value and the frequency of this current. **[12.6 A; 55 Hz]**

2. Four coils are connected in series. Each has induced in it a sinusoidal e.m.f. of 100 V, 50 Hz and there is a phase difference of 14 electrical degrees between one coil and the next. What is the total e.m.f. generated in the circuit ? **[384 V]**

3. The instantaneous voltage across each of the four coils connected in series is given by
$v_1 = 100 \sin 471t$; $v_2 = 250 \cos 471\, t$; $v_3 = 150 \sin (471\, t + \pi/6)$; $v_4 = 200 \sin (471\, t - \pi/4)$

Determine the total p.d. expressed in similar form to those given. What will be the resultant p.d. If v_2 is reversed in sign ? **[v = 414 sin (471 t + 26.5°) ; v = 486 sin (471 t–40°)]**

4. An alternating voltage of $v = 100 \sin 376.8\, t$ is applied to a circuit consisting of a coil having a resistance of 6Ω and an inductance of 21.22 mH.

(*a*) express the current flowing in the circuit in the form $i = I_m \sin (376.8\, t \pm \phi)$

(*b*) If a moving-iron voltmeter, a wattmeter and a frequency meter are connected in the circuit, what would be the respective readings on the instruments ? **[i = 10 sin (376.8 t–53.1°) ; 70.7 V ; 300 W ; 60 Hz]**

5. Three circuits *A*, *B* and *C* are connected in series across a 200-*V* supply. The voltage across circuit *A* is 50 *V* lagging the supply voltage by 45° and the voltage across circuit *C* is 100 V leading the supply voltage by 30°. Determine graphically or by calculation, the voltage across circuit *B* and its phase displacement from the supply voltage. **[79.4 V ; 10°38′ lagging]**

6. Three alternating currents are given by
$i_1 = 141 \sin (\omega t + \pi/4)$ $\quad i_2 = 30 \sin (\omega t + \pi/2)$ $\quad i_3 = 20 \sin (\omega t - \pi/6)$

and are fed into a common conductor. Find graphically or otherwise the equation of the resultant current and its r.m.s. value. **[i = 167.4 sin (ωt + 0.797), I_{rms} = 118.4 A]**

7. Four e.m.fs $e_1 = 100 \sin \omega t$, $e_2 = 80 \sin (\omega t - \pi/6)$, $e_3 = 120 \sin (\omega t + \pi/4)$ and $e_4 = 100 \sin (\omega t - 2\pi/3)$ are induced in four coils connected in series so that the vector sum of four e.m.fs. is obtained. Find graphically or by calculation the resultant e.m.f. and its phase difference with (*a*) e_1 and (*b*) e_2.

If the connections to the coil in which the e.m.f. e_2 is induced are reversed, find the new resultant e.m.f. **[208 sin (ωt–0.202 (*a*) 11°34′ lag (*b*) 18′26′ lead ; 76 sin (ωt + 0.528)**

8. Draw to scale a vector diagram showing the following voltages :

$v_1 = 100 \sin 500\,t$; $v_2 = 200 \sin (500\,t + \pi/3)$; $v_3 = -50 \cos 500\,t$; $v_4 = 150 \sin (500\,t - \pi/4)$

Obtain graphically or otherwise, their vector sum and express this in the form $V_m \sin (500\,t \pm \phi)$, using v_1 as the reference vector. Give the r.m.s. value and frequency of the resultant voltage. **[360.5 sin (500 t + 0.056) ; 217 V ; 79.6 Hz]**

11.28. A.C. through Resistance, Inductance and Capacitance

We will now consider the phase angle introduced between an alternating voltage and current when the circuit contains resistance only, inductance only and capacitance only. *In each case, we will assume that we are given the alternating voltage of equation* $e = E_m \sin \omega t$ and will proceed to find the equation and the phase of the alternating current produced in each case.

11.29. A.C. through Pure Ohmic Resistance Alone

The circuit is shown in Fig. 11.54. Let the applied voltage be given by the equation.

$$v = V_m \sin \theta = V_m \sin \omega t \quad \ldots(i)$$

Let R = ohmic resistance ; i = instantaneous current

Obviously, the applied voltage has to supply ohmic voltage drop only. Hence for equilibrium

$$v = iR ;$$

Putting the value of 'v' from above, we get $V_m \sin \omega t = iR$; $i = \frac{V_m}{R} \sin \omega t$ ...(*ii*)

Current 'i' is maximum when $\sin \omega t$ is unity. ∴ $I_m = V_m/R$ Hence, equation (*ii*) becomes, $i = I_m \sin \omega t$ ...(*iii*)

Comparing (*i*) and (*ii*), we find that the alternating voltage and current are in phase with each other as shown in Fig. 11.55. It is also shown vectorially by vectors V_R and I in Fig. 11.54.

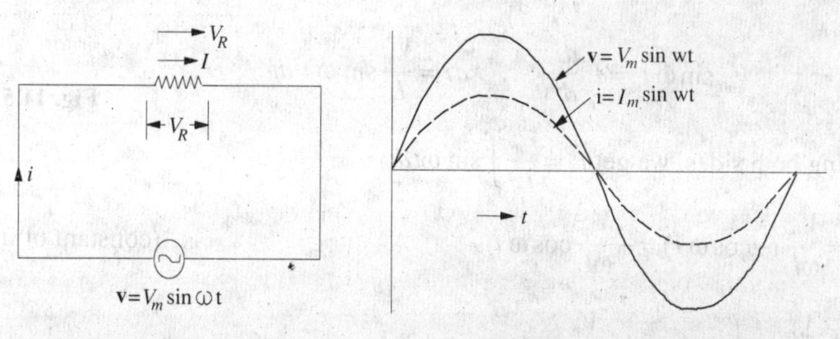

Fig. 11.54 Fig. 11.55

Power. Instantaneous power, $p = vi = V_m I_m \sin^2 \omega t$ ...(Fig. 11.56)

$$= \frac{V_m I_m}{2}(1 - \cos 2\omega t) = \frac{V_m I_m}{2} - \frac{V_m I_m}{2} \cos 2\omega t$$

Power consists of a constant part $\frac{V_m I_m}{2}$ and a fluctuating part $\frac{V_m I_m}{2} \cos 2\omega t$ of frequency double that of voltage and current waves. For a complete cycle, the average value of $\frac{V_m I_m}{2} \cos 2\omega t$ is zero.

Hence, power for the whole cycle is

$$P = \frac{V_m I_m}{2} = \frac{V_m}{\sqrt{2}} \times \frac{I_m}{\sqrt{2}}$$

or $P = V \times I$ watt

where V = r.m.s. value of applied voltage.

I = r.m.s. value of the current.

It is seen from Fig. 11.56 that no part of the power cycle becomes negative at any time. In other words, in a purely resistive circuit, power is never zero. This is so because the instantaneous values of voltage and current are always either both positive or negative and hence the product is always positive.

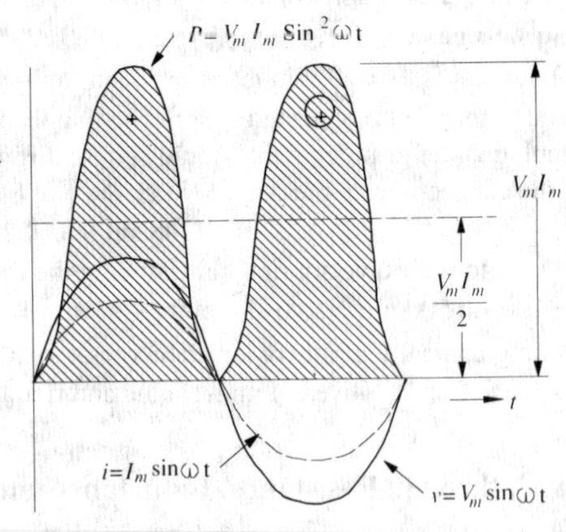

Fig. 11.56

Example 11.34. *A 60-Hz voltage of 115 V (r.m.s.) is impressed on a 100 ohm resistance :*

(i) Write the time equations for the voltage and the resulting current. Let the zero point of the voltage wave be at t = 0 (ii) Show the voltage and current on a time diagram. (iii) Show the voltage and current on a phasor diagram. **[Elect Technology. Hyderabad Univ. 1992]**

Solution. (*i*) $V_{max} = \sqrt{2}\,V = \sqrt{2} \times 115 = 163$ **V**

$I_{max} = V_{max}/R = 163/100 = 1.63$ A; $\phi = 0$; $\omega = 2\pi f = 2\pi \times 60 = 377$ rad/s

The required equations are : $v(t) = 1.63 \sin 377\, t$ and $i(t) = 1.63 \sin 377\, t$

(*ii*) and (*iii*) These are similar to those shown in Fig. 11.54 and 11.55

11.30. A.C. Through Pure Inductance Alone

Whenever an alternating voltage is applied to a purely inductive coil*, a back e.m.f. is produced due to the self-inductance of the coil. The back e.m.f., at every step, opposes the rise or fall of current through the coil. As there is no ohmic voltage drop, the applied voltage has to overcome this self-induced e.m.f. only. So at every step

$$v = L\frac{di}{dt}$$

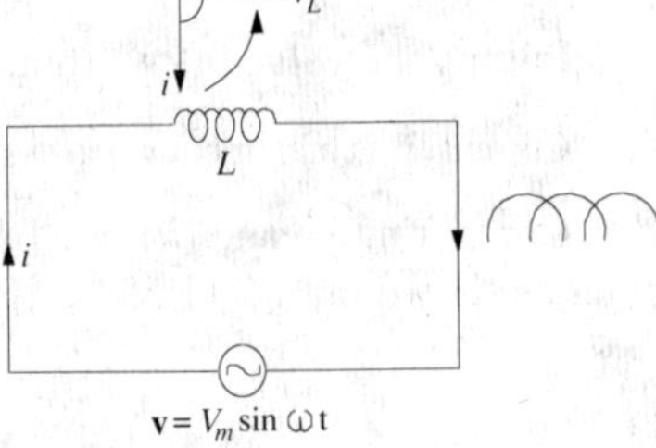

Fig. 11.57

Now $$v = V_m \sin \omega t$$

$$\therefore \quad V_m \sin \omega t = L\frac{di}{dt} \quad \therefore \quad di = \frac{V_m}{L} \sin \omega t\, dt$$

Integrating both sides, we get $i = \frac{V_m}{L}\int \sin \omega t\, dt$

$$= \frac{V_m}{\omega L}(-\cos \omega t) = -\frac{V_m}{\omega L}\cos \omega t \qquad \text{...(constant of integration = 0)}$$

$$\therefore \quad = \frac{V_m}{\omega L}\sin\left(\omega t - \frac{\pi}{2}\right) = \frac{V_m}{X_L}\sin(\omega t - \pi/2) \qquad ...(ii)$$

Max. value of i is $I_m = \frac{V_m}{\omega L}$ when $\sin\left(\omega t - \frac{\pi}{2}\right)$ is unity.

Hence, the equation of the current becomes $i = I_m \sin(\omega t - \pi/2)$.

*By purely inductive coil is meant one that has no ohmic resistance and hence no I^2R loss. Pure inductance is actually not attainable, though it is very nearly approached by a coil wound with such thick wire that its resistance is negligible. If it has some actual resistance, then it is represented by a separate equivalent resistance joined in series with it.

So, we find that if applied voltage is represented by $v = V_m \sin \omega t$, then current flowing in a *purely* inductive circuit is given by $i = I_m \sin\left(\omega t - \frac{\pi}{2}\right)$

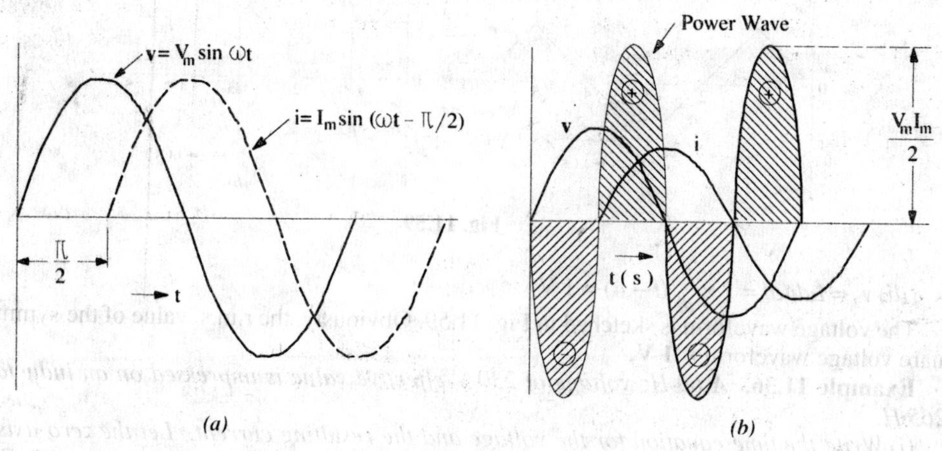

Fig. 11.58

Clearly, the current lags behind the applied voltage by a quarter cycle (Fig. 11.58) or the phase difference between the two is $\pi/2$ with voltage leading. Vectors are shown in Fig. 11.57 where voltage has been taken along the reference axis. We have seen that $I_m = V_m / \omega L = V_m / X_L$ Here 'ωL' plays the part of 'resistance'. It is called the (inductive) *reactance* X_L of the coil and is given in ohms if L is in henry and ω is in radian/second.

Now, $X_L = \omega L = 2\pi f L$ ohm. It is seen that X_L depends directly on frequency of the voltage. Higher the value of f, greater the reactance offered and vice versa.

Power

Instantaneous power $= v_i = V_m I_m \sin \omega t \sin\left(\omega t - \frac{\pi}{2}\right) = -V_m I_m \sin \omega t . \cos \omega t^* = -\frac{V_m I_m}{2} \sin 2\omega t$

Power for whole cycle is $P = -\frac{V_m I_m}{2} \int_0^{2\pi} \sin 2\omega t \, dt = 0$

It is also clear from Fig. 11.58 (*b*) that the average demand of power from the supply for a complete cycle is zero. Here again it is seen that power wave is a sine wave of frequency double that of the voltage and current waves. The maximum value of the instantaneous power is $V_m I_m / 2$.

Example 11.35. *Through a coil of inductance 1 henry, a current of the wave-form shown in Fig. 11.59 (a) is flowing. Sketch the wave form of the voltage across the inductance and calculate the r.m.s. value of the voltage.* **(Elect. Technology, Indore Univ. 1985)**

Solution. The instantaneous current $i(t)$ is given by

(*i*) $0 < t < 1$ second, here slope of the curve is $1/1 = 1$.

$\therefore i = 1 \times t = t$ ampere

(*ii*) $1 < t < 3$ second, here slope is $\frac{1-(-1)}{2} = 1$

$\therefore t = 1 - (1)(t-1) = 1 - (t-1) = (2-t)$ ampere

(*iii*) $3 < t < 4$ second, here slope is $\frac{-1-0}{1} = -1$

(*b*) $i = -1 - (-1)(t-3) = (t-4)$ ampere

---

*Or $p = \frac{1}{2} E_m I_m$ [cos 90° – cos (2 ω t – 90°)]. The constant component $= \frac{1}{2} E_m I_m \cos 90° = 0$. The pulsating component is $-\frac{1}{2} E_m I_m \cos(2\omega t - 90°)$ whose average value over one complete cycle is zero.

The corresponding voltages are (*i*) $v_1 = Ldi/dt = 1 \times 1 = 1\text{V}$ (*ii*) $v_2 = Ldi/dt = 1 \times \frac{d}{dt}(2-t) = -1$ V

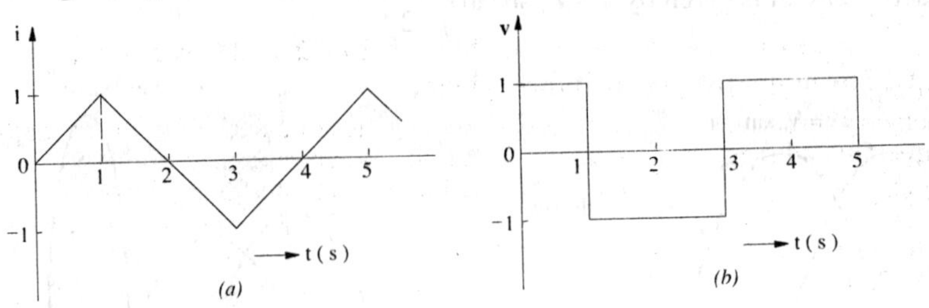

Fig. 11.59

(*iii*) $v_3 = Ldi/dt = 1 \times \frac{d}{dt}(t-4) = 1$ V

The voltage waveform is sketched in Fig. 11.59. Obviously, the r.m.s. value of the symmetrical square voltage waveform is **1 V.**

Example 11.36. *A 60-Hz voltage of 230-V effective value is impressed on an inductance of 0.265 H.*

(*i*) *Write the time equation for the voltage and the resulting current. Let the zero axis of the voltage wave be at t = 0.* (*ii*) *Show the voltage and current on a phasor diagram.* (*iii*) *Find the maximum energy stored in the inductance.* **(Elect. Engineering, Bhagalpur Univ. 1985)**

Solution. $V_{max} = \sqrt{2}\,V = \sqrt{2} \times 230$ V, $f. = 60$ Hz.

$\omega = 2\pi f = 2\pi \times 60 = 377$ rad/s, $X_L = \omega L = 377 \times 0.265 = 100\ \Omega$

(*i*) The time equation for voltage is $v(t) = 230\sqrt{2} \sin 377\,t$;

$I_{max} = V_{max}/X_L = 230\sqrt{2}/100 = 2.3\sqrt{2}$, $\phi = 90°$ (lag)

∴ Current equation is $i(t) = 2.3\sqrt{2} \sin(377t - \pi/2)$ or $= 2.3\sqrt{2} \cos 377\,t$.

(*ii*) It is shown in Fig. 11.56. (*iii*) $E_{max} = \frac{1}{2} LI_{max}^2 = \frac{1}{2} \times 0.265 \times (2.3\sqrt{2})^2 =$ **1.4 J.**

11.31. Complex Voltage Applied to Pure Inductance

In Art. 11.30, the applied voltage was a pure sine wave (*i.e.* without harmonics) given by

$$v = V_m \sin \omega t.$$

The current was given by $i = I_m \sin(\omega t - \pi/2)$

Now, it the applied voltage has a complex form and is (say) given by*

$$v = V_{1m} \sin \omega t + V_{3m} \sin 3\omega t + V_{5m} \sin 5\omega t$$

then the reactances offered to the fundamental voltage wave and the harmonics would be different.

For the fundamental wave, $X_1 = \omega L$. For 3rd harmonic ; $X_3 = 3\omega L$. For 5th harmonic ; $x_5 = 5\omega L$.

Hence, the current would be given by the equation.

$$i = \frac{V_{1m}}{\omega L} \sin\left(\omega t - \frac{\pi}{2}\right) + \frac{V_{3m}}{3\omega L} \sin\left(3\omega t - \frac{\pi}{2}\right) + \frac{V_{5m}}{5\omega L} \sin\left(5\omega t - \frac{\pi}{2}\right)$$

Obviously, the harmonics in the current wave are much smaller than in the voltage wave. For example, the 5th harmonic of the current wave is only 1/5th of the harmonic in the voltage wave. It means that the self-inductance of a coil has the effect of 'smoothening' the current waveform when the voltage waveform is complex *i.e.* contains harmonics.

Example 11.37. *The voltage applied to a purely inductive coil of self-inductance 15.9 mH is given by the equation, v = 100 sin 314t + 75 sin 942 t + 50 sin 1570 t. Find the equation of the resulting current wave.*

Solution. Here $\omega = 314$ rad/s ∴ $X_1 = \omega L = (15.9 \times 10^{-3}) \times 314 = 5\ \Omega$

$$X_3 = 3\omega L = 3 \times 5 = 15\ \Omega;\ X_5 = 5\omega L = 5 \times 5 = 25\ \Omega$$

---

*It is assumed that the harmonics have no individual phase differences.

Hence, the current equation is

$$i = (100/5)\sin(314t - \pi/2) + (75/15)\sin(942t - \pi/2) + (50/25)\sin(1570t - \pi/2)$$

or $\quad i = \mathbf{20\ sin\ (314 - \pi/2) + 5\ sin\ (942t - \pi/2) + 2\ sin\ (1570\ t\ - \pi/2)}$

11.32. A.C. Through Pure Capacitance Alone

When an alternating voltage is applied to the plates of a capacitor, the capacitor is charged first in one direction and then in the opposite direction. When reference to Fig. 11.60, let

$v =$ p.d. developed between plates at any instant

$q =$ charge on plates at that instant.

Then $\quad q = C\,v \qquad$ ...where C is the capacitance

$= C\,V_m \sin \omega t \qquad$ ...putting the value of v.

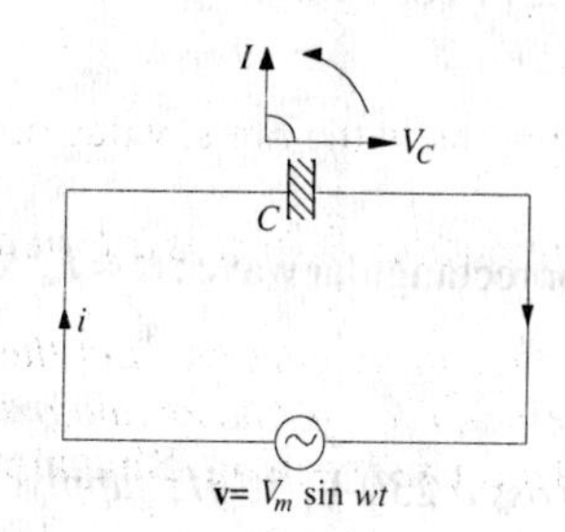

Fig. 11.60

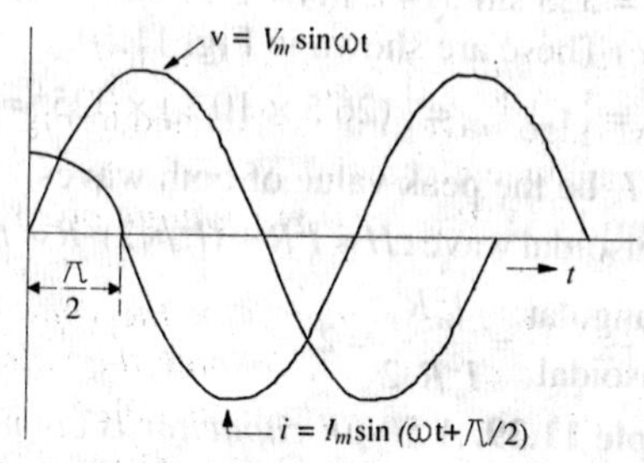

Fig. 11.61

Now, current i is given by the rate of flow of charge.

$$\therefore \quad i = \frac{dq}{dt} = \frac{d}{dt}(Cv_m \sin \omega t) = \omega C V_m \cos \omega t \text{ or } i = \frac{V_m}{1/\omega C}\cos \omega t = \frac{V_m}{1/\omega C}\sin\left(\omega t + \frac{\pi}{2}\right)$$

$$\text{Obviously,} \quad I_m = \frac{V_m}{1/\omega C} = \frac{V_m}{X_C} \qquad \therefore \quad i = I_m \sin\left(\omega t + \frac{\pi}{2}\right)$$

The denominator $X_C = 1/\omega C$ is known as capacitive *reactance* and is in ohms if C is in farad and ω in radian/second. It is seen that if the applied voltage is given by $v = V_m \sin \omega t$, then the current is given by $i = I_m \sin(\omega t + \pi/2)$.

Hence, we find that the current in a pure capacitor leads its voltage by a quarter cycle as shown in Fig. 11.61 or phase difference between its voltage and current is π/2 with the current leading. Vector representation is given in Fig. 11.61. Note that V_C is taken along the reference axis.

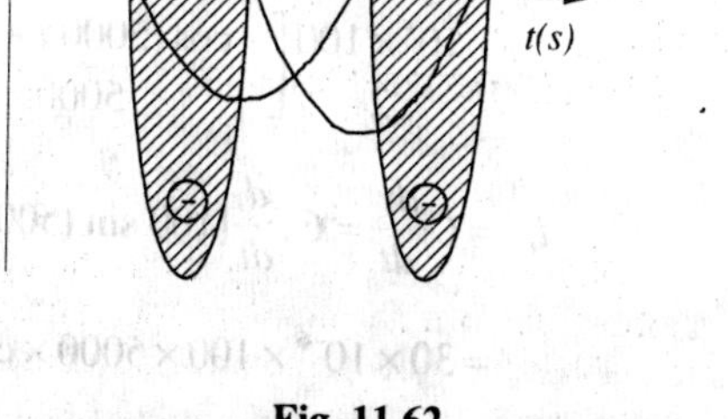

Power

Instantaneous power

$$p = vi = V_m \sin \omega t . I_m \sin(\omega t + 90°)$$

$$= V_m I_m \sin \omega t \cos \omega t^* = \tfrac{1}{2} V_m I_m \sin 2\omega t$$

Power for the whole cycle

$$= \frac{1}{2} V_m I_m \int_0^{2\pi} \sin 2\,\omega t \; dt = 0$$

Fig. 11.62

This fact is graphically illustrated in Fig. 11.62. We find that in a purely capacitive circuit**, the average demand of power from supply is zero (as in a purely inductive circuit). Again, it is seen

*Or power $p = \frac{1}{2}E_m I_m [\cos 90° - \cos(2\omega t - 90°)]$. The constant component is again zero. The pulsating component averaged over one complete cycle is zero.

that power wave is a sine wave of frequency double that of the voltage and current waves. The maximum value of the instantaneous power is $V_m I_m/2$.

Example 11.38. *A 50-Hz voltage of 230 volts effective value is impressed on a capacitance of 26.5 μF. (a) Write the time equations for the voltage and the resulting current. Let the zero axis of the voltage wave be at t = 0. (b) Show the voltage and current on a time diagram.*

(c) Show the voltage and current on a phasor diagram. (d) Find the maximum energy stored in the capacitance. Find the relative heating effects of two current waves of equal peak value, the one sinusoidal and the other rectangular in waveform. **(Elect. Technology, Allahabad Univ. 1991)**

Solution. $V_{max} = 230\sqrt{2} = 325$ V

$\omega = 2\pi \times 50 = 314$ rad/s ; $X_C = 1/\omega C = 10^6/314 \times 26.5 = 120\ \Omega$

$I_{max} = V_{max}/X_C = 325/120 = 2.71$ A, $\phi = 90°$ (lead)

(*a*) $v(t) = 325 \sin 314\,t$; $i(t) = 2.71 \sin(314t + \pi/2) = 2.71 \cos 314t$.

(*b*) & (*c*) These are shown in Fig. 11.59.

(*d*) $E_{max} = \frac{1}{2} CV^2_{max} = \frac{1}{2}(26.5 \times 10^{-6}) \times 325^2 = \mathbf{1.4\ J}$

(*e*) Let I_m be the peak value of both waves.

For sinusoidal wave : $H \propto I^2 R \propto (I_m/\sqrt{2})^2 R \propto I_m^2 R/2$ For rectangular wave : $H \propto I_m^2 R$ –Art. 12.15.

$$\frac{H \text{ rectangular}}{H \text{ sinusoidal}} = \frac{I_m^2 R}{I_m^2 R/2} = 2$$

Example 11.39. *A 50-μF capacitor is connected across a 230-V, 50-Hz supply. Calculate (a) the reactance offered by the capacitor (b) the maximum current and (c) the r.m.s. value of the current drawn by the capacitor.*

Solution. (*a*) $X_C = \frac{1}{\omega C} = \frac{1}{2\pi f_C} = \frac{1}{2\pi \times 50 \times 50 \times 10^{-6}} = \mathbf{63.6\ \Omega}$

(*c*) Since 230 V represents the r.m.s. value,

$\therefore$ $I_{r.m.s.} = 230/X_C = 230/63.6 = \mathbf{3.62\ A}$ (*b*) $I_m = I_{r.m.s.} \times \sqrt{2} = 3.62 \times \sqrt{2} = \mathbf{5.11\ A}$

Example 11.40. *The voltage applied across 3-branched circuit of Fig. 11.23 is given by v = 100 sin (5000t + π/4). Calculate the branch currents and total current.*

Solution. The total instantaneous current is the vector sum of the three branch currents.

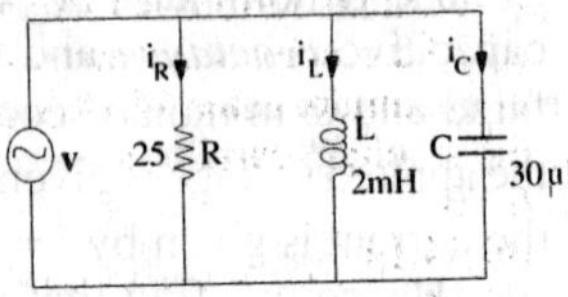

$i_r = i_R + i_L + i_C$. Now $i_R = v/R = 100 \sin(5000\,t + \pi/4)/25$

$= 4 \sin(5000\,t + \pi/4)$

Fig. 11.63

$$i_L = \frac{1}{L}\int v\,dt = \frac{10^3}{2}\int 100 \sin\left(5000t + \frac{\pi}{4}\right) dt$$

$$= \frac{10^3 \times 100}{2}\left[\frac{-\cos(5000\,t + \pi/4)}{5000}\right] = -10 \cos(5000\,t + \pi/4)$$

$$i_C = C\frac{dv}{dt} = C.\frac{d}{dt}[100 \sin(5000\,t + \pi/4)]$$

$$= 30 \times 10^{-6} \times 100 \times 5000 \times \cos(5000\,t + \pi/4) = 15 \cos(5000\,t + \pi/4)$$

$$i_T = 4 \sin(5000\,t + \pi/4) - 10 \cos(5000\,t + \pi/4) + 15 \cos(5000t + \pi/4)$$

$$= \mathbf{4 \sin(5000\,t + \pi/4) + 5 \cos(5000\,t + \pi/4)}$$

---

*By pure capacitor is meant one that has neither resistance nor dielectric loss. If there is loss in a capacitor, then it may be represented by loss in (*a*) high resistance joined in parallel with the pure capacitor or (*b*) by a comparatively low resistance joined in series with the pure capacitor. But out of the two alternatives, usually (*a*) is chosen (Art. 13.8).

OBJECTIVE TESTS–11

1. An a.c. current given by $i = 14.14 \sin(\omega t + \pi/6)$ has an rms value of —— amperes.
(*a*) 10
(*b*) 14.14
(*c*) 1.96
(*d*) 7.07
and a phase of —— degrees.
(*e*) 180
(*f*) 30
(*g*) –30
(*h*) 210

2. If $e_1 = A \sin \omega t$ and $e_2 = B \sin(\omega t - \phi)$, then
(*a*) e_1 lags e_2 by θ
(*b*) e_2 lags e_1 by θ
(*c*) e_2 leads e_1 by θ
(*d*) e_1 is in phase with e_2

3. From the two voltage equations $e_A = E_m \sin 100\pi t$ and $e_B = E_m \sin(100\pi t + \pi/6)$, it is obvious that
(*a*) *A* leads *B* by 30°
(*b*) B achieves its maximum value 1/600 second before *A* does.
(*c*) *B* lags behind *A*
(*d*) *A* achieves its zero value 1/600 second before *B*.

4. The r.m.s. value of a half-wave rectified current is 10 A, its value for full-wave rectification would be —— amperes.
(*a*) 20
(*b*) 14.14
(*c*) $20/\pi$
(*d*) $40/\pi$

5. A resultant current is made of two components : a 10 *A* d.c. component and a sinusoidal component of maximum value 14.14A. The average value of the resultant current is —— amperes.
(*a*) 0
(*b*) 24.14
(*c*) 10
(*d*) 4.14
and r.m.s. value is —— amperes.
(*e*) 10
(*f*) 14.14
(*g*) 24.14
(*h*) 100

6. The r.m.s. value of sinusoidal ac current is equal to its value at an angle of —— degree
(*a*) 60 (*b*) 45 (*c*) 30 (*d*) 90

7. Two sinusoidal currents are given by the equations : $i_1 = 10 \sin(\omega t + \pi/3)$ and $i_2 = 15 \sin(\omega t - \pi/4)$. The phase difference between them is —— degrees.
(*a*) 105 (*b*) 75 (*c*) 15 (*d*) 60

8. A sine wave has a frequency of 50 Hz. Its angular frequency is —— radian/second.
(*a*) $50/\pi$ (*b*) $50/2\pi$ (*c*) 50π (*d*) 100π

9. An a.c. current is given by $i = 100 \sin 100$. It will achieve a value of 50 A after —— second.
(*a*) 1/600 (*b*) 1/300 (*c*) 1/1800 (*d*) 1/900

10. The reactance offered by a capacitor to alternating current of frequency 50 Hz is 10 Ω. If frequency is increased to 100 Hz reactance becomes —— ohm.
(*a*) 20 (*b*) 5 (*c*) 2.5 (*d*) 40

11. A complex current wave is given by $i = 5 + 5 \sin 100\pi t$ ampere. Its average value is —— ampere.
(*a*) 10 (*b*) 0 (*c*) $\sqrt{50}$ (d) 5

12. The current through a resistor has a waveform as shown in Fig. 11.64. The reading shown by a

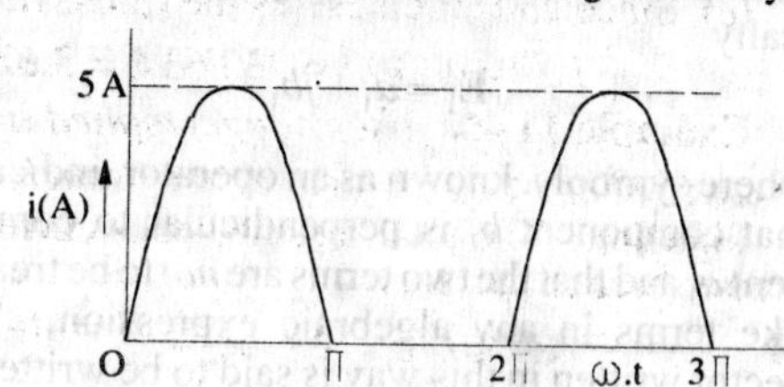

Fig. 11.64

moving coil ammeter will be —— ampere.
(*a*) $5/\sqrt{2}$ (*b*) $2.5/\sqrt{2}$ (*c*) $5/\pi$ (*d*) 0

(Principles of Elect. Engg. Delhi Univ. July 1984)

Answers

1. *a*, *f* **2.** *b* **3.** *b* **4.** *b* **5.** *c*, *f* **6.** *b* **7.** *a* **8.** *d* **9.** *a* **10.** *b* **11.** *d* **12.** *c*

12 COMPLEX NUMBERS

12.1. Mathematical Representation of Vectors

There are various forms or methods of representing vector quantities, all of which enable those operations which are carried out graphically in a phasor diagram, to be performed analytically. The various methods are :

(*i*) *Symbolic Notation*. According to this method, a vector quantity is expressed algebraically in terms of its rectangular components. Hence, this form of representation is also known as Rectangular or Cartesian form of notation or representation.

(*ii*) *Trigonometrical Form* (*iii*) *Exponential Form* (*iv*) *Polar Form*.

12.2. Symbolic Notation

A vector can be specified in terms of its X-component and Y-component. For example, the vector OE_1 (Fig. 12.1) may be completely described by stating that its horizontal component is a_1 and vertical component is b_1. But instead of stating this verbally, we may express symbolically

$$\mathbf{E}_1 = a_1 + jb_1$$

where symbol j, known as an operator, indicates that component b_1 is perpendicular to component a_1 and that the two terms are *not* to be treated like terms in any algebraic expression. The vector written in this way is said to be written in '*complex form*'. In Mathematics, a_1 is known as real component and b_1 as imaginary component but in electrical engineering, these are known as *in phase* (or active) and *quadrature* (or reactive) components respectively.

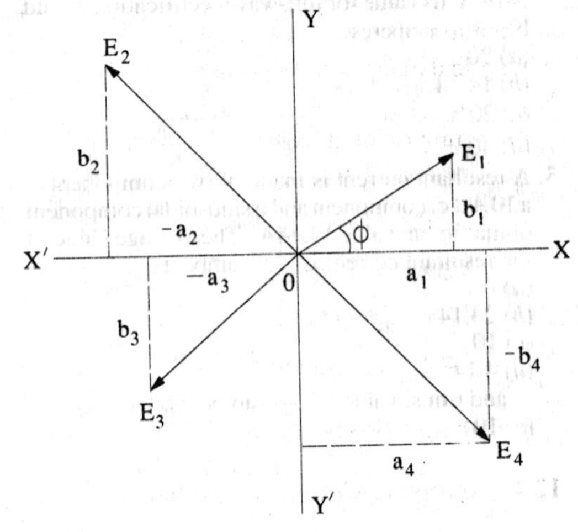

Fig. 12.1

The other vectors OE_2, OE_3 and OE_4 can similarly, be expressed in this form.

$$\mathbf{E}_2 = -a_2 + jb_2\ ;\ \mathbf{E}_3 = -a_3 - jb_3\ ;\ \mathbf{E}_4 = +a_4 - jb_4$$

It should be noted that in this book, a vector quantity would be represented by letters in heavy type and its numerical or scalar value by the same letter in ordinary type.* Other method adopted for indicating a vector quantity is to put an arrow above the letter such as $\vec{E}$

The numerical value of vector $\mathbf{E}_1$ is $\sqrt{a_1^2 + b_1^2}$. Its angle with X-axis is given by $\phi = \tan^{-1}(b_1/a_1)$.

*The magnitude of a vector is sometimes called 'modulus' and is represented by |E| or E.

12.3. Significance of Operator *j*

The letter *j* used in the above expression is a symbol of an operation. Just as symbols ×, + √, ∫ etc. are used with numbers for indicating certain operations to be performed on those numbers, similarly, symbol *j* is used to indicate the counter-clockwise rotation of a vector through 90°. It is assigned a value of $\sqrt{(-1)}$*. The double operation of *j* on a vector rotates it counter-clockwise through 180° and hence reverses its sense because

$$j\,j = j^2 = \sqrt{(-1)^2} = -1$$

When operator *j* is operated on vector **E**, we get the new vector *j***E** which is displaced by 90° in counter-clockwise direction from **E** (Fig. 12.2). Further application of *j* will give j^2**E** = –**E** as shown.

If the operator *j* is applied to the vector j^2**E**, the result is j^3**E** = –*j***E**. The vector j^3**E** is 270° counter-clockwise from the reference axis and is directly opposite to *j***E**. If the vector j^3**E** is, in turn, operated on by *j*, the result will be

$$j^4\mathbf{E} = [\sqrt{(-1)}]^4\mathbf{E} = \mathbf{E}$$

Hence, it is seen that successive applications of the operator *j* to the vector **E** produce successive 90° steps of rotation of the vector in the counter-clockwise direction without in any way affecting the magnitude of the vector.

It will also be seen from Fig. 12.2 that the application of –*j* to **E** yields – *j***E** which is a vector of identical magnitude but rotated 90° *clockwise* from **E**.

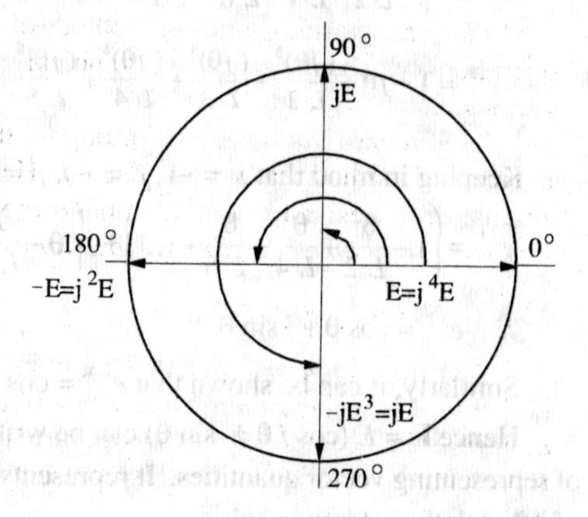

Fig. 12.2

Summarising the above, we have

j = 90° ccw rotation $= \sqrt{(-1)}$

j^2 = 180° ccw rotation $= [\sqrt{(-1)}]^2 = -1;$

j^3 = 270° ccw rotation $= [\sqrt{(-1)}]^3 = -\sqrt{(-1)} = -j$

j^4 = 360° ccw rotation $= [\sqrt{(-1)}]^4 = +1;$

j^5 = 450° ccw rotation $= [\sqrt{(-1)}]^5 = \sqrt{(-1)} = j$

It should also be noted that $\frac{1}{j} = \frac{j}{j^2} = \frac{j}{-1} = -j$

12.4. Conjugate Complex Numbers

Two complex numbers are said to be conjugate if they differ only in the algebraic sign of their quadrature components. Accordingly, the numbers ($a + jb$) and ($a - jb$) are conjugate. The sum of two conjugate numbers gives in-phase (or active) component and their difference gives quadrature (or reactive) component.

12.5. Trigonometrical Form of Vector

From Fig. 12.3, it is seen that *X-component of* **E** is $E\cos\theta$ and *Y*-component is E sign θ. Hence, we can represent the vector **E** in the form: **E** = $E(\cos\theta + j\sin\theta)$

---

*In Mathematics, $\sqrt{(-1)}$ is denoted by *i* but in electrical engineering *j* is adopted because letter *i* is reserved for representing current. This helps to avoid confusion.

This is equivalent to the rectangular form $\mathbf{E} = a + jb$ because $a = E \cos 0$ and $b = E \sin \theta$. In general, $\mathbf{E} = E (\cos \theta \pm j \sin \theta)$.

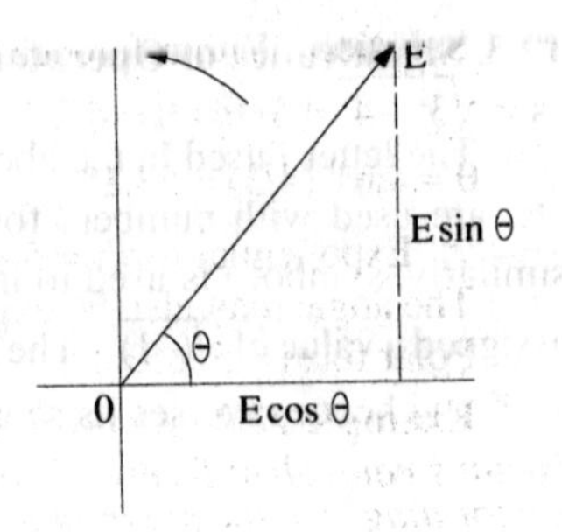

Fig. 12.3

12.6. Exponential Form of Vector

It can be proved that $e^{\pm j\theta} = (\cos \theta \pm j \sin \theta)$

This equation is known as Euler's equation after the famous mathematician of 18th century : Leonard Euler.

This equation follows directly from an inspection of Maclaurin* series expansions of sin θ, cos θ and $e^{j\theta}$.

When expanded into series form :

$$\cos\theta = 1 - \frac{\theta^2}{L2} + \frac{\theta^4}{L4} - \frac{\theta^6}{L6} + \ldots \text{and } \sin\theta = \theta - \frac{\theta^3}{L3} + \frac{\theta^5}{L5} - \frac{\theta^7}{L7} + \ldots$$

$$e^{j\theta} = 1 + j\theta + \frac{(j\theta)^2}{L2} + \frac{(j\theta)^3}{L3} + \frac{(j\theta)^4}{L4} + \frac{(j\theta)^5}{L5} + \frac{(j\theta)^6}{L6} + \ldots$$

Keeping in mind that $j^2 = -1, j^3 = -j, j^4 = 1, j^5 = -j, j^6 = -1$, we get

$$e^{j\theta} = \left(1 - \frac{\theta^2}{L2} + \frac{\theta^4}{L4} - \frac{\theta^6}{L6} + \ldots\right) + j\left(\theta - \frac{\theta^3}{L3} + \frac{\theta^5}{L5} - \frac{\theta^7}{L7} \ldots\right)$$

$$\therefore \quad e^{j\theta} = \cos\theta + j\sin\theta$$

Similarly, it can be shown that $e^{-j\theta} = \cos \theta - j \sin \theta$

Hence $\mathbf{E} = E (\cos j\, \theta \pm \sin \theta)$ can be written as $\mathbf{E} = Ee^{\pm j\theta}$. This is known as exponential form of representing vector quantities. It represents a vector of numerical value E and having phase angle of $\pm\theta$ with the reference axis.

12.7. Polar Form of Vector Representation

The expression $E (\cos \theta + j \sin \theta)$ is written in the simplified form of $E \angle \theta$. In this expression, E represents the magnitude of the vector and θ its inclination (in ccw direction) with the X-axis. For angles in clockwise direction the expression becomes $E \angle -\theta$. In general, the expression is written as $E \angle \pm\theta$. *It may be pointed out here that* $E \angle \pm\theta$ *is simply a short-hand or symbolic style of writing* $Ee^{\pm j\theta}$. Also, the form is purely conventional and does not possess the mathematical elegance of the various other forms of vector representation given above.

Summarizing, we have the following alternate ways of representing vector quantities

(*i*) Rectangular form (or complex form) $\mathbf{E} = a + jb$

(*ii*) Trigonometrical form $\mathbf{E} = E(\cos \theta \pm j \sin \theta)$

(*iii*) Exponential form $\mathbf{E} = Ee^{\pm j\theta}$

(*iv*) Polar form (conventional) $\mathbf{E} = E \angle \pm\theta$.

Example 12.1. *Write the equivalent exponential and polar forms of vector 3 + j4. How will you illustrate the vector means of diagram ?*

---

*Functions like cos θ, sin θ and $e^{j\theta}$ etc. can be expanded into series form with the help of Macalurin's Theorem. The theorem states : $f(\theta) = f(0) + \frac{f'(0)\theta}{1} + \frac{f''(0)\theta^2}{\angle 2} + \frac{f'''(0)\,\theta^3}{\angle 3} + \ldots$ where $f(\theta)$ is function of θ which is to be expanded, $f(0)$ is the value of the function when $\theta = 0$, $f'(0)$ is the value of first derivative of $f(\theta)$ when $\theta = 0$, $f''(0)$ is the value of second derivative of function $f(\theta)$ when $\theta = 0$ etc.

Solution. With reference to Fig. 12.4, magnitude of the vector is $= \sqrt{3^2+4^2} = 5$. $\tan\theta = 4/3$

$\therefore\ \theta = \tan^{-1}(4/3) = 53.1°$

$\therefore$ Exponential form $= 5\,e^{j53.1°}$

The angle may also be expressed in radians.

Polar form $= 5\,\angle 53.1°$.

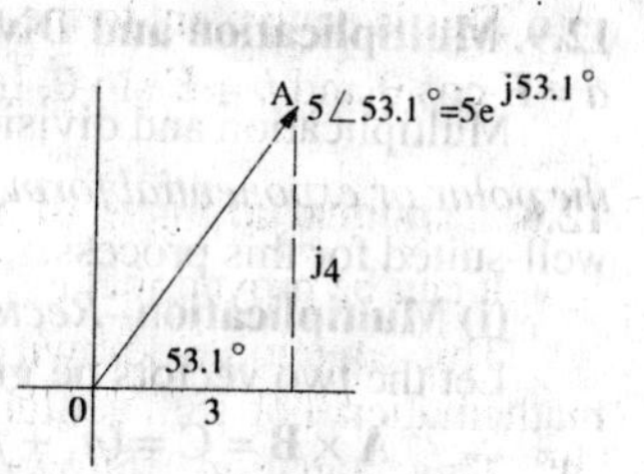

Fig. 12.4

Example 12.2. *A vector is represented by 20 $e^{-j2\pi/3}$. Write the various equivalent forms of the vector and illustrate by means of a vector diagram, the magnitude and position of the above vector.*

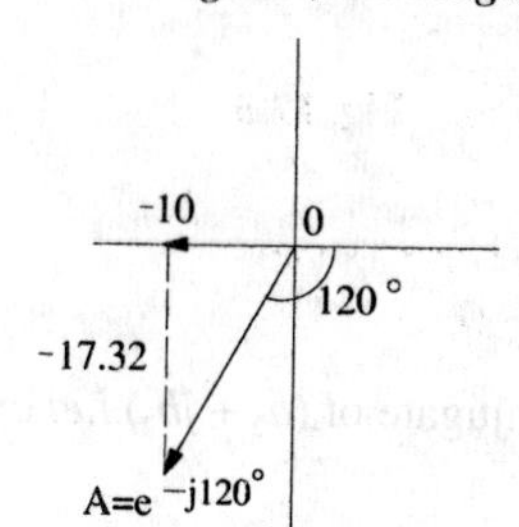

Fig. 12.5

Solution. The vector is drawn in a direction making an angle of $2\pi/3 = 120°$ in the clockwise direction (Fig. 12.5). The clockwise direction is taken because the angle is negative.

(i) Rectangular Form $(a) = 20\cos(-120°) = -10$; $b = 20\sin(-120°) = -17.32$

$\therefore$ Expression is $= (-10 - j17.32)$

(ii) Polar Form is $20\,\angle -120°$

12.8. Addition and Subtraction of Vector Quantities

Rectangular form is best suited for addition and subtraction of vector quantities. Suppose we are given two vector quantities $\mathbf{E}_1 = a_1 + jb_1$ and $\mathbf{E}_2 = a_2 + jb_2$ and it is required to find their sum and difference.

Addition. $\mathbf{E} = \mathbf{E}_1 + \mathbf{E}_2 = a_1 + jb_1 + a_2 + jb_2 = (a_1 + a_2) + j(b_1 + b_2)$

The magnitude of resultant vector **E** is $\sqrt{(a_1+a_2)^2 + (b_1+b_2)^2}$

The position of **E** with respect to *X*-axis is $\theta = \tan^{-1}\left(\dfrac{b_1+b_2}{a_1+a_2}\right)$

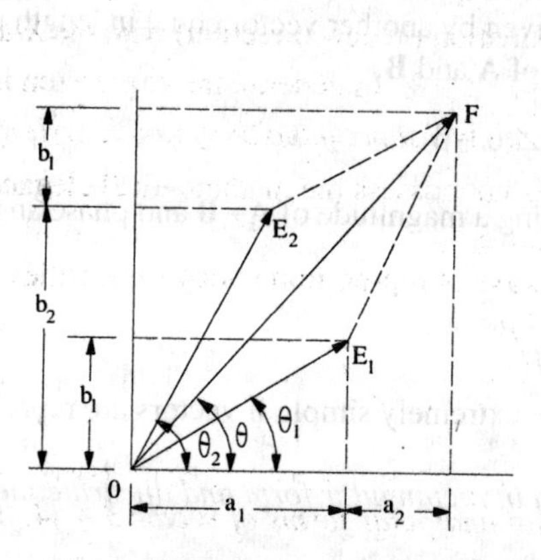

Fig. 12.6

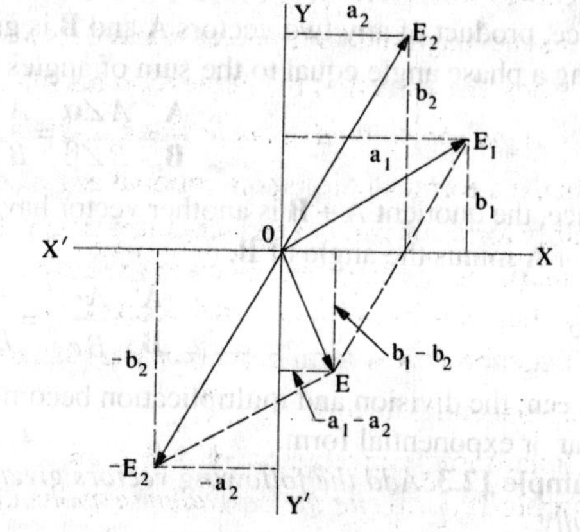

Fig. 12.7

A graphic representation of the addition process is shown in Fig. 12.6.

Subtraction. $\mathbf{E} = \mathbf{E}_1 - \mathbf{E}_2 = (a_1 + jb_1) - (a_2 + jb_2) = (a_1 - a_2) + j(b_1 - b_2)$

Magnitude of $\mathbf{E} = \sqrt{(a_1-a_2)^2 + (b_1-b_2)^2}$

Its position with respect to x – axis is given by the angle $\theta = \tan^{-1}\left(\dfrac{b_1-b_2}{a_1-a_2}\right)$.

The graphic representation of the process of subtraction is shown in Fig. 12.7.

12.9. Multiplication and Division of Vector Quantities

Multiplication and division of vectors becomes very simple and easy *if they are represented in the polar or exponential form.* As will be shown below, the rectangular form of representation is not well-suited for this process.

(i) Multiplication–*Rectangular form*

Let the two vectors be given by $\mathbf{A} = a_1 + jb_1$ and $\mathbf{B} = a_2 + jb_2$

$$\therefore \quad \mathbf{A} \times \mathbf{B} = \mathbf{C} = (a_1 + jb_1)(a_2 + jb_2) = a_1a_2 + j^2b_1b_2 + j(a_1b_2 + b_1a_2)$$

$$= (a_1a_2 - b_1b_2) + j(a_1b_2 + b_1a_2) \qquad (\because j^2 = -1)$$

The magnitude of $\mathbf{C} = \sqrt{[(a_1a_2 - b_1b_2)^2 + (a_1b_2 + b_1a_2)^2]}$

Its angle with respect to X-axis is given by $\theta = \tan^{-1}\left(\frac{a_1b_2 + b_1a_2}{a_1a_2 - b_1b_2}\right)$

(ii) Division – *Rectangular Form*: $\frac{\mathbf{A}}{\mathbf{B}} = \frac{a_1 + jb_1}{a_2 + jb_2} = \frac{(a_1 + jb_1)(a_2 - jb_2)}{(a_2 + jb_2)(a_2 - jb_2)}$

Both the numerator and denominator have been multiplied by the conjugate of $(a_2 + jb_2)$ *i.e.* by $(a_2 - jb_2)$

$$\therefore \quad \frac{\mathbf{A}}{\mathbf{B}} = \frac{(a_1a_2 + b_1b_2) + j(b_1a_2 - a_1b_2)}{a_2^2 + b_2^2} = \frac{a_1a_2 + a_1b_2}{a_2^2 + b_2^2} + j\frac{b_1a_2 - a_1b_2}{a_2^2 + b_2^2}$$

The magnitude and the angle with respect to X-axis can be found in the same way as given above.

As will be noted, both the results are somewhat awkward but unfortunately, there is no easier way to perform multiplication in rectangular form.

(iii) Multiplication–*Polar Form*

Let $\quad \mathbf{A} = a_1 + jb_1 = A\angle\alpha = A\,e^{j\alpha} \quad$ where $\alpha = \tan^{-1}(b_1/a_1)$

$$\mathbf{B} = a_2 + jb_2 = B\angle\beta = B\,e^{j\beta} \quad \text{where} \quad \beta = \tan^{-1}(b_2/a_2)$$

$$\therefore \quad \mathbf{AB} = A\angle\alpha \times B\angle\beta = AB\angle(\alpha + \beta)^* \text{ or } \mathbf{AB} = Ae^{j\alpha} \times Be^{j\beta} = ABe^{j(\alpha+\beta)}$$

Hence, product of any two vectors **A** and **B** is given by another vector equal in length to $\mathbf{A} \times \mathbf{B}$ and having a phase angle equal to the sum of angles of **A** and **B**.

$$\frac{\mathbf{A}}{\mathbf{B}} = \frac{A\angle\alpha}{B\angle\beta} = \frac{A}{B}\angle(\alpha - \beta)$$

Hence, the quotient $\mathbf{A} \div \mathbf{B}$ is another vector having a magnitude of $\mathbf{A} \div \mathbf{B}$ and phase angle equal to angle of **A** minus the angle of **B**.

Also $$\frac{\mathbf{A}}{\mathbf{B}} = \frac{Ae^{j\alpha}}{Be^{j\beta}} = \frac{A}{B}e^{j(\alpha-\beta)}$$

As seen, the division and multiplication become extremely simple if vectors are represented in their polar or exponential form.

Example 12.3. *Add the following vectors given in rectangular form and illustrate the process graphically.*

$$^*\mathbf{A} = A(\cos\alpha + j\sin\alpha) \text{ and } \mathbf{B} = B(\cos\beta + j\sin\beta)$$

$$\therefore \quad \mathbf{AB} = AB(\cos\alpha\cos\beta + j\sin\alpha\cos\beta + j\cos\alpha\sin\beta + j^2\sin\alpha\sin\beta)$$

$$= AB[\cos\alpha\cos\beta - \sin\alpha\sin\beta) + j(\sin\alpha\cos\beta + \cos\alpha\sin\beta)]$$

$$= AB[\cos(\alpha + \beta) + j\sin(\alpha + \beta)] = AB\angle(\alpha + \beta)$$

$A = 16 + j12, \quad B = -6 + j10.4$

Solution. $\mathbf{A} + \mathbf{B} = C = (16 + j12) + (-6 + j10.4) = 10 + j\,22.4$

$\therefore$ Magnitude of $\mathbf{C} = \sqrt{(10^2 + 22.4^2)} = 25.5$ units

Slope of $\mathbf{C} = \theta = \tan^{-1}\left(\dfrac{22.4}{10}\right) = 65.95°$

The vector addition is shown in Fig. 12.8.

$\alpha = \tan^{-1}(12/16) = 36.9°$

$\beta = \tan^{-1}(-10.4/6) = -240°$ or $120°$

The resultant vector is found by using parallelogram law of vectors (Fig. 12.8).

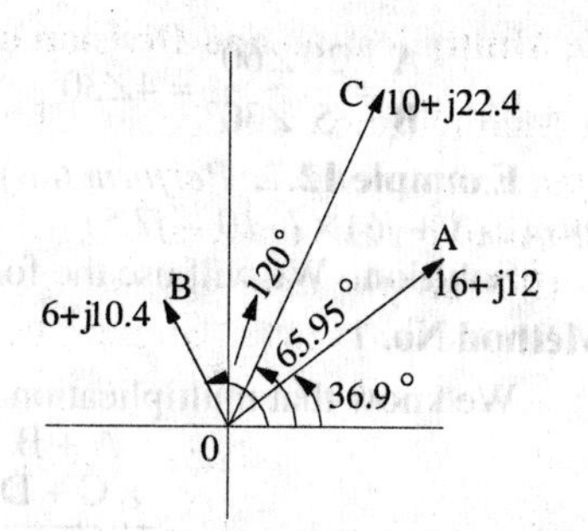

Fig. 12.8

Example 12.4. *Perform the following operation and express the final result in the polar form :*
$5\angle 30° + 8\angle -30°$ **(Elect. Engg. & Electronics Bangalore Univ. 1989)**

Solution. $5\angle 30° = 5\,(\cos 30° + j\sin 30°) = 4.33 + j2.5$

$8\angle -30° = 8[\cos(-30°) + j\sin(-30°)] = 8(0.866 - 0.5) = 6.93 - j4$

$\therefore \quad 5\angle 30° + 8\angle -30° = 4.33 + j2.5 + 6.93 - j4 = 11.26 - j1.5 = \sqrt{11.26^2 + 1.5^2}\angle \tan^{-1}(-1.5/11.26) = 11.35$

$\tan^{-1}(-0.1332) = 11.35 \quad \angle 7.6°$

Example 12.5. *Subtract the following given vectors from one another :*

$A = 30 + j\,52$ *and* $B = -39.5 - j\,14.36$

Solution. $\mathbf{A} - \mathbf{B} = \mathbf{C} = (30 + j\,52) - (-39.5 - j\,14/36) = 69.5 + j\,66.36$

$\therefore$ Magnitude of $\mathbf{C} = \sqrt{(66.5^2 + 66.36^2)} = 96$

Slope of $\mathbf{C} = \tan^{-1}(66.36/69.5) = 43.6° \quad \therefore \quad \mathbf{C} = 96\angle 43.6°$.

Similarly $\mathbf{B} - \mathbf{A} = -69.5 - j\,66.36 = 96\angle 223.6°$* or $= 96\angle -136.4°$

Example 12.6. *Given the following two vectors :*

$A = 20\angle 60°$ *and* $B = 5\angle 30°$

Perform the following indicated operations and illustrate graphically (i) A × B and (ii) A/B.

Solution. (*i*) $\mathbf{A} \times \mathbf{B} = \mathbf{C} = 20\angle 60° \times 5\angle 30° = 100\angle 90°$

Vectors are shown in Fig. 12.9.

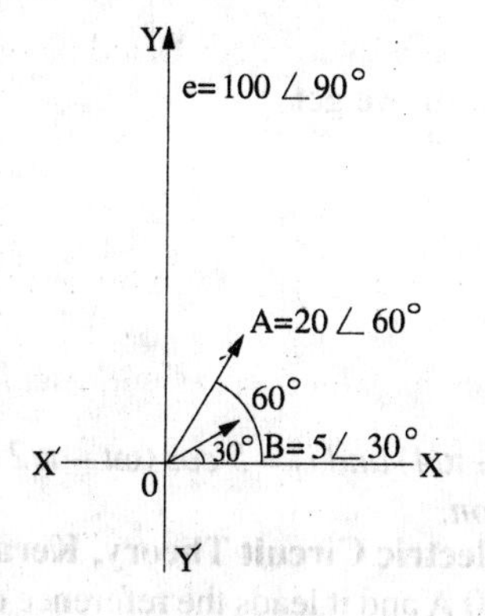

Fig. 12.9

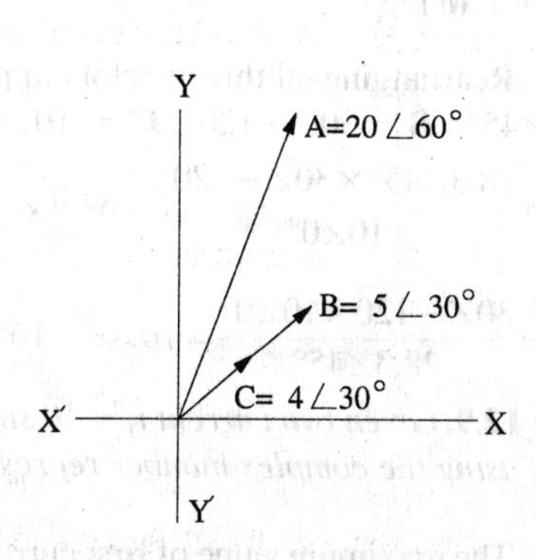

Fig. 12.10

*$\tan\theta = 66.36/69.5$ or $\theta = \tan^{-1}(66.36/69.5) = 43.6°$. Since both components are negative, the vector lies in third quadrant. Hence, the angle as measured from +ve direction of X-axis and in the CCW directions is $= (180 + 43.6) = 223.6°$.

(ii) $\frac{\mathbf{A}}{\mathbf{B}}=\frac{20\angle 60^\circ}{5\angle 30^\circ}=4\angle 30^\circ$ —Fig. 12.10

Example 12.7. *Perform the following operation and the final result may be given in the polar form : (8 + j6) × (–10 – j7.5)* **(Elect. Engg. & Electronics Bangalore Univ. 1990)**

Solution. We will use the following two methods to solve the above question.

Method No. 1

We know that multiplication of (*A* + *B*) and (*C* + *D*) can be found as under :

$$\begin{array}{r} A+B \\ \times C+D \\ \hline CA+CB \\ +\ DA+DB \\ \hline CA+CD+DA+DB \\ \hline \end{array}$$

Similarly, the required multiplication can be carried out as follows :

$$\begin{array}{l} 8+j6 \\ \times -10-j7.5 \\ \hline -80-j\,60 \\ -j60-j^2\,45 \\ \hline -80-j120+45 \end{array}$$

or $-35-j120=\sqrt{(-35)^2+(-120)^2}\;\angle \tan^{-1}(120/35)=125\tan^{-1}3.42=125\angle 73.8^\circ$

Since both the components of the vector are negative, it obviously lies in the third quadrant. As measured from the *X*-axis in the CCW direction, its angle is = 180° + 73.8° = 253.8°. Hence, the product vector can be written as 125∠53.8°.

Method No. 2 $8+j\,6=10\angle 36.9^\circ,\ -10-j\,7.5=12.5\tan^{-1}0.75=12.5\angle 36.9^\circ.$

Again as explained in Method 1 above, the actual angle of the vector is 180° + 36.9° = 216.9°

$\therefore\ -10-j7.5=12.5\angle 216.9\ \therefore\ \ 10\angle 36.9^\circ\times 12.5\angle 216.9^\circ=125<253.8^\circ$

Example 12.8. *The following three vectors are given : -*

A = 20 + *j*20, *B* = 30 ∠–120 and c = 10 + *j*0

Perform the following indicated operations :

(i) $\frac{AB}{C}$ *and* (ii) $\frac{BC}{A}$.

Solution. Rearranging all three vectors in polar form, we get

$\mathbf{A}=28.3\angle 45^\circ,\ \mathbf{B}=30\angle -120^\circ,\ \mathbf{C}=10\angle 0^\circ$

(i) $\frac{\mathbf{AB}}{\mathbf{C}}=\frac{28.3\angle 45^\circ\times 30\angle -120^\circ}{10\angle 0^\circ}=84.9\angle -75^\circ$

(ii) $\frac{\mathbf{BC}}{\mathbf{A}}=\frac{30\angle -120\times 10\angle 0^\circ}{28.3\angle 45^\circ}=10.6\angle -165^\circ$

Example 12.9. *Given two current $i_1 = 10\sin(\omega t+\pi/4)$ and $i_2 = 5\cos(\omega t-\pi/2)$, find the r.m.s. value of $i_1 + i_2$ using the complex number representation.*

[Electric Circuit Theory, Kerala Univ. 1985]

Solution. The maximum value of first current is 10 A and it leads the reference quantity by 45°. The second current can be written as

$$i_2=5\cos(\omega t-\pi/2)=5\sin([90+(\omega t-\pi/2)]=5\sin\omega t$$

Hence, its maximum value is 5A and is in phase with the reference quantity.

$\therefore\ \ \mathbf{I}_{m1}=10(\cos 45^\circ+j\sin 45^\circ)=(7.07+j\,7.07)$

$\mathbf{I}_{m2}=5(\cos 0^\circ+j\sin 0^\circ)=(5+j\,0)$

The maximum value of resultant current is

$$\mathbf{I}_m = (7.07 + j\,7.07) + (5 + j\,0) = 12.07 + j\,7.07 = 14 \angle 30.4°$$

$\therefore$ R.M.S. value $= 14/\sqrt{2} = \mathbf{10\ A}$

12.10. Power and Roots of Vectors

(a) Powers

Suppose it is required to find the cube of the vector $3 \angle 15°$. For this purpose, the vector has to be multiplied by itself three times.

$\therefore (3 \angle 15)^3 = 3 \times 3 \times 3.\ \angle(15° + 15° + 15°) = 27 \angle 45°$. In general, $\mathbf{A}^n = A^n \angle n\alpha$

Hence, *n*th power of vector **A** is a vector whose magnitude is A^n and whose phase angle with respect to *X*-axis is $n\alpha$.

It is also clear that $\mathbf{A}^n\mathbf{B}^n = A^nB^n \angle(n\alpha + n\beta)$

(b) Roots

It is clear that $\sqrt[3]{(8 \angle 45°)} = 2 \angle 15°$

In general, $\sqrt[n]{\mathbf{A}} = \sqrt[n]{A} \angle \alpha/n$

Hence, *n*th root of a vector **A** is a vector whose magnitude is $n\sqrt{A}$ and whose phase angle with respect to *X-axis is* α/n.

12.11. The 120° Operator

In three-phase work where voltage vectors are displaced from one another by 120°, it is convenient to employ an operator which rotates a vector through 120° forward or backwards without changing its length. This operator is '*a*'. Any operator which is multiplied by '*a*' remains unchanged in magnitude but is rotated by 120° in the counter-clockwise (ccw) direction.

$$\therefore \quad \alpha = 1 \angle 120°$$

This, when expressed in the cartesian form, becomes

$$a = \cos 120° + j \sin 120° = -0.5 + j\,0.866$$

Similarly, $a^2 = 1 \angle 120° \times 1 \angle 120° = 1 \angle 240° = \cos 240° + j \sin 240° = -0.5 - j\,0.866$

Hence, operator 'a^2' will rotate the vector in ccw by 240°. This is the same as rotating the vector in *clockwise* direction by 120°.

$\therefore a^2 = 1 \angle -120°$. Similarly, $a^3 = 1 \angle 360° = 1$*

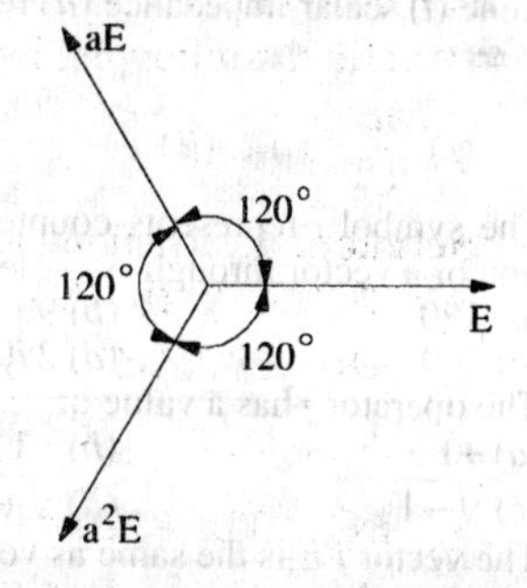

Fig. 12.11

As shown in Fig. 12.11, the 3-phase voltage vectors with standard phase sequence may be represented as E, a^2E and aE or as E, $E(-0.5 - j\,0.866)$ and $E\,(-0.5 + j.0.866)$

It is easy to prove that

(*i*) $a^2 + a = -1$ (*ii*) $a^2 + a + 1 = 0$ (*iii*) $a^3 + a^2 + a = 0$

Note. We have seen in Art. 12.3 that operator $-j$ turns a vector through $-90°$ *i.e.* through 90° in *clockwise* direction. But it should be clearly noted that operator '$-a$' *does not* turn a vector through $-120°$. Rather '$-a$' turns a vector through $-60°$ as shown below.

Example 12.10. *Evaluate the following expressions in the polar form*
(*i*) $a^2 - 1$, (*ii*) $1 - a - a^2$ (*iii*) $2a^2 + 3 + a$ (*iv*) ja

[Elect. Meas and Meas. Inst., Madras Univ. 1984]

Solution. (*i*) $a^2 = a \times a = 1\angle 120° \times 1\angle 120° = 1\angle 240° = 1\angle -120° = -\frac{1}{2} - j\frac{\sqrt{3}}{2}$

*Numerically, *a* is equivalent to the cube root of unity.

$$\therefore\quad a^2-1 = -\frac{1}{2}-j\frac{\sqrt{3}}{2}-1 = -\frac{3}{2}-j\frac{\sqrt{3}}{2}\quad \sqrt{3}\left(-\frac{\sqrt{3}}{2}-j\frac{1}{2}\right) = \sqrt{3}\angle 30° + 180° = \sqrt{3}\angle 210°$$

(*ii*) $$a = 1\angle 120° = -\frac{1}{2}+j\frac{\sqrt{3}}{2}\ ;\ a^2 = 1\angle 240° = -\frac{1}{2}-j\frac{\sqrt{3}}{2}$$

$$\therefore\quad 1-a-a^2 = 1+\frac{1}{2}-j\frac{\sqrt{3}}{2}+\frac{1}{2}+j\frac{\sqrt{3}}{2} = 2+j0 = 2\angle 0°$$

(*iii*) $$2a^2 = 2\left(-\frac{1}{2}-j\frac{\sqrt{3}}{2}\right) = -1-j\sqrt{3}$$

$$2a = 2\left(-\frac{1}{2}+j\frac{\sqrt{3}}{2}\right) = -1+j\sqrt{3}$$

$$\therefore\quad 2a^2+3+2a = 3-1-j\sqrt{3}-1+j\sqrt{3} = 1\angle 0°$$

(*iv*) $$ja = j\times a = 1\angle 90° \times 1\angle 120° = 1\angle 210°$$

Tutorial Problem No. 12.1

1. Perform the following indicated operations :
(*a*) $(60 + j80) + (30 - j40)$ (*b*) $(12 - j6) - (40 - j20)$ (*c*) $(6 + j8)(3 - j4)$ (*d*) $(16 + j8) \div (3 - j4)$
[(*a*) (90 + j40) (*b*) (–28 + j14) (*c*) (50 + j0 (*d*) (–0.56 + j1.92)

2. Two impedances $\mathbf{Z}_1 = 2 + j6\ \Omega$ and $\mathbf{Z}_2 = 6 - 12\ \Omega$ are connected in a circuit so that they are additive. Find the resultant impedance in the polar form. **[10 ∠ –36.9°]**

3. Express in rectangular form and polar form a vector, the magnitude of which is 100 units and the phase of which with respect to reference axis is
(*a*) + 30° (*b*) + 180° (*c*) –60° (*d*) + 120° (*e*) –120° (*f*) –210°.
[(*a*) 86.6 + j50 ∠30° (*b*) (–100 +j0), 100 ∠180 ° (*c*) 50–j86.6, 100 ∠ -60° (*d*) (–50 +j86.6), 100 ∠ -120°] (*e*) (–50–*j*86.6), 100 ∠ –120° (*f*) (–50 + j86.6), 100 ∠ –210°]

4. In the equation $\mathbf{V}_m = \mathbf{V} - \mathbf{ZI}$, $\mathbf{V} = 100\angle 0°$ volts, $\mathbf{Z} = 10\angle 60°\ \Omega$ and $\mathbf{I} = 8\angle -30°$ amperes. Express $\mathbf{V}_m$ in polar form. **[50.5 ∠ –52°]**

5. A voltage $\mathbf{V} = 150 + j180$ is applied across an impedance and the current flowing is found to be $\mathbf{I} = 5 - j4$. Determine (*i*) scalar impedance (*ii*) resistance (*iii*) reactance (*iv*) power consumed.
[(*i*) 3.73 Ω (*ii*) 0.75 Ω (*iii*) 36.6 Ω (*iv*) 30 W]

Objective Tests–12

1. The symbol *j* represents counterclockwise rotation of a vector through —— degrees.
(*a*) 180 (*b*) 90
(*c*) 360 (*d*) 270

2. The operator *j* has a value of
(*a*) +1 (*b*) –1
(*c*) $\sqrt{-1}$ (*d*) $\sqrt{+1}$

3. The vector j^5E is the same as vector
(*a*) jE (*b*) j^2E
(*c*) j^3E (*d*) j^4E

4. The conjugate of $(-a + jb)$ is
(*a*) $(a - jb)$ (*b*) $(-a - jb)$
(*c*) $(a + jb)$ (*d*) $(jb - a)$

5. The operator '$-a$' turns a vector through —— degrees.
(*a*) –120 (*b*) 120
(*c*) 60 (*d*) –60

6. The polar form of the expression *ja* is
(*a*) 2 ∠0° (*b*) $\sqrt{3}\angle 210°$
(*c*) 1 ∠210° (*d*) 1∠0°

Answers

1. *b* **2.** *c* **3.** *a* **4.** *b* **5.** *d* **6.** *c*

13 SERIES A.C. CIRCUITS

13.1. A.C. Through Resistance and Inductance

A pure resistance R and a pure inductive coil of inductance L are shown connected in series in Fig. 13.1.

Let V = r.m.s. value of the applied voltage, I = r.m.s. value of the resultant current

$V_R = IR$ – voltage drop across R (in phase with I), $V_L = I.X_L$ –voltage drop across coil (ahead of I by 90°)

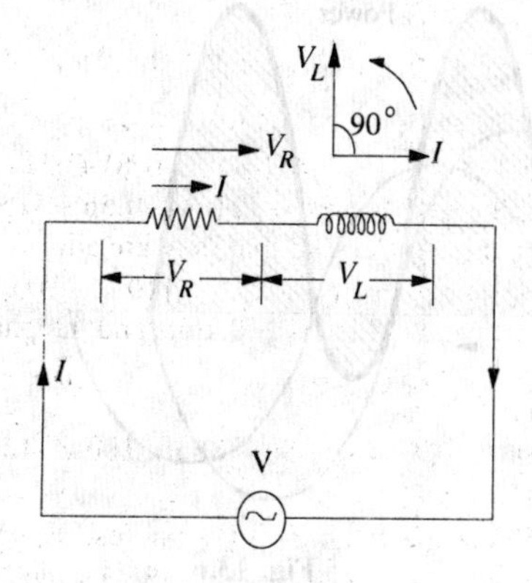

Fig. 13.1

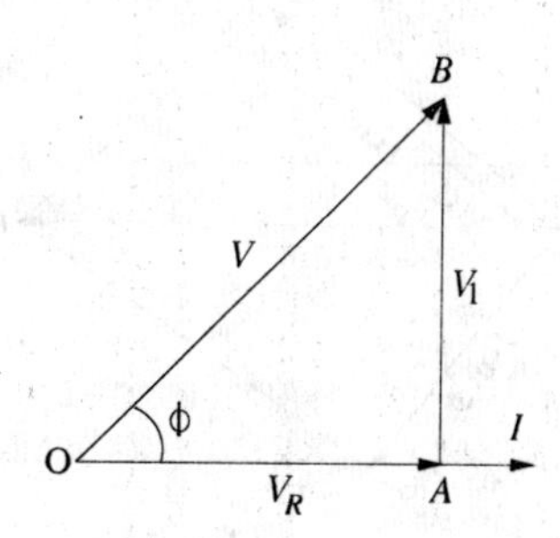

Fig. 13.2

These voltage drops are shown in voltage triangle OAB in Fig. 13.2. Vector OA represents ohmic drop V_R and AB represents inductive drop V_L. The applied voltage V is the vector sum of the two *i.e.* OB.

$$\therefore \quad V = \sqrt{(V_R^2 + V_L^2)} = \sqrt{[(IR)^2 + (I.X_L)^2]} = I\sqrt{(R^2 + X_L^2)} = \frac{V}{\sqrt{(R^2 + X_L^2)}}$$

The quantity $\sqrt{(R^2 + X_L^2)}$ is known as the *impedance* (Z) of the circuit. As seen from the impedance triangle ABC (Fig. 13.3) $Z^2 = R^2 + X_L^2$

i.e. $\quad (Impedance)^2 = (resistance)^2 + (reactance)^2$

From Fig. 13.2, it is clear that the applied voltage V leads the current I by an angle ϕ such that

$$\tan\phi = \frac{V_L}{V_R} = \frac{I.X_L}{I.R} = \frac{X_L}{R} = \frac{\omega L}{R} = \frac{\text{reactance}}{\text{resistance}} \quad \therefore \quad \phi = \tan^{-1}\frac{X_L}{R}$$

The same fact is illustrated graphically in Fig. 13.4.

In other words, current I lags behind the applied voltage V by an angle ϕ.

Hence, if applied voltage is given by $\upsilon = V_m \sin \omega t$, then current equation is

$$i = I_m \sin(\omega t - \phi) \text{ where } I_m = V_m / Z$$

Power

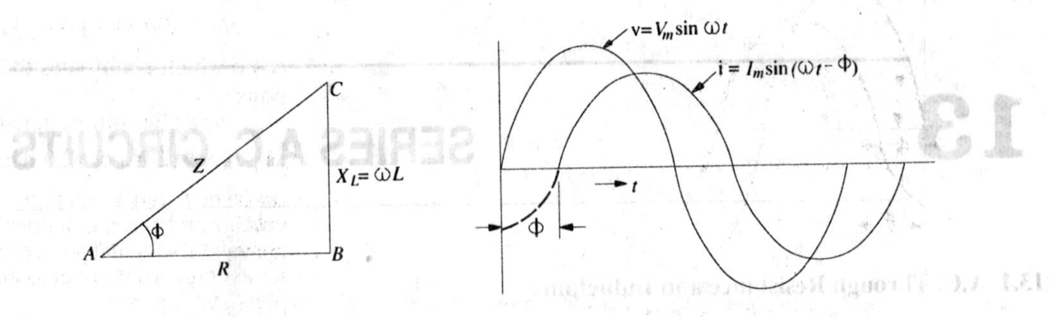

Fig. 13.3 **Fig. 13.4**

In Fig. 13.5, I has been resolved into its two mutually perpendicular components, $I \cos \phi$ along the applied voltage V and $I \sin \phi$ in quadrature (*i.e.* perpendicular) with V.

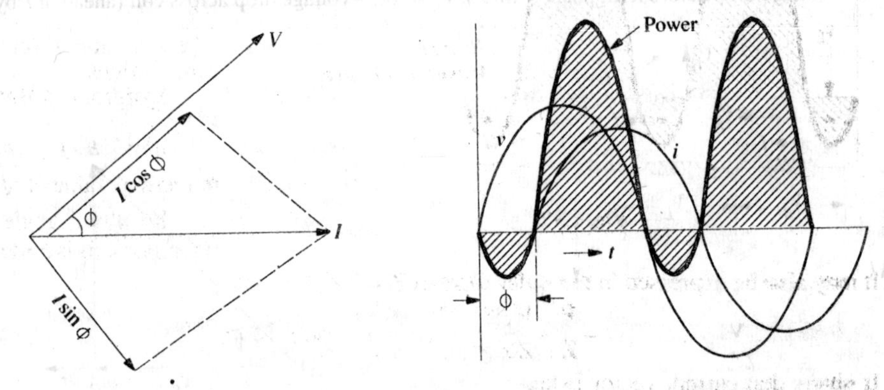

Fig. 13.5 **Fig. 13.6**

The mean power consumed by the circuit is given by the product of V and *that component of the current I which is in phase with V.*

So $P = V \times I \cos \phi$ = r.m.s. voltage × r.m.s. current × cos φ

The term 'cos φ' is called the power factor of the circuit.

Remember that in an a.c. circuit, the product of r.m.s. volts and r.m.s. amperes gives volt-amperes (VA) and **not** *true power in watts.* True power (W)= volt-amperes (VA) × power factor.

or **Watts = VA × cos φ***

It should be noted that power consumed is due to ohmic resistance only because pure inductance does not consume any power.

Now $P = VI \cos \phi = VI \times (R/Z) = (V/Z) \times I \, . \, R = I^2 R$ ($\because$ $\cos \phi = R/Z$) or $P = I^2 R$ watt

Graphical representation of the power consumed is shown in Fig. 14.6.

Let us calculate power in terms of instantaneous values.

Instantaneous power is $= v\,i = V_m \sin \omega t \times I_m \sin(\omega t - \phi) = V_m I_m \sin \omega t \sin(\omega t - \phi)$

$$= \tfrac{1}{2} V_m I_m [\cos \phi - \cos(2\omega t - \Phi)]$$

---

*While dealing with large supplies of electric power, it is convenient to use kilowatt as the unit

kW = kVA × cos φ

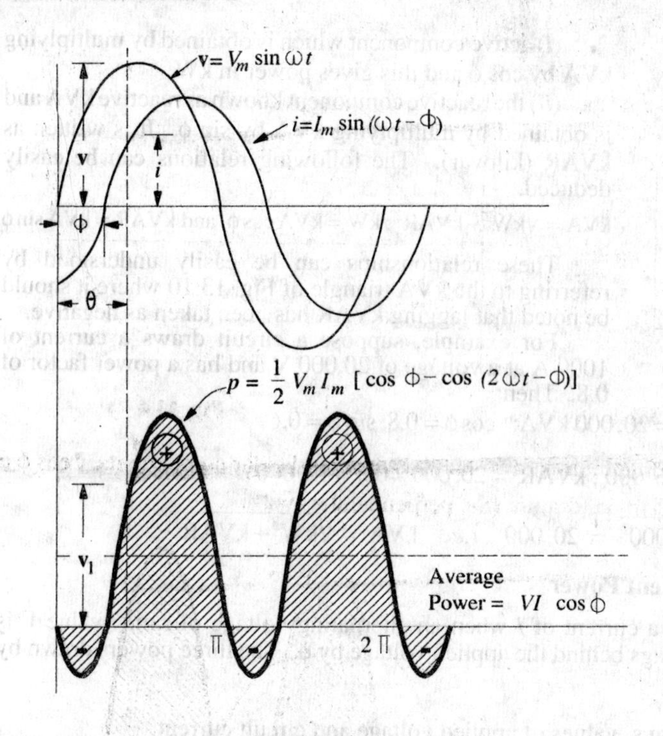

Fig. 13.7

Obviously, this power consists of two parts (Fig. 13.7).

(*i*) a constant part $\frac{1}{2} V_m I_m \cos\phi$ which contributes to real power.

(*ii*) a pulsating component $\frac{1}{2} V_m I_m \cos(2\omega t - \phi)$ which has a frequency twice that of the voltage and current. It does not contribute to actual power since its average value over a complete cycle is zero.

Hence, average power consumed $= \frac{1}{2} V_m I_m \cos\phi = \frac{V_m}{\sqrt{2}} \cdot \frac{I_m}{\sqrt{2}} \cos\phi = V I \cos\phi$, where *V* and *I* represent the r.m.s. values.

Symbolic Notation. $Z = R + jX_L$

Impedance vector has numerical value of $\sqrt{(R^2 + X_L^2)}$

Its phase angle with the reference axis is $\phi = \tan^{-1}(X_L/R)$

It may also be expressed in the polar form as $\mathbf{Z} = Z\angle\phi°$

(*i*) Assuming $\mathbf{V} = V\angle 0°$; $\mathbf{I} = \frac{\mathbf{V}}{\mathbf{Z}} = \frac{V\angle 0°}{Z\angle\phi°} = \frac{V}{Z}\angle -\phi°$ (Fig. 13.8)

It shows that current vector is lagging behind the voltage vector by $\phi°$. The numerical value of current is *V/Z*.

(*ii*) However, if we assume that $\mathbf{I} = I\angle 0$, then

$\mathbf{V} = \mathbf{IZ} = I\angle 0° \times Z\angle\phi° = IZ\angle\phi°$

It shows that voltage vector is $\phi°$ ahead of current vector in ccw direction as shown in Fig. 13.9.

V = V ∠ 0°
0
ϕ
I = I ∠ −ϕ

V = V ∠ ϕ
ϕ
0
I = I ∠ 0°

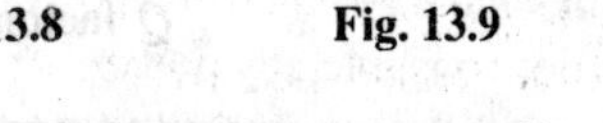

Fig. 13.8 **Fig. 13.9**

13.2. Power Factor

It may be defined as

(*i*) cosine of the angle of lead or lag

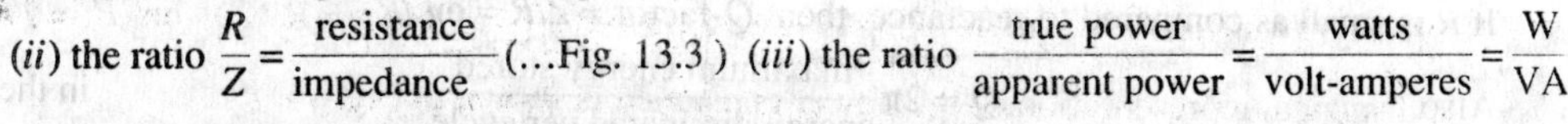

(*ii*) the ratio $\frac{R}{Z} = \frac{\text{resistance}}{\text{impedance}}$ (...Fig. 13.3) (*iii*) the ratio $\frac{\text{true power}}{\text{apparent power}} = \frac{\text{watts}}{\text{volt-amperes}} = \frac{\text{W}}{\text{VA}}$

13.3. Active and Reactive Components of Circuit Current I

Active component is that which is in phase with the applied voltage *V i.e. I* cos ϕ. It is also known as 'wattful' component.

Reactive component is that which is in quadrature with *V i.e. I* sin ϕ. It is also known as 'wattless' or 'idle' component.

It should be noted that the product of volts and amperes in an a.c. circuit gives voltamperes (VA). Out of this, the actual power is *VA* cos ϕ = ***W*** and reactive power is *VA* sin ϕ. Expressing the values in kVA, we find that it has two rectangular components :

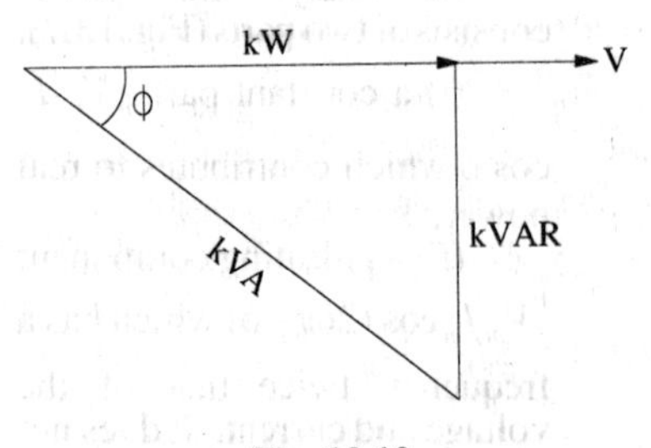

Fig. 13.10

(*i*) active component which is obtained by multiplying kVA by cos ϕ and this gives power in kW.

(*ii*) the reactive component known as reactive kVA and is obtained by multiplying kVA by sin ϕ. It is written as kVAR (kilovar). The following relations can be easily deduced.

$$kVA = \sqrt{kW^2 + kVAR^2}; kW = kVA \cos\phi \text{ and } kVAR = kVA \sin\phi$$

These relationships can be easily understood by referring to the kVA triangle of Fig. 13.10 where it should be noted that lagging kVAR has been taken as negative.

For example, suppose a circuit draws a current of 1000 A at a voltage of 20,000 V and has a power factor of 0.8. Then

input $= 1,000 \times 20,000/1000 = 20,000$ kVA; $\cos\phi = 0.8; \sin\phi = 0.6$

Hence $kW = 20,000 \times 0.8 = 16,000$; $kVAR = 20,000 \times 0.6 = 12,000$

Obviously, $\sqrt{16000^2 + 12000^2} = 20,000$ *i.e.* $kVA = \sqrt{kW^2 + kVAR^2}$

13.4. Active, Reactive and Apparent Power

Let a series R–L circuit draw a current of I when an alternating voltage of r.m.s. value V is applied to it. Suppose that current lags behind the applied voltage by ϕ. The three powers drawn by the circuit are as under :

(*i*) **apparent power (S)**

It is given by the product of r.m.s. values of applied voltage and circuit current.

$\therefore \quad S = VI = (IZ) \, . \, I = I^2Z$ volt-amperes (VA)

(*ii*) **active power (P** or **W)**.

It is the power which is actually dissipated in the circuit resistance. $P = I^2R = VI \cos\phi$ watts

(*iii*) **reactive power (Q)**

It is the power developed in the inductive reactance of the circuit.

$Q = I^2X_L = I^2 . \, Z \sin\phi = I.(IZ) \, . \, \sin\phi = VI \sin\phi$ volt-amperes-reactive (VAR)

These three powers are shown in the power triangle of Fig. 13.11 from where it can be seen that

$$S^2 = P^2 + Q^2 \text{ or } S = \sqrt{P^2 + Q^2}$$

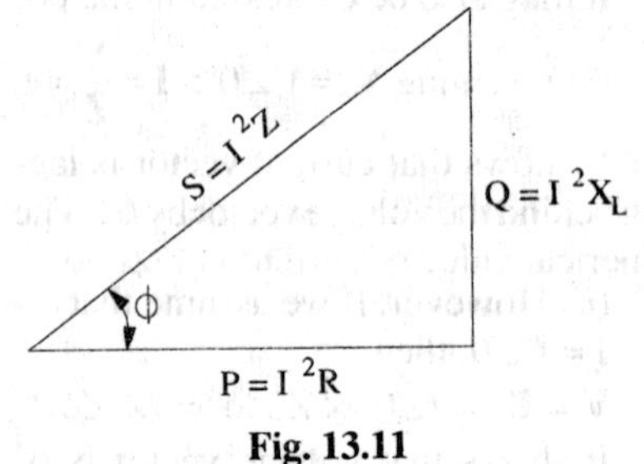

Fig. 13.11

13.5. Q-factor of a Coil

Reciprocal of power factor is called the Q-factor of a coil or its figure of merit. It is also known as quality factor of the coil.

$$Q \text{ factor} = \frac{1}{\text{power factor}} = \frac{1}{\cos\phi} = \frac{Z}{R}$$

If R is small as compared to reactance, then Q-factor $= Z/R = \omega L/R$

Also, $$Q = 2\pi \frac{\text{maximum energy stored}}{\text{energy dissipated per cycle}} \text{ in the coil}$$

Example 13.1. *In a series circuit containing pure resistance and a pure inductance, the current and the voltage are expressed as :*

$$i(t) = 5 \sin(314t + 2\pi/3) \text{ and } v(t) = 15 \sin(314t + 5\pi/6)$$

(a) What is the impedance of the circuit ? (b) What is the value of the resistance ? (c) What is the inductance in henrys ? (d) What is the average power drawn by the circuit ? (e) What is the power factor ? **[Elect. Technology, Indore Univ. 1986]**

Solution. Phase angle of current $= 2\pi/3 = 2 \times 180°/3 = 120°$ and phase angle of voltage $= 5\pi/6 = 5 \times 180°/6 = 150°$. Also, $Z = V_m/I_m = 3\Omega$.

Hence, current *lags* behind voltage by 30°. It means that it is an *R–L* circuit. Also $314 = 2\pi f$ or $f = 50$ Hz. Now, $R/Z = \cos 30° = 0.866$; $R = 2.6\ \Omega$; $X_L/Z = \sin 30° = 0.5$

$\therefore\ X_L = 1.5\ \Omega$. $314\,L = 1.5$, $L = 4.78$ mH

(*a*) $Z = \mathbf{3\ \Omega}$ (*b*) $R = \mathbf{2.6\ \Omega}$ (*c*) $L = \mathbf{4.78\ mH}$

(*d*) $P = I^2R = (5/\sqrt{2})^2 \times 2.6 = \mathbf{32.5\ W}$ (*e*) p.f. = cos 30° = **0.866 (lag)**.

Example 13.2. *In a circuit, the equations for instantaneous voltage and current are given by* $v = 141.4 \sin\left(\omega t - \frac{2\pi}{3}\right)$, *volt and* $i = 7.07 \sin\left(\omega t - \frac{\pi}{2}\right)$, *amp, where* $\omega = 314$ *rad/sec.*

(i) Sketch a neat phasor diagram for the circuit. (ii) Use polar notation to calculate impedance with phase angle. (iii) Calculate average power & power factor. (iv) Calculate the instantaneous power at the instant t = 0. **(F.Y. Engg. Pune Univ. Nov. 1988)**

Solution. (*i*) From the voltage equation, it is seen that the voltage lags behind the reference quantity by $2\pi/3$ radian or $2 \times 180/3 = 120°$. Similarly, current lags behind the reference quantity by $\pi/2$ radian or $180/2 = 90°$. Between themselves, voltage lags behind the current by $(120 - 90) = 30°$ as shown in Fig. 13.12(*b*).

(*ii*) $V = V_m/\sqrt{2} = 141.4/\sqrt{2} = 100$ V ; $I = I_m/\sqrt{2} = 7.07/\sqrt{2} = 5$ A.

$$\therefore\quad \dot{V} = 100\angle -120° \text{ and } I = 5\angle -90° \quad \therefore\quad Z = \frac{100\angle -120°}{5\angle -90°} = 20\angle -30°$$

(*iii*) Average power $= VI\cos\phi = 100 \times 5 \times \cos 30° = \mathbf{433\ W}$

(*iv*) At $t = 0$; $v = 141.4 \sin(0 - 120°) = -122.45$ V ; $i = 7.07 \sin(0 - 90°) = -7.07$ A.

$\therefore$ instantaneous power at $t = 0$ is given by $vi = (-122.45) \times (-7.07) = \mathbf{865.7\ W}$.

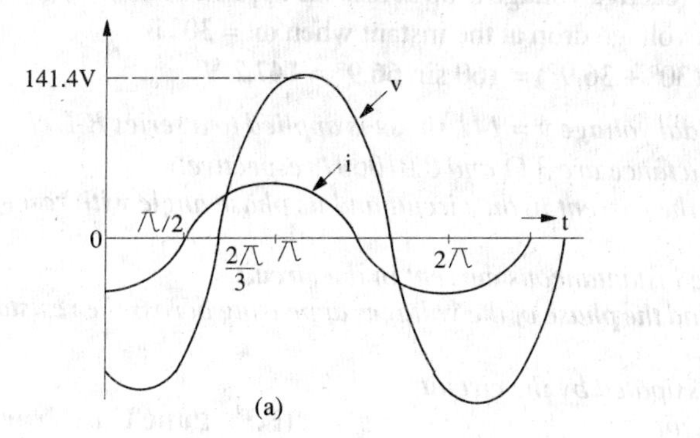

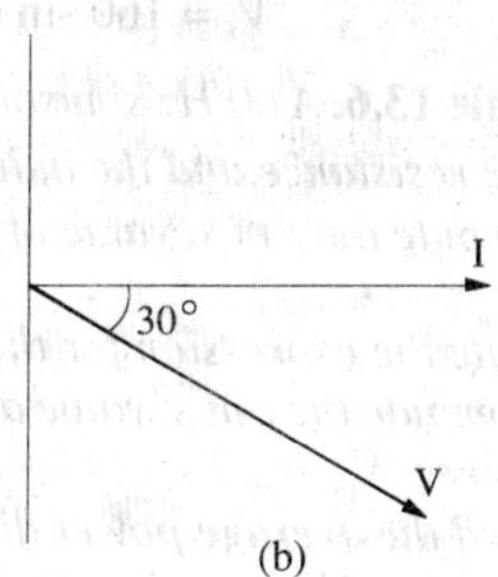

Fig. 13.12

Example 13.3. *The potential difference measured across a coil is 4.5 V, when it carries a direct current of 9 A. The same coil when carries an alternating current of 9 A at 25 Hz, the potential difference is 24 V. Find the current, the power and the power factor when it is supplied by 50 V, 50 Hz supply.* **(F.Y. Pune Univ. May 1989)**

Solution. Let *R* be the *d.c.* resistance and *L* be the inductance of the coil.

$\therefore\ R = V/I = 4.5/9 = 0.5\ \Omega$;

With a.c. current of 25 Hz, $Z = V/I = 24/9 = 2.66\ \Omega$

$$X_L = \sqrt{Z^2 - R^2} = \sqrt{2.66^2 - 0.5^2} = 2.62\ \Omega. \text{ Now, } X_L = 2\pi \times 25 \times L;\ L = 0.0167\ \Omega$$

At 50 Hz

$X_L = 2.62 \times 2 = 5.24\ \Omega$; $Z = \sqrt{0.5^2 + 5.24^2} = 5.26\ \Omega$

Current $I = 50/5.26 = \mathbf{9.5\ A}$; Power $= I^2R = 9.5^2 \times 0.5 = \mathbf{45\ W}$.

Example 13.4. *In a particular R-L series circuit a voltage of 10 V at 50 Hz produces a current of 700 mA while the same voltage at 75 Hz produces 500 mA. What are the values of R and L in the circuit ?* **(Network Analysis A.M.I.E. Sec. B, S 1990)**

Solution. (*i*) $Z = \sqrt{R^2+(2\pi\times 50\,L)^2} = \sqrt{R^2+98696\,L^2}; V = IZ$ or $10 = 700\times 10^{-3}\sqrt{(R^2+98696L^2)}$

$$\sqrt{(R^2+98696\,L^2)} = 10/700\times 10^{-3} = 100/7 \text{ or } R^2+98696\,L^2 = 10000/49 \quad \ldots(i)$$

(*ii*) In the second case $Z = \sqrt{R^2+(2\pi\times 75L)^2} = \sqrt{(R^2+222066L^2)}$

$\therefore\ 10 = 500\times 10^{-3}\sqrt{(R^2+222066L^2)}$ *i.e.* $\sqrt{(R^2+222066L^2)} = 20$ or $R^2+222066L^2 = 400$ (*ii*)

Substracting Eq. (*i*) from (*ii*), we get
$222066\,L^2 - 98696\,L^2 = 400 - (10000/49)$ or $123370\ L^2 = 196$ or $L = 0.0398\,H =$ **40 mH.**
Substituting this value of L in Eq. (*ii*), we get, $R^2 + 222066\,(0.398)^2 = 400\ \therefore\ R =$ **6.9Ω.**

Example 13.5. *A series circuit consists of a resistance of 6 Ω and an inductive reactance of 8 Ω. A potential difference of 141.4 V (r.m.s.) is applied to it. At a certain instant the applied voltage is + 100 V, and is increasing. Calculate at this instant, (i) The current (ii) the voltage drop across the resistance and (iii) Voltage drop across inductive reactance.* **(F.E. Pune Univ. May 1989)**

Solution. $Z = R + jX = 6 + j8 = 10\angle 53.1°$

It shows that current lags behind the applied voltage by 53.1°. Let V be taken as the reference quantity. Then, $v = (141.4\times\sqrt{2})\sin\omega t = 200\sin\omega t$; $i = (V_m/Z)\sin\ (\omega t) - 30° = 20\sin(\omega t - 53.1°)$.

(*i*) When the voltage is +100 V and increasing ; $100 = 200\sin\omega t$; $\sin\omega t = 0.5$; $\omega t = 30°$

At this instant, the current is given by $i = 20\sin(30°-53.1°) = -20\sin 23.1° =$ **−7.847 A.**

(*ii*) drop across resistor $= iR = -7.847\times 6 = -47$ V.

(*iii*) Let us first find the equation of the voltage drop V_L across the inductive reactance. Maximum value of the voltage drop $= I_m X_L = 20\times 8 = 160$ V. It leads the current by 90°. Since current itself lags the applied voltage by 53.1°, the reactive voltage drop across the applied voltage by (90°−53.1°) = 36.9°. Hence, the equation of this voltage drop at the instant when $\omega t = 30°$ is

$$V_L = 160\sin(30° + 36.9°) = 160\sin 66.9° = \mathbf{147.2\ V.}$$

Example 13.6. *A 60 Hz sinusoidal voltage $v = 141\sin\omega t$ is applied to a series R-L circuit. The values of the resistance and the inductance are 3 Ω and 0.0106 H respectively.*

(*i*) *Compute the r.m.s. value of the current in the circuit and its phase angle with respect to the voltage.*

(*ii*) *Write the expression for the instantaneous current in the circuit.*

(*iii*) *Compute the r.m.s. value and the phase of the voltages appearing across the resistance and the inductance.*

(*iv*) *Find the average power dissipated by the circuit.*

(*v*) *Calculate the p.f. of the circuit.* **(F.E. Pune Univ. Nov. 1989)**

Solution. $V_m = 141$ V; $V = 141/\sqrt{2} = 100$ V $\therefore\ V = 100 + j0$

$$X_L = 2\pi\times 60\times 0.0106 = 4\ \Omega;\ Z = 3 + j4 = 5\angle 53.1°$$

(*i*) $I = V/Z = 100\angle 0°/5\angle 53.1° = 20\angle -53.1°$

Since angle is minus, the current lags behind the voltage by 53.1

(*ii*) $I_m = \sqrt{2}\times 20 = 28.28$; $\therefore\ i = 28.28\ \sin(\omega t - 53.1°)$

(*iii*) $VR = IR = 20\angle -53.1°\times 3 = 60\angle -53.1°$ volt.

$V_L = jIX_L = 1\angle 90°\times 20\angle -53.1°\times 4 = 80\angle 36.9°$

(*iv*) $P = VI\cos\phi = 100\times 20\times\cos 53.1° =$ **1200 W.**

(*v*) p.f. $= \cos\phi = \cos 53.1° =$ **0.6.**

Example 13.7. *In a given R-L circuit, R = 3.5 Ω and L = 0.1 H. Find (i) the current through the circuit and (ii) power factor if a 50-Hz voltage V = 220 ∠30° is applied across the circuit.*

Solution. The vector diagram is shown in Fig. 13.13.

$$X_L = 2\pi fL = 2\pi\times 50\times 0.1 = 31.42\ \Omega$$

$$Z = \sqrt{(R^2+X_L^2)} = \sqrt{3.5^2+31.42^2} = 31.6\ \Omega$$

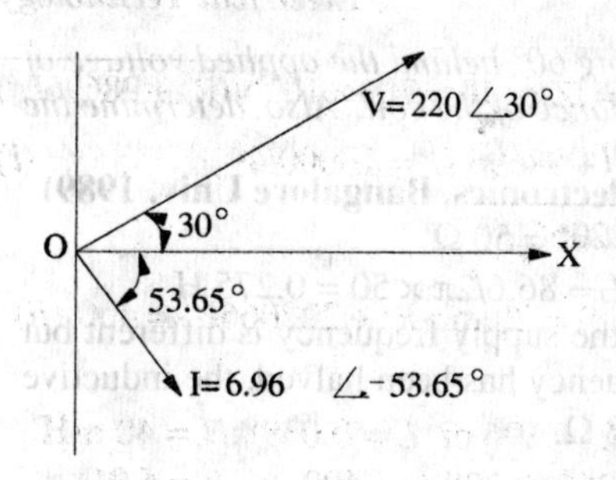

Fig. 13.13

$\therefore\ \mathbf{Z} = 31.6\angle\tan^{-1}(31.42/3.5) = 31.6\angle 83.65°$

(*i*) $\mathbf{I} = \dfrac{\mathbf{V}}{\mathbf{Z}} = \dfrac{220\angle 30°}{31.6\angle 83.65°} = 6.96\angle -53.65°$

(*ii*) Phase angle between voltage and current is
$= 53.65° + 30° = 83.65°$ with current lagging.

p.f. $= \cos 83.65° = \mathbf{0.11}$ **(lag).**

Example 13.8. *In an alternating circuit, the impressed voltage is given by V = (100–j50) volts and the current in the circuit is I = (3 – j4) A. Determine the real and reactive power in the circuit.* **(Electrical Engg., Calcutta Univ. 1991)**

Solution. Power will be found by the conjugate method. Using current conjugate, we have

$$P_{VA} = (100 - j50)(3 + j4) = 300 + j400 - j150 + 200 = 500 + j250$$

Hence, real power is **500 W** and reactive power or VAR is **250**. Since the second term in the above expression is positive, the reactive volt-amperes of 250 are inductive.*

Example 13.9. *In the circuit of Fig. 14.14, applied voltage V is given by (0+j10) and the current is (0.8 + j0.6) A. Determine the values of R and X and also indicate if X is inductive or capacitive.* **(Elect. Technology, Nagpur Univ. 1991)**

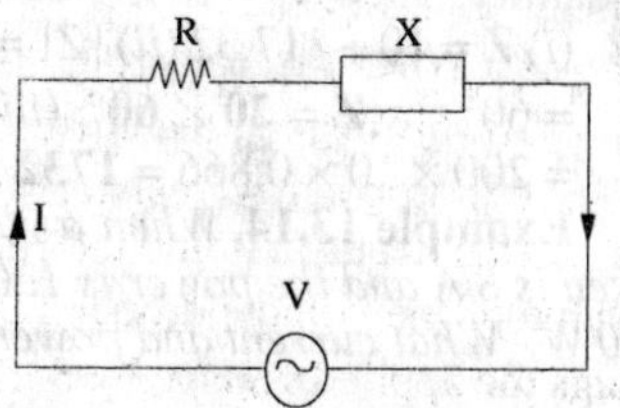

Fig. 13.14

Solution. $\mathbf{V} = 0 + j10 = 10\angle 90°\ ;\ \mathbf{I} = 0.8 + j0.6 = 1\angle 36.9°$

As seen, *V* leads the reference quantity by 90° whereas *I* leads by 36.9°. In other words, *I lags* behind the applied voltage by (90°–36.9°) = 53.1°.

Hence, the circuit of Fig. 13.14 is an *R-L* circuit.

Now, $\mathbf{Z} = \mathbf{V}/\mathbf{I} = 10\angle 90°/1\angle 36.9° = 10\angle 53.1° = 6 + j8$

Hence, $R = \mathbf{6\ \Omega}$ and $X_L = \mathbf{8\ \Omega}$

Example 13.10. *A two-element series circuit is connected across an a.c. source e = 200√2 sin (ωt + 20°) V. The current in the circuit then is found to be i = 10√2 cos (314t – 25°) A. Determine the parameters of the circuit.* **(Electromechanic Allahabad Univ. 1991)**

Solution. The current can be written as $i = 10\sqrt{2}\sin(314t - 25° + 90°) = 10\sqrt{2}\sin(314t + 65°)$. It is seen that applied voltage leads by 20° and current leads by 65° with regards to the reference quantity, their mutual phase difference is = 65°–(20°) = 45°. Hence, p.f. = cos 45° = $\mathbf{1/\sqrt{2}}$ **(lead).**

Now, $V_m = 200\sqrt{2}$ and $I_m = 10\sqrt{2}$ $\quad\therefore\ Z = V_m/I_m = 200\sqrt{2}/10\sqrt{2} = 20\ \Omega$

$$R = Z\cos\phi = 20/\sqrt{2}\ \Omega = \mathbf{14.1\ \Omega}\ ;\ X_c = Z\sin\phi = 20/\sqrt{2} = \mathbf{14.1\ \Omega}$$

Now, $f = 314/2\pi = 50$ Hz. Also, $X_c = 1/2\pi fC\ \therefore\ C = 1/2\pi\times 50\times 14.1 = \mathbf{226\ \mu F}$

Hence, the given circuit is an *R–C* circuit.

Example 13.11. *Transform the following currents to the time domain : (i) 6 – j8 (ii) – 6 + j8 (iii) –j5.*

Solution. (*i*) Now, (6–*j*8) when expressed in the polar form is $\sqrt{6^2+8^2}\ \angle -\tan^{-1}8/6 = 10\angle -53.1°$. The time domain representation of this current is; $i(t) = 10\sin(\omega t - 53.1°)$

(*ii*) $-6 + j8 = \sqrt{6^2+8^2}\ \angle\tan^{-1}8/-6 = 10\angle 126.9°$

$\therefore\ i(t) = 10\sin(\omega t + 126.9°)$

(*iii*) $-j5 = 10\angle -90°\ \therefore\ i(t) = 10\sin(\omega t - 90°)$

*If voltage conjugate is used, then capacitive VARs are positive and inductive VARs negative. If current conjugate is used, then capacitive VARs are negative and inductive VARs are positive.

Example 13.12. *A choke coil takes a current of 2 A lagging 60° behind the applied voltage of 200 V at 50 Hz. Calculate the inductance, resistance and impedance of the coil. Also, determine the power consumed when it is connected across 100-V, 25-Hz supply.*

(Elect. Engg. & Electronics, Bangalore Univ. 1989)

Solution. (*i*) $Z_{coil} = 200/2 = 100\ \Omega$; $R = Z\cos\phi = 100\cos 60° = 50\ \Omega$

$X_L = Z\sin\phi = 100\sin 60° = 86.6\ \Omega\ X_L = 2\pi fL = 86.6\quad \therefore L = 86.6/2\pi\times 50 = 0.275$ H

(*ii*) Now, the coil will have different impedance because the supply frequency is different but its resistance would remain the same *i.e.* 50 Ω. Since the frequency has been halved, the inductive reactance of the coil is also halved *i.e.* it becomes 86.6/2 = 43.3 Ω.

$$Z_{coil} = \sqrt{50^2 + 43.3^2} = 66.1\ \Omega$$

$$I = 100/66.1 = 1.5\text{ A, p.f. } \cos\phi = 50/66.1 = 0.75$$

Power consumed by the coil = $VI\cos\phi = 100\times 1.5\times 0.75 =$ **112.5 W**

Example 13.13. *An inductive circuit draws 10 A and 1 kW from a 200-V, 50 Hz a.c. supply. Determine :-*

(i) the impedance in cartesian form (a + jb) (ii) the impedance in polar form Z∠θ (iii) the power factor (iv) the reactive power (v) the apparent power.

Solution. $Z = 200/10 = 20\ \Omega$; $P = I^2R$ or $1000 = 10^2\times R$; $R = 10\ \Omega$; $X_L = \sqrt{20^2 - 10^2} = 17.32\ \Omega$.

(*i*) $\mathbf{Z} = 10 + j\,17.32$ (*ii*) $|Z| = \sqrt{(10^2 + 17.32^2)} = 20\ \Omega$; $\tan\phi = 17.32/10 = 1.732$; $\phi = \tan^{-1}(1.732)$ $= 60°\quad \therefore \mathbf{Z} = \mathbf{20\angle 60°}$. (*iii*) p.f. = $\cos\phi = \cos 60° =$ **0.5 lag** (*iv*) reactive power = $VI\sin\phi$ $= 200\times 10\times 0.866 =$ **1732 VAR** (*v*) apparent power = $VI = 200\times 10 =$ **2000 VA.**

Example 13.14. *When a voltage of 100 V at 50 Hz is applied to a choking coil A, the current taken is 8 A and the power is 120 W. When applied to a coil B, the current is 10 A and the power is 500 W. What current and power will be taken when 100 V is applied to the two coils connected in series ?*

(Elements of Elect. Engg., Bangalore Univ. 1985)

Solution. $Z_1 = 100/8 = 12.5\ \Omega$; $P = I^2.R_1$ or $120 = 8^2\times R_1$; $R_1 = 15/8\Omega$

$$X_1 = \sqrt{Z_1^2 - R_1^2} = \sqrt{12.5^2 - (15/8)^2} = 12.36\ \Omega$$

$$Z_2 = 100/10 = 10\ \Omega;\ 500 = 10^2\times R_2 \text{ or } R_2 = 5\ \Omega$$

$$X_2 = \sqrt{10^2 - 5^2} = 8.66\ \Omega$$

When Joined in Series

$$R = R_1 + R_2 = (15/8) + 5 = 55/8\ \Omega;\ X = 12.36 + 8.66 = 21.02\ \Omega$$

$$Z = \sqrt{(55/8)^2 + (21.02)^2} = 22.1\ \Omega,\quad I = 100/22.1 = 4.52\text{ A}, P = I^2R = 4.52^2\times 55/8 = \mathbf{140\ W}$$

Example 13.15. *A coil takes a current of 6 A when connected to a 24-V d.c. supply. To obtain the same current with a 50-Hz a.c. supply, the voltage required was 30 V.*

Calculate (i) the inductance of the coil (ii) the power factor of the coil.

(F.Y. Engg. Pune Univ. 1989)

Solution. It should be kept in mind that coil offers only resistance to direct voltage whereas it offers impedance to an alternating voltage.

$\therefore\quad R = 24/6 = 4\ \Omega$; $Z = 30/6 = 5\ \Omega$

(*i*) $\therefore\quad X_L = \sqrt{Z^2 - R^2} = \sqrt{5^2 - 4^2} = 3\ \Omega$ (*iii*) p.f. $= \cos\phi = R/Z = 4/5 =$ **0.8 (lag)**

Example 13.16. *A resistance of 20 ohm, inductance of 0.2 H and capacitance of 150 μF are connected in series and are fed by a 230 V, 50 Hz supply Find X_L, X_C, Z, Y, p.f., active power and reactive power.*

(Elect. Science-I, Allahabad Univ. 1992)

Solution. $X_L = 2\pi fL = 2\pi\times 50\times 0.2 = 62.8\ \Omega$; $X_C = 1/2\pi fC$

$$= 10^{-6}|\,2\pi\times 50\times 150 = 21.2\ \Omega;\ X = (X_L - X_C) = 41.6\ \Omega;$$

$$Z = \sqrt{R^2 + X^2} = \sqrt{20^2 + 41.6^2} = 46.2\ \Omega;\ I = V/Z = 230/46.2 = 4.98\text{ A.}$$

Also, $Z = R + jX = 20 + j41.6 = 46.2\angle 64.3°$ ohm

$\therefore Y = 1/Z = 1/46.2\angle 64.3° = 0.0216\angle -64.3°$ siemens

p.f. $= \cos 64.3° = 0.4336$ (lag)

Active power $= VI \cos \phi = 230 \times 4.98 \times 0.4336 = 497$ W

Reactive power $= VI \sin \phi = 230 \times 4.98 \times \sin 64.3° = 1031$ VAR

Example 13.17. *A 120-V, 60-W lamp is to be operated on 220-V, 50-Hz supply mains. Calculate what value of (a) non-inductive resistance (b) pure inductance would be required in order that lamp is run on correct voltage. Which method is preferable and why ?*

Solution. Rated current of the bulb = 60/120 = 0.5 A

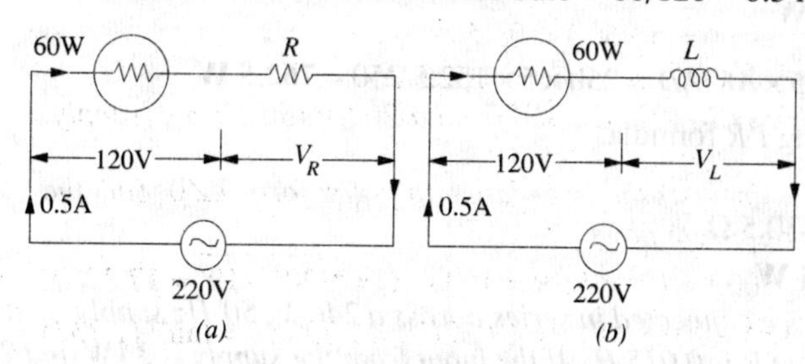

Fig. 13.15

(*a*) Resistor has been shown connected in series with the lamp in Fig. 13.15 (*a*).

P.D. across R is

$V_R = 220 - 120 = 100$ V

It is in phase with the applied voltage. $\therefore R = 100/0.5 =$ **200 Ω**

(*b*) P.D. across bulb = 120 V

P.D. across L is $V_L = \sqrt{(220^2 - 120^2)}$ = 184.4 V

(Remember that V_L is in quadrature with V_R – the voltage across the bulb).

Now $V_L = 0.5 \times X_L$ or $184.4 = 0.5 \times L \times 2\pi \times 50$ $\therefore$ $L = 184.4/0.5 \times 314 =$ **1.17 H**

Method (*b*) is preferable to (*a*) because in method (*b*), there is no loss of power. Ohmic resistance of 200 Ω itself dissipates large power (*i.e.* 100 × 0.5 = **50 W**).

Example 13.18. *A non-inductive resistor takes 8 A at 100 V. Calculate the inductance of a choke coil of negligible resistance to be connected in series in order that this load may be supplied from 220-V, 50-Hz mains. What will be the phase angle between the supply voltage and current ?*

(Elements of Elect. Engg.-I, Bangalore Univ. 1987)

Solution. It is a case of pure resistance in series with pure inductance as shown in Fig. 13.16 (*a*).

Here $V_R = 100\text{V}, V_L = \sqrt{220^2 - 100^2} = 196$ V

Now, $V_L = I.X_L$

or $196 = 8 \times 2\pi \times 50 \times L$ $\therefore$ $L =$ **0.078 H**

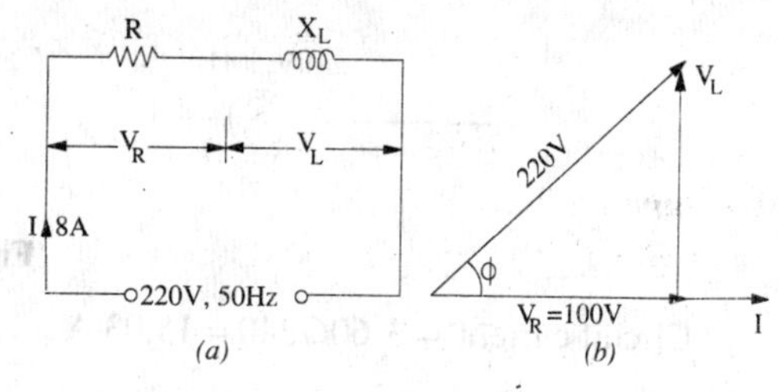

Fig. 13.16

Example 13.19. *A current of 5 A flows through a non-inductive resistance in series with a choking coil when supplied at 250-V, 50-Hz. If the voltage across the resistance is 125 V and across the coil 200 V, calculate*

(a) impedance, reactance and resistance of the coil (b) the power absorbed by the coil and (c) the total power. Draw the vector diagram. **(Elect. Engg., Madras Univ. 1988)**

Solution. As seen from the vector diagram of Fig. 13.17 (*b*).

$BC^2 + CD^2 = 200^2$ ...(*i*) $\qquad (125 + BC)^2 + CD^2 = 250^2$ ...(*ii*)

Substracting Eq. (*i*) from (*ii*), we get, $(125 + BC)^2 - BC^2 = 250^2 - 200^2$

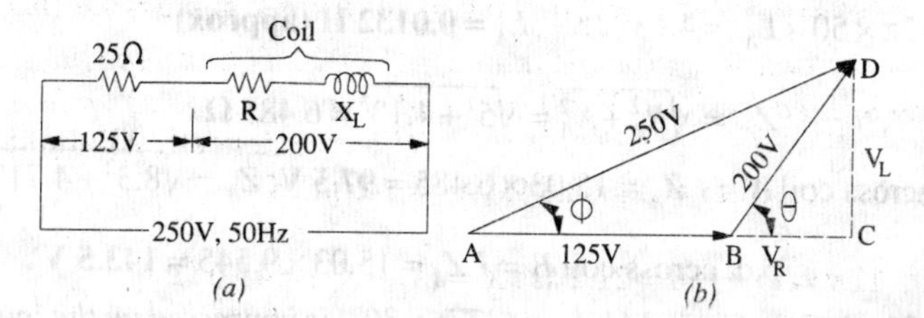

Fig. 13.17

$\therefore \quad BC = 27.5\text{V}; \; CD = \sqrt{200^2 - 27.5^2} = 198.1\text{V}$

(*i*) Coil impedance $= 200/5 = \mathbf{40\ \Omega}$

$$V_R = IR = BC \text{ or } 5R = 27.5 \quad \therefore \quad R = 27.5/5 = \mathbf{5.5\ \Omega}$$

Also $\quad V_L = I.X_L = CD = 198.1 \quad \therefore \quad X_L = 198.1/5 = \mathbf{39.62\ \Omega}$

or $\quad X_L = \sqrt{40^2 - 5.5^2} = \mathbf{39.62\ \Omega}$

(*ii*) Power absorbed by the coil is $= I^2R = 5^2 \times 5.5 = 137.5$ W

Also $P = 200 \times 5 \times 27.5/200 = \mathbf{137.5\ W}$

(*iii*) Total power $= VI \cos\phi = 250 \times 5 \times AC/AD = 250 \times 5 \times 152.5/250 = \mathbf{762.5\ W}$

The power may also be calculated by using I^2R formula.

Series resistance $= 125/5 = 25\ \Omega$

Total *circuit* resistance $= 25 + 5.5 = 30.5\ \Omega$

$\therefore$ Total power $= 5^2 \times 30.5 = \mathbf{762.5\ W}$

Example 13.20. *Two coils A and B are connected in series across a 240-V, 50-Hz supply. The resistance of A is 5 Ω and the inductance of B is 0.015 H. If the input from the supply is 3 kW and 2 kVAR, find the inductance of A and the resistance of B. Calculate the voltage across each coil.*

(Elect. Technology Hyderabad Univ. 1991)

Solution. The kVA triangle is shown in Fig. 13.18 (*b*) and the circuit in Fig. 13.18 (*a*). The circuit kVA is given by, kVA $= \sqrt{(3^2 + 2^2)} = 3.606$ or VA = 3,606 voltmeters

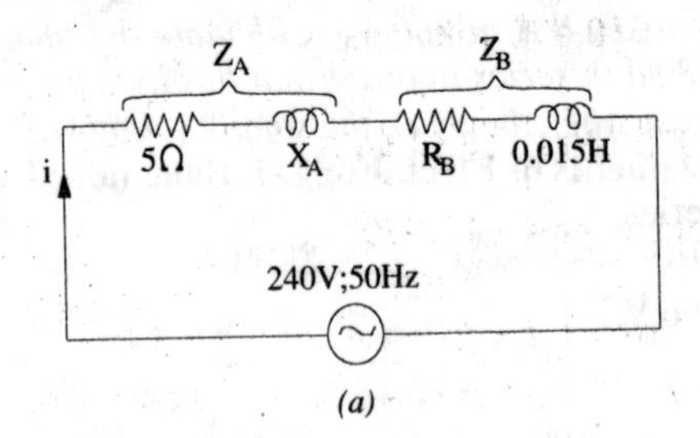

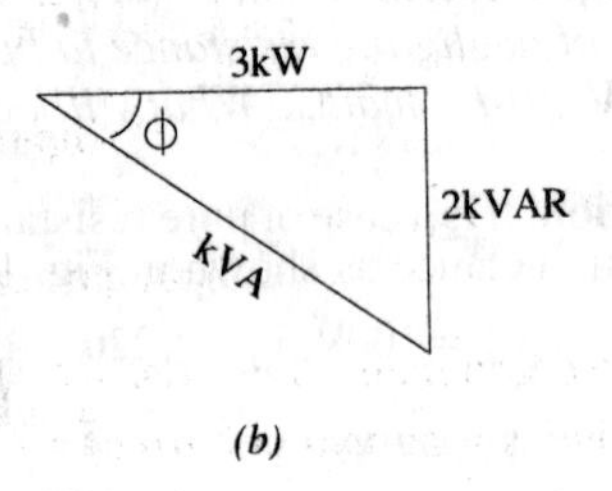

Fig. 13.18

Circuit current $= 3{,}606/240 = 15.03$ A $\quad \therefore \quad 15.03^2\,(R_A + R_B) = 3{,}000$

$$\therefore \quad R_A + R_B = 3{,}000/15.03^2 = 13.3\ \Omega \quad \therefore \quad R_B = 13.3 - 5 = \mathbf{8.3\ \Omega}$$

Now, impedance of the whole circuit is given by Z = 240/15.03 = 15.97 Ω

$$\therefore \quad X_A + X_B = \sqrt{Z^2 - (R_A + R_B)^2} = \sqrt{15.97^2 - 13.3^2} = 8.84\ \Omega$$

Now $\quad X_B = 2\pi \times 50 \times 0.015 = 4.713\ \Omega \quad \therefore \quad X_A = 8.843 - 4.713 = 4.13\ \Omega$

or $\quad 2\pi \times 50 \times L_A = 4.13 \quad \therefore \quad L_A = \mathbf{0.0132\ H\ (approx)}$

Now $\quad Z_A = \sqrt{R_A^2 + X_A^2} = \sqrt{5^2 + 4.13^3} = 6.485\ \Omega$

P.D. across coil $A = I.Z_A = 15.03 \times 6.485 = \mathbf{97.5\ V}$; $Z_B = \sqrt{8.3^2 + 4.713^2} = 9.545\ \Omega$

$\therefore \quad$ p.d. across coil $B = I.Z_B = 15.03 \times 9.545 = \mathbf{143.5\ V}$

Example 13.21. *An e.m.f. $e_0 = 141.4 \sin(377t + 30°)$ is impressed on the impedance coil having a resistance of 4 Ω and an inductive reactance of 1.25 Ω, measured at 25 Hz. What is the equation*

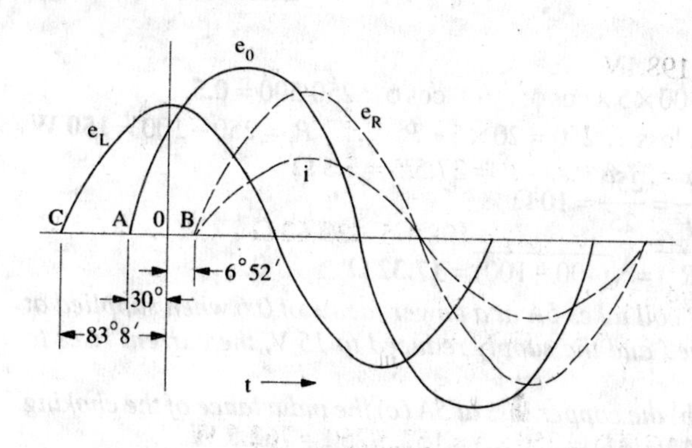

Fig. 13.19

of the current ? Sketch the waves for i, e_R, e_L and e_0.

Solution. The frequency of the applied voltage is $f = 377/2\pi = 60$ Hz

Since coil reactance is 1.25 Ω at 25 Hz, its value at 60 Hz = 1.25 × 60/25 = 3 Ω

Coil impedance, $Z = \sqrt{4^2 + 3^2} = 5\ \Omega$; $\phi = \tan^{-1}(3/4) = 36°52'$

It means that circuit current lags behind the applied voltage by 36° 52′. Hence, equation of the circuit current is $i = (141.4/5)\sin(377t + 30° - 36°52') = 28.3 \sin(377t - 6°52')$

Since, resistive drop is in phase with current, its equation is, $e_R = iR = 113.2 \sin(377t - 6°52')$

The inductive voltage drop *leads the current* by 90°, hence its equation is

$$e_L = iX_L = 3 \times 28.3 \sin(377t - 6°52' + 90°) = 54.9 \sin(377t + 83°8')$$

The waves for i, e_R, e_L and e_0 have been drawn in Fig. 13.19.

Example 13.22. *A single-phase, 7.46 kW motor is supplied from a 400-V, 50-Hz a.c. mains. If its efficiency is 85% and power factor 0.8 lagging, calculate (a) the kVA input (b) the reactive components of input current and (c) kVAR.*

Solution. Efficiency $= \dfrac{\text{output in watts}}{\text{input in watts}}$ $\therefore\ 0.85 = \dfrac{7.46 \times 1000}{VI \times \cos\phi} = \dfrac{7{,}460}{VI \times 0.8}$

$$\therefore \quad VI = \frac{7460}{0.85 \times 0.8} = 10{,}970 \text{ voltamperes}$$

(*a*) ∴ Input = 10,970/1000 = **10.97 kVA**

(*b*) Input current $I = \dfrac{\text{voltamperes}}{\text{volts}} = \dfrac{10{,}970}{400} = 27.43$ A

Active component of current = $I\cos\phi$ = **27.43 × 0.8 = 21.94 A**

Reactive component of current = $I \sin\phi$ = 27.43 × 0.6 = **16.46 A** (∵ $\sin\phi = 0.6$)

(Reactive component = $\sqrt{27.43^2 - 21.94^2}$ = 16.46 A)

(*c*) kVAR = kVA $\sin\phi$ = 10.97 × 0.6 = **6.58** (or kVAR = $VI \sin\phi \times 10^{-3} = 400 \times 16.46 \times 10^{-3} = 6.58$)

13.6. Power in an Iron-cored Choking Coil

Total power, P taken by an iron-cored choking coil is used to supply (*i*) power loss in ohmic resistance *i.e.* I^2R. (*ii*) iron-loss in core, P_i

$\therefore\ P^* = I^2R + P_i$ or $\dfrac{P}{I^2} = R + \dfrac{P_1}{I^2}$ is known as the **effective resistance** of the choke.

∴ effective resistance = true resistance + equivalent resistance $\dfrac{P_i}{I^2}$ $\therefore\ R_{eff} = \dfrac{P}{I^2} = R + \dfrac{P_i}{I^2}$

Example 13.23. *An iron-cored choking coil takes 5A when connected to a 20-V d.c. supply and takes 5A at 100 V a.c. and consumes 250 W. Determine (a) impedance (b) the power factor (c) the iron loss (d) inductance of the coil.* **(Elect. Engg & M.A.S I. June, 1991)**

*At higher frequencies like radio frequencies, there is skin-effect loss also.

Solution. (*a*) $Z = 100/5 = \mathbf{20\ \Omega}$

(*b*) $P = VI\cos\phi$ or $250 = 100 \times 5 \times \cos\phi$ $\therefore$ $\cos\phi = 250/500 = \mathbf{0.5}$

(*c*) Total loss = loss in resistance + iron loss $\therefore$ $250 = 20 \times 5 + P_i$ $\therefore$ $P_i = 250 - 100 = \mathbf{150\ W}$

(*d*) Effective resistance of the choke is $\dfrac{P}{I^2} = \dfrac{250}{25} = 10\ \Omega$

$$\therefore \quad X_L = \sqrt{(Z^2 - R^2)} = \sqrt{(400 - 100)} = \mathbf{17.32\ \Omega}$$

Example 13.24. *An iron-cored choking coil takes 5A at a power factor of 0.6 when supplied at 100-V, 50 Hz. When the iron core is removed and the supply reduced to 15 V, the current rises to 6 A at power factor of 0.9.*

Determine (a) the iron loss in the core (b) the copper loss at 5A (c) the inductance of the choking coil with core when carrying a current of 5A.

Solution. When core is removed, then $Z = 15/6 = 2.5\ \Omega$

True resistance, $R = Z\cos\phi = 2.5 \times 0.9 = 2.25\ \Omega$

With Iron Core

Power input $= 100 \times 5 \times 0.6 = 300$ W

Power wasted in the true resistance of the choke when current is 5A $= 5^2 \times 2.25 = 56.2$ W

(*a*) Iron loss $= 300 - 56.2 =$ **244 W (approx)** (*b*) Cu loss at 5A = **56.2 W**

(*c*) $Z = 100/5 = 20\ \Omega$; $X_L = Z\sin\phi = 20 \times 0.8 = 16\ \Omega$ $\therefore$ $2\pi \times 50 \times L = 16$ $\therefore$ $L = \mathbf{0.0509\ N}$

Tutorial Problem No. 13.1

1. The voltage applied to a coil having $R = 200\ \Omega$, $L = 638$ mH is represented by $e = 200 \sin 100\pi t$. Find a corresponding expression for the current and calculate the average value of the power taken by the coil. **[$i = 0.707 \sin (100\pi t - \pi/4)$; 50 W]** (*I.E.E. London*)

2. The coil having a resistance of 10 Ω and an inductance of 0.2 *H* is connected to a 100-V, 50-Hz supply. Calculate (*a*) the impedance of the coil (*b*) the reactance of the coil (*c*) the current taken and (*d*) the phase difference between the current and the applied voltage. **[(*a*) 63.5 Ω (*b*) 62.8 Ω (*c*) 1.575 A (*d*) 80°57′]**

3. An inductive coil having a resistance of 15 Ω takes a current of 4 A when connected to a 100-V, 60 Hz supply. If the coil is connected to a 100-V, 50 Hz supply, calculate (*a*) the current (*b*) the power (*c*) the power factor. Draw to scale the vector diagram for the 50-Hz conditions, showing the component voltages. **[(*a*) 4.46 A (*b*) 298 W (*c*) 0.669]**

4. When supplied with current at 240-V, single-phase at 50 Hz, a certain inductive coil takes 13.62 A. If the frequency of supply is changed to 40 Hz, the current increases to 16.12 A. Calculate the resistance and inductance of the coil. **[17.2 Ω, 0.05 H]** (*London Univ.*)

5. A voltage $v(t) = 141.4 \sin (314\ t + 10°)$ is applied to a circuit and a steady current given by $i(t) = 14.4 \sin (314\ t - 20°)$ is found to flow through it. Determine (*i*) the p.f. of the circuit and (*ii*) the power delivered to the circuit. **[0.866 (lag) ; 866 W]**

6. A circuit takes a current of 8 A at 100 V, the current lagging by 30° behind the applied voltage. Calculate the values of equivalent resistance and reactance of the circuit. **[10.81 Ω ; 6.25 Ω]**

7. Two inductive impedances *A* and *B* are connected in series. *A* has $R = 5\ \Omega$, $L = 0.01$ H; *B* has $R = 3\ \Omega$, $L = 0.02$ H. If a sinusoidal voltage of 230 V at 50 Hz is applied to the whole circuit calculate (*a*) the current (*b*) the power factor (*c*) the voltage drops. Draw a complete vector diagram for the circuit. **[(*a*) 18.6 (*b*) 0.648 (*c*) V_A = 109.5 V, V_B = 129.5 V]** (*I.E.E. London*)

8. A coil has an inductance of 0.1 H and a resistance of 30 Ω at 20°C. Calculate (*i*) the current and (*ii*) the power taken from 100-V, 50-Hz mains when the temperature of the coil is 60° C, assuming the temperature coefficient of resistance to be 0.4% per °C from a basic temperature of 20° C. **[(*i*) 2.13 A (*ii*) 158.5 W]** (*London Univ.*)

9. An air-cored choking coil takes a current of 2 A and dissipates 200 W when connected to a 200-V, 50-Hz mains. In other coil, the current taken is 3 A and the power 270 W under the same conditions. Calculate the current taken and the total power consumed when the coils are in series and connected to the same supply. **[1.2 A, 115 W]** (*City and Guilds, London*)

10. A circuit consists of a pure resistance and a coil in series. The power dissipated in the resistance is 500 W and the drop across it is 100 V. The power dissipated in the coil is 100 W and the drop across it is 50 V. Find the reactance and resistance of the coil and the supply voltage. **[9.168 Ω ; 4 Ω ; 128.5 V]**

11. A choking coil carries a current of 15 A when supplied from a 50-Hz, 230-V supply. The power in the circuit is measured by a wattmeter and is found to be 1300 watt. Estimate the phase difference between the current and p.d. in the circuit. **[0.3768]** (*I.E.E. London*)

12. An ohmic resistance is connected in series with a coil across 230-V, 50-Hz supply. The current is 1.8 A and p.ds. across the resistance and coil are 80 V and 170 V respectively. Calculate the resistance and inductance of the coil and the phase difference between the current and the supply voltage.

[61.1 Ω , 0.229 H, 34°20′] (*App. Elect. London Univ.*)

13. A coil takes a current of 4 A when 24 V d.c. are applied and for the same power on a 50-Hz a.c. supply, the applied voltage is 40. Explain the reason for the difference in the applied voltage. Determine (*a*) the reactance (*b*) the inductance (*c*) the angle between the applied p.d. and current (*d*) the power in watts.

[(*a*) 8 Ω (*b*) 0.0255 H (*c*) 53°7′ (*d*) 96 W]

14. An inductive coil and a non-inductive resistance *R* ohms are connected in series across an a.c. supply. Derive expressions for the power taken by the coil and its power factor in terms of the voltage across the coil, the resistance and the supply respectively. If *R* = 12 Ω and the three voltages are in order, 110 V, 180 V and 240 V, calculate the power and the power factor of the coil.

[546 W; 0.331]

15. Two coils are connected in series. With 2 A d.c. through the circuit, the p.ds. across the coils are 20 and 30 V respectively. With 2 A a.c. at 40 Hz, the p.ds. across the coils are 140 and 100 V respectively. If the two coils in series are connected to a 230-V, 50-Hz supply, calculate (*a*) the current (*b*) the power (*c*) the power factor.

[(*a*) 1.55 A (*b*) 60 W (*c*) 0.1684]

16. It is desired to run a bank of ten 100-W, 100-V lamps in parallel from a 230-V, 50-Hz supply by inserting a choke coil in series with the bank of lamps. If the choke coil has a power factor of 0.2, find its resistance, reactance and inductance.

[R = 4.144 Ω, X = 20.35 Ω, L = 0.065 H] (*London Univ.*)

17. At a frequency for which ω = 796, an e.m.f. of 6 V sends a current of 100 mA through a certain circuit. When the frequency is raised so that ω = 2866, the same voltage sends only 50 mA through the same circuit. Of what does the circuit consist ?

[R = 52 Ω, L = 0.0378 H in series] (*I.E.E. London*)

18. An iron-cored electromagnet has a d.c. resistance of 7.5 Ω and when connected to a 400-V 50-Hz supply, takes 10 A and consumes 2 kW. Calculate for this value of current (*a*) power loss in iron core (*b*) the inductance of coil (*c*) the power factor (*d*) the value of series resistance which is equivalent to the effect of iron loss.

[1.25 kW, 0.11 H, 0.5 ; 12.5 Ω] (*I.E.E. London*)

13.7. A.C. Through Resistance and Capacitance

The circuit is shown in Fig. 13.20 (*a*). Here $V_R = IR$ = drop across *R* –in phase when *I*

$V_C = IX_C$ = drop across capacitor–lagging *I* by π/2

As capacitive reactance X_C is taken negative, V_C is shown along negative direction of *Y*-axis in the voltage triangle [Fig. 13.20 (*b*)].

Now $V = \sqrt{V_R^2 + (-V_C)^2} = \sqrt{(IR)^2 + (-IX_C)^2} = I\sqrt{R^2 + X_C^2}$ or $I = \dfrac{V}{\sqrt{R^2 + X_C^2}} = \dfrac{V}{Z}$

The denominator is called the *impedance* of the circuit. So, $Z = \sqrt{R^2 + X_C^2}$

Impedance triangle is shown in Fig. 13.20 (*c*)

From Fig. 13.20 (*b*) it is found that *I* leads *V* by angle φ such that tan φ = $-X_C/R$

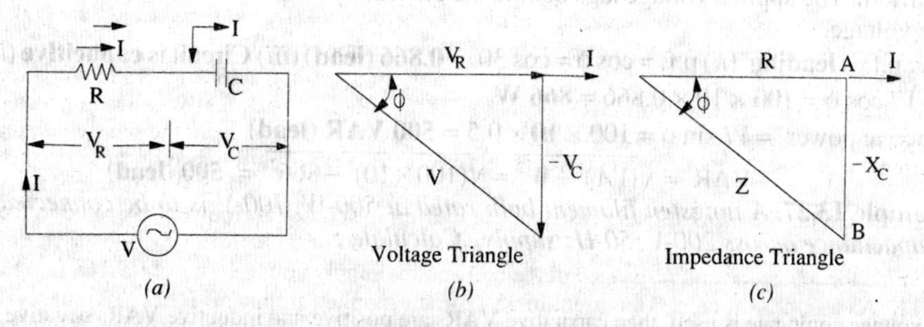

Fig. 13.20

Hence, it means that if the equation of the applied alternating voltage is $v = V_m \sin \omega t$, the equation of the resultant current in the *R-C* circuit is $i = I_m \sin(\omega t + \phi)$ so that current *leads* the applied voltage by an angle ϕ. This fact is shown graphically in Fig. 13.21.

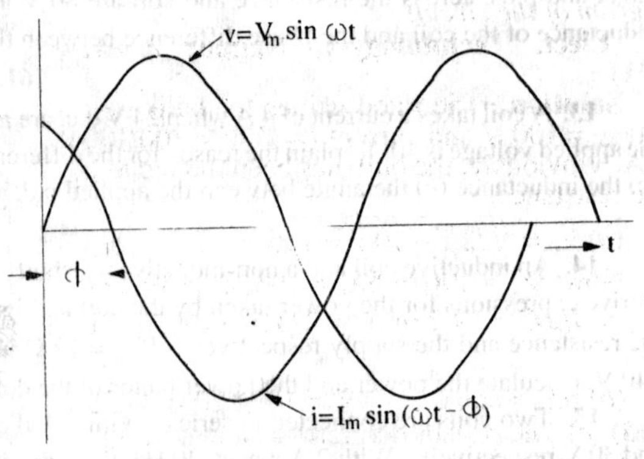

Fig. 13.21

Example 13.25. *An a.c. voltage (80 + j60) volts is applied to a circuit and the current flowing is (–4 + j 10) amperes. Find (i) impedance of the circuit (ii) power consumed and (iii) phase angle.*

[Elect. Technology, Indore Univ. 1989]

Solution. $\mathbf{V} = (80 + j60) = 100 \angle 36.9°$;

$\mathbf{I} = -4 + j10 = 10.77 \angle \tan^{-1}(-2.5) = 10.77 \angle(180° - 68.2°) = 10.77 \angle 111.8°$

(*i*) $\mathbf{Z} = \mathbf{V}/\mathbf{I} = 100\angle 36.9°/10.77\angle 111.8° = 9.28\angle -74.9°$

$= 9.28(\cos 74.9° - j \sin 74.9°) = 2.42 - j8.96\ \Omega$

Hence $R = 2.42\ \Omega$ and $X_C = 8.96\ \Omega$ capacitive

(*ii*) $P = I^2R = 10.77^2 \times 2.42 = \mathbf{2.81\ W}$

(*iii*) phase angle between voltage and current = 74.9° with current *leading* as shown in Fig. 13.22.

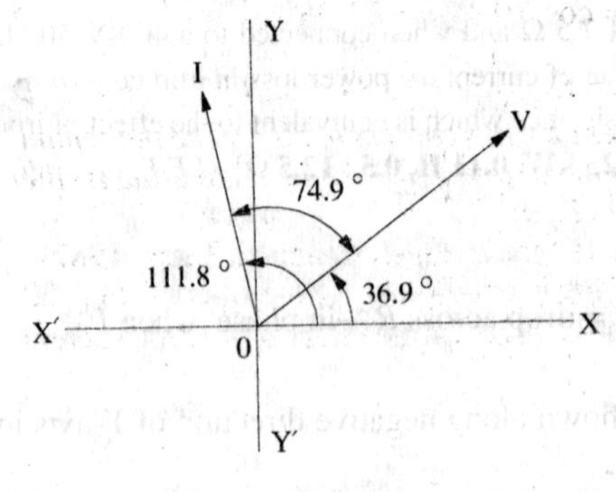

Fig. 13.22

Alternative Method for Power

The method of conjugates will be used to determine the real power and reactive volt-ampere. It is a convenient way of calculating these quantities when both voltage and current are expressed in cartesian form. If the conjugate of *current* is multiplied by the voltage in cartesian form, the result is a complex quantity, the real part of which gives the real power and *j* part of which gives the reactive volt-amperes (*VAR*). It should, however, be noted that real power as obtained by this method of conjugates is the same regardless of whether *V* or *I* is reversed although sign of voltamperes will depend on the choice of *V* or *I*.

Using current conjugate, we get $P_{VA} = (80 + j60)(-4 - j10) = 280 - j1040$*

∴ Power consumed = **280 W**

Example 13.26. *In a circuit, the applied voltage is 100 V and is found to lag the current of 10 A by 30°. (i) Is the p.f. lagging or leading ? (ii) What is the value of p.f. ?*

(iii) Is the circuit inductive or capacitive ? (iv) What is the value of active and reactive power in the circuit ? **(Basic Electricity, Bombay Univ. 1985)**

Solution. The applied voltage lags behind the current which, in other words, means that current leads the voltage.

(*i*) ∴ p.f. is **leading** (*ii*) p.f. = $\cos\phi = \cos 30° = \mathbf{0.866}$ **(lead)** (*iii*) Circuit is **capacitive** (*iv*) Active power = $VI\cos\phi = 100 \times 10 \times 0.866 = \mathbf{866\ W}$

Reactive power $= VI \sin\phi = 100 \times 10 \times 0.5 = \mathbf{500\ VAR}$ **(lead)**

or $\text{VAR} = \sqrt{(VA)^2 - W^2} = \sqrt{(100 \times 10)^2 - 866^2} = \mathbf{500}$ **(lead)**

Example 13.27. *A tungsten filament bulb rated at 500-W, 100-V is to be connected to series with a capacitance across 200-V, 50-Hz supply. Calculate :*

*If voltage conjugate is used, then capacitive VARs are positive and inductive VARs negative. If current conjugate is used, then capacitive VARs are negative and inductive VARs are positive.

(a) the value of capacitor such that the voltage and power consumed by the bulb are according to the rating of the bulb. (b) the power factor of the current drawn from the supply. (c) draw the phasor diagram of the circuit.

(Elect. Technology-1, Nagpur Univ. 1991)

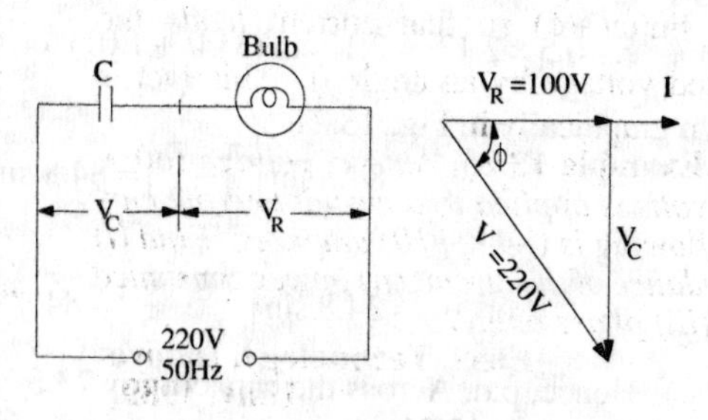

Fig. 13.23

Solution. The rated values for bulb are : voltage = 100 V and current $I = W/V = 500/100 = 5$A. Obviously, the bulb has been treated as a pure resistance.

(a) $V_C = \sqrt{220^2 - 100^2} = 196$ V

Now, $IX_C = 196$ or $5X_C = 196$, $X_C = 39.2\ \Omega$

$\therefore\ I/\omega C = 39.2$ or $C = 1/314 \times 39.2 =$ **81 μF**

(b) p.f. = cos ϕ = $V_R/V = 100/220 =$ **0.455 (lead)** *(c)* The phasor diagram is shown in Fig. 13.23.

Example 13.28. *A pure resistance of 50 ohms is in series with a pure capacitance of 100 microfarads. The series combination is connected across 100-V, 50-Hz supply. Find (a) the impedance (b) current (c) power factor (d) phase angle (e) voltage across resistor (f) voltage across capacitor. Draw the vector diagram.* **(Elect. Engg-1, JNT Univ. Warrangel 1985)**

Solution. $X_C = 10^6\pi | 2\pi \times 50 \times 100 = 32\ \Omega;\ R = 50\ \Omega$

(a) $Z = \sqrt{50^2 + 32^2} = \mathbf{59.4\ \Omega}$ *(b)* $I = V/Z = 100/59.4 = \mathbf{1.684\ A}$

(c) p.f. $= R/Z = 50/59.4 =$ **0.842 (lead)** *(d)* $\phi = \cos^{-1}(0.842) = \mathbf{32°36'}$

(e) $V_R = IR = 50 \times 1.684 = \mathbf{84.2\ V}$ *(f)* $V_C = IX_C = 32 \times 1.684 = \mathbf{53.9\ V}$

Example 13.29. *A 240-V, 50-Hz series R-C circuit takes an r.m.s current of 20 A. The maximum value of the current occurs 1/900 second before the maximum value of the voltage. Calculate (i) the power factor (ii) average power (iii) the parameters of the circuit.*

(Elect. Engg-I, Calcutta Univ. 1987)

Solution. Time-period of the alternating voltage is 1/50 second. Now a time interval of 1/50 second corresponds to a phase difference of 2π radian or 360°. Hence, a time interval of 1/900 second corresponds to a phase difference of 360 × 50/900 = 20°.

Hence, current leads the voltage by 20°

(i) power factor = cos 20° = **0.9397 (lead)**

(ii) average power = 240 × 20 × 0.9397 = **4,510 W**

(iii) Z = 240/20 = 12Ω ; R = Z cos Φ = 12 × 0.9397 = **11.28 Ω**

X_C = Z sin Φ = 12 × sin 20° = 12 × 0.342 = 4.1 Ω

C = $10^6/2\pi \times 50 \times 4.1$ = **775 μF**

Example 13.30. *A voltage v = 100 sin 314t is applied to a circuit consisting of a 25 Ω resistor and an 80 μF capacitor in series. Determine : (a) an expression for the value of the current flowing at any instant (b) the power consumed (c) the p.d. across the capacitor at the instant when the current is one-half of its maximum value.*

Solution. $X_C = 1/(314 \times 80 \times 10^{-6}) = 39.8\ \Omega$, $Z = \sqrt{25^2 + 39.8^2} = 47\ \Omega$;

$I_m = V_m/Z = 100/47 = 2.13$ A,

$\phi = \tan^{-1}(39.8/25) = 57°\ 52' = 1.01$ radian (lead)

(a) Hence, equation for the instantaneous current

i = **2.13 sin (314 t + 1.01)** *(b)* Power = $I^2R = (2.13/\sqrt{2})^2 \times 25 =$ **56.7 W** *(c)* The voltage across the capacitor lags the circuit current by π/2 radians. Hence, its equation is given by

$$v_c = V_{cm} \sin\left(314t + 1.01 - \frac{\pi}{2}\right) \text{ where } V_{cm} = I_m \times X_C = 2.13 \times 39.8 = 84.8 \text{ V}$$

Now, when i is equal to half the maximum current (say, in the positive direction) then

$i = 0.5 \times 2.13$ A

$$\therefore \; 0.5 \times 2.13 = 2.13 \sin(314t + 1.01) \text{ or } 314t + 1.01 = \sin^{-1}(0.5) = \frac{\pi}{6} \text{ or } \frac{5\pi}{6} \text{ radian}$$

$$\therefore \; v_C = 84.8 \sin\left(\frac{\pi}{6} - \frac{\pi}{2}\right) = 84.8 \sin(-\pi/3) = -73.5 \text{ V}$$

$$\text{or } \upsilon_c = 34.8 \sin\left(\frac{5\pi}{6} - \frac{\pi}{2}\right) = 84.4 \sin \pi|3 = 73.5\ V$$

Hence, p.d. Across the capacitor is **73.5** V

Example 13.31. *A capacitor and a non-inductive resistance are connected in series to a 200-V, single-phase supply. When a voltmeter having a non-inductive resistance of 13,500 Ω is connected across the resistor, it reads 132 V and the current then taken from the supply is 22.35 mA.*

Indicate on a vector diagram, the voltages across the two components and also the supply current (a) when the voltmeter is connected and (b) when it is disconnected.

Solution. The circuit and vector diagrams are shown in Fig. 13.24 (*a*) and (*b*) respectively.

(*a*) $V_C = \sqrt{200^2 - 132^2} = \mathbf{150\ V}$.

It is seen that $\phi = \tan^{-1}(150/132) = 49°$ in Fig. 13.24 (*b*). Hence

(*i*) Supply voltage lags behind the current by 49°. (*ii*) V_R leads supply voltage by 49° (*iii*) V_C lags behind the supply voltage by (90°–49°) = 41°

The supply current is, as given, equal to **22.35 mA**. The value of unknown resistance *R* can be found as follows :

Current through voltmeter = 132/13,500 = 9.78 mA

∴ Current through *R* = 22.35 – 9.78 = 12.57 mA ∴ $R = 132/12.57 \times 10^{-3} = 10{,}500\ \Omega$;

$X_C = 150/22.35 \times 10^{3} = 6.711\ \Omega$

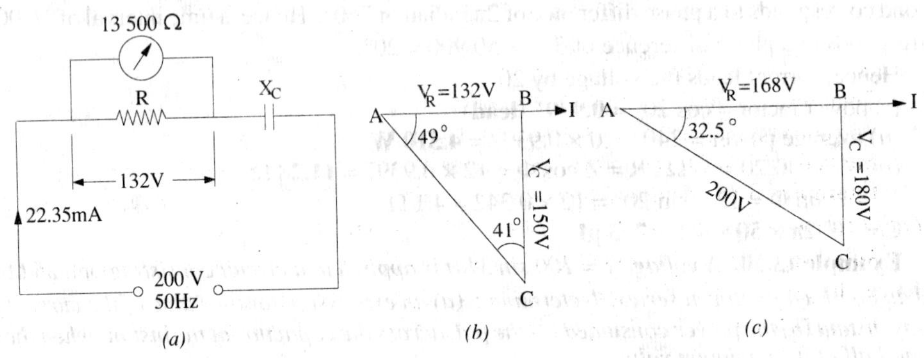

Fig. 13.24

(*b*) **When voltmeter is disconnected,** $Z = \sqrt{R^2 \times X_L^2} = \sqrt{10{,}500^2 + 6{,}730^2} = 12{,}500\ \Omega$

Supply current $= 200/12{,}461 = \mathbf{16.0\ mA}$

In this case, $V_R = 16.0 \times 10^{-3} \times 10{,}500 = \mathbf{168\ V}$

$V_C = 16.0 \times 10^{-3} \times 6711 = \mathbf{107.4\ V}$; $\tan\phi = 107.4/168$

$\therefore \; \phi = 32.5°$

In this case, the supply voltage lags the circuit current by 32.5° as shown in Fig. 13.24(*c*).

Example 13.32. *It is desired to operate a 100-W, 120-V electric lamp at its current rating from a 240-V, 50-Hz supply. Give details of the simplest manner in which this could be done using (a) a resistor (b) a capacitor and (c) an inductor having resistance of 10 Ω. What power factor would be presented to the supply in each case and which method is the most economical of power.*

(Principles of Elect. Engg.-I, Jadavpur Univ. 1985)

Solution. Rated current of the bulb is = 100/120 = 5/6 A

The bulb can be run at its correct rating by any one of the three methods shown in Fig. 13.25.

(*a*) With reference to Fig. 13.25 (*a*), we have

P.D. Across $R = 240 - 120 = 120$ V

$\therefore\ R = 120/(5/6) = \mathbf{144\ \Omega}$

Power factor of the circuit is **unity.** Power consumed = 240 × 5/6 = **200 W**

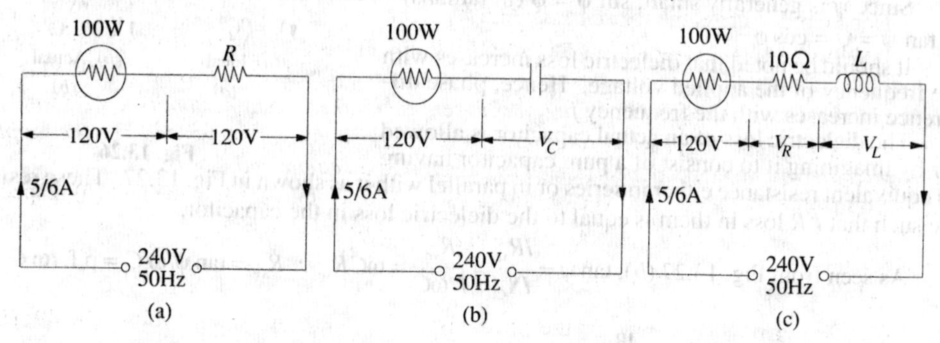

Fig. 13.25

(*b*) Referring to Fig. 13.25 (*b*), we have

$V_C = \sqrt{240^2 - 120^2} = 207.5$ V; $X_C = 207.5/(5/6) = 249\ \Omega$

$\therefore\quad 1/314C = 249$ or $C = \mathbf{12.8\ \mu F}$; p.f. $= \cos\phi = 120/240 = \mathbf{0.5\ (lead)}$

Power consumed $= 240 \times (5/6) \times 0.5 = \mathbf{100\ W}$

(*c*) The circuit connections are shown in Fig. 13.25 (*c*)

$$V_R = (5/6) \times 10 = 25/3\text{ V} \quad \therefore\quad V_L = \sqrt{240^2 - \left(120 + \frac{25}{3}\right)^2} = 203\text{ V}$$

$\therefore\quad 314L \times (5/6) = 203 \quad \therefore\ L = \mathbf{0.775\ H}$

Total resistive drop = 120 + (25/3) = 128.3 V ; cos φ = 128.3/240 = **0.535 (lag)**

Power consumed = 240 × (5/6) × 0.535 = **107 W**

Method (*b*) is most economical because it involves least consumption of power.

Example 13.33. *A two-element series circuit consumes 700 W and has a p.f. = 0.707 leading. If applied voltage is v = 141.1 sin (314 t + 30°), find the circuit constants.*

Solution. The maximum value of voltage is 141.4 V and it leads the reference quantity by 30°. Hence, the given sinusoidal voltage can be expressed in the phasor form as

$\mathbf{V} = (141.4/\sqrt{2})\angle 30° = 100\angle 30°$ Now, $P = VI\cos\phi \quad \therefore\ 700 = 100 \times I \times 0.707; I = 10$ A.

Since p.f. $= 0.707$(lead) ; $\phi = \cos^{-1}(0.707) = 45°$(lead).

It means that current leads the given voltage by 45° for it *leads the common reference quantity by (30° + 45°) = 75.°* Hence, it can be expressed as **I** = 10∠75°

$$\mathbf{Z} = \frac{\mathbf{V}}{\mathbf{I}} = \frac{100\angle 30°}{10\angle 75°} = 10\angle -45° = 7.1 - j7.1 \quad \therefore\ R = \mathbf{7.1\ \Omega}$$

Since $X_C = 7.1 \quad \therefore\quad 1/314C = 7.1; \quad \therefore\ C = \mathbf{450\ \mu F}$

13.8. Dielectric Loss and Power Factor of a Capacitor

An ideal capacitor is one in which there are no losses and whose current leads the voltage by

90° as shown in Fig. 13.26 (*a*). In practice, it is impossible to get such a capacitor although close approximation is achieved by proper design. In every capacitor, there is always some dielectric loss and hence it absorbs some power from the circuit. Due to this loss, the phase angle is somewhat less than 90° [Fig. 13.26 (*b*)]. In the case of a capacitor with a poor dielectric, the loss can be considerable and the phase angle much less than 90°. This dielectric loss appears as heat. By *phase difference* is meant the difference between the ideal and actual phase angles. As seen from Fig. 13.26 (*b*), the phase difference ψ is given by $\psi = 90 - \phi$ where ϕ is the actual phase angle, $\sin\psi = \sin(90 - \phi) = \cos\phi$ where $\cos\phi$ is the power factor of the capacitor.

Since ψ is generally small, $\sin\psi = \psi$ (in radians)

$\therefore \tan\psi = \psi = \cos\phi$.

It should be noted that dielectric loss increases with the frequency of the applied voltage. Hence, phase difference increases with the frequency *f*.

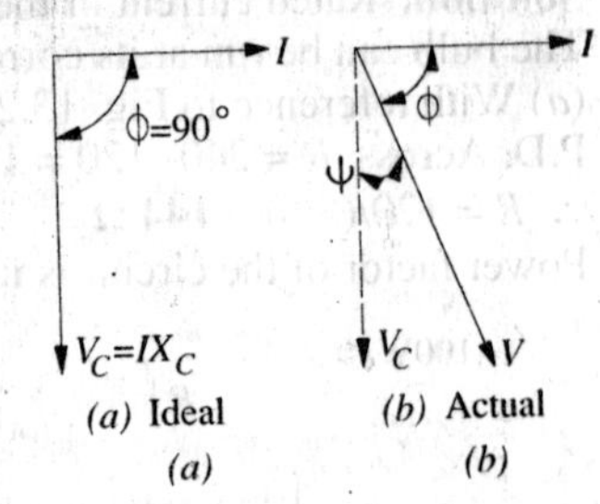

(*a*) (*b*)

Fig. 13.26

The dielectric loss of an actual capacitor is allowed for by imagining it to consist of a pure capacitor having an equivalent resistance either in series or in parallel with it as shown in Fig. 13.27. These resistances are such that I^2R loss in them is equal to the dielectric loss in the capacitor.

As seen from Fig. 13.27 (*b*), $\tan\psi = \dfrac{IR_{se}}{IX_C} = \dfrac{R_{se}}{I/\omega C} = \omega CR_{se}$ $\therefore R_{se} = \tan\psi/\omega C = \text{p.f.}/\omega C$

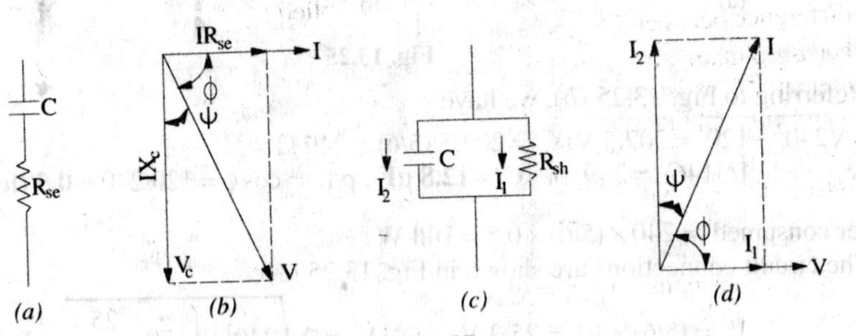

Fig. 13.27

Similarly, as seen from Fig. 14.26 (*d*), $\tan\psi = \dfrac{I_1}{I_2} = \dfrac{V/R_{sh}}{V/X_C} = \dfrac{X_C}{R_{sh}} = \dfrac{I}{\omega CR_{sh}}$

$$R_{sh} = \frac{1}{\omega C.\tan\psi} = \frac{1}{\omega C \times \text{power factor}} = \frac{1}{\omega C \times \text{p.f.}}$$

The power loss in these resistances is $P = V^2/R_{sh} = \omega CV^2 \tan\psi = \omega CV^2 \times \text{p.f.}$

or $= I^2R_{se} = (I^2 \times \text{p.f.})/\omega C$

where p.f. stands for the power factor of the capacitor.

Note. (*i*) In case, ψ is not small, then as seen from Fig. 14.26 (*b*) $\tan\phi = \frac{X_C}{R_{se}}$ (Ex. 13.34.) $R_{se} = X_C/\tan\phi$

From Fig. 13.27 (*d*), we get $\tan\phi = \dfrac{I_2}{I_1} = \dfrac{V/X_C}{V/R_{sh}} = \dfrac{R_{sh}}{X_C}$ $\therefore R_{sh} = X_C\tan\phi = \tan\phi/\omega C$

(*ii*) It will be seen from above that both R_{se} and R_{sh} vary inversely as the frequency of the applied voltage. In other words, the resistance of a capacitor decreases in proportion to the increase in frequency.

$$\frac{R_{se1}}{R_{se2}} = \frac{f_2}{f_1}$$

Example 13.34. *A capacitor has a capacitance of 10 μF and a phase difference of 10°. It is inserted in series with a 100 Ω resistor across a 220-V, 50-Hz line. Find (i) the increase in resistance due to the insertion of this capacitor (ii) power dissipated in the capacitor and (iii) circuit power factor.*

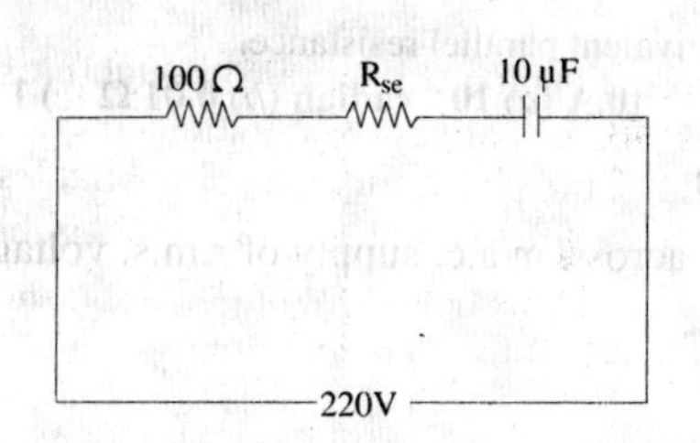

Fig. 13.28

Solution. $X_C = \dfrac{10^6}{2\pi\times 50\times 10} = 318.3\ \Omega$

The equivalent series resistance of the capacitor is

$$R_{se} = X_C/\tan\phi$$

Now $\phi = 90 - \psi = 90° - 10° = 80°$

$\tan\phi = \tan 80° = 5.671$

$\therefore\quad R_{se} = 318.3/5.671 = 56.1\ \Omega$

(*i*) Hence, resistance of the circuit increases by **56.1 Ω**.

(*ii*) $Z = \sqrt{(R+R_{se})^2+X_C^2} = \sqrt{156.1^2+318.3^2} = 354.4\ \Omega;\ I = 220/354 = \mathbf{0.62\ A}$

Power dissipated in the capacitor $= I^2R_{se} = 0.62^2\times 56.1 = \mathbf{21.6\ W}$

(*iii*) Circuit power factor is $= (R+R_{se})/Z = 156.1/354.4 =$ **0.44 (lead)**

Example 13.35. *Dielectric heating is to be employed to heat a slab of insulating material 2 cm thick and 150 sq. cm in area. The power required is 200 W and a frequency of 30 MHz is to be used. The material has a relative permittivity of 5 and a power factor of 0.05. Determine the voltage necessary and the current which will flow through the material. If the voltage were to be limited to 600-V, to what value would the frequency have to be raised ?*

(Elect. Engg. AMIETE (New Scheme) June 1992)

Solution. The capacitance of the parallel-plate capacitor formed by the insulating slab is

$$C = \frac{\varepsilon_0\varepsilon_r A}{d} = \frac{8.854\times 10^{-12}\times 5\times 150\times 10^{-4}}{2\times 10^{-2}} = 33.2\times 10^{-12}\ \text{F}$$

As shown in Art. 13.8 $R_{sh} = \dfrac{1}{\omega C\times p.f.} = \dfrac{1}{(2\pi\times 30\times 10^6)\times 33.2\times 10^{-12}\times 0.05} = 3196\ \Omega$

Now, $P = V^2/R_{sh}$ or $V = \sqrt{P\times R_{sh}} = \sqrt{200\times 3196} = 800\ \text{V}$

Current $I = V/X_C = \omega CV = (2\pi\times 30\times 10^6)\times 33.2\times 10^{-12}\times 800 = 5\ \text{A}$

Now, as seen from above $P = \dfrac{V^2}{R_{sh}} = \dfrac{V^2}{1/\omega C\times \text{p.f.}} = V^2\omega C\times\text{p.f.}$ or $P\propto V^2 f$

$$\therefore\quad 800^2\times 30 = 600^2\times f \text{ or } f = \left(\frac{800}{600}\right)^2\times 30 = \mathbf{53.3\ MHz}$$

Tutorial Problem No. 13.2

1. A capacitor having a capacitance of 20 μF is connected in series with a non-inductive resistance of 120 Ω across a 100-V, 50-Hz supply. Calculate (*a*) voltage (*b*) the phase difference between the current and the supply voltage (*c*) the power. Also draw the vector diagram. **[(*a*) 0.501 A (*b*) 52.9° (*c*) 30.2 W]**

2. A capacitor and resistor are connected in series to an a.c. supply of 50 V and 50 Hz. The current is 2 A and the power dissipated in the circuit is 80 W. Calculate the resistance of the resistor and the capacitance of the capacitor. **[20 Ω ; 212 μF]**

3. A voltage of 125 *V* at 50 Hz is applied to a series combination of non-inductive resistor and a lossless capacitor of 50 μF. The current is 1.25 A. Find (*i*) the value of the resistor (*ii*) power drawn by the network (*iii*) the power factor of the network. Draw the phasor diagram for the network.

[(*i*) 77.3 Ω (*ii*) 121 W (*iii*) 0.773 (lead)] (*Electrical Technology-1, Osmania Univ. Dec. 1979*)

4. A black box contains a two-element series circuit. A voltage (40–*j*30) drives a current of (40–*j*3)A in the circuit. What are the values of the elements ? Supply frequency is 50 Hz.

[R = 1.05 ; C = 4750 μF] (*Elect. Engg. & Electronics Bangalore Univ. 1986*)

5. Following readings were obtained from a series circuit containing resistance and capacitance :

$V = 150\ V;\ I = 2.5\ A;\ P = 37.5\ W, f = 60$ Hz

Calculate (*i*) power factor (*ii*) effective resistance (*iii*) capacitive reactance and (*iv*) capacitance.

[(*i*) 0.1 (*ii*) 6 Ω (*iii*) 59.7 Ω (*iv*) 44.4 μF]

6. An alternating voltage of 10 volt at a frequency of 159 kHz is applied across a capacitor of 0.01 μF. Calculate the current in the capacitor. If the power dissipated within the dielectric is 100μW, calculate (*a*) loss angle (*b*) the equivalent series resistance (*c*) the equivalent parallel resistance.

[0.A (*a*) 10^{-4} radian (*b*) 0.01 Ω (*c*) 1 MΩ]

13.9. Resistance, Inductance and Capacitance in Series

The three are shown in Fig. 13.29 (*a*) joined in series across an a.c. supply of r.m.s. voltage *V*.

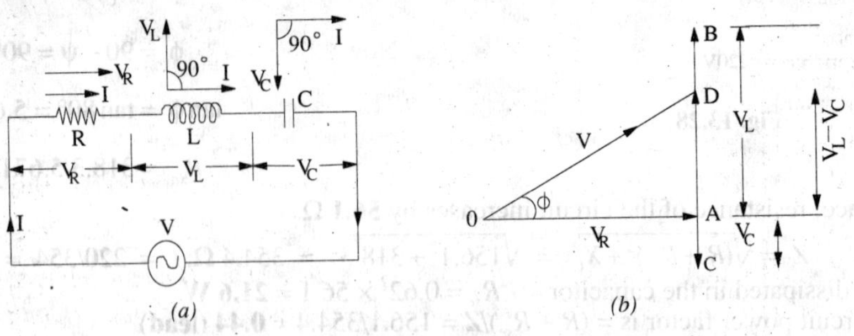

Fig. 13.29

Let $V_R = IR$ = voltage drop across *R* – in phase with *I*

$V_L = I.X_L$ = voltage drop across *L* – leading *I* by π/2

$V_C = I.X_C$ = voltage drop across *C* – lagging *I* by π/2

In voltage triangle of Fig. 13.29 (*b*), *OA* represents V_R, *AB* and *AC* represent the inductive and capacitive drops respectively. It will be seen that V_L and V_C are 180° out of phase with each other *i.e.* they are in direct opposition to each other.

Subtracting *BD* (= *AC*) from *AB*, we get the net reactive drop $AD = I(X_L - X_C)$

The applied voltage *V* is represented by *OD* and is the vector sum of *OA* and *AD*

$$\therefore \quad OD = \sqrt{OA^2 + AD^2} \quad \text{or} \quad V = \sqrt{(IR)^2 + (IX_l - IX_C)^2} = I\sqrt{R^2 + (X_L - X_C)^2}$$

or
$$I = \frac{V}{\sqrt{R^2 + (X_L - X_C)^2}} = \frac{V}{\sqrt{R^2 + X^2}} = \frac{V}{Z}$$

The term $\sqrt{R^2 + (X_L - X_C)^2}$ is known as the impedance of the circuit. Obviously,

$$(impedance)^2 = (resistance)^2 + (net\ reactance)^2$$

or
$$Z^2 = R^2 + (X_L - X_C)^2 = R^2 + X^2$$

where *X* is the net reactance (Fig. 14.29).

Phase angle ϕ is given by $\tan\phi = (X_L - X_C)/R = X/R$ = net reactance/resistance

Power factor is $\cos\phi = \dfrac{R}{Z} = \dfrac{R}{\sqrt{R^2 + (X_L - X_C)^2}} = \dfrac{R}{\sqrt{R^2 + X^2}}$

Hence, it is seen that if the equation of the applied voltage is $v = V_m \sin\omega t$, then equation of the resulting current in an *R-L-C* circuit is given by $i = I_m \sin(\omega t \pm \phi)$

The +ve sign is to be used when current leads *i.e.* $X_C > X_L$.

The –ve sign is to be used when current lags *i.e.* when $X_L > X_C$.

In general, the current lags or leads the supply voltage by an angle ϕ such that $\tan\phi = X/R$

Using symbolic notation, we have (Fig. 13.31), $\mathbf{Z} = R + j(X_L - X_C)$

Numerical value of impedance

$$Z = \sqrt{R^2 + (X_L - X_C)^2}$$

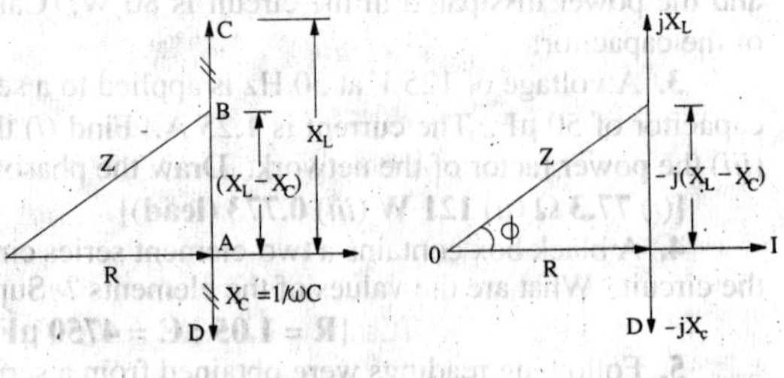

Fig. 13.30 **Fig. 13.31**

Its phase angle is $\Phi = \tan^{-1}[X_1 - X_C)/R]$

$\mathbf{Z} = Z\angle\tan^{-1}[(X_L - X_C)/R] = Z\angle\tan^{-1}(X/R)$

If $\mathbf{V} = V\angle 0$, then, $\mathbf{I} = \mathbf{V}/\mathbf{Z}$

Summary of Results of Series AC Circuits

Type of Impedance	*Value of Impedance*	*Phase angle for current*	*Power factor*
Resistance only	R	0°	1
Inductance only	ωL	90° lag	0
Capacitance only	$1/\omega C$	90° lead	0
Resistance and Inductance	$\sqrt{[R^2+(\omega L)^2]}$	$0 < \phi < 90°$ lag	1>p.f.>0 lag
Resistance and Capacitance	$\sqrt{[R^2+(-1/\omega C)^2]}$	$0 < \phi < 90°$ lead	1 > p.f. > 0 lead
R–L–C	$\sqrt{[R^2+(\omega L \sim 1/\omega C)^2]}$	between 0° and 90° lag or lead	between 0 and unity lag or lead

Example 13.36. *A resistance of 20 Ω, an inductance of 0.2 H and a capacitance of 100 μF are connected in series across 220-V, 50-Hz mains. Determine the following (a) impedance (b) current (c) voltage across R, L and C (d) power in watts and VA (e) p.f. and angle of lag.*

(*Elect. Engg. A.M.Ae. S.I. 1992*)

Solution. $X_C = 0.2 \times 314 = 63\ \Omega \quad C = 10\mu F = 100 \times 10^{-6} = 10^{-4}$ farad

$$X_C = \frac{1}{\omega C} = \frac{1}{314 \times 10^{-4}} = 32\ \Omega,\ X = 63 - 32 = \mathbf{31\ \Omega}\ (inductive)$$

$(a)\ Z = \sqrt{(20^2 + 31^2)} = \mathbf{37\ \Omega} \qquad (b)\ I = 220/37 = \mathbf{6\ A\ (approx)}$

$(c)\ V_R = I \times R = 6 \times 20 = \mathbf{120\ V};\quad V_L = 6 \times 63 = \mathbf{378\ V},\ V_C = 6 \times 32 = \mathbf{192\ V}$

(d) Power in VA $= 6 \times 220 = \mathbf{1320}$

Power in watts $= 6 \times 220 \times 0.54 = \mathbf{713\ W}$

(e) p.f. $= \cos\phi = R/Z = 20/37 = \mathbf{0.54}$; $\phi = \cos^{-1}(0.54) = \mathbf{57°18'}$

Example 13.37. *A voltage e(t) = 100 sin 314 t is applied to a series circuit consisting of 10 ohm resistance, 0.0318 henry inductance and a capacitor of 63.6 μF. Calculate (i) expression for i(t) (ii) phase angle between voltage and current (ii) power factor (iv) active power consumed (v) peak value of pulsating energy.* **(Elect. Technology, Indore Univ. 1985)**

Solution. Obviously, $\omega = 314$ rad/s ; $X_L = \omega L = 314 \times 0.0318 = 10\ \Omega$

$X_C = 1/\omega C = 1/314 \times 63.6 \times 10^{-6} = 50\ \Omega$; $X = X_L - X_C = (10 - 50) = -40\ \Omega$ (capacitive)

$$\mathbf{Z} = 10 - j40 = 41.2\ \angle -76°;\ \mathbf{I} = \frac{\mathbf{V}}{\mathbf{Z}} = \frac{(100/\sqrt{2})\angle 0°}{41.2\angle -76°} = 1.716\angle 76°$$

$I_m = I \times \sqrt{2} = 1.716 \times \sqrt{2} = 2.43$ A

(*i*) $i(t) = 2.43 \sin(314\,t + 76°)$

(*ii*) $\phi = \mathbf{76°}$ with current leading

(*iii*) p.f. $= \cos\phi = \cos 76° = \mathbf{0.24\ (lead)}$

(*iv*) Active power, $P = VI\cos\phi$

$= (100/\sqrt{2})\,(2.43/\sqrt{2}) \times 0.24 = \mathbf{29.16\ W}$

(*v*) As seen from Fig. 13.32, peak value of

pulsating energy is $\dfrac{V_m I_m}{2} + \dfrac{V_m I_m}{2}\cos\phi$

$$= \frac{V_m I_m}{2}(1 + \cos\phi) = \frac{100 \times 2.43}{2}(1 + 0.24) = \mathbf{151\ W}$$

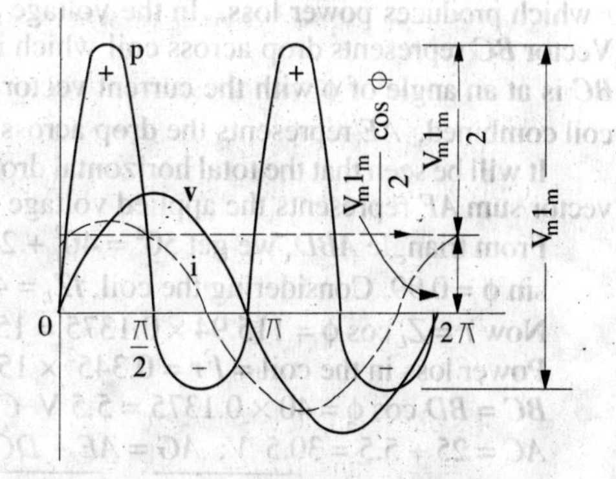

Fig. 13.32

Example 13.38. *Two impedances Z_1 and Z_2 when connected separately across a 230-V, 50-Hz supply consumed 100 W and 60 W at power factors of 0.5 lagging and 0.6 leading respectively. If these impedances are now connected in series across the same supply, find :*

(i) total power absorbed and overall p.f. (ii) the value of the impedance to be added in series so as to raise the overall p.f. to unity. **(Elect. Circuits-I, Bangalore Univ. 1987)**

Solution. Inductive Impedance $V_1 I_1 \cos\phi_1$ = power ; $230 \times I_1 \times 0.5 = 100$; $I_1 = 0.87$ A

Now $I_1^2 R_1$ = power or $0.87^2 R_1 = 100$; $R_1 = 132\ \Omega$; $Z_1 = 230/0.87 = 264\ \Omega$

$$X_L = \sqrt{Z_1^2 - R_1^2} = \sqrt{264^2 - 132^2} = 229\ \Omega$$

Capacitive Impedance $I_2 = 60/230 \times 0.6 = 0.434\ A$; $R_2 = 60/0.434^2 = 318\ \Omega$

$$Z_2 = 230/0.434 = 530\ \Omega\ ;\ X_C = \sqrt{530^2 - 318^2} = 424\ \Omega \text{ (capacitive)}$$

When Z_1 and Z_2 are connected in series

$$R = R_1 + R_2 = 132 + 318 = 450\ \Omega\ ;\ X = 229 - 424 = -195\ \Omega \text{ (capacitive)}$$

$$Z = \sqrt{R^2 + X^2} = \sqrt{450^2 + (-195)^2} = 490\ \Omega, I = 230/490 = 0.47\ A$$

(*i*) Total power absorbed = $I^2R = 0.47^2 \times 450$ = **99 W,** $\cos\phi = R/Z = 450/490$ = **0.92 (lead)**

(*ii*) Power factor will become unity when the net capacitive reactance is neutralised by an equal inductive reactance. The reactance of the required series pure inductive coil is **195 Ω.**

Example 13.39. *A resistance R, an inductance L = 0.01 H and a capacitance C are connected in series. When a voltage v = 400 cos (300t–10°) volts is applied to the series combination, the current flowing is 10 $\sqrt{2}$ cos (3000t–55°) amperes. Find R and C.*

(Elect. Circuits Nagpur Univ. 1992)

Solution. The phase difference between the applied voltage and circuit current is (55° – 10°) = 45° with current lagging. The angular frequency is ω =3000 radian/second. Since current lags, $X_L > X_C$.

Net reactance $X = (X_L - X_C)$. Also $X_L = \omega L = 3000 \times 0.01 = 30\ \Omega$

$$\tan\phi = X/R \quad \text{or} \quad \tan 45° = X/R \quad \therefore X = R \text{ Now, } Z = \frac{V_m}{I_m} = \frac{400}{10\sqrt{2}} = 28.3\ \Omega$$

$$Z^2 = R^2 + X^2 = 2R^2 \quad \therefore \quad R = Z/\sqrt{2} = 28.3/\sqrt{2} = 20\ \Omega; X = X_L - X_C = 30 - X_C = 20$$

$$X_C = 10\Omega \quad \text{or} \quad \frac{1}{\omega C} = 10 \quad \text{or} \quad \frac{1}{3000\ C} = \text{ or } C = \mathbf{33.3\ \mu F}$$

Example 13.40. *A non-inductive resistor is connected in series with a coil and a capacitor. The circuit is connected to a single-phase a.c. supply. If the voltages are as indicated in Fig. 13.33 when current flowing through the circuit is 0.345 A, find the applied voltage and the power loss in coil.*

(Elect. Engg. Pune Univ. 1988)

Solution. It may be kept in mind that the coil has not only inductance *L* but also some resistance *r* which produces power loss. In the voltage vector diagram, *AB* represents drop across *R* = 25 V. Vector *BC* represents drop across coil which is due to *L* and *r*. Which value is 40 V and the vector *BC* is at an angle of φ with the current vector. *AD* represents 50 V which is the drop across *R* and coil combined. *AE* represents the drop across the capacitor and leads the current by 90°.

It will be seen that the total horizontal drop in the circuit is *AC* and the vertical drop is *AG*. Their vector sum *AF* represents the applied voltage V.

From triangle *ABD*, we get $50^2 = 40^2 + 25^2 + 2 \times 25 \times 40 \times \cos\phi \therefore \cos\phi = 0.1375$ and $\sin\phi = 0.99$. Considering the coil, $IZ_L = 40 \therefore Z_L = 40/0.345 = 115.94\ \Omega$

Now $r = Z_L \cos\phi = 115.94 \times 0.1375 = 15.94\ \Omega$

Power loss in the coil = $I^2 r = 0.345^2 \times 15.94$ = **1.9 W**

$BC = BD \cos\phi = 40 \times 0.1375 = 5.5$ V $CD = BD \sin\phi = 40 \times 0.99 = 39.6$ V

$AC = 25 + 5.5 = 30.5$ V; $AG = AE - DC = 55 - 39.6 = 15.4$ V

$$AF = \sqrt{AC^2 + CF^2} = \sqrt{30.5^2 + 15.4^2} = \mathbf{34.2\ V}$$

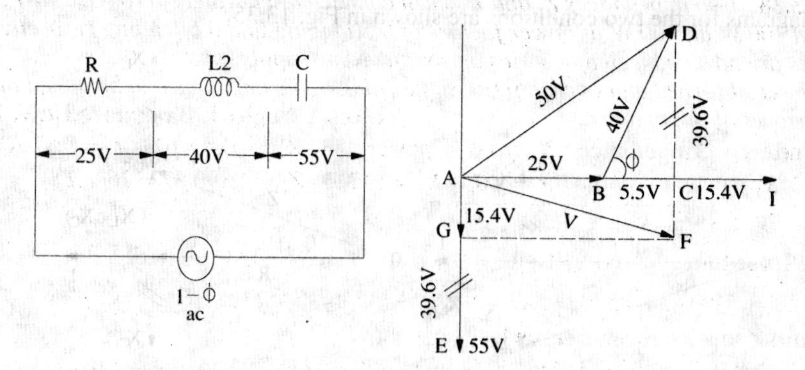

Fig. 13.33

Example 13.41. *A 4.7 H inductor which has a resistance of 20 Ω, a 4-μF capacitor and a 100-Ω non-inductive resistor are connected in series to a 100-V, 50-Hz supply. Calculate the time interval between the positive peak value of the supply voltage and the next peak value of power.*

Solution. Total resistance = 120 Ω ; $X_L = 2\pi \times 50 \times 4.7 = 1477\ \Omega$

$X_C = 10^6/2\pi \times 50 \times 4 = 796\ \Omega$; $X = 1477 - 798 = 681\ \Omega$; $Z = \sqrt{120^2 + 681^2} = 691.3\ \Omega$

$\cos \phi = R/Z = 120/691.3 = 0.1736$; $\phi = 80°$

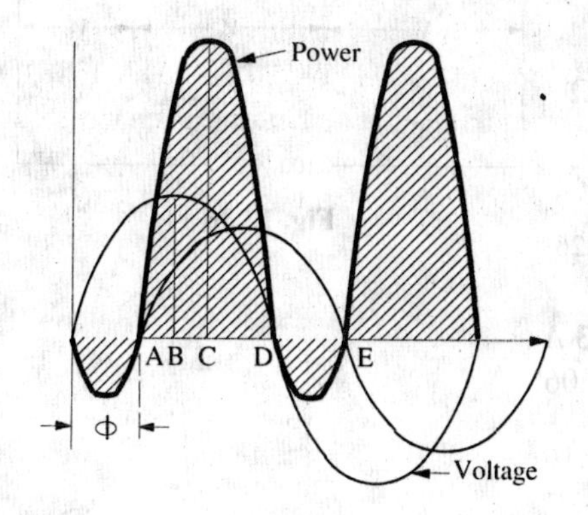

Fig. 13.34

Now, as seen from Fig. 13.34, the angular displacement between the peak values of supply voltage and power cycles is $BC = \phi/2$ because $AB = 90 - \phi$ and $AD = 180 - \phi$.

Hence $AC = 90 - \phi/2$

$\therefore \quad BC = AC - AB = (90 - \phi/2) - (90 - \phi) = \phi/2$

Angle difference = $\phi/2 = 80°/2 = 40°$

Since a full cycle of 360° corresponds to a time interval of 1/50 second

$\therefore$ 40° angular interval $= \dfrac{40}{50 \times 360} = \mathbf{2.22\ ms.}$

Example 13.42. *A coil is in series with a 20 μF capacitor across a 230-V, 50-Hz supply. The current taken by the circuit is 8 A and the power consumed is 200 W. Calculate the inductance of the coil if the power factor of the circuit is (i) leading (ii) lagging.*

Sketch a vector diagram for each condition and calculate the coil power factor in each case.

(Elect. Engg.-I Nagpur Univ. 1993)

Solution. (*i*) Since power factor is leading, net reactance $X = (X_C - X_L)$ as shown in Fig. 13.35(*a*).

$$I^2R = 200 \text{ or } 8^2 \times R = 200 ; \therefore R = 200/64 = 25/8\ \Omega = 3.125\ \Omega$$

$$Z = V/I = 230/8 = 28.75\ \Omega ; X_C = 10^6/2\pi \times 50 \times 20 = 159.15\ \Omega$$

$$R^2 + X^2 = 28.75^2 \quad \therefore \quad X = 28.58\ \Omega \quad \therefore (X_C - X_L) = 28.58 \text{ or } 159.15 - X_L = 28.58$$

$$\therefore X_L = 130.57\ \Omega \text{ or } 2\pi \times 50 \times L = 130.57 \quad \therefore \quad L = 0.416\ H$$

If θ is the p.f. angle of the coil, then $\tan \theta = R/X_L = 3.125/130.57 = 0.024$; $\theta = 1.37°$, p.f. of the coil = 0.9997

(*ii*) When power factor is lagging, net reactance is $(X_L - X_C)$ as shown in Fig. 13.36(*b*).

$\therefore X_L - 159.15 = 28.58$ or $X_L = 187.73\ \Omega \therefore 187.73 = 2\pi \times 50 \text{ x } L$ or $L = 0.597$ H.

In this case, $\tan \theta = 3.125/187.73 = 0.0167$; $\theta = 0.954° \therefore \cos \theta = 0.9998$.

The vector diagrams for the two conditions are shown in Fig. 13.35.

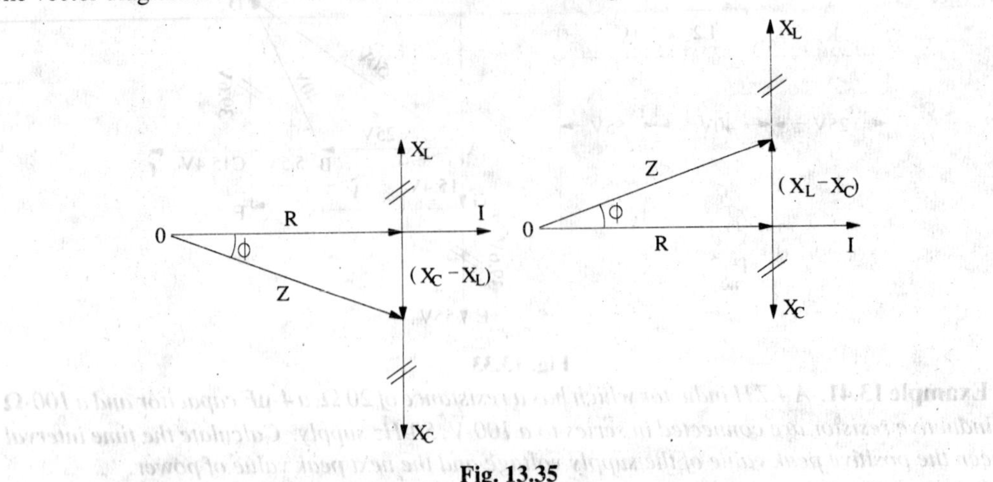

Fig. 13.35

Example 13.43. *In Fig. 13.36, calculate (i) current (ii) voltage drops V_1, V_2, and V_3 and (iii) power absorbed by each impedance and total power absorbed by the circuit. Take voltage vector along the reference axis.*

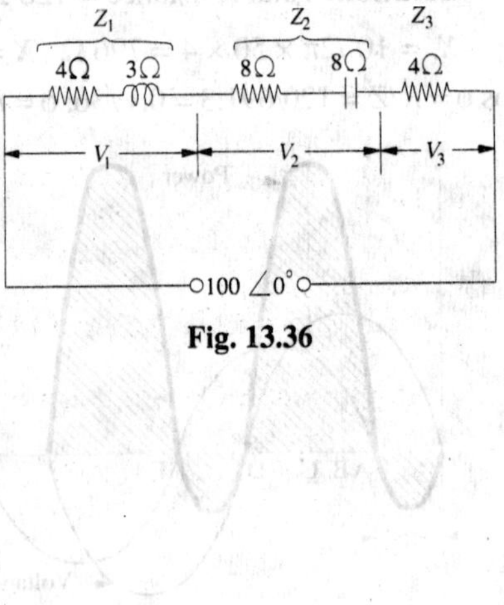

Fig. 13.36

Solution. $\mathbf{Z}_1 = (4 + j3)\ \Omega$; $\mathbf{Z}_2 = (6 - j8)\ \Omega$; $\mathbf{Z}_3 = (4 + j0)\ \Omega$

$$\mathbf{Z} = \mathbf{Z}_1 + \mathbf{Z}_2 + \mathbf{Z}_3 = (4 + j3) + (6 - j8) + (4 + j0) = (14 - j5)\Omega$$

Taking $\mathbf{V} = V\angle 0° = 100\angle 0° = (100 + j0)$

$$\therefore \mathbf{I} = \frac{\mathbf{V}}{\mathbf{Z}} = \frac{100}{(14 - j5)} = \frac{100(14 + j5)}{(14 - j5)(14 + j5)} = 6.34 + j2.26$$

(*i*) Magnitude of the current $= \sqrt{(6.34^2 + 2.26^2)} = \mathbf{6.73\ A}$

(*ii*) $\mathbf{V}_1 = \mathbf{IZ}_1 = (6.34 + j2.26)\,(4 + j3) = 18.58 + j28.06$

$\mathbf{V}_2 = \mathbf{IZ}_2 = (6.34 + j2.26)\,(6 - j8) = 56.12 - j37.16$

$\mathbf{V}_3 = \mathbf{IZ}_3 = (6.34 + j2.26)\,(4 + j0) = 25.36 + j9.04,$

$\mathbf{V} = 100 + j0$ (check)

(*iii*) $P_1 = 6.73^2 \times 4 = 181.13$ W,

$P_2 = 6.73^2 \times 6 = 271.74$ W, $P_3 = 6.73^2 \times 4 = 181.13$ W,

Total = 634 W

Otherwise $\mathbf{P}_{VA} = (100 + j0)(6.34 - j2.26)$ (using current conjugate)

$= 634 - j226 \therefore$

real power = 634 W (as a check)

Example 13.44. *Draw a vector for the circuit shown in Fig. 13.37 indicating the resistance and reactance drops, the terminal voltages V_1 and V_2 and the current. Find the values of (i) the current I (ii) V_1 and V_2 and (iii) p.f.* **(Elements of Elect Engg-I, Bangalore Univ. 1988)**

Solution. $L = 0.05 + 0.1 = 0.15$ H ; $X_L = 314 \times 0.5 = 47.1\ \Omega$

$X_C = 10^6/314 \times 50 = 63.7\ \Omega$; $X = 47.1 - 63.7 = -16.6\ \Omega, R = 30\ \Omega,$

$Z = \sqrt{30^2 + (-16.6)^2} = 34.3\ \Omega$

(*i*) $I = V/Z = 200/34.3 =$ **5.83 A**

(*ii*) $X_{L1} = 314 \times 0.05$

$= 15.7\ \Omega$

$Z_1 = \sqrt{10^2 + 15.7^2} = 18.6\ \Omega$

$V_1 = IZ_1 = 5.83 \times 18.6$

$=$ **108.4 V**

$\phi_1 = \cos^{-1}(10/18.6) = 57.5°$(lag)

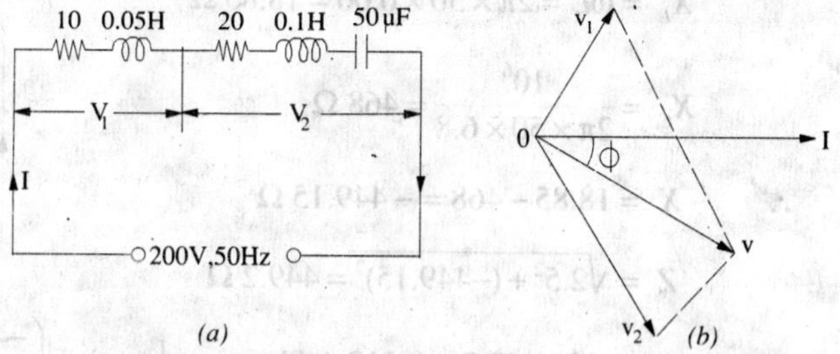

(*a*) (*b*)

Fig. 13.37

$X_{L2} = 314 \times 0.1 = 31.4\ \Omega, X_C = -63.7\ \Omega, X = 31.4 - 63.7 = -32.3\ \Omega, Z_2 = \sqrt{20^2 + (-32.3)^2} =$ **221 V**

$\phi_2 = \cos^{-1}(20/38) = 58.2$ (lead)

(*iii*) combined p.f. $= \cos\phi = R/Z = 30/34.3 =$ **0.875 (lead)**

Example 13.45. *In a circuit, the applied voltage is found to lag the current by 30°.*

(a) Is the power factor lagging or leading ? (b) What is the value of the power factor ? (c) Is the circuit inductive or capacitive ?

In the diagram of Fig. 13.38, the voltage drop across Z_1 is (10+j0) volts. Find out (i) the current in the circuit (ii) the voltage drops across Z_2 and Z_3 (iii) the voltage of the generator.

(Elect. Engg.-1, Bombay Univ. 1991)

Solution. (*a*) Power factor is *leading* because current leads the voltage.

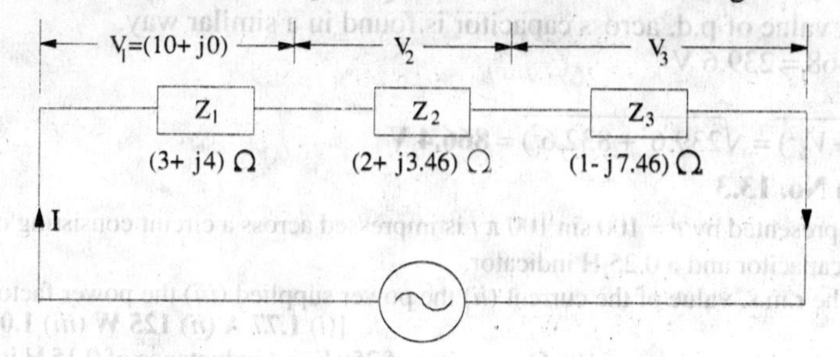

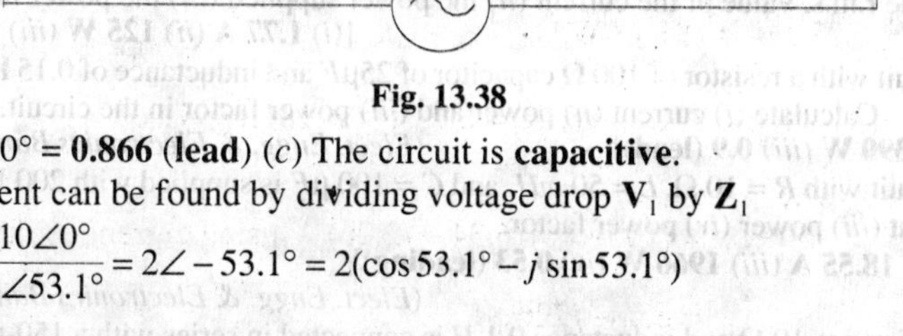

Fig. 13.38

(*b*) p.f. = cos 30° = **0.866 (lead)** (*c*) The circuit is **capacitive.**

(*i*) Circuit current can be found by dividing voltage drop $\mathbf{V}_1$ by $\mathbf{Z}_1$

$$\mathbf{I} = \frac{10 + j0}{3 + j4} = \frac{10\angle 0°}{5\angle 53.1°} = 2\angle -53.1° = 2(\cos 53.1° - j\sin 53.1°)$$

$$= 2(0.6 - j0.8) = 1.2 - j1.6$$

$$\mathbf{Z}_2 = 2 + j3.46;\ \mathbf{V}_2 = \mathbf{IZ}_2 = (1.2 - j1.6)(2 + j3.46) = (7.936 + j0.952) \text{ volt}$$

$$\mathbf{V}_3 = (1.2 - j1.6)(1 - j7.46) = (-10.74 - j10.55) \text{ volt}$$

(*ii*) $$\mathbf{V} = \mathbf{V}_1 + \mathbf{V}_2 + \mathbf{V}_3 = (10 + j0) + (7.936 + j0.952) + (-10.74 - j10.55)$$

$$= (7.2 - j9.6) = 12\angle -53.1°$$

Incidentally, it shows that current **I** and voltage **V** are in phase with each other.

Example 13.46. *A 230-V, 50-Hz alternating p.d. supplies a choking coil having an inductance of 0.06 henry in series with a capacitance of 6.8 μF, the effective resistance of the circuit being 2.5 Ω. Estimate the current and the angle of the phase difference between it and the applied p.d. If the p.d. has a 10% harmonic of 5 times the fundamental frequency, estimate (a) the current due to it and (b) the p.d. across the capacitance.* **(Electrical Network Analysis, Nagpur Univ 1993)**

Solution. Fundamental Frequency

$$X_L = \omega L = 2\pi \times 50 \times 0.06 = 18.85\ \Omega$$

$$X_C = \frac{10^6}{2\pi \times 50 \times 6.8} = 468\ \Omega$$

$$\therefore \quad X = 18.85 - 468 = -449.15\ \Omega$$

$$Z = \sqrt{2.5^2 + (-449.15)^2} = 449.2\ \Omega$$

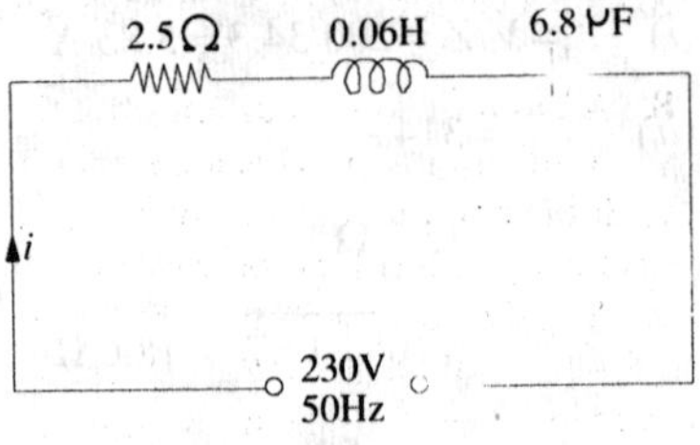

2.5Ω 0.06H 6.8 μF i 230V 50Hz

Fig. 13.39

Current $I_f = 230/449.2 = 0.512\ A$ Phase angle $= \tan^{-1}\left(\dfrac{-449.2}{2.5}\right) = -89°.42'$

∴ current leads p.d. by 89° 42′.

Fifth Harmonic Frequency $X_L = 18.85 \times 5 = 94.25\ \Omega$; $X_C = 468 \div 5 = 93.6\ \Omega$

$X = 94.25 - 93.6 = 0.65\ \Omega$; $\mathbf{Z} = \sqrt{2.5^2 + 0.65^2} = 2.585\ \Omega$ Harmonic p.d. $= 230 \times 10/100 = 23$ V

∴ Harmonic current $I_h = 23/2.585 = 8.893$ A

P.D. Across capacitor at harmonic frequency is, $V_h = 8.893 \times 93.6 = 832.6$ V

The total current flowing through the circuit, due to the complex voltage wave form, is found from the fundamental and harmonic components thus. Let,

I = the r.m.s. value of total circuit current,

I_f = r.m.s. value of fundamental current,

I_h = r.m.s. value of fifth harmonic current,

(*a*) $\therefore \ I = \sqrt{I_f^2 + I_h^2} = \sqrt{0.512^2 + 8.893^2} = \mathbf{8.9\ A}$

(*b*) The r.m.s. value of p.d. across capacitor is found in a similar way.

$V_f = 0.512 \times 468 = 239.6$ V

$$\therefore \ V = \sqrt{(V_f^2 + V_h^2)} = \sqrt{239.6^2 + 832.6^2)} = \mathbf{866.4\ V}$$

Tutorial Problem No. 13.3

1. An e.m.f. represented by $e = 100 \sin 100\ \pi\ t$ is impressed across a circuit consisting of 40-Ω resistor in series with a 40-μ*F* capacitor and a 0.25 H indicator.

Determine (*i*) the r.m.s. value of the current (*ii*) the power supplied (*iii*) the power factor.

[(*i*) **1.77 A** (*ii*) **125 W** (*iii*) **1.0**] (*London Univ.*)

2. A series circuit with a resistor of 100 Ω capacitor of 25μ*F* and inductance of 0.15 H is connected across 220-V, 60-Hz supply. Calculate (*i*) current (*ii*) power and (*iii*) power factor in the circuit.

[(*i*) **1.97A;** (*ii*) **390 W** (*iii*) **0.9 (lead)**] (*Elect. Engg. & Electronics Bangalore Univ. 1985*)

3. A series circuit with $R = 10\ \Omega$, $L = 50\ mH$ and $C = 100\ \mu F$ is supplied with 200 *V*/50 Hz. Find (*i*) the impedance (*ii*) current (*iii*) power (*iv*) power factor.

[(i) **18.94 Ω** (*ii*) **18.55 A** (*iii*) **1966 W** (*iv*) **0.53 (leading)**]

(*Elect. Engg. & Electronics Bangalore Univ. 1986*)

4. A coil of resistance 10 Ω and inductance 0.1 *H* is connected in series with a 150-μ*F* capacitor across a 200-V, 50-Hz supply. Calculate (*a*) the inductive reactance, (*b*) the capacitive reactance, (*c*) the impedance (*d*) the current, (*e*) the power factor (*f*) the voltage across the coil and the capacitor respectively.

[(*a*) **31.4 Ω** (*b*) **21.2 Ω** (*c*) **14.3 Ω** (*d*) **14 A** (*e*) **0.7 lag** (*f*) **460 V, 297 V**]

5. A circuit is made up of 10 Ω resistance, 12 mH inductance and 281.5 μ*F* capacitance in series. The supply voltage is 100 V (constant). Calculate the value of the current when the supply frequency is (*a*) 50 Hz and (*b*) 150 Hz.

[8 A leading ; 8 A lagging]

6. A coil having a resistance of 10 Ω and an inductance of 0.2 H is connected in series with a capacitor of 50.7 μ*F*. The circuit is connected across a 100-V, 50-Hz a.c. supply. Calculate (*a*) the current flowing (*b*) the voltage across the capacitor (*c*) the voltage across the coil. Draw a vector diagram to scale.

[(*a*) **10 A** (*b*) **628 V** (*c*) **635 V**]

7. A coil is in series with a 20 μF capacitor across a 230-V, 50-Hz supply. The current taken by the circuit is 8 A and the power consumed is 200 W. Calculate the inductance of the coil if the power factor of the circuit is (*a*) leading and (*b*) lagging.

Sketch a vector diagram for each condition and calculate the coil power factor in each case.

[0.415 H; 0.597 H; 0.0238 ; 0.0166]

8. A circuit takes a current of 3 A at a power factor of 0.6 lagging when connected to a 115-V, 50-Hz supply. Another circuit takes a current of 5 A at a power factor of 0.707 leading when connected to the same supply. If the two circuits are connected in series across a 230-V, 50-Hz supply, calculate

(*a*) the current (*b*) the power consumed and (*c*) the power factor. [(*a*) **5.5 A** (*b*) **1.188 kW** (*c*) **0.939 lag**]

9. A coil of insulated wire of resistance 8 ohms and inductance 0.03 H is connected to an a.c. supply at 240 V, 50-Hz. Calculate (*a*) the current, the power and power factor (*b*) the value of a capacitance which, when connected in series with the above coil, causes no change in the values of current and power taken from the supply. [(*a*) **19.4 A, 3012 W, 0.65 lag** (*b*) **168.7 μF**] (*London Univ.*)

10. A series circuit, having a resistance of 10 Ω, an inductance of 0.025 H and a variable capacitance is connected to a 100-V, 25-Hz single-phase supply. Calculate the capacitance when the value of the current is 8 A. At this value of capacitance, also calculate (*a*) the circuit impedance (*b*) the circuit power factor and (*c*) the power consumed. **[556 μF** (*a*) **1.5 Ω** (*b*) **0.8 leading** (*c*) **640 W]**

11. An alternating voltage is applied to a series circuit consisting of a resistor and iron-cored inductor and a capacitor. The current in the circuit is 0.5 A and the voltages measured are 30 V across the resistor, 48 V across the inductor, 60 V across the resistor and inductor and 90 V across the capacitor. Find (*a*) the combined copper and iron losses in the inductor (*b*) the applied voltage, [(*a*) **3.3 W** (*b*) **56 V**] (*City & Guilds, London*)

12. When an inductive coil is connected across a 250-V, 50-Hz supply, the current is found to be 10 A and the power absorbed 1.25 kW. Calculate the impedance, the resistance and the inductance of the coil.

A capacitor which has a reactance twice that of the coil, is now connected in series with the coil across the same supply. Calculate the p.d. Across the capacitor. **[25 Ω ; 12.5 Ω ; 68.7 mH; 433 V]**

13. A voltage of 200 V is applied to a series circuit consisting of a resistor, an inductor and a capacitor. The respective voltages across these components are 170, 150 and 100 V and the current is 4 A. Find the power factor of the inductor and of the circuit. **[0.16 ; 0.97]**

14. A pure resistance *R*, a choke coil and a pure capacitor of 50 μF are connected in series across a supply of V volts, and carry a current of 1.57 A. Voltage across *R* is 30 V, acros choke coil 50 V and across capacitor 100 V. The voltage across the combination of *R* and choke coil is 60 volt. Find the supply voltage *V*, the power loss in the choke, frequency of the supply and power factor of the complete circuit. Draw the phasor diagram.

[60.7 V; 6.5 W; 0.562 lead] [*F.E. Pune Univ. No. 1986*)

13.10. Resonance in R-L-C Circuits

We have seen from Art. 13.9 that net reactance in an *R-L-C* circuit of Fig. 13.40 (*a*) is

$$X = X_L - X_C \text{ and } Z = \sqrt{[R^2 + (X_L - X_C)^2]} = \sqrt{R^2 + X^2}$$

Let such a circuit be connected across an a.c. source of constant voltage *V* but of frequency varying from zero to infinity. There would be a certain frequency of the applied voltage which would make X_L equal to X_C in magnitude. In that case, $X = 0$ and $Z = R$ as shown in Fig. 13.40 (c). Under this condition, the circuit is said to be in electrical resonance.

As shown in Fig. 13.40 (c), $V_L = I \cdot X_L$ and $V_C = I \cdot X_C$ and the two are equal in magnitude but opposite in phase. Hence, they cancel each other out. The two reactances taken together act as a short-circuit since no voltage develops across them. Whole of the applied voltage drops across *R* so that $V = V_R$. The circuit impedance $Z = R$. The phasor diagram for series resonance is shown in Fig. 13.40 (*d*).

Calculation of Resonant Frequency

The frequency at which the net reactance of the series circuit is zero is called the resonant frequency f_0. Its value can be found as under : $X_L - X_C = O$ or $X_L = X_C$ or $\omega_0 L = 1/\omega_0 C$

$$\text{or } \omega_0^2 = \frac{1}{LC} \text{ or } (2\pi f_0)^2 = \frac{1}{LC} \text{ or } f_0 = \frac{1}{2\pi\sqrt{LC}}$$

If *L* is in henry and *C* in farad, then f_0 is given in *Hz*.

When a series *R-L-C* circuit is in resonance, it possesses minimum impedance $Z = R$. Hence, circuit current is maximum, it being limited by value of *R* alone. The current $I_0 = V/R$ and is in phase with V.

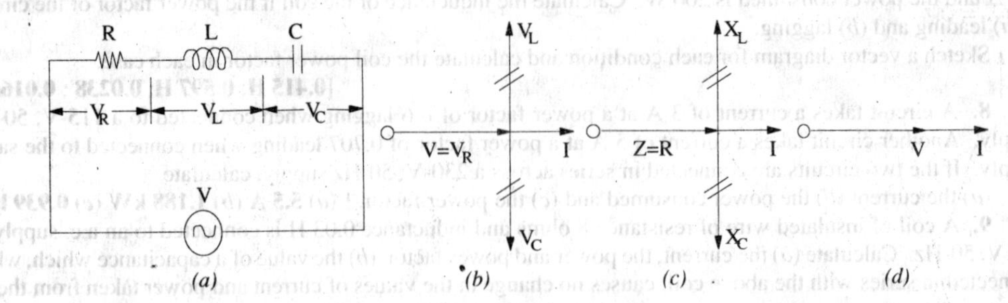

Fig. 13.40

Since circuit current is maximum, it produces large voltage drops across *L* and *C*. But these drops being equal and opposite, cancel each other out. Taken together, *L* and *C* from part of a circuit across which no voltage develops, however, large the current flowing. If it were not for the presence of *R*, such a resonant circuit would act like a short-circuit to currents of the frequency to which it resonates. Hence, a series resonant circuit is sometimes called *acceptor* circuit and the series resonance is often referred to as voltage resonance.

In fact, at resonance the series *RLC* circuit is reduced to a purely resistive circuit, as shown in Fig. 13.40.

Incidentally, it may be noted that if X_L and X_C are shown at any frequency *f*, that the value of the resonant frequency of such a circuit can be found from the relation $f_o = f\sqrt{X_C/X_L}$.

Summary

When an *R-L-C* circuit is in resonance

1. net reactance of the circuit is zero *i.e.* $(X_L \sim X_C) = 0$. or $X = 0$.
2. circuit impedance is minimum *i.e.* $Z=R$. Consequently, circuit admittance is maximum.
3. circuit current is maximum and is given by $I_0 = V/Z_0 = V/R$.
4. power dissipated is maximum *i.e.* $P_0 = I_0^2 R = V^2/R$.
5. circuit power factor angle $\theta = 0$. Hence, power factor $\cos\theta = 1$.
6. although $V_L = V_C$ yet V_{coil} is greater than V_C because of its resistance.
7. at resonance, $\omega^2 LC = 1$
8. $Q = \tan\theta = \tan 0° = 0$*.

13.11. Graphical Representation of Resonance

Suppose an alternating voltage of constant magnitude, but of varying frequency is applied to an *R-L-C* circuit. The variations of resistance, inductive reactance X_L and capacitive reactance X_C with frequency are shown in Fig. 13.41(*a*).

(*i*) *Resistance*: It is independnet of *f*, hence, it is represented by a straight line.

(*ii*) *Inductive Reactance*: It is given by $X_L = \omega L = 2\pi fL$. As seen, X_L is directly proportional to *f* *i.e.* X_L increases linearly with *f*. Hence, its graph is a straight line passing through the origin.

(*iii*) *Capacitive Reactance*: It is given by $X_C = 1/\omega C = 1/2\pi fC$. Obviously, it is inversely proportional to *f*. Its graph is a rectangular hyperbola which is drawn in the fourth quadrant because X_C is regarded negative. It is asymptotic to the horizontal axis at high frequencies and to the vertical axis at low frequencies.

(*iv*) *Net Reactance*: It is given by $X = X_L \sim X_C$. Its graph is a hyperbola (not rectangular) and crosses the X-axis at point *A* which represents resonant frequency f_0.

(*v*) *Circuit Impedance* It is given by $Z = \sqrt{[R^2 + (X_L \sim X_C)^2]} = \sqrt{R^2 + X^2}$

*However, value of Q_0 is as given in Art 13.5, 13.9 and 13.17.

At low frequencies Z is large because X_C is large. Since $X_C > X_L$, the net circuit reactance X is capacitive and the p.f. is leading [Fig. 13.41 (*b*)]. At high frequencies, Z is again large (because X_L is large) but is inductive because $X_L > X_C$. Circuit impedance has minimum values at f_0 given by $Z = R$ because $X = 0$.

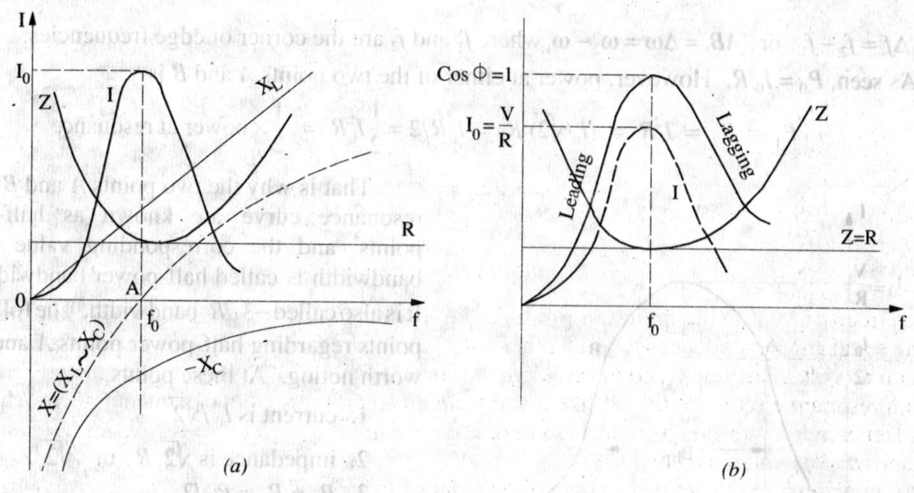

Fig. 13.41

(*vi*) *Current* I_0

It is the reciprocal of the circuit impedance. When Z is low, I_0 is high and vice versa. As seen, I_0 has low value on both sides of f_0 (because Z is large there) but has maximum value of $I_0 = V/R$ at resonance. Hence, maximum power is dissipated by the series circuit under resonant conditions. At frequencies below and above resonance, current decreases as shown in Fig. 13.41 (b). Now, $I_0 = V/R$ and $I = V/Z = V/\sqrt{(R^2+X^2)}$. Hence, $I/I_0 = R/Z = R/\sqrt{(R^2+X^2)}$ where X is the net circuit reactance at any frequency f.

(*vii*) *Power Factor*

As pointed out earlier, X is capacitive below f_0. Hence, current leads the applied voltage. However, at frequencies above f_0, X is inductive. Hence, the current lags the applied voltage as shown in Fig. 13.41 The power factor has maximum value of unity at f_0.

13.12. Resonance Curve

The curve, between circuit current and the frequency of the applied voltage, is known as resonance curve. The shapes of such a curve, for different values of R are shown in Fig. 13.42. For smaller values of R, the resonance curve is sharply peaked and such a circuit is said to be sharply resonant or highly selective. However, for larger values of R, resonance curve is flat and is said to have poor selectivity. The ability of a resonant circuit to discriminate between one particular frequency and all others is called its selectivity. The selectivities of different resonant circuits are compared in terms of their half-power bandwidths (Art. 13.13).

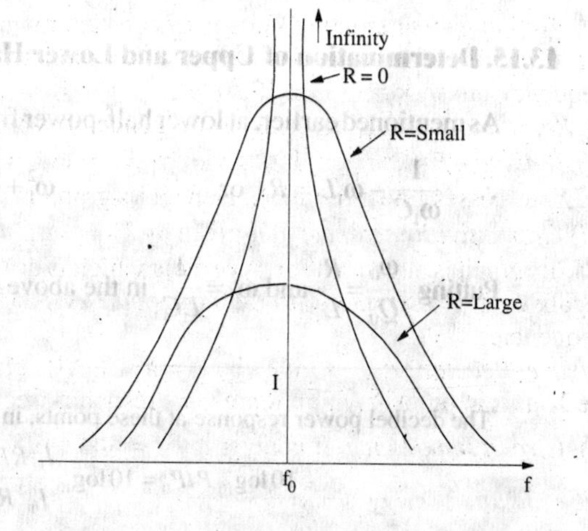

Fig. 13.42

13.13. Half-power Bandwidth of a Resonant-Circuit

As discussed earlier, in an R-L-C circuit, the maximum current at resonance is solely determined by circuit resistance R ($\because X = 0$) but at off-resonance frequencies, the current amplitude depends on Z (where $X \neq 0$). The half-wave bandwidth of a circuit is given by the band of frequencies which lies between two points on either side of f_0 where current falls to $I_0/\sqrt{2}$. Narrower the bandwidth, higher the selectivity of the circuit and vice versa. As shown in Fig. 13.43 the half-power bandwidth AB is given by

$AB = \Delta f = f_2 - f_1$ or $AB = \Delta\omega = \omega_2 - \omega_1$ where f_1 and f_2 are the corner or edge frequencies.

As seen, $P_0 = I_0^2 R$. However, power at either of the two points A and B is

$$P_1 = P_2 = I^2 R = (I_0/\sqrt{2})^2 R = I_0^2 R/2 = \frac{1}{2} I_0^2 R = \frac{1}{2} \times \text{power at resonance}$$

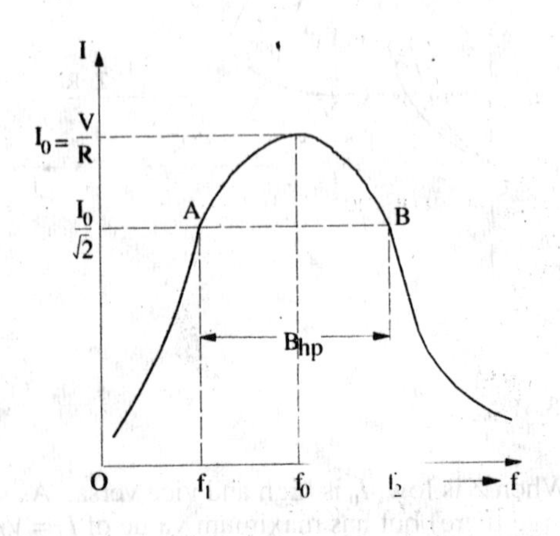

Fig. 13.43

That is why the two points A and B on the resonance curve are known as half-power points* and the corresponding value of the bandwidth is called half-power bandwidth B_{hp}. It is also called $-3\,dB$* bandwidth. The following points regarding half-power points A and B are worth noting. At these points,

1. current is $I_0/\sqrt{2}$
2. impedance is $\sqrt{2}.R$ or $\sqrt{2}.Z_0$
3. $P_1 = P_2 = P_0/2$
4. the circuit phase angle is $\theta = \pm 45°$
5. $Q = \tan\theta = \tan 45° = 1$
6. $B_{hp} = f_2 - f_1 = f_0/Q_0 = \sqrt{f_1 f_2}/Q_0 = R/2\pi L$.

It is interesting to note that B_{hp} is independent of the circuit capacitance.

13.14. Bandwidth B at any Off-resonance Frequency

It is found that the bandwidth of a given *R-L-C* circuit at ANY off-resonance frequencies f_1 and f_2 is given by $B = f_0\, Q/Q_0 = \sqrt{f_1 f_2}$; $. Q/Q_0 = f_2 - f_1$
where f_1 and f_2 are any frequencies (not necessarily half-power frequencies) below and above f_0.
Q = tangent of the circuit phase angle at the off-resonance frequencies f_1 and f_2.

$$Q_0 = \text{quality factor at resonance} = \frac{\omega_0 L}{R} = \frac{1}{R}\sqrt{\frac{L}{C}}$$

13.15. Determination of Upper and Lower Half-power Frequencies

As mentioned earlier, at lower half-power frequencies, $\omega_1 < \omega_0$ so that $\omega_1 L < 1/\omega_1 C$ and $\phi = -45°$

$$\therefore \quad \frac{1}{\omega_1 C} - \omega_1 L = R \quad \text{or} \quad \omega_1^2 + \frac{R}{L}\omega_1 - \frac{1}{LC} = 0$$

Putting $\frac{\omega_0}{Q_0} = \frac{R}{L}$ and $\omega_0^2 = \frac{1}{LC}$ in the above equation, we get $\omega_1^2 + \frac{\omega_0}{Q_0}\omega_1 - \omega_0^2 = 0$

*The decibel power response at these points, in terms of the maximum power at resonance, is

$$10\log_{10} P/P_0 = 10\log_{10}\frac{I_m^2 R/2}{I_m^2 R} = 10\log_{10}\frac{1}{2} = -10\log_{10} 2 = -3dB$$

Hence, the half-power points are also referred to as -3 dB points.

The positive solution of the above equation is, $\omega_1 = \omega_0 \left[\sqrt{\left(1+\frac{1}{4Q_0^2}\right)} - \frac{1}{2Q_0}\right]$

Now at the upper half-power frequency, $\omega_2 > \omega_0$ so that $\omega_2 > 1/\omega_2 C$ and $\phi = +45°$

$$\therefore \quad \omega_2 L - \frac{1}{\omega_2 C} = R \text{ or } \omega_2^2 - \frac{\omega_0}{Q_0}\omega_2 - \omega_0^2 = 0$$

The positive solution of the above equation is $\omega_2 = \omega_0\left[\sqrt{\left(1+\frac{1}{4Q_0^2}\right)} + \frac{1}{2Q_0}\right]$

In case $Q_0 > 10$; then the term $1/4\, Q_0^2$ is negligible as compared to 1.

Hence, in that case $\quad \omega_1 \cong \omega_0\left(1 - \frac{1}{2Q_0}\right)$ and $\omega_2 \cong \omega_0\left(1+\frac{1}{2Q_0}\right)$

Incidentally, it may be noted from above that $\omega_2 - \omega_1 = \omega_0 / Q_0$

13.16. Values of Edge Frequencies

Let us find the values of ω_1 and ω_2, $I_0 = V/R$ ... at resonance

$$I = \frac{V}{[R^2 + (\omega L - 1/\omega C)^2]^{1/2}} \text{ ...at any frequency}$$

At points A and B, $I = \frac{I_0}{\sqrt{2}} = \frac{I}{\sqrt{2}} \cdot \frac{V}{R}$

$$\therefore \frac{1}{\sqrt{2}} \cdot \frac{V}{R} = \frac{V}{[(R^2 + (\omega L - 1/\omega C)^2]^{1/2}} \text{ or } R = \pm(\omega L - 1/\omega C) = \pm X$$

It shows that at half-power points, net reactance is equal to the resistance.

(Since resistance equals reactance, p.f. of the circuit at these points is = $1/\sqrt{2}$ *i.e.* 0.707, though leading at point A and lagging at point B).

Hence $\quad R^2 = (\omega L - 1/\omega C)^2 \therefore \omega = \pm\frac{R}{2L} \pm \sqrt{\frac{R^2}{4L^2} + \frac{1}{LC}} = \pm\alpha \pm \sqrt{\alpha^2 + \omega_0^2}$

where $\quad \alpha = \frac{R}{2L}$ and $\omega_0 = \frac{1}{\sqrt{LC}}$

Since $R^2/4L^2$ is much less than $1/\sqrt{LC}$ $\quad \therefore \quad \omega = \pm\frac{R}{2L} \pm \frac{1}{\sqrt{LC}} = \pm\frac{R}{2L} \pm \omega_0$

Since only positive values of ω_0 are considered, $\omega = \omega_0 \pm R/2L = \omega_0 \pm \alpha$

$$\therefore \quad \omega_1 = \omega_0 - \frac{R}{2L} \text{ and } \omega_2 = \omega_0 + \frac{R}{2L}$$

$$\therefore \quad \Delta\omega = \omega_2 - \omega_1 = \frac{R}{L} \text{ rad/s and } \Delta f = f_2 - f_1 = \frac{R}{2\pi L} \text{ Hz} = \frac{f_0}{Q_0} \text{ Hz}$$

Also $\quad f_1 = f_0 - \frac{R}{4\pi L} Hz$ and $f_2 = f_0 + \frac{R}{4\pi L}$ Hz

It is obvious that f_0 is the *centre* frequency between f_1 and f_2.

Also, $\omega_1 = \omega_0 - \frac{1}{2}\Delta\omega$ and $\omega_2 = \omega_0 + \frac{1}{2}\Delta\omega$

As stated above, bandwidth is a measure of circuit's selectivity. Narrower the bandwidth, higher the selectivity and vice versa.

13.17. Q-Factor of a Resonant Series Circuit

The Q-factor of an R-L-C series circuit can be defined in the following different ways.

(*i*) it is given by the voltage magnification produced in the circuit at resonance.

We have seen that at resonance, current has maximum/value $I_0 = V/R$. Voltage across either coil or capacitor = $I_0 X_{L0}$ or $I_0 X_{C0}$ supply voltage $V = I_0 R$

$$\therefore \text{ Voltage magnification } = \frac{V_{L0}}{V} = \frac{I_0 X_{L0}}{I_0 R} = \frac{\text{reactive power}}{\text{active power}} = \frac{X_{L0}}{R} = \frac{\omega_0 L}{R} = \frac{\text{reactance}}{\text{resistance}}$$

or
$$= \frac{V_{C0}}{V} = \frac{I_0 X_{C0}}{I_0 R} = \frac{\text{reactive power}}{\text{active power}} = \frac{X_{C0}}{R} = \frac{\text{reactance}}{\text{resistance}} = \frac{1}{\omega_0 CR}$$

$$\therefore Q-\text{factor}, Q_0 = \frac{\omega_0 L}{R} = \frac{2\pi f_0 L}{R} = \tan\phi \qquad \ldots(i)$$

where ϕ is power factor of the coil.

(*ii*) The Q-factor may also be defined as under.

$$\text{Q-factor} = 2\pi \frac{\text{maximum stored energy}}{\text{energy dissipated per cycle}} \qquad \ldots\text{in the circuit}$$

$$= 2\pi \frac{\frac{1}{2} L I_0^2}{I^2 R T_0} = 2\pi \frac{\frac{1}{2} L (\sqrt{2} I)^2}{I^2 R (1/f_0)} = \frac{I^2 2\pi f_0 L}{I^2 R} = \frac{\omega_0 L}{R} = \frac{1}{\omega_0 CR} \qquad \ldots(T_0 = 1/f_0)$$

$$\text{In other words, } Q_0 = \frac{\text{energy stored}^*}{\text{energy lost}} \qquad \ldots\text{in the circuit}$$

(*iii*) We have seen above that resonant frequency, $f_0 = \dfrac{1}{2\pi\sqrt{(LC)}}$ or $2\pi f_0 = \dfrac{1}{\sqrt{(LC)}}$

Substituting this value in Eq. (*i*) above, we get the Q – factor, $Q_0 = \dfrac{1}{R}\sqrt{\left(\dfrac{L}{C}\right)}$

(*iv*) In the case of series resonance, higher Q-factor means not only higher voltage magnification but also higher selectivity of the tuning coil. In fact, Q-factor of a resonant series circuit may be written as

$$Q_0 = \frac{\omega_0}{\text{bandwidth}} = \frac{\omega_0}{\Delta\omega} = \frac{\omega_0}{R/L} = \frac{\omega_0 L}{R} = \frac{L}{R\sqrt{LC}} = \frac{1}{R}\sqrt{\frac{L}{C}} \qquad \ldots\text{as before}$$

Obviously, Q-factor can be increased by having a coil of large inductance but of small ohmic resistance.

(*v*) In summary, we can say that

$$Q_0 = \frac{\omega_0 L}{R} = \frac{1}{\omega_0 CR} = \frac{1}{R}\sqrt{\frac{L}{C}} = \sqrt{\frac{X_{L0} X_{C0}}{R}} = \frac{f_0}{B_{hp}} = \frac{\omega_0}{\omega_2 - \omega_1} = \frac{f_0}{f_2 - f_1}$$

13.18. Circuit Current at Frequencies Other Than Resonant Frequencies

At resonance, $I_0 = V/R$

At any other frequency above the resonant frequency, the current is given by I

$$= \frac{V}{Z} = \frac{V}{\sqrt{R^2 + (\omega L - 1/\omega C)^2}}$$

This current lags behind the applied voltage by a certain angle ϕ

---

*The author often jokingly tells students in his class that these days the quality of a person is also measured in terms of a quality factor given by

$$Q = \frac{\text{money earned}}{\text{money spent}}$$

Obviously, a person should try to have a high a quality factor as possible by minimising the denominator and/or maximizing the numerator.

$$\therefore \quad \frac{I}{I_0} = \frac{V}{\sqrt{R^2+(\omega L - 1/\omega C)^2}} \times \frac{R}{V} = \frac{1}{\sqrt{1+\frac{1}{R^2}(\omega L - 1/\omega C)^2}} = \frac{1}{\left[1+\left(\frac{\omega_0 L}{R}\right)^2\left(\frac{\omega}{\omega_0}-\frac{\omega_0}{\omega}\right)^2\right]^{1/2}}$$

Now, $\omega_0 L/R = Q_0$ and $\omega/\omega_0 = f/f_0$ hence, $\dfrac{I}{I_0} = \dfrac{1}{\left[1+Q_0^2\left(\frac{f}{f_0}-\frac{f_0}{f}\right)^2\right]^{1/2}}$

13.19. Relation Between Resonant Power P_0 and Off-resonant Power P

In a series RLC resonant circuit, current is maximum *i.e.* I_0 at the resonant frequency f_0. The maximum power P_0 is dissipated by the circuit at this frequency where X_L equals X_C. Hence, circuit impedance $Z_0 = R$.

$$\therefore \quad P_0 = I_0^2 R = (V/R)^2 \times R = V^2/R$$

At any other frequency either above or below f_0 the power is (Fig. 13.44).

$$P = I^2 R = \left(\frac{V}{Z}\right)^2 \times R = \frac{V^2 R}{Z^2} = \frac{V^2 R}{R^2+X^2} = \frac{V^2 R}{R^2+X^2R^2/R^2}$$

$$= \frac{V^2 R}{R^2+R^2Q^2} = \frac{V^2 R}{R^2(1+Q^2)} = \frac{V^2}{R(1+Q^2)} = \frac{P_0}{(1+Q^2)}$$

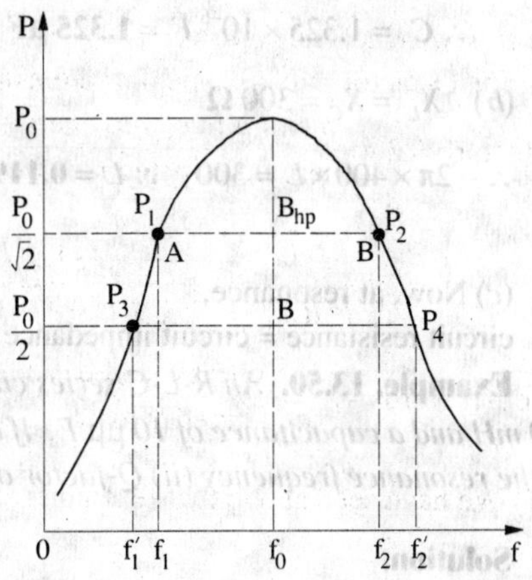

Fig. 13.44

The above equation shows that at any frequency other than f_0, the circuit power P is reduced by a factor of $(1 + Q^2)$ where Q is the tangent of the circuit phase angle (and not Q_0). At resonance, circuit phase angle $\theta = 0$, and $Q = \tan\theta = 0$. Hence, $P = P_0 = V^2/R$ (values of Q_0 are given in Art.)

Example 13.47. *For a series R.L.C circuit the inductor is variable. Source voltage is 200 $\sqrt{2}$ sin 100 πt. Maximum current obtainable by varying the inductance is 0.314 A and the voltage across the capacitor then is 300 V. Find the circuit element values.*

(Circuit and Field Theory, A.M.I.E. Sec B, 1993)

Solution. Under resonant conditions, $I_m = V/R$ and $V_L = V_C$.

$\therefore \quad R = V/I_m = 200/0.314 = 637\ \Omega$, $V_C = I_m \times X_{CD} = I_m/\omega_0 C$

$\therefore \quad C = I_m/\omega_0 V_C = 0.314/100\ \pi \times 300 = 3.33\ \mu F$.

$V_L = I_m \times X_L = I_m\, \omega_0 L$; $L = V_L/\omega_0 I_m = 300/100\ \pi \times 0.314 =$ **3.03 H**

Example 13.48. *A coil having an inductance of 50 mH and resistance 10 Ω is connected in series with a 25 μF capacitor across a 200 V ac supply. Calculate (a) resonance frequency of the circuit (b) current flowing at resonance and (c) value of Q_0 by using different data.*

[Elect. Engg. A.M.Ae, S.I, June 1991]

Solution. (*a*) $f_0 = \dfrac{1}{2\pi\sqrt{LC}} = \dfrac{1}{2\pi\sqrt{50\times10^{-3}\times25\times10^{-6}}} = 142.3$ Hz

(*b*) $I_0 = V/R = 200/10 = 20$ A

(*c*) $Q_0 = \dfrac{\omega_0 L}{R} = \dfrac{2\pi\times142.3\times50\times10^{-3}}{10} = 4.47$

$$Q_0 = \frac{1}{\omega_0 CR} = \frac{1}{2\pi\times142.3\times25\times10^{-6}\times10} = 4.47$$

$$Q_0 = \frac{1}{R}\sqrt{\frac{L}{C}} = \frac{1}{10}\sqrt{\frac{50\times10^{-3}}{25\times10^{-6}}} = 4.47$$

Example 13.49. *A 20-Ω resistor is connected in series with an inductor, a capacitor and an ammeter across a 25-V variable frequency supply. When the frequency is 400-Hz, the current is at its maximum value of 0.5 A and the potential difference across the capacitor is 150 V. Calculate*

(a) the capacitance of the capacitor

(b) the resistance and inductance of the inductor.

Solution. Since current is maximum, the circuit is in resonance.

$X_C = V_C/I = 150/0.5 = 300\ \Omega$

(a) $X_C = 1/2\pi fC$ or $300 = 1/2\pi \times 400 \times C$

$\therefore C = 1.325 \times 10^{-6}\ F = \mathbf{1.325\ \mu F}$

(b) $X_L = X_C = 300\ \Omega$

$\therefore \quad 2\pi \times 400 \times L = 300 \quad \therefore L = \mathbf{0.119\ H}$

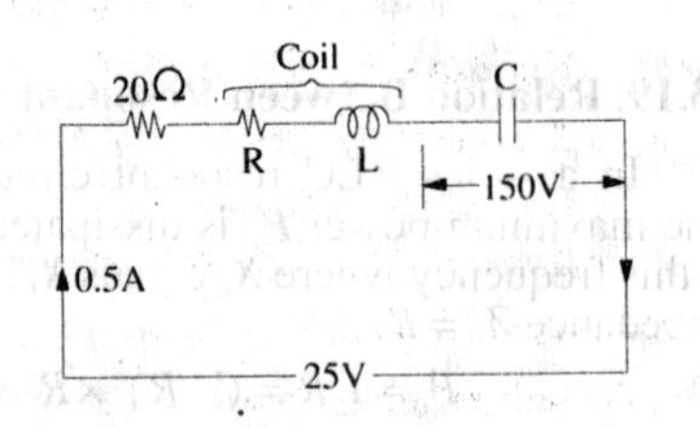

Fig. 13.45

(c) Now, at resonance,

circuit resistance = circuit impedance or $20 + R = V/I = 25/0.5 \quad \therefore \quad R = 30\ \Omega$ ...Fig. 13.45

Example. 13.50. *An R-L-C series circuit consists of a resistance of 1000 Ω, an inductance of 100 mH and a capacitance of 10 μμ F. If a voltage of 100 V is applied across the combination, find (i) the resonance frequency (ii) Q-factor of the circuit and (iii) the half-power points.*

(Elect. Circuit Analysis, Bombay Univ. 1985)

Solution.

$$(i)\ f_0 = \frac{1}{2\pi\sqrt{10^{-1} \times 10^{-11}}} = \frac{10^6}{2\pi} = \mathbf{159\ kHz}$$

$$(ii)\ Q = \frac{1}{R}\sqrt{\frac{L}{C}} = \frac{1}{1000} \times \sqrt{\frac{10^{-1}}{10^{-11}}} = \mathbf{100}$$

$$(iii)\ f_1 = f_0 - \frac{R}{4\pi L} = 159 \times 10^3 - \frac{1000}{4\pi \times 10^{-1}} = \mathbf{158.2\ kHz,}$$

$$f_2 = f_0 + \frac{R}{4\pi L} = 159 \times 10^3 + \frac{1000}{4\pi \times 10^{-1}} = \mathbf{159.8\ kHz}$$

Example 13.51. *A series R-L-C circuit consists of R = 1000 Ω, L = 100 mH and C = 10 picofarads. The applied voltage across the circuit is 100 V.*

(i) Find the resonant frequency of the circuit.

(ii) Find the quality factor of the circuit at the resonant frequency.

(iii) At what angular frequencies do the half power points occur ?

(iv) Calculate the bandwidth of the circuit. **(Networks-I, Delhi Univ. Jan. 1986)**

Solution. (i) $$f_0 = \frac{1}{2\pi\sqrt{LC}} = \frac{1}{2\pi\sqrt{100 \times 10^{-3} \times 10 \times 10^{-12}}} = 159.15\ \text{kHz}$$

$$(ii)\quad Q_0 = \frac{1}{R}\sqrt{\frac{L}{C}} = \frac{1}{1000} \times \sqrt{\frac{1000 \times 10^{-3}}{10 \times 10^{-12}}} = 100$$

$$(iii)\quad B_{hp} = \frac{R}{2\pi L} = \frac{1000}{2\pi \times 100 \times 10^{-3}} = 1591.5\ \text{Hz}$$

Also. $B_{hp} = f_0/Q_0 = 159.15k\text{Hz}/100 = 1.5915\ \text{kHz} = \mathbf{1591.5\ Hz}$ ...as above

$$(iv)\quad \omega_1 = \omega_0\left(1-\frac{1}{2Q_0}\right) = 2\pi\times 159.15\left(1-\frac{1}{2\times 100}\right) = 994.969 \text{ radia/sec.}$$

$$\omega_2 = \omega_0\left(1+\frac{1}{2Q}\right) = 2\pi\times 159.15\left(1+\frac{1}{2\times 100}\right) = 1004.969 \text{ rad/sec}$$

(*iv*) Band width = $(\omega_2 - \omega_1)$ = 1004.969 – 994.969 = 10.00 rad/sec.

Example 13.52. *An R-L-C series resonant circuit has the following parameters :*
Resonance frequency = 5000/2π Hz; impedance at resonance = 56 Ω and Q-factor = 25.
Calculate the capacitance of the capacitor and the inductance of the inductor.
Assuming that these values are independent of the frequency, find the two frequencies at which the circuit impedance has a phase angle of π/4 radian.

Solution. Here $\omega_0 = 2\pi f_0 = 2\pi\times 5000/2\pi = 5000$ rad/s

Now, $$Q = \frac{\omega_0 L}{R} \quad \text{or} \quad 25 = \frac{5000L}{56} \quad \text{or} \quad L = 0.28 \text{ H}$$

Also at resonance $\omega_0 L = 1/\omega_0 C$ or $5000\times 0.28 = 1/5000\times C$ ∴ $C = 0.143\ \mu F$

The circuit impedance has a phase shift of 45° and the two half-power frequencies which can be found as follows :

$$BW = \frac{f_0}{Q} = \frac{5000/2\pi}{25} = 31.83 \text{ Hz}$$

Therefore lower half-power frequency = $(f_0 - 31.83/2)$ = 5000/2π – 15.9 = **779.8 Hz.**
Upper half-power frequency = $(f_0 + 31.83/2)$ = 5000/2π + 15.9 = **811.7 Hz.**

Example 13.53. *An R-L-C series circuit is connected to a 20-V variable frequency supply. If R = 20 Ω , L = 20 mH and C = 0.5 μF, calculate the following :*
(a) resonant frequency f_0 (b) resonant circuit Q_0 using L/C ratio (c) half-power bandwidth using f_0 and Q_0 (d) half-power bandwidth using the general formula for any bandwidth (e) half-power bandwidth using the given component values (f) maximum power dissipated at f_0.

Solution. (*a*) $f_0 = 1/2\pi\sqrt{LC} = 1/2\pi\sqrt{(20\times 10^{-3}\times 0.5\times 10^{-6}} =$ **1591 HZ**

$$(b)\ Q_0 = \frac{1}{R}\cdot\sqrt{\frac{L}{C}} = \frac{1}{20}\cdot\sqrt{\frac{20\times 10^{-3}}{0.5\times 10^{-6}}} = \mathbf{10}$$

(*c*) $B_{hp} = f_0/Q_0$ = 1591/10 = **159.1 Hz**

(*d*) $B_{hp} = f_0\, Q/Q_0$ = 1591 × tan 45°/10 = **159.1 Hz.**

It is so because the power factor angle at half-power frequencies is ± 45°.

(*e*) B_{hp} = R/2π L = 20/2π × 20 × 10^{-3} = **159.1 Hz**

(*f*) $f_0 = V^2/R = 20^2/20$ = **20 W**

Example 13.54. *An inductor having a resistance of 25Ω and a Q_0 of 10 at a resonant frequency of 10 kHz is fed from a 100∠0° supply. Calculate*
(a) Value of series capacitance required to produce resonance with the coil
(b) the inductance of the coil (c) Q_0 using the L/C ratio (d) voltage across the capacitor (e) voltage across the coil.

Solution. (*a*) $X_{L0} = Q_0 R$ = 10 × 25 = 250 Ω.

Now, $X_{C0} = X_{L0}$ = 250 Ω.

Hence, $C = 1/2\pi f_0\times X_{C0} = 1/2\pi\times 10^4\times 250 = 63.67\times 10^{-9} F$ = **63.67 nF**

(b) $L = X_{L0}/2\pi f_0 = 250/2\pi\times 10^4 = \mathbf{3.98\ mH}$

(c) $$Q_0 = \frac{1}{R}\cdot\sqrt{\frac{L}{C}}$$

Now, $$\frac{L}{C} = \frac{3.98\times 10^{-3}}{63.67\times 10^{-9}} = 6.25\times 10^4$$

$$\therefore \quad Q_0 = \frac{1}{25}\times\sqrt{6.25\times 10^4} = \mathbf{10}\ \text{(verification)}$$

(d) $V_{CO} = -jQ_0V = -j100\angle 0°\times 10 = -j1000V = \mathbf{-100\angle -90°V}$

(e) Since $V_{LO} = V_{CO}$ in magnitude, hence, $V_{LO} = +j1000$ V

$= 1000\angle 90°V$; Also, $V_R = V = 100\angle 0°$

Hence, $V_{coil} = V_R + V_{LO}$

$= 100 + 1000\angle 90° = 100 + j1000 = \mathbf{1005\angle 84.3°}$

Example 13.55. *A series L.C circuit has L = 100 pH, C = 2500 μF and Q = 70. Find (a) resonant frequency f_0 (b) half-power points and (c) bandwidth.*

Solution. (a) $f_0 = \dfrac{1}{2\pi\sqrt{LC}} = \dfrac{10^9}{2\pi\sqrt{100\times 2500}} = \mathbf{318.3\ kHz}$

(b) $f_2 - f_1 = \Delta f = f_0/Q = 318.3/70 = \mathbf{4.55\ kHz}$

(c) $f_1f_2 = f_0^2 = 318.3^2;\ f_2 - f_1 = 4.55$ kHz

Solving for f_1 and f_2, we get, f_1 = **316.04 kHz** and f_2 = 320.59 kHz.

Note. Since Q is very high, there would be negligible error in assuming that the half-power points are equidistant from the resonant frequency.

Example 13.56. *A series R-L-C circuit is excited from a constant-voltage variable frequency source. The current in the circuit becomes maximum at a frequency of 600/2π Hz and falls to half the maximum value at 400/2π Hz. If the resistance in the circuit is 3 Ω, find L and C.*

(Grad. I.E.T.E. Summer 1991)

Solution. Current at resonance is $I_0 = V/R$

Actual current at any other frequency is $I = \dfrac{V}{\sqrt{R^2 + \left(\omega L - \frac{1}{\omega C}\right)^2}}$

$$\therefore \quad \frac{I}{I_0} = \frac{V}{\left[R^2 + \left(\omega L - \frac{1}{\omega C}\right)^2\right]^{1/2}}\cdot\frac{R}{V} = \frac{1}{\left[1 + \frac{1}{R^2}\left(\omega L - \frac{1}{\omega C}\right)^2\right]^{1/2}} = \frac{1}{\left[1 + \left(\frac{\omega_0 L}{R}\right)^2\left(\frac{\omega}{\omega_0} - \frac{\omega_0}{\omega}\right)^2\right]^{1/2}}$$

Now $Q = \dfrac{\omega_0 L}{R}$ and $\dfrac{\omega}{\omega_0} = \dfrac{f}{f_0}$, hence $\dfrac{I}{I_0} = \dfrac{1}{\left[1 + Q^2\left(\frac{f}{f_0} - \frac{f_0}{f}\right)^2\right]^{1/2}}$

In the present case, $f_0 = 600/2\pi$ Hz, $f = 400/2\pi$ Hz and $I/I_0 = 1/2$

$$\therefore \quad \frac{1}{2} = \frac{1}{\left[1 + Q^2\left(\frac{400}{600} - \frac{600}{400}\right)^2\right]^{1/2}} = \frac{1}{\left[1 + Q^2\left(\frac{2}{2} - \frac{3}{2}\right)^2\right]^{1/2}}$$

or $$\frac{1}{4} = \frac{1}{1 + 25Q^2/36} \quad \therefore \quad Q = 2.08$$

Now, $Q = \dfrac{1}{\omega_0 RC}$ or $2.08 = \dfrac{1}{600 \times 3 \times C}$ $\therefore C = 267 \times 10^{-6} F = \mathbf{267\ \mu F}$

Also $Q = \omega_0 L/R$ $\therefore$ $2.08 = \dfrac{600L}{R} = \dfrac{600L}{3}$ $\therefore$ $L = \mathbf{10.4\ mH}$

Example 13.57. *Discuss briefly the phenomenon of electrical resonance in simple R-L-C circuits.*

A coil of inductance L and resistance R in series with a capacitor is supplied at constant voltage from a variable-frequency source. Call the resonance frequency ω_0 and find, in terms of L, R and ω_0, the values of that frequency at which the circuit current would be half as much as at resonance.

(Basic Electricity, Bomaby Univ. 1985)

Solution. For discussion of resonance, please refer to Art. 13.10.

The current at resonance is maximum and is given by $I_0 = V/R$. Current at any other frequency

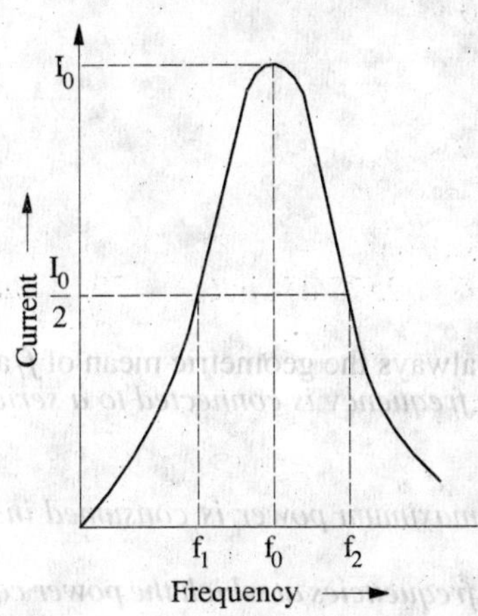

Fig. 13.46

is $I = \dfrac{V}{[R^2 + (\omega L - 1/\omega C)^2]^{1/2}}$

$$\therefore \frac{I_0}{I} = \frac{[R^2 + (\omega L - 1/\omega C)^2]^{1/2}}{R}$$

$$\text{or } N = \left[1 + \frac{1}{R^2}\left(\omega L - \frac{1}{\omega C}\right)^2\right]^{1/2}$$

$$\text{Now } Q = \frac{\omega_0 L}{R} = \frac{1}{\omega_0 CR}$$

$$\therefore R = \frac{\omega_0 L}{Q} = \frac{1}{\omega_0 CQ}$$

Substituting this value in the above equation, we get

$$N = \left[1 + Q^2\left(\frac{f}{f_0} - \frac{f_0}{f}\right)^2\right]^{1/2}$$

or $$\left(\frac{f}{f_0} - \frac{f_0}{f}\right)^2 = \frac{N^2 - 1}{Q^2} \quad \therefore \quad \frac{\sqrt{(N^2-1)}}{Q} = \pm\left(\frac{f}{f_0} - \frac{f_0}{f}\right)$$

or $$\frac{\sqrt{(N^2-1)}}{Q} = \frac{f_2}{f_0} - \frac{f_0}{f_2} = \frac{f_0}{f_1} - \frac{f_1}{f_0}$$

where $f_2 > f_0$ and $f_1 < f_0$ are the two frequencies at which the current has fallen to $1/N$ of the resonant value.

In the present case, $N = 2$ (Fig. 13.46) $\therefore$ $\dfrac{f_2}{f_0} - \dfrac{f_0}{f_2} = \dfrac{\sqrt{3}}{Q}$ and $\dfrac{f_0}{f_1} - \dfrac{f_1}{f_0} = \dfrac{\sqrt{3}}{Q}$

From these equations, f_1 and f_2 may be calculated.

Example 13.58. *A coil of inductance 9H and resistance 50 Ω in series with a capacitor is supplied at constant voltage from a variable frequency source. If the maximum current of 1A occurs at 75 Hz, find the frequency when the current is 0.5 A.*

(Principles of Elect. Engg. Delhi Univ. 1987)

Solution. Here, $N = I_0/I = I/0.5 = 2$; $Q = \omega_0 L/R = 2\pi \times 75 \times 9/50 = 84.8$

Let f_1 and f_2 be the frequencies at which current falls to half its maximum value at resonance frequency. Then, as seen from above,

$$\frac{f_0}{f_1} - \frac{f_1}{f_0} = \frac{\sqrt{3}}{Q} \quad \text{or} \quad \frac{75}{f_1} - \frac{f_1}{75} = \frac{\sqrt{3}}{84.4}$$

or $(75^2 - f_1^2)/75f_1 = 0.02$ or $f_1^2 + 1.5f_1 - 5625 = 0$ or $f_1 = \mathbf{74.25\ Hz}$

Also $\dfrac{f_2}{75} - \dfrac{75}{f_2} = \dfrac{\sqrt{3}}{84.4}$ or $f_2^2 - 1.5f_2 - 5625 = 0$ or $f_2 = \mathbf{75.75\ Hz.}$

Example 13.59. *Using the data given in Ex. 13.45 find the following when the power drops to 4 W on either side of the maximum power at resonance.*

(a) circuit Q (b) circuit phase angle φ (c) 4-W bandwidth B

(d) lower frequency f_1 (e) upper frequency f_2.

(f) resonance frequency using the value of f_1 and f_2.

Solution. *(a)* $P = \dfrac{P_0}{(1 + Q_0^2)}$ $\therefore$ $Q = \sqrt{(P_0/P_1) - 1} = \sqrt{(20/4) - 1} = \mathbf{2}$

(b) $\tan(\pm\theta) = 2j \pm\theta/\tan^{-1}2 = \pm\,\mathbf{63.4°}$

(c) $B_{hp} = \dfrac{f_0 Q}{Q_0} = \dfrac{1591 \times 2}{10} = 318.2\text{ Hz}$

(d) $f_1 = f_0 - B/2 = 1591 - (318.2/2) = 1431.9\text{ Hz}$

(e) $f_2 = f_0 + B/2 = 1591 + (318.2/2) = 1750.1\ Hz$

(f) $f_0 = \sqrt{f_1 f_2} = \sqrt{1431.9 \times 1750.2} = \mathbf{1591\ Hz.}$

It shows that regardless of the bandwidth magnitude, f_0 is always the geometric mean of f_l and f_2.

Example 13.60. *A constant e.m.f. source of variable frequency is connected to a series R.L.C circuit of Fig. 13.47.*

(a) Show the nature of the frequency – V_R graph

(b) Calculate the following *(i) frequency at which maximum power is consumed in the 2 Ω resistor*

(ii) Q-factor of the circuit at the above frequency (iii) frequencies at which the power consumed in 2Ω resistor is one-tenth of its maximum value.

(Network Analysis A.M.I.E Sec. B.W. 1989)

Solution. *(a)* The graph of angular frequency ω versus voltage drop across *R i.e.* V_R is shown in Fig. 13.47. It is seen that as frequency of the applied voltage increases, V_R increases till it reaches its maximum value when the given *RLC* circuit becomes purely reactive *i.e.* when $X_L = X_C$ (Art. 13.10). *(b) (i)* Maximum power will be consumed in the 2 Ω resistor when maximum current flows in the circuit under resonant condition.

For resonance $\omega_0 L = 1/\omega_0 X_C$ or $\omega_0 = 1/\sqrt{LC} = 1/\sqrt{40 \times 10^{-6} \times 160 \times 10^{-12}} = 10^9/80$ rad/s

$\therefore$ $f_0 = \omega_0/2\pi = 10^9/2\pi \times 80 = \mathbf{1.989\ MHz}$

(ii) Q-factor, $Q_0 = \dfrac{\omega_0 L}{R} = \dfrac{10^9 \times 40 \times 10^{-6}}{80 \times 2} = \mathbf{250}$

(iii) Maximum current $I_0 = V/R$ (Art. 13.10). Current at any other frequency is $I = \dfrac{V}{\sqrt{R^2 + (\omega L - 1/\omega C)^2}}$

Power at any frequency $I^2R = \dfrac{V^2}{R^2 + (\omega L - 1/\omega C)^2} \cdot R$

Maximum power $I_0^2 R = \left(\dfrac{V}{R}\right)^2 \cdot R$

Hence, the frequencies at which power consumed would be one-tenth of the maximum power will be given by the relation. $\dfrac{1}{10} \cdot \left(\dfrac{V}{R}\right)^2 R = \dfrac{V^2}{R^2 + (\omega L - 1/\omega C)^2} \cdot R$

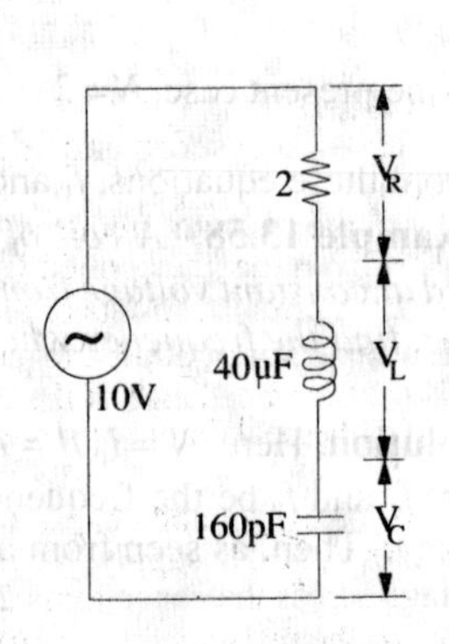

Fig. 13.47

or cross multiplying, we get $R^2 + (\omega L - 1/\omega C)^2 = 10R^2$ or $(\omega L - 1/\omega)C^2 = 9R^2$

$$\therefore \quad (\omega L - 1/\omega C) = \pm 3R \text{ and } \omega_1 L - 1/\omega_1 C = 3R$$

$$\text{and } \omega_2 L - 1/\omega_2 C = -3R$$

Adding the above two equations, we get

$$(\omega_1 + \omega_2)L - \frac{1}{C}\left(\frac{\omega_2 + \omega_1}{\omega_1 . \omega_2}\right) = 0 \text{ or } \omega_1\omega_2 = \frac{1}{LC} = \omega_0^2$$

Subtracting the same two equations, we have $L(\omega_1 - \omega_2) \frac{1}{C}\left(\frac{1}{\omega_2} - \frac{1}{\omega_1}\right) = 6R\,(\omega_1 - \omega_2) + \frac{1}{LC}$

$$\left(\frac{\omega_1 - \omega_2}{\omega_1 \cdot \omega_2}\right) = \frac{6R}{L}$$

Substituting the value of $1/LC = \omega^2 = \omega_1\omega_2$, we get

$$(\omega_1 - \omega_2) + \omega_1\omega_2\left(\frac{\omega_1 - \omega_2}{\omega_1\omega_2}\right) = \frac{6R}{L} \text{ or } (\omega_1 - \omega_2) = \Delta\omega = \frac{3R}{L}$$

Now, $\omega_2 = \omega_0 + \frac{\Delta\omega}{2} = \omega_0 + \frac{1.5R}{L}$ and $\omega_1 = \omega_0 - \frac{1.5R}{L}$

$$\therefore f_2 = f_0 + 1.5R/2\pi L = 1.989 \times 10^6 + 1.5 \times 2/2\pi \times 40 \times 10^{-6} = 1.989 \times 10^6 + 0.0119 \times 10^6 = \mathbf{2\ MHz}$$

$$f_1 = f_0 - 1.5R/2\pi L = 1.989 \times 10^6 - 0.0119 \times 10^6 = \mathbf{1.977\ MHz}$$

Example 13.61. *Show that in an R-L-C circuit, the resonant frequency ω_0 is the geometric mean of the lower and upper half-power frequencies ω_l and ω_2 respectively.*

Solution. As stated earlier, at lower half-power radiant frequency ω_1; $X_C > X_L$ and at frequencies higher than half-power frequencies $X_L > X_C$. However, the difference between the two equals R.

$$\therefore \quad \text{at } \omega_1, X_C - X_L = R \text{ or } 1/\omega_1 C - \omega_1 L = R \quad \text{...(i)}$$

$$\text{At } \omega_2, X_L - X_C = R \text{ or } \omega_2 L - 1/\omega_2 C = R$$

Multiplying both sides of Eq. (i) by C and substituting $\omega_0^2 = 1\sqrt{LC}$, we get

$$\frac{1}{\omega_1} - \frac{\omega_1}{\omega_0^2} = \frac{\omega_2}{\omega_0^2} - \frac{1}{\omega_2} \text{ or } \frac{1}{\omega_1} + \frac{1}{\omega_2} = \frac{\omega_1 + \omega_2}{\omega_0^2} \text{ or } \omega_0 = \sqrt{\omega_1\omega_2}$$

Example 13.62. *Prove that in a series R-L-C circuit, $Q_0 = \omega_0 L/R = f_0/bandwidth = f_0/BW$.*

Solution. As has been proved in Art. 14.13, at half power frequencies, net reactance equals resistance. Moreover, at ω_1, capacitive reactance exceeds inductive reactance whereas at ω_2, inductive reactance exceeds capacitive reactance.

$$\therefore \quad \frac{1}{2\pi f_1 C} - 2\pi f_1 L = R \text{ or } f_1 = \frac{-R + \sqrt{R^2 + 4L/C}}{4\pi L}$$

$$\text{Also} \quad 2\pi f_2 L - \frac{1}{2\pi f_2 C} = \text{ or } f_2 = \frac{R + \sqrt{R^2 + 4L/C}}{4\pi L}$$

Now, $BW = f_2 - f_1 = R/2\pi L$. Hence, $Q_0 = f_0/BW = 2\pi f_0 L/R = \omega_0 L/R$

Tutorial Problem No. 13.4

1. An a.c. series circuit has a resistance of 10 Ω, an inductance of 0.2 H and a capacitance of 60 μF. Calculate

(*a*) the resonant frequency (*b*) the current and (c) the power at resonance.

Given that the applied voltage is 200 V. **[46 Hz ; 20 A; 4 kW]**

2. A circuit consists of an inductor which has a resistance of 10 Ω and an inductance of 0.3 H, in series with a capacitor of 30μ*F* capacitance. Calculate

(*a*) the impedance of the circuit to currents of 40 Hz (*b*) the resonant frequency

(*c*) the peak value of stored energy in joules when the applied voltage is 200 V at the resonant frequency. **[58.31 Ω ; 53 Hz; 120 J]**

3. A resistor and a capacitor are connected in series with a variable inductor. When the circuit is connected to a 240-V, 50-Hz supply, the maximum current given by varying the inductance is 0.5 A. At this current, the voltage across the capacitor is 250 *V*. Calculate the values of

(*a*) the resistance (*b*) the capacitance (*c*) the inductance. **[480 Ω, 6.36 μF; 1.59 H]**

Neglect the resistance of the inductor

4. A circuit consisting of a coil of resistance 12 Ω and inductance 0.15 H in series with a capacitor of 12μF is connected to a variable frequency supply which has a constant voltage of 24 V.

Calculate (*a*) the resonant frequency (*b*) the current in the circuit at resonance (*c*) the voltage across the capacitor and the coil at resonance. **[(*a*) 153 Hz (*b*) 2 A (*c*) 224 V]**

5. A resistance, a capacitor and a variable inductance are connected in series across a 200-V, 50-Hz supply. The maximum current which can be obtained by varying the inductance is 314 mA and the voltage across the capacitor is then 300 V. Calculate the capacitance of the capacitor and the values of the inductance and resistance. **[3.33 μF, 3.04 H, 637 Ω]** (*I.E.E. London*)

6. A 250-V circuit, consisting of a resistor, an inductor and a capacitor in series, resonates at 50 Hz. The current is then 1A and the p.d. across the capacitor is 500 V. Calculate (*i*) the resistance (*ii*) the inductance and (*iii*) the capacitance. Draw the vector diagram for this condition and sketch a graph showing how the current would vary in a circuit of this kind if the frequency were varied over a wide range, the applied voltage remaining constant. **[(*i*) 250 Ω (*ii*) 0.798 H (*iii*) 12.72 μF]** (*City & Guilds, London*)

7. A resistance of 24 Ω, a capacitance of 150 μF and an inductance of 0.16 H are connected in series with each other. A supply at 240 V, 50 Hz is applied to the ends of the combination. Calculate (*a*) the current in the circuit (*b*) the potential differences across each element of the circuit (*c*) the frequency to which the supply would need to be changed so that the current would be at unity power-factor and find the current at this frequency.
[(*a*) 6.37 A (*b*) V_R = 152.8 V, V_C = 320 V, V_C = 123.3 V (*c*) 32 Hz ; 10 A] (*London Univ.*)

8. A series circuit consists of a resistance of 10 Ω, an inductance of 8 mH and a capacitance of 500 μμ*F*. A sinusoidal E.M.F. of constant amplitude 5 V is introduced into the circuit and its frequency varied over a range including the resonant frequency.

At what frequencies will the current be (*a*) a maximum (*b*) one-half the-maximum ?
[(*a*) 79.6 kHz (*b*) 79.872 kHz, 79.528 kHz] (*App. Elect. London Univ.*)

9. A circuit consists of a resistance of 12 ohms, a capacitance of 320 μF and an inductance of 0.08 H, all in series. A supply of 240 V, 50 Hz is applied to the ends of the circuit. Calculate :
(*a*) the current in the coil.
(*b*) the potential differences across each element of the circuit.
(*c*) the frequency at which the current would have unity power-factor.
[(*b*) 12.4 A (*b*) 149 V, 311 V (*c*) 32 Hz] (*London Univ.*)

10. A series circuit consists of a reactor of 0.1 henry inductance and 5 ohms resistance and a capacitor of 25.5 μF capacitance.

Find the resonance frequency and the percentage change in the current for a divergence of 1 percent from the resonance frequency.
[100 Hz, 1.96 % at 99 Hz; 4.2% at 101 Hz] (*City and Guilds, London*)

Objective Tests–13

1. In a series *R-L* circuit, V_L —— V_R by —— degrees.
(*a*) lags, 45 (*b*) lags, 90
(*c*) leads, 90 (*d*) leads, 45

2. The voltage applied across an *R–L* circuit is equal to —— of V_R and V_L.
(*a*) arithmetic sum (*b*) algebraic sum
(*c*) phasor sum (*d*) sum of the squares.

3. The power in an a.c. circuit is given by
(*a*) $VI \cos \phi$ (*b*) $VI \sin \phi$
(*c*) $I^2 Z$ (*d*) $I^2 X_L$

4. The p.f. of an *R–C* circuit is
(*a*) often zero
(*b*) between zero and 1
(*c*) always unity
(*d*) between zero and -1.0

5. Which phasor diagram of Fig. 13.48 is correct for a series *R–C* circuit ?

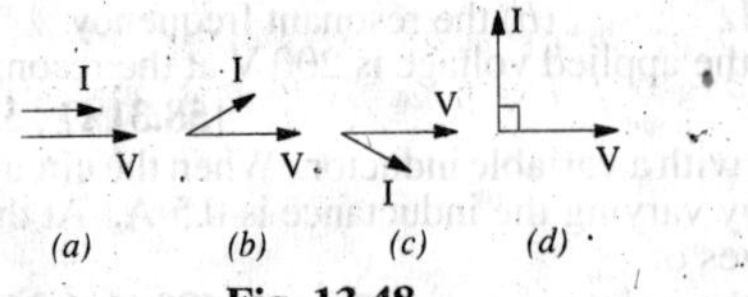

Fig. 13.48

6. In an *R-L-C* circuit, $v(t)$ = 20 sin (314*t* + 5π/6) and $i(t)$ = 10 sin (314 *t* + 2π/3). The p.f. of the circuit is
(*a*) 0.5 lead (*b*) 0.866 lag
(*c*) 0.866 lead (*d*) 0.5 lag
and power drawn is —— watt.
(*e*) 200 (*f*) 86.6
(*g*) 173.2 (*h*) 50

7. The input of an a.c. circuit having p.f. of 0.8 lagging is 20 kVA. The power drawn by the circuit is —— kW.
(*a*) 12 (*b*) 20
(*c*) 16 (*d*) 8

8. The power factor of an a.c. circuit is given by
(*a*) cosine of the phase angle
(*b*) tangent of the phase angle
(*c*) the ratio R/X_L
(*d*) the ratio X_L/Z

9. In a series *R-L-C* circuit, *R* = 100 Ω, X_L = 300 Ω and X_C = 200 Ω. The phase angle ϕ of the circuit is —— degrees.
(*a*) 0 (*b*) 90
(*c*) 45 (*d*) –45

10. The phase angle of a series *R-L-C* circuit is leading if

(*a*) $X_L = 0$ (*b*) $R = 0$
(*c*) $X_C > X_L$ (*d*) $X_C < X_L$

11. In an a.c. circuit, the ratio of kW/kVA represents
(*a*) power factor
(*b*) load factor
(*c*) form facor
(*d*) diversity factor

12. If p.f. of a circuit is unity, its reactive power is
(*a*) a maximum
(*b*) equal to I^2R
(*c*) zero
(*d*) a negative quantity

13. In a series *R-L-C* circuit, resonance occurs when
(*a*) $R = X_L \sim X_C$
(*b*) $X_L = X_C$
(*c*) $X_L = 10\ X_C$ or more
(*d*) net $X > R$

14. The p.f. of a series *R-L-C* circuit at its half-power points is
(*a*) unity (*b*) lagging
(*c*) leading (*d*) either (*b*) or (*c*)

15. A resonance curve for a series circuit is a plot of frequency versus ——.
(*a*) voltage (*b*) impedance
(*c*) current (*d*) reactance

16. At half-power points of a resonance curve, the current is —— times the maximum current.
(*a*) 2 (*b*) $\sqrt{2}$
(*c*) $1\sqrt{2}$ (*d*) 1/2

17. Higher the *Q* of a series circuit,
(*a*) greater its bandwidth
(*b*) sharper its resonance
(*c*) broader its resonance curve
(*d*) narrower its passband

18. As the *Q*-factor of a circuit——, its selectivity becomes ——.
(*a*) increases, better
(*b*) increases, worse
(*c*) decreases, better
(*d*) decreases, narrower

19. An *R-L-C* circuit has a resonance frequency of 160 kHz and a *Q*-factor of 100. Its bandwidth is
(*a*) 1.6 kHz
(*b*) 0.625 kHz
(*c*) 16 MHz
(*d*) none of the above

20. An *R-L-C* circuit has $R = 10\ \Omega$, $X_L = 20\ \Omega$ and $X_C = 30\ \Omega$. The impedance of the circuit is given by the expression
(*a*) $\mathbf{Z} = 10 + j\,20$ (*b*) $\mathbf{Z} = 10 + j\,50$
(*c*) $\mathbf{Z} = 10\ \ j\,20$ (*d*) $\mathbf{Z} = -10 + j\,20$

21. The power factor of an a.c. circuit is equal to
(*a*) cosine of the phase angle
(*b*) sine of the phase angle
(*c*) unity for a resistive circuit
(*d*) unity for a reactive circuit
(*Network Theory Nagpur Univ. 1993*)

22. The power factor of an ordinary electric bulb is ——.
(*a*) zero (*b*) unity
(*c*) slightly more than zero
(*d*) slightly less than unity
(*Prinicples of Elect. Engg. Delhi Univ. July 1984*)

23. In a parallel *R-L* circuit if I_R is the current in the resistor and I_L is the current in the inductor, then
(*a*) I_R lags I_L by 90°
(*b*) I_R leads I_L by 270°
(*c*) I_L leads I_R by 270°
(*d*) I_L lags I_R by 90°
(*Principles of Elect. Engg. Delhi Univ. 1984*)

24. In a parallel resonant circuit there is practically no difference between the condition for unity power factor and the condition for maximum impedance so long as *Q* is
(*a*) very small of the order of 5
(*b*) small of the order of 20
(*c*) large of the order of 1000
(*Principls of Elect. Engg. Delhi Univ. 1988*)

25. A series *R–L–C* circuit will have unity power factor if operated at a frequency of
(*a*) $1/LC$
(*b*) $1/\omega\sqrt{LC}$
(*c*) $1/\omega^2 LC$
(*d*) $1/2\,\pi\sqrt{LC}$
(*Principles of Elect. Engg. Delhi Univ. July 1984*)

26. A two terminal black box contains one of the *R-L-C* elements. The black box is connected to a 220 volts A.C. supply. The current through the source is *I*. When a capacitance of 0.1 F is inserted in series between the source and the box the current through the source is 2I. The element is
(*a*) a resistance
(*b*) an inductance
(*c*) a capacitance
(*d*) it is not possible to determine the element
(*Network Theory Nagpur Univ. 1993*)

Answers

1. *c* **2.** *c* **3.** *a* **4.** *b* **5.** *b* **6.** *b*, *f* **7.** *c* **8.** *a* **9.** *c* **10.** *c* **11.** *a* **12.** *c* **13.** *b* **14.** *d* **15.** *c*
16. *c* **17.** *d* **18.** *a* **19.** *a* **20.** *c* **21.** *a* **22.** *b* **23.** *d* **24.** *c* **25.** *d* **26** *d*.

14 PARALLEL A.C. CIRCUITS

14.1. Solving Parallel Circuits

When impedances are joined in parallel, there are three methods available to solve such circuits : *(a) Vector or Phasor Method (b) Admittance Method and (c) Vector Algebra.*

14.2. Vector or Phasor Method

Consider the circuit shown in Fig. 14.1. Here, two reactors *A* and *B* have been joined in parallel across an r.m.s supply of *V* volts. The voltage across two parallel branches *A* and *B* is the same, but currents through them are different.

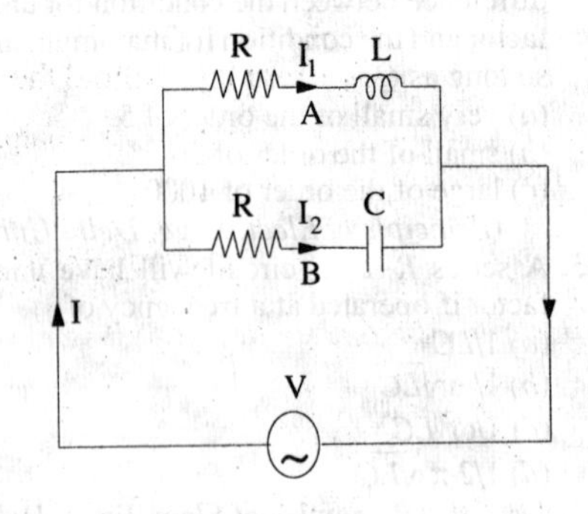

Fig. 14.1

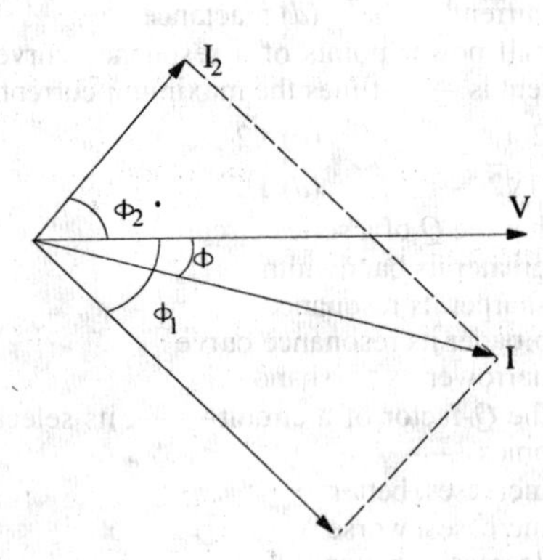

Fig. 14.2

For Branch A, $Z_1 = \sqrt{(R_1^2 + X_L^2)}$; $I_1 = V/Z_1$; $\cos\phi_1 = R_1/Z_1$ or $\phi_1 = \cos^{-1}(R_1/Z_1)$

Current I_1 lags behind the applied voltage by ϕ_1 (Fig. 14.2).

For Branch B, $Z_2 = \sqrt{(R_2^2 + X_c^2)}$; $I_2 = V/Z_2$; $\cos\phi_2 = R_2/Z_2$ or $\phi_2 = \cos^{-1}(R_2/Z_2)$

Current I_2 leads V by ϕ_2 (Fig. 14.2).

Resultant Current I

The resultant circuit current *I* is the vector sum of the branch currents I_1 and I_2 and can be found by (*i*) using parallelogram law of vectors, as shown in Fig. 14.2 or (*ii*) resolving I_2 into their *X*-and *Y*-components (or active and reactive components respectively) and then by combining these components, as shown in Fig. 14.3. Method (*ii*) is preferable, as it is quick and convenient.

With reference to Fig. 14.3 (*a*), we have

Sum of the active components of I_1 and I_2

$= I_1 \cos\phi_1 + I_2 \cos\phi_2$

Sum of the reactive components of I_1 and $I_2 = I_2 \sin\phi_2 - I_1 \sin\phi_1$

If I is the resultant current and ϕ its phase, then its active and reactive components must be equal to these X-and Y-components respectively [Fig. 14.3 (*b*)].

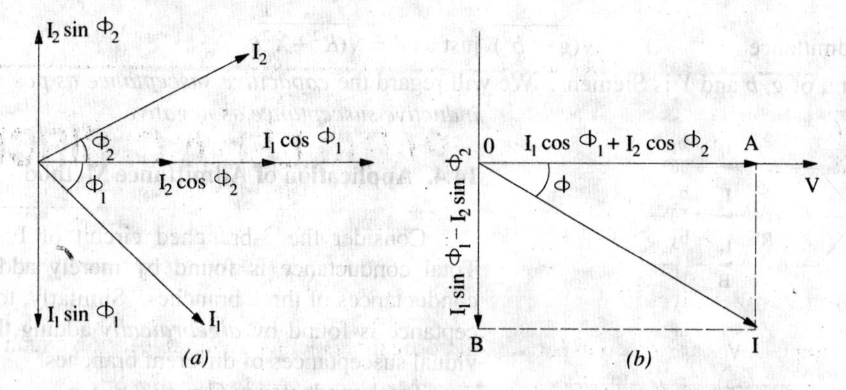

Fig. 14.3

$$I\cos\phi = I_1\cos\phi_1 + I_2\cos\phi_2 \text{ and } I\sin\phi = I_2\sin\phi_2 - I_1\sin\phi_1$$

$$\therefore \quad I = \sqrt{[(I_1\cos\phi_1 + I_2\cos\phi_2)^2 + (I_2\sin\phi_2 - I_1\sin\phi_1)^2]} \text{ and } \tan\phi = \frac{I_2\sin\phi_2 - I_1\sin\phi_1}{I_1\cos\phi_1 + I_2\cos\phi_2} = \frac{Y-\text{component}}{X-\text{component}}$$

If $\tan\phi$ is positive, then current leads and if $\tan\phi$ is negative, then current lags behind the applied voltage V. Power factor for the whole circuit is given by $\cos\phi = \dfrac{I_1\cos\phi_1 + I_2\cos\phi_2}{1} = \dfrac{X\text{-comp.}}{I}$.

14.3. Admittance Method

Admittance of a circuit is defined as *the reciprocal of its impedance*. Its symbol is Y.

$$\therefore \quad Y = \frac{1}{Z} = \frac{I}{V} \quad \text{or} \quad Y = \frac{\text{r.m.s. amperes}}{\text{r.m.s. volts}}$$

Its unit is Siemens (S). A circuit having an impedance of one ohm has an admittance of one Siemens. The old unit was mho (ohm spelled backwards).

As the impedance Z of a circuit has two components X and R (Fig. 14.4), similarly, admittance Y also has two components as shown in Fig. 14.5. The X-component is known as *conductance* and Y-component as *susceptance*.

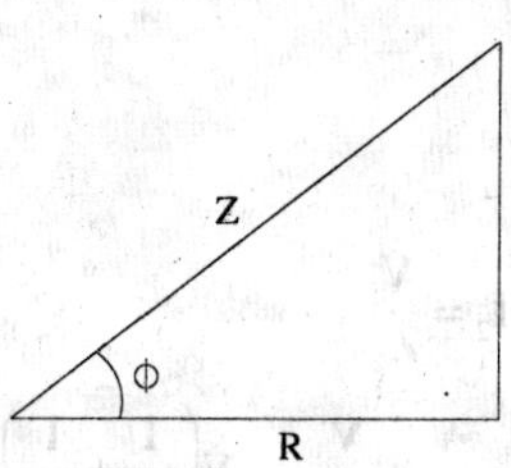

Obviously, conductance $g = Y\cos\phi$

Z X R φ

Fig. 14.4

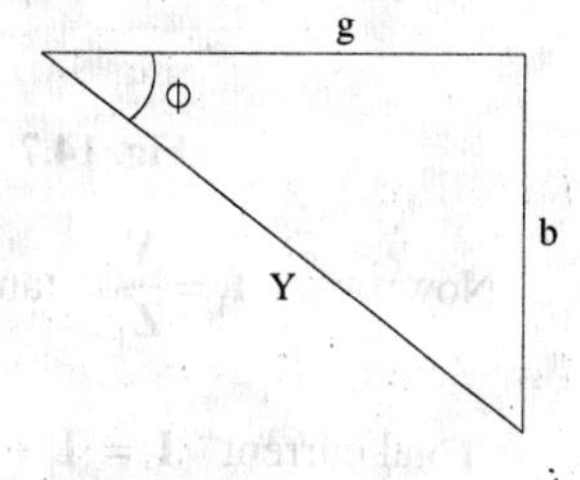

Fig. 14.5

$$\text{or} \quad g = \frac{1}{Z}\cdot\frac{R}{Z} \text{(from Fig. 14.4)}$$

$$\therefore \quad g^* = \frac{R}{Z^2} = \frac{R}{R^2+X^2}$$

---

*In the special case when $X = 0$, then $g = 1/R$ *i.e.* conductance becomes reciprocal of resistance, not otherwise.

Similarly, susceptance b = Y sin $\phi = \frac{1}{Z} \cdot \frac{X}{Z}$. $\therefore$ $b^* = X/Z^2 = X/(R^2 + X^2)$ (from Fig. 14.5)

The admittance $Y = \sqrt{(g^2 + b^2)}$ just as $Z = \sqrt{(R^2 + X^2)}$

The unit of g, b and Y is Siemens. We will regard the *capacitive susceptance as positive and inductive susceptance as negative.*

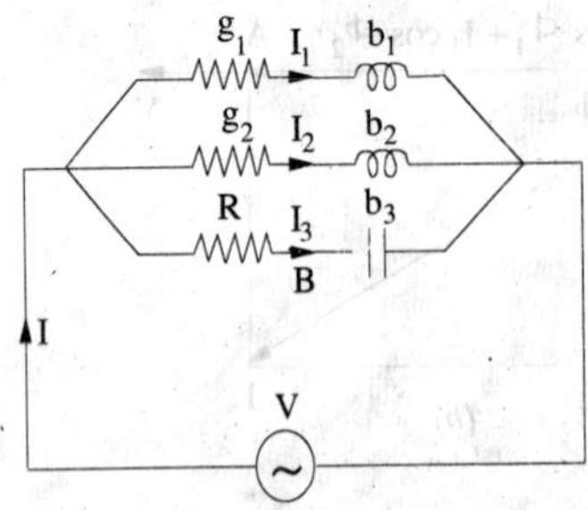

Fig. 14.6

14.4. Application of Admittance Method

Consider the 3-branched circuit of Fig. 14.6. Total conductance is found by merely adding the conductances of three branches. Similarly, total susceptance is found by *algebraically* adding the individual susceptances of different branches.

Total conductance $G = g_1 + g_2 + g_3$ .....

Total susceptance $B = (-b_1) + (-b_2) + b_3$ .....

(algebraic sum)

$\therefore$ Total admittance $Y = \sqrt{(G^2 + B^2)}$

Total current $I = VY$; Power factor $\cos\phi = G/Y$

14.5. Complex or Phasor Algebra

Consider the parallel circuit shown in Fig. 14.7. The two impedances, $\mathbf{Z}_1$ and $\mathbf{Z}_2$, being in parallel, have the same p.d. across them.

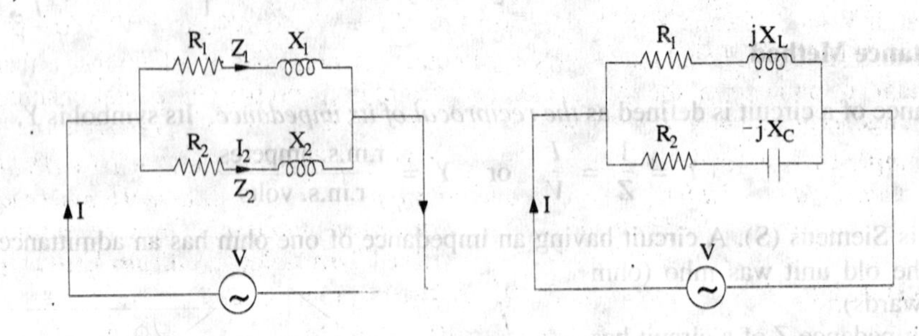

Fig. 14.7 Fig. 14.8

Now $\mathbf{I}_1 = \frac{\mathbf{V}}{\mathbf{Z}_1}$ and $\mathbf{I}_2 = \frac{\mathbf{V}}{\mathbf{Z}_2}$

Total current $\mathbf{I} = \mathbf{I}_1 + \mathbf{I}_2 = \frac{\mathbf{V}}{\mathbf{Z}_1} + \frac{\mathbf{V}}{\mathbf{Z}_2} = \mathbf{V}\left(\frac{\mathbf{1}}{\mathbf{Z}_1} + \frac{\mathbf{1}}{\mathbf{Z}_3}\right) = \mathbf{V}(\mathbf{Y}_1 + \mathbf{Y}_2) = \mathbf{VY}$

where $\mathbf{Y}$ = total admittance $= \mathbf{Y}_1 + \mathbf{Y}_2$

It should be noted that *admittances* are *added* for parallel branches, whereas for branches in series, it is the *impedances* which are *added. However, it is important to remember that since both admittances and impedances are complex quantities, all additions must be in complex form.* Simple arithmetic additions must not be attempted.

---

*Similarly, in the special case when $R = 0$, $b = 1/X$ *i.e.* susceptance becomes reciprocal of reactance, not otherwise.

Considering the two parallel branches of Fig. 14.8, we have

$$\mathbf{Y}_1 = \frac{1}{\mathbf{Z}_1} = \frac{1}{R_1 + jX_L} = \frac{(R_1 - jX_L)}{(R_1 + jX_L)(R_1 - jX_L)}$$

$$= \frac{R_1 - jX_L}{R_1^2 + X_L^2} = \frac{R_1}{R_1^2 + X_L^2} - j\frac{X_L}{R_1^2 + X_L^2} = g_1 - jb_1$$

where $g_1 = \frac{R_1}{R_1^2 + X_L^2}$ – conductance of upper branch,

$b_1 = -\frac{X_L}{R_1^2 + X_L^2}$ – susceptance of upper branch

Similarly, $\mathbf{Y}_2 = \frac{1}{\mathbf{Z}_2} = \frac{1}{R_2 - jX_C}$

$$= \frac{R_2 + jX_C}{(R_2 - jX_C)(R_2 + jX_C)} = \frac{R_2 + jX_C}{R_2^2 + X_C^2} = \frac{R_2}{R_2^2 + X_C^2} + j\frac{X_C}{R_2^2 + X_C^2} = g_2 + jb_2$$

Total admittance $\mathbf{Y} = \mathbf{Y}_1 + \mathbf{Y}_2 = (g_1 - jb_1) + (g_2 + jb_2) = (g_1 + g_2) - j(b_1 - b_2) = G - jB$

$$\mathbf{Y} = \sqrt{[(g_1 + g_2)^2 + (b_1 - b_2)^2]}\ ; \quad \phi = \tan^{-1}\left(\frac{b_1 - b_2}{g_1 + g_2}\right)$$

The polar form for admittance is $\mathbf{Y} = Y \angle \phi°$ where ϕ is as given above.

$$Y = \sqrt{G^2 + B^2} \angle \tan^{-1}(B/G)$$

Total current $\quad \mathbf{I} = \mathbf{VY}\ ;\ \mathbf{I}_1 = \mathbf{VY}_1$ and $\mathbf{I}_2 = \mathbf{VY}_2$

If $\quad \mathbf{V} = V\angle 0°$ and $\mathbf{Y} = Y\angle\phi$ then $\mathbf{I} = \mathbf{VY} = V\angle 0° \times Y\angle\phi = VY\angle\phi$

In general, if $\mathbf{V} = V\angle\alpha$ and $\mathbf{Y} = Y\angle\beta$, then $\mathbf{I} = \mathbf{VY} = V\angle\alpha \times Y\angle\beta = VY\angle\alpha + \beta$

Hence, it should be noted that when vector voltage is multiplied by admittance either in complex (rectangular) or polar form, the result is vector current in its proper phase relationship with respect to the voltage, *regardless of the axis to which the voltage may have been referred to.*

Example 14.1. *Two circuits, the impedances of which are given by $Z_1 = 10 + j15$ and $Z_2 = 6 - j8$ ohm are connected in parallel. If the total current supplied is 15 A, what is the power taken by each branch ? Find also the p.f. of individual circuits and of combination. Draw vector diagram.*

(Elect. Technology, Vikram Univ. Ujjain 1989)

Solution. Let $\quad \mathbf{I} = 15\angle 0°\ ;\ \mathbf{Z}_1 = 10 + j15 = 18\angle 57°$

$\mathbf{Z}_2 = 6 - j8 = 10\angle -53.1°$

Total impedance, $\mathbf{Z} = \frac{\mathbf{Z}_1\mathbf{Z}_2}{\mathbf{Z}_1 + \mathbf{Z}_2} = \frac{(10 + j15)(6 - j8)}{16 + j7}$

$= 9.67 - j3.6 = 10.3\angle -20.4°$

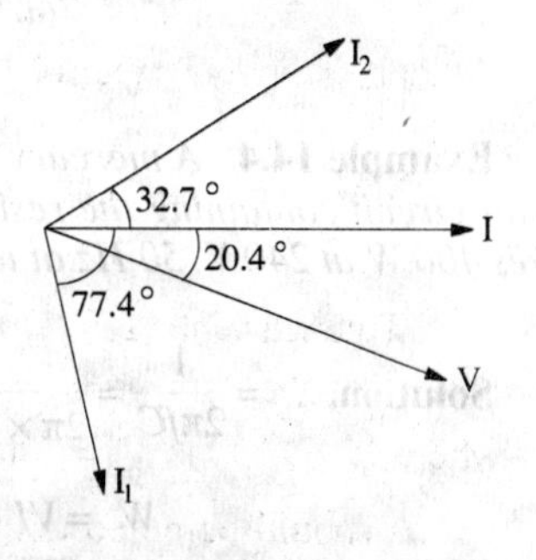

Fig. 14.9

Applied voltage is given by

$\mathbf{V} = \mathbf{IZ} = 15\angle 0° \times 10.3\ \angle -20.4° = 154.5\angle -20.4°$

$\mathbf{I}_1 = \mathbf{V}/\mathbf{Z}_1 = 154.5\angle -20.4°/18\angle 57° = 8.58\angle -77.4°$

$\mathbf{I}_2 = \mathbf{V}/\mathbf{Z}_2 = 154.5\angle -20.4°/10\angle -53.1° = 15.45\angle 32.7°$

We could also find branch currents as under :

$$\mathbf{I}_1 = \mathbf{I}.\ \mathbf{Z}_2/(\mathbf{Z}_1 + \mathbf{Z}_2) \text{ and } \mathbf{I}_2 = \mathbf{I}.\ \mathbf{Z}_1/(\mathbf{Z}_1 + \mathbf{Z}_2)$$

It is seen from phasor diagram of Fig. 14.9 that $\mathbf{I}_1$ lags behind $\mathbf{V}$ by (77.4°–20.4°) = 57° and $\mathbf{I}_2$ leads it by (32.7° + 20.4°) = 53.1°.

$\therefore \quad P_1 = I_1^2R_1 = 8.58^2 \times 10 =$ **736 W** ; p.f. = cos 57° = **0.544 (lag)**

$P_2 = I_2^2R_2 = 15.45^2 \times 6 =$ **1432 W** ; p.f. = cos 53.1° = **0.6**

Combined p.f. = cos 20.4° = **0.937 (lead)**

Example 14.2. *Two impedances Z_1 = (8 + j6) and Z_2 = (3–j4) are in parallel. If the total current of the combination is 25 A, find the current taken and power consumed by each impedance.*

(F.Y. Engg. Pune Univ. May 1988)

Solution. $Z_1 = (8 + j6) = 10\angle 36.87°; Z_2 = (3 - j4) = 5\angle -53.1°$

$$Z = \frac{Z_1 Z_2}{Z_1 + Z_2} = \frac{(10\angle 36.87°)(5\angle -53.1°)}{(8 + j6) + (3 - j4)} = \frac{50\angle -16.23°}{11 + j2} = \frac{50\angle -16.23°}{11.18\angle 10.3°} = 4.47\angle -26.53°$$

Let $I = 25\angle 0°;\ V = IZ = 25\angle 0° \times 4.47\angle -26.53° = 111.75\angle -26.53°$

$I_1 = V/Z_1 = 111.75\angle -26.53°/10\angle 36.87° = 11.175\angle -63.4°$

$I_2 = 111.75\angle -26.53/5\angle -53.1° = 22.35\angle 26.57°$

Now, the phase difference between V and I_1 is 63.4° – 26.53° = 36.87° with current lagging. Hence, $\cos\phi_1 = \cos 36.87° = 0.8$.

Power consumed in $\mathbf{Z}_1 = VI_1 \cos\phi = 11.175 \times 111.75 \times 0.8$ = **990 W**.

Similarly, $\phi_2 = 26.57 - (-26.53) = 53.1°$; $\cos 53.1° = 0.6$.

Power consumed in $\mathbf{Z}_2 = VI_2 \cos\phi_2 = 111.75 \times 22.35 \times 0.6$ = **1499 W**

Example 14.3. *Refer to the circuit of Fig. 14.10(a) and determine the resistance and reactance of the lagging coil load and the power factor of the combination when the currents are as indicated.*

(Elect. Engg. A.M.Ae. S.I. Dec. 1989)

Solution. As seen from the $\Delta\,ABC$ of Fig. 14.10 (*b*).

$5.6^2 = 2^2 + 4.5^2 + 2 \times 2 \times 4.5 \times \cos\theta$, ∴ $\cos\theta = 0.395$, $\sin\theta = 0.919$. $Z = 300/4.5 = 66.67\ \Omega$,

$R = Z\ \cos\theta = 66.67 \times 0.919 = 61.3\ \Omega$

p.f = $\cos\phi = AC/AD = (2 + 4.5 \times 0.395)/5.6$ = **0.67 (lag)**

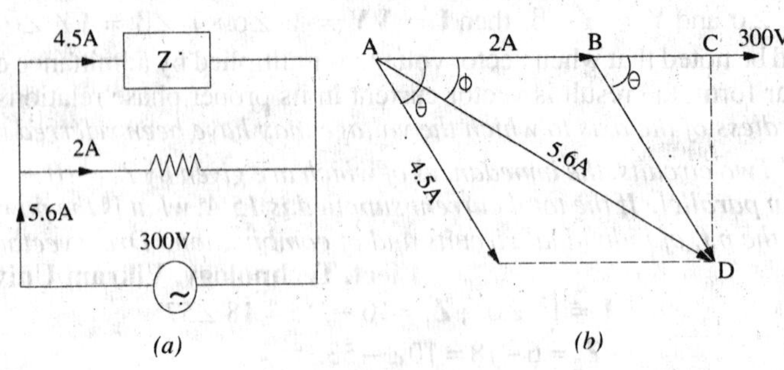

Fig. 14.10

Example 14.4. *A mercury vapour lamp unit consists of a 25μf condenser in parallel with a series circuit containing the resistive lamp and a reactor of negligible resistance. The whole unit takes 400 W at 240 V, 50 Hz at unity p.f. What is the voltage across the lamp ?*

(F.Y. Engg. Pune Univ. Nov. 1987)

Solution. $X_C = \frac{1}{2\pi fC} = \frac{1}{2\pi \times 50 \times (25 \times 10^{-6})} = 127.3\ \Omega$, ∴ $I_C = \frac{240}{127.3} = 1.885$ A

$$W = VI\cos\phi = VI \ \therefore \quad I = W/V = 400/240 = 1.667 \text{ A}$$

In the vector diagram of Fig. 14.10 (*b*) I_C leads V by 90° and current I_1 in the series circuit lags V by ϕ_1 where ϕ_1 is the power factor angle of the series circuit. The vector sum of I_C and ϕ_1 gives the total current I. As seen $\tan\phi_1 = I_C/I = 1.885/1.667 = 1.13077$. Hence, ϕ_1 = 48.5° lag. The applied voltage V is the vector sum of the drop across the resistive lamp which is in phase with I_1 and drop across the coil which leads I_1 by 90°.

Voltage across the lamp $= V\cos\Phi_1 = 340 \times \cos 48.5 = 240 \times 0.662$ = **159 V**.

Example 14.5. *The currents in each branch of a two-branched parallel circuit are given by the expressions* $i_a = 7.07 \sin(314t - \pi/4)$ *and* $i_b = 21.2 \sin(314\,t + \pi/3)$

The supply voltage is given by the expression v = 354 sin 314t. Derive a similar expression for the supply current and calculate the ohmic value of the components, assuming two pure components in each branch. State whether the reactive components are inductive or capacitive.

(Elect. Engineering., Calcutta Univ. 1991)

Solution. By inspection, we find that i_a lags the voltage by $\pi/4$ radian or 45° and i_b leads it by $\pi/3$ radian or 60°. Hence, branch *A* consists of a resistance in series with a pure inductive reactance. Branch *B* consists of a resistance in series with pure capacitive reactance as shown in Fig. 14.11 (*a*).

Maximum value of current in branch *A* is 7.07 A and in branch *B* is 21.2 A. The resultant current can be found vecctorially As seen from vector diagram,

X-comp = 21.2 cos 60° + 7.07 cos 45° = 15.6 A

Y-comp = 21.2 sin 60°−7.07 sin 45° = 13.36 A

Maximum value of the resultant current is = $\sqrt{15.6^2 + 13.36^2} = 20.55$ A

$\phi = \tan^{-1}(13.36/15.6) = \tan^{-1}(0.856) = 40.5°$ (lead)

Hence, the expression for the supply current is **i = 20.55 sin (314t + 40.5°)**

$Z_A = 354/7.07 = 50\ \Omega$; $\cos\phi_A = \cos 45°$

$= 1/\sqrt{2}$. $\sin\phi_A = \sin 45° = 1/\sqrt{2}$

$R_A = Z_A \cos\phi_A = 50 \times 1/\sqrt{2} = \mathbf{35.4\ \Omega}$

$X_L = Z_A \sin\phi_A = 50 \times 1/\sqrt{2} = \mathbf{35.4\ \Omega}$

$Z_B = 354/20.2 = \mathbf{17.5\ \Omega}$

$R_B = 17.5 \times \cos 60° = \mathbf{8.75\ \Omega}$

$X_C = 17.5 \times \sin 60° = \mathbf{15.16\ \Omega}$

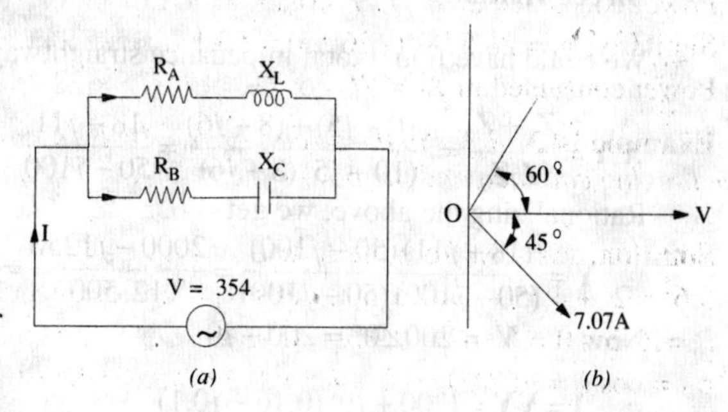

Fig. 14.11

Example 14.5(a). *A total current of 10 A flows through the parallel combination of three impedances : (2−j5) Ω, (6 + j3) Ω and (3 + j4) Ω. Calculate the current flowing through each branch. Find also the p.f. of the combination.*

(Elect. Engg., -I, Delhi Univ. 1987)

Solution. Let $\mathbf{Z}_1 = (2 - j5),\quad \mathbf{Z}_2 = (6 + j3),\quad \mathbf{Z}_3 = (3 + j4)$

$$\mathbf{Z}_1\mathbf{Z}_2 = (2 - j5)(6 + j3) = 27 - j24.\ \mathbf{Z}_2\mathbf{Z}_3 = (6 + j3)(3 + j4) = 6 + j33$$

$$\mathbf{Z}_3\mathbf{Z}_1 = (3 + j4)(2 - j5) = 26 - j7;\ \mathbf{Z}_1\mathbf{Z}_2 + \mathbf{Z}_2\mathbf{Z}_3 + \mathbf{Z}_3\mathbf{Z}_1 = 59 + j2$$

With reference to Art, 1.25

$$\mathbf{I}_1 = \mathbf{I}.\frac{\mathbf{Z}_2\mathbf{Z}_3}{\mathbf{Z}_1\mathbf{Z}_2 + \mathbf{Z}_2\mathbf{Z}_3 + \mathbf{Z}_3\mathbf{Z}_1} = (10 + j0) \times \frac{6 + j33}{59 + j2} = 1.21 + j5.55$$

$$\mathbf{I}_2 = \mathbf{I}.\frac{\mathbf{Z}_3\mathbf{Z}_1}{\Sigma\mathbf{Z}_1\mathbf{Z}_2} = (10 + j0) \times \frac{26 - j7}{59 + j2} = 4.36 - j1.33$$

$$\mathbf{I}_3 = \mathbf{I}.\frac{\mathbf{Z}_1\mathbf{Z}_2}{\Sigma\mathbf{Z}_1\mathbf{Z}_2} = (10 + j0) \times \frac{27 - j24}{59 + j2} = 4.43 - j4.22$$

Now,
$$\mathbf{Z} = \frac{\mathbf{Z}_1\mathbf{Z}_2\mathbf{Z}_3}{\mathbf{Z}_1\mathbf{Z}_2 + \mathbf{Z}_2\mathbf{Z}_3 + \mathbf{Z}_3\mathbf{Z}_1} = \frac{(2 - j5)(6 + j33)}{59 + j2} = 3.01 + j0.51$$

$$\mathbf{V} = 10\angle 0° \times 3.05\angle 9.6° = 30.5\ \angle 9.6°$$

Combination p.f. = cos 9.6° = **0.986 (lag)**

Example 14.6. *Two impedances given by $Z_1 = (10 + j5)$ and $Z_2 = (8 + j6)$ are joined in parallel and connected across a voltage of v = 200 + j0. Calculate the circuit current, its phase and the branch currents. Draw the vector diagram.* **(Electrotechnics -I, M.S. Univ. Baroda 1985)**

Solution. The circuit is shown in Fig. 14.12.

Branch A, $\mathbf{Y}_1 = \dfrac{1}{\mathbf{Z}_1} = \dfrac{1}{(10+j5)}$

$$= \frac{10-j5}{(10+j5)(10-j5)} = \frac{10-j5}{100+25}$$

$$= 0.08 - j0.04 \text{ Siemens}$$

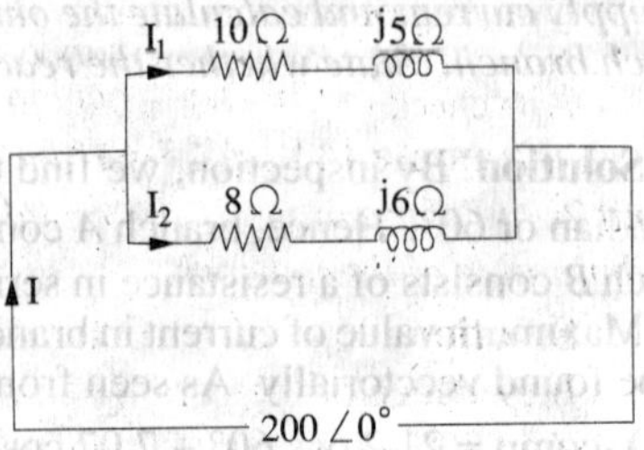

Branch B, $\mathbf{Y}_2 = \dfrac{1}{\mathbf{Z}_2} = \dfrac{1}{(8+j6)}$

$$= \frac{8-j6}{(8+j6)(8-j6)} = \frac{8-j6}{64+36} = 0.08 - j0.06 \text{ Siemens}$$

Fig. 14.12

$$\mathbf{Y} = (0.08 - j0.04) + (0.08 - j0.06) = 0.16 - j0.1 \text{ Siemens}$$

Direct Method

We could have found total impedance straightway like this : $\dfrac{1}{\mathbf{Z}} = \dfrac{1}{\mathbf{Z}_1} + \dfrac{1}{\mathbf{Z}_2} = \dfrac{\mathbf{Z}_1 + \mathbf{Z}_2}{\mathbf{Z}_1\mathbf{Z}_2}$

$$\therefore \mathbf{Y} = \frac{\mathbf{Z}_1 + \mathbf{Z}_2}{\mathbf{Z}_1\mathbf{Z}_2} = \frac{(10+j5)+(8+j6)}{(10+j5)(8+j6)} = \frac{18+j11}{50+j100}$$

Rationalizing the above, we get

$$\mathbf{Y} = \frac{(18+j11)(50-j100)}{(50+j100)(50-j100)} = \frac{2000-j1250}{12,500} = 0.16 - j0.1 \text{(same as before)}$$

Now $\quad \mathbf{Y} = 200\angle 0° = 200 + j0$

$$\therefore \quad \mathbf{I} = \mathbf{VY} = (200+j0)(0.16-j0.1)$$

$$= 32 - j20 = 37.74\angle -32° \quad \text{...polar form}$$

Power factor = cos 32° = 0.848

$\mathbf{I}_1 = \mathbf{VY}_1 = (200 + j0)(0.08 - j0.04)$

$= 16 - j8 = 17.88 \angle -26°32'$

It lags behind the applied voltage by 26°32′.

$\mathbf{I}_2 = \mathbf{VY}_2 = (200 + j0)(0.08 - j0.06)$

$= 16 - j12 = 20 \angle -36°46'$

It lags behind the applied voltage by 36°46′. The vector diagram is shown in Fig. 14.13.

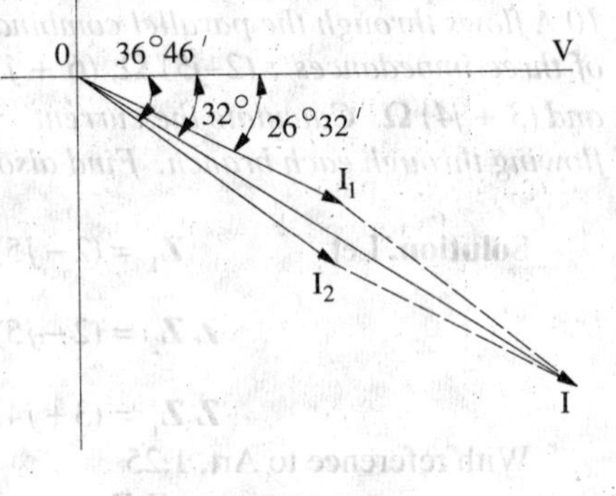

Fig. 14.13

Example 14.7. *Explain the term admittance.*

Two impedances Z_1 = (6–j8) ohm and Z_2 = (16 + j12) ohm are connected in parallel. If the total current of the combination is (20 + j10) amperes, find the complexor for power taken by each impedance. Draw and explain the complete phasor diagram.

(Basic Electricity, Bombay Univ 1986)

Solution. Let us first find out the applied voltage, $\mathbf{Y} = \mathbf{Y}_1 + \mathbf{Y}_2 = \dfrac{1}{6-j8} + \dfrac{1}{16+j12}$

$$= (0.06 + j0.08) + (0.04 - j0.03) = 0.1 + j0.05 = 0.1118\angle 26°34'$$

$$\mathbf{I} = 20 + j10 = 22.36\angle 26°34'$$

Now $\mathbf{I} = \mathbf{VY} \therefore \quad \mathbf{V} = \dfrac{I}{Y} = \dfrac{22.36\angle 26°34'}{0.1118\angle 26°34'} = 200\angle 0°,$

$\mathbf{I}_1 = \mathbf{VY}_1 = (200 + j0)(0.06 + j0.08) = 12 + j16\text{Â}, \quad \mathbf{I}_2 = 200(0.04 - j0.03) = 8 - j6 \text{ A}$

Using the method of conjugates and taking voltage conjugate, the complexor power taken by each branch can be found as under :

$\mathbf{P}_1 = (200-j0)(12+j16) = 2400 + j3200$; $\mathbf{P}_2 = (200-j0)(8-j6) = 1600-j1200$

Drawing of phasor diagram is left to the reader.

Note. Total voltamperes = $4000 + j2000$

As a check, $\mathbf{P} = \mathbf{VI} = 200(20 + j10) = 4000 + j2000$

Example 14.8. *A 15-mH inductor is in series with a parallel combination of an 80 Ω resistor and 20 μF capacitor. If the angular frequency of the applied voltage is ω = 1000 rad/s, find the admittance of the network.* **(Basic Circuit Analysis Osmania Univ. Jan/Feb 1992)**

Solution. $X_L = \omega L = 1000 \times 15 \times 10^{-3} = 15\Omega$; $X_C = 1/\omega C = 10^6/1000 \times 20 = 50\ \Omega$

Impedance of the parallel combination is given by

$Z_p = 80 || -J50 = -j4000/(80 - j50) = 22.5 - j36,$

Total impedance $= j15 + 22.5 - j36 = 22.5 - j21$

Admittance $Y = \dfrac{1}{Z} = \dfrac{1}{22.5 - j21} = 0.0238 - j0.022$ siemens

Example 14.9. *An impedance (6 + j8) is connected across 200-V, 50-Hz mains in parallel with another circuit having an impedance of (8 - j6) Ω. Calculate (a) the admittance, the conductance, the susceptance of the combined circuit (b) the total current taken from the mains and its p.f.*

(Elect. Engg-AMAE, S.I. 1992)

Solution. $\mathbf{Y}_1 = \dfrac{1}{6+j8} = \dfrac{6-j8}{6^2+8^2} = 0.06 - j0.08$ Siemens, $\mathbf{Y}_2 = \dfrac{1}{8-j6} = \dfrac{8+j6}{100} = 0.08 + j0.06$ Siemens

(*a*) Combined admittance is $\mathbf{Y} = \mathbf{Y}_1 + \mathbf{Y}_2 = 0.14 - j0.02 = \mathbf{0.1414 \angle -8.°8'}$ **Siemens**

Conductance, G = **0.14 Siemens** ; Susceptance, B = **−0.02 Siemens** (inductive)

(*b*) Let $\mathbf{V} = 200\angle 0°$; $\mathbf{I} = \mathbf{VY} = 200 \times 0.1414\angle -8°8' V = 28.3\angle -8°8'$

p.f. $= \cos 8°8' = \mathbf{0.99}$ **(lag)**

Example 14.10. *If the voltmeter in Fig. 14.14 reads 60 V, find the reading of the ammeter.*

Solution. $I_2 = 60/4 = 15$A. Taking it as reference quantity, we have $\mathbf{I}_2 = 15 \angle 0$.

Obviously, the applied voltage is

$\mathbf{V} = 15\angle 0° \times (4 - j4) = 84.8\angle -45°$

$\mathbf{I}_1 = 84.8\angle -45°/(6 + j3) = 84.8\angle -45 \div 6.7\angle 26.6$

$= 12.6\angle -71.6° = (4 - j12)$

$\mathbf{I} = \mathbf{I}_1 + \mathbf{I}_2 = (15 + j0) + (4 - j12) = 19 - j12 = 22.47\angle -32.3°$

Hence, ammeter reads **22.47**.

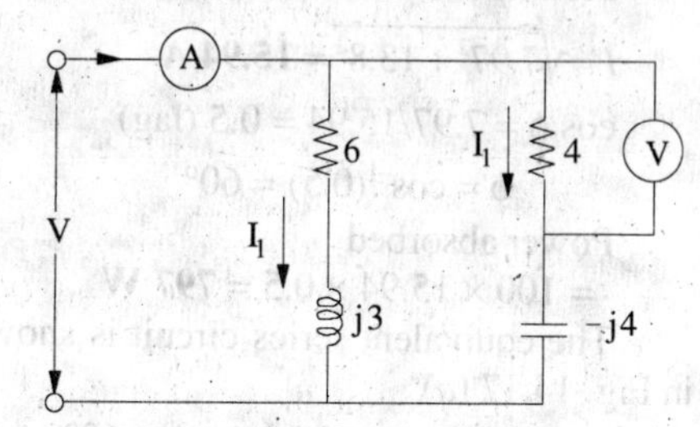

Fig. 14.14

Example 14.11. *Find the reading of the ammeter when the voltmeter across the 3 ohm resistor in the circuit of Fig. 14.15 reads 45 V*

(Elect. Engg. & Electronics Bangalore Univ. 1988)

Solution. Obviously $I_1 = 45/3 = 15$ A. If we take it as reference quantity, $I_1 = 3\angle 0°$.

Now, $Z_1 = 3 - j3 = 4.24 \angle -45°$.

Hence, $V = I_1 Z_1 = 15 \angle 0° \times 4.24 \angle -45° = 63.6 \angle -45°$.

$I_2 = \dfrac{V}{Z_2} = \dfrac{63.6\angle -45°}{5 + j2} = \dfrac{63.6\angle -45°}{5.4\angle 21.8°}$

$= 11.77\angle -66.8° = 4.64 - j10.8$

$I = I_1 + I_2 = 19.64 - j\ 10.8 = \mathbf{22.4\angle 28.8°}$

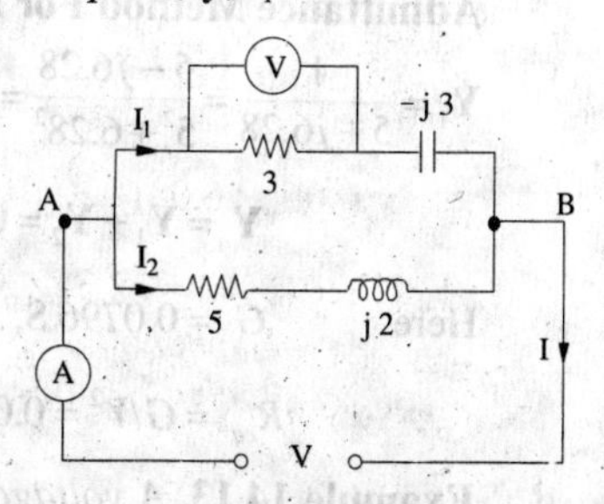

Fig. 14.15

Example 14.12. *A coil having a resistance of 5 Ω and an inductance of 0.02 H is arranged in parallel with another coil having a resistance of 1 Ω and an inductance of 0.08 H. Calculate* *the current through the combination and the power absorbed when a voltage of 100 V at 50 Hz is applied. Estimate the resistance of a single coil which will take the same current at the same power factor.*

Solution. The circuit and its phasor diagram are shown in Fig. 14.16.

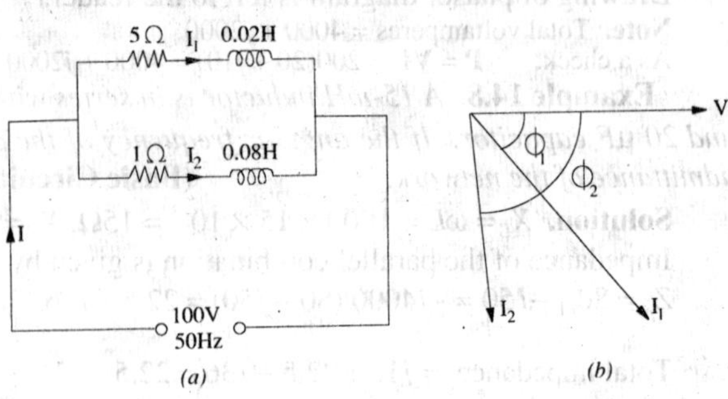

Fig. 14.16

Branch No. 1

$$X_1 = 314 \times 0.02 = 6.28\ \Omega$$

$$Z_1 = \sqrt{5^2 + 6.28^2} = 8\ \Omega$$

$$I_1 = 100/8 = 12.5\ \text{A}$$

$$\cos\phi_1 = R_1/Z_1 = 5/8$$

$$\sin\phi_1 = 6.28/8$$

Branch No. 2

$X_2 = 314 \times 0.08 = 25.12\ \Omega,\ Z_2 = \sqrt{1^2 + 25.12^2} = 25.14\ \Omega,\ I_2 = 100/25.14 = 4\ \text{A}$

$\cos\phi_2 = 1/25.14$ and $\sin\phi_2 = 25.12/25.14$

X–components of I_1 and $I_2 = I_1 \cos\phi_1 + I_2 \cos\phi_2 = (12.5 \times 5/8) + (4 \times 1/25.14) = 7.97$ A

Y-components of I_1 and $I_2 = I_1 \sin\phi_1 + I_2 \sin\phi_2 = (12.5 \times 6.28/8) + (4 \times 25.12/25.14) = 13.8$ A

$I = \sqrt{7.97^2 + 13.8^2} = \mathbf{15.94\ A}$

$\cos\phi = 7.97/15.94 = 0.5$ (lag)

$\phi = \cos^{-1}(0.5) = 60°$

Power absorbed

$= 100 \times 15.94 \times 0.5 = \mathbf{797\ W}$

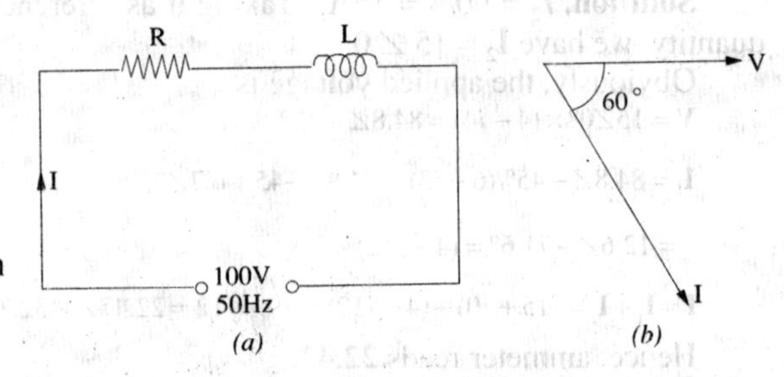

Fig. 14.17

The equivalent series circuit is shown in Fig. 14.17 (*a*).

$V = 100$ V; $I = 15.94$ A ; $\phi = 60°$

$Z = 100/15.94 = 6.27\ \Omega$;

$R = Z\cos\phi = 6.27 \times \cos 60° = \mathbf{3.14\ \Omega}$

$X = Z\sin\phi = 6.27 \times \sin 60° = \mathbf{5.43\ \Omega}$

Admittance Method For Finding Equivalent Circuit

$$\mathbf{Y}_1 = \frac{1}{5 + j6.28} = \frac{5 - j6.28}{5^2 + 6.28^2} = 0.078 - j0.098\ \text{S},\ \mathbf{Y}_2 = \frac{1}{1 + j25.12} = \frac{1 - j25.12}{1^2 + 25.12^2} = 0.00158 - j0.0397\ \text{S},$$

$$\mathbf{Y} = \mathbf{Y}_1 + \mathbf{Y}_2 = 0.0796 - j0.138 = 0.159\angle -60°$$

Here $\quad G = 0.0796$ S, $\quad B = -0.138$ S, $\quad Y = 0.159\ \Omega$

$\therefore \quad R_{eq} = G/Y^2 = 0.0796/0.159^2 = \mathbf{3.14\ \Omega}, X_{eq} = B/Y^2 = 0.138/0.159^2 = \mathbf{5.56\ \Omega}$

Example 14.13. *A voltage of 200 $\angle 53°8'$ is applied across two impedances in parallel. The values of impedances are (12 + j16) and (10 – j20). Determine the kVA, kVAR and kW in each branch and the power factor of the whole circuit.* **(Elect. Technology, Indore Univ. 1986)**

Solution. The circuit is shown in Fig. 14.18.

$\mathbf{Y}_A = 1/(12 + j16) = (12 - j16)/[(12 + j16)(12 - j16)]$

$= (12 - j16)/400 = 0.03 - j0.04$ mho

$\mathbf{Y}_B = 1/(10 - j20) = 10 + j20/[(10 - j20)(10 + j20)]$

$= \dfrac{10 + j20}{500} = 0.02 + j0.04$ mho

Now $V = 200\angle 53°8' = 200(\cos 53°8' + j \sin 53°8')$

$= 200(0.6 + j0.8) = 120 + j160$ volt

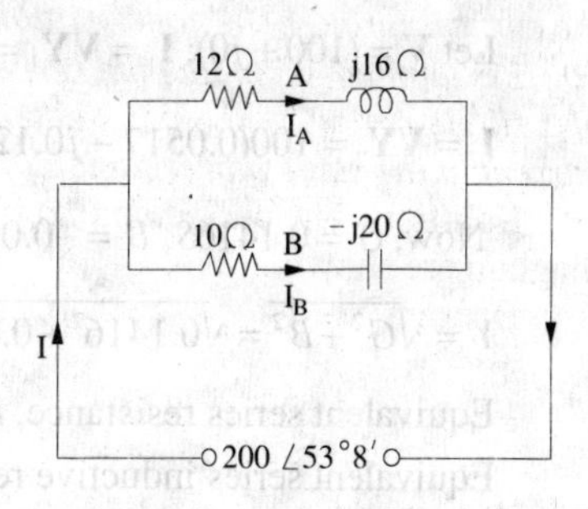

Fig. 14.18

$\mathbf{I}_A = \mathbf{VY}_A = (120 + j160)(0.03 - j0.04) = (10 + j0)$ ampere (along the reference axis)

$\therefore$ $\mathbf{I}_B = \mathbf{VY}_B = (120 + j160)(0.02 + j0.04) = -4.0 + j8$ ampere (leading)

Power Calculations

Power etc. can be calculated by the method of conjugates as explained in Ex. 14.3.

Branch A

The current conjugate of $(10 + j0)$ is $(10 - j0)$

$\therefore$ $\mathbf{VI}_A = (120 + j160)(10 - j0) = 1200 + j1600$ $\therefore$ kW = 1200/1000 = **1.2**

$\therefore$ kVAR = 1600/1000 = **1.6**. The fact that it is positive merely shows that reactive volt-amperes are due to a lagging current*. kVA = $\sqrt{(1.2^2 + 1.6^2)}$ = **2**

Branch B

The current conjugate of $(-4.0 + j8)$ is $(-4.0 - j8)$

$\therefore$ $\mathbf{VI}_B = (120 + j160)(-4 - j8) = 800 - j1600$

$\therefore$ kW = 800/1000 = **0.8** $\therefore$ kVAR = −1600/1000 = **−1.6**

The negative sign merely indicates that reactive volt-amperes are due to leading current

$\therefore$ $kVA = \sqrt{[0.8^2 + (-1.6)^2]} = \mathbf{1.788}$

$\mathbf{Y} = \mathbf{Y}_A + \mathbf{Y}_B = (0.03 - j0.04) + (0.02 + j0.04) = 0.05 + j0$

$\mathbf{I} = \mathbf{VY} = (120 + j160)(0.05 + j0) = 6 + j8 = 10\angle 53°8'$

or $\mathbf{I} = \mathbf{I}_A + \mathbf{I}_B = (10 + j0) + (-4 + j8) = 6 + j8$ (same as above)

Circuit p.f. $= \cos 0° = 1$ ($\because$ current is in phase with voltage)

Example 14.14. *An impedance $Z_1 = (8 - j5)\ \Omega$ is in parallel with an impedance $Z_2 = (3 + j7)\ \Omega$. If 100 V are impressed on the parallel combination, find the branch currents I_1, I_2 and the resultant current. Draw the corresponding phasor diagram showing each current and the voltage drop across each parameter. Calculate also the equivalent resistance, reactance and impedance of the whole circuit.* **(Elect. Technology-I, Gwalior Univ. 1988)**

Solution. Admittance Method

$\mathbf{Y}_1 = 1/(8 - j5) = (0.0899 + j0.0562)$ S

$\mathbf{Y}_2 = 1/(3 + j7) = (0.0517 - j0.121)$ S, $\mathbf{Y} = \mathbf{Y}_1 + \mathbf{Y}_2 = (0.1416 - j0.065)$ S

---

*If voltage conjugate is used, then capacitive VARs are positive and inductive VARs negative.

Let $\mathbf{V} = (100 + j0)$; $\mathbf{I}_1 = \mathbf{VY}_1 = 100(0.0899 + j0.0562) = 8.99 + j5.62$

$\mathbf{I}_2 = \mathbf{VY}_2 = 100(0.0517 - j0.121) = 5.17 - j12.1$; $\mathbf{I} = \mathbf{VY} = 100(0.1416 - j0.065) = 14.16 - j6.5$

Now, $G = 0.1416S, B = -0.065$ S(inductive;)

$Y = \sqrt{G^2 + B^2} = \sqrt{0.1416^2 + 0.065^2} = 0.1558$ S

Equivalent series resistance, $R_{eq} = G/Y^2 = 0.1416/0.1558^2 = 5.38\ \Omega$

Equivalent series inductive reactance $X_{eq} = B/Y^2 = 0.065/0.1558^2 = 2.68\ \Omega$

Equivalent series impedance $\mathbf{Z} = 1/\mathbf{Y} = 1/0.1558 = \mathbf{6.42}\ \Omega$

Impedance Method

$\mathbf{I}_1 = \mathbf{V}/\mathbf{Z}_1 = (100 + j0)/(8 - j5) = 8.99 + j5.62$

$\mathbf{I}_2 = \mathbf{V}/\mathbf{Z}_2 = 100/(3 + j7) = 5.17 - j12.1$

$$\mathbf{Z} = \frac{\mathbf{Z}_1\mathbf{Z}_2}{\mathbf{Z}_1 + \mathbf{Z}_2} = \frac{(8 - j5)(3 + j7)}{(11 + j2)} = \frac{59 + j41}{(11 + j2)} = 5.848 + j2.664 = 6.426 = \angle 24.5°,$$

$\mathbf{I} = 100/6.426\angle 24.5° = 15.56\angle -24.5° = 14.16 - j6.45$

As seen from the expression for Z, equivalent series resistance is **5.848** Ω and inductive reactance is **2.664 ohm.**

Example 14.15. *The impedances $Z_1 = 6 + j8$, $Z_2 = 8 - j6$ and $Z_3 = 10 + j0$ ohms measured at 50 Hz, form three branches of a parallel circuit. This circuit is fed from a 100 volt, 50-Hz supply. A purely reactive (inductive or capacitive) circuit is added as the fourth parallel branch to the above three-branched parallel circuit so as to draw minimum current from the source. Determine the value of L or C to be used in the fourth branch and also find the minimum current.*

(Electrical Circuits, South Gujarat Univ. 1986)

Solution. Total admittance of the 3-branched parallel circuit is

$$\mathbf{Y} = \frac{1}{6 + j8} + \frac{1}{8 - j6} + \frac{1}{10 + j0} = 0.06 - j0.08 + 0.08 + j0.06 + 0.1 = 0.24 - j0.02$$

Current taken would be minimum when net susceptance is zero. Since combined susceptance is inductive, it means that we must add capacitive susceptance to neutralize it. Hence, we must connect a pure capacitor in parallel with the above circuit such that its susceptance equals $+ j0.02$ S

$\therefore\ I/X_C = 0.02$ or $2\pi fC = 0.02$; $C = 0.02/314 = \mathbf{63.7\ \mu F}$

Admittance of four parallel branches = $(0.24 - j0.02) + j0.02 = 0.24$ S

$\therefore$ Minimum current drawn by the circuit = $100 \times 0.24 = \mathbf{24\ A}$

Example 14.16. *The total effective current drawn by parallel circuit of Fig. 14.19 is 20 A. Calculate (i) VA (ii) VAR and (iii) watts drawn by the circuit.*

Solution. The combined impedance of the circuit is

$$\mathbf{Z} = \frac{\mathbf{Z}_1\mathbf{Z}_2}{\mathbf{Z}_1 + \mathbf{Z}_2} = \frac{10(6 - j8)}{(16 - j8)} = (5 - j2.5) \text{ ohm}$$

(*iii*) Power = $I^2R = 20^2 \times 5 = \mathbf{2000\ W}$ (*ii*) $Q = I^2X = 20^2 \times 2.5$ = **1000 VAR (leading)** (*i*) $S = P + jQ = 2000 + j1000$ $= 2236\angle 27°$; $S = \mathbf{2236\ VA}$

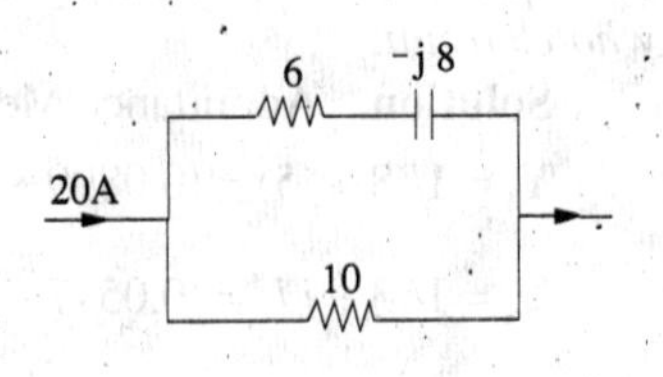

Fig. 14.19

Example 14.17. *Calculate (i) total current and (ii) equivalent impedance for the four-branched circuit of Fig. 14.20.*

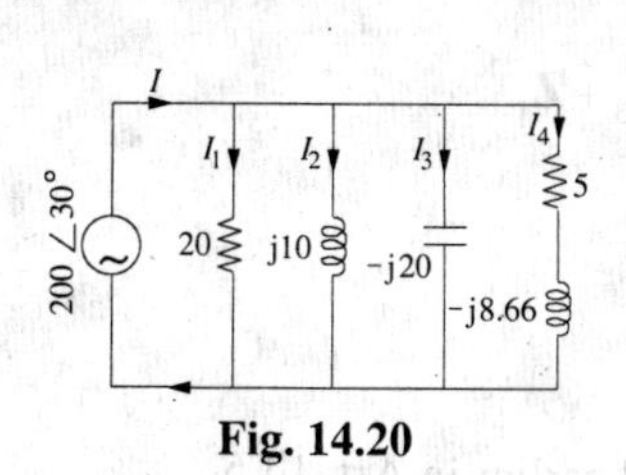

Fig. 14.20

Solution. $\mathbf{Y}_1 = 1/20 = 0.05$ S, $\mathbf{Y}_2 = 1/j10 = -j0.1$ S; $\mathbf{Y}_3 = 1/-j20 = j0.05$ S ; $\mathbf{Y}_4 = 1/5 - j8.66 = 1/10 \angle 60°$
$= 0.1 \angle -60° = (0.05 - j0.0866)$ S
$Y = \mathbf{Y}_1 + \mathbf{Y}_2 + \mathbf{Y}_3 + \mathbf{Y}_4 = (0.1 - j0.1366)$ S
$= 0.169 \angle -53.8°$ S

(i) $\mathbf{I} = \mathbf{VY} = 200 \angle 30° \times 0.169 \angle -53.8° = 33.8 \angle -23.8°$ A

(ii) $\mathbf{Z} = 1/\mathbf{Y} = 1/0.169 \angle -53.8° = 5.9 \angle 53.8° \Omega$

Example 14.18. *The power consumed by both branches of the circuit shown in Fig. 14.21 is 2200 W. Calculate power of each branch and the reading of the ammeter.*

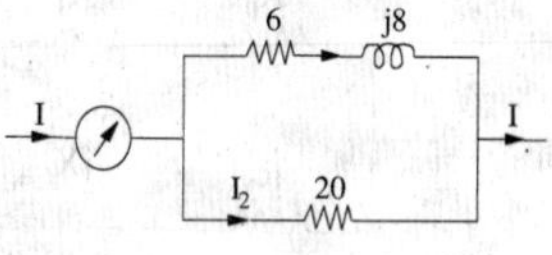

Solution. $\mathbf{I}_1 = \mathbf{V}/\mathbf{Z}_1$

$= \mathbf{V}/(6 + j8) = \mathbf{V}/10\angle 53.1° \quad \mathbf{I}_2 = \mathbf{V}/\mathbf{Z}_2 = \mathbf{V}/20$

Fig. 14.21

$\therefore I_1/I_2 = 20/10 = 2, P_1 = I_1^2 R_1$ and $P_2 = I_2^2 R_2$

$$\therefore \frac{P_1}{P_2} = \frac{I_1^2 R_1}{I_2^2 R_2} = 2^2 \times \left(\frac{6}{20}\right) = \frac{6}{5}$$

Now, $P = P_1 + P_2$ or $\dfrac{P}{P_2} = \dfrac{P_1}{P_2} + 1 = \dfrac{6}{5} + 1 = \dfrac{11}{5}$

or $P_2 = 2200 \times \dfrac{5}{11} = 1000$ W $\therefore P_1 = 2200 - 1000 = 1200$ W

Since $P_1 = I_1^2 R_1$ or $1200 = I_1^2 \times 6$; $I_1 = 14.14$ A

If $\mathbf{V} = V\angle 0°$, then $\mathbf{I}_1 = 14.14\angle -53.1° = 8.48 - j11.31$

Similarly, $P_2 = I_2^2 R_2$ or $1000 = I_2^2 \times 20$; $I_2 = 7.07$ A or $\mathbf{I}_2 = \mathbf{7.07\angle 0°}$

Total current $\mathbf{I} = \mathbf{I}_1 + \mathbf{I}_2 = (8.48 - j11.31) + 7.07 = 15.55 - j11.31 = \mathbf{19.3 \angle -36°}$

Hence, ammeter reads **19.3 A**

14.6. Series-parallel Circuits

(i) By Admittance Method

In such circuits, the parallel circuit is first reduced to an equivalent series circuit and then, as usual, combined with the rest of the circuit. For a parallel circuit

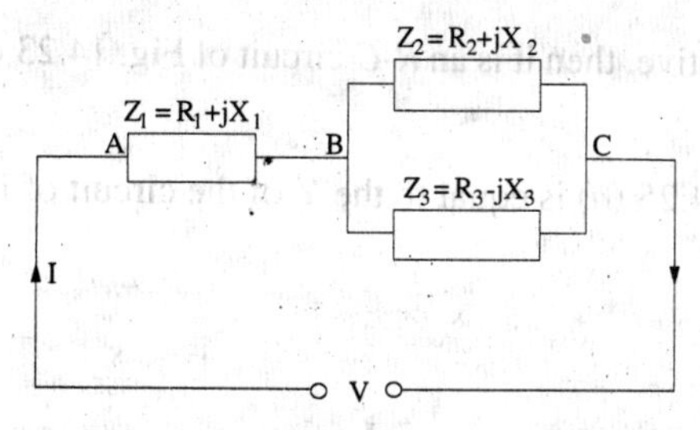

Fig. 14.22

Equivalent series resistance $R_{eq} = Z \cos = \dfrac{1}{Y} \cdot \dfrac{G}{Y} = \dfrac{G}{Y^2}$

—See Ex. 14.14

Equivalent series reactance $X_{eq} = Z \sin \phi = \dfrac{I}{Y} \cdot \dfrac{B}{Y} = \dfrac{B}{Y^2}$

(ii) By Symbolic Method

Consider the circuit of Fig. 14.22. First, equivalent impedance of parallel branches is calculated and it is then added to the series impedance to get the total circuit impedance. The circuit current can be easily found.

$$Y_2 = \frac{1}{R_2 + jX_2}; \mathbf{Y}_3 = \frac{1}{R_3 - jX_3}$$

$$\therefore \quad \mathbf{Y}_{23} = \frac{1}{R_2 + jX_2} + \frac{1}{R_3 - jX_3}$$

$$\therefore \quad \mathbf{Z}_{23} = \frac{1}{\mathbf{Y}_{23}}; \mathbf{Z}_1 = R_1 + jX_1; \mathbf{Z} = \mathbf{Z}_{23} + \mathbf{Z}_1$$

$$\therefore \quad \mathbf{I} = \frac{\mathbf{V}}{\mathbf{Z}} \quad \text{(See Ex. 14.21)}$$

14.7. Series Equivalent of a Parallel Circuit

Consider the parallel circuit of Fig. 14.23 (*a*). As discussed earlier in Art. 14.5.

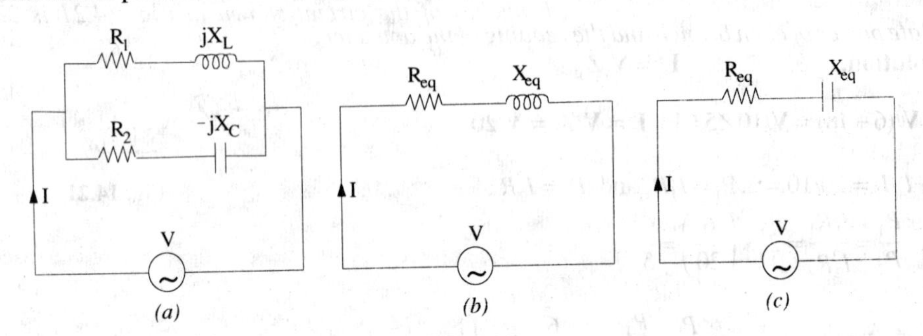

Fig. 14.23

$$\mathbf{Y}_1 = \frac{R_1}{R_1^2 + X_L^2} - j\frac{X_L}{R_1^2 + X_L^2} = g_1 - jb_1\,;\ \mathbf{Y}_2 = \frac{R_2}{R_2^2 + X_C^2} + j\frac{X_2}{R_2^2 + X_C^2} = g_2 + jb_2$$

$$\mathbf{Y} = \mathbf{Y}_1 + \mathbf{Y}_2 = g_1 - jb_1 + g_2 + jb_2 = (g_1 + g_2) + j(b_2 - b_1) = G + jB = \sqrt{(G^2 + B^2)} \angle \tan^{-1}(B/G)$$

As seen from Fig. 14.24.

$$R_{eq} = Z\cos\phi = \frac{1}{Y}\cdot\frac{G}{Y} = \frac{G}{Y^2}$$

$$X_{eq} = Z\sin\phi = \frac{1}{Y}\cdot\frac{B}{Y} = \frac{B}{Y^2}$$

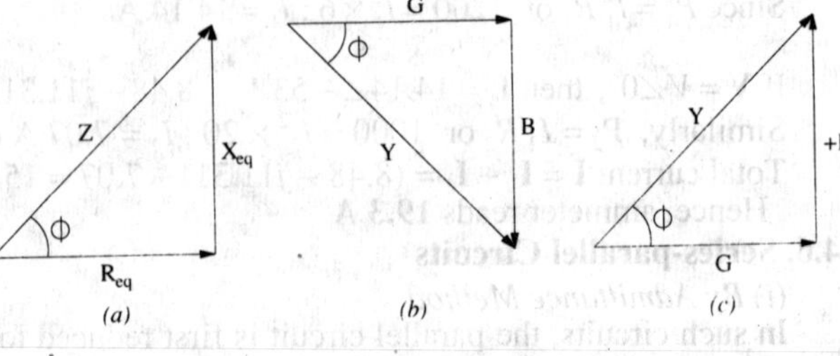

Fig. 14.24

Hence, equivalent series circuit is as shown in Fig. 14.23 (*b*) or (*c*) depending on whether net susceptance *B* is negative (inductive) or positive (capacitive). If *B* is negative, then it is an *R-L* circuit of Fig. 14.23 (*b*) and if *B* is positive, then it is an *R-C* circuit of Fig. 14.23 (*c*).

14.8. Parallel Equivalent of a Series Circuit

The two circuits will be equivalent if **Y** of Fig. 14.25 (*a*) is equal to the **Y** of the circuit of Fig. 14.25 (*b*).

Series Circuit

$$\mathbf{Y}_S = \frac{1}{R_S + jX_S}$$

$$= \frac{R_S - jX_S}{(R_S + jX_S)(R_S - jX_S)}$$

$$= \frac{R_S - jX_S}{R_S^2 + X_S^2} = \frac{R_S}{R_S^2 + X_S^2} - j\frac{X_S}{R_S^2 + X_S^2}$$

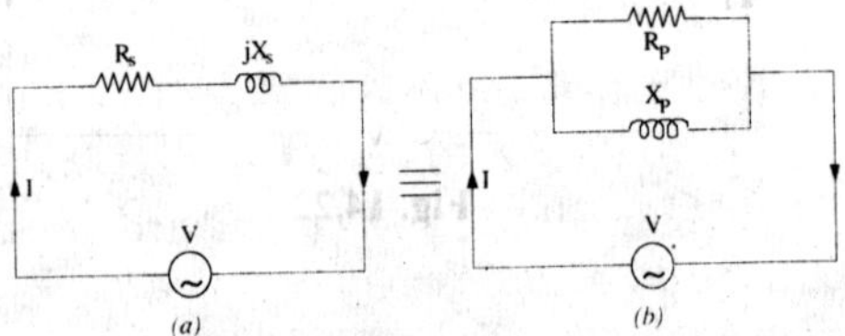

Fig. 14.25

Parallel Circuit

$$\mathbf{Y}_P = \frac{1}{R_P + j0} + \frac{1}{0 + jX_P} = \frac{1}{R_P} + \frac{1}{jX_P} = \frac{1}{R_P} - \frac{j}{X_p}$$

$$\therefore \quad \frac{R_s}{R_S^2+X_S^2} - j\frac{X_S}{R_S^2+X_S^2} = \frac{1}{R_P} - \frac{j}{X_P} \quad \therefore \quad \frac{1}{R_P} = \frac{R_S}{R_S^2+X_S^2} \text{ or } R_P = R_S + \frac{X_S^2}{R_S} = R_S\left(1+\frac{X_S^2}{R_S^2}\right)$$

Similarly $X_P = X_S + \dfrac{R_S^2}{X_S} = X_S\left(1+\dfrac{R_S^2}{X_S^2}\right)$

Example 14.19. *The admittance of a circuit is (0.03–j0.04) Siemens. Find the values of the resistance and inductive reactance of the circuit if they are joined (a) in series and (b) in parallel.*

Solution. (*a*) $\mathbf{Y} = 0.03 - j0.04$

$$\therefore \quad \mathbf{Z} = \frac{1}{\mathbf{Y}} = \frac{1}{0.03 - j0.04} = \frac{0.03 + j0.04}{0.03^2 + 0.04^2} = \frac{0.03 + j0.04}{0.0025} = 12 + j16$$

Hence, if the circuit consists of a resistance and inductive reactance in series, then resistance is **12Ω** and inductive reactance is 16 Ω as shown in Fig. 14.26.

(*b*) Conductance = 0.03 mho ∴ Resistance = 1/0.03 = 33.3 Ω

Susceptance (inductive) = 0.04 S ∴ inductive reactance = 1/0.04 = 25 Ω

Hence, if the circuit consists of a resistance connected in parallel with an inductive reactance, then resistance is **33.3 Ω** and inductive reactance is **25 Ω** as shown in Fig. 14.27.

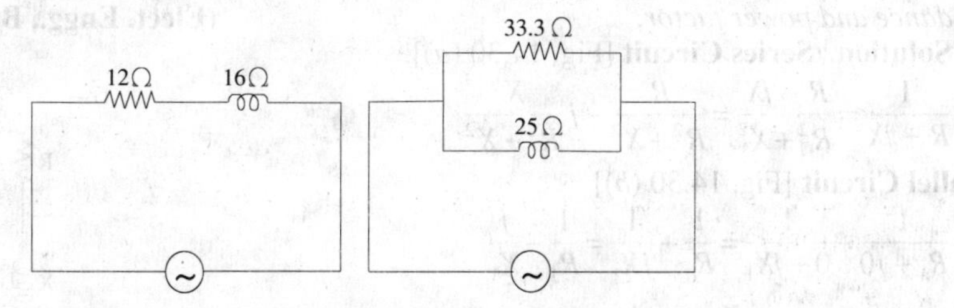

Fig. 14.26 Fig. 14.27

Example 14.20. *A circuit connected to a 115-V, 50-Hz supply takes 0.8 A at a power factor of 0.3 lagging. Calculate the resistance and inductance of the circuit assuming (a) the circuit consists of a resistance and inductance in series and (b) the circuit consists of a resistance and inductance in parallel.* **(Elect. Engg. -I, Sardar Patel Univ. 1988)**

Solution. Series Combination

$Z = 115/0.8 = 143.7\Omega$; $\cos\phi = R/Z = 0.3$ ∴ $R = 0.3 \times 143.7 = \mathbf{43.1\ \Omega}$

Now $X_L = \sqrt{Z^2 - R^2} = \sqrt{143.7^2 - 43.1^2} = 137.1\ \Omega$

∴ $L = 137.1/2\pi \times 50 = \mathbf{0.436\ H}$

Parallel Combination

Active component of current (drawn by resistance)

$= 0.8 \cos\phi = 0.8 \times 0.3 = 0.24$ A $\quad R = 115/0.24 = \mathbf{479\ \Omega}$

Quadrature component of current (drawn by inductance) = $0.8 \sin\phi = 0.8\sqrt{1 - 0.3^2} = 0.763$ A

∴ $X_L = 115/0.763\ \Omega$ ∴ $L = 115/0.763 \times 2\pi \times 50 = \mathbf{0.48\ H}$

Example 14.21. *The active and lagging reactive components of the current taken by an a.c. circuit from a 250-V supply are 50 A and 25 A respectively. Calculate the conductance, susceptance, admittance and power factor of the circuit. What resistance and reactance would an inductive coil have if it took the same current from the same mains at the same power factor ?*

(Elect. Technology, Sumbal Univ. 1987)

Solution. The circuit is shown in Fig. 14.28.

Resistance = 250/50 = 5Ω ; Reactance = 250/25 = 10 Ω

∴ Conductance $g = 1/5 = \mathbf{0.2\ S}$, Susceptance $b = -1/10 = \mathbf{-0.1\ S}$

Admittance $Y = \sqrt{g^2 + b^2} = \sqrt{0.2^2 + (-0.1)^2} = \sqrt{0.05} = \mathbf{0.224\ S}$

$\therefore$ $\mathbf{Y} = 0.2 - j0.1 = 0.224 \angle -26° 34'$. Obviously, the total current lags the supply voltage by 26°34′, p.f. = cos 26°34′ = **0.894 (lag)**

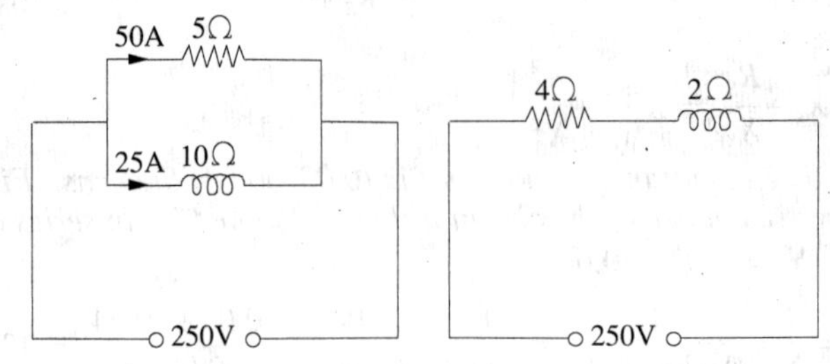

Fig. 14.28 **Fig. 14.29**

Now $\mathbf{Z} = \frac{1}{\mathbf{Y}} = \frac{1}{0.2 - j0.1} = \frac{0.2 + j0.1}{0.05} = 4 + j2$

Hence, resistance of the coil = **4 Ω** Reactance of the coil = **2 Ω** (Fig. 14.29).

Example 14.22. *The series and parallel circuits shown in Fig. 15.30 have the same impedance and the same power factor. If R = 3 Ω and X = 4 Ω, find the values of R_1 and X_1. Also, find the impedance and power factor.* **(Elect. Engg., Bombay Univ. 1988)**

Solution. Series Circuit [Fig. 14.30 (*a*)].

$$\mathbf{Y}_S = \frac{1}{R + jX} = \frac{R - jX}{R^2 + X^2} = \frac{R}{R^2 + X^2} - j\frac{X}{R^2 + X^2}$$

Parallel Circuit [Fig. 14.30 (*b*)]

$$\mathbf{Y}_P = \frac{1}{R_1 + j0} + \frac{1}{0 + jX_1} = \frac{1}{R_1} + \frac{1}{jX_1} = \frac{1}{R_1} - \frac{j}{X_1}$$

$$\therefore \frac{R}{R^2 + X^2} - j\frac{X}{R^2 + X^2} = \frac{1}{R_1} - \frac{j}{X_1}$$

$$\therefore \quad R_1 = R + X^2/R \quad \text{and} \quad X_1 = X + R^2/X$$

$$\therefore \quad R_1 = 3 + (16/3) = \mathbf{8.33\Omega}; \qquad X_1 = 4 + (9/4) = \mathbf{6.25\ \Omega}$$

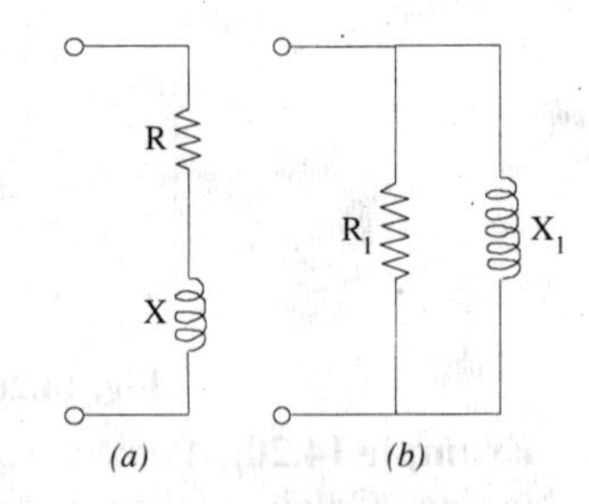

Fig. 14.30

Impedance = 3 + *j*4 = 5 ∠ 53.1° ; Power factor = cos 53.1° = **0.6 (lag)**

Example 14.23. *Find the value of the resistance R and inductance L which when connected in parallel will take the same current at the same power factor from 400-V, 50-Hz mains as a coil of resistance R_1 = 8 Ω and an inductance L_1 = 0.2 H from the same source of supply.*

Show that when the resistance R_1 of the coil is small as compared to its inductance L_1, then R and L are respectively equal to $\omega^2 L_1^2/R_1$ and L_1. **(Elect. Technology, Utkal Univ. 1985)**

Solution. As seen from Art. 14.8.

$$R = R_1 + X_1^2/R_1 \qquad \ldots(i)$$

$$X = X_1 + R_1^2/X_1 \qquad \ldots(ii)$$

Now $R_1 = 8\ \Omega \quad X_1 = 2\pi \times 50 \times 0.2 = 62.8\ \Omega$

$$\therefore \quad R = 8 + (62.8^2/8) = \mathbf{508\ \Omega}$$

$$X = 62.8 + (64/62.8) = \mathbf{63.82\ \Omega}$$

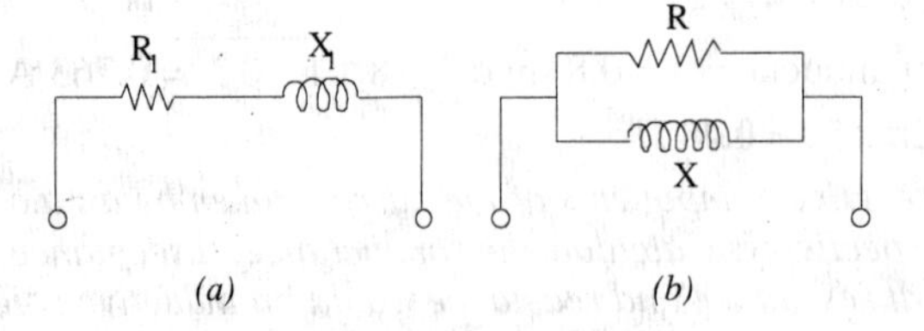

Fig. 14.31

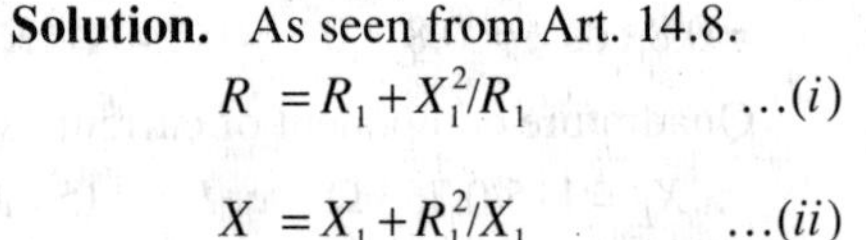

From (*i*), it is seen that if R_1 is negligible, then $R = X_1^2/R_1 = \omega^2 L_1^2/R_1$

Similarly, from (*ii*) we find that the term R_1^2/X_1 is negligible as compared to X_1,

$\therefore X = X_1$ or $L = L_1$

Example 14.24. *Determine the current drawn by the following circuit (Fig. 14.32) when a voltage of 200 V is applied across the same. Draw the phasor diagram.*

Solution. As seen from the figure

$$\mathbf{Z}_2 = 10 - j12 = 15.6\angle -50.2°;\ \mathbf{Z}_3 = 6 + j10 = 11.7\angle 58°$$

$$\mathbf{Z}_1 = 4 + j6 = 7.2\angle 56.3°; \qquad \mathbf{Z}_{BC} \frac{(10 - j12)(6 + j10)}{16 - j2} = 10.9 + j3.1 = 11.3\angle 15.9°$$

$$\mathbf{Z} = \mathbf{Z}_1 + \mathbf{Z}_{BC} = (4 + j6) + (10.9 + j3.1) = 14.9 + j9.1 = 17.5\angle 31.4°$$

Assuming $\mathbf{V} = 200\angle 0°;\ \mathbf{I} = \dfrac{\mathbf{V}}{\mathbf{Z}} = \dfrac{200\angle 0°}{17.5\angle 31.4°} = 11.4\angle -31.4°$

For drawing the phasor diagram, let us find the following quantities :

(*i*) $\mathbf{V}_{AB} = \mathbf{IZ}_1 = 11.4\angle -31.4° \times 7.2\angle 56.3° = 82.2\angle 24.9°$

$\mathbf{V}_{BC} = \mathbf{I.Z}_{BC} = 11.4\angle -31.4° \times 11.3\angle 15.9° = 128.8\angle -15.5°$

$$\mathbf{I}_2 = \frac{\mathbf{V}_{AB}}{\mathbf{Z}_2} = \frac{128.8\angle -15.5°}{15.6\angle -50.2°} = 8.25\angle 34.7°$$

$$\mathbf{I}_3 = \frac{128.8\angle -15.5°}{11.7\angle 59°} = 15.1\angle -74.5°$$

Various currents and voltages are shown in their phase relationship in Fig. 14.32 (*b*).

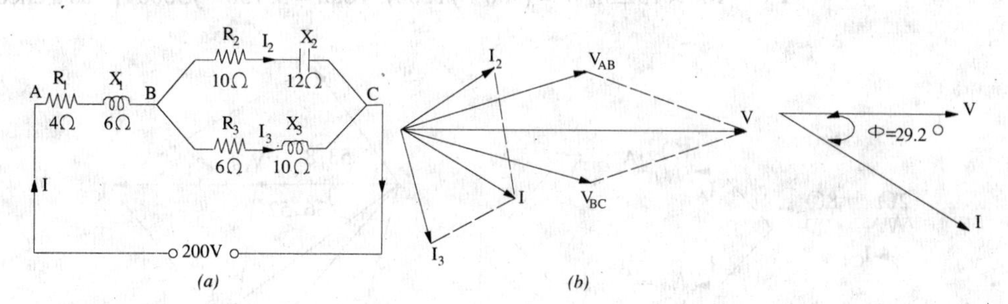

Fig. 14.32 **Fig. 14.33**

Example 14.25. *For the circuit shown in Fig. 14.33(A), find (i) total impedance (ii) total current (iii) total power absorbed and power-factor. Draw a vector diagram.*

(Elect. Tech. Osmania Univ. Jan/Feb 1992)

Solution. $Z_{BC} = (4 + j8) \| (5 - j8) = 9.33 + j0.89$

(*i*) $Z_{AC} = 3 + j6 + 9.33 + j0.89 = 12.33 + j6.89 = 14.13\angle 29.2°$

(*ii*) $I = 100/14.13\angle 29.2°$

$= 7.08\angle -29.2°$

(*iii*) $\phi = 29.2°;\ \cos\phi = 0.873;\ P = VI\cos\phi$

$= 100 \times 7.08 \times 0.873 = 618$ W.

Fig. 14.33(A)

Example 14.26. *In a series-parallel circuit, the parallel branches A and B are in series with C. The impedances are : $Z_A = (4 + j3)$; $Z_B = (4 - j16/3)$; $Z_C = (2 + j8)$ ohm.*

If the current $I_C = (25 + j0)$, draw the complete phasor diagram determining the branch currents and voltages and the total voltage. Hence, calculate the complex power (the active and reactive powers) for each branch and the whole circuit. **(Basic Electricity, Bombay Univ. 1986)**

Solution. The Circuit is shown in Fig. 14.34(*a*)

$\mathbf{Z_A} = (4 + j3) = 5\angle 36°52'$; $\quad \mathbf{Z}_B = (4 - j16/3) = 20/3\angle -53°8'$; $\mathbf{Z}_C = (2 + j8) = 8.25\angle 76°$;

$\mathbf{I}_C = (25 + j0) = 25\angle 0°$; $\mathbf{V}_C = \mathbf{I}_C\mathbf{Z}_C = 206\angle 76°$

$$\mathbf{Z_{AB}} = \frac{(4+j3)(4-j16/3)}{(8-j7/3)} = \frac{(32-j28/3)}{(8-j7/3)} = 4 + j0 = 4\angle 0°$$

$$\mathbf{V_{AB}} = \mathbf{I}_C\,Z_{AB} = 25\angle 0° \times 4\angle 0° = 100\angle 0°$$

$$\mathbf{Z} = \mathbf{Z}_C + \mathbf{Z}_{AB} = (2+j8) + (4+j0) = (6+j8) = 10\angle 53°8'; \mathbf{V} = \mathbf{I}_C\mathbf{Z} = 25\angle 0° \times 10\angle 53°8' = 250\angle 53°8'$$

$$\mathbf{I}_A = \frac{\mathbf{V}_{AB}}{\mathbf{Z}_A} = \frac{100\angle 0°}{5\angle 36°52'} = 20\angle -36°52'; \mathbf{I_B} = \frac{\mathbf{V_{AB}}}{\mathbf{Z}_B} = \frac{100\angle 0°}{(20/3)\angle -53°8'} = 15\angle 53°8'$$

Various voltages and currents are shown in Fig. 14.34 (*b*). Powers would be calculated by using voltage conjugates.

Power for whole circuit is $\mathbf{P} = \mathbf{VI}_C = 250\angle -53°8' \times 25\angle 0° = 6{,}250\angle -53°8'$

$$= 6250(\cos 53°8' - j\sin 53°8') = 3750 - j5000$$

$$P_C = 25 \times 206\angle -76° = 5150(\cos 76° - j\sin 76°) = 1250 - j5000$$

$$\mathbf{P}_A = 100 \times 20\angle -36°52' = 2000\angle -36°52' = 1600 - j1200$$

$$\mathbf{P}_B = 100 \times 15\angle 53°8' = (900 + j1200);\ \text{Total} = 3{,}750 - j5000°\ |\ (\text{as a check})$$

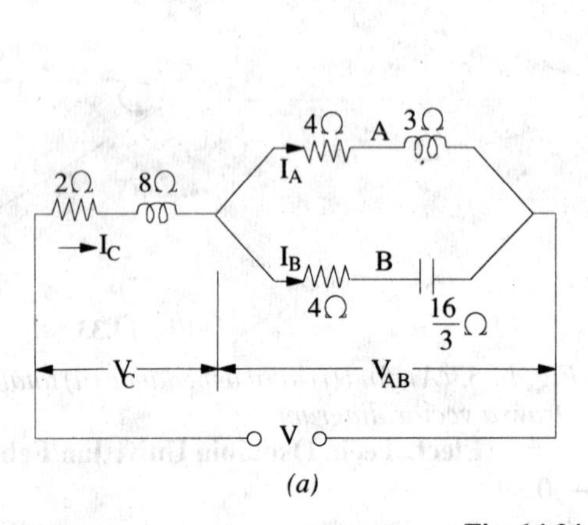

(*a*)

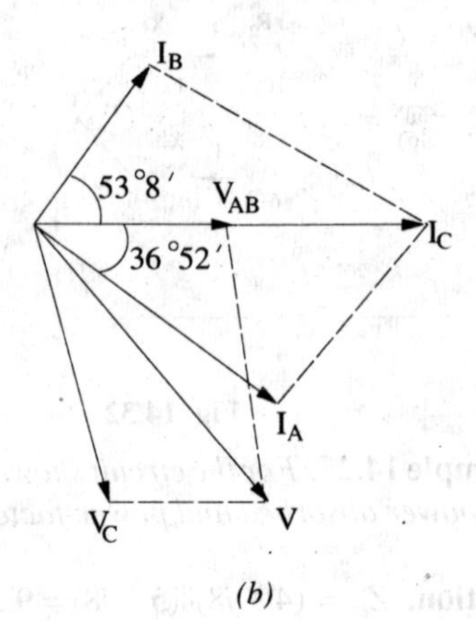

(*b*)

Fig. 14.34

Example 14.27. *Find the value of the power developed in each arm of the series-parallel circuit shown in Fig. 14.35.*

Solution. In order to find the circuit current, we must first find the equivalent impedance of the whole circuit.

$$Z_{AB} = (5 + j12) \,||\, (-j20) = \frac{(5+j12)(-j20)}{5 + j12 - j20} = \frac{13\angle 67.4° \times 20\angle -90°}{9.43\angle -58°}$$

$$= 27.57\angle 35.4° = (22.47 + j15.97)$$

$$Z_{AC} = (10 + j0) + (22.47 + j15.97) = (32.47 + j14.97)$$

$$= 36.2\angle 26.2°$$

$$I = \frac{V}{Z} = \frac{50\angle 0°}{36.2\angle 26.2°} = 1.38\angle -26.2° \text{A}$$

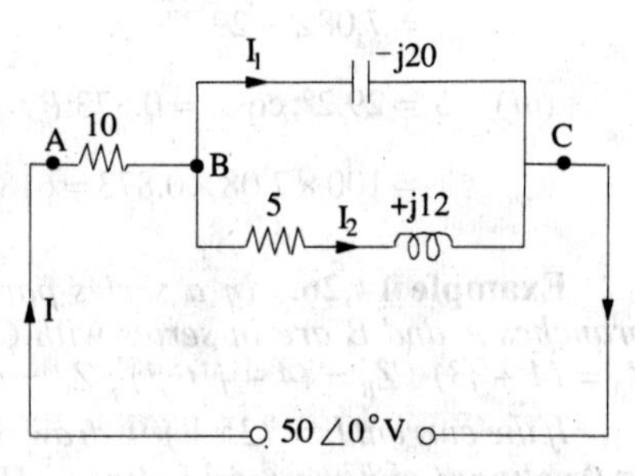

Fig. 14.35

Power developed in 10 Ω resistor = $I^2R = 1.38^2 \times 10 =$ **19 W**.

Potential difference across 10 Ω resistor is

$$IR = 1.38\angle-26.2° \times 10 = 13.8\angle-26.2° = (12.38 - j\,6.1)$$

$$V_{BC} = \text{supply voltage - drop across 10 Ω resistor}$$

$$= (50 + j0) - (12.38 - j6.1) = (37.62 + j6.1) = 38.1\angle 9.21°$$

$$I_2 = \frac{V_{BC}}{(5 + j12)} = \frac{38.1\angle 9.21°}{13\angle 67.4°} = 2.93\angle-58.2°$$

Power developed = $I_2^2 \times 5 = 2.93^2 \times 5 =$ **43 W**

No power is developed in the capacitor branch because it has no resistance.

Example 14.28. *In the circuit shown in Fig. 14.36 determine the voltage at a frequency of 50 Hz to be applied across AB in order that the current in the circuit is 10 A. Draw the phasor diagram.*

(Elect. Engg. & Electronics Bangalore Univ. 1988)

Solution. $X_{L1} = 2\pi \times 50 \times 0.05 = 15.71\ \Omega$; $X_{L2} = 2\pi \times 50 \times 0.02 = 6.28\ \Omega$;

$X_C = 1/2\pi \times 50 \times 400 \times 10^{-6} = 7.95\ \Omega$

$$\mathbf{Z}_1 = R_1 + jX_{L1} = 10 + j15.71 = 18.6\angle 57°33'$$

$$\mathbf{Z}_2 = R_2 + jX_{L2} = 5 + j6.28 = 8\angle 51°30'$$

$$\mathbf{Z}_3 = R_3 - jX_C = 10 - j7.95 = 12.77\angle 38°30'$$

$$\mathbf{Z}_{BC} = Z_2 || Z_3 (5 + j6.28) || (10 - j7.95) = 6.42 + j2.25 = 6.8\angle 19°18'$$

$$Z = Z_1 + Z_{BC} = (10 + j15.71) + (6.42 + j2.25) = 16.42 + j17.96 = 24.36\angle 47°36'$$

Let I = 10∠0°; ∴ $V = IZ = 10\angle 0° \times 24.36\angle 47°36 = 243.6\angle 47°36'$

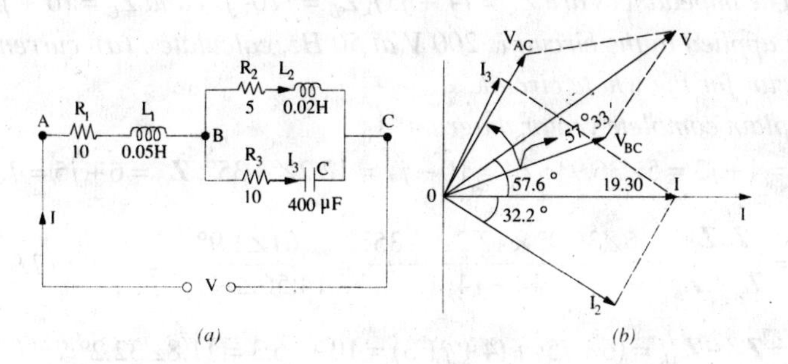

Fig. 14.36

$$V_{BC} = IZ_{BC} = 10\angle 0° \times 6.8\angle 19°18' = 68\angle 19°18';\ I_2 = \frac{V_{BC}}{Z_2} = \frac{68\angle 19°18'}{8\angle 51°30'} = 8.5\angle-32°12'$$

$$I_3 = \frac{V_{BC}}{Z_3} = \frac{68\angle 19°18'}{12.77\angle-38°30'} = 5.32\angle 57°48';\ V_{AC} = IZ_1 = 10\angle 0° \times 18.6\angle 57°33' = 186\angle 57°33'$$

The phasor diagram is shown in Fig. 14.36.

Example 14.29. *Determine the average power delivered to each of the three boxed networks in the circuit of Fig. 14.37.* **(Basic Circuit Analysis Osmania Univ. Jan/Feb 1992)**

Solution. $Z_1 = 6 - j8 = 10\angle-53.13°$; $Z_2 = 2 + j14 = 14.14\angle 81.87°$; $Z_3 = 6 - j8 = 10\angle-53.13°$

$$\dot{Z}_{23} - \frac{Z_2 Z_3}{Z_2 + Z_3} = 14.14\angle -8.13° = 14 - j2$$

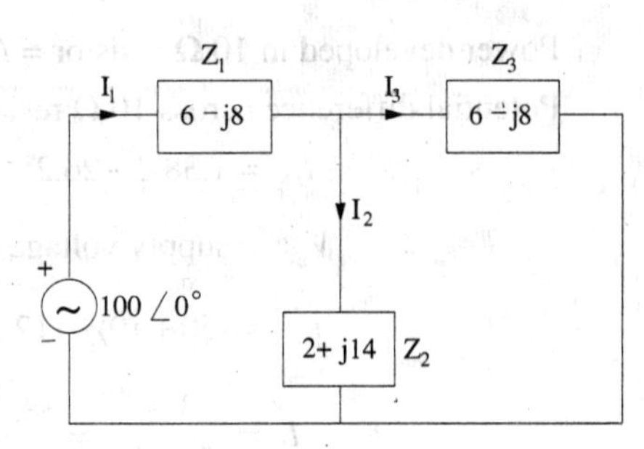

Fig. 14.37

Drop across two parallel impedances is given by

$$V_{23} = 100\frac{14 - j2}{(6 - j8) + (14 - j2)} = 63.2\angle 18.43° = 60 + j20$$

$$V_1 = 100\frac{10\angle -53.13°}{6 - j8 + (14 - j2)} = 44.7\angle -26.57° = 40 - j20$$

$$I_1 = \frac{44.7\angle -26.57°}{10\angle -53.13°} = 4.47\angle 26.56°$$

$$I_2 = \frac{63.2\angle 18.43°}{14.14\angle 81.87°} = 4.47\angle -63.44°$$

$$I_3 = \frac{63.2\ \angle 18.43°}{10\angle -53.13°} = 6.32\angle 71.56°;$$

$$P_1 = V_1 I_1 \cos\phi_1 = 44.7 \times 4.47 \times \cos 53.13° = 120\ \text{W}$$

$$P_2 = V_2 I_2 \cos\phi_2 = 63.2 \times 4.47 \times \cos 81.87° = 40\ \text{W};$$

$$P_3 = V_3 I_3 \cos\phi_3 = 63.2 \times 6.32 \times \cos 53.13° = \mathbf{240\ W}, \text{Total} = 400\ \text{W}$$

As a check, power delivered by the 100-V source is,

$P = VI_1 \cos\phi = 100 \times 4.47 \times \cos 26.56° = 400$ W

Example 14.30. *In a series-parallel circuit of Fig. 15.38 (a), the parallel branches A and B are in series with C. The impedances are $Z_A = (4 + j3)$, $Z_B = (10 - j7)$ and $Z_C = (6 + j5)\Omega$.*

If the voltage applied to the circuit is 200 V at 50 Hz, calculate : (a) current I_A, I_B and I_C; (b) the total power factor for the whole circuit.

Draw and explain complete vector diagram.

Solution. $\mathbf{Z}_A = 4 + j3 = 5\angle 36.9°$; $\mathbf{Z}_B = 10 - j7 = 12.2\angle -35°$; $\mathbf{Z}_C = 6 + j5 = 7.8\angle 39.8°$

$$\mathbf{Z}_{AB} = \frac{\mathbf{Z}_A \mathbf{Z}_B}{\mathbf{Z}_A + \mathbf{Z}_B} = \frac{5\angle 36.9° \times 12.2\angle -35°}{14 - j4} = \frac{61\angle 1.9°}{14.56\angle -16°} = 4.19\angle 17.9° = 4 + j1.3$$

$$\mathbf{Z} = \mathbf{Z}_C + \mathbf{Z}_{AB} = (6 + j5) + (4 + j1.3) = 10 + j6.3 = 11.8\angle 32.2°$$

Let $\mathbf{V} = 200\angle 0°$; $\mathbf{I}_C = (\mathbf{V}/\mathbf{Z}) = (200/11.8)\angle 32.2° = 16.35\ \angle\ 32.2°$

$$\mathbf{I}_A = \mathbf{I}_C \cdot \frac{\mathbf{Z}_B}{\mathbf{Z}_A + \mathbf{Z}_B} = 16.95°\angle -32.2° \times \frac{12.2\angle -35°}{14.56\angle -16°} = 14.0\angle -51.2°$$

$$\mathbf{I}_B = \mathbf{I}_C \cdot \frac{\mathbf{Z}_A}{\mathbf{Z}_A + \mathbf{Z}_B} = 14.95\angle -32.2° \times \frac{5\angle 36.9°}{14.56\angle -16°} = 5.82\angle 20.7°$$

The phase angle between V and total circuit current I_C is 32.2°. Hence, p.f. for the whole circuit is = cos 32.2° = **0.846 (lag)**

For drawing the phasor diagram of Fig. 14.38 (*b*) following quantities have to be calculated :

$$\mathbf{V}_{CA} = \mathbf{I}_C \mathbf{Z}_C = 17.85\angle -32.2^\circ \times 7.8\angle 39.8^\circ = 132.2\angle 7.6^\circ$$

$$\mathbf{V}_{AB} = \mathbf{I}_C \mathbf{Z}_{AB} = 17.85\angle -32.2^\circ \times 4.19\angle 17.9^\circ = 71\angle -14.3^\circ$$

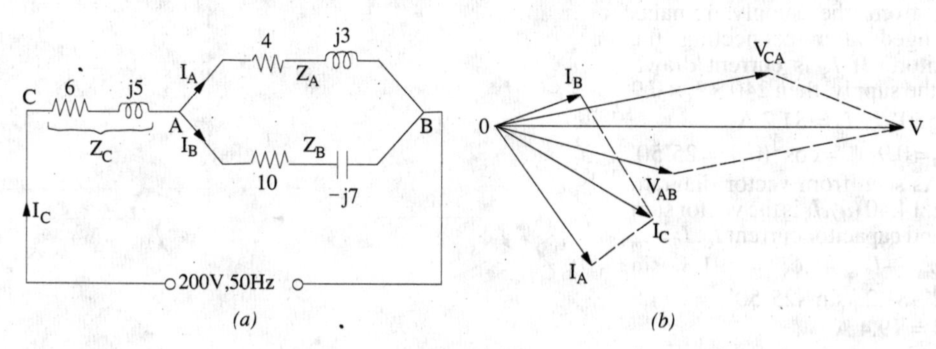

Fig. 14.38

The circuit and phasor diagrams are shown in Fig. 14.38.

Example 14.31. *A fluorescent lamp taking 80 W at 0.7 power factor lagging from a 230-V, 50-Hz supply is to be corrected to unity power factor. Determine the value of the correcting apparatus required.*

Solution. Power taken by the 80-W lamp circuit can be found from the following equation,

$$230 \times I \times 0.7 = 80 \quad \therefore \quad I = 80/230 \times 0.7 = 0.5 \text{ A}$$

Reactive component of the lamp current is = $I \sin\phi = 0.5\sqrt{1 - 0.7^2} = 0.357$ A

The power factor of the lamp circuit may be raised to unity by connecting a suitable capacitor across the lamp circuit. The leading reactive current drawn by it should be just equal to 0.357 A. In that case, the two will cancel out leaving only the in-phase component of the lamp current.

$$I_C = 0.357 \text{ A} \qquad X_C = 230/0.357 = \mathbf{645\ \Omega}$$

Now $\quad X_C = I/\omega C \quad \therefore \quad 645 = 1/2\pi \times 50 \times C; \quad C = \mathbf{4.95\ \mu F}$

Example 14.32. *For the circuit shown in Fig. 14.39, calculate I_1, I_2 and I_3. The values marked on the inductance and capacitance give their reactances.*

(Elect. Science-I Allahabad Univ. 1992)

Solution. $Z_{BC} = Z_2 || Z_3 = \dfrac{(4+j2)(1-j5)}{(4+j2)+(1-j5)} = \dfrac{14-j18}{5-j3} = \dfrac{(14-j18)(5+j3)}{5^2+3^2} = 3.65 - j1.41 = 3.9\angle 21.2^\circ$

$$Z = Z_1 + Z_{BC} = (2+j3) + (3.65 - j1.41) = 5.65 + j1.59 = 5.82\angle 74.3^\circ$$

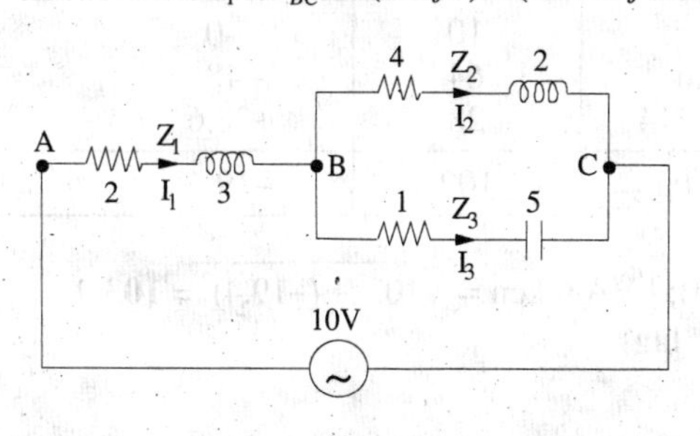

Fig. 14.39

Let $V = 10\angle 0^\circ$; $I_1 = V/Z = 10\angle 0^\circ / 5.82\angle 74.3^\circ$
$= \mathbf{1.72\angle -74.3^\circ}$

$V_{BC} = I_1 Z_{BC} = 1.72\angle -74.3^\circ \times 3.9\angle 21.2^\circ$
$= 6.7\angle -53.1$

Now, $Z_2 = 4 + j2 = 4.47\angle 63.4^\circ$;
$Z_3 = 1 - j5 = 5.1\angle -11.3^\circ$

$I_2 = V_{BC}/Z_2 = 6.7\angle -53.1^\circ / 4.47\angle 63.4^\circ$
$= \mathbf{1.5\angle 10.3^\circ}$;

$I_3 = V_{BC}/Z_3 = 6.7\angle -53.1^\circ / 5.1\angle -11.3^\circ = \mathbf{1.3\angle -41.8^\circ}$

Example 14.33. *A workshop has four 240-V, 50-Hz single-phase motors each developing 3.73 kW having 85% efficiency and operating at 0.8 power factor. Calculate the values of capacitor that would be required to change the power factor of the total load on the supply to (a) 0.9 lagging and (b) 0.9 leading. For each case, sketch a vector diagram and find the value of the supply current.*

Solution. Total motor power input $= 4 \times 3730/0.85 = 17{,}550$ W

Motor current $\quad I_m = 17{,}550/240 \times 0.8 = 91.3$ A

Motor p.f. $= \cos \phi_m = 0.8 \quad \therefore \quad \phi_m = \cos^{-1}(0.8) = 36°52'$

(*a*) Since capacitor does not consume any power, the power taken from the supply remains unchanged after connecting the capacitor. If I_S is current drawn from the supply, then $240 \times I_S \times 0.9 = 17{,}550 \therefore I_S = 81.2$ A, $\cos\phi_S = 0.9$; $\phi_S = \cos^{-1}(0.9) = 25°50'$

As seen from vector diagram of Fig. 14.40 (*a*), I_S is the vector sum of I_m and capacitor current I_C. $I_C = I_m \sin \phi_m - I_S \sin \phi_S = 91.3 \sin 36°52' - 81.2 \sin 25°50' = 54.8 - 35.4 = 19.4$ A.

Now $I_C = \omega VC$

or $19.4 = 240 \times 2\pi \times 50 \times C$

$\therefore \quad C = 257 \times 10^{-6} \text{F} = \mathbf{257 \mu F}$

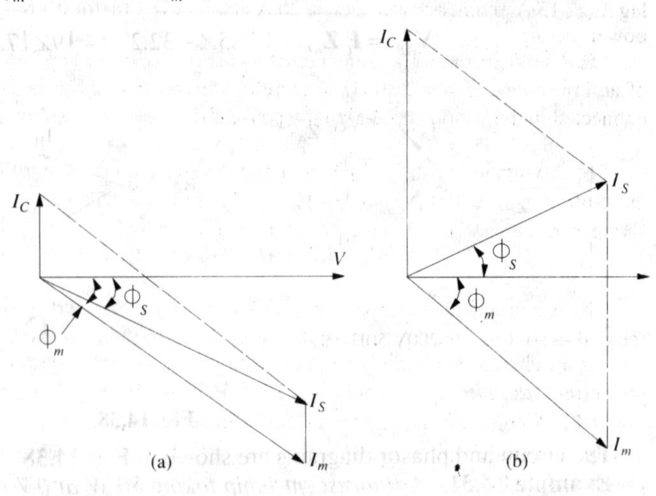

Fig. 14.40

(*b*) In this case, I_S leads the supply voltage as shown in Fig. 14.40 (*b*).

$$I_C = I_m \sin \phi_m + I_S \sin \phi_S = 54.8 + 35.4 = 90.2 \text{ A}$$

Now $I_C = \omega VC \quad \therefore \quad 90.2 = 240 \times 2\pi \times 50 \times C$

$\therefore \quad C = 1196 \times 10^{-6}$ F $= \mathbf{1196 \; \mu F}$

The line or supply current is, as before, 81.2 A (leading).

Example 14.34. *The load taken from a supply consists of (a) lamp load of 10 kW at unity power factor (b) motor load of 80 kVA at 0.8 power factor (lag) and (c) motor load of 40 kVA at 0.7 power factor leading. Calculate the total load taken from the supply in kW and in kVA and the power factor of the combined load.*

Solution. Since it is more convenient to adopt the tabular method for such questions, we will use the same as illustrated below. We will tabulate the kW, kVA and and kVAR (whether leading or lagging) of each load. The lagging kVAR will be taken as negative and leading kVAR as positive.

Load	*kVA*	*cos φ*	*sin φ*	*kW*	*kVAR*
(a)	10	1	0	10	0
(b)	80	0.8	0.6	64	−48
(c)	40	0.7	0.714	28	+28.6
			Total	102	−19.4

Total kW = **102** ; Total kVAR = −19.4 (lagging); kVA taken = $\sqrt{102^2 + (-19.4)^2} = \mathbf{103.9}$

Power factor = kW/kVA = 102/103.9 = **0.9822 (lag)**

Tutorial Problem No. 14.1

1. A capacitor of 50 μF capacitance is connected in parallel with a reactor of 22 Ω resistance and 0.07 henry inductance across 200-V, 50-Hz mains. Calculate the total current taken. Draw the vector diagram in explanation. **[4.76 A lagging, 17°12′]** (*City & Guilds, London*)

2. A non-inductive resistor is connected in series with a capacitor of 100 μF capacitance across 200-V, 50-Hz mains. The p.d. measured across the resistor is 150 V. Find the value of resistance and the value of current taken from the mains if the resistor were connected in parallel-with the capacitor instead of in series. **[R = 36.1 Ω; 8.37 Ω]** (*City & Guilds, London.*)

3. An impedance of (10 + *j*15) Ω is connected in parallel with an impedance of (6−*j*8) Ω. The total current is 15 A. Calculate the total power. **[2036 W]** (*City & Guilds, London*)

4. The load on a 250-V supply system is : 12 A at 0.8 power factor lagging; 10 A at 0.5 power factor lagging ; 15 A at unity power factor; 20 A at 0.6 power factor leading. Find (*i*) the total lead in kVA and (*ii*) its power factor. [(*i*) **10.4 kVA** (*ii*) **1.0**] (*City & Guilds London*)

5. A voltage, having a frequency of 50 Hz and expressed by **V** = 200 + *j*100 is applied to a circuit consisting of an impedance of 50 ∠ 30° Ω in parallel with a capacitance of 10 μF. Find (*a*) the reading on a ammeter connected in the supply circuit (*b*) the phase difference between the current and the voltage.

[(*a*) **4.52** (*b*) **26.6° lag**] (*London University*)

6. A voltage of 200° ∠30°V is applied to two circuits *A* and *B* connected in parallel. The current is *A* is 20 ∠ 60° A and that in *B* is 40∠ –30° A. Find the kVA and kW in each branch circuit and the main circuit. Express the current in the main circuit in the form *A* + *jB*.

[$\mathbf{kVA_A = 4, kVA_B = 8, kV_A = 12, kW_A = 3.46, kW_B = 4, kW = 7.46. I = 44.64 - j2.68}$]

(*City & Guilds, London*)

7. A coil having an impedance of (8 + *j*6) Ω is connected across a 200-V supply. Express the current in the coil in (*i*) polar and (*ii*) rectangular co-ordinate forms.

If a capacitor having a susceptance of 0.1 S is placed in parallel with the coil, find (*iii*) the magnitude of the current taken from the supply. [(*i*) **20 ∠ 36.8° A** (*ii*) **16–j12 A** (*iii*) **17.9 A**] (*City & Guilds, London*)

8. A coil-*A* of inductance 80 mH and resistance 120 Ω, is connected to a 230-V, 50 Hz single-phase supply. In parallel with it is a 16 μ F capacitor is series with a 40 Ω non-inductive resistor *B*. Determine (*i*) the power factor of the combined circuit and (*ii*) the total power taken from the supply.

[(*i*) **0.945 lead** (*ii*) **473 W**] (*London University*)

9. A choking coil of inductance 0.08 H and resistance 12 ohm, is connected in parallel with a capacitor of 120 μF. The combination is connected to a supply at 240 V, 50 Hz. Determine the total current from the supply and its power factor. Illustrate your answers with a phasor diagram.

[**3.94 A, 0.943 lag**] (*London University*)

10. A choking coil having a resistance of 20 Ω and an inductance of 0.07 henry is connected with a capacitor of 60 μF capacitance which is in series with a resistor of 50 Ω. Calculate the total current and the phase angle when this arrangement is connected to 200-V, 50 Hz mains.

[**7.15 A , 24°39′ lag**] (*City & Guilds, London*)

11. A coil of resistance 15 Ω and inductance 0.05 H is connected in parallel with a non-inductive resistance of 20 Ω. Find (*a*) the current in each branch (*b*) the total current (*c*) the phase angle of whole arrangement for an applied voltage of 200 V at 50 Hz. [**9.22 A ; 10 A ; 22.1°**]

12. A sinusoidal 50-Hz voltage of 200 V (r.m.s.) supplies the following three circuits which are in parallel : (*a*) a coil of inductance 0.03 H and resistance 3 Ω (*b*) a capacitor of 400 μF in series with a resistance of 100 Ω (*c*) a coil of inductance 0.02 H and resistance 7 Ω in series with a 300 μF capacitor. Find the total current supplied and draw a complete vector diagram. [**29.4 A**] (*Sheffield Univ. U.K.*)

13. A 50-Hz, 250-V single-phase power line has the following loads placed across it in parallel : 4 kW at a p.f. of 0.8 lagging ; 6 kVA at a p.f. of 0.6 lagging; 5 kVA which includes 1.2 kVAR leading. Determine the overall p.f. of the system and the capacitance of the capacitor which, if connected across the mains would restore the power factor to unity. [**0.884 lag ; 336 μF**]

14. Define the terms admittance, conductance and susceptance with reference to alternating current circuits. Calculate their respective values for a circuit consisting of resistance of 20 Ω, in series with an inductance of 0.07 H when the frequency is 50 Hz. [**0.0336 S, 0.0226 S, 0.0248 S**] (*City & Guilds, London*)

15. Explain the terms admittance, conductance, susceptance as applied to a.c. circuits. One branch *A*, of a parallel circuit consists of a coil, the resistance and inductance of which are 30 Ω and 0.1 *H* respectively. The other branch *B*, consists of a 100 μF capacitor in series with a 20 Ω resistor. If the combination is connected to 240-V, Hz mains, calculate (*i*) the line current and (*ii*) the power. Draw to scale a vector diagram of the supply current and the branch-circuit currents. [(*i*) **7.38 A** (*ii*) **1740 W**] (*City & Guilds, London*)

16. Find the value of capacitance which when placed in parallel with a coil of resistance 22 Ω and inductance of 0.07 H, will make it resonate on a 50-Hz circuit. [**72.33 μF**] (*City & Guilds, London*)

17. A parallel circuit has two branches. Branch *A* consists of a coil of inductance 0.2 H and a resistance of 15 Ω ; branch *B* consists of a 30 μF capacitor in series with a 10 Ω resistor. The circuit so formed is connected to a 230-V, 50-Hz supply. Calculate (*a*) current in each branch (*b*) line current and its power factor (*c*) the constants of the simplest series circuit which will take the same current at the same power factor as taken by the two branches in parallel. [**3.57 A, 2.16 A ; 1.67 A, 0.616 lag, 8.48 Ω, 0.345 H**]

18. A 3.73 kW, 1-phase, 200-V motor runs at an efficiency of 75% with a power factor of 0.7 lagging. Find (*a*) the real input power (*b*) the kVA taken (*c*) the reactive power and (*d*) the current. With the aid of a vector diagram, calculate the capacitance required in parallel with the motor to improve the power factor to 0.9 lagging. The frequency is 50 Hz. [**4.97 kW ; 7.1 kVA ; 5.07 kVAR ; 35.5 A : 212 μF**].

19. The impedances of two parallel circuits can be represented by $(20+j\,15)$ and $(10-j\,60)\ \Omega$ respectively. If the supply frequency is 50 Hz, find the resistance and the inductance or capacitance of each circuit. Also derive a symbolic expression for the admittance of the combined circuit and thence find the phase angle between the applied voltage and the resultant current. State whether this current is leading or lagging relatively to the voltage. **[20 Ω ; 0.0478 H ; 10 Ω; 53 μF ; (0.0347 – j 0.00778) S ; 12°38′ lag]**

20. One branch *A* of a parallel circuit consists of a 60-μF capacitor. The other branch *B* consists of a 30 Ω resistor in series with a coil of inductance 0.2 *H* and negligible resistance. A 140 Ω resistor is connected in parallel with the coil. Sketch the circuit diagram and calculate (*i*) the current in the 30 Ω resistor and (*ii*) the line current if supply voltage is 230-*V* and the frequency 50 Hz. [(*i*) **3.1 ∠–44°A** (*ii*) **3.1 ∠ 45°A**]

21. A coil having a resistance of 45 Ω and an inductance of 0.4 *H* is connected in parallel with a capacitor having a capacitance of 20 μ*F* across a 230-V, 50-Hz system. Calculate (*a*) the current taken from the supply (*b*) the power factor of the combination and (*c*) the total energy absorbed in 3 hours.

[(*a*) **0.615** (*b*) **0.951** (*c*) **0.402 kWh**] (*London Univ.*)

22. A series circuit consists of a resistance of 10 Ω and reactance of 5 Ω. Find the equivalent values of conductance and susceptance in parallel. **[0.08 S, 0.04 S]**

23. An alternating current passes through a non-inductive resistance *R* and an inductance *L* in series. Find the value of the non-inductive resistance which can be shunted across the inductance without altering the value of the main current. [$\omega^2 L^2/2R$] (*Elect. Meas. London Univ.*)

24. A p.d. of 200 V at 50 Hz is maintained across the terminals of a series-parallel circuit, of which the series branch consists of an inductor having an inductance of 0.15 H and a resistance of 30 Ω, one the parallel branches consists of 100-μ*F* capacitor and the other consists of a 40-Ω resistor.

Calculate (*a*) the current taken by the capacitor (*b*) the p.d. across the inductor and (*c*) the phase difference of each of these quantities relative to the supply voltage. Draw a vector diagram representing the various voltages and currents. [(*a*) **2.95A** (*b*) **210 V** (*c*) **7.25°, 26.25°**] (*City & Guilds, London*)

25. A coil (*A*) having an inductance of 0.2 *H* and resistance of 3.5 Ω is connected in parallel with another coil (*B*) having an inductance of 0.01 H and a resistance of 5 Ω. Calculate (*i*) the current and (*ii*) the power which these coils would take from a 100-V supply system having a frequency of 50-Hz. Calculate also (*iii*) the resistance and (*iv*) the inductance of a single coil which would take the same current and power.

[(*i*) **29.9 A** (*ii*) **2116 W** (*iii*) **2.365 Ω** (*iv*) **0.00752 H**] (*London Univ.*)

26. Two coils, one (*A*) having *R* = 5 Ω, *L* = 0.031 *H* and the other (*B*) having *R* = 7 Ω ; *L* = 0.023 *H*, are connected in parallel to an a.c. supply at 200 *V*, 50 Hz. Determine (*i*) the current taken by each coil and also (*ii*) the resistance and (*iii*) the inductance of a single coil which will take the same total current at the same power factor as the two coils in parallel. [(*i*) $\mathbf{I_A}$ **= 18.28 A,** $\mathbf{I_B}$ **= 19.9 A** (*ii*) **3.12 Ω** (*iii*) **0.0137 H**] (*London Univ.*)

27. Two coils are connected in parallel across 200-V, 50-Hz mains. One coil takes 0.8 kW and 1.5 kVA and the other coil takes 1.0 kW and 0.6 kVAR. Calculate (*i*) the resistance and (*ii*) the reactance of a single coil which would take the same current and power as the original circuit.

[(*i*) **10.65 Ω** (*ii*) **11.08 Ω**] (*City & Guilds, London*)

28. An a.c. circuit consists of two parallel branches, one (*A*) consisting of a coil, for which *R* = 20 Ω and *L* = 0.1 H and the other (*B*) consisting of a 40-Ω non-inductive resistor in series with a 60-μ*F* capacitor. Calculate (*i*) the current in each branch (*ii*) the line current (*iii*) the power, when the circuit is connected to 230-V mains having a frequency of 50 Hz. Calculate also (*iv*) the resistance and (*b*) the inductance of a single coil which will take the same current and power from the supply.

[(*i*) **6.15 A, 3.46 A** (*ii*) **5.89 A** (*iii*) **1235 W** (*iv*) **35.7 Ω** (*b*) **0.0509 H**)] (*London Univ.*)

29. One branch (*A*) of a parallel circuit, connected to 230-V, 50-Hz mains consists of an inductive coil (*L* = 0.15 H, *R* = 40 Ω and the other branch (*B*) consists of a capacitor (*C* = 50 μ*F*) in series with a 45 Ω resistor. Determine (*i*) the power taken (*ii*) the resistance and (*iii*) the reactance of the equivalent series circuit.

[(*i*) **946 W** (*ii*) **55.4 Ω** (*iii*) **4.6 Ω**] (*London Univ.*)

14.9. Resonance in Parallel Circuits

We will consider the practical case of a coil in parallel with a capacitor, as shown in Fig. 14.41. Such a circuit is said to be in electrical resonance when the reactive (or wattless) component of line current becomes zero. The frequency at which this happens is knwon as *resonant* frequency.

The vector diagram for this circuit is shown in Fig. 14.41 (*b*).

Net reactive or wattless component $= I_C - I_L \sin \phi_L$

As at resonance, its value is zero, hence $I_C - I_L \sin \phi_L = 0$ or $I_L \sin \phi_L = I_C$

Now, $I_L = V/Z$; $\sin \phi_L = X_L$ and $I_C = V/X_C$

Hence, condition for resonance becomes

Fig. 14.41

$$\frac{V}{Z} \times \frac{X_L}{Z} = \frac{V}{X_C} \quad \text{or} \quad X_L \times X_C = Z^2$$

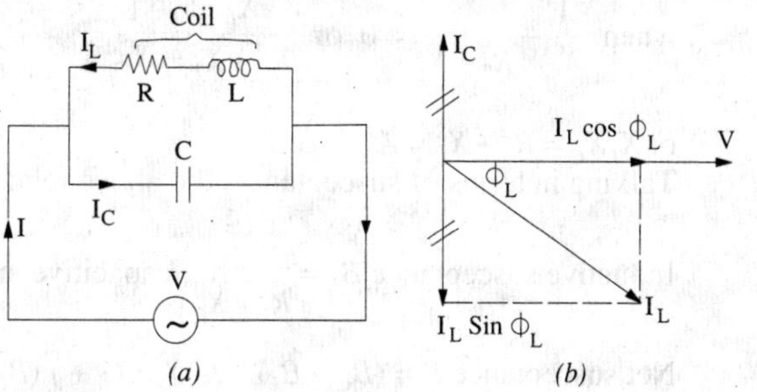

$$\text{Now, } X_L = \omega L, X_C = \frac{1}{\omega C} \quad \therefore \quad \frac{\omega L}{\omega C} = Z^2 \text{ or } \frac{L}{C} = Z^2 \quad ...(i)$$

$$\text{or } \frac{L}{C} = R^2 + X_L^2 = R^2 + (2\pi f L)^2$$

$$\text{or } (2\pi f_0 L)^2 = \frac{L}{C} - R^2 \text{ or } 2\pi f_0 = \sqrt{\frac{1}{LC} - \frac{R^2}{L^2}} \text{ or } f_0 = \sqrt{\frac{1}{LC} - \frac{R^2}{L^2}}$$

This is the resonant frequency and is given in Hz if R is in ohm, L is in henry and C is in farad.

If R is negligible, then $f_0 = \dfrac{1}{2\pi\sqrt{(LC)}}$ ...same as for series resonance

Current at Resonance

As shown in Fig. 14.41(b), since wattless component of the current is zero, the circuit current is $I = I_L \cos \phi_L = \dfrac{V}{Z} - \dfrac{R}{Z}$ or $I = \dfrac{VR}{Z^2}$.

Putting the value of $Z^2 = L/C$ from (*i*) above, we get $I = \dfrac{VR}{L/C} = \dfrac{V}{L/CR}$

The denominator L/CR is known as the *equivalent* or *dynamic impedance* of the parallel circuit at resonance. It should be noted that impedance is 'resistive' only. Since current is minimum at resonance, L/CR must, therefore, represent the maximum impedance of the circuit. In fact, parallel resonance is a condition of maximum impedance or minimum admittance.

Current at resonance is minimum, hence such a circuit (when used in radio work) is sometimes known as *rejector* circuit because it rejects (or takes minimum current of) that frequency to which it resonates. This resonance is often referred to as current resonance also because the current circulating *between* the two branches is many times greater than the line current taken from the supply.

The phenomenon of parallel resonance is of great practical importance because it forms the basis of tuned circuits in Electronics.

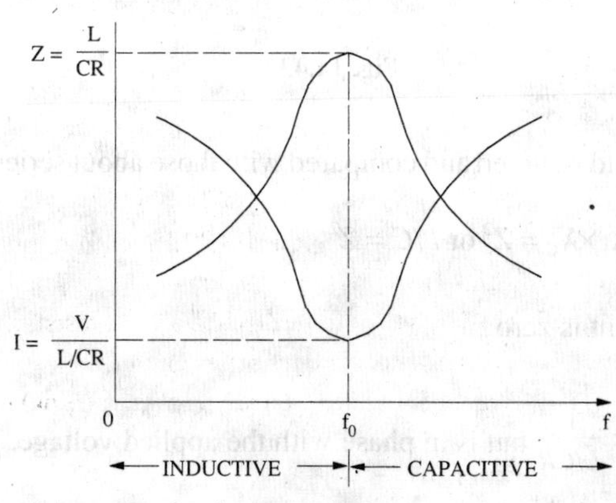

Fig. 14.42

The variations of impedance and current with frequency are shown in Fig. 14.42. As seen, at resonant frequency, impedance is maximum and equals L/CR. Consequently, current at resonance is minimum and is $= V/(L/CR)$. At off-resonance frequencies, impedance decreases and, as a result, current increases as shown.

Alternative Treatment

$$\mathbf{Y}_1 = \frac{1}{R + jX_L} = \frac{R}{R^2 + X_L^2} - j\frac{X_L}{R^2 + X_L^2}; \ \mathbf{Y}_2 = \frac{1}{-jX_C} = \frac{j}{X_C}$$

$$\mathbf{Y} = \frac{R}{R^2 + X_L^2} + j\left(\frac{1}{X_C} - \frac{X_L}{R^2 + X_L^2}\right)$$

Now, circuit would be in resonance when *j*-component of the complex admittance is zero *i.e.*

when $\frac{1}{X_C} - \frac{X_L}{R^2 + X_L^2} = 0$ or $\frac{X_L}{R^2 + X_L^2} = \frac{1}{X_C}$

or $X_L X_C = R^2 + X_L^2 = Z^2$ —as before

Talking in terms of susceptance, the above relations can be put as under :

Inductive susceptance $B_L = \frac{X_L^2}{R^2 + X_L^2}$; capacitive susceptance $B_C = \frac{1}{X_C}$

Net susceptance $B = (B_C - B_L)$ $\therefore$ $Y = G + j(B_C - B_L) = G + jB$.

The parallel circuit is said to be in resonance when $B = 0$.

$$\therefore \quad B_C - B_L = 0 \quad \text{or} \quad \frac{1}{X_C} = \frac{X_L}{R^2 + X_L^2}$$

The rest procedure is the same as above. It may be noted that at resonance, the admittance equals the conductance.

14.10. Graphic Representation of Parallel Resonance

We will now discuss the effect of variation of frequency on the susceptance of the two parallel branches. The variations are shown in Fig. 14.43.

(*i*) **Inductive susceptance** ; $b = -I/X_L = -1/2\pi fL$

It is inversely proportional to the frequency of the applied voltage. Hence, it is represented by a rectangular hyperbola drawn in the fourth quadrant ($\because$ it is assumed negative).

(*ii*) **Capacitive susceptance**

$b = 1/X_C = \omega C = 2\pi f C$

It increases with increase in the frequency of the applied voltage. Hence, it is represented by a straight line drawn in the first quadrant (it is assumed positive).

(*iii*) **Net susceptance B**

It is the difference of the two susceptances and is represented by the dotted hyperbola. At point *A*, net susceptance is zero, hence admittance is minimum (and equal to *G*). So at point *A*, line current is minimum.

Obviously, below resonant frequency (corresponding to point *A*), inductive susceptance predominates, hence line current lags behind the applied voltage. But for frequencies above the resonant frequency, capacitive susceptance predominates, hence line current leads.

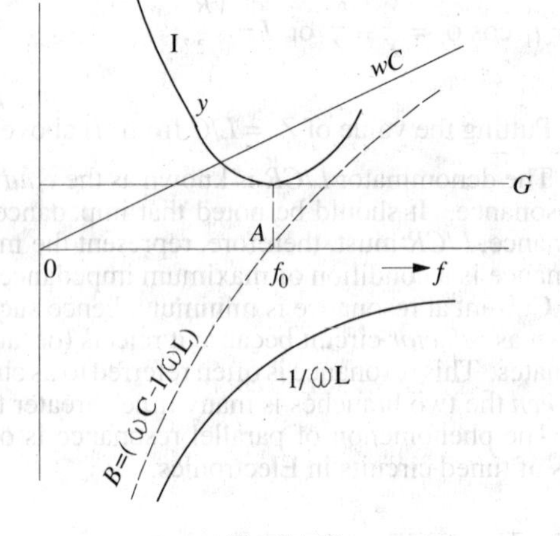

Fig. 14.43

14.11. Points to Remember

Following points about parallel resonance should be noted and compared with those about series resonance. At resonance.

1. net susceptance is zero *i.e.* $1/X_C = X_L/Z^2$ or $X_L \times X_C = Z^2$ or $L/C = Z^2$
2. the admittance equals conductance
3. reactive or wattless component of line current is zero.
4. dynamic impedance = L/CR ohm.
5. line current at resonance is minimum and $= \frac{V}{L/CR}$ but is in phase with the applied voltage.
6. power factor of the circuit is unity.

14.12. Bandwidth of a Parallel Resonant Circuit

The bandwidth of a parallel circuit is defined in the same way as that for a series circuit. This circuit also has upper and lower half-power frequencies where power dissipated is half of that at resonant frequency.

At bandwidth frequencies, the net susceptance B equals the conductance. Hence, at f_2, $B = B_{C2} - B_{L2} = G$. At f_1, $B = B_{L1} - B_{C1} = G$. Hence, $Y = \sqrt{G^2 + B^2} = \sqrt{2}.G$ and $\phi = \tan^{-1}(B/G) = \tan^{-1}(1) = 45°$.

However, at off-resonance frequencies, $Y > G$ and $B_C \neq B_L$ and the phase angle is greater than zero.

Comparison of Series and Parallel Resonant Circuits

item	*series circuit (R–L–C)*	*parallel circuit (R–L and C)*
Impedance at resonance	Minimum	Maximum
Current at resonance	Maximum = V/R	Minimum = $V/(L/CR)$
Effective impedance	R	L/CR
Power factor at resonance	Unity	Unity
Resonant frequency	$1/2\pi\sqrt{(LC)}$	$\frac{1}{2\pi}\sqrt{\left(\frac{1}{LC} - \frac{R^2}{L^2}\right)}$
It magnifies	Voltage	Current
Magnification is	$\omega L/R$	$\omega L/R$

14.13. Q-factor of a Parallel Circuit

It is defined as *the ratio of the current circulating between its two branches to the line current drawn from the supply or simply, as the current magnification.* As seen from Fig. 14.44, the circulating current between capacitor and coil branches is I_C.

Hence $Q\text{-factor} = I_C/I$

Now $I_C = V/X_C = V/(1/\omega C) = \omega CV$

and $I = V/(L/CR)$

$$\therefore\ Q - \text{factor} = \omega CV \div \frac{V}{L/CR} = \frac{\omega L}{R} = \frac{2\pi fL}{R}$$

$= \tan\phi$ (same as for series circuit)

where ϕ is the power factor angle of the *coil*.

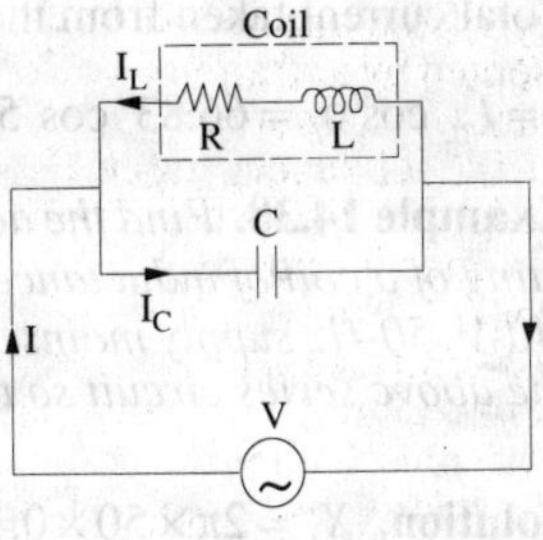

Fig. 14.44

Now, resonant frequency when R is negligible is,

$$f_0 = \frac{1}{2\pi\sqrt{(LC)}}$$

Putting this value above, we get, $Q\text{-factor} = \frac{2\pi L}{R} \times \frac{1}{2\pi\sqrt{(LC)}} = \frac{1}{R}\sqrt{\frac{L}{C}}$

It should be noted that in series circuits, Q-factor gives the voltage magnification, whereas in parallel circuits, it gives the current magnification.

Again, $$Q = 2\pi\,\frac{\text{maximum stored energy}}{\text{energy dissipated/cycle}}$$

Example 14.35. *A capacitor is connected in parallel with a coil having L = 5.52 mH and R = 10 Ω, to a 100-V, 50-Hz supply. Calculate the value of the capacitance for which the current taken from the supply is in phase with voltage.* **(Elect. Machines, A.M.I.E. Sec B, 1992)**

Solution. At resonance, $L/C = Z^2$ or $C = L/Z^2$

$X_L = 2\pi \times 50 \times 5.52 \times 10^{-3} = 1.734\ \Omega$, $Z^2 = 10^2 + 1.734^2 = 10.1\ \Omega$

$C = 5.52 \times 10^{-3}/10.1 =$ **54.6 μF**

Example 14.36. *Calculate the impedance of the parallel-tuned circuit as shown in Fig. 14.45 at a frequency of 500 kHz and for bandwidth of operation equal to 20 kHz. The resistance of the coil is 5 Ω.* **(Circuit and Field Theory, A.M.I.E. Sec B, 1993)**

Solution. At resonance, circuit impedance is L/CR. We have been given the value of R but that of L and C has to be found from the given data.

$$BW = \frac{R}{2\pi L} \quad 20 \times 10^3 = \frac{5}{2\pi \times L} \quad \text{or } L = 39 \mu\text{H}$$

$$f_0 = \frac{1}{2\pi}\sqrt{\frac{1}{LC} - \frac{R^2}{L^2}} = \frac{1}{2\pi}\sqrt{\frac{1}{39 \times 10^{-6} C} - \frac{5^2}{(39 \times 10^{-6})^2}}$$

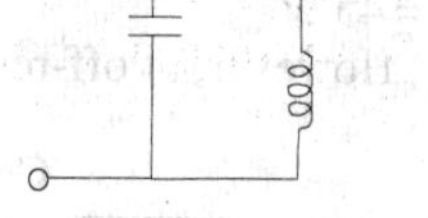

$\therefore$ $C = 2.6 \times 10^{-9}$F, $Z = L/CR = 39 \times 10^{-6}/2.6 \times 10^{-9} \times 5 = \mathbf{3 \times 10^3\ \Omega}$

Fig. 14.45

Example 14.37. *An inductive circuit of resistance 2 ohm and inductance 0.01 H is connected to a 250-V, 50-Hz supply. What capacitance placed in parallel will produce resonance ? Find the total current taken from the supply and the current in the branch circuits.* **(Elect. Engineering, Kerala Univ. 1987)**

As seen from Art. 14.9, at resonance, $C = L/Z^2$

Now, $R = 2\Omega;\ X_L = 314 \times 0.01 = 3.14\ \Omega;\ Z = \sqrt{2^2 + 3.14^2} = 3.74\ \Omega$

$$C = 0.01/3.74^2 = 714 \times 10^{-6}\text{F} = \mathbf{714\ \mu F}\ ;\ I_{RL} = 250/3.74 = \mathbf{66.83\ A}$$

$$\tan\phi_L = 3.14/2 = 1.57\ ;\ \phi_L = \tan^{-1}(1.57) = 57.5°$$

Hence, current in R-L branch lags the applied voltage by 57.5°

$$\therefore \quad I_C = \frac{V}{X_C} = \frac{V}{I/\omega C} = \omega VC = 250 \times 314 \times 714 \times 10^{-6} = \mathbf{56.1\ A}$$

This current leads the applied voltage by 90°.

Total current taken from the supply under resonant condition is

$$I = I_{RL}\cos\phi_L = 66.83\ \cos 57.5° = 66.83 \times 0.5373 = \mathbf{35.9\ A}\left(\text{or } I = \frac{V}{L/CR}\right)$$

Example 14.38. *Find the active and reactive components of the current taken by a series circuit consisting of a coil of inductance 0.1 henry and resistance 8 Ω and a capacitor of 120 μF connected to a 240-V, 50-Hz supply mains. Find the value of the capacitor that has to be connected in parallel with the above series circuit so that the p.f. of the entire circuit is unity.* **(Elect. Technology, Mysore Univ. 1986)**

Solution. $X_L = 2\pi \times 50 \times 0.1 = 31.4\ \Omega,\ X_C = I/\omega C = 1/2\pi \times 50 \times 120 \times 10^{-6} = 26.5\ \Omega$

$X = X_L - X_C = 31.4 - 26.5 = 5\ \Omega,\ Z = \sqrt{(8^2 + 5^2)} = 9.43\ \Omega;\ I = V/Z = 240/9.43 = 25.45$ A

$\cos\phi = R/Z = 8/9.43 = 0.848, \sin\phi = X/Z = 5/9.43 = 0.53$

active component of current $= I\cos\phi = 25.45 \times 0.848 = 21.58$ A

reactive component of current $= I\sin\phi = 25.45 \times 0.53 = 13.49$ A

Let a capacitor of capacitance C be joined in parallel across the circuit.

$$\mathbf{Z}_1 = R + jX = 8 + j5; \mathbf{Z}_2 = -jX_C;$$

$$\mathbf{Y} = \mathbf{Y}_1 + \mathbf{Y}_2 = \frac{1}{\mathbf{Z}_1} + \frac{1}{\mathbf{Z}_2} = \frac{1}{8 + j5} + \frac{1}{-jX_C}$$

$$= \frac{8 - j5}{89} + \frac{j}{X_C} = 0.0899 - j0.056 + \frac{j}{X_C}; = 0.0899 + j(1/X_C - 0.056)$$

For p.f. to be unity, the j-component of **Y** must be zero.

$$\therefore \quad \frac{1}{X_C} - 0.056 = 0 \text{ or } 1/X_C = 0.056 \text{ or } \omega C = 0.056 \text{ or } 2\pi \times 50C = 0.056$$

$$\therefore \quad C = 0.056/100\,\pi = 180 \times 10^{-6}F = \mathbf{180\ \mu F}$$

Example 14.39. *A coil of resistance 20 Ω and inductance 200 μH is in parallel with a variable capacitor. This combination is in series with a resistor of 8000 Ω. The voltage of the supply is 200 V at a frequency of 10^6 Hz. Calculate*

(i) the value of C to give resonance *(ii) the Q of the coil*

(iii) the current in each branch of the circuit at resonance.

Solution. The circuit is shown in Fig. 14.46.

$X_L = 2\pi fL = 2\pi \times 10^6 \times 200 \times 10^{-6} = 1256\ \Omega$

Since coil resistance is negligible as compared to its reactance, the resonant frequency is given by

$$f_0 = \frac{1}{2\pi\sqrt{LC}}$$

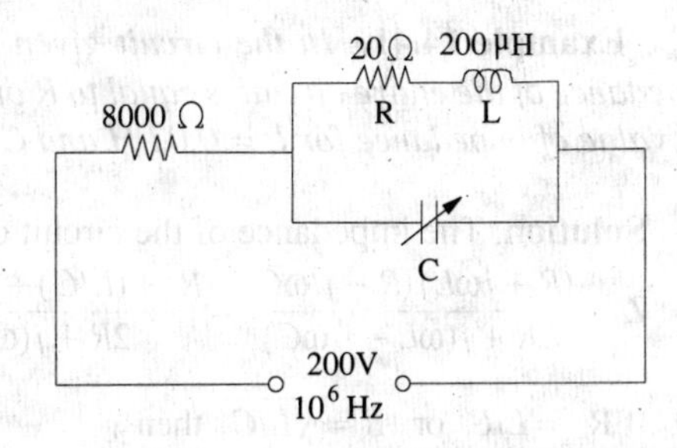

Fig. 14.46

$$\therefore \qquad 10^6 = \frac{1}{2\pi\sqrt{200 \times 10^{-6} \times C}}$$

$$(i)\ \therefore \qquad C = \mathbf{125\ \mu\mu F}$$

$$(ii) \qquad Q = \frac{2\pi fL}{R} = \frac{2\pi \times 10^6 \times 200 \times 10^{-4}}{20} = \mathbf{62.8}$$

(*iii*) Dynamic resistance of the circuit is

$$= \frac{L}{CR} = \frac{200 \times 10^{-6}}{125 \times 10^{-12} \times 20} = 80,000\ \Omega$$

Total equivalent resistance of the tuned circuit is 80,000 + 8,000 = 88,000 Ω

$\therefore$ Current $I = 200/88,000 = 2.27$ mA

p.d. across tuned circuit = current × dynamic resistance $= 2.27 \times 10^{-3} \times 80,000 = 181.6$ V

$$\text{Current through inductive branch} = \frac{181.6}{\sqrt{10^2 + 1256^2}} = 0.1445A = \mathbf{144.5\ mA}$$

Current through capacitor branch

$$= \frac{V}{1/\omega C} = \omega VC = 181.6 \times 2\pi \times 10^6 \times 125 \times 10^{-12} = \mathbf{142.7\ mA}$$

Note. It may be noted in passing that current in each branch is nearly 62.8 (*i.e.* Q-factor) times the resultant current taken from the supply.

Example 14.40. *Impedances Z_2 and Z_3 in parallel are in series with an impedance Z_1 across a 100-V, 50-Hz a.c. supply. Z_1 = (6.25 + j 1.25) ohm; Z_2 = (5 + j0) ohm and Z_3 = (5−j X_C) ohm. Determine the value of capacitance of X_C such that the total current of the circuit will be in phase with the total voltage. When is then the circuit current and power ?*

(Elect. Engg-I, Nagpur Univ, 1992)

Solution. $\mathbf{Z}_{23} = \dfrac{5(5 - jX_c)}{(10 - jX_C)}$

$$= \frac{25 - j5X_C}{(10 - jX_C)} \times \frac{10 + jX_C}{10 + jX_C} = \frac{250 + 5X_C^2}{100 + X_C^2} - j\frac{25X_C}{100 + X_C^2}$$

$$\mathbf{Z} = 6.25 + j1.25 + \frac{250 + 5X_C^2}{100 + X_C^2} - j\frac{25X_C}{100 + X_C^2}$$

$$= \left(6.25 + \frac{250 + 5X_C^2}{100 + X_C^2}\right) - j\left(\frac{25X_C}{100 + X_C^2} - \frac{5}{4}\right)$$

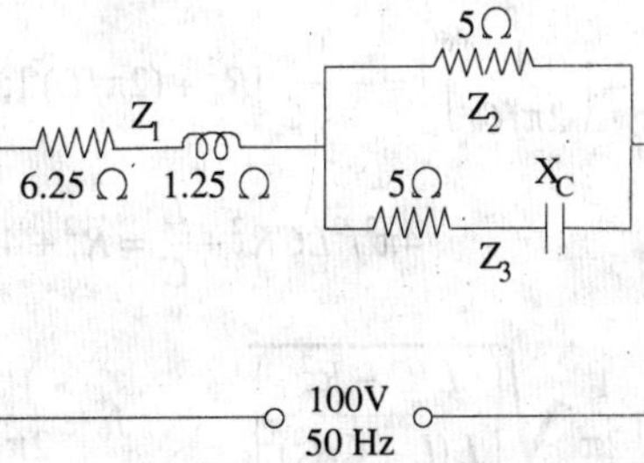

Fig. 14.47

Power factor will be unity or circuit current will be in phase with circuit voltage if the j term in the above equation is zero.

$$\therefore \quad \left(\frac{25X_C}{100+X_C^2}-\frac{5}{4}\right)=0 \text{ or } X_C=10 \quad \therefore \quad 1/\omega C = 10 \text{ or } C = 1/314\times 10 = \mathbf{318\ \mu F}$$

Substituting the value of $X_C = 10\ \Omega$ above, we get

$$Z = 10 - j0 = 10\angle 0° \text{ and } I = 100/10 = \mathbf{10\ A}\text{ ; Power } I^2R = 10^2\times 10 = \mathbf{1000\ W}$$

Example 14.41. *In the circuit given below, if the value of $R = \sqrt{L/C}$, then prove that the impedance of the entire circuit is equal to R only and is independent of the frequency of supply. Find the value of impedance for L = 0.02 H and C = 100 µF.*

(Communication System, Hyderabad Univ. 1991)

Solution. The impedance of the circuit of Fig. 14.48 is

$$\mathbf{Z} = \frac{(R+j\omega L)(R-j/\omega C)}{2R+j(\omega L-1/\omega C)} = \frac{R^2+(L/C)+jR(\omega L-1/\omega C)}{2R+j(\omega L-1/\omega C)}$$

If $R^2 = L/C$ or $R=\sqrt{L/C}$, then

$$\mathbf{Z} = \frac{R^2+R^2+jR(\omega L-1/\omega C).}{2R+j(\omega L-1/\omega C)}$$

$$= R\left[\frac{2R+j(\omega L-1/\omega C)}{2R+j(\omega L-1/\omega C)}\right] \text{ or } \mathbf{Z}=R$$

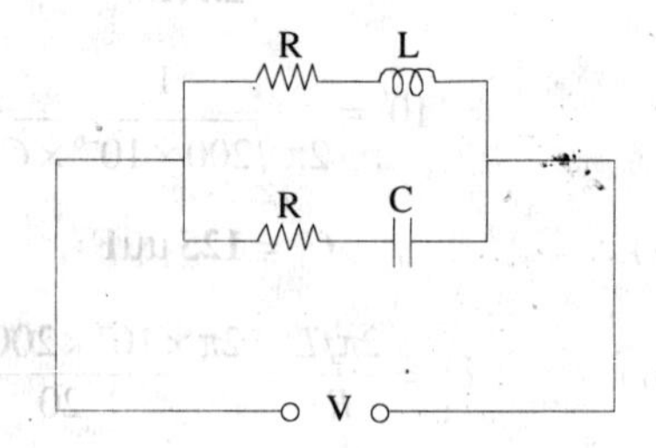

Fig. 14.48

Now, $R = \sqrt{\frac{L}{C}} = \sqrt{\frac{0.02}{100\times 10^{-6}}} = \mathbf{14.14\ \Omega}$

Example 14.42. *Derive an expression for the resonant frequency of the parallel circuit shown in Fig. 14.49.*

(Electrical Circuit, Nagpur Univ, 1993)

Solution. As stated in Art. 14.9, for resonance of a parallel circuit, total circuit susceptance should be zero. Susceptance of the *R-L* branch is

$$B_1 = -\frac{X_L}{R_1^2+X_L^2}$$

Similarly, susceptance of the *R-C* branch is

$$B_2 = \frac{X_C}{R_2^2+X_C^2}$$

Net susceptance is $B = -B_1 + B_2$

For resonance $B = 0$ or $0 = -B_1 + B_2$ $\therefore$ $B_1 = B_2$

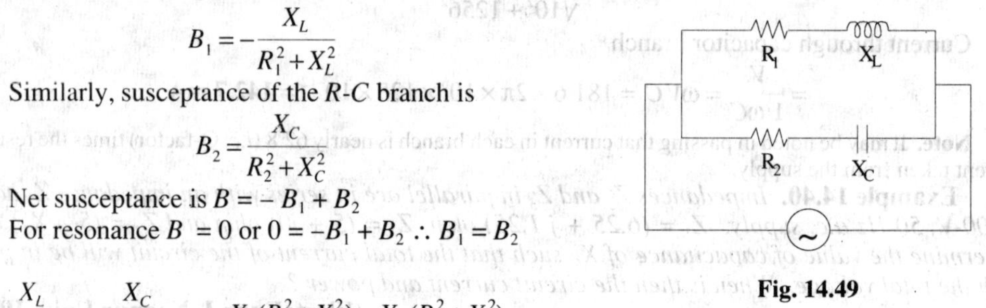

Fig. 14.49

$$\text{or} \quad \frac{X_L}{R_1^2+X_L^2} = \frac{X_C}{R_2^2+X_C^2} \quad \text{or} \quad X_L(R_2^2+X_C^2) = X_C(R_1^2+X_L^2)$$

$$2\pi fL\left[R_2^2+\left(\frac{1}{2\pi fC}\right)^2\right] = \frac{1}{2\pi fC}[R_1^2+(2\pi fL)^2]; \ 4\pi^2f^2LC\left[R_2^2+\left(\frac{1}{2\pi fC}\right)^2\right] = [R_1^2+(2\pi fL)^2]$$

$$\therefore \quad 4\pi^2f^2LCR_2^2+\frac{L}{C} = R_1^2+4\pi^2f^2L^2; \quad 4\pi^2f^2[L(L-CR_2^2)] = \frac{L}{C}-R_1^2$$

$$\therefore \quad f_0 = \frac{1}{2\pi}\sqrt{\left(\frac{L/C-R_1^2}{L(L-CR_2^2)}\right)} \quad \therefore \quad f_0 = \frac{1}{2\pi}\sqrt{\left(\frac{L-CR_1^2}{LC(L-CR_2^2)}\right)}; \ \omega_0 = \frac{1}{\sqrt{LC}}\sqrt{\left(\frac{L-CR_1^2}{L-CR_2^2}\right)}$$

Note. If both R_1 and R_2 are negligible, then $f_0 = \frac{1}{2\pi\sqrt{LC}}$ —as in Art. 14.9

Example 14.43. *Calculate the resonant frequency of the network shown in Fig. 14.50.*

Solution. Total impedance of the network between terminals *A* and *B* is

$$Z_{AB} = (R_1 || jX_L) + [R_2 || (-jX_C)] = \frac{jR_1X_L}{R_1 + jX_L} + \frac{R_2(-jX_C)}{R_2 - jX_C} = \frac{jR_1\omega L}{R_1 + j\omega L} - \frac{jR_2/\omega C}{R_2 - j/\omega C}$$

$$= \frac{R_1\omega^2L^2}{R_1^2 + \omega^2L^2} + \frac{R_2}{\omega C(R_2^2 + 1/\omega^2C^2)} + j\left[\frac{R_1^2\omega L}{R_1^2 + \omega^2L^2} - \frac{R_2^2}{\omega C(R_2^2 + 1/\omega^2C^2)}\right]$$

At resonance, $\omega = \omega_0$ and the j term of Z_{AB} is zero.

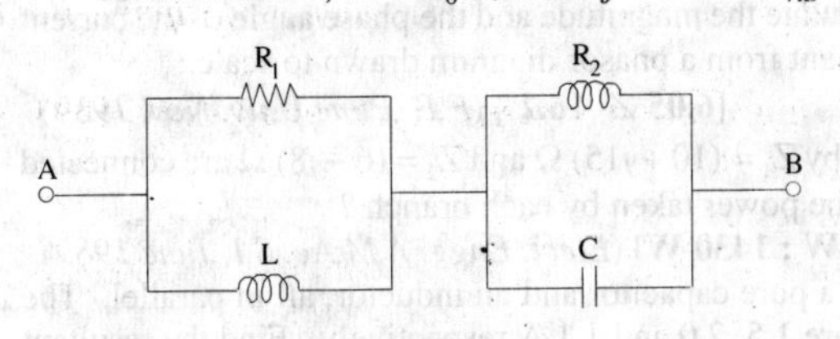

Fig. 14.50

$$\therefore \quad \frac{R_1^2\omega_0 L}{R_1^2 + \omega_0^2L^2} - \frac{R_2^2}{\omega_0 C(R_2^2 + 1/\omega_0^2C^2)} = 0$$

$$\text{or} \quad \frac{R_1^2\omega_0 L}{R_1^2 + \omega_0^2L^2} = \frac{R_2^2\omega_0 C}{R_2^2\omega_0^2C^2 + 1}$$

Simplifying the above, we get

$$\omega_0^2 = \frac{G_2^2 - C/L}{LC(G_1^2 - C/L)} \quad \text{where} \quad G_1 = \frac{1}{R_1} \text{ and } G_2 = \frac{1}{R_2}$$

The resonant frequency of the given network in Hz is

$$f_0 = \frac{\omega_0}{2\pi} = \frac{1}{2\pi}\sqrt{\frac{G^2 - C/L}{LC(G_1^2 - C/L)}}$$

Example 14.44. *Compute the value of C which results in resonance for the circuit shown in Fig. 14.51 when f = 2500/π Hz.*

Solution. $\mathbf{Y}_1 = 1/(6 + j8)$

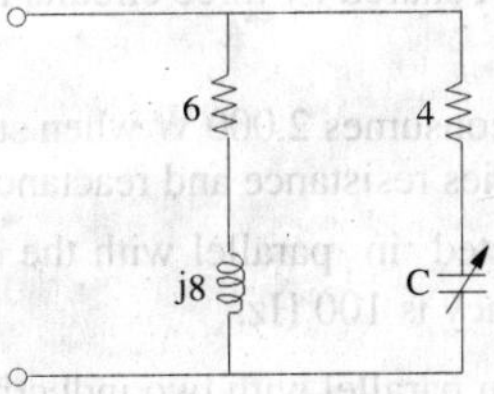

$$\mathbf{Y}_2 = 1(4 - jX_C)$$

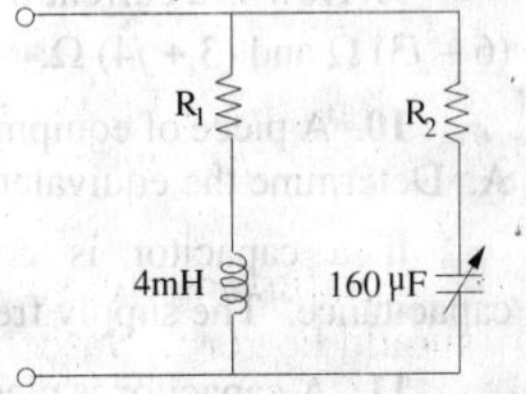

$$\mathbf{Y} = \mathbf{Y}_1 + \mathbf{Y}_2 = \frac{1}{6 + j8} = \frac{1}{4 + jX_C}$$

$$= \left(0.06 + \frac{4}{16 + X_C^2}\right) + j\left(\frac{X_C}{16 + X_C^2} - 0.08\right)$$

Fig. 14.51 **Fig. 14.52**

For resonance, j part of admittance is zero *i.e.* the complex admittance is a real number.

$$\therefore \quad X_C/(16 + X_C^2) - 0.08 = 0 \text{ or } 0.08X_C^2 - X_C + 1.28 = 0$$

$$\therefore \quad X_C = 11.05 \text{ or } 1.45 \quad \therefore \quad 1/\omega C = 11.05 \text{ or } 1.45$$

(*i*) $1/5000C = 11.05$ or $C =$ **18 μF** (*ii*) $1/5000\,C = 1.45$ or $C =$ **138 μF**

Example 14.45. *Find the values of R_1 and R_2 which will make the circuit of Fig. 14.52 resonate at all frequencies.*

Solution. As seen from Example 14.42, the resonant frequency of the given circuit is

$$\omega_0 = \frac{1}{\sqrt{LC}}\sqrt{\left(\frac{L - CR_1^2}{L - CR_2^2}\right)}$$

Now, ω_0 can assume any value provided $R_1^2 = R_2^2 = L/C$.

In the present case, $L/C = 4 \times 10^{-3}/160 \times 10^{-6} = 25$. Hence, $R_1 = R_2 = \sqrt{25} = $ **5 ohm.**

Tutorial Problem No. 14.2

1. A resistance of 20 Ω and a coil of inductance 31.8 mH and negligible resistance are connected in parallel across 230 V, 50 Hz supply. Find (*i*) The line current (*ii*) power factor and (*iii*) The power consumed by the circuit. [(*i*) **25.73 A** (*ii*) **0.44 T lag** (*iii*) **Z 246 W**] (*F.E. Pune Univ. May 1989*)

2. Two impedances $Z_1 = (150 + j157)$ ohm and $Z_2 = (100 + j\,110)$ ohm are connected in parallel across a 220-V, 50-Hz supply. Find the total current and its power factor.

[**24∠ −47° A; 0.68 (lag)**] (*Elect. Engg. & Electronics Bangalore Univ. 1988*)

3. Two impedances (14 + j5) Ω and (18 + j10) Ω are connected in parallel across a 200-V, 50-Hz supply. Determine (*a*) the admittance of each branch and of the entire circuit ; (*b*) the total current, power, and power factor and (*c*) the capacitance which when connected in parallel with the original circuit will make the resultant power factor unity.

[(*a*) **0.0634 –J0.0226), (0.0424–J0.023) (0.1058–j0.0462 S)** (*b*) **23.1 A, 4.232 kW, 0.915** (*c*) **147 μF.**]

4. A parallel circuit consists of two branches A and B. Branch A has a resistance of 10 Ω and an inductance of 0.1 H in series. Branch B has a resistance of 20 Ω and a capacitance of 100 μF in series. The circuit is connected to a single-phase supply of 250 V, 50 Hz. Calculate the magnitude and the phase angle of the current taken from the supply. Verify your answer by measurement from a phasor diagram drawn to scale.

[**6.05 ∠–15.2°**] (*F.E. Pune Univ. Nov. 1989*)

5. Two circuits, the impedances of which are given by $Z_1 = (10 + j15)$ Ω and $Z_2 = (6 - j8)$ Ω are connected in parallel. If the total current supplied is 15 A, what is the power taken by each branch ?

[**737 W ; 1430 W**] (*Elect. Engg. A.M.Ae. S.I. June 1989*)

6. A voltage of 240 V is applied to a pure resistor, a pure capacitor and an inductor, all in parallel. The resultant current is 2.3 A, while the component currents are 1.5, 2.0 and 1.1 A respectively. Find the resultant power factor and the power factor of the inductor. [**0.88 ; 0.5**]

7. Two parallel circuits comprise respectively (*i*) a coil of resistance 20 Ω and inductance 0.07 H and (*ii*) a capacitance of 60 μ*F* in series with a resistance of 50 Ω. Calculate the current in the mains and the power factor of the arrangement when connected across a 200-V, 50-Hz supply.

[**7.05 A; 0.907 lag.**] (*Elect. Engg. & Electronics, Bangalore Univ. 1987*)

8. Two circuits having the same numerical ohmic impedance are joined in parallel. The power factor of one circuit is 0.8 lag and that of the other 0.6 lag. Find the power factor of the whole circuit.

[**0.707**] (*Elect. Engg. Pune Univ. 1988*)

9. How is a current of 10 A shared by three circuits in parallel, the impedances of which are (2–j5) Ω, (6 + j3) Ω and (3 + j4) Ω. [**5.68 A; 4.57 A; 6.12 A**]

10. A piece of equipment consumes 2,000 W when supplied with 110 V and takes a lagging current of 25 A. Determine the equivalent series resistance and reactance of the equipment.

If a capacitor is connected in parallel with the equipment to make the power factor unity, find its capacitance. The supply frequency is 100 Hz. [**3.2 Ω, 3.02 Ω, 248 μF**] (*Sheffield Univ. U.K.*)

11. A capacitor is placed in parallel with two inductive loads, one of 20 A at 30° lag and one of 40° A at 60° lag. What must be the current in the capacitor so that the current from the external circuit shall be at unity power factor ? [**44.5 A**] (*City & Guilds, London*)

12. An air-cored choking coil is subjected to an alternating voltage of 100 V. The current taken is 0.1 A and the power factor 0.2 when the frequency is 50 Hz. Find the capacitance which, if placed in parallel with the coil, will cause the main current to be a minimum. What will be the impedance of this parallel combination (*a*) for currents of frequency 50 (*b*) for currents of frequency 40 ?

[**3.14 μF** (*a*) **5000 Ω** (*b*) **1940 Ω**] (*London Univ.*)

13. A circuit, consisting of a capacitor in series with a resistance of 10 Ω, is connected in parallel with a coil, having L = 55.2 mH and R = 10 Ω, to a 100-V, 50-Hz supply. Calculate the value of the capacitance for which the current taken from the supply is in phase with the voltage. Show that for the particular values given, the supply current is independent of the frequency. [**153 μF**] (*London Univ.*)

14. In a series-parallel circuit, the two parallel branches A and B are in series with C. The impedances are Z_A = (10–j8) Ω, Z_B = (9–j6) Ω and Z_C = (100 + j 0). Find the currents I_A and I_B and the phase difference between them. Draw the phasor diagram.

[$\mathbf{I_A}$ **= 12.71 ∠– 30° 58′** $\mathbf{I_B}$ **= 15∠–35°56′ ; 4°58′**] (*Elect. Engg. & Electronics Bangalore Univ. 1985*)

15. For the series-parallel circuit shown in Fig. 14.53 calculate (*a*) impedance between points *B* and *C* (*b*) total impedance of the circuit between points *A* and *C* and (*c*) the circuit current.

[(*a*) **(0.57–j0.25)** Ω; (*b*) **(0.97 + j0.55)** Ω, (*c*) **(77.8–j44.6) A, 89.7 ∠–29.8 A**]

16. Calculate the value of the current flowing in the series-parallel circuit of Fig. 14.54.

[**(2.02–j3.07) A; 3.68 ∠–56.54° A**]

17. Calculate the amount of power developed in each arm of the series parallel circuit shown in Fig. 14.55.

[**zero ; 400 W; 400 W**]

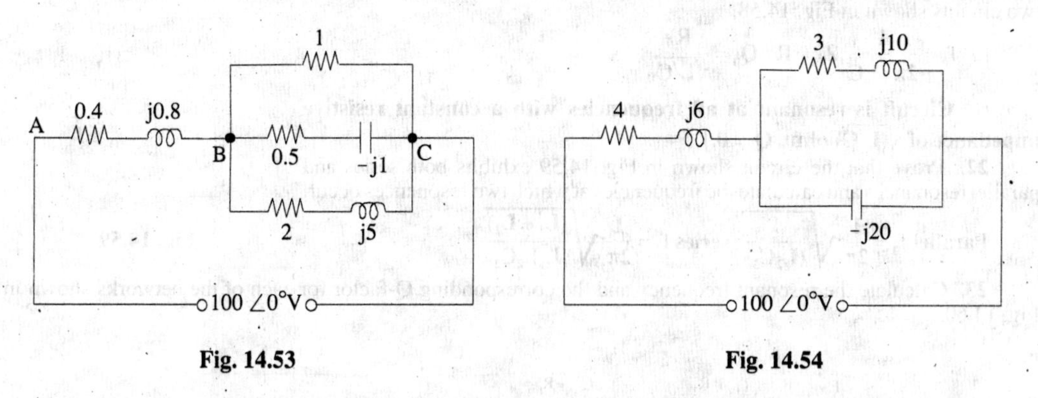

Fig. 14.53 **Fig. 14.54**

18. Calculate the power developed in each branch series of the parallel circuit shown in Fig. 14.56.
[238.4 W; 205.7 W; zero]

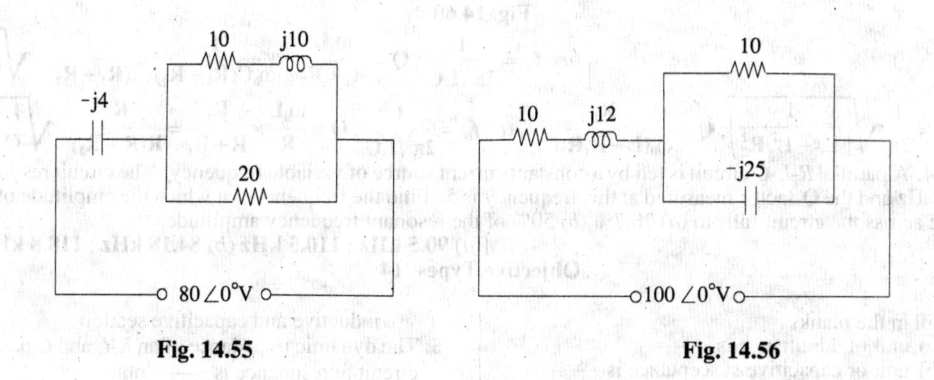

Fig. 14.55 **Fig. 14.56**

19. Find the equivalent series circuit of the 4-branch parallel circuit shown in Fig. 14.57.
[(4.41 + j2.87) Ω [A resistor of 4.415 Ω is series with a 4.57 mH inductor.]

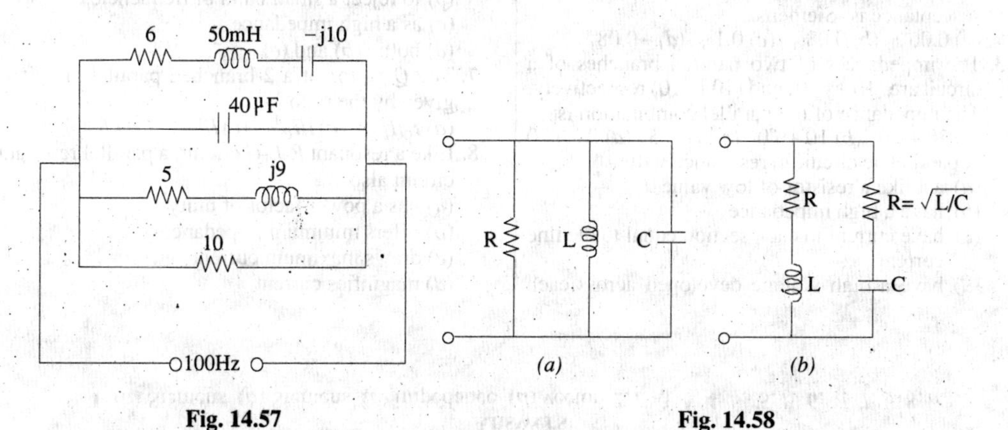

Fig. 14.57 **Fig. 14.58**

20. A coil of 20 Ω resistance has an inductance of 0.2 H and is connected in parallel with a 100-μF capacitor. Calculate the frequency at which the circuit will act as a non-inductive resistance of *R* ohms. Find also the value of *R*. **[31.8 Hz; 100 Ω]**

21. Calculate the resonant frequency, the impedance at resonance and the Q-factor at resonance for the two circuits shown in Fig. 14.58.

$$(a)\ f_0 = \frac{1}{2\pi\sqrt{LC}};Z_0 = R\ ;\ Q_0 = \frac{R}{\sqrt{L/C}}$$

(*b*) **Circuit is resonant at all frequencies with a constant resistive impedance of $\sqrt{(L/C)}$ ohm, Q = 0.]**

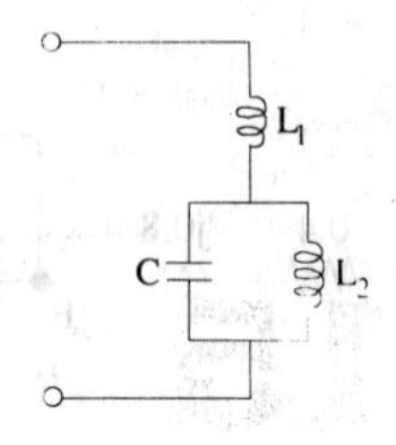

Fig. 14.59

22. Prove that the circuit shown in Fig. 14.59 exhibits both series and parallel resonances and calculate the frequencies at which two resonances occur.

$$\textbf{Parallel } f_0 = \frac{1}{2\pi}\sqrt{\frac{1}{(L_2C_2)}}\ ;\ \textbf{series } f_0 = \frac{1}{2\pi}\sqrt{\frac{(L_1+L_2)}{L_1L_2C_2}}$$

23. Calculate the resonant frequency and the corresponding Q-factor for each of the networks shown in Fig. 14.60.

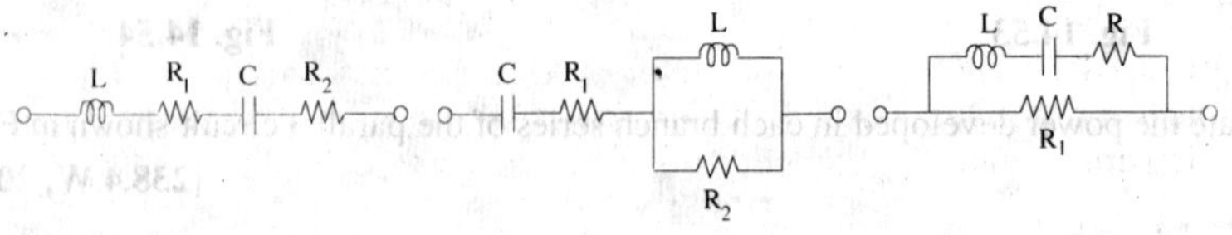

(*a*) (*b*) (*c*)

Fig. 14.60

$$\left[(a)\ f_0 = \frac{1}{2\pi\sqrt{LC}};\ Q = \frac{\omega_0 L}{R_1+R_2} = \frac{1}{\omega_0 C(R_1+R_2)} = \frac{1}{(R_1+R_2)}\cdot\sqrt{\left(\frac{L}{C}\right)}\right.$$

$$(b)\ f_0 = \frac{1}{2\pi}\sqrt{\left(\frac{1}{L_C - L^2/R_2^2}\right)};\ Q = \frac{R_2}{\omega_0(L+R_1R_2C)}\quad (c)\ f_0 = \frac{1}{2\pi\sqrt{LC}};\ Q = \frac{\omega_0 L}{R}\times\frac{R_1}{R+R_1} = \frac{R_1}{R(R+R_1)}\cdot\left.\sqrt{\frac{L}{C}}\right]$$

24. A parallel *R-L-C* circuit is fed by a constant current source of variable frequency. The circuit resonates at 100 kHz and the Q-factor measured at this frequency is 5. Find the frequencies at which the amplitude of the voltage across the circuit falls to (*a*) 70.7% (*b*) 50% of the resonant frequency amplitude.

[(*a*) 90.5 kHz ; 110.5 kHz (*b*) 84.18 kHz ; 118.8 kHz]

Objective Types–14

1. Fill in the blanks
   (*a*) unit of admittance is ——
   (*b*) unit of capacitive susceptance is ——
   (*c*) admittance equals the reciprocal of ——
   (*d*) admittance is given by the —— sum of conductance and susceptance.
2. An *R-L* circuit has **Z** = (6 + *j*8) ohm. Its susceptance is -Siemens.
   (*a*) 0.06 (*b*) 0.08 (*c*) 0.1 (*d*) −0.08
3. The impedances of two parallel branches of a circuit are (10 + *j* 10) and (10 − *j* 10) respectively. The impedance of the parallel combination is
   (*a*) 20 + *j* 0 (*b*) 10 + *j* 0 (*c*) 5 − *j* 5 (*d*) 0 − *j* 20
4. A parallel ac circuit in resonance will
   (*a*) act like a resistor of low value
   (*b*) have a high impedance
   (*c*) have current in each section equal to the line current
   (*d*) have a high voltage developed across each inductive and capacitive section
5. The dynamic impedance of an *R-L* and *C* parallel circuit at resonance is —— ohm.
   (*a*) *C/LR* (*b*) *L/CR* (*c*) *LC/R* (*d*) *R/LC*
6. A parallel resonant circuit can be used
   (*a*) to amplify certain frequencies
   (*b*) to reject a small band of frequencies
   (*c*) as a high impedance
   (*d*) both (*b*) and (*c*)
7. The *Q*-factor of a 2-branched parallel circuit is given by the ratio
   (*a*) I_C/I_L (*b*) I/I_C (*c*) I/I_L (*d*) *L/C*
8. Like a resonant *R-L-C* circuit, a parallel resonant circuit also
   (*a*) has a power factor of unity
   (*b*) offers minimum impedance
   (*c*) draws maximum current
   (*d*) magnifies current

ANSWERS

1. (*a*) Siemens (*b*) Siemens (*c*) impedance (*d*) vector **2.** *d* **3.** *b* **4.** *b*. **5.** *b* **6.** *d* **7.** *b* **8.** *a*.

15 A.C. NETWORK ANALYSIS

15.1. Introduction

We have already discussed various d.c. network theorems in Chapter 2 of this book. The same laws are applicable to a.c. networks except that instead of resistances, we have impedances and instead of taking algebraic sum of voltages and currents we have to take the phasor sum.

15.2. Kirchhoff's Laws

The statements of Kirchhoff's laws are similar to those given in Art. 2.2 for d.c. networks except that instead of algebraic sum of currents and voltages, we take phasor or vector sums for a.c. networks.

1. *Kirchhoff's Current Law.* According to this law, in any electrical network, the phasor sum of the currents meeting at a junction is zero.

In other words, $\Sigma I = 0$ ...at a junction

Put in another way, it simply means that in any electrical circuit the phasor sum of the currents flowing towards a junction is equal to the phasor sum of the currents going away from that junction.

2. *Kirchhoff's Voltage Law.* According to this law, the phasor sum of the voltage drops across each of the conductors in any closed path (or mesh) in a network plus the phasor sum of the e.m.fs. connected in that path is zero.

In other words, $\Sigma IR + \Sigma \text{e.m.f.} = 0$ ... round a mesh

Example 15.1. *Use Kirchhoff's laws to find the current flowing in each branch of the network shown in Fig. 15.1.*

Solution. Let the current distribution be as shown in Fig. 15.1 (*b*). Starting from point *A* and applying KVL to closed loop *ABEFA*, we get

$$-10(x + y) - 20\,x + 100 = 0 \quad \text{or} \quad 3\,x + y = 10 \qquad ....(i)$$

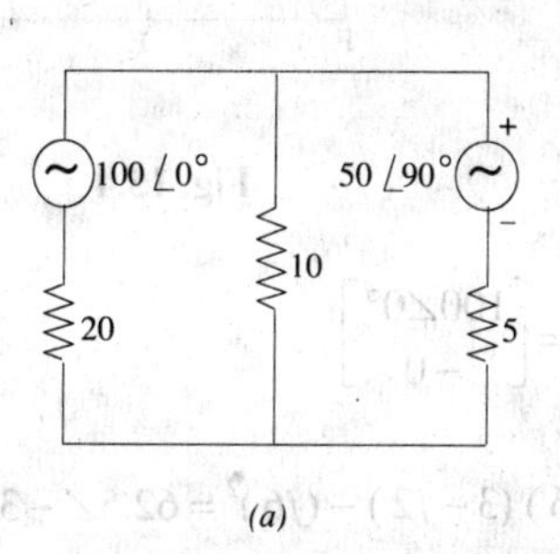

(*a*)

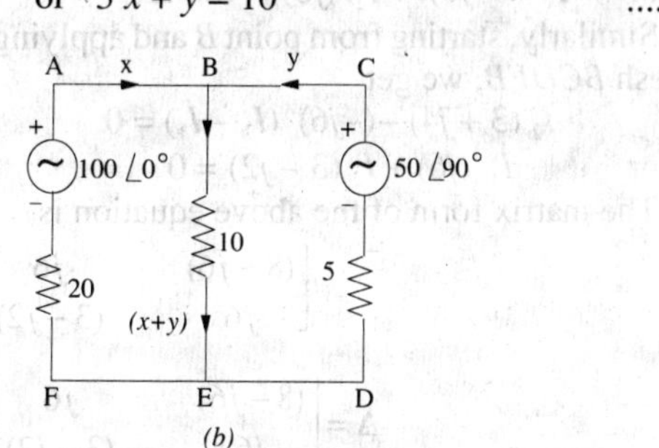

(*b*)

Fig. 15.1

Similarly, considering the closed loop BCDEB and starting from point *B*, we have

$$-50\,\angle 90° + 5y + 10\,(x + y) = 0 \quad \text{or} \quad 2x + 3y = j10 \qquad ...(ii)$$

Multiplying Eq. (*i*) by 3 and subtracting it from Eq. (*ii*), we get

$$7x = 30 - j10 \quad \text{or} \quad x = 4.3 - j1.4 = 4.52\,\angle{-18°}$$

Substituting this value of *x* in Eq. (*i*), we have

$$y = 10 - 3x = 5.95\,\angle\,\mathbf{119.15°} = -2.9 + j5.2$$

$$\therefore \quad x + y = 4.3 - j1.4 - 2.9 + j5.2 = 1.4 + j3.8$$

Tutorial Problem No. 15.1

1. Using Kirchhoff's Laws, calculate the current flowing through each branch of the circuit shown in Fig. 15.2. **[$I = 0.84\angle 47.15°$A; $I_1 = 0.7\angle -88.87°$A; $I_2 = 1.44\angle 67.12°$A]**

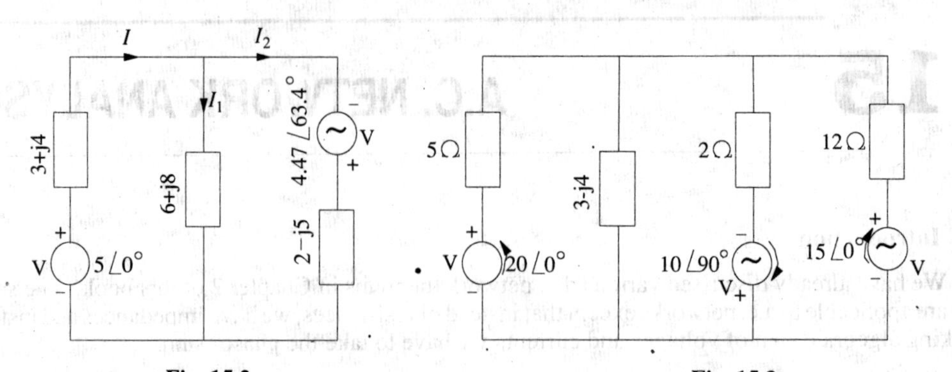

Fig. 15.2 **Fig. 15.3**

2. Use Kirchhoff's laws to find the current flowing in the capacitive branch of Fig. 15.3. **[5.87 A]**

15.3. Mesh Analysis

It has already been discussed in Art. 2.3. Sign convention regarding the voltage drops across various impedances and the e.m.f.s is the same as explained in Art. 2.3. The circuits may be solved with the help of *KVL* or by the use of determinants and Cramer's rule or with the help of impedance matrix $[\mathbf{Z_m}]$.

Example 15.2. *Find the power output of the voltage source in the circuit of Fig. 15.4. Prove that this power equals the power in the circuit resistors.*

Solution. Starting from point *A* in the clockwise direction and applying KVL to the mesh *ABEFA*, we get

$$-8\,I_1 - (-j6)\,(I_1 - I_2) + 100\angle 0° = 0$$

or $\quad I_1\,(8 - j6) + I_2\,(j6) = 100\angle 0° \quad \ldots(i)$

Similarly, starting from point *B* and applying KVL to mesh *BCDEB*, we get

$$-I_2\,(3 + j4) - (-j6)\,(I_2 - I_1) = 0$$

or $\quad I_1\,(j6) + I_2\,(3 - j2) = 0 \quad \ldots(ii)$

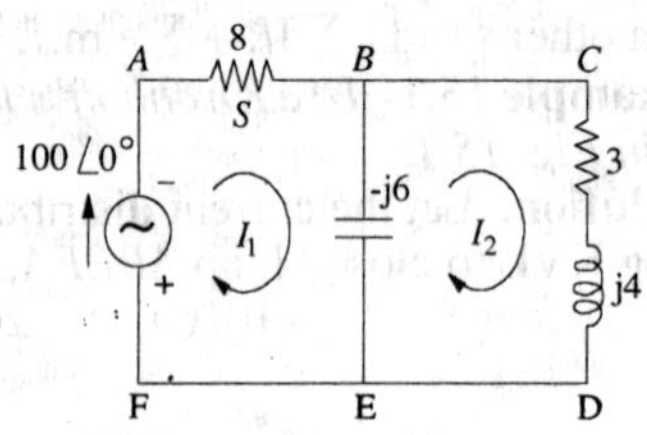

Fig. 15.4

The matrix form of the above equation is

$$\begin{bmatrix} (8-j6) & j6 \\ j6 & (3-j2) \end{bmatrix} = \begin{bmatrix} I_1 \\ I_2 \end{bmatrix} = \begin{bmatrix} 100\angle 0° \\ -0 \end{bmatrix}$$

$$\Delta = \begin{vmatrix} (8-j6) & j6 \\ j6 & (3-j2) \end{vmatrix} = (8-j6)(3-j2) - (j6)^2 = 62.5\angle -39.8°$$

$$\Delta_1 = \begin{vmatrix} 100\angle 0° & j6 \\ 0 & (3-j2) \end{vmatrix} = (300 - j200) = 360\angle -26.6°$$

$$\Delta_2 = \begin{vmatrix} (8-j6) & 100\angle 0° \\ j6 & 0 \end{vmatrix} = 600\angle 90°$$

$$I_1 = \frac{\Delta_1}{\Delta} = \frac{360\angle -26.6°}{62.5\angle -39.8°} = 5.76\angle 13.2°\,;\; I_2 = \frac{\Delta_2}{\Delta} = \frac{600\angle 90°}{62.5\angle -39.8°} = 9.6\angle 129.8°$$

Example 15.3. *Using Maxwell's loop current method, find the value of current in each branch of the network shown in Fig. 15.5(a).*

Solution. Let the currents in the two loops be I_1 and I_2 flowing in the clockwise direction as shown in Fig. 15.5 (*b*) Applying KVL to the two loops, we get

Loop No. 1

$$25-I_1\ (40+j50) - (-j100)\ (I_1-I_2) = 0$$

$$\therefore \quad 25-I_1\ (40-j50) - j100\, I_2 = 0 \qquad ...(i)$$

Loop No. 2

$$-60I_2 - (-j100)\ (I_2-I_1) = 0$$

$$\therefore \quad -j100\, I_1 - I_2\,(60-j100) = 0$$

$$\therefore \quad I_2 = \frac{-j100I_1}{(60-j100)} = \frac{100\angle-90° I_1}{116.62\angle-59°} = 0.8575\angle 31° I_1 \qquad ...(ii)$$

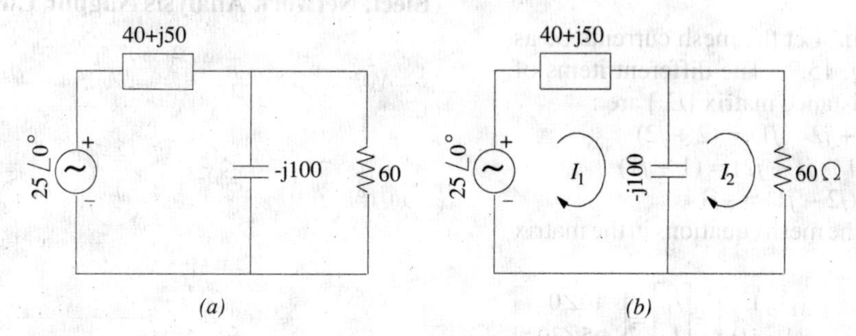

Fig. 15.5

Substituting this value of I_2 in (*i*) above, we get $25 - I_1\,(40 - j50) - j100 \times 0.8575\ \angle 31°\ I_1 = 0$

or $$25 - 40\,I_1 + j50\,I_1 - 85.75\ \angle 59°\ I_1 = 0 \quad (j100 = 100\ \angle\ 90°)$$

or $$25 - I_1\,(84.16 + j23.5\,) = 0.$$

$$\therefore \quad I_1 = \frac{25}{(84.16+j23.5)} = \frac{25}{87.38\angle 15.6°} = 0.286\angle-15.6° \text{ A}$$

Also, $$I_2 = 0.8575\ \angle - 31°\ I_1 \times 0.286\ \angle - 15.6° = 0.2452\ \angle - 46.6° \text{A}$$

Current through the capacitor = $(I_1-I_2) = 0.286\ \angle{-}\ 15.6° - 0.2452\ \angle 46.6° = 0.107 + j0.1013 = 0.1473\ \angle 43.43°$ A.

Example 15.4. *Write the three mesh current equations for the network shown in Fig. 15.6.*

Solution. While moving along I_1, if we apply KVL, we get

$$-(-j10)\,I_1 - 10(I_1-I_2) - 5\ (I_1-I_3) = 0$$

or $$I_1\,(15-j10) - 10\,I_2 - 5\,I_3 = 0 \qquad ...(i)$$

In the second loop, current through the a.c. source is flowing upwards indicating that its upper end is positive and lower is negative. As we move along I_2, we go from the positive terminal of the voltage source to its negative terminal. Hence, we experience a decrease in voltage which as per Art. would be taken as negative.

$$-j5\ I_2 - 10\ \angle 30° - 8(I_2 - I_3) - 10\ (I_2 - I_1) = 0$$

or $$10\ I_1 - I_2\,(18 + j5) + 8\,I_3 = 10\ \angle 30° \qquad ...(ii)$$

Similarly, from third loop, we get

$$-20\ \angle 0° - 5(I_3 - I_1) - 8\ (I_3 - I_2) - I_3\,(3 + j4) = 0$$

or $$5\ I_1 + 8\,I_2 - I_3\,(\,16 + j4) = 20\ \angle 0° \qquad ...(iii)$$

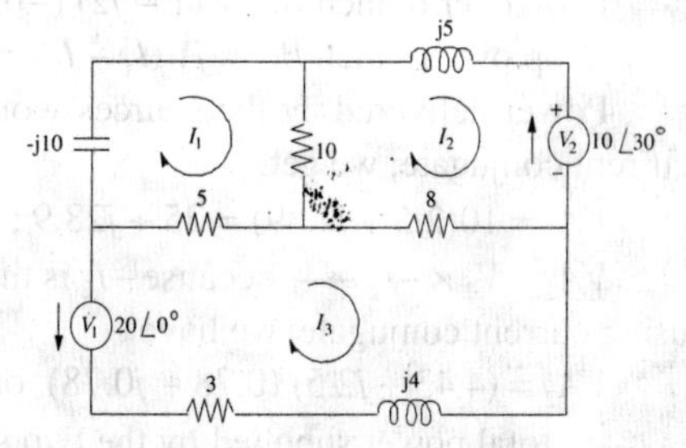

Fig. 15.6

The values of the three currents may be calculated with the help of Cramer's rule. However, the same values may be found with the help of mesh impedance $[Z_m]$ whose different items are as under :

$$Z_{11} = -j10 + 10 + 5 = (15 - j10);\ Z_{22} = (18 + j5)$$

$$Z_{33} = (16 + j5); Z_{12} = Z_{21} = -10; Z_{23} = Z_{32} = -8$$

$$Z_{13} = Z_{31} = -5; E_1 = 0 ; E_2 = -10\angle 30°; \quad E_3 = -20\angle 0°$$

Hence, the mesh equations for the three currents in the matrix form are as given below :

$$\begin{bmatrix} (15-j10) & -10 & -5 \\ -10 & (18+j5) & -8 \\ -5 & -8 & (16+j5) \end{bmatrix} = \begin{bmatrix} I_1 \\ I_2 \\ I_3 \end{bmatrix} = \begin{bmatrix} 0 \\ -10\angle 30° \\ -20\angle 0° \end{bmatrix}$$

Example 15.5. *For the circuit shown in Fig.15.7 determine the branch voltage and currents and power delivered by the source using mesh analysis.*

(Elect. Network Analysis Nagpur Univ. 1993)

Solution. Let the mesh currents be as shown in Fig. 15.7. The different items of the mesh resistance matrix $[E_m]$ are :

$Z_{11} = (2 + j1 + j2 - j1) = (2 + j2)$

$Z_{22} = (-j2 + 1 - j1 + j2) = (1 - j1)$

$Z_{12} = Z_{21} = -(j2 - j1) = -j1$

Hence, the mesh equations in the matrix form are

$$\begin{bmatrix} (2+j2) & -j1 \\ -j1 & (1-j1) \end{bmatrix} = \begin{bmatrix} I_1 \\ I_2 \end{bmatrix} = \begin{bmatrix} 10\angle 0° \\ -5\angle 30° \end{bmatrix}$$

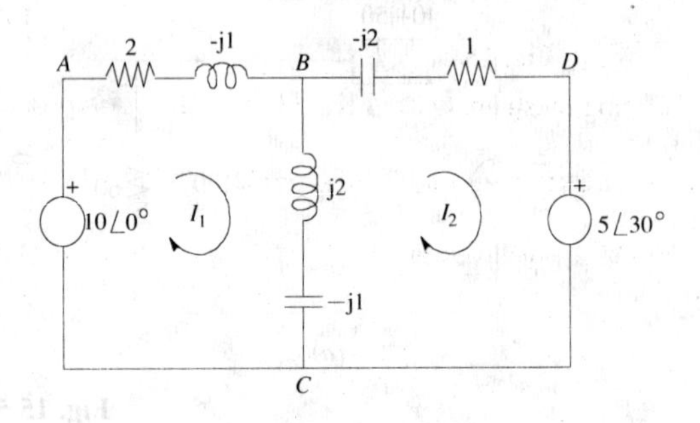

Fig. 15.7

$$\therefore \quad \Delta = (2 + j2)\,(1 - j1) + 1 = 5$$

$$\Delta_1 = \begin{bmatrix} 10 & -j1 \\ -(4.43+j2.5) & (1-j1) \end{bmatrix} = 10(1 - j1) - j1(4.43 + j2.5) = 12.5 - j14.43 = 19.1\angle -49.1°$$

$$\Delta_2 = \begin{bmatrix} (2+j2) & 10 \\ -j1 & -(4.43+j2.5) \end{bmatrix} = (2 + j2)\,(4.43 + j2.5) + j10 = -3.86 - j3.86 = 5.46\angle -135° \text{ or } \angle 225°$$

$$I_1 = \Delta_1/\Delta = 19.1\angle -49.1°/5 = 3.82\angle -49.1° = 2.5 - j2.89$$

$$I_2 = \Delta_2/\Delta = 5.46\angle -135°/5 = 1.1\angle -135° = -0.78 - j0.78$$

Current through branch $BC = I_1 - I_2 = 2.5 - j2.89 + 0.78 + j0.78 = 3.28 - j2.11 = 3.49\angle -32.75°$

Drop over branch $AB = (2 + j1)\,(2.5 - j2.89) = 7.89 - j3.28$

Drop over branch $BD = (1 - j2)\,(-0.78 - j0.78) = -2.34 + j0.78$

Drop over branch $BC = j1\,(I_1 - I_2) = j1\,(3.28 - j2.11) = 2.11 + j3.28$

Power delivered by the sources would be found by using conjugate method (Art.). Using current conjugate, we get

$VA_1 = 10(2.5 + j2.89) = 25 + j28.9$; $\therefore\ W_1 = 25$ W

$VA_2 = V_2 \times -I_2$ — because $-I_2$ is the current coming out of the second voltage source. Again, using current conjugate, we have

$VA_2 = (4.43 + j2.5)\,(0.78 - j0.78)$ or $W_2 = 4.43 \times 0.78 + 2.5 \times 0.78 = 5.4$ W

$\therefore$ total power supplied by the two sources = 25+5.4 = 30.4 W

Incidentally, the above fact can be verified by adding up the powers dissipated in the three branches of the circuit. It may be noted that there is no power dissipation in the branch BC.

Power dissipated in branch $AB = 3.82^2 \times 2 = 29.2$ W

Power dissipated in branch $BD = 1.1^2 \times 1 = 1.21$ W

Total power dissipated = 29.2 + 1.21 = 30.41 W.

Tutorial Problems No. 15.2

1. Using mesh analysis, find current in the capacitor of Fig. 15.8. **[$13.1 \angle 70.12°$ A]**

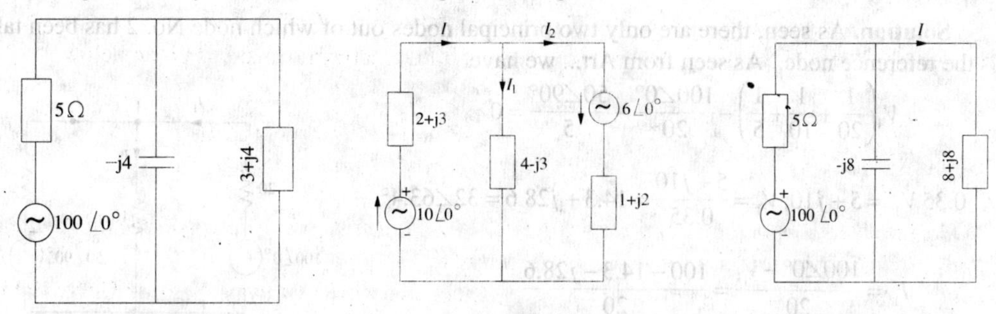

Fig. 15.9 **Fig. 15.10**

Fig. 15.8

2. Using mesh analysis or Kirchhoff's laws, determine the values of I, I_1 and I_2 (in Fig. 15.9)
[$I = 2.7 \angle -58.8°$ A; $I_1 = 0.1 \angle 97°$ A; $I_2 = 2.8 \angle -59.6°$ A]

3. Using mesh current analysis, find the value of current I and active power output of the voltage source in fig 15.10). **[$7 \angle - 50°$ A ; 645 W]**

4. Find the mesh currents I_1, I_2 and I_3 for the circuit shown in Fig. 15.11. All resistances and reactances are in ohms **[$I_1 = (1.168 + j1.281)$; $I_2 = (0.527 - j0.135)$; $I_3 = (0.718 + J0.412)$]**

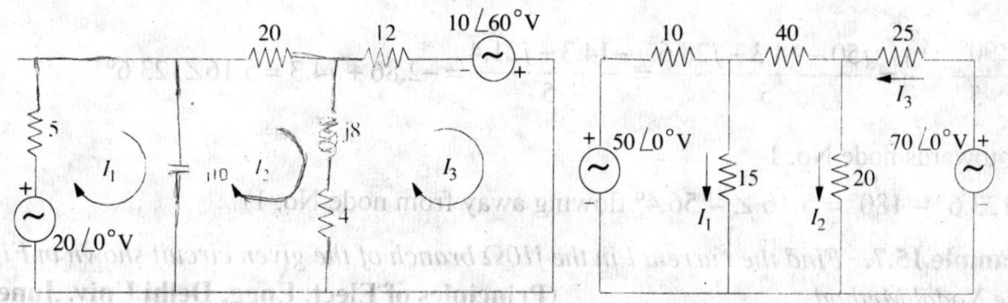

Fig. 15.11 **Fig. 15.12**

5. Find the values of branch currents I_1. I_2 and I_3 in the circuit shown in Fig. 15.12 by using mesh analysis. All resistances are in ohms. **[$I_1 = 2.008 \angle 0°$; $I_2 = 1.545 \angle 0°$; $I_3 = 1.564 \angle 0°$)**

6. Using mesh-current analysis, determine the current I_1, I_2 and I_3 flowing in the branches of the networks shown in Fig. 15.13. **[$I_1 = 8.7 \angle -1.37°$ A; $I_2 = 3 \angle -48.7°$A ; $I_3 = 7 \angle 17.25°$ A]**

7. Apply mesh-current analysis to determine the values of current I_1 to I_5 in different branches of the circuit shown in Fig. 15.14.
[$I_1 = 2.4 \angle 52.5°$A ; $I_2 = 1.0 \angle 46.18°$ A ; $I_3 = 1.4 \angle 57.17°$A ; $I_4 = 0.86 \angle 166.3°$ A ; $I_5 = 1.0 \angle 83.7°$ A]

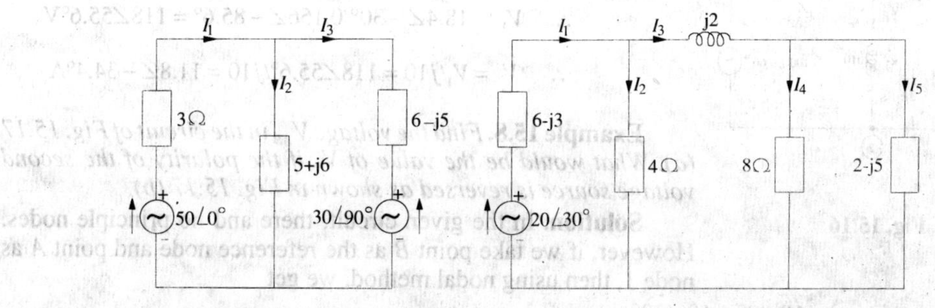

Fig. 15.13 **Fig. 15.14**

15.4. Nodal Analysis

This method has already been discussed in details in Art. 15. --. This technique is the same although we have to deal with circuit impedances rather than resistances and take phasor sum of voltages and currents rather than algebraic sum.

Example 15.6. *Use Nodal analysis to calculate the current flowing in each branch of the network shown in Fig. 15.15.*

Solution. As seen, there are only two principal nodes out of which node No. 2 has been taken as the reference node. As seen from Art... we have

$$V_1\left(\frac{1}{20}+\frac{1}{10}+\frac{1}{5}\right)-\frac{100\angle 0^\circ}{20}-\frac{50\angle 90^\circ}{5}=0$$

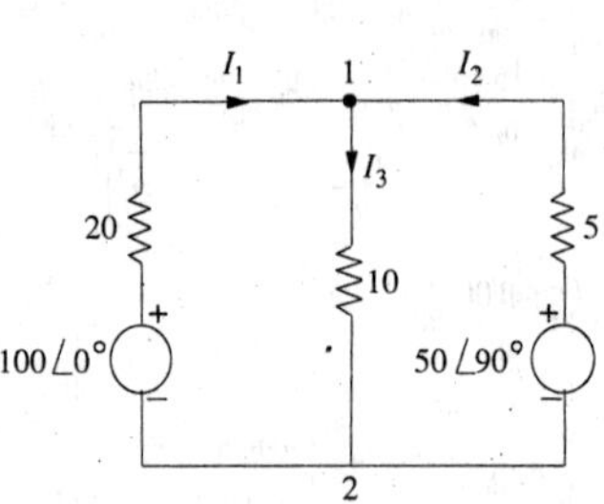

Fig. 15.15

$$\therefore\ 0.35\,V_1 = 5+j10;\ V_1=\frac{5+j10}{0.35}=14.3+j28.6=32\angle 63.4^\circ$$

$$\therefore\quad I_1=\frac{100\angle 0^\circ - V_1}{20}=\frac{100-14.3-j28.6}{20}$$

$$=4.3-j1.4=4.5\angle -18^\circ \text{ flowing towards node No. 1}$$

(or 4.5 ∠ –18° + 180° = 45∠162° flowing away from node No. 1)

$$I_3=\frac{V_1}{10}=\frac{32\angle 63.4^\circ}{10}=3.2\angle 63.4^\circ = 1.4+j2.9 \text{ flowing from node No. 1 to node No. 2}$$

$$I_2=\frac{50\angle 90^\circ - V_1}{5}=\frac{j50-14.3-j28.6}{5}=\frac{-14.3+j21.4}{5}=-2.86+j4.3=5.16\angle 123.6^\circ$$

flowing towards node No. 1

∠ 123.6° – 180° = 5.16 ∠ – 56.4° flowing away from node No. 1)

Example 15.7. *Find the current I in the j10Ω branch of the given circuit shown in Fig. 15.16 using the Nodal Method.* **(Principles of Elect. Engg. Delhi Univ. June 1985)**

Solution. There are two principal nodes out of which node No. 2 has been taken as the reference node. As per Art.

$$V_1\left(\frac{1}{6+j8}+\frac{1}{6-j8}+\frac{1}{j10}\right)-\frac{100\angle 0^\circ}{6+j8}\ \frac{100\angle -60^\circ}{6-j8}=0$$

$$V_1(0.06-j0.08+0.06+j0.08-j0.1)=6-j8+9.93-j1.2=18.4\angle -30^\circ$$

$$\therefore\quad V_1(0.12-j0.1)=18.4\angle 30^\circ \text{ or } V_1\times 0.156\angle -85.6^\circ = 18.4\angle -30^\circ$$

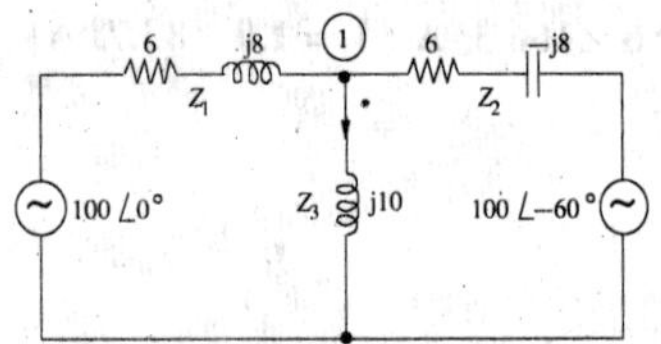

Fig. 15.16

$$\therefore\quad V_1=18.4\angle -30^\circ/0.156\angle -85.6^\circ = 118\angle 55.6^\circ \text{V}$$

$$\therefore\quad V=V_1/j10=118\angle 55.6^\circ/j10=11.8\angle -34.4^\circ \text{A}$$

Example 15.8. *Find the voltage V_{AB} in the circuit of Fig. 15.17 (a). What would be the value of V_1 if the polarity of the second voltage source is reversed as shown in Fig. 15.17 (b).*

Solution. In the given circuit, there and no principle nodes. However, if we take point *B* as the reference node and point *A* as node 1, then using nodal method, we get

$$V_1\left(\frac{1}{10}+\frac{1}{8+j4}\right)-\frac{10\angle 0^\circ}{10}-\frac{10\angle 30^\circ}{8+j4}=0$$

$$V_1\times 0.2\angle -14.1^\circ = 1+1.116+j0.066=4.48\angle 1.78^\circ$$

$$\therefore\ V_1=4.48\angle 1.78^\circ/0.2\angle -14.1^\circ = 22.4\angle 1^{\cdot}.88^\circ$$

When source polarity is Reversed

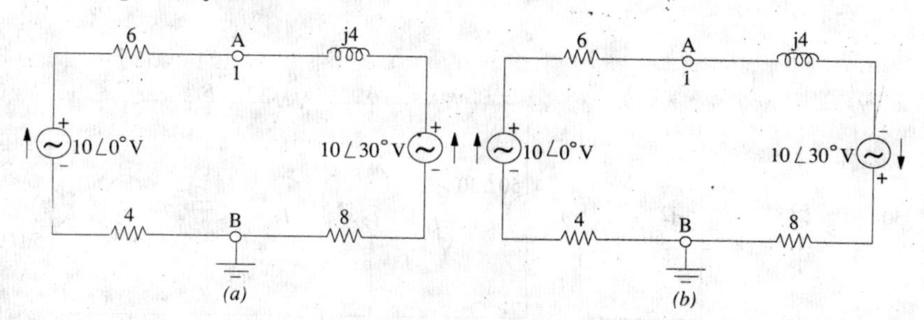

Fig. 15.17

$$V_1\left(\frac{1}{10}+\frac{1}{8+j4}\right)-\frac{10\angle 0°}{10}+\frac{10\angle 30°}{8+j4}=0 \text{ or } = V_1 = 0.09\angle 223.7°$$

Example 15.9. *Write the nodal equations for the network shown in Fig. 15.18.*

Solution. Keeping in mind the guidance given in Art. 2.10, it would be obvious that since current of the second voltage source is flowing away from node 1, it would be taken as negative. Hence, the term containing this source will become positive because it has been reversed twice. As seen, node 3 has been taken as the reference node. Considering node 1, we have

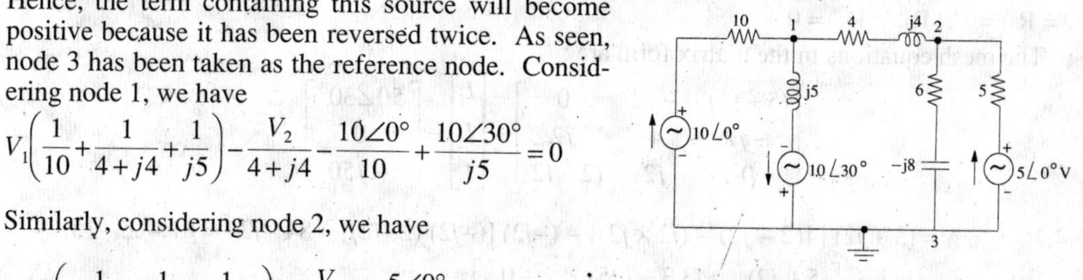

Fig. 15.18

$$V_1\left(\frac{1}{10}+\frac{1}{4+j4}+\frac{1}{j5}\right)-\frac{V_2}{4+j4}-\frac{10\angle 0°}{10}+\frac{10\angle 30°}{j5}=0$$

Similarly, considering node 2, we have

$$V_2\left(\frac{1}{4+j4}+\frac{1}{5}+\frac{1}{6-j8}\right)-\frac{V_1}{4+j4}-\frac{5\angle 0°}{5}=0,$$

Example 15.10. *In the network of Fig. 15.19 determine the current flowing through the branch of 4 Ω resistane using nodal analysis.* **(Network Analysis Nagpur Univ. 1993)**

Solution. We will find voltages V_A and V_B by using Nodal analysis and then find the current through 4 Ω resistor by dividing their difference by 4.

$$V_A\left(\frac{1}{5}+\frac{1}{4}+\frac{1}{j2}\right)-\frac{V_B}{4}-\frac{50\angle 30°}{5}=0 \quad \text{...for node A}$$

$$\therefore \quad V_A(9-j10)-5V_B = 200\angle 30° \quad \text{...(i)}$$

Similarly, from node B, we have

$$V_B\left(\frac{1}{4}+\frac{1}{2}+\frac{1}{-j2}\right)-\frac{V_A}{4}-\frac{50\angle 90°}{2}=0 \therefore V_B(3+j2)-V_A = 100\angle 90° = j100 \quad \text{...(ii)}$$

V_A can be eliminated by multiplying. Eq. (*ii*) by (9–j10) and adding the result.

$$\therefore \quad V_B(42-j12) = 1173+j1000 \text{ or } V_B = \frac{1541.4\angle 40.40°}{43.68\angle -15.9°} = 35.29\angle 56.3° = 19.58+j29.36$$

Substituting this value of V_B in Eq. (*ii*), we get

$$V_A = V_B(3+j2)-j100 = (19.58+j29.36)(3+j2)-j100 = j27.26$$

$$\therefore \quad V_A - V_B = j27.26-19.58-j29.36 = -19.58-j2.1 = 19.69\angle 186.12°$$

$$\therefore \quad I_2 = (V_A-V_B)/4 = 19.69\angle 186.12°/4 = 4.92\angle 186.12°$$

For academic interest only, we will solve the above question with the help of following two methods :

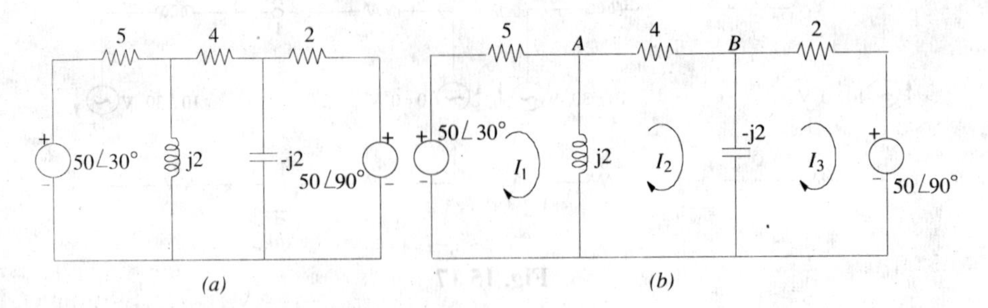

Fig. 15.19

Solution by using Mesh Resistance Matrix

Let the mesh currents I_1, I_2 and I_3 be as shown in Fig. 15.19 (*b*). The different items of the mesh resistance matrix $[R_m]$ are as under :

$R_{11} = (5 + j2)$; $R_{22} = 4$; $R_{33} = (2 - j2)$; $R_{12} = R_{21} = -j2$;

$R_{23} = R_{32} = j2$; $R_{31} = R_{13} = 0$

The mesh equations in the matrix form are :

$$\begin{bmatrix} (5+j2) & -j2 & 0 \\ -j2 & 4 & j2 \\ 0 & j2 & (2-j2) \end{bmatrix} \begin{bmatrix} I_1 \\ I_2 \\ I_3 \end{bmatrix} = \begin{bmatrix} 50\angle 30° \\ 0 \\ -j50 \end{bmatrix}$$

$$\Delta = (5 + j2)\,[4(2 - j2) - (j2 \times j2)] - (-j2)\,[(-j2)\,(2-j2)] = 84 - j24 = 87.4\angle -15.9°$$

$$\Delta_2 = \begin{bmatrix} (5+j2) & (43.3+j25) & -0 \\ -j2 & 0 & j2 \\ 0 & -j50 & (2-j2) \end{bmatrix} = (5+j2)\,[-j2(-j50)] +$$

$$j2[43.3+25)\,(2-j2)] = -427 + j73 = 433\angle 170.3°$$

$$\therefore \quad I_2 = \Delta_2/\Delta = 433\angle 170.3°/87.4\angle -15.9° = 4.95\angle 186.2°$$

Solution by using Thevenin's Theorem

When the 4 Ω resistor is disconnected, the given figure becomes as shown in Fig. 15.20 (*a*). The voltage V_A is given by the drop across *j*2 reactance. Using the voltage-divider rule, we have

$$V_A = 50\angle 30° \times \frac{j2}{5+j2} = 18.57\angle 98.2° = -2.65 + j18.38$$

Similarly,
$$V_B = 50\angle 90° \frac{-j2}{2-j2} = 35.36\angle 45° = 25 + j25$$

$$\therefore \quad V_{th} = V_A - V_B = -2.65 + 18.38 - 25 - j25 = 28.43\angle 193.5°$$

$$R_{th} = 5||j2 + 2||(-j2) = \frac{j10}{5+j2} + \frac{-j4}{2-j2} + 1.689 + j0.72$$

The Thevenin's equivalent circuit consists of a voltage source of 28.43 ∠ 193.5° V and an impedance of (1.689 + *j*0.72) Ω as shown in Fig. 15.20 (*c*). Total resistance is 4 + (1.689 + *j*0.72) = 5.689 + j 0.72 = 5.73∠7.2°. Hence, current through the 4 Ω resistor is = 28.43 ∠193.5°/5.73∠7.20° = 4.96 ∠ 186.3°.

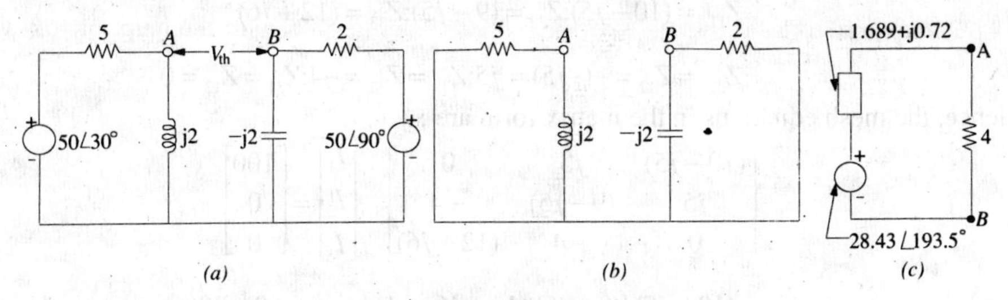

Fig. 15.20

Note. The slight variations in the answers are due to the approximations made during calculations.

Example 15.11. *Using any suitable method, calculate the current through 4 ohm resistance of the network shown in Fig. 15.21.* **(Network Analysis AMIE Sec. B Summer 1990)**

Solution. We will solve this question with the help of (*i*) Kirchhoff's laws (*ii*) Mesh analysis and (*iii*) Nodal analysis.

(*i*) Solution by using Kirchhoff's Laws

Let the current distribution be as shown in Fig. 15.21(*b*). Using the same sign convention as given in Art. we have

First Loop $-10(I_1 + I_2 + I_3) - (-j5)\, I_1 + 100 = 0$

or $I_1\,(10 - j5) + 10I_2 + 10I_3 = 100$ ...(*i*)

Second Loop $-5(I_2 + I_3) - 4I_2 + (-j\,5)\, I_1 = 0$

or $j5I_1 + 9\,I_2 + 5I_3 = 0$ ...(ii)

Third Loop $-I_3\,(8 + j6) + 4I_2 = 0$

or $0I_1 + 4I_2 - I_3\,(8 + j6) = 0$ ...(*iii*)

The matrix form of the above three equations is

$$\begin{bmatrix} (10-j5) & -10 & 10 \\ j5 & 9 & 5 \\ 0 & 4 & -(8+j6) \end{bmatrix} \begin{bmatrix} I_1 \\ I_2 \\ I_3 \end{bmatrix} = \begin{bmatrix} 100 \\ 0 \\ 0 \end{bmatrix}$$

$$\Delta = (10-j5)\,[-9(8+j6)-20] - j5[-10(8+j6)-40]$$

$$= -1490 + j520 = 1578\angle 160.8°$$

Since we are interested in finding I_2 only, we will calculate the value of Δ_2.

$$\Delta_2 = \begin{bmatrix} (10-j5) & 100 & 10 \\ j5 & 0 & 5 \\ 0 & 0 & -(8+j6) \end{bmatrix}$$

$$= -j5(-800 - j600) = -3000 + j4000 = 5000\angle 126.9°$$

$$I_2 = \frac{\Delta_2}{\Delta} = \frac{500.0\angle 126.9°}{1578\angle 160.8°} = 3.17\angle -33.9°\text{ A}$$

(*ii*) Solution by using Mesh Impedance Matrix

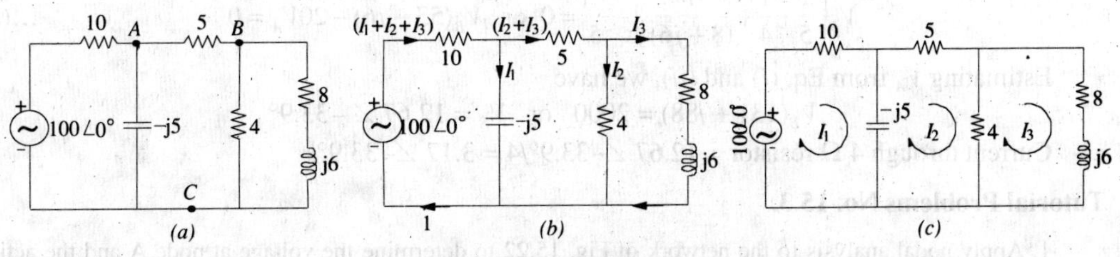

Fig. 15.21

Let the mesh currents I_1, I_2 and I_3 be as shown in Fig. 15.21 (*c*). From the inspection of Fig. 15.21 (*c*), the different items of the mesh impedance matrix $[Z_m]$ are as under:

$$Z_{11} = (10 - j5); Z_{22} = (9 - j5); Z_{33} = (12 + j6)$$

$$Z_{21} = Z_{12} = -(-j5) = j5; Z_{23} = Z_{32} = -4; Z_{31} = Z_{13} = 0$$

Hence, the mesh equations in the matrix form are :

$$\begin{bmatrix} (10-j5) & j5 & 0 \\ j5 & (9-j5) & -4 \\ 0 & -4 & (12+j6) \end{bmatrix} \begin{bmatrix} I_1 \\ I_2 \\ I_3 \end{bmatrix} = \begin{bmatrix} 100 \\ 0 \\ 0 \end{bmatrix}$$

$$\therefore \quad \Delta = (10 - j5)[(9 - j5)(12 + j6) - 16] - j5(j60 - 30)$$

$$= 1490 - j520 = 1578\angle 19.2°$$

It should be noted that the current passing through 4 Ω resistance is the vector difference ($I_2 - I_3$). Hence, we will find I_2 and I_3 only.

$$\Delta_2 = \begin{bmatrix} (10-j5) & 100 & 0 \\ j5 & 0 & -4 \\ 0 & 0 & (12+j6) \end{bmatrix} = -j5(1200 + j600) = 3000 - j6000 = 6708\angle -63.4°$$

$$\Delta_2 = \begin{bmatrix} (10-j5) & j5 & 100 \\ j5 & (9-j5) & 0 \\ 0 & -4 & 0 \end{bmatrix} = -j5(400) = -j2000 = 2000\angle -90°$$

$$\therefore \quad I_2 = \frac{\Delta_2}{\Delta} = \frac{6708\angle -63.4°}{1578\angle -19.2°} = 4.25\angle -44.2° = 3.05 - j2.96$$

$$I_3 = \frac{\Delta_3}{\Delta} = \frac{2000\angle -90°}{1578\angle -19.2°} = 1.27\angle -70.8° = 0.42 - j1.2$$

Current $\quad (I_2 - I_3) = 2.63 - j1.76 = 3.17\angle -33.9°$

(*iii*) **Solution by Nodal Analysis**

The current passing through 4 Ω resistance can be found by finding the voltage V_B of node B with the help of Nodal analysis. For this purpose point C in Fig. 15.21 (*a*) has been taken as the reference node.. Using the Nodal technique as explained in Art. we have

$$V_A\left(\frac{1}{10} + \frac{1}{5} + \frac{1}{-j5}\right) - \frac{V_B}{5} - \frac{100\angle 0°}{10} = 0 \qquad \text{...for node A}$$

$$V_A(3 + j2) - 2V_B = 100 \qquad \text{...}(i)$$

Similarly, for node B, we have

$$V_B\left(\frac{1}{5} + \frac{1}{4} + \frac{1}{(8+j6)}\right) - \frac{V_A}{5} = 0 \text{ or } V_B(53 - j6) - 20V_A = 0 \qquad \text{...}(ii)$$

Estimating V_A from Eq. (*i*) and (*ii*), we have

$$V_B(131 + j88) = 2000 \quad \text{or} \quad V_B = 12.67\angle -33.9°$$

Current through 4 Ω resistor = 12.67 ∠−33.9°/4 = 3.17 ∠−33.9°

Tutorial Problems No. 15.3.

1. Apply nodal analysis to the network of Fig. 15.22 to determine the voltage at node A and the active power delivered by the voltage source. **[8∠3.7° V; 9.85 W]**

2. Using nodal analysis, determine the value of voltages at models 1 and 2 in Fig. 15.23. **[V_1 = 88.1 ∠33.88° A; V_2 = 58.7 ∠72.34°A]**

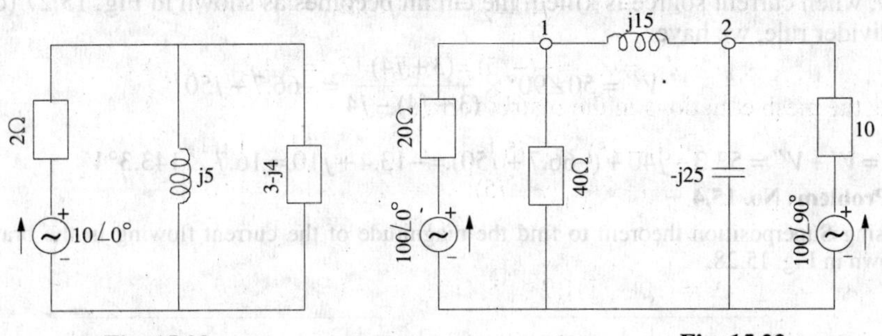

Fig. 15.22 **Fig. 15.23**

3. Using Nodal analysis, find the nodal voltages V_1 and V_2 in the circuit shown in Fig. 15.24. All resistances are given in terms of siemens **[V_1 = 1.64 V ; V_2 = 0.38 V]**

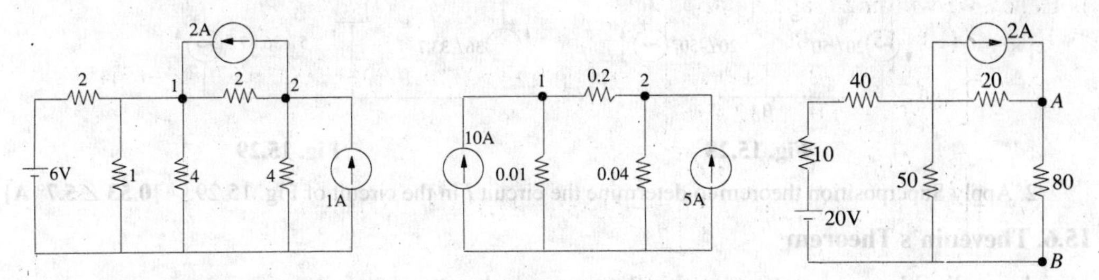

Fig. 15.24 **Fig. 15.25** **Fig. 15.26**

4. Find the values of nodal voltages V_1 and V_2 in the circuit of Fig. 15.25. Hence, find the current going from nodel 1 to node 2. All resistances are given in siemens. **[V_1 = 327 V; V_2 = 293.35 V; 6.73 A]**

5. Using Nodal analysis, find the voltage across points *A* and *B* in the circuit of Fig. 15.26: Check your answer by using mesh analysis. **[32 V]**

15.5. Superposition Theorem

As applicable to a.c. networks, it states as follows:

In any network made up of linear impedances and containing more than one source of e.m.f., the current flowing in any branch is the phasor sum of the currents that would flow in that branch if each source were considered separately, all other e.m.f. sources being replaced for the time being, by their respective internal impedances (if any).

Note. It may be noted that independent sources can be 'killed' *i.e.* removed leaving behind their internal impedances (if any) but dependent sources should not be killed.

Example 15.12. *Use Superposition theorem to find the voltage V in the network shown in Fig. 15.27.*

Solution. When the voltage source is killed, the circuit becomes as shown in the Fig. 15.27 (*b*). Using current-divider rule,

$$I = 10\angle 0° \times \frac{-j4}{(3+j4)-j4}, \quad \text{Now, } V' = I(3+j4)$$

$$\therefore \quad V' = 10\frac{-j4(3+j4)}{3} = 53.3 - j40$$

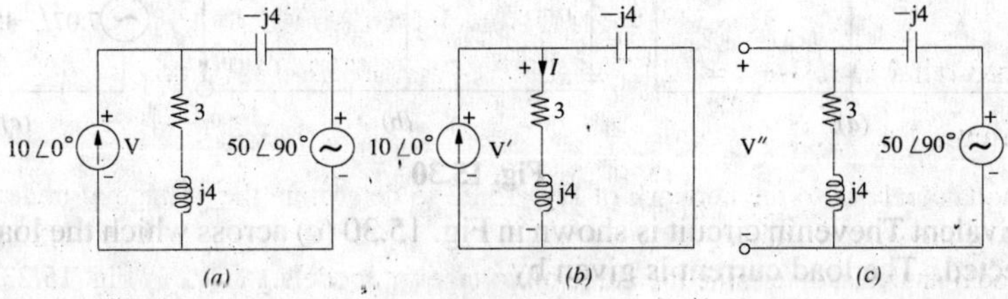

(*a*) (*b*) (*c*)

Fig. 15.27

Now, when current source is killed, the circuit becomes as shown in Fig. 15.27 (c). Using the voltage-divider rule, we have

$$V'' = 50\angle 90^\circ \times \frac{(3+j4)}{(3+j4)-j4} = -66.7 + j50$$

$\therefore$ drop $V = V' + V'' = 53.3 - j40 + (-66.7 + j50) = -13.4 + j10 = 16.7 \angle 143.3^\circ\ V$

Tutorial Problems No. 15.4

1. Using Superposition theorem to find the magnitude of the current flowing in the branch *AB* of the circuit shown in Fig. 15.28. **[2.58 A]**

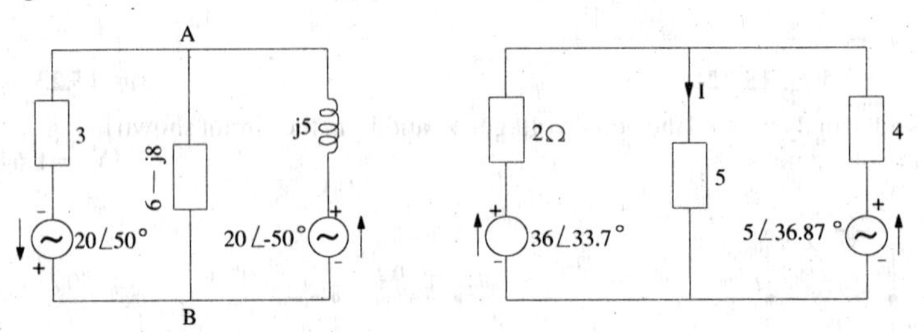

Fig. 15.28 **Fig. 15.29**

2. Apply Superposition theorem to determine the circuit *I* in the circuit of Fig. 15.29. **[0.53 ∠ 5.7° A]**

15.6. Thevenin's Theorem

As applicable to a.c. networks, this theorem may be stated as follows :

The current through a load impedance Z_L connected across any two terminals *A* and *B* of a linear network is given by $V_{th}/(Z_{th} + Z_L)$ where V_{th} is the open-circuit voltage across *A* and *B* and Z_{th} is the internal impedance of the network as viewed from the open-circuited terminals *A* and *B* with all voltage sources replaced by their internal impedances (if any) and current sources by infinite impedance.

Example 15.13. *In the network shown in Fig. 15.30.*

$Z_1 = (8 + j8)\ \Omega$; $Z_2 = (8 - j8)\ \Omega$; $Z_3 = (2 + j20)$; $V = 10\ \angle 0^\circ$ *and* $Z_L = j10\ \Omega$

Find the current through the load Z_L *using Thevenin's theorem.*

Solution. When the load impedance Z_L is removed, the circuit becomes as shown in Fig. 15.30(*b*). The open-circuit voltage which appears across terminals *A* and *B* represents the Thevenin voltage V_{th}. This voltage equals the drop across Z_2 because there is no current flow through Z_3.

Current flowing through Z_1 and Z_2 is

$$I = V(Z_1 + Z_2) = 10\ \angle 0^\circ\ [(8 + j8) + (8 - j8)] = 10\ \angle 0^\circ/16 = 0.625\ \angle 0^\circ$$

$$\therefore \quad V_{th} = IZ_2 = 0.625\ (8 - j\,8) = (5 - j5) = 7.07\ \angle -45^\circ$$

The Thevenin impedance Z_{th} is equal to the impedance as viewed from open terminals *A* and *B* with voltage source shorted.

$$\therefore \quad Z_{th} = Z_3 + Z_1 \parallel Z_2 = (2 + j20) + (8 + j8) \parallel (8 - j8) = (10 + j20)$$

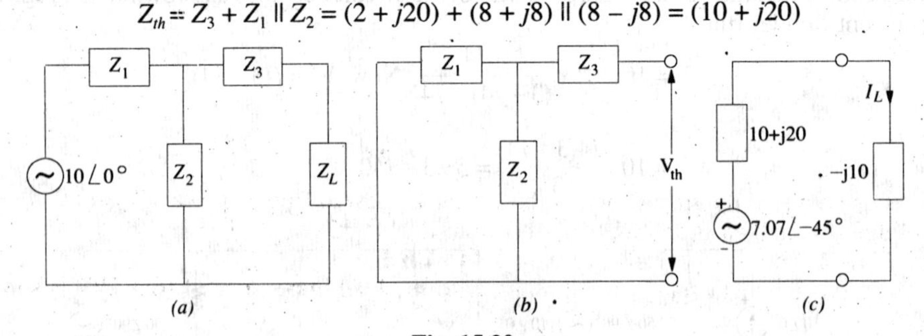

Fig. 15.30

The equivalent Thevenin circuit is shown in Fig. 15.30 (*c*) across which the load impedance has been reconnected. The load current is given by

$$\therefore \quad I_L = \frac{V_{th}}{Z_{th}+Z_L} = \frac{(5-j5)}{(10+j20)+(-j\,10)} = \frac{-j}{2}$$

Example 15.13 A. *Find the Thevenin equivalent circuit at terminals AB of the circuit given in Fig. 15.31 (a).*

Solution. For finding $V_{th} = V_{AB}$, we have to find the phasor sum of the voltages available on the way as we go from point B to point A because V_{AB} means voltage of point A with respect to that of point B (Art.). The value of current $I = 100\ \angle 0°/(6 - j8) = (6 + j8)$ A.

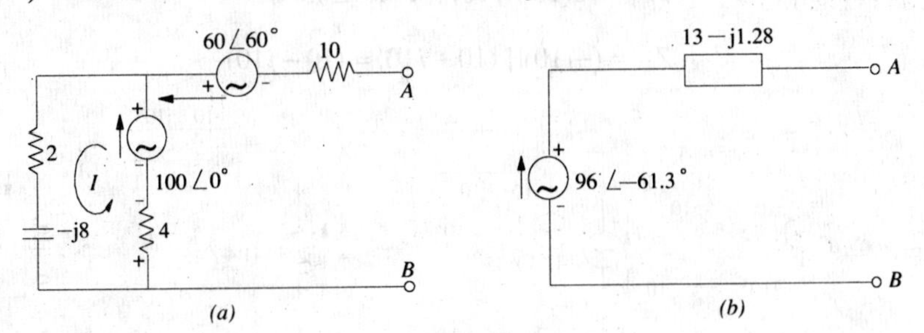

Fig. 15.31

Drop across 4 Ω resistor = $4(6 + j8) = (24 + j32)$

$$\therefore \quad V_{th} = V_{AB} = -(24 + j32) + (100 + j0) - 60(0.5 + j0.866)$$
$$= 46 - j84 = 96\angle -61.3°$$
$$Z_{AB} = Z_{th} = [10 + [4 \parallel (2 - j8)] = (13 - j1.28)$$

The Thevenin equivalent circuit is shown in Fig. 15.31 (*b*).

Example 15.14. *Find the Thevenin's equivalent of the circuit shown in Fig. 15.32 and hence calculate the value of the current which will flow in an impedance of (6 + j30) Ω connected across terminals A and B. Also calculate the power dissipated in this impedance.*

Solution. Let us first find the value of V_{th} *i.e.* the Thevenin voltage across open terminals A and B. With terminals A and B open, there is no potential drop across the capacitor. Hence, V_{th} is the drop across the pure inductor $j3$ ohm.

$$\text{Drop across the inductor} = \frac{10+j0}{(4+j3)} \times j3 = \frac{j30}{4+j3} = \frac{j30(4-j3)}{4^2+3^2} = (3.6+j4.8)V$$

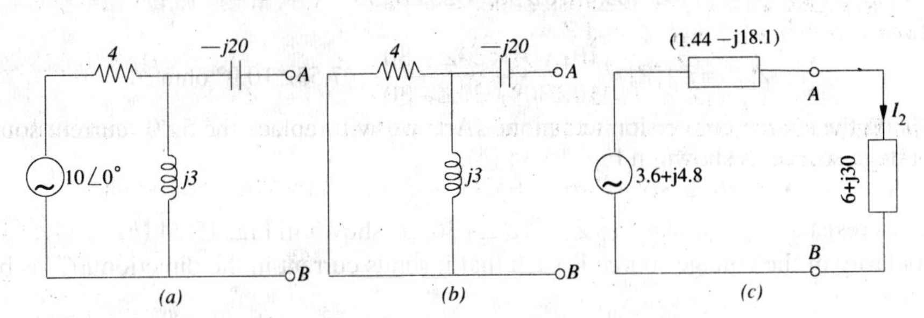

Fig. 15.32

Let us now find the impedance of the circuit as viewed from terminals A and B after replacing the voltage source by a short circuit as shown in Fig. 15.32 (*a*).

$Z_{th} = -j20 + 4 \parallel j3 = -j20 + 1.44 + j1.92 = 1.44 - j18.1$

The equivalent Thevenin circuit along with the load impedance of $(6 + j30)$ is shown in Fig. 15.32 (*c*).

$$\text{Load current} = \frac{(3.6+j4.8)}{(1.44-j18.1)+(6+j30)} = \frac{(3.6+j4.8)}{(7.44+j11.9)} = \frac{6\angle 53.1°}{14\angle 58°} = 0.43\angle -4.9°$$

The current in the load is 0.43 A and lags the supply voltage by 4.9°

Power in the load impedance is $0.43^2 \times 6$ = **1.1 W**

Example 15.15. *Using Thevenin's theorem, calculate the current flowing through the load connected across terminals A and B of the circuit shown in Fig. 15.33 (a). Also calculate the power delivered to the load.*

Solution. The first step is to remove the load from the terminals *A* and *B*. $V_{th} = V_{AB}$ = drop across $(10 + j10)$ ohm with *A* and *B* open.

Circuit current *I* $= \dfrac{100}{-j10 + 10 + j10} = 10\angle 0°$

$\therefore$ $$V_{th} = 10(10 + j10) = 141.4\angle 45°$$

$$Z_{th} = (-j10) || (10 + j10) = (10 - j10)$$

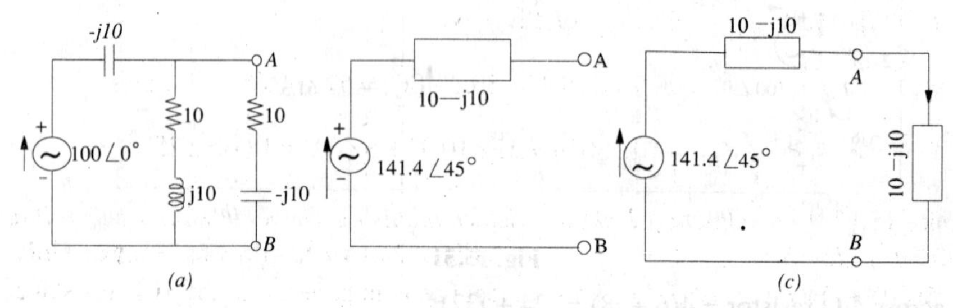

Fig. 15.33

The equivalent Thevenin's source is shown in Fig. 15.33 (*b*). Let the load be re-connected across *A* and *B* as shown in Fig. 15.33 (*c*).

$$\therefore \quad I_L = \frac{141.4\angle 45°}{(10 - j10) + (10 - j10)} = \frac{141.4\angle 45°}{20 - j20} = \frac{141.4\angle 45°}{28.3°\angle -45°} = 5\angle 90°$$

Power delivered to the load = $I_L^2 R_L = 5^2 \times 10 = 250$ W

Example 15.16. *Find the Thevenin's equivalent across terminals A and B of the networks shown in Fig. 15.34(a).*

Solution. The solution of this circuit involves the following steps :

(*i*) Let us find the equivalent Thevenin voltage V_{CD} and Thevenin impedance Z_{CD} as viewed from terminals *C* and *D*.

$$V_{CD} = V\frac{Z_2}{Z_1 + Z_2} = \frac{100\angle 0° \times 20\angle -30°}{10\angle 30° + 20\angle -30°} = 75.5\angle 19.1°V$$

$$Z_{CD} = Z_1 || Z_2 = \frac{10\angle 30° \times 20\angle -30°}{10\angle 30° + 20\angle -30°} = 7.55\angle 10.9° \text{ ohm}$$

(*ii*) Using the source conversion technique (Art) we will replace the $5\angle 0°$ current source by a voltage source as shown in Fig. 15.34 (*b*).

$V_{EC} = 5\angle 0° \times 10\angle -30° = 50\angle -30°$

Its series resistance is the same as $Z_3 = 10\angle -30°$ as shown in Fig. 15.34 (*b*).

The polarity of the voltage source is such that it sends current in the direction *EC*, as before.

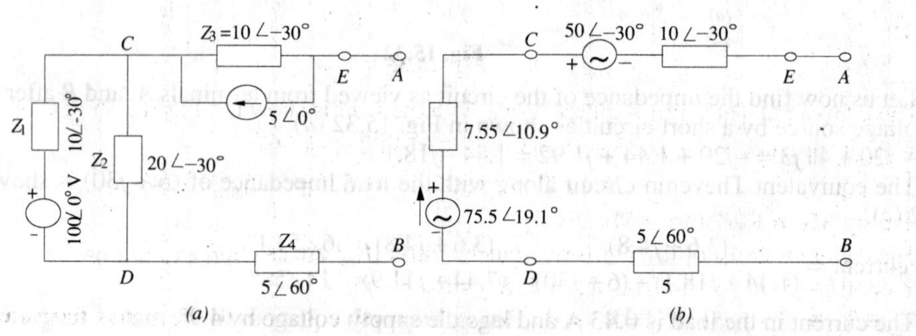

Fig. 15.34

(*iii*) From the above information, we can find V_{th} and Z_{th}

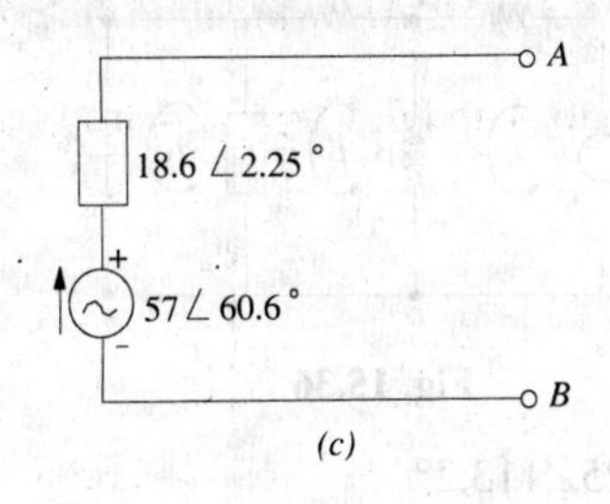

Fig. 15.34

$$V_{th} = V_{CD} = 75.5\angle 19.1° - 50\angle -30° = 57\angle 60.6°$$

$$Z_{th} = Z_{CD} = 10\angle -30° + 7.55\angle 10.9° + 5\angle 60° = 18.6\angle 2.25°$$

The Thevenin equivalent will respect to the terminals *A* and *B* is shown in Fig. 15.34(*c*).

For finding V_{AB} *i.e.* voltage at point *A* will respect to point *B*, we start from point *B* in Fig. 15.34(*b*) and go to point *A* and calculate the phasor sum of the voltages met on the way.

$$\therefore \quad V_{AB} = 75.1\angle 19.1° - 50\angle -30° = 57\angle 60.6°$$

$$Z_{AB} = 10\angle 30° + 7.55\angle 10.9° + 5\angle 60° = 18.6\angle 2.25°$$

Example 15.17. *For the network shown, determine using Thevenin's theorem, voltage across capacitor in . Fig. 15.35.* **(Elect. Network Analysis Nagpur Univ. 1993)**

$Z_{CD} = j5 \| (10 + j5) = 1.25 + j3.75$. This impedance is in series with the 10Ω resistance. Using voltage divider rule, the drop over Z_{CD} is

Solution. When load of –j5 Ω is removed the circuit becomes as shown in Fig. 15.35(b). Thevenin is voltage is given by the voltage drop produced by 100–V source over (5 + *j*5) impedance. It can be calculated as under.

$$V_{CD} = 100\frac{(1.25 + j3.75)}{10 + (1.25 + j3.75)} = \frac{125 + j375}{11.25 + j3.75}$$

This V_{CD} is applied across j5 reactance as well as across the series combination of 5Ω and (5 + *j*5) Ω. Again, using voltage-divider rule for V_{CD}, we get

$$V_{AB} = V_{th} = V_{CD} \times \frac{5 + j5}{10 \times j5} = \frac{(125 + j375)}{11.25 + j3.75} \times \frac{5 + j5}{10 + j5} = 21.1\angle 71.57° = 6.67 + j20$$

As looked into terminals *A* and *B*, the equivalent impedance is given by

$$R_{AB} = R_{th} = (5 + j5) \| (5 + 10 \| j5) = (5 + j5) \| (7 + j4) = 3 + j2.33$$

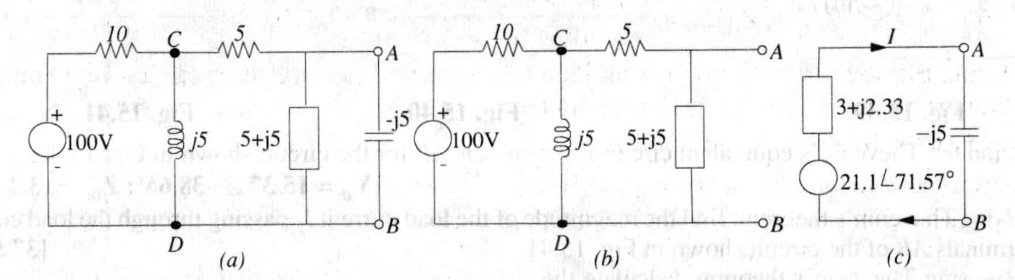

Fig. 15.35

The equivalent Thevenin's source along with the load is shown in Fig. 15.35 (*c*).

Total impedance = 3 + *j*2.33 – *j*5 = 3 – *j*2.67 = 4.02 ∠–41.67°

$$\therefore \quad I = 21.1\angle 71.57°/4.02\angle -41.67° = 5.25\angle 113.24°$$

Solution by Mesh Resistance Matrix

The different items of the mesh resistance matrix $[R_m]$ are as under:

$R_{11} = 10 + j5$; $R_{22} = 10 + j10$; $R_{33} = 5$; $R_{12} = R_{21} = -j5$;

$R_{23} = R_{32} = -(5 + j5)$; $R_{31} = R_{13} = 0$. Hence, the mesh equations in the matrix form are as given below

$$\begin{bmatrix}(10+j5) & -j5 & 0\\ -j5 & (10+j110) & -(5+j5)\\ 0 & -(5+j5) & 5\end{bmatrix}\begin{bmatrix}I_1\\ I_2\\ I_3\end{bmatrix}=\begin{bmatrix}100\\ 0\\ 0\end{bmatrix}$$

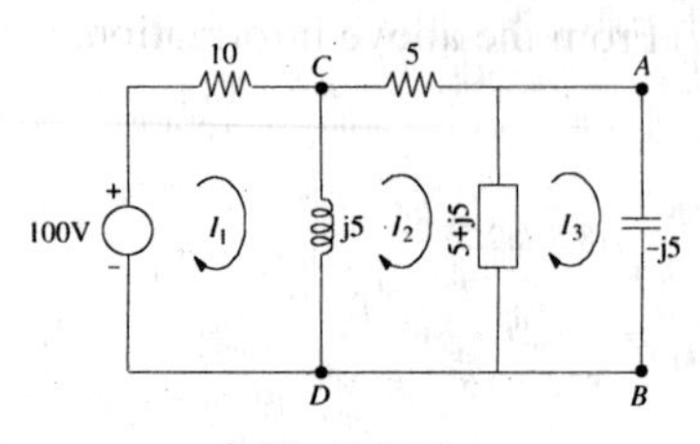

Fig. 15.36

$$\therefore\ \Delta=(10+j5)[5(10+j10)-(5+j5)(5+j5)] +j5(-j25)=625+j250=673\angle 21.8°$$

$$\begin{vmatrix}(10+j5) & -j5 & 100\\ -j5 & (10+j10) & 0\\ 0 & -(5+j5) & 0\end{vmatrix}=j5\,(500+j500)=3535\angle 135°$$

$$\therefore\qquad I_3=\Delta_3/\Delta=3535\angle 135°/673\angle 21.8°=5.25\angle 113.2°$$

Tutorial Problems No. 15.5

1. Determine the Thevenin's equivalent circuit with respect to terminals *AB* of the circuit shown in Fig. 15.37. **[V_{th} = 14.3 ∠6.38°, Z_m = (4 + j0.55) Ω]**

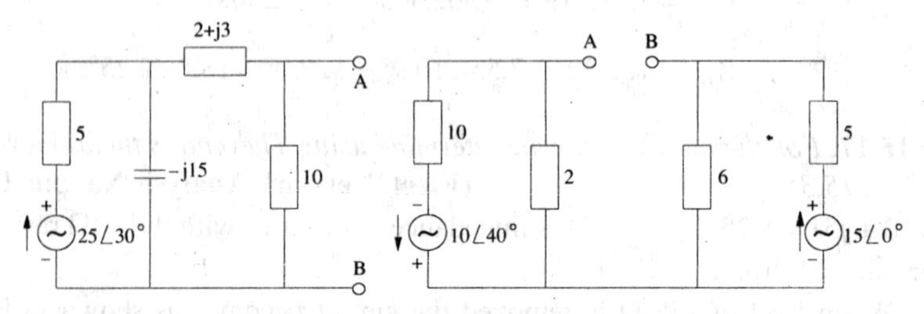

Fig. 15.37

2. Determine Thevenin's equivalent circuit with respect to terminals *AB* in Fig. 15.38. **[V_{th} = 9.5 ∠ 6.46° ; Z_{th} = 4.4 ∠0°]**

3. The e.m.fs. of two voltage sources shown in Fig. 15.39 are in phase with each other. Using Thevenin's theorem, find the current which will flow in a 16 Ω resistor connected across terminals *A* and *B*. **[V_{th} = 100 V ; Z_{th} = (48 + j32) ; I = 1.44 ∠–26.56 °]**

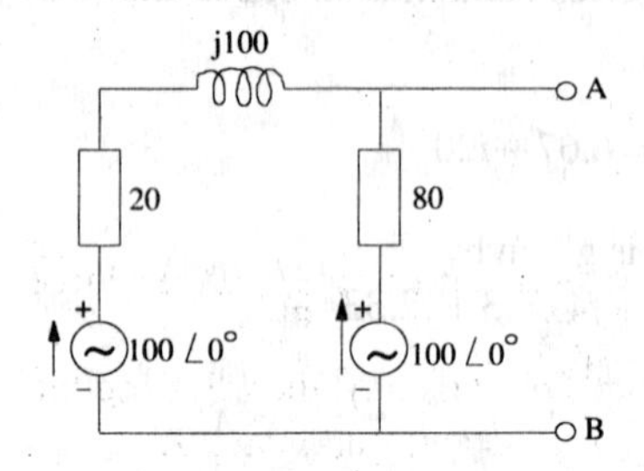

Fig. 15.39

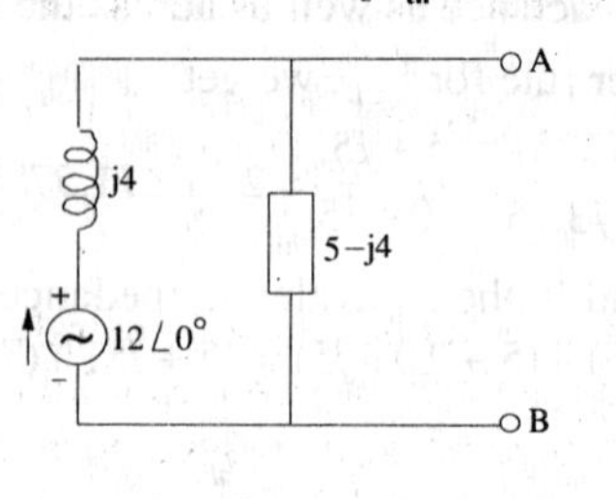

Fig. 15.40

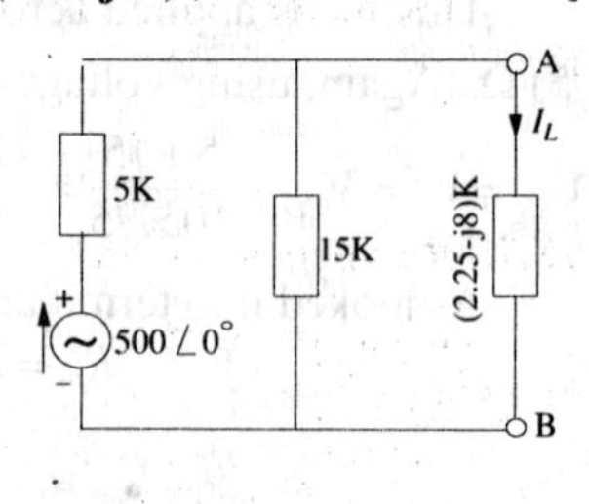

Fig. 15.41

4. Find the Thevenin's equivalent circuit for terminals *AB* for the circuit shown in Fig. 15.40. **[V_{th} = 15.37 ∠–38.66°; Z_{th} = (3.2 + j4) Ω]**

5. Using Thevenin's theorem, find the magnitude of the load current I_L passing through the load connected across terminals *AB* of the circuit shown in Fig. 15.41. **[37.5 mA]**

6. By using Thevenin's theorem, calculate the current flowing through the load connected across terminals *A* and *B* of the circuit shown in Fig. 15.42. All resistances and reactances are in ohms. **[V_{th} = 56.9 ∠ 50.15° ; 3.11 ∠85.67°]**

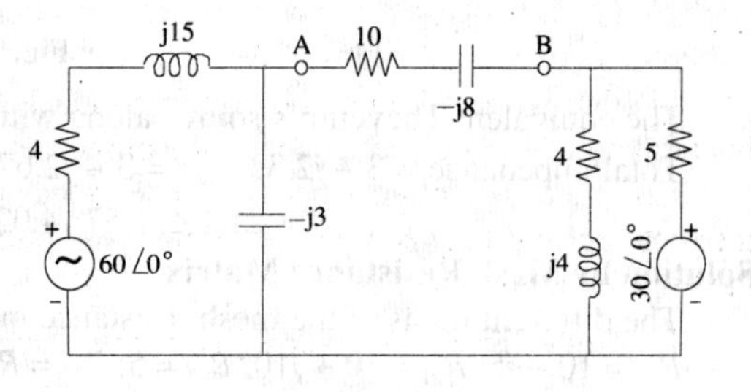

Fig. 15.42

7. Calculate the equivalent Thevenin's source with respect to the terminals *A* and *B* of the circuit shown in Fig. 15.43. **[V_{th} = (6.34 + j2.93) V ; Z_{th} = (3.17 – j5.07) Ω**

8. What is the Thevenin's equivalent source with respect to the terminals *A* and *B* of the circuit shown in Fig. 15.44? **[V_{th} = (9.33 + j8) V ; Z_{th} = (8–j11) Ω]**

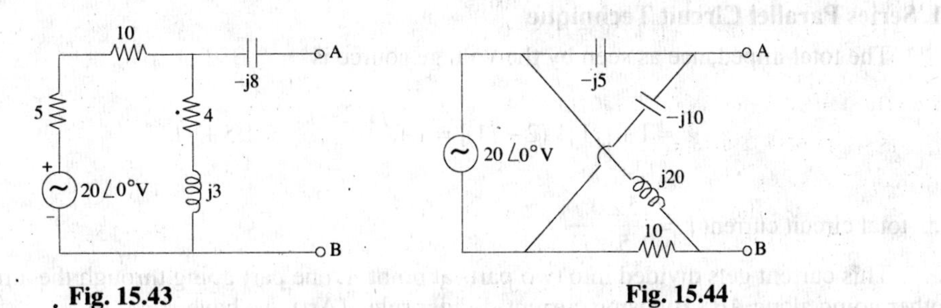

Fig. 15.43 Fig. 15.44

9. What is the Thevenin's equivalent source with respect to terminals A and B of the circuit shown in Fig. 15.45 ? Also, calculate the value of impedance which should be connected across AB for MPT. All resistances and reactances are in ohms. **[V_{th} = (16.87 + j15.16) V; Z_{th} = (17.93 − j1.75) Ω ; (17.93 + j1.75) Ω]**

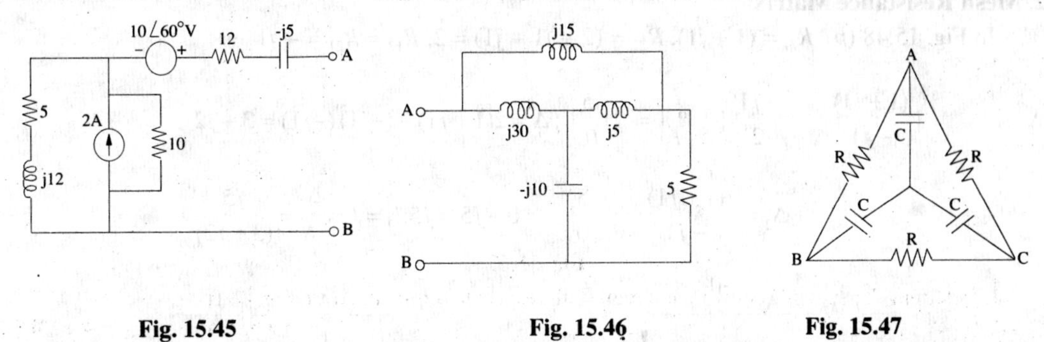

Fig. 15.45 Fig. 15.46 Fig. 15.47

10. Find the impedance of the network shown in Fig. 15.46, when viewed from the terminals A and B. All resistances and reactances are in ohms. **[(4.435 + j6.878)]**

11. Find the value of the impedance that would be measured across terminals BC of the circuit shown in Fig. 15.47.

$$\left[\frac{2R}{9+\omega^2C^2R^2}(3-j\omega CR)\right]$$

15.7. Reciprocity Theorem

This theorem applies to networks containing linear bilateral elements and a single voltage source or a single current source. This theorem may be stated as follows :

If a voltage source in branch A of a network causes a current of l in branch B, then shifting the voltage source (but not its impedance) to branch B will cause the same current I in branch A.

It may be noted that currents in other branches will generally not remain the same. A simple way of stating the above theorem is that if an ideal voltage source and an ideal ammeter are interchanged, the ammeter reading would remain the same. The ratio of the input voltage in branch A to the output current in branch B is called the transfer impedance.

Similarly, if a current source between nodes I and 2 causes a potential difference of V between nodes 3 and 4, shifting the current source (but not its admittance) to nodes 3 and 4 causes the same voltage V between nodes 1 and 2.

In other words, the interchange of an ideal current source and an ideal voltmeter in any linear bilateral network does not change the voltmeter reading.

However, the voltages between other nodes would generally not remain the same. The ratio of the input current between one set of nodes to output voltage between another set of nodes is called the transfer admittance.

Example 15.18. *Verify Reciprocity theorem for V & I in the circuit shown in Fig. 15.48.* **(Elect. Network Analysis, Nagpur Univ. 1993)**

Solution. We will find the value of the current I as read by the ammeter first by applying series parallel circuits technique and then by using mesh resistance matrix (Art)

1. Series Parallel Circuit Technique

The total impedance as seen by the voltage source is

$$= 1 + [j1 \,||\, (2 - j1)] = 1 + \frac{j1(2 - j1)}{2} = 1.5 + j1$$

$\therefore$ total circuit current $i = \dfrac{5\angle 0°}{1.5 + j1}$

This current gets divided into two parts at point A, one part going through the ammeter and the other going along AB. By using current-divider rule. (Art), we have

$$I = \frac{5}{1.5 + j1} \times \frac{j1}{(2 + j1 - j1)} = \frac{j5}{3 + j2}$$

2. Mesh Resistance Matrix

In Fig. 15.48 (*b*), $R_{11} = (1 + j1)$, $R_{22} = (2 + j1 - j1) = 2$; $R_{12} = R_{21} = -j1$

$$\therefore \begin{vmatrix} (1+j1) & -j1 \\ -j1 & 2 \end{vmatrix} \begin{vmatrix} I_1 \\ I_2 \end{vmatrix} = \begin{vmatrix} 5 \\ 10 \end{vmatrix}; \Delta = 2(1 + j1) - (-j1)(-j1) = 3 + j2$$

$$\Delta_2 = \begin{vmatrix} (1+j1) & 5 \\ -j1 & 0 \end{vmatrix} = 0 + j5 = j5; I_2 = I = \frac{\Delta_2}{\Delta} = \frac{j5}{(3 + j2)}$$

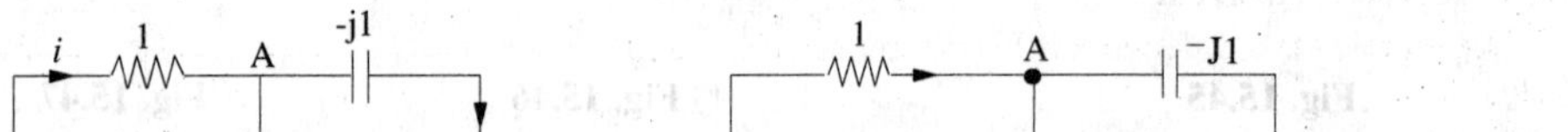

Fig. 15.48

As shown in Fig. 15.48 (*c*), the voltage source has been interchanged with the ammeter. The polarity of the voltage source should be noted in particular. It looks as if the voltage source has been pushed along the wire in the counterclockwise direction to its new position, thus giving the voltage polarity as shown in the figure. We will find the value of I in the new position of the ammeter by using the same two techniques as above.

1. Series Parallel Circuit Technique

As seen by the voltage source from its new position, the total circuit impedance is

$$= 2(2 - j1) + j1 \,||\, 1 = \frac{3 + j2}{1 + j1}$$

The total circuit current $i = 5 \times \dfrac{1+j1}{3+j2}$

This current i gets divided into two parts at point B as per the current-divider rule.

$$\therefore \qquad I = \frac{5(1+j1)}{3+j2} \times j\frac{1}{1+j1} = \frac{j5}{3+j2}$$

2. Mesh Resistance Matrix

As seen from Fig. 15.48(*d*).

$$\begin{vmatrix} (1+j1) & -j1 \\ -j1 & 2 \end{vmatrix} \begin{vmatrix} I_1 \\ I_2 \end{vmatrix} = \begin{vmatrix} 0 \\ 5 \end{vmatrix}; \Delta = 2(1+j1)+1 = 3+j2$$

$$\Delta = \begin{vmatrix} 0 & -j1 \\ 5 & 2 \end{vmatrix} = j5; I = I_1 = \frac{\Delta_1}{\Delta} = \frac{j5}{3+j2}$$

The reciprocity theorem stands verified from the above results.

Tutorial problem No. 15.6

1. State reciprocity theorem. Verify it for the circuit Fig. 15.49, with the help of any suitable current through any element.

(Elect. Network Analysis Nagpur Univ. 1993)

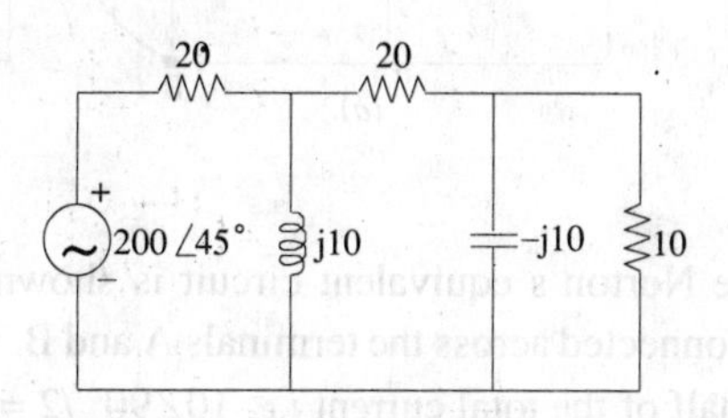

Fig. 15.49

15.8. Norton's Theorem

As applied to a.c. networks, this theorem can be stated as under :

Any two terminal active linear network containing voltage sources and impedances when viewed from its output terminals is equivalent to a constant current source and a parallel impedance. The constant current is equal to the current which would flow in a short-circuit placed across the terminals and the parallel impedance is the impedance of the network when viewed from open-circuited terminals after voltage sources have been replaced by their internal impedances (if any) and current sources by infinite impedance.

Example 15.19. *Find the Norton's equivalent of the circuit shown in Fig. 15.50. Also find the current which will flow through an impedance of (10−j20) Ω across the terminals A and B.*

Solution. As shown in Fig. 15-50 (*b*), the terminals *A* and *B* have been short-circuited.

$\therefore \qquad I_{SC} = I_N = 25/(10 + j20) = 25/22.36 \angle 63.4° = 1.118 \angle{-63.4°}$

When voltage source is replaced by a short, then the internal resistance of the circuit, as viewed from open terminals *A* and *B*, is $R_N = (10 + j20)\ \Omega$. Hence, Norton's equivalent circuit becomes as shown in Fig. 15.50(*c*).

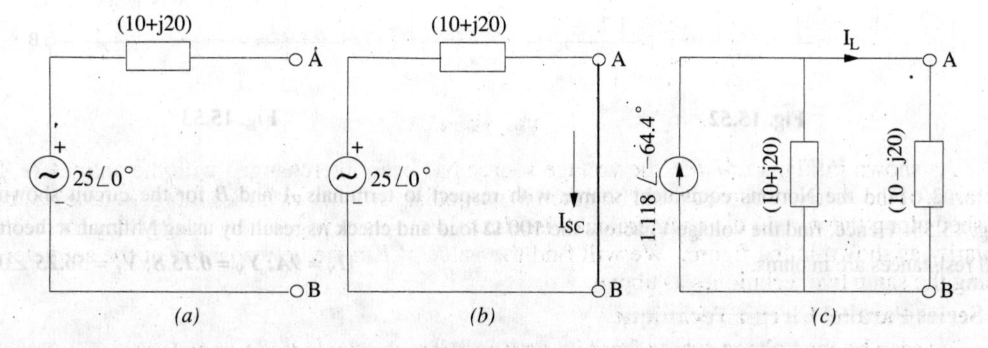

Fig. 15.50

When the load impedance of (10 − *j*20) is applied across the terminals *A* and *B*, current through it can be found with the help of current-divider rule.

$$\therefore \quad I_L = 1.118\angle-63.4° \times \frac{(10+j20)}{(10+j20)+(10-j20)} = 1.25 \text{ A}$$

Example 15.20. *Use Norton's theorem to find current in the load connected across terminals A and B of the circuit shown in Fig. 15.51(a).*

Solution. The first step is to short-circuit terminals *A* and *B* as shown in Fig. 15.51(*a*).* The short across *A* and *B* not only short-circuits the load but the (10 + *j* 10) impedance as well.

$$\therefore \quad I_N = 100\angle 0° /(-j10) = j10 = 10\angle 90°$$

Since the impedance of the Norton and Thevenin equivalent circuits is the same, $Z_N = 10 - j10$.

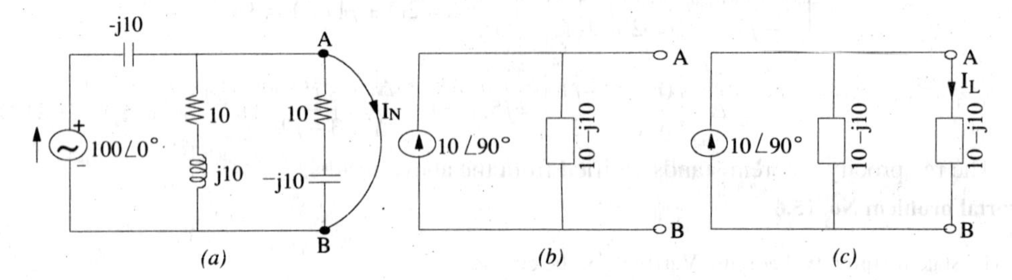

Fig. 15.51

The Norton's equivalent circuit is shown in Fig. 15.51 (*b*). In Fig. 15.51(*c*), the load has been reconnected across the terminals A and B. Since the two impedances are equal, current through each is half of the total current *i.e.* 10∠90° /2 = 5 ∠90°.

Tutorial Problems No. 15.7

1. Find the Norton's equivalent source with respect to terminals *A* and *B* of the networks shown in Fig. 15.51(*a*) and (*b*). All resistances and reactances are expressed in siemens in Fig. 15.51 (a) and in ohms in Fig. 15.52.

[(*a*) $\mathbf{I_N}$ = −(2.1−j3) A; 1/$\mathbf{Z_N}$ = (0.39 + j0.3)S (*b*) $\mathbf{I_N}$ = (6.87 + j0.5) A; 1/$\mathbf{Z_N}$ = (3.17 + j1.46)S]

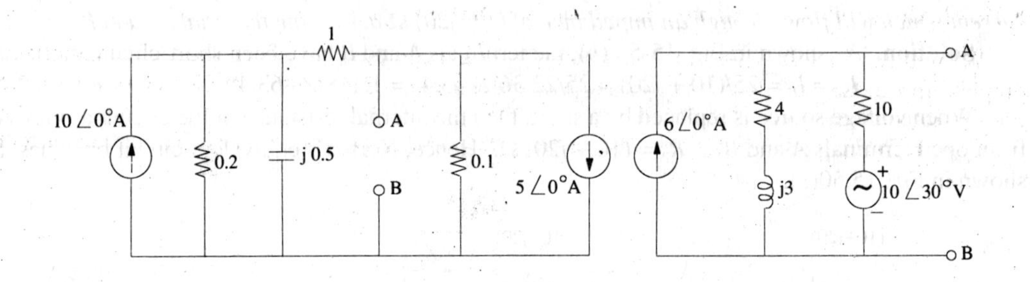

Fig. 15.52 **Fig. 15.53**

2. Find the Nortons equivalent source with respect to terminals *A* and *B* for the circuit shown in Fig. 15.53. Hence, find the voltage V_L across the 100 Ω load and check its result by using Millman's theorem. All resistances are in ohms. **[I_N = 9A; Y_N = 0.15 S; V_L = 56.25 ∠0°]**

---

*For finding I_N, we may or may not remove the load from the terminals (because, in either case, it would be short-circuited) but for finding R_N, it has to be removed as in the case of Thevenin's theorem.

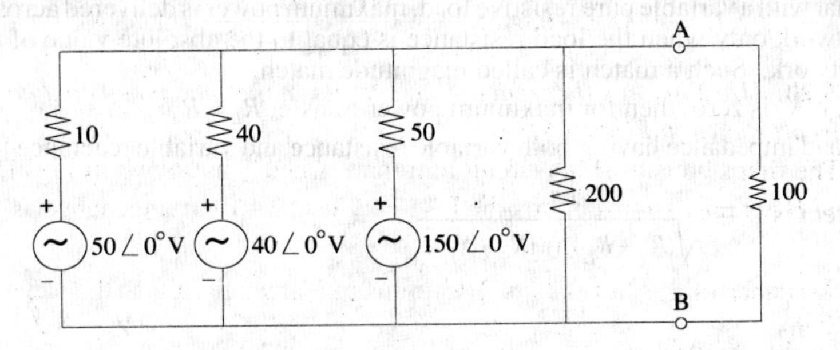

Fig. 15.54

3. Find the Norton's equivalent network at terminals AB of the circuit shown in Fig. 15.55.

[I_{SC} = 2.17∠–44° A; Z_N = (2.4 + j1.47)Ω]

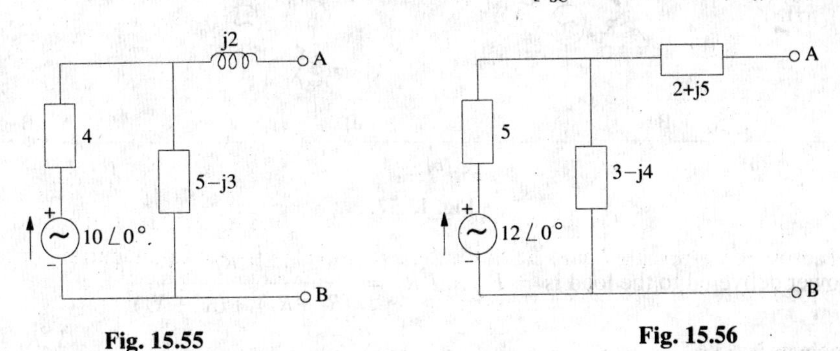

Fig. 15.55 **Fig. 15.56**

4. What is the Norton equivalent circuit at terminals AB of the network shown in Fig. 15.56.

[I_{SC} = 1.15 ∠–66.37°; Z_N = (4.5 + j3.75) Ω]

15.9. Maximum Power Transfer Theorem

As explained earlier in Art. this theorem is particularly useful for analysing communication networks where the goal is transfer of maximum power between two circuits and not highest efficiency.

15.10. Maximum Power Transfer Theorems – General Case

We will consider the following maximum power transfer theorems when the source has a fixed complex impedance and delivers power to a load consisting of a variable resistance or a variable complex impedance.

Case 1. When load consists only of a variable resistance R_L [Fig. 15.57 (*a*)]. The circuit current is

$$I = \frac{V_g}{\sqrt{(R_g + R_L)^2 + X_g^2}}$$

Power delivered to R_L is $P_L = \dfrac{V_g^2 R_L}{(R_g + R_L)^2 + X_g^2}$

To determine the value of R_L for maximum transfer of power, we should set the first derivative dP_L/dR_L to zero.

$$\frac{dP_L}{dR_L} = \frac{d}{dR_L}\left[\frac{V_g^2 R_L}{(R_g + R_L)^2 + X_g^2}\right] = V_g^2\left\{\frac{[(R_g + R_L)^2 + X_g^2] - R_L(2)(Rg + R_L)}{[(R_g + R_L)^2 + X_g^2]^2}\right\} = 0$$

or $$R_g^2 + 2R_gR_L + R_L^2 + X_g^2 - 2R_LR_g - 2R_L^2 = 0 \text{ and } R_g^2 + X_g^2 = R_L^2$$

∴ $$R_L = \sqrt{R_g^2 + X_g^2} = |\mathbf{Z}_g|$$

It means that with a variable pure resistive load, maximum power is delivered across the terminals of an active network only when the load resistance is equal to the absolute value of the impedance of the active network. Such a match is called magnitude match.

Moreover, if X_g is zero, then for maximum power transfer $R_L = R_g$

Case 2. Load impedance having both variable resistance and variable reactance [Fig. 15.57(*b*)].

The circuit current is $I = \dfrac{V_g}{\sqrt{(R_g+R_L)^2+(X_g+X_L)^2}}$

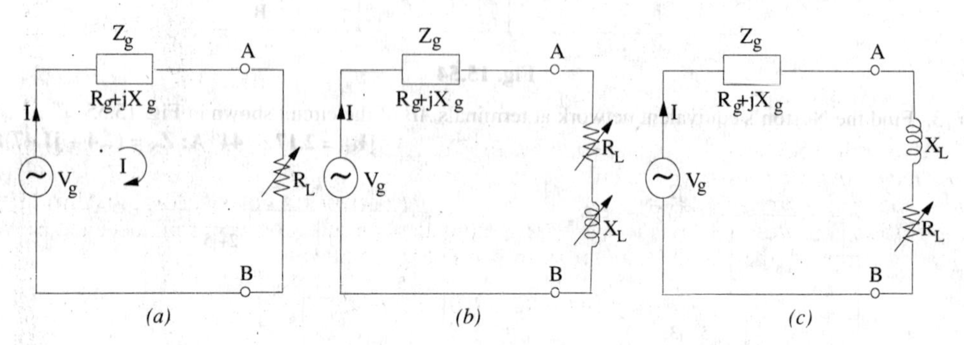

Fig. 15.57

The power delivered to the load is $= P_L = I^2R_L = \dfrac{V_g^2R_L}{(R_g+R_L)^2+(X_g+X_L)^2}$

Now, if R_L is held fixed, P_L is maximum when $X_g = -X_L$. In that case $P_{L\max} = \dfrac{V_g^2R_L}{(R_g+R_L)^2}$

If on the other hand, R_L is variable then, as in **Case** 1 above, maximum power is delivered to the load when $R_L = R_g$. In that case if $R_L = R_g$ and $X_L = -X_g$, then $\mathbf{Z}_L = \mathbf{Z}_g$. Such a match is called conjugate match.

From the above, we come to the conclusion that in the case of a load impedance having both variable resistance and variable reactance, maximum power transfer across the terminals of the active network occurs when $\mathbf{Z}_L$ equals the complex conjugate of the network impedance $\mathbf{Z}_g$ *i.e.* the two impedances are conjugately matched.

Case 3. Z_L with variable resistance and fixed reactance [(Fig. 15.57(*c*)]. The equations for current I and power P_L are the same as in **Case** 2 above except that we will consider X_L to remain constant. When the first derivative of P_L with respect to R_L is set equal to zero, it is found that

$$R_L^2 = R_g^2 + (X_g + X_L)^2 \text{ and } R_L = |\mathbf{Z}_g + jX_L|$$

Since Z_g and X_L are both fixed quantities, these can be combined into a single impedance. Then with R_L variable, **Case** 3 is reduced to **Case** 1 and the maximum power transfer takes place when R_L equals the absolute value of the network impedance.

Summary

The above facts can be summarized as under :

1. When load is purely resistive and adjustable, MPT is achieved when $R_L = |Z_g| = \sqrt{R_g^2+X_g^2}$.
2. When both load and source impedances are purely resistive (*i.e.* $X_L = X_g = 0$), MPT is achieved when $R_L = R_g$.
3. When R_L and X_L are both independently adjustable, MPT is achieved when $X_L = -X_g$ and $R_L = R_g$.
4. When X_L is fixed and R_L is adjustable, MPT is achieved when $R_L = \sqrt{[R_g^2+(X_g+X_L)^2]}$

Example 15.21. *In the circuit of Fig. 15.58, which load impedance of p.f. = 0.8 lagging when connected across terminals A and B will draw the maximum power from the source. Also find the power developed in the load and the power loss in the source.*

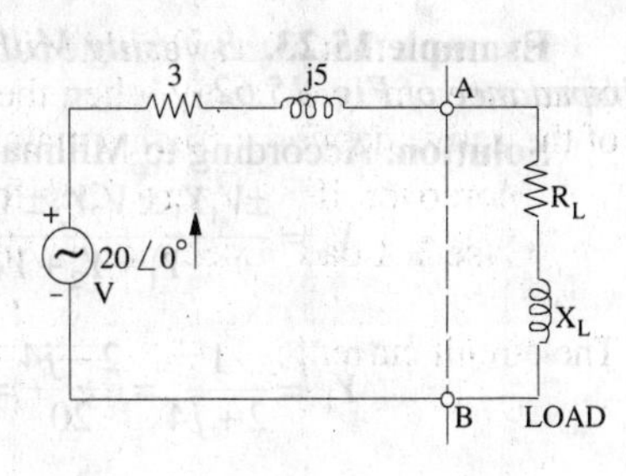

Fig. 15.58

Solution. For maximum power transfer $|Z_L| = |Z_1|$ $\sqrt{(3^2+5^2)} = 5.83\ \Omega$.

For p.f. = 0.8, $\cos\phi = 0.8$ and $\sin\phi = 0.6$.

$\therefore\ R_L = Z_L \cos\phi = 5.83 \times 0.8 = 4.66\ \Omega$. $X_L = Z_L \sin\phi = 5.83 \times 0.6 = 3.5\ \Omega$.

Total circuit impedance $Z = \sqrt{[(R_1+R_L)^2+(X_1+X_L)^2]}$

$$=\sqrt{[(3+4.66)^2+(5+3.5)^2]} = 11.44\ \Omega$$

$$\therefore \qquad I = V/Z = 20/11.44 = 1.75\text{ A.}$$

Power in the load = $I^2R_L = 1.75^2 \times 4.66 = 14.3$ W
Power loss in the source = $1.75^2 \times 3 = 9.2$ W.

Example 15.22. *In the network shown in Fig. 15.59 find the value of load to be connected across terminals AB consisting of variable resistance R_L and capacitive reactance X_C which would result in maximum power transfer.* **(Network Analysis, Nagpur Univ. 1993)**

Solution. We will first find the Thevenin's equivalent circuit between terminals *A* and *B*. When the load is removed, the circuit becomes as shown in Fig. 15.59 (b).

$$V_{th} = \text{drop across } (2+j10) = 50\angle 45° \times \frac{2+j10}{7+j10}$$

$$= 41.8\angle 68.7° = 15.2 + j38.9$$

$$R_{th} = 5\ ||\ (2+j10) = 4.1\angle 23.7° = 3.7 + j1.6$$

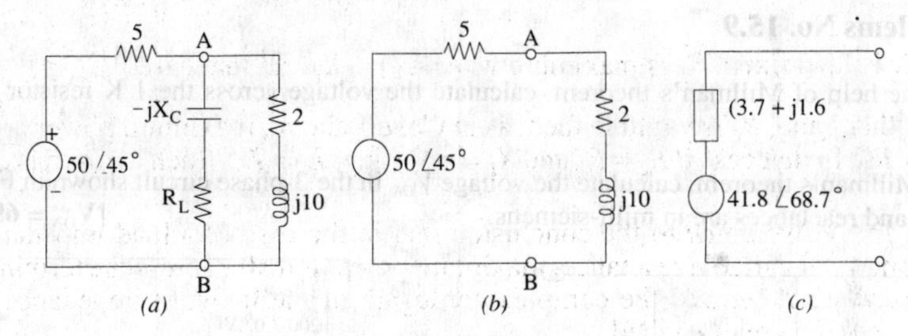

Fig. 15.59

The Thevenin's equivalent source is shown in Fig. 15.59 (*c*)
Since for MPT, a conjugate match is required hence, $X_C = 1.6\ \Omega$ and $R_L = 3.7\ \Omega$.

Tutorial Problem No. 15.8

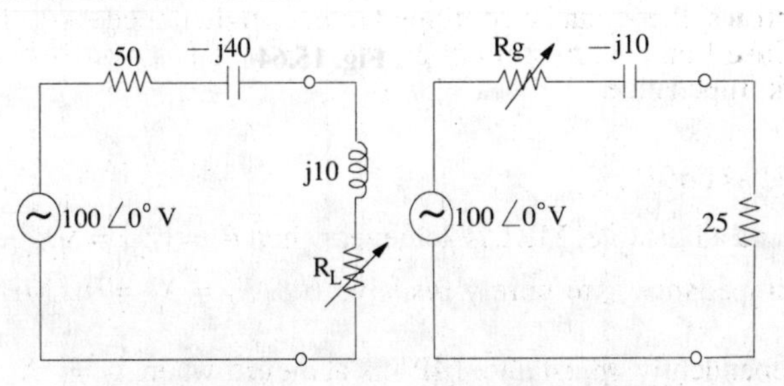

Fig. 15.60 **Fig. 15.61**

1. In the circuit of Fig. 15.60 the load consists of a fixed inductance having a reactance of $j10\ \Omega$ and a variable load resistor R_L. Find the value of R_L for MPT and the value of this power. **[58.3 Ω ; 46.2 W]**

2. In the circuit of Fig. 15.61, the source resistance R_g is variable between 5 Ω and 50 Ω but R_L has a fixed value of 25 Ω. Find the value of R_g for which maximum power is dissipated in the load and the value of this power. **[5 Ω ; 250 W]**

15.11. Millman's Theroem

It permits any number of parallel branches consisting of voltage sources and impedances to be reduced to a single equivalent voltage source and equivalent impedance. Such multi-branch circuits are frequently encountered in both electronics and power applications.

Example 15.23. *By using Millman's theorem, calculate node voltage V and current in the j6 impedance of Fig. 15.62.*

Solution. According to Millman's theorem as applicable to voltage sources.

$$V = \frac{\pm V_1Y_1 \pm V_2Y_2 \pm V_3Y_3 \pm \ldots \pm V_nY_n}{Y_1 + Y_2 + Y_3 + \ldots + Y_n}$$

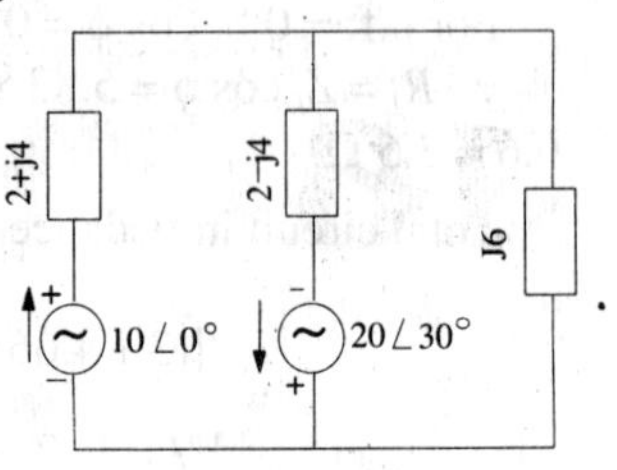

Fig. 15.62

$$Y_1 = \frac{1}{2+j4} = \frac{2-j4}{20} = 0.01 - j0.02 = 0.022\angle -63.4°$$

$$Y_2 = \frac{1}{2-j4} = \frac{2+j4}{20} = 0.01 + j0.02 = 0.022\angle 63.4°$$

$$Y_3 = \frac{1}{j6} = 0.167\angle -90° = 0 - j0.167$$

$$Y_1 + Y_2 + Y_3 = 0.02 - j0.167$$

In the present case, $V_3 = 0$ and also V_2Y_2 would be taken as negative because current due to V_2 flows away from the node.

$$\therefore \quad V = \frac{V_1Y_1 - V_2Y_2}{Y_1 + Y_2 + Y_3} = \frac{10\angle 0° \times 0.22\angle -63.4° - 20\angle 30° \times 0.022\angle 63.4°}{0.02 - j0.167}$$

$$= 3.35\angle 177°$$

Current through $j\,6$ impedance $= 3.35\angle 177°/6\angle 90° = 0.56\angle 87°$

Tutorial Problems No. 15.9

1. With the help of Millman's theorem, calculate the voltage across the 1 K resistor in the circuit of Fig. 15.63. **[2.79 V]**

2. Using Millman's theorem, calculate the voltage V_{ON} in the 3-phase circuit shown in Fig. 15.64. All load resistances and reactances are in milli-siemens. **[V_{ON} = 69.73 ∠113.53°]**

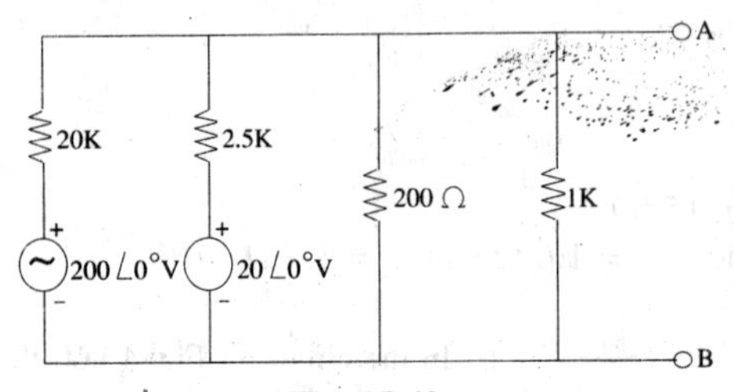

Fig. 15.63

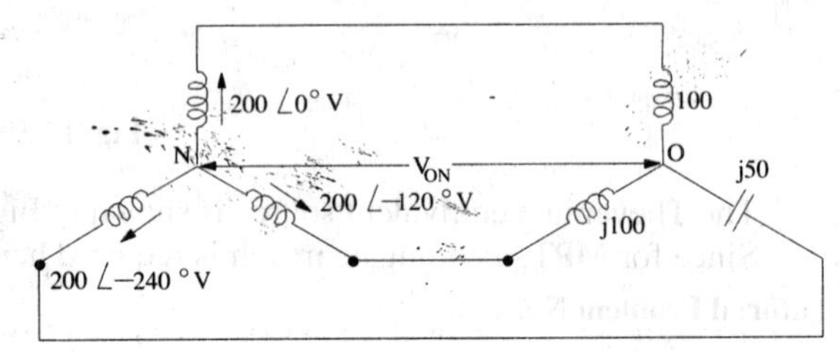

Fig. 15.64

16 A.C. BRIDGES

16.1. A.C. Bridges

Resistances can be measured by direct-current Wheatstone bridge, shown in Fig. 16.1 (*a*) for which the condition of balance is that

$$\frac{R_1}{R_2} = \frac{R_4}{R_3} \quad \text{or} \quad R_1 R_3 = R_2 R_4^*$$

Inductances and capacitances can also be measured by a similar four-arm bridge, as shown in Fig. 16.1 (*b*); instead of using a source of direct current, alternating current is employed and galvanometer is replaced by a vibration galvanometer (for commercial frequencies or by telephone detector if frequencies are higher (500 to 2000 Hz).

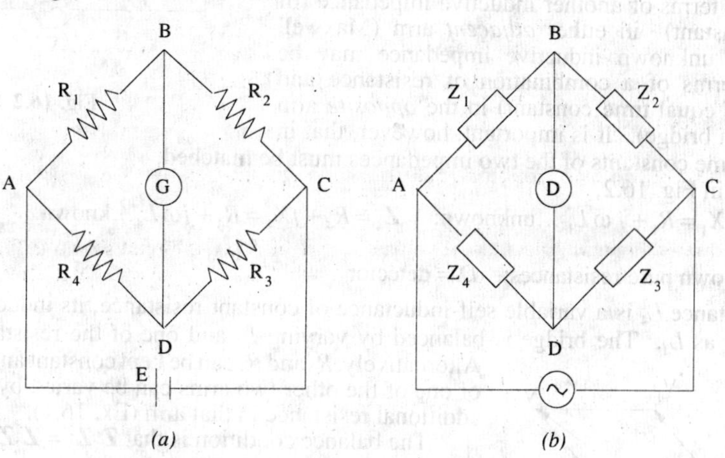

Fig. 16.1

The condition for balance is the same as before but instead of resistances, impedances are used *i.e.*

$$\mathbf{Z_1/Z_2 = Z_4/Z_3} \quad \text{or} \quad \mathbf{Z_1Z_3 = Z_2Z_4}$$

But there is one important difference *i.e.* not only should there be balance for the magnitudes of the impedances but also a phase balance. Writing the impedances in their polar form, the above condition becomes

$$Z_1\angle\phi_1 . Z_3\angle\phi_3 = Z_2\angle\phi_2 \cdot Z_4\angle\phi_4 \text{ or } Z_1Z_3\angle\phi_1 + \phi_3 = Z_2Z_4\angle\phi_2 + \phi_4$$

Hence, we see that, in fact, there are two balance conditions which must be satisfied simultaneously in a four-arm a.c. impedance bridge.

(*i*) $Z_1Z_3 = Z_2Z_4$ ...for magnitude balance

(*ii*) $\phi_1 + \phi_3 = \phi_2 + \phi_4$ ...for phase angle balance

*Products of opposite arm resistances are equal.

In this chapter, we will consider a few of the numerous bridge circuits used for the measurement of self-inductance, capacitance and mutual inductance, choosing as examples some bridges which are more common.

16.2. Maxwell's Inductance Bridge

The bridge circuit is used for medium inductances and can be arranged to yield results of considerable precision. As shown in Fig. 16.2, in the two arms, there are two pure resistances so that for balance relations, the phase balance depends on the remaining two arms. If a coil of an unknown impedance Z_1 is placed in one arm, then its positive phase angle ϕ_1 can be compensated for in either of the following two ways :

(*i*) A known impedance with an equal positive phase angle may be used in either of the *adjacent* arms (so that $\phi_1 = \phi_2$ or $\phi_1 = \phi_4$), remaining two arms have zero phase angles (being pure resistances). Such a network is known as Maxwell's a.c. bridge or L_1/L_4 bridge.

(*ii*) Or an impedance with an equal *negative* phase angle (*i.e.* capacitance) may be used in *opposite* arm (so that $\phi_1 + \phi_3 = 0$). Such a network is known as Maxwell-Wien bridge (Fig. 16.5) or Maxwell's L/C bridge.

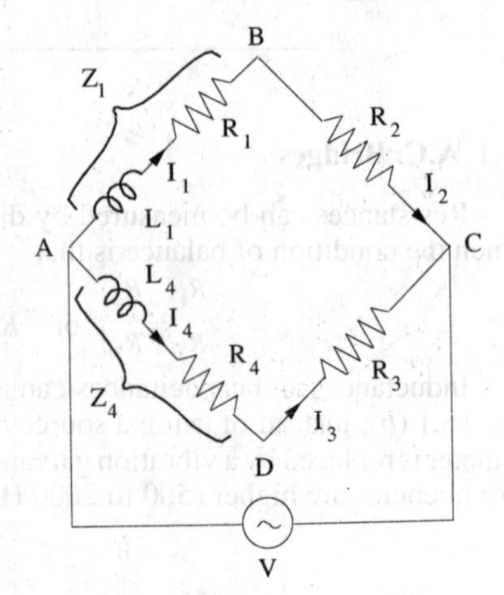

Fig. 16.2

Hence, we conclude that an inductive impedance may be measured in terms of another inductive impedance (of equal time constant) in either *adjacent* arm (Maxwell bridge) or the unknown inductive impedance may be measured in terms of a combination of resistance and capacitance (of equal time constant) in the *opposite* arm (Maxwell-Wien bridge). It is important, however, that in each case the time constants of the two impedances must be matched.

As shown in Fig. 16.2,

$\mathbf{Z}_1 = R_1 + jX_1 = R_1 + j\,\omega\,L_1$ ...unknown; $\quad \mathbf{Z}_4 = R_4 + jX_4 = R_4 + j\omega\,L_4$...known

R_2, R_3 = known pure resistances; $\quad D$ = detector

The inductance L_4 is a variable self-inductance of constant resistance, its inductance being of the same order as L_1. The bridge is balanced by varying L_4 and one of the resistances R_2 or R_3. Alternatively, R_2 and R_3 can be kept constant and the resistance of one of the other two arms can be varied by connecting an additional resistance in that arm (Ex. 16.1).

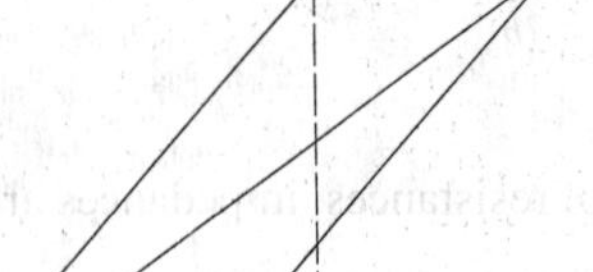

Fig. 16.3

The balance condition is that $\mathbf{Z}_1\mathbf{Z}_3 = \mathbf{Z}_2\mathbf{Z}_4$

$\therefore \quad (R_1 + j\omega L_1)R_3 = (R_4 + j\omega L_4)\,R_2$

Equating the real and imaginary parts on both sides, we have

$R_1R_3 = R_2R_4$ or $R_1/R_4 = R_2/R_3$

(*i.e.* products of the resistances of opposite arms are equal).

and $\quad \omega L_1 R_3 = \omega L_4 R_2 \quad$ or $\quad L_1 = L_4 \dfrac{R_2}{R_3}$

We can also write that $\quad L_1 = L_4 \cdot \dfrac{R_1{}^*}{R_4}$

Hence, the unknown self-inductance can be measured in terms of the known inductance L_4 and the two resistors. Resistive and reactive terms balance independently and the conditions are independent of frequency. This bridge is often used for measuring the iron losses of the transformers at audio frequency.

---

* Or $\dfrac{L_1}{R_1} = \dfrac{L_4}{R_4}$ *i.e.*, the time constants of the two coils are matched.

The balance condition is shown vectorially in Fig. 16.3. The currents I_4 and I_3 are in phase with I_1 and I_2. This is, obviously, brought about by adjusting the impedances of different branches, so that these currents lag behind the applied voltage V by the same amount. At balance, the voltage drop V_1 across branch 1 is equal to that across branch 4 and $I_3 = I_4$. Similarly, voltage drop V_2 across branch 2 is equal to that across branch 3 and $I_1 = I_2$.

Example 16.1. *The arms of an a.c. Maxwell bridge are arranged as follows : AB and BC are non-reactive resistors of 100 Ω each, DA is a standard variable reactor L_1 of resistance 32.7 Ω and CD comprises a standard variable resistor R in series with a coil of unknown impedance. Balance was obtained with L_1 = 47.8 mH and R = 1.36 Ω. Find the resistance and inductance of the coil.*

(Elect. Inst. & Meas. Nagpur Univ. 1993)

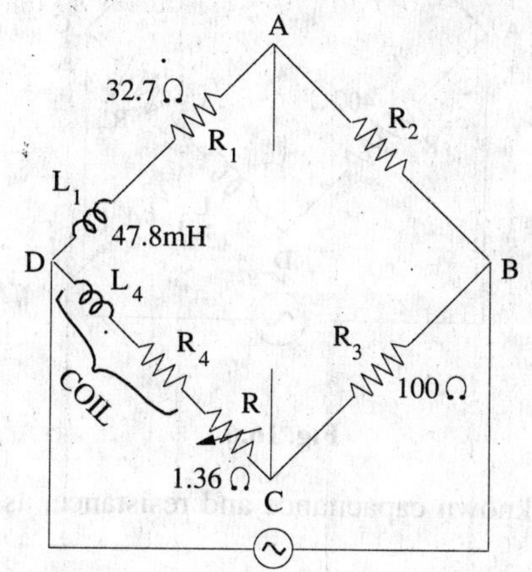

Fig. 16.4

Solution. The a.c. bridge is shown in Fig. 16.4.

Since the products of the resistances of opposite arms are equal

$$\therefore \quad 32.7 \times 100 = (1.36 + R_4)100$$

$$\therefore \quad 32.7 = 1.36 + R_4 \text{ or } R_4 = 32.7 - 1.36 = \mathbf{31.34\ \Omega}$$

Since $L_1 \times 100 = L_4 \times 100 \ \therefore \ L_4 = L_1 = \mathbf{47.8\ mH}$

or because time constants are the same, hence

$L_1/32.7 = L_4/(31.34 + 1.36) \quad \therefore \quad L_4 = 47.8$ mH

16.3. Maxwell-Wien Bridge or Maxwell's L/C Bridge

As referred to in Art. 16.2, the *positive* phase angle of an inductive impedance may be compensated by the *negative* phase angle of a capacitive impedance put in the *opposite* arm. The unknown inductance then becomes known in terms of this capacitance.

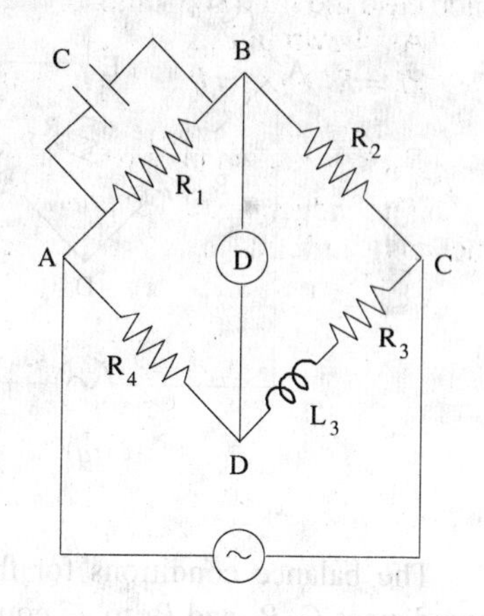

Fig. 16.5

Let us first find the combined impedance of arm 1.

$$\frac{1}{\mathbf{Z}_1} = \frac{1}{R_1} + \frac{1}{-jX_C} = \frac{1}{R_1} + \frac{j}{X_C} = \frac{1}{R_1} + j\omega C = \frac{1 + j\omega CR_1}{R_1}$$

$$\therefore \mathbf{Z}_1 = \frac{R_1}{1 + j\omega CR_1}; \ \mathbf{Z}_2 = R_2$$

$$\mathbf{Z}_3 = R_3 + j\omega L_3 \text{ and } Z_4 = R_4$$

Balance condition is $\mathbf{Z}_1\mathbf{Z}_3 = \mathbf{Z}_2\mathbf{Z}_4$

or
$$\frac{R_1(R_3 + j\omega L_3)}{1 + j\omega CR_1} = R_2R_4 \text{ or } R_1R_3 + j\omega L_3R_1 = R_2R_4 + j\omega CR_1R_2R_4$$

Separating the reals and imaginaries, we get

$$R_1R_3 = R_2R_4 \text{ and } L_3R_1 = CR_1R_2R_4; \ R_3 = \frac{R_2R_4}{R_1} \text{ and } L_3 = CR_2R_4$$

Example 16.2. *The arms of an a.c. Maxwell bridge are arranged as follows.: AB is a non-inductive resistance of 1,000 Ω, in parallel with a capacitor of capacitance 0.5 μF, BC is a non-inductive resistance of 600 Ω, CD is an inductive impedance (unknown) and DA is a non-inductive resistance of 400 Ω. If balance is obtained under these conditions, find the value of the resistance* *and the inductance of the branch CD.*

[Elect. & Electronic Meas. Madras Univ. 1986)

Solution. The bridge is shown in Fig. 16.6. The conditions of balance have already been derived in Art. 16.3 above.

Since $R_1 R_3 = R_2 R_4 \quad \therefore \quad R_3 = R_2 R_4 / R_1$

$$\therefore \quad R_3 = \frac{600 \times 400}{1000} = \mathbf{240\ \Omega}$$

Also $L_3 = C R_2 R_4$

$$= 0.5 \times 10^{-6} \times 400 \times 600$$

$$= 12 \times 10^{-2} = \mathbf{0.12\ H}$$

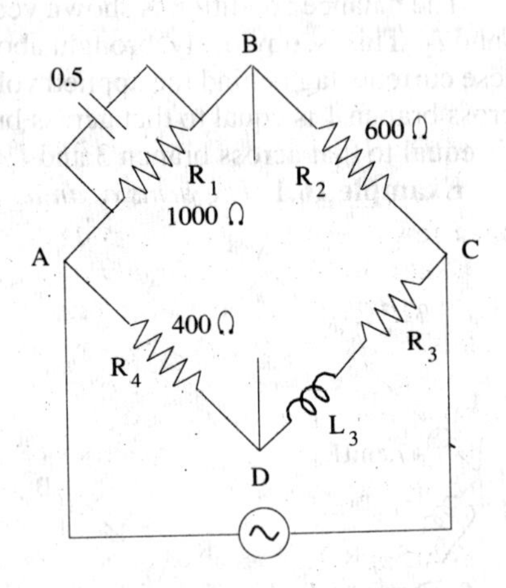

Fig. 16.6

16.4. Anderson Bridge

It is a very important and useful modification of the Maxwell-Wien bridge described in Art. 16.3. In this method, the unknown inductance is measured in terms of a known capacitance and resistance, as shown in Fig. 16.7.

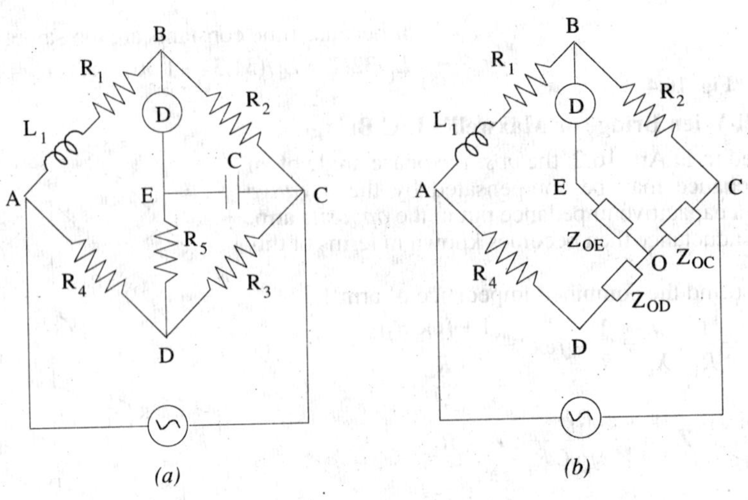

Fig. 16.7

The balance conditions for this bridge may be easily obtained by converting the mesh of impedances C, R_5 and R_3 to an equivalent star with star point O by Δ / Y transformation. As seen from Fig. 16.7 (*b*).

$$\mathbf{Z}_{OD} = \frac{R_3 R_5}{(R_3 + R_5 + 1/j\,\omega\,C)}\,;\; \mathbf{Z}_{OC} = \frac{R_3 | j\omega C}{(R_3 + R_5 + 1/j\omega C)} = \mathbf{Z}_3$$

With reference to Fig. 16.7 (*b*) it is seen that

$$\mathbf{Z}_1 = (R_1 + j\omega L_1)\,;\; \mathbf{Z}_2 = R_2\,;\; \mathbf{Z}_3 = \mathbf{Z}_{OC} \text{ and } \mathbf{Z}_4 = R_4 + \mathbf{Z}_{OD}$$

For balance $\mathbf{Z}_1 \mathbf{Z}_3 = \mathbf{Z}_2 \mathbf{Z}_4 \quad \therefore \quad (R_1 + j\omega L_1) \times \mathbf{Z}_{OC} = R_2 (R_4 + \mathbf{Z}_{OD})$

$$\therefore \quad (R_1 + j\omega L_1) \frac{R_3 / j\omega C}{(R_3 + R_5 + I / j\omega C)} = R_2 \left(R_4 + \frac{R_3 R_5}{R_3 + R_5 + 1/j\omega C} \right)$$

Further simplification leads to $R_2 R_3 R_4 + R_2 R_4 R_5 - j\dfrac{R_2 R_4}{\omega C} + R_2 R_3 R_5 = -j\dfrac{R_1 R_3}{\omega C} + \dfrac{R_3 L_1}{C}$

$$\therefore \quad \frac{-jR_2 R_4}{\omega C} = -\frac{jR_1 R_3}{\omega C} \quad \text{or} \quad R_1 = R_2 R_4 / R_3$$

$$\text{Also} \quad \frac{R_3 L_1}{\omega C} = R_2 R_3 R_4 + R_2 R_3 R_5 + R_2 R_4 R_5 \quad \therefore \quad L_1 = CR_2\left(R_4 + R_5 + \frac{R_4 R_5}{R_3}\right)$$

This method is capable of precise measurements of inductances over a wide range of values from a few micro-henrys to several henrys and is one of the commonest and the best bridge methods.

Example 16.3. *An alternating current bridge is arranged as follows : The arms AB and BC consists of non-inductive resistances of 100-ohm each, the arms BE and CD of non-inductive variable resistances, the arm EC of a capacitor of 1 μ F capacitance, the arm DA of an inductive resistance. The alternating current source is connected to A and C and the telephone receiver to E and D. A balance is obtained when resistances of arms CD and BE are 50 and 2,500 ohm respectively. Calculate the resistance and inductance of arm DA.*

Draw the vector diagram showing voltage at every point of the network.

(Elect. Measurements. Pune Univ. 1985)

Solution. The circuit diagram and voltage vector diagram are shown in Fig. 16.8. As seen, I_2 is vector sum of I_C and I_3. Voltage $V_2 = I_2 R_2 = I_C X_C$. Also, vector sum of V_1 and V_2 is V as well as that of V_3 and V_4. I_C is at right angles to V_2.

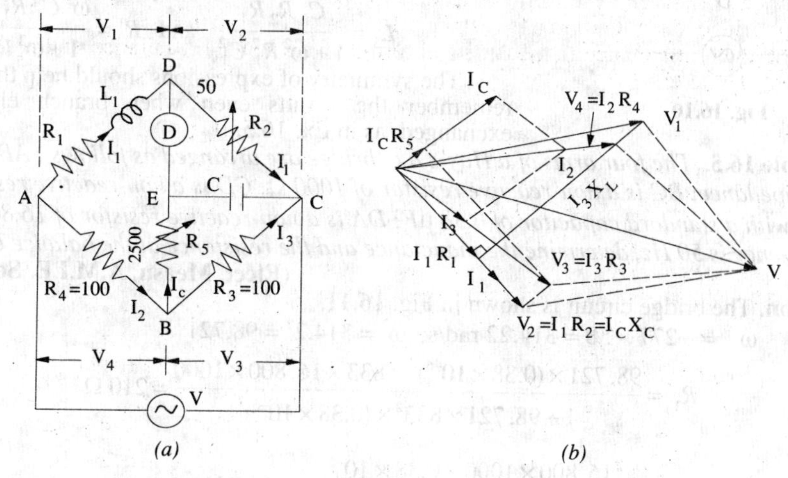

Fig. 16.8

Similarly, V_3 is the vector sum of V_2 and $I_C R_5$.

As shown in Fig. 16.8, $R_1 = R_2 . R_4/R_3 = 50 \times 100/100 = \mathbf{50\ \Omega}$

The inductance is given by $L = CR_2 (R_4 + R_5 + R_4 R_5/R_3)$

$\therefore \quad L = 1{\times}10^{-6} \times 50\ (100 + 2500 + 100 \times 2500/100) =$ **0.2505 H**

Example 16.4. *Fig. 16.9 gives the connection of Anderson's bridge for measuring the inductance L_1 and resistance R_1 of a coil. Find R_1 and L_1 if balance is obtained when $R_3 = R_4 = 2000$ ohms, $R_2 = 1000$ ohms $R_5 = 200$ ohms and $C = 1$ μF. Draw the vector diagram for the voltages and currents in the branches of the bridge at balance.*

(Elect. Measurements, AMIE Sec. B Summer 1990)

Solution. $R_1 = R_2 R_4/R_3 = 1000 \times 2000/2000 = \mathbf{1000\ \Omega}$

$$L_1 = CR_2\left(R_4 + R_5 + \frac{R_4 R_5}{R_3}\right)$$

$$= 1 \times 10^{-6} \times 1000\left(2000 + 200 + \frac{2000 \times 200}{2000}\right) = \mathbf{2.4\ H}$$

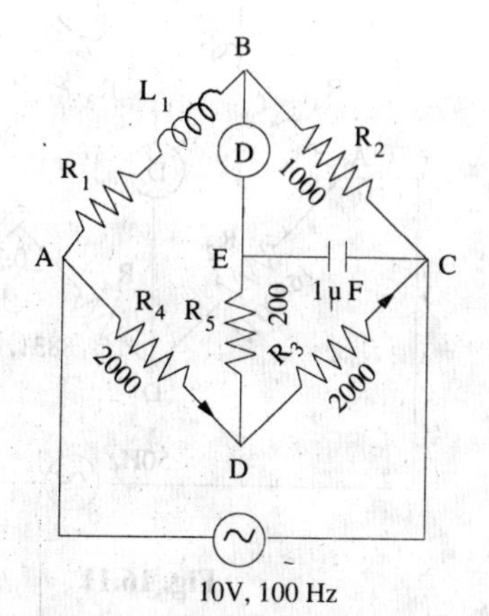

Fig. 16.9

16.5. Hay's Bridge

It is also a modification of the Maxwell-Wien bridge and is particularly useful if the phase angle of the inductive impedance $\phi_m = \tan^{-1}(\omega L/R)$ is large. The network is shown in Fig. 16.10. It is seen that, in this case, a comparatively smaller series resistance R_1 is used instead of a parallel resistance (which has to be of a very large value).

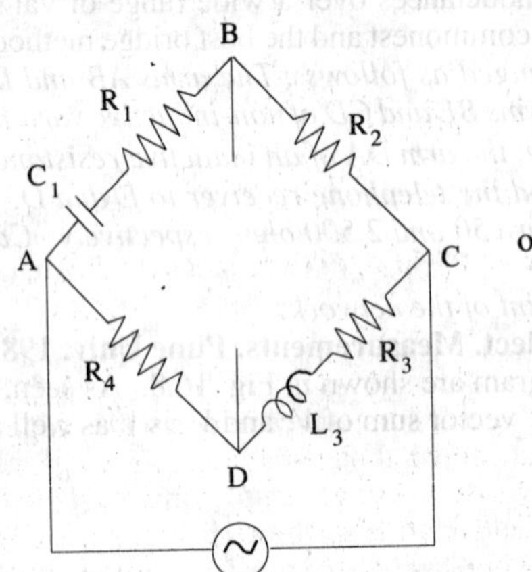

Fig. 16.10

Here $\mathbf{Z}_1 = R_1 - \dfrac{j}{\omega C_1}$; $\mathbf{Z}_2 = R_2$

$\mathbf{Z}_3 = R_3 + j\omega L_3$; $\mathbf{Z}_4 = R_4$

Balance condition is $\mathbf{Z}_1\mathbf{Z}_3 = \mathbf{Z}_2\mathbf{Z}_4$

or $$\left(R_1 - \frac{j}{\omega C_1}\right)(R_3 + j\omega L_3) = R_2 R_4$$

Separating the reals and the imaginaries, we obtain

$$R_1 R_3 + \frac{L_3}{C_1} = R_2 R_4 \text{ and } \omega L_3 R_1 - \frac{R_3}{\omega C_1} = 0$$

Solving these simultaneous equations, we get

$$L_3 = \frac{C_1 R_2 R_4}{1 + \omega^2 R_1^2 C_1^2} \text{ and } R_3 = \frac{\omega^2 C_1^2 R_1 R_2 R_4}{1 + \omega^2 R_1^2 C_1^2}$$

The symmetry of expressions should help the readers to remember the results even when branch elements are exchanged, as in Ex. 16.5.

Example 16.5. *The four arms of a Hay's a.c. bridge are arranged as follows : AB is a coil of unknown impedance; BC is a non-reactive resistor of 1000 Ω; CD is a non-reactive resistor of 833 Ω in series with a standard capacitor of 0.38 μF; DA is a non-reactive resistor of 16,800 Ω. If the supply frequency is 50 Hz, determine the inductance and the resistance at the balance condition.*

(Elect. Measu. A.M.I.E. Sec B, 1992)

Solution. The bridge circuit is shown in Fig. 16.11.

$$\omega = 277 \times 50 = 314.22 \text{ rad/s}; \ \omega^2 = 314.2^2 = 98,721$$

$$R_1 = \frac{98,721 \times (0.38 \times 10^{-6})^2 \times 833 \times 16,800 \times 1000}{1 + 98,721 \times 833^2 \times (0.38 \times 10^{-6})^2} = 210\ \Omega$$

$$L_1 = \frac{16,800 \times 1000 \times 0.38 \times 10^{-6}}{1 + 98,721 \times 833^2 \times (0.38 \times 10^{-6})^2} = 6.38 \text{ H}$$

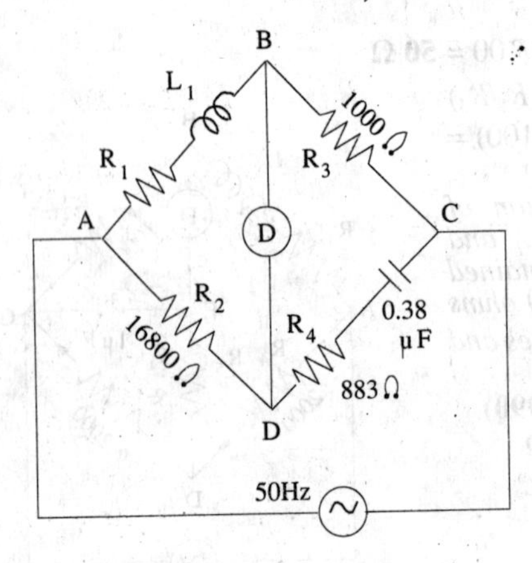

Fig. 16.11

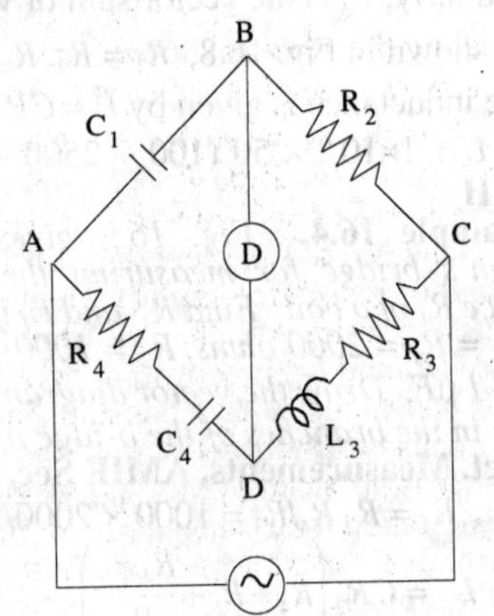

Fig. 16.12

16.6. The Owen Bridge

The arrangement of this bridge is shown in Fig. 16.12. In this method, also, the inductance is determined in terms of resistance and capacitance. This method has, however, the advantage of being useful over a very wide range of inductances with capacitors of reasonable dimensions.

Balance condition is $\mathbf{Z}_1 \mathbf{Z}_3 = \mathbf{Z}_2 \mathbf{Z}_4$

Here $\mathbf{Z}_1 = -\dfrac{j}{\omega C_1}$; $\mathbf{Z}_2 = R_2$; $\mathbf{Z}_3 = R_3 + j\omega L_3$; $\mathbf{Z}_4 = R_4 - \dfrac{j}{\omega C_4}$

$$\therefore \qquad -\frac{j}{\omega C_1}(R_3 + j\omega L_3) = R_2\left(R_4 - \frac{j}{\omega C_4}\right)$$

Separating the reals and imaginaries, we get $R_3 = R_2 \dfrac{C_1}{C_4}$ and $L_3 = C_1 R_2 R_4$

Since ω does not appear in the final balance equations, hence the bridge is unaffected by frequency variations and wave-form.

16.7. Heaviside-Campbell Equal Ratio Bridge

It is a mutual inductance bridge and is used for measuring self-inductance over a wide range in terms of mutual inductometer readings. The connections for Heaviside's bridge employing a standard variable mutual inductance are shown in Fig. 16.13. The primary of the mutual inductometer is inserted in the supply circuit and the secondary having self-inductance L_2 and resistance R_2 is put in arm 2 of the bridge. The unknown inductive impedance having self inductance of L_1 and resistance R_1 is placed in arm 1. The other two arms have pure resistances of R_3 and R_4.

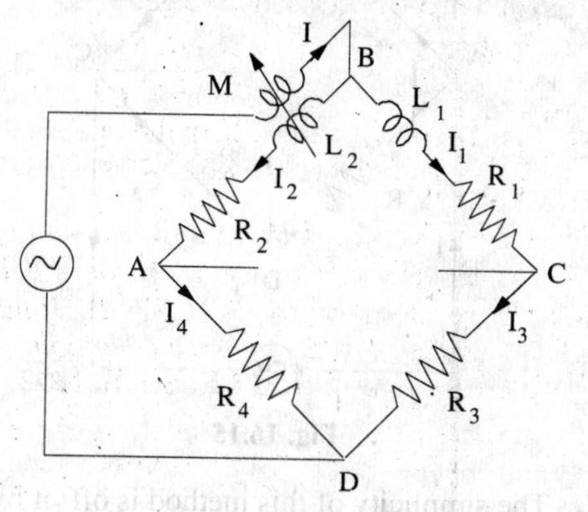

Fig. 16.13

Balance is obtained by varying mutual inductance M and resistances R_3 and R_4.

For balance, $I_1 R_3 = I_2 R_4 \ldots(i)$

$I_1(R_1 + j\omega L_1) = I_2(R_2 + j\omega L_2) + j\omega M I \ldots(ii)$

Since $I = I_1 + I_2$, hence putting the value of I in equation (*ii*), we get

$$I_1' [R_1 + j\omega (L_1 - M)] = I_2[R_2 + j\omega (L_2 + M)] \ldots(iii)$$

Dividing equation (*iii*) by (*i*), we have $\dfrac{R_1 + j\omega (L_1 - M)}{R_3} = \dfrac{R_2 + j\omega (L_2 + M)}{R_4}$

$$\therefore \qquad R_3 [R_2 + j\omega (L_2 + M)] = R_4 [R_1 + j\omega (L_1 - M)]$$

Equating the reals and imaginaries, we have $R_2 R_3 = R_1 R_4$ ...(*iv*)

Also, $R_3 (L_2 + M) = R_4 (L_1 - M)$. If $R_3 = R_4$, then $L_2 + M = (L_1 - M)$ $\therefore$ $L_1 - L_2 = 2M$ ...(*v*)

This bridge, as modified by Campbell, is shown in Fig. 16.14. Here $R_3 = R_4$. A balancing coil or a test coil of self-inductance equal to the self-inductance L_2 of the secondary of the inductometer and of resistance slightly greater than R_2 is connected in series with the unknown inductive impedance (R_1 and L_1) in arm 1. A non-inductive resistance box along with a constant-inductance rheostat are also introduced in arm 2 as shown.

Balance is obtained by varying M and r. Two readings are taken; one when Z_1 is in circuit and second when Z_1 is removed or short-circuited across its terminals.

With unknown impedance Z_1 still in circuit, suppose for balance the values of mutual inductance and r are M_1 and r_1. With Z_1 short-circuited, let these values be M_2 and r_2. Then

$\therefore$ $L_1 = 2(M_1 - M_2)$ and $R_1 = r_1 - r_2$

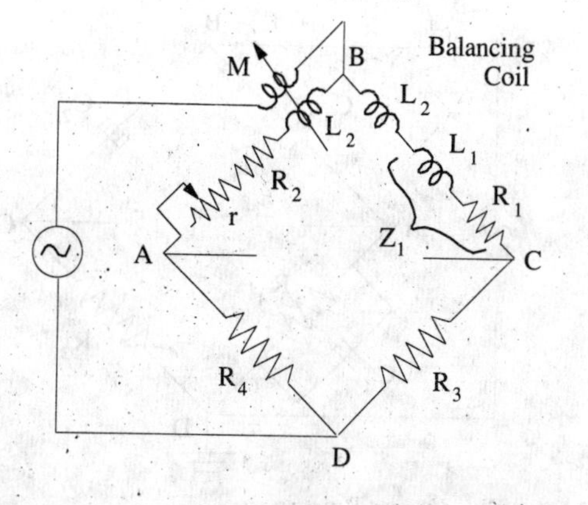

Fig. 16.14

By this method, the self-inductance and resistance of the leads are eliminated.

Example 16.6. *The inductance of a coil is measured by using the Heaviside-Campbell equal ratio bridge. With the test coil short-circuited, balance is obtained when adjustable non-reactive resistance is 12.63 Ω and mutual inductometer is set at 0.1 mH. When the test coil is in circuit, balance is obtained when the adjustable resistance is 25.9 Ω and mutual inductometer is set at 15.9 mH. What is the resistance and inductance of the coil ?*

Solution. With reference to Art. 16.7 and Fig. 16.14, $r_1 = 25.9\ \Omega$, $M_1 = 15.9$ mH

With test coil short-circuited $r_2 = 12.63\ \Omega$; $M_2 = 0.1$ mH

$L_1 = 2(M_1 - M_2) = 2(15.9 - 0.1) = \mathbf{31.6\ mH}$

$R_1 = r_1 - r_2 = 25.9 - 12.63 = \mathbf{13.27\ \Omega}$

16.8. Capacitance Bridges

We will consider only De Sauty bridge method of comparing two capacitances and Schering bridge used for the measurement of capacitance and dielectric loss.

16.9. De Sauty Bridge

With reference to Fig. 16.15, let

C_2 = capacitor whose capacitance is to be measured

C_3 = a standard capacitor

R_1, R_2 = non-inductive resistors

Balance is obtained by varying either R_1 or R_2.

For balance, points B and D are at the same potential.

$$\therefore \quad I_1 R_1 = I_2 R_2 \text{ and } \frac{-j}{\omega C_2} \cdot I_1 = \frac{-j}{\omega C_3} \cdot I_2$$

Dividing one equation by the other, we get

$$\frac{R_1}{R_2} = \frac{C_2}{C_3}; \quad C_2 = C_3 \frac{R_1}{R_2}$$

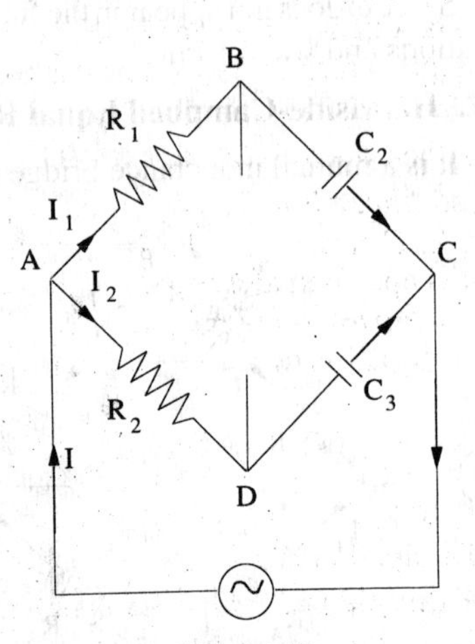

Fig. 16.15

The bridge has maximum sensitivity when $C_2 = C_3$. The simplicity of this method is offset by the impossibility of obtaining a perfect balance if both the capacitors are not free from the dielectric loss. A perfect balance can only be obtained if air capacitors are used.

16.10. Schering Bridge

It is one of the very important and useful methods of measuring the capacitance and dielectric loss of a capacitor. In fact, it is a device for comparing an imperfect capacitor C_2 in terms of a loss-free standard capacitor C_1 [Fig. 16.16 (*a*)]. The imperfect capacitor is represented by its equivalent loss-free capacitor C_2 in series with a resistance r [Fig. 16.16 (*b*)].

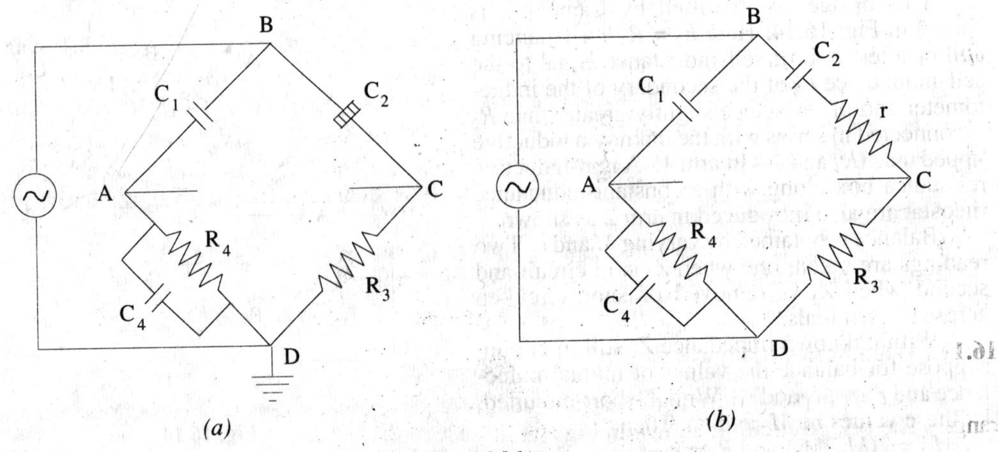

Fig. 16.16

For high voltage applications, the voltage is applied at the junctions shown in the figure. The junction between arms 3 and 4 is earthed. Since capacitor impedances at lower frequencies are much higher than resistances, most of the voltage will appear across capacitors. Grounding of the junction affords safety to the operator form the high-voltage hazards while making balancing adjustment in arms 3 and 4.

Now $\quad \mathbf{Z}_1 = \dfrac{-j}{\omega C_1}$; $\mathbf{Z}_2 = r - \dfrac{j}{\omega C_2}$; $\mathbf{Z}_3 = R_3$; $\mathbf{Z}_4 = \dfrac{1}{(1/R_4) + j\omega C_4} = \dfrac{R_4}{1 + j\omega C_4 R_4}$

For balance, $$\mathbf{Z}_1\mathbf{Z}_3 = \mathbf{Z}_2\mathbf{Z}_4$$

or $\quad \dfrac{-jR_3}{\omega C_1} = \left(r - \dfrac{j}{\omega C_2}\right)\left(\dfrac{R_4}{1 + j\omega C_4 R_4}\right)$ or $\dfrac{-j R_3}{\omega C_1}(1 + \omega_a C_4 R_4) = R_4\left(r - \dfrac{j}{\omega C_2}\right)$

Separating the reals and imaginaries, we have $C_2 = C_1(R_4/R_3)$ and $r = R_3 . (C_4/C_1)$.

The quality of a capacitor is usually expressed in terms of its phase defect angle or dielectric loss angle which is defined as the angle by *which current departs from exact quadrature from the applied voltage i.e.* the complement of the phase angle. If ϕ is the actual phase angle and δ the defect angle, then $\phi + \delta = 90°$. For small values of δ, tan δ = sin δ = cos ϕ (approximately). Tan δ is usually called the *dissipation factor* of the *R–C* circuit. For low power factors, therefore, dissipation factor is approximately equal to the power factor.

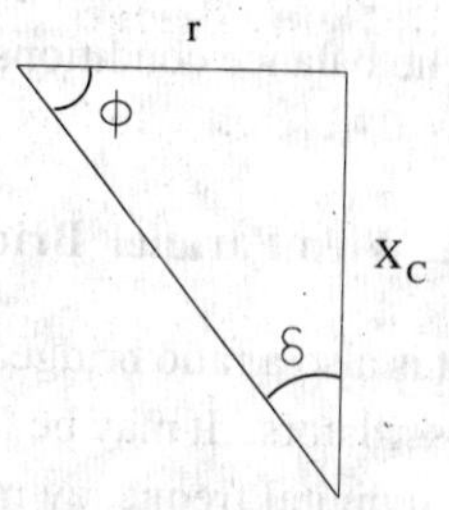

As shown in Fig. 16.17,

Dissipation factor = power factor = tan δ

$$= \frac{r}{X_C} = \frac{r}{1/\omega C_2} = \omega r C_2$$

Putting the value of rC_2 from above,

Dissipation factor = $\omega r C_2 = \omega C_4 R_4$ = power factor.

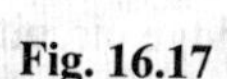

Fig. 16.17

Example 16.7. *In a test on a bakelite sample at 20 kV, 50 Hz by a Schering bridge, having a standard capacitor of 106 μF, balance was obtained with a capacitance of 0.35 μF in parallel with a non-inductive resistance of 318 ohms, the non-inductive resistance in the remaining arm of the bridge being 130 ohms. Determine the capacitance, the p.f. and equivalent series resistance of the specimen. Derive any formula used. Indicate the precautions to be observed for avoiding errors.*

(Elect. Engg. Paper I Indian Engg. Services 1991)

Solution. Here C_1 = 106 pF, C_4 = 0.35 μF, R_4 = 318 Ω, R_3 = 130 Ω.

$$C_2 = C_1.(R_4|R_3) = 106 \times 318/130 = \mathbf{259.3\ pF}$$

$$r = R_3 \cdot (C_4/C_1) = 130 \times 0.35 \times 10^{-6}/106 \times 10^{-12} = \mathbf{0.429\ M\Omega}$$

$$\text{p.f.} = \omega_r C_2 = (2\pi \times 50) \times 0.429 \times 10^{+6} \times 259.3 \times 10^{-12} = \mathbf{0.035}$$

Example 16.8. *A losy capacitor is tested with a Schering bridge circuit. Balance obtained with the capacitor under test in one arm, the succeeding arms being, a non-inductive resistor of 100 Ω, a non-reactive resistor of 309 Ω in parallel with a pure capacitor of 0.5 μF and a standard capacitor of 109 μμF. The supply frequency is 50 Hz. Calculate from the equation at balance the equivalent series capacitance and power factor (at 50 Hz) of the capacitor under test.*

(Measu. & Instru., Nagpur Univ. 1992)

Solution. Here, we are given

$$C_1 = 109\ \text{pF} ;\ R_3 = 100\ \Omega ;\ C_4 = 0.5\ \mu\text{F} ;\ R_4 = 309\ \Omega$$

Equivalent capacitance is $\quad C_2 = 109 \times 309/100 = \mathbf{336.8\ pF}$

$$p.f. = \omega C_4 R_4 = 314 \times 0.5 \times 10^{-6} \times 309 = \mathbf{0.0485}$$

16.11. Wien Series Bridge

It is a simple ratio bridge and is used for audio-freqeuncy measurement of capacitors over a wide range. The bridge circuit is shown in Fig. 16.18.

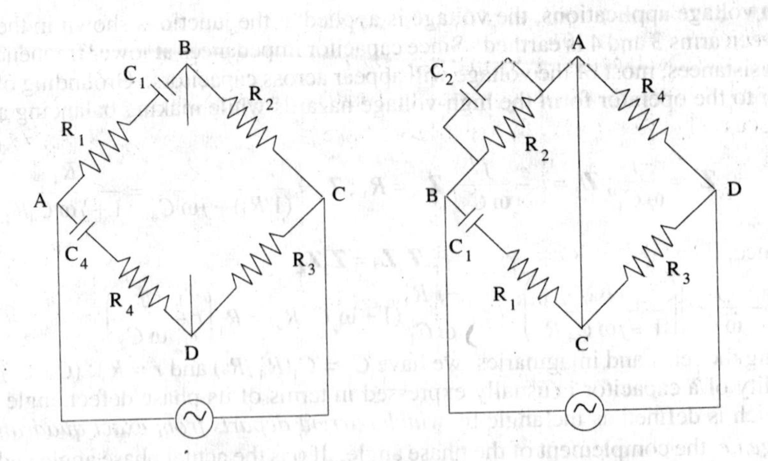

Fig. 16.18 Fig. 16.19

The balance conditions may be obtained in the usual way. For balance

$$R_1 = R_2 R_4/R_3 \text{ and } C_1 = C_4 (R_3/R_2)$$

16.12. Wien Parallel Bridge

It is also a ratio bridge used mainly as the feedback network in the wide-range audio-frequency *R-C* oscillators. It may be used for measuring audio-frequencies although it is not as accurate as the modern digital frequency meters.

The bridge circuit is shown in Fig. 16.19. In the simple theory of this bridge, capacitors C_1 and C_2 are assumed to be loss-free and resistances R_1 and R_2 are separate resistors.

The usual relationship for balance gives

$$R_4\left(R_1 - \frac{j}{\omega C_1}\right) = R_3\left(\frac{R_2}{1 + j\omega C_2 R_2}\right) \text{ or } R_4\left(R_1 - \frac{j}{\omega C_1}\right)(1 + j\omega C_2 R_2) = R_2 R_3$$

Separating the real and imaginary terms, we have

$$R_1 R_4 + R_2 R_4 \frac{C_2}{C_1} = R_2 R_3 \text{ or } \frac{C_2}{C_1} = \frac{R_3}{R_4} - \frac{R_1}{R_2} \quad \ldots(i)$$

and

$$\omega C_2 R_1 R_2 R_4 - \frac{R_4}{\omega C_1} = 0 \text{ or } \omega^2 = \frac{1}{R_1 R_2 C_1 C_2} \quad \ldots(ii)$$

or

$$f = \frac{1}{2\pi\sqrt{R_1 R_2 C_1 C_2}} \text{ Hz}$$

Note. Eq. (*ii*) may be used to find angular frequency ω of the source if other terms are known. For such purposes, it is convenient to make $C_1 = 2C_2$, $R_3 = R_4$ and $R_2 = 2R_1$. In that case, the bridge has equal ratio arms so that Eq. (*i*) will always be satisfied. The bridge is balanced simultaneously by adjusting R_2 and R_1 (though maintaining $R_2 = 2R_1$). Then, as seen from Eq. (*ii*) above

$$\omega^2 = 1/(R_1.2R_1.2C_2.C_2) \text{ or } \omega = 1/(2R_1C_2)$$

Example 16.9. *The arms of a four-arm bridge ABCD, supplied with a sinusoidal voltage, have the following values :*

AB : 200 ohm resistance in parallel with 1 μF capacitor; BC : 400 ohm resistance ; CD : 1000 ohm resistance and DA : resistance R in series with a 2μF capacitor.

Determine (i) the value of R and (ii) the supply frequency at which the bridge will be balanced.
(Elect. Meas. A.M.I.E. Sec. 1991)

Solution. The bridge circuit is shown in Fig. 16.20.

(*i*) As discussed in Art. 16.12, for balance we have

$$\frac{C_2}{C_1} = \frac{R_3}{R_4}\frac{R_1}{R_2} \quad \text{or} \quad \frac{2}{1} = \frac{1000}{4000}\frac{R_1}{200}$$

$$\therefore \quad R_1 = 200 \times 0.5 = \mathbf{100\ \Omega}$$

(*ii*) The frequency at which bridge is balanced is given by

$$f = \frac{1}{2\pi\sqrt{R_1 R_2 C_1 C_2}} \text{ Hz}$$

$$= \frac{10^6}{2\pi\sqrt{100 \times 200 \times 1 \times 2}} = \mathbf{796\ Hz}$$

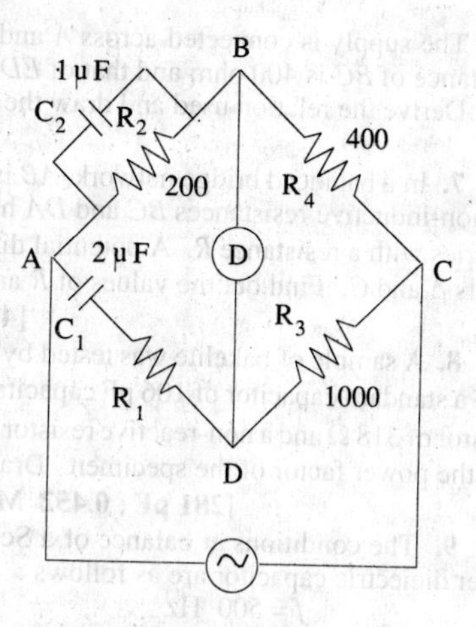

Fig. 16.20

Tutorial Problems No. 16.1

1. In Anderson a.c. bridge, an impedance of inductance *L* and resistance *R* is connected between *A* and *B*. For balance following data is obtained. An ohmic resistance of 1000 Ω each in arms *AD* and *CD*, a non-inductive resistance of 500 Ω in *BC*, a pure resistance of 200 Ω between points *D* and *E* and a capacitor of 2μF between *C* and *E*. The supply is 10 volt (A.C.) at a frequency of 100 Hz and is connected across points *A* and *C*. Find *L* and *R*. **[1.4 H ; 500 Ω]**

2. A balanced bridge has the following components connected between its five nodes, A, B, C, D and *E* :

Between *A* and *B* :	1,000 ohm resistance;	Between *B* and *C*:	1,000 ohm resistance
Between *C* and *D* :	an inductor;	Between *D* and *A* :	218 ohm resistance
Between *A* and *E* :	469 ohm resistance;	Between *E* and *B*:	10μF capacitance
Between *E* and *C* :	a detector ;	Between *B* and *D* :	a power supply (a.c.)

Derive the equations of balance and hence deduce the resistance and inductance of the inductor.
[R = 218 Ω , L = 7.89 H] *(Elect. Theory and Meas, London Univ.)*

3. An a.c. bridge is arranged as follows : The arms *AB* and *BC* consist of non-inductive resistance of 100 Ω, the arms, *BE* and *CD* of non-inductive variable resistances, the arm *EC* of a capacitor of 1 μF capacitance, the arm *DA* of an inductive resistance. The a.c. source is connected to *A* and *C* and the telephone receiver to *E* and *D*. A balance is obtained when the resistances of the arms *CD* and *BE* are 50 Ω and 2500 Ω respectively.

Calculate the resistance and the inductance of the arm *DA*. What would be the effect of harmonics in the waveform of the alternating current source ? **[50 Ω ; 0.25 H]**

4. For the Anderson's bridge of Fig. 16.21, the values are under balance conditions. Determine the values of unknown resistance *R* and inductance *L*. **[R = 500 Ω ; L = 1.5 H]**
(Elect. Meas & Inst. Madras Univ. Nov. 1978)

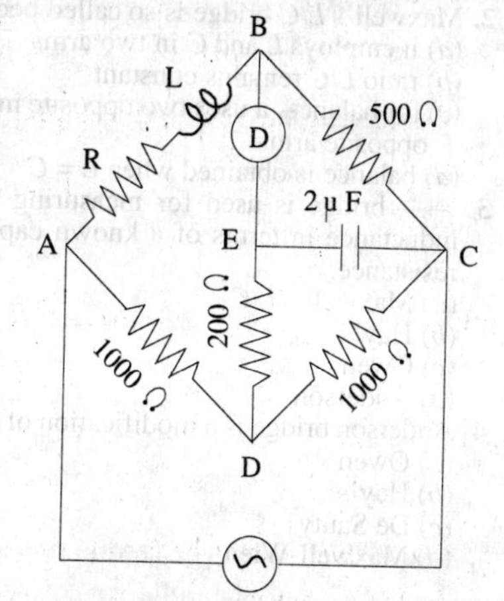

Fig. 16.21

5. An Anderson's bridge is arranged as under and balanced for the following values of the bridge components :

Branch *AB* – unknown coil of inductance *L* and resistance *R*
Branch *BC* –non-inductive resistance of 500 Ω
Branches *AD* & *CD* –non-inductive resistance of 100 Ω each
Branch *DE* – non-inductive resistance of 200 Ω
Branch *EB* – vibration galvanometer
Branch *EC*– 2.0 μF capacitance
Between *A* and *C* is 10 V, 100-Hz a.c. supply. Find the values of *R* and *L* of the unknown coil.
[R = 500 Ω ; L = 0.5 H] *(Elect. Meas & Meas. Inst., Gujarat Univ. Oct. 1977)*

6. An a.c. Anderson bridge is arranged as follows :
(*i*) branches *BC* and *ED* are variable nonreactive resistors
(*ii*) branches *CD* and *DA* are nonreactive resistors of 200 ohm each
(*iii*) branch *CE* is a loss-free capacitor of 1 μF capacitance.

The supply is connected across A and C and the detector across B and E. Balance is obtained when the resistance of BC is 400 ohm and that of ED is 500 ohm. Calculate the resistance and inductance of AB. Derive the relation used and draw the vector diagram for balanced condition of the bridge.

[400 Ω; 0.48 H] (*Elect. Measurements, Poona Univ. May 1979*)

7. In a balanced bridge network, AB is a resistance of 500 ohm in series with an inductance of 0.18 henry, the non-inductive resistances BC and DA have values of 1000 ohm and arm CD consists of a capacitance of C in series with a resistance R. A potential difference of 5 volts at a frequency $5000/2\pi$ is the supply between the points A and C. Find out the values of R and C and draw the vector diagram.

[472 Ω ; 0.235 μF] (*Elect. Measurements, Poona Univ. April 1979*)

8. A sample of bakelite was tested by the Schering bridge method at 25 kV, 50-Hz. Balance was obtained with a standard capacitor of 106 pF capacitance, a capacitor of capacitance 0.4 μF in parallel with a non-reactive resistor of 318 Ω and a non-reactive resistor of 120 Ω. Determine the capacitance, the equivalent series resistance and the power factor of the specimen. Draw the phas or diagram for the balanced bridge.

[281 pF ; 0.452 M Ω; 0.04] (*Elect. Measurements-II ; Bangalore Univ. Jan. 1981*)

9. The conditions at balance of a Schering bridge set up to measure the capacitance and loss angle of a paper dielectric capacitor are as follows :

$f = 500$ Hz

Z_1 = a pure capacitance of 0.1 μF

Z_2 = a resistance of 500 Ω shunted by a capacitance of 0.0033 μF

Z_3 = pure resistance of 163 Ω

Z_4 = the capacitor under test

Calculate the approximate values of the loss resistance of the capacitor assuming –

(*a*) series loss resistance (*b*) shunt loss resistance.

[5.37 Ω, 197,000 Ω] (*London Univ.*)

OBJECTIVE TESTS–16

1. Maxwell-Wien bridge is used for measuring
(*a*) capacitance
(*b*) dielectric loss
(*c*) inductance
(*d*) phase angle

2. Maxwell's L/C bridge is so called because
(*a*) it employs L and C in two arms
(*b*) ratio L/C remains constant
(*c*) for balance, it uses two opposite impedances in opposite arms
(*d*) balance is obtained when $L = C$

3. —— bridge is used for measuring an unknown inductance in terms of a known capacitance and resistance.
(*a*) Maxwell's L/C
(*b*) Hay's
(*c*) Owen
(*d*) Anderson

4. Anderson bridge is a modification of —— bridge.
(*a*) Owen
(*b*) Hay's
(*c*) De Sauty
(*d*) Maxwell-Wien

5. Hay's bridge is particularly useful for measuring
(*a*) inductive impedance with large phase angle
(*b*) mutual inductance
(*c*) self inductance
(*d*) capacitance and dielectric loss

6. The most useful ac bridge for comparing capacitances of two air capacitors is —— bridge.
(*a*) Schering
(*b*) De Sauty
(*c*) Wien series
(*d*) Wien parallel

7. Heaviside-Campbell Equal Ratio bridge is used for measuring
(*a*) self-inductance in terms of mutual inductance
(*b*) capacitance in terms of inductance
(*c*) dielectric loss of an imperfect capacitor
(*d*) phase angle of a coil

8. The capacitance and dielectric loss of a capacitor is generally measured with the help of —— bridge.
(*a*) De Sauty
(*b*) Schering
(*c*) Wien Series
(*d*) Anderson

Answers

1. *c* **2.** *c* **3.** *d* **4.** *d* **5.** *a* **6.** *b* **7.** *a* **8.** *b* **9.** *a* **10.** *c*

17 A.C. FILTER NETWORKS

17.1. Introduction

The reactances of inductors and capacitors depend on the frequency of the a.c. signal applied to them. That is why these devices are known as frequency-selective. By using various combinations of resistors, inductors and capacitors, we can make circuits that have the property of passing or rejecting either low or high frequencies or bands of frequencies. These frequency-selective networks, which alter the amplitude and phase characteristics of the input a.c. signal, are called filters. Their performance is usually expressed in terms of how much attenuation a band of frequencies experiences by passing through them. Attenuation is commonly expressed in terms of decibels (dB).

17.2. Applications

A.C. filters find application in audio systems and television etc. Bandpass filters are used to select frequency ranges corresponding to desired radio or television station channels. Similarly, bandstop filters are used to reject undesirable signals that may contaminate the desirable signal. For example, low-pass filters are used to eliminate undesirable hum in d.c. power supplies.

No loudspeaker is equally efficient over the entire audible range of frequencies. That is why high-fidelity loudspeaker systems use a combination of low-pass, high-pass and bandpass filters (called crossover networks) to separate and then direct signals of appropriate frequency range to the different loudspeakers making up the system. Fig. 17.1 shows the output circuit of a high-fidelity

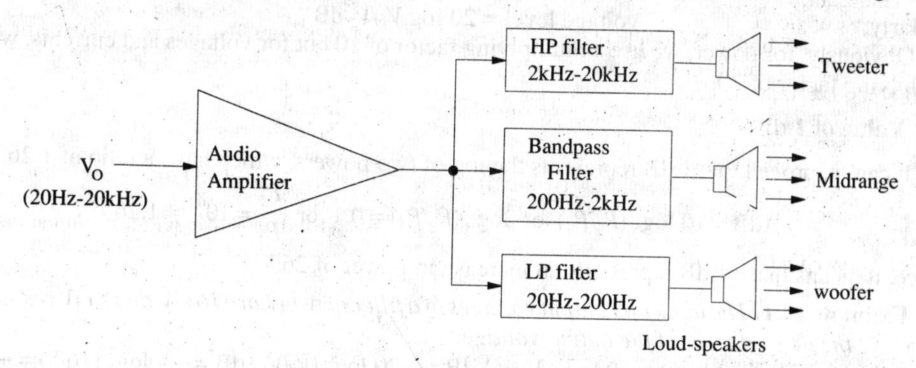

Fig. 17.1

audio amplifier, which uses three filters to separate, the low, mid-range and high frequencies, for feeding them to individual loudspeakers, best able to reproduce them.

17.3. Different Types of Filters

A.C. filter networks are divided into two major categories : (*i*) active networks and (*ii*) passive networks.

Active filter networks usually contain transistors and/or operational amplifiers in combination with *R*, *L* and *C* elements to obtain the desired filtering effect. These will not be discussed in this book. We will consider passive filter networks only which usually consist of series-parallel combinations of *R*, *L* and *C* elements. There are four types of such networks, as described below :

1. Low-Pass Filter. As the name shows, it allows only low frequencies to pass through, but attenuates (to a lesser or greater extent) all higher frequencies. The maximum frequency which it allows to pass through, is called cutoff frequency f_c (also called break frequency). There are R_L and R_C low-pass filters.

2. High–Pass Filter. It allows signals with higher frequencies to pass from input to output while rejecting lower frequencies. The minimum frequency it allows to pass is called cutoff frequency f_c There are R_L and R_C high-pass filters.

3. Bandpass Filter. It is a resonant circuit which is tuned to pass a certain band or range of frequencies while rejecting all frequencies below and above this range (called passband).

4. Bandstop Filter. It is a resonant circuit that rejects a certain band or range of frequencies while passing all frequencies below and above the rejected band. Such filters are also called wavetraps, notch filters or band-elimination, band-separation or band-rejection filters.

17.4. Octaves and Decades of Frequency

A filter's performance is expressed in terms of the number of decibels the signal is increased or decreased per frequency octave or frequency decade. An octave means a doubling or halving of a frequency whereas a decade means tenfold increase or decrease in frequency.

17.5. The Decibel System

These system of logarithmic measurement is widely used in audio, radio, TV and instrument industry for comparing two voltages, currents or power levels. These levels are measured in a unit called bel (B) or decibel (dB) which is 1/10th of a bel.

Suppose we want to compare the output power P_0 of a filter with its input power P_i. The power level change is $= 10 \log_{10}{}^{th} P_0/P_i$ dB

It should be noted that dB is the unit of power change (*i.e.* increase or decrease) and not of power itself. Moreover, 20 dB is not twice as much power as 10 dB.

However, when voltage and current levels are required, then the expressions are :

$$\text{Current level} = 20 \log_{10}(I_0/I_i) \text{ dB}$$

Similarly, $\quad$ voltage level $= 20 \log V_0/V_i$ dB

Obviously, for power, we use a multiplying factor of 10 but for voltages and currents, we use a multiplying factor of 20.

17.6. Value of 1 dB

It can be proved that 1 dB represents the log of two powers, which have a ratio of 1.26.

$$1\text{dB} = 10 \log_{10}(P_2/P_1) \text{ or } \log_{10}(P_2/P_1) = 0.1 \text{ or } \frac{P_2}{P_1} = 10^{0.1} = 1.26$$

Hence, it means that +1 dB represents an increase in power of 26%.

Example 17.1. *The input and output voltages of a filter network are 16 mV and 8 mV respectively. Calculate the decibel level of the output voltage.*

Solution. Decibel level $= 20 \log_{10}(V_0/V_i)$ dB $= -20 \log_{10}(V_i/V_0)$ dB $= -20\log_{10}(16/8) = -6$ dB

Whenever voltage ratio is less than 1, its log is negative which is often difficult to handle. In such cases, it is best to invert the fraction and then make the result negative, as done above.

Example 17.2. *The output power of a filter is 100 mW when the signal frequency is 5 kHz. When the frequency is increased to 25 kHz, the output power falls to 50 mW. Calculate the dB change in power.*

Solution. The decibel change in power is

$= 10 \log_{10}(50/100) = -10 \log_{10}(100/50) = -10 \log_{10}2 = -10 \times 0.3 = -3$ dB

Example 17.3. *The output voltage of an amplifier is 10 V at 5 kHz and 7.07 V at 25 kHz. What is the decibel change in the output voltage ?*

Solution. Decibel change $= 20 \log_{10}(V_0/V_i) = 20 \log_{10}(7.07/10) = -20 \log_{10}(10/7.07)$

$$= -20 \log_{10}(1.4/4)$$
$$= -20 \times 0.15 = -3 \text{ dB}$$

17.7. Low-Pass RC Filter

A simple low-pass *RC* filter is shown in Fig. 17.2 (*a*). As stated earlier, it permits signals of low frequencies upto f_c to pass through while attenuating frequencies above f_c. The range of frequencies upto f_c is called the passband of the filter. Fig. 17.2 (*b*) shows the frequency response curve of such a filter. It shows how the signal output voltage V_0 varies with the signal frequency. As seen at f_c, output signal voltage is reduced to 70.7% of the input voltage. The output is said to be −3 dB at f_c. Signal outputs beyond f_c roll-off or attenuate at a fixed rate of −6 dB/octave or −20 dB/decade. As seen from the frequency-phase response curve of Fig. 17.2 (*c*), the phase angle between V_0 and V_i is 45° at cutoff frequency f_c.

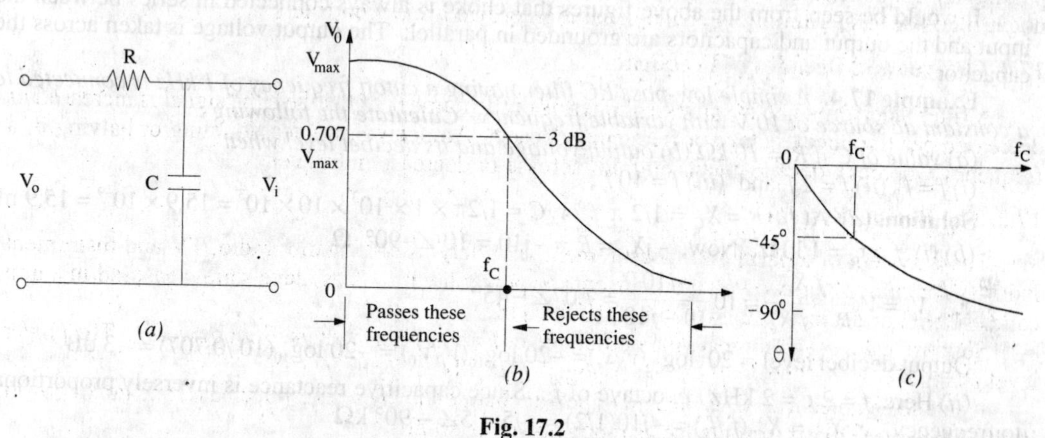

Fig. 17.2

By definition cutoff frequency f_c occurs where (*a*) $V_0 = 70.7\%\ V_i$ *i.e.* V_0 is −3 dB down from V_i (*b*) $R = X_C$ and $V_R = V_C$ in magnitude. (*c*) The impedance phase angle $\theta = -45°$. The same is the angle between V_0 and V_i.

As seen, the output voltage is taken across the capacitor. Resistance *R* offers fixed opposition to frequencies but the reactance offered by capacitor *C* decreases with increase in frequency. Hence, low-frequency signal develops over *C* whereas high-frequency signals are grounded. Signal frequencies above f_c develop negligible voltage across *C*. Since *R* and *C* are in series, we can find the low-frequency output voltage V_0 developed across *C* by using the voltage-divider rule.

$$\therefore \qquad V_0 = V_i \frac{-j\,X_C}{R - j\,X_C} \quad \text{and} \quad f_c = \frac{1}{2\,\pi\,C\,R}$$

17.8. Other Types of Low-Pass Filters

There are many other types of low-pass filters in which instead of pure resistance, series chokes are commonly used alongwith capacitors.

(*i*) **Inverted −L Type.** It is shown in Fig. 17.3 (*a*). Here, inductive reactance of the choke blocks higher frequencies and *C* shorts them to ground. Hence, only low frequencies below f_c (for which *X* is very low) are passed without significant attenuation.

(*ii*) **T-Type.** It is shown in Fig. 17.3 (*b*). In this case, a second choke is connected on the output side which improves the filtering action.

(*iii*) **π-Type.** It is shown in Fig. 17.3 (*c*). The additional capacitor further improves the filtering action by grounding higher frequencies.

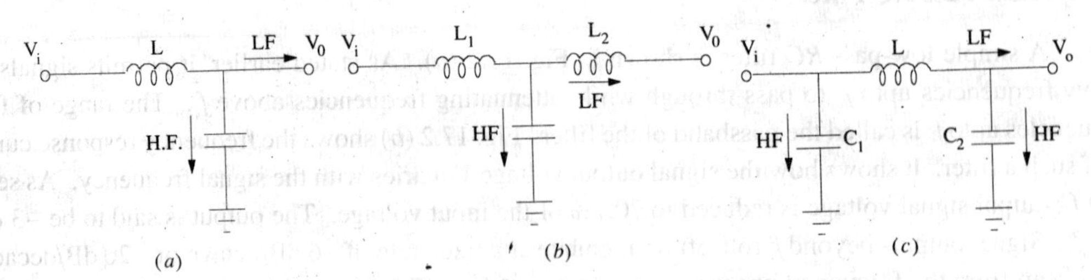

Fig. 17.3

It would be seen from the above figures that choke is always connected in series between the input and the output and capacitors are grounded in parallel. The output voltage is taken across the capacitor.

Example 17.4. *A simple low-pass RC filter having a cutoff frequency of 1 kHz is connected to a constant ac source of 10 V with variable frequency. Calculate the following :*

(a) value of C if R = 10 kΩ (b) output voltage and its decibel level when

(i) $f = f_c$ *(ii)* $f = 2 f_c$ and *(iii)* $f = 10 f_c$.

Solution. *(a)* At f_c, $R = X_C = 1/2\pi f_c$ or $C = 1/2\pi \times 1 \times 10^3 \times 10 \times 10^3 = 15.9 \times 10^{-9} = 15.9$ nF

(b) (i) $f = f_C = 1$ kHz. Now, $-jX_C = R = -j10 = 10 \angle -90°\ \Omega$

$$\therefore \quad V_0 = V_i \frac{-j X_C}{R - j X_C} = 10 \frac{-j10}{10 - j10} = 7.07\angle -45°$$

Output decibel level = $20 \log_{10}(v_0/v_i) = -20 \log_{10}(V_i/V_0) = -20 \log_{10}(10/0.707) = -3$ dB

(ii) Here, $f = 2 f_c = 2$ kHz *i.e.* octave of f_c. Since capacitive reactance is inversely proportional to frequency, $\therefore X_{C2} = X_{C1}(f_1/f_2) = -j10(1/2) = -j5 = 5\angle -90°$ kΩ

$$\therefore V_0 = \frac{5\angle -90°}{10 - j5} = \frac{5°\angle -90°}{11.18\angle -26.6°} = 4.472\angle -63.4°$$

Decibel level = $-20\log_{10}(V_i/V_0) = -20 \log_{10}(10/4.472) = -6.98$ dB

(iii) $X_{C3} = X_{C1}(f_1/f_3) = -j10\ (1/10) = j1 = 1\angle -90°$ kΩ

$$\therefore \quad V_0 = 10 \frac{10\angle -90°}{10 - j1} = 1\angle -84.3°$$

Decibel level = $-20 \log_{10}(10/1) = -20$ dB

17.9. Low–Pass RL Filter

It is shown in Fig. 17.4 (*a*). Here, coil offers high reactance to high frequencies and low reactance to low frequencies. Hence, low frequencies upto f_c can pass through the coil without much opposition. The output voltage is developed across *R*. Fig. 17.4 (*b*) shows the frequency-output response curve of the filter. As seen at f_c, $V_0 = 0.707\ V_i$ and its attenuation level is -3 *dB* with respect to V_0 *i.e.* the voltage at $f = 0$.

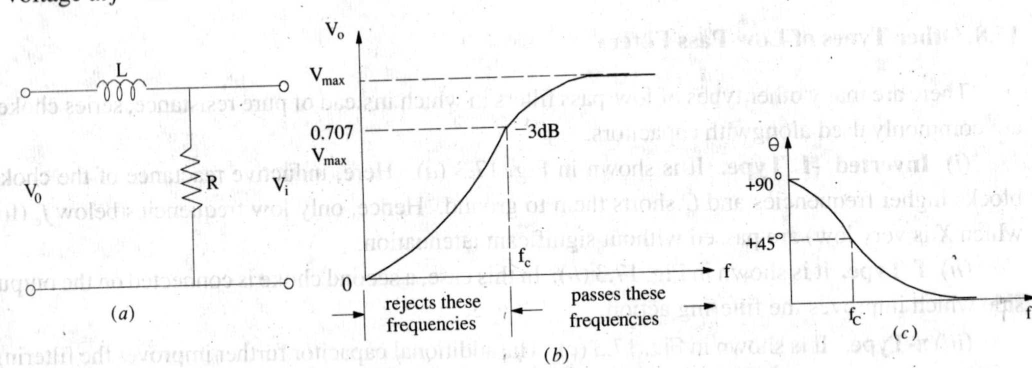

Fig. 17.4

However, it may be noted that being an *RL* circuit, the impedance phase angle is +45° (and not −45° as in low-pass *RC* filter). Again at f_c, $R = X_L$.

Using the voltage-divider rule, the output voltage developed across *R* is given by

$$V_0 = V_i \frac{R}{R + jX_L} \quad \text{and} \quad f_c = \frac{R}{2\pi L}$$

Example 17.5. *An ac signal having constant amplitude of 10 V but variable frequency is applied across a simple low-pass RL circuit with a cutoff frequency of 1 kHz. Calculate (a) value of L if R = 1 k Ω (b) output voltage and its decibel level when (i) $f = f_c$ (ii) $f = 2f_c$ and (iii) $f = 10 f_c$.*

Solution. (*a*) $L = R/2\pi f_c = 1 \times 10^3/2\pi \times 10^3 = 159.2$ mH

(*b*) (*i*) $f = f_c = 1k\ Hz$; $jX_L = R = j1$; $V_0 = 10 \dfrac{1}{(1+j1)} = 7.07 \angle -45°$ V

Decibel decrease = $-20 \log_{10}(V_i/V_b) = -20 \log_{10}10/7.07 = -3$ dB

(*ii*) $f = 2f_C = 2$ kHz. Since X_L varies directly with

$$f, X_{L2} = X_{L1}(f_2/f_1) = 1 \times 2/1 = 2\ k\Omega$$

$$\therefore \quad V_0 = 10 \frac{1}{(1+j2)} = \frac{10}{2.236\angle 63.4°} = 4.472\angle -63.4°$$

Decibel decrease = $-20 \log_{10}(10/4.472) = -6.98$ dB

(*iii*) $f = 10 f_c = 10$ kHz ; $X_{L3} = 1 \times 10/1 = 10\ \Omega$, $V_0 = 10 \dfrac{1}{(1+j10)} = 1\angle -84.3°$

Decibel decrease = $-20 \log_{10}(10/1) = -20$ dB

17.10. High-Pass RC Filter

It is shown in Fig. 17.5 (*a*). Lower frequencies experience considerable reactance by the capacitor and are not easily passed. Higher frequencies encounter little reactance and are easily passed. The high frequencies passing through the filter develop output voltage V_0 across *R*. As seen from the frequency response of Fig. 13.5 (*b*), all frequencies above f_c are passed whereas those below it are attenuated. As before, f_c corresponds to −3 dB output voltage or half-power point. At f_c, $R = X_C$ and the phase angle between V_0 and V_i is +45° as shown in Fig. 17.5 (*c*). It may be noted that high-pass *RC* filter can be obtained merely by interchanging the positions of *R* and *C* in the low-pass *RC* filter of Fig. 17.5 (*a*).

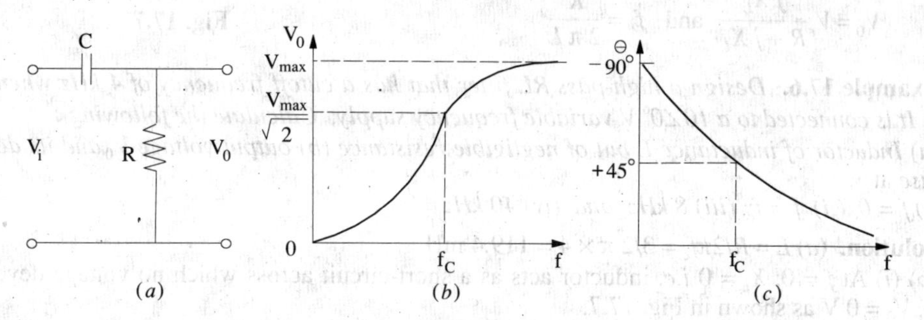

Fig. 17.5

Since *R* and *C* are in series across the input voltage, the voltage drop across *R*, as found by the voltage-divider rule, is

$$V_0 = V_i \frac{R}{R - jX_C} \quad \text{and} \quad f_c = \frac{1}{2\pi CR}$$

A very common application of the series capacitor high-pass filter is a coupling capacitor between two audio amplifier stages. It is used for passing the amplified audio signal from one stage to the next and simultaneously block the constant d.c. voltage.

Other high-pass *RC* filter circuits exist besides the one shown in Fig. 17.5 (*a*). These are shown in Fig. 17.6.

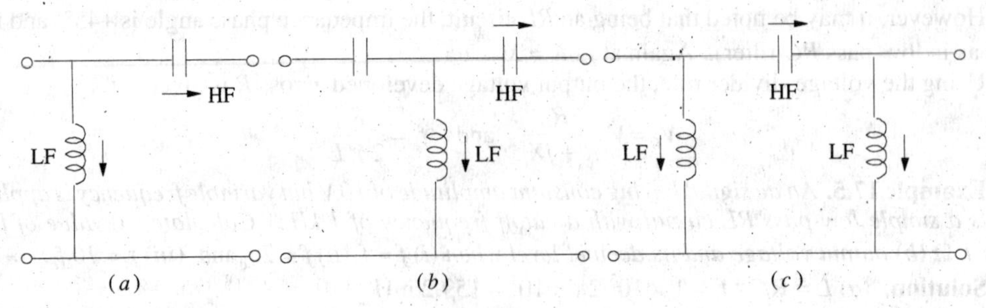

Fig. 17.6

(*i*) **Inverted-L Type.** It is so called because the capacitor and inductor form an upside down *L*. It is shown in Fig. 17.6 (*a*). At lower frequencies, X_C is large but X_L is small. Hence, most of the input voltage drops across X_C and very little across X_L. However, when the frequency is increased, X_C becomes less but X_L is increased thereby causing the output voltage to increase. Consequently, high frequencies are passed while lower frequencies are attenuated.

(*ii*) **T-Type.** It uses two capacitors and a choke as shown in Fig. 17.6 (*b*). The additional capacitor improves the filtering action.

(*iii*) **π-Type.** It uses two inductors which shunt out the lower frequencies as shown in Fig. 17.6 (*c*).

It would be seen that in all high-pass filter circuits, capacitors are in series between the input and output and the coils are grounded. In fact, capacitors can be viewed as shorts to high frequencies but as open to low frequencies. Opposite is the case with chokes.

17.11. High-Pass RL Filter

It is shown in Fig. 17.7 and can be obtained by 'swapping' positions of *R* and *L* in the low-pass *RL* circuit of Fig. 17.4 (*a*). Its response curves are the same as for high-pass *RC* circuit and are shown in Fig. 17.5 (*b*) and (*c*).

As usual, its output voltage equals the voltage which drops across X_L. It is given by

$$V_0 = V_i \frac{j X_L}{R + j X_L} \quad \text{and} \quad f_c = \frac{R}{2\pi L}$$

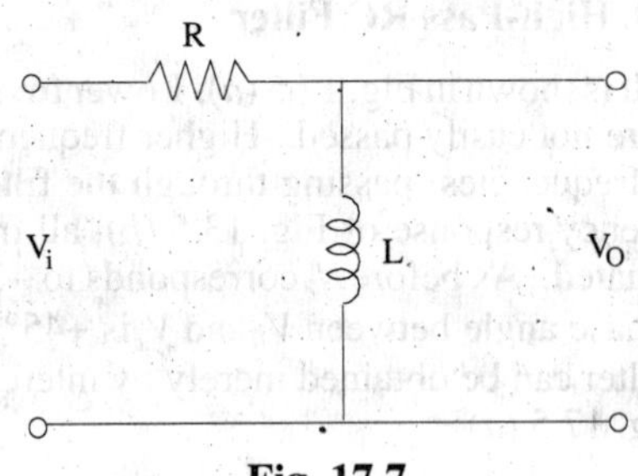

Fig. 17.7

Example 17.6. *Design a high-pass RL filter that has a cutoff frequency of 4 kHz when R = 3 k Ω. It is connected to a 10∠0° V variable frequency supply. Calculate the following :*

(a) Inductor of inductance L but of negligible resistance (b) output voltage V_0 and its decibel decrease at

(i) f = 0 (ii) $f = f_c$ (iii) 8 kHz and (iv) 40 kHz

Solution. (*a*) $L = R/2\pi f_c = 3/2\,\pi \times 4 = 119.4$ mH

(*b*) (*i*) At $f = 0$; $X_L = 0$ *i.e.* inductor acts as a short-circuit across which no voltage develops. Hence, $V_0 = 0$ V as shown in Fig. 17.7.

(*ii*) $f = f_c = 4$ kHz ; $X_L = R$. ∴ $jX_L = j3 = 3\angle 90°$ kΩ

$$\therefore \quad V_0 = V_i \frac{jX_L}{R + jX_L} = 10\angle 0° \frac{3\angle 90°}{3 + j3} = \frac{30\angle 90°}{4.24\angle 45°} = 7.07\angle -45° \text{ V}$$

Decibel decrease $= -20 \log_{10}(10/7.07) = -3$ dB

(*iii*) $f = 2 f_c = 8$ kHz. Here, $X_{L2} = 2 \times j3 = j6$ kΩ

$$\therefore \quad V_0 = 10\angle 0° \frac{6\angle 90°}{3 + j6} = \frac{60\angle 90°}{6.7\angle 63.4°} = 8.95\angle 26.6° \text{ V}$$

Decibel decrease $= -20 \log_{10}(10/8.95) = -0.96$ dB

(*iv*) $f = 10f_c = 40$ kHz; $X_{L3} = 10 \times j3 = j30$ kΩ

$$\therefore \quad V_0 = 10\frac{j30}{3 + j30} = \frac{300\angle 90°}{30.15\angle 84.3°} = 9.95\angle 5.7° \text{ V}$$

Decibel decrease = $-20 \log_{10}(10/9.95) = 0.04$ dB

As seen from Fig. 17.7, as frequency is increased, V_0 is also increased.

17.12. R-C Bandpass Filter

It is a filter that allows a certain band of frequencies to pass through and attenuates all other frequencies below and above the passband. This passband is known as the band width of the filter. As seen, it is obtained by cascading a high-pass *RC* filter to a low-pass *RC* filter. It is shown in Fig. 17.8 alongwith its response curve. The passband of this filter is given by the band of frequencies lying between f_{c1} and f_{c2}. Their values are give by

$$f_{c1} = 1/2\pi\, C_1R_1 \quad \text{and} \quad f_{c2} = 1/2\pi C_2R_2$$

The ratio of the output and input voltages is given by

$$\frac{V_0}{V_i} = \frac{R_1}{R_1 - jX_{C1}} \quad \text{.....from } f_1 \text{ to } f_{C1};$$

$$= \frac{-jX_{C2}}{R_2 - jX_{C2}} \quad \text{.....from } f_{C2} \text{ to } f_2$$

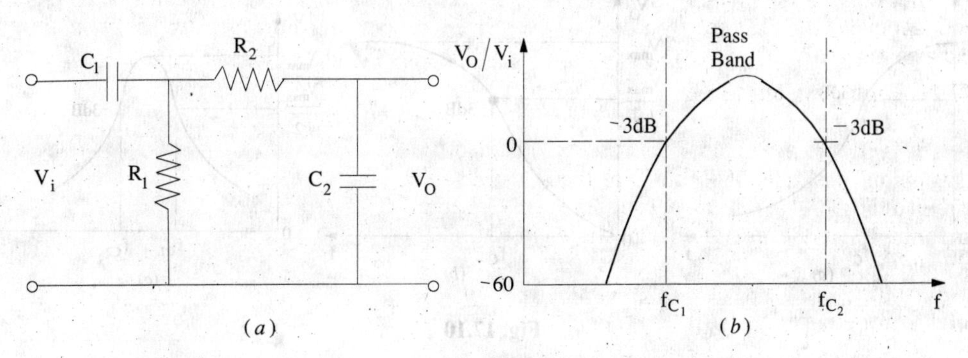

Fig. 17.8

17.13. R-C Bandstop Filter

It is a series combination of low-pass and high-pass RC filters as shown in Fig. 17.9 (*a*). In fact, it can be obtained by reversing the cascaded sequence of the RC bandpass filter. As stated earlier, this filter attenuates a single band of frequencies and allows those on either side to pass through. The stopband is represented by the group of frequencies that lie between f_1 and f_2 where response is below − 60 dB.

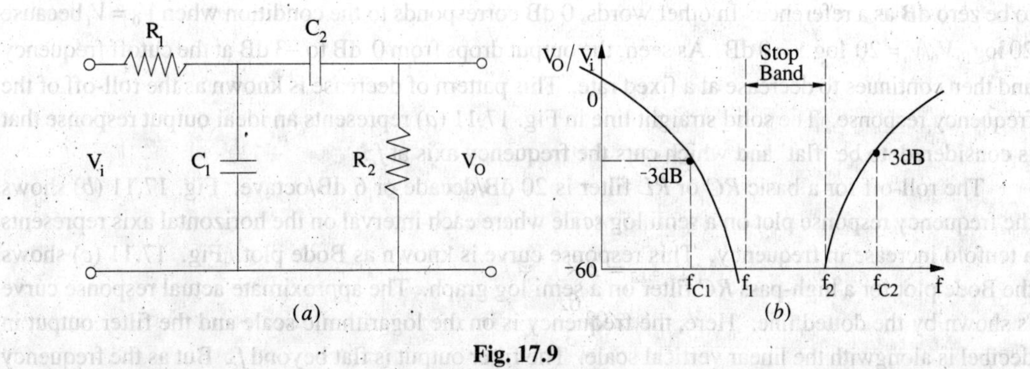

Fig. 17.9

For frequencies from f_{C1} to f_1, the following relationships hold good :

$$\frac{V_0}{V_i} = \frac{-j X_{C1}}{(R_1 - j X_{C1})} \quad \text{and} \quad f_{C1} = \frac{1}{2\pi C_1 R_1}$$

For frequencies from f_2 to f_{C2}, the relationships are as under :

$$\frac{V_0}{V_i} = \frac{R_2}{(R_2 - j X_{C2})} \quad \text{and} \quad f_{C2} = \frac{1}{2\pi C_2 R_2}$$

In practices, several low-pass *RC* filter circuits cascaded with several high-pass *RC* filter circuits which provide almost vertical roll-offs and rises. Moreover, unlike *RL* filters, *RC* filters can be produced in the form of large-scale integrated circuits. Hence, cascading is rarely done with *RL* circuits.

17.14. The –3 dB Frequencies

The output of an a.c. filter is said to be down 3 dB or –3 dB at the cutoff frequencies. Actually at this frequency, the output voltage of the circuit is 70.7% of the maximum input voltage as shown in Fig. 17.10 (*a*) for low-pass filter and in Fig. 17.10 (*b*) and (*c*) for high-pass and bandpass filters respectively. Here, maximum voltage is taken as the 0 dB reference.

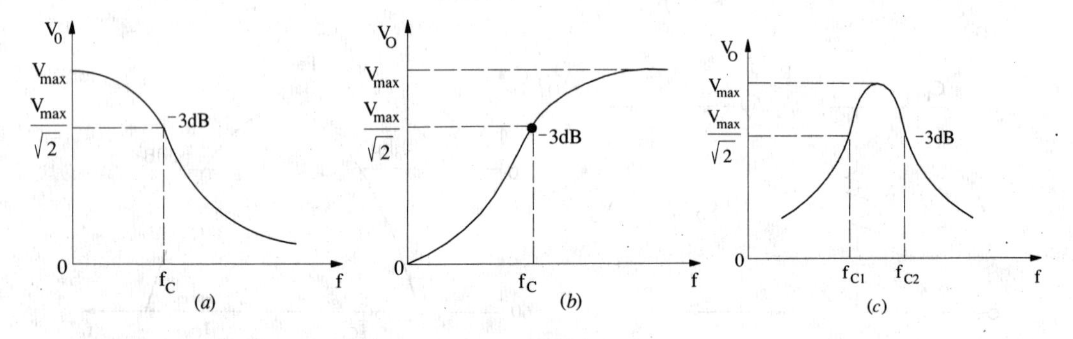

Fig. 17.10

It can also be shown that the power output at the cutoff frequency is 50% of that at zero frequency in the case of low-pass and high-pass filters and of that at f_0 in case of resonant-circuit filter.

17.15. Roll-off of the Response Curve

Gradual decreasing of the output of an a.c. filter is called roll-off. The dotted curve in Fig. 17.11 (*a*) shows an actual response curve of a low-pass *RC* filter. The maximum output is defined to be zero dB as a reference. In other words, 0 dB corresponds to the condition when $V_0 = V_i$ because $20 \log_{10} V_0/V_i = 20 \log 1 = 0$ dB. As seen, the output drops from 0 dB to –3 dB at the cutoff frequency and then continues to decrease at a fixed rate. This pattern of decrease is known as the roll-off of the frequency response. The solid straight line in Fig. 17.11 (*a*) represents an ideal output response that is considered to be 'flat' and which cuts the frequency axis at f_c.

The roll-off for a basic *RC* or *RL* filter is 20 dB/decade or 6 dB/octave. Fig. 17.11 (*b*) shows the frequency response plot on a semi log scale where each interval on the horizontal axis represents a tenfold increase in frequency. This response curve is known as Bode plot. Fig. 17.11 (*c*) shows the Bode plot for a high-pass *RC* filter on a semi log graph. The approximate actual response curve is shown by the dotted line. Here, the frequency is on the logarithmic scale and the filter output in decibel is alongwith the linear vertical scale. The filter output is flat beyond f_c. But as the frequency is reduced below f_c, the output drops at the rate of –20 dB/decade.

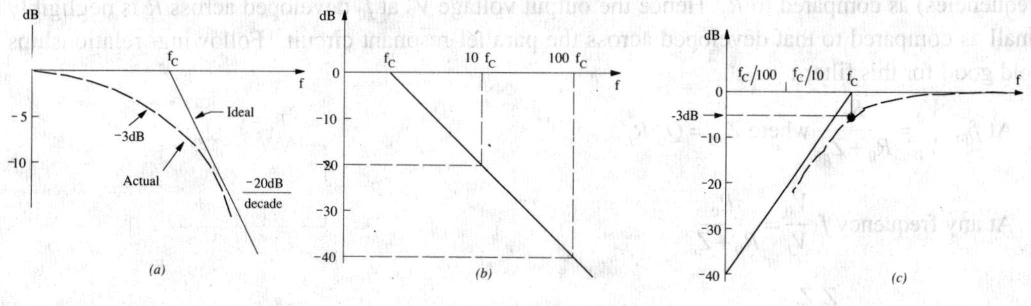

Fig. 17.11

17.16. Bandstop and Bandpass Resonant Filter Circuits

Frequency resonant circuits are used in electronic systems to make either bandstop or bandpass filters because of their characteristic Q-rise to either current or voltage at the resonant frequency. Both series and parallel resonant circuits are used for the purpose. It has already been discussed in Chap. No. 7 that

(*i*) a series resonant circuit offers minimum impedance to input signal and provides maximum current. Minimum impedance equals R because $X_L = X_C$ and maximum current $I = V/R$.

(*ii*) a parallel circuit offers maximum impedance to the input signal and provides minimum current. Maximum impedance offered is $= L/CR$ and minimum current $I = V/(L/CR)$.

17.17. Series-and Parallel-Resonant Bandstop Filters

The series resonant bandstop filter is shown in Fig. 17.12 (*a*) where the output is taken across the series resonant circuit. Hence, at resonant frequency f_0, the output circuit 'sees' a very low resistance R over which negligible output voltage V_0 is developed. That is why there is a sharp resonant dip in the response curve of Fig. 17.12 (*b*). Such filters are commonly used to reject a particular frequency such as 50-cycle hum produced by transformers or inductors or turntable rumble in recording equipment.

For the series-resonant bandstop filter shown in Fig. 17.12 (*a*), the following relationships hold good :

$$\text{At } f_0, \frac{V_0}{V_i} = \frac{R_L}{(R_L + R_S)};\ Q_0 = \frac{\omega_0 L}{(R + R_S)} \quad \text{and} \quad B_{hp} = \frac{1/2\pi\sqrt{LC}}{Q_0}$$

$$\text{At any other frequency } f, \quad \frac{V_0}{V_i} = \frac{R_L + j(X_L - X_C)}{(R_L + R_S) + j(X_L - X_C)}$$

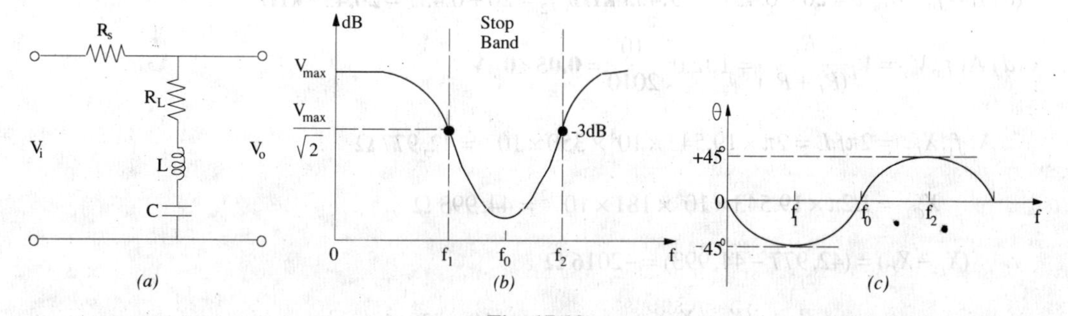

Fig. 17.12

17.18. Parallel-Resonant Bandstop Filter

In this filter, the parallel-resonant circuit is in series with the output resistor R as shown in Fig. 17.13. At resonance, the parallel circuit offers extremely high impedance to f_0 (and nearby frequencies) as compared to R. Hence the output voltage V_0 at f_0 developed across R is negligibly small as compared to that developed across the parallel-resonant circuit. Following relationships hold good for this filter :

At f_0, $\frac{V_0}{V_i}=\frac{R_0}{R_0+Z_{p0}}$ where $Z_{p0}=Q_0^2 R_L$

At any frequency f; $\frac{V_0}{V_i}=\frac{R_0}{R_0+Z_p}$

where $Z_p=\frac{Z_L Z_C}{R_L+j(X_L-X_C)}$

Also $Q_0=\omega_0 L/R_L$ and $B_{hp}=(1/2\pi\sqrt{LC})/Q_0$

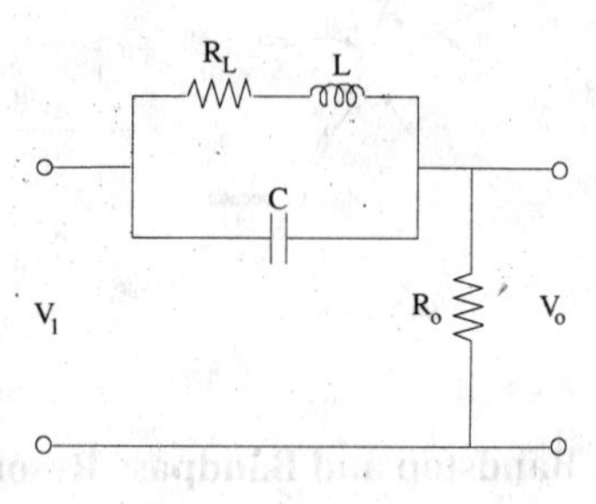

Fig. 17.13

It should be noted that the same amplitude phase response curves apply both to the series resonant and parallel-resonant bandstop filters. Since X_C predominates at lower frequencies, phase angle θ is negative below f_0. Above f_0, X_L predominates and the phase current leads. At cutoff frequency f_1, $\theta=-45°$ and at other cutoff frequency f_2, $\theta=+45°$ as in the case of any resonant circuit.

Example 17.7. *A series-resonant bandstop filter consist of a series resistance of 2 kΩ across which is connected a series-resonant circuit consisting of a coil of resistance 10 Ω and inductance 350 mH and a capacitor of capacitance 181 pF. If the applied signal voltage is 10∠0° of variable frequency, calculate*

(a) resonant frequency f_0 (b) half-power bandwidth B_{hp} (c) edge frequencies f_1 and f_2 (d) output voltage at frequencies f_0, f_1 and f_2

Solution. We are given that $R_S=2\ k\Omega$; $R=10\ \Omega$; $L=350$ mH; $C=181$ pF.

$$(a)\ f_0=1/2\pi\sqrt{LC}=1/2\pi\sqrt{350\times10^{-3}\times181\times10^{-12}}=20\text{ kHz}$$

$$(b)\ Q_0=\omega_0 L/(R_S+R_L)=2\pi\times20\times10^3\times350\times10^{-3}/2010=21.88$$

$$B_{hp}=f_0/Q_0=0.914\text{ kHz}$$

$$(c)\ f_1=f_0-B_{hp}/2=20-0.457=\mathbf{19.453\ kHz};\ f_2=20+0.457=\mathbf{20.457\ kHz}$$

$$(d)\text{ At } f_0, V_0=V_i\frac{R_L}{(R_L+R_S)}=10\angle0°\frac{10}{2010}=\mathbf{0.05\angle0°\ V}$$

$$\text{At } f_1, X_{L1}=2\pi f_1 L=2\pi\times19.543\times10^3\times350\times10^{-3}=42,977\ \Omega$$

$$X_{C1}=1/2\pi\times19.543\times10^3\times181\times10^{-12}=44,993\ \Omega$$

$$\therefore\quad (X_L-X_C)=(42,977-44,993)=-2016\ \Omega$$

$$V_{01}=V_i\frac{R_L+j(X_L-X_C)}{(R_S+R_L)+j(X_L-X_C)}=10\angle0°\times\frac{10-j2016}{2010-j2016}$$

$$=\frac{20160\angle-89.7°}{2847\angle-45°}=\mathbf{7.07\angle-44.7°\ V}$$

At $f_2, X_{L2} = X_{L1}(f_2/f_1) = 42977 \times 20.457/19.453$

$= 44,987\ \Omega; X_{C2} = X_{C1}(f_1/f_2) = 44,993 \times 19.543/20.457$

$= 42,983\ \Omega;\ (X_{L2} - X_{C2}) = 44,987 - 42,983 = 2004\ \Omega$

$$\therefore \quad V_0 = 10\angle 0° \frac{10 + j2004}{2010 + j2004} = \frac{2004\angle 89.7°}{2837\angle 44.9°} = \mathbf{7.07\angle 44.8°\ V}$$

17.19. Series-Resonant Bandpass Filter

As shown in Fig. 17.14 (*a*), it consists of a series-resonant circuit shunted by an output resistance R_0. It would be seen that this filter circuit can be produced by 'swapping' a series resonant bandstop filter. At f_0, the series resonant impedance is very small and equals R_L which is negligible as compared to R_0. Hence, output voltage is maximum at f_0 and falls to 70.7% at cutoff frequency f_1 and f_2 as shown in the response curve of Fig. 17.14 (*b*). The phase angle is positive for frequencies above f_0 and negative for frequencies below f_0 as shown in Fig. 17.14 (*c*) by the solid curve.

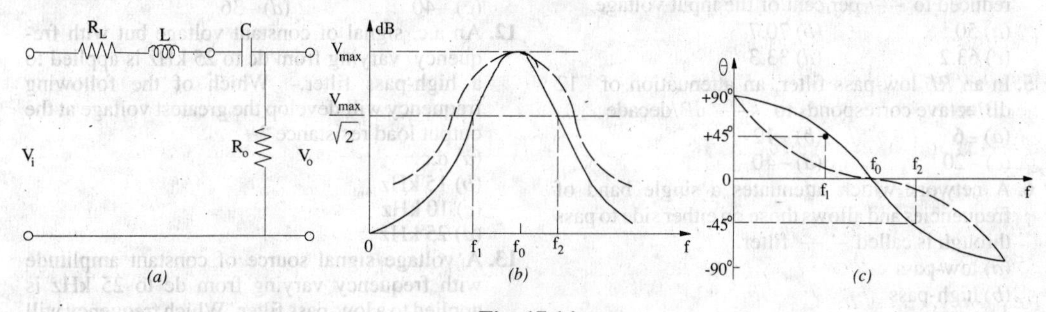

Fig. 17.14

Following relationships hold good for this filter circuit.

$$\text{At } f_0, \frac{V_0}{V_i} = \frac{R_0}{(R_L + R_0)};\ Q_0 = \frac{\omega_0 L}{(R_L + R_0)} \text{ and } B_{hp} = \frac{1/2\pi\sqrt{LC}}{Q_0}$$

17.20. Parallel-Resonant Bandpass Filter

It can be obtained by transposing the circuit elements of a bandstop parallel-resonant filter. As shown in Fig. 17.15, the output is taken across the two-branch parallel-resonant circuit. Since this circuit offers maximum impedance at resonance, this filter produces maximum output voltage V_0 at f_0. The amplitude-response curve of this filter is similar to that of the series-resonant bandpass filter discussed above [Fig. 17.14 (*b*)]. The dotted curve in Fig. 17.14(*c*) represents the phase relationship between the input and output voltages of this filter. The following relationships apply to this filter :

$$\text{At } f_0, \frac{V_0}{V_i} = \frac{R_0}{(R_0 + Z_{p0})} \quad \text{where } Z_{p0} = R_{p0} = Q_r^2 R_L$$

$$\text{and } Q_0 = \frac{R_{p0}}{X_{CO}} \text{ and } B_{hp} = \frac{1/2\pi\sqrt{LC}}{Q_0}$$

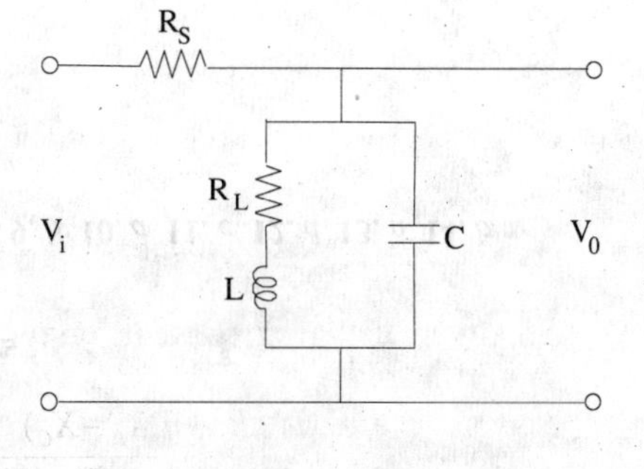

Fig. 17.15

$$\text{At any frequency } f, \frac{V_0}{V_i} = \frac{Z_p}{R_0 + Z_p} \quad \text{where } Z_p = \frac{Z_L(-jX_C)}{R_L + j(X_L - X_C)}$$

OBJECTIVE TEST-17

1. The decibel is a measure of
   (*a*) power
   (*b*) voltage
   (*c*) current
   (*d*) power level
2. When the output voltage level of a filter decreases by –3 dB, its absolute value changes by a factor of
   (*a*) $\sqrt{2}$ (*b*) $1/\sqrt{2}$
   (*c*) 2 (*d*) 1/2
3. The frequency corresponding to half-power point on the response curve of a filter is known as —— frequency.
   (*a*) cutoff (*b*) upper
   (*c*) lower (*d*) roll-off
4. In a low-pass filter, the cutoff frequency is represented by the point where the output voltage is reduced to —— per cent of the input voltage.
   (*a*) 50 (*b*) 70.7
   (*c*) 63.2 (*d*) 33.3
5. In an *RL* low-pass filter, an attenuation of –12 dB/octave corresponds to —— dB/decade.
   (*a*) –6 (*b*) –12
   (*c*) –20 (*d*) –40
6. A network which attenuates a single band of frequencies and allows those on either side to pass through is called —— filter.
   (*a*) low-pass
   (*b*) high-pass
   (*c*) bandstop
   (*d*) bandpass
7. In a simple high-pass *RC* filter, if the value of capacitance is doubled, the cutoff frequency is
   (*a*) doubled
   (*b*) halved
   (*c*) tripled
   (*d*) quadrupled
8. In a simple high-pass *RL* filter circuit, the phase difference between the output and input voltages at the cutoff frequency is —— degrees.
   (*a*) –90 (*b*) 45
   (*c*) –45 (*d*) 90
9. In a simple low-pass *RC* filter, attenuation is –3 dB at f_c. At $2f_c$, attenuation is –6 dB. At $10f_c$, the attenuation would be —— dB.
   (*a*) –30 (*b*) –20
   (*c*) –18 (*d*) –12
10. An a.c. signal of constant voltage 10 V and variable frequency is applied to a simple high-pass RC filter. The output voltage at ten times the cutoff frequency would be —— volt.
    (*a*) 1
    (*b*) 5
    (*c*) $10/\sqrt{2}$
    (*d*) $10\sqrt{2}$
11. When two simple low-pass filters having same values of *R* and *C* are cascaded, the combined filter will have a roll-off of —— dB/decade.
    (*a*) –20 (*b*) –12
    (*c*) –40 (*d*) –36
12. An a.c. signal of constant voltage but with frequency varying from dc to 25 kHz is applied to a high-pass filter. Which of the following frequency will develop the greatest voltage at the output load resistance ?
    (*a*) d.c.
    (*b*) 15 kHz
    (*c*) 10 kHz
    (*d*) 25 kHz
13. A voltage signal source of constant amplitude with frequency varying from dc to 25 kHz is applied to a low-pass filter. Which frequency will develop greatest voltage across the output load resistance ?
    (*a*) d.c.
    (*b*) 10 kHz
    (*c*) 15 kHz
    (*d*) 25 kHz
14. The output of a filter drops from 10 to 5 V as the frequency is increased from 1 to 2 kHz. The dB change in the output voltage is
    (*a*) –3 dB/decade
    (*b*) –6 dB/octave
    (*c*) 6 dB/octave
    (*d*) –3 dB/octave

Answers

1. *d* **2.** *b* **3.** *a* **4.** *b* **5.** *d* **6.** *c* **7.** *b* **8.** *b* **9.** *b* **10.** *a* **11.** *c* **12.** *d* **13.** *a* **14.** *b*

18 CIRCLE DIAGRAMS

18.1. Circle Diagram of a Series Circuit

Circle diagrams are helpful in analysing the operating characteristics of circuits, which, under some conditions, are used in representing transmission lines and a.c. machinery (like induction motor etc.).

Consider a circuit having a constant reactance but variable resistance varying from zero to infinity and supplied with a voltage of constant magnitude and frequency (Fig. 18.1).

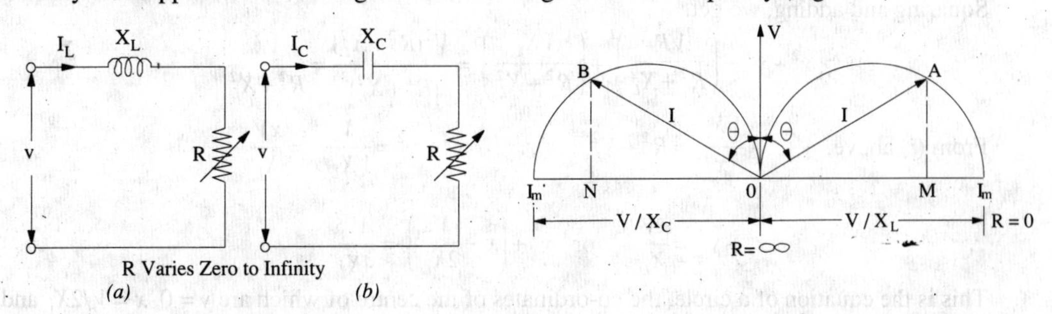

Fig. 18.1

Fig. 18.2

If $R = 0$, then $I = V/X_L$ or V/X_C and has maximum value. It will lag or lead the voltage by 90° depending on whether the reactance is inductive or capacitive. In Fig. 18.2, angle θ represents the phase angle. If R is now increased from its zero value, then I and θ will both decrease. In the limiting case, when $R = \infty$, then $I = 0$ and $\theta = 0°$. It is found that the locus of end point of current vector OA or OB represents a semi-circuit with diameter equal to V/X as shown in Fig. 18.2. It can be proved thus :

$I = V/Z$ and $\sin\theta = X/Z$ or $Z = X/\sin\theta$ $\therefore$ $I = V\sin\theta/X$

For constant value of V and X, the above is the polar equation of a circle of diameter V/X. This equation is plotted in Fig. 18.2. Here, OV is taken as reference vector. It is also seen that for inductive circuit, the current semi-circle is on the right-hand side of reference vector OV so that current vector OA lags by θ°. The current semi-circle for R-C circuit is drawn on the left-hand side of OV so that current vector OB lends OV by θ°. It is obvious that $AM = I\cos\theta$, hence AM represents, on a suitable scale, the power consumed by the R-L circuit. Similarly, BN represents the power consumed by the R–C circuit.

18.2. Rigorous Mathematical Treatment

We will again consider both R-L and R–C circuits. The voltage drops across R and X_L (or X_C) will be 90° out of phase with each other. Hence, for any given value of resistance, the vector diagram for the two voltage drops (*i.e.* IR and IX) is a rightangled triangle having applied voltage as the hypotenuse.

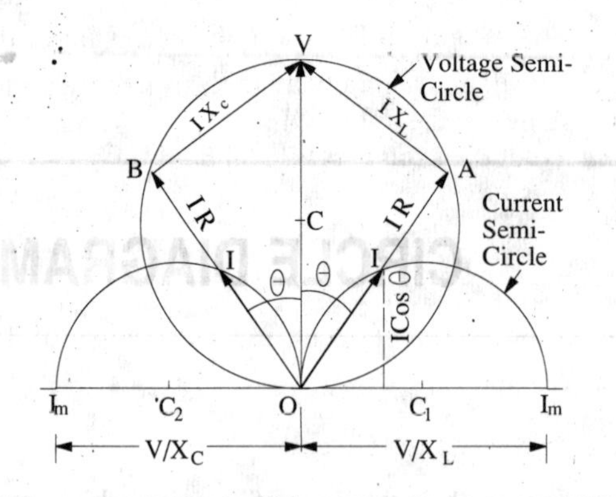

Fig. 18.3

For a constant applied voltage and reactance, the vector diagrams for different values of R are represented by a series of right-angled triangles having common hypotenuse as shown in Fig. 18.3. The locus of the apex of the right-angled voltage triangles is a semicircle described on the hypotenuse. The voltage semi–circle for R-L circuit (OAV) is on the right and for R-C circuit (OBV) on the left of the reference vector OV as shown in Fig. 18.3.

The foci of end points of current vectors are also semi-circles as shown but their centres lie on the opposite sides of and in an axis perpendicular to the reference vector OV.

(*i*) **R-L Circuit** [Fig. 18.1 (*a*)]

The co-ordinates of point A with respect to the origin O are

$$y = I\cos\theta = \frac{V}{Z}\cdot\frac{R}{Z} = V\frac{R}{Z^2} = V\frac{R}{R^2+X_L^2};\ x = I\sin\theta = \frac{V}{Z}\cdot\frac{X_L}{Z} = V\cdot\frac{X_L}{Z^2} = V\frac{X_L}{R^2+X_L^2} \quad \ldots(i)$$

Squaring and adding, we get

$$x^2+y^2 = \left(\frac{VR}{R^2+X_L^2}\right)^2 + \left(\frac{VX_L}{R^2+X_L^2}\right)^2 = \frac{V^2(R^2+X_L^2)}{(R^2+X_L^2)^2} = \frac{V^2}{R^2+X_L^2}$$

From (*i*) above, $\quad \frac{VX_L}{x} = R^2+X_L^2 \qquad \therefore \qquad x^2+y^2 = \frac{V^2}{VX_L/x} = \frac{xV}{X_L}$

$$\therefore \qquad x^2+y^2 = \frac{xV}{X_L} \quad \text{or} \quad y^2+\left(x-\frac{V}{2X_L}\right)^2 = \frac{V^2}{4X_L^2}$$

This is the equation of a circle, the co-ordinates of the centre of which are $y = 0$, $x = V/2X_L$ and whose radius is $V/2X_L$.

(*ii*) **R-C Circuit.** In this case it can be similarly proved that the locus of the end point of current vector is a semi-circle. The equation of this circle is

$$y^2+\left(x+\frac{V}{2\,X_C}\right)^2 = \frac{V^2}{4\,X_C^2}$$

The centre has co-ordinates of $y = 0$, $x = -V/2X_C$.

18.3. Constant Resistance But Variable Reactance

Fig. 18.4 shows two circuits having constant resistance but variable reactance X_L or X_C which vary from zero to infinity. When $X_L = 0$, current is maximum and equals V/R. For other values,

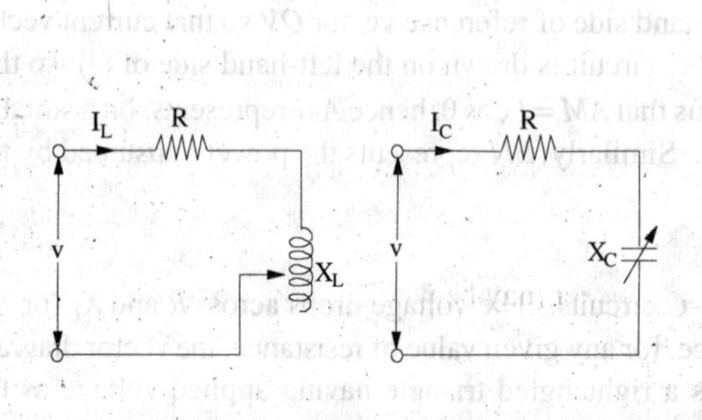

Fig. 18.4

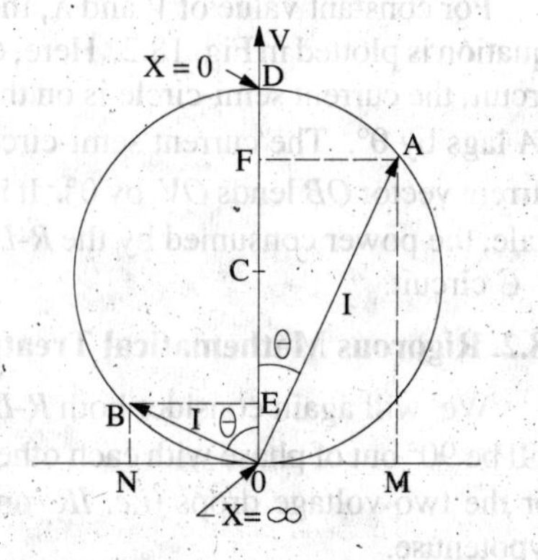

Fig. 18.5

$I = V/\sqrt{R^2 + X_L^2}$. Current becomes zero when $X_L = \infty$. As seen from Fig. 18.5, the end point of the current vector describes a semi-circle with radius $OC = V/2R$ and centre lying in the reference sector *i.e.* voltage vector OV. For R-C circuit, the semi-circle lies to the left of OV.

As before, it may be proved that the equation of the circle shown in Fig. 18.5 is

$$x^2 + \left(y - \frac{V}{2R}\right)^2 = \frac{V^2}{4R^2}$$

The co-ordinates of the centre are $x = 0$, $y = V/2R$ and radius $= V/2R$.

As before, power developed would be maximum when current vector makes an angle of 45° with the voltage vector OV. In that case, current is $I_m/\sqrt{2}$ and $P_m = VI_m/2$.

16.4. Properties of Constant Reactance But Variable Resistance Circuit

From the circle diagram of Fig. 18.3, it is seen that circuits having a constant reactance but variable resistance or *vice-versa* have the following properties :

(*i*) the current has limiting value

(*ii*) the power supplied to the circuit has a limiting value also

(*iii*) the power factor corresponding to maximum power supply is 0.707 (= cos 45°)

Obviously, the maximum current in the circuit is obtained when $R = 0$.

$$\therefore \quad I_m = V/X_L = V/\omega L \qquad \text{...for } R\text{-}L \text{ circuit}$$

$$= -V/X_C = -\omega VC \qquad \text{...for } R\text{-}C \text{ circuit}$$

Now, power P taken by the circuit is $VI \cos\theta$ and if V is constant, then $P \propto I \cos\theta$. Hence, the ordinates of current semi-circles are proportional to $I \cos\theta$. The maximum ordinate possible in the semi-circle represents the maximum power taken by the circuit. The maximum ordinate passes through the centre of semi-circle so that current vector makes an angle of 45° with both the diameter and the voltage vector OV. Obviously, power factor corresponding to maximum power intake is cos 45° = 0.707.

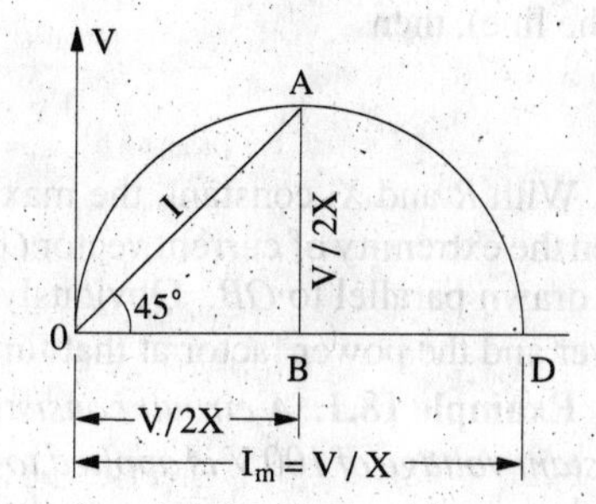

Fig. 18.6

Maximum power,

$$P_m = V \times AB = V \times \frac{I_m}{2} = \frac{1}{2} VI_m$$

Now, for R-L circuit, $I_m = V/X_L$

$$\therefore \quad P_m = \frac{V^2}{2X_L} = \frac{V^2}{2\omega L}$$

$$\text{For } R\text{-}C \text{ circuit} \quad P_m = \frac{V^2}{2X_C} = \frac{V^2 \omega C}{2}$$

As said above, at maximum power, $\theta = 45°$, hence vector triangle for voltages is an isosceles triangle which means that voltage drops across resistance and reactance are each equal to 0.707 of supply voltage *i.e.* $V/\sqrt{2}$. As current is the same, for maximum power, resistance equals reactance *i.e.* $R = X_L$ (or X_C).

Hence, the expression representing maximum power may be written as $P_m = V^2/2R$.

18.5. Simple Transmission Line Circuit

In Fig. 18.7 (*a*) is shown a simple transmission circuit having negligible capacitance and reactance. R and X_L represent respectively the resistance and reactance of the line and R represents load resistance.

If R and X_L are constant, then as R_L is varied, the current AM follows the equation $I = (V/X) \sin \theta$ (Art. 18.1). The height AM in Fig. 18.7 (*b*) represents the power consumed by the circuit but, in the present case, this power is consumed both in R and R_L. The power absorbed by each resistance can be represented on the circle diagram.

In Fig. 18.7 (*b*), OB represents the line current when $R_L = 0$. The current $OB = V/\sqrt{(R^2 + X_L^2)}R^2 + X_L^2)$ and power factor is $\cos \theta_1$. The ordinate BN then represents on a different scale the power dissipated in R only. OA represents current when R_L has some finite value *i.e.* $OA = V/\sqrt{(R + R_L)^2 + X_L^2}$. The ordinate AM represents total power dissipated, out of which ME is consumed in R and AE in R_L.

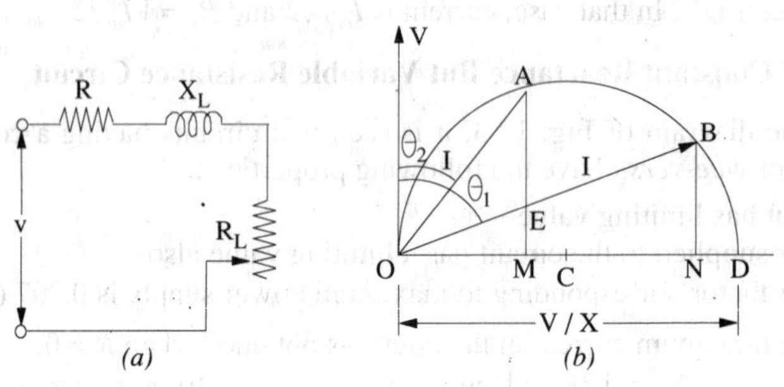

Fig. 18.7

In fact, if $OA^2 \times R_L (= AE)$ is considered to be the output of the circuit (the power transmitted by the line), then

$$\eta = \frac{AE}{AM}$$

With R and X_L constant, the maximum power that can be transmitted by such a circuit occurs when the extremity of current vector OA coincides with the point of tangency to the circle of a straight line drawn parallel to OB. Obviously, V times AE under these conditions represents the maximum power and the power factor at that time is $\cos \theta_2$.

Example 18.1. *A circuit consists of a reactance of 5 Ω in series with a variable resistance. A constant voltage of 100 V is applied to the circuit. Show that the current locus is circular. Determine (a) the maximum power input to the circuit (b) the corresponding current, p.f. and value of the resistance.* **(Electrical Science II, Allahabad Univ. 1992)**

Solution. For the first part, please refer to Art. 18.1

(*a*) $I_m = V/X = 100/5 = 20$ A ; $P_m = \frac{1}{2} VI_m = \frac{1}{2} \times 100 \times 20 =$ **1000 W**

(*b*) At maximum power input, current is = OA (Fig. 18.6)

$\therefore\ OA = I_m/\sqrt{2} = 20/\sqrt{2} = 14.14$ A ; *p.f.* = cos 45° = **0.707** ; $R = X =$ **5 Ω**

Example 18.2. *If a coil of unknown resistance and reactance is connected in series with a 100-V, 50-Hz supply, the current locus diagram is found to have a diameter of 5 A and when the value of series resistor is 15 Ω, the power dissipated is maximum. Calculate the reactance and resistance of the coil and the value of the maximum power in the circuit and the maximum current.*

Solution. Let the unknown resistance and reactance of the coil be R and X respectively

Diameter $= V/X \quad \therefore\ 5 = 100/X$ or $X = 20\ \Omega$

Power is maximum when total resistance = reactance

or $\quad 15 + R = 20 \quad \therefore \quad R =$ **5 Ω**

Maximum power $P_m = V^2/2X = 100^2/2 \times 20 =$ **250 W**

Maximum current $I_m = 100/\sqrt{(20^2 + 5^2)} =$ **4.85 A**

Example 18.3. *A constant alternating sinusoidal voltage at constant frequency is applied across a circuit consisting of an inductance and a variable resistance in series. Show that the locus diagram of the current vector is a semi-circle when the resistance is varied between zero and infinity.*

If the inductance has a value of 0.6 henry and the applied voltage is 100 V at 25 Hz, calculate (a) the radius of the arc (in amperes) and (b) the value of variable resistance for which the power taken from the mains is maximum and the power factor of the circuit at the value of this resistance.

Solution. $X_L = \omega L = 0.6 \times 2\pi \times 25 = 94.26\ \Omega$

(*a*) Radius = $V/2X_L$ = 100/2 x 94.26 = **0.531 A** (Art. 18.1)

(*b*) R = **94.26** Ω–for maximum power; Power factor = **0.707.**

Example 18.4. *A resistor of 10 Ω is connected in series with an inductive reactor which is variable between 2 Ω and 20 Ω. Obtain the locus of the current vector when the circuit is connected to a 250-V supply. Determine the value of the current and the power factor when the reactance is (i) 5 Ω (ii) 10 Ω (iii) 15 Ω.* **(Basic Electricity, Bombay Univ. 1987)**

Solution. As discussed in Art. 18.3, the end point of current vector describes a semi-circle whose diameter (Fig. 18.8) equals V/R = 250/10 = 25 A and whose centre lies to right side of the vertical voltage vector OV.

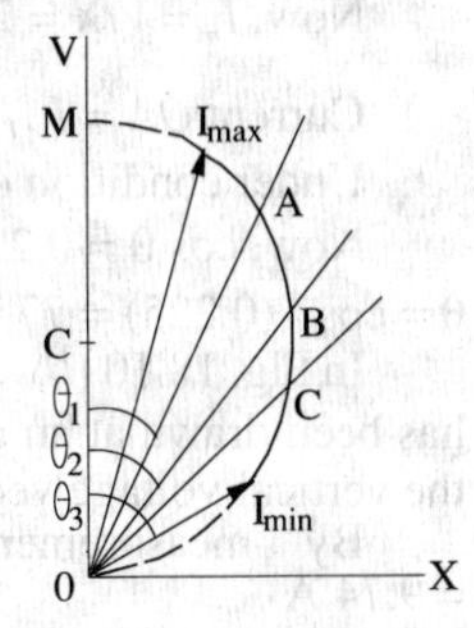

Fig. 18.8

$$I_{max} = 250/\sqrt{10^2+2^2} = 24 \text{ A}; \theta = \tan^{-1}(2/10) = 11.3°;$$

$$I_{min} = 250/\sqrt{10^2+20^2} = 11.2 \text{ A}; \theta = \tan^{-1}(20/10) = 63.5°$$

$(i)\ \theta_1 = \tan^{-1}(5/10) = 26.7°$, p.f. = cos 26.7° = **0.89**

$I = OA =$ **22.4 A**

$(ii)\ \theta_2 = \tan^{-1}(10/10) = 45°$, p.f. = cos 45° = **1**;

$I = OB =$ **17.7 A**

$(iii)\ \theta_3 = \tan^{-1}(15/10) = 56.3°$; p.f. = cos 56.3° = **0.55;** $I = OC$ = **13.9 A.**

Example 18.5. *A voltage of 100 sin 10,000 t is applied to a circuit consisting of a 1 μF capacitor in series with a resistance R. Determine the locus of the tip of the current phasor when R is varied from 0 to ∞. Take the applied voltage as the reference phasor.*

(Network Theory and Design, AMIETE 1990)

Solution. As seen from Art. 18.2 the locus of the tip of the current phasor is a circle whose equation

is
$$y^2 + \left(x + \frac{V}{2X_C}\right)^2 = \frac{V}{4X_C^2}$$

We are given that $V = V_m/\sqrt{2} = 100/\sqrt{2} = 77.7$ V

ω = 10,000 rad/s; $X_C = 1/\omega \times C = 1/10{,}000 \times 1 \times 10^{-6} = 100\ \Omega$ C, $(V/2X_C)^2 = (77.7/2 \times 100)^2$ = 0.151

$\therefore\ y^2 + (x + 0.389)^2 = 0.151$

Example 18.6. *Prove that polar locus of current drawn by a circuit of constant resistance and variable capacitive reactance is circular when the supply voltage and frequency are constant.*

If the constant resistance is 10 Ω and the voltage is 100 V, draw the current locus and find the values of the current and p.f. when the reactance is (i) 5.77 Ω (ii) 10 Ω and (iii) 17.32 Ω. Explain when the power will be maximum and find its value. **(Electromechanics, Allahabad Univ. 1992)**

Solution. For the first part, please refer to Art. 18.3. The current semicircle will be drawn on the vertical axis with a radius $OM = V/2R$ = 100/2 × 10 = 5 A as shown in Fig. 18.9 (*b*).

(*i*) $\theta_1 = \tan^{-1}(5.77/10) = 30°$; $\cos\theta_1$ = **0.866 (lead)**; current = OA = **8.66 A**

(*ii*) $\theta_2 = \tan^{-1}(10/10) = 45°$; $\cos\theta_2$ = **0.707 (lead)** current = OB = **7.07 A**

(*iii*) $\theta_3 = \tan^{-1}(17.32/10) = 60°$; $\cos\theta_3$ = **0.5 (lead)** current = OC = **5 A**

Power would be maximum for point B where θ = 45°; $I_m = V/R$ = 100/10 = 10 A

$P_m = V \times OB \times \cos 45° = V \times I_m \cos 45° \times \cos 45° = \frac{1}{2}VI_m = \frac{1}{2} \times 100 \times 10 =$ **500 W**

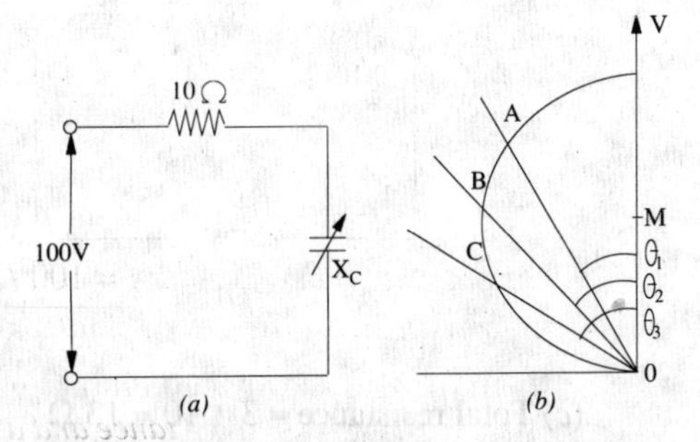

Fig. 18.9

Example 18.7. *Prove that the polar locus of the current drawn by a circuit of constant reactance and variable resistance is circular when the supply voltage and frequency are constant.*

If the reactance of such a circuit is 25 Ω and the voltage 250, draw the said locus and locate there on the point of maximum power and for this condition, find the power, current, power factor and resistance.

Locate also the point at which the power factor is 0.225 and for this condition find the current, power and resistance. **(Basic Electricity, Bombay Univ. 1985)**

Solution. For the first part, please refer to Art. 18.3.

Radius of the current semi-circle is = $V/2X = 250/2 \times 25 = 5$ A. As discussed in Art. 18.3, point A [Fig. 18.10 (*a*)] corresponds to maximum power.

Now, $I_m = V/X = 250/25 = 10$ A; $P_m = \frac{1}{2} VI_m = \frac{1}{2} \times 250 \times 10 =$ **1250 W**

Current $OA = I_m / \sqrt{2} = 10/\sqrt{2} =$ **7.07 A** ; *p.f.* = cos 45° = **0.707.**

Under condition of maximum power, $R = X =$ **25 Ω.**

Now, cos θ = 0.225 ;

$\theta = \cos^{-1}(0.225) = 77°$

In Fig. 18.10 (*b*), current vector *OA* has been drawn at an angle of **77°** with the vertical voltage vector *OV*.

By measurement, current *OA* = 9.74 A

By calculation, *OA*, = I_m cos 13° = 10 × 0.974 = **9.74 A**

Power = *VI* cos θ = 250 × 9.74 × 0.225 = **548 W**

Fig. 18.10

$P = I^2R$; $R = P/I^2 = 548/9.74^2 =$ **5.775 Ω.**

Example 18.8. *A non-inductive resistance R, variable between 0 and 10 Ω, is connected in series with a coil of resistance 3 Ω and reactance 4 Ω and the circuit supplied from a 240-V a.c. supply. By means of a locus diagram, determine the current supplied to the circuit when R is (a) zero (b) 5 Ω and (c) 10 Ω. By means of the symbolic method, calculate the value of the current when R = 5 Ω.*

Solution. The locus of the current vector is a semi-circle whose centre is (0, *V*/2*X*) and whose radius is obviously equal to *V*/2*X*. Now, $V/2X = 240/2 \times 4 = 30$ A.

Hence, the semi-circle is drawn as shown in Fig. 18.11 (*b*).

(*a*) Total resistance = 3 Ω and *X* = 4 Ω ∴ $\tan\theta_1 = 4/3$ ∴ $\theta_1 = 53°8'$

Hence, current vector *OA* is draw making an angle of 53°8′ with vector *OV*. Vector *OA* measures **49 A.**

(*b*) total resistance = 3 + 5 = 8 Ω

Reactance = 4 Ω ; $\tan\theta_2 = 4/8 = 0.5$ ∴ $\theta_2 = 26°34'$

Current vector *OB* is drawn at an angle of 26°34′ with *OV*. It measures **27 A (approx.)**

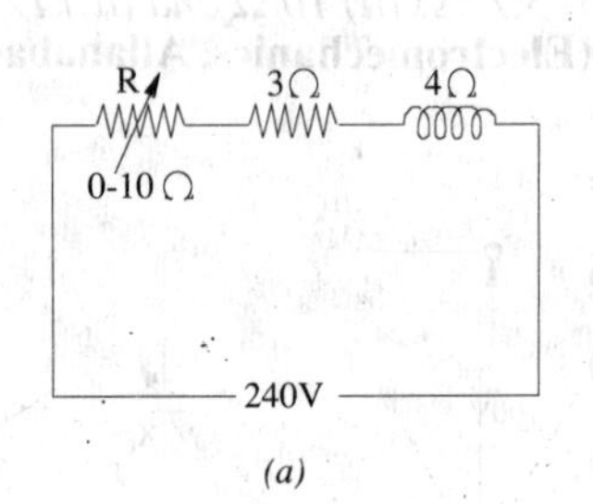

(*a*)

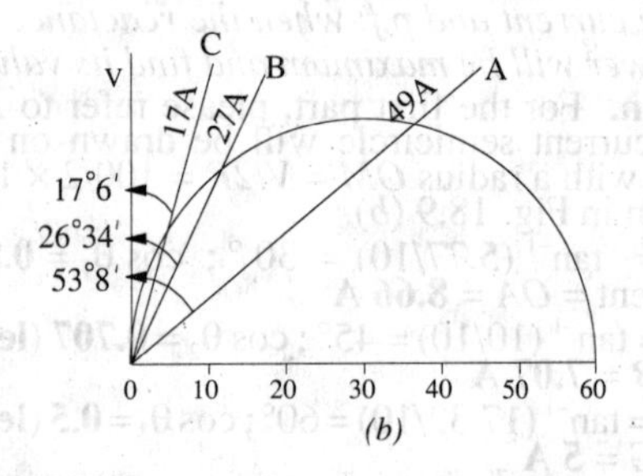

(*b*)

Fig. 18.11

(*c*) Total resistance = 3 + 10 = 13 Ω

Reactance = 4 Ω; $\tan\theta_3 = 4/13$ ∴ $\theta_3 = 17°6'$

Current vector OC is drawn at an angle of 17°6′ with vector OA. It measures **17 A.**

Symbolic Method

$$\mathbf{I} = \frac{240 + j0}{(5+3)+j4} = \frac{240}{8+j4} = \frac{240}{8.96\angle 26.5°} = 26.7\angle 26.5°$$

Note. There is difference in the magnitudes of the currents and the angles as found by the two different methods. It is so because one has been found exactly by mathematical calculations, whereas the other has been measured from the graph.

Example 18.9. *A circuit consisting of a 50-Ω resistor in series with a variable reactor is shunted by a 100 Ω resistor. Draw the locus of the extremity of the total current vector to scale and determine the reactance and current corresponding to the minimum overall power factor, the supply voltage being 100 V.* **(Elect. Engg-I Nagpur Univ. 1992)**

Solution. The parallel circuit is shown in Fig. 18.12 (*a*).

The resistive branch draws a fixed current I_2 = 100/100 = 1A. The current I_1 drawn by the reactive branch is maximum when $X_L = 0$ and its maximum value is = 100/50 = 2 A and is in phase with voltage.

In the locus diagram of Fig. 18.12 (*b*), the diameter OA of the reactive current semi-circle is = 2 A. OB is the value of I_1 for some finite value of X_L. $O'O$ represents I_2. Being in phase with voltage, it is drawn in phase with voltage vector OV. Obviously, $O'B$ represents total circuit current, being the vector sum of I_1 and I_2.

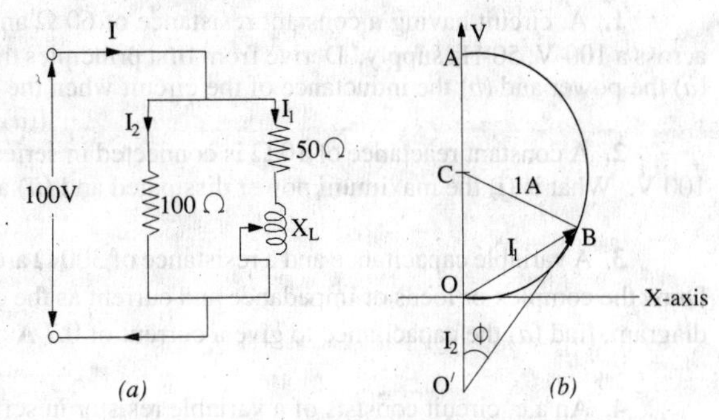

Fig. 18.12

The minimum power factor which corresponds to maximum phase difference between $O'B$ and $O'V$ occurs when $O'B$ is tangential to the semi-circle. In that case, $O'B$ is perpendicular to BC. It means that $O'BC$ is a right-angled triangle.

Now, $\sin\phi = BC/O'C = 1/(1+1) = 0.5$; $\phi = 30°$

∴ Minimum $p.f. = \cos 30° = 0.866$ (lag)

Current corresponding to minimum $p.f.$ is $O'B = O'C \cos\phi = 2 \times 0.866 =$ **1.732 A.**

Now, Δ OBC is an equilateral triangle, hence $I_1 = OB = 1$ A. Considering reactive branch, $Z = 100/1 = 100\ \Omega$, $X_L = \sqrt{100^2 - 50^2} = \mathbf{88.6\ \Omega}$

Example 18.10. *A coil of resistance 60 Ω and inductance 0.4 H is connected in series with a capacitor of 17.6 μF across a variable frequency source which is maintained at a fixed potential of 120 V. If the frequency is varied through a range of 40Hz to 80 Hz, draw the complete current locus and calculate the following :*

(i) the resonance frequency (ii) the current and power factor at 40 Hz and
(iii) the current and power factor at 80 Hz. **(Elect. Circuits, South Gujarat Univ. 1989)**

Solution. (*i*) $f_0 = 10^3/2\pi\sqrt{0.4 \times 17.6} = \mathbf{60\ Hz.}$

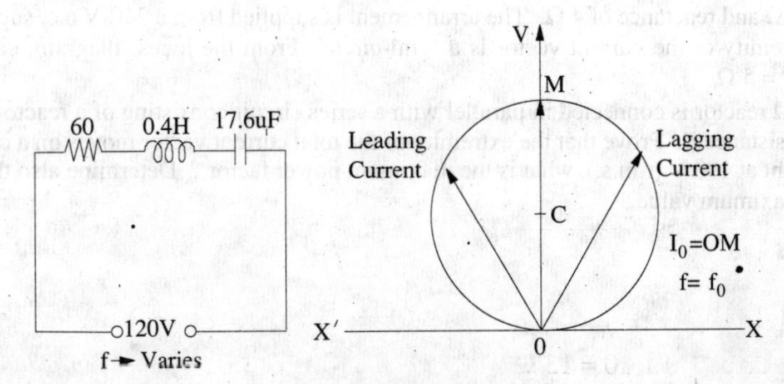

Fig. 18.13

(*ii*) **f = 40 Hz**

$X_L = 2\pi \times 40 \times 0.4 = 100\ \Omega;\ X_C = 10^6/2\pi \times 40 \times 17.6 = 226\ \Omega$

$X = 100 - 226 = -126\ \Omega$ (capacitive); $I = 120/\sqrt{60^2 + (-126)^2} =$ **0.86 A**

p.f. = $\cos\theta = R/Z = 60/139.5 =$ **0.43 (lead)**

(*iii*) **f = 80 Hz**

$X_L = 100 \times 2 = 200\ \Omega$; $X_C = 226/2 = 113\ \Omega$; $X = 200 - 113 = 87\ \Omega$ (inductive)

$Z = \sqrt{R^2 + X^2} = \sqrt{60^2 + 87^2} = 105.3\ \Omega$

$I = 120/105.3 =$ **1.14 A** ; p.f. = $\cos\theta = 60/105.3 =$ **0.57 (lag)**

Tutorial Problems No. 18.1

1. A. circuit having a constant resistance of 60 Ω and a variable inductance of 0 to 0.4 H is connected across a 100-V, 50-Hz supply. Derive from first principles the locus of the extremity of the current vector. Find (*a*) the power and (*b*) the inductance of the circuit when the power factor is 0.8.

[(*a*) **107 W** (*b*) **0.143 H**] (*App. Elect. London Univ.*)

2. A constant reactance of 10 Ω is connected in series with a variable resistor and the applied voltage is 100 V. What is (*i*) the maximum power dissipated and (*ii*) at what value of resistance does it occur ?

[(*i*) **500 W** (*ii*) **10 Ω**] (*City & Guilds London*)

3. A variable capacitance and a resistance of 300 Ω are connected in series across a 240-V, 50-Hz supply. Draw the complex or locus of impedance and current as the capacitance changes from 5 μF to 30 μ*F*. From the diagram, find (*a*) the capacitance to give a current of 0.7 A and (*b*) the current when the capacitance is 10 μ*F*.

[**19.2 μF, 0.55 A**] (*London Univ.*)

4. An a.c. circuit consists of a variable resistor in series with a coil, for which $R = 20\ \Omega$ and $L = 0.1$ H. Show that when this circuit is supplied at constant voltage and frequency and the resistance is varied between zero and infinity, the locus diagram of the current vector is a circular arc. Calculate when the supply voltage is 100 V and the frequency 50 Hz (*i*) the radius (in amperes) of the arc (*ii*) the value of the variable resistor in order that the power taken from the mains may be a maximum. [(*i*) **1.592 A** (*ii*) **11.4 Ω**] (*London Univ.*)

5. A circuit consists of an inductive coil ($L = 0.2\ H, R = 20\ \Omega$) in series with a variable resistor (0–200 Ω). Draw to scale the locus of the current vector when the circuit is connected to 230-V, 50-Hz supply mains and the resistor is varied between 0 and 200 Ω. Determine (*i*) the value of the resistor which will give maximum power in the circuit (*ii*) the power when the resistor is 150 Ω.

[(*i*) **42.8 Ω** (*ii*) **275 W**] (*London Univ.*)

6. A 15 μF capacitor, an inductive coil ($L = 0.135$ H, $R = 50\ \Omega$) and a variable resistor are in series and connected to a 230-V, 50-Hz supply.

Draw to scale the vector locus of the current when the variable resistor is varied between 0 and 500 Ω. Caluclate (*i*) the value of the variable resistor when the power is a maximum (*ii*) the power under these conditions. [(*i*) **120 Ω** (*ii*) **155.5 W**] (*London Univ.*)

7. As a.c. circuit supplied at 100 V, 50-Hz consists of a variable resistor in series with a fixed 100 μF capacitor.

Show that the extremity of the current vector moves on a circle. Determine the maximum power dissipated in the circuit and the corresponding power factor and the value of the resistor. [**157 W ; 0.707 : 31.8 Ω**]

8. A variable non-inductive resistor *R* of maximum value 10 Ω is placed in series with a coil which has a resistance of 3 Ω and reactance of 4 Ω. The arrangement is supplied from a 240-V a.c. supply. Show that the locus of the extremity of the current vector is a semi-circle. From the locus diagram, calculate the current supplied when $R = 5\ \Omega$. [**26.7 A**]

9. A 20-Ω reactor is connected in parallel with a series circuit consisting of a reactor of reactance 10 Ω and a variable resistance *R*. Prove that the extremity of the total current vector moves on a circle. If the supply voltage is constant at 100 V (r.m.s.), what is the maximum power factor ? Determine also the value of *R* when the p.f. has its maximum value. [**0.5 ; 17.3 Ω**]

19 POLYPHASE CIRCUITS

19.1. Generation of Polyphase Voltage

The kind of alternating currents and voltages discussed in chapters 12 to 15 are known as single-phase voltage and current, because they consist of a single alternating current and voltage wave. A single-phase alternator was diagrammatically depicted in Fig. 11.1 (*b*) and it was shown to have one armature winding only. But if the number of armature windings is increased, then it becomes polyphase alternator and it produces as many independent voltage waves as the number of windings or phases. These windings are displaced from one another by equal angles, the values of these angles being determined by the number of phases or windings. In fact, the word 'poly-phase' means poly (*i.e.* many or numerous) and phases (*i.e.* winding or circuit).

In a two-phase alternator, the armature windings are displaced 90 electrical degrees apart. A 3-phase alternator, as the name shows, has three independent armature windings which are 120 electrical degrees apart. Hence, the voltages induced in the three windings are 120° apart in time-phase. With the exception of two-phase windings, it can be stated that, in general, the electrical displacement between different phases is $360/n$ where n is the number of phases or windings.

Three-phase systems are the most common, although, for certain special jobs, greater number of phases is also used. For example, almost all mercury-arc rectifiers for power purposes are either six-phase or twelve-phase and most of the rotary converters in use are six-phase. All modern generators are practically three-phase. For transmitting large amounts of power, three-phase is invariably used. The reasons for the immense popularity of three-phase apparatus are that (*i*) it is more efficient (*ii*) it uses less material for a given capacity and (*iii*) it costs less than single-phase apparatus etc.

In Fig. 19.1 is shown a two-pole, stationary-armature, rotating-field type three-phase alternator. It has three armature coils *aa'*, *bb'* and *cc'* displaced 120° apart from one another. With the position and clockwise rotation of the poles as indicated in Fig. 19.1, it is found that the e.m.f. induced in conductor '*a*' of coil *aa'* is maximum and its direction* is *away* from the reader. The e.m.f. in conductor '*b*' of coil *bb'* would be maximum and away from the reader when the *N*-pole has turned through 120° *i.e.* when *N-S* axis lies along *bb'*. It is clear that the induced e.m.f. in conductor '*b*' reaches its maximum value 120° later than the maximum value in conductor '*a*'. In the like manner, the maximum e.m.f. induced (in the direction away from the reader) in conductor '*c*' would occur 120° later than that in '*b*' or 240° later than that in '*a*'.

Thus the three coils have three e.m.fs. induced in them which are similar in all respects except that they are 120° out of time phase with one another as pictured in Fig. 19.3. Each voltage wave is assumed to be sinusoidal and having maximum value of E_m.

In practice, the space on the armature is completely covered and there are many slots per phase per pole.

---

*The direction is found with the help of Fleming's Right-hand rule. But while applying this rule, it should be remembered that the relative motion of the conductor with respect to the field is anticlockwise although the motion of the field with respect to the conductor is clockwise as shown. Hence, thumb should point to the left.

Fig. 19.2 illustrates the relative positions of the windings of a 3-phase, 4-pole alternator and Fig. 19.4 shows the developed diagram of its armature windings. Assuming full-pitched winding and the direction of rotation as shown, phase '*a*' occupies the position under the centres of *N* and *S*-poles. It starts at S_a and ends or finishes at F_a.

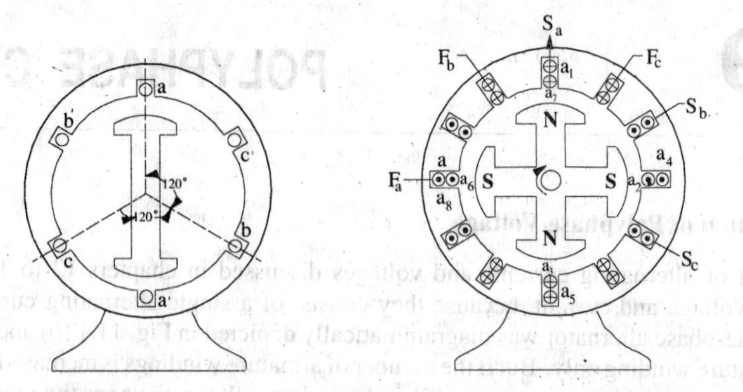

Fig. 19.1 **Fig. 19.2**

The second phase '*b*' starts at S_b which is 120 electrical degrees apart from the start of phase '*a*', progresses round the armature clockwise (as does '*a*') and finishes at F_b. Similarly, phase '*c*' starts at S_c, which is 120 electrical degrees away from S_b, progresses round the armature and finishes at F_c. As the three circuits are exactly similar but are 120 electrical degrees apart, the e.m.f. waves generated in them (when the field rotates) are displaced from each other by 120°. Assuming these waves to be sinusoidal and counting the time from the instant when the e.m.f. in phase '*a*' is zero, the instantaneous values of the three e.m.fs. will be given by curves of Fig. 19.3.

Their equations are :

$$e_a = E_m \sin \omega t \qquad \ldots(i)$$

$$e_b = E_m \sin(\omega t - 120°) \qquad \ldots(ii)$$

$$e_c = E_m \sin(\omega t - 240°) \qquad \ldots(iii)$$

As shown in Art. 11.23. alternating voltages may be represented by revolving vectors which indicate their maximum values (or r.m.s. values if desired). The actual values of these voltages vary from peak positive to zero and to peak negative values in one revolution of the vectors. In Fig. 19.5 are shown the three vectors representing the r.m.s voltages of the three phases E_a, E_b and E_c (in the present case $E_a = E_b = E_c = E$, say).

It can be shown that the sum of the three phase e.m.fs. is zero in the following three ways :

(*i*) The sum of the above three equations (*i*), (*ii*) and (*iii*) is zero as shown below :

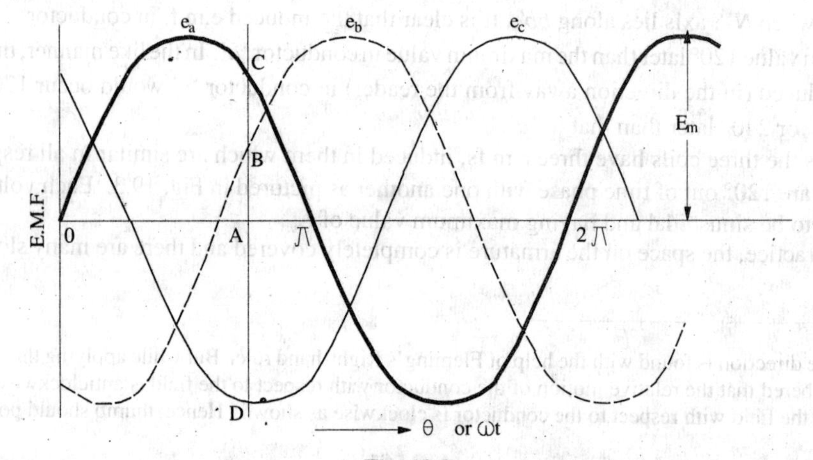

Fig. 19.3

Resultant instantaneous e.m.f. $= e_a + e_b + e_c$

$$= E_m \sin \omega t + E_m \sin (\omega t - 120°) + E_m (\omega - 240)$$

$$= E_m [\sin \omega t + 2 \sin (\omega t - 180°) \cos 60°]$$

$$= E_m [\sin \omega t - 2 \sin \omega t \cos 60] = 0$$

(*ii*) The sum of ordinates of three e.m.f. curves of Fig. 19.3 is zero. For example, taking ordinates *AB* and *AC* as positive and *AD* as negative, it can be shown by actual measurement that

$$AB + AC + (-AD) = 0$$

(*iii*) If we add the three vectors of Fig. 19.5 either vectorially or by calculation, the result is zero.

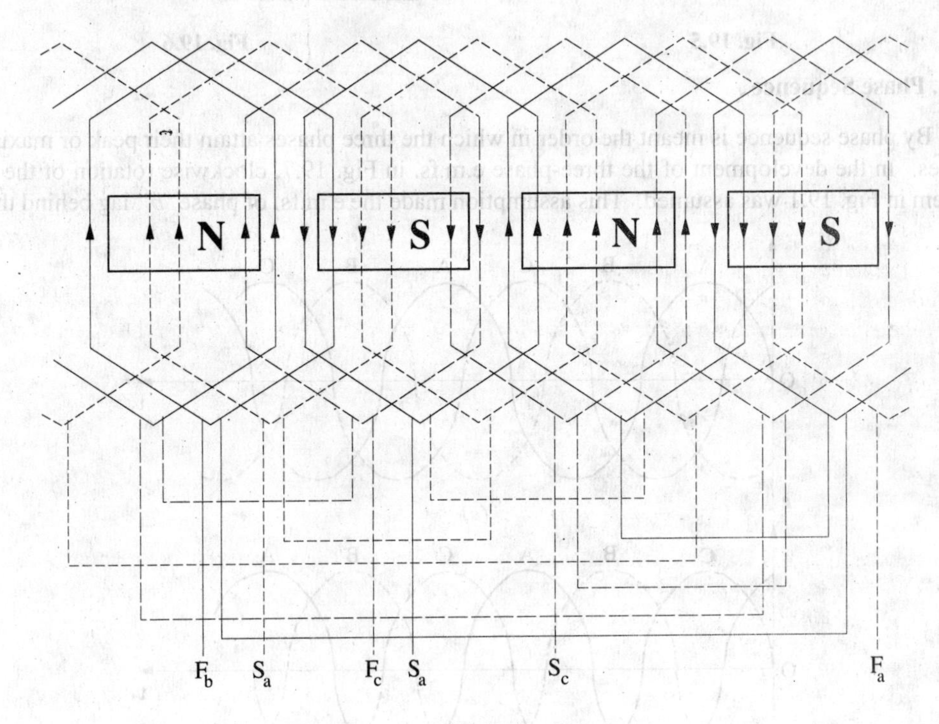

Fig. 19.4

Vector Addition

As shown in Fig. 19.6, the resultant of E_a and E_b is E_r and its magnitude is $2E \cos 60° = E$ where $E_a = E_b = E_c = E$.

This resultant E_r is equal and opposite to E_c. Hence, their resultant is zero.

By Calculation

Let us take E_a as reference voltage and assuming clockwise phase sequence

$\mathbf{E}_a = E \angle 0° = E + j0$

$\mathbf{E}_b = E \angle -120° = E(-0.5 - j\, 0.866)$

$\mathbf{E}_c = E \angle\ 240° = E \angle 120° = E\,(-0.05 + j\, 0.866)$

$\therefore\ \mathbf{E}_a + \mathbf{E}_b + \mathbf{E}_c = (E + j0) + E(-0.5 - 0.866) + E(-0.5 + j0.866) = 0$

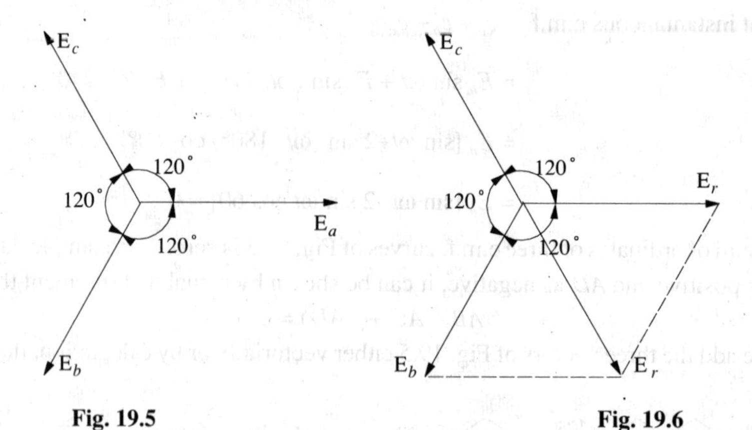

Fig. 19.5 Fig. 19.6

19.2. Phase Sequence

By phase sequence is meant the order in which the three phases attain their peak or maximum values. In the development of the three-phase e.m.fs. in Fig. 19.7, clockwise rotation of the field system in Fig. 19.1 was assumed. This assumption made the e.m.fs. of phase '*b*' lag behind that of

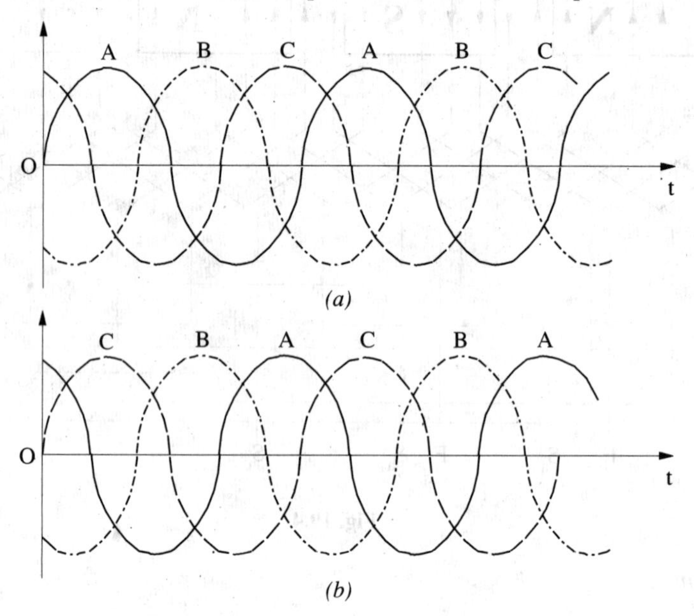

Fig. 19.7

'*a*' by 120° and in a similar way, made that of '*c*' lag behind that of '*b*' by 120° (or that of '*a*' by 240°). Hence, the order in which the e.m.fs. of phases *a*, *b* and *c* attain their maximum values is *a b c*. *It is called the phase order* or *phase sequence* $a \rightarrow b \rightarrow c$ as illustrated in Fig. 19.7 (a).

If, now, the rotation of the field structure of Fig. 19.1 is reversed *i.e.* made anticlockwise, then the order in which the three phases would attain their corresponding maximum voltages would also be reversed. The phase sequence would become $a \rightarrow c \rightarrow b$. This means that e.m.f. of phase '*c*' would now lag behind that of phase '*a*' by 120° instead of 240° as in the previous case as shown in Fig. 19.7(*b*). By repeating the letters, this phase sequence can be written as *acbacba* which is the same thing as *cba*. Obviously, a three-phase system has only two possible sequences : *abc* and *cba* (*i.e. abc* read in the reverse direction).

19.3. Phase Sequence At Load

In general, the phase sequence of the voltages applied to load is determined by the order in which the 3-phase lines are connected. The phase sequence can be reversed by interchanging any pair of lines. In the case of an induction motor, reversal of sequence results in the reversed direction of motor rotation. In the case of 3-phase unbalanced loads, the effect of sequence reversal is, in general, to cause a completely different set of values of the currents. Hence, when working on such systems, it is essential that phase sequence be clearly specified otherwise unnecessary confusion will arise. Incidentally, reversing the phase sequence of a 3-phase generator which is to be paralleled with a similar generator can cause extensive damage to both the machines.

Fig. 19.8 illustrates the fact that by interchanging any two of the three cables the phase sequence *at the load* can be reversed though sequence of 3-phase supply remains the same *i.e. abc*. It is customary to define phase sequence at the load by reading repetitively from top to bottom. For example, load phase sequence in Fig. 19.8 (*a*) would be read as *abcabcabc*– or simply *abc*. The changes are as tabulated below :

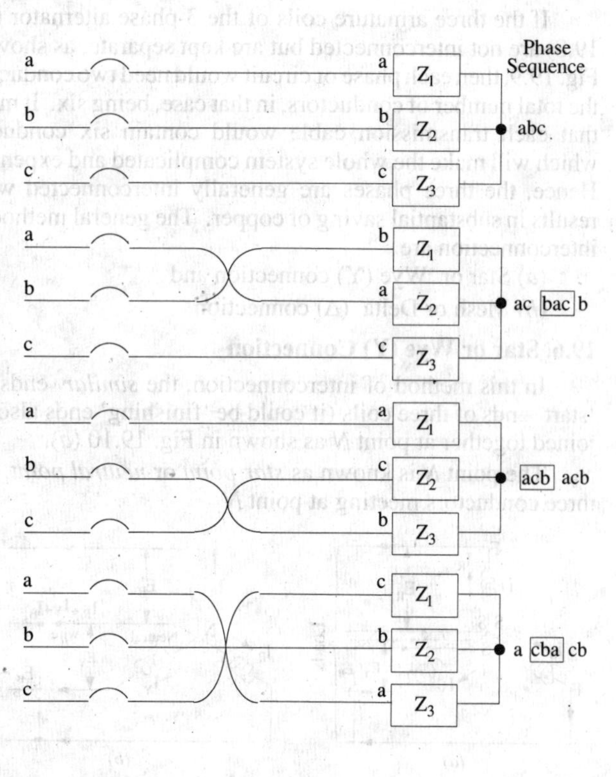

Fig. 19.8

Cables Interchanged	*Phase Sequence*
a and *b*	*b a* c b a c b a *c* – or *c b a*
b and *c*	*a* c b a c b a *c b* – or *c b a*
c and *a*	c b a c b a *c b a* – or *c b a*

19.4. Numbering of Phases

The three phases may be numbered 1, 2, 3 or *a, b, c* or as is customary, they may be given three colours. The colours used commercially are red, yellow (or sometimes white) and blue. In this case, the sequence is *RYB*.

Obviously, in any three-phase system, there are two possible sequences in which the three coil or phase voltages may pass through their maximum values *i.e.* red → yellow → blue (*RYB*) or red → blue → yellow (*RBY*). By convention, sequence *RYB* is taken as positive and *RBY* as negative.

19.5. Interconnection of Three Phases

If the three armature coils of the 3-phase alternator (Fig. 19.8) are not interconnected but are kept separate, as shown in Fig. 19.9, then each phase or circuit would need two conductors, the total number of conductors, in that case, being six. It means that each transmission cable would contain six conductors which will make the whole system complicated and expensive. Hence, the three phases are generally interconnected which results in substantial saving of copper. The general methods of interconnection are

(*a*) Star or Wye (Y) connection and

(*b*) Mesh or Delta (Δ) connection.

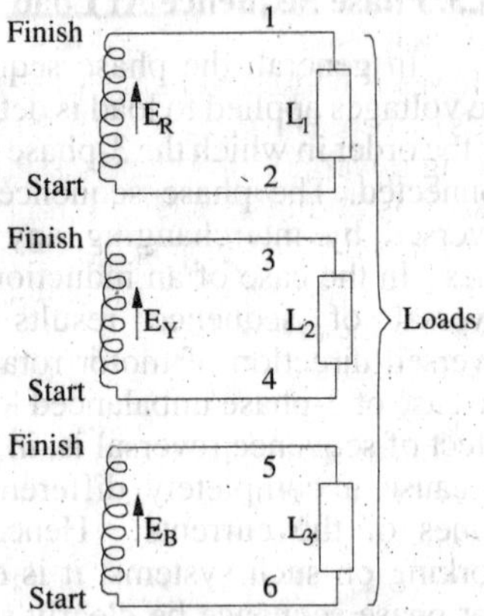

Fig. 19.9

19.6. Star or Wye (Y) Connection

In this method of interconnection, the *similar** ends say, 'start' ends of three coils (it could be 'finishing' ends also) are joined together at point *N* as shown in Fig. 19.10 (*a*).

The point *N* is known as *star point* or *neutral point*. The three conductors meeting at point *N* are replaced by a single conductor known as *neutral conductor* as shown in Fig. 19.10 (*b*). Such an interconnected system is known as four-wire, 3-phase system and is diagrammatically shown in Fig. 19.10 (*b*). If this three-phase voltage system is applied across a balanced symmetrical load, the neutral wire will be carrying three currents which are exactly equal in magnitude but are 120° out of phase with each other. Hence, their vector sum is zero.

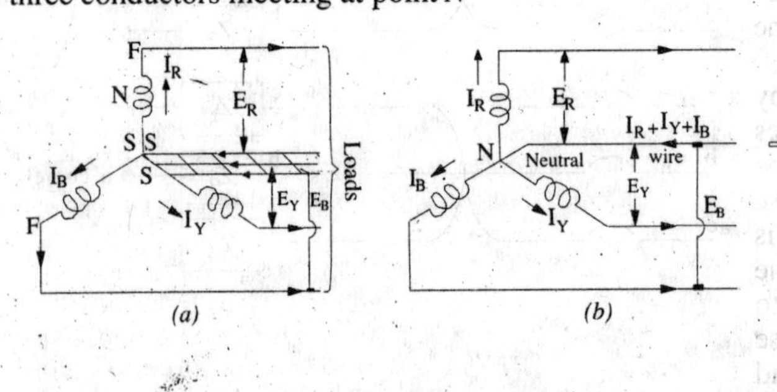

Fig. 19.10

i.e. $\mathbf{I_R} + \mathbf{I_Y} + \mathbf{I_B} = 0$ ...vectorially

The neutral wire, in that case, may be omitted although its retention is useful for supplying lighting loads at low voltages (Ex. 19.22). The p.d. between any terminal (or line) and neutral (or star) point gives the *phase* or *star* voltage. But the p.d. between any two lines gives the line-to-line voltage or simply line voltage.

19.7. Values of Phase Currents

When considering the distribution of current in a 3-phase system, it is extremely important to bear in mind that :

(*i*) the arrow placed alongside the currents I_R, I_Y and I_B flowing in the three phases [Fig. 19.10 (*b*)] indicate the directions of currents when they are assumed to be *positive* and not the directions at a particular instant. It should be clearly understood *that at no instant will all the three currents flow in the same direction either outwards or inwards.* The three arrows indicate that first the current flows outwards in phase *R*, then after a phase-time of 120°, it will flow outwards from phase *Y* and after a further 120°, outwards from phase *B*.

(*ii*) the current flowing outwards in one or two conductors is always equal to that flowing inwards in the remaining conductor or conductors. In other words, *each conductor in turn, provides a return path for the currents of the other conductors.*

---

*As an aid to memory, remember that first letter *S* of Similar is the same as that of Star.

In Fig. 19.11 are shown the three phase currents, having the same peak value of 20 A but displaced from each other by 120°. At instant '*a*', the currents in phases *R* and *B* are each + 10A (*i.e.* flowing outwards) whereas the current in phase *Y* is –20A (*i.e.* flowing inwards). In other words, at the instant '*a*', phase *Y* is acting as return path for the currents in phases *R* and *B*. At instant *b*, I_R = +15 A and I_Y = +5A but I_B = –20A which means that now phase *B* is providing the return path.

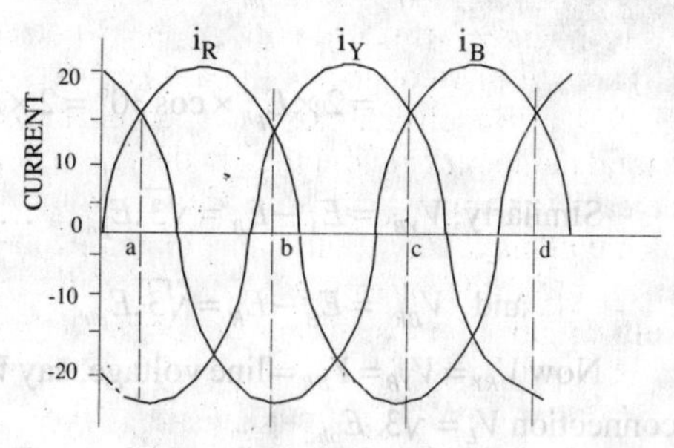

Fig. 19.11

At instant *c*, I_Y = +15 *A* and I_B = + 5A and I_R = –20A.

Hence, now phase *R* carries current inwards whereas *Y* and *B* carry current outwards. Similarly at point *d*, I_R = 0, I_B = 17.3 A and I_Y = –17.3 A. In other words, current is flowing outwards from phase *B* and returning via phase *Y*.

In addition, it may be noted that although the distribution of currents between the three lines is continuously changing, yet at any instant the algebraic sum of the *instantaneous* values of the three currents is zero *i.e.* $i_R + i_Y + i_B = 0$ –algebraically.

19.8. Voltages and Currents in Y-Connection

The voltage induced in each winding is called the *phase* voltage and current in each winding is likewise known as *phase* current. However, the voltage available between any pair of terminals (or outers) is called *line* voltage (V_L) and the current flowing in each *line* is called line current (I_L).

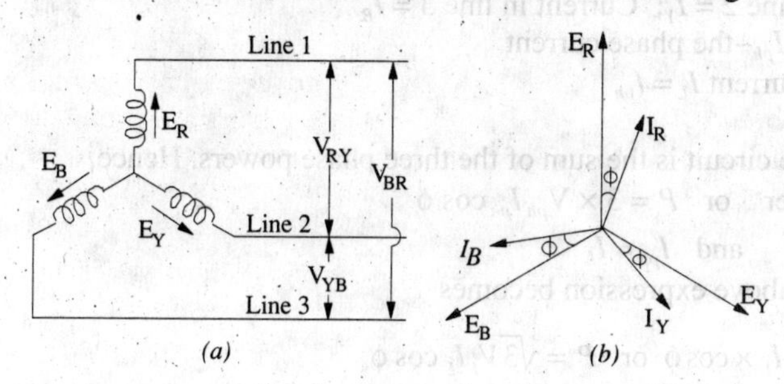

Fig. 19.12

As seen from Fig. 19.12 (*a*), in this form of interconnection, there are two phase windings between each pair of terminals but since their *similar* ends have been joined together, they are in opposition. Obviously, the *instantaneous* value of p.d. between any two terminals is the *arithmetic difference* of the two phase e.m.fs. concerned. However, the r.m.s. value of this p.d. is given by the *vector difference* of the two phase e.m.fs.

The vector diagram for phase voltages and currents in a star connection is shown in Fig. 19.12 (*b*) where a balanced system has been assumed.* It means that $E_R = E_Y = E_B = E_{ph}$ (phase e.m.f.).

Line voltage V_{RY} between line 1 and line 2 is the vector difference of E_R and E_Y.

Line voltage V_{YB} between line 2 and line 3 is the vector difference of E_Y and E_B.

Line voltage V_{BR} between line 3 and line 1 is the vector difference of E_B and E_R.

(*a*) Line Voltages and Phase Voltages

The p.d. between lines 1 and 2 is $V_{RY} = E_R - E_Y$ ...vector difference.

Hence, V_{RY} is found by compounding E_R and E_Y reversed and its value is given by the diagonal of the parallelogram of Fig. 19.13. Obviously, the angle between E_R and E_Y reversed is 60°. Hence if $E_R = E_Y = E_B$ = say, E_{ph} –the phase e.m.f., then

---

*A balanced system is one in which (*i*) the voltages in all phases are equal in magnitude and differ in phase from one another by equal angles, in this case, the angle = 360/3 = 120° (*ii*) the currents in the three phases are equal in magnitude and also differ in phase from one another by equal angles.

A 3-phase balanced load is that in which the loads connected across three phases are identical.

$$V_{RY} = 2 \times E_{ph} \times \cos(60°/2)$$

$$= 2 \times E_{ph} \times \cos 30° = 2 \times E_{ph} \times \frac{\sqrt{3}}{2} = \sqrt{3}E_{ph}$$

Similarly, $V_{YB} = E_Y - E_B = \sqrt{3}.E_{ph}$ ...vector difference

and $V_{BR} = E_B - E_R = \sqrt{3}.E_{ph}$

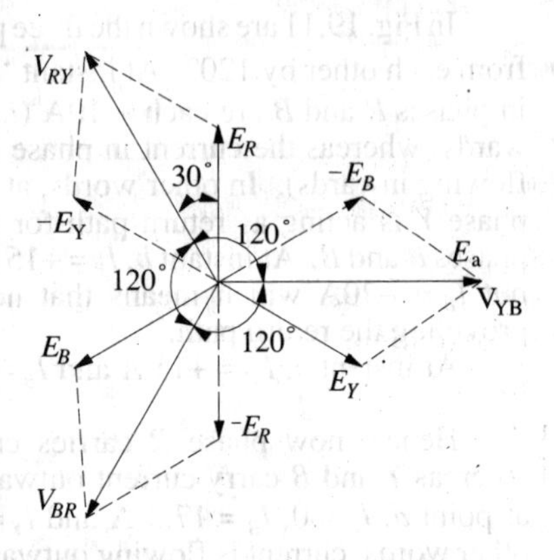

Fig. 19.13

Now $V_{RY} = V_{YB} = Y_{BR}$ = line voltage, say V_L. Hence, in star connection $V_L = \sqrt{3}.\ E_{ph}$

It will be noted from Fig. 19.13 that

1. Line voltages are 120° apart.
2. Line voltages are 30° ahead of their respective *phase* voltages.
3. The angle between the line currents and the corresponding line voltages is (30 + ϕ) with current lagging.

(*b*) Line Currents and Phase Currents

It is seen from Fig. 19.12 (*a*) that each line is in series with its individual phase winding, hence the line current in each line is the same as the current in the phase winding to which the line is connected.

Current in line 1 = I_R ; Current in line 2 = I_Y; Current in line 3 = I_B

Since $I_R = I_Y = I_B$ = say, I_{ph}–the phase current

$\therefore$ line current $I_L = I_{ph}$

(*c*) Power

The total active or true power in the circuit is the sum of the three phase powers. Hence, total active power = 3 × phase power or $P = 3 \times V_{ph} I_{ph} \cos\phi$

Now $V_{ph} = V_L/\sqrt{3}$ and $I_{ph} = I_L$

Hence, in terms of line values, the above expression becomes

$$P = 3 \times \frac{V_L}{\sqrt{3}} \times I_L \times \cos\phi \quad \text{or} \quad P = \sqrt{3}V_L I_L \cos\phi$$

It should be particularly noted that ϕ is the angle between *phase* voltage and *phase* current and **not** between the line voltage and line current.

Similarly, total reactive power is given by $\mathbf{Q} = \sqrt{3}\, V_L I_L \sin\phi$

By convention, reactive power of a coil is taken as positive and that of a capacitor as negative. The total apparent power of the three phases is

$$S = \sqrt{3}\, V_L I_L \qquad \text{Obviously, } S = \sqrt{P^2 + Q^2} \qquad \text{–Art. 13.4}$$

Example 19.1. *A balanced star-connected load of (8 + j6) Ω per phase is connected to a balanced 3-phase 400-V supply. Find the line current, power factor, power and total volt-amperes.*

(Elect. Engg., Bhagalpur Univ. 1985)

Solution. $Z_{ph} = \sqrt{8^2 + 6^2} = 10\ \Omega$

$$V_{ph} = 400/\sqrt{3} = 231 \text{ V}$$

$$I_{ph} = V_{ph}/Z_{ph} = 231/10 = 23.1 \text{ A}$$

(*i*) $I_L = I_{ph} = \mathbf{23.1\ A}$

(*ii*) p.f. = $\cos\phi = R_{ph}/Z_{ph} = 8/10 = \mathbf{0.8\ (lag)}$

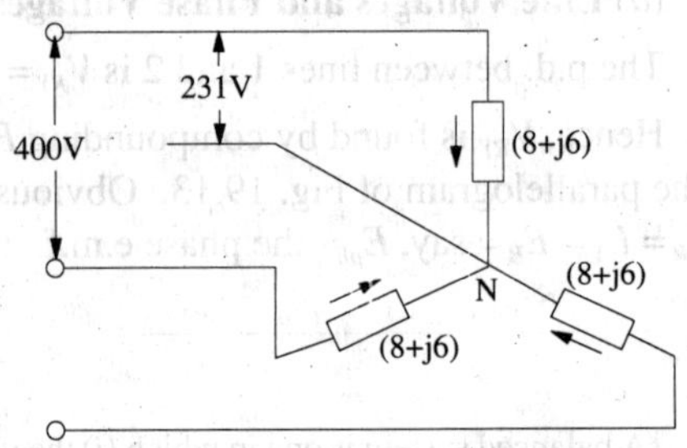

Fig. 19.14

(*iii*) Power $P = \sqrt{3}\, V_L I_l \cos\phi$

$$= \sqrt{3} \times 400 \times 23.1 \times 0.8 = \mathbf{12{,}800\ W} [\text{Also, } P = 3I_{ph}^2 R_{ph} = 3(23.1)^2 \times 8 = 12,800W]$$

(*iv*) Total volt-amperes, $S = \sqrt{3}\, V_L I_L = \sqrt{3} \times 400 \times 23.1 = \mathbf{16{,}000\ VA}$

Example 19.2. *Phase voltages of a star connected alternator are $E_R = 231 \angle 0°$ V; ; $E_Y = 231 \angle -120°$ V: and $E_B = 231 \angle +120$ °V;. What is the phase sequence of the system ? Compute the line voltages = E_{RY} and E_{YB}.* **(Elect. Mechines AMIE Sec. B Winter 1990)**

Solution. The phase voltage $E_B = 231 \angle 120°$ can be written as $E_B = 231 \angle -240°$. Hence, the three voltages are: $E_R = 231 \angle 0°$, $E_Y = 231 \angle -120°$ and $E_B = 231 \angle -240°$. It is seen that E_R is the reference voltage, E_Y lags behind it by 120° whereas E_B lags behind it by 240°. Hence, phase sequence is *RYB*. Moreover, it is a symmetrical 3-phase voltage system.

$$\therefore \qquad E_{RY} = E_{YB} = \sqrt{3} \times 231 = \mathbf{400\ V}$$

Example 19.3. *Three equal star-connected inductors take 8 kW at a power factor 0.8 when connected across a 460 V, 3-phase, 3-wire supply. Find the circuit constants of the load per phase.* **(Elect. Machines AMIE Sec. B 1992)**

Solution. $P = \sqrt{3}\, V_L I_L \cos\phi$ or

$8000 = \sqrt{3} \times 460 \times I_L \times 0.8$

$\therefore \quad I_L = 12.55$ A $\therefore\ I_{ph} = 12.55\ A;\ V_{ph} = V_L/\sqrt{3} = 460/\sqrt{3} = 265$ V

$I_{ph} = V_{ph}/Z_{ph};\ \therefore\ Z_{ph} = V_{ph}/I_{ph} = 265/12.55 = 21.1\ \Omega$

$R_{ph} = Z_{ph} \cos\phi = 21.1 \times 0.8 = 16.9\ \Omega$;

$X_{ph} = Z_{ph} \sin\phi$

$= 21.1 \times 0.6 = 12.66\ \Omega$

The circuit is shown in Fig. 19.15.

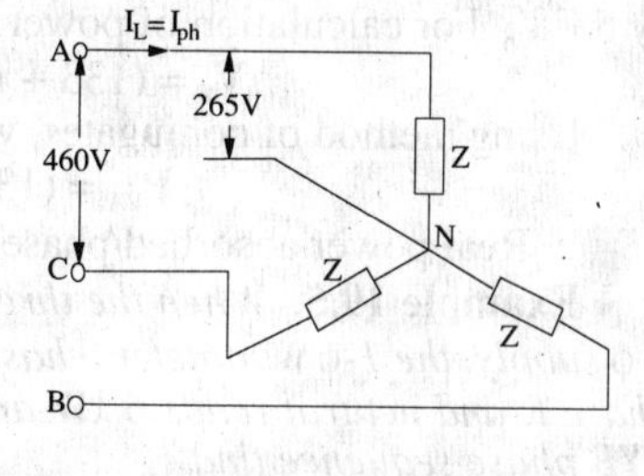

Fig. 19.15

Example 19.4. *Given a balanced 3-φ, 3-wire system with Y-connected load for which line voltage is 230 V and impedance of each phase is (6 + j8) ohm. Find the line current and power absorbed by each phase.* **(Elect. Engg -II Pune Univ. 1991)**

Solution. $Z_{ph} = \sqrt{6^2 + 8^2} = 10\ \Omega;\ V_{ph} = V_L/\sqrt{3} = 230/\sqrt{3} = 133$ V

$$\cos\phi = R/Z = 6/10 = 0.6;\ I_{ph} = V_{ph}/Z_{ph} = 133/10 = 13.3\ \text{A}$$

$$\therefore \qquad I_L = I_{ph} = \mathbf{13.3\ A}$$

Power absorbed by each phase $= I_{ph}^2 R_{ph} = 13.3^2 \times 6 = \mathbf{1061\ W}$

Solution by Symbolic Notation

In Fig. 19.16 (*b*), V_R, V_Y and V_B are the phase voltage whereas I_R, I_Y and I_B are phase currents. Taking V_R as the reference vector, we get

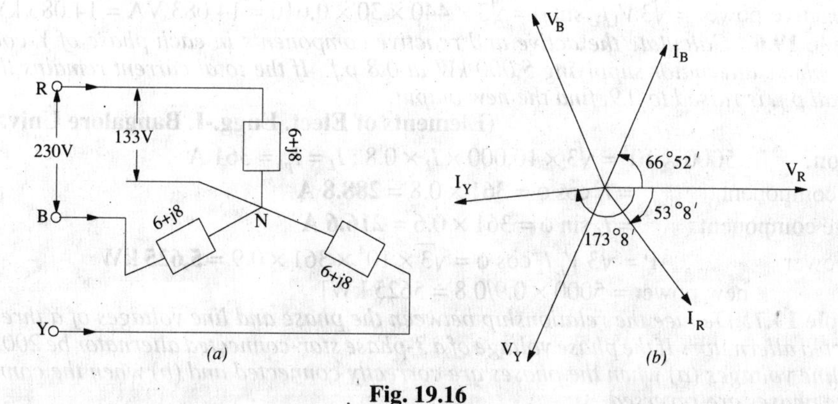

Fig. 19.16

$$\mathbf{V}_R = 133\angle 0° = 133 + j0 \text{ volt}$$
$$\mathbf{V}_Y = 133\angle -120° = 133(-0.5 - j0.866) = (-66.5 - j115) \text{ volt.}$$
$$\mathbf{V}_B = 133\angle 120° = 133(-0.5 + j0.866) = (-66.5 + j115) \text{ volt}$$
$$\mathbf{Z} = 6 + j8 = 10\angle 53°8'; \mathbf{I}_R = \frac{\mathbf{V}_R}{\mathbf{Z}} = \frac{133\angle 0°}{10\angle 53°8'} = 13.3\angle -53°8'$$

This current lags behind the reference voltage by 53°8′ [Fig. 19.16 (*b*)]

$$\mathbf{I}_Y = \frac{\mathbf{V}_Y}{\mathbf{Z}} = \frac{133\angle -120°}{10\angle 53°8'} = 13.3\angle -173°8'$$

It lags behind the reference vector *i.e.* V_R by 173°.8′ which amounts to lagging behind its phase voltage V_Y by 53°.8′.

$$\mathbf{I}_B = \frac{\mathbf{V}_B}{\mathbf{Z}} = \frac{133\angle 120°}{10\angle 53°8'} = 13.3\angle 66°52'$$

This current leads V_R by 66°52′ which is the same thing as *lagging* behind its phase voltage by 53°8′. For calculation of power, consider *R*-phase

$$\mathbf{V}_R = (133 + j0);\ \mathbf{I}_R = 13.3(0.6 - j0.8) = (7.98 - j10.64)$$

Using method of conjugates, we get

$$\mathbf{P}_{VA} = (133 - j0)(7.98 - j10.64) = 1067 - j1415$$

∴ Real power absorbed/phase = 1067 W –as before

Example 19.5. *When the three identical star-connected coils are supplied with 440 V, 50 Hz, 3-ϕ supply, the 1-ϕ wattmeter whose current coil is connected in line R and pressure coil across the phase R and neutral reads 6 kW and the ammeter connected in R-phase reads 30 Amp. Assuming RYB phase sequence find :*

(i) resistance and reactance of the coil
(ii) the power factor of the load
(iii) reactive power of 3–ϕ load.

(Elect. Engg.-I Nagpur Univ. 1993)

Solution. $V_{ph} = 440/\sqrt{3} = 254$ V; $I_{ph} = 30$ A (Fig. 19.17.)

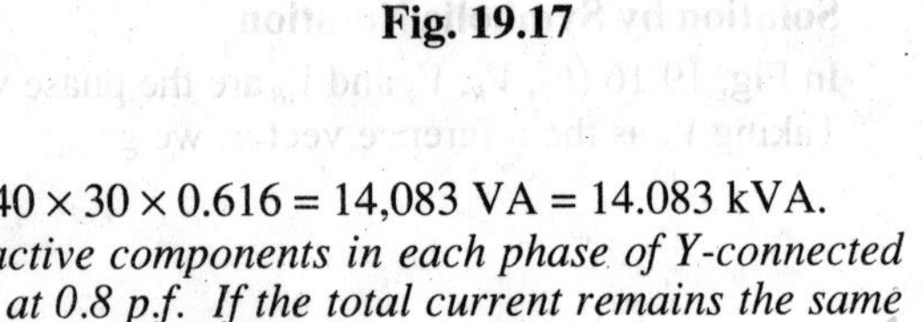

Fig. 19.17

Now, $V_{ph} I_{ph} \cos\phi = 6000$; $254 \times 30 \times \cos\phi = 6000$

∴ $\cos\phi = 0.787$; $\phi = 38.06°$ and $\sin\phi = 0.616°$;

$Z_{ph} = V_{ph}/I_{ph} = 254/30 = 8.47\ \Omega$

(*i*) Coil resistance $R = Z_{ph} \cos\phi = 8.47 \times 0.787 = 6.66\ \Omega$

$X_L = Z_{ph} \sin\phi = 8.47 \times 0.616 = 5.22\ \Omega$

(*ii*) p.f. = cos ϕ = 0.787 (lag)

(*iii*) Reactive power = $\sqrt{3}\, V_L I_L \sin\phi = \sqrt{3} \times 440 \times 30 \times 0.616 = 14{,}083$ VA = 14.083 kVA.

Example 19.6. *Calculate the active and reactive components in each phase of Y-connected 10,000 V, 3-phase alternator supplying 5,000 kW at 0.8 p.f. If the total current remains the same when the load p.f. is raised to 0.9, find the new output.*

(Elements of Elect. Engg.-I, Bangalore Univ. 1985)

Solution. $5000 \times 10^3 = \sqrt{3} \times 10{,}000 \times I_L \times 0.8$; $I_L = I_{ph} = 361$ A

active component $= I_L \cos\phi = 361 \times 0.8 =$ **288.8 A**

reactive component $= I_L \sin\phi = 361 \times 0.6 =$ **216.6 A**

New power $\mathrm{P} = \sqrt{3}\, V_L I_L \cos\phi = \sqrt{3} \times 10^4 \times 361 \times 0.9 =$ **5,625 kW**

[or new power = 5000 × 0.9/0.8 = 5625 kW]

Example 19.7. *Deduce the relationship between the phase and line voltages of a three-phase star-connected alternator. If the phase voltage of a 3-phase star-connected alternator be 200 V, what will be the line voltages (a) when the phases are correctly connected and (b) when the connections to one of the phases are reversed.*

Solution. (*a*) When phases are correctly connected, the vector diagram is as shown in Fig. 19.12 (*b*). As proved in Art. 19.7

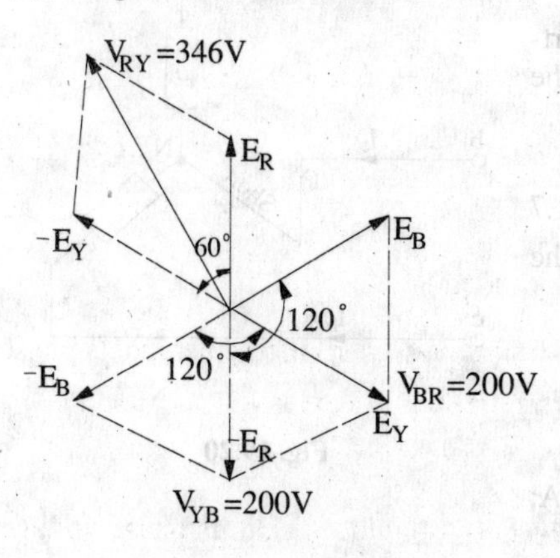

Fig. 19.18

$$V_{RY} = V_{YB} = V_{BR} = \sqrt{3}.E_{ph}$$

Each line voltage $= \sqrt{3} \times 200 =$ **346 V**

(*b*) Suppose connections to *B*-phase have been reversed. Then voltage vector diagram for such a case is shown in Fig. 19.18. It should be noted that E_B has been drawn in the reversed direction, so that angles between the three-phase voltages are 60° (instead of the usual 120°)

$$V_{RY} = E_R - E_Y \quad \text{...vector difference}$$

$$= 2 \times E_{ph} \times \cos 30° = \sqrt{3} \times 200 = \mathbf{346\ V}$$

$$V_{YB} = E_Y - E_B \quad \text{...vector difference}$$

$$= 2 \times E_{ph} \times \cos 60° = 2 \times 200 \times \frac{1}{2} = \mathbf{200\ V}$$

$$V_{BR} = E_B - E_R \quad \text{...vector difference} = 2 \times E_{ph} \times \cos 60° = 2 \times 200 \times \frac{1}{2} = \mathbf{200\ V}$$

Example 19.8. *In a 4-wire, 3-phase system, two phases have currents of 10A and 6A at lagging power factors of 0.8 and 0.6 respectively while the third phase is open-circuited. Calculate the current in the neutral and sketch the vector diagram.*

Solution. The circuit is shown in Fig. 19.19 (*a*).

$$\phi_1 = \cos^{-1}(0.8) = 36°54'; \ \phi_2 = \cos^{-1}(0.6) = 53°6'$$

Let V_R be taken as the reference vector. Then

$\mathbf{I}_R = 10\angle -36°54' = (8 - j6)$ and $\mathbf{I}_y = 6\angle -173°6' = (-6 - j0.72)$

The neutral current I_N, as shown in Fig. 19.16 (*b*), is the sum of these two currents.

$$\therefore \quad \mathbf{I}_N = (8 - j6) + (-6 - j0.72) = 2 - j6.72 = 7\angle -73°26'$$

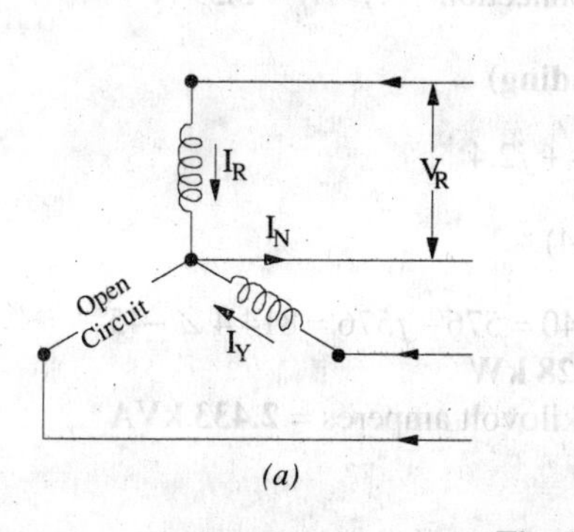

(*a*)

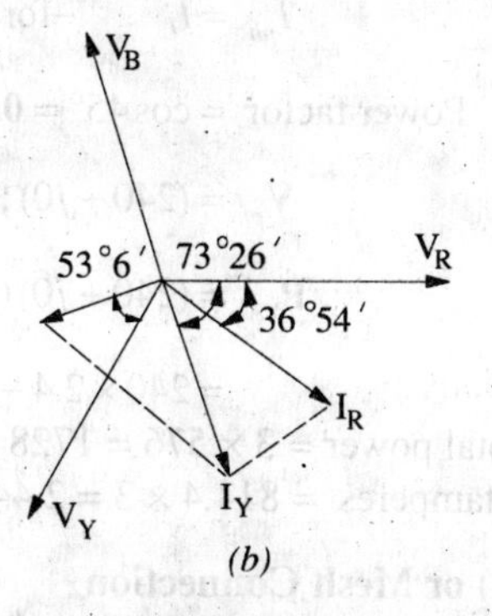

(*b*)

Fig. 19.19

Example 19.9. *Three equal star-connected inductors take 8 kW at power factor 0.8 when connected to a 460-V, 3-phase, 3-wire supply. Find the line currents if one inductor is short-circuited.*

Solution. Since the circuit is balanced, the three line voltages are represented by

$V_{ab} = 460\angle 0°$; $V_{bc} = 460\angle -120°$ and $V_{ca} = 460\angle 120°$

The phase impedance can be found from the given data :

$$8000 = \sqrt{3} \times 460 \times I_L \times 0.8 \qquad \therefore \quad I_L = I_{ph} = 12.55 \text{ A}$$

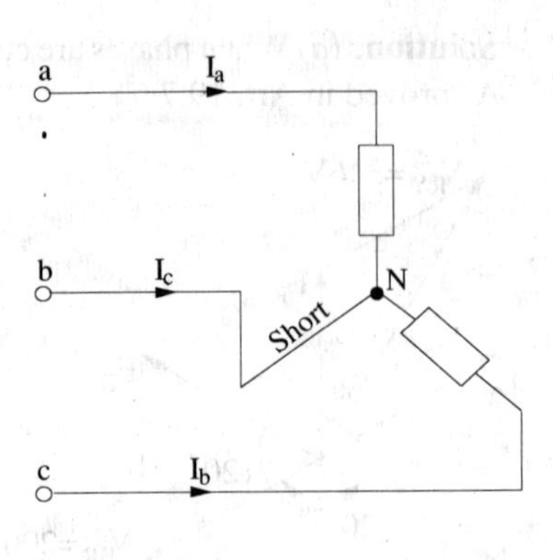

Fig. 19.20

$Z_{ph} = V_{ph}/I_{ph} = 460/\sqrt{3} \times 12.55 = 21.2\ \Omega$;

$\therefore\ Z_{ph} = 21.2\angle 36.9°$ because $\phi = \cos^{-1}(0.8) = 36.9°$

As shown in the Fig. 19.20, the phase *c* has been short-circuited. The line current $I_a = V_{ac}/Z_{ph} = -V_{ca}/Z_{ph}$ because the current enters at point *a* and leaves from point *c*.

$\therefore\ I_a = -460\angle 120°/21.2\angle 36.9° = 21.7\angle 83.1°$

Similarly, $I_b = V_{bc}/Z_{ph} = 460\angle 120°/21.2\angle 36.9° = 21.7\angle -156.9°$. The current I_c can be found by applying KVL to the neutral point N.

$$\therefore\quad I_a + I_b + I_c = 0 \quad \text{or} \quad I_c = -I_a - I_b$$

$$\therefore\quad I_C = 21.7\angle 83.1° - 21.7\angle -156.9° = 37.3\angle 53.6°$$

Hence, the magnitudes of the three currents are : 21.7 A; 21.7 A; 37.3 A

Example 19.9 (a). *Each phase of a star-connected load consists of a non-reactive resistance of 100 Ω in parallel with a capacitance of 31.8 μF.*

Calculate the line current, the power absorbed, the total kVA and the power factor when connected to a 416-V, 3-phase, 50-Hz supply.

Solution. The circuit is shown in Fig. 14.20.

$V_{ph} = (416/\sqrt{3})\angle 0° = 240\angle 0° = (240 + j0)$

Admittance of each phase is

$$\mathbf{Y}_{ph} = \frac{1}{R} + j\omega C = \frac{1}{100} + j314 \times 31.8 \times 10^{-6} = 0.01 + j0.01$$

$$\therefore\ \mathbf{I}_{ph} = \mathbf{V}_{ph} \cdot \mathbf{Y}_{ph} = 240(0.01 + j0.01)$$

$$= 2.4 + j2.4 = 3.39\angle 45°$$

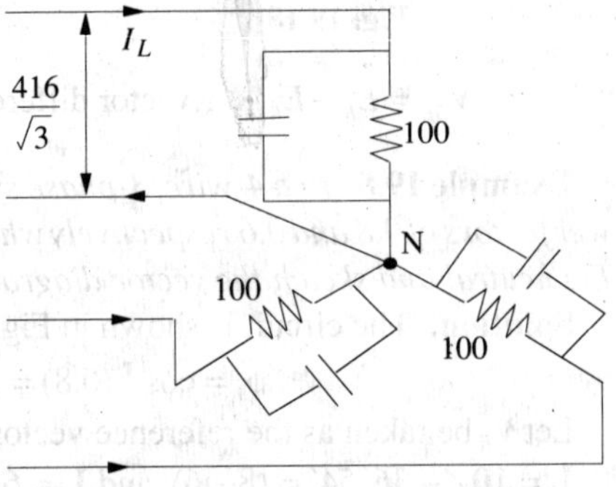

Fig. 19.21

Since $\quad I_{ph} = I_L \quad$ –for a star connection $\quad \therefore\ I_L = \mathbf{3.39\ A}$

Power factor $= \cos 45° = \mathbf{0.707\ (leading)}$

Now, $\quad \mathbf{V}_{ph} = (240 + j0)\ ;\ I_{ph} = 2.4 + j2.4$

$$\therefore\ \mathbf{P}_{VA} = (240 + j0)(2.4 + j2.4)$$

$$= 240 \times 2.4 - j2.4 \times 240 = 576 - j576 = 814.4\angle -45° \quad \text{...per phase}$$

Hence, total power = 3 × 576 = 1728 W = **1.728 kW**

Total voltamperes = 814.4 × 3 = 2,443 VA ; kilovolt amperes = **2.433 kVA**

19.9. Delta (Δ) or Mesh Connection

In this form of interconnection the *dissimilar** ends of the three phase windings are joined together *i.e.* the 'starting' end of one phase is joined to the 'finishing' end of the other phase and so on as shown in Fig. 19.22 (*a*). In other words, the three windings are joined in series to form a closed mesh as shown in Fig. 19.22 (*b*).

---

*As an aid to memory, remember that first letter *D* of Dissimilar is the same as that of Delta.

Three leads are taken out from the three junctions as shown and outward directions are taken as positive.

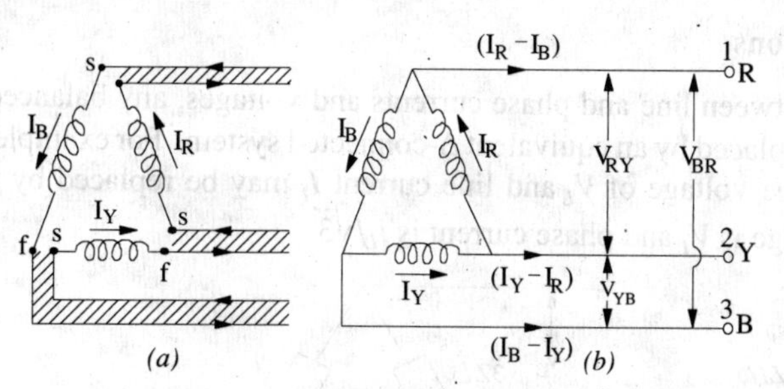

Fig. 19.22

It might look as if this sort of interconnection results in short-circuiting the three windings. However, if the system is balanced then sum of the three voltages round the closed mesh is zero, hence no current of fundamental frequency can flow around the mesh when the terminals are open. It should be clearly understood that at any instant, the e.m.f in one phase is equal and opposite to the resultant of those in the other two phases.

This type of connection is also referred to as 3-phase, 3-wire system.

(*i*) **Line Voltages and Phase Voltages**

It is seen from Fig. 19.22 (*b*) that there is only one phase winding completely included between any pair of terminals. Hence, in Δ-connection, the voltage between any pair of lines is equal to the phase voltage of the phase winding connected between the two lines considered. Since phase sequence is *R Y B*, the voltage having its positive direction from *R* to *Y* leads by 120° on that having its positive direction from *Y* to *B*. Calling the voltage between lines 1 and 2 as V_{RY} and that between lines 2 and 3 as V_{YB}, we find that V_{RY} leads V_{YB} by 120°. Similarly, V_{YB} leads V_{BR} by 120° as shown in Fig. 19.23. Let $V_{RY} = V_{YB} = V_{BR}$ = line voltage V_L. Then, it is seen that $V_L = V_{ph}$.

(*ii*) **Line Currents and Phase Currents**

It will be seen from Fig. 19.22 (*b*) that current in each line is the *vector difference* of the two phase currents flowing through that line. For example

$$\left.\begin{array}{l}\text{Current in line 1 is } I_1 = I_R - I_B \\ \text{Current in line 2 is } I_2 = I_Y - I_R \\ \text{Current in line 3 is } I_3 = I_B - I_Y\end{array}\right\} \text{vector difference}$$

Current in line No. 1 is found by compounding I_R and I_B reversed and its value is given by the diagonal of the parallelogram of Fig. 19.23. The angle between I_R and I_B reversed (*i.e.* $-I_B$) is 60°. If $I_R = I_Y$ = phase current I_{ph} (say), then

Current in line No. 1 is

$I_1 = 2 \times I_{ph} \times \cos(60°/2) = 2 \times I_{ph} \times \sqrt{3}/2 = \sqrt{3}\, I_{ph}$

Current in line No. 2 is

$I_2 = I_Y - I_R$ ...vector difference $= \sqrt{3}\, I_{ph}$

and current in line No. 3 is

$I_3 = I_B - I_Y$ ...vector difference $= \sqrt{3}.I_{ph}$

Since all the line currents are equal in magnitude *i.e.*

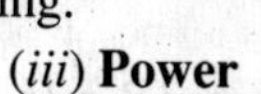

$I_1 = I_2 = I_3 = I_L$

$\therefore\ I_L = \sqrt{3} I_{ph}$

With reference to Fig. 19.23, it should be noted that

1. line currents are 120° apart
2. line currents are 30° behind the respective phase currents
3. the angle between the line currents and the corresponding line voltages is (30° + φ) with the current lagging.

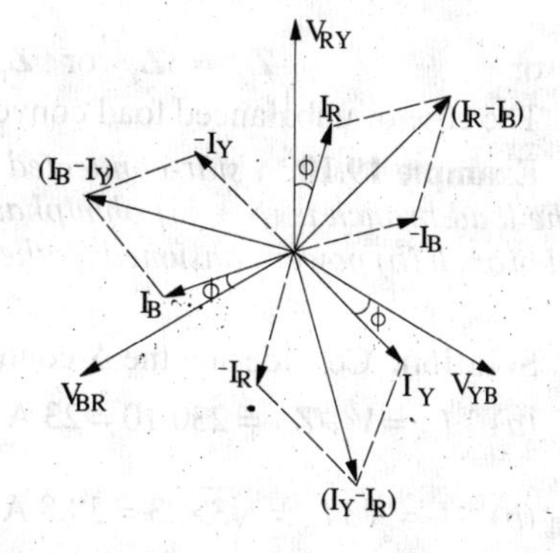

Fig. 19.23

(*iii*) **Power**

Power/phase = $V_{ph}\, I_{ph} \cos \phi$; Total power = $3 \times V_{ph} I_{ph} \cos \phi$. However, $V_{ph} = V_L$ and $I_{ph} = I_L/\sqrt{3}$

Hence, in terms of line values, the above expression for power becomes

$$P = 3 \times V_L \times \frac{I_L}{\sqrt{3}} \times \cos \phi = \sqrt{3}\, V_L I_L \cos \phi$$

where ϕ is the phase power factor angle.

19.10. Balanced Y/Δ and Δ/Y Conversions

In view of the above relationship between line and phase currents and voltages, any balanced Y-connected system may be completely replaced by an equivalent Δ-connected system. For example, a 3-phase, Y-connected system having the voltage of V_L and line current I_L may be replaced by a Δ-connected system in which phase voltage is V_L and phase current is $I_L/\sqrt{3}$.

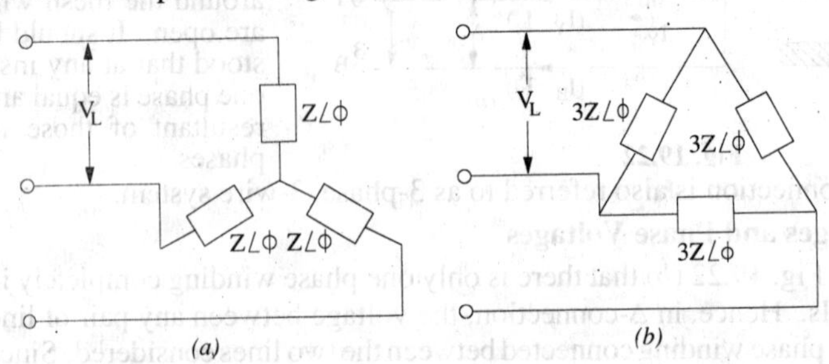

Fig. 19.24

Similarly, a balanced Y-connected load having equal branch impedances each of $Z \angle \phi$ may be replaced by an equivalent Δ-connected load whose each phase impedance is $3Z \angle \phi$. This equivalence is shown in Fig. 19.24.

For a balanced star-connected load, let

V_L = line voltage; I_L = line current ; Z_Y = impedance/phase

$\therefore \quad V_{ph} = V_L/\sqrt{3}, \quad I_{ph} = I_L; \quad Z_Y = V_L/(\sqrt{3}\, I_L)$

Now, in the equivalent Δ-connected system, the line voltages and currents must have the same values as in the Y-connected system, hence we must have

$$V_{ph} = V_L, \quad I_{ph} = I_L/\sqrt{3} \quad \therefore \quad Z_\Delta = V_L/(I_L/\sqrt{3}) = \sqrt{3}\, V_L/I_L = 3Z_Y$$

$$\therefore \quad Z_\Delta \angle\phi = 3Z_Y \angle\phi \qquad (\because V_L/I_L = \sqrt{3}\, Z_Y)$$

$$\text{or} \qquad \mathbf{Z}_\Delta = 3\mathbf{Z}_Y \quad \text{or} \quad \mathbf{Z}_Y = \mathbf{Z}_\Delta/3$$

The case of unbalanced load conversion is considered later. (Art. 19.34)

Example 19.10. *A star-connected alternator supplies a delta connected load. The impedance of the load branch is (8 + j6) ohm/phase. The line voltage is 230 V. Determine (a) current in the load branch (b) power consumed by the load (c) power factor of load (d) reactive power of the load.*

(Elect. Engg. A.M.Ae. S.I. June 1991)

Solution. Considering the Δ-connected load, we have $Z_{ph} = \sqrt{8^2 + 6^2} = 10\ \Omega$; $V_{ph} = V_L = 230$ V

(*a*) $I_{ph} = V_{ph}/Z_{ph} = 230/10 = 23$ A

(*b*) $I_L = \sqrt{3}\, I_{ph} = \sqrt{3} \times 23 = 39.8$ A; $\quad P = \sqrt{3}\, V_L I_L \cos\phi = \sqrt{3} \times 230 \times 39.8 \times 0.8 = 12{,}684$ W

(*c*) p.f. $= \cos\phi = R|Z = 8/10 = 0.8$ (lag)

(*d*) Reactive power $Q = \sqrt{3}\, V_L I_L \sin\phi = \sqrt{3} \times 230 \times 39.8 \times 0.6 = 9513$ W

Example 19.11. *A 220-V, 3-φ voltage is applied to a balanced delta-connected 3-φ load of phase impedance (15 + j20) Ω*

(*a*) *Find the phasor current in each line.*

(*b*) *What is the power consumed per phase ?*

(*c*) *What is the phasor sum of the three line currents ? Why does it have this value ?*

(Elect. Circuits and Instruments, B.H.U. 1985)

Solution. The circuit is shown in Fig. 19.25.

$V_{ph} = V_L = 220$ V ; $Z_{ph} = \sqrt{15^2 + 20^2} = 25\ \Omega$, $I_{ph} = V_{ph}/Z_{ph} = 220/25 = 8.8$ A

(*a*) $I_L = \sqrt{3}\, I_{ph} = \sqrt{3} \times 8.8 =$ **15.24 A** (*b*) $P = I_{ph}^2 R_{ph} = 8.8^2 \times 15 =$ **462 W**

(*c*) Phasor sum would be zero because the three currents are equal in magnitude and have a mutual phase difference of 120°.

Solution by Symbolic Notation

Taking V_{RY} as the reference vector, we have [Fig. 19.25 (*b*)]

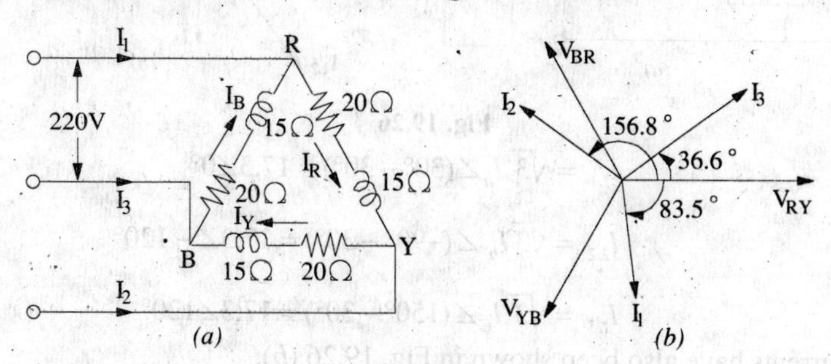

Fig. 19.25

$\mathbf{V}_{RY} = 220\angle 0°$; $\mathbf{V}_{YB} = 220\angle -120°$

$\mathbf{V}_{BR} = 220°\angle 120°$; $\mathbf{Z} = 15 + j20 = 25\angle 53°8'$

$$\mathbf{I}_R = \frac{\mathbf{V}_{RY}}{\mathbf{Z}} = \frac{220\angle 0°}{25\angle 53°8'} = 8.8\angle -53°8' = (5.28 - j7.04)\text{ A}$$

$$\mathbf{I}_Y = \frac{\mathbf{V}_{YB}}{\mathbf{Z}} = \frac{220\angle -120°}{25\angle 53°8'} = 8.8\angle -173°8' = (-8.75 - j1.05)\text{ A}$$

$$\mathbf{I}_B = \frac{\mathbf{V}_{BR}}{\mathbf{Z}} = \frac{220\angle 120°}{25\angle 53°8'} = 8.8\angle 66°52' = (3.56 + j8.1)$$

(*a*) Current in line No. 1 is

$\mathbf{I}_1 = \mathbf{I}_R - \mathbf{I}_B = (5.28 - j7.04) - (3.56 + j8.1) = (1.72 - j15.14) = 15.23\angle -83.5°$

$\mathbf{I}_2 = \mathbf{I}_Y - \mathbf{I}_R = (-8.75 - j1.05) - (5.28 - j7.04) = (-14.03 + j6.0) = 15.47\angle 156.8°$

$\mathbf{I}_3 = \mathbf{I}_B - \mathbf{I}_Y = (3.56 + j8.1) - (-8.75 - j1.05) = (12.31 + j9.15) = 15.26\angle 36.6°$

(*b*) Using conjugate of voltage, we get for *R*-phase

$\mathbf{P}_{VA} = \mathbf{V}_{RY} \cdot \mathbf{I}_R = (220 - j0)(5.28 - j7.04) = (1162 - j1550)$ voltampere

Real power per phase = **1162 W**

(*c*) Phasor sum of three line currents

$= \mathbf{I}_1 + \mathbf{I}_2 + \mathbf{I}_3 = (1.72 - j15.14) + (-14.03 + j6.0) + (12.31 + j9.15) = 0$

As expected, phasor sum of 3 line currents drawn by a balanced load is zero because these are equal in magnitude and have a phase difference of 120° amongst themselves.

Example 19.12. *A 3-φ, Δ-connected alternator drives a balanced 3-φ load whose each phase current is 10 A in magnitude. At the time when $I_a = 10\angle 30°$, determine the following, for a phase sequence of abc.*

(i) Polar expression for I_b and I_c and (ii) polar expressions for the three line currents. Show the phase and line currents on a phasor diagram.

Solution. (*i*) Since it is a balanced 3-phase system, I_b lags I_a by 120° and I_c lags I_a by 240° or leads it by 120°.

$$\therefore \quad I_b = I_a\angle -120° = 10\angle(30° - 120°) = 10\angle -90°$$

$$I_c = I_a\angle 120° = 10\angle(30° + 120°) = 10\angle 150°$$

The 3-phase currents have been represented on the phasor diagram of Fig. 19.26 (*b*).

As seen from Fig. 19.26 (*b*), the line currents lag behind their nearest phase currents by 30°.

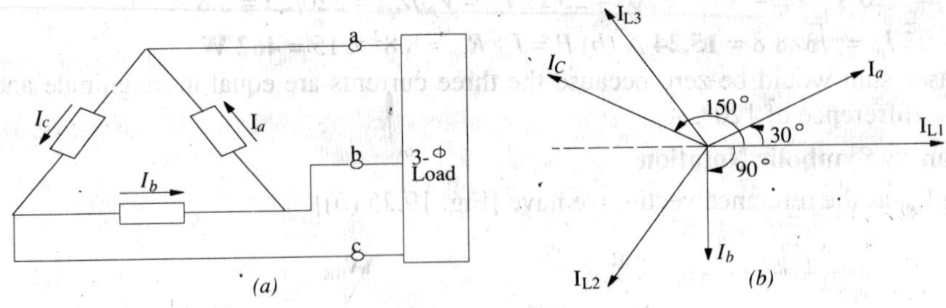

Fig. 19.26

$$\therefore \quad I_{L1} = \sqrt{3}.I_a\angle(30° - 30°) = 17.3\angle 0°$$

$$I_{L2} = \sqrt{3}\,I_b\,\angle(-90° - 30°) = 17.3\angle -120°$$

$$I_{L3} = \sqrt{3}.I_C\angle(150° - 30°) = 17.3\angle 120°$$

These line currents have also been shown in Fig. 19.26 (*b*).

Example 19.13. *Three similar coils, each having a resistance of 20 ohms and an inductance of 0.05 H are connected in (i) star (ii) mesh to a 3-phase, 50-Hz supply with 400-V between lines. Calculate the total power absorbed and the line current in each case. Draw the vector diagram of current and voltages in each case.* **(Elect. Technology, Punjab Univ. 1990)**

Solution. $X_L = 2\pi \times 50 \times 0.05 \cong 15\ \Omega$, $Z_{ph} = \sqrt{15^2 + 20^2} = 25\ \Omega$

(*i*) **Star Connection.** [Fig. 19.27 (*a*)]

$V_{ph} = 400/\sqrt{3} = 231$ V ; $I_{ph} = V_{ph}/Z_{ph} = 231/25 = 9.24\ \Omega$

$I_L = I_{ph} =$ **9.24 A** ; $P = \sqrt{3} \times 400 \times 9.24 \times (20/25) =$ **5120 W**

(*ii*) **Delta Connection** [Fig. 19.27 (*b*)]

$V_{ph} = V_L = 400$ V ; $I_{ph} = 400/25 = 16$ A; $I_L = \sqrt{3}I_{ph} = \sqrt{3} \times 16 =$ **27.7 A**

$P = \sqrt{3} \times 400 \times 27.7 \times (20/25) =$ **15,360 W**

Note. It may be noted that line current as well as power are three times the star values.

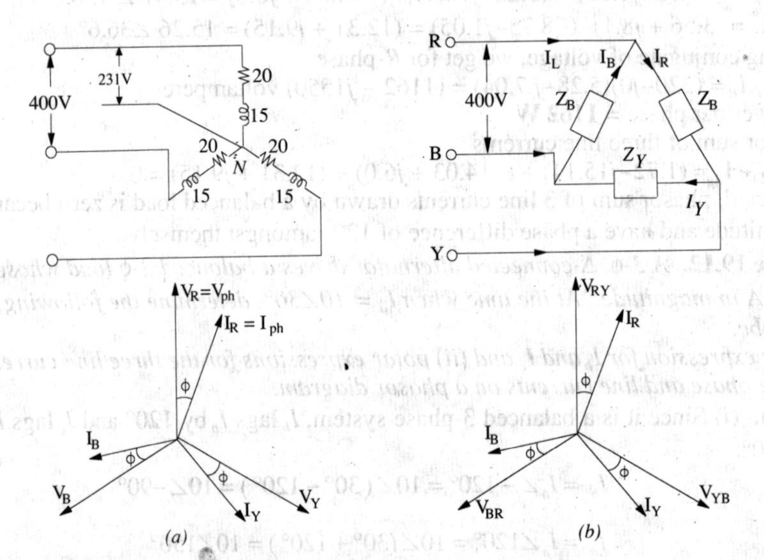

Fig. 19.27

Example 19.14. *A Δ-connected balanced 3-phase load is supplied from a 3-phase, 400-V supply. The line current is 20 A and the power taken by the load is 10,000 W. Find (i) impedance in each branch (ii) the line current, power factor and power consumed if the same load is connected in star.* **(Electrical Machines, A.M.I.E. Sec. B. 1992)**

Solution. Delta Connection.

$$V_{ph} = V_L = 400\text{ V}; I_L = 20 A; I_{ph} = 20/\sqrt{3}\text{ A}$$

(*i*) ∴ $$Z_{ph} = \frac{400}{20|\sqrt{3}} = 20\sqrt{3} = 34.64\ \Omega$$

Now $P = \sqrt{3}V_L I_L \cos\phi$ ∴ $\cos\phi = 10,000/\sqrt{3}\times 400\times 20 = \mathbf{0.7217}$

(*ii*) **Star Connection**

$$V_{ph} = \frac{400}{\sqrt{3}},\ I_{ph} = \frac{400/\sqrt{3}}{20\sqrt{3}} = \frac{20}{3}\text{ A},\ I_L = I_{ph} = \mathbf{\frac{20}{3}}\text{ A}$$

Power factor remains the same since impedance is the same.

Power consumed = $\sqrt{3}\times 400\times(20/3)\times\ 0.7217 = \mathbf{3,330\ W}$

Note. The power consumed is 1/3 of its value for Δ-connection.

Example 19.15. *Three similar resistors are connected in star across 400-V, 3-phase lines. the line current is 5 A. Calculate the value of each resistor. To what value should the line voltage be changed to obtain the same line current with the resistors delta-connected.*

Solution. Star Connection

$I_L = I_{ph} = 5$ A; $V_{ph} = 400/\sqrt{3} = 231$ V ∴ $R_{ph} = 231/5 = \mathbf{46.2\ \Omega}$

Delta Connection

$I_L = 5$ A ...(given) $I_{ph} = 5/\sqrt{3}$ A; $R_{ph} = 46.2\ \Omega$ ...found above

$V_{ph} = I_{ph}\times R_{ph} = 5\times 46.2/\sqrt{3} = \mathbf{133.3\ V}$

Note. Voltage needed is 1/3 rd the star value.

Example 19.16. *A balanced delta connected load, consisting of three coils, draws 10 $\sqrt{3}$ A at 0.5 power factor from 100 V, 3-phase supply. If the coils are re-connected in star across the same supply, find the line current and total power consumed.*

(Elect. Technology, Punjab Univ. Nov. 1988)

Solution. Delta Connection

$$V_{ph} = V_L = 100\text{V};\ I_L = 10\sqrt{3}\ A;\ I_{ph} = 10\sqrt{3}|\sqrt{3} = 10\text{ A}$$

$$Z_{ph} = V_{ph}/I_{ph} = 100/10 = 10\ \Omega;\ \cos\phi = 0.5(\text{given});\ \sin\phi = 0.866$$

$$\therefore R_{ph} = Z_{ph}\cos\phi = 10\times 0.5 = \mathbf{5\ \Omega};\ X_{ph} = Z_{ph}\sin\phi = 10\times 0.866 = \mathbf{8.66\ \Omega}$$

Incidentally, total power consumed = $\sqrt{3}\,V_L I_L\cos\phi = \sqrt{3}\times 100\times 10\sqrt{3}\times 0.5 = 1500$ W

Star Connection

$$V_{ph} = V_L|\sqrt{3} = 100|\sqrt{3};\ Z_{ph} = 10\ \Omega;\ I_{ph} = V_{ph}/Z_{ph} = 100|\sqrt{3}\times 10 = \mathbf{10|\sqrt{3}\ A}$$

Total power absorbed = $\sqrt{3}\times 100\times(10\sqrt{3})\times 0.5 = \mathbf{500\ W}$

It would be noted that the line current as well as the power absorbed are one-third of that in the delta connection.

Example 19.17. *Three identical impedances are connected in delta to a 3 φ supply of 400 V. The line current is 35 A and the total power taken from the supply is 15 kW. Calculate the resistance and reactance values of each impedance.* **(Elect. Technology, Punjab Univ. Dec. 1989)**

Solution. $V_{ph} = V_L = 400$ V; $I_L = 35$ A ∴ $I_{ph} = 35/\sqrt{3}$ A

$$Z_{ph} = V_{ph}/I_{ph} = 400\times\sqrt{3}/35 = 19.8\ \text{A}$$

Now, Power $P = \sqrt{3}V_L I_L \cos\phi \therefore \cos\phi = \frac{P}{\sqrt{3}V_L I_L} = \frac{15,000}{\sqrt{3}\times 400\times 35} = 0.619$; But $\sin\phi = 0.786$

$\therefore \quad R_{ph} = Z_{ph}\cos\phi = 19.8\times 0.619 = \mathbf{12.25\ \Omega}$; $X_{ph} = Z_{ph}\sin\phi$ and $X_{ph} = 19.8\times 0.786 = \mathbf{15.5\ \Omega}$

Example 19.18. *Three 100 Ω non-inductive resistances are connected in (a) star (b) delta across a 400-V, 50-Hz, 3-phase mains. Calculate the power taken from the supply system in each case. In the event of one of the three resistances getting open-circuited, what would be the value of total power taken from the mains in each of the two cases ?*

(Elect. Engg. A.M.Ae. S.I June, 1993)

Solution. (*i*) **Star Connection**

$V_{ph} = 400/\sqrt{3}$ V

$P = \sqrt{3}\ V_L I_L \cos\phi$

$= \sqrt{3}\times 400\times 4\times 1/\sqrt{3} = \mathbf{1600\ W}$

(*ii*) **Delta Connection**

$V_{ph} = 400$V; $R_{ph} = 100\ \Omega$

$I_{ph} = 400/100 = 4$ A

$I_L = 4\times\sqrt{3}$ A

$P = \sqrt{3}\times 400\times 4\times\sqrt{3}\times 1$

$= \mathbf{4800\ W}$

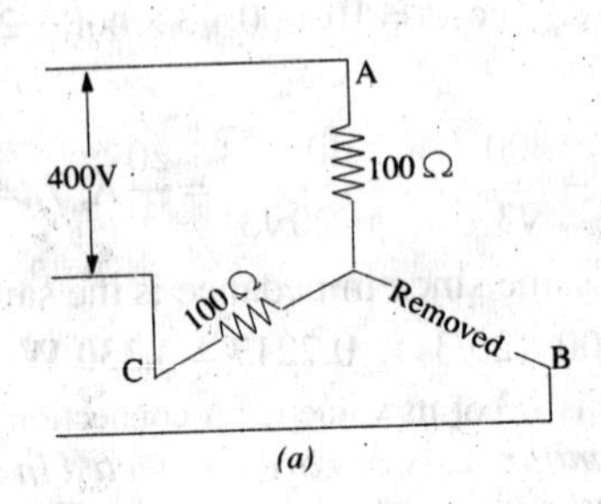

(*a*)

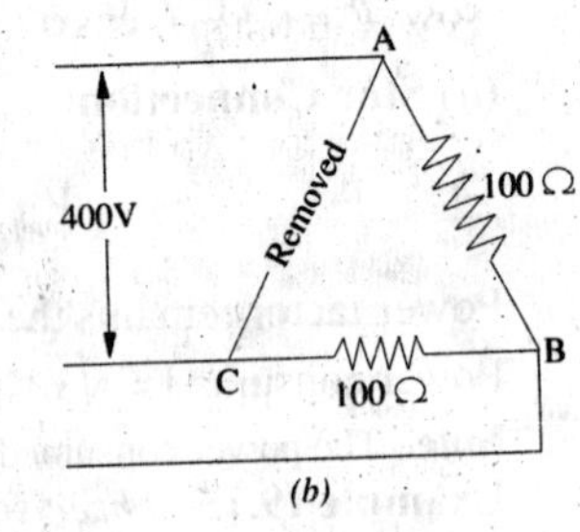

(*b*)

Fig. 19.28

When one of the resistors is disconnected

(*i*) **Star Connection** [Fig. 19.28 (*a*)]

The circuit no longer remains a 3-phase circuit but consists of two 100 Ω resistors in series across a 400-V supply. Current in lines *A* and *C* is = 400/200 = 2 A

Power absorbed in both = 400 × 2 = **800 W**

Hence, by disconnecting one resistor, the power consumption is reduced by half.

(*ii*) **Delta Connection** [Fig. 19.28 (*b*)]

In this case, currents in *A* and *C* remain as usual 120° out of phase with each other.

Current in each phase = 400/100 = 4 A

Power consumption in both = $2\times 4^2\times 100 = \mathbf{3200\ W}$

(or $P = 2\times 4\times 400 = 3200$ W)

In this case, when one resistor is disconnected, the power consumption is reduced by one-third.

Example 19.19. *A 200-V, 3-φ voltage is applied to a balanced Δ-connected load consisting of the groups of fifty 60-W, 200-V lamps. Calculate phase and line currents, phases voltages, power consumption of all lamps and of a single lamp included in each phase for the following cases.:*

(a) under normal conditions of operation

(b) after blowout in line R′R (c) after blowout in phase YB

Neglect impedances of the line and internal resistances of the sources of e.m.f.

Solution. The load circuit is shown in Fig. 19.29 where each lamp group is represented by two lamps only. It should be kept in mind that lamps remain at the line voltage of the supply irrespective of whether the Δ-connected load is balanced or not.

(*a*) **Normal operating conditions** [Fig. 19.29 (*a*)]

Since supply voltage equals the rated voltage of the bulbs, the power consumption of the lamps equals their rated wattage.

Power consumption/lamp = 60 W ; Power consumption/phase = 50 × 60 = **3,000 W**

Phase current =3000/200 = **15 A** ; Line current = $15\times\sqrt{3} = \mathbf{26\ A}$

(*b*) **Line Blowout** [Fig. 19.29 (*b*)]

When blowout occurs in line R, the lamp group of phase Y-B remains connected across line voltage $V_{YB} = V_{Y'B'}$. However, the lamp groups of other two phases get connected in series across the same voltage V_{YB}. Assuming that lamp resistances remain constant, voltage drop across $YR = V_{YR}$ 200/2 = **100 V** and that across RB = **100 V**.

Hence, phase currents are as under :

$$I_{YB} = 3000/200 = \mathbf{15\ A}, I_{YB} = I_{RB} = 15/2 = \mathbf{7.5\ A}$$

The line currents are :

$$I'_R{}_R = 0, I'_Y{}_Y = I'_B{}_B = I_{YB} + I_{YR} - 15 + 7{,}5 = \mathbf{22.5\ A}$$

Power in phase $YR = 100 \times 7.5 = 750$ W; Power/lamp = 750/50 = **15 W**
Power in phase $YB = 200 \times 15 = 3000$ W ; Power/lamp = 3000/50 = **60 W**
Power in phase $RB = 100 \times 7.5 = 750$ W ; Power/lamp = 750/50 = **15 W**

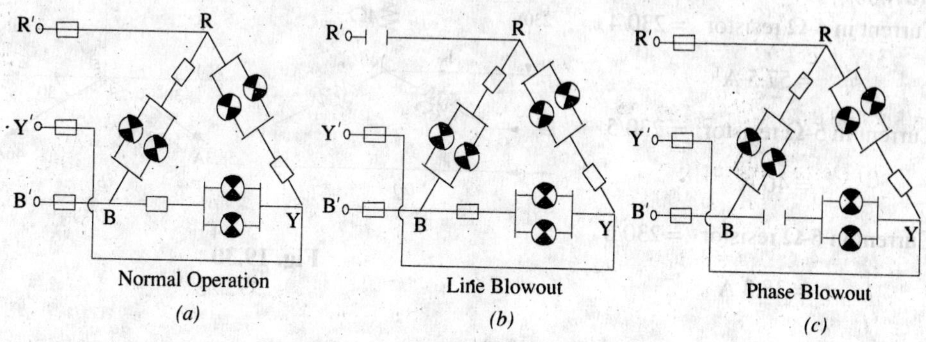

Fig. 19.29

(*c*) **Phase Blowout** [Fig. 19.29 (*c*)]

When fuse in phase Y-B blows out, the phase voltage becomes zero (though voltage across the open remains 200 V). However, the voltage across the other two phases remains the same as under normal operating conditions.

Hence, different phase currents are :

$$I_{RY} = \mathbf{15\ A}, I_{BR} = \mathbf{15\ A}, I_{YB} = \mathbf{0}$$

The line currents become

$$I'_R{}_R = 15\sqrt{3} = \mathbf{26\ A}\ ; I'_Y{}_Y = \mathbf{15\ A}, I'_B{}_B = \mathbf{15\ A}$$

Power in phase $RY = 200 \times 15$ = **3000 W**, Power/lamp = 3000/50 = **60 W**
Power in phase $RB = 200 \times 15$ = **3000 W**, Power/lamp = 3000/50 = **60 W**
Power in phase YB = 0 ; power/lamp = 0.

Example 19.20. *The load connected to a 3-phase supply comprises three similar coils connected in star. The line currents are 25 A and the kVA and kW inputs are 20 and 11 respectively. Find the line and phase voltages, the kVAR input and the resistance and reactance of each coil.*

If the coils are now connected in delta to the same three-phase supply, calculate the line currents and the power taken.

Solution. Star Connection

$$\cos\phi = \text{kW/kVA} = 11/20 \qquad I_L = 25\text{ A} \qquad P = 11\text{ kW} = 11{,}000\text{ W}$$

Now $\quad P = \sqrt{3}\, V_L I_L \cos\phi \quad \therefore \quad 11{,}000 = \sqrt{3} \times V_L \times 25 \times 11/20$

$\therefore \quad V_L = \mathbf{462\ V}\ ; \quad V_{ph} = 462/\sqrt{3} = \mathbf{267\ V}$

$$\text{kVAR} = \sqrt{\text{kVA}^2 - \text{kW}^2} = \sqrt{20^2 - 11^2} = \mathbf{16.7}\ ; Z_{ph} = 267/25 = 10.68$$

$\therefore \quad R_{ph} = Z_{ph} \times \cos\phi = 10.68 \times 11/20 = \mathbf{5.87\ \Omega}$

$\therefore \quad X_{ph} = Z_{ph} \times \sin\phi = 10.68 \times 0.838 = \mathbf{8.97\ \Omega}$

Delta Connection

$$V_{ph} = V_L = 462 \text{ V} \quad \text{and} \quad Z_{ph} = 10.68\ \Omega$$

$$\therefore \quad I_{ph} = 462/10.68 \text{ A} \qquad I_L = \sqrt{3} \times 462/10.68 = \mathbf{75\ A}$$

$$P = \sqrt{3} \times 462 \times 75 \times 11/20 = \mathbf{33{,}000\ W}$$

Example 19.21. *A 3-phase, star-connected system with 230 V between each phase and neutral has resistances of 4, 5 and 6 Ω respectively in the three phases. Estimate the current flowing in each phase and the neutral current. Find the total power absorbed.* **(I.E.E. London)**

Solution. Here, V_{ph} = 230 V [Fig. 19.30 (*a*)].

Current in 4-Ω resistor = 230/4 = 57.5 A

Current in 5-Ω resistor = 230/5 = 46 A

Current in 6-Ω resistor = 230/6 = 38.3 A

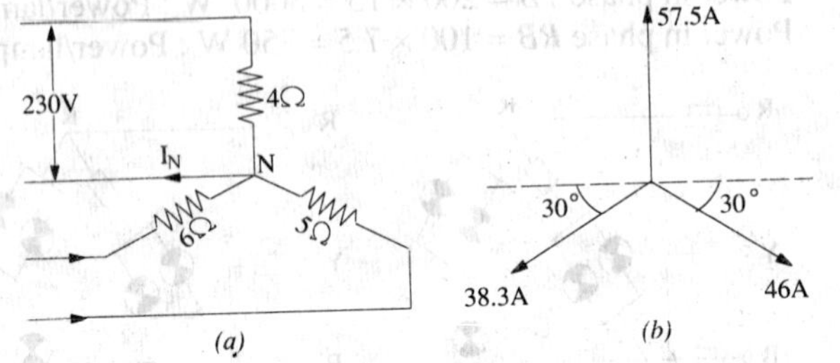

Fig. 19.30

These currents are mutually displaced by 120°. The neutral current I_N is the vector sum* of these three currents. I_N can be obtained by splitting up these three phase currents into their *X*-components and *Y*-components and then by combining them together.

X-component = 46 cos 30° − 38.3 cos 30° = 6.64 A

Y-component = 57.5 − 46 sin 30° − 38.3 sin 30° = 15.3 A $\therefore$ $I_N = \sqrt{6.64^2 + 15.3^2} = \mathbf{16.71\ A}$

The power absorbed = 230 (57.5 + 46 + 38.3) = **32.610 W**

Example 19.22. *A 3-phase, 4-wire system supplies power at 400 V and lighting at 230 V. If the lamps in use require 70, 84 and 33 A in each of the three lines, what should be the current in the neutral wire ? If a 3-phase motor is now started, taking 200 A from the line at a power factor of 0.2, what would be the current in each line and the neutral current ? Find also the total power supplied to the lamps and the motor.* **(Elect. Technology, Aligarh Univ. 1985)**

Solution. The lamp and motor connections are shown in Fig. 19.31.

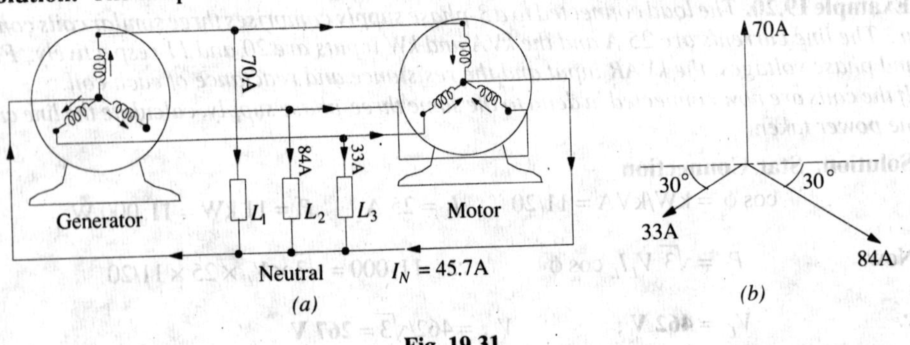

Fig. 19.31

*Some writers disagree with this statement on the ground that according to Kirchhoff's Current Law, at any junction, $\mathbf{I_N} + \mathbf{I_R} + \mathbf{I_Y} + \mathbf{I_B} = 0 \quad \therefore \quad \mathbf{I_N} = -(\mathbf{I_R} + \mathbf{I_Y} + \mathbf{I_B})$

Hence, according to them, numerical value of $\mathbf{I_N}$ is the same but its phase is changed by 180°.

When motor is not started

The neutral current is the vector sum of lamp currents. Again, splitting up the currents into their *X*- and *Y*-components, we get

X-component = 84 cos 30° − 33 cos 30° = 44.2 A

Y-component = 70 − 84 sin 30° − 33 sin 30° = 11.5 A

$\therefore \quad I_N = \sqrt{44.2^2 + 11.5^2} = \mathbf{45.7\ A}$

When motor is started

A 3-phase motor is a balanced load. Hence, when it is started, it will change the line currents but being a balanced load, it contributes nothing to the neutral current. Hence, *the neutral current remains unchanged even after starting the motor*.

Now, the motor takes 200 A from the lines. It means that each line will carry motor current (which lags) as well as lamp current (which is in phase with the voltage). The current in each line would be the vector of sum of these two currents.

Motor p.f. = cos ϕ = 0.2 ; sin ϕ = 0.9799 ...from tables

Active component of motor current 200 × 0.2 = 40 A

Reactive component of motor current 200 × 0.9799 = 196 A

(*i*) Current in first line $= \sqrt{(40+70)^2 + 196^2} = \mathbf{224.8\ A}$

(*ii*) Current in second line $= \sqrt{(40+84)^2 + 196^2} = \mathbf{232\ A}$

(*iii*) Current in third line $= \sqrt{(40+33)^2 + 196^2} = \mathbf{210.6\ A}$

Power supplied to lamps $= 230(33+84+70) = \mathbf{43{,}000\ W}$

Power supplied to motor $= \sqrt{3} \times 200 \times 400 \times 0.2 = \mathbf{27{,}700\ W}$

19.11. Star and Delta connected Lighting Loads

In Fig. 19.32 (*a*) is shown a *Y*-connected lighting network in a three storey house. For such a load, it is essential to have neutral wire in order to ensure uniform distribution of load among the three phases despite random switching on and off or burning of lamps. It is seen from Fig. 19.32 (*a*),

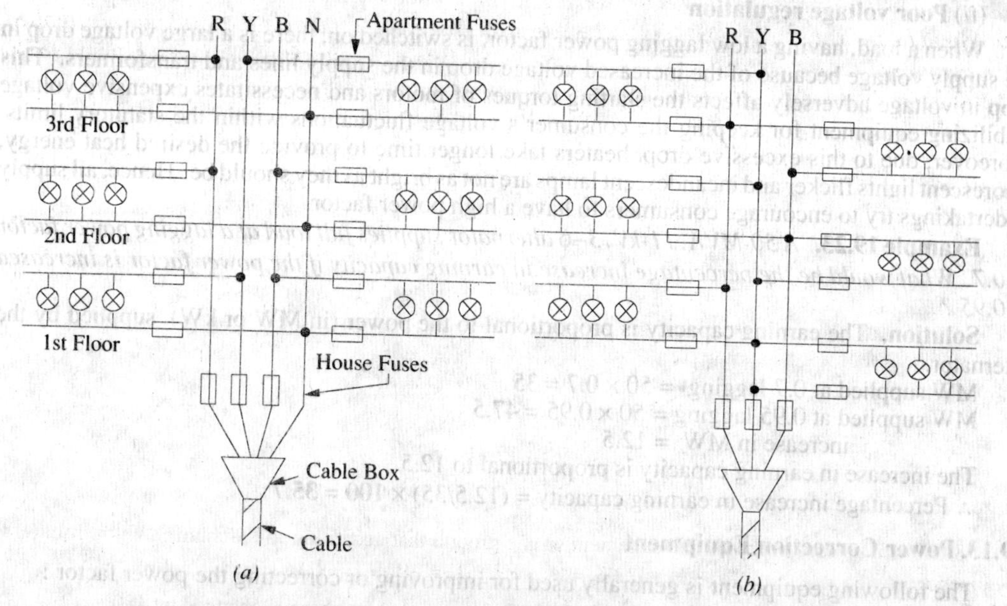

Fig. 19.32

that network supplies two flats on each floor of the three storey residence and there is balanced distribution of lamp load among the three phases. There are house fuses at the cable entry into the building which protect the two mains against short-circuits in the main cable. At the flat entry, there are apartment (or flat) fuses in the single-phase supply which protect the two mains and other flats in the same building from short-circuits in a given building. There is no fuse (or switch) on the neutral wire of the mains because blowing of such a fuse (or disconnection of such a switch) would mean a break in the neutral wire. This would result in unequal voltages across different groups of lamps in case they have different power ratings or number. Consequently, filaments in one group would burn dim whereas in other groups they would burn too bright resulting in their early burn-out.

The house-lighting wire circuit for Δ-connected lamps is shown in Fig. 19.32(*b*).

19.12. Power Factor Improvement

The heating and lighting loads supplied from 3-phase supply have power factors, ranging from 0.95 to unity. But motor loads have usually low lagging power factors, ranging from 0.5 to 0.9. Single-phase motors may have as low a power factor as 0.4 and electric welding units have even lower power factors of 0.2 or 0.3.

The power factor is given by $\cos\phi = \dfrac{\text{kW}}{\text{kVA}}$ or $\text{kVA} = \dfrac{\text{kW}}{\cos\phi}$

In the case of single-phase supply, $\text{kVA} = \dfrac{VI}{1000}$ or $I = \dfrac{1000\ \text{kVA}}{V}$ $\therefore\ I \propto \text{kVA}$

In the case of 3-phase supply $\text{kVA} = \dfrac{\sqrt{3}\,V_L I_L}{1000}$ or $I_L = \dfrac{1000\ \text{kVA}}{\sqrt{3} \times V_L}$ $\therefore\ I_L \propto \text{kVA}$

In each case, the kVA is directly proportional to current. The chief disadvantage of a low p.f. is that the current required for a given power, is very high. This fact leads to the following undesirable results.

(*i*) Large kVA for given amount of power

All electric machinery, like alternators, transformers, switchgears and cables are limited in their current-carrying capacity by the permissible temperature rise, which is proportional to I^2. Hence, they may all be fully loaded with respect to their rated kVA, without delivering their full power. Obviously, it is possible for an existing plant of a given kVA rating to increase its earning capacity (which is proportional to the power supplied in kW) if the overall power factor is improved *i.e.* raised.

(*ii*) Poor voltage regulation

When a load, having a low lagging power factor, is switched on, there is a large voltage drop in the supply voltage because of the increased voltage drop in the supply lines and transformers. This drop in voltage adversely affects the starting torques of motors and necessitates expensive voltage stabilizing equipment for keeping the consumer's voltage fluctuations within the statutory limits. Moreover, due to this excessive drop, heaters take longer time to provide the desired heat energy, fluorescent lights flicker and incandescent lamps are not as bright as they should be. Hence, all supply undertakings try to encourage consumers to have a high power factor.

Example 19.23. *A 50-MVA, 11-kV, 3−φ alternator supplies full load at a lagging power factor of 0.7. What would be the percentage increase in earning capacity if the power factor is increased to 0.95 ?*

Solution. The earning capacity is proportional to the power (in MW or kW) supplied by the alternator.

MW supplied at 0.7 lagging $= 50 \times 0.7 = 35$
MW supplied at 0.95 lagging $= 50 \times 0.95 = 47.5$
increase in MW $= 12.5$

The increase in earning capacity is proportional to 12.5

$\therefore$ Percentage increase in earning capacity $= (12.5/35) \times 100 = \mathbf{35.7}$

19.13. Power Correction Equipment

The following equipment is generally used for improving or correcting the power factor :

(*i*) Synchronous Motors (or capacitors)

These machines draw leading kVAR when they are over-excited and, especially, when they are running idle. They are employed for correcting the power factor in bulk and have the special advantage that the amount of correction can be varied by changing their excitation.

(*ii*) **Static Capacitors**

They are installed to improve the power factor of a group of a.c. motors and are practically loss-free (*i.e.* they draw a current leading in phase by 90°). Since their capacitances are not variable, they tend to over-compensate on light loads, unless arrangements for automatic switching off the capacitor bank are made.

(*iii*) **Phase Advancers**

They are fitted with individual machines.

However, it may be noted that the economical degree of correction to be applied in each case, depends upon the tariff arrangement between the consumers and the supply authorities.

Example 19.24. *A 3-phase, 37.3 kW, 440-V, 50-Hz induction motor operates on full load with an efficiency of 89% and at a power factor of 0.85 lagging. Calculate the total kVA rating of capacitors required to raise the full-load power factor ot 0.95 lagging. What will be the capacitance per phase if the capacitors are (a) delta-connected and (b) star-connected?*

Solution. It is helpful to approach such problems from the 'power triangle' rather than from vector diagram viewpoint.

Motor power input $P = 37.3/0.89 = 41.91$ kW

Power factor 0.85 (lag)

$\cos\phi_1 = 0.85 : \phi_1 = \cos^{-1}(0.85) = 31.8°$; $\tan\phi_1 = \tan 31.8° = 0.62$

Motor kVAR$_1$ = $P \tan\phi_1 = 41.91 \times 0.62 = 25.98$

Power factor 0.95 (lag)

Motor power input $P = 41.91$ kW ....as before

It is the same as before because capacitors are loss-free *i.e.* they do not absorb any power.

$\cos\phi_2 = 0.95 \quad \therefore \phi_2 = 18.2°$; $\tan 18.2° = 0.3288$

Motor kVAR$_2$ = $P \tan\phi_2 = 41.91 \times 0.3288 = 13.79$

The difference in the values of kVAR is due to the capacitors which supply *leading* kVAR to partially neutralize the *lagging* kVAR of the motor.

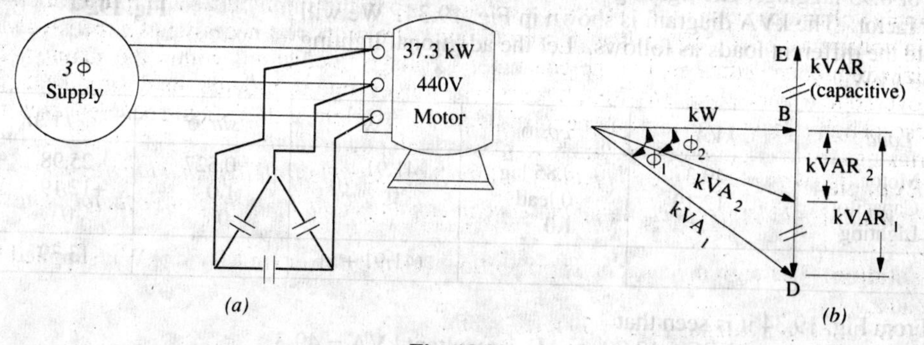

Fig. 19.33

∴ leading kVAR supplied by capacitors is

= kVAR$_1$ − kVAR$_2$ = 25.98−13.79 = **12.19** ...*CD* in Fig. 19.33. (*b*)

Since capacitors are loss-free, their kVAR is the same as kVA

∴ kVA/capacitor = 12.19/3 = 4.063 ∴ VAR/capacitor = 4,063

(*a*) In Δ-connection, voltage across each capacitor is 440 V

Current drawn by each capacitor $I_C = 4063/440 = 9.23$ A

Now, $I_C = \frac{V}{X_C} = \frac{V}{1/\omega C} = \omega VC$

$\therefore$ $C = I_C/\omega V = 9.23/2\pi \times 50 \times 440 = 66.8 \times 10^{-6}\text{F} = \mathbf{66.8\ \mu F}$

(*b*) In star connection, voltage across each capacitor is = $440/\sqrt{3}$ volt

Current drawn by each capacitor, $I_C = \frac{4063}{440/\sqrt{3}} = 16.0$ A

Now, $I_C = \frac{V}{X_C} = \omega VC$ or $16 = \frac{440}{\sqrt{3}} \times 2\pi \times 50 \times C$

$\therefore$ $C = 200.4 \times 10^{-6}\text{F} = \mathbf{200.4\ \mu F}$

Note. Star value is three times the delta value.

Example 19.25. *If the motor of Example 19.24 is supplied through a cable of resistance 0.04 Ω per core, calculate*

(i) the percentage reduction in cable Cu loss and

(ii) the additional balanced lighting load which the cable can supply when the capacitors are connected.

Solution. Original motor $\text{kVA}_1 = P/\cos\phi_1 = 41.91/0.85 = 49.3$

Original line current, $I_{L1} = \frac{kVA_1 \times 1000}{\sqrt{3} \times 440} = \frac{49.3 \times 1000}{\sqrt{3} \times 440} = 64.49$ A

$\therefore$ Original Cu loss/conductor $= 64.69^2 \times 0.04 = 167.4$ W

From Fig. 19.34, it is seen that the new kVA *i.e.* kVA_2 when capacitors are connected is given by $\text{kVA}_2 = \text{kW}/\cos\phi_2 = 41.91/0.95 = 44.12$

New line current $I_{L2} = \frac{44,120}{\sqrt{3} \times 440} = 57.89$ A

New Cu loss $= 57.89^2 \times 0.04 = 134.1$ W

(*i*) $\therefore$ percentage reduction $= \frac{167.4 - 134.1}{167.4} \times 100 = \mathbf{19.9}$

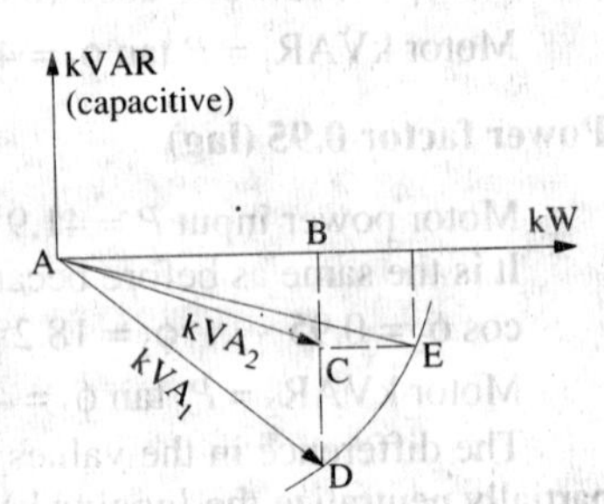

Fig. 19.34

The total kVA which the cable can supply is 49.3 kVA. When the capacitors are connected, the kVA supplied is 44.12 at a power factor of 0.95 lagging. The lighting load will be assumed at unity power factor. The kVA diagram is shown in Fig. 19.34. We will tabulate the different loads as follows. Let the additional lighting load be x kW.

Load	*kVA*	*cos φ*	*kW*	*sin φ*	*kVAR*
Motor	49.3	0.85 lag	41.91	0.527	−25.98
Capacitors	12.19	0 lead	0	1.0	+12.19
Lighting	—	1.0	x	0	0
			$(41.91 + x)$		−13.79

From Fig. 19.34 it is seen that

$AF = 41.91 + x$ and $EF = 13.79$ AE = resultant kVA = 49.3

Also $AF^2 + EF^2 = AE^2$ or $(41.91 + x)^2 + 13.79^2 = 49.3^2$ $\therefore$ $x = \mathbf{5.42\ kW}$

Example 19.26. *Three impedance coils, each having a resistance of 20 Ω and a reactance of 15 Ω, are connected in star to a 400-V, 3-φ, 50-Hz supply. Calculate (i) the line current (ii) power supplied and (iii) the power factor.*

If three capacitors, each of the same capacitance, are connected in delta to the same supply so as to form parallel circuit with the above impedance coils, calculate the capacitance of each capacitor to obtain a resultant power factor of 0.95 lagging.

Solution. $V_{ph} = 100|\sqrt{3}\,V, Z_{ph} = \sqrt{20^2 + 15^2} = 25\ \Omega$

$$\cos\phi_1 = R_{ph}|Z_{ph} = 20/25 = 0.8 \text{ lag} ; \sin\phi_1 = 0.6 \text{ lag}$$

where ϕ_1 is the power factor angle of the coils.

When capacitors are not connected

(i) $I_{ph} = 400/25 \times \sqrt{3} = 9.24$ A $\qquad \therefore I_L = \mathbf{9.24\ A}$

(ii) $P = \sqrt{3}\,V_L I_L \cos\phi_1 = \sqrt{3} \times 400 \times 9.24 \times 0.8 = \mathbf{5.120\ W}$

(iii) Power factor = **0.8** (lag)

$\therefore$ Motor $VAR_1 = \sqrt{3}\,V_L I_L \sin\phi_1 = \sqrt{3} \times 400 \times 9.24 \times 0.6 = 3{,}840$

When capacitors are connected

Power factor, $\cos\phi_2 = 0.95$, $\phi_2 = 18.2°$; $\tan 18.2° = 0.3288$

Since capacitors themselves do not absorb any power, power remains the same *i.e.* 5,120 W even when capacitors are connected. The only thing that changes is the VAR.

Now $VAR_2 = P \tan\phi_2 = 5120 \times 0.3288 = 1684$

Leading VAR supplied by the three capacitors is

$= VAR_1 - VAR_2 = 3840 - 1684 = 2156$ = BD or CE in Fig. 19.35 *(b)*

VAR/capacitor = 2156/3 = 719

For delta connection, voltage across each capacitor is 400 V $\therefore I_c = 719/400 = 1.798$ A

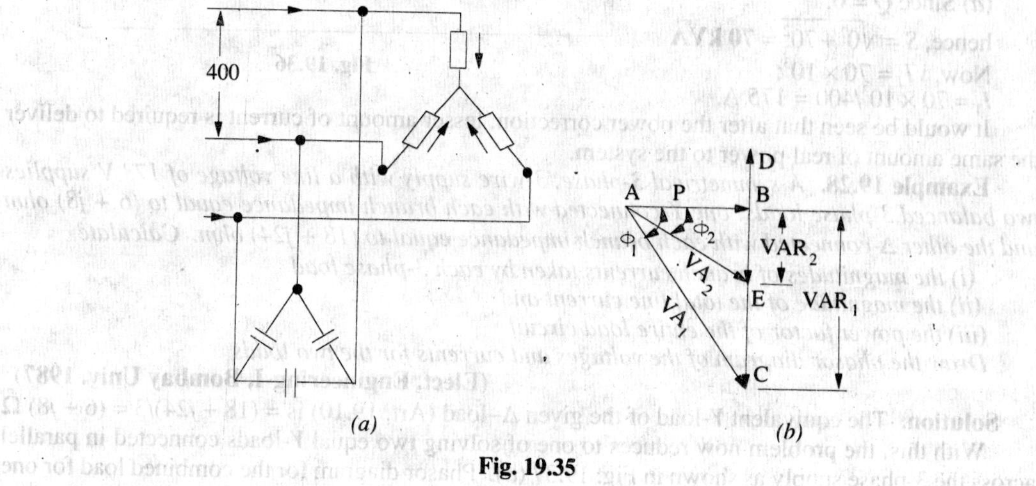

Fig. 19.35

Also $\quad I_C = \dfrac{V}{1/\omega C} = \omega VC \ \therefore C = 1.798/2\pi \times 50 \times 400 = 14.32 \times 10^{-6}\ \text{F} = \mathbf{14.32\ \mu F}$

19.14. Parallel Loads

A combination of balanced 3-phase loads connected in parallel may be solved by any one of the following three methods :

1. All the given loads may be converted into equivalent Δ-loads and then combined together according to the law governing parallel circuits.
2. All the given loads may be converted into equivalent *Y*-loads and treated as in (1) above.
3. The third method, which-requires less work, is to work in terms of volt-amperes. The special advantage of this approach is that voltmeters can be added regardless of the kind of connection involved. The real power of various loads can be added arithmetically and VARs may be added algebraically so that total voltamperes are given by

$$VA = \sqrt{W^2 + VAR^2} \quad \text{or} \quad S = \sqrt{P^2 + Q^2}$$

where P is the power in watts and Q represents reactive voltamperes.

Example 19.27. *For the power distribution system shown in Fig. 19.36, find*

(*a*) *total apparent power, power factor and magnitude of the total current I_T, without the capacitor in the system*

(*b*) *the capacitive kVARs that must be supplied by C to raise the power factor of the system to unity*

(*c*) *the capacitance C necessary to achieve the power correction in part (b) above*

(*d*) *total apparent power and supply current I_T after the power factor correction.*

Solution. (*a*) We will take the inductive *i.e.* lagging kVARs as negative and capacitive *i.e.* leading kVARs as positive.

Total $Q = -16 + 6 - 12 = -22$ kVAR (lag); Total $P = 30 + 4 + 36 = 70$ kW

$\therefore$ apparent power $S = \sqrt{(-22)^2 + 70^2} =$ **73.4 kVA**; p.f. $= \cos\phi = P/S = 70/73.4 =$ **0.95**

$S = VI_T$ or $73.4 \times 10^3 = 400 \times I_T \; \therefore \; I_T =$ **183.5 A**

(*b*) Since total lagging kVARs are **−22**, hence, for making the power factor unity, 22 leading kVARs must be supplied by the capacitor to neutralize them. In that case, total $Q = 0$ and $S = P$ and p.f. is unity.

(*c*) If I_C is the current drawn by the capacitor, then $22 \times 10^3 = 400 \times I_C$

Now, $I_C = V/X_C = V\omega C$

$= 400 \times 2\pi \times 50 \times C$

$\therefore \; 22 \times 10^3 = 400 \times (400 \times 2\pi \times 50 \times C)$;

$\therefore C =$ **483 μF**

(*d*) Since $Q = 0$,

hence, $S = \sqrt{0^2 + 70^2} =$ **70 kVA**

Now, $VI_T = 70 \times 10^3$;

$I_T = 70 \times 10^3/400 = 175$ A.

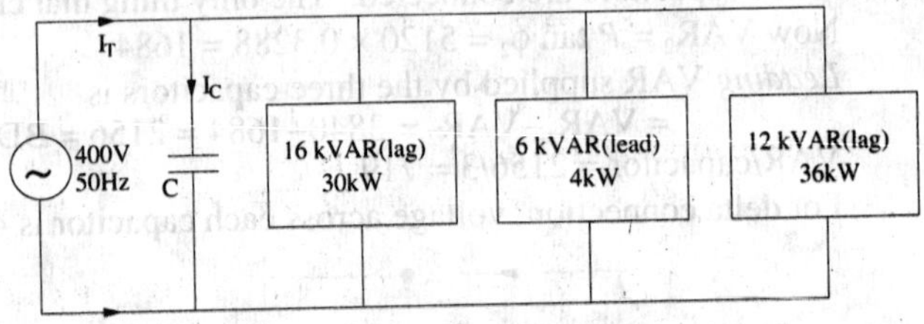

Fig. 19.36

It would be seen that after the power correction, lesser amount of current is required to deliver the same amount of real power to the system.

Example 19.28. *A symmetrical 3-phase, 3-wire supply with a line voltage of 173 V supplies two balanced 3-phase loads; one Y-connected with each branch impedance equal to (6 + j8) ohm and the other Δ-connected with each branch impedance equal to (18 + j24) ohm. Calculate*

(*i*) *the magnitudes of branch currents taken by each 3-phase load*

(*ii*) *the magnitude of the total line current and*

(*iii*) *the power factor of the entire load circuit*

Draw the phasor diagram of the voltages and currents for the two loads.

(Elect. Engineering-I, Bombay Univ. 1987)

Solution. The equivalent *Y*-load of the given Δ–load (Art. 19.10) is $= (18 + j24)/3 = (6 + j8)\ \Omega$

With this, the problem now reduces to one of solving two equal *Y*-loads connected in parallel across the 3-phase supply as shown in Fig. 19.37 (*a*). Phasor diagram for the combined load for one phase only is given in Fig. 19.37 (*b*).

Combined load impedance

$= (6 + j8)/2 = 3 + j4$

$= 5\angle 53.1°$ ohm

$V_{ph} = 173/\sqrt{3} = 100$ V

Let $\quad V_{ph} = 100\angle 0°$

$$\therefore \quad I_{ph} = \frac{100\angle 0°}{5\angle 53.1°} = 20\angle -53.1°$$

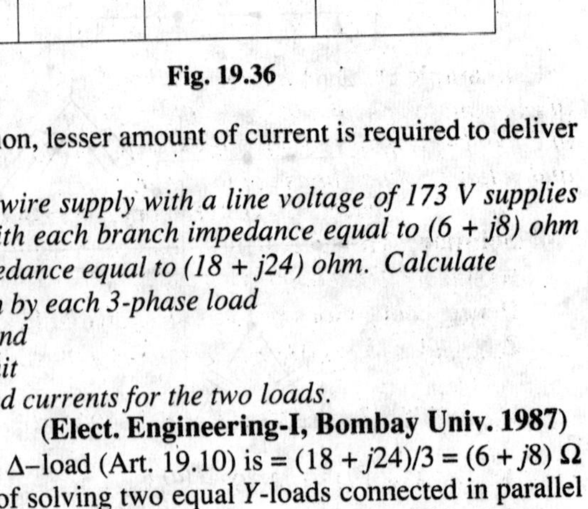

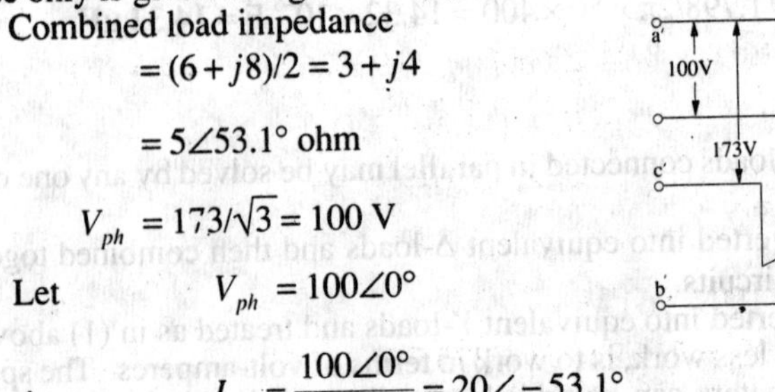

Fig. 19.37

Current in each load $= 10\angle -53.1$ A

(*i*) branch current taken by each load is **10 A** (*ii*) line current is **20 A**

(*iii*) combined power factor $= \cos 53.1° =$ **0.6 (lag).**

Example 19.29. *Three identical impedances of 30 ∠30° ohms are connected in delta to a 3-phase, 3-wire, 208 V volt abc system by conductors which have impedances of (0.8 + j 0.63) ohm. Find the magnitude of the line voltage at the load end.* **(Elect. Engg. Punjab Univ. May 1990)**

Solution. The equivalent Z_Y of the given Z_Δ is $30\angle 30/3 = 10\angle 30° = (8.86 + j5)$. Hence, the load connections become as shown in Fig. 19.38.

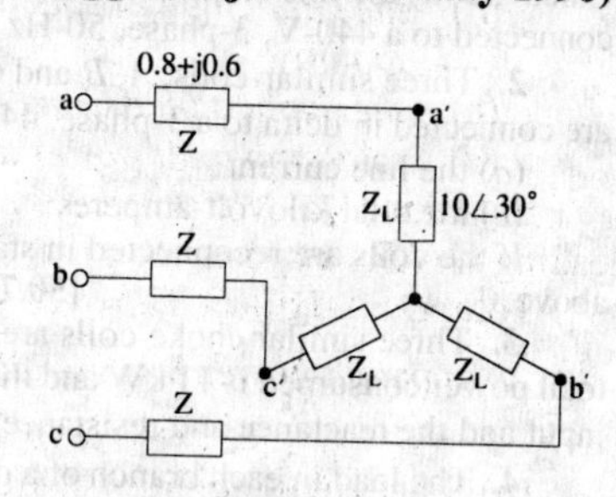

Fig. 19.38

$$Z_{an} = (0.8 + j0.6) + (8.86 + j5)$$

$$= 9.66 + j5.6 = 11.16\angle 30.1°$$

$$V_{an} = V_{ph} = 208/\sqrt{3} = 120\text{V}$$

Let $V_{an} = 120\angle 0°$

$\therefore$ $I_{an} = 120\angle 0°/11.16\angle 30.1° = 10.75\angle -30.1°$

Now, $Z_{aa'} = 0.8 + j0.6 = 1\angle 36.9°$

Voltage drop on line conductors is

$$V_{aa}' = I_{an} Z_{aa}' = 10.75\angle -30.1° \times 1\angle 36.9° = 10.75\angle 6.8° = 10.67 + j1.27$$

$\therefore$ $V_{a'n}' = V_{an} - V_{aa}' = (120 + j0) - (10.67 + j1.27) = \mathbf{109.3\angle 2.03°}$

Example 19.30. *A balanced delta-connected load having an impedance $Z_L = (300 + j210)$ ohm in each phase is supplied from 400-V, 3-phase supply through a 3-phase line having an impedance of $Z_s = (4+j8)$ ohm in each phase. Find the total power supplied to the load as well as the current and voltage in each phase of the load.*

(Elect. Circuit Theory, Kerala Univ. 1988)

Solution. The equivalent *Y*-load of the given Δ-load is $= (300 + j210)/3 = (100 + j70)\ \Omega$

Hence, connections become as shown in Fig. 19.39.

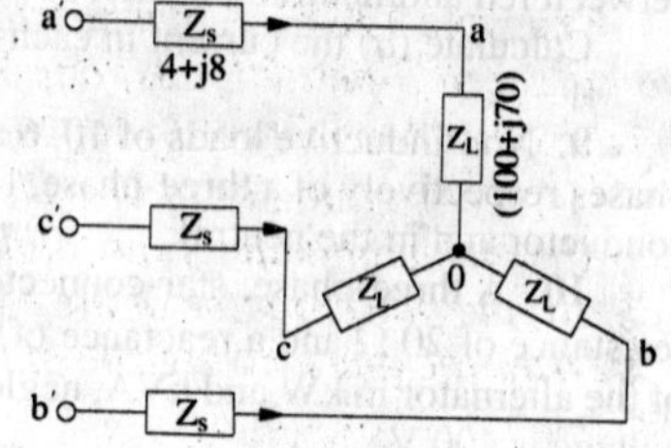

Fig. 19.39

$\mathbf{Z}_{a0}' = (4 + j8) + (100 + j70) = 104 + j78 = 130\angle 36.9°$

$V_{a0}' = 400/\sqrt{3} = 231V$ Let $\mathbf{V}_{a0}' = 231\angle 0°$

$\mathbf{I}_{a0}' = 231\angle 0°/130\angle 36.9° = 1.78\angle -36.9°$

Now, $\mathbf{Z}_{a0}' = (4 + j8) = 8.94\angle 63.4°$

Line drop $\mathbf{V}_{aa}' = \mathbf{I}_{aa}' \mathbf{Z}_{aa}' = 1.78\angle -36.9° \times 8.94\angle 63.4° = 15.9\angle 26.5° = 14.2 + j7.1$

$$\mathbf{V}_{a0} = \mathbf{V}_{a0}' - \mathbf{V}_{aa}' = (231 + j0) - (14.2 + j7.1)$$

$$= (216.8 - j7.1) = 216.9\angle -1°52'$$

Phase voltage at load end, V_{a0} = **216.9 V**

Phase current at load end, I_{a0} = **1.78 A**

Power supplied to load = $3 \times 1.78^2 \times 100$ = **951 W**

Incidentally, line voltage at load end $V_{ac} = 216.9 \times \sqrt{3} = 375.7$ V*

***It should be noted that total line drop is not the numerical sum of the individual line drops because they are 120° out of phase with each other. By a laborious process $\mathbf{V}_{ac} = \mathbf{V}_{ac}' - \mathbf{V}_{aa}' - \mathbf{V}_{ac}'$.**

Tutorial Problem No. 19.1

1. Each phase of a delta-connected load comprises a resistor of 50 Ω and capacitor of 50 μF in series. Calculate (*a*) the line and phase currents (*b*) the total power and (*c*) the kilovoltamperes when the load is connected to a 440-V, 3-phase, 50-Hz supply. [(*a*) **9.46 A ; 5.46 A** (*b*) **4480 W** (*c*) **7.24 kVA**]

2. Three similar-coils, *A*, *B* and *C* are available. Each coil has 9 Ω resistance and 12 Ω reactance. They are connected in delta to a 3-phase, 440-V, 50-Hz supply. Calculate for this load :

(*a*) the line current (*b*) the power factor
(*c*) the total kilovolt-amperes (*d*) the total kilowatts

If the coils are reconnected in star, calculate for the new load the quantities named at (*a*), (*b*), (*c*) and (*d*) above. [**50.7 A ; 0.6 ; 38.6 kVA; 23.16 kW; 16.9 A ; 0.6; 12.867 kVA ; 7.72 kW**]

3. Three similar choke coils are connected in star to a 3-phase supply. If the line currents are 15 A, the total power consumed is 11 kW and the volt-ampere input is 15 kVA, find the line and phase voltages, the VAR input and the reactance and resistance of each coil. [**577.3 V ; 333.3 V ; 10.2 kVAR ; 15.1 Ω ; 16.3 Ω**]

4. The load in each branch of a delta-connected balanced 3-φ circuit consits of an inductance of 0.0318 H in series with a resistance of 10 Ω. The line voltage is 400 V at 50 Hz. Calculate (*i*) the line current and (*ii*) the total power in the circuit. [(*i*) **49 A** (*ii*) **24 kW**] (*London Univ.*)

5. A 3-phase, delta-connected load, each phase of which has $R = 10\ \Omega$ and $X = 8\ \Omega$, is supplied from a star-connected secondary winding of a 3-phase transformer each phase of which gives 230 V. Calculate

(*a*) the current in each phase of the load and in the secondary windings of the transformer
(*b*) the total power taken by the load
(*c*) the power factor of the load. [(*a*) **31.1 A ; 54 A** (*b*) **29 kW** (*c*) **0.78**]

6. A 3-phase load consists of three similar inductive coils, each of resistance 50 Ω and inductance 0.3 H. The supply is 415 V, 50 Hz, Calculate (*a*) the line current (*b*) the power factor and (*c*) the total power when the load is (*i*) star-connected and (*ii*) delta-connected.
[(*i*) **2.25 A, 0.47 lag, 762 W** (*ii*) **6.75 A, 0.47 lag, 2280 W**] (*London Univ.*)

7. Three 20 Ω non-inductive resistors are connected in star across a three phase supply the line voltage of which is 480 V. Three other equal non-inductive resistors are connected in delta across the same supply so as to take the same-line current. What are the resistance values of these other resistors and what is the current -flowing through each of them ? [**60 Ω; 8 A**] (*Sheffield Univ. U.K.*)

8. A 415-V, 3-phase, 4-wire system supplies power to three non-inductive loads. The loads are 25 kW between red and neutral, 30 kW between yellow and neutral and 12 kW between blue and neutral.

Calculate (*a*) the current in each-line wire and (*b*) the current in the neutral conductor.
[(*a*) **104.2 A, 125 A, 50 A** (*b*) **67 A**] (*London Univ.*)

9. Non-inductive loads of 10, 6 and 4 kW are connected between the neutral and the red, yellow and blue phases respectively of a three-phase, four-wire system. The line voltage is 400 V. Find the current in each line conductor and in the neutral. [(*a*) **43.3 A, 26 A, 17.3 A, 22.9 A**] (*App. Elect. London Univ.*)

10. A three-phase, star-connected alternator supplies a delta-connected load, each phase of which has a resistance of 20 Ω and a reactance of 10 Ω. Calculate (*a*) the current supplied by the alternator (*b*) the output of the alternator in kW and kVA, neglecting the losses in the lines between the alternator and the load. The line voltage is 400 V. [(*a*) **30.95 A** (*b*) **19.2 kW, 21.45 kVA**]

11. Three non-inductive resistances, each of 100 Ω, are connected in star to 3-phase, 440-V supply. Three equal choking coils each of reactance 100 Ω are also connected in delta to the same supply. Calculate :

(*a*) line current
(*b*) p.f. of the system. [(*a*) **8.04 A** (*b*) **0.3156**] (*I.E.E. London*)

12. In a 3-phase, 4-wire system, there is a balanced 3-phase motor load taking 200 kW at a power factor of 0.8 lagging, while lamps connected between phase conductors and the neutral take 50, 70 and 100 kW respectively. The voltage between phase conductors is 430 V. Calculate the current in each phase and in the neutral wire of the feeder supplying the load. [**512 A, 5.87 A, 699 A ; 213.3 A**] (*Elect. Power, London Univ.*)

13. A 440-V, 50-Hz induction motor takes a line current of 45 A at a power factor of 0.8 (lagging). Three Δ-connected capacitors are installed to improve the power factor to 0.95 (lagging). Calculate the kVA of the capacitor bank and the capacitance of each capacitor. [**11.45 kVA, 62.7 μF**] (*I.E.E. London*)

14. Three resistances, each of 500 Ω, are connected in star to a 400-V, 50-Hz, 3-phase supply. If three capacitors, when connected in delta to the same supply, take the same line currents, calculate the capacitance of each capacitor and the line current. [**2.123 μF, 0.653 A**] (*London Univ.*)

15. A factory takes the following balanced loads from a 440-V, 3-phase, 50-Hz supply :

(*a*) a lighting load of 20 kW
(*b*) a continuous motor load of 30 kVA at 0.5 p.f. lagging.
(*c*) an intermittent welding load of 30 kVA at 0.5 p.f. lagging.

Calculate the kVA rating of the capacitor bank required to improve the power factor of loads (*a*) and (*b*) together to unity. Give also the value of capacitor required in each phase if a star-connected bank is employed.

What is the new overall p.f. if, after correction has been applied, the welding load is switched on.

[30 kVAR ; 490μ F ; 0.945 kg]

16. A three-wire, three-phase system, with 400 V between the line wires, supplies a balanced delta-connected load taking a total power of 30 kW at 0.8 power factor lagging. Calculate (*i*) the resistance and (*ii*) the reactance of each branch of the load and sketch a vector diagram showing the line voltages and line currents. If the power factor of the system is to be raised to 0.95 lagging by means of three delta-connected capacitors, calculate (*iii*) the capacitance of each branch assuming the supply frequency to be 50 Hz.

[(*i*) **10.24 A** (*ii*) **7.68 Ω** (*iii*) **83.2 μF**] (*London Univ.*)

19.15. Power Measurement in 3-phase Circuits

Following methods are available for measuring power in a 3-phase load.

(*a*) **Three Wattmeter Method**

In this method, three wattmeters are inserted one in each phase and the algebraic sum of their readings gives the total power consumed by the 3-phase load.

(*b*) **Two Wattmeter Method**

(*i*) This method gives true power in the 3-phase circuit without regard to balance or wave form provided in the case of *Y*-connected load. The neutral of the load is isolated from the neutral of the source of power. Or if there is a neutral connection, the neutral wire should not carry any current. This is possible only if the load is perfectly balanced and there are no harmonics present of triple frequency or any other multiples of that frequency.

(*ii*) This method can also be used for 3-phase, 4-wire system in which the neutral wire carries the neutral current. In this method, the current coils of the wattmeters are supplied from current transformers inserted in the principal line wires in order to get the correct magnitude and phase differences of the currents in the current coils of the wattmeter, because in the 3-phase, 4-wire system, the sum of the instantaneous currents in the principal line wires is not necessarily equal to zero as in 3-phase 3-wire system.

(*c*) **One Wattmeter Method**

In this method, a single wattmeter is used to obtain the two readings which are obtained by two wattmeters by the two-wattmeter method. This method can, however, be used only when the load is balanced.

19.16. Three Wattmeter Method

A wattmeter consists of (*i*) a low resistance current coil which is inserted in series with the line carrying the current and (*ii*) a high resistance pressure coil which is connected across the two points whose potential difference is to be measured.

A wattmeter shows a reading which is proportional to the product of the current through its current coil, the p.d. across its potential or pressure coil and cosine of the angle between this voltage and current.

As shown in Fig. 19.40 in this method three wattmeters are inserted in each of the three phases of the load whether Δ-connected or *Y*-connected. The current coil of each wattmeter carries the current of one phase only and the pressure coil measures the phase-voltage of this phase. Hence, each wattmeter measures the power in a single phase. The algebraic sum of the readings of three wattmeters must give the total power in the load.

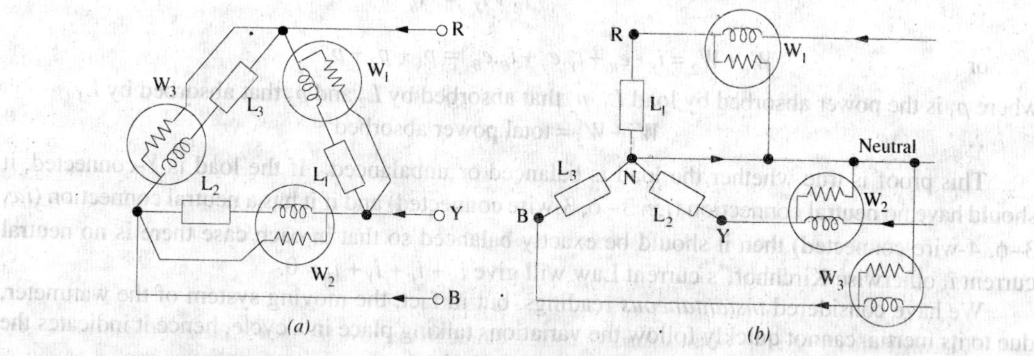

Fig. 19.40

The difficulty with this method is that under ordinary conditions it is not generally feasible to break into the phases of a delta-connected load nor is it always possible, in the case of a *Y*-connected load, to get at the neutral point which is required for connections as shown in Fig. 19.40 (*b*). However, it is not necessary to use three wattmeters to measure power, two wattmeters can be used for the purpose as shown below.

19.17. Two Wattmeter Method–*Balanced* or *Unbalanced Load*

As shown in Fig. 19.41, the current coils of the two wattmeters are inserted in *any two* lines and the potential coil of each joined to the third line. It can be proved that the sum of the instantaneous powers indicated by W_1 and W_2 gives the instantaneous power absorbed by the three loads L_1, L_2 and L_3. A star-connected load is considered in the following discussion although it can be equally applied to Δ-connected loads because a Δ-connected load can always be replaced by an equivalent *Y*-connected load.

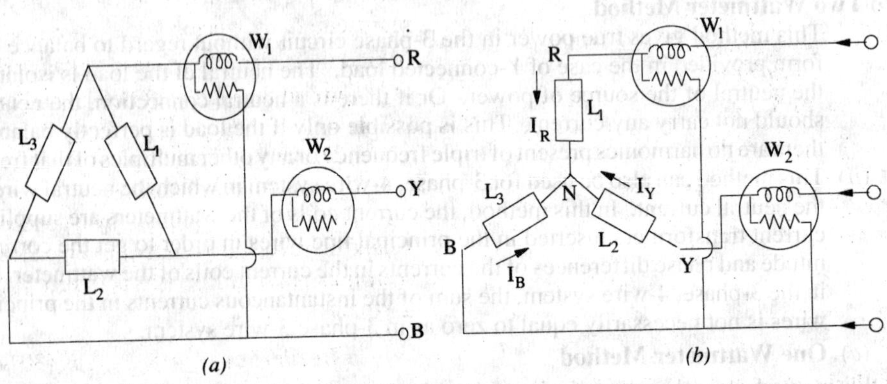

Fig. 19.41

Now, before we consider the currents through and p.d. across each wattmeter, it may be pointed out that *it is important to take the direction of the voltage through the circuit the same as that taken for the current when establishing the readings of the two wattmeters.*

Instantaneous current through W_1 $= i_R$

p.d. across $W_1 = e_{RB}$ $= e_R - e_B$

p.d across power read by W_1 $= i_R\,(e_R - e_B)$

Instantaneous current through W_2 $= i_Y$

Instantaneous p.d. across W_2 $= e_{YB} = (e_Y - e_B)$

Instantaneous power read by W_2 $= i_Y\,(e_Y - e_B)$

$$\therefore \quad W_1 + W_2 = i_R(e_R - e_B) + i_Y(e_Y - e_B) = i_R e_R + i_Y e_Y - e_B(i_R + i_Y)$$

Now, $i_R + i_Y + i_B = 0$ ...Kirchhoff's Current Law

$$\therefore \quad i_R + i_Y = -i_B$$

or $$W_1 + W_2 = i_R \cdot e_R + i_Y . e_y + i_B . e_B = p_1 + p_2 + p_3$$

where p_1 is the power absorbed by load L_1, p_2 that absorbed by L_2 and p_3 that absorbed by L_3

$$\therefore \quad W_1 + W_2 = \text{total power absorbed}$$

This proof is true whether the load is balanced or unbalanced. If the load is *Y*-connected, it should have no neutral connection (*i.e.* 3– φ, 3-wire connected) and if it has a neutral connection (*i.e.* 3–φ, 4-wire connected) then it should be exactly balanced so that in each case there is no neutral current i_N otherwise Kirchhoff's current Law will give $i_N + i_R + i_Y + i_B = 0$.

We have considered *instantaneous* readings, but in fact, the moving system of the wattmeter, due to its inertia, cannot quickly follow the variations talking place in a cycle, hence it indicates the *average* power.

$$\therefore \qquad W_1 + W_2 = \frac{1}{T}\int_0^T i_R\, e_{RB} dt + \frac{1}{T}\int_0^T i_Y e_{YB}\, dt$$

19.18. Two Wattmeter Method–*Balanced Load*

If the load is balanced, then power factor of the load can also be found from the two wattmeter readings. The Y-connected load in Fig. 19.41 (*b*) will be assumed inductive. The vector diagram for such a balanced Y-connected load is shown in Fig. 19.42. We will now consider the problem in terms of r.m.s. values instead of instantaneous values.

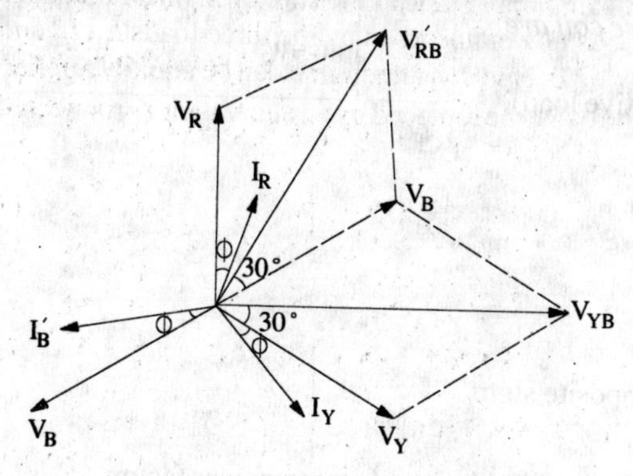

Fig. 19.42

Let V_R, V_Y and V_B be the r.m.s. values of the three phase voltages and I_R, I_Y and I_B the r.m.s. values of the currents. Since these voltages and currents are assumed sinusoidal, they can be represented by vectors, the currents lagging behind their respective phase voltages by ϕ.

Current through wattmeter W_1 [Fig. 19.41 (*b*)] is $= I_R$.

P.D. across voltage coil of W_1 is

$V_{RB} = V_R - V_B$ ...vectorially

This V_{RB} is found by compounding V_R and V_B reversed as shown in Fig. 19.42. It is seen that phase difference between V_{RB} and I_R $= (30° - \phi)$.

$\therefore$ Reading of $W_1 = I_R V_{RB} \cos (30° - \phi)$

Similarly, as seen from Fig. 19.41.(*b*). Current through $W_2 = I_Y$

P.D. across $W_2 = V_{YB} = V_Y - V_B$ ....vectorially

Again, V_{YB} is found by compounding V_Y with V_B reversed as shown in Fig. 19.42. The angle between I_Y and V_{YB} is $(30° + \phi)$. Reading of $W_2 = I_Y V_{YB} \cos (30° + \phi)$

Since load is balanced, $V_{RB} = V_{YB}$ = line voltage V_L ; $I_Y = I_R$ = line current, I_L

$\therefore \quad W_1 = V_L I_L \cos (30° - \Phi)$ and $W_2 = V_L I_L \cos (30° + \phi)$

$\therefore W_1 + W_2 = V_L I_L \cos (30° - \phi) + V_L I_L (\cos (30° + \phi)$

$= V_L I_L [\cos 30° \cos \phi + \sin 30° \sin \phi + \cos 30° \cos \phi - \sin 30° \sin \phi]$

$= V_L I_L (2 \cos 30° \cos \phi) = \sqrt{3}\, V_L I_L \cos \phi$ = total power in the 3-phase load

Hence, the sum of the two wattmeter readings gives the total power consumption in the 3-phase load.

It should be noted that phase sequence of RYB has been assumed in the above discussion. Reversal of phase sequence will interchange the readings of the two wattmeters.

19.19. Variations in Wattmeter Readings

It has been shown above that for a *lagging* power factor

$W_1 = V_L I_L \cos (30° - \varphi)$ and $W_2 = V_L I_L \cos (30° + \phi)$

From this it is clear that individual readings of the wattmeters not only depend *on the load but upon its power factor also.* We will consider the following cases :

(*a*) When $\phi = 0$ *i.e.* power factor is unity (*i.e.* resistive load) then,

$$W_1 = W_2 = V_L I_L \cos 30°$$

Both wattmeters indicate equal and positive *i.e.* up-scale readings.

(*b*) When $\phi = 60°$ *i.e.* power factor = 0.5 (lagging)

Then $W_2 = V_L I_L \cos (30° + 60°) = 0$. Hence, the power is measured by W_1 alone.

ϕ	$0°$	$60°$	$90°$
$\cos\phi$	1	0.5	0
W_1	+ve	+ve	+ve
W_2	+ve	0	−ve
	$W_1 = W_2$		$W_1 = W_2$

(c) When $90° > \phi > 60°$ *i.e.* $0.5 > \text{p.f.} > 0$, then W_1 is still positive but reading of W_2 is reversed* because the phase angle between the current and voltage is more than 90°. For getting the total power, the reading of W_2 is to be subtracted from that of W_1.

Under this condition, W_2 will read 'down scale' *i.e.* backwards. Hence, to obtain a reading on W_2 it is necessary to reverse either its pressure coil or current coil, usually the former. *All readings taken after reversal of pressure coil are to be taken as negative.*

(d) When $\phi = 90°$ (*i.e.* pure inductive or capacitive load), then

$$W_1 = V_L I_L \cos(30° - 90°) = V_L I_L \sin 30°;$$

$$W_2 = V_L I_L \cos(30° + 90°) = -V_L I_L \sin 30°$$

As seen, the two readings are equal but of opposite sign.

$$\therefore \quad W_1 + W_2 = 0$$

• The above facts have been summarised in the above table for a lagging power factor.

19.20. Leading Power Factor

In the above discussion, lagging angles are taken positive. Now, we will see how wattmeter readings are changed if the power factor becomes leading. For $\phi = +60°$ (lag), W_2 is zero. But for $\phi = -60°$ (lead), W_1 is zero. So we find that for angles of lead, the reading of the two wattmeters are interchanged. Hence, for a *leading* power factor

$$W_1 = V_L I_L \cos(30° + \phi) \quad \text{and} \quad W_2 = V_L I_L \cos(30° - \phi)$$

19.21. Power Factor–*Balanced load*

In case the load is balanced (and currents and voltages are sinusoidal) and for a *lagging* power factor :

$$W_1 + W_2 = V_L I_L \cos(30° - \phi) + V_L I_L \cos(30° + \phi) = \sqrt{3} V_L I_L \cos\phi \qquad ...(i)$$

Similarly $W_1 - W_2 = V_L I_L \cos(30° - \phi) - V_L I_L \cos(30° + \phi)$

$$= V_L I_L (2 \times \sin\phi \times 1/2) = V_L I_L \sin\phi \qquad ...(ii)$$

Dividing (*ii*) by (*i*), we have $\tan\phi = \dfrac{\sqrt{3}(W_1 - W_2)**}{(W_1 + W_2)}$ ...(*iii*)

Knowing $\tan\phi$ and hence ϕ, the value of power factor $\cos\phi$ can be found by consulting the trigonometrical tables. It should, however, be kept in mind that if W_2 reading has been taken after reversing the pressure coil *i.e.* if W_2 is negative, then the above relation becomes

---

*For a leading p.f., conditions are just the opposite of this. In that case, W_1 reads negative (Art. 19.22).

**For a leading power factor, this expression becomes

$$\tan\phi = -\sqrt{3}\left(\frac{W_1 - W_2}{W_1 + W_2}\right) \qquad \text{....Art. 19.22}$$

$$\tan \phi = \sqrt{3}\ \frac{W_1 - (-W_2)}{W_1 + (-W_2)} = \sqrt{3}\ \frac{W_1 + W_2}{W_1 - W_2}$$

Obviously, in this expression, only *numerical* values of W_1 and W_2 should be substituted.

We may express power factor in terms of the ratio of the two wattmeters as under :

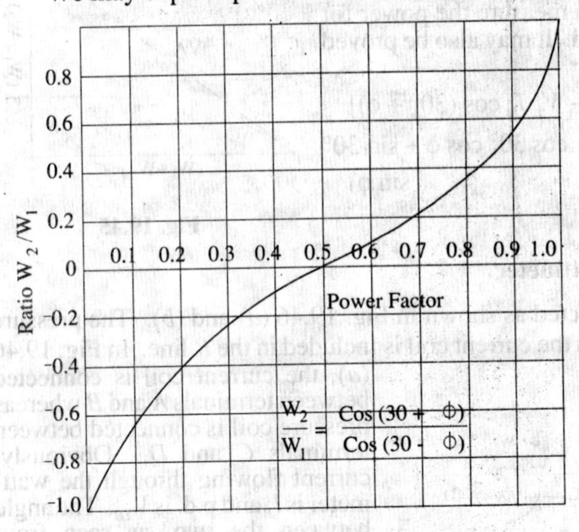

Fig. 19.43

Let $\frac{\text{smaller reading}}{\text{larger reading}} = \frac{W_2}{W_1} = r$

Then from equation (*iii*) above,

$$\tan \phi = \frac{\sqrt{3}[1 - (W_2/W_1)]}{1 + (W_2/W_1)} = \frac{\sqrt{3}\,(1 - r)}{1 + r}$$

Now $\sec^2\phi = 1 + \tan^2\phi$

or $$\frac{1}{\cos^2\phi} = 1 + \tan^2\phi$$

$$\therefore \qquad \cos\phi = \frac{1}{\sqrt{1 + \tan^2\phi}}$$

$$= \frac{1}{\sqrt{1 + 3\left(\frac{1-r}{1+r}\right)^2}}$$

$$= \frac{1 + r}{2\sqrt{1 - r + r^2}}$$

If r is plotted against $\cos\phi$, then a curve called watt-ratio curve is obtained as shown in Fig. 19.43.

19.22. Balanced Load – *leading power factor*

In this case, as seen from Fig. 19.44

$$W_1 = V_L I_L \cos(30 + \phi)$$

and $$W_2 = V_L I_L \cos(30 - \phi)$$

$$\therefore \quad W_1 + W_2 = \sqrt{3}\, V_L I_L \cos\phi - \text{ as found above}$$

$$W_1 - W_2 = -V_L I_L \sin\phi$$

$$\therefore \quad \tan\phi = -\frac{\sqrt{3}\,(W_1 - W_2)}{(W_1 + W_2)}$$

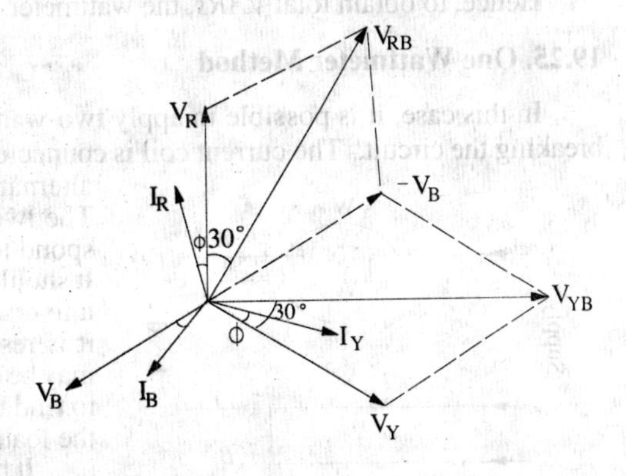

Fig. 19.44

Obviously, if $\phi > 60°$, then phase angle between V_{RB} and I_R becomes more than 90°. Hence, W_1 reads 'down-scale' *i.e.* it indicates negative reading. However, W_2 gives positive reading even in the extreme case when $\phi = 90°$.

19.23. Reactive Voltamperes with Two Wattmeters

We have seen that $\tan\phi = \frac{\sqrt{3}\,(W_1 - W_2)}{(W_1 + W_2)}$

Since the tangent of the angle of lag between phase current and phase voltage of a circuit is always equal to the ratio of the reactive power to the active power (in watts), it is clear that $\sqrt{3}$ (W_1-W_2) represents the reactive power (Fig. 19.45). Hence, for a balanced load, the reactive power is given by $\sqrt{3}$ times the difference of the readings of the two wattmeters used to measure the power for a 3-phase circuit by the two wattmeter method. It may also be proved mathematically as follows :

$$= \sqrt{3}\,(W_1-W_2) = \sqrt{3}\,[V_L I_L \cos(30° - \phi) - V_L I_L \cos(30° + \phi)]$$

$$= \sqrt{3}\,V_L I_L (\cos 30° \cos\phi + \sin 30° \sin\phi - \cos 30° \cos\phi + \sin 30° \sin\phi)$$

$$= \sqrt{3}\,V_L I_L \sin\phi$$

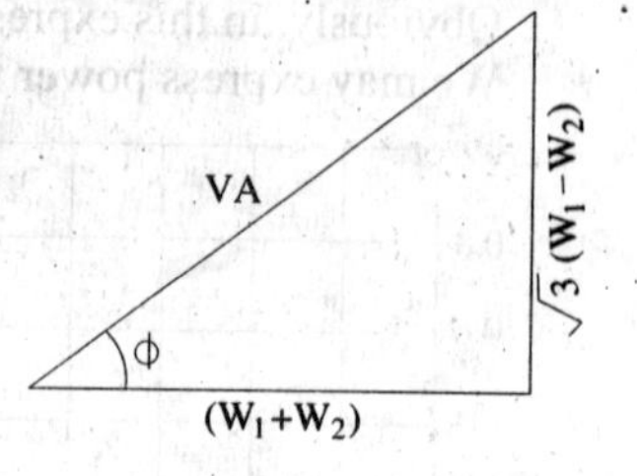

Fig. 19.45

19.24. Reactive Voltamperes with One Wattmeter

For this purpose, the wattmeter is connected as shown in Fig. 19.46 (*a*) and (*b*). The pressure coil is connected across *Y* and *B* lines whereas the current coil is included in the *R* line. In Fig. 19.46 (*a*), the current coil is connected between terminals *A* and *B* whereas pressure coil is connected between terminals *C* and *D*. Obviously, current flowing through the wattmeter is I_R and p.d. is V_{YB}. The angle between the two, as seen from vector diagram of Fig. 19.42, is $(30 + 30 + 30 - \phi) = (90 - \phi)$

Hence, reading of the wattmeter is

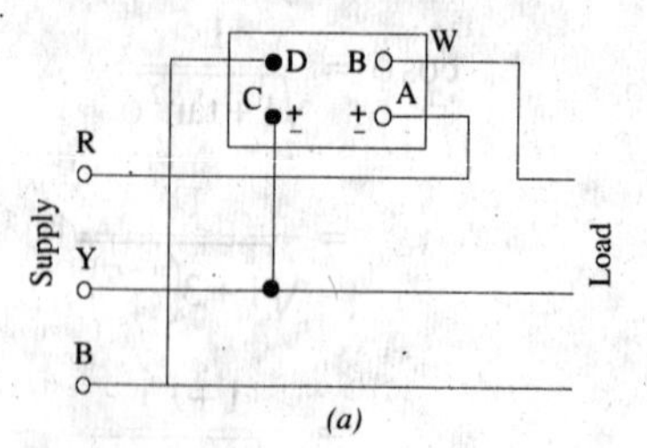

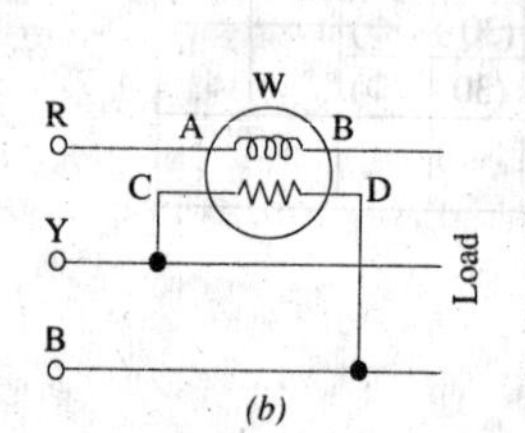

Fig. 19.46

$$W = V_{YB} I_R \cos(90 - \phi) = V_{YB} I_R \sin\phi$$

For a balanced load, V_{YB} equals the line voltage V_L and I_R equals the line current I_L, hence

$$W = V_L I_L \sin\phi$$

We know that the total reactive voltamperes of the load are $Q = \sqrt{3}\,V_L I_L \sin\phi$.

Hence, to obtain total *VARs*, the wattmeter reading must be multiplied by a factor of $\sqrt{3}$.

19.25. One Wattmeter Method

In this case, it is possible to apply two-wattmeter method by means of one wattmeter without breaking the circuit. The current coil is connected in any one line and the pressure coil is connected alternately between this and the other two lines (Fig. 19.47). The two readings so obtained, for a balanced load, correspond to those obtained by normal two wattmeter method. It should be kept in mind that this method is not of as much universal application as the two wattmeter method because it is restricted to fairly balanced loads only. However, it may be conveniently applied, for instance, when it is desired to find the power input to a factory motor in order to check the load upon the motor.

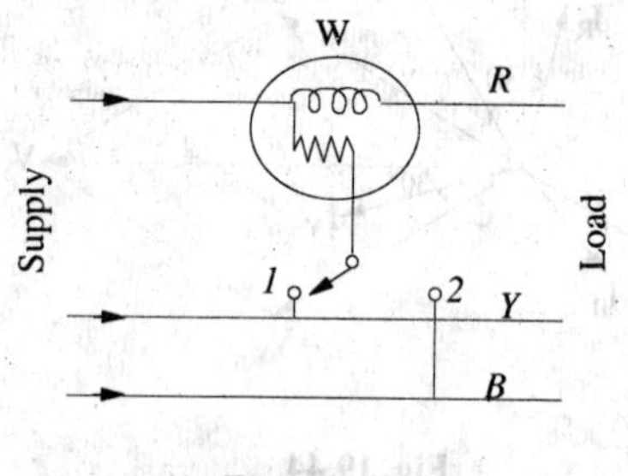

Fig. 19.47

It may be pointed out here that the two wattmeters used in the two-wattmeter method (Art. 19.17) are usually combined into a single instrument in the case of switchboard wattmeter which is then known as a polyphase wattmeter. The combination is affected by arranging the two sets of coils in such a way as to operate on a single moving system resulting in an indication of the total power on the scale.

19.26. Copper Required for Transmitting Power under Fixed Conditions

The comparison between 3-phase and single-phase systems will be done on the basis of a fixed amount of power transmitted to a fixed distance with the same amount of loss and at the same maximum voltage between conductors. In both cases, the weight of copper will be directly proportional to the number of wires (since the distance is fixed) and inversely proportional to the resistance of each wire. We will assume the same power factor and same voltage.

$$P_1 = VI_1 \cos \phi \text{ and } P_3 = \sqrt{3}\, VI_3 \cos \phi$$

where I_1 = r.m.s. value of currrent in 1-phase system

I_3 = r.m.s. value of line current in 3-phase system

$$\because \quad P_1 = P_2 \quad \therefore \quad VI_1 \cos \phi = \sqrt{3}\, VI_3 \cos \phi \qquad \therefore I_1 = \sqrt{3}\, I_3$$

also $$I_1^2 R_1 \times 2 = I_3^2 R_3 \times 3 \text{ or } \frac{R_1}{R_3} = \frac{3I_3^2}{2I_1^2}$$

Substituting the value of I_1, we get $\dfrac{R_1}{R_3} = \dfrac{3I_3^2}{3I_3^2 \times 2} = \dfrac{1}{2}$

$$\therefore \quad \frac{\text{copper 3-phase}}{\text{copper 1-phase}} = \frac{\text{No. of wires 3-phase}}{\text{No. of wires 1-phase}} \times \frac{R_1}{R_3} = \frac{3}{2} \times \frac{1}{2} = \frac{3}{4}$$

Hence, we find that for transmitting the same amount of power over a fixed distance with a fixed line loss, we need only three-fourths of the amount of copper that would be required for a single phase or to put it in another way, one-third more copper is required for a 1-phase system than would be necessary for a three phase system.

Example 19.31. *Phase voltage and current of a star-connected inductive load is 150 V and 25 A. Power factor of load is 0.707 (lag). Assuming that the system is 3-wire and power is measured using two wattmeters, find the readings of wattmeters.*

(Elect. Instrument & Measurements Nagpur Univ. 1993)

Solution. $V_{ph} = 150$ V; $V_L = 150 \times \sqrt{3}$ V; $I_{ph} = I_L = 25$ A

Total power $= \sqrt{3}\, V_L I_L \cos \phi = \sqrt{3} \times 150 \times \sqrt{3} \times 25 \times 0.707 = 7954$ W

$\therefore \; W_1 + W_2 = 7954$ W ...(*i*)

$\cos \phi = 0.707$; $\phi = \cos^{-1}(0.707) = 45°$; $\tan 45° = 1$

Now, for a lagging power factor, $\tan \phi = \sqrt{3}\,(W_1 - W_2)/(W_1 + W_2)$ or $1 = \sqrt{3}(W_1 - W_2)/7954$

$\therefore \; (W_1 - W_2) = 4592$ W. ...(*ii*)

From (*i*) and (*ii*) above, we get, $\mathbf{W_1 = 6273}$ **W**; $\mathbf{W_2 = 1681}$ **W**.

Example 19.32. *In a balanced 3-phase 400-V circuit, the line current is 115.5 A. When power is measured by two wattmeter method, one meter reads 40 kW and the other zero. What is the power factor of the load ? If the power factor were unity and the line current the same, what would be the reading of each wattmeter ?*

Solution. Since $W_2 = 0$, the whole power is measured by W_1. As per Art. 19.18, in such a situation, p.f. = 0.5. However, it can be calculated as under.

Since total power is 40 kW, $\therefore \; 40{,}000 = \sqrt{3} \times 400 \times 115.5 \times \cos \phi$; $\cos \phi = 0.5$

If the power factor is unity with line currents remaining the same, we have

$$\tan \phi = \frac{\sqrt{3}(W_1 - W_2)}{(W_1 + W_2)} = 0 \text{ or } W_1 = W_2$$

Also, $(W_1 + W_2) = \sqrt{3} \times 400 \times 115.5 \times 1 = 80000$ W = 80 kW

As per Art. 19.19, at unity p.f., $W_1 = W_2$. Hence, each wattmeter reads = 80/2 = **40 kW**.

Example 19.33. *The input power to a three-phase motor was measured by two wattmeter method. The readings were 10.4 KW and –3.4 KW and the voltage was 400 V. Calculate (a) the power factor (b) the line current.*

(Elect. Engg. A.M.Ae, S.I. June 1991)

Solution. As given in Art. 19.21, when W_2 reads negative, then we have

$\tan \phi = \sqrt{3}\,(W_1 + W_2)\,/\,(W_1 - W_2)$. Substituting numerical values of W_1 and W_2, we get

$\tan \phi = \sqrt{3}\,(10.4 + 3.4)/(10.4 - 3.4) = 1.97$; $\phi = \tan^{-1}(1.97) = 63.1°$

(*a*) p.f. = cos ϕ = cos 63.1° = **0.45 (lag)**

(*b*) W = 10.4 – 3.4 = 7 KW = 7,000 W

$7000 = \sqrt{3}\, I_L \times 400 \times 0.45\ ;\ I_L = \mathbf{22.4\ A}$

Example 19.33 (A). *A three-phase, three-wire, 100-V, ABC system supplies a balanced delta-connected load with impedance of 20∠ 40° ohm.*

(a) Determine the phase and line currents and draw the phase or diagram (b) Find the wattmeter readings when the two wattmeter method is applied to the system.

(Elect. Machines, A.M.I.E. Sec B, 1989)

Solution. (*a*) The phase or diagram is shown in Fig. 19.48 (*b*).

Let V_{AB} = 100 ∠0°. Since phase sequence is *ABC*, V_{BC} = 100 ∠–120° and V_{CA}=100 ° 120∠ °

$$\text{Phase current } I_{AB} = \frac{V_{AB}}{Z_{AB}} = \frac{100\angle 0^\circ}{20\angle 45^\circ} = 5\angle -45^\circ$$

$$I_{BC} = \frac{V_{BC}}{Z_{BC}} = \frac{100\angle -120^\circ}{20\angle 45^\circ} = 5\angle -165^\circ, I_{CA} = \frac{V_{CA}}{Z_{CA}} = \frac{100\angle 120^\circ}{20\angle 45^\circ} = 5\angle 75^\circ$$

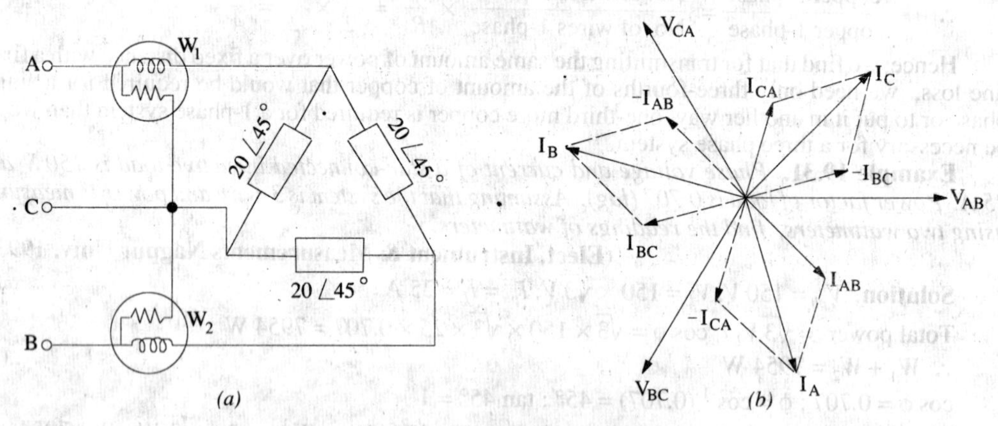

Fig. 19.48

Applying *KCL* to junction A, we have

$$I_A + I_{CA} - I_{AB} = 0 \quad \text{or} \quad I_A = I_{AB} - I_{CA}$$

∴ Line current I_A = 5 ∠-45° –5∠ 75° = 8.66 ∠–75°

Since the system is balanced, I_B will lag I_A by 120° and I_C will lag I_A by 240°.

$\therefore I_B = 8.66\angle(75^\circ - 120^\circ) = 8.66\angle -195^\circ; I_C = 8.66\angle(-75^\circ - 240^\circ) = 8.66\angle -315^\circ = 8.66\angle 45^\circ$

(*b*) As shown in Fig. 19.48 (*b*), reading of wattmeter W_1 is $W_1 = V_{AC}\ I_C \cos\varphi$. Phasor V_{AC} is the reverse of phasor V_{CA}. Hence, V_{AC} is the reverse of phasor V_{CA}. Hence, V_{AC} lags the reference vector by 60° whereas I_A lags by 75°. Hence, phase difference between the two is (75°–60°) = 15°

$\therefore\ W_1 = 100 \times 8.66 \times \cos\ 15^\circ = 836.5$ W

Similarly $W_2 = V_{BC}\, I_B \cos\phi = 100 \times 8.66 \times \cos 75^\circ = 224.1$ W

$\therefore\ W_1 + W_2 = 836.5 + 224.1 = 1060.6$ W

Resistance of each delta branch = 20 cos 45° = 14.14 Ω

Total power consumed = $3\, I^2 R = 3 \times 5^2 \times 14.14 = 1060.6$ W

Hence, it proves that the sum of the two wattmeter readings gives the total power consumed.

Example 19.34. *A 3-phase, 500-V motor load has a power factor of 0.4. Two wattmeters connected to measure the power show the input to be 30 kW. Find the reading on each instrument.*

(Electrical Meas. Nagpur Univ. 1991)

Solution. As seen from Art. 19.21

$$\tan\phi = \frac{\sqrt{3}\,(W_1 - W_2)}{W_1 + W_2} \qquad \ldots(i)$$

Now, $\cos\phi = 0.4;\ \phi = \cos^{-1}(0.4) = 66.6°;\quad \tan 66.6° = 2.311$

$$W_1 + W_2 = 30 \qquad \ldots(ii)$$

Substituting these values in equation (*i*) above, we get

$$2.311 = \frac{\sqrt{3}\,(W_1 - W_2)}{30} \quad \therefore \quad W_1 - W_2 = 40 \qquad \ldots(iii)$$

From Eq. (*ii*) and (*iii*), we have $W_1 = \mathbf{45\ kW}$ and $W_2 = \mathbf{-5\ kW}$

Since W_2 comes out to be negative, second wattmeter reads 'down scale.' Even otherwise it is obvious that p.f. being less than 0.5, W_2 must be negative (Art. 19.19).

Example 19.34 (a). *The power in a 3-phase circuit is measured by two wattmeters. If the total power is 100 kW and power factor is 0.66 leading, what will be the reading of each wattmeter ? Give the connection diagram for the wattmeter circuit. For what p.f. will one of the wattmeters read zero?*

Solution. $\phi = \cos^{-1}(0.66) = 48.7°\ ;\ \tan\phi = 1.1383$

Since p.f. is leading,

$$\therefore \tan\phi = -\frac{\sqrt{3}\,(W_1 - W_2)}{W_1 + W_2} \quad \therefore \quad 1.1383 = -\sqrt{3}(W_1 - W_2)/100$$

$$\therefore \quad W_1 - W_2 = -65.7 \text{ and } W_1 + W_2 = 100 \quad \therefore \quad W_1 = \mathbf{17.14\ kW};\quad W_2 = \mathbf{82.85\ kW}$$

Connection diagram is similar to that shown in Fig. 19.41(*b*). One of the wattmeters will read zero when p.f. = **0.5**

Example 19.35. *Two wattmeters are used for measuring the power input and the power factor of an over-excited synchronous motor. If the readings of the meters are (−2.0 kW) and (+ 7.0 kW) respectively, calculate the input and power factor of the motor.*

(Elect. Technology, Punjab Univ. June, 1991)

Solution. Since an over-excited synchronous motor runs with a leading p.f., we should use the relationship derived in Art. 19.22.

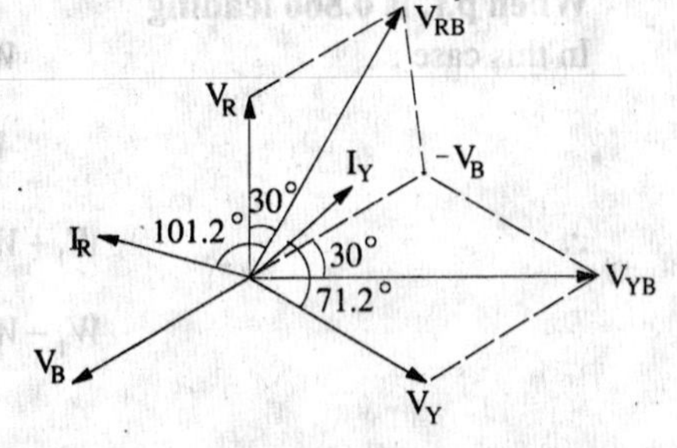

Fig. 19.49

$$\tan\phi = \frac{\sqrt{3}\,(W_1 - W_2)}{W_1 + W_2}$$

Moreover, as explained in the same article, it is W_1 that gives negative reading and not W_2.

Hence, $W_1 = -2$ kW

$$\therefore \quad \tan\phi = -\frac{\sqrt{3}(-2-7)}{-2+7} = \sqrt{3} \times \frac{9}{5} = 3.1176$$

$$\therefore \quad \phi = \tan^{-1}(3.1176) = 71.2°(\text{lead})$$

$$\therefore \quad \cos\phi = \cos 71.2° = \mathbf{0.3057\ (lead)} \text{ and}$$

Input $= W_1 + W_2 = -2 + 7 = \mathbf{5\ kW}$

Example 19.36. *A 440-V, 3-phase, delta-connected induction motor has an output of 14.92 kW at a p.f. of 0.82 and efficiency 85%. Calculate the readings on each of the two wattmeters connected to measure the input. Prove any formula used.*

If another star-connected load of 10 kW at 0.85 p.f. lagging is added in parallel to the motor, what will be the current drawn from the line and the power taken from the line ?

(Elect. Technology-I, Bombay Univ. 1986)

Solution. Motor input = 14,920/0.85 = 17,600 W $\quad\therefore\quad W_1 + W_2 = 17.6$ kW $\qquad\ldots(i)$

$$\cos\phi = 0.82\ ;\ \phi = 34.9°,\quad \tan 34.9° = 0.6976 \quad \therefore\ 0.6976 = \sqrt{3}\,\frac{W_1 - W_2}{17.6}$$

$$\therefore \quad W_1 - W_2 = 7.09 \text{ kW} \qquad \ldots(ii)$$

From (i) and (ii) above, we get W_1 = **12.35 kW** and W_2 = **5.26 kW**

$$\text{Motor kVA, } S_m = \frac{\text{motor kW}}{\cos\phi_m} = \frac{17.6}{0.82} = 21.46 \quad \therefore \quad \mathbf{S}_m = 21.46\angle -34.9° = (17.6 - j12.28) \text{ kVA}$$

Load p.f. = 0.85 $\therefore$ $\phi = \cos^{-1}(0.85) = 31.8°$ Load kVA, $S_Y = 10/0.85 = 11.76$

$$\therefore \quad \mathbf{S}_Y = 11.76\angle -31.8° = (10 - j6.2) \text{ kVA}$$

$$\text{Combined kVA,} \quad \mathbf{S} = \mathbf{S}_m + \mathbf{S}_Y = (27.6 - j18.48) = 33.2\angle -33.8° \text{ kVA}$$

$$I = \frac{S}{\sqrt{3}.V} = \frac{33.2\times10^3}{\sqrt{3}\times440} = \mathbf{43.56\ A}$$

Power taken = **27.6 kW**

Example 19.37. *The power input to a synchronous motor is measured by two wattmeters both of which indicate 50 kW. If the power factor of the motor be changed to 0.866 leading, determine the readings of the two wattmeters, the total input power remaining the same. Draw the vector diagram for the second condition of the load.* **(Elect. Technology, Nagpur Univ. 1992)**

Solution. In the first case both wattmeters read equal and positive. Hence motor must be running at unity power factor (Art. 19.22).

When p.f. is 0.866 leading

In this case ;

$$W_1 = V_L I_L \cos(30° + \phi);$$

$$W_2 = V_L I_L \cos(30° - \phi)$$

$$\therefore \quad W_1 + W_2 = \sqrt{3} V_L I_L \cos\phi$$

$$W_1 - W_2 = -V_L I_L \sin\phi$$

$$\therefore \quad \tan\phi = -\frac{\sqrt{3}(W_1 - W_2)}{(W_1 + W_2)}$$

$$\phi = \cos^{-1}(0.866) = 30°$$

$$\tan\phi = 1/\sqrt{3}$$

$$\therefore \quad \frac{1}{\sqrt{3}} = \frac{-\sqrt{3}(W_1 - W_2)}{100}$$

$$\therefore \quad W_1 - W_2 = -100/3$$

$$\text{and} \quad W_1 + W_2 = 100$$

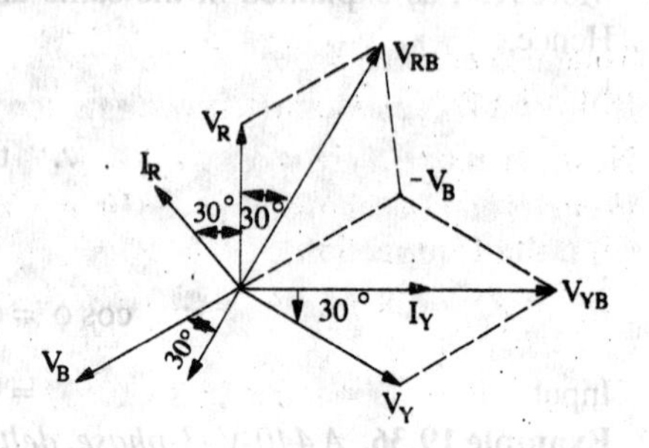

Fig. 19.50

$$\therefore \quad 2W_1 = 200/3;\ W_1 = \mathbf{33.33\ kW};\ W_2 = \mathbf{66.67\ kW}$$

For connection diagram, please refer to Fig. 19.41. The vector or phasor diagram is shown in Fig. 19.50.

Example 19.37 (a). *A star-connected balanced load is supplied from a 3-φ balanced supply with a line voltage of 416 volts at a frequency of 50 Hz. Each phase of the load consists of a resistance and a capacitor joined in series and the reading on two wattmeters connected to measure the total power supplied are 782 W and 1980 W, both positive. Calculate*

(i) power factor of circuit (ii) the line current (iii) the capacitance of each capacitor.

(Elect. Engg. I Nagpur Univ. 1993)

Solution. (*i*) As seen from Art. 19.21 $\tan\phi = -\dfrac{\sqrt{3}(W_1 - W_2)}{(W_1 + W_2)} = -\dfrac{\sqrt{3}(782 - 1980)}{(782 + 1980)} = 0.75;$
$\phi = 36.9°, \cos\phi = 0.8$

(*ii*) $\sqrt{3} \times 416 \times I_L \times 0.8 = 2762,\ I_L = 4.8$ A

(*iii*) $Z_{ph} = V_{ph}/I_{ph} = (416/\sqrt{3})/4.8 = 50\ \Omega,\ X_C = Z_{ph}\sin\phi = 50 \times 0.6 = 30\ \Omega$

Now, $X_C = 1/2\pi fC = 1/2\phi \times 50 \times C = 106 \times 10^{-6}$F

Example 19.38. *Each phase of a 3-phase, Δ-connected load consists of an impedance Z = 20 ∠ 60° ohm. The line voltage is 440 V at 50 Hz. Compute the power consumed by each phase impedance and the total power. What will be the readings of the two wattmeters connected ?*

(Elect. and Mech. Technology, Osmania Univ 1980)

Solution. $Z_{ph} = 20\ \Omega;\ V_{ph} = V_L = 440\text{V};\ I_{ph} = V_{ph}/Z_{ph} = 440/20 = 22\text{A}$

Since $\phi = 60°; \cos\phi = \cos 60° = 0.5°;\quad R_{ph} = Z_{ph} \times \cos 60° = 20 \times 0.5 = 10\ \Omega$

$\therefore$ Power/phase $= I_{ph}^2 R_{ph} = 22^2 \times 10 = \mathbf{4{,}840\ W}$

Total power $= 3 \times 4{,}840 = \mathbf{14{,}520\ W}$ [or $P = \sqrt{3} \times 440 \times (\sqrt{3} \times 22) \times 0.5 = 14{,}520$ W]

Now, $W_1 + W_2 = 14{,}520.$

Also $\tan\phi = \sqrt{3} \cdot \dfrac{W_1 - W_2}{W_1 + W_2} \quad \therefore \quad \tan 60° = \sqrt{3} = \sqrt{3} \cdot \dfrac{W_1 - W_2}{14{,}520}$

$\therefore \quad W_1 - W_2 = 14{,}520.$ Obviously, $W_2 = 0$

Even otherwise it is obvious that W_2 should be zero because p.f. = cos 60° = 0.5 (Art. 19.19).

Example 19.39. *Three identical coils, each having a reactance of 20 Ω and resistance of 20 Ω are connected in (a) star (b) delta across a 440-V, 3-phase line. Calculate for each method of connection the line current and readings on each of the two wattmeters connected to measure the power.*

(Electro-mechanics Allahabad Univ. 1992)

Solution. (*a*) **Star Connection**

$$Z_{ph} = \sqrt{20^2 + 20^2} = 20\sqrt{2} = 28.3\ \Omega;\ V_{ph} = 440/\sqrt{3} = 254\text{ V}$$

$$I_{ph} = 254/28.3 = 8.97\ A;\ I_L = \mathbf{8.97\ A};\ \cos\phi = R_{ph}/Z_{ph} = 20/28.3 = 0.707$$

Total power taken $= \sqrt{3}\, V_L I_L \cos\phi = \sqrt{3} \times 440 \times 8.97 \times 0.707 = 4830$ W

If W_1 and W_2 are wattmeter readings, then $W_1 + W_2 = 4830$ W ...(*i*)

Now, $\tan\phi = 20/20 = \sqrt{3}\,(W_1 - W_2)/(W_1 + W_2)\ ;\ W_1 - W_2)\ ; = 2790$ W ...(*ii*)

From (*i*) and (*ii*) above, $W_1 = \mathbf{3810\ W}$; $W_2 = \mathbf{1020\ W}$

(*b*) **Delta Connection**

$$Z_{ph} = 28.3\ \Omega, V_{ph} = 440\ V, I_{ph} = 440/28.3 = 15.5\ A;\ I_L = 15.5 \times \sqrt{3} = 28.8\text{ A}$$

$P = \sqrt{3} \times 440 \times 28.8 \times 0.707 = 14{,}490$ W (it is 3 times the Y – power)

$\therefore \quad W_1 + W_2 = 14{,}490$ W ...(*iii*)

$\tan\phi = 20/20 = \sqrt{3}\,(W_1 - W_2)/14{,}490;\ W_1 - W_2 = 8370$ ...(*iv*)

From Eq. (*iii*) and (*iv*), we get, $W_1 = \mathbf{11{,}430\ W}$; $W_2 = \mathbf{3060\ W}$

Note: These readings are 3-times the Y-readings.

Example 19.40. *Three identical coils are connected in star to a 200-V, three-phase supply and each takes 500 W. The power factor is 0.8 lagging. What will be the current and the total power if the same coils are connected in delta to the same supply ? If the power is measured by two wattmeters, what will be their readings ? Prove any formula used.* **(Elect. Engg. A.M.Ae. S.I. Dec. 1991)**

Solution. When connected in star as shown in Fig. 19.51(*a*), $V_{ph} = 200/\sqrt{3} = 115.5$ V

Now, $V_{ph} I_{ph} \cos\phi$ – power per phase or $115.5 \times I_{ph} \times 0.8 = 500$

$\therefore \quad I_{ph} = 5.41$ A; $Z_{ph} = V_{ph}/I_{ph} = 115.5/5.41 = 21.34\ \Omega$

$R = Z_{ph} \cos\phi = 21.34 \times 0.8 = 17\Omega$; $X_L = Z_{ph} \sin\phi = 21.34 \times 0.6 = 12.8\ \Omega$

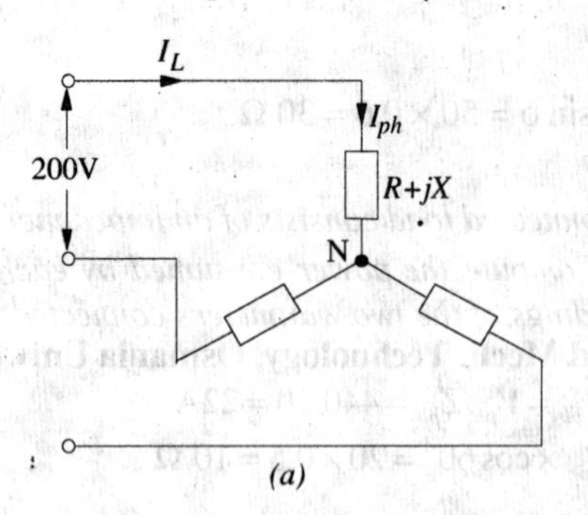

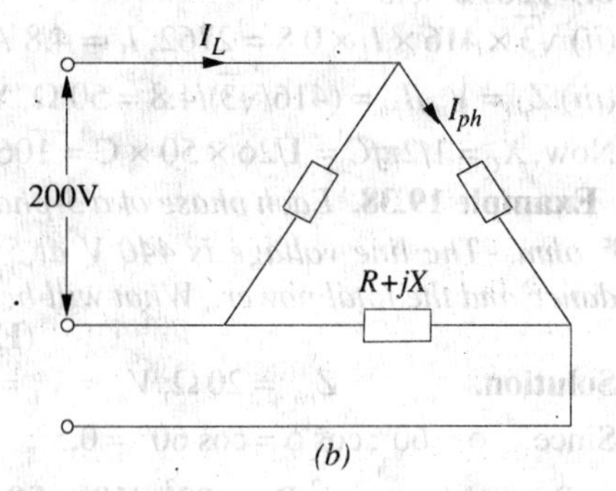

Fig. 19.51

The same three coils have been connected in delta in Fig. 19.51(*b*). Here, $V_{ph} = V_L = 200$ V.

$$I_{ph} = 200/21.34 = 9.37 \text{ A}; I_L = \sqrt{3}\, I_{ph} = 9.37 \times 1.732 = \mathbf{16.23\ A}$$

Total power consumed $= \sqrt{3} \times 200 \times 16.23 \times 0.8 =$ **4500 W**

It would be seen that when the same coils are connected in delta, they consume three times more power than when connected in star.

Wattmeter Readings

Now, $W_1 + W_2 = 4500$; $\tan\phi = \dfrac{\sqrt{3}\,(W_1 - W_2)}{(W_1 + W_2)}$

$$\phi = \cos^{-1}(0.8) = 36.87°; \tan\phi = 0.75$$

$$0.75 = \frac{\sqrt{3}\,(W_1 - W_2)}{4500} \quad \therefore \quad (W_1 - W_2) = 1950 \text{ W}$$

$$\therefore \quad W_1 = (4500 + 1950)/2 = 3225 \text{ W}; W_2 = \mathbf{1275\ W}.$$

Example 19.41(A). *A 3-phase, 3-wire, 415-V system supplies a balanced load of 20 A at a power factor 0.8 lag. The current coil of wattmeter 1 is in phase R and of wattmeter 2 in phase B. Calculate (i) the reading on 1 when its voltage coil is across R and Y (ii) the reading on 2 when its voltage coil is across B and Y and (iii) the reading on 1 when its voltage coil is across Y and B. Justify your answer with relevant phasor diagram.* **(Elect, Machines, A.M.I.E. Sec B, 1991)**

Solution. (*i*) As seen from phasor diagram of Fig. 19.51(A)

$$W_1 = V_{RY} I_A \cos(30 + \phi) = \sqrt{3} \times 415 \times 20 \times \cos(36.87° + 30°) = \mathbf{5647\ W}$$

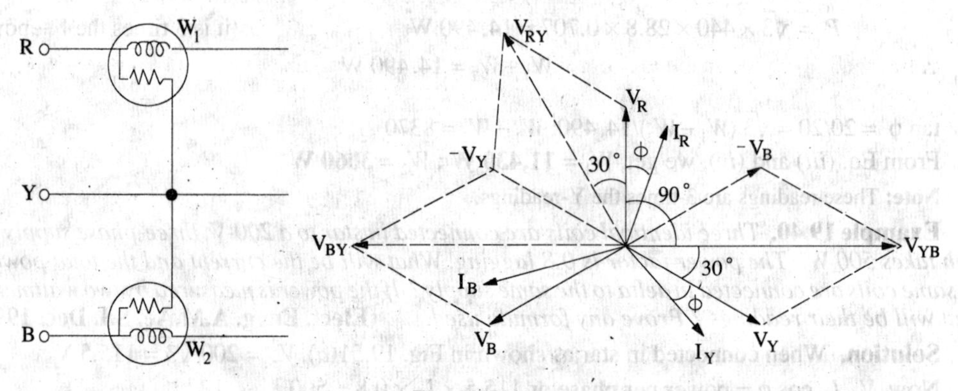

Fig. 19.51(*A*)

(*ii*) Similarly, $W_2 = V_{BY} I_B \cos(30-\phi)$

It should be noted that voltage across W_2 is V_{BY} and not V_{YB}. Moreover, $\phi = \cos^{-1}(0.8) = 36.87°$.

$$\therefore \quad W_2 = \sqrt{3} \times 415 \times 20 \times \cos(30° - 36.87°) = \mathbf{14{,}275\ W}$$

(*iii*) Now, phase angle between I_R and V_{YB} is $(90°-\phi)$

$$\therefore \quad W_2 = V_{YB} I_R \cos(90°-\phi) = \sqrt{3} \times 415 \times 20 \times \sin 36.87° = 8626\ \text{VAR}$$

Example 19.41(*B*). *A wattmeter reads 5.54 kW when its current coil is connected in R phase and its voltage coil is connected between the neutral and the R phase of a symmetrical 3-phase system supplying a balanced load of 30 A at 400 V. What will be the reading on the instrument if the connections to the current coil remain unchanged and the voltage coil be connected between B and Y phasés ? Take phase sequence RYB. Draw the corresponding phasor diagram.*

(Elect. Machines, A.M.I.E.Sec. B, 1992)

Solution. As seen from Fig. 19.50(B).

$W_1 = V_R I_R \cos\phi$ or $5.54 \times 10^3 = (400/\sqrt{3}) \times 30 \times \cos\phi$; $\therefore \cos\phi = 0.8, \sin\phi = 0.6$

In the second case (Fig. 19.50(*B*))

$$W_2 = V_{YB} I_R \cos(90° - \phi) = 400 \times 30 \times \sin\phi = 400 \times 30 \times 0.6 = \mathbf{7.2\ kW}$$

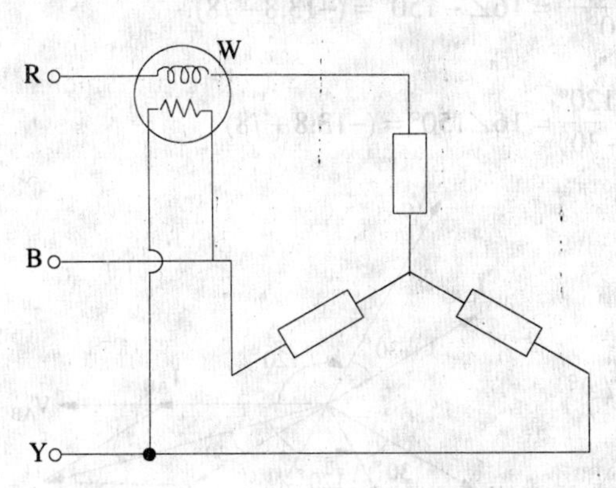

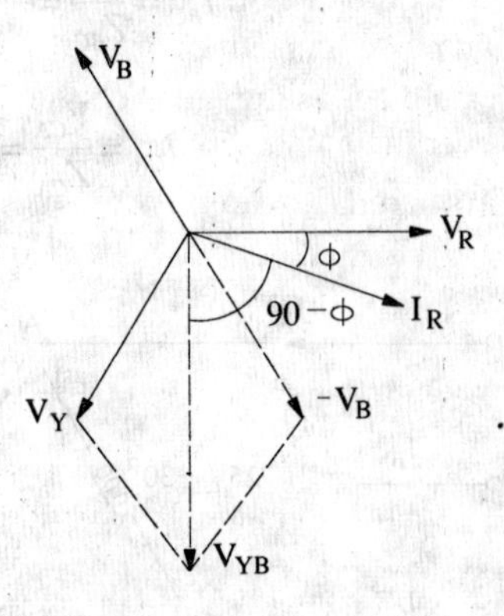

Fig. 19.51(*B*)

Example 19.41(*C*). *A 3-phase, 3-wire balanced load with a lagging power factor is supplied at 400 V (between lines). A 1-phase wattmeter (scaled in kW) when connected with its current coil in the R-line and voltage coil between R and Y lines gives a reading of 6 kW. When the same terminals of the voltage coil are switched over to Y- and B-lines, the current-coil connections remaining the same, the reading of the wattmeter remains unchanged. Calculate the line current and power factor of the load. Phase sequence is R→ Y→ B.* **(Elect. Engg–1, Bombay Univ. 1985)**

Solution. The current through the wattmeter is I_R and p.d. across its pressure coil is V_{RY}. As seen from the phasor diagram of Fig. 19.52, the angle between the two is $(30° + \phi)$.

$$\therefore \quad W_1 = V_{RY} I_R \cos(30° + \phi) = V_L I_L \cos(30° + \phi) \quad \ldots(i)$$

In the second case, current is I_R but voltage is V_{YB}. The angle between the two is $(90° - \phi)$

$$\therefore \quad W_2 = V_{YB} I_R \cos(90° - \phi) = V_L I_L \cos(90° - \phi)$$

Since $\quad W_1 = W_2$ we have

$$V_L I_L \cos(30° + \phi) = V_L I_L \cos(90° - \phi)$$

$$\therefore \quad 30° + \phi = 90° - \phi$$

or $\quad 2\phi = 60° \quad \therefore \phi = 30°$

$\therefore$ load power factor $= \cos 30° = \mathbf{0.866\ (lag)}$

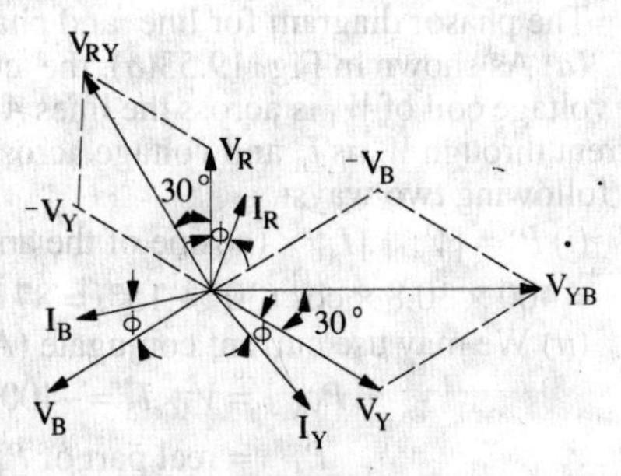

Fig. 19.52

Now $\quad W_1 = W_2 = 6\ \text{kW}$.

Hence, from (*i*) above, we get

$$6000 = 400 \times I_L \cos 60°; \; I_L = \mathbf{30\ A}$$

Example 19.42. *A 3-phase, 400 V circuit supplies a Δ-connected load having phase impedances of $Z_{AB} = 25\angle 0°$; $Z_{BC} = 25\angle 30°$ and $V_{CA} = 25\angle -30°$.*

Two wattmeters are connected in the circuit to measure the load power. Determine the wattmeter readings if their current coils are in the lines (a) A and B (b) B and C and (c) C and A. The phase sequence is ABC. Draw the connections of the wattmeter for the above three cases and check the sum of the two wattmeter readings against total power consumed.

Solution. Taking V_{AB} as the reference voltage, we have $Z_{AB} = 400\angle 0°$; $Z_{BC} = 400\angle -120°$ and $Z_{CA} = 400\angle 120°$

The three phase currents can be found as follows :

$$I_{AB} = \frac{V_{AB}}{Z_{AB}} = \frac{400\angle 0°}{25\angle 0°} = 16\angle 0° = (16 + j0)$$

$$I_{BC} = \frac{V_{BC}}{Z_{BC}} = \frac{400\angle -120°}{25\angle 30°} = 16\angle -150° = (-13.8 - j8)$$

$$I_{CA} = \frac{V_{CA}}{Z_{CA}} = \frac{400\angle 120°}{25\angle -30°} = 16\angle 150° = (-13.8 + j8)$$

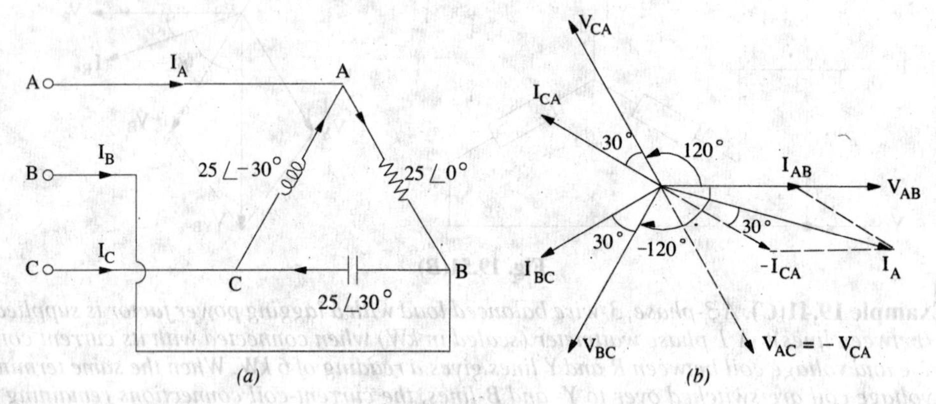

Fig. 19.53

The line currents I_A, I_B and I_C can be found by applying KCL at the three nodes *A*, *B* and *C* of the load.

$I_A = I_{AB} + I_{AC} = I_{AB} - I_{CA} = (16 + j0) - (-13.8 + j8) = 29.8 - j8 = 30.8\angle -15°$

$I_B = I_{BC} - I_{AB} = (-13.8 - j8) - (16 + j0) = -29.8 - j8 = 30.8\angle -165°$

$I_C = I_{CA} - I_{BC} = (-13.8 + j8) - (-13.8 - j8) = 0 + j16 = 16\angle 90°$

The phasor diagram for line and phase currents is shown in Fig. 19.53(*a*) and (*b*).

(*a*) As shown in Fig. 19.53(*a*), the current coils of the wattmeters are in the line *A* and *B* and the voltage coil of W_1 is across the lines *A* and *C* and that of W_2 is across the lines *B* and *C*. Hence, current through W_1 is I_A and voltage across it is V_{AC}. The power indicated by W_1 may be found in the following two ways:

(*i*) $P_1 = |V_{AC}|.|I_A| \times$ (cosine of the angle between V_{AC} and I_A).

$= 400 \times 30.8 \times \cos(30° + 15°) = 8710$ W

(*ii*) We may use current conjugate (Art.) for finding the power

$$P_{VA} = V_{AC}.I_A^* = -400\angle 120° \times 30.8\angle 15°$$

$$\therefore \quad P_1 = \text{real part of } P_{VA} = -400 \times 30.8 \times \cos 135° = 8710 \text{ W}$$

$$P_2 = \text{real part of } [V_{BC}^* Z_B] = 400\angle 120° \times 30.8\angle -165°$$

$$= 400 \times 30.8 \times \cos(-45°) = 8710 \text{ W}$$

$\therefore \quad P_1 + P_2 = 8710 + 8710 = 17{,}420$ W.

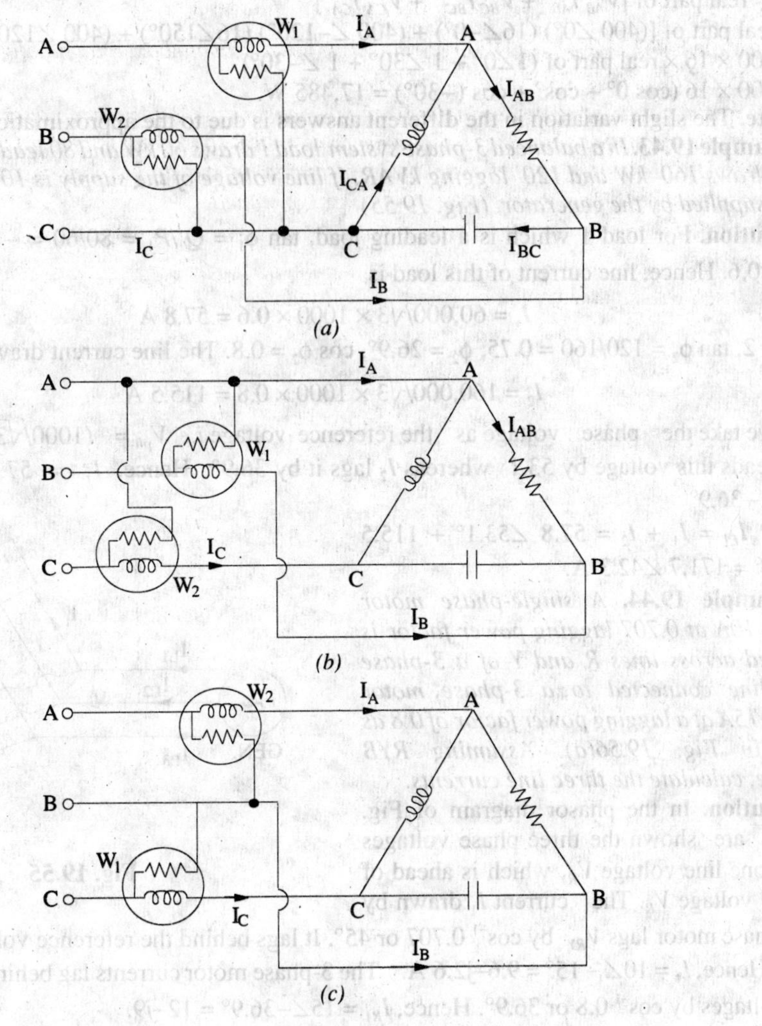

Fig. 19.54

(*b*) As shown in Fig. 19.54(*b*), the current coils of the wattmeters are in the lines *B* and *C* whereas voltage coil of W_1 is across the lines *B* and *A* and that of W_2 is across lines *C* and *A*.

(*i*) $\therefore \; P_1 = |V_{BA}| \cdot |I_B|$ (cosine of the angle between V_{BA} and I_B)

$= 400 \times 30.8 \times \cos 15° = 11{,}900$ W

(*ii*) Using voltage conjugate (which is more convenient in this case), we have

$P_{VA} = V_{BA}^{*} \cdot I_B = -400\angle 0° \times 30.8 \angle \times -165°$

$\therefore P_1 =$ real part of $P_{VA} = -400 \times 30.8 \times \cos(-165°) = 11{,}900$ W

$P_2 =$ real part of $[V_{CA}{}^{*}I_C] = [400\angle -120° \times 16\angle 90°] = 400 \times 16 \times \cos(-30°) = 5{,}540$ W.

$\therefore P_1 + P_2 = 11{,}900 + 5{,}540 = 17{,}440$ W.

(*c*) As shown in Fig. 19.54(*c*), the current coils of the wattmeters are in the lines *C* and *A* whereas the voltage coil of W_1 is across the lines *C* and *B* and that of W_2 is across the lines *A* and *B*.

(*i*) $P_1 =$ real part of $[V_{CB}{}^{*}I_C] = [(-400\angle 120°) \times 16\angle 90°]$

$= -400 \times 16 \times \cos \angle 210° = 5540$ W

$P_2 =$ real part of $[V_{AB}{}^{*}I_A] = [400\angle °0 \times 30.8 \angle -15° = 400 \times 30.8 \times \cos \angle -15° = 11{,}900$ W

$\therefore \; P_1 + P_2 = 5{,}540 + 11{,}900 = 17{,}440$ W

Total power consumed by the -phase load can be found directly as under :-

P_T = real part of $[V_{AB} I_{AB}^* + V_{BC} I_{BC}^* + V_{CA} I_{CA}^*]$

= real part of $[(400 \angle 0°)(16\angle -0°) + (400 \angle -120°)(16\angle 150°) + (400 \angle 120°)(16 \angle -150°)]$

= $400 \times 16 \times$ real part of $(1\angle 0° + 1 \angle 30° + 1 \angle -30°)$

= 400×16 (cos 0° + cos° + cos (−30°) = 17,485 W.

Note. The slight variation in the different answers is due to the approximation made.

Example 19.43. *In a balanced 3-phase system load 1 draws 60 kW and 80 leading kVAR whereas load 2 draws 160 kW and 120 lagging kVAR. If line voltage of the supply is 1000 V, find the line current supplied by the generator. (Fig. 19.55)*

Solution. For load 1 which is a leading load, tan $\phi_1 = Q_1/P_1$ = 80/60 = −1.333; ϕ_1 = 53.1°, cos ϕ_1 = 0.6. Hence, line current of this load is

$$I_1 = 60{,}000/\sqrt{3} \times 1000 \times 0.6 = 57.8 \text{ A}$$

For load 2, tan ϕ_2 = 120/160 = 0.75; ϕ_2 = 26.9°, cos ϕ_2 = 0.8. The line current drawn by this load is

$$I_2 = 160{,}000/\sqrt{3} \times 1000 \times 0.8 = 115.5 \text{ A}$$

If we take the phase voltage as the reference voltage *i.e.* $V_{ph} = (1000/\sqrt{3}) \angle 0° = 578\angle 0°$; then I_1 leads this voltage by 53.1° whereas I_2 lags it by 36.9°. Hence, $I_1 = 57.8\angle 53.1°$ and $I_2 = 115.5 \angle -36.9°$

$\therefore \quad I_{L1} = I_1 + I_2 = 57.8 \angle 53.1° + 115.5 \angle -36.9° = 171.7 \angle 42.3°$A.

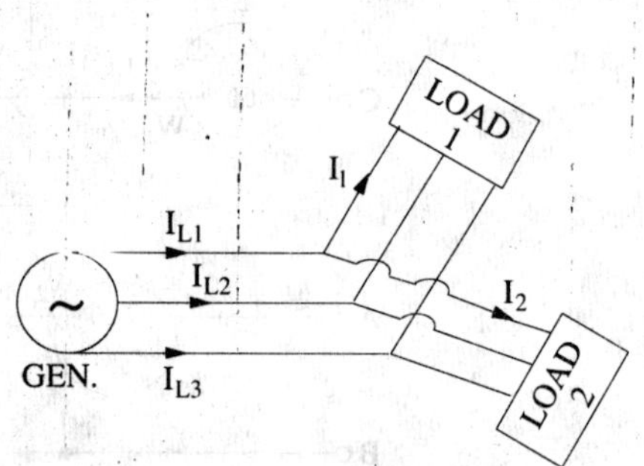

Fig. 19.55

Example 19.44. *A single-phase motor drawing 10A at 0.707 lagging power factor is connected across lines R and Y of a 3-phase supply line connected to a 3-phase motor drawing 15A at a lagging power factor of 0.8 as shown in Fig. 19.56(a). Assuming RYB sequence, calculate the three line currents.*

Solution. In the phasor diagram of Fig. 19.56(*b*) are shown the three phase voltages and the one line voltage V_{RY} which is ahead of its phase voltage V_R. The current I_1 drawn by single-phase motor lags V_{RY} by cos^{-1} 0.707 or 45°. It lags behind the reference voltage V_R by 15° as shown. Hence, $I_1 = 10\angle -15° = 9.6 - j2.6$ A. The 3-phase motor currents lag behind their respective phase voltages by cos^{-1} 0.8 or 36.9°. Hence, $I_{R1} = 15\angle -36.9° = 12 - j9$.

$$I_{Y1} = 1.5 \angle(-120° - 36.9°) = 15\angle -156.9° = -13.8 - j5.9$$

$$I_B = 15\angle(120° - 36.9°) = 15\angle 83.1°$$

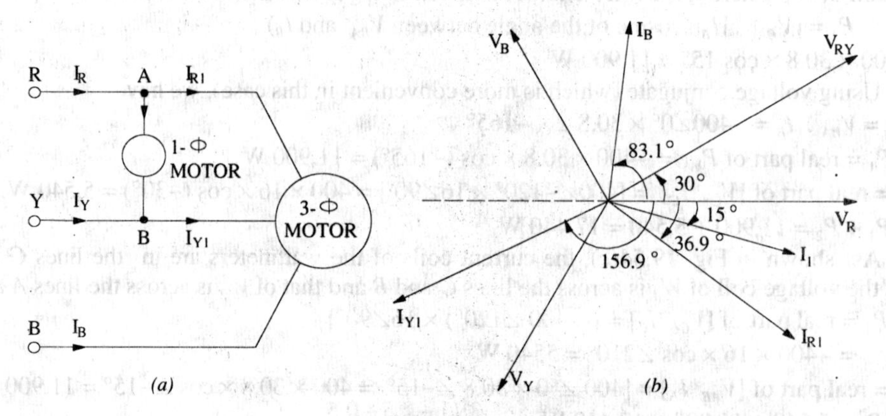

Fig. 19.56

Applying Kirchhoff's laws to point A of Fig. 19.56(*a*), we get

$$I_R = I_1 + I_{R1} = 9.6 - j2.6 + 12 - j9 = 21.6 - j11.6 = 24.5 \angle -28.2°$$

Similarly, applying KCL to point *B*, we get

$$I_Y + I_1 = I_{Y1} \text{ or } I_Y = I_{Y1} - I_1 = -13.8 - j5.9 - 9.6 + j2.6 = -23.4 - j3.3 = 23.6 \angle -172°.$$

Example 19.45. *A 3-ϕ, 434-V, 50-Hz, supply is connected to a 3-ϕ, Y-connected induction motor and synchronous motor. Impedance of each phase of induction motor is (1.25 + j2.17) Ω. The 3-ϕ synchronous motor is over-excited and it draws a current of 120 A at 0.87 leading p.f. Two wattmeters are connected in usual manner to measure power drawn by the two motors. Calculate (i) reading on each wattmeter (ii) combined power factor.* **(Elect. Technology, Hyderabad Univ. 1992)**

Solution. It will be assumed that the synchronous motor is *Y*-connected. Since it is over-excited it has a leading p.f. The wattmeter connections and phasor diagrams are as shown in Fig. 19.57.

$$\mathbf{Z}_1 = 1.25 + j2.17 = 2.5\angle 60°$$

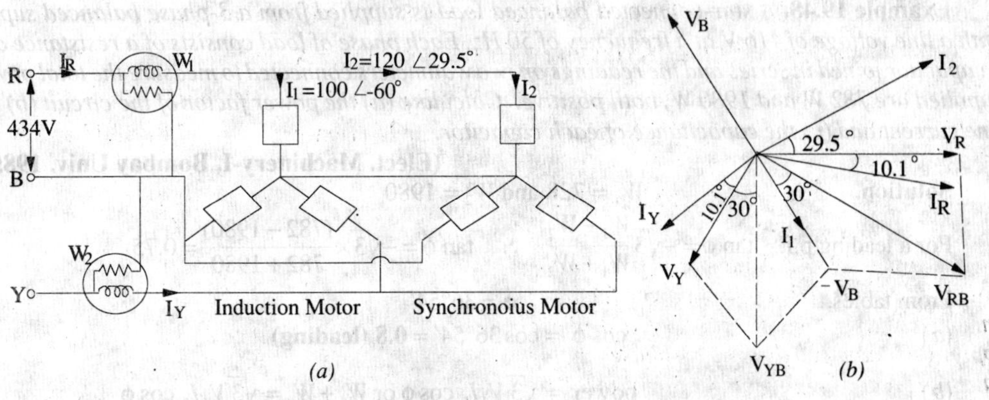

Fig. 19.57

Phase voltage in each case $= 434/\sqrt{3} = 250$ V

$I_1 = 250/2.5 = 100$ A lagging the reference vector V_R by 60°. Current $I_2 = 120$ A and leads V_R by an angle $= \cos^{-1}(0.87) = 29.5°$

$$\therefore \quad \mathbf{I}_1 = 100\angle -60° = 50 - j86.6; \; \mathbf{I}_2 = 120\angle 29.5° = 104.6 + j59°$$

$$\mathbf{I}_R = \mathbf{I}_1 + \mathbf{I}_2 = 154.6 - j27.6 = 156.8\angle -10.1°$$

(*a*) As shown in Fig. 19.57 (*b*), I_R lags V_R by 10.1°. Similarly, I_Y lags V_Y by 10.1°.

As seen from Fig. 19.57 (*a*), current through W_1 is I_R and voltage across it is $V_{RB} = V_R - V_B$. As seen, $V_{RB} = 434$ V lagging by 30°. Phase difference between V_{RB} and I_R is $= 30 - 10.1 = 19.9°$.

$\therefore$ reading of $W_1 = 434 \times 156.8 \times \cos 19.9° =$ **63,970 W**

Current I_Y is also (like I_R) the vector sum of the line currents drawn by the two motors. It is equal to 156.8 A and lags behind its respective phase voltage V_Y by 10.1°. Current through W_2 is I_Y and voltage across it is $\mathbf{V}_{YB} = \mathbf{V}_Y - \mathbf{V}_B$. As seen, $V_{YB} = 434$ V. Phase difference between V_{YB} and $I_Y = 30° + 10.1° = 40.1°$ (lag).

$\therefore$ reading of $W_2 = 434 \times 156.8 \times \cos 40.1° =$ **52,050 W**

(*b*) Combined p.f. $= \cos 10.1° =$ **0.9845 (lag)**

Example 19.46. *Power in a balanced 3-phase system is measured by the two-wattmeter method and it is found that the ratio of the two readings is 2 to 1. What is the power factor of the system ?* **(Elect. Science-I, Allahabad Univ. 1991)**

Solution. We are given that $W_1 : W_2 \cdot\cdot\; 2 : 1$. Hence, $W_2/W_2 = r = 1/2 = 0.5$. As seen from Art. 19.21.

$$\cos\phi = \frac{1+r}{2\sqrt{1-r+r^2}} = \frac{1+0.5}{2\sqrt{1-0.5+0.5^2}} = \mathbf{0.866\ lag}$$

Example 19.47. *A synchronous motor absorbing 50 kW is connected in parallel with a factory load of 200 kW having a lagging power factor of 0.8. If the combination has a power factor of 0.9 lagging, find the kVAR supplied by the motor and its power factor.*

(Elect. Machines, A.M.I.E. Sec B, 1989)

Solution. Load kVA = 200/0.8 = 250

Load kVAR = 250 × 0.6 = 150 (lag) [cos ϕ = 0.8 sin ϕ = 0.6]

Total combined load = 50 + 200 = 250 kW

kVA of combined load = 250/0.9 = 277.8

Combined kVAR = 277.8 × 0.4356 = 121 (inductive) (combined cos ϕ = 0.9, sin ϕ = 0.4356)

Hence, leading kVAR supplied by synch, motor = 150 − 121 = 29 (capacitive)

kVA of motor alone = $\sqrt{(kW^2 + kVAR^2)} = \sqrt{50^2 + 29^2} = 57.8$

p.f. of motor = kW/kVA = 50/57.8 = 0.865 (leading)

Example 19.48. *A star-connected balanced load is supplied from a 3-phase balanced supply with a line voltage of 416 V at a frequency of 50 Hz. Each phase of load consists of a resistance and a capacitor joined in series and the readings on two wattmeters connected to measure the total power supplied are 782 W and 1980 W, both positive. Calculate (a) the power factor of the circuit (b) the line current and (c) the capacitance of each capacitor.*

(Elect. Machinery-I, Bombay Univ. 1985)

Solution. $W_1 = 728$ and $W_2 = 1980$

For a leading p.f. $\tan\phi = -\sqrt{3}\dfrac{W_1 - W_2}{W_1 + W_2}$ $\therefore$ $\tan\phi = -\sqrt{3} \times \dfrac{(782 - 1980)}{782 + 1980} = 0.75$

From tables, $\phi = 36°54'$

(*a*) $\therefore \cos\phi = \cos 36°54' = \mathbf{0.8\ (leading)}$

(*b*) power $= \sqrt{3}\, V_L I_L \cos\phi$ or $W_1 + W_2 = \sqrt{3}\, V_L I_L \cos\phi$

or $(782 + 1980) = \sqrt{3} \times 416 \times I_L \times 0.8$ $\therefore$ $I_L = I_{ph} = \mathbf{4.8\ A}$

(*c*) Now $V_{ph} = 416/\sqrt{3}$ V $\therefore$ $Z_{ph} = 416/\sqrt{3} \times 4.8 = 50\ \Omega$

$\therefore$ $Z_{ph} = 50\angle -36°54' = 50(0.8 - j0.6) = 40 - j30$

Capacitive reactance X_C = 30; or $\dfrac{1}{2\pi \times 50 \times C} = 30$ $\therefore C = \mathbf{106\ \mu F}$

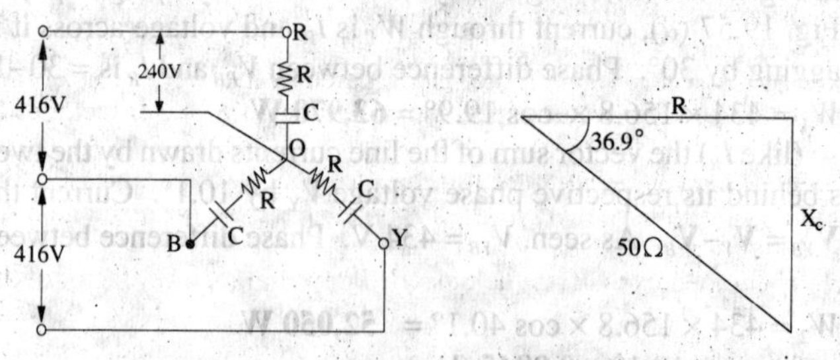

Fig. 19.58

Example 19.49. *The two wattmeters A and B, give readings as 5000 W and 1000 W respectively during the power measurement of 3-ϕ, 3-wire, balanced load system. (a) Calculate the power and power factor if (i) both meters read direct and (ii) one of them reads in reverse. (b) If the voltage of the circuit is 400 V, what is the value of capacitance which must be introduced in each phase to cause the whole of the power to appear on A. The frequency of supply is 50 Hz.*

(Elect. Engg.-I, Nagpur Univ. 1992)

Solution. (*a*) (*i*) *Both Meters Read Direct*

$W_1 = 5000$ W; $W_2 = 1000$ W; $\therefore W_1 + W_2 = 6000$ W; $W_1 - W_2 = 4000$ W

$$\tan \phi = \sqrt{3}(W_1 - W_2)/(W_1 + W_2) = \sqrt{3} \times 4000/6000 = 1.1547$$

$\therefore$ $\phi = \tan^{-1}(1.1547) = 49.1°$;p.f. $= \cos 49.1° = 0.655$ (lag)

Total power = 5000 + 1000 = 6000 W

(*ii*) One Meter Reads in Reverse

In this case, $\tan \phi = \sqrt{3}(W_1 + W_2)/(W_1 - W_2) = \sqrt{3} \times 6000/4000 = 2.598$

$\therefore$ $\phi = \tan^{-1}(2.598) = 68.95°$; p.f. $= \cos 68.95° = 0.36$ (lag)

Total power = $W_1 + W_2 = 5000 - 1000 = 4000$ W ...Art.

(*b*) The whole of power would be measured by wattmeter W_1 if the load power factor is 0.5 (lagging) or less. It means that in the present case p.f. of the load will have to be reduced from 0.655 to 0.5. In other words, capacitive reactance will have to be introduced in each phase of the load in order to partially neutralize the inductive reactance.

Now, $\sqrt{3}\, V_L I_L \cos \phi = 6000$ or $\sqrt{3} \times 400\, I_L \times 0.655 = 6000$

$\therefore$ $I_L = 13.2$A; $\therefore I_{ph} = 13.2/\sqrt{3} = 7.63$ A

$$Z_{ph} = V_{ph}/I_{ph} = 400/7.63 = 52.4\ \Omega$$

$$X_L = Z_{ph} \sin \phi = 52.4 \times \sin 49.1° = 39.6\ \Omega$$

When p.f. = 0.5

$\sqrt{3} \times 400 \times I_L \times 0.5 = 6000; I_L = 17.32A; I_{ph} = 17.32/\sqrt{3} = 10A; Z_{ph} = 400/10 = 40\Omega$

$\cos \phi = 0.5$; $\phi = 60$; $\sin 60° = 0.886$; $X = Z_{ph} \sin \phi = 40 \times 0.886 = 35.4\ \Omega$

$\therefore$ $X = X_L - X_C = 35.4$ or $39.6 - X_C = 35.4$; $\therefore X_C = 4.2\ \Omega$.

If C is the required capacitance, then $4.2 = 1/2\pi \times 50 \times C$; $\therefore C = 758\ \mu$F.

Tutorial Problems No. 19.2

1. Two wattmeters connected to measure the input to a balanced three-phase circuit indicate 2500 W and 500 W respectively. Find the power factor of the circuit (*a*) when both readings are positive and (*b*) when the latter reading is obtained after reversing the connections to the current coil of one instrument.
[(*a*) **0.655** (*b*) **0.3591**] (*City & Guilds, London*)

2. A 400-V, 3-phase induction motor load takes 900 kVA at a power factor of 0.707. Calculate the kVA rating of the capacitor bank to raise the resultant power factor of the installation of 0.866 lagging.

Find also the resultant power factor when the capacitors are in circuit and the motor load has fallen to 300 kVA at 0.5 power factor. [**269 kVA, 0.998 leading**] (*City & Guilds, London*)

3. Two wattmeters measure the total power in three-phase circuits and are correctly connected. One reads 4,800 W while other reads backwards. On reversing the latter, it reads 400 W. What is the total power absorbed by the circuit and the power factor? [**4400 W; 0.49**] (*Sheffield Univ. U.K.*)

4. The power taken by a 3-phase, 400-V motor is measured by the two wattmeter method and the readings of the two wattmeters are 460 and 780 watts respectively. Estimate the power factor of the motor and the line current. [**0.913, 1.96 A**] (*City & Guilds, London*)

5. Two wattmeters, W_1 and W_2 connected to read the input to a three-phase induction motor running unloaded, indicate 3 kW and 1 kW respectively. On increasing the load, the reading on W_1 increases while that on W_2 decreases and eventually reverses.

Explain the above phenomenon and find the unloaded power and power factor of the motor.
[**2 kW, 0.287 lag**] (*London Univ.*)

6. The power flowing in a 3-ϕ, 3-wire, balanced-load system is measured by the two wattmeter method. The reading on wattmeter *A* is 5,000 W and on wattmeter *B* is –1,000 W

(*a*) What is the power factor of the system ?

(*b*) If the voltage of the circuit is 440, what is the value of capacitance which must be introduced into each phase to cause the whole of the power measured to appear on wattmeter *A*?
[**0.359; 5.43 Ω**] (*Meters and Meas. Insts. A.M.I.E.E. London*)

7. Two wattmeters are connected to measure the input to a 400 V; 3-phase, connected motor outputting 24.4 kW at a power factor of 0.4 (lag) and 80% efficiency. Calculate the

(*i*) resistance and reactance of motor per phase.

(*ii*) reading of each wattmeters. [(*i*) **2.55 Ω; 5.85 Ω**; (*ii*) **34,915 W; – 4850 W**]
(*Elect. Machines, A.M.I.E. Sec. B, 1993*)

8. The readings of the two instruments connected to a balanced three-phase load are 128 W and 56 W. When a resistor of about 25 Ω is added to each phase, the reading of the second instrument is reduced to zero. State, giving reasons, the power in the circuit before the resistors were added. **[72 W]** (*London Univ.*)

9. A balanced star-connected load, each phase having a resistance of 10 Ω and inductive reactance of 30 Ω is connected to 400-*V*, 50-Hz supply. The phase rotation is read, yellow and blue. Wattmeters connected to read total power have their current coils in the red and blue lines respectively. Calculate the reading on each wattmeter and draw a vector diagram in explanation. **[2190 W, – 583 W]** (*London Univ.*)

10. A 7.46 kW induction motor runs from a 3-phase, 400-V supply. On no-load, the motor takes a line current of 4 A at a power factor of 0.208 lagging. On full load, it operates at a power factor of 0.88 lagging and an efficiency of 89 per cent. Determine the readings on each of the two wattmeters connected to read the total power on (*a*) no load and (*b*) full load. **[1070 W, –494 W ; 5500 W ; 2890 W]**

11. A balanced inductive load, connected in star across 415-V, 50- Hz, three-phase mains, takes a line current of 25A. The phase sequence is *RYB*. A single-phase wattmeter has its current coil connected in the *R* line and its voltage coil across the lines *YB*. With these connections, the reading is 8 kW. Draw the vector diagram and find (*i*) the kW (*ii*) the kVAR (*iii*) the kVA and (*iv*) the power factor of the load.

[(*i*) **11.45 kW** (*ii*) **13.87 kVAR** (*iii*) **18 kVA** (*iv*) **0.637**] (*City & Guilds, London*)

19.27. Double Subscript Notation

In symmetrically-arranged networks, it is comparatively easier and actually more advantageous, to use single-subscript notation. But for unbalanced 3-phase circuits, it is essential to use double subscript notation, in order to avoid unnecessary confusion which is likely to result in serious errors.

Suppose, we are given two coils whose induced e.m.fs. are 60° out of phase with each other [Fig. 19.59 (*a*)]. Next, suppose that it is required to connect these coils in additive series *i.e.* in such

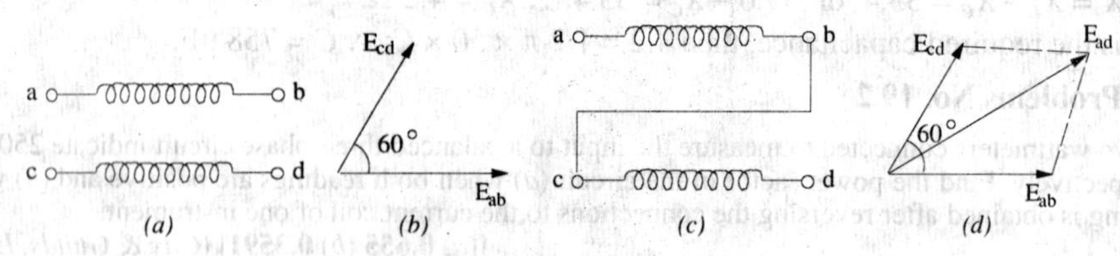

Fig. 19.59

a way that their e.m.fs. add at an angle of 60°. From the information given, it is impossible to know whether to connect terminal '*a*' to terminal '*c*' or to terminal '*d*'. But if additionally it were given that e.m.f. from terminal '*c*' to terminal '*d*' is 60° out of phase with that from terminal '*a*' to terminal '*b*', then the way to connect the coils is definitely fixed, as shown in Fig. 19.59 (*b*) and 19.60 (*a*). The double-suscript notation is obviously very convenient in such cases. The order in which these subscripts are written indicates the direction along which the voltage acts (or current flows). For example the e.m.f. '*a*' to '*b*' [Fig. 19.59(*a*)], may be written as E_{ab} and that from '*c*' to '*d*' as E_{cd}. The e.m.f. between '*a*' and '*d*' is E_{ad} where $\mathbf{E}_{ad} = \mathbf{E}_{ab} + \mathbf{E}_{cd}$ and is shown in Fig. 19.59 (*b*).

Example 19.59. *If in Fig. 19.60 (a), terminal 'b' is connected to 'd', find E_{ac} if E = 100 V.*

Solution. Vector diagram is shown in Fig. 19.60 (*b*)

Obviously, $\mathbf{E}_{ac} = \mathbf{E}_{ab} + \mathbf{E}_{dc} = \mathbf{E}_{ab} + (-\mathbf{E}_{cd})$

Hence, $\mathbf{E}_{cd}$ is reversed and *added* to $\mathbf{E}_{ab}$ to get $\mathbf{E}_{ac}$ as shown in Fig. 19.60 (*b*). The magnitude of resultant vector is

$$E_{ac} = 2 \times 100 \cos 120°/2 = 100 \text{ V}; \mathbf{E}_{ac} = 100\angle-60°$$

Example 19.51. *In Fig. 19.60 (a) with terminal 'b' connected to 'd', find E_{ca}.*

Solution. $\mathbf{E}_{ca} = \mathbf{E}_{cd} + \mathbf{E}_{ba} = \mathbf{E}_{cd} + (-\mathbf{E}_{ab})$

As shown in Fig. 19.61, vector $\mathbf{E}_{ab}$ is reversed and then combined with $\mathbf{E}_{cd}$ to get $\mathbf{E}_{ca}$.

Magnitude of $\mathbf{E}_{ca}$ is given by 2 × 100 × cos 60° = 100 *V* but it leads $\mathbf{E}_{ab}$ by 120°.

$$\therefore \qquad \mathbf{E}_{ca} = 100\angle 120°$$

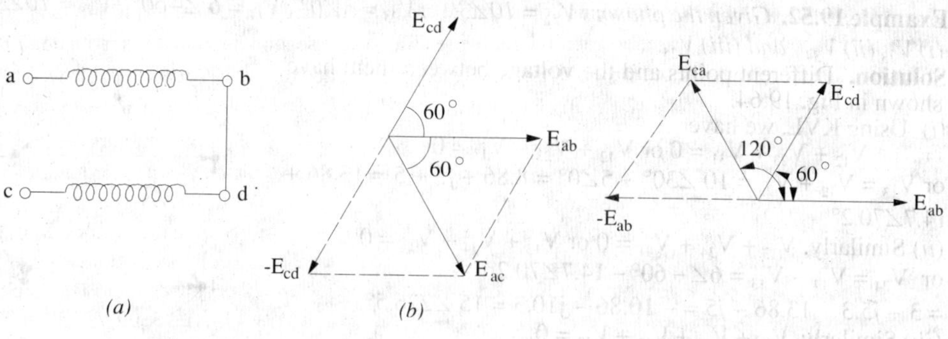

Fig. 19.60 **Fig. 19.61**

In Fig. 19.62 (*b*) is shown the vector diagram of the e.m.fs induced in the three phases 1, 2, 3 (or *R*, *Y*, *B*) of a 3-phase alternator [Fig. 19.62 (*a*)]. According to double subscript notation, each phase e.m.f. may be written as $\mathbf{E}_{01}$, $\mathbf{E}_{02}$ and $\mathbf{E}_{03}$, the order of the subscripts indicating the direction in which the e.m.fs. act. It is seen that while passing from phase 1 to phase 2 through the *external* circuit, we are in opposition to $\mathbf{E}_{02}$.

$$\mathbf{E}_{12} = \mathbf{E}_{20} + \mathbf{E}_{01} = (-\mathbf{E}_{02}) + \mathbf{E}_{01} = \mathbf{E}_{01} - \mathbf{E}_{02}$$

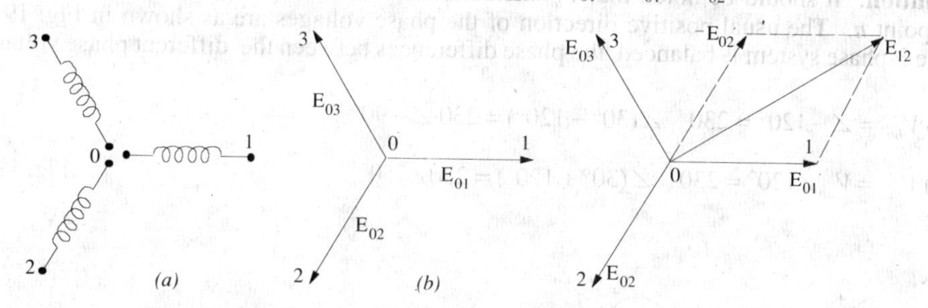

Fig. 19.62 **Fig. 19.63**

It means that for obtaining $\mathbf{E}_{12}$, $\mathbf{E}_{20}$ has to be reversed to obtain $-\mathbf{E}_{02}$ which is then combined with $\mathbf{E}_{01}$ to get $\mathbf{E}_{12}$ (Fig. 19.51). Similarly,

$$\mathbf{E}_{23} = \mathbf{E}_{30} + \mathbf{E}_{02} = (-\mathbf{E}_{03}) + \mathbf{E}_{02} = \mathbf{E}_{02} - \mathbf{E}_{03}$$

$$\mathbf{E}_{31} = \mathbf{E}_{10} + \mathbf{E}_{03} = (-\mathbf{E}_{01}) + \mathbf{E}_{03} = \mathbf{E}_{03} - \mathbf{E}_{01}$$

By now it should be clear that double-subscript notation is based on lettering every junction and terminal point of diagrams of connections and on the use of two subscripts with all vectors representing voltage or current. The subscripts on the vector diagram, taken from the diagram of connections, indicate that the positive direction of the current or voltage is from the first subscript to the second. For example, according to this notation $\mathbf{I}_{ab}$ represents a current whose +*ve* direction is from *a* to *b* in the branch *ab* of the circuit in the diagram of connections. In the like manner, $\mathbf{E}_{ab}$ represents the e.m.f. which produces this current. Further, $\mathbf{I}_{ba}$ will represent a current flowing from *b* to *a*, hence its vector will be drawn equal to but in a direction opposite to that of $\mathbf{I}_{ab}$ *i.e.* $\mathbf{I}_{ab}$ and $\mathbf{I}_{ba}$ differ in phase by 180° although they do not differ in magnitude.

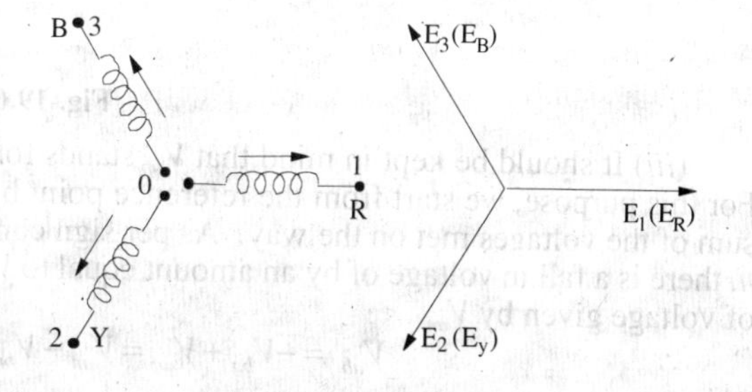

Fig. 19.63 (a)

In single subscript notation (*i.e.* the one in which single subscript is used) the +*ve* directions are fixed by putting arrows on the circuit diagrams as shown in Fig. 19.63 (*a*). According to this notation

$$\mathbf{E}_{12} = -\mathbf{E}_2 + \mathbf{E}_1 = \mathbf{E}_1 - \mathbf{E}_2\ ;\ \mathbf{E}_{23} = -\mathbf{E}_3 + \mathbf{E}_2 = \mathbf{E}_2 - \mathbf{E}_3 \text{ and } \mathbf{E}_{31} = -\mathbf{E}_1 + \mathbf{E}_3 = \mathbf{E}_3 - \mathbf{E}_1$$

or $$\mathbf{E}_{RY} = \mathbf{E}_R - \mathbf{E}_Y\ ;\ \mathbf{E}_{YB} = \mathbf{E}_Y - \mathbf{E}_B\ ;\ \mathbf{E}_{BR} = \mathbf{E}_B - \mathbf{E}_R$$

Example 19.52. *Given the phasors* $V_{12} = 10\angle 30°$; $V_{23} = 5\angle 0°$; $V_{14} = 6\angle -60°$; $V_{45} = 10\angle 90°$. *Find (i)* V_{13} *(ii)* V_{34} *and (iii)* V_{25}.

Solution. Different points and the voltage between them have been shown in Fig. 19.64.

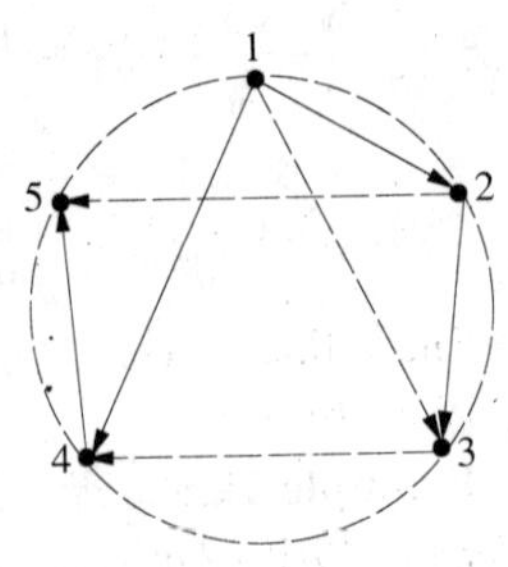

Fig. 19.64

(*i*) Using KVL, we have

$$V_{12} + V_{23} + V_{31} = 0 \text{ or } V_{12} + V_{23} - V_{13} = 0$$

or $V_{13} = V_{12} + V_{23} = 10\angle 30° + 5\angle 0° = 8.86 + j5 + 5 = 13.86 + j5 = 14.7\angle 70.2°$

(*ii*) Similarly, $V_{13} + V_{34} + V_{41} = 0$ or $V_{13} + V_{34} - V_{14} = 0$

or $V_{34} = V_{14} - V_{13} = 6\angle -60° - 14.7\angle 70.2°$

$= 3 - j5.3 - 13.86 - j5 = -10.86 - j10.3 = 15\angle 226.5°$

(*iii*) Similarly, $V_{23} + V_{34} + V_{45} + V_{52} = 0$

or $V_{23} + V_{34} + V_{45} - V_{25} = 0$

or $V_{25} = V_{23} + V_{34} + V_{45} = 5\angle 0° + 15\angle 226.5° + 10\angle 90°$

$= 5 - 10.86 - j10.3 + j10 = -5.86 - j0.3 = 5.86 - \angle 2.9°$.

Example 19.53. *In a balanced 3-phase Y-connected voltage source having phase sequence abc,* $V_{an} = 230\angle 30°$. *Calculate analytically (i)* V_{bn} *(ii)* V_{cn} *(iii)* V_{ab} *(iv)* V_{bc} *and (v)* V_{ca}. *Show the phase and line voltages on a phasor diagram.*

Solution. It should be noted that V_{an} stands for the voltage of terminal a with respect to the neutral point *n*. The usual positive direction of the phase voltages are as shown in Fig. 19.65(*a*). Since the 3-phase system is balanced, the phase differences between the different phase voltages are 120°.

$$(i)\ V_{bn} = \angle -120° = 230\ \angle(30° - 120°) = 230\angle -90°$$

$$(ii)\ V_{cn} = V_{an}\angle 120° = 230\ \angle(30° + 120°) = 230\angle 150° \qquad \text{...Fig. 19.65}(b)$$

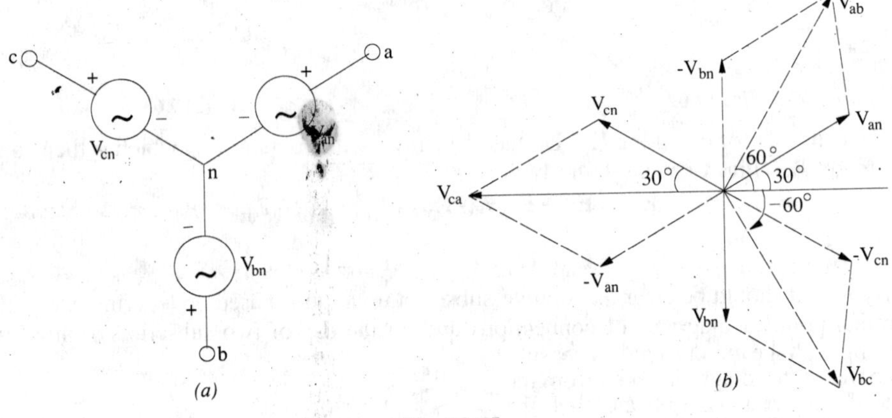

Fig. 19.65

(*iii*) It should be kept in mind that V_{ab} stands for the voltage of point *a* with respect to point *b*. For this purpose, we start from the reference point b in Fig. 19.65(*a*) and go to point *a* and find the sum of the voltages met on the way. As per sign convention given in Art, 19.27 as we go from *b* to *n*, there is a fall in voltage of by an amount equal to V_{bn}. Next as we go from *n* to *a*, there is increase of voltage given by V_{an}.

$$\therefore \quad V_{ab} = -V_{bn} + V_{an} = V_{an} - V_{bn} = 230\angle 30° - 230\angle -90°$$

$$= 230(\cos 30° + j\sin 30°) - 230(0 - j\sin 90°)$$

$$= 230\left(\frac{\sqrt{3}}{2} + j\frac{1}{2}\right) + j230 = 230\left(\frac{\sqrt{3}}{2} + j\frac{3}{2}\right)$$

$$= 230\sqrt{3}\left(\frac{1}{2} + j\frac{\sqrt{3}}{2}\right) = 400\angle 60°$$

$(iv)\ V_{bc} = V_{bn} - V_{cn} = 230\angle -90° - 230\angle 150° = -j230 - 230$

$$\left(-\frac{\sqrt{3}}{2} + j\frac{1}{2}\right) = 230\sqrt{3}\left(\frac{1}{2} - j\frac{\sqrt{3}}{2}\right) = 400\angle -60°$$

$$(v) V_{ca} = V_{cn} - V_{an} = 230\angle 150° - 230\angle 30° = \left(-\frac{\sqrt{3}}{2} + j\frac{1}{2}\right) - 230\left(\frac{\sqrt{3}}{2} + j\frac{1}{2}\right) = -400 = 400\angle 180°$$

These line voltages along with the phase voltages have been shown in the phasor diagram of Fig. 19.65(*b*).

Example 19.54. *Three non-inductive resistances, each of 100 Ω are connected in star to a 3-phase, 440-V supply. Three equal choking coils are also connected in delta to the same supply ; the resistance of one coil being equal to 100 Ω. Calculate (a) the line current and (b) the power factor of the system.* **(Elect. Technology-II, Sumbal Univ. 1987)**

Solution. The diagram of connections and the vector diagram of the *Y*-and Δ-connected impedances are shown in Fig. 19.66.

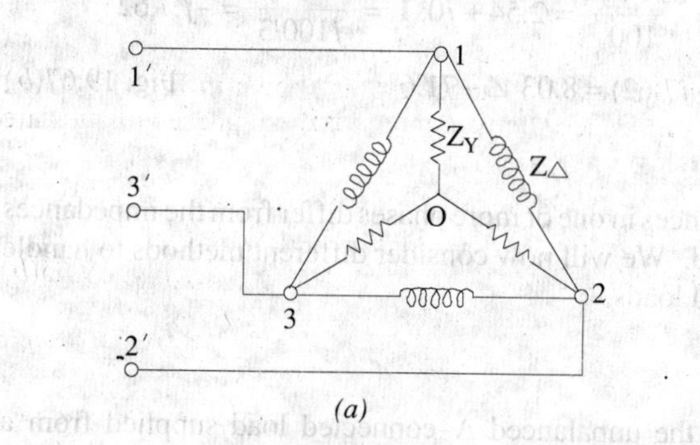

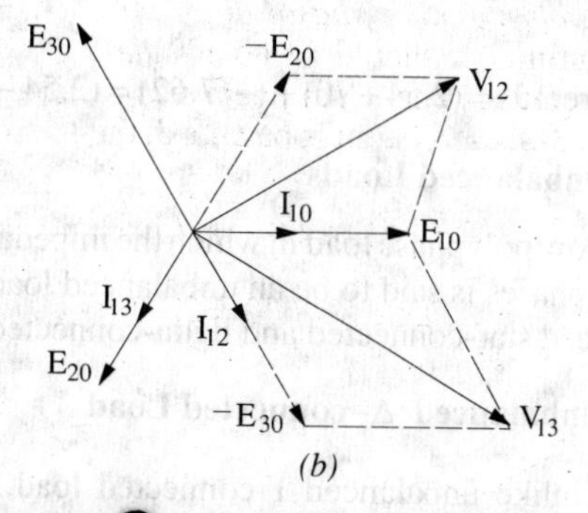

Fig. 19.66

The voltage E_{10} between line 1 and neutral is taken along the *X*-axis. Since the load is balanced, it will suffice to determine the current in one line only. Applying Kirchhoff's Law to junction 1, we have $\mathbf{I}_1{}'_1 = \mathbf{I}_{10} + \mathbf{I}_{12} + \mathbf{I}_{13}$

Let us first get the vector expressions for $\mathbf{E}_{10}$, $\mathbf{E}_{20}$ and $\mathbf{E}_{30}$.

$$\mathbf{E}_{10} = \frac{440}{\sqrt{3}}(1 + j0) = 254 + j0,\ \mathbf{E}_{20} = 254\left(-\frac{1}{2} - j\frac{\sqrt{3}}{2}\right) = -127 - j220$$

$$\mathbf{E}_{30} = 254\left(-\frac{1}{2} + j\frac{\sqrt{3}}{2}\right) = -127 + j220$$

Let us now derive vector expressions for $\mathbf{V}_{12}$ and $\mathbf{V}_{31}$.

$$\mathbf{V}_{10} = \mathbf{E}_{10} + \mathbf{E}_{02} = \mathbf{E}_{10} - \mathbf{E}_{20} = (254 + j0) - (-127 - j220) = 381 + j220$$

$$\mathbf{V}_{13} = \mathbf{E}_{10} + \mathbf{E}_{03} = \mathbf{E}_{10} - \mathbf{E}_{30} = (254 + j0) - (-127 + j220) = 381 - j220$$

$$\mathbf{I}_{10} = \frac{\mathbf{E}_{10}}{\mathbf{Z}_Y} = \frac{254 + j0}{100} = 2.54 + j0,\ \mathbf{I}_{12} = \frac{\mathbf{V}_{13}}{\mathbf{Z}_\Delta} = \frac{381 + j220}{j100} = 2.2 - j3.81 = 4.4\angle -60°$$

$$\mathbf{I}_{13} = \frac{V_{13}}{Z_\Delta} = \frac{381 - j220}{j100} = -2.2 - j3.81 = 4.4\angle -120°$$

$(a)\ \mathbf{I}_1{}'_1 = (2.54 + j0) + (2.2 - j3.81) + (-2.2 - j3.81) = (2.54 - j7.62) = 8.03\angle -71$°

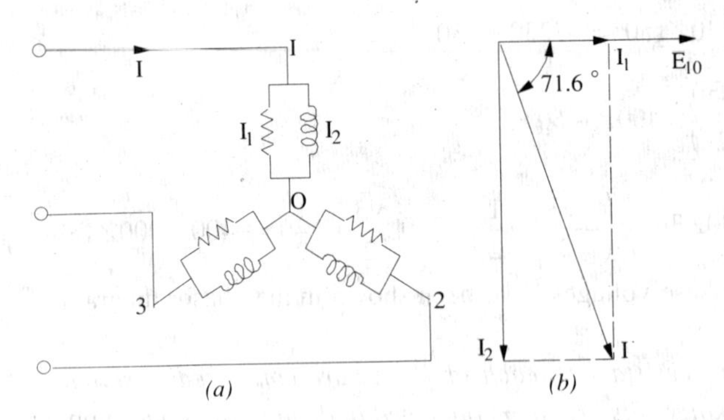

Fig. 19.67

(*b*) p.f. = cos 71.6° = **0.316 (lag)**

Alternative Method

This question may be easily solved by Δ/Y conversion. The star equivalent of the delta reactance is 100/3 Ω per phase.

As shown in Fig. 19.67, there are now two parallel circuits across each phase, one consisting of a resistance of 100 Ω and the other of a reactance of 100/3 Ω.

Taking E_{10} as the reference vector, we have

$$\mathbf{E}_{10} = (254 + j0)$$

$$\mathbf{I}_1 = \frac{254 + j0}{100} = 2.54 + j0; \ \mathbf{I}_2 = \frac{254 + j0}{j100/3} = -j.7.62$$

Line current $\mathbf{I} = (2.54 + j0) + (-j7.62) = (2.54 - j7.62) = 8.03 \angle -71.6°$ ...Fig. 19.67(*b*)

19.28. Unbalanced Loads

Any polyphase load in which the impedances in one or more phases differ from the impedances of other phases is said to be an unbalanced load. We will now consider different methods to handle unbalanced star-connected and delta-connected loads.

19.29. Unbalanced Δ–connected Load

Unlike unbalanced *Y*-connected load, the unbalanced Δ–connected load supplied from a balanced 3-phase supply does not present any new problems because the voltage across each load phase is fixed. It is independent of the nature of the load and is equal to line voltage. In fact, the problem resolves itself into three independent single-phase circuits supplied with voltages which are 120° apart in phase.

The different phase currents can be calculated in the usual manner and the three line currents are obtained by taking the vector difference of phase currents in pairs.

If the load consists of three different pure resistances, then trigonometrical method can be used with advantage, otherwise symbolic method may be used.

Example 19.55. *A 3-phase, 3-wire, 240 volt, CBA system supplies a delta-connected load in which $Z_{AB} = Z_{AB} = 25 \angle 90°$, $Z_{BC} = 15 \angle Z_{CA} = 20 \angle 0°$ ohms, Find the line currents and total power.* **(Advanced Elect. Machines AMIE Sec. B, Summer 1991)**

Solution. As explained in Art. 19.2, a 3-phase system has only two possible sequences : *ABC* and *CBA*. In the *ABC* sequence, the voltage of phase *B* lags behind voltage of phase *A* by 120° and that of phase *C* lags behind phase *A* voltage by 240°. In the *CBA* phase which can be written as $A \rightarrow C \rightarrow B$, voltage of *C* lags behind voltage *A* by 120° and that of *B* lags behind voltage *A* by 240°. Hence, the phase voltage which can be written as

$$E_{AB} = E\angle 0° \ ; \ E_{BC} = E \angle -120°$$

$$\text{and } E_{CA} = E\angle -240° \text{ or } E_{CA} = \angle 120°$$

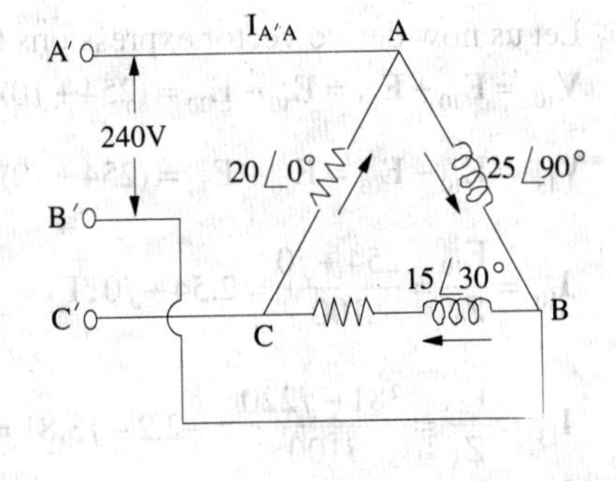

Fig. 19.68

$$\therefore \qquad I_{AB} = \frac{E_{AB}}{Z_{AB}} = \frac{240\angle 0°}{25\angle 90°} = 9.6\angle -90° = -j9.6 \text{ A}$$

$$I_{BC} = \frac{E_{BC}}{Z_{BC}} = \frac{240\angle 120°}{15\angle 30°} = 16\angle 90° = j16 \text{ A}$$

$$I_{CA} = \frac{E_{CA}}{Z_{CA}} = \frac{240\angle -120°}{20\angle 0°} = 12\angle -120° = 12(0.5 - j0.866) = (-6 - j10.4) \text{ A}$$

The circuit is shown in Fig. 19.68.

Line current $I_{A'A} = I_{AB} + I_{AC} = I_{AB} - I_{CA} = -j9.6 - (-6 - j10.4) = 6 + j0.8$

Line current $I_{B'B} = I_{BC} - I_{AB} = j16 - (-j9.6) = j\,25.6$ A

$I_{C'C} = I_{CA} - I_{BC} = (-6 - j\,10.4) - j\;16 = (-6 - j26.4)$ A

Now, $R_{AB} = 0$; $R_{BC} = 15 \cos 30 = 13\ \Omega$; $R_{CA} = 20\ \Omega$

Power

$W_{AB} = 0$; $W_{BC} = I_{BC}^2 R_{BC} = 16^2 \times 13 = 3328$ W; $W_{CA} = I_{CA}^2 \times R_{CA} = 27^2 \times 20 = 14{,}580$ W.

Total Power = 3328 + 14580 = 17,908 W.

Example 19.56. *In the network of Fig. 19.69, E_{na} = 230 ∠0° and the phase sequence is abc. Find the line currents I_a, I_b and I_c as also the phase currents I_{AB}, I_{BC} and I_{CA}.*

E_{na}, E_{nb}, E_{nc} is a balanced three-phase voltage system with phase sequence abc.

(Network Theory, Nagpur Univ. 1993)

Solution. Since the phase sequence is *abc,* the generator phase voltages are :

$$E_{na} = 230\angle 0° \;;\; E_{nb} = 230\angle -120° \;;\; E_{nc} = 230\angle 120°$$

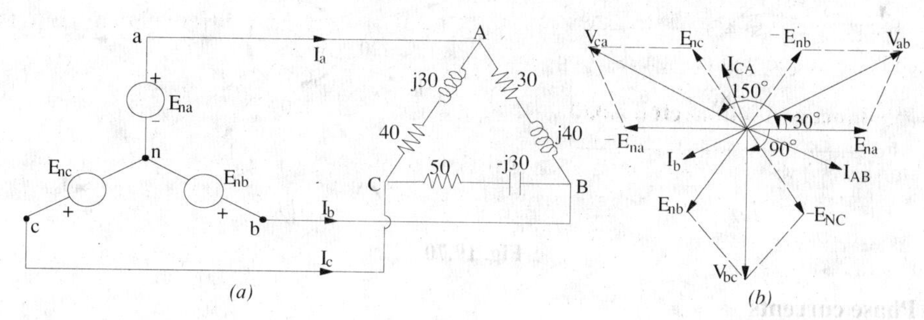

Fig. 19.69

As seen from the phasor diagram of Fig. 19.69 (*b*), the line voltages are as under :-

$$V_{ab} = E_{na} - E_{nb} \;;\; V_{bc} = E_{nb} = E_{nb} - E_{nc} \;;\; V_{ca} = E_{nc} - E_{na}$$

$\therefore \qquad V_{ab} = \sqrt{3} \times 230\ \sqrt{30}° = 400\angle 30°$ *i.e.* it is ahead of the reference generator phase voltage E_{na} by 30°.

$V_{bc} = \sqrt{3} \times 230\angle -90° = 400\angle -90°$. This voltage is 90° behind E_{na} but 120° behind V_{ab}.

$V_{ca} = \sqrt{3} \times 230\ \angle 150° = 400\ \angle 150°$ or $\angle -210°$. This voltage leads reference voltage E_{na} by 150° but leads V_{ab} by 120°.

These voltages are applied across the unbalanced Δ-connected lead as shown in Fig. 19.69 (*a*).

$$Z_{AB} = 30 + j40 = 50\angle 53.1°; Z_{BC} = 50 - j30 = 58.3\angle -31°,$$

$$Z_{CA} = 40 + j30 = 50\angle 36.9°$$

$$I_{AB} = \frac{V_{ab}}{Z_{AB}} = \frac{400\angle 30°}{50\angle 53.1°} = 8\angle -23.1° = 7.36 - j3.14$$

$$I_{BC} = \frac{V_{bc}}{Z_{BC}} = \frac{400\angle -90°}{58.3\angle -31°} = 6.86\ \angle -59° = 3.53 - j5.88$$

$$I_{CA} = \frac{V_{ca}}{Z_{CA}} = \frac{400\angle 150°}{50\angle 36.9°} = 8\angle 113.1° = -3.14 + j7.36$$

$$I_a = I_{AB} - I_{CA} = 7.36 - j3.14 + 3.14 - j7.36 = 10.5 - j10.5 = 14.85\ \angle -45°$$

$$I_b = I_{BC} - I_{AB} = 3.53 - j5.88 - 7.36 + j3.14 = -3.83 - j2.74 = 4.71\angle 215.6°$$

$$I_C = I_{CA} - I_{BC} = -3.14 + j7.36 - 3.53 + j5.88 = -6.67 + j13.24 = 14.8\angle 116.7°$$

Example 19.57. *For the unbalanced Δ-connected load of Fig. 19.55 (a), find the phase currents, line currents and the total power consumed by the load when phase sequence is (a) abc and (b) acb.*

Solution. (*a*) **Phase sequence abc** (Fig. 19.70).

Let $\mathbf{V}_{ab} = 100\angle 0° = 100 + j0$

$$\mathbf{V}_{bc} = 100\ \angle -120° = 100\left(-\frac{1}{2} - j\frac{\sqrt{3}}{2}\right) = -50 - j86.6$$

$$\mathbf{V}_{ca} = 100\ \angle\ 120° = 100\left(-\frac{1}{2} + j\frac{\sqrt{3}}{2}\right) = -50 + j86.6$$

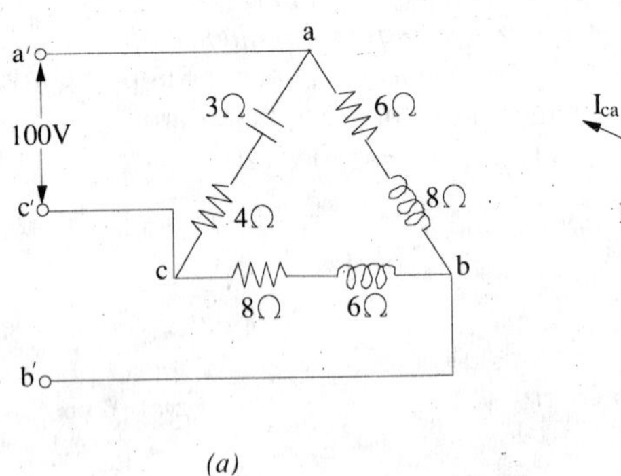

(*a*)

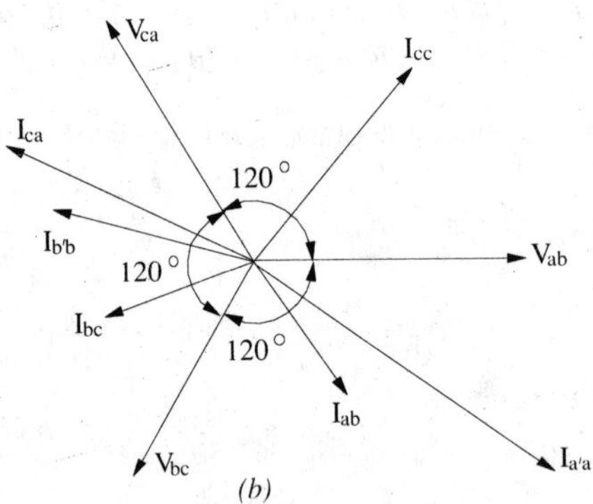

(*b*)

Fig. 19.70

(*i*) **Phase currents**

Phase current, $\mathbf{I} = \frac{\mathbf{V}_{ab}}{\mathbf{Z}_{ab}} = \frac{100 + j0}{6 + j8} = 6 - j8 = 10\angle -53°8'$

Similarly, $\mathbf{I}_{bc} = \frac{\mathbf{V}_{bc}}{\mathbf{Z}_{bc}} = \frac{-30 - j86.6}{8 + j6} = -9.2 - j3.93 = 10\angle -156°52'$

$$\mathbf{I}_{ca} = \frac{\mathbf{V}_{ca}}{\mathbf{Z}_{ca}} = \frac{-50 + j86.6}{4 - j3} = -18.39 + j7.86 = 20\angle 156°\ 52'$$

(*ii*) **Line Currents**

Line Current $\mathbf{I}_{a'a} = \mathbf{I}_{ab} + \mathbf{I}_{ac} = \mathbf{I}_{ab} - \mathbf{I}_{ca} = (6 - j8) - (-18.39 + j7.86)$

$$= 24.39 - j15.86 = 29.1\angle -33°2'$$

Similarly, $\mathbf{I}_{b'b} = I_{bc} + I_{ba} = \mathbf{I}_{bc} - \mathbf{I}_{ab} = (-9.2 - j3.93) - (6 - j8) = -15.2 + j4.07 = 15.73\angle 165°30'$

$$\mathbf{I}_{c'c} = \mathbf{I}_{ca} + \mathbf{I}_{cb} = \mathbf{I}_{ca} - \mathbf{I}_{bc} = (-18.39 + j7.86) - (-9.2 - j3.93)$$

$$= 9.19 + j11.79 = 14.94\angle 52°3'$$

Check $\Sigma\mathbf{I} = 0 + j0$

(*iii*) **Power**

$$W_{ab} = I_{ab}^2 R_{ab} = 10^2 \times 6 = 600 \text{ W}$$

$$W_{bc} = I_{bc}^2 R_{bc} = 10^2 \times 8 = 800 \text{ W}$$

$$W_{ca} = I_{ca}^2 R_{ca} = 20^2 \times 4 = 1600 \text{ W}$$

$$\text{Total} \quad = 3000 \text{ W}$$

(*b*) **Phase sequence acb** (Fig. 19.71)

Here, $\mathbf{V}_{ab} = 100 \angle 0° = 100 + j9$

$\mathbf{V}_{bc} = 100\angle 120° = -50 + 86.6$

$\mathbf{V}_{ca} = 100 \angle -120° = -50 - j86.6$

(*i*) **Phase Currents**

$$\mathbf{I}_{ab} = \frac{100}{6 + j8} = 6 - j8 = 10\angle -53°8'$$

$$\mathbf{I}_{bc} = \frac{-50 + j86.6}{8 + j6}$$

$$= (1.2 + j9.93) = 10\angle 83°8'$$

$$I_{ca} = \frac{-50 - j86.6}{4 - j3}$$

$$= (2.4 - j19.86) = 20 - 83°8'$$

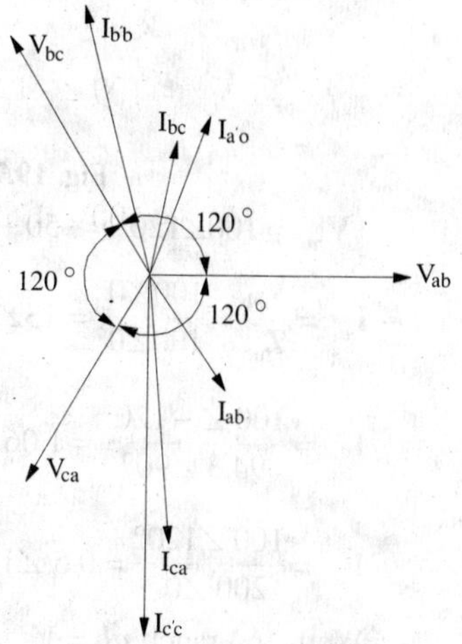

Fig. 19.71

(*ii*) **Line Currents**

$$\mathbf{I}_{a'a} = \mathbf{I}_{ab} + \mathbf{I}_{ac} = \mathbf{I}_{ab} - \mathbf{I}_{ca}$$

$$= (6 - j8) - (2.4 - j19.86) = (3.6 + j11.86) = 12.39 \angle 73°6'$$

$$\mathbf{I}_{b'b} = (1.2 + j9.93) - (6 - j8) = (-4.8 + j17.93) = 18.56\angle 105°$$

$$\mathbf{I}_{c'c} = (2.4 - j19.86) - (1.2 + j9.93) = (1.2 - j29.79) = 29.9\angle -87°42'$$

It is seen that $\Sigma\mathbf{I} = 0 + j0$

(*iii*) **Power**

$$W_{ab} = 10^2 \times 6 = 600 \text{ W}$$
$$W_{bc} = 10^2 \times 8 = 800 \text{ W}$$
$$W_{ca} = 20^2 \times 4 = 1600 \text{ W}$$

Total = 3000 W —as before

It will be seen that the effect of phase reversal on an unbalanced Δ-connected load is as under:

(*i*) phase currents change in angle only, their magnitudes remaining the same

(*ii*) consequently, phase powers remain unchanged

(*iii*) line currents change both in magnitude and angle.

The adjoining tabulation emphasizes the effect of phase sequence on the line currents drawn by an unbalanced 3-phase load.

Line	*Ampere Sequence a b c*	*Sequence c b a*
a	$29.1\angle -33°2'$	$12.39\angle 73.1°$
b	$15.73\angle 165°$	$18.56\angle 105°$
c	$14.94\angle 52°3'$	$29.9\angle -87.7°$

Example 19.58. *A balanced 3-phase supplies an unbalanced 3-phase*

delta-connected load made up of to resistors 100 Ω and 200 Ω and a reactor having an inductance of 0.3 H with negligible resistance. V_L = 100 V at 50 Hz. Calculate (a) the total power in the system and (b) the total reactive voltamperes. **(Elect. Engineering-I, Madras Univ. 1988)**

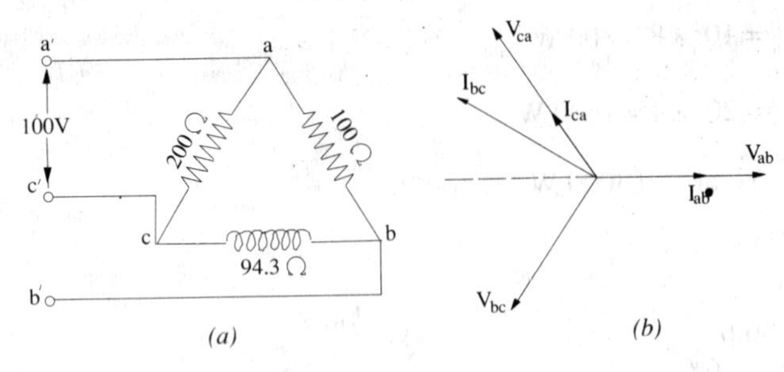

Fig. 19.72

Solution. The Δ-connected load and its phasor diagram are shown in Fig. 19.72 (*a*).

$$X_L = \omega L = 314.2 \times 0.2 = 94.3\ \Omega$$

$$\text{Let } \mathbf{V}_{ab} = 100\angle 0° = 100 + j0$$

$$\mathbf{V}_{bc} = 100\angle -120° = -50 - j86.6$$

$$\mathbf{V}_{ca} = 100\angle 120° = -50 + j86.6$$

$$\mathbf{I}_{ab} = \frac{\mathbf{V}_{ab}}{\mathbf{Z}_{ab}} = \frac{100\angle 0°}{100\angle 0°} = 13\angle 0° = 1 + j0$$

$$\mathbf{I}_{bc} = \frac{100\ \angle -120°}{94.3\ \angle 90°} = 1.06\ \angle -210° = -0.92 + j0.53$$

$$\mathbf{I}_{ca} = \frac{100\ \angle 120°}{200\ \angle 0°} = 0.5\ \angle 120° = -0.25 + j0.43$$

Watts in branch $ab = V_{ab}^2/R_{ab} = 100^2/100 = 100$ W ; VARs = 0

Watts in branch bc = 0; VRAs = 100 × 1.06 = 106 (lag)

Watts in branch $ca = V_{ca}^2/R_{ca} = 100^2/200 = 50$ W; VRAs = 0

(*a*) Total power = 100 + 50 = **150 W** ; VARs = **106 (lag)**

19.30. Four-wire Star-connected Unbalanced Load

It is the simplest case of an unbalanced load and may be treated as three separate single-phase systems with a common return wire. It will be assumed that impedance of the line wires and source phase windings is zero. Should such an assumption be unacceptable, these impedances can be added to the load impedances. Under these conditions, source and load line terminals are at the same potential.

Consider the following two cases :

(*i*) **Neutral wire of zero impedance**

Because of the presence of neutral wire (assumed to be having zero impedance), the star points of the generator and load are tied together and are at the same potential. Hence, the voltages across the three load impedances are *equalized* and each is equal to the voltage of the corresponding phase of the generator. In other words, due to the provision of the neutral, each phase voltage is a *forced* voltage so that the three phase voltages are balanced when line voltages are balanced even though phase impedances are unbalanced. However, it is worth noting that a break or open ($Z_N = \infty$) in the neutral wire of a 3-phase, 4-wire system with *unbalanced* load always causes large (in most cases inadmissible) changes in currents and phase voltages. It is because of this reason that no fuses and circuit breakers are ever used in the neutral wire of such a 3-phase system.

The solution for currents follows a pattern similar to that for the unbalanced delta.

Obviously, the vector sum of the currents in the three lines is not zero but is equal to neutral current.

(*ii*) **Neutral wire with impedance Z_N**

Such a case can be easily solved with the help of Node-pair Voltage method as detailed below*. Consider the general case of a *Y* - to – *Y* system with a neutral wire of impedance Z_N as shown in Fig. 19.73 (*a*). As before, the impedance of line wires and source phase windings would be assumed to be zero so that the line and load terminals, *R*, *Y*, *B* and *R′*, *Y′*, *B′* are at the same respective potentials.

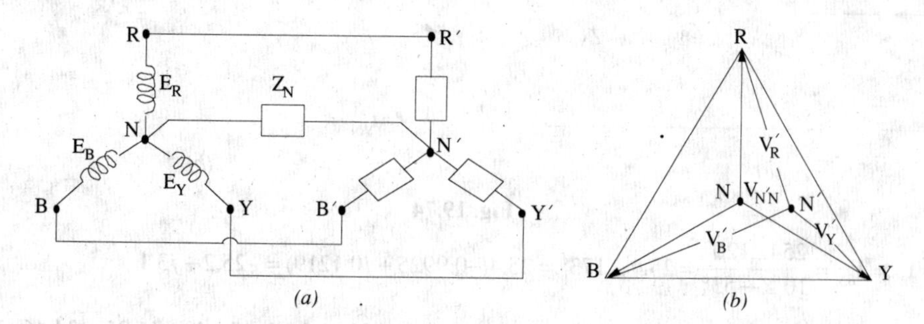

Fig. 19.73

According to Node-pair Voltage method, the above star-to-star system can be looked upon as multi-mesh network with a single pair of nodes *i.e.* neutral points *N* and *N′*. The node potential *i.e.* the potential difference between the supply and local neutrals is given by

$$\mathbf{V}_{N'N} = \frac{\mathbf{E}_R\mathbf{Y}_R + \mathbf{E}_Y\mathbf{Y}_Y + \mathbf{E}_B\mathbf{Y}_B}{\mathbf{Y}_R + \mathbf{Y}_Y + \mathbf{Y}_B + \mathbf{Y}_N}$$

where $\mathbf{Y}_R$, $\mathbf{Y}_Y$ and $\mathbf{Y}_B$ represent the *load phase* admittances. Obviously, the load neutral *N′* does not coincide with source neutral *N*. Hence, load phase voltages are no longer equal to one another even when phase voltages are as seen from Fig. 19.73 (*b*).

The load phase voltage are given by

$$\mathbf{V}_{R}' = \mathbf{E}_R - V_{N'N};\ \mathbf{V}_Y' = \mathbf{E}_Y - \mathbf{V}_{N'N} \text{ and } \mathbf{V}_{B'} = \mathbf{E}_B - \mathbf{V}_{N'N}$$

The phase currents are

$$\mathbf{I}_R = \mathbf{V}_R'\mathbf{Y}_R,\quad \mathbf{I}_Y = \mathbf{V}_Y'\mathbf{Y}_Y \text{ and } \mathbf{I}_B = \mathbf{V}_B'\mathbf{Y}_B$$

The current in the neutral wire is $\mathbf{I}_N = \mathbf{V}_N\mathbf{Y}_N$

Note. In the above calculations, $I_R = I_R' = I_{RR}'$.
Similarly, $I_Y = I_{Y'} = I_{YY}'$ and $I_B' = I_{BB}'$.

Example 19.59. *A 3-phase, 4-wire system having a 254-V line-to-neutral has the following loads connected between the respective lines and neutral ; Z_R = 10 ∠ 0° ohm; Z_Y = 10 ∠ 37° ohm and Z_B = 10 ∠–53° ohm. Calculate the current in the neutral wire and the power taken by each load when phase sequence is (i) RYB and (ii) RBY.*

Solution. (*i*) **Phase sequence RYB** (Fig. 19.74).

$$\mathbf{V}_{RN} = 254\angle 0°;\qquad \mathbf{V}_{YN} = 254\angle -120°;\qquad \mathbf{V}_{BN} = 254\angle 120°$$

$$\mathbf{I}_R = \mathbf{I}_{RN} = \frac{\mathbf{V}_{RN}}{\mathbf{R}_R} = \frac{254\angle 0°}{10\angle 0°} = 25.4\angle 0°$$

$$\mathbf{I}_Y = \mathbf{I}_{YN} = \frac{254\angle -120°}{10\angle 37°} = 25.4\angle -157° = 25.4(-0.9205 - j0.3907) = -23.38 - j9.95$$

*This method is similar to Millman's Theorem of Art. 19.32.

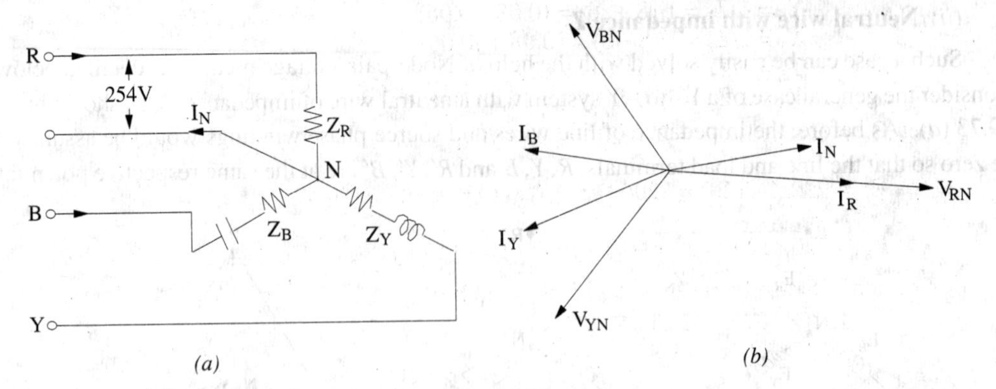

Fig. 19.74

$$\mathbf{I}_B = \mathbf{I}_{BN} = \frac{254\angle 120°}{10\ \angle -53°} = 25.4\ \angle 173° = 25.4(-0.9925 + j0.1219) = -25.2 + j3.1$$

$$\mathbf{I}_N = -(\mathbf{I}_R + \mathbf{I}_Y + \mathbf{I}_B) = -[25.4 + (-23.38 - j9.55) + (-25.21 + j3.1)] = 23.49 + j6.85 = \mathbf{24.46\ \angle 16°15'}$$

Now, $R_R = 10\ \Omega;\ R_Y = 10 \cos 37° = 8\ \Omega;\ R_B = 10 \cos 53° = 6\ \Omega$

$$W_R = 25.4^2 \times 10 = \mathbf{6.452\ W};\ W_Y = 25.4^2 \times 8 = \mathbf{5,162\ W}$$

$$W_B = 25.4^2 \times 6 = \mathbf{3,871\ W}$$

(*ii*) **Phase sequence RBY** [Fig. 19.75]

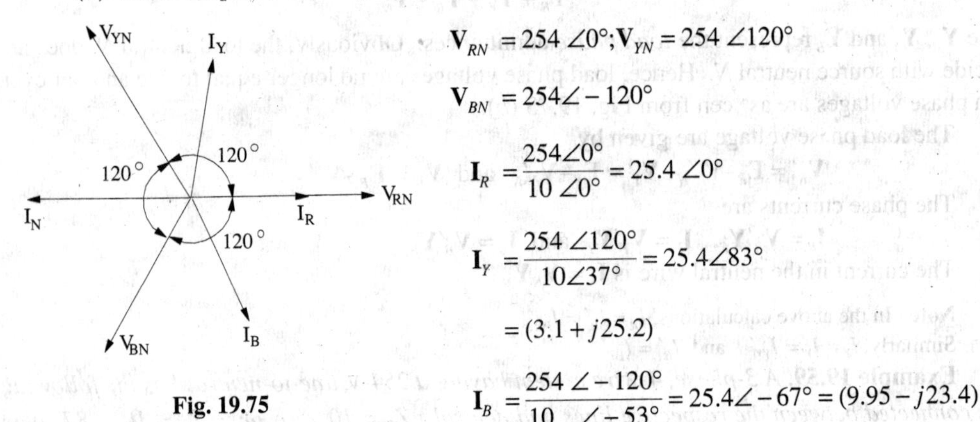

Fig. 19.75

$$\mathbf{V}_{RN} = 254\ \angle 0°;\ \mathbf{V}_{YN} = 254\ \angle 120°$$

$$\mathbf{V}_{BN} = 254\angle -120°$$

$$\mathbf{I}_R = \frac{254\angle 0°}{10\ \angle 0°} = 25.4\ \angle 0°$$

$$\mathbf{I}_Y = \frac{254\ \angle 120°}{10\angle 37°} = 25.4\angle 83°$$

$$= (3.1 + j25.2)$$

$$\mathbf{I}_B = \frac{254\ \angle -120°}{10\quad \angle -53°} = 25.4\angle -67° = (9.95 - j23.4)$$

$$\therefore\ \mathbf{I}_N = -(\mathbf{I}_R + \mathbf{I}_Y + \mathbf{I}_B) = -(38.45 + j1.8) = -38.45 - j1.8 = 38.5\angle -177.3°$$

Obviously, power would remain the same because magnitude of branch currents is unaltered. From the above, we conclude that phase reversal in the case of a 4-wire unbalanced load supplied from a balanced voltage system leads to the following changes :

(*i*) it changes the angles of phase currents but not their magnitudes.

(*ii*) however, power remains unchanged.

(*iii*) it changes the magnitude as well as angle of the neutral current I_N.

Example. 19.60. *A 3-φ, 4-wire, 380-V supply is connected to an unbalanced load having phase impedances of : $Z_R = (8 + j6)\ \Omega$, $Z_Y = (8 - j6)\ \Omega$ and $Z_B = 5\Omega$. Impedance of the neutral wire is $Z_N = (0.5 + j1)\ \Omega$.*

Ignoring the impedances of line wires and internal impedances of the e.m.f. sources, find the phase currents and voltages of the load.

Solution. This question will be solved by using Node-pair Voltage method discussed in Art. 19.30. The admittances of the various branches connected between nodes N and N' are

$$\mathbf{Y}_R = 1/\mathbf{Z}_R = 1/(8 + j6) = (0.08 - j0.06)$$
$$\mathbf{Y}_Y = 1/\mathbf{Z}_Y = 1/(8 - j6) = (0.08 + j0.06)$$
$$\mathbf{Y}_B = 1/\mathbf{Z}_B = 1/(5 + j0) = 0.2$$
$$\mathbf{Y}_N = 1/\mathbf{Z}_N = 1/(0.5 + j1) = (0.4 - j0.8)$$

Let $\mathbf{E}_R = (380/\sqrt{3})\angle 0° = 220\angle 0° = 220 + j0$

$$\mathbf{E}_Y = 220\angle -120° = 220(-0.5 - j0.866) = -110 - j190$$

$$\mathbf{E}_B = 220\angle 120° = 220(-0.5 + j0.866) = -110 + j190$$

The node voltage between N' and N is given by

$$\mathbf{V}_N'N = \frac{\mathbf{E}_R\mathbf{Y}_R + \mathbf{E}_Y\mathbf{Y}_Y + \mathbf{E}_B\mathbf{Y}_B}{\mathbf{Y}_R + \mathbf{Y}_Y + \mathbf{Y}_B + \mathbf{Y}_N}$$

$$= \frac{200(0.08 - j0.06) + (-110 - j190)(0.08 + j0.06) + (-110 + j190) \times 0.2}{(0.08 - j0.06) + (0.08 + j0.06) + 0.2 + (0.4 - j0.8)}$$

$$= \frac{-1.8 + j3}{0.76 - j0.8} = -3.41 + j0.76$$

The three load phase voltages are as under

$\mathbf{V}_R' = \mathbf{E}_R - \mathbf{V}_{N'N} = 220 + 3.41 - j0.76 = 223.41 - j0.76$

$\mathbf{V}_Y' = \mathbf{E}_Y - \mathbf{V}_{N'N} = (-110 - j90) + 3.41 - j0.76 = -106.59 - j190.76$

$\mathbf{V}_B' = \mathbf{E}_B - \mathbf{V}_{N'N} = (-110 + j190) + 3.41 - j0.76 = -106.59 + j189.24$

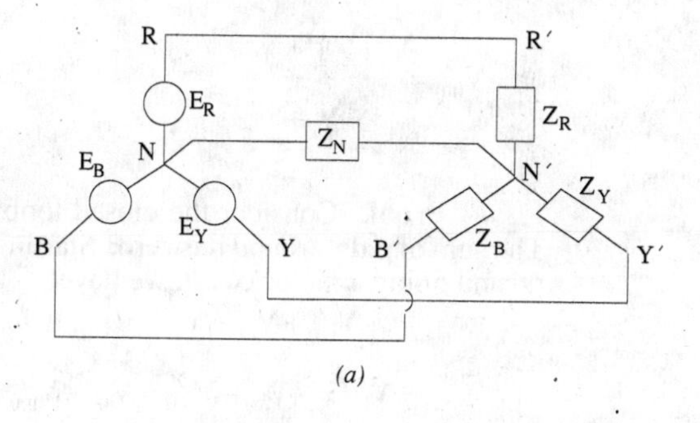

(a)

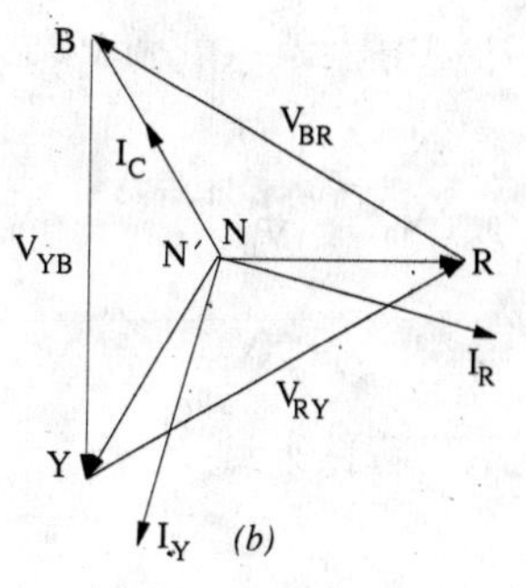

(b)

Fig. 19.76

The phase currents are as under :

$\mathbf{I}_R' = \mathbf{V}_R'\mathbf{Y}_R = (223.41 - j0.76)(0.08 - j0.06) = 17.83 - j13.1 = 22.1 \angle -36.3°$ A

$\mathbf{I}_Y' = \mathbf{V}_Y'\mathbf{Y}_Y = (-106.59 - j190.76)(0.08 + j0.06) = 2.92 - j21.66 = 21.86 \angle 82.4$ A

$\mathbf{I}_B' = \mathbf{V}_B'\mathbf{Y}_B = (-106.59 + j190.76) \times 0.2 = -21.33 + j37.85 = 43.45 \angle 119.4°$ A

$\mathbf{I}_N = \mathbf{V}_N\mathbf{Y}_N = (-3.41 + j0.76)(0.4 - j0.8) = -0.76 + j3.03 = 3.12\angle 104.1°$ A

These voltage and currents are shown in the phasor diagram of Fig. 19.76 (*b*) where displacement of the neutral point has not been shown due to low value of $\mathbf{V}_{N'N}$.

Note. It can be shown that $I_N = I_{R'} + I_{Y'} + I_B'$

19.31. Unbalanced Y-connected Load Without Neutral

When a star-connected load is unbalanced and it has no neutral wire, then its star point is isolated from the star point of the generator. The potential of the load star-point is different from that of the generator star-point. The potential of the former is subject to variations according to the imbalance of the load and under certain conditions of loading, the potentials of the two star - point may differ considerably. Such an isolated load star-point or neutral point is called a 'floating' neutral point because its potential is always changing and is not fixed.

All *Y*-connected unbalanced loads supplied from polyphase systems have floating neutral points without a neutral wire. Any unbalancing of the load causes variations not only of the potential of star-point but also of the voltages across the different branches of the load. *Hence, in that case, phase voltage of the load is not* $1/\sqrt{3}$ *of the line voltage.* In general, the phase voltages are not only unequal in magnitude but also subtend angles other than 120° with one another.

There are many methods to tackle such unbalanced *Y*-connected loads having isolated neutral points.

1. By converting the *Y*-connected load to an equivalent unbalanced Δ-connected load by using *Y*-Δ conversion theorem. The equivalent Δ-connection can be solved as shown in Fig. 19.80. The line currents so calculated are equal in magnitude and phase to those taken by the original unbalanced *Y*-connected load.
2. By applying Kirchhoff's Laws.
3. By applying Millman's Theorem.
4. By using Maxwell's Mesh or Loop Current Method.

19.32. Millman's Theorem

Fig. 19.77 shows a number of linear bilateral admittances, $\mathbf{Y}_1, \mathbf{Y}_2, \ldots \mathbf{Y}_n$ connected to a common point or node *O′*. The voltages of the free ends of these admittances with respect to another common point *O* are $\mathbf{V}_{10}, \mathbf{V}_{20} \ldots \mathbf{V}_{no}$. Then, according to this theorem, the voltage of *O′* with respect to *O* is

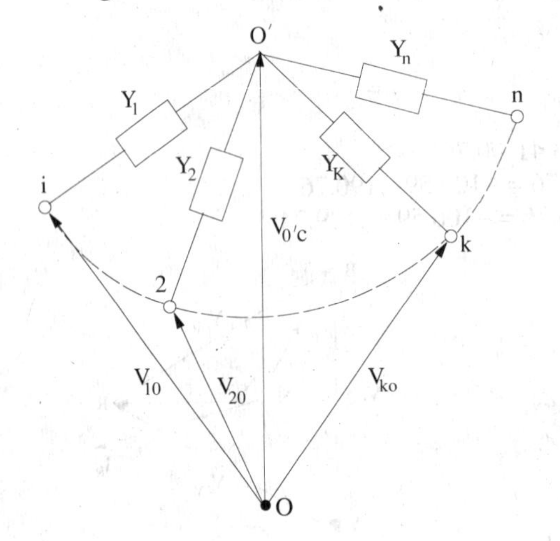

Fig. 19.77

given by $\mathbf{V}_{0'0} = \dfrac{\mathbf{V}_{10}\mathbf{Y}_1 + \mathbf{V}_{20}\mathbf{Y}_2 + \mathbf{V}_{30}\mathbf{Y}_3 + ..\mathbf{V}_{no}\mathbf{Y}_n}{\mathbf{Y}_1 + \mathbf{Y}_2 + \mathbf{Y}_3 + \ldots\mathbf{Y}_n}$

or $\mathbf{V}_{0'0} = \dfrac{\sum_{k=1}^{n} \mathbf{V}_{k0}\mathbf{Y}_k}{\sum_{k=1}^{n} \mathbf{Y}_k}$

Proof. Consider the closed loop *0′Ok*. The sum of p.ds. around it is zero. Starting from *O′* and going anticlockwise, we have

$$\mathbf{V}_{0'0} + \mathbf{V}_{ok} + \mathbf{V}_{ko}' = 0$$

$$\therefore \quad \mathbf{V}_{ko}' = -\mathbf{V}_{ok} - \mathbf{V}_{0'0} = \mathbf{V}_{ko} - \mathbf{V}_{0'0}$$

Current through $\mathbf{Y}_k$ is $\mathbf{I}_{ko}' = \mathbf{V}_{ko}'\,\mathbf{Y}_k = (\mathbf{V}_{ko} - \mathbf{V}_{0'0})\,\mathbf{Y}_k$

By Kirchhoff's Current Law, sum of currents meeting at point *0′* is zero.

$$\therefore \quad \mathbf{I}_{10}' + \mathbf{I}_{20}' + \ldots\ldots\mathbf{I}_{ko}'\ldots\ldots + \mathbf{I}_{no}' = 0$$

$$(\mathbf{V}_{10} - \mathbf{V}_{0'0})\mathbf{Y}_1 + (\mathbf{V}_{20} - \mathbf{V}_{0'0})\,\mathbf{Y}_2 + \ldots\ldots\ (\mathbf{V}_{K0} - \mathbf{V}_{0'0})\,\mathbf{Y}_K + \ldots. = 0$$

$$\text{or} \quad \mathbf{V}_{10}\mathbf{Y}_1 + \mathbf{V}_{20}\,\mathbf{Y}_2 + \ldots\ldots + \mathbf{V}_{K0}\mathbf{Y}_k + \ldots\ldots = \mathbf{V}_{0'0}(\mathbf{Y}_1 + \mathbf{Y}_2 + \ldots. + \mathbf{Y}_K + \ldots)$$

$$\therefore \quad V_{0'0} = \frac{\mathbf{V}_{10}\mathbf{Y}_1 + \mathbf{V}_{20}\mathbf{Y}_2 + \ldots.}{\mathbf{Y}_1 + \mathbf{Y}_2 + \ldots.}$$

19.33. Application of Kirchhoff's Laws

Consider the unbalanced *Y*-connected load of Fig. 19.78. Since the common point of the three load impedances is not at the potential of the neutral, it is marked 0′ instead of *N**. Let us assume

* For the sake of avoiding printing difficulties, we will take the load star point as 0 instead of 0′ for this article.

the phase sequence $\mathbf{V}_{ab}$, $\mathbf{V}_{bc}$, $\mathbf{V}_{ca}$ *i.e.* $\mathbf{V}_{ab}$ leads $\mathbf{V}_{bc}$ and $\mathbf{V}_{bc}$ leads $\mathbf{V}_{ca}$. Let the three branch impedance be $\mathbf{Z}_{ca}$, $\mathbf{Z}_{ab}$ and $\mathbf{Z}_{oc}$. However, since double subscript notation is not necessary for a *Y*-connected impedances in order to indicate to which phase it belongs, single-subscript notation may be used with advantage. Therefore $\mathbf{Z}_{oa}$, $\mathbf{Z}_{ob}$, $\mathbf{Z}_{oc}$ can be written as $\mathbf{Z}_a$, $\mathbf{Z}_b$, $\mathbf{Z}_c$ respectively. It may be pointed out that double-subscript notation is essential for mesh-connected impedances in order to make them definite.

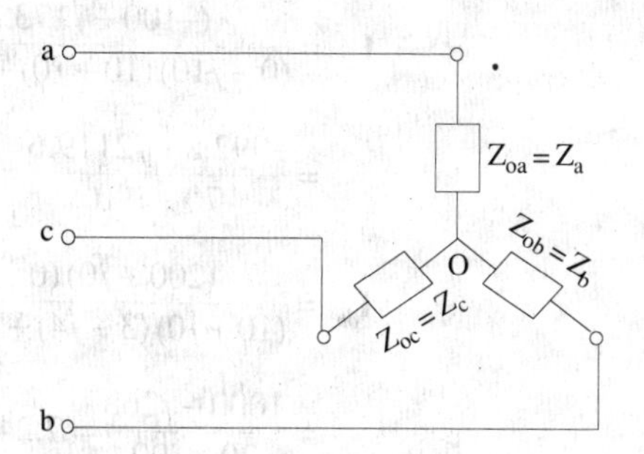

Fig. 19.78

From Kirchhoff's laws, we obtain

$$\mathbf{V}_{ab} = \mathbf{I}_{ao}\mathbf{Z}_a + \mathbf{I}_{ob}\mathbf{Z}_b \quad \ldots(1)$$

$$\mathbf{V}_{bc} = \mathbf{I}_{bo}\mathbf{Z}_b + \mathbf{I}_{oc}\mathbf{Z}_c \quad \ldots(2)$$

$$\mathbf{V}_{ca} = \mathbf{I}_{co}\mathbf{Z}_C + \mathbf{I}_{oa}\mathbf{Z}_a \quad \ldots(3)$$

and $\mathbf{I}_{ao} + \mathbf{I}_{bo} + \mathbf{I}_{co} = 0$ –point law ...(4)

Equations (1), (3) and (4) can be used for finding $\mathbf{I}_{bo}$.

Adding (1) and (3), we get

$$\mathbf{V}_{ab} + \mathbf{V}_{ca} = \mathbf{I}_{ao}\mathbf{Z}_a + \mathbf{I}_{ob}\mathbf{Z}_b + \mathbf{I}_{co}\mathbf{Z}_c + \mathbf{I}_{oc}\mathbf{Z}_a$$

$$= \mathbf{I}_{ob}\mathbf{Z}_b + \mathbf{I}_{co}\mathbf{Z}_c + \mathbf{I}_{oa}\mathbf{Z}_a - \mathbf{I}_{oa}\mathbf{Z}_a = \mathbf{I}_{ob}\mathbf{Z}_b + \mathbf{I}_{CO}\mathbf{Z}_C \quad \ldots(5)$$

Substituting $\mathbf{I}_{oa}$ from equation (4) in equation (3), we get

$$\mathbf{V}_{ca} = \mathbf{I}_{co}\mathbf{Z}_c + (\mathbf{I}_{bo} + \mathbf{I}_{co})\mathbf{Z}_a = \mathbf{I}_{co}(\mathbf{Z}_c + \mathbf{Z}_a) + \mathbf{I}_{bo}\mathbf{Z}_a \quad \ldots(6)$$

Putting the value of $\mathbf{I}_{co}$ from equation (5) in equation (6), we have

$$\mathbf{V}_{ca} = (\mathbf{Z}_c + \mathbf{Z}_a)\frac{(\mathbf{V}_{ab} + \mathbf{V}_{ca}) - \mathbf{I}_{ob}\mathbf{Z}_b}{\mathbf{Z}_c} + \mathbf{I}_{bo}\mathbf{Z}_a$$

$$\mathbf{V}_{ca}\mathbf{Z}_c = -\mathbf{I}_{ob}\mathbf{Z}_b\mathbf{Z}_c - \mathbf{I}_{ob}\mathbf{Z}_b\mathbf{Z}_a + \mathbf{I}_{bo}\mathbf{Z}_a\mathbf{Z}_c + \mathbf{V}_{ab}\mathbf{Z}_c + \mathbf{V}_{ca}\mathbf{Z}_c + \mathbf{V}_{ab}\mathbf{Z}_a + \mathbf{V}_{ca}\mathbf{Z}_a$$

$$\therefore \quad \mathbf{I}_{ob} = \frac{(\mathbf{V}_{ab} + \mathbf{V}_{ca})\mathbf{Z}_a + \mathbf{V}_{ab}\mathbf{Z}_a}{\mathbf{Z}_a\mathbf{Z}_b + \mathbf{Z}_b\mathbf{Z}_c + \mathbf{Z}_c\mathbf{Z}_a} \quad \text{Since } \mathbf{V}_{ab} + \mathbf{V}_{bc} + \mathbf{V}_{ca} = 0 \therefore \mathbf{I}_{ob} = \frac{\mathbf{V}_{ab}\mathbf{Z}_c - \mathbf{V}_{bc}\mathbf{Z}_a}{\mathbf{Z}_a\mathbf{Z}_b + \mathbf{Z}_b\mathbf{Z}_c + \mathbf{Z}_c\mathbf{Z}_a} \quad \ldots(7)$$

From the symmetry of the above equation, the expressions for the other branch currents are,

$$\mathbf{I}_{oc} = \frac{\mathbf{V}_{bc}\mathbf{Z}_a - \mathbf{V}_{ca}\mathbf{Z}_b}{\mathbf{Z}_a\mathbf{Z}_b + \mathbf{Z}_b\mathbf{Z}_c + \mathbf{Z}_c\mathbf{Z}_a} \quad \ldots(8) \quad \mathbf{I}_{oa} = \frac{\mathbf{V}_{ca}\mathbf{Z}_b - \mathbf{V}_{ab}\mathbf{Z}_c}{\mathbf{Z}_a\mathbf{Z}_b + \mathbf{Z}_b\mathbf{Z}_c + \mathbf{Z}_c\mathbf{Z}_a} \quad \ldots(9)$$

Note. Obviously, the three line currents can be written as

$$\mathbf{I}_{ao} = -\mathbf{I}_{oa} = \frac{\mathbf{V}_{ab}\mathbf{Z}_c - \mathbf{V}_{ca}\mathbf{Z}_b}{\Sigma\mathbf{Z}_a\mathbf{Z}_b} \quad \ldots(10) \quad \mathbf{I}_{bo} = -\mathbf{I}_{ob} = \frac{\mathbf{V}_{bc}\mathbf{Z}_a - \mathbf{V}_{ab}\mathbf{Z}_c}{\Sigma\mathbf{Z}_a\mathbf{Z}_b} \quad \ldots(11) \quad \mathbf{I}_{co} = -\mathbf{I}_{oc} = \frac{\mathbf{V}_{ca}\mathbf{Z}_b - \mathbf{V}_{bc}\mathbf{Z}_a}{\Sigma\mathbf{Z}_a\mathbf{Z}_b} \quad \ldots(12)$$

Example 19.61. *If in the unbalanced Y-connected load of Fig. 19.78, Z_a = (10 + j0), Z_b = (3 + j4) and Z_c = (0 –j10) and the load is put across a 3-phase, 200-V circuit with balanced voltages, find the three line currents and voltages across each branch impedance. Assume phase sequence of V_{ab}, V_{bc}. V_{ca}.*

Solution. Take $\mathbf{V}_{ab}$ along the axis of reference. The vector expressions for the three voltages are

$$\mathbf{V}_{ab} = 200 + j0$$

$$\mathbf{V}_{bc} = 200\left(-\frac{1}{2} - j\frac{\sqrt{3}}{2}\right) = -100 - j173.2; \; \mathbf{V}_{ca} = 200\left(-\frac{1}{2} + j\frac{\sqrt{3}}{2}\right) = -100 + j173.2$$

From equation (9) given above

$$\mathbf{I}_{oa} = \frac{(-100+j173.2)(3+j4)-(200+j0)(0-j10)}{(0-j10)(10+j0)+(10+j0)(3+j4)+(3+j4)(0-j10)}$$

$$= \frac{-992.8+j2119.6}{70-j90} = -20.02+j4.54$$

$$\mathbf{I}_{ob} = \frac{(200+j0)(0-j10)-(-100-j173.2)(10+j0)}{(10+j0)(3+j4)+(3+j4)(0-j10)+(0-j10)(10+j0)}$$

$$= \frac{1000-j268}{70-j90} = 7.24+j5.48$$

Now, $\mathbf{I}_{oc}$ may also be calculated in the same way or it can be found easily from equation (4) of Art. 19.33.

$$\mathbf{I}_{oc} = \mathbf{I}_{ao}+\mathbf{I}_{bo} = -\mathbf{I}_{oa}-\mathbf{I}_{ob} = 20.02-j4.54-7.24-j5.48 = 12.78-j10.02$$

Now $\quad \mathbf{V}_{oa} = \mathbf{I}_{oa}\mathbf{Z}_a = (-20.02+j4.54)(10+j0) = 200.2+j45.4$

$$\mathbf{V}_{ob} = \mathbf{I}_{ob}\mathbf{Z}_b = (7.24+j5.48)(3+j4) = -0.2+j45.4$$

$$\mathbf{V}_{oc} = \mathbf{I}_{oc}\mathbf{Z}_c = (12.78-j10.02)(0.-j10) = -100.2-j127.8$$

As a check, we may combine $\mathbf{V}_{oa}$, $\mathbf{V}_{ob}$ and $\mathbf{V}_{oc}$ to get line voltages which should be equal to the applied line voltages. In passing from *a* to *b* through the circuit *internally,* we find that we are in opposition to $\mathbf{V}_{oa}$ but in the same direction as the positive direction of $\mathbf{V}_{ob}$.

$$\mathbf{V}_{ab} = \mathbf{V}_{ao}+\mathbf{V}_{ob} = -\mathbf{V}_{oa}+\mathbf{V}_{ob} = -(-200.2+j45.4)+(-0.2+j45.4) = 200+j0$$

$$\mathbf{V}_{bc} = \mathbf{V}_{bo}+\mathbf{V}_{oc} = -\mathbf{V}_{ob}+\mathbf{V}_{oc} = -(-0.2+j45.4)+(-100.2-j127.8) = -100-j173.2$$

$$\mathbf{V}_{ca} = \mathbf{V}_{co}+\mathbf{V}_{oa} = -\mathbf{V}_{oc}+\mathbf{V}_{oa} = -(-100.2-j127.8)+(-200.2+j45.4) = -100+j173.2$$

19.34. Delta/Star and Star/Delta Conversions

Let us consider the unbalanced Δ-connected load of Fig. 19.79 (*a*) and *Y*-connected load of Fig. 19.79 (*b*). If the two systems are to be equivalent, then the impedances between corresponding pairs of terminals must be the same.

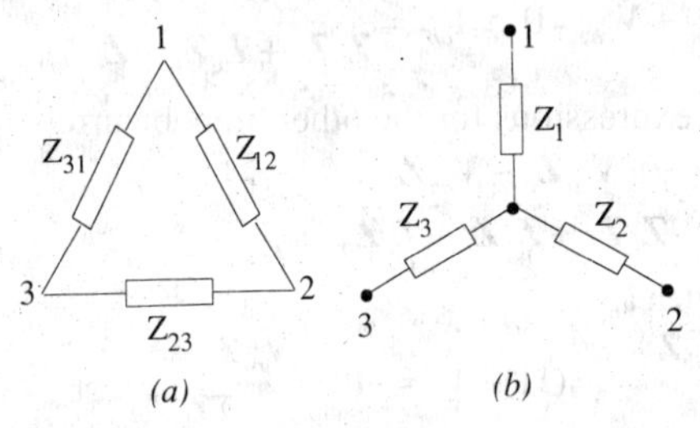

Fig. 19.79

(*i*) Delta/Star Conversion

For *Y*-load, total impedance between terminals 1 and 2 is = $\mathbf{Z}_1 + \mathbf{Z}_2$ (it should be noted that double subscript notation of $\mathbf{Z}_{01}$ and $\mathbf{Z}_{02}$ has been purposely avoided).

Considering terminals 1 and 2 of Δ-load, we find that there are two parallel paths having impedances of $\mathbf{Z}_{12}$ and $(\mathbf{Z}_{31} + \mathbf{Z}_{23})$. Hence, the equivalent impedance between terminals 1 and 2 is given by

$$\frac{1}{\mathbf{Z}} = \frac{1}{\mathbf{Z}_{12}} + \frac{1}{\mathbf{Z}_{23}+\mathbf{Z}_{31}} \quad \text{or} \quad \mathbf{Z} = \frac{\mathbf{Z}_{12}(\mathbf{Z}_{23}+\mathbf{Z}_{31})}{\mathbf{Z}_{12}+\mathbf{Z}_{23}+\mathbf{Z}_{31}}$$

Therefore, for equivalence between the two systems $\mathbf{Z}_1+\mathbf{Z}_2 = \dfrac{\mathbf{Z}_{12}(\mathbf{Z}_{23}+\mathbf{Z}_{31})}{\mathbf{Z}_{12}+\mathbf{Z}_{23}+\mathbf{Z}_{31}}$ ...(1)

Similarly $\quad \mathbf{Z}_2+\mathbf{Z}_3 = \dfrac{\mathbf{Z}_{23}(\mathbf{Z}_{31}+\mathbf{Z}_{12})}{\mathbf{Z}_{12}+\mathbf{Z}_{23}+\mathbf{Z}_{31}}$ ...(2) $\quad \mathbf{Z}_3+\mathbf{Z}_1 = \dfrac{\mathbf{Z}_{31}(\mathbf{Z}_{12}+\mathbf{Z}_{23})}{\mathbf{Z}_{12}+\mathbf{Z}_{23}+\mathbf{Z}_{31}}$ ...(3)

Adding equation (3) to (1) and substracting equation (2), we get

$$2\,\mathbf{Z}_1 = \frac{\mathbf{Z}_{12}(\mathbf{Z}_{23}+\mathbf{Z}_{31})+\mathbf{Z}_{31}(\mathbf{Z}_{12}+\mathbf{Z}_{23})-\mathbf{Z}_{23}(\mathbf{Z}_{31}+\mathbf{Z}_{12})}{\mathbf{Z}_{12}+\mathbf{Z}_{23}+\mathbf{Z}_{31}} = \frac{2\mathbf{Z}_{12}\mathbf{Z}_{31}}{\mathbf{Z}_{12}+\mathbf{Z}_{23}+\mathbf{Z}_{31}}$$

$$\therefore \qquad \mathbf{Z}_1 = \frac{\mathbf{Z}_{12}\mathbf{Z}_{31}}{\mathbf{Z}_{12}+\mathbf{Z}_{23}+\mathbf{Z}_{31}} \qquad \ldots(4)$$

The other two results may be written down by changing the subscripts cyclically

$$\therefore \qquad \mathbf{Z}_2 = \frac{\mathbf{Z}_{23}\mathbf{Z}_{12}}{\mathbf{Z}_{12}+\mathbf{Z}_{23}+\mathbf{Z}_{31}}; \quad \ldots(5) \quad \mathbf{Z}_3 = \frac{\mathbf{Z}_{31}\mathbf{Z}_{23}}{\mathbf{Z}_{12}+\mathbf{Z}_{23}+\mathbf{Z}_{31}} \qquad \ldots(6)$$

The above expression can be easily obtained by remembering that (Art. 2.19)

$$\text{Star } \mathbf{Z} = \frac{\text{Product of } \Delta\ \mathbf{Z}\text{'s connected to the same terminals}}{\text{Sum of } \Delta\ \mathbf{Z}\text{'s}}$$

It should be noted that all **Z**' are to be expressed in their complex form.

(*iii*) **Star/Delta Conversion**

The equations for this conversion can be obtained by rearranging equations (4), (5) and (6). Rewriting these equations, we get

$$\mathbf{Z}_1(\mathbf{Z}_{12}+\mathbf{Z}_{23}+\mathbf{Z}_{31}) = \mathbf{Z}_{12}\mathbf{Z}_{31} \qquad \ldots(7)$$

$$\mathbf{Z}_2(\mathbf{Z}_{12}+\mathbf{Z}_{23}+\mathbf{Z}_{31}) = \mathbf{Z}_{23}\mathbf{Z}_{12} \qquad \ldots(8)$$

$$\mathbf{Z}_3(\mathbf{Z}_{12}+\mathbf{Z}_{23}+\mathbf{Z}_{31}) = \mathbf{Z}_{31}\mathbf{Z}_{23} \qquad \ldots(9)$$

Dividing equation (7) by (9), we get $\dfrac{\mathbf{Z}_1}{\mathbf{Z}_3} = \dfrac{\mathbf{Z}_{12}}{\mathbf{Z}_{23}} \quad \therefore \quad \mathbf{Z}_{23} = \mathbf{Z}_{12}\dfrac{\mathbf{Z}_3}{\mathbf{Z}_1}$

Dividing equation (8) by (9), we get $\dfrac{\mathbf{Z}_2}{\mathbf{Z}_3} = \dfrac{\mathbf{Z}_{12}}{\mathbf{Z}_{31}} \quad \therefore \quad \mathbf{Z}_{31} = \mathbf{Z}_{12}\dfrac{\mathbf{Z}_3}{\mathbf{Z}_2}$

Substituting these values in equation (7), we have $\mathbf{Z}_1\left(\mathbf{Z}_{12}+\mathbf{Z}_{12}\dfrac{\mathbf{Z}_3}{\mathbf{Z}_1}+\mathbf{Z}_{31}\right) = \mathbf{Z}_{12}\cdot\mathbf{Z}_{12}\dfrac{\mathbf{Z}_3}{\mathbf{Z}_2}$

or $$\mathbf{Z}_1\mathbf{Z}_{12}\left(1+\frac{\mathbf{Z}_3}{\mathbf{Z}_1}+\frac{\mathbf{Z}_{31}}{\mathbf{Z}_{12}}\right) = \mathbf{Z}_{12}\mathbf{Z}_{12}\frac{\mathbf{Z}_3}{\mathbf{Z}_2} \quad \therefore \quad \mathbf{Z}_1\mathbf{Z}_2+\mathbf{Z}_2\mathbf{Z}_3+\mathbf{Z}_3\mathbf{Z}_1 = \mathbf{Z}_{12}\times\mathbf{Z}_3$$

$$\mathbf{Z}_{12} = \frac{\mathbf{Z}_1\mathbf{Z}_2+\mathbf{Z}_2\mathbf{Z}_3+\mathbf{Z}_3\mathbf{Z}_1}{\mathbf{Z}_3} \quad \text{or} \quad \mathbf{Z}_{12} = \mathbf{Z}_1+\mathbf{Z}_2+\frac{\mathbf{Z}_1\mathbf{Z}_2}{\mathbf{Z}_3}$$

Similarly, $$\mathbf{Z}_{23} = \mathbf{Z}_2+\mathbf{Z}_3+\frac{\mathbf{Z}_2\mathbf{Z}_3}{\mathbf{Z}_1} = \frac{\mathbf{Z}_1\mathbf{Z}_2+\mathbf{Z}_2\mathbf{Z}_3+\mathbf{Z}_3\mathbf{Z}_1}{\mathbf{Z}_1}$$

$$\mathbf{Z}_{31} = \mathbf{Z}_3+\mathbf{Z}_1+\frac{\mathbf{Z}_3\mathbf{Z}_1}{\mathbf{Z}_2} = \frac{\mathbf{Z}_1\mathbf{Z}_2+\mathbf{Z}_2\mathbf{Z}_3+\mathbf{Z}_3\mathbf{Z}_1}{\mathbf{Z}_2}$$

As in the previous case, it is to be noted that all impedances must be expressed in their complex form.

Another point wor noting is that *the line currents of this equivalent delta are the currents in the phases of the Y-connected load.*

Example 19.62. *An unbalanced star-connected load has branch impedances of* $Z_1 = 10\angle 30°\ \Omega$, $Z_2 = 10\angle{-45°}\ \Omega$, $Z_3 = 20\angle 60°\ \Omega$ *and is connected across a balanced 3-phase, 3-wire supply of 200 V. Find the line currents and the voltage across each impedance using Y/Δ conversion method.*

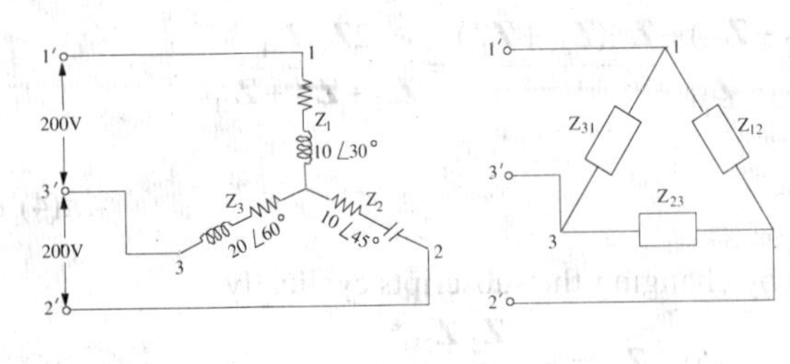

Fig. 19.80

Solution. The unbalanced *Y*-connected load and its equivalent Δ-connected load are shown in Fig. 19.80.

Now $\mathbf{Z}_1\mathbf{Z}_2 + \mathbf{Z}_2\mathbf{Z}_3 + \mathbf{Z}_3\mathbf{Z}_1$ = (10 ∠ 30°) (10 ∠ − 45°) + (10 ∠ −45°) (20 ∠ 60°) + (20 ∠ 60°)(10 ∠30°) = 100 ∠ −15° + 200 ∠ 15° + 200 ∠ 90°

Converting these into their cartesian form, we get

$$= 100[\cos(-15°) - j\sin 15°] + 200(\cos 15° + j\sin 15°) + 200(\cos 90° + j\sin 90°)$$

$$= 96.6 - j25.9 + 193.2 + j51.8 + 0 + j200 = 289.8 + j225.9 = 368\angle 38°$$

$$\mathbf{Z}_{12} = \frac{\mathbf{Z}_1\mathbf{Z}_2 + \mathbf{Z}_2\mathbf{Z}_3 + \mathbf{Z}_3\mathbf{Z}_1}{\mathbf{Z}_3} = \frac{368\angle 38°}{20\angle 60°} = 18.4\angle -22° = 17.0 - j6.9$$

$$\mathbf{Z}_{23} = \frac{368\angle 38°}{10\angle -30°} = 36.8\angle 8° = 36.4 + j5.1$$

$$\mathbf{Z}_{31} = \frac{368\angle 38°}{10\angle -45°} = 36.8\angle 83° = 4.49 + j36.5$$

Assuming clockwise phase sequence of voltages $\mathbf{V}_{12}$, $\mathbf{Z}_{23}$ and $\mathbf{V}_{31}$, we have

$$\mathbf{V}_{12} = 200\angle 0°, \mathbf{V}_{23} = 200\angle -120°, \mathbf{V}_{31} = 200\angle 120°$$

$$\mathbf{I}_{12} = \frac{\mathbf{V}_{12}}{\mathbf{Z}_{12}} = \frac{200\ \angle 0°}{18.4\ \angle -22°} = 10.86\ \angle 22° = 10.07 + j4.06$$

$$\mathbf{I}_{23} = \frac{\mathbf{V}_{23}}{\mathbf{V}_{23}} = \frac{200\angle -120°}{36.8\angle 8°} = 5.44\angle -128° = -3.35 - j4.29$$

$$\mathbf{I}_{31} = \frac{\mathbf{V}_{31}}{\mathbf{Z}_{31}} = \frac{200\angle 120°}{36.8\angle 83°} = 5.44\angle 37° = 4.34 + j3.2$$

Line current = $\mathbf{I}'_1 = \mathbf{I}_{12} + \mathbf{I}_{13} = \mathbf{I}_{12} - \mathbf{I}_{31}$

= (10.07 + *j*4.06) −(4.34 + *j*3.2) = 5.73 + *j*0.86 = 5.76 ∠8° 32′

$\mathbf{I}'_2$ = $\mathbf{I}_{23} - \mathbf{I}_{12}$ = (−3.35−*j*4.29) −(10.07 + *j*4.06) = −13.42−*j*8.35 = 15.79 ∠ − 148°6′

$\mathbf{I}'_3$ = $\mathbf{I}_{31} - \mathbf{I}_{23}$ = (4.34 + *j*3.2) −(−3.35−*j*4.29) = 7.69 + *j*7.49 = 10.73∠ ° 44°16′

These are currents in the *phases of the Y-connected unbalanced load.* Let us find voltage drop across each star-connected branch impedance.

Voltage drop across $\mathbf{Z}_1 = \mathbf{V}_{10} = \mathbf{I}'_1\ \mathbf{Z}_1$ = 5.76 ∠8°32′.10 ∠30° = 57.6∠38°32′

Voltage drop across $\mathbf{Z}_2 = \mathbf{V}_{20} = \mathbf{I}'_2\mathbf{Z}_2$ = 15.79 ∠ − 148°6.10 ∠ −45° = 157.9∠−193°6′

Voltage drop across $\mathbf{Z}_3 = \mathbf{V}_{30} = \mathbf{I}'_3\mathbf{Z}_3$ = 10.73 ∠ 44°16′.20 ° ∠60° = 214.6 ∠ 104°16′

Example 19.63. *A 300-V (line) 3-phase supply feeds a star-connected load consisting of non-inductive resistors of 15, 6 and 10 Ω connected to the R, Y and B lines respectively. The phase sequence is RYB. Calculate the voltage across each resistor.*

Solution. The *Y*-connected unbalanced load and its equivalent Δ-connected load are shown in Fig. 19.81. Using *Y*/Δ conversion method we have

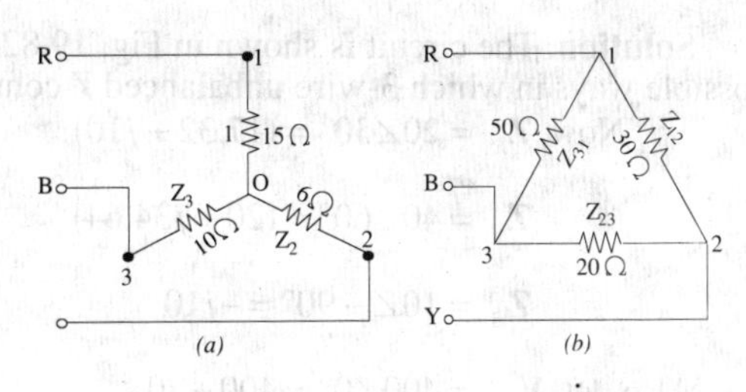

Fig. 19.81

$$\mathbf{Z}_{12} = \frac{\mathbf{Z}_1\mathbf{Z}_2 + \mathbf{Z}_2\mathbf{Z}_3 + \mathbf{Z}_3\mathbf{Z}_1}{\mathbf{Z}_3}$$

$$= \frac{90 + 60 + 150}{10} = 30\ \Omega$$

$$\mathbf{Z}_{23} = 300/15 = 20\ \Omega$$

$$\mathbf{Z}_{31} = 300/6 = 50\ \Omega$$

Phase current $I_{RY} = V_{RY}/Z_{12} = 300/30 = 10$ A

Similarly $I_{YB} = V_{YB}/Z_{23} = 300/20 = 15$ A

$I_{BR} = V_{BR}/Z_{31} = 300/50 = 6$ A

Each current is in phase with its own voltage because the load is purely resistive.

The line currents for the delta connection are obtained by compounding these phase currents in pairs, either trigonometrically or by phasor algebra. Using phasor algebra and choosing $\mathbf{V}_{RY}$ as the reference axis, we get

$\mathbf{I}_{RY} = 10 + j0$; $\mathbf{I}_{YB} = 15(-\frac{1}{2} - j\sqrt{3}/2) = -7.5 - j13.0$; $\mathbf{I}_{BR} = 6(-\frac{1}{2} + j\sqrt{3}/2) = -3.0 + j5.2$

Line currents for delta-connection [Fig. 19.66 (*b*)] are

$\mathbf{I}_R = \mathbf{I}_{RY} + \mathbf{I}_{RB} = \mathbf{I}_{RY} - \mathbf{I}_{BR} = (10 + j0) - (-3 + j5.2) = 13 - j5.2$ or 14 A in magnitude

$\mathbf{I}_Y = \mathbf{I}_{YR} + \mathbf{I}_{YB} = \mathbf{I}_{YB} - \mathbf{I}_{RY} = (-7.5 - j13.0) - (10 + j0) = -17.5 - j13$ or 21.8 A in magnitude

$\mathbf{I}_B = \mathbf{I}_{BR} + \mathbf{I}_{BY} = \mathbf{I}_{BR} - \mathbf{I}_{YB} = (-3.0 + j5.2) - (-7.5 - j13.0) = 4.5 + j18.2 = 18.7$ A magnitude

These line currents for Δ-connection are the phase currents for *Y*-connection. Voltage drop across each limb of *Y*-connected load is

$\mathbf{V}_{RN} = \mathbf{I}_R\mathbf{Z}_l = (13 - j5.2)(15 + j0) = 195 - 78$ volt or 210 **V**

$\mathbf{V}_{YN} = \mathbf{I}_Y\mathbf{Z}_2 = (-17.5 - j13.0)(6 + j0) = -105 - j78$ volt or 131 V

$\mathbf{V}_{BN} = \mathbf{I}_B\mathbf{Z}_3 = (4.5 + j18.2)(10 + j0) = 45 + j182$ volt or 187 V

As a check, it may be verified that the difference of phase voltages taken in pairs should give the three line voltages. Going through the circuit *internally*, we have

$\mathbf{V}_{RY} = \mathbf{V}_{RN} + \mathbf{V}_{NY} = \mathbf{V}_{RN} - \mathbf{V}_{YN} = (195 - j78) - (105 - j78) = 300\angle 0°$

$\mathbf{V}_{YB} = \mathbf{V}_{YN} - \mathbf{V}_{BN} = (-105 - j78) - (45 + j182) = -150 - j260 = 300\angle -120°$

$\mathbf{V}_{BR} = \mathbf{V}_{BN} - \mathbf{V}_{RN} = (45 + j182) - (195 - j78) = -150 + j260 = 300\angle 120°$

This question could have been solved by direct geometrical methods as shown in Ex. 19.50.

Example 19.64. *A Y-connected load is supplied from a 400-V, 3-phase, 3-wire symmetrical system RYB. The branch circuit impedances are*

$$Z_R = 10\sqrt{3} + j10\ ;\ Z_Y = 20 + j20\sqrt{3}\ ;\ Z_B = 0 - j10$$

Determine the current in each branch. Phase sequence is RYB.

(Network Analysis, Nagpur Univ. 1993)

Solution. The circuit is shown in Fig. 19.82. The problem will be solved by using all the four possible ways in which 3-wire unbalanced *Y* connected load can be handled.

Now, $\mathbf{Z}_R = 20\angle 30° = (17.32 + j10)$

$\mathbf{Z}_Y = 40\angle 60° = (20 + j34.64)$

$\mathbf{Z}_B = 10\angle -90° = -j10$

Also, let, $\mathbf{V}_{RY} = 400\angle 0° = 400 + j0$

$\mathbf{V}_{RB} = 400\angle -120° = -200 - j346$

$\mathbf{V}_R = 400\angle 120° = -200 + j346$

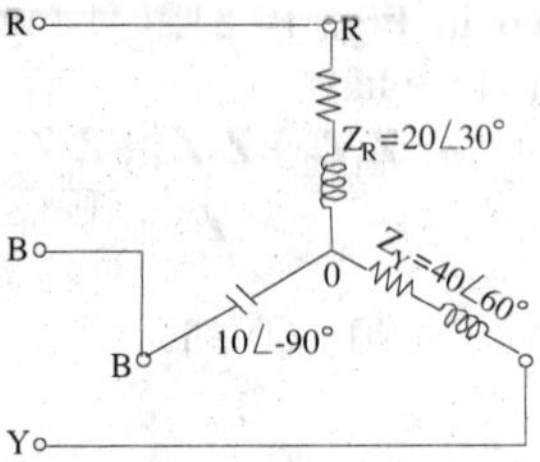

Fig. 19.82

(*a*) **By applying Kirchhoff's Laws**

With reference to Art. 19.33, it is seen that

$$\mathbf{I}_{RO} = \mathbf{I}_R = \frac{\mathbf{V}_{RY}\mathbf{Z}_B - \mathbf{V}_{BR}\mathbf{Z}_Y}{\mathbf{Z}_R\mathbf{Z}_Y + \mathbf{Z}_Y\mathbf{Z}_B + \mathbf{Z}_B\mathbf{Z}_R};\ \mathbf{I}_{YO} = \mathbf{I}_Y = \frac{\mathbf{V}_{YB}\mathbf{Z}_R - \mathbf{V}_{RY}\mathbf{Z}_B}{\mathbf{Z}_R\mathbf{Z}_Y + \mathbf{Z}_Y\mathbf{Z}_B + \mathbf{Z}_B\mathbf{Z}_R};\ \mathbf{I}_{BO} = \mathbf{I}_B = \frac{\mathbf{V}_{BR}\mathbf{Z}_Y - \mathbf{V}_{YB}\mathbf{Z}_R}{\mathbf{Z}_R\mathbf{Z}_Y + \mathbf{Z}_Y\mathbf{Z}_B + \mathbf{Z}_B\mathbf{Z}_R}$$

Now, $\mathbf{Z}_R\mathbf{Z}_Y + \mathbf{Z}_Y\mathbf{Z}_B + \mathbf{Z}_B\mathbf{Z}_R$

$$= 20\angle 30°.40\angle 60° + 40\angle 60°.10\angle -90° + 10\angle -90°.20\angle 30°$$
$$= 800\angle 90° + 400\angle -30° + 200\angle -60° = 446 + j426 = 617\angle 43.7°$$

$$\mathbf{V}_{RY}\mathbf{Z}_B - \mathbf{V}_{BR}\mathbf{Z}_Y = 400 \times 10\angle -90° - 400\angle 120° \cdot 40\angle 60°$$

$$= 16{,}000 - j4000 = 16{,}490\angle -14°3'$$

$$\therefore \quad \mathbf{I}_R = \frac{16{,}490\angle -14°3'}{617\angle 43.7°} = 26.73\angle -57°45'$$

$$\mathbf{V}_{YB}\mathbf{Z}_R - \mathbf{V}_{RY}\mathbf{Z}_B = 400\angle -120°.20\angle 30° - 400.10\angle -90° = -j4000 = 4000\angle -90°$$

$$\mathbf{I}_Y = \frac{4000\angle -90°}{617\angle 43.7°} = 6.48\angle -133.7°$$

$$\mathbf{V}_{BR}\mathbf{Z}_Y - \mathbf{V}_{YB}\mathbf{Z}_R = 400\angle 120°.40\angle 60° - 400\angle -120°20 \cdot \angle 30°$$

$$= -16{,}000 + j8{,}000 = 17{,}890\angle 153°26'$$

$$\therefore \quad \mathbf{I}_B = \frac{17{,}890\angle 153°26'}{617\angle 43.7°} = 29\angle 109°45'$$

(*b*) **By Star/Delta Conversion** (Fig. 19.83)

The given star may be converted into the equivalent delta with the help of equations given in Art. 19.34.

$$\mathbf{Z}_{RY} = \frac{\mathbf{Z}_R\mathbf{Z}_Y + \mathbf{Z}_Y\mathbf{Z}_B + \mathbf{Z}_B\mathbf{Z}_R}{\mathbf{Z}_B} = \frac{617\angle 43.7°}{10\angle -90°} = 61.73\angle 133.7°$$

$$\mathbf{Z}_{YB} = \frac{\Sigma\mathbf{Z}_R\mathbf{Z}_Y}{\mathbf{Z}_R} = \frac{617\angle 43.7°}{20\angle 30°} = 30.87\angle 13.7°$$

$$\mathbf{Z}_{BR} = \frac{\Sigma\mathbf{Z}_R\mathbf{Z}_Y}{\mathbf{Z}_Y} = \frac{617\angle 43.7°}{40\angle 60°} = 15.43\angle -16.3°$$

$$\mathbf{I}_{RY} = \frac{\mathbf{V}_{RY}}{\mathbf{Z}_{RY}} = \frac{400}{61.73\angle 133.7°} = 6.48\angle -133.7° = (-4.47 - j4.68)$$

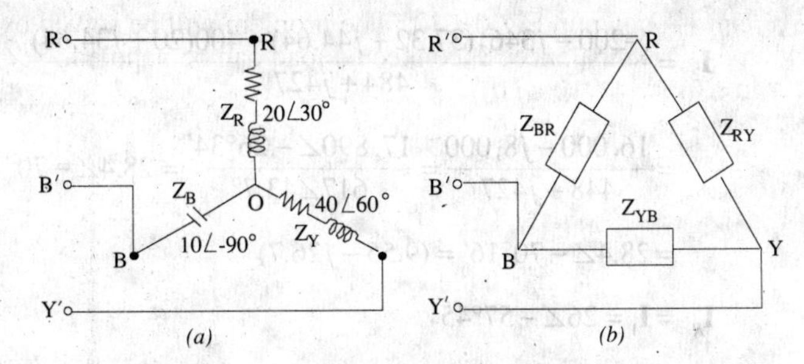

Fig. 19.83

$$\mathbf{I}_{YB} = \frac{\mathbf{V}_{YB}}{\mathbf{Z}_{YB}} = \frac{400\angle-120°}{30.87\angle13.7°} = 12.95\angle-133.7° = (-8.95 - j9.35)$$

$$\mathbf{I}_{BR} = \frac{\mathbf{V}_{BR}}{\mathbf{Z}_{BR}} = \frac{400\angle120°}{15.43\angle-16.3°} = 25.9\angle136.3° = (-18.7 + j17.9)$$

$$\mathbf{I}_{R'R} = \mathbf{I}_{RY} - \mathbf{I}_{BR} = 14.23 - j22.58 = 26.7\angle-57°48'$$

$$\mathbf{I}_{Y'Y} = \mathbf{I}_{YB} - \mathbf{I}_{RY} = -4.48 - j4.67 = 6.47\angle-134°6'$$

$$\mathbf{I}_{B'B} = \mathbf{I}_{BR} - \mathbf{I}_{YB} = -9.85 + j27.25 = 29\angle109°48'$$

$$\Sigma\mathbf{I} = (0 + j0) \quad \text{—as a check}$$

As explained in Art. 19.34, these line currents of the equivalent delta represent the phase currents of the star-connected load of Fig. 19.83 (*a*).

Note. Minor differences are due to accumulated errors.

(*c*) **By Using Maxwell's Loop Current Method**

Let the loop or mesh currents be as shown in Fig. 19.84. It may be noted that

$$\mathbf{I}_R = \mathbf{I}_1\,;\ \mathbf{I}_Y = \mathbf{I}_2 - \mathbf{I}_1 \ \text{ and } \ \mathbf{I}_B = -\mathbf{I}_2$$

Considering the drops across R and Y-arms, we get

$$\mathbf{I}_1\mathbf{Z}_R + \mathbf{Z}_Y(\mathbf{I}_1 - \mathbf{I}_2) = V_{RY}$$

or

$$\mathbf{I}_1(\mathbf{Z}_R + \mathbf{Z}_Y) - \mathbf{I}_2\mathbf{Z}_Y = \mathbf{V}_{RY} \qquad \ldots(i)$$

Similarly, considering the legs Y and B, we have

$$\mathbf{Z}_Y(\mathbf{I}_2 - \mathbf{I}_1) + \mathbf{Z}_B\mathbf{I}_2 = V_{YB}$$

or

$$-\mathbf{I}_1\mathbf{Z}_Y + \mathbf{I}_2(\mathbf{Z}_B + \mathbf{Z}_Y) = \mathbf{V}_{YB} \qquad \ldots(ii)$$

Solving for $\mathbf{I}_1$ and $\mathbf{I}_2$, we get

$$\mathbf{I}_1 = \frac{\mathbf{V}_{RY}(\mathbf{Z}_Y + \mathbf{Z}_B) + \mathbf{Z}_{YB}\mathbf{V}_Y}{(\mathbf{Z}_R + \mathbf{Z}_Y)(\mathbf{Z}_Y + \mathbf{Z}_B) - \mathbf{Z}_Y^2};$$

$$\mathbf{I}_2 = \frac{\mathbf{V}_{YB}(\mathbf{Z}_R + \mathbf{Z}_Y) + \mathbf{V}_{RY}\mathbf{Z}_Y}{(\mathbf{Z}_R + \mathbf{Z}_Y)(\mathbf{Z}_Y + \mathbf{Z}_B) - \mathbf{Z}_Y^2}$$

$$\mathbf{I}_1 = \frac{400(20 + j24.64) + 400\angle-120°.40\angle60°}{(37.32 + j44.64)(20 + j24.64) - 1600\angle120°}$$

$$= \frac{16,000 - j4,000}{448 + j427} = \frac{16,490\angle-14°3'}{617\angle43.7°}$$

$$= 26\angle-57°45' = (13.9 - j22)$$

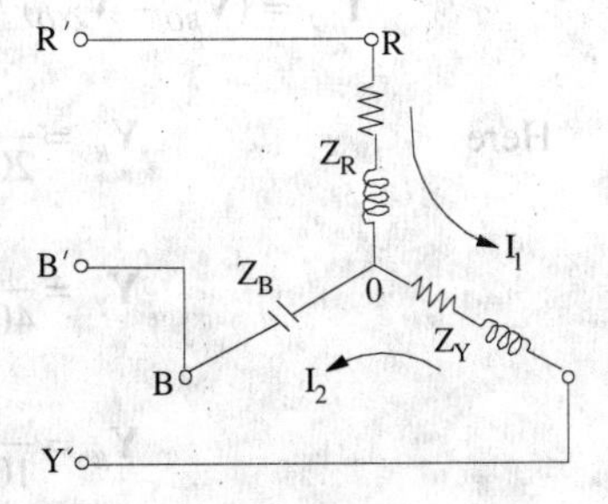

Fig. 19.84

$$\mathbf{I}_2 = \frac{(-200 - j346)(37.32 + j44.64) + 400(20 + j34.64)}{484 + j427}$$

$$= \frac{16{,}000 - j8{,}000}{448 + j427} = \frac{17{,}890\angle -26°34'}{617\angle 43.7°} = 28.4\angle -70°16'$$

$$= 28.4\angle -70°16' = (9.55 - j26.7)$$

$\therefore$
$$\mathbf{I}_R = \mathbf{I}_1 = 26\angle -57°45'$$

$$\mathbf{I}_Y = \mathbf{I}_2 - \mathbf{I}_1 = (9.55 - j26.7) - (13.9 - j22) = -4.35 - j4.7 = 6.5\angle -134°$$

$$\mathbf{I}_B = -\mathbf{I}_2 = -28.4\angle -70°16' = 28.4\angle 109°44'$$

(*d*) **By Using Millman's Theorem**

According to this theorem, the voltage of the *load* star point O' with respect to the star point or neutral O of the generator or supply (normally zero potential) is given by

$$\mathbf{V}_{O'O} = \frac{\mathbf{V}_{RO}\mathbf{Y}_R + \mathbf{V}_{YO}\mathbf{Y}_Y + \mathbf{V}_{BO}\mathbf{Y}_B}{\mathbf{Y}_R + \mathbf{Y}_Y + \mathbf{Y}_B}$$

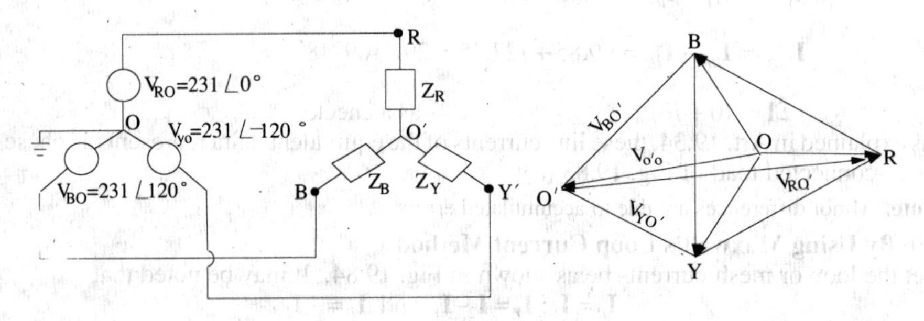

Fig. 19.85

where $\mathbf{V}_{RO}$, $\mathbf{V}_{YO}$ and $\mathbf{V}_{BO}$ are the phase voltages of the generator or 3-phase supply.

As seen from Fig. 19.85, voltage across each phase of the load is*

$\mathbf{V}_{RO'} = \mathbf{V}_{RO} - \mathbf{V}_{O'O}$ $\quad \mathbf{V}_{YO'} = \mathbf{V}_{YO} - \mathbf{V}_{O'O}$ $\quad \mathbf{V}_{BO'} = \mathbf{V}_{BO} - \mathbf{V}_{O'O}$

Obviously, $\mathbf{I}_{RO'} = (\mathbf{V}_{RO} - \mathbf{V}_{O'O})\,\mathbf{Y}_R$; $\mathbf{I}_{YO'} = (\mathbf{V}_{YO} - \mathbf{V}_{O'O})\,\mathbf{Y}_Y$ and

$$\mathbf{I}_{BO'} = (\mathbf{V}_{BO} - \mathbf{V}_{O'O})\,\mathbf{Y}_B$$

Here
$$\mathbf{Y}_R = \frac{1}{20\angle 30°} = 0.05\angle -30° = (0.0433 - j0.025)$$

$$\mathbf{Y}_Y = \frac{1}{40\angle 60°} = 0.025\angle -60° = (0.0125 - j0.0217)$$

$$\mathbf{Y}_B = \frac{1}{10\angle -90°} = 0.1\angle 90° = 0 + j0.1$$

---

*Incidentally, it may be noted that the p.d. between load neutral and supply neutral is given by

$$\mathbf{V}_{O'O} = -\frac{\mathbf{V}_{RO'} + \mathbf{V}'_{YO} + \mathbf{V}_{BO'}}{3}$$

$\therefore \qquad \mathbf{Y}_R + \mathbf{Y}_Y + \mathbf{Y}_B = 0.0558 + j0.0533 = 0.077\angle 43.7°$

Let $\qquad \mathbf{V}_{RO} = \frac{400}{\sqrt{3}}\angle 0° \quad = (231 + j0)$

$$\mathbf{V}_{BO} = 231\angle -120° = -115.5 - j200$$

$$\mathbf{V}_{BO} = 231\angle 120° = -115.5 + j200$$

$$\mathbf{V}_{O'O} = \frac{231.\ 0.05\angle -30° + 231\angle -120°.\ 0.025\angle -60° + 231\angle 120°.\ 0.1\angle 90°}{0.077\angle 43.7°}$$

$$= \frac{-15.8 - j17.32}{0.077\angle 43.7°} = \frac{23.5\angle -132.4°}{0.077\angle 43.7°} = 305\angle -176.1° = (-304.5 - j20.8)$$

$$\mathbf{V}_{RO}' = \mathbf{V}_{RO} - \mathbf{V}_{O'O} = 231 - (-304.5 - j20.8) = 535.5 + j20.8 = 536\angle 2.2°$$

$$\mathbf{V}_{YO}' = (-115.5 - j200) - (-304.5 - j20.8) = 189 - j179 = 260\angle -43°27'$$

$$\mathbf{V}_{BO}' = (-115.5 + j200) - (-304.5 - j20.8) = 189 + j221 = 291\angle 49°27'$$

$\therefore \qquad \mathbf{I}_{RO}' = 536\ \angle 2.2° \times 0.05\angle -30° = 26.5\angle -27.8°$

$$\mathbf{I}_{YO}' = 260\angle -43°27' \times 0.025\angle -60° = 6.5\angle -103°27'$$

$$\mathbf{I}_{BO}' = 291\angle 49°27' \times 0.1\angle 90° = 29.1\ \angle 139°27'$$

Note. As seen from above, $\mathbf{V}_{RO}' = \mathbf{V}_{RO}' - \mathbf{V}_{O'O}$

Substituting the value of $\mathbf{V}_{O'O}$, we have

$$\mathbf{V}_{RO}' = \mathbf{V}_{RO} - \left(\frac{\mathbf{V}_{RO}\mathbf{Y}_R + \mathbf{V}_{YO}\mathbf{Y}_Y + \mathbf{V}_{BO}\mathbf{Y}_B}{\mathbf{V}_R + \mathbf{V}_Y + \mathbf{V}_B}\right)$$

$$= \frac{(\mathbf{V}_{RO} - \mathbf{V}_{YO})\mathbf{Y}_Y + (\mathbf{V}_{RO} - \mathbf{V}_{BO})\mathbf{Y}_B}{\mathbf{Y}_R + \mathbf{Y}_Y + \mathbf{Y}_B}$$

$$= \frac{\mathbf{V}_{RY}\mathbf{Y}_Y + \mathbf{V}_{RB}\mathbf{Y}_B}{\mathbf{Y}_R + \mathbf{Y}_Y + \mathbf{Y}_B}$$

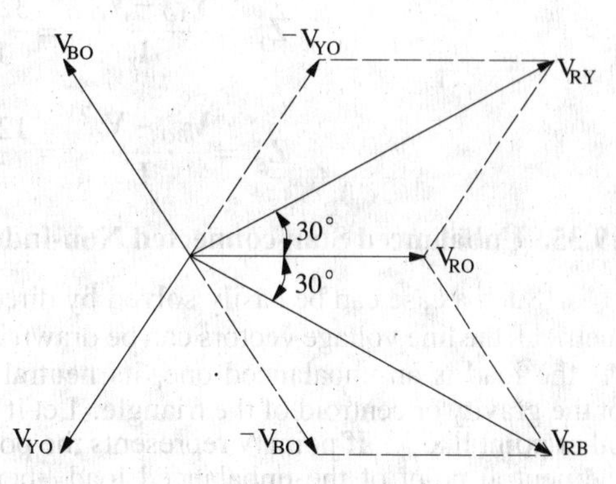

Fig. 19.86

Since $\mathbf{V}_{RO}$ is taken as the reference vector, then as seen from Fig. 19.86.

$\mathbf{V}_{RY} = 400\angle 30°$ and $\mathbf{V}_{RB} = 400\angle -30°$

$\therefore \qquad \mathbf{V}_{RO}' = \frac{400\angle 30° \times 0.025\angle -60° + 400\angle -30° \times 0.1\angle 90°}{0.077\angle 43.7°}$

$$= \frac{28.6 + j29.64}{0.077\angle 43.7°} = \frac{41\angle 46°}{0.077\angle 43.7°} = 532.5\angle 2.3°$$

$$\mathbf{I}_{RO}' = \mathbf{V}_{RO}'\mathbf{Y}_R = 532.5\angle 2.3° \times 0.05 \times \angle -30° = 26.6\angle -27.7°$$

Similarly, $\mathbf{V}_{YO}'$ and $\mathbf{V}_{BO}'$ may be found and $\mathbf{I}_Y$ and $\mathbf{I}_B$ calculated thereform.

Example 19.65. *Three impedances, Z_R, Z_Y and Z_B are connected in star across a 440-V, 3-phase supply. If the voltage of star-point relative to the supply neutral is 200 ∠ 150° volt and Y and B line currents are 10 ∠−90° A and 20∠90° A respectively, all with respect to the voltage between the supply neutral and the R line, calculate the valves of Z_R, Z_Y and Z_B.*

(Elect Circuit; Nagpur Univ. 1991)

Solution. Let O and O' be the supply and load neutrals respectively. Also, let,

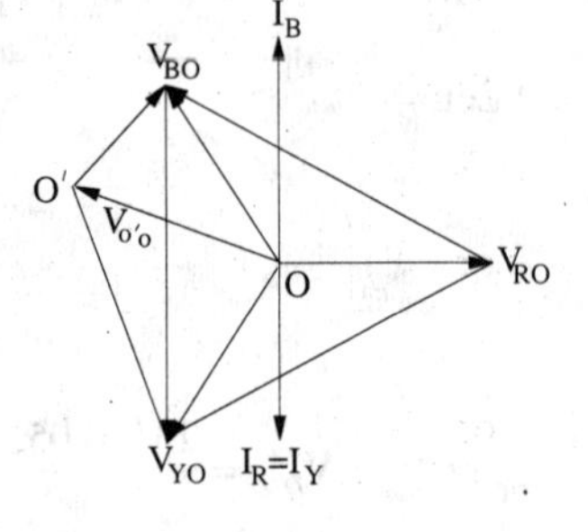

Fig. 19.87

$$\mathbf{V}_{RO} = \frac{440}{\sqrt{3}} \angle 0° = 254 \angle 0° = 254 + j0$$

$$\mathbf{V}_{YO} = 254 \angle -120° = -127 - j220$$

$$\mathbf{V}_{BO} = 254 \angle 120° = -127 + j220$$

$$\mathbf{I}_Y = 10 \angle -90° = -j10; \mathbf{I}_B = 20 \angle 90° = j20$$

$$\mathbf{I}_R = -(\mathbf{I}_Y + \mathbf{I}_B) = -j10$$

Also,
$$\mathbf{V}_{O'O} = 200 \angle 150° = -173 + j100$$

$$\mathbf{V}_{RO} - \mathbf{V}_{O'O} = 254 - (-173 + j100) = 427 - j100 = 438.5 \angle -13.2°$$

$$\mathbf{V}_{YO} - \mathbf{V}_{O'O} = (-127 - j220) - (-173 + j100) = 46 - j320 = 323 \angle -81.6°$$

$$\mathbf{V}_{BO} - \mathbf{V}_{O'O} = (-127 + j220) - (-173 + j100) = 46 + j120 = 128.6 \angle 69°$$

As seen from Art. 19.32.

$$\mathbf{Z}_R = \frac{\mathbf{V}_{RO} - \mathbf{V}_{O'O}}{\mathbf{I}_R} = \frac{438.5 \angle -13.2°}{10 \angle -90°} = 43.85 \angle 76.8°$$

$$\mathbf{Z}_Y = \frac{\mathbf{V}_{YO} - \mathbf{V}_{O'O}}{\mathbf{I}_Y} = \frac{323 \angle -81.6°}{10 \angle -90°} = 32.3 \angle 8.4°$$

$$\mathbf{Z}_B = \frac{\mathbf{V}_{BO} - \mathbf{V}_{O'O}}{\mathbf{I}_B} = \frac{128.6 \angle 69°}{20 \angle 90°} = 6.43 \angle -21°$$

19.35. Unbalanced Star-connected Non-inductive Load

Such a case can be easily solved by direct geometrical method. If the supply system is symmetrical, the line voltage vectors can be drawn in the form of an equilateral triangle *RYB* (Fig. 19.88). As the load is an unbalanced one, its neutral point will not, obviously, coincide with the centre of the gravity or centroid of the triangle. Let it lie at any other point like N. If point N represents the potential of the neutral point of the unbalanced load, then vectors drawn from N to points, R, Y and B represent the voltages across the branches of the load. These voltages can be represented in their rectangular co-ordinates with respect to the rectangular axis drawn through N. It is seen that taking co-ordinates of N as (0, 0), the co-ordinates of point R are $[(V/2 - x), -y]$

of point Y are $[-(V/2 + x), -y]$

and of point B are $[-x, (\sqrt{3}V/2 - y)]$

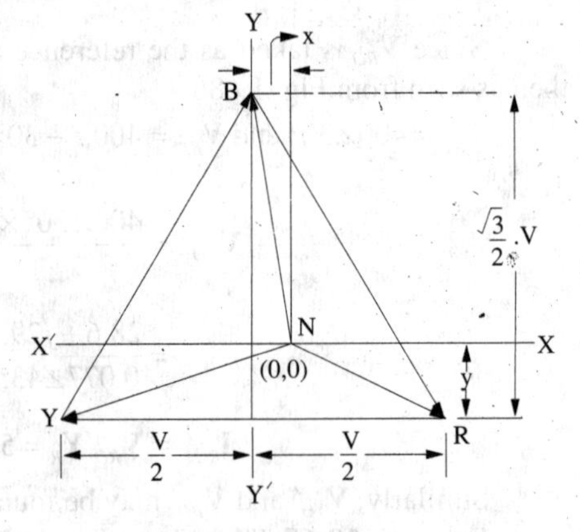

Fig. 19.88

$$\mathbf{V}_{RN} = \left(\frac{V}{2} - x\right) - jy; \mathbf{V}_{YN} = -\left(\frac{V}{2} + x\right) - jy$$

$$\mathbf{V}_{BN} = -x + j\left(\frac{\sqrt{3}V}{2} - y\right)$$

Let R_1, R_2 and R_3 be the respective branch impedances, $\mathbf{Y}_1$, $\mathbf{Y}_2$ and $\mathbf{Y}_3$ the respective admittances and $\mathbf{I}_R$, $\mathbf{I}_Y$ and $\mathbf{I}_B$ the respective currents in them.

Then $$\mathbf{I}_R = \mathbf{V}_{RN}/\mathbf{R}_1 = \mathbf{V}_{RN}\mathbf{Y}_1$$

Similarly, $\mathbf{I}_Y = \mathbf{V}_{YN}\ \mathbf{Y}_2$ and $\mathbf{I}_B = \mathbf{V}_{BN}\mathbf{Y}_3$. Since $\mathbf{I}_R + \mathbf{I}_Y + \mathbf{I}_B = 0$

$$\therefore \quad \mathbf{V}_{RN}\mathbf{Y}_1 + \mathbf{V}_{YN}\mathbf{Y}_2 + \mathbf{V}_{BN}\mathbf{Y}_3 = 0$$

or $$Y_1\left[\left(\frac{V}{2}-x\right)-jy\right]+Y_2\left[-\left(\frac{V}{2}+x\right)-jy\right]+Y_3\left[-x+j\left(\frac{\sqrt{3}\,V}{2}-y\right)\right]=0$$

or $$-x(Y_1+Y_2+Y_3)+\frac{V}{2}(Y_1-Y_2)+j\left[Y_3\frac{\sqrt{3}\,V}{2}-y(Y_1+Y_2+Y_3)\right]=0$$

$$\therefore \quad -x(Y_1+Y_2+Y_3)+\frac{V}{2}(Y_1-Y_2)=0 \quad \therefore x=\frac{V}{2}\frac{(Y_1-Y_2)}{Y_1+Y_2+Y_3}$$

Also $$Y_3\frac{\sqrt{3}\,V}{2}-y(Y_1+Y_2+Y_3)=0 \quad \therefore y=\frac{\sqrt{3}\,V}{2}\frac{Y_3}{(Y_1+Y_2+Y_3)}$$

Knowing the values of x, the values of $\mathbf{V}_{RN}$, $\mathbf{V}_{YN}$ and $\mathbf{V}_{BN}$ and hence, of $\mathbf{I}_R$, $\mathbf{I}_Y$ and $\mathbf{I}_B$ can be found as illustrated by Ex. 19.66.

Example 19.66. *Three non-inductive resistances of 5, 10 and 15 Ω are connected in star and supplied from a 230-V symmetrical 3-phase system. Calculate the line currents (magnitudes).*

(Principles of Elect. Engg. Jadavpur Univ. 1987)

Solution.

(*a*) **Star/Delta Conversion Method**

The Y-connected unbalanced load and its equivalent Δ-connected load are shown in Fig. 19.89 (*a*) and (*b*) respectively. Using *Y*/Δ conversion, we have

$$\mathbf{Z}_{12}=\frac{\mathbf{Z}_1\mathbf{Z}_2+\mathbf{Z}_2\mathbf{Z}_3+\mathbf{Z}_3\mathbf{Z}_1}{\mathbf{Z}_3}=\frac{50+150+75}{15}=\frac{55}{3}\,\Omega$$

$$\mathbf{Z}_{23}=275/5=55\,\Omega \quad \text{and} \quad \mathbf{Z}_{31}=275/10=27.5\,\Omega$$

Phase current $\mathbf{I}_{RY}=\mathbf{V}_{RY}/\mathbf{Z}_{12}=230/(55/3)=12.56$ A

Similarly, $\mathbf{I}_{YB}=V_{YB}/Z_{23}=230/55=4.18$ A; $\mathbf{I}_{BR}=\mathbf{V}_{BR}/\mathbf{Z}_{31}=230/27.5=8.36$ A

The line currents for Δ-connection are obtained by compounding the above phase currents trigonometrically or vectorially. Choosing vector addition and taking $\mathbf{V}_{RY}$ as the reference vector, we get ; $\mathbf{I}_{RY}=(12.56+j0)$

$$\mathbf{I}_{YB}=4.18\left(-\frac{1}{2}-j\frac{\sqrt{3}}{2}\right)=-2.09-j3.62$$

$$\mathbf{I}_{BR}=8.36\left(-\frac{1}{2}+j\frac{\sqrt{3}}{2}\right)=-4.18+j7.24$$

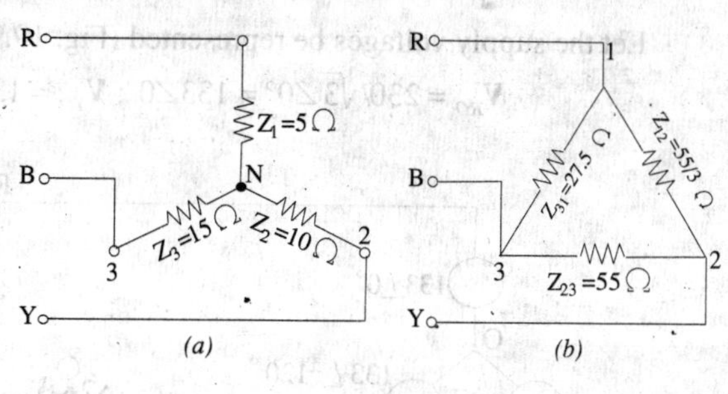

Fig. 19.89

Hence, line currents for Δ-connection of Fig. 19.74 (*b*) are

$$\mathbf{I}_R=\mathbf{I}_{RY}+\mathbf{I}_{RB}=\mathbf{I}_{RY}-\mathbf{I}_{BR}$$

$$=(12.56+j0)-(-4.18+j7.24)=16.74-j7.24 \text{ or } 18.25\text{ A} - \text{in magnitude}$$

$$\mathbf{I}_Y=\mathbf{I}_{YR}+\mathbf{I}_{YB}=\mathbf{I}_{YB}-\mathbf{I}_{RY}$$

$$=(-2.09-j3.62)-(12.56+j0)=-14.65-j3.62 \text{ or } 15.08\text{ A} - \text{in magnitude}$$

$$\mathbf{I}_B=\mathbf{I}_{BR}+\mathbf{I}_{BY}=\mathbf{I}_{BR}-\mathbf{I}_{YB}$$

$$=(-4.18+j7.24)-(-2.09-j3.62)=-2.09+j10.86 \text{ or } 11.06\text{ A} - \text{in magnitude}$$

(*b*) **Geometrical Method**

Here, $R_1 = 5\ \Omega, R_2 = 10\ \Omega$ and $R_3 = 15\ \Omega$ $Y_1 = 1/5$ S ; $Y_2 = 1/10$ S ; $Y_3 = 1/15$ S

As found above in Art. 19.35 $x = \frac{V}{2}(Y_1 - Y_2)/(Y_1 + Y_2 + Y_3)$

$$= \frac{230}{2}\left(\frac{1}{5} - \frac{1}{10}\right) / \left(\frac{1}{5} + \frac{1}{10} + \frac{1}{15}\right) = 31.4$$

$$y = \left(\frac{\sqrt{3}V}{2}.Y_3\right) / (Y_1 + Y_2 + Y_3) = (\sqrt{3} \times 115 \times 1/15)/(11/30) = 36.2$$

$$\mathbf{V}_{RN} = \left(\frac{V}{2} - x\right) - jy = (115 - 31.4) - j36.2 = 83.6 - j36.2$$

$$\mathbf{V}_{YN} = -\left(\frac{V}{2} + x\right) - jy = -146.4 - j36.2$$

$$\mathbf{V}_{BN} = -x + j\left(\frac{\sqrt{3}V}{2} - y\right) = -31.4 + j163$$

$$\mathbf{I}_R = \mathbf{V}_{RN}\mathbf{Y}_1 = (83.6 - j36.2) \times 1/5 = 16.72 - j7.24$$

$$\mathbf{I}_Y = \mathbf{V}_{YN}\mathbf{Y}_2 = (-146.4 - j36.2) \times 1/10 = -14.64 - j3.62$$

$$\mathbf{I}_B = \mathbf{V}_{BN}\mathbf{Y}_3 = (-31.4 + j163) \times 1/15 = -2.1 + j10.9$$

These are the same currents as found before.

(*c*) **Solution by Millman's Theorem**

$\mathbf{Y}_R = 1/5\angle 0°$; $\mathbf{Y}_Y = 1/10\angle 0°$; $\mathbf{Y}_B = 1/15\angle 0°$ and $\mathbf{Y}_R + \mathbf{Y}_Y + \mathbf{Y}_B = 11/30\ \angle 0°$ Siemens

Let the supply voltages be represented (Fig. 19.90) by

$$\mathbf{V}_{RO} = 230/\sqrt{3}\angle 0° = 133\angle 0°;\ \mathbf{V}_{YO} = 133\angle -120°;\ \mathbf{V}_{BO} = 133\angle 120°$$

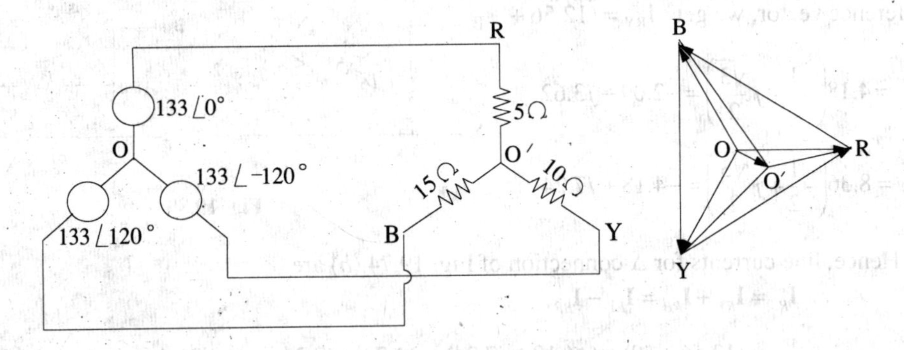

Fig. 19.90

The p.d. between load and supply neutrals is

$$\mathbf{V}_{o'o} = \frac{133/5 + (133/10)\angle -120° + (133/15)\angle 120°}{30/11\angle 0°}$$

$$= 42.3 - j10.4 = 43.6\angle -13.8°$$

$$\mathbf{V}_{RO}' = 133-(42.3-j10.4) = 90.7+j10.4$$

$$\mathbf{V}_{YO}' = 133\angle-120° - (42.3\,j-10.4)$$

$$= (-66.5-j115)-(42.3-j10.4) = -108.8-j104.6$$

$$\mathbf{V}_{BO}' = 133\angle 120° - \mathbf{V}_{O'O} = (-66.5+j115)-(42.3-j10.4) = -108.8+j125.4$$

$$\mathbf{I}_R = \mathbf{V}_{RO}'\mathbf{Y}_R = 1/5.(90.7+j10.4) = 18.1+j2.1 \text{ or } 18.22 \text{ A in magnitude}$$

$$\mathbf{I}_Y = -10.88-j10.5 \text{ or } 15.1 \text{ A in magnitude}$$

$$\mathbf{I}_B = -7.5+j8.4 \text{ or } 11.7 \text{ A in magnitude}$$

Example 19.67. *The unbalanced circuit of Fig. 19.91 (a) is connected across a symmetrical 3-phase supply of 400-V. Calculate the currents and phase voltages. Phase sequence is RYB.*

Solution. The line voltages are represented by the sides of an equilateral triangle *ABC* in Fig. 19.91 (*b*). Since phase impedances are unequal, phase voltages are unequal and are represented by lengths *NA*, *NB* and *NC* where *N* is the neutral point which is shifted from its usual position. *CM* and *ND* are drawn perpendicular to horizontal side *AB* Let co-ordinates of point *N* be (0, 0). Obviously, $AM = BM = 200$ V, $CM = \sqrt{3} \times 200 = 346$ V. Let $DM = x$ volts and $ND = y$ volts.

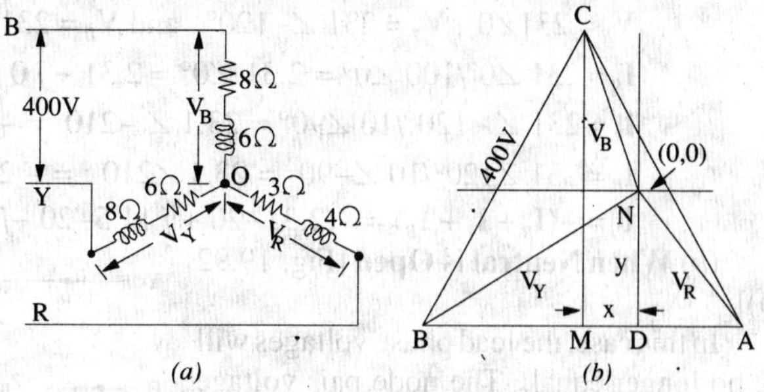

(a) (b)

Fig. 19.91

Then, with reference to point *N*, the vector expressions for phase voltages are

$$\mathbf{V}_R = (200-x)-jy, \mathbf{V}_Y = -(200+x)-jy; \mathbf{V}_B = -x+j(346-y)$$

$$\mathbf{I}_R = \frac{\mathbf{V}_R}{\mathbf{Z}_R} = \frac{(200-x)-jy}{3+j4} \times \frac{3-j4}{3-j4} = (24-0.12x-0.16y)+j(-32+0.16x-0.12y)$$

$$\mathbf{I}_Y = \frac{\mathbf{V}_Y}{\mathbf{Z}_Y} = \frac{-(200+x)-jy}{6+j8} \times \frac{6-j8}{6-j8}$$

$$= (-12-0.06x-0.08y)+j(16+0.08x-0.06y)$$

$$\mathbf{I}_B = \frac{\mathbf{V}_B}{\mathbf{Z}_B} = \frac{-x+j(346-y)}{8+j6} \times \frac{8-j6}{8-j6}$$

$$= (20.76-0.08x-0.06y)+j(27.68+0.06x-0.08y)$$

Now, $\mathbf{I}_R + \mathbf{I}_Y + \mathbf{I}_B = 0$

$\therefore$ $(32.76-0.26x - 0.3y) + j(11.68 + 0.3x - 0.26y) = 0$

Obviously, the real component as well as the *j*-component must be zero.

$\therefore$ $32.76-0.26 \times -0.\,3y = 0$ and $11.68 + 0.3x-0.26y = 0$

Solving these equations for *x* and *y*, we have $x = 31.9$ V and $y = 81.6$ V

$$\mathbf{V}_R = (200-31.9)-j81.6 = 168-j81.6 = 186.7\angle-25.9°$$

$$\mathbf{V}_Y = -(200+31.9)-j81.6 = -231.9-j81.6 = 245.8\angle 199.4°$$

$$\mathbf{V}_B = -31.9+j(346-81.6) = -31.9+j261.1 = 266.3\angle 83.1°$$

Substituting these values of *x* and *y* in the expressions for currents, we get

$$\mathbf{I}_R = (24-0.12\times 31.9 - 0.16\times 81.6) + j(-32 + 0.16\times 31.9 - 0.12\times 81.6)$$
$$= 7.12 - j36.7$$

Similarly $\mathbf{I}_Y = -20.44 + j13.65$; $\mathbf{I}_B = 13.3 + j23.06$

$\Sigma I = (0. + j0)$ – as a check

Example 19.68. *A 3-ϕ, 4-wire, 400-V symmetrical system supplies a Y-connected load having following branch impedances :*

$Z_R = 100\ \Omega$, $Z_Y = j10\ \Omega$ and $Z_B = -j10\ \Omega$

Compute the values of load phase voltages and currents and neutral current. Phase sequence is RYB.

How will these values change in the event of an open in the neutral wire ?

Solution. (*a*) **When Neutral Wire is Intact.** [Fig 19.92 (*a*)]. As discussed in Art. 19.30, the *load* phase voltages would be the same as *supply* phase voltages despite imbalance in the load. The three load phase voltages are :

$$\mathbf{V}_R = 231\angle 0°,\ \mathbf{V}_Y = 231\angle -120° \text{ and } \mathbf{V}_B = 231\angle 120°$$
$$\mathbf{I}_R = 231\angle 0°/100\angle 0° = 2.31\angle 0° = 2.31 + j\,0$$
$$\mathbf{I}_Y = 231\angle -120°/10\angle 90° = 23.1\angle -210° = -20 + j\,11.5$$
$$\mathbf{I}_B = 231\angle 120°/10\angle -90° = 23.1\angle 210° = -20 - j\,11.5$$
$$\mathbf{I}_N = -(\mathbf{I}_R + \mathbf{I}_Y + \mathbf{I}_B) = -(2.31 - 20 + j\,11.5 - 20 - j\,11.5) = 37.7 \text{ A}$$

(*b*) **When Neutral is Open** [Fig. 19.92 (*b*)]

In this case, the load phase voltages will be no longer equal. The node pair voltage method will be used to solve the question. Let the supply phase voltages be given by

$$\mathbf{E}_R = 231\angle 0°,\ \mathbf{E}_Y = 231\angle -120° = -115.5 - j200$$
$$\mathbf{E}_B = 231\angle 120° = -115.5 + j200$$

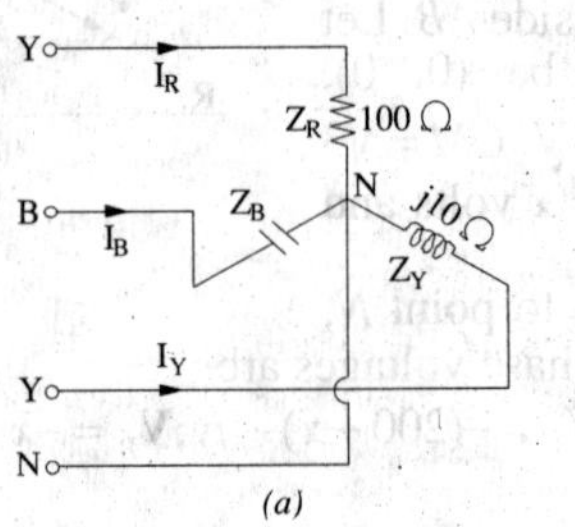

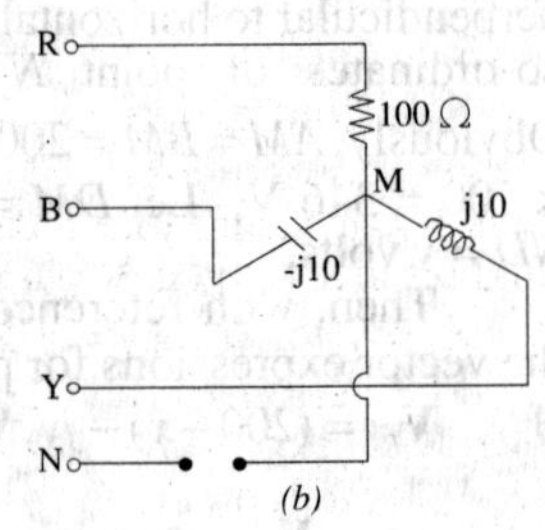

Fig. 19.92

$$\mathbf{Y}_R = 1/100 = 0.01;\ \mathbf{Y}_Y = 1/j10 = -j0.1 \text{ and } \mathbf{Y}_B = 1/-j10 = j0.1$$

$$\mathbf{V}_{N'N} = \frac{231\times 0.01 + (-j0.1)(-115.5 - j200) + j0.1(-115.5 + j200)}{0.01 + (-j0.1) + j0.1} = -3769 + j0$$

The load phase voltages are given by

$$\mathbf{V}_R' = \mathbf{E}_R - \mathbf{V}_{N'N} = (231 + j0) - (-3769 + j0) = 4000 \text{ V}$$

$$\mathbf{V}_Y' = \mathbf{E}_Y - \mathbf{V}_{N'N} = -115.5 - j200 - (-3769 + j0) = (3653.5 - j200)$$

$$\mathbf{V}_B' = \mathbf{E}_B - \mathbf{V}_{N'N} = -115.5 + j200 - (-3769 + j0) = (3653.5 + j0)$$

$$\mathbf{I}_R = \mathbf{V}_R'\mathbf{Y}_R = 4000\times 0.01 = -40 \text{ A}$$

$$\mathbf{I}_Y = (-j0.1)(3653.5 - j200) = (20 - j3653.5)$$

$$\mathbf{I}_B = (j0.1)(3653.5 + j200) = -20 + j3653.5$$

Obviously, the neutral current will just not exist.

Note. As hinted in Art. 19.30 (*i*), the load phase voltages and currents become abnormally high.

Example 19.69. *For the circuit shown in Fig. 19.93 find the readings on the two wattmeters W_a and W_c.*

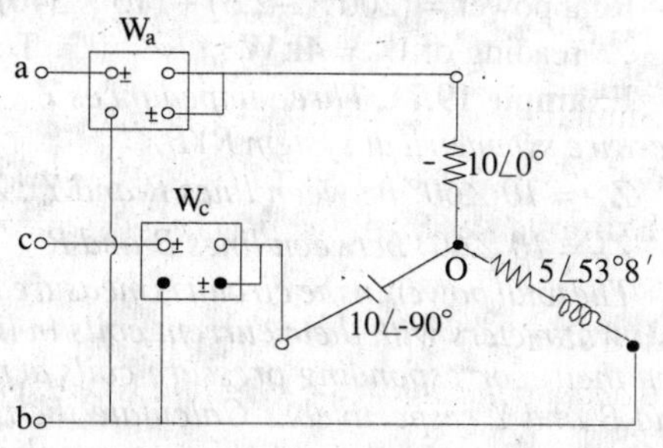

Fig. 19.93

Solution. The three line currents for this problem have already been determined in Example 19.43.

$$\mathbf{I}_{ao} = 20.02 - j\,4.54$$
$$\mathbf{I}_{bo} = -7.24 - j\,5.48$$
$$\mathbf{I}_{co} = -12.78 + j10.12$$

The line voltages are given by

$$\mathbf{V}_{ab} = 200 + j0$$
$$\mathbf{V}_{bc} = -100 - j\,173.2$$
$$\mathbf{V}_{ca} = -100 + j173.2$$

Wattmeter W_a carries a current of using $\mathbf{I}_{ao} = 20.02 - j\,4.54$ and has voltage $\mathbf{V}_{ab}$ impressed across its pressure coil. Power can be found by using current conjugate.

$$\mathbf{P}_{VA} = (200 + j\,0)\,(20.02 + j\,4.54) = (200)(20.02) + j(200)(4.54)$$

Actual power $= 200 \times 20.02 = 4004$ W $\therefore W_a = \mathbf{4004\ W}$

The other wattmeter W_c carries current of $\mathbf{I}_{co} = -12.78 + j\,10.02$ and has a voltage $\mathbf{V}_{cb} = -\mathbf{V}_{bc} = 100 + j\,173.2$ impressed across it. By the same method, wattmeter reading is

$$W_c = (100 \times -12.78) + (173.2 \times 10.02) = -1278 + 1735.5 = \mathbf{457.5\ W}$$

Example 19.70. *Three resistors 10, 20 and 20 Ω are connected in star to the terminals A, B and C of a 3−ϕ, 3 wire supply through two single-phase wattmeters for measurement of total power with current coils in lines A and C and pressure coils between A and B and C and B. Calculate (i) the line currents (ii) the readings of each wattmeter.*

The line voltage is 400-V. **(Electrical Engineering -I, Bombay Univ. 1987)**

Solution. Let $\mathbf{V}_{AB} = 400\angle 0°$; $\mathbf{V}_{BC} = 400\angle -120°$ and $\mathbf{V}_{CA} = 400\angle 120°$

As shown in Fig. 19.94, current through wattmeter W_1 is $\mathbf{I}_{AO}$ or $\mathbf{I}_A$ and that through W_2 is $\mathbf{I}_{CO}$ or $\mathbf{I}_C$ and the voltages are $\mathbf{V}_{AB}$ and $\mathbf{V}_{CB}$ respectively. Obviously,

$$\mathbf{Z}_A = 10\angle 0°,\ \mathbf{Z}_B = 20\angle 0°,\ \mathbf{Z}_C = 20\angle 0°$$

The currents $\mathbf{I}_A$ and $\mathbf{I}_C$ may be found be applying either Kirchhoff's laws (Art. 19.33) or Maxwell's Mesh Method. Both methods will be used for illustration.

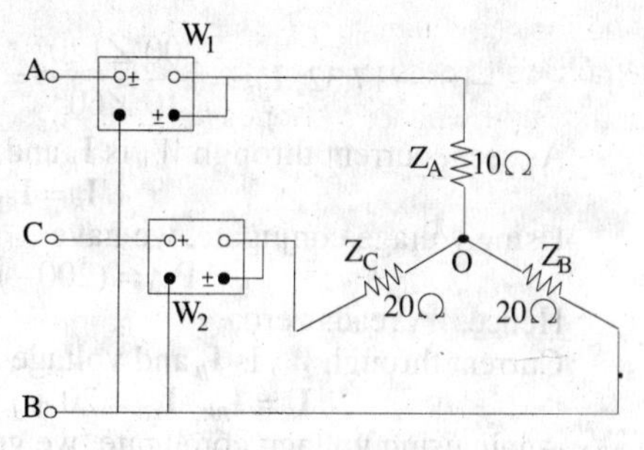

Fig. 19.94

(*a*) From Eq. (10), (11) and (12) of Art. 19.33, we have

$$\mathbf{I}_A = \frac{400 \times 20 - 20(-200 + j346)}{(10 \times 20) + (20 \times 20) + (20 \times 10)}$$

$$= \frac{12{,}000 - j6{,}920}{800} = 15 - j8.65 \text{ A}$$

$$\mathbf{I}_C = \frac{20(-200 + j346) - 10(-200 - j346)}{800} = \frac{-2000 + j10{,}380}{800} = -2.5 + j13$$

(*b*) From Eq. (*i*) and (*ii*) of solved Example 17.48 (*c*) we get

$$\mathbf{I}_A = \mathbf{I}_1 = \frac{400 \times 40 + 20(-200 - j346)}{30 \times 40 - 20^2} = 15 - j8.65 \text{ A}$$

$$\mathbf{I}_C = -\mathbf{I}_2 = \frac{30 \times (-200 - j346) + 400 \times 20}{800} = -25 + j13$$

As seen, wattmeter W_1 carries current $\mathbf{I}_A$ and has a voltage $\mathbf{V}_{AB}$ impressed across its pressure coil. Power may be found by using voltage conjugate.

$$\mathbf{P}_{VA} = (400 - j0)(15 - j8.65) = 6000 - j\,3{,}460$$

$\therefore$ reading of $W_1 = 6000$ W $= \mathbf{6\ kW}$

Similarly, W_2 carries $\mathbf{I}_C$ and has voltage $\mathbf{V}_{CB}$ impressed across its pressure coil. Now, $\mathbf{V}_{CB} = -\mathbf{V}_{BC} - (200 + j\,346)$. Using voltage conjugate, we get

$$\mathbf{P}_{VA} = (200 - j346)(-2.5 + j13)$$

Real power = (200 × –2.5) + (13 × 346) = 4000 W

∴ reading of W_2 = **4kW** Total power = 10 kW

Example 19.71. *Three impedances Z_A, Z_B and Z_C are connected in delta to a 200-V, 3-phase three-wire symmetrical system RYB.*

Z_A = 10 ∠60° between lines R and Y ; Z_B = 10 ∠ 0° between lines Y and B

Z_C = 10 ∠60° between lines B and R

The total power in the circuit is measured by means of two wattmeters with their current coils in lines R and B and their corresponding pressure coils across R and Y and B and Y respectively. Calculate the reading on each wattmeter and the total power supplied. Phase sequence RYB.

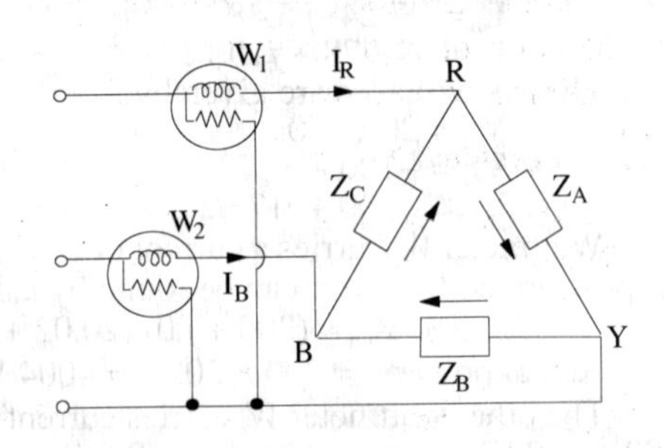

Fig. 19.95

Solution. The wattmeter connections are shown in Fig. 19.95.

$$\mathbf{V}_{RY} = 200\angle 0° = 200 + j0$$

$$\mathbf{V}_{YB} = 200\angle -120° = -100 - j173.2$$

$$\mathbf{V}_{BR} = 200\angle 120° = -100 + j173.2$$

$$\mathbf{I}_{BR} = \frac{200\angle 0°}{10\angle 60°} = 20\angle -60° = 10 - j17.32\ \Omega$$

$$\mathbf{I}_{YB} = \frac{200\angle -120°}{10\angle 0°} = 20\angle -120°$$

$$= -10 - j17.32;\ \mathbf{I}_{BR} = \frac{200\angle 120°}{10\angle 60°} = 20\angle 60° = 10 + j17.32$$

As seen, current through W_1 is $\mathbf{I}_R$ and voltage across its pressure coil is $\mathbf{V}_{RY}$.

$$\mathbf{I}_R = \mathbf{I}_{RY} - \mathbf{I}_{BR} = -j34.64\text{ A}$$

Using voltage conjugate, we have

$$\mathbf{P}_{VA} = (200 - j0)(-j34.64) = 0 - j6{,}928$$

Hence, W_1 reads zero.

Current through W_2 is $\mathbf{I}_B$ and voltage across its pressure coil is $\mathbf{V}_{BY}$

$$\mathbf{I}_B = \mathbf{I}_{BR} - \mathbf{I}_{YB} = 20 + j34.64\ ;\ \mathbf{V}_{BY} = -\mathbf{V}_{YB} = 100 + j173.2$$

Again using voltage conjugate, we get

$$\mathbf{P}_{VA} = (100 - j173.2)(20 + j34.64)$$

$$= 8000 + j0$$

∴ reading of W_2 = **8000 W**

19.36. Phase Sequence Indicators

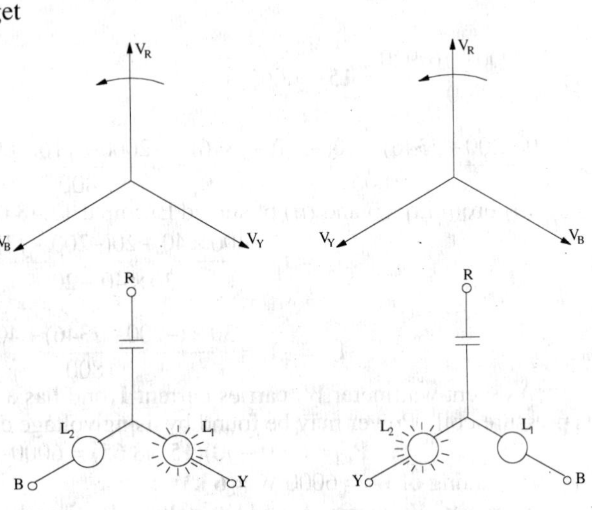

Fig. 19.96

In unbalanced 3-wire star-connected loads, phase voltages change considerably if the phase sequence of the supply is reversed. One or the other load phase voltage becomes dangerously large which may result in damage to the equipment. Some phase voltage becomes too small which is equally deterimental to some types of electrical equipment. Since phase voltage depends on phase sequence, this fact has been made the basis of several types of phase sequence indicators.* A simple phase sequence indicator may be made by connecting two suitable incandescent lamps and a capacitor in a *Y*-connection as shown in Fig. 19.96. It will be found that for phase sequence *RYB*, lamp L_1 will glow because its phase voltage will be large whereas L_2 will not glow because of low voltage across it.

When, phase sequence is *RBY*, opposite conditions develop so that this time L_2 glows but not L_1.

Another method of determining the phase sequence is by means of a small 3-phase motor. Once its direction of rotation with a known sequence is found, the motor may be used thereafter for determining an unknown sequence.

Tutorial Problem No. 19.3

1. Three impedances $\mathbf{Z}_1$, $\mathbf{Z}_2$ and $\mathbf{Z}_3$ are mesh-connected to a symmetrical 3-phase, 400-V, 50-Hz supply of phase sequence $R \rightarrow Y \rightarrow B$.

$\mathbf{Z}_1 = (10 + j0)$ ohm –between *R* and *Y* lines
$\mathbf{Z}_2 = (8 + j6)$ ohm –between *Y* and *B* lines
$\mathbf{Z}_3 = (5 - j5)$ ohm –between *B* and *R* lines

Calculate the phase and line currents and total power consumed.
[40 A, 40 A, 56.6 A ; 95.7 A, 78.4 A, 35.2 A ; 44.8 kW]

2. A symmetrical 3-ϕ, 380-V supply feeds a mesh-connected load as follows :
Load *A* : 19 kVA at p.f. 0.5 lag ; Load *B* : 20 kVA at p.f. 0.8 lag : Load *C* : 10 kVA at p.f. 0.9 lead
Determine the line currents and their phase angles for RYB sequence.
[74.6 $\angle$–51° A, 98.6 $\angle$ 172.7° A ; 68.3 $\angle$ 41.8° A]

3. Determine the line currents in an unbalanced *Y*-connected load supplied from a symmetrical 3-ϕ, 440-V, 3-wire system. The branch impedances of the load are : $\mathbf{Z}_1 = 5 \angle 30°$ ohm, $\mathbf{Z}_2 = 10 \angle 45°$ ohm and $\mathbf{Z}_3 = 10 \angle 45°$ ohm and $\mathbf{Z}_3 = 10 \angle 60°$ ohm. The sequence is *RYB*. **[35.7 A, 32.8 A; 27.7 A]**

4. A 3-ϕ, *Y*-connected alternator supplies an unbalanced load consisting of three impedances $(10 + j20)$, $(10 - j20)$ and 10 Ω respectively, connected in star. There is no neutral connection. Calculate the voltage between the star point of the alternator and that of the load. The phase voltage of the alternator is 230 V.
[–245.2 V]

5. Non-reactive resistors of 10, 20 and 25 Ω are star-connected to the *R*, *Y* and *B* phases of a 400-V, symmetrical system. Determine the current and power in each resistor and the voltage between star point and neutral. Phase sequence, *RYB*. **[16.5 A, 2.72 kW ; 13.1 A, 3.43 kW; 11.2 A, 3.14 kW ; 68 V]**

6. Determine the line current in an unbalanced, star-connected load supplied from a symmetrical 3-phase, 440-V system. The branch impedances of the load are $\mathbf{Z}_R = 5 \angle 30°\ \Omega$, $Z_Y = 10\angle 45°\ \Omega$ and $Z_B = 10 \angle 60°\ \Omega$. The phase sequence is *RYB*. **[35.7 A, 32.8 A, 27.7 A]**

7. Three non-reactive resistors of 3, 4 and 5 Ω respectively are star-connected to a 3-phase, 400-V symmetrical system, phase sequence *RYB*. Find (*a*) the current in each resistor (*b*) the power dissipated in each resistor (*c*) the phase angles between the currents and the corresponding line voltages (*d*) the star-point potential. Draw to scale the complete vector diagram.
[(a) 66.5 A, 59.5 A, 51.8 A (*b*) 13.2, 14.15, 13.4 kW (*c*) 26°24′, 38°10′, 25° 20′ (*d*) 34 V]

8. An unbalanced *Y*-connected load is supplied from a 400-V, 3-ϕ, 3-wire symmetrical system. The branch circuit impedances and their connections are $(2 + j2)\ \Omega$, *R* to *N* ; $(3 - j3)\ \Omega$, *Y* to *N* and $(4 + j1)\ \Omega$, *B* to *N* of the load. Calculate (*i*) the value of the voltage between lines *Y* and *N* and (*ii*) the phase of this voltage relative to the voltage between lines *R* and *Y*. Phase sequence *RYB*.
[(*i*) (–216–j 135.2) or 225.5 V (*ii*) 2° or –178°]

9. A star-connection of resistors $R_a = 10\ \Omega$; $R_b = 20\ \Omega$ and $R_e = 40\ \Omega$ is made to the terminals *A*, *B* and *C* respectively of a symmetrical 400-*V*, 3 ϕ supply of phase sequence $A \rightarrow B \rightarrow C$. Find the branch voltages and currents and star-point voltage to neutral.
**[$V_A = 148.5 + j28.6$; $I_A = 14.85 + j2.86$; $V_B = -198 - j171.4$; $I_B = -9.9 - j8.57$
$V_C = -198 + j228.6$; $I_C = -4.95 + j5.71$. $V_N = 82.5 - j28.6$ (to be subtracted from supply voltage)]**

*It may, however, be noted that phase sequence of *currents* in an unbalanced load is not necessarily the same as the *voltage* phase sequence. Unless indicated otherwise, voltage phase sequence is implied.

10. Three non-reactive resistances of 5, 10 and 5 ohm are star-connected across the three lines of a 230-V, 3-phase, 3-wire supply. Calculate the line currents. **[(18.1 + J2.1) A ; (−10.9−j 10.45) A ; (−7.3 + j8.4) A]**

11. A 3-ϕ, 400-V symmetrical supply feeds a star-connected load consisting of non-reactive resistors of 3,4 and 5 Ω connected to the *R*, *Y* and *B* lines respectively. The phase sequence is *RYB*. Calculate (*i*) the load star point potential (*ii*) current in each resistor and power dissipated in each resistor.

[(*i*) 34.5 V (*ii*) 66.4 A, 5 9.7 A, 51.8 A (*iii*) 13.22 kW, 14.21 kW, 13.42 kW]

12. A 20-Ω resistor is connected between lines *R* and *Y*, a 50-Ω resistor between lines *Y* and *B* and a 10-Ω resistor between lines *B* and *R* of a 415-V, 3-phase supply. Calculate the current in each line and the reading on each of the two wattmeters connected to measure the total power, the respective current coils of which are connected in lines *R* and *Y*. **[(25.9 −j9) ; (−24.9 − j 7.2); (−1.04 + j16.2) ; 8.6 kW ; 7.75 kW]**

13. A three-phase supply, giving sinusoidal voltage of 400 V at 50 Hz is connected to three terminals marked *R*, *Y* and *B*. Between *R* and *Y* is connected a resistance of 100 Ω, between *Y* and *B* an inductance of 318 mH and negligible resistance and between *B* and *R* a capacitor of 31.8 μF. Determine(*i*) the current flowing in each line and (*ii*) the total power supplied. Determine (*iii*) the resistance of each phase of a balanced star-connected, non-reactive load, which will take the same total power when connected across the same supply.

[(*i*) 7.73 A, 7.73 A, (*ii*) 1,600 W (*iii*) 100 Ω (*London Univ.*)

14. An unbalanced, star-connected load is fed from a symmetrical 3-phase system. The phase voltages across two of the arms of the load are $V_B = 295 \angle 97° 30'$ and $V_R = 206 \angle -25°$. Calculate the voltage between the star-point of the load and the supply neutral. **[52.2 ∠−49. 54′]**

15. A symmetrical 440-V, 3-phase system supplies a star-connected loa with the following branch impedances : $Z_R = 100\ \Omega$, $Z_Y = j5\ \Omega$, $Z_B = -j5\ \Omega$. Calculate the voltage drop across each branch and the potential of the neutral point to earth. The phase sequence is RYB. Draw the vector diagram.

[8800 ∠−30°, 8415 ∠ − 31.5°, 8420 −∠−28.5°, 8545 ∠150°]

16. Three star-connected impedances, $Z_1 = (20 + j37.7)\ \Omega$ per phase are in parallel with three delta-connected impedances, $Z_2 = (30 - j159.3)\Omega$ per phase. The line voltage is 398 V. Find the line current, power factor, power and reactive volt-amperes taken by the combination.

[3.37 ∠−10.4° ; 0.984 lag; 2295 W; 420 VAR.]

17. A 3-phase, 440-V, delta-connected system has the loads : branch *RY*, 20 KW at power factor. 1.0: branch *YB*, 30 kVA at power factor 0.8 lagging; branch *BR*, 20 kVA at power factor 0.6 leading. Find the line currents and readings on watt-meters whose current coils are in phases *R* and *B*.

[90.5 ∠ 176.5° ; 111.4 ∠14°; 36.7 ∠−119; ° 39.8 kW; 16.1 kW]

18. A 415 V, 50 Hz, 3-phase supply of phase sequence RYB is connected to a delta connected load in which branch *RY* consists of $R_1 = 100\ \Omega$, branch *YB* conists of $R_2 = 20\ \Omega$ in series with $X_2 = 60\ \Omega$ and branch *BR* consists of a capacitor $C = 30\ \mu F$. Take V_{RY} as the reference and calculate the line currents. Draw the complete phasor diagrams. **(Elect. Machines, A.M.I.E. Sec. B, 1989)**

[$I_R = 7.78 \angle 14.54°$, $I_Y = 10.66 \angle 172.92°$, $I_B = 4.46 \angle -47°$]

19. Three resistances of 5, 10 and 15 Ω are connected in delta across a 3-phase supply. Find the values of the three resistors, which if connected in star across the same supply, would take the same line currents.

If this star-connected load is supplied from a 4-wire, 3-phase system with 260 *V* between lines, calculate the current in the neutral.

[2.5 Ω, 1.67 Ω, 5 Ω ; 52 A] (*London Univ.*)

20. The impedances of the three phase of a star-connected load (no neutral wire) are (5 + *j*20), (12 + *j*0) and (1−*j*10) in that order. The voltage is 400 volts. Find the line currents.

[(0.5 − j 29.65), (16.24−j11.5) ; (−16.74 + j41.15)] (*Elect. Meas. London Univ.*)]

21. Three voltmeters having resistances of 10, 10, and 5 kΩ respectively, are connected in star to a balanced 3-phase 4-wire supply. The line voltage is 440 V. Determine the readings of the three voltmeters.

[190 V, 290 V, 290 V] (*I.E.C. London*)

22. A 440-V, symmetrical, 3-ϕ supply feeds a *Y*-connected load consisting of three non-inductive resistances of 10, 5, and 12 Ω connected to the *R*, *Y* and *B* phases respectively.

Calculate the line currents and the voltage across each resistor. Phase sequence *RYB*.

[28.9 A, 36.5 A, 25.4 V, 290 V, 182 V, 304 V (*London University*)

23. A symmetrical 3-ϕ, 440-V system supplies a *Y*-connected load, of which the branch resistances are *A* = 10 Ω, *B* = 13 Ω, *C* = 15 Ω. Calculate the voltage to earth of the load star-point, assuming the neutral of the supply to be earthed. Phase sequence *A*, *B*, *C*. **[31 V]** (*I.E.E. London*)

24. Three non-reactive resistors *A*, *B*, *C* of 5Ω, 10 Ω and 15Ω respectively are star-connected across the lines of a 230 V, 3−ϕ, 3-wire supply. Calculate the line currents.

[$I_A = 15.16$ A, $I_B = 11.1$ A, $I_C = 18.25$ A] (*City & Guilds, London*)

25. The impedances, in ohms, of a three-phase star-connected load are ; $\mathbf{Z}_1 = 10 + j0$; $\mathbf{Z}_2 = 10 + j5$; $\mathbf{Z}_3 = 5 - j10$. The load is connected to a three phase 50-Hz supply having a balanced line voltage of 440 V. Calculate the reading of single-phase wattmeter which has its current coil in line 1 and its voltage coil between lines 1 and 3. Assume the phase sequence to be 1, 2, 3. Draw a vector diagram showing the relationships of the voltages and currents. **[11.75 kW]** (*London Univ.*)

26. An unbalanced star-connected load is supplied from a symmetrical 400-V, three-phase, three-wire system. The impedances are ; $(1 + j2)\ \Omega$ between line *R* and neutral point of load : $(2 + j3)\ \Omega$. between line *Y* and neutral point of load ; $(3 - j3)\ \Omega$, between line *B* and neutral point of load. Calculate the current in line. *B*. Phase sequence *RYB*. **[35.5 A]** (*London Univ.*)

27. An unbalanced star-connected load is supplied from symmetrical three-phase mains. The load impedance are Z_a, Z_b, Z_c and the positive phase sequences is *a*, *b*, *c*. Calculate the value of the positive phase-sequence current when the line voltage of the three-phase balanced supply is 400 *V* and $\mathbf{Z}_a = (0 + j10)\ \Omega$, $\mathbf{Z}_b = (0 - j5)\ \Omega$ and $\mathbf{Z}_c = (10 + j0)\ \Omega$. **[36.6 A]** (*Elect. Theory and Meas. London Univ.*)

28. The impedances are mesh-connected to a symmetrical three-phase, 40-V, 50, Hz supply of phase sequence *RYB*.

$\mathbf{Z}_1 = (5 + j10)\ \Omega$ is connected between lines *R* and *Y*

$\mathbf{Z}_2 = (5 + j5)\ \Omega$ is connected between lines *Y* and *B*

$\mathbf{Z}_3 = (6 - j4)\ \Omega$ is connected between lines *B* and *R*.

Two wattmeters connected to measure input have their current coils in lines *R* and *Y* respectively. Calculate the line currents and the wattmeter readings.

[95.5 A, 79.4 A, 43.4 A ; 39.8 kW, 9.6 kW] (*London Univ.*)

29. In Fig. 19.97 three non-reactive resistors, $R_1 = 10\Omega$, $R_2 = 20\ \Omega$ and $R_3 = 30\ \Omega$ are connected in star to a symmetrical 3-phase supply of line voltage 400 V and phase sequence *RYB*. The wattmeter *W* is connected as shown, with the current coil in the *R* line and the voltage circuit connected between lines *R* and *Y*. Calculate the currents in the three lines and the reading of the wattmeter. **[15.9 A, 13.1 A, 9.8A, 5.82 kW]** (*I.E.E. London*)

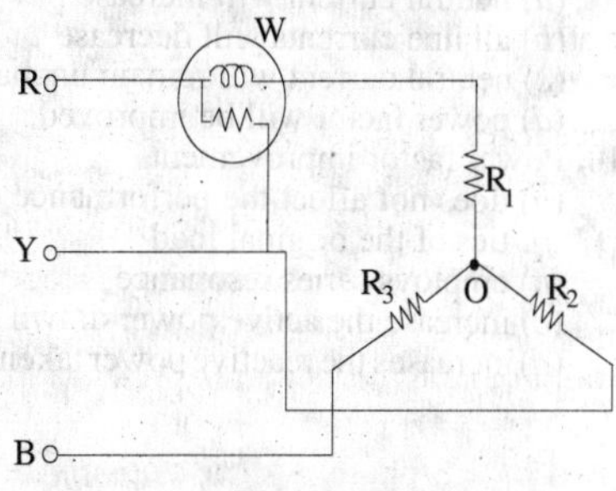

Fig. 19.97

30. A 3φ- *Y*-connected load is connected between three line terminals, *R, Y, B,* the impedances (in ohms) of the load being: $(1 + j2)$ between *R* and the star point, $(2 + j3)$ between *Y* and the star point and $(3 + j4)$ between *B* and the star point. The phase sequence is *RYB*. If the line voltage is 400 V, calculate the voltage between *R* and the star point. **[181 V)** (*London Univ.*)

31. A 3-phase 3-wire star-connected load consists of a capacitive reactance of 100 Ω in the red phase and a resistor of 100 Ω in each of the other phases. The line voltage is 440 V and the phase sequence is *RYB*. If the power is measured by the 2-wattmeter method, the current coils being connected in the red and yellow leads respectively, determine the reading of each instrument. **[–86.5 W, 1,640 W]** (*I.E.E. London*)

32. The impedances, in ohms, of the branches of a three-phase star-connected load are as follows : $\mathbf{Z}_R = (20 + j0)$, $\mathbf{Z}_Y = (10 + j10)$; $\mathbf{Z}_B = (0 - j25)$. If this load, with its neutral point isolated, is connected to a symmetrical three-phase, four-wire system, with 400 *V* between the line wires, determine the p.d. between the neutral point of the supply system and the neutral point of the load. Phase sequence *RYB*. **[267.5 V]** (*London Univ.*)

Objective Tests–19

1. Electric power is almost exclusively generated, transmitted and distributed, by three phase system because it
(*a*) is more efficient
(*b*) uses less material for a given capacity
(*c*) costs less than single-phase apparatus
(*d*) all of the above.

2. The voltages induced in the three windings of a 3-phase alternator are —— degree apart in time phase.
(*a*) 120 (*b*) 60
(*c*) 90 (*d*) 30

3. If positive phase sequence of a 3-phase *load* is *abc*, the negative sequence would be
(*a*) *bac* (*b*) *cba* (*c*) *acb*
(*d*) all of the above

4. In the balanced 3-phase voltage system generated by a *Y*-connected alternator, V_{YB} lags E_R by —— electrical degrees.
(*a*) 90 (*b*) 120
(*c*) 60 (*d*) 30

5. The power taken by a 3-φ load is given by the expression
(*a*) $3\ V_L I_L \cos\phi$
(*b*) $\sqrt{3}\ V_L I_L \cos\phi$
(*c*) $3\ V_L I_L \sin\phi$
(*d*) $\sqrt{3}\ V_L I_L \sin\phi$

6. In a balanced 3-phase voltage generator, the different phase voltages reach their maximum values —— degree apart.
(*a*) 120 (*b*) 60
(*c*) 240 (*d*) 30

7. If the *B*-phase of a 3-phase, *Y*-connected alternator become reverse connected by mistake, it will not affect
(*a*) V_{YB} (*b*) V_{RY}
(*c*) V_{BR} (*c*) V_{BY}

8. Three equal impedances are first connected in star across a balanced 3-phase supply. If connected in delta across the same supply.
(*a*) phase current will be tripled
(*b*) phase current will be doubled
(*c*) line current will become on-third
(*d*) power consumed will increase three-fold.

9. A 3-phase, 4-wire, 230/440-V system is supplying lamp load at 230 V. If a 3-phase motor is now switched on across the same supply, then
(*a*) neutral current will increase
(*b*) all line currents will decrease
(*c*) neutral current will remain unchanged
(*d*) power factor will be improved.

10. Power factor improvement
(*a*) does not affect the performance characteristics of the original load
(*b*) employs series resonance
(*c*) increase the active power drawn by the load
(*d*) increases the reactive power taken by the load.

11. The chief disadvantage of a low power factor is that
(*a*) more power is consumed by the load
(*b*) current required for a given load power is higher.
(*c*) active power developed by a generator exceeds its rated output capacity
(*d*) heat generated is more than the desired amount

12. In the 2-wattmeter method of measuring 3-phase power, the two wattmeters indicate equal and opposite readings when load power factor angle is —— degrees lagging.
(*a*) 60 (*b*) 0
(*c*) 30 (*d*) 90

13. When phase sequence at the 3-phase load is reversed
(*a*) phase powers are changed
(*b*) phase currents are changed
(*c*) phase currents change in angle but not in magnitude
(*d*) total power consumed in changed.

14. Phase reversal of a 4-wire unbalanced load supplied from a balanced 3-phase supply changes.
(*a*) magnitude of phase currents
(*b*) magnitudes as well as phase angle of neutral current
(*c*) the power consumed
(*c*) only the magnitude of neutral current.

Answers

1. *d* **2.** *a* **3.** *d* **4.** *a* **5.** *b* **6.** *b* **7.** *b* **8.** *d* **9.** *c* **10.** *a* **11.** *b* **12.** *d* **13.** *c* **14.** *b* **15.** *b* **16.** *a*

20 HARMONICS

20.1. Fundamental Wave and Harmonics

Upto this stage, while dealing with alternating voltages and currents, it had been assumed that they have sinusoidal waveform or shape. Such a waveform is an ideal one and much sought after by the manufacturers and designers of alternators. But it is nearly impossible to realize such a waveform in practice. All the alternating waveforms deviate, to a greater or lesser degree, from this ideal sinusoidal shape. Such waveforms are referred to as *non-sinusoidal* or *distorted* or *complex waveforms*.

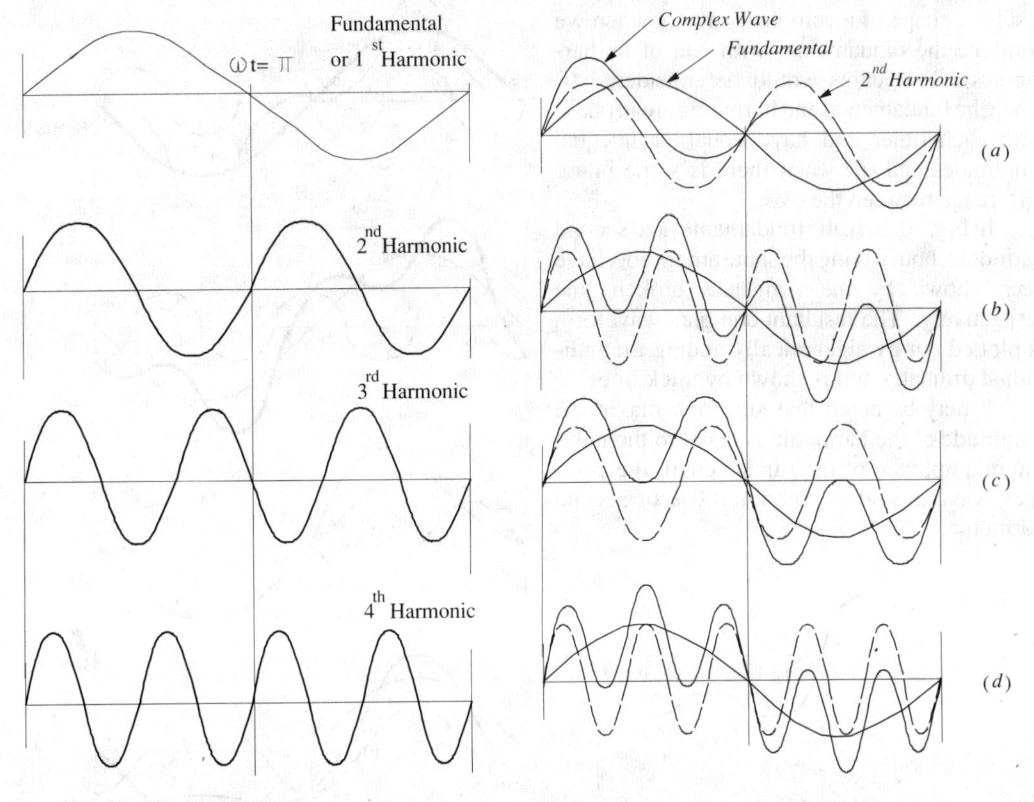

Fig. 20.1 Fig. 20.2

Complex waveforms are produced due to the superposition of sinusoidal waves of different frequencies. Such waves occur in speech, music, TV, rectifier outputs and many other applications of electronics. On analysis, it is found that a complex wave essentially consists of

(*a*) *a fundamental* wave – it has the lowest frequency, say '*f*'

(*b*) a number of other sinusoidal waves whose frequencies are an integral multiple of the fundamental or basic frequency like 2*f*, 3*f* and 4*f* etc.

The fundamental and its higher multiples form a *harmonic* series.

As shown in Fig. 20.1, fundamental wave itself is called the *first* harmonic. The *second* harmonic has frequency *twice* that of the fundamental, the *third* harmonic has frequency *thrice* that of the fundamental and so on.

Waves having frequencies of $2f$, $4f$ and $6f$ etc. are called *even harmonics* and those having frequencies of $3f$, $5f$ and $7f$ etc. are called *odd* harmonics. Expressing the above in angular frequencies, we may say that successive odd harmonics have frequencies of 3ω, 5ω and 7ω etc. and even harmonics have frequencies of 2ω, 4ω and 6ω etc.

As mentioned earlier, harmonics are introduced in the output voltage of an alternator due to many reasons such as the irregularities of the flux distribution in it. Considerations of waveform and form factor are very important in the transmission of a.c. power but they are of much greater importance in radio work where the intelligibility of a signal is critically dependent on the faithful transmission of the harmonic structure of sound waves. In fact, it is only the rich harmonic content of the consonants and lesser but still plentiful harmonic content of vowels which helps the ear to distinguish a well-regulated speech from a more rhythmical succession of musical sounds.

20.2. Different Complex Waveforms

Let us now find out graphically what the resultant shape of a complex wave is when we combine the fundamental with one of its harmonics. Two cases would be considered (*i*) when the fundamental and harmonic are in phase with each other and have equal or unequal amplitudes and (*ii*) when there is some phase difference between the two.

In Fig. 20.2(*a*), the fundamental and second harmonic, both having the same amplitude, have been shown by the firm and broken line respectively. The resultant complex waveform is plotted out by algebraically adding the individual ordinates and is shown by thick line.

It may be noted that since the maximum amplitude of the harmonic is equal to the maximum amplitude of the fundamental, the complex wave is said to contain 100% of second harmonic.

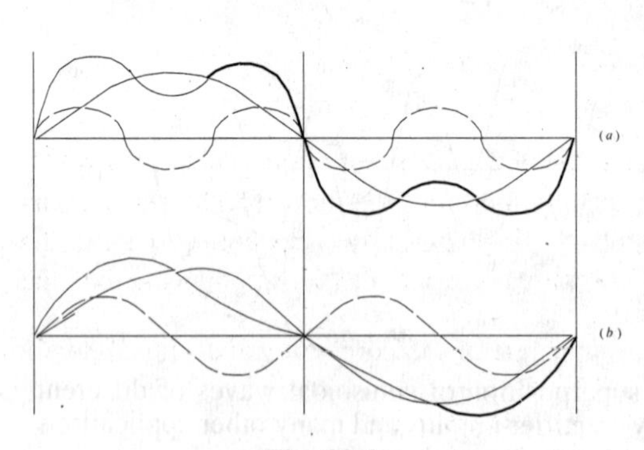

Fig. 20.3

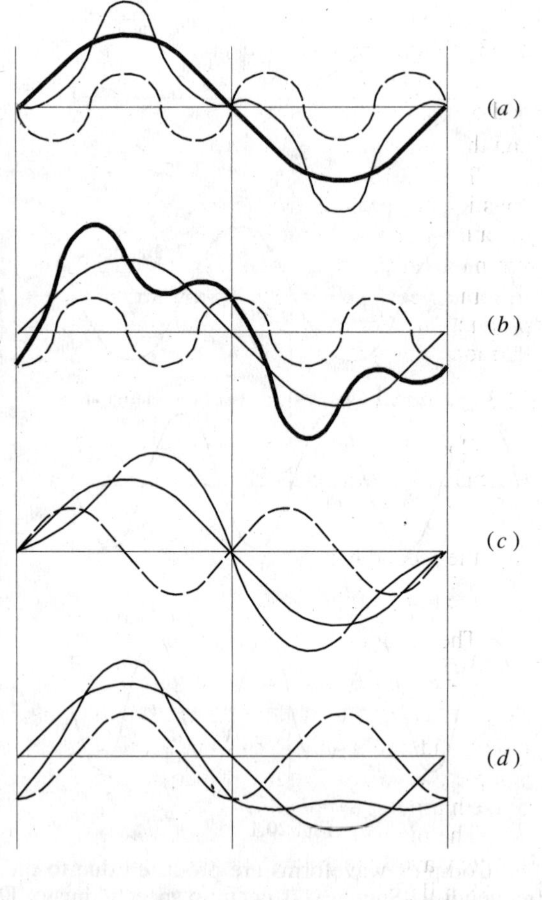

Fig. 20.4

The complex wave of Fig. 20.2(*b*) is made up of the fundamental and 4th harmonic, that of Fig. 20.2(*c*) consists of the fundamental and 3rd harmonic whereas that shown in Fig. 20.2(*d*) is made up of the fundamental and 5th harmonic. Obviously, in all these cases, there is no phase difference between the fundamental and the harmonic.

Fig. 20.2(*a*) and (*c*) have been reconstructed as Fig. 20.3(*a*) and (*b*) respectively with the only difference that in this case, the amplitude of harmonic is half that of the fundamental *i.e.* the harmonic content is 50%.

The effect of phase difference between the fundamental and the harmonic on the shape of the resultant complex wave has been illustrated in Fig. 20.4.

Fig. 20.4 (*a*) shows the fundamental and second harmonic with phase difference of $\pi/2$ and Fig. 20.4(*b*) shows the same with a phase difference of π. In Fig. 20.4(*c*) and (*d*) are shown the fundamental and third harmonic with a phase difference of $\pi/2$ and π respectively. In all these figures, the amplitude of the harmonic has been taken equal to half that of the fundamental.

A careful examination of the above figures leads us to the following conclusions :

1. With *odd* harmonics, the positive and negative halves of the complex wave are symmetrical whatever the phase difference between the fundamental and the harmonic. In other words, the first and third quarters (*i.e.* ωt from 0 to $\pi/2$ and ωt from π to $3\pi/2$) and the second and fourth quarters (*i.e.* ωt from $\pi/2$ to π and ωt from $3\pi/2$ to 2π) are respectively similar.

2. (*i*) When *even* harmonics are present and their phase difference with the fundamental is 0 or π, then the first and fourth quarters of the complex wave are of the same phase but inverted and the same holds good for the second and third quarters.

(*ii*) When *even* harmonics are present and their phase difference with the fundamental is $\pi/2$ or $3\pi/2$, then there is no symmetry as shown in Fig. 20(*a*).

3. It may also be noted that the resultant displacement of the complex wave (whether containing odd or even harmonics) is zero at $\omega t = 0$ only when the phase difference between the fundamental and the harmonics is either 0 or π.

The above conclusions are of great help in analysing a complex waveform into its harmonic constituents because a visual inspection of the complex wave enables us to rule out the presence of certain harmonics. For example, if the positive and negative half-cycles of a complex wave are symmetrical (*i.e.* the wave is symmetrical about $\omega t = 0$), then we need not look for even harmonics. In some cases, we *may* be able to forecast the types of harmonics to be expected from their mode of production. For example, in alternators which are symmetrically designed, we should expect *odd* harmonics only.

20.3. General Equation of a Complex Wave

Consider a complex wave which is built up of the fundamental and a few harmonics, each of which has its own peak value and phase angle. The fundamental may be represented by

$$e_1 = E_{1m} \sin(\omega t + \Psi_1)$$

the second harmonic by $\quad e_2 = E_{2m} \sin(2\omega t + \Psi_2)$

the third harmonic by $\quad e_3 = E_{3m} \sin(3\omega t + \Psi_3)$ and so on.

The equation for the instantaneous value of the complex wave is given by

$$e = e_1 + e_2 + ...e_n = E_{1m} \sin(\omega t + \Psi_1) \times E_{2m} \sin(2\omega t + \Psi_2) + ... + E_{nm} \sin(n\,\omega t + \Psi_n)$$ where

E_{1m}, E_{2m} and E_{nm} etc. denote the maximum values or the amplitudes of the fundamental, second harmonic and *n*th harmonic etc. and Ψ_1, Ψ_2 and Ψ_n represent the phase differences with respect to the complex wave* (*i.e.* angle between the zero value of complex wave and the corresponding zero value of the harmonic).

The number of terms in the series depends on the shape of the complex wave. In relatively simple waves, the number of terms in the series would be less, in others, more.

Similarly, the instantaneous value of the complex wave is given by

$$i = I_{1m} \sin(\omega t + \phi_1) + I_{2m} \sin(2\omega t + \phi_2) + ... + I_{nm} \sin(n\omega t + \phi_n)$$

---

*We could also express these phase anlges with respect to the fundamental wave instead of the complex wave.

Obviously $(\Psi_1 - \phi_1)$ is the phase difference between the harmonic voltage and current for the fundamental, $(\Psi_2 - \phi_2)$ for the second harmonic and $(\Psi_n - \Phi_n)$ for the nth harmonic.

20.4. R.M.S. Value of a Complex Wave

Let the equation of the given complex current wave be

$$i = I_{1m}\sin(\omega t + \phi_1) + I_{2m}\sin(2\omega t + \phi_2) \;.....\; I_{nm}\sin(n\omega t + \phi_n)$$

Its r.m.s. value is given by $I = \sqrt{\text{average value of } i^2 \text{over whole cycle}}$

Now $\quad i^2 = [I_{1m}\sin(\omega t + \phi_1) + I_{2m}\sin(2\omega t + \phi_1) + ... I_{nm}\sin(n\omega t + \phi_n)]^2$

$$= I_{1m}^2\sin^2(\omega t + \phi_1) + I_{2m}^2\sin^2(2\omega t + \phi_2) + ... I_{nm}^2\sin^2(n\omega t + \phi_n)]^2$$
$$+ 2I_{1m}I_{2m}\sin(\omega t + \phi_1)\sin(2\omega t + \phi_2) + 2I_{1m}I_{3m}\sin(\omega t + \Phi_1)\sin(\omega t + \Phi_2) + ...$$

The right-hand side of the above equation consists of two types of terms

(*i*) harmonic self-products, the general expression for which is $I_{pm}^2\sin^2(p\omega t + \phi_p)$ for the pth harmonic and

(*ii*) the products of different harmonics of the general form $2I_{pm}I_{qm}\sin(p\omega t + \phi_p)\sin(q\omega t + \phi_q)$

The average value of i^2 is the sum of the average values of these individual terms in the above equation. Let us now find the average value of the general term $I_{pm}^2\sin^2(p\omega t + \phi_p)$ over a whole cycle.

$$\text{Average value} = \frac{1}{2\pi}\int_0^{2\pi} I_{pm}^2\sin^2(p\omega t + \phi_p)d(\omega t) = \frac{I_{pm}^2}{2\pi}\int_0^{2\pi}\sin^2(p\theta + \phi_p)d\theta$$

$$= \frac{I_{pm}^2}{2\pi}\int_0^{2\pi}\left(\frac{1 - \cos 2(p\theta + \phi_p)}{2}\right)d\theta = \frac{I_{pm}^2}{4\pi}\left|\theta - \sin^2(p\theta + \phi_p)\right|_0^{2\pi}$$

$$= \frac{I_{pm}^2}{4\pi}\times 2\pi = \frac{I_{pm}^2}{2}$$

From this result, we can generalize that

$$\text{Average value of } I_{1m}^2\sin^2(\omega t + \phi_1) = \frac{I_{1m}^2}{2}$$

$$\text{Average value of } I_{2m}^2\sin^2(2\omega t + \phi_2) = \frac{I_{2m}^2}{2}$$

$$\text{Average value of } I_{nm}^2\sin^2(n\omega t + \phi_n) = \frac{I_{nm}^2}{2} \text{ and so on.}$$

Now, the average value of the product terms is

$$= \frac{1}{2\pi}\int_0^{2\pi} I_{pm}I_{qm}\sin(p\omega t + \phi_p)\sin(q\omega t + \phi_q)d(\omega t)$$

$$= \frac{I_{pm}I_{qm}}{2\pi}\int_0^{2\pi}\sin(p\theta + \phi_p)\sin(q\theta + \phi_q)d\theta = 0$$

$$\therefore \quad \text{Average value of } i^2 = \frac{I_{1m}^2}{2} + \frac{I_{2m}^2}{2} + ... + \frac{I_{nm}^2}{2}$$

$$\therefore \quad \text{r.m.s. value}, I = \sqrt{\text{average value of } i^2} = \sqrt{\frac{I_{1m}^2}{2} + \frac{I_{2m}^2}{2} + ...\frac{I_{nm}^2}{2}} \qquad ...(i)$$

$$= 0.707\sqrt{I_{1m}^2 + I_{2m}^2 + ... + I_{nm}^2}$$

Equation (*i*) above may also be put in the form

$$I = \sqrt{\left(\frac{I_{1m}}{\sqrt{2}}\right)^2 + \left(\frac{I_{2m}}{\sqrt{2}}\right)^2 + \ldots + \left(\frac{I_{nm}}{\sqrt{2}}\right)^2} = \sqrt{I_1^2 + I_2^2 + \ldots I_n^2}$$

where $I_1 = I_{1m}/\sqrt{2}$ –r.m.s. value of fundamental

$I_2 = I_{2m}/\sqrt{2}$ –r.m.s. value of 2nd harmonic

$I_n = I_{nm}/\sqrt{2}$ –r.m.s. value of nth harmonic

Similarly, the r.m.s. value of a complex voltage wave is

$$E = 0.707\sqrt{E_{1m}^2 + E_{2m}^2 + \ldots E_{nm}^2} = \sqrt{E_1^2 + E_2^2 + \ldots E_n^2}$$

Hence, the **rule** *is that the r.m.s. value of the complex current (or voltage) wave is given by the square-root of the sum of the squares of the r.m.s. values of its individual components.*

Note. If complex current wave contains a d.c. component of constant value I_D, then its equation is given by

$$i = I_D + I_{1m}\sin(\omega t + \phi_1) + I_{2m}\sin(2\omega t + \phi_2) + \ldots I_{nm}\sin(n\omega t + \phi_n)$$

r.m.s. value, $I = \sqrt{I_D^2 + (I_{1m}/\sqrt{2})^2 + (I_{2m}/\sqrt{2})^2 \ldots (I_{nm}/\sqrt{2}^2)} = I_D^2 + I_1^2 + I_2^2 + \ldots I_n^2$

20.5. Form Factor of a Complex Wave

In general, it may be defined as $k_f = \dfrac{\text{R.M.S. value}}{\text{average value}}$

A general expression for form factor in some simple cases may be found as under :

(*i*) Sine Series. Suppose the equation of a complex voltage wave is

$$v = V_{1m}\sin\omega t \pm V_{3m}\sin 3\omega t \pm V_{5m}\sin 5\omega t$$

$$= V_{1m}\sin\theta \pm V_{3m}\sin 3\theta \pm V_{5m}\sin 5\theta \quad \text{where} \quad \omega = 2\pi/T.$$

Obviously, zeros occurs at $t = 0$ or at $\theta = 0°$ and $\theta = 180°$ or $t = T/2$.

Mean value over half-cycle is

$$V_{av} = \frac{1}{\pi}\int_0^{\pi} v\,d\theta$$

$$= \frac{1}{\pi}\left[V_{1m}\int_0^{\pi}\sin\theta.d\theta \pm V_{3m}\int_0^{\pi}\sin 3.d\theta \pm V_{5m}\int_0^{\pi}\sin 5\theta.d\theta\right] = \frac{2}{\pi}\left(\frac{V_{1m}}{1} \pm \frac{V_{3m}}{3} \pm \frac{V_{5m}}{5}\right)$$

As found in Art. 20.4,

$$V = (V_1^2 + V_3^2 + V_5^2)^{1/2} = \frac{1}{\sqrt{2}}(V_{1m}^2 + V_{3m}^2 + V_{5m}^2)^{1/2}$$

$$\therefore \quad k_f = \frac{(1/\sqrt{2})(V_{1m}^2 + V_{3m}^2 + V_{5m}^2)^{1/2}}{\left(\frac{2}{\pi}\right)\left(V_{1m} \pm \frac{V_{3m}}{3} \pm \frac{V_{5m}}{5}\right)}$$

(ii) Cosine Series. Consider the following cosine series :

$$v = V_{1m}\cos\omega t \pm V_{3m}\cos 3\omega t \pm V_{5m}\cos 5\omega t$$

$$= V_{1m}\cos\theta \pm V_{3m}\cos 3\theta \pm V_{5m}\cos 5\theta$$

Obviously, in this case, zeros occur at $\theta = \pm\,\pi/2$ or 90°. Moreover, positive and negative half-cycles are symmetrical.

$$\therefore \quad V_{av} = \frac{1}{\pi}\int_{-\pi/2}^{\pi/2}(V_{1m}\cos\theta \pm V_{3m}\cos 3\theta \pm V_{5m}\cos 5\theta)d\theta = \frac{2}{\pi}\left(V_{1m} \pm \frac{V_{3m}}{3} \pm \frac{V_{5m}}{5}\right)$$

$$\therefore \quad k_f = \frac{(1/\sqrt{2})(V_{1m}^2 + V_{3m}^2 + V_{5m}^2)^{1/2}}{\left(\frac{2}{\pi}\right)\left(V_{1m} \mp \frac{V_{3m}}{3} \mp \frac{V_{5m}}{5}\right)}$$

Example 20.1. *A voltage given by v = 50 + 24 sin ωt–20 sin 2ωt is applied across the circuit shown in Fig. 20.5. What would be the readings of the instruments if ω = 10,000 rad/s. A_1 is thermoelectric ammeter, A_2 a moving-coil ammeter an V an electrostatic voltmeter.*

Solution. It may be noted that the thermoelectric ammeter and the electrostatic voltmeter record the r.m.s. values of the current and voltage respectively. But the moving coil ammeter records the average values. Since the average values of the sinusoidal waves are zero, hence the moving coil ammeter reads the d.c. component of the current only. The d.c. will pass only through the inductive branch and not through the capacitive branch.

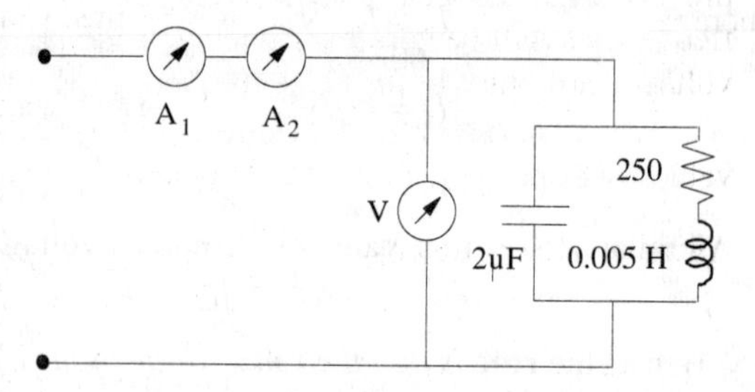

Fig. 20.5

$$\therefore \quad I_{DC} = \frac{V_{DC}}{R} = \frac{50}{250} = \mathbf{0.2\ A}$$

Equivalent impedance of the circuit at fundamental frequency is

$$\mathbf{Z}_1 = \frac{(R + jX_{L1})(-jX_{C1})}{R + jX_{L1} - jX_{C1}} = \frac{(250 + j50)(-j50)}{250 + j(50 - 50)} = \frac{2,500 - j12,500}{250} = 10 - j50 = 51\angle -78°42'$$

∴ r.m.s. fundamental current $I_1 = I_{1m}/\sqrt{2} = 24/51 \times \sqrt{2} = 0.33$ A

Equivalent impedance of the circuit at the second harmonic is

$$\mathbf{Z}_2 = \frac{(R + jX_{L2})(-jX_{C2})}{R + jX_{L2} - jX_{C2}} = \frac{(250 + 100)(-j25)}{200 + j75} = 31.5\angle -88°43'$$

∴ r.m.s. value of second harmonic current

$$I_2 = I_{2m}/\sqrt{2} = 20/31.5 \times \sqrt{2} = 0.449 \text{ A}$$

r.m.s. current in the circuit is

$$I = \sqrt{I_{DC}^2 + I_1^2 + I_2^2} = \sqrt{0.2^2 + 0.33^2 + 0.449^2} = 0.593 \text{ A}$$

Hence, the reading of the thermoelectric ammeter is **0.593 A**

The voltmeter reading $V = \sqrt{50^2 + (24/\sqrt{2})^2 + (20/\sqrt{2})^2} = \mathbf{54.66}$

Example 20.2. *Draw one complete cycle of the following wave*

i = 100 sin ωt + 40 sin 5ωt

Determine the average value, the r.m.s. value and form factor of the wave.

(Elect. Engineering, Osmania Univ. 1985)

Solution.
$$I_{av} = \frac{2}{\pi}\left(\frac{I_{2m}}{1} + \frac{I_{5m}}{5}\right) = \frac{2}{\pi}\left(\frac{100}{1} + \frac{40}{5}\right) = \mathbf{68.7\ A}$$

$$I = \sqrt{\frac{I_{1m}^2}{2} + \frac{I_{5m}^2}{2}} = \frac{1}{\sqrt{2}}(100^2 + 40^2)^{1/2} = \mathbf{76.2\ A}$$

Form factor
$$= \frac{I}{I_{av}} = \frac{76.2}{68.7} = \mathbf{1.109}$$

20.6. Power Supplied by a Complex Wave

Let the complex voltage be represented by the equation

$$e = E_{1m} \sin \omega t + E_{2m} \sin 2\omega t + \ldots E_{nm} \sin n\omega t$$

be applied to a circuit. Let the equation of the resultant current wave be

$$i = I_{1m} \sin(\omega t + \phi_1) + I_{2m} \sin(2\omega t + \phi_2) + \ldots I_{nm} \sin(n\omega t + \phi_n)$$

The instantaneous value of the power in the circuit is $p = ei$ watt

For obtaining the value of this product, we will have to multiply every term of the voltage wave, in turn, by every term in the current wave. The average power supplied during a cycle would be equal to the sum of the average values over one cycle of each individual product term. However, as proved in Art. 20.4 earlier, the average value of all product terms involving harmonics of different frequencies will be zero over one cycle, so that we need consider only the products of current and voltage harmonics of the same frequency.

Let us consider a general term of this nature *i.e.* $E_{nm} \sin n\omega t \times I_{nm} \sin (n\omega t - \phi_n)$ and find its average value over one cycle of the fundamental.

$$\text{Average value of power} = \frac{1}{2\pi}\int_0^{2\pi} E_{nm} I_{nm} \sin n\omega t \sin (n\omega t - \phi_n)\, d(\omega t)$$

$$= \frac{E_{nm} I_{nm}}{2\pi}\int_0^{2\pi} \sin n\theta \sin (n\theta - \phi_n) d\theta$$

$$= \frac{E_{nm} I_{nm}}{2\pi}\int_0^{2\pi} \frac{\cos \phi_n - \cos(2n\theta - \phi_n)}{2} d\theta$$

$$= \frac{E_{nm} I_{nm} \cos \phi}{2} = \frac{E_{nm}}{\sqrt{2}} \cdot \frac{I_{nm}}{\sqrt{2}} \cdot \cos \phi_n = E_n I_n \cos \phi_n$$

where E_n and I_n are the r.m.s. values of the voltage and current respectively. Hence, total average power supplied by a complex wave *is the sum of the average power supplied by each harmonic component acting independently.*

$\therefore$ Total power is $P = E_1 I_1 \cos \phi_1 + E_2 I_2 \cos \phi_2 + ... E_n I_n \cos \phi_n$

The overall power factor is given by

$$\text{p.f.}^* = \frac{\text{total watts}}{\text{total voltamperes}} = \frac{E_1 I_1 \cos \phi_1 + E_2 I_2 \cos \phi_2 + \ldots}{E \times I}$$

where E = r.m.s. value of the complex voltage wave

I = r.m.s. value of the complex current wave

Example 20.3. *A single-phase voltage source 'e' is given by*

$$e = 141 \sin \omega t + 42.3 \sin 3\omega t + 28.8 \sin 5\omega t$$

The corresponding current in the load circuit is given by

$$i = 16.5 \sin (\omega t + 54.5°) + 8.43 \sin (3\omega t - 38°) + 4.65 \sin (5\omega t - 34.3°)$$

Find the power supplied by the source.

(Electrical Circuits, Nagpur Univ. 1991)

Solution. In problems of such type, it is best to deal with each harmonic separately

$$\text{Power at fundamental} = E_1 I_1 \cos \phi_1 = \frac{E_{1m}}{\sqrt{2}} \cdot \frac{I_{1m}}{\sqrt{2}} \cos \phi_1 = \frac{E_{1m} I_{1m}}{2} \cos \phi_1 = \frac{141 \times 16.5}{2} \cos 54.5° = 675.5 \text{ W}$$

$$\text{Power at 3rd harmonic} = \frac{E_{3m} I_{3m}}{2} \cos \phi_3 = \frac{42.3 \times 8.43}{2} \cos 38° = 140.5 \text{ W}$$

$$\text{Power at 5th harmonic} = \frac{28.8 \times 4.65}{2} \cos 34.3° = 55.5 \text{ W}$$

Total power supplied $= 675.5 + 140.5 + 55.5 =$ **871.5 W**

---

*When harmonics are present, it is obvious that the overall p.f. of the circuit cannot be stated lagging or leading. It is simply the ratio of power in watts to voltamperes.

Example 20.4. *A complex voltage is given by e = 60 sin ωt + 24 sin (3ωt + ω/6) + 12 sin (5ωt + π/3) is applied across a certain circuit and the resulting current is given by*

i = 0.6 sin (ωt −2π/10) + 0.12 sin (3ωt−2π/24) + 0.1 sin (5π−3π/4)

Find (i) r.m.s. value of current and voltage (ii) total power supplied and (iii) the overall power factor.

Solution. In such problems where harmonics are involved, it is best to deal with each harmonic separately.

$$\text{Power at fundamental} = E_1 I_1 \cos\phi_1 = \frac{E_{1m} I_{1m}}{2}\cos\phi_1 = \frac{60 \times 0.6}{2} \times \cos 36° = 14.56 \text{ W}$$

$$\text{Power at 3rd harmonic} = \frac{E_{3m} I_{3m}}{2}\cos 45° = \frac{24 \times 0.12}{2} \times 0.707 = 1.02 \text{ W}$$

$$\text{Power at 5th harmonic} = \frac{E_{5m} I_{5m}}{2}\cos 75° = \frac{12 \times 0.1}{2} \times 0.2588 = 0.16 \text{ W}$$

$$(i) \text{ R.M.S. current} \quad I = \sqrt{I_1^2 + I_3^2 + I_5^2} = \sqrt{\left(\frac{I_{1m}}{\sqrt{2}}\right)^2 + \left(\frac{I_{3m}}{\sqrt{2}}\right)^2 + \left(\frac{I_{5m}}{\sqrt{2}}\right)^2}$$

$$(ii) \text{ Total power} = 14.56 + 1.02 + 0.16 = \mathbf{15.74\ W} = \sqrt{\frac{0.6^2}{2} + \frac{0.12^2}{2} + \frac{0.1^2}{2}} = \mathbf{0.438\ A}$$

$$\text{R.M.S. volts,} \quad E = \sqrt{\frac{60^2}{2} + \frac{24^2}{2} + \frac{12^2}{2}} = \mathbf{46.5\ V}$$

$$(iii) \text{ Overall p.f.} = \frac{\text{watts}}{\text{voltamperes}} = \frac{15.74}{46.5 \times 0.438} = \mathbf{0.773}$$

20.7. Harmonics in Single-phase A.C. Circuits

If an alternating voltage, containing various harmonics, is applied to a single-phase circuit containing linear circuit elements, then the current so produced also contains harmonics. Each harmonic voltage will produce its own current independent of others. By the principle of superposition, the combined current can be found. We will now consider some of the well-known elements like pure resistance, pure inductance and pure capacitance and then various combinations of these. In each case, we will assume that the applied complex voltage is represented by

$$e = E_{1m} \sin \omega t + E_{2m} \sin 2\omega t + \ldots\ldots + E_{nm} \sin n\omega t$$

(a) Pure Resistance

Let the circuit have a resistance of R which is independent of frequency.

The instantaneous current i_1 due to fundamental voltage is

$$i_1 = \frac{E_{1m} \sin \omega t}{R}$$

Similarly, $$i_2 = \frac{E_{2m} \sin 2\omega t}{R} \quad \text{for 2nd harmonic}$$

and $$i_n = \frac{E_{nm} \sin n\omega t}{R} \quad \ldots\text{for } n\text{th harmonic}$$

Total current $$i = i_1 + i_2 + \ldots.. + i_n$$

$$= \frac{E_{1m} \sin \omega t}{R} + \frac{E_{2m} \sin 2\omega t}{R} + \ldots.. + \frac{E_{nm} \sin n\omega t}{R}$$

$$= I_{1m} \sin \omega t + I_{2m} \sin 2\omega t + \ldots\ldots + I_{nm} \sin n\omega t$$

It shows that

(*i*) the waveform of the resulting current is similar to that of the applied voltage *i.e.* the two waves are identical.

(*ii*) the percentage of harmonic content in the current wave is the same as in the applied voltage.

(b) Pure Inductance

Let the inductance of the circuit be L henry whose reactance varies directly as the frequency of the applied voltage. Its reactance for the fundamental would be $X_1 = \omega L$; for the second harmonic, $X_2 = 2\omega L$, for the third harmonic, $X_3 = 3\omega L$ and for the nth harmonic $X_n = n\omega L$. However, for every harmonic term, the current will lag behind the voltage by 90°.

Current due to fundamental, $i_1 = \dfrac{E_{1m}}{\omega L} \sin(\omega t - \pi/2)$

Current due to 2nd harmonic, $i_2 = \dfrac{E_{2m}}{2\omega L} \sin(2\omega t - \pi/2)$

Current due to 3rd harmonic, $i_3 = \dfrac{E_{3m}}{3\omega L} \sin(3\omega t - \pi/2)$

Current due to nth harmonic, $i_n = \dfrac{E_{nm}}{n\omega L} \sin(n\omega t - \pi/2)$

$\therefore$ Total current $i = i_1 + i_2 + \ldots\ldots i_n$

$$= \frac{E_{1m}}{\omega L} \sin(\omega t - \pi/2) + \frac{E_2 m}{2\omega L} \sin(2\omega t - \pi/2) + \ldots + \frac{E_{nm}}{n\omega L} \sin(n\omega t - \pi/2)$$

It can be seen from the above equation that

(*i*) the waveform of the current differs from that of the applied voltage.

(*ii*) for the nth harmonic, the percentage harmonic content in the current-wave is $1/n$ of the corresponding harmonic content in the voltage wave. It means that in an inductive circuit, the current waveform shows less distortion than the voltage waveform. In this case, current more nearly approaches a sine wave than it does in a circuit containing resistance.

(c) Pure Capacitance

In this case,

$X_1 = \dfrac{1}{\omega C}$ – for fundamental ; $X_2 = \dfrac{1}{2\omega C}$ –for 2nd harmonic

$X_3 = \dfrac{1}{3\omega C}$ –for 3rd harmonic ; $X_n = \dfrac{1}{n\omega C}$ – for nth harmonic

$$i_1 = \frac{E_{1m}}{1/\omega C} \sin(\omega t + \pi/2) = \omega C E_{1n} \sin(\omega t + \pi/2)$$

$$i_2 = \frac{E_{2m}}{1/2\omega C} \sin(2\omega t + \pi/2) = 2\omega C E_{2m} \sin(2\omega t + \pi/2)$$

$$i_n = \frac{E_{nm}}{1/n\omega C} \sin(n\omega t + \pi/2) = n\,\omega C E_{nm} \sin(n\omega t + \pi/2)$$

For every harmonic term, the current will lead the voltage by 90°.

Now $i = i_1 + i_2 + \ldots\ldots + i_n$

$$= \omega C E_{1m} \sin(\omega t + \pi/2) + 2\omega C E_{2m} \sin(2\omega t + \pi/2) + \ldots + n\omega C E_{nm} \sin(n\omega t + \pi/2)$$

This equation shows that

(*i*) the current and voltage waveforms are dissimilar.

(*ii*) percentage harmonic content of the current is larger than that of the applied voltage wave. For example, for nth harmonic, it would be n times larger.

(*iii*) as a result, the current wave is more distorted than the voltage wave.

(*iv*) effect of capacitor on distrotion is just the reverse of that of inductance.

Example 20.5. *A complex wave of r.m.s. value 240 V has 20% 3rd harmonic content, 5% 5th harmonic content and 2% 7th harmonic content. Find the r.m.s. values of the fundamental and of each harmonic.* **(Elect. Circuits, Gujarat Univ. 1985)**

Solution. Let V_1, V_3, V_5 and V_7 be the r.m.s. values of the fundamental and harmonic voltages. Then

$$V_3 = 0.2\,V_1;\; V_5 = 0.05\,V_1 \text{ and } V_7 = 0.02\,V_1$$

$$240 = (V_1^2 + V_3^2 + V_5^2 + V_7^2)^{1/2}$$

$$\therefore \quad 240 = [V_1^2 + (0.2V_1)^2 + (0.05V_1)^2 + (0.02V_1)^2]^{1/2}$$

$$\therefore \quad V_1 = \mathbf{235\ V};\; V_3 = 0.2 \times 235 = \mathbf{47\ V}$$

$$V_5 = 0.05 \times 235 = \mathbf{11.75\ V};\; V_7 = 0.02 \times 235 = \mathbf{4.7\ V}$$

Example 20.6. *Derive an expression for the power, power factor and r.m.s. value for a complex wave.*

A voltage e = 250 sin ωt + 50 sin (3ωt + π/3) + 20 sin (5ωt + 5π/6) is applied to a series circuit of resistance 20Ω and inductance 0.05 H. Derive (a) an expression for the current (b) the r.m.s. value of the current and for the voltage (c) the total power supplied and (d) the power factor. Take ω = 314 rad/s. **(Electrical Circuits, Nagpur Univ. 1991)**

Solution. For Fundamental

$$X_1 = \omega L = 314 \times 0.05 = 15.7\,\Omega;\; \mathbf{Z}_1 = 20 + j15.7 = 25.4\angle 38.1°\,\Omega$$

For Third Harmonic

$$X_3 = 3\,\omega L = 3 \times 15.7 = 47.1\,\Omega;\; \mathbf{Z}_3 = 20 + j47.1 = 51.2\angle 67°\,\Omega$$

For Fith Harmonic

$$X_5 = 5\,\omega L = 5 \times 15.7 = 78.5\,\Omega;\; \mathbf{Z}_5 = 20 + j78.5 = 81\angle 75.7°\,\Omega$$

(*a*) Expression for the current is

$$i = \frac{250}{25.4}\sin(\omega t - 38.1°) + \frac{50}{51.2}\sin(3\omega t + 60° - 67°) + \frac{20}{81}\sin(5\,\omega t + 150° - 75.7°)$$

$$\therefore \quad i = 9.84\sin(\omega t - 38.1°) + 0.9\sin(3\omega t - 7°) + 0.25\sin(5\omega t + 74.3°)$$

(*b*) *R.M.S. current*
$$I = \sqrt{\frac{I_{1m}^2}{2} + \frac{I_{3m}^2}{2} + \frac{I_{5m}^2}{2}}$$

$$I^2 = \frac{9.84^2}{2} + \frac{0.9^2}{2} + \frac{0.25^2}{2} = 48.92$$

$$\therefore \quad I = \sqrt{48.92} = \mathbf{6.99\ A}$$

R.M.S. voltage
$$V = \sqrt{\frac{250^2}{2} + \frac{50^2}{2} + \frac{20^2}{2}} = \mathbf{180.8\ V}$$

(*c*) Total power $= I^2R = 48.92 \times 20 = \mathbf{978\ W}$

(*d*) Power factor $= \dfrac{\text{watts}}{\text{VA}} = \dfrac{978}{180.8 \times 6.99} = \mathbf{0.773}$

Example 20.7. *An r.m.s. current of 5A, which has a third harmonic content, is passed through a coil having a resistance of 1 Ω and an inductance of 10 mH. The r.m.s voltage across the coil is 20 V. Calculate the magnitude of the fundamental and harmonic components of current if the fundamental frequency is 300/2π Hz. Also, find the power dissipated.*

Solution. (*i*) **Fundamental Frequency**

$\omega = 300$ rad/s ; $X_L = 300 \times 10^{-2} = 3\ \Omega$ $\therefore$ $\mathbf{Z}_1 = 1 + j3 = 3.16\angle 71.6°$ ohm

If V_1 is the r.m.s. value of the fundamental voltage across the coil, then

$$V_1 = I_1 Z_1 = 3.16\, I_1$$

(*ii*) **Third Harmonic**

$X_3 = 3 \times 3 = 9\ \Omega$; $\mathbf{Z}_3 = 1 + j9 = 9.05\angle 83.7°$ ohm ; $V_3 = I_3 Z_3 = 9.05\, I_3$

Since r.m.s. current of the complex wave is 5A and r.m.s. voltage drop 20 V

$$5 = \sqrt{I_1^2 + I_3^2} \text{ and } 20 = \sqrt{V_1^2 + V_3^2}$$

Substituting the values of V_1 and V_3, we get, $20 = [(3.16\, I_1)^2 + (9.05\, I_3)^2]^{1/2}$

Solving for I_1 and I_3, we have $I_1 =$ **4.8 A** and $I_3 =$ **1.44 A**

Power dissipated $= I^2 R = 5^2 \times 1 =$ **25 W**

Example 20.8. *An e.m.f. represented by the equation e = 150 sin 314t + 50 sin 942t is applied to a capacitor having a capacitance of 20 μF. What is the r.m.s. value of the charging current ?*

Solution. For Fundamental

$X_{C1} = 1/\omega C = 10^6/20 \times 314 = 159\ \Omega$; $I_{1m} = E_{1m}/X_{C1} = 150/159 = 0.943$ A

For Third Harmonic

$X_{C3} = 1/3\omega C = 159/3 = 53\ \Omega$ $\therefore$ $I_{3m} = E_{3m}/X_{C3} = 50/53 = 0.943$ A

r.m.s. value of charging current,

$$I = \sqrt{\frac{I_{1m}^2}{2} + \frac{I_{3m}^2}{2}} = \sqrt{\frac{0.943^2}{2} + \frac{0.943^2}{2}} \quad \text{or} \quad I = \mathbf{0.943\ A}$$

Example 20.9. *The voltage given by v = 100 cos 314t + 50 sin (1570t–30°) is applied to a circuit consisting of a 10 Ω resistance, a 0.02 H inductance and a 50 μF capacitor. Determine the instantaneous current through the circuit. Also find the r.m.s. value of the voltage and current.*

Solution. For Fundamental

$\omega = 314$ rad/s; $X_L = 314 \times 0.02 = 6.28\ \Omega$

$X_C = 10^6/314 \times 50 = 63.8\ \Omega$; $X = X_L - X_C = 6.28 - 63.8 = -57.32\ \Omega$

$Z = \sqrt{10^2 + (-57.32)^2} = 58.3\ \Omega$; $I_{1m} = 100/58.3 = 1.71$ A

$\phi_1 = \tan^{-1}(-57.32/10) = -80.2°$ (lead) ; $i_1 = 1.71 \cos(314t + 80.2°)$

For Fifth Harmonic

Inductive reactance $= 5X_L = 5 \times 6.28 = 31.4\ \Omega$

Capacitive reactance $= X_C/5 = 63.8/5 = 12.76\ \Omega$

Net reactance $= 31.4 - 12.76 = 18.64\ \Omega$

$$Z = \sqrt{10^2 + 18.64^2} = 21.2\ \Omega$$

$$I_{5m} = 50/21.2 = 2.36\text{ A}; \ \phi_5 = \tan^{-1}(18.64/10) = 61.8°\text{(lag)}$$

$$i_5 = 2.36 \sin(1570t - 30° - 61.8°) = 2.36 \sin(1570t - 91.8°)$$

Hence, total instantaneous current is

$$i = i_1 + i_5 = 1.71 \cos(314t + 80.2°) + 2.36 \sin(1570t - 91.8°)$$

$$R.M.S.\ volt = \sqrt{\frac{100^2}{2} + \frac{50^2}{2}} = \mathbf{79.2\ V}$$

$$R.M.S.\ current = \sqrt{\frac{1.71^2}{2} + \frac{2.36^2}{2}} = \mathbf{2.06\ A}$$

Example 20.10. *A 6.36 μF capacitor is connected in parallel with a resistance of 500 Ω and the combination is connected in series with a 500-Ω resistor. The whole circuit is connected across an a.c. voltage given by e = 300 sin ω t + 100 sin (3ω t + π/6).*

If ω = 314 rad/s, find

(i) power dissipated in the circuit

(ii) an expression for the voltage across the series resistor

(iii) the percentage harmonic content in the resultant current.

Solution. For Fundamental

$$X_{C1} = \frac{1}{\omega C} = \frac{10^6}{314 \times 6.36} = 500\ \Omega$$

The impedance of the whole series-parallel circuit is given by

$$\mathbf{Z}_1 = 500 + \frac{500(-j500)}{500 - j500} = 750 - j250 = 791\angle -18.4°$$

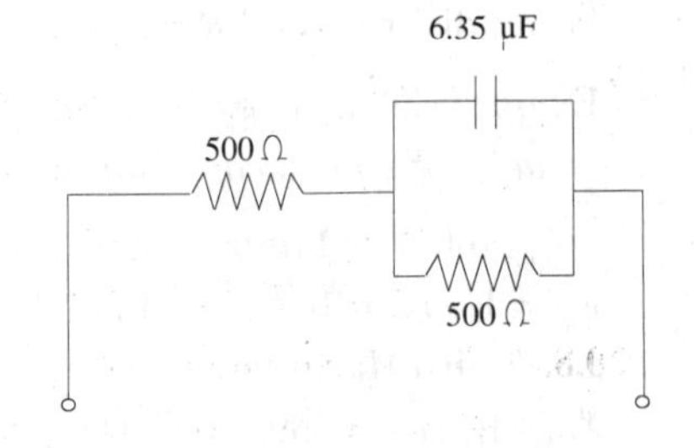

Fig. 20.6

For Third Harmonic

$$X_{C3} = 1/3\omega C = 500/3 = 167\ \Omega$$

$$\therefore \quad \mathbf{Z}_3 = 500 + \frac{500(-j167)}{500 - j167} = 550 - j150 = 570\angle -15.3°$$

$$\therefore \quad i = \frac{300}{791}\sin(\omega t + 18.4°) + \frac{100}{570}\sin(3\omega t + 45.3°)$$

$$= 0.379 \sin(\omega t + 18.4°) + 0.175 \sin(3\omega t + 45.3)$$

$$(i)\ \text{Power dissipated} = \frac{E_{1m}I_{1m}}{2}\cos\phi_1 + \frac{E_{3m}I_{3m}}{2}\cos\phi_3$$

$$= \frac{300 \times 0.379}{2} \times \cos 18.4° + \frac{100 \times 0.175}{2}\cos 15.3 = \mathbf{62.4\ W}$$

(*ii*) The voltage drop across the series resistor would be

$E_R = iR = 500[0.379 \sin(\omega t + 18.4°) + 0.175 \sin(3\omega t + 45.3°)]$

$e_R = \mathbf{189.5\ sin\ (\omega t + 18.4°) + 87.5\ sin\ (3\omega t + 45.3°)}$

(*iii*) The percentage harmonic content of the current is = 87.5/189.5) × 100 = 46.2%

Example 20.11. *An alternating voltage of v = 1.0 sin 500t + 0.5 sin 1500t is applied across a capacitor which can be represented by a capacitance of 0.5 μF shunted by a resistance of 4,000 Ω. Determine*

(i) the r.m.s value of the current (ii) the r.m.s value of the applied voltage

(iii) the p.f. of the circuit.

(Circuit Theory and Components, Madras Univ. 1981)

Solution. For Fundamental [Fig. 20.7 (*a*)]

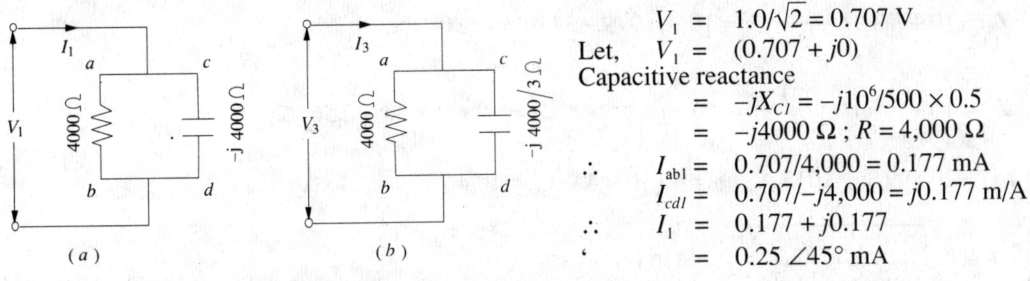

Fig. 20.7

$$V_1 = 1.0/\sqrt{2} = 0.707 \text{ V}$$

Let, $V_1 = (0.707 + j0)$

Capacitive reactance

$$= -jX_{C1} = -j10^6/500 \times 0.5$$
$$= -j4000\ \Omega\ ;\ R = 4,000\ \Omega$$
$$\therefore \quad I_{ab1} = 0.707/4,000 = 0.177 \text{ mA}$$
$$I_{cd1} = 0.707/-j4,000 = j0.177 \text{ m/A}$$
$$\therefore \quad I_1 = 0.177 + j0.177$$
$$= 0.25\angle 45° \text{ mA}$$

Hence, I_1 leads the fundamental voltage by 45°.

$P_{ab1} = 0.707 \times 0.177 = 0.125$ mW ; $P_{cd1} = 0$

For Third Harmonic [Fig. 20.7 (*b*)]

$\mathbf{V}_3 = 0.5/\sqrt{2} = 0.3535\ \angle 0°$; $R = 4,000\ \Omega : \mathbf{X}_{C3} = -j4,000/3\ \Omega$

$\mathbf{I}_{ab3} = 0.3535/4000 = 0.0884\ \angle 0°$ mA ; $\mathbf{I}_{cd3} = 0.3535/-j(4,000/3) = j0.265$ mA

$\mathbf{I}_3 = 0.0884 + j0.265 = 0.28\ \angle\ 71.6°$ mA

$P_{ab3} = 0.3535 \times 0.0884 = 0.0313$ mW ; $P_{cd3} = 0$

(*i*) *R.M.S.* current $= \sqrt{I_1^2 + I_3^2} = \sqrt{0.25^2 + 0.28^2} = \mathbf{0.374\ mA}$

(*ii*) *R.M.S.* voltage $= \sqrt{(1/\sqrt{2})^2 + (0.5/\sqrt{2})^2} = \mathbf{0.79\ V}$

(*iii*) Power factor = watts/voltamperes

Wattage $= (0.125 + 0.0313) \times 10^{-3} = 0.1563 \times 10^{-3}$ W

Volt-amperes $= 0.79 \times 0.374 = 0.295$; p.f. $= 0.1563 \times 10^{-3}/0.295 = \mathbf{0.0005}$

20.8. Selective Resonance Due to Harmonics

When a complex voltage is applied across a circuit containing both inductance and capacitance, it may happen that the circuit resonates at one of the harmonic frequencies of the applied voltage. This phenomenon is known as *selective resonance*.

If it is a series circuit, then large currents would be produced at resonance, even though the applied voltage due to this harmonic may be small. Consequently, it would result in large harmonic voltage appearing across both the capacitor and the inductance.

If it is a parallel circuit, then at resonant frequency, the resultant current drawn from the supply would be minimum.

It is because of the possibility of such selective resonance happening that every effort is made to eliminate harmonics in supply voltage.

However, the phenomenon of selective resonance has been usefully employed in some wave analysers for determining the harmonic content of alternating waveforms. For this purpose, a variable inductance, a variable capacitor, a variable non-inductive resistor and a fixed non-inductive resistance or shunt for an oscillograph are connected in series and connected to show the wave-form of the voltage across the fixed non-inductive resistance. The values of inductance and capacitance are adjusted successively to give resonance for the first, third, fifth and seventh harmonics and a record of the waveform is obtained by the oscillograph. A quick inspection of the shape of the waveform helps to detect the presence or absence of a particular harmonic.

Example 20.12. *An e.m.f. e = 200 sin ωt + 40 sin 3 ωt + 10 sin 5 ωt is impressed on a circuit comprising of a resistance of 10 Ω, a variable inductor and a capacitance of 30 μF, all connected in series. Find the value of the inductance which will give resonance with triple frequency component of the pressure and estimate the effective p.f. of the circuit, ω = 300 radian/second.*

(Elect. Engg.I, Bombay Univ. 1985)

Solution. For resonance at third harmonic

$3\omega L = 1/3\omega C \quad \therefore\ L = 1/9\omega^2 C = 10^6/9 \times 300^2 \times 30 = \mathbf{0.041\ H}$

$$\mathbf{Z}_1 = 10 + j\left(300 \times 0.041 - \frac{10^6}{300 \times 30}\right) = 10 + j(12.3 - 111.1) = 10 - j98.8 = 99.3\ \angle -84.2°$$

$$\mathbf{Z}_3 = 10 + j\left(3\omega L - \frac{1}{3\omega C}\right) = 10 + j(36.9 - 37.0) = 10\angle 0°$$

$$\mathbf{Z}_5 = 10 + j\left(5\omega L - \frac{1}{5\omega L}\right) = 10 + j(61.5 - 22.2) = 10 + j39.3 = 40.56\ \angle 75.7°$$

$$I_{1m} = 200/99.3 = 2.015\ \text{A}\ ;\ I_{3m} = 40/10 = 4\text{A}\ ;\ I_{5m} = 10/40.56 = 0.246\ \text{A}$$

$$I = \sqrt{\frac{2.015^2}{2} + \frac{4^2}{2} + \frac{0.246^2}{2}} = \sqrt{10.06} = 3.172\ \text{A}$$

$$V = \sqrt{\frac{200^2}{2} + \frac{40^2}{2} + \frac{10^2}{2}} = 144.5\ \text{V}\ ;\ \text{Power} = I^2R = 10.06 \times 10 = 100.6\ \text{W}$$

Volt-amperes = VI = 144.5 × 3.172 = 458 VA ; Power factor = 100.6/458 = **0.22**

Example 20.13. *A coil having R = 100 Ω and L = 0.1 H is connected in series with a capacitor across a supply, the voltage of which is given by e = 200 sin 314t + 5 sin 3454t. What capacitance would be required to produce resonance with the 11th harmonic. Find (a) the equation of the current and (b) the r.m.s. value of the current, if this capacitance is in circuit.*

Solution. For series resonance, $X_L = X_C$

Since resonance is required for 11th harmonic whose frequency is 3454 rad/s, hence

$$3454\ L = \frac{1}{3454C}\ ;\ C = \frac{1}{3454^2 \times 0.1}\ \text{farad} = \mathbf{0.838\ \mu F}$$

(a) For Fundamental

Inductive reactance $= \omega L = 314 \times 0.1 = 31.4\ \Omega$

Capacitive reactance $= 1/\omega C = 10^6/0.838 \times 314 = 3796\ \Omega$

∴ Net reactance $= 3796 - 31.4 = 3765\Omega$; Resistance = 100 Ω

∴ $Z_1 = \sqrt{100^2 + 3765^2} = 3767\ \Omega$; $\tan \phi_1 = 3765/100 = 37.65$

∴ $\phi_1 = 88°28'$ (leading) = 1.546 radian

Now $E_{1m} = 200\ V$; $Z_1 = 3767\ \Omega$ ∴ $I_{1m} = 200/3767 = 0.0531$ A

Eleventh Harmonic

Net reactance = 0; Impedance $= Z_{11} = 100\ \Omega$

∴ Current I_{11m} = 5/100 = 0.05 A ; $\phi_{11} = 0$ ...at resonance

Hence, the equation of the current is

$$i = \frac{200}{3767}(\sin 314t + 1.546) + \frac{5}{100}\sin(3454\,t + 0)$$

$$i = \mathbf{0.0531\ sin\ (314t + 1.546) + 0.05\ sin\ 3454\ t}$$

(*b*) $I = \sqrt{(0.0531)^2/2 + (0.05)^2/2} = \mathbf{0.052\ A}$

20.9. Effect of Harmonics on Measurement of Inductance and Capacitance

Generally, with the help of ammeter and voltmeter readings, the value of impedance, inductance and capacitance of a circuit can be calculated. But while dealing with complex voltages, the use of instrument readings does not, in general, give correct values of inductance and capacitance except in the case of a circuit containing only pure resistance. It is so because, in the case of resistance, the voltage and current waveforms are similar and hence the values of r.m.s. volts and r.m.s amperes (as read by the voltmeter and ammeter respectively) would be the same whether they are sinusoidal or non-sinusoidal (*i.e.* complex).

(i) Effect on Inductance

Let L be the inductance of a circuit and E and I the r.m.s. values of the applied voltage and current as read by the instruments connected in the circuit. For a complex voltage

$$E = 0.707\sqrt{(E_{1m}{}^2 + E_{3n}{}^2 + E_{5m}{}^2 + ...)}$$

Hence $$I = 0.707\sqrt{\left[\left(\frac{E_{1m}}{\omega L}\right)^2 + \left(\frac{E_{3m}}{3\omega L}\right)^2 + \left(\frac{E_{5m}}{5\omega L}\right)^2 + ...\right]}$$

$$= \frac{0.707}{\omega L}\sqrt{\left(E_{1m}{}^2 + \frac{1}{9}E_{3m}{}^2 + \frac{1}{25}E_{5m}{}^2 + ...\right)}$$

$\therefore$ $$L = \frac{0.707}{\omega I}\sqrt{\left(E_{1m}{}^2 + \frac{1}{9}E_{3m}{}^2 + \frac{1}{25}E_{5m}{}^2 + ...\right)}$$

For calculating the value of L from the above expression, it is necessary to know the absolute value of the amplitudes of several harmonic voltages. But, in practice, it is more convenient to deal with relative values than with absolute values. For this purpose, let us multiply and divide the right-hand side of the above expression by E but write the E *in the denominator* in its form $0.707\sqrt{(E_{1m}{}^2 + E_{3m}{}^2 + E_{5m}{}^2 + ...)}$

$\therefore$ $$L = \left[\frac{0.707}{\omega I}\sqrt{(E_{1m}{}^2 + 1/9 \,.\, E_{3m}{}^2 + 1/25 \,.\, E_{5m}{}^2 + ...)}\right] \times \left[\frac{E}{0.707\sqrt{(E_{1m}{}^2 + E_{3m}{}^2 + E_{5m}{}^2 + ...)}}\right]$$

or $$L = \frac{E}{\omega I}\sqrt{\left[\frac{E_{1m}{}^2 + 1/9 \,.\, E_{3m}{}^2 + 1/25 \,.\, E_{5m}{}^2 + ...)}{(E_{1m}{}^2 + E_{3m}{}^2 + E_{5m}{}^2 + ...)}\right]} = \frac{E}{\omega I}\sqrt{\left[\frac{1 + 1/9 \,.\, (E_{3m}/E_{1m})^2 + 1/25 \,.\, (E_{5m}/E_{1m})^2 + ...}{1 + (E_{3m}/E_{1m})^2 + (E_{5m}/E_1)^2 + ...}\right]}$$

If the effect of harmonics were to be neglected, then the value of the inductance would appear to be $E/\omega I$ but the true or actual value is less than this. The apparent value has to be multiplied by the quantity under the radical to get the true value of inductance when harmonics are present.

The quantity under the radical is called the correction factor *i.e.*

True inductance (L) = Apparent inductance (L') $\times$ correction factor

(ii) Effect on Capacitance

Let the capacitance of the circuit be C farads and E and I the instrument readings for voltage and current. Since the instruments read r.m.s. values, hence, as before,

$$E = 0.707\sqrt{(E_{1m}{}^2 + E_{3m}{}^2 + E_{5m}{}^2 + ...)}$$

Hence $$I = 0.707\sqrt{\left[\left(\frac{E_{1m}}{1/\omega C}\right)^2 + \left(\frac{E_{3m}}{1/3\omega C}\right)^2 + \left(\frac{E_{5m}}{1/5\omega C}\right)^2 + ...\right]}$$

$$= 0.707\sqrt{(\omega C E_{1m})^2 + (3\omega C E_{3m})^2 + (5\omega C\, E_{5m})^2 + ...}$$

$$= 0.707\ \omega C\sqrt{E_{1m}{}^2 + 9E_{3m}0^2 + 25E_{sm}{}^2 + \ .....}$$

$\therefore$ $$C = \frac{I}{0.707\omega\sqrt{(E_{1m}{}^2 + 9E_{3m}{}^2 + 25E_{5m}{}^2 + ...)}}$$

Again, we will multiply and divide the right-hand side E but in this case, we will write E in the *numerator* in its form $[0.707\sqrt{(E_{1m}{}^2 + E_{3m}{}^2 + E_{5m}{}^2 + ...)}]$

$\therefore$ $$C = \left[\frac{1}{0.707\ \omega E\sqrt{(E_{1m}{}^2 + 9E_{3m}{}^2 + 25E_{5m}{}^2 + ...)}}\right] \times \left[0.707\sqrt{(E_{1m}{}^2 + E_{3m}{}^2 + E_{5m}{}^2 + ...)}\right]$$

$$=\frac{1}{\omega E}\sqrt{\left[\frac{E_{1m}^2+E_{3m}^2+E_{5m}^2\ldots}{E_{1m}^2+9E_{3m}^2+25E_{5m}^2+\ldots}\right]}=\frac{1}{\omega E}\sqrt{\left[\frac{1+(E_{3m}/E_{1m})^2+(E_{5m}/E_{1m})^2\ldots\ldots}{1+9(E_{3m}/E_{1m})^2+25(E_{5m}/E_{1m})^2+\ldots\ldots}\right]}$$

Again, if the effects of harmonics were neglected, the value of capacitance would appear to be $I/\omega E$ but its true value is less than this. For getting the true value, this apparent value will have to be multiplied by the quantity under the radical (which, therefore, is referred to as correction factor).*

∴ True capacitance (C) = Apparent capacitance (C')× correction factor

Example 20.14. *A current of 50-Hz containing first, third and fifth harmonics of maximum values 100, 15 and 12 A respectively, is sent through an ammeter and an inductive coil of negligibly small resistance. A voltmeter connected to the terminals shows 75 V. What would be the current indicated by the ammeter and what is the exact value of the inductance of the coil in henrys ?*

Solution. The r.m.s. current is

$$I = 0.707\sqrt{I_{1m}^2+I_{3m}^2+I_{5m}^2} = 0.707\sqrt{(100^2+15^2+12^2)} = 72 \text{ A}$$

Hence, current indicated by the ammeter is **72 A**

Now $E = 0.707\sqrt{(E_{1m}^2+E_{3m}^2+E_{5m}^2)}$

Also $I_{1m} = \frac{E_{1m}}{\omega L}$; $I_{3m} = \frac{E_{3m}}{3\omega L}$; $I_{5m} = \frac{E_{5m}}{5\omega L}$

∴ $E_{1m} = I_{1m}.\omega L$; $E_{3m} = I_{3m}\, 3\omega L$; $E_{5m} = I_{5m}.5\omega L$

∴ $E = 0.707\sqrt{(I_{1m}\omega L)^2+(I_{3m}3\,\omega L)^2+(I_{5m}5\omega L)^2} = 0.707\omega L\sqrt{I_{1m}^2+9I_{3m}^2+25I_{5m}^2}$

∴ $75 = 0.707\, L\times 2\pi\times 50\sqrt{100^2+9\times 15^2+25\times 12^2}$ ∴ $L = \mathbf{0.0027\ H}$

Note. Apparent inductance $L' = \frac{E}{\omega I} = \frac{75}{2\pi\times 50\times 72} = 0.00331$ H

Example 20.15. *The capacitance of a 20 μF capacitor is checked by direct connection to an alternating voltage which is supposed to be sinusoidal, an electrostatic voltmeter and a dynamometer ammeter being used for measurement. If the voltage actually follows the law,*

$$e = 100 \sin 250t + 20 \sin (500t - \phi) + 10 \sin (750t - \phi)$$

Calculate the value of capacitance as obtained from the direct ratio of the instrument readings.

Solution. True value, $C = 20\ \mu F$

Apparent value C' = value read by the instruments

Now, $C = C' \times$ correction factor.

Let us find the value of correction factor.

Here $E_{1m} = 100; E_{2m} = 20$ and $E_{3m} = 10$

∴ Correction factor $=\sqrt{\left[\frac{E_{1m}^2+E_{2m}^2+E_{3m}^2}{E_{1m}^2+4E_{2m}^2+9E_{5m}^2}\right]}=\sqrt{\left[\frac{100^2+20^2+10^2}{100^2+4\times 20^2+9\times 10^2}\right]}=0.9166$

$20 = C'\times 0.9166$ ∴ $C' = 20/0.9166 = \mathbf{21.82}\ \mu$**F.**

20.10 Harmonics in Different Three-phase Systems

In three-phase systems, harmonics may be produced in the same way as in single-phase systems.

*It may be noted that this correction factor is different from that in the case of pure inductance.

Hence, for all calculations they are treated in the same manner *i.e.* each harmonic is treated separately. Usually, even harmonics are absent in such systems. But care must be exercised when dealing with odd, especially, third harmonics and all multiples of 3rd harmonic (also called the triple-*n* harmonics).

(a) Expressions for Phase E.M.Fs.

Let us consider a 3-phase alternator having identical phase windings (*R*, *Y* and *B*) in which harmonics are produced. The three phase e.m.fs. would be represented in their proper phase sequence by the equation.

$$e_R = E_{1m}(\omega t + \Psi_1) + E_{3m}(3\omega t + \Psi_3) + E_{5m}\sin(5\omega t + \Psi_5) + \ldots\ldots$$

$$e_Y = E_{1m}\sin\left(\omega t - \frac{2\pi}{3} + \Psi_1\right) + E_{3m}\sin\left\{3\left(\omega t - \frac{2\pi}{3}\right) + \Psi_3\right\} + E_{5m}\sin\left\{5\left(\omega t - \frac{2\pi}{3}\right) + \Psi_5\right\}$$

$$e_B = E_{1m}\sin\left(\omega t - \frac{4\pi}{3} + \Psi_1\right) + E_{3m}\sin\left\{3\left(\omega t - \frac{4\pi}{3}\right) + \Psi_3\right\} + E_{5m}\sin\left\{5\left(\omega t - \frac{4\pi}{3}\right) + \Psi_5\right\}$$

On simplification, these become

$$e_R = E_{1m}\sin(\omega t + \Psi_1) + E_{3m}(3\omega t + \Psi_3) + E_{5m}\sin(5\omega t + \Psi_5) + \ldots\ldots \quad \text{—as before}$$

$$e_Y = E_{1m}\sin\left(\omega t - \frac{2\pi}{3} + \Psi_1\right) + E_{3m}\sin(3\omega t - 2\pi + \Psi_3) + E_{5m}\sin\left(5\omega t - \frac{10\pi}{3} + \Psi_5\right) + \ldots\ldots$$

$$= E_{1m}\sin\left(\omega t - \frac{2\pi}{3} + \Psi_1\right) + E_{3m}\sin(3\omega t + \Psi_3) + E_{5m}\sin\left(5\omega t - \frac{4\pi}{3} + \Psi_5\right) + \ldots\ldots$$

$$e_B = E_{1m}\sin\left(\omega t - \frac{4\pi}{3} + \Psi_1\right) + E_{3m}\sin(3\omega t + \Psi_3) + E_{5m}\sin\left(5\omega t - \frac{2\pi}{3} + \Psi_5\right) + \ldots\ldots$$

From these expressions, it is clear that

(*i*) All third harmonics are equal in all phases of the circuit *i.e.* they are in time phase.

(*ii*) Fifth harmonics in the three phases have a negative phase sequence of *R, B, Y* because the fifth harmonic of blue phase reaches its maximum value before that in the yellow phase.

(*iii*) All harmonics which are not multiples of three, have a phase displacement of 120° so that they can be dealt with in the usual manner.

(*iv*) At any instant, all the e.m.fs. have the same direction which means that in the case of a *Y*-connected system they are directed either away from or towards the neutral point and in the case of Δ-connected system, they flow in the same direction.

Main points can be summarized as below :

(*i*) all triple-*n* harmonics *i.e.* 3rd, 9th, 15th etc. are in phase,

(*ii*) the 7th, 13th and 19th harmonics have positive phase rotation of *R*, *Y* , *B*.

(*iii*) the 5th, 11th and 17th harmonics have a negative phase sequence of *R*, *B*, *Y*.

(b) Line Voltage for a Star-connected System

In this system, the line voltages will be the *difference* between successive phase voltages and hence will contain no third harmonic terms because they, being identical in each phase, will cancel out. The fundamental will have a line voltage $\sqrt{3}$ times the phase voltage. Also, fifth harmonic has line voltage $\sqrt{3}$ its phase voltage.

But it should be noted that in this case the r.m.s. value of the line voltage will be less than $\sqrt{3}$ times the r.m.s. value of the phase voltage due to the absence of third harmonic term from the line voltage. It can be proved that for any line voltage.

$$\text{Line value} = \sqrt{3}\sqrt{\frac{E_1^2 + E_5^2 + E_7^2}{E_1^2 + E_3^2 + E_5^2 + E_7^2}}$$

where E_1, E_3 etc. are r.m.s. values of the phase e.m.fs.

(c) Line Voltage for a Δ–connected System

If the windings of the alternator are delta-connected, then the resultant e.m.f. acting round the closed mesh would be the sum of the phase e.m.fs. The sum of these e.m.fs. is zero for fundamental, 5th, 7th, 11th etc. harmonics. Since the third harmonics are in phase, there will be a resultant third harmonic e.m.f. of three times the phase value acting round the closed mesh. It will produce a circulating current whose value will depend on the impedance of the windings at the third harmonic frequency. It means that the third harmonic e.m.f. would be short-circuited by the windings with the result that there will be no third harmonic voltage across the lines. The same is applicable to all triple-*n* harmonic voltages. Obviously, the line voltage will be the phase voltage but without the triple-*n* terms.

Example 20.16. *A 3-φ generator has a generated e.m.f. of 230 V with 15 per cent third harmonic and 10 per cent fifth harmonic content. Calculate*

(i) the r.m.s. value of line voltage for Y-connection.

(ii) the r.m.s. value of line voltage for Δ-connection.

Solution. Let E_1, E_3, E_5 be the r.m.s. values of the phase e.m.fs. Then

$$E_3 = 0.15\,E_1 \text{ and } E_5 = 0.1\,E_1$$

$$\therefore \quad 230 = \sqrt{E_1^2 + (0.15E_1)^2 + (0.1E_1)^2}$$

$$E_1 = 226 \text{ V} \quad \therefore E_3 = 0.15 \times 226 = 34 \text{ V} \quad \text{and } E_5 = 0.1 \times 226 = 22.6 \text{ V}$$

(*i*) r.m.s. value of the fundamental line voltage $= \sqrt{3} \times 226 = 392$ V

r.m.s. value of third harmonic line voltage $= 0$

r.m.s. value of 5th harmonic line voltage $= \sqrt{3} \times 22.6 = \mathbf{39.2\ V}$

$\therefore$ r.m.s. value of line voltage $V_L = \sqrt{392^2 + 39.2^2} = \mathbf{394\ V}$

(*ii*) In Δ- connection, again the third harmonic would be absent from the line voltage

$\therefore$ r.m.s. value of line voltage $V_L = \sqrt{226^2 + 22.6^2} = \mathbf{227.5\ V}$

(d) Circulating Current in Δ–connected Alternator

Let the three symmetrical phase e.m.fs. of the alternator be represented by the equations,

$$e_R = E_{1m}\sin(\omega t + \Psi_1) + E_{3m}(3\omega t + \Psi_3) + E_{5m}\sin(5\omega t + \Psi_5) + \ldots..$$

$$e_Y = E_{1m}\sin(\omega t + \Psi_1 - 2\pi/3) + E_{3m}\sin(3\omega t + \Psi_3) + E_{5m} + \sin(5\omega t + \Psi_5 - 4\pi/3) \ldots...$$

$$e_B = E_{1m}\sin(\omega t + \Psi_1 - 4\pi/3) + E_{3m}\sin(3\omega t + \Psi_3) + E_{5m}\sin(5\,\omega t + \Psi_5 - 2\pi/3) + \ldots.$$

The resultant e.m.f. acting round the Δ-connected windings of the armature is the sum of these e.m.fs. Hence it is given by $e = e_R + e_Y + e_B$

$$\therefore \quad e = 3E_{3m}\sin(3\omega t + \phi_3) + 3E_{9m}\sin(9\,\omega t + \Psi_9) + 3E_{15}\sin(15\omega t + \Psi_{15}) + \ldots.$$

If *R* and *L* represent respectively the resistance and inductance per phase of the armature winding, then the circulating current due to the resultant e.m.f. is given by

$$i_i = \frac{3E_{3m}\sin(3\omega t + \Psi_3)}{3\sqrt{(R^2 + 9L^2\omega^2)}} + \frac{3E_9m\sin(9\omega t + \Psi_9)}{3\sqrt{(R^2 + 81L^2\omega^2)}} + \frac{3E_{15m}\sin(15\omega t + \Psi_{15})}{3\sqrt{(R^2 + 225L^2\omega^2)}} + \ldots.$$

$$= \frac{E_{3m}\sin(3\omega t + \Psi_3)}{\sqrt{(R^2 + 9L^2\omega^2)}} + \frac{E_{9m}\sin(9\omega t + \Psi_9)}{\sqrt{(R^2 + 81L^2\omega^2)}} + \frac{E_{15m}\sin(15\omega t + \Psi_{15})}{\sqrt{(R^2 + 225L^2\omega^2)}}$$

The r.ms. value of the current is given by

$$I_C = 0.707\sqrt{[E_{3m}^2/(R^2 + 9L^2\omega^2) + E_{9m}^2/(R^2 + 81L^2\omega^2) + (E_{15m}^2/(R^2 + 225L^2\omega^2) + \ldots..]}$$

(e) Three-phase Four-wire System

In this case, there will be no third harmonic component in line voltage. For the 4-wire system, each phase voltage (*i.e.* line to neutral) may contain a third harmonic component. If it is actually present, then current will flow in the Y-connected load. In case load is balanced, the resulting third harmonic line currents will all be in phase so that neutral wire will have to carry three times the third harmonic line current. There will be no current in the neutral wire either at fundamental frequency, or any harmonic frequency other than the triple-n frequency.

20.11. Harmonics in Single and 3-phase Transformers

The flux density in transformer cores is usually maintained at a fairly high value in order to keep the required volume of iron to the minimum. However, due to the non-linearity of magnetisation curve, some third harmonic distortion is always produced. Also, there is usually a small percentage of fifth harmonic. The magnetisation current drawn by the primary contains mainly third harmonic whose proportion depends on the size of the primary applied voltage. Hence, the flux is sinusoidal.

In the case of three-phase transformers, the production of harmonics will be affected by the method of connection and the type of construction employed.

(a) Primary Windings Δ-CONNECTED

Each primary phase can be considered as separately connected across the sinusoidal supply.

(*i*) The core flux will be sinusoidal which means that magnetizing current will contain 3rd harmonic component in addition to relatively small amounts of other harmonics of higher order.

(*ii*) In each phase, these third harmonic currents will be in phase and so produce a circulating current round the mesh with the result that there will be no third harmonic component in the line current.

b) Primary Windings Connected in 4-wire Star

Each phase of the primary can again be considered as separately connected across a sinusoidal supply.

(*i*) The flux in the transformer core would be sinusoidal and so would be the output voltage.

(*ii*) The magnetizing current will contain 3rd harmonic component. This component being in phase in each winding will, therefore, return through the neutral wire.

(c) Primary Windings Connected 3-wire Star

Since there is no neutral wire, there will be no return path for the 3rd harmonic component of the magnetizing current. Hence, there will exist a condition of forced magnetization so that core flux must contain third harmonic component which is in phase in each limb of the transformer core. Although there will be a magnetic path for these fluxes in the case of shell type 3-phase transformer, yet in the case of three-limb core type transformer, the third harmonic component of the flux must return *via* the air. Because of the high reluctance magnetic path in such transformers, the third harmonic flux is reduced to a very small value. However, if the secondary of the transformer is delta-connected, then a third harmonic circulating current would be produced. This current would be in accordance with Lenz's law tend to oppose the very cause producing it *i.e.* it would tend to minimize the third harmonic component of the flux.

Should the third and fifth secondary be Y-connected, then provision of an additional Δ-connected winding, in which this current can flow, becomes necessary. This tertiary winding additionally serves the purpose of preserving magnetic equilibrium of the transformer in the case of unbalanced loads. In this way, the output voltage from the secondary can be kept reasonably sinusoidal.

Example 20.17. *Determine whether the following two waves are of the same shape*

$$e = 10 \sin(\omega t + 30°) - 50 \sin(3\omega t - 60°) + 25 \sin(5\omega t + 40°)$$
$$i = 1.0 \sin(\omega t - 60°) + 5 \sin(3\omega t - 150°) + 2.5 \cos(5\omega t - 140°)$$

(Principles of Elect. Engg - II Jadavpur Univ. 1989)

Solution. Two waves possess the same waveshape

(*i*) if they contain the same harmonics

(*ii*) if the ratio of the corresponding harmonics to their respective fundamentals is the same (*iii*) if the harmonics are similarly spaced with respect to their fundamentals.

In other words,

(*a*) the ratio of the magnitudes of corresponding harmonics must be constant and

(*b*) with fundamentals in phase, the corresponding harmonics of the two waves must be in phase.

The test is applied first by checking the ratio of the corresponding harmonics and then coinciding the fundamentals by shifting one wave. If the phase angles of the corresponding harmonics are the same, then the two waves have the same shape.

In the present case, condition (*i*) is fulfilled because the voltage and current waves contain the same hermonies, *i.e*.. third and fifth.

Secondly, the ratio of the magnitudes of corresponding current and voltage harmonics is the same *i.e* 1/10.

Now, let the fundamental of the current wave be shifted ahead by 90° so that it is brought in phase with the fundamental of the voltage wave. It may be noted that the third and fifth harmonics of the current wave will be shifted by 3 × 90° = 270° and 5 × 90° = 450° respectively. Hence, the current wave becomes

$$i' = 1.0 \sin (\omega t - 60° + 90°) + 5 \sin (3\,\omega t - 150° + 270°) + 2.5 \cos (5\omega t - 140° + 450°)$$
$$= 1.0 \sin (\omega t + 30°) + 5 \sin (3\omega t + 120°) + 2.5 \cos (5\omega t + 310°)$$
$$= 1.0 \sin (\omega t + 30°) - 5 \sin (3\omega t - 60°) + 2.5 \sin (5\omega t + 40°)$$

It is seen that now the corresponding harmonics of the voltage and current waves are in phase. Since all conditions are fulfilled, the two waves are of the same waveshape.

Tutorial Problem No. 20.1

1. A series circuit consists of a coil of inductance 0.1 H and resistance 25 Ω and a variable capacitor. Across this circuit is applied a voltage whose instantaneous value is given by

$v = 100 \sin \omega t + 20 \sin (3\omega t - 45) + 5 \sin (5\omega t - 30)$ where $\omega = 314$ rad/s

Determine the value of *C* which will produce response at third harmonic frequency and with this value of *C*, find

(*a*) an expression for the current in the circuit (*b*) the r.m.s. value of this current (*c*) the total power absorbed.

[11.25 μF, (*a*) i = 0.398 sin (ωt + 84.3) + 0.8 sin (3ωt + 45) + 0.485 sin (5ωt + 106) (*b*) 0.633 A (*c*) 10 W]

2. A voltage given by $v = 200 \sin 314t + 50 \sin (942\,t + 45°)$ is applied to a circuit consisting of a resistance of 20 Ω, an inductance of 20 mH and a capacitance of 56.3 μF all connected in series.

Calculate the r.m.s values of the applied voltage and current. Find also the total power absorbed by the circuit. **[146 V ; 3.16 A 200 W]**

3. A voltage given by $v = 100 \sin \omega t + 8 \sin 3\omega t$ is applied to a circuit which has a resistance of 1Ω, an inductance of 0.02 H and a capacitance of 60μF. A hot-wire ammeter is connected in series with the circuit and a hot-wire voltmeter is connected to the terminals. Calculate the ammeter and voltmeter readings and the power supplied to the circuit. **[71 V ; 5.18 A ; 26.8 W]**

4. A certain coil has a resistance of 20 Ω and an inductance of 0.04 H. If the instantaneous current flowing in it is represented by $i = 5 \sin 300\,t + 0.8 \sin 900\,t$ amperes, derive an expression for the instantaneous value of the voltage applied across the ends of the coil and calculate the r.ms. value of that voltage.

[V = 117 sin (300t + 0.541) + 33 sin (900 t + 1.06) ; 0.86 V]

5. A voltage given by the equation $v = \sqrt{2}\; 100 \sin 2\pi \times 50t + \sqrt{2}.20 \sin 2\pi.\; 150t$ is applied to the terminals of a circuit made up of a resistance of 5 Ω, an inductance of 0.0318 H and a capacitor of 12.5 μF all in series. Calculate the effective current and the power supplied to the circuit. **[0.547 A ; 1.5 W]**

6. An alternating voltage given by the expression $v = 1{,}000 \sin 314t + 100 \sin 942t$ is applied to a circuit having a resistance of 100 Ω and an inductance of 0.5 H. Calculate r.m.s. value of the current and p.f.

[3.81 A ; 0.535]

7. The current in a series circuit consisting of a 159μF capacitor, a reactor with a resistance of 10Ω and an inductance of 0.0254 H is given by $i = \sqrt{2}(8 \sin \omega\, t + 2 \sin 3\omega\, t)$ amperes. Calculate the power input and the power factor. Given $\omega = 100\pi$ radian/second. **[680 W ; 0.63]**

8. If the terminal voltage of a circuit is $100 \sin \omega t + 50 \sin (3\omega t + \pi/4)$ and the current is $10 \sin (\omega t + \pi\; 3) + 5 \sin 3\omega t$, calculate the power consumption. **[522.6 W]**

9. A single-phase load takes a current of $4 \sin \omega t + \pi/6) + 1.5 \sin (3\omega t + \pi/3)$ A from a source represented by $360 \sin \omega t$ volts. Calculate the power dissipated by the circuit and the circuit power factor.

[623.5 W ; 0.837]

10. An e.m.f. given by $e = 100 \sin \omega t + 40 \sin (3\omega t - \pi/6) + 10 \sin (5\omega t - \pi 3)$ volts is applied to a series circuit having a resistance of 100 Ω, an inductance of 40.6 mH and a capacitor of 10μF. Derive an expression for the current in the circuit. Also, find the r.m.s. value of the current and the power dissipated in the circuit. Take $\omega = 314$ rad/second.

[0.329 A, 10.8 W]

11. A p.d. of the form $\upsilon = 400 \sin \omega t + 30 \sin 3 \omega t$ is applied to a rectifier having a resistance of 50 Ω in one direction and 200 in the reverse direction. Find the average and effective values of the current and the p.f. of the circuit.

[1.96 A, 4.1 A, 0.51]

12. A coil having $R = 2$ Ω and $L = 0.01$ H carries a current given by $i = 50 + 20 \sin 300\, t$

A moving-iron ammeter, a moving-coil voltmeter and a dynameter wattmeter, are used to indicate current, voltage and power respectively. Determine the readings of the instruments and the equation for the p.d.

[121.1 V ; 52 A ; 5.4 kW, v = 100 + 72 sin (300 t + 0.982)]

13. Two circuits having impedances at 50Hz of $(10 + j6)$ Ω and $(10 - j6)$ respectively are connected in parallel across the terminals of an a.c. system, the waveform of which is represented by $v = 100 \sin \omega t + 35 \sin 3\omega t + 10 \sin 5\omega t$, the fundamental frequency being 50 Hz, Determine the ratio of the readings of two ammeters, of negligible resistance, connected one in each circuit.

[6.35 ; 6.72]

14. Explain what is meant by harmonic resonance in a.c. circuits.

A current having an instantaneous value of $2 (\sin \omega t + \sin 3\omega t)$ amperes is passed through a circuit which consists of a coil of resistance R and inductance L in series with a capacitor C. Derive an expression for the value of ω at which the r.m.s circuit voltage is a minimum. Determine this voltage if the coil has inductance 0.1 H and resistance 150 Ω and the capacitance is 10μ F. Determine also the circuit voltage at the fundamental resonant frequency.

[$\omega = 1/\sqrt{(LC)}$; 378 V ; 482 V]

15. An r.m.s. current of 5 A which has a third-harmonic content, is passed through a coil having a resistance of 1Ω and an inductance of 10 mH. The r.m.s. voltage across the coil is 20 V. Calculate the magnitudes of the fundamental and harmonic components of current if the fundamental frequency is 300/2π Hz Also, find the power dissipated.

[4.8 A ; 1.44 A; 25 W]

16. Derive a general expression for the form factor of a complex wave containing only odd-order harmonics. Hence, calculate the form factor of the alternating current represented by

$i = 2.5 \sin 157t + 0.7 \sin 471t + 0.4 \sin 785t$

[1.038]

OBJECTIVE TESTS – 20

1. Non-sinusoidal waveforms are made up of

(*a*) different sinusoidal waveforms
(*b*) fundamental and even harmonics
(*c*) fundamental and odd harmonics
(*d*) even and odd harmonics only.

2. The positive and negative halves of a complex wave are symmetrical when

(*a*) it contains even harmonics
(*b*) phase difference between even harmonics and fundamental is 0 or π
(*c*) it contains odd harmonics
(*d*) phase difference between even harmonics and fundamental is either π/2 or 3π/2.

3. The r.m.s value of the complex voltage given by $v = 16\sqrt{2} \sin \omega t + 12\sqrt{2} \sin 3\omega t$ is

(*a*) $20\sqrt{2}$ (*b*) 20
(*c*) $28\sqrt{2}$ (*d*) 192

4. In a 3-phase system, —— th harmonic has negative phase sequence of RBY.

(*a*) 9 (*b*) 13
(*c*) 5 (*d*) 15

5. A complex current wave is given by the equation $i = 14 \sin \omega t + 2 \sin 5\omega t$. The r.m.s value of the current is —— ampere.

(*a*) 16 (*b*) 12
(*c*) 10 (*d*) 8

6. When a pure inductive coil is fed by a complex voltage wave, its current wave

(*a*) has larger harmonic content
(*b*) is more distorted
(*c*) is identical with voltage wave
(*d*) shows less distortion.

7. A complex voltage wave is applied across a pure capacitor. As compared to the fundamental voltage, the reactance offered by the capacitor to the third harmonic voltage would be

(*a*) nine times (*b*) three times
(*b*) one-third (*d*) one-ninth

8. Which of the following harmonic voltage components in a 3-phase system would be in phase with each other ?

(*a*) 3rd, 9th, 15th etc.
(*b*) 7th, 13th, 19th etc.
(*c*) 5th, 11th, 17th etc.
(*d*) 2nd, 4th, 6th etc.

ANSWERS

1. *a* 2. *c* 3. *b* 4. *c* 5. *c* 6. *d* 7. *c* 8. *a*

21 FOURIER SERIES

21.1. Harmonic Analysis

By harmonic analysis is meant the process of determining the magnitude, order and phase of the several harmonics present in a complex periodic wave.

For carrying out this analysis, the following methods are available which are all based on Fourier theorem :

(*i*) Analytical Method– the standard Fourier Analysis

(*ii*) Graphical Method– (*a*) by Superposition Method (Wedgemore' Method) (*b*) Twenty four Ordinate Method

(*iii*) Electronic Method– by using a special instrument called 'harmonic analyser'

We will consider the first and third methods only.

21.2. Periodic Functions

A function $f(t)$ is said to be periodic if $f(t+T) = f(t)$ for all values of t where T is some positive number. This T is the interval between two successive repetitions and is called the period of $f(t)$. A sine wave having a period of $T = 2\pi/\omega$ is a common example of periodic function.

21.3. Trigonometric Fourier Series

Suppose that a given function $f(t)$ satisfies the following conditions (known as Dirichlet conditions) :

1. $f(t)$ is periodic having a period of T.
2. $f(t)$ is single-valued everywhere.
3. In case it is discontinuous, $f(t)$ has a finite number of discontinuities in any one period.
4. $f(t)$ has a finite number of maxima and minima in any one period.

The function $f(t)$ may represent either a voltage or current waveform. According to Fourier theorem, this function $f(t)$ may be represented in the trigonometric form by the infinite series.

$$f(t) = a_0 + a_1 \cos \omega_0 t + a_2 \cos 2\omega_0 t + a_3 \cos 3\omega_0 t + ... + a_n \cos n\omega_0 t$$
$$+ b_1 \sin \omega_0 t + b_2 \sin 2\omega_0 t + b_3 \sin 3\,\omega_0 t + ... + b_n \sin n\omega_0 t$$
$$= a_0 + \sum_{n=1}^{\infty} (a_n \cos n\omega_0 t + b_n \sin n\omega_0 t) \quad ....(i)$$

Putting $\omega_0 t = \theta$, we can write the above equation as under

$$f(\theta) = a_0 + a_1 \cos\theta + a_2 \cos 2\theta + a_3 \cos 3\theta + ... + a_n \cos n\theta \quad + b_1 \sin\theta + b_2 \sin 2\theta + b_3 \sin 3\theta + ... + b_n \sin n\theta$$
$$= a_0 + \sum_{n=1}^{\infty} (a_n \cos n\theta + b_n \sin n\theta) \quad ...(ii)$$

Since $\omega_0 = 2\,\pi/T$, Eq. (*i*) above can be written as

$$f(t) = a_0 + \sum_{n=1}^{\infty} \left(a_n \cos \frac{2\pi n}{T} t + b_n \sin \frac{2\pi n}{T} t \right) \quad ...(iii)$$

where ω_0 is the fundamental angular frequency, T is the period and a_0, a_n and b_n are constants which depend on n and $f(t)$. The process of determining the values of the constants a_0, a_n and b_n is called Fourier Analysis. Also, $\omega_0 = 2\,\pi\, T = 2\,\pi f_0$ where f_0 is the fundamental frequency.

It is seen from the above Fourier Series that the periodic function consists of sinusoidal components of frequency 0, ω_0, 2 ω_0 .....n ω_0. This representation of the function $f(t)$ is in the frequency domain. The first component a_0 with zero frequency is called the *dc* component. The sine and cosine terms represent the harmonics. The number n represents the order of the harmonics.

When $n = 1$, the component $(a_1 \cos \omega_0 t + b_1 \sin \omega_0 t)$ is called the first harmonic or the fundamental component of the waveform.

When $n = 2$, the component $(a_2 \cos 2\, \omega_0 t + b_2 \sin 2\, \omega_0 t)$ is called the second harmonic of the waveform.

The nth harmonic of the waveform is represented by $(a_n \cos n\, \omega_0 t + b_n \sin n\, \omega_0 t)$. It has a frequency of $n\omega_0$ *i.e.* n times the frequency of the fundamental component.

21.4. Alternate Forms of Trigonometric Fourier Series

Eq. (*i*) given above can be written as

$f(t) = a_0 + (a_1 \cos \omega_0 t + b_1 \sin \omega_0 t) + (a_2 \cos 2\, \omega_0 t + b_2 \sin 2\omega_0 t) + ... +(a_n \cos n\omega_0 t + b_n \sin n\, \omega_0 t)$

Let, $a_n \cos n\omega_0 t + b_n \sin n\omega_0 t = A_n \cos (n\omega_0 t - \phi_n)$

$= A_n \cos n\omega_0 t \cos \phi_n + A_n \sin n\omega_0 t \sin \phi_n$

$\therefore \quad a_n = A_n \cos \phi_n$ and $b_n = A_n \sin \phi_n$

$\therefore \quad A_n \sqrt{a_n^2 + b_n^2}$ and $\phi_n = \tan^{-1} b_n / a_n$

Similarly, let $(a_n \cos n\, \omega_0 t + b_n \sin n\, \omega_0 t = A_n \sin (n\omega_0 t + \Psi_n)$

$= A_n \sin n\, \omega_0 t \cos \psi_n + A_n \cos n\, \omega_0 t \sin \phi_n$

As seen from Fig. 21.1, $b_n = A_n \cos \Psi_n$ and $a_n = A_n \sin \Psi_n$

$\therefore \quad A_n = \sqrt{a_n^2 + b_n^2}$ and $\Psi = \tan^{-1} a_n / b_n$

The two angles ϕ_n and ψ_n are complementary angles.

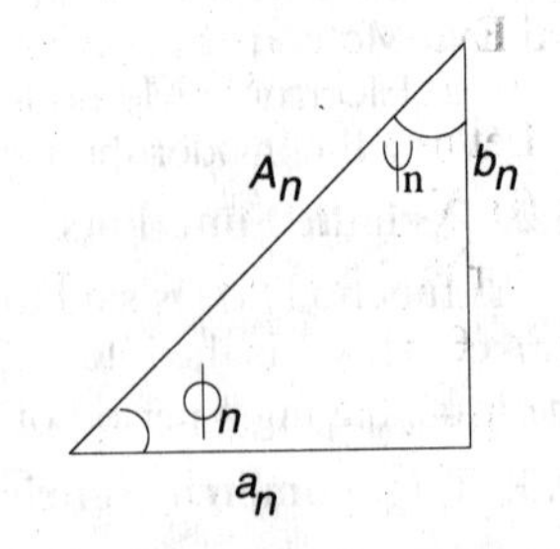

Fig. 21.1

Hence, the Fourier series given in Art. 21.2 may be put in the following two alternate forms

$$f(t) = A_0 + \sum_{n=1}^{\infty} A_n \cos (n\omega_0 t - \phi_n)$$

or

$$f(t) = A_0 + \sum_{n=1}^{\infty} A_n \sin (n\omega_0 t + \psi_n)$$

21.5. Certain Useful Integral Calculus Theorems

The Fourier coefficients or constants $a_0, a_1, a_2 ..... a_n$ and $b_1, b_2 ..... b_n$ can be evaluated by integration process for which purpose the following theorems will be used.

(*i*) $\int_0^{2\pi} \sin n\theta\, d\theta = \frac{1}{n} | \cos n\theta |_0^{2\pi} = \frac{1}{n}(1-1) = 0$

(*ii*) $\int_0^{2\pi} \cos n\, \theta\, d\, \theta = -\frac{1}{n} | \sin n\theta |_0^{2\pi} = -\frac{1}{n}(0-0) = 0$

(*iii*) $\int_0^{2\pi} \sin^2 n\theta\, d\, heta = \frac{1}{2}\int_0^{2\pi} (1 - \cos 2n\theta) d\, \theta = \frac{1}{2} | \theta - \frac{1}{2n} \sin 2n\theta |_0^{2\pi} = \pi$

(*iv*) $\int_0^{2\pi} \cos^2 n\, \theta\, d\, \theta = \frac{1}{2}\int_0^{2\pi} (\cos 2n\theta + 1) d\, \theta = \frac{1}{2} | \frac{1}{2n} \sin 2n\theta + \theta |_0^{2\pi} = \pi$

(v) $\int_0^{2\pi} \sin m\theta \cos n\theta \, d\theta = \frac{1}{2}\int_0^{2\pi} \{\sin (m+n)^\theta + \sin (m-n)\theta\} d\theta$

$$= \frac{1}{2} \left| -\frac{1}{m+n}\cos(m+n)\theta - \frac{1}{m-n}\cos(m-n)\theta \right|_0^{2\pi} = 0$$

(vi) $\int_0^{2\pi} \cos m\theta \cos n\theta \, d\theta = \frac{1}{2}\int_0^{2\pi} \{\cos(m+n)\,\theta + \cos(m-n)\theta\} d\theta$

$$= \frac{1}{2} \left| \frac{1}{m+n}\sin(m+n)\,\theta + \frac{1}{m-n}\sin(m-n)\theta \right|_0^{2\pi} = 0 \ldots \text{for } n \neq m$$

(vii) $\int_0^{2\pi} \sin m\theta \sin n\theta \, d\theta = \frac{1}{2}\int_0^{2\pi} \{\cos(m-n)\,\theta - \cos(m+n)\theta\} d\theta$

$$= \frac{1}{2} \left| \frac{1}{m-n}\sin(m-n)\,\theta - \frac{1}{m+n}\sin(m+n)\theta \right|_0^{2\pi} = 0 \ldots \text{for } n \neq m$$

where m and n are any positive integers.

21.6. Evaluation of Fourier Constants

Let us now evaluate the constants a_0, a_n and b_n by using the above integral calculus theorems

(i) Value of a_0

For this purpose we will integrate both sides of the series given below over one period *i.e.* for $\theta = 0$ to $\theta = 2\pi$.

$$f(\theta) = a_0 + a_1 \cos\theta + a_2 \cos 2\theta + \ldots\ldots + a_n \cos n\theta$$
$$+ b_1 \sin\theta + b_2 \sin 2\theta + \ldots\ldots + b_n \sin n\theta$$

$$\therefore \int_0^{2\pi} f(\theta)\, d\theta = \int_0^{2\pi} a_0 d\,\theta + a_1 \int_0^{2\pi} \cos\theta\, d\theta + a_2 \int_0^{2\pi} \cos 2\theta\, d\theta + \ldots\ldots + a_n \int_0^{2\pi} \cos n\theta\, d\theta$$

$$+ b_1 \int_0^{2\pi} \sin\theta\, d\theta + b_2 \int_0^{2\pi} \sin 2\theta\, d\,\theta + \ldots\ldots + b_n \int_0^{2\pi} \sin n\theta\, d\theta$$

$$= a_0 |\theta|_0^{2\pi} + 0 + 0 + \ldots\ldots 0 + 0 + 0 + \ldots\ldots + 0 = 2\pi a_0$$

$$\therefore \qquad a_0 = \frac{1}{2\pi}\int_0^{2\pi} f(\theta) d\theta \quad \text{or} \quad = \frac{1}{2\pi}\int_{-\pi}^{\pi} f(\theta) d\theta$$

= mean value of $f(\theta)$ between the limits 0 to 2 π *i.e.* over one cycle or period.

Also, $a_0 = \frac{1}{2\pi}(\text{net area})_0^{2\pi}$

If we take the periodic function as $f(t)$ and integrate over period T (which corresponds to 2π),

we get $\quad a_0 = \frac{1}{T}\int_0^T f(t)dt = \frac{1}{T}\int_{-T/2}^{T/2} f(t)dt = \frac{1}{T}\int_{t_1}^{t_1+T} f(t)dt$

where t_1 can have any value.

(ii) Value of a_n

For finding the value of a_n, multiply both sides of the Fourier Series by cos $n\theta$ and integrate between the limits $\theta = 0$ to/2π

$$\therefore \int_0^{2\pi} f(\theta)\cos n\theta\, d\theta = a_0\int_0^{2\pi}\cos n\theta\, d\theta + a_1\int_0^{2\pi}\cos\theta\cos n\,\theta d\,\theta$$

$$+a_2\int_0^{2\pi}\cos 2\theta\cos n\,\theta d\,\theta + a_n\int_0^{2\pi}\cos^2 n\theta\, d\theta +$$

$$b_1\int_0^{2\pi}\sin\theta\cos n\theta\, d\theta + b_2\int_0^{2\pi}\sin 2\theta\cos n\theta\, d\theta + \ldots\ldots + b_n\int_0^{2\pi}\sin n\theta\cos n\theta\, d\theta$$

$$= 0+0+0+\ldots\ldots + a_n\int_0^{2\pi}\cos^2 n\theta\, d\theta + 0 + 0 + \ldots\ldots + 0 = a_n\int_0^{2\pi}\cos^2 n\theta\, d\theta = \pi a_n$$

$$\therefore a_n = \frac{1}{\pi}\int_0^{2\pi} f(\theta)\cos n\theta\, d\theta = 2\times\frac{1}{2\pi}\int_0^{2\pi} f(\theta)\cos n\theta\, d\theta$$

= 2 × average value of $f(\theta)$ cos $n\theta$ over one cycle of the fundamental.

$$\text{Also, } a_n = \frac{1}{\pi}\int_{-\pi}^{\pi} f(\theta)\cos n\theta\, d\theta = 2\times\frac{1}{2\pi}\int_{-\pi}^{\pi} f(\theta)\cos n\theta\, d\theta$$

If we take periodic function as $f(t)$, then different expressions for a_n are as under.

$$a_n = \frac{2}{T}\int_0^{T} f(t)\cos\frac{2\pi n}{T}t\, dt = \frac{2}{T}\int_{-T/2}^{T/2} f(t)\cos\frac{2\pi n}{T}t\, dt$$

Giving different numerical values to n, we get

a_1 = 2 × average value of $f(\theta)$ cos θ over one cycle ---$n = 1$

a_2 = 2 × average value of $f(\theta)$ cos 2θ over one cycle etc. ---- $n = 2$

(iii) Value of b_n

For finding its value, multiply both sides of the Fourier Series of Eq. (*i*) by sin $n\theta$ and integrate between limits θ = 0 to θ = 2π.

$$\therefore \int_0^{2\pi} f(\theta)\sin n\theta\, d\theta = a_0\int_0^{2\pi}\sin n\theta d\theta + a_1\int_0^{2\pi}\cos\theta\sin n\theta d\,\theta$$

$$+a_2\int_0^{2\pi}\cos 2\theta\sin n\theta\, d\theta + \ldots\ldots + a_n\int_0^{2\pi}\cos n\theta\sin n\theta\, d\theta$$

$$+b_1\int_0^{2\pi}\sin\theta\sin n\theta\, d\theta + b_2\int_0^{2\pi}\sin 2\theta\sin n\theta\, d\theta + \ldots\ldots + b_n\int_0^{2\pi}\sin^2 n\theta\, d\theta$$

$$= 0+0+0+\ldots\ldots + 0 + 0 + \ldots\ldots + b_n\int_0^{2\pi}\sin^2 n\theta\, d\theta = b_n\int_0^{2\pi}\sin^2 n\theta\, d\theta = b_n\pi$$

$$\therefore \int_0^{2\pi} f(\theta)\sin n\theta d\theta = b_n\times\pi$$

$$\therefore b_n = \frac{1}{\pi}\int_0^{2\pi} f(\theta)\sin n\theta\, d\theta = 2\times\frac{1}{2\pi}\int_0^{2\pi} f(\theta)\sin n\theta\, d\theta$$

= 2 × average value of $f(\theta)$ sin n θ over one cycle of the fundamental.

∴ b_1 = 2 × average value of $f(\theta)$ sin θ over one cycle ... $n = 1$

b_2 = 2 × average value of $f(\theta)$ sin 2 θ over one cycle ...$n = 2$

Also, $b_n = \frac{2}{T}\int_0^T f(t) \sin \frac{2\pi n}{T} t\, dt + \frac{2}{T}\int_{-T/2}^{T/2} f(t) \sin \frac{2\pi n}{T} t\, dt$

$= \frac{2}{T}\int_0^T f(t) \sin n\, \omega_0 t\, dt = \frac{2}{T}\int_{-T/2}^{T/2} f(t) \sin n\, \omega_0\, t\, dt$

Hence, for Fourier analysis of a periodic function, the following procedure should be adopted :

(*i*) Find the term a_0 by integrating both sides of the equation representing the periodic function between limits 0 to 2π or 0 to T or $-T/2$ to $T/2$ or t_1 to $(t_1 + T)$.

$$\therefore \quad a_0 = \frac{1}{2\pi}\int_0^{2\pi} f(\theta)\, d\theta = \frac{1}{T}\int_0^T f(t)\, dt = \frac{1}{T}\int_{-T/2}^{T/2} f(t) dt = \frac{1}{T}\int_{t_1}^{t_1+T} f(t) dt$$

= average value of the function over one cycle.

(*ii*) Find the value of a_n by multiplying both sides of the expression for Fourier series by $\cos n\,\theta$ and then integrating it between limits 0 to 2π or 0 to T or $-T/2$ to $T/2$ or t_1 to $(t_1 + T)$.

$$\therefore a_n = \frac{1}{\pi}\int_0^{2\pi} f(\theta) \cos n\theta\, d\theta = 2 \times \frac{1}{2\pi}\int_0^{2\pi} f(\theta) \cos n\theta\, d\theta$$

Since $\pi = T/2$, we have

$$a_n = \frac{2}{T}\int_0^T f(t) \cos \frac{2\pi n}{T} t\, dt = \frac{2}{T}\int_{-T/2}^{T/2} f(t) \cos \frac{2\pi n}{T} t\, dt = \frac{2}{T}\int_{t_1}^{t_1+T} f(t) \cos \frac{2\pi n}{T} t\, dt$$

$$= \frac{2}{T}\int_0^T f(t) \cos n\omega_0\, t\, dt = \frac{2}{T}\int_{-T/2}^{T/2} f(t) \cos n\omega_0\, t\, dt = \frac{2}{T}\int_{t_1}^{t_1+T} f(t) \cos n\, \omega_0\, t\, dt$$

$= 2 \times$ average value of $f(\theta) \cos n\theta$ over one cycle of the fundamental.

Values of a_1, a_2, a_3 etc. can be found from above by putting $n = 1, 2, 3$ etc.

(*iii*) Similarly, find the value of b_n by multiplying both sides of Fourier series by $\sin n\,\theta$ and integrating it between the limits 0 to 2π or 0 to T or $-T/2$ to $T/2$ or t_1 to $(t_1 + T)$.

$$\therefore \quad b_n = \frac{1}{\pi}\int_0^{2\pi} f(\theta) \sin n\theta = 2 \times \frac{1}{2\pi}\int_0^{2\pi} f(\theta) \sin n\,\theta\, d\theta$$

$$= \frac{2}{T}\int_0^T f(t) \sin \frac{2\pi n}{T} t\, dt = \frac{2}{T}\int_{-T/2}^{T/2} f(t) \sin \frac{2\pi n}{T} t\, dt = \frac{2}{T}\int_{t_1}^{t_1+T} f(t) \sin \frac{2\pi n}{t} t\, dt$$

$$= \frac{2}{T}\int_0^T f(t) \sin n\, \omega_0\, t\, dt = \frac{2}{T}\int_{-T/2}^{T/2} f(t) \sin n\omega_0\, t\, dt = \frac{2}{T}\int_{t_1}^{t_1+T} f(t) \sin n\, \omega_0\, t\, dt$$

= 2x average value of f (θ) sin $n\,\theta$ of $f(t) \sin \frac{2\pi n}{T} t$ or $f(t) \sin n\, \omega_0\, t$ over one cycle of the fundamental.

Values of b_1, b_2, b_3 etc. can be found from above by putting $n = 1, 2, 3$ etc.

21.7. Different Types of Functional Symmetries

A non-sinusoidal wave can have the following types of symmetry :

1. Even Symmetry

The function $f(t)$ is said to possess even symmetry if $f(t) = f(-t)$.

It means that as we travel equal amounts in time to the left and right of the origin (*i.e.* along the + X–axis and –X–axis), we find the function to have the same value. For example in Fig. 21.2 (*a*), points A and B are equidistant from point O. Here the two function values are equal and positive. At points C and D, the two values of the function are again equal, though negative. Such a function is symmetric with respect to the vertical axis. Examples of even function are : t^2, cos $3t$, $\sin^2 5t\,(2 + t^2 + t^4)$ and a constant A because the replacement of t by $(-t)$ does not change the value of any of these functions. For example, $\cos \omega t = \cos(-\omega t)$.

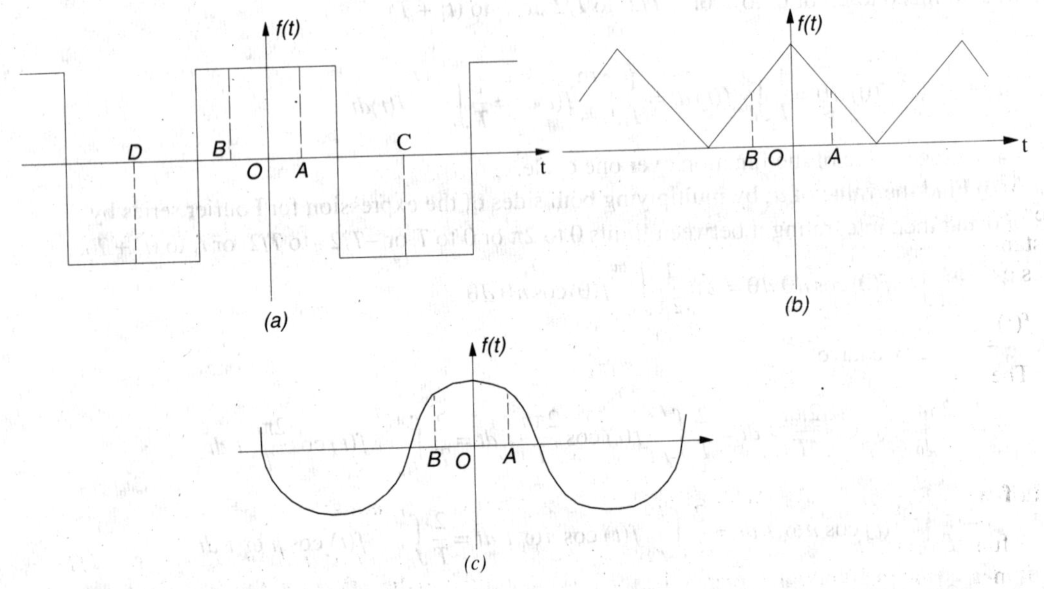

Fig. 21.2

This type of symmetry can be easily recognised graphically because mirror symmetry exists about the vertical or $f(t)$ axis. The function shown in Fig. 21.2 has even symmetry because when folded along vertical axis, the portions of the graph of the function for positive and negative time fit exactly, one on top of the other.

The effect of the even symmetry on Fourier series is that the constant $b_n = 0$ *i.e.* the wave has no sine terms. In general, $b_1, b_2, b_3 \ldots\ldots b_n = 0$. The Fourier series of an even function contains only a constant term and cosine terms *i.e.*

$$f(t) = a_0 + \sum_{n=1}^{\infty} a_n \cos n\,\omega_0 t = a_0 + \sum_{n=1}^{\infty} a_n \cos \frac{2\pi n}{T} t$$

The value of a_n may be found be integrating over any half-period.

$$\therefore \quad a_n = \frac{2}{\pi}\int_0^{\pi} f(\theta)\cos n\theta\, d\theta = \frac{4}{T}\int_0^{T/2} f(t)\cos n\omega t\, dt$$

2. Odd Symmetry

A function $f(t)$ is said to possess odd symmetry if $f(-t) = -f(t)$.

It means that as we travel an equal amount in time to the left or right from the origin, we find the function to be the same except for a reversal in sign. For example, in Fig. 21.3 the two points A and B are equidistant from point O. The two function values at A and B are equal in magnitude but opposite in sign. In other words, if we replace t by $(-t)$, we obtained the negative of the given function. The X-axis divides an odd function into two halves with equal areas above and below the X-axis. Hence, $a_0 = 0$.

Examples of odd functions are : t, $\sin t$, $t \cos 50\,t\,(t + t^3 + t^5)$ and $t\sqrt{(1 + t^2)}$

A sine function is an odd function because sin $(-\omega t) = -\sin \omega t$.

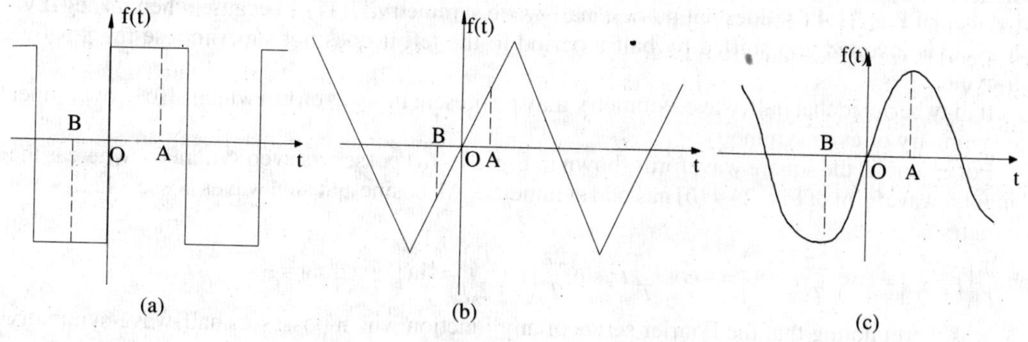

Fig. 21.3

An odd function has symmetry about the origin rather than about the $f(t)$ axis which was the case for an even function. The effect of odd symmetry on a Fourier series is that it contains no constant term or consine term. It means that $a_0 = 0$ and $a_n = 0$ *i.e.* $a_1, a_2, a_3 .... a_n = 0$. The Fourier series expansion contains only sine terms.

$$\therefore \quad f(t) = \sum_{n=1}^{\infty} b_n \sin n\omega_0 t$$

The value of b_n may be found by integrating over any half-period.

$$\therefore \qquad b_n = \frac{2}{\pi}\int_0^{\pi} f(\theta) \sin n\theta \, d\theta = \frac{4}{T}\int_0^{T/2} f(t) \sin n\omega t \, dt$$

3. Half-wave Symmetry or Mirror Symmetry or Rotational Symmetry

A function $f(t)$ is said to possess half-wave symmetry if $f(t) = -f(t \pm T/2)$ or $-f(t) = f(t \pm T/2)$.

It means that the function remains the same if it is shifted to the left or right by half a period and then flipped over (*i.e.* multiplied by -1) in respect to the t–axis or horizontal axis. It is called mirror symmetry because the negative portion of the wave is the mirror image of the positive portion of the wave displaced horizontally a distance $T/2$.

In other words, a waveform possesses half symmetry only when we invert its negative half-cycle and get an exact duplicate of its positive half-cycle. For example, in Fig. 21.4 (*a*) if we invert the negative half-cycle, we get the dashed *ABC* half-cycle which is exact duplicate of the positive half-cycle. Same is the case with the waveforms of Fig. 21.4 (*b*) and Fig. 21.4 (*c*). In case of doubt, it is helpful to shift the inverted half-cycle by a half-period to the left and see if it super-imposes the

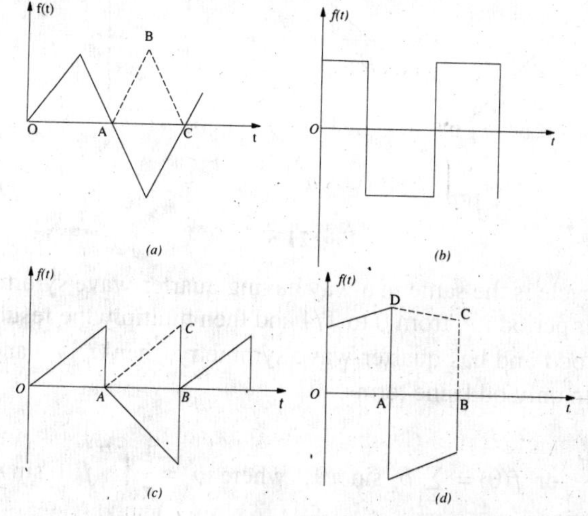

Fig. 21.4

positive half-cycle. If it does so, there exists half-wave symmetry otherwise not. It is seen that the waveform of Fig. 21.4 (*d*) does not possess half-wave symmetry. It is so because when its negative half-cycle is inverted and shifted by half a period to the left it does not superimpose the positive half-cycle.

It may be noted that half-wave symmetry may be present in a waveform which also shows either odd symmetry or even symmetry:

For example, the square waveform shown in Fig. 21.4(*a*) possesses even symmetry whereas the triangular waveform of Fig. 24.4 (*b*) has odd symmetry. All cosine and sine waves possess half-wave symmetry because

$$\cos\frac{2\pi}{T}\left(t\pm\frac{T}{2}\right)=\cos\left(\frac{2\pi}{T}t\pm\pi\right)=-\cos\frac{2\pi}{T}t\,;\ \sin\frac{2\pi}{T}\left(t\pm\frac{T}{2}\right)=\sin\left(\frac{2\pi}{T}t\pm\pi\right)=-\sin\frac{2\pi}{T}t$$

It is worth noting that the Fourier series of any function which possesses half-wave symmetry has zero average value and contains only odd harmonics and is given by

$$f(t)=\sum_{\substack{n=1\\ odd}}^{\infty}\left(a_n\cos\frac{2\pi n}{T}t+b_n\sin\frac{2\pi n}{T}t\right)=\sum_{\substack{n=1\\ odd}}^{\infty}(a_n\cos n\omega_0 t+b_n\sin n\omega_0 t)=\sum_{\substack{n=1\\ odd}}^{\infty}(a_n\cos n\theta+b_n\sin n\theta)$$

$$\text{where, } a_n=\frac{4}{T}\int_0^{T/2}f(t)\cos\frac{2\pi n}{T}dt=\frac{2}{\pi}\int_0^{\pi}f(\theta)\cos n\theta\,d\theta \quad \ldots n \text{ odd}$$

$$b_n=\frac{4}{T}\int_0^{T/2}f(t)\sin\frac{2\pi n}{T}dt=\frac{2}{\pi}\int_0^{\pi}f(\theta)\sin n\theta\,d\theta \quad \ldots n \text{ odd}$$

4. Quarter-wave Symmetry

An odd or even function with rotational symmetry is said to possess quarter-wave symmetry.

Fig. 21.5 (*a*) possesses half-wave symmetry as well as odd symmetry. The wave shown in Fig. 21.5 (*b*) has both half-wave symmetry and even symmetry.

The mathematical test for quarter-wave symmetry is as under :

Odd quarter-wave $f(t) = -f(t + T/2)$ and $f(-t) = -f(t)$

Even quarter-wave $f(t) = -f(t + T/2)$ and $f(t) = f(-t)$

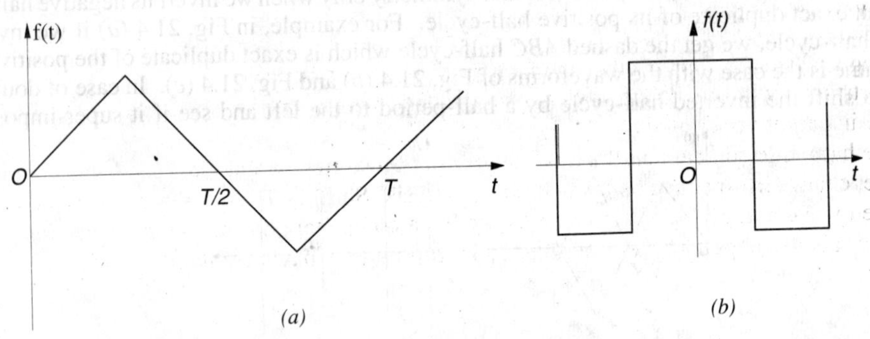

Fig. 21.5

Since each quarter cycle is the same in a way having quarter-wave symmetry, it is sufficient to integrate over one quarter period *i.e.* from 0 to T/4 and then multiply the result by 4.

(*i*) If $f(t)$ or $f(\theta)$ is odd and has quarter-wave symmetry, then a_0 is 0 and a_n is 0. Hence, the Fourier series will contain only odd sine terms.

$$\therefore\ f(t)=\sum_{\substack{n=1\\ odd}}^{\infty}b_n\sin\frac{2\pi nt}{T}\ \text{ or }\ f(\theta)=\sum_{\substack{n=1\\ odd}}^{\infty}b_n\sin n\theta,\ \text{ where }\ b_n=\frac{1}{\pi}\int_0^{2\pi}f(\theta)\sin n\theta\,d\theta$$

It may be noted that in the case of odd quarter-wave symmetry, the integration may be carried over a quarter cycle.

$$\therefore \quad a_n = \frac{4}{\pi}\int_0^{\pi/2} f(\theta)\cos n\,\theta\, d\,\theta \qquad \text{....}n \text{ odd}$$

$$= \frac{8}{T}\int_0^{T/4} f(t)\sin n\omega\, t dt \qquad \text{....}n \text{ odd}$$

(ii) If $f(t)$ or $f(\theta)$ is even and, additionally, has quarter-wave symmetry, then a_0 is 0 and b_n is 0. Hence, the Fourier series will contain only odd cosine terms.

$$\therefore \quad f(\theta) = \sum_{\substack{n=1\\ odd}}^{\infty} a_n \cos n\theta\, d\theta = \sum_{\substack{n=1\\ odd}}^{\infty} a_n \cos n\omega_0 t\, dt \text{ ; where } a_n = \frac{1}{\pi}\int_0^{2\pi} f(\theta)\cos n\theta\, d\theta$$

In this case a_n may be found by integrating over any quarter period.

$$a_n = \frac{4}{\pi}\int_0^{\pi/2} f(\theta)\sin n\theta\, d\theta \qquad \text{...}n \text{ odd}$$

$$= \frac{8}{T}\int_0^{T/4} f(t)\cos n\omega t\, dt \qquad \text{...}n \text{ odd}$$

21.8. Line or Frequency Spectrum

A plot which shows the amplitude of each frequency component in a complex waveform is called the line spectrum or frequency spectrum (Fig. 21.6). The amplitude of each frequency component is indicated by the length of the vertical line located at the corresponding frequency. Since the spectrum represents frequencies of the harmonics as discrete lines of appropriate height, it is also called a discrete spectrum. The lines decrease rapidly for waves having convergent series. Waves with discontinuities such as the sawtooth and square waves have spectra whose amplitudes decrease slowly because their series have strong high harmonics. On the other hand, the line spectra of waveforms without discontinuities and with a smooth appearance have lines which decrease in height very rapidly.

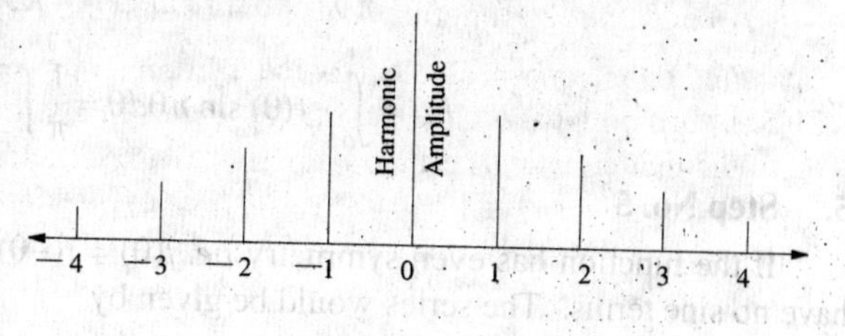

Fig. 21.6

The harmonic content and the line spectrum of a wave represent the basic nature of that wave and never change irrespective of the method of analysis. Shifting the zero axis changes the symmetry of a given wave and gives its trigonometric series a completely different appearance but the same harmonics always appear in the series and their amplitude remains constant.

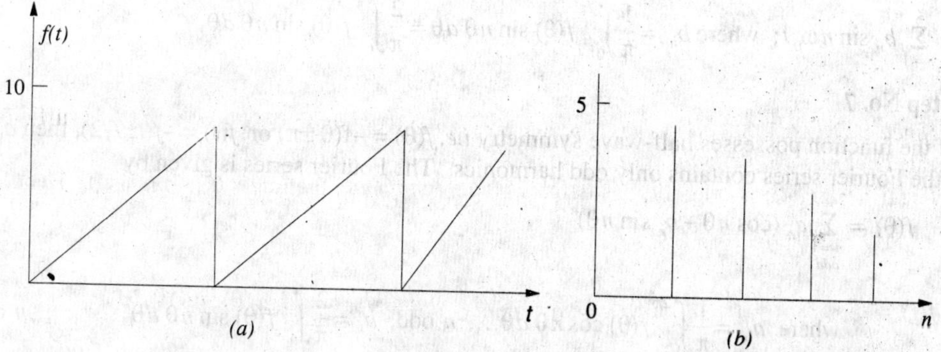

Fig. 21.7

Fig. 21.7 shows a smooth wave alongwith its line spectrum. Since there are only sine terms in its trigonometric series (apart from $a_0 = \pi$), the harmonic amplitudes are given by b_n.

21.9 Procedure for Finding the Fourier Series of a Given Function

It is advisable to follow the following steps :

1. **Step No.1**

If the function is defined by a set of equations, sketch it approximately and examine for symmetry.

2. **Step No. 2**

Whatever be the period of the given function, take it as 2π (Ex. 20.6) and find the Fourier series in the form

$$f(\theta) = a_0 + a_1 \cos\theta + a_2 \cos 2\theta + ... + a_n \cos n\theta + b_1 \sin\theta + b_2 \sin 2\theta + ... + b_n \sin n\theta$$

3. **Step No. 3**

The value of the constant a_0 can be found in most cases by inspection. Otherwise it can be found as under :

$$a_0 = \frac{1}{2\pi}\int_0^{2\pi} f(\theta)d\theta = \frac{1}{2\pi}\int_{-\pi}^{\pi} f(\theta)d\theta$$

4. **Step No. 4**

If there is no symmetry, then a_0 is found as above whereas the other two fourier constants can be found by the relation.

$$a_n = \frac{1}{\pi}\int_0^{2\pi} f(\theta)\cos n\theta d\theta = \frac{1}{\pi}\int_{-\pi}^{\pi} f(\theta)\cos n\,\theta\,d\theta$$

$$b_n = \int_0^{2\pi} f(\theta)\sin n\theta d\theta = \frac{1}{\pi}\int_{-\pi}^{\pi} f(\theta)\sin n\theta\,d\theta$$

5. **Step No. 5**

If the function has even symmetry *i.e.* $f(\theta) = f(-\theta)$, then $b_n = 0$ so that the Fourier series will have no sine terms. The series would be given by

$$f(\theta) = a_0 + \sum_{n=1}^{\infty} a_n \cos n\theta\,d\theta \text{ where } a_n = \frac{1}{\pi}\int_0^{2\pi} f(\theta)\cos n\theta = \frac{2}{\pi}\int_0^{\pi} f(\theta)\cos n\theta d\theta$$

6. **Step No. 6**

If the given function has odd symmetry *i.e.* $f(-\theta) = -f(\theta)$ then $a_0 = 0$ and $a_n = 0$. Hence, there would be no cosine terms in the Fourier series which accordingly would be given by

$$f(\theta) = \sum_{n=1}^{\infty} b_n \sin n\omega_0 t;\ \text{where } b_n = \frac{1}{\pi}\int_0^{2\pi} f(\theta)\sin n\theta\,d\theta = \frac{2}{\pi}\int_0^{\pi} f(\theta)\sin n\theta\,d\theta$$

7. **Step No. 7**

If the function possesses half-wave symmetry *i.e.* $f(\theta) = -f(\theta \pm \pi)$ or $f(t) = -f(t \pm T/2)$, then a_0 is 0 and the Fourier series contains only odd harmonics. The Fourier series is given by

$$f(\theta) = \sum_{\substack{n=1 \\ odd}}^{\infty} a_n (\cos n\theta + b_n \sin n\theta)$$

$$\text{where } a_n = \frac{1}{\pi}\int_0^{2\pi} f(\theta)\cos n\theta\,d\theta \text{.... } n \text{ odd, } b_n = \frac{2}{\pi}\int_0^{\pi} f(\theta)\sin n\theta\,d\theta \qquad \text{...}n \text{ odd}$$

8. Step No. 8

If the function has even quarter-wave symmetry then $a_0 = 0$ and $b_n = 0$. It means the Fourier series will contain no sine terms but only odd cosine terms. It would be given by

$$f(\theta) = \sum_{\substack{n=1 \\ odd}}^{\infty} a_n \cos n\theta; \text{ where } a_n = \frac{1}{\pi}\int_0^{2\pi} f(\theta)\cos n\theta\, d\theta = \frac{2}{\pi}\int_0^{\pi} f(\theta)\cos n\theta\, d\theta = \frac{4}{\pi}\int_0^{\pi/2} f(\theta)\cos n\theta\, d\theta \;\ldots n \text{ odd}$$

9. Step No. 9

If the function has odd quarter-wave symmetry, then $a_0 = 0$ and $a_n = 0$. The Fourier series will contain only odd sine terms (but no cosine terms).

$$\therefore f(\theta) = \sum_{\substack{n=1 \\ odd}}^{\infty} b_n \sin n\theta; \text{ where } b_n = \frac{1}{\pi}\int_0^{2\pi} f(\theta)\sin n\theta\, d\theta = \frac{2}{\pi}\int_0^{\pi} f(\theta)\sin n\theta\, d\theta = \frac{4}{\pi}\int_0^{\pi/2} f(\theta)\sin n\theta\, d\theta \;\ldots n \text{ odd}$$

10. Step No. 10

Having found the coefficients, the Fourier series as given in step No. 2 can be written down.

11. Step No. 11

The different harmonic amplitudes can be found by combining similar sine and cosine terms *i.e.*

$$A_n = \sqrt{a_n^2 + b_n^2}$$

where A_n is the amplitude of the n_{th} harmonic.

Table No. 21.1

	Wave from	*Appearance*	*Equation*
A.	Sine wave	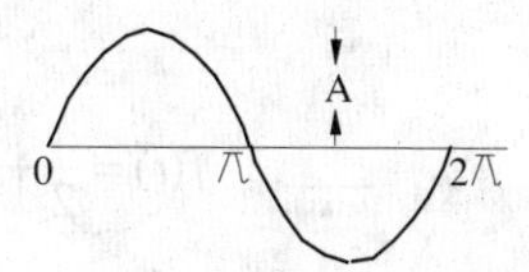	$f(t) = A \sin \omega t$
B.	Half-wave rectified sine wave	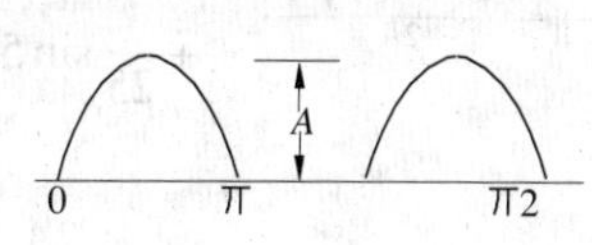	$f(t) = A\left(\frac{1}{\pi} + \frac{1}{2}\sin \omega t - \frac{2}{3\pi}\cos 2\omega t - \frac{2}{15\pi}\cos 4\omega t - \frac{2}{35\pi}\cos 6\,\omega t - \frac{2}{63\pi}\cos 8\omega t \ldots\ldots\right)$
C.	Full-wave rectified sine wave	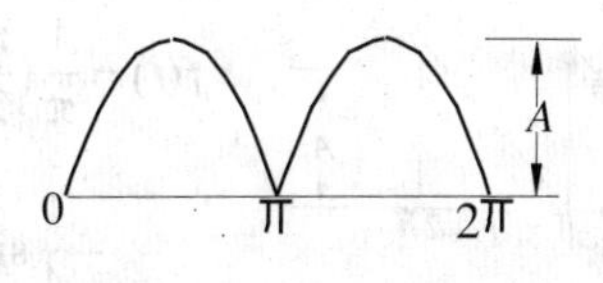	$f(t) = \frac{2A}{\pi}\left(1 - \frac{2}{3}\cos 2\omega t - \frac{2}{15}\cos 4\omega t - \frac{2}{35}\cos 6\omega\, t \ldots.\right)$
D.	Rectangular or square wave	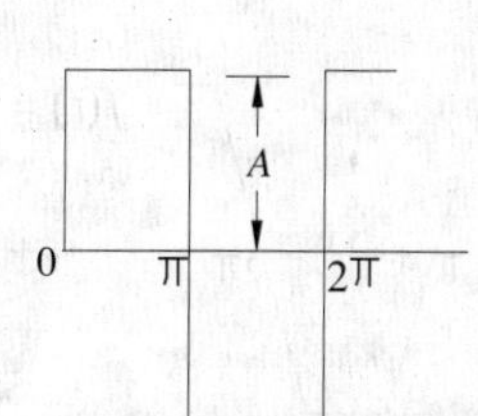	$f(t) = \frac{4A}{\pi}\left(\sin \omega\, t + \frac{1}{3}\sin 3\omega\, t + \frac{1}{5}\sin 5\omega\, t + \frac{1}{7}\sin 7\omega\, t + \ldots.\right)$

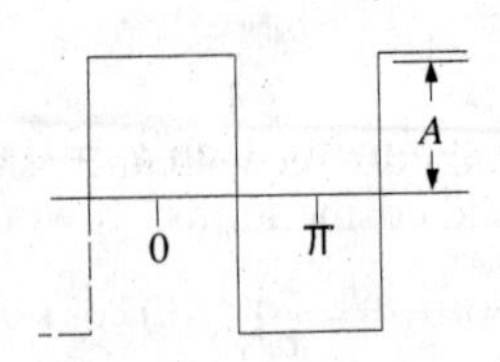

$$f(t)=\frac{4A}{\pi}(\cos\omega t-\frac{1}{3}\cos 3\omega t+\frac{1}{5}\cos 5\omega t.......)$$

E. Rectangular or square wave pulse

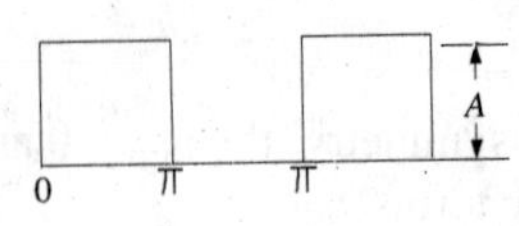

$$f(t)=\frac{A}{2}+\frac{2A}{\pi}(\sin\omega t+\frac{1}{3}\sin 3\omega t+\frac{1}{5}\sin 5\omega t+\frac{1}{7}\sin 7\omega t+....)$$

F. Triangular wave

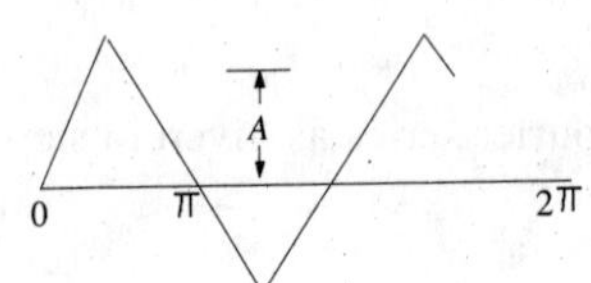

$$f(t)=\frac{8A}{\pi^2}(\sin\omega t-\frac{1}{9}\sin 3\omega t+\frac{1}{25}\sin 5\omega t-\frac{1}{49}\sin 7\omega t+.....)$$

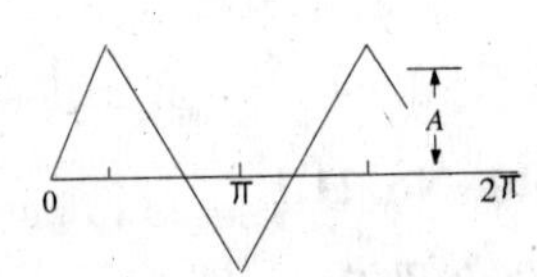

$$f(t)=\frac{8A}{\pi^2}(\cos\omega t+\frac{1}{9}\cos 3\omega t+\frac{1}{25}\cos 5\omega t+\frac{1}{49}\cos 7\omega t+....)$$

G. Triangular pulse

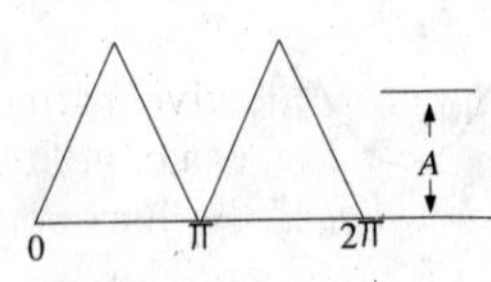

$$f(t)=\frac{A}{2}+\frac{4A}{\pi^2}(\sin\omega t-\frac{1}{9}\sin 3\omega t+\frac{1}{25}\sin 5\omega t-\frac{1}{49}\sin 7\omega t+....)$$

H. Sawtooth wave

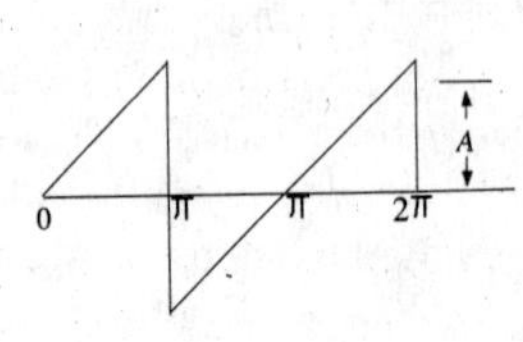

$$f(t)=\frac{2A}{\pi}(\sin\omega t-\frac{1}{2}\sin 2\omega t+\frac{1}{3}\sin 3\omega t-\frac{1}{4}\sin 4\omega t+....)$$

I. Sawtooth pulse

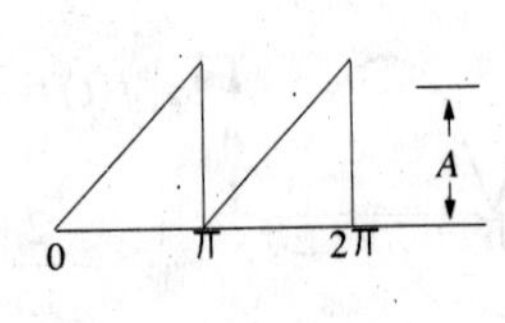

$$f(t)=\frac{A}{\pi}(\frac{\pi}{2}-\sin\omega_0 t-\frac{1}{2}\sin 2\omega t-\frac{1}{3}\sin 3\omega t-\frac{1}{4}\sin 4\omega t)$$

J. Trapezoidal wave

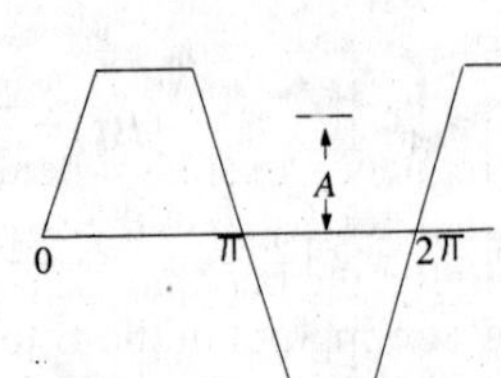

$$f(t)=\frac{6\sqrt{3}.A}{\pi^2}(\sin\omega t-\frac{1}{25}\sin 5\omega t+\frac{1}{49}\sin 7\omega t.....)$$

21.10. Wave Analyzer

A wave analyzer is an instrument designed to measure the individual amplitude of each harmonic

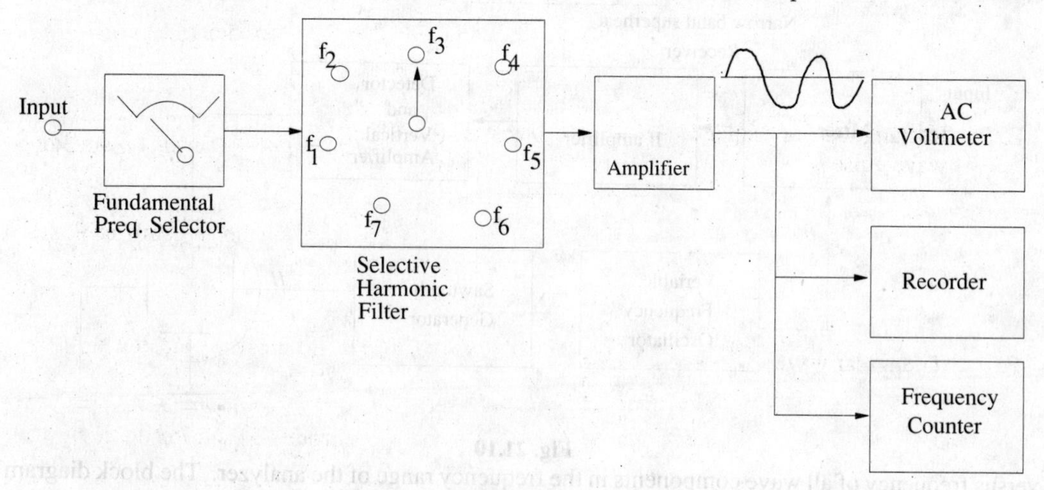

Fig. 21.8

component in a complex waveform. It is the simplest form of analysis in the frequency domain and can be performed with a set of tuned filters and a voltmeter. That is why such analyzers are also called frequency-selective voltmeters. Since such analyzers sample the frequency spectrum successively, *i.e*. one after the other, they are called non-real-time analyzers.

The block diagram of a simple wave analyzer is shown in Fig. 21.8. It consists of a tunable fundamental frequency selector that detects the fundamental frequency f_1 which is the lowest frequency contained in the input waveform.

Once tuned to this fundamental frequency, a selective harmonic filter enables switching to multiples of f_1. After amplification, the output is fed to an a.c. voltmeter, a recorder and a frequency counter. The voltmeter reads the r.m.s amplitude of the harmonic wave, the recorder traces its waveform and the frequency counter gives its frequency. The line spectrum of the harmonic component can be plotted from the above data.

For higher frequencies (MHz) heterodyne wave analyzers are generally used. Here, the input complex wave signal is heterodyned to a higher intermediate frequency (IF) by an internal local oscillator. The output of the IF amplifier is rectified and is applied to a dc voltmeter called heterodyned - tuned voltmeter.

The block diagram of a wave analyzer using the heterodyning principle is shown in Fig. 21.9.

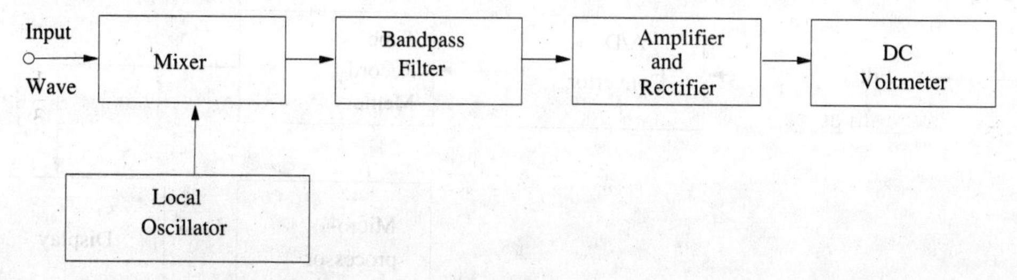

Fig. 21.9

The signal from the internal, variable-frequency oscillator heterodynes with the input signal in a mixer to produce output signal having frequencies equal to the sum and difference of the oscillator frequency f_0 and the input frequency f_i. Generally, the bandpass filter is tuned to the 'sum frequency' which is allowed to pass through. The signal coming out of the filter is amplified, rectified and then applied to a dc voltmeter having a decibel-calibrated scale. In this way, the peak amplitudes of the fundamental component and other harmonic components can be calculated.

21.11. Spectrum Analyzer

It is a real-time instrument *i.e.* it simultaneously displays on a CRT, the harmonic peak values

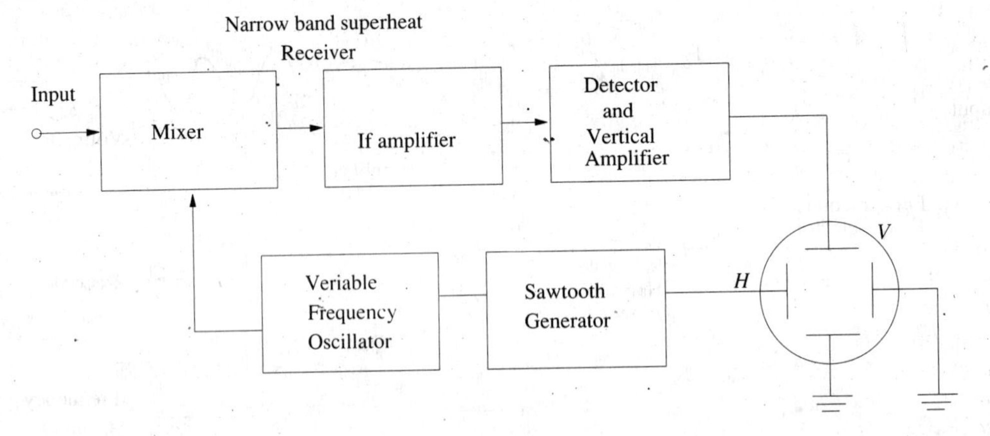

Fig. 21.10

versus frequency of all wave components in the frequency range of the analyzer. The block diagram of such an analyzer is shown in Fig. 21.10.

As seen, the spectrum analyzer uses a CRT in combination with a narrow-band superheterodyne receiver. The receiver is tuned by varying the frequency of the voltage-tuned variable-frequency oscillator which also controls the sawtooth generator that sweeps the horizontal time base of the CRT deflection plates. As the oscillator is swept through its frequency band by the sawtooth generator, the resultant signal mixes and beats with the input signal to produce an intermediate frequency (IF) signal in the mixer. The mixer output occurs only when there is a corresponding harmonic component in the input signal which matches with the sawtooth generator signal. The signals from the IF amplifier are detected and further amplified before applying them to the vertical deflection plates of the CRT. The resultant output displayed on the CRT represents the line spectrum of the input complex or nonsinusoidal waveform.

21.12. Fourier Analyzer

A Fourier analyzer uses digital signal processing technique and provides information regarding the contents of a complex waye which goes beyond the capabilities of a spectrum analyzer. These analyzers are based on the calculation of the discrete Fourier transform using an algorithm called the fast Fourier transformer. This algorithm calculates the amplitude and phase of each harmonic component from a set of time domain samples of the input complex wave signal.

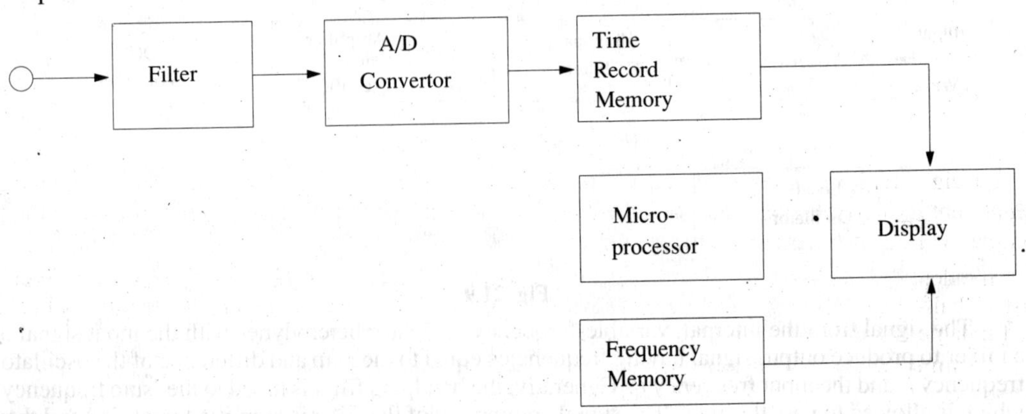

Fig. 21.11

A basic block diagram of a Fourier analyzer is shown in Fig. 21.11. The complex wave signal applied to the instrument is first filtered to remove out-of-band frequency components. Next, the signal is applied to an analog/digital (A/D) convertor which samples and digitizes it at regular time intervals until a full set of samples (called a time record) has been collected. The microprocessor then performs a series of computations on the time data to obtain the frequency-domain results *i.e.* amplitude versus frequency relationships. These results which are stored in memory can be displayed on a CRT or recorded permanently with a recorder or plotter.

Since Fourier analyzers are digital instruments, they can be easily interfaced with a computer or other digital systems. Moreover, as compared to spectrum analyzers, they provide a higher degree of accuracy, stability and repeatability.

21.13. Harmonic Synthesis

It is the process of building up the shape of a complex waveform by adding the instantaneous values of the fundamental and harmonics. It is a graphical procedure based on the knowledge of the different components of a complex waveform.

Example 21.1. *A complex voltage waveform contains a fundamental voltage of r.m.s value 220 V and frequency 50 Hz alongwith a 20% third harmonic which has a phase angle lagging by 3π/4 radian at t = 0. Find an expression representing the instantaneous complex voltage v. Using harmonic synthesis, also sketch the complex waveform over one cycle of the fundamental.*

Solution. The maximum value of the fundamental voltage is = $220\times\sqrt{2}$ = 310 V. Its angular velocity is $\omega = 2\pi \times 50 = 100\pi$ rad/s. Hence, the fundamental voltage is represented by 310 sin 100π *t*.

The amplitude of the third harmonic = 20% of 310 = 62 V. Its frequency is 3 × 50 = 150 Hz. Hence, its angular frequency is = 2π × 150 = 300 π rad/s. Accordingly, the third harmonic voltage can be represented by the equation 62 sin (300 π t−3π/4). The equation of the complex voltage is given by $\upsilon = 310 \sin 100\,\pi t + 62 \sin (300\,\pi\, t - 3\pi/4)$

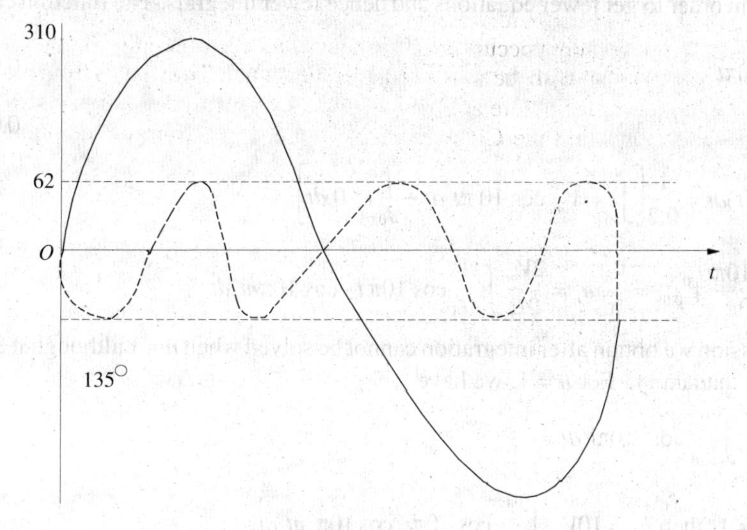

Fig. 21.12

In Fig. 21.12 are shown one cycle of the fundamental and three cycles of the third harmonic component initially lagging by 3 π/4 radian or 135°. By adding ordinates at different intervals, the complex voltage waveform is built up as shown.

Incidentally, it would be seen that if the negative half-cycle is reversed, it is identical to the postive half-cycle. This is a feature of waveforms possessing half-wave symmetry which contain the fundamental and odd harmonics.

Example 21.2. *For the nonsinusoidal wave shown in Fig. 21.13, determine (a) Fourier coefficients a_0, a_3 and b_4 and (b) frequency of the fourth harmonic if the wave has a period of 0.02 second.*

Solution. The function f(θ) has a constant value from θ = 0 to θ = 4π/3 radian and 0 value from θ − 4π /3 radian to θ = 2 π radian.

$$(a)a_0 = \frac{1}{2\pi}(\text{net area per cycle})_0^{2\pi} = \frac{1}{2\pi}\left(6\times\frac{4\pi}{3}\right) = 4$$

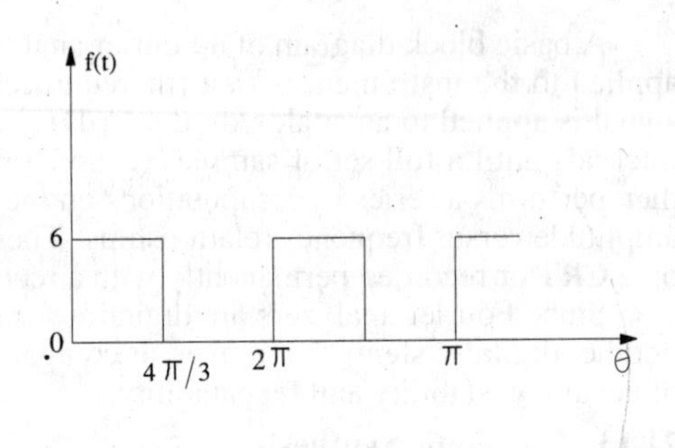

Fig. 21.13

$$a_3 = \frac{1}{\pi}\int_0^{2\pi} f(\theta)\cos 3\theta\, d\theta = \frac{1}{\pi}\int_0^{4\pi/3} 6\cos 3\theta\, d\theta$$

$$= \frac{6}{\pi}\left|\frac{\sin 3\theta}{3}\right|_0^{4\pi/3} = \frac{2}{\pi}(\sin 4\pi) = 0$$

$$b_4 = \frac{1}{\pi}\int_0^{2\pi} f(\theta)\sin 4\theta\, d\theta = \frac{1}{\pi}\int_0^{4\pi/3} 6\sin 4\theta\, d\theta$$

$$= \frac{6}{\pi}\left|\frac{-\cos 4\theta}{4}\right|_0^{4\pi/3} = -\frac{3}{2\pi}\left(\frac{\cos 16\pi}{3} - \cos 0\right) = \frac{-3}{2\pi}(-0.5-1) = \frac{9}{4\pi}$$

(*b*) Frequency of the fourth harmonic $= 4 f_0 = 4/T = 4/0.02 = 200$ Hz.

Example 21.3. *Find the Fourier series of the "half sinusoidal" voltage waveform which represents the output of a half-wave rectifier. Sketch its line spectrum.*

Solution. As seen from Fig. 21.14 (*a*), $T = 0.2$ second, $f_0 = 1/T = 1/0.2 = 5$ Hz and $\omega_0 = 2\pi, f_0 = 10\pi$ rad/s. Moreover, the function has even symmetry. Hence, the Fourier series will contain no sine terms because $b_n = 0$.

The limits of integration would not be taken from $t = 0$ to $t = 0.2$ second, but from $t = -0.5$ to $t = 0.15$ second in order to get fewer equations and hence fewer integrals. The function can be written as

$$f(t) = V_m \cos 10\pi t \qquad -0.05 < t < 0.05$$

$$= 0 \qquad 0.05 < t < 0.15$$

$$a_0 = \frac{1}{T}\int_{-0.05}^{0.15} f(t)dt = \frac{1}{0.2}\left[\int_{-0.05}^{0.05} V_m \cos 10\,\pi t\; dt + \int_{0.05}^{0.15} 0.dt\right]$$

$$= \frac{V_m}{0.2}\left|\frac{\sin 10\pi t}{10\pi}\right|_{-0.05}^{0.05} = \frac{V_m}{\pi}\ ; \quad a_n = \frac{2V_m}{0.2}\int_{-0.05}^{0.05} \cos 10\pi t \cdot \cos 10\pi nt\; dt$$

The expression we obtain after integration cannot be solved when $n = 1$ although it can be solved when n is other than unity. For $n = 1$, we have

$$a_1 = 10V_m \int_{-0.05}^{0.05} \cos^2 10\pi t\; dt = \frac{V_m}{2}$$

When $n \neq 1$, then $a_n = 10V_m \int_{-0.05}^{0.05} \cos 10\pi t . \cos 10\pi\; nt\; dt$

$$= \frac{10V_m}{2}\int_{-0.05}^{0.05} [\cos 10\,\pi(1+n)t\, \cos 10\,\pi(1-n)t]\, dt = \frac{2V_m}{\pi}\cdot\frac{\cos(\pi\, n/2)}{(1-n^2)} \qquad \ldots n \neq 1$$

$$a_2 = \frac{2V_m}{\pi}\cdot\frac{\cos\pi}{1-4} = \frac{2Vm}{\pi}\cdot\frac{-1}{-3} = \frac{2V_m}{3\pi}\ ; \ a_3 = \frac{2V_m}{\pi}\cdot\frac{\cos 3\pi/2}{1-3^2} = 0\ ; \ a_4 = \frac{2V_m}{\pi}\cdot\frac{\cos 2\pi}{1-4^2} = -\frac{2V_m}{15\pi}$$

$$a_5 = 0;\ a_6 = \frac{2V_m}{\pi}\cdot\frac{\cos 3\pi}{1-6^2} = \frac{2V_m}{35\pi} \text{ and so on}$$

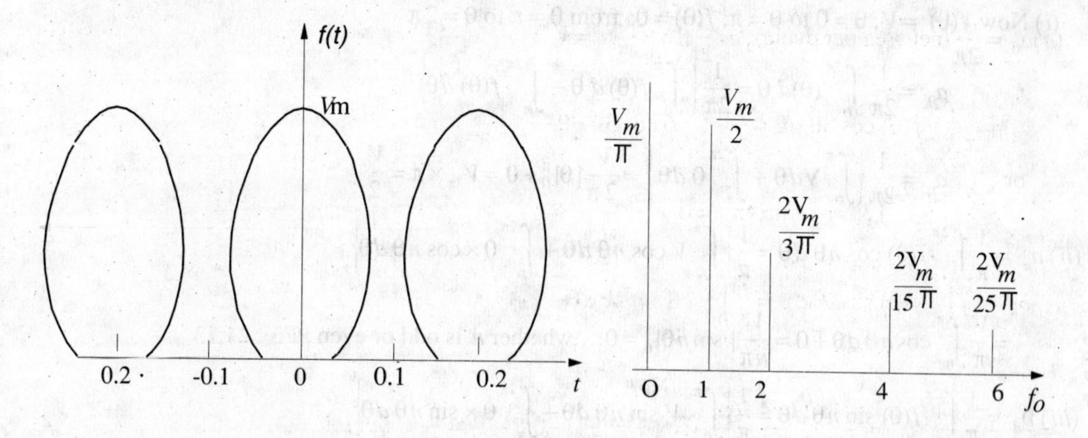

Fig. 21.14

Substituting the value of a_0, a_1, a_2, a_4 etc. in the standard Fourier series expression given in Art. 20.3, we have

$$f(t) = a_0 + a_1 \cos 2\omega_0 t + a_2 \cos {}_2\omega 0\, t + a_4 \cos 4\omega_0 t + a_6 \cos 6\omega_0 t + \ldots\ldots$$

$$= \frac{V_m}{\pi} + \frac{V_m}{2} \cos 10\pi t + \frac{2V_m}{3\pi} \cos 20\pi t - \frac{2V_m}{15\pi} \cos 40\pi t + \frac{2V_m}{35} \cos 60\pi t \ldots\ldots$$

$$= V_m \left(\frac{1}{\pi} + \frac{1}{2} \cos \omega_0 t + \frac{2}{3\pi} \cos 2\omega_0 t\, \pi - \frac{2}{15\pi} \cos 4\omega_0 t + \frac{2}{35\pi} \cos 6\omega_0 t \ldots\ldots \right)$$

The line spectrum which is a plot of the harmonic amplitudes versus frequency is given in Fig. 21.14(*b*).

Example 21.4. *Determine the Fourier series for the square voltage pulse shown in Fig. 21.15 (a) and plot its line spectrum.* **(Network Theory, Nagpur Univ. 1992)**

Solution. The wave represents a periodic function of θ or ωt or $\left(\frac{2\pi t}{T}\right)$ having a period extending over 2 π radians or *T* seconds. The general expression for this wave can be written as

$$f(\theta) = a_0 + a_1 \cos \theta + a_2 \cos 2\theta + a_3 \cos 3\theta + \ldots.$$

$$\ldots\ldots + b_1 \sin \theta + b_2 \sin 2\theta + b_3 \sin 3\theta + \ldots.$$

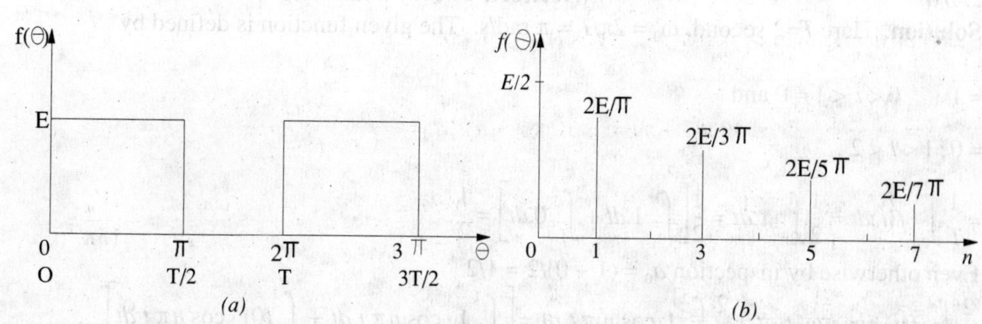

f(Θ)
E
0 π 2π 3π Θ
O T/2 T 3T/2
(a)

f(Θ)
E/2
2E/π
2E/3π
2E/5π
2E/7π
0 1 3 5 7 n
(b)

Fig. 21.15

(*i*) Now, $f(\theta) = V; \theta = 0$ to $\theta = \pi$; $f(\theta) = 0$, from $\theta = \pi$ to $\theta = 2\pi$

$$\therefore \quad a_0 = \frac{1}{2\pi}\int_0^{2\pi}(\theta)d\,\theta = \frac{1}{2\pi}\left\{\int_0^{\pi} f(\theta)\,d\,\theta + \int_{\pi}^{2\pi} f(\theta)\,d\theta\right\}$$

$$\text{or} \quad a_0 = \frac{1}{2\pi}\left\{\int_0^{\pi} V d\theta + \int_{\pi}^{2\pi} 0\,d\theta\right\} = \frac{V}{2\pi}|\theta|_0^{\pi} + 0 = V_{2\pi} \times \pi = \frac{V}{2}$$

$$(ii)\ a_n = \frac{1}{\pi}\int_0^{2\pi} f(\theta)\cos n\theta\,d\theta = \frac{1}{\pi}\left\{\int_0^{\pi} V\cos n\theta\,d\theta + \int_{\pi}^{2\pi} 0\times\cos n\theta\,d\theta\right\}$$

$$= \frac{V}{\pi}\int_0^{\pi}\cos n\theta\,d\theta + 0 = \frac{V}{n\pi}|\sin n\theta|_0^{\pi} = 0 \text{....whether } n \text{ is odd or even}$$

$$(iii)\ b_n = \frac{1}{\pi}\int_0^{2\pi} f(\theta)\sin n\theta\,d\theta = \frac{1}{\pi}\left\{\int_0^{\pi} V\sin n\theta\,d\theta + \int_{\pi}^{2\pi} 0\times\sin n\theta\,d\theta\right\}$$

$$= \frac{V}{\pi}\int_0^{\pi}\sin n\theta d\theta + 0 = \frac{V}{\pi}\left|\frac{-\cos n\theta}{n}\right|_0^{\pi} = \frac{V}{n\pi}(-\cos n\pi + 1)$$

Now, when n is odd, $(1 - \cos n\pi) = 2$ but when n is even, $(1 - \cos n\pi) = 0$.

$$\therefore\ b_1 = \frac{2V}{\pi} \ \ldots n = 1;\ b_2 = \frac{V}{2\pi}\times 0 = 0 \ \ldots n = 2;\ b_3 = \frac{V}{3\pi}\times 2 = \frac{2V}{3\pi} \ \ldots n = 3 \text{ and so on.}$$

Hence, substituting the values of a_0, a_1, a_2 etc. and b_1, b_2 etc. in the above given Fourier series, we get

$$f(\theta) = \frac{V}{2} + \frac{2V}{\pi}\sin\theta + \frac{2V}{3\pi}\sin 3\theta + \frac{2V}{5\pi}\sin 5\theta + \ldots = \frac{E}{2} + \frac{2V}{\pi}\left(\sin\omega_0 t + \frac{1}{3}\sin 3\omega_0 t + \frac{1}{5}\sin 5\omega_0 t + \ldots\right)$$

It is seen that the Fourier series contains a constant term $V/2$ and odd harmonic components whose amplitudes are as under :

Amplitude of fundamental or first harmonic $= \frac{2V}{\pi}$.

Amplitude of second harmonic $= \frac{2V}{3\pi}$

Amplitude of third harmonic $= 2\frac{EV}{5\pi}$ and so on.

The plot of harmonic amplitude versus the harmonic frequencies (called line spectrum) is shown in Fig. 21.15 (*b*).

Example 21.5. *Obtain the Fourier series for the square wave pulse train indicated in Fig. 21.16.* **(Network Theory and Design, AMIETE June, 1990)**

Solution. Here T=2 second, $\omega_0 = 2\pi/T = \pi$ rad/s. The given function is defined by

$$f(t) = 1 \qquad 0 < t < 1 = 0 \text{ and}$$

$$= 0;\ 1 < t < 2$$

$$a_0 = \frac{1}{T}\int_0^T f(t)dt = \frac{1}{2}\int_0^2 1.dt = \frac{1}{2}\left[\int_0^1 1.dt + \int_1^2 0.dt\right] = \frac{1}{2}$$

Even otherwise by inspection $a_0 = (1 + 0)/2 = 1/2$

$$a_n = \frac{2}{T}\int_0^T f(t)\cos n\omega_0 t\,dt = \frac{2}{2}\int_0^2 1.\cos n\pi t\,dt = \left[\int_0^1 1\cdot\cos n\pi t\,dt + \int_1^2 (0)\cdot\cos n\pi t\,dt\right]$$

$$= \int_0^1 \cos n\pi t \, dt = \left| \frac{\sin n\pi t}{n\pi} \right|_0^1 = 0$$

$$b_n = \frac{2}{T}\int_0^T f(t) \sin n\omega_0 t \, dt = \left[\int_0^1 1 . \sin n\pi t \, dt \right.$$

$$\left. + \int_1^2 (0) \sin n\pi t dt \right] = \int_0^1 \sin n\pi t \, dt$$

$$= \left| \frac{-\cos n\pi t}{n\pi} \right|_0^1 = \frac{1 - \cos n\pi}{n\pi}$$

$\therefore \quad b_n = 2/n\pi$ ...when n is odd; $= 0$ ...when n is even

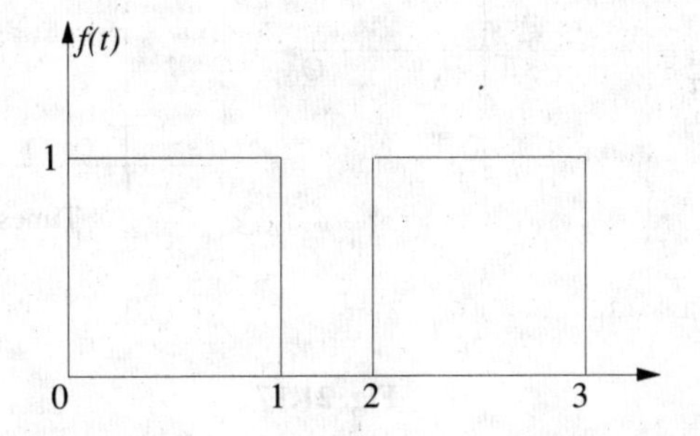

Fig. 21.16

$$\therefore \quad f(t) = a_0 + \frac{2}{\pi}\sum_{\substack{n=1 \\ odd}}^{\infty} \frac{1}{n} \sin 3n\,\pi\, t = \frac{1}{2} + \frac{2}{\pi}\left(\sin \pi\, t + \frac{1}{3} \sin 3\,\pi t + \frac{1}{5} \sin 5\pi\, t \, etc. \right)$$

Example 21.6. *Find the trigonometric Fourier series for the square voltage waveform shown in Fig. 21.17(a) and sketch the line spectrum.*

Solution. The function shown in Fig. 21.17(*a*) is an odd function because at any time $f(-t) = -f(t)$. Hence, its Fourier series will contain only sine terms *i.e.* $a_n = 0$. The function also possesses half-wave symmetry, hence, it will contain only odd harmonics.

As seen from Art. 21.7(2) the Fourier series for the above wave is given by

$$f(\theta) = \sum_{\substack{n=1 \\ odd}}^{\infty} b_n \sin n\theta t \qquad \text{where } b_n = \frac{1}{\pi}\int_0^{2\pi} f(\theta) \sin n\theta \, d\theta$$

$$= \frac{1}{\pi}\left\{ \int_0^{\pi} V \sin n\,\theta \, d\theta + \int_{\pi}^{2\pi} -V \sin n\theta \, d\theta \right\} = \frac{V}{\pi n} \left| -\cos n\theta \right|_0^{\pi} + \frac{V}{\pi n} \left| \cos n\,\theta \right|_{\pi}^{2\pi}$$

$$= \frac{V}{\pi n}\{(-\cos n\pi + \cos 0) - \cos \pi n)\}$$

$$= \frac{V}{\pi\, n}\{(1 - \cos n\pi) + (1 - \cos n\pi)\} = \frac{2V}{\pi n}(1 - \cos n\pi)$$

Now, $1 - \cos n.\pi = 2$ when n is odd

and $= 0$ when n is even

$$\therefore \quad b_1 = \frac{2V}{\pi} \times 2 = \frac{4V}{\pi} \text{ ... putting } n = 1 \text{ ; } b_2 = 0 \text{ ... putting } n = 2$$

$$b_3 = \frac{2V}{\pi 3} \times 2 = \frac{4V}{3\pi} \text{ ...putting } n = 3 \text{ ; } b_4 = 0\text{... putting } n = 4$$

$$b_5 = \frac{2V}{\pi .5} \times 2 = \frac{4V}{5\pi} \text{ ... putting } n = 5 \text{ and so on.}$$

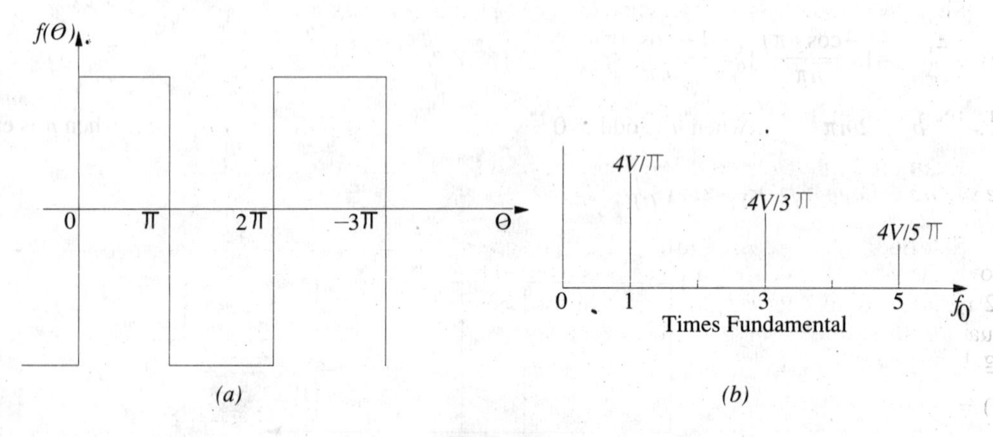

Fig. 21.17

Hence, the Fourier series for the given waveform is

$$f(\theta) = \frac{4V}{\pi} \sin \theta + \frac{4V}{3\pi} \sin 3\theta + \frac{4V}{5\pi} \sin 5\theta + \ldots$$

$$= \frac{4V}{\pi}\left(\sin \omega_0 t + \frac{1}{3} \sin 3\omega_0 t + \frac{1}{5} \sin 5\omega_0 t + \ldots.\right) = \frac{4v}{\pi}\left(\sin \frac{2\pi}{T} t + \frac{1}{3} \sin \frac{6\pi}{T} t + \frac{1}{5} \sin \frac{10\pi}{T} t + \ldots.\right)$$

The line spectrum of the function is shown in Fig. 21.17(*b*). It would be seen that the harmonic amplitudes decrease as $1/n$, that is, the third harmonic amplitude is 1/3 as large as the fundamental, the fifth harmonic is 1/5 as large and so on.

Example 21.7. *Determine the Fourier series for the square voltage waveform shown in Fig. 21.17(a). Plot its line spectrum.*

Solution. This is the same question as given in Ex. 21.6 but has been repeated to illustrated a slightly different technique. As seen from Fig. 21.17 (*a*) $T = 2\pi$, hence, $\omega_0 = 2\pi f_0 = 2\pi/T = 2\pi/\pi = 1$. Over one period the function can be defined as

$$f(t) = V \quad 0 < t < 5$$

$$= -V, \quad \pi < t < 2\pi$$

$$b_n = \frac{1}{\pi}\int_0^{2\pi} f(t) \sin n\omega_0 t \, dt = \frac{1}{\pi}\int_0^{2\pi} f(t) \sin nt \, dt = \frac{1}{\pi}\left[\int_0^{\pi} f(t) \sin nt \, dt + \frac{1}{\pi}\int_{\pi}^{2\pi} f(t) \sin nt \, dt\right]$$

$$= \frac{1}{\pi}\int_0^{\pi} V \sin nt \, dt + \frac{1}{\pi}\int_{\pi}^{2\pi} (-V) \sin nt \, dt = \frac{V}{\pi}\left|\frac{-\cos nt}{n}\right|_0^{\pi} + \frac{V}{\pi}\left|\frac{\cos nt}{n}\right|_{\pi}^{2\pi}$$

$$= -\frac{V}{n\pi}(\cos n\pi - \cos 0) + \frac{V}{n\pi}(\cos 2n\pi - \cos n\pi)$$

Since cos 0 is 1 and cos $2n\pi = 1$ $\therefore$ $b_n = \frac{2V}{n\pi}(1 - \cos n\pi)$

When n is even, $\cos n\pi = 1$ $\therefore$ $b_n = 0$

When n is odd, $\cos n\pi = -1$ $\therefore$ $b_n = \frac{2V}{n\pi}(1+1) = \frac{4V}{n\pi}$ $n = 1, 3, 5....$

Substituting the value of b_n, the Fourier series become

$$f(t) = \sum_{\substack{n=1\\odd}}^{\infty} \frac{4V}{n\pi}\sin nt = \frac{4V}{\pi}\sum_{\substack{n=1\\odd}}^{\infty}\frac{1}{n}\sin nt = \frac{4V}{\pi}\left(\sin t + \frac{1}{3}\sin 3t + \frac{1}{5}\sin 5t + ...\right)$$

Since $\omega_0 = 1$, the above expression in general terms becomes

$$f(t) = \frac{4V}{\pi}\left(\sin\omega_0 t + \frac{1}{3}\sin 3\omega_0 t + \frac{1}{5}\sin 5\omega_0 t + ...\right)$$

The line spectrum is as shown in Fig. 21.17(*b*).

Example 21.8. *Determine the Fourier series of the square voltage waveform shown in Fig. 20.18.*

Solution. As compared to Fig. 21.17(*a*) given above, the vertical axis of figure has been shifted by $\pi/2$ radians. Replacing t by $(t + \pi/2)$ in the above equation, the Fourier series of the waveform shown in Fig. becomes

$$f(t) = \frac{4V}{\pi}\left[\sin\left(t+\frac{\pi}{2}\right) + \frac{1}{3}\sin 3\left(t+\frac{\pi}{2}\right) + \frac{1}{5}\sin 5\left(t+\frac{\pi}{2}\right) + ...\right]$$

$$= \frac{4V}{\pi}\left(\cos t - \frac{1}{3}\cos 3t + \frac{1}{5}\cos 5t....\right)$$

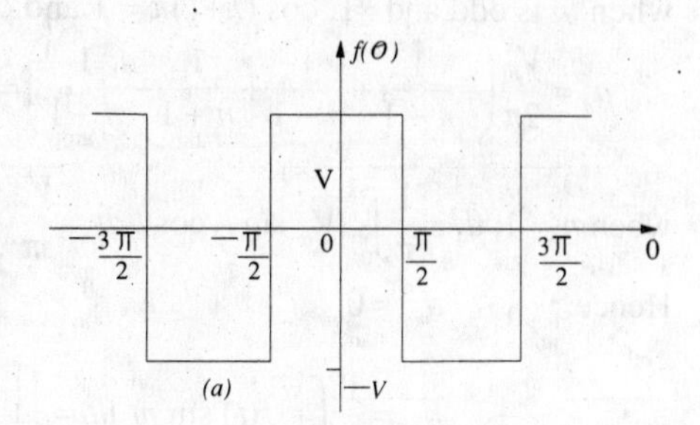

Fig. 21.18

Example 21.9. *Determine the trigonometric Fourier series for the half-wave rectified sine wave form shown in Fig. 21.19(a) and sketch its line spectrum.*

Solution. The given waveform shows no symmetry, hence its series would contain both sine and cosine terms. Moreover, its average value is not obtainable by inspection, hence it will have to be found by integration.

Here, $T = 2\pi$, $\omega_0 = 2\pi/T = 1$. Hence, equation of the given waveform is $V = V_m \sin\omega t = V_m \sin t$.

The given waveform is defined by

$$f(t) = V_m \sin t, 0 < t < \pi = 0 \quad \pi < t < 2\pi$$

$$a_0 = \frac{1}{2\pi}\int_0^{2\pi} f(t)\,dt = \frac{1}{2\pi}[\int_0^{\pi} V_m \sin t\,dt + \int_{\pi}^{2\pi} dt = \frac{1}{2\pi}\int_0^{\pi} V_m \sin t\,dt$$

$$= \frac{V_m}{2\pi}| -\cos t|_0^{\pi} = -\frac{V_m}{2\pi}(\cos\pi - \cos 0) = \frac{V_m}{\pi}$$

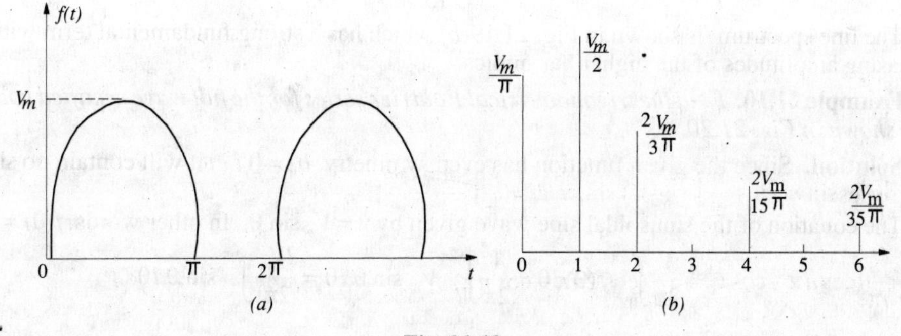

Fig. 21.19

$$a_n = \frac{1}{\pi}\int_0^{2\pi} f(t)\cos nt\,dt = \frac{1}{\pi}\int_0^{\pi} V_m \sin t \cos nt\,dt$$

$$= \frac{V_m}{2\pi}\int_0^{\pi}[\sin(n+1)t - \sin(n-1)t]\,dt = \frac{V_m}{2\pi}\left|\frac{-\cos(n+1)t}{n+1} + \frac{\cos(n-1)t}{n-1}\right|_0^{\pi}$$

$$= \frac{V_m}{2\pi}\left[-\frac{\cos(n+1)\pi}{n+1} + \frac{\cos(n-1)\pi}{n-1} + \frac{1}{n+1} - \frac{1}{n-1}\right]$$

when n is even, $\cos(n+1)\pi = -1$ and $\cos(n-1)\pi = -1$

$$\therefore\; a_n = \frac{V_m}{2\pi}\left[\frac{1}{n+1} - \frac{1}{n-1} + \frac{1}{n+1} - \frac{1}{n-1}\right] = -\frac{2V_m}{\pi(n^2-1)} \qquad \ldots n = 2, 4, 6 \text{ etc.}$$

when n is odd and $\neq 1$, $\cos(n+1)\pi = 1$ and $\cos(n-1)\pi = 1$

$$\therefore\; a_n = \frac{V_m}{2\pi}\left(-\frac{1}{n+1} + \frac{1}{n-1} + \frac{1}{n+1} - \frac{1}{n-1}\right) = 0 \qquad \ldots n = 3, 5, 7 \text{ etc.}$$

$$\text{when } n = 1,\; a_1 = \frac{1}{\pi}\int_0^{\pi} V_m \sin t.\cos t\,dt = \frac{V_m}{\pi}\int_0^{\pi}\sin t\cos t\,dt = \frac{V_m}{2\pi}\int_0^{\pi}\sin 2t\,dt = 0$$

Hence, $\qquad a_n = 0 \qquad \ldots n = 1, 3, 5\ldots.$

$$b_n = \frac{1}{\pi}\int_0^{2\pi} f(t)\sin nt\,dt = \frac{1}{\pi}\left[\int_0^{\pi} V_m \sin t \sin nt\,dt + \int_{\pi}^{2\pi}(0)\sin nt\,dt\right]$$

$$= \frac{V_m}{\pi}\int_0^{\pi}\sin t\sin nt\,dt = 0 \text{ for } n = 2, 3, 4, 5 \text{ etc.}$$

However, the expression is indeterminate for $n = 1$ so that b_1 has to be evaluated separately.

$$b_1 = \frac{1}{\pi}\int_0^{\pi} V_m \sin t.\sin t\,dt = \frac{V_m}{\pi}\int_0^{\pi}\sin^2 t\,dt = \frac{V_m}{2}$$

The required Fourier series for the half-wave rectified voltage waveform is

$$f(t) = \frac{V_m}{\pi} + \frac{V_m}{2}\sin t - \frac{2V_m}{\pi}\sum_{\substack{n=1\\ \text{even}}}^{\infty}\left(\frac{\cos nt}{n^2-1}\right)$$

$$= \frac{V_m}{\pi}\left(1 + \frac{\pi}{2}\sin t - \frac{2}{3}\cos 2t - \frac{2}{15}\cos 4t - \frac{2}{35}\cos 6t\ldots.\right)$$

$$= \frac{V_m}{\pi}\left(1 + \frac{\pi}{2}\sin\omega_0 t - \frac{2}{3}\cos 2\,\omega_0 t - \frac{2}{15}\cos\omega_0 t - \frac{2}{35}\cos 6\omega_0 t - \ldots.\right)$$

$$= \frac{V_m}{T}\left(1 + \frac{\pi}{2}\sin\theta - \frac{2}{3}\cos 2\theta - \frac{2}{15}\cos 4\theta - \frac{2}{35}\cos 6\theta - \ldots\right)$$

The line spectrum is shown in Fig. 21.19(*b*) which has a strong fundamental term with rapidly decreasing amplitudes of the higher harmonics.

Example 21.10. *Find the trigonometrical Fourier series for the full wave rectified voltage sine wave shown in Fig. 21.20.*

Solution. Since the given function has even symmetry, $b_n = 0$ *i.e.* it will contain no sine terms in its series.

The equation of the sinusoidal sine wave given by $v = V_m \sin\theta$. In other words, $f(\theta) = V_m \sin\theta$.

$$a_0 = \frac{1}{2\pi}\int_0^{2\pi} f(\theta)d\theta = \frac{1}{2\pi}\int_0^{2\pi} V_m \sin\theta d\theta = \frac{2V_m}{2\pi}\int_0^{\pi}\sin\theta d\theta$$

It is so because the two parts 0 −π and π −2 are identical

$$\therefore \quad a_0 = \frac{V_m}{\pi} | -\cos\theta|_0^{\pi} = \frac{2V_m}{\pi}$$

$$a_n = \frac{1}{\pi}\int_0^{2\pi} f(\theta)\cos n\theta\, d\theta = \frac{2V_m}{\pi}\int_0^{\pi} \sin\theta \cos n\theta\, d\theta$$

Now, $\sin A \cos B = \frac{1}{2}[\sin(A+B) + \sin(A-B)]$

$$\therefore \quad a_n = \frac{2V_m}{\pi}\int_0^{\pi} [\sin(1+n)\theta + \sin(1-n)\theta]\, d\theta$$

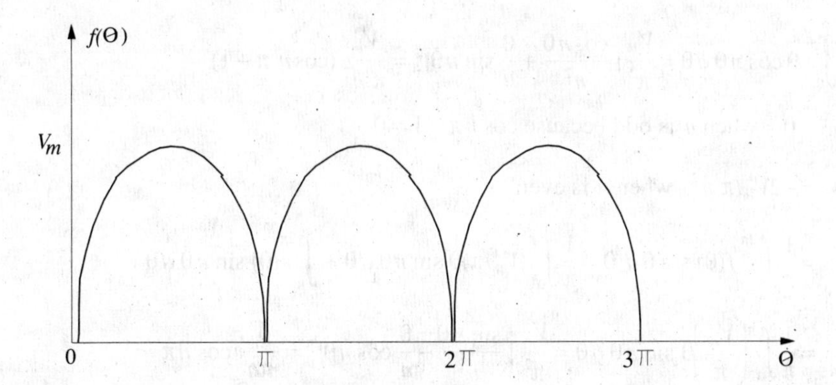

Fig. 21.20

$$= -\frac{V_m}{\pi}\left[\frac{\cos(1+n)\theta}{(1+n)} + \frac{\cos(1-n)\theta}{(1-n)}\right.$$

$$= 0 = \frac{-V_m}{\pi}\left[\frac{\cos(1+n)\pi}{1+n} + \frac{\cos(1-n)\pi}{1-n} - \frac{1}{1+n} - \frac{1}{1-n}\right] \text{...when } n \text{ is odd}$$

However, when n is even, then

$$a_n = \frac{-V_m}{\pi}\left[-\frac{1}{1+n} - \frac{1}{1-n} - \frac{1}{1+n} - \frac{1}{1-n}\right] = \frac{2V_m}{\pi}\left[\frac{1}{1+n} + \frac{1}{1-n}\right] = \frac{-4V_m}{\pi(n^2-1)}$$

$$\therefore \quad f(\theta) = a_0 - \frac{4V_m}{\pi}\sum_{\substack{n=2\\ \text{even}}}^{\infty} \frac{\cos 2\theta}{(n^2-1)}$$

$$\therefore \quad f(\theta) = \frac{2V_m}{\pi} - \frac{4V_m}{\pi}\left(\frac{1}{3}\cos^2\theta - \frac{1}{15}\cos 4\theta + \frac{1}{35}\cos 6\theta + \ldots\right)$$

Example 21.11. *Determine the Fourier series for the waveform shown in Fig. 21.21 (a) and sketch its line spectrum.*

Solution. It is seen from Fig. 21.21 (*a*) that the waveform equation is $f(\theta) = (V_m/\pi)\theta$. The given function $f(\theta)$ is defined by

$$f(\theta) = \left(\frac{V_m}{\pi}\right)\theta \qquad 0 < \theta < \pi$$

$$= 0 \quad \pi < \theta < 2\pi$$

Since the function possesses neither even nor odd symmetry, it will contain both sine and cosine terms.

Average value of the wave over one cycle is $V_m/4$ or $a_0 = V_m/4$. It is so because the average value over the first half cycle is $V_m/2$ and over the second half cycle is 0 hence, the average value for full cycle is $= \dfrac{(V_m/2)+0}{2} = \dfrac{V_m}{4}$

$$a_n = \frac{1}{\pi}\int_0^{2\pi} f(\theta) n\theta\, d\theta = \frac{1}{\pi}\left[\int_0^{\pi}(V_m/\pi)\theta \cos n\theta\, d\theta + \int_{\pi}^{2\pi}(0)\cos n\theta\, d\theta\right]$$

$$= \frac{V_m}{\pi^2}\int_0^{\pi}\theta\cos n\theta\, d\theta = \frac{V_m}{\pi^2}\left|\frac{\cos n\theta}{n^2} + \frac{\theta}{n}\sin n\theta\right|_0^{\pi} = \frac{V_m}{\pi^2 n^2}(\cos n\pi - 1)$$

$$\therefore \quad a_n = 0 \quad \text{when } n \text{ is odd because } \cos n\pi - 1 = 0$$

$$= -2V_m/\pi^2 n^2 \quad \text{when } n \text{ is even}$$

$$b_n = \frac{1}{\pi}\int_0^{2\pi} f(\theta)\sin\theta\, d\theta = \frac{1}{\pi}\int_0^{\pi}(V_m/\pi)\theta\sin n\theta\, d\theta + \int_{\pi}^{2\pi}(0)\sin n\theta\, d\theta$$

$$= \frac{1}{\pi}\int_0^{\pi}\left(\frac{V_m}{\pi}\right)\theta\sin n\theta\, d\theta = \frac{V_m}{\pi^2}\left|\frac{\sin n\theta}{n^2} - \frac{\theta}{n}\cos n\theta\right|_0^{\pi} = \frac{-V_m}{\pi n}\cos n\pi$$

$\therefore \; b_n = -V_m|\pi n$ *when n* is even $b_n = +V_m|\pi_n$ when n is odd

Substituting the values of various constants in the general expression for Fourier series, we get

$$f(\theta) = \frac{V_m}{4} - \frac{2V_m}{\pi^2}\cos\theta - \frac{2V_m}{(3\pi)^2}\cos 3\theta - \frac{2V_m}{(5\pi)^2}\cos 5\theta \ldots.$$

$$+ \frac{V_m}{\pi}\sin\theta - \frac{V_m}{2\pi}\sin 2\theta + \frac{V_m}{3\pi}\sin 3\theta \ldots.$$

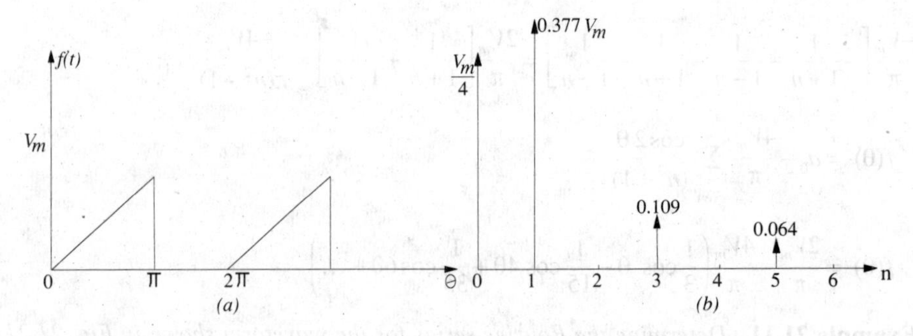

Fig. 21.21

The amplitudes of even harmonics are given directly by b_n but amplitudes of odd harmonics are given by $A_n = \sqrt{a_n^2 + b_n^2}$ (Art. 21.4).

For example,

$$A_1 = \sqrt{(2V_m/\pi^2)^2 + (V_m/\pi)^2} \qquad = 0.377\ V_m$$

$$A_3 = \sqrt{\left(\frac{2V_m}{(3\pi)^2}\right)^2 + \left(\frac{V_m}{3\pi}\right)^2} \qquad = 0.109\ V_m$$

$$A_5 = \left(\frac{2V_m}{(5\pi)^2}\right)^2 + \left(\frac{V_m}{5\pi}\right)^2 \qquad = 0.064\ V_m \text{ and so on.}$$

The line spectrum is as shown in Fig. 21.21(*b*).

Example 21.12. *Find the Fourier series for the sawtooth waveform shown in Fig. 21.22(a). Sketch its line spectrum.*

Solution. Using by the relation $y = mx$, the equation of the function becomes $f(t) = 1.\ t$ or $f(t) = t$.

$T = 2$, $\omega_0 = 2\pi/t = 2\pi/2 = \pi$

By inspection it is clear that $a_0 = 2/2 = 1$

$$a_n = \frac{2}{T}\int_0^T f(t) \cos n\ \omega_0\ t = \int_0^2 t.\ \cos n\pi\ t\ dt$$

Since we have to find the integral of two functions, we use the technique of integration by parts *i.e.*

$$\int u\ v\ dx = u\int v\ dx - \int\left(\frac{du}{dx}\int v\ dx\right)dx$$

$$\therefore\quad a_n = t\int_0^2 \cos n\pi\ t\ dt - \int_0^2\left(1.\int_0^2 \cos n\pi\ t\ dt\right)dt$$

$$= \left|\frac{t}{n\pi}\sin n\pi\ t\right|_0^2 + \left|\frac{\cos n\pi t}{(n\pi)^2}\right|_0^2 = 0 + \frac{1}{(n\pi)^2}(\cos 2n\ \pi - \cos 0)$$

since $\cos 2n\pi = \cos 0$ for all values of n, hence $a_n = 0$

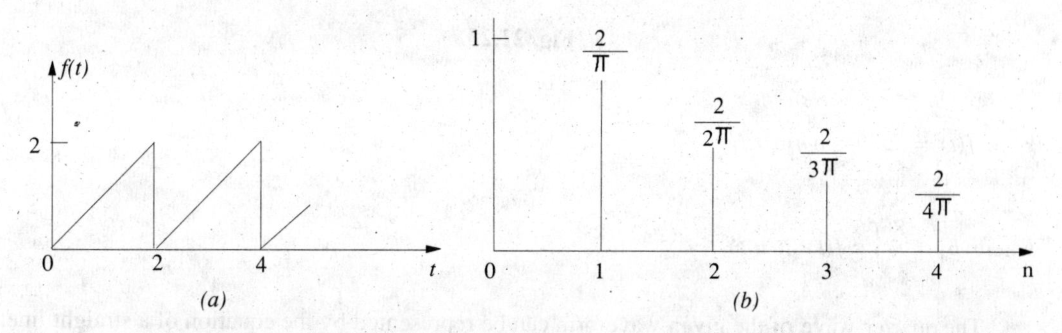

Fig. 21.22

$$b_n = \frac{2}{T}\int_0^T f(t) \sin n\ \omega_0\ t\ dt = \int_0^T t \sin n\pi\ t\ dt$$

Employing integration by parts we get

$$b_n = t\int_0^2 \sin n\pi t \, dt - \int_0^2\left(1.\int_0^2 \sin n\pi t \, dt\right) dt$$

$$= | t.\frac{-\cos n\,\pi\,t}{n\pi}|_0^2 - \int_0^2 \frac{-\cos n\pi t}{n\pi} dt =)\frac{-t}{n\pi}.\cos n\pi\, t|_0^2 + |\frac{\sin n\pi t}{(n\pi)^2}|_0^2 = \frac{\sin 2n\,\pi}{(n\pi)^2} - \frac{2}{n\pi}\cos 2n\,\pi$$

The sine term is 0 for all values of n because sign of any multiple of 2π is 0. Since value of cosine term is 1 for any multiple of 2π, we have $b_n = -2/n\,\pi$.

$$\therefore\ f(t) = a_0 + \sum_{n=1}^{\infty} b_n \sin n\omega_0 t = a_0 - \frac{2}{\pi}\sum_{n=1}^{\infty}\frac{1}{n}\sin n\pi t$$

$$= 1 - \frac{2}{\pi}\left(\sin \pi t + \frac{1}{2}\sin 2\pi t + \frac{1}{3}\sin 3\pi t + \ldots\right)$$

The line spectrum showing the amplitudes of various harmonics is shown in Fig. 21.22(*b*).

Example 21.13. *Determine the trigonometric Fourier series of the triangular waveform shown in Fig. 21.23.*

Solution. Since the waveform possesses odd symmetry, hence $a_0 = 0$ and $a_n = 0$ *i.e.* there would be no cosine terms in the series. Moreover, the waveform has half-wave symmetry. Hence, series will have only odd harmonics. In the present case, there would be only odd sine terms. Since the waveform possesses quarter-wave symmetry, it is necessary to integrate over only one quarter period of finding the Fourier coefficients.

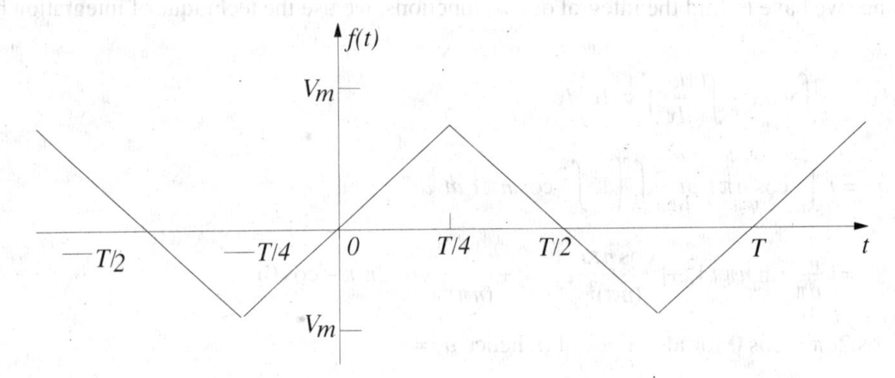

Fig. 21.23

$$f(t) = \sum_{\substack{n=1 \\ \text{odd}}}^{\infty} b_n \sin n\omega_0 t$$

$$\text{where } b_n = \frac{8}{T}\int_0^{t/4} f(t) \sin n\,\omega_0 t$$

The quarter-wave of the given waveform can be represented by the equation of a straight line. Slope of the straight line is = $V_m/(T/40) = 4Vm/T$.

Hence, using $Y = mx$, we have

$$f(t) = \left(\frac{4V_m}{T}\right)t \quad 0 < t < T/4 \quad \therefore\ b_n = \frac{8}{T}\int_0^{T/4}\left(\frac{4V_m}{T}\right)t.\sin n\omega_0 t\, dt = \frac{32V_m}{T^2}\int_0^{T/4} t\cdot \sin n\,\omega_0 t\, dt$$

Using the theorem of integration by parts, we have

$$b_n = \frac{32V_m}{T^2}\left[t\int_0^{T/4} \sin n\omega_0 t\ dt - \int_0^{T/4}\left(1.\int_0^{T/4} \sin n\omega_0 t\ dt\right)dt\right]$$

$$= \frac{32V_m}{T^2}[)t.\frac{-\cos n\ \omega_0 t}{n\omega_0}|_0^{T/4} - \int_0^{T/4}\frac{-\cos n\ \omega_0 t}{n\ \omega_0}dt]$$

$$= \frac{32V_m}{T^2}\left[\left(\frac{T}{4}.\frac{-\cos n\omega_0 T/4}{n\omega_0}\right) + \left|\frac{\sin n\omega_0 t}{(n\omega_0)^2}|^{T/4}\right] = \frac{32V_m}{T^2}\left[\left(\frac{T}{4}.\frac{\cos n\omega_0 T/4}{n\omega_0}\right) + \left(\frac{\sin n\omega_0 T/4}{(n\omega_0)^2}\right)\right]$$

Now, $\omega_0 = 2\pi/T$ or $\omega_0 T = 2\pi$ $\therefore$ $n\omega 0\ T/4 = n\pi/2$

$\therefore$ $\cos n\ \omega_0\ T/4 = \cos n\pi/2 = 0$ when n is odd

$$\therefore \quad b_n = \frac{32V_m}{n^2\omega_0^2 T^2}\sin n\ \omega_0\frac{T}{4} = \frac{32V_m}{n^2(2\pi)^2}\sin\frac{n\pi}{2} = \frac{8V_m}{n^2\pi^2}\sin\frac{n\pi}{2} \quad \ldots n \text{ odd}$$

$$\therefore \quad b_n = \frac{8V_m}{n^2\pi^2} \wedge \ldots n = 1, 5, 9, 13 \ldots b_n = \frac{-8V_m}{n^2\pi^2} \quad \ldots n = 3, 7, 11, 15, \ldots$$

Substituting this value of b_n, the Fourier series for the given waveform becomes ·

$$f(t) = \frac{8V_m}{\pi^2}\left(\sin\omega_0 t - \frac{1}{3^2}3\omega_0 t + \frac{1}{5^2}\sin 5\omega_0 t - \frac{1}{7^2}\sin 7\omega_0 t + \ldots\right)$$

Example 21.14. *Determine the Fourier series of the triangular waveform shown in Fig. 21.24.*

Solution. Since the function has even symmetry, $b_n = 0$. Moreover, it also has half-wave symmetry, hence, $a_0 = 0$. The Fourier series can be written as

$$f(t) = \sum_{n=1}^{\infty} a_n \cos n\omega_0 t \text{ where } a_n = \frac{2}{T}\int_0^T f(t)\cos n\omega_0 t\ dt$$

The function is given by the relation

$$f(t) = \frac{-4V_m}{T}\left(t - \frac{T}{4}\right)$$

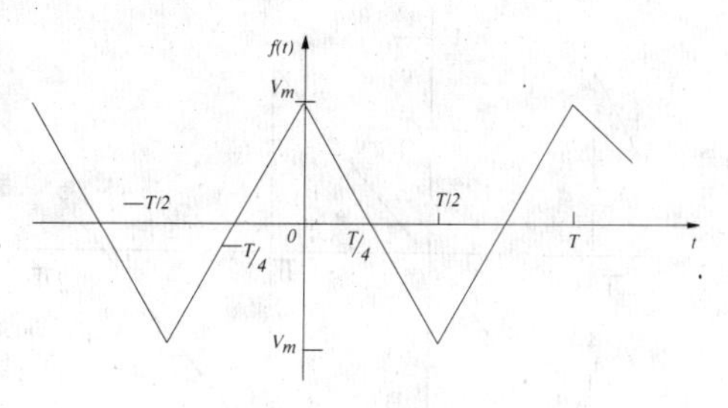

Fig. 21.24

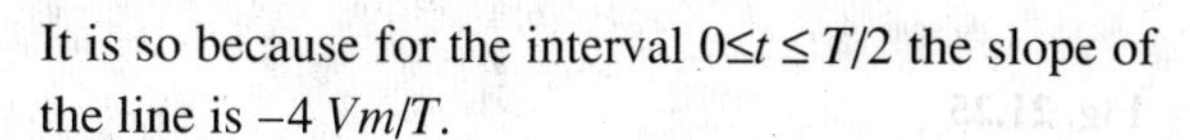

It is so because for the interval $0 \le t \le T/2$ the slope of the line is $-4\ Vm/T$.

$$\therefore \quad a_n = \frac{2}{T}\int_0^T f(t)\cos n\omega_0 t\ dt = \frac{4}{T}\int_0^{T/2} f(t)\cos n\omega_0 t\ dt$$

$$= \frac{-16V_m}{T^2}\int_0^{T/2}\left(t - \frac{T}{4}\right)\cos n\ \omega_0 t\ dt = \frac{-16V_m}{T^2}\int_0^{T/2} t\cdot\cos n\ \omega_0 t\ dt + \frac{4V_m}{T}\int_0^{T/2}\cos n\omega_0 t\ dt$$

$$= \frac{-16V_m}{T^2}|\frac{1}{n^2\omega_0^2}.\cos n\omega_0 t + \frac{t}{n\omega_0}\sin n\ \omega_0 t|_0^{T/2} + \frac{4V}{T}|\frac{\sin n\ \omega_0 t}{n\ \omega_0}|_0^{T/2}$$

Substituting $\omega = 2\pi/T$, we get

$$a_n = \frac{-16V_m}{T^2}\left[\frac{T^2}{4\pi^2 n^2}(\cos n\ \pi - 1) + \frac{T^2}{4\pi n}(\sin n\ \pi)\right] + \frac{2V_m}{\pi n}.\sin n\pi$$

Now, $\sin n\ \pi = 0$ for all values of n, $\cos n\pi - 1$ when n is even and -1 when n is odd.

$$\therefore\ a_n = \frac{8V_m}{\pi^2 n^2} \qquad \ldots n \text{ odd only}$$

$$\therefore\ f(t) = \sum_{\substack{n=1\\odd}}^{\infty} \frac{8V_m}{\pi^2 n^2} \cos n\omega_0 t = \frac{8V_m}{\pi^2} \sum_{\substack{n=1\\odd}}^{\infty} \frac{\cos n\omega_0 t}{n^2}$$

$$= \frac{8V_m}{\pi^2}\left(\cos \omega_0 t + \frac{1}{9}\cos 3\omega_0 t + \frac{1}{25}\cos 5\omega_0 t + \frac{1}{49}\cos 7\omega_0 t + \ldots\right)$$

Alternative Solution

We can deduce the Fourier series from Fig. of Ex. 21.11 by shifting the vertical axis by $\pi/2$ radians to the right. Replacing t by $(t + \pi/2)$ in the Fourier series of Ex. 21.11, we get

$$f(t) = \frac{8V_m}{\pi^2}\left[\sin \omega_0\left(t + \frac{\pi}{2}\right) - \frac{1}{3^2}\sin 3\omega_0\left(t + \frac{\pi}{2}\right) + \frac{1}{5^2}\sin 5\omega_0\left(t + \frac{\pi}{2}\right) - \frac{1}{7^2}\sin 7\omega_0\left(t + \frac{\pi}{2}\right) + \ldots\right]$$

$$= \frac{8V_m}{\pi^2}\left(\cos \omega_0 t + \frac{1}{9}\cos 3\omega_0 t + \frac{1}{25}\cos 5\omega_0 t + \frac{1}{49}\cos 7\omega_0 t + \ldots\right)$$

Example 21.15. *Obtain the Fourier series representation of the sawtooth waveform shown in Fig. 21.25(a) and plot its spectrum.*

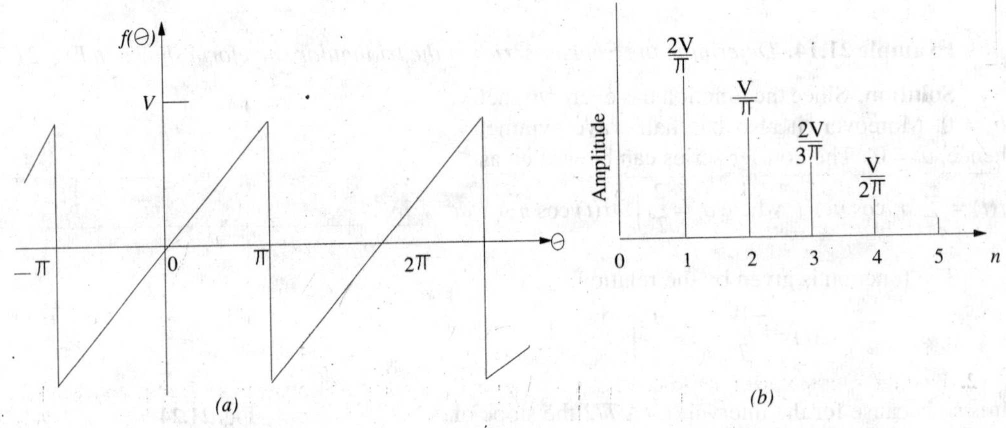

Fig. 21.25

Solution. By inspection we know that the average value of the wave is zero over a cycle because the height of the curve below and above the X-axis is the same hence, $a_0 = 0$. Moreover, it has odd symmetry so that $a_n = 0$ *i.e.* there would be no cosine terms. The series will contain only sine terms.

$$\therefore\ f(\theta) = \sum_{n=1}^{\infty} b_n \sin n\theta \text{ where } b_n = \frac{1}{\pi}\int_0^{2\pi} f(\theta) \sin n\theta\, d\theta$$

The slope of curve is $m = V/\pi$

$\therefore$ we get, $f(\theta) = (V/\pi)\,\theta$.

If we are the limit of integration from $-\pi$ to $+\pi$ then

$$b_n = \frac{1}{\pi}\int_{-\pi}^{\pi}\left(\frac{V}{\pi}\right)\theta \sin n\theta\, d\theta = \frac{V}{\pi^2}\left|\frac{1}{n^2}\sin n\theta - \frac{\theta}{n}\cos n\theta\right|_{-\pi}^{+\pi} = \frac{-2V}{n\pi}\cos n\pi$$

The above result has been obtained by making use of integration by parts as explained earlier. cos n π is positive when n is even and is negative when n is odd and thus the signs of the coefficients alternate. The required Fourier series is

$$f(\theta) = \frac{2V}{\pi}\left(\sin\theta - \frac{1}{2}\sin 2\theta + \frac{1}{3}\sin 3\theta - \frac{1}{4}\sin 4\theta + \ldots\right)$$

$$\text{or } f(t) = \frac{2V}{\pi}\left(\sin\omega_0 t - \frac{1}{2}\sin 2\omega_0 t + \frac{1}{3}\sin 3\omega_0 t - \frac{1}{4}\sin 4\omega_0 t + \ldots\right)$$

As seen, the coefficients decrease as $1/n$ so that the series converges slowly as shown by the line spectrum of Fig. 21.25(*b*).

The amplitudes of the fundamental or first harmonic, second harmonic, third harmonic and fourth harmonic are $(2/\pi)$, $(2V/2\pi)$, $(2V/3\pi)$ and $(2V/4\pi)$ respectively.

Tutorial Problems No. 21.1

1. Determine the Fourier series for the triangular waveform shown in Fig. 21.26(*a*).

(Network Theory and Design, AMIETE June 1990)

$$\left[\frac{1}{2} - \frac{4}{\pi^2}\cos\omega_0 t + \frac{1}{3^2}\cos 3\omega_0 t + \frac{1}{5^2}\cos 5\omega_0 t + \ldots\right]$$

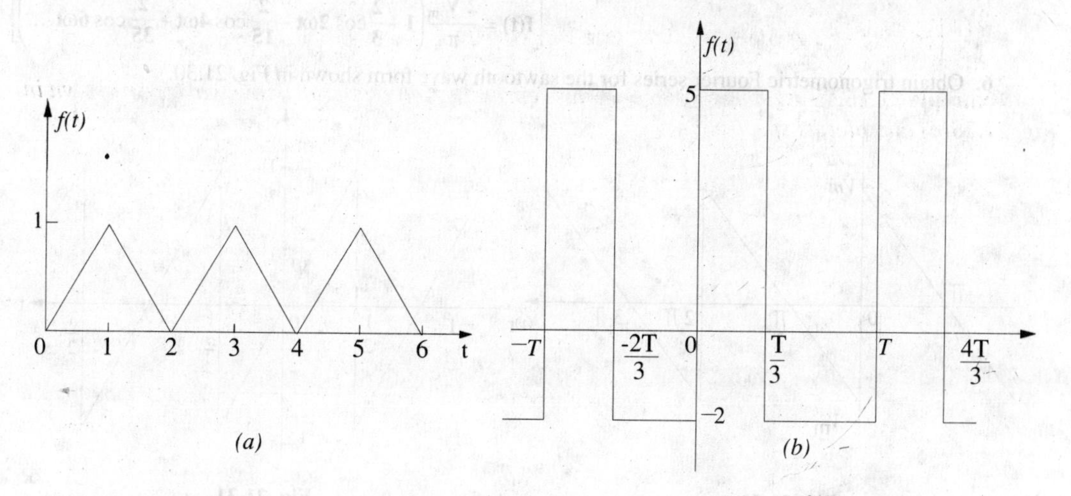

Fig. 21.26

2. Find the values of the Fourier coefficients a_0, a_n and b_n for the function given in Fig. 21.26(*b*).

$$\left[a_0 = \frac{2}{3};\ a_n = \frac{7}{n\pi}\sin\frac{2\pi n}{3};\ b_n = \frac{7}{n\pi}\left(1 - \cos\frac{2\pi n}{3}\right)\right]$$

3. Determine the trigonometric series of the triangular waveform shown in Fig. 21.27. Sketch its line spectrum.

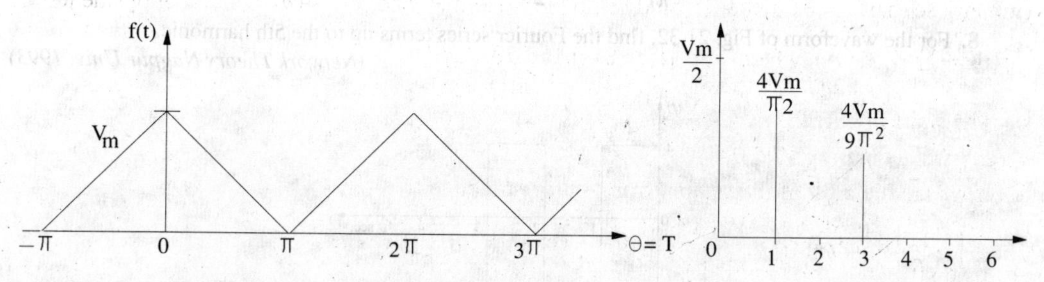

Fig. 21.27

$$\left[\frac{V_m}{2} + \frac{4V_m}{\pi^2}\left(\cos\omega_0 t + \frac{1}{3^2}\cos 3\omega_0 t + \frac{1}{5^2}\cos 5\omega_0 t + \ldots\right)\right]$$

4. Determine the Fourier series for the sawtooth waveform shown in Fig. 21.28.

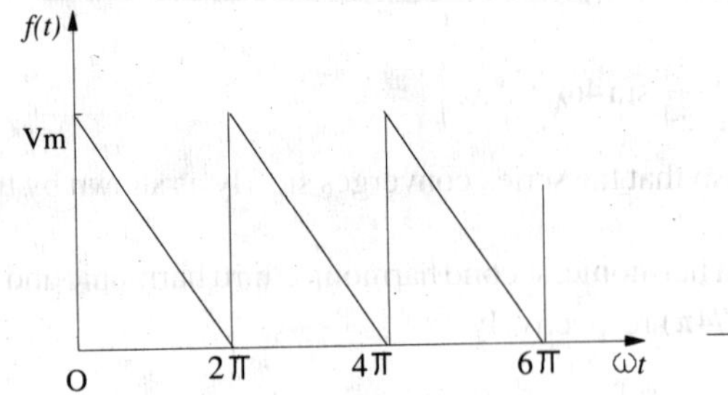

Fig. 21.28

Fig. 21.29

$$\left[f(t) = \frac{V_m}{2} + \frac{V_m}{\pi}\left(\sin \omega_0 t + \frac{1}{2}\sin 2\omega_0 t + \frac{1}{3}\sin 3\omega_0 t + \ldots \right) \right]$$

5. Represent the full-wave rectified voltage sine waveform shown in Fig. 21.29 by a Fourier series.

$$\left[f(t) = \frac{2\,V_m}{\pi}\left(1 + \frac{2}{3}\cos 2\omega t - \frac{2}{15}\cos 4\omega t + \frac{2}{35}\cos 6\omega t \ldots \right) \right]$$

6. Obtain trigonometric Fourier series for the sawtooth wave form shown in Fig. 21.30.

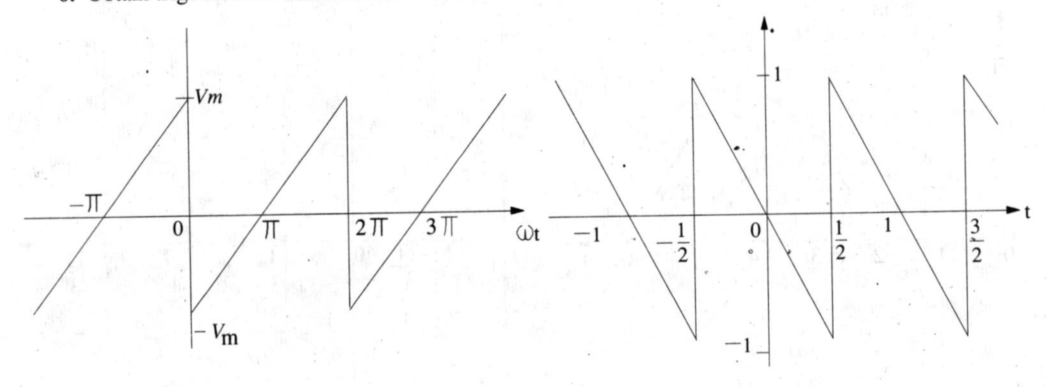

Fig. 21.30

Fig. 21.31

$$\left[f(t) = \frac{-2V_m}{\pi}\left(\sin \omega_0 t + \frac{1}{2}\sin 2\omega_0 t + \frac{1}{3}\sin 3\omega_0 t + \frac{1}{4}\sin 4\omega_0 t + \ldots \right) \right]$$

7. Find the Fourier series for the sawtooth waveform shown in Fig. 21.31.

$$\left[f(t) = \frac{2}{\pi}\left(-\sin 2\pi t + \frac{1}{2}\sin 4\pi t - \frac{1}{3}\sin 6\pi t + \ldots \right) \right]$$

8. For the waveform of Fig. 21.32, find the Fourier series terms up to the 5th harmonic.

(Network Theory Nagpur Univ. 1993)

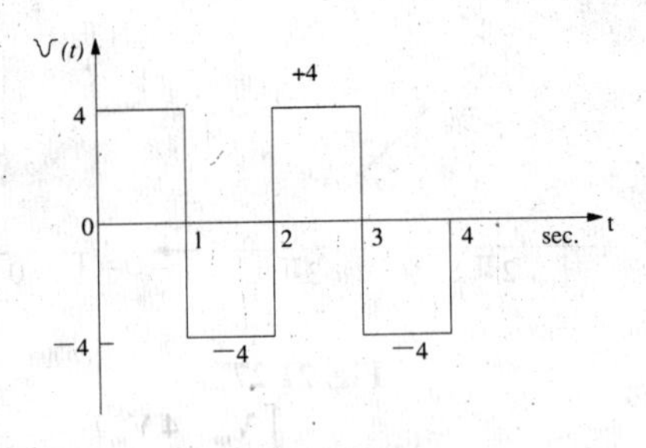

Fig. 21.32

$$\left[\mathbf{V(t)} = \frac{\mathbf{16}}{\pi}\left(\mathbf{\sin t} + \frac{\mathbf{2}}{\mathbf{3}} \mathbf{\sin 3t} \right) + \frac{\mathbf{1}}{\mathbf{5}} \mathbf{\sin 5t} + \ldots \right]$$

9. Determine Fourier series of a repetitive triangular wave as shown in Fig. 21.33.

(*a*) What is the magnitude of d.c. component ?

(*b*) What is the fundamental frequency ?

(*c*) What is the magnitude of the fundamental ?

(*d*) Obtain its frequency spectrum. (*Network Theory Nagpur Univ. 1993*)

[(*a*) 5V (*b*) 1 Hz (*c*) 10/π volt]

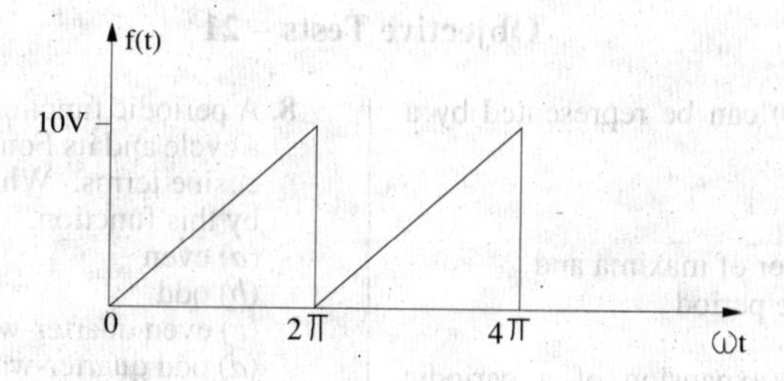

Fig. 21.33

10. Determine the Fourier series of voltage response obtained at the o/p of a half wave rectifier shown in Fig. 21.34. Plot the discrete spectrum of the waveform. (*Elect. Network Analysis Nagpur Univ. 1993*)

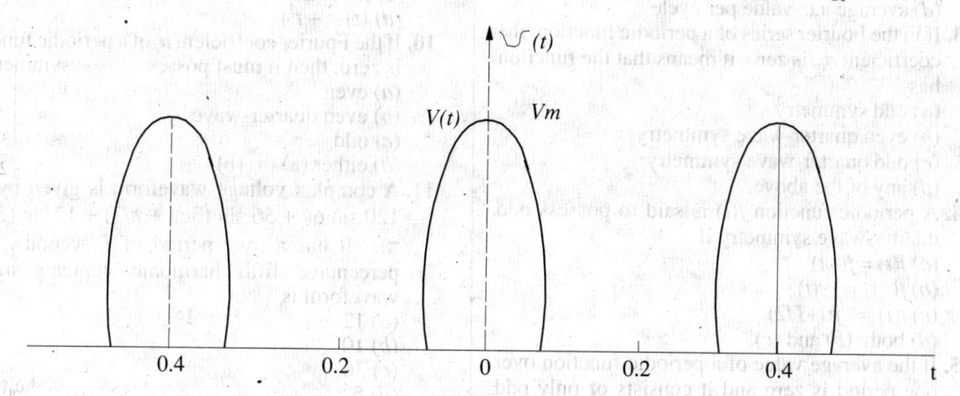

Fig. 21.34

$$[v(t) = \frac{V_m}{\pi} + \frac{V_m}{2} \cos \omega_0 t + \frac{2V_m}{3\pi} \cos 10\pi t - \frac{2V_m}{15} \cos 20\pi t + \frac{2V_m}{35\pi} \cos 30\pi t \ldots]$$

11. Determine the Fourier coefficients and plot amplitude and phase spectra for(*Network Analysis Nagpur Univ. 1993*)

$$[\mathbf{a_3 = 0, b_n = 0, a_1 = \frac{4 V_m}{\pi}, a_2 = 0}$$

$$\mathbf{a_3 = \frac{-4 V_m}{3\pi}, a_4 = 0 a_5 - \frac{4 V_m}{\pi}}]$$

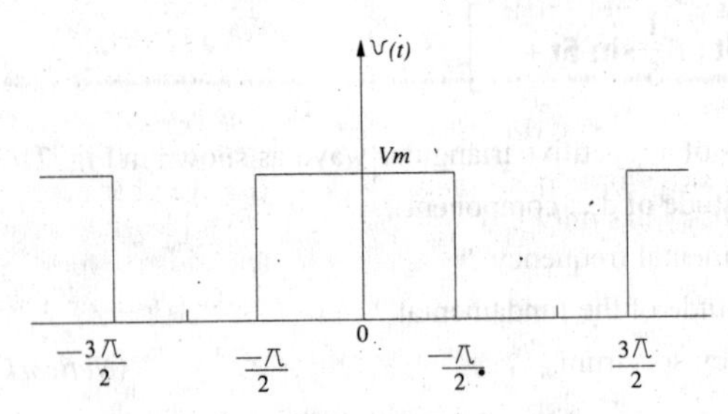

Fig. 21.35

Objective Tests – 21

1. A given function $f(t)$ can be represented by a Fourier series if it
   (a) is periodic
   (b) is single-valued
   (c) has a finite number of maxima and minima in any one period
   (d) all of the above.
2. In a Fourier series expansion of a periodic function, the coefficient a_0 represents its
   (a) net area per cycle
   (b) d.c. value
   (c) averge value over half cycle
   (d) average a.c. value per cycle
3. If in the Fourier series of a periodic function, the coefficient a_0 is zero, it means that the function has
   (a) odd symmetry
   (b) even quarter-wave symmetry
   (c) odd quarter-wave symmetry
   (d) any of the above.
4. A periodic function $f(t)$ is said to possess odd quarter-wave symmetry if
   (a) $f(t) = f(-t)$
   (b) $f(-t) = -f(t)$
   (c) $f(t) = -f(t+T/2)$
   (d) both (b) and (c).
5. If the average value of a periodic function over one period is zero and it consists of only odd harmonics then it must be possessing —— symmetry.
   (a) half-wave
   (b) even quarter-wave
   (c) odd quarter-wave
   (d) odd.
6. If in the Fourier series of a periodic function, the coefficient $a_0 = 0$ and $a_n = 0$, then it must be having —— symmetry.
   (a) odd
   (b) odd quarter-wave
   (c) even
   (d) either (a) or (b).
7. In the case of a periodic function possessing half-wave symmetry, which Fourier coefficient is zero ?
   (a) a_n (b) b_n
   (c) a_0 (d) none of the above.
8. A periodic function has zero average value over a cycle and its Fourier series consist of only odd cosine terms. What is the symmetry possessed by this function.
   (a) even
   (b) odd
   (c) even quarter-wave
   (d) odd quarter-wave
9. Which of the following periodic function possesses even symmetry ?
   (a) cos 3 t
   (b) sin t
   (c) t. cos 50 t
   (d) $(t+t^2+t^5)$.
10. If the Fourier coefficient b_n of a periodic function is zero, then it must possess —— symmetry.
    (a) even
    (b) even quarter-wave
    (c) odd
    (d) either (a) or (b).
11. A complex voltage waveform is given by $V = 120 \sin \omega t + 36 \sin (3\omega t + \pi/2) + 12 \sin (5\omega t + \pi)$. It has a time period of T seconds. The percentage fifth harmonic contents in the waveform is
    (a) 12
    (b) 10
    (c) 36
    (d) 5.
12. In the waveform of Q. 11 above, the phase displacement of the third harmonic represents a time interval of —— seconds.
    (a) T/12
    (b) T/3
    (c) 3T
    (d) T/36.
13. When the negative half-cycle of a complex waveform is reversed, it becomes identical to its positive half-cycle. This feature indicates that the complex waveform is composed of
    (a) fundamental
    (b) odd harmonics
    (c) even harmonics
    (d) both (a) and (b)
    (e) both (a) and (c).
14. A periodic waveform possessing half-wave symmetry has no

(*a*) even harmonics
(*b*) odd harmonics
(*c*) sine terms
(*d*) cosine terms

15. The Fourier series of a waveform possessing even quarter-wave symmetry has only
(*a*) even harmonics
(*b*) odd cosine terms
(*c*) odd sine terms
(*d*) both (*b*) and (*c*).

16. The Fourier series of a waveform possessing odd quarter-wave symmetry contains only
(*a*) even harmonics
(*b*) odd cosine terms
(*c*) odd sine terms
(*d*) none of the above

ANSWERS

1. (*d*) **2.** (*b*) **3.** (*d*) **4.** (*d*) **5.** (*a*) **6.** (*d*) **7.** (*c*) **8.** (*c*) **9.** (*a*) **10.** (*d* **11.** (*b*)
12. (*a*) **13.** (*d*) **14.** (*a*) **15.** (*b*) **16.** (*c*)

22 TRANSIENTS

22.1. Introduction

It is quite an easy job to calculate the steady current, which flows in a circuit, when it is connected to a d.c. generator or a battery. Similarly, the alternating current which flows in a circuit when connected to an alternator can also be calculated by the various methods discussed in Chapters 13 and 14. These currents are known as *steady* currents because in such cases, it is assumed that (*i*) the circuit components are constant and (*ii*) the circuit has been connected to the generator long enough for any disturbance produced on initial switching, to resolve itself.

In general, transient disturbances are produced whenever

(*a*) an apparatus or circuit is *suddenly* connected to or disconnected from the supply,

(*b*) a circuit is shorted and

(*c*) there is a *sudden* change in the applied voltage from one finite value to another.

We will now discuss the transients produced whenever different circuits are suddenly switched on or off from the supply voltage. In each case, we will assume that the resultant current consists of two parts (*i*) a final steady-stage or normal current and (*ii*) a transient current superimposed on the steady-stage current.

It is essential to remember that the transient currents are not driven by any part of the applied voltage but are entirely associated with the changes in the stored energy in inductors and capacitors. Since there is no stored energy in resistors, *there are no transients in pure resistive circuits.*

22.2. Types of Transients

There are *single-energy* transients and *double-energy* transients. Single-energy transients are those in which only one form of energy, either electromagnetic or electrostatic is involved as in *R-L* and *R-C* circuits.

However, double-energy transients are those in which both electromagnetic or electrostatic is involved as in *R-L-C* circuits. Transient disturbances may be further classified as follows :

(a) Initiation Transients : These are produced when a circuit, which is originally dead, is energised.

(b) Subsidence Transients : These are produced when an energised circuit is rapidly de-energised and reaches an eventual steady-stage of zero current or voltage, as in the case of short-circuiting an *R-L* or *R-C* circuit suddenly.

(c) Transition Transients : These are due to sudden but energtic changes from one steady state to another.

(d) Complex Transients : These are produced in a circuit which is simultaneously subjected to two transients due to two independent disturbances or when the disturbing force producing the transients is itself variable.

(e) Relaxation Transients : In these transients, the transition occurs cyclically towards states, which when reached, become unstable themselves.

A distinction may also be made between free and forced transients which are produced due to the applied voltage being itself transient.

22.3. Important Differential Equations

Some of the important differential equations, used in the treatment of single and double energy transients, are given below. We will consider both first-order and second-order differential equations.

1. First Order Equations

(*i*) Let $\frac{dy}{dx} + ay = 0$ where a is a constant.

Its solution is $y = k\,e^{-ax}$ where k is the constant of integration whose value can be found from the boundary conditions *i.e.* conditions prevalent at the instant when the voltage to a circuit is applied or excluded.

(*ii*) If $\frac{dy}{dx} + ay = b$ where a and b are constants, then solution is $y = \frac{b}{a} + k\,e^{-ax}$

The value of k can again be found from boundary conditions.

(*iii*) If $\frac{dy}{dx} + Ay = B$

where A and B are not constants but are functions of x, then the solution is given by

$$y = e^{-\int A.dx} \int e^{\int A.dx} B\,dx + k e^{-\int A.dx}$$

If $A = a$ = constant, then the above equation simplifies to

$$y = e^{-ax} \int e^{ax} B\,dx + k e^{-ax}$$

2. Second Order Equations

(*i*) Suppose $\frac{d^2y}{dx^2} + a\frac{dy}{dx} + by = 0$ where a and b are constants, then the solution is

$$y = k_1 e^{\lambda_1 x} + k_2 e^{\lambda_2 x}$$

where λ_1 and λ_2 are constants of integration and whose values are,

$$\lambda_1 = -\frac{a}{2} - \sqrt{\left(\frac{a^2}{4} - b\right)} \quad \text{and} \quad \lambda_2 = -\frac{a}{2} + \sqrt{\left(\frac{a^2}{4} - b\right)}$$

(*a*) If $a^2/4 > b$, the roots are real and the above solution can be applied without any difficulty.

(*b*) If $a^2/4 < b$, the radicals contain a negative quantity. In that-case, the solution is given by

$$y = e^{-\frac{1}{2}ax}\,(k_3 \sin \lambda_0 x + k_4 \cos \lambda_0 x)$$

where k_3 and k_4 are the new c onstants of integration and

$$\lambda_0 = \sqrt{b - \frac{a^2}{4}}$$

(*c*) If $a^2/4 = b$, then both roots are equal and each is $= -a/2$.

Hence, in this case, the solution becomes $y = k_5\,e^{\lambda t} + k_6 t.e^{\lambda t}$

(*ii*) Let $\frac{d^2y}{dx^2} + a\frac{dy}{dx} + by = c$

where a, b and c are constants. In this case also, the solution will again depend on the root as discussed above.

$$y = k_1 e^{\lambda_1 x} + k_2 e^{\lambda_2 x} + c/b$$

(*iii*) (*a*) Let the differential equation be given by

$$\frac{d^2y}{dx^2} + a\frac{dy}{dx} + by = u$$

where a, b and c are constants but u is a particular function of the variable x. The solution of such an equation consists of a *particular integral* and a complementary function.

(*b*) Let y be a sinusoidal function of x, then

$$\frac{d^2y}{dx^2} + a\frac{dy}{dx} + by = c \sin \omega t$$

In this case, particular integral is

$$y_1 = \frac{-c}{\sqrt{a^2\omega^2 + (\omega^2 - b)^2}} \cos\left[\omega x - \tan^{-1}\left(\frac{\omega^2 - b}{a\omega}\right)\right]$$

The complementary function is given by

$$y_2 = k_1 e^{\lambda_1 x} + k_2 e^{\lambda_2 x} \text{ where } \lambda_1 = -\frac{a}{2} - \sqrt{\left(\frac{a^2}{4} - b\right)} \text{ and } \lambda_2 = -\frac{a}{2} + \sqrt{\left(\frac{a^2}{4} - b\right)}$$

The complete solution for the above equation is $y = y_1 + y_2$

Further treatment is the same as for case 2 (*i*) above.

22.4. Transients in R-L Circuits (D.C.)

If i, I_s and i_t be the resultant current, final steady-state current and transient current respectively in *R-L* circuit of Fig. 22.1 (*a*), then by superimposition, the equation for the resultant current, for the duration of initiation transient, is

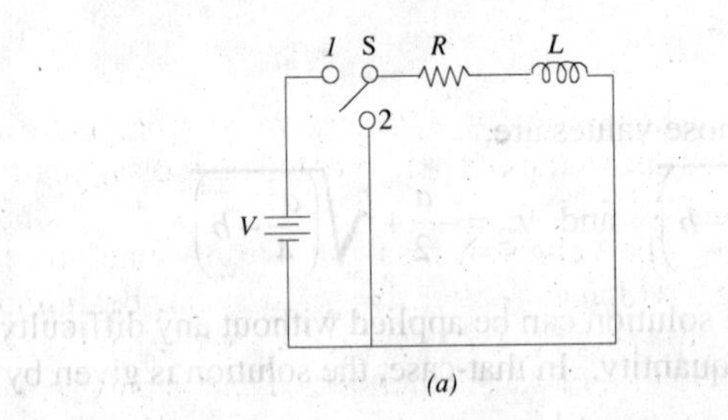

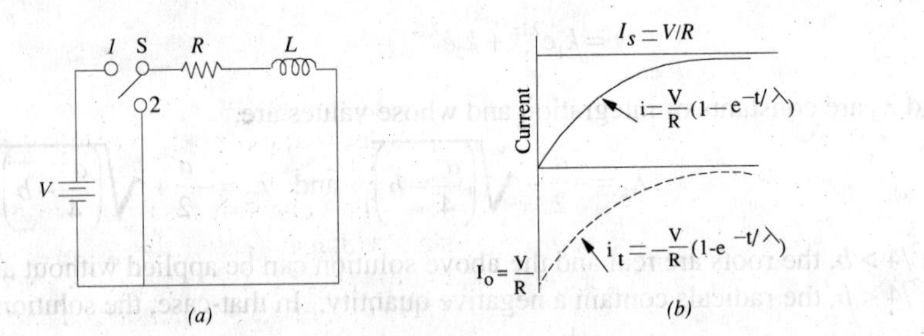

Fig. 22.1

$$i = I_s + i_t \qquad \ldots(i)$$

Since the applied voltage V drives the steady-state current, hence

$$I_s = V/R$$

Since the transient current i_t is not associated with any voltage,

$$\therefore \quad i_t R + L\frac{d_{it}}{dt} = 0 \qquad \ldots(ii)$$

$$\text{or} \quad \frac{di_t}{i_t} = -\frac{R}{L}dt \qquad \ldots(iii)$$

$$\text{or} \quad \int\frac{di_t}{i_t} = -\frac{R}{L}\int dt \qquad \therefore \quad \text{logh } i_t = -\frac{R}{L}t + K^* \qquad \ldots(iv)$$

where K is the constant of integration whose value may be found from the initial conditions.

Now when $t = 0$, $i_t = I_0$ (say). Then from Eq. (*iv*) above we get, logh $I_0 = 0 + K$

Putting this value of K in Eq. (*iv*), we have

$$\text{logh } i_t - \text{logh } I_0 = -\frac{R}{L}t \text{ or logh } i_t/I_0 = -\frac{R}{L}t = \frac{-t}{\lambda} \qquad \therefore \quad i_t = I_0\, e^{-t/\lambda} \quad \ldots(v)$$

where $\lambda = L/R$ is called the *time-constant* of the circuit. Its reciprocal R/L is called the *damping coefficient* of the circuit. The current decreases exponentially as shown in Fig. 22.1 (*b*). From Eq. (*i*) and (*v*), we have

$$i = I_s + I_0\, e^{-t/\lambda} \qquad \ldots(vi)$$

If the time is reckoned when the voltage V is applied, so that when $t = 0$, $i = 0$, then from equation (*vi*), we get

$$0 = I_s + i_0\, e^{-0} = I_s + I_0 \qquad \therefore \qquad I_0 = -I_s = -\frac{V}{R}$$

In that case, Eq. (*vi*) becomes

$$i = \frac{V}{R} - \frac{V}{R}e^{-t/\lambda} \qquad \ldots(vii)$$

$$= \frac{V}{R}(1 - e^{-t/\lambda}) \qquad \ldots(viii)$$

Curves for I_s and i_t have been plotted in Fig. 22.1 (*b*). The curve for resultant current has been obtained by the superposition of steady-state current I_s ($= V/R$) and transient current

$$i_t = \frac{V}{R}e^{-t/\lambda}$$

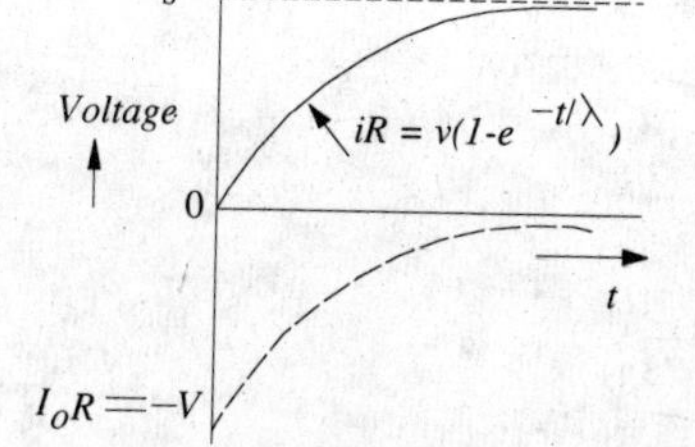

Fig. 22.2

Theoretically, the transient current i_t takes infinite time to die off but, in practice, it disappears in a very short time.

The values of resultant, steady-state and transient voltages across the resistor can be found by multiplying Eq. (*vii*) by R and are shown in Fig. 22.2. The e.m.f. of self-induction–Ldi_t/dt is only transient in nature and equals $i_t R$ as seen from Eq. (*ii*) above.

22.5. Short Circuit Current

After some time, the transient current would disappear and the only current flowing in the circuit

---

*$\log_e x$ is written as logh x. Obviously, logh $x = 2.3 \log_{10} x$.

would be the steady-state current $I_s = V/R$. Let the *R-L* circuit be closed upon itself *i.e.* be short-circuited by shifting the switch [Fig. 22.1 (*a*)] to position 2. Since the voltage *V* has been excluded from the circuit, the trapped current I_s will immediately cease to be a steady-state current, but on the other hand, will become the initial value I_0 of a new subsidence transient current i_t. If time is measured at the instant of short-circuit, so that when $t = 0$, the current is $I_s = V/R$, then Eq. (*v*) becomes

$$i_t = \frac{V}{R}e^{-t/\lambda} \qquad \ldots(ix)$$

This equation has been plotted in Fig. 22.3. The only voltage acting in the circuit is that due to self-induction *i.e.* $-L\,di_t/dt$ which equals $i_t R$.

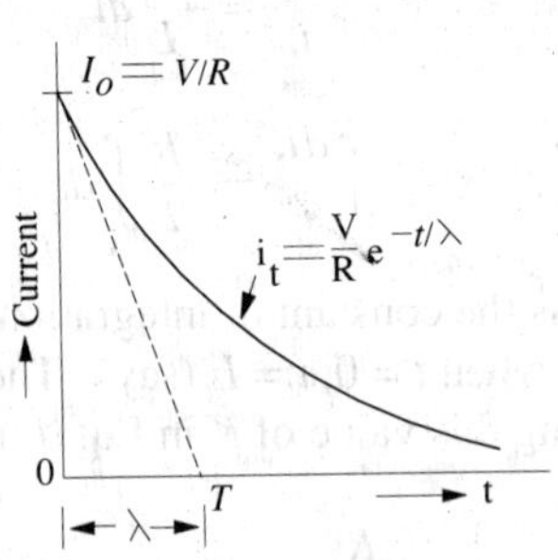

Fig. 22.3

22.6. Time Constant

The time constant of a circuit is defined as the time it would take for the transient current to decrease to zero, if the decrease were linear instead of being exponential.

In other words, it is the time during which the transient current would have decreased to zero, had it maintained its initial rate of decrease.

The initial rate of decrease can be found by differentiating Eq. (*vi*) and putting $t = 0$

$$\therefore \quad \frac{di_t}{dt} = \frac{I_0}{\lambda}e^{-t/\lambda} \qquad \therefore \quad \left(\frac{di_t}{dt}\right)_{t=0} = -\frac{I_0}{\lambda}$$

If the rate of decrease were constant throughout and equal to $-I_0/\lambda$, then the straight line showing the relation between i_t and t would be given by

$$i_t = -\frac{I_0}{\lambda}t$$

The time-period would be equal to the sub-tangent *OT* drawn to the exponential curve of Fig. 2.3 at $i_t = I_0$ *i.e.* at the beginning of the curve.

If we put $t = \lambda$ in Eq. (*v*), then $i_t = I_0\,e^{-1} = I_0/e = I_0/2.718 = 0.37\ I_0$

Hence, time period of a circuit is the time during which the transient current decreases to 0.37 of its initial value.

Example 22.1. *A coil having a resistance of 30Ω and an inductance of 0.09 H is connected across a battery of 20 V. Plot the current and its two components. Assume that t = 0 when the circuit is completed.* **(Electromechanic Allahabad Univ. 1992)**

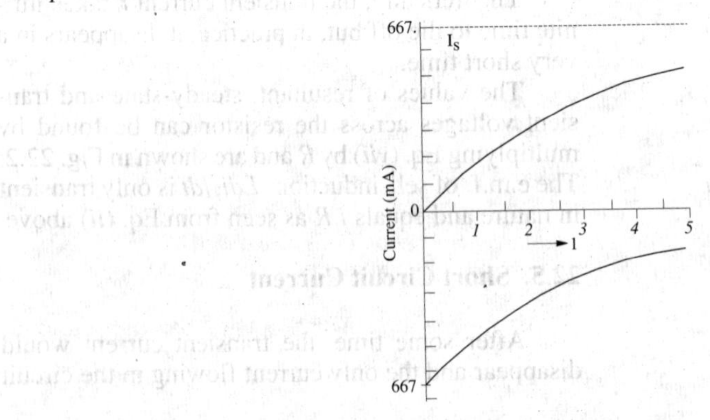

Fig. 22.4

Solution. The two components of the circuit current are (*i*) steady current $I_s = V/R = 20/30 = 2/3\ A$ $= 667\ mA$ and (*ii*) transient current $i_t = -(V/R)e^{-t/\lambda}$

Total current is $i = I_s + i_t$. Let us find the value of transient current after various time intervals. In the present case, $\lambda = L/R = 0.09/30 = 0.003$ second = 3 millisecond.

The values of i_t and i at various times are tabulated below. Value of $i = I_s + i_t$

t (ms)	$e^{-t/\lambda}$	i_t (mA)	i	t (ms)	$e^{-t/\lambda}$	i_t (mA)	i
0.0	1.000	−667	0	2.5	0.435	−290	377
0.5	0.847	−565	102	3.0	0.368	−244	423
1.0	0.716	−477	190	3.5	0.311	−208	459
1.5	0.606	−405	262	4.0	0.264	−176	491
2.0	0.514	−344	323	4.5	0.223	−148	519

It is seen that whereas transient current decreases exponentially, total circuit current increases exponentially as expected (Fig. 22.4).

Example 22.2. *A circuit of resistance 10 Ω and inductance 0.1 H in series has a direct voltage of 200 V suddenly applied to it. Find the voltage drop across the inductance at the instant of switching on and at 0.01 second. Find also the flux-linkages at these instants.*

(Basic Electricity, Bombay Univ., 1985)

Solution. (i) Switching instant

At the instant of switching on, $i = 0$, so that $iR = 0$, hence all applied voltage must drop across the inductance only. Therefore, voltage drop across inductance = **200 V**.

Since at this instant $i = 0$, there are no flux-linkages of the coil.

(ii) When t = 0.01 second

As time passes, current grows so that the applied voltage is partly dropped across the resistance and partly across the coil. Let us first find iR drop for which purpose, we need the value of i at $t = 0.01$ second.

Now, time period of the circuit is $\lambda = L/R = 0.1/10 = 0.01$ second. Since the given time happens to be equal to time constant,

$\therefore \quad i = (200/10) \times 0.632 = 12.64$ A ; $iR = 12.64 \times 10 =$ **126.4 V**

Drop across inductance $= \sqrt{200^2 - 126.4^2} =$ **155 V**

Now, $\quad L = N\Phi/i \quad$ or $\quad N\Phi = Li$

$\therefore \quad$ Flux-linkages $Li = 0.1 \times 12.64 =$ **1.264 Wb-turns.**

Example 22.3. *A coil of 10 H inductance and 5 Ω resistance is connected in parallel with a 20 Ω resistor across a 100-V d.c. supply which is suddenly disconnected. Find*

(a) the initial rate of change of current after switching.

(b) the voltage across the 20 Ω resistor initially and after 0.3 s.

(c) the voltage across the switch contacts at the instant of separation and

(d) the rate at which the coil is losing stored energy 0.3 second after switching.

Solution. (*a*) Since the steady-state current is zero, $i = I_0\, e^{-t/\lambda}$

Now, when $t = 0$, current is = 100/5 = 20 A. It means the current flowing through the coil immediately before opening the switch is 20 A.

$\therefore \quad I_0 = 20A$

Hence, the above equation becomes $i = 20\, e^{-t/\lambda}$

Now $\lambda = L/R = 10/25 = 1/2.5$ $\therefore\ i = 20e^{-2.5t}$

$$\left(\frac{di}{dt}\right)_{t=0} = (-20 \times 2.5e^{-2.5t})_{t=0} = \mathbf{-50\ A/s}$$

The negative sign merely shows that the current is decreasing.

(*b*) After the supply has been disconnected, the current through the 20-Ω resistor is *i* since it is in series with the coil.

Initial p.d. across the 20 Ω resistor = (current at $t = 0$) × 20 = 20 × 20 = **400 V**

Current through the resistor after 0.3 second = $20e^{-2.5 \times 0.3} = 9.45$A

∴ Voltage across the resistor after 0.3 second

= (current at $t = 0.3$ second) × 20 = 9.45 × 20 = **189 V**

(*c*) The e.m.f. induced in the coil at break tends to maintain the current through it in the original direction. Hence, the direction of the current through 20-Ω resistor is upwards so that the p.d. across the switch contacts will be the *sum* of supply voltage and the voltage across 20 Ω resistor.

∴ Initial voltage across switch contacts = 400 + 100 = **500 V**.

(*d*) The rate of loss of energy = power = induced e.m.f. in coil × current (after 0.3 s)

$$= L\left(\frac{di}{dt}\right)_{t=0.3} \times i \text{ (after 0.3 second)}$$

Now, after 0.3 second, $i = 9.45$ A

Value of di/dt after 0.3 second = $-20 \times 2.5 \times e^{-0.75} = -23.6$ A/second

∴ Rate of loss of energy = $-10 \times 23.6 \times 9.45 = -2{,}230$ joule/second

22.7. Transients in R-L Circuits (A.C)

Let a voltage given by $v = V_m \sin(\omega t + \Psi)$ be *suddenly* applied across an *R-L* circuit [Fig. 22.5 (*a*)] at a time when $t = 0$. It means that the voltage is applied when it is passing through the value $V_m \sin \Psi$. Since the contact may be closed at any point of the cycle, angle Ψ may have any value lying between zero and 2π radians. The resultant current, as before, is given by

$$i = i_s + i_t$$

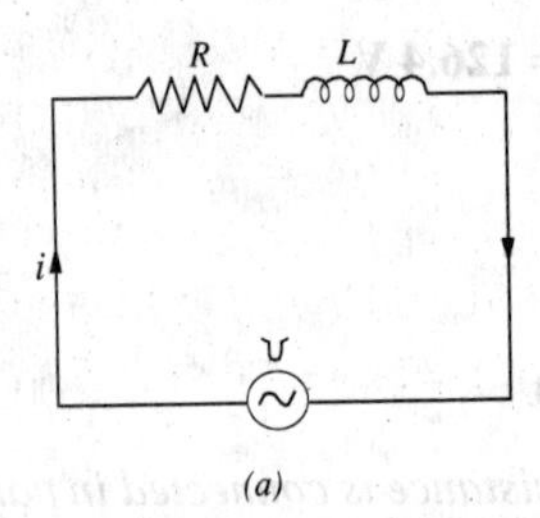

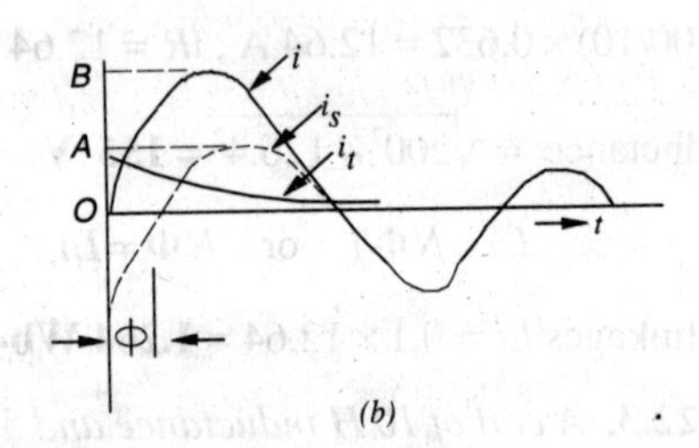

Fig. 22.5

The value of steady-state current is found by the normal circuit theory. The peak steady-state current is given by

$$I_m = \frac{V_m}{\sqrt{R^2 + X_L^2}} = \frac{V_m}{Z}$$

where $\sqrt{R^2 + X_L^2}$ is the impedance of the circuit. This current lags behind the applied voltage by an angle ϕ such that $\tan\phi = X_L/R$ or $\phi = \tan^{-1}(X_L/R)$

Hence, the equation for the instantaneous value of the steady-state current becomes

$$i_s = I_m \sin(\omega t + \Psi - \phi)$$

As before, the transient current is given by

$$i_t = I_0\, e^{-t/\lambda} \quad \therefore \quad i = I_m \sin(\omega t + \Psi - \phi) + I_0\, e^{-t/\lambda} \qquad ...(i)$$

Now, when $t = 0$, $i = 0$, hence putting these values in Eq. (*i*) above, we get

$$0 = I_m \sin(\Psi - \phi) + I_0 \quad \therefore \quad I_0 = -I_m \sin(\Psi - \phi)$$

Hence Eq. (*i*) can be written as

$$i = I_m \sin(\omega t + \Psi - \phi) - I_m \sin(\Psi - \phi) e^{-t/\lambda} \qquad ...(ii)$$

From the above, it is seen that the value of I_0 and hence the size of the transient current depends on angle Ψ *i.e.* it depends on the instant in the cycle at which the circuit is closed. We will consider the following three cases :

Case 1

When $t = 0$, let the voltage pass through its zero value and become positive *i.e.* let $\Psi = 0$. In that case, putting this value of Ψ in Eq. (*ii*), we get

$$i = I_m \sin(\omega t - \phi) - I_m \sin(-\phi) e^{-t/\lambda} = I_m[\sin(\omega t - \phi) + \sin\phi\, e^{-t/\lambda}]$$

This is shown in Fig. 22.5 (*b*). It is seen that maximum instantaneous peak current *OB* is larger than the normal peak current *OA*.

Case 2

Let $t = 0$ when voltage is passing through its value $V_m \sin\phi$ so that $\Psi = \phi$ or $\Psi - \phi = 0$

In that case, $I_0 = 0$, there is no transient current at the time of switching on (*i.e.* $i_t = 0$). It corresponds to the contacts closing at the instant when the steady state current itself is zero.

Case 3

When $t = 0$, let the voltage be passing through

$$V_m \sin\left(\phi \pm \frac{\pi}{2}\right) \; i.e.\; \Psi = \phi \pm \frac{\pi}{2} \text{ and } \Psi - \phi = \pm\pi/2$$

In this case, the transient [as found from Eq. (*ii*)] would be given by

$$i_t = -I_m \sin\left(\pm\frac{\pi}{2}\right) e^{-t/\lambda} = \mp I_m e^{-t/\lambda}$$

Under these conditions, the transient would have its maximum possible initial value.

Example 22.4. *A 1.0 H choke has a resistance of 50 Ω. This choke is supplied with an a.c. voltage given by e = 141 sin 314 t. Find the expression for the transient component of the current flowing through the choke after the voltage is suddenly switched on.*

(Principles of Elect. Engg-11, Jadavpur Univ. 1986)

Solution. The equation of the transient component of the current is (Art. 22.7 Case 1)

$$i_t = I_m \sin\phi\, e^{-t/\lambda}$$

Here, $\lambda = L/R = 1/50 = 0.02$ second ; $\mathbf{Z} = 50 + j\,314 = 318 \angle 80.95°$

$I_m = V_m/Z = 141/318 = 0.443$ A; $\sin 80.95° = 0.9875$

$$\therefore \quad i_t = 0.443 \times 0.9875 e^{-t/0.02} = 0.4376^{-t/0.02}$$

Example 22.5. *A 50-Hz sinusoidal voltage of maximum value of 400 V is applied to a series circuit of resistance 10 Ω and inductance 0.1 H. Find an expression for the value of the current at any instant after the voltage is applied, assuming that voltage is zero at the instant of applic ation. Calculate its value 0.02 second after switching on* **(Electric Circuita, Punjab Univ. 1990)**

Solution. In such cases, as seen from Art. 22.7 (Case 1), the current consists of a steady-state component and a transient component. The equation of the resultant current is

$$i = I_m \sin(\omega t - \phi) + I_m \sin\phi\, e^{-t/\lambda}$$

steady-state current   transient current

where $I_m = V_m/Z$; $\phi = \tan^{-1}(X_L/R)$; $\lambda = L/R$ second

$R = 10\ \Omega$; $X_L = 314 \times 0.1 = 31.4\ \Omega$; $\mathbf{Z} = 10 + j\,31.4 = 33 \angle 72.3°$

$I_m = 400/33 = 12.1$ A ; $\phi = 72.3° = 1.262$ rad.

$\sin\phi = \sin 72.3° = 0.9527$; $\lambda = 0.1/10 = 1/100$ second

$$i = 12.1 \quad \{\sin(314\,t - 1.262) + 0.9527\, e^{-100t}\}$$

Substituting $t = 0.02$ second, we get

$$i = 12.1\,\{\sin(314 \times 0.02 - 1.262\} + 0.9527\, e^{-2}\}$$

$$= 12.1\,(\sin 5.02 + 0.9527\, e^{-2}) = 12.1\,(\sin 288° + 0.9527\, e^{-2})$$

$$= 12.1\,(-\sin 72° + 0.9527 \times 0.1353) = 12.1\,(-0.9511 + 0.1289) = \mathbf{-9.95\ A}$$

Example 22.6. *An alternating voltage $v = 400 \sin(314\ t + \Psi)$ is suddently applied across a coil of resistance 0.2 Ω and inductance 6.36 mH. Determine the first peak value of the resultant current when the transient current has maximum value.*

Solution. Obviously, $\omega = 314$ rad/s

$$X_L = \omega L = 314 \times 6.36 \times 10^{-3} = 2\ \Omega$$

Coil impedance $\mathbf{Z} = 0.2 + j2 \sim 2 \angle 84.3°$

Max. value of steady-state current = 400/2 = 200 A

As seen from Art. 22.7, the maximum value of transient current will occur when

$\Psi = \phi \pm \pi/2$ where $\phi = 84.3°$ *i.e.* the phase angle of the current w.r.t. voltage

$\therefore\ \Psi = 84.3° - 90° = -5.7°$

$\therefore$ resultant current, $i = 400 \sin(314\,t - 90°) + I_0\, e^{-31.4t}$

Now, at $t = 0$, $i = 0$ $\quad\therefore\quad 0 = 400 \sin(-90°) + I_0 \quad\therefore\quad I_0 = 400$ A

Hence, the above equation becomes

$$i = 400 \sin(\omega t - 90°) + 400 e^{-31.4t}$$

The procedure for determining an exact solution for the first peak of the resultant current is first to differentiate the above expression, next to equate the result to zero and then to solve the resulting expression graphically for t. However, sufficiently accurate result can be obtained by determining the instant at which steady-state current reaches its first positive peak value and then to add to it the value of the transient current at this instant. The first peak value of seady-state current occurs when

$$(314\,t - 90°) = \pi/2 \text{ rad} ; \quad i.e. \text{ when } t = \pi/314 = 0.01 \text{ second}$$

At this time, $\quad i_t = 400\, e^{-0.314} = 292$ A

$\therefore$ resultant current i at this time = 200 + 292 = **492 A**

22.8. Transients in R-C Series Circuits (D.C.)

When a d.c. voltage V is *suddently* applied to an R-C series circuit (Fig. 22.6), the voltage v_c across the capacitor rises from zero value to the steady-state value V. If v_c is the voltage across capacitor, v_{tc} the transient voltage, then

$$v_c = V + v_{ct} \qquad ...(i)$$

The charging current is maximum at the beginning but then is reduced to zero so that there is no steady-state current but a transient one.

Since the transient current is not associated with any applied voltage, hence

$$i_t R + v_{ct} = 0 \qquad ...(ii)$$

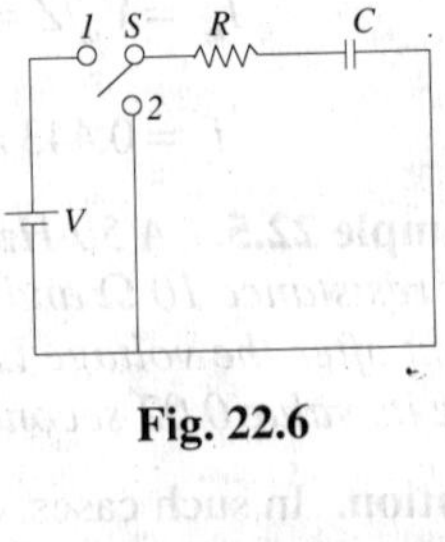

1 S R C
2
V

Fig. 22.6

Now, capacitor voltage $v_{ct} = q_t / C$

Hence, Eq. (*ii*) becomes

$$i_t R + \frac{q_t}{C} = 0$$

or $R\cdot\frac{di_t}{dt} + \frac{1}{C}\cdot\frac{dq_t}{dt} = 0$ or $\frac{di_t}{dt} = -\frac{1}{CR}\frac{dq_t}{dt} = -\frac{1}{CR}i_t$ $(\because \; dq_t/dt = i_t)$

$\therefore \quad \frac{di_t}{i_t} = -\frac{d_t}{CR};$ As before $i_t = I_0 e^{-t/CR} = I_0 e^{-t/\lambda}$

where $CR = \lambda$ = time constant. The reciprocal I/CR is known as damping coefficient.

(i) Charging Current

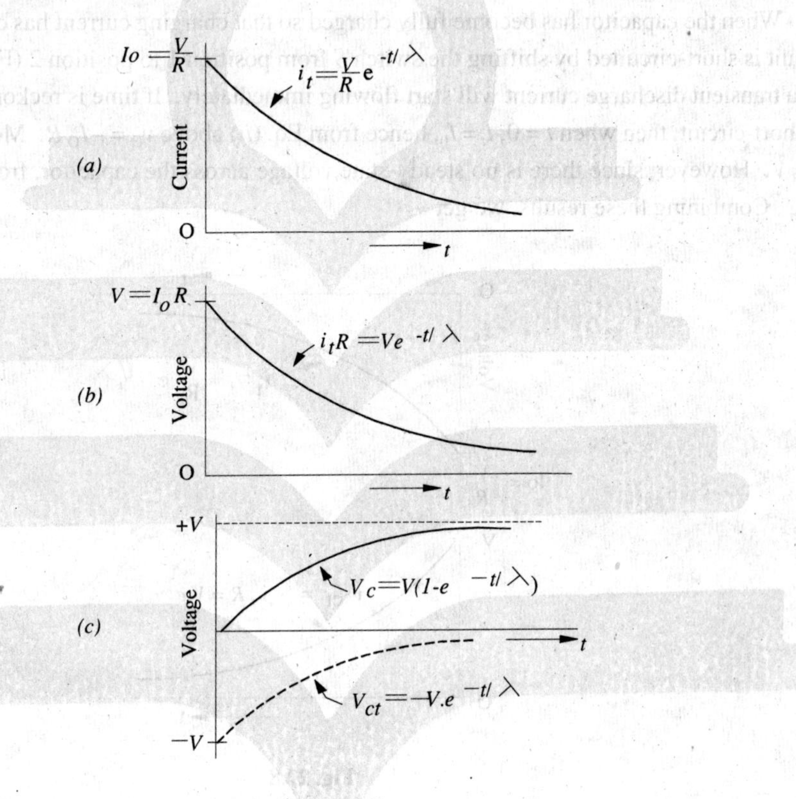

Fig. 22.7

When $t = 0$, transient current $i_t = I_0$, so that from Eq. (*ii*) $v_{ct} = -I_0R$. Moreover, when $t = 0$, $v_c = 0$, hence from Eq. (*i*), $v_{ct} = -V$

Combining these results, we get

$$I_0 = V/R$$

$$\therefore \quad i_t = I_0 e^{-t/\lambda} = \frac{V}{R}e^{-t/\lambda}$$

This is plotted in Fig. 22.7 (*a*)

The transient voltage across the resistor R is given by

$$i_t R = \frac{V}{R}e^{-t/\lambda} \times R$$

$$= V e^{-t/\lambda} \qquad \text{...Fig. 19.7 } (b)$$

From Eq. (*ii*) the value of transient voltage across the capacitor is $v_{ct} = -i_t R$

Hence, Eq. (*i*) becomes

$$v_c = V - i_t R = V - Ve^{-t/\lambda}$$

or $$v_c = V(1 - e^{-t/\lambda}) \qquad \ldots(iii)$$

The voltage across the capacitor v_c which is the sum of the transient voltage v_{ct} and steady-state voltage V has been plotted in Fig. 22.7 (*c*).

The charge across the capacitor is given by

$q = v_c = CV(1 - e^{-t/\lambda})$ or $q = Q(1 - e^{-t/\lambda})$ $(\therefore \quad Q = CV)$

(ii) Discharge Current

When the capacitor has become fully charged so that charging current has ceased, then the *R-C* circuit is short-circuited by shifting the switch *S* from position 1 to position 2 (Fig. 22.6). On doing so, a transient discharge current will start flowing immediately. If time is reckoned from the instant of short-circuit, then when $t = 0$, $i_t = I_0$, hence from Eq. (*ii*) above $v_{ct} = -I_O R$. Moreover, when $t = 0$, $v_c = V$. However, since there is no steady-state voltage across the capacitor, from Eq. (*i*), we get $v_c = v_{ct}$. Combining these results, we get

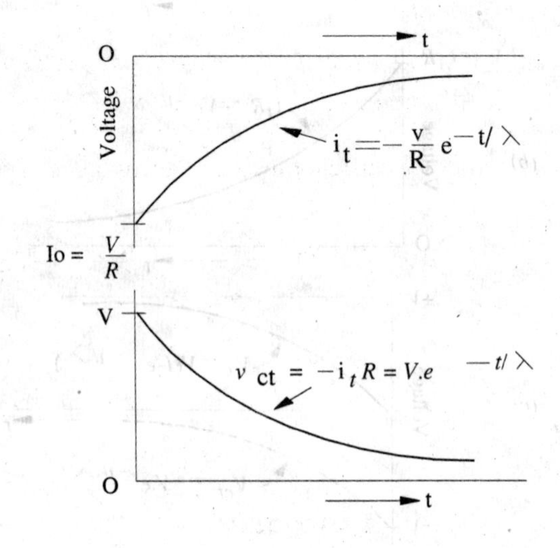

Fig. 22.8

$$I_0 = -V/R$$

$$\therefore \qquad i_t = -\frac{V}{R} e^{-t/\lambda}$$

It is plotted in Fig. 22.8 (*a*). The negative sign shows that discharge current flows in a direction opposite to that in which the charging current flows. That is why the curve has been drawn below the *X*-axis. It may be noted that the only voltage in the circuit is v_{ct} which equals $-i_t R$.

Example 22.7. *In a simple saw-tooth generator circuit with the thyrdtron switches on at 150 V and switches off at 10 V. If this circuit is supplied with 250 V d.c. source; find the time period of saw-tooth wave. The resistance and capacitance have the values of 10 kΩ and 1 μF respectively.* **(Principles of Elect. Engg-II, Jadavpur Univ. 1987)**

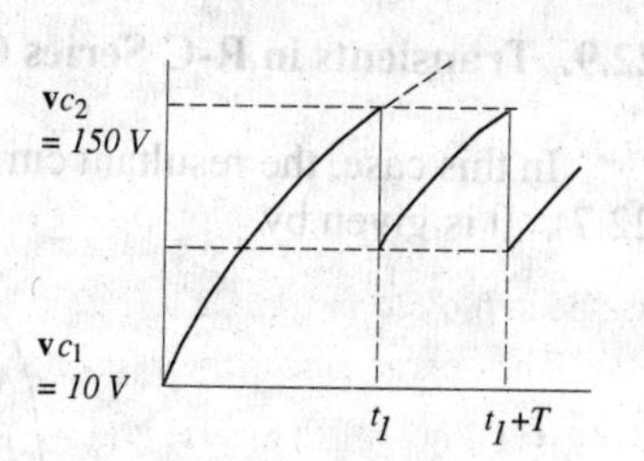

Fig. 22.9

Solution. With reference to Fig. 22.9, let
V = applied voltage
v_{c1} = switching-off voltage of the thyratron = 10 V
v_{c2} = switching-on voltage of the thyratron = 150 V
Now, $v = V(1 - e^{-t/\lambda}$
$\therefore v_c 1 = V(1 - e^{-t/\lambda})$...(i)

$V\,(1 - e^{(-t_1 + T)}/\lambda$ ....(ii)

where T is the time-period of the saw-tooth wave.
From Eq. (i) and (ii), we get

$$T = \lambda \log_e (V - v_{c1})/(V - v_{c2})$$

Now $$\lambda = CR = 10^4 \times 10^{-6} = 10^{-2} \text{ second}$$

$$V - v_{c1} = 250 - 10 = 240 \text{ V} ; V - v_{c2} = 250 - 150 = 100 \text{ V}$$

$$\therefore \quad \frac{V - v_{c1}}{V - v_{c2}} = \frac{240}{100} = 2.4 \qquad \therefore T = 10^{-2} \log_e{}^{2.4} = \mathbf{0.00875 \text{ second}}$$

Example 22.8. *A simple neon-tube time base for a cathode-ray oscillography employs a 300 kΩ and a 0.016 μF capacitor. The striking and extinction voltages of the neon-tube are 170 V and 140 V respectively. Calculate the frequency of the time base if the supply voltage is 200 V.*

Solution. The voltage across the capacitor increases according to the equation

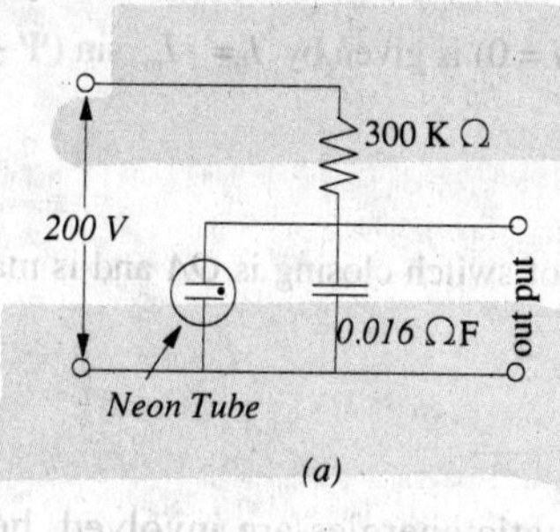

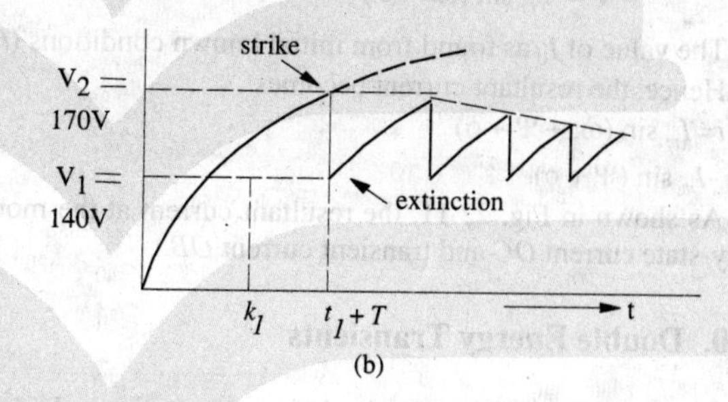

Fig. 22.10

$$v_c = V(1 - e^{-t/CR})$$

It is shown in Fig. 22.10 (b)

$$\therefore v_{c1} = V\left(1 - e^{-t_1/\lambda}\right) \quad ...(i)$$

$$v_{c2} = V\left(1 - e^{(-t_1 + T)/\lambda}\right) \quad ...(ii)$$

From eq. (i) and (ii), we get

$$T = \lambda \,\text{logh}\,(V - v_{C1}) / (V - v_{C2})$$

Now $$\lambda = CR = 0.016 \times 10^{-6} \times 300 \times 10^3 = 4.8 \times 10^{-3} \text{second}$$

$$V - v_{c1} = 200 - 140 = 60 \text{ V and } V - v_{c2} = 200 - 170 = 30 \text{ V}$$

$$\therefore \quad T = 4.8 \times 10^{3} \log 60/30 = 1/300 \text{ socond}$$

$\therefore$ Frequency of time base = $1/T$ = **300 Hz**

22.9. Transients in R-C Series Circuits (A.C.)

In this case, the resultant current can be determined in the same way as for an $R-L$ circuit (Art. 22.7). It is given by

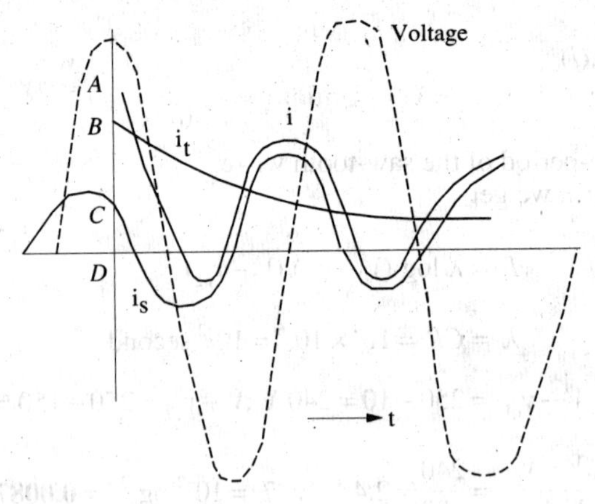

Fig. 22.11

$$i = i_s + i_t = I_m \sin(\omega t + \Psi + \phi) + I_0 e^{-t/\lambda}$$

where $I_m = V_m / \sqrt{R^2 + X_C^2}$

and $v = V_m \sin(\omega t + \Psi)$

The value of I_0 as found from initial known conditions ($t = 0$, $i = 0$) is given by $I_0 = -I_m \sin(\Psi + \phi)$

Hence, the resultant current becomes

$i = I_m \sin(\omega t + \Psi + \phi)$

$-I_m \sin(\Psi + \phi) e^{-t/\lambda}$

As shown in Fig. 22.11, the resultant current at the moment of switch closing is OA and is made up of steady-state current OC and transient current OB.

22.10. Double Energy Transients

In an R-L-C circuit, both electromagnetic and electrostatic energies are involved, hence any sudden change in the conditions of the circuit involves the redistribution of these two forms of energy. The transient currents produced due to this redistribution are known as double-energy transients. The transient current produced may be unidirectional or a decaying oscillatory current.

In an R-L-C circuit, the transient voltages across the three circuit parameters are $i_t R$, $(L\, di_t/dt)$ and q_t/C. Hence, the equation of the transient voltage is

$$i_t R + L\frac{di_t}{dt} + \frac{q_t}{C} = 0 \qquad ...(i)$$

Differentiating the above equation and putting i_t for dq_t/dt, we get

$$\frac{d^2 i_t}{dt^2} + \frac{R}{L}\frac{di_t}{dt} + \frac{1}{LC} i_t = 0 \qquad ...(ii)$$

This is a linear differential equation of the second order with constant coefficient like 2 (*i*) given in Art. No. 22.3. Its solution is given by

$$i_t = k_1 e^{\lambda_1 t} + k_2 e^{\lambda_2 t} \qquad ...(iii)$$

where k_1 and k_2 are constants whose values are found from the boundary conditions. The values of λ_1 and λ_2 are given by

$$\lambda_1 = -\frac{R}{2L} - \sqrt{\frac{R^2}{4L^2} - \frac{1}{LC}} \quad \text{and} \quad \lambda_2 = -\frac{R}{2L} + \sqrt{\frac{R^2}{4L^2} = -\frac{1}{LC}}$$

Depending on the value of λ_1 and λ_2, four different conditions of the circuit are distinguishable. We will now examine these four conditions in the case of an *R-L-C* circuit.

Case 1. Loss-free Circuit, R = 0 ***i.e.*** **Undamped**

In this case,

$$\lambda_1 = \sqrt{-\frac{1}{LC}} = j\frac{1}{\sqrt{LC}} = -j\omega \quad \text{and} \quad \lambda_2 = -\sqrt{-\frac{1}{LC}} = -j\frac{1}{\sqrt{LC}} = -j\omega$$

Hence, Eq. (*iii*) given above becomes

$$i_t = k_1 e^{j\omega t} + k_2 e^{-j\omega t} = k_1 (\cos \omega t + j \sin \omega t) + k_2 (\cos \omega t - j \sin \omega t)$$

$$= (k_1 + k_2) \cos \omega t + j\,(k_1 - k_2) \sin \omega t$$

or $$i_t = A \cos \omega t + B \sin \omega t \qquad \text{...}(iv)$$

where $$A = k_1 + k_2 \text{ and } B = j(k_1 - k_2)$$

Eq. (*iv*) can be still further simplified to

$$i_t = I_m \sin(\omega t + \phi) \qquad \text{...}(v)$$

where $$I_m = \sqrt{A^2 + B^2} \quad \text{and} \quad \phi = \tan^{-1}(A/B)$$

As seen from Eq. (*v*), the transient current in this case is sinusoidal wave of constant peak value and frequency $f = 1/2\pi\sqrt{LC}$ as shown in Fig. 22.12 (*a*). The values of two constant terms I_m and ϕ can be determined from any *two* known initial circuit conditions which are (*i*) the initial current in the inductance and (*ii*) the initial voltage across the capacitor.

Case 2. Low-loss Circuit : $\frac{R^2}{4L^2} < \frac{1}{LC}$ *i.e.* **Under-damped**

In this case, λ_1 and λ_2 would be conjugate complex numbers because the term under the square root sign in each case would be negative.

$$\therefore \quad \lambda_1 = \frac{R}{2L} + j\sqrt{\frac{1}{LC} - \frac{R^2}{4L^2}}$$

If $$a = \frac{R}{2L} \quad \text{and} \quad \omega = \sqrt{\frac{1}{LC} - \frac{R^2}{4L^2}} \quad \text{then } \lambda_1 = -a + j\omega \quad \text{and} \quad \lambda_2 = -a - j\omega$$

Putting these values in equation (*v*), we get

$$i_t = k_1 e^{(-a + j\omega)t} + k_2 e^{(-a - j\omega)t} = e^{-at}\,(k_1 e^{j\omega t} + k_2 e^{-j\omega t})$$

This equation can be reduced, as before, to the form

$$i_t = I_m e^{-at} \sin(\omega t + \phi) \qquad \text{...}(vi)$$

where I_m and ϕ are constants as before. Equation (*vi*) represents damped transient oscillatory current as shown in Fig. 22.12 (*b*).

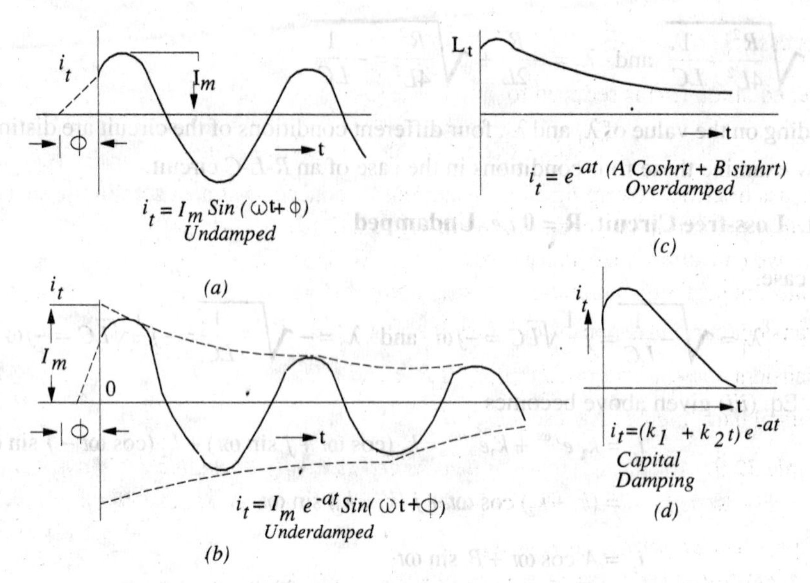

Fig. 22.12

The exponential term e^{-at} which accounts for the decay of oscillations, is called the decay or damping factor or merely *decrement*. It makes each current peak a definite fraction less than that preceding it. The logarithm to the Naperian base '*e*' of the ratio of peaks one cycle apart in time is $a/f = R/2fL$ and is referred to as *logarithmic decrement*. The frequency of damped oscillations is given by

$$f = \sqrt{\frac{1}{LC} - \frac{R^2}{4L^2}} \text{ and is called the natural frequency of the circuit}$$

if $\quad \frac{R^2}{4L^2} < \frac{1}{LC}$, then $f = \frac{1}{2\pi\sqrt{LC}}$

Case 3. High - loss Circuit : $\frac{R^2}{4L^2} > \frac{1}{LC}$ i.e. overdamped

In this case, λ_1 and λ_2 will be pure numbers.

$$\lambda_1 - \frac{R}{2L} + \sqrt{\frac{R^2}{4L^2} - \frac{1}{LC}} = -a + \gamma \text{ and } \gamma_2 = -a - \gamma$$

$$\therefore \quad i_t = k_1 e^{(-a+\gamma)t} + k_2 e^{(-a-\gamma)t} = e^{-at}\,(k_1 e^{\gamma t} + k_2 e^{-\gamma t})$$

Now $\quad e^{rt} = \sin h\ \gamma t + \cosh \gamma t$

and $\quad e^{-\gamma t} = \cosh \gamma t - \sinh \gamma t$

$\therefore \quad i_t = e^{-at}\{(k_1 + k_2)\cosh \gamma t + (k_1 - k_2)\sinh \gamma t\}$

or $\quad i_t = e^{-at}\,(A \cosh \gamma t + B \sinh \gamma t)$

A typical curve of this equation is shown in Fig. 22.12(*c*).

Case 4. $\frac{R^2}{4L^2} = \frac{1}{LC}$ *i.e.* Critical Damping

In this case, $\lambda_1 = \lambda_2 = -\frac{R}{2L}$

Hence, equation (*iii*) is reduced to

$$i_t = (k_1 + k_2 t)\, e^{-\frac{R}{2L}t} \quad \text{or} \quad i_t = (k_1 + k_2\, t) e^{-at}$$

It is a case of critical damping because current is reduced to almost zero in the shortest possible time. The above equation has been plotted in Fig. 22.12 (*d*).

Hence, we can summarize as follows:

1. Transient current is an undamped sine wave if $R = 0$
2. Transient current is non-osicillatory if $R < 2\sqrt{L/C}$
3. Transient current is non-oscillatory if $R \geq 2\sqrt{L/C}$
4. Critical damping occurs if $R = 2\sqrt{L/C}$

Example 22.9. *A 5-μF capacitor is discharged suddenly through a coil having an inductance of 2H and a resistance of 200 Ω. The capacitor is initially charged to a voltage of 10 V. Find*

(*a*) *an expression for the current*
(*b*) *the additional resistance required to give critical damping.*

Solution. Since there is no battery or generator in the circuit (Fig. 22.13), the steady-state current must be zero. It means that resultant current is simply the transient current.

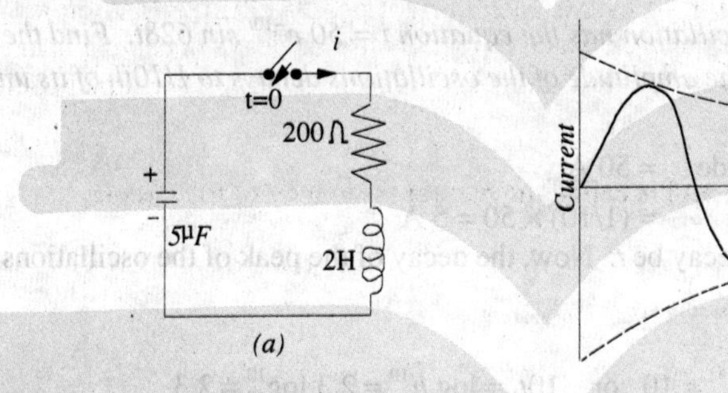

Fig. 22.13

Value of $2\sqrt{L/C} = 2\sqrt{2/5 \times 10^{-6}}$

$= 1265\ \Omega$

Since $R < 2\sqrt{L/C}$, the circuit is originally oscillatory.

(*a*) the expression for the transient current, therefore, is

$$i_t = I_m e^{-at} \sin(\omega t + \phi)$$

where $a = R/2L = 200/2 \times 2 = 50$

$$\omega = \sqrt{\frac{1}{LC} - \frac{R^2}{4L^2}} = \sqrt{100{,}000 - 2500} = 312.3 \text{ rad/s}$$

$$\therefore \quad i_t = I_m e^{-50t} \sin(312.3t + \phi) = i \quad \ldots(i)$$

Two initial conditions are known from which I_m and ϕ can be found (*a*) at $t = 0$; $i = 0$ and (*b*) at $t = 0$; $v_c = 10$ V. Applying condition (*a*) to Eq. (*i*), we get

$$0 = I_m \sin\phi, \text{ hence } \phi = 0 \quad \therefore \quad i = I_m e^{-50t} \sin 312.3t \quad \ldots(ii)$$

Now, at $t = 0$, the voltage across the inductance must be 10 V because the current in the resistance is zero.

i.e. $(L\,di/dt)_{t=0} = 10$ V $\quad\therefore\quad (di/dt)_{t=0} = 10/L = 5$ A/s ...(*iii*)

Now, from equation (*ii*), we have

$$\frac{di}{dt} = -50\,I_m e^{-50t}\sin 312.3t + 312.3\,I_m e^{-50t}\cos 312.3t \quad ...(iv)$$

Putting $t = 0$, it becomes $(di/dt)_{t=0} = 312.3\,I_m$

From equation (*iii*), we have

$312.3\,I_m = 5 \quad\therefore\quad I_m = 5/312.3 = 0.016$ A

Hence, the general expression for the current becomes

$$\mathbf{i = 0.016\,e^{-50t}\,\sin\,312.3t}$$

It is roughly plotted (the first few cycles only) in Fig. 22.13 (*b*).

(*b*) Critical damping is achieved when $R = 2\sqrt{L/C}$

$$\therefore \quad R = 2\sqrt{2/5\times10^{-6}} = 1265\ \Omega$$

$\therefore$ Additional resistance reqd. $= 1265 - 200 =$ **1065 Ω**

Example 22.10. *A damped oscillation has the equation $i = 50\,e^{-10t}\sin 628t$. Find the number of oscillations which occur before the amplitude of the oscillations decays to 1/10th of its undamped value.*

Solution. Undamped amplitude $= 50$ A

1/10th amplitude $= (1/10)\times 50 = 5$ A

Let the time required for this decay be t. Now, the decay of the peak of the oscillations is given by the term $50e^{-10t}$.

$$\therefore \quad 5 = 50\,e_1^{-10t} \quad\therefore\quad e_1^{10t_1} = 10 \text{ or } 10t_1 = \log h^{10} = 2.3\log_{10}^{10} = 2.3$$

$$\therefore \quad t_1 = 0.23 \text{ second}$$

Frequency of oscillations $= 628/2\pi = 100$ Hz.

Hence, the number of oscillations which occur before the amplitude falls to 1/10th of its undamped value is $= 0.23\times 100 =$ **23**

Example 22.11. *If, in Fig. 22.14, a break occurs at a point marked X, what would be the voltage across the break ? It may be assumed that prior to the break, steady conditions existed in the circuit.*

Solution. Steady-state current through the inductance $= 120/60 = 2$ A

Energy stored in the inductor prior to the break

$$= \frac{1}{2}LI^2 = \frac{1}{2}\times 12\times10^{-3}\times 4 = 24\times10^{-3}\text{ J}$$

Energy initially stored in the capacitor

$$= \frac{1}{2}CV^2 = \frac{1}{2}\times10^{-8}\times120^2 = 72\times10^{-6}\text{ J} = 0 \text{ -practically}$$

When the break occurs, the energy stored in the inductor is transferred to the capacitor. If loss of energy during first transfer is neglected, then maximum energy stored in the capacitor is

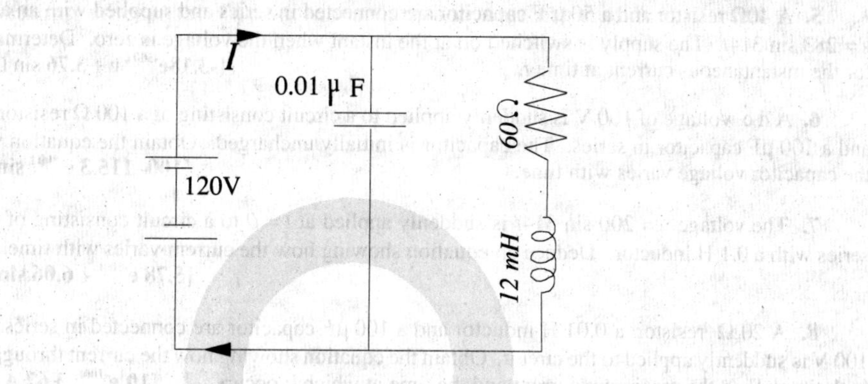

Fig. 22.14

$= 24 \times 10^{-3}$ J $\quad \therefore \quad \frac{1}{2} C V_m^2 = 24 \times 10^{-3}$

$\therefore \quad V_m = \sqrt{2 \times 24 \times 10^{-3} \times 10^8} = 2,190$ V

Maximum voltage across the break is = 2190 + 120 = 2310 V

The voltage would be oscillatory because the energy alternates between the inductor and capacitor.

Frequency of voltage oscillation is

$$f = 1/2\pi\sqrt{LC} = 10^5/2\pi \times \sqrt{1.2} = 14,530 \text{ Hz}$$

Decay or damping factor = e^{-at} Art. 22.10, Case 2

Here, $a = R/2L = 60/2 \times 12 \times 10^{-3} = 2500$ $\quad \therefore$ damping factor = $e^{-2500\,t}$

Hence, voltage across the break is

$$= 120 + 2190\, e^{-2500t} \sin 2\pi \times 14,500t = 120 + 2190\, e^{-2500t} \sin 91,290t.$$

Tutorial Problem No : 22.1

1. Deduce an expression for the growth of current in an inductive circuit.

A 15-H inductance coil of 10 Ω resistance is suddenly connected to a 20 V d.c. supply. Calculate:

(*a*) the initial rate of change of current (*b*) the current after 2 second

(*c*) the rate of change of current after 2 second

(*d*) the energy stored in the magnetic field in this time

(*e*) the energy lost as heat in this time (*f*) the time constant.

[(*a*) **1.33 A/s** (*b*) **1.47 A** (*c*) **0.352 A/s** (*d*) **16.3 joules** (*e*) **19.5 joules** (*f*) **1.5 s**]

2. A circuit consisting of a 20 Ω resistor in series with a 0.2 H inductor is supplied from 200 V (r.m.s) 50 Hz a.c. mains. Deduce equations showing how the current varies with time if the supply is suddenly switched on (*a*) at the instant when the voltage is zero (*b*) at the instant when the voltage is a maximum.

[(*a*) $\mathbf{4.11\, e^{-100} + 4.32 \sin (314\, t - 72°16')}$ **A** (*b*) $\mathbf{-1.245\, e^{-100t} + 4.32 \cos (314t - 72°16')}$ **A**]

3. A circuit consisting of a 20 Ω resistor, 20 mH inductor and a 100 μF capacitor in series is connected to a 200 V, d.c. supply. The capacitor is initially uncharged. Determine the equation relating the instantaneous current to the time and find the maximum instantaneous curren

[$\mathbf{(20\, e^{-500t} \sin 500\, t)}$ **A; 6.44 A**]

4. Find an expression for the value of current at any instant after a sinusoidal voltage of amplitude 600 V at 50 Hz is applied to a series circuit of resistance 10 Ω and inductance 0.1 H, assuming that voltage is zero at the instant of switching. Also, find the value of transeint current at $t = 0.02$ second.

[**−15.14 A ; 2.17 A**] (*Electric Circuits and Fields, Gujarat Univ. 1978*)

5. A 40Ω resistor and a 50 μ F capacitor are connected in series and supplied with an alternating voltage $v = 283 \sin 314t$. The supply is switched on at the instant when the voltage is zero. Determine the expression for the instantaneous current at time t. **[$-3.18e^{-500\,t}\ v + 3.76 \sin (314\,t + 57° 50')$]**

6. A d.c. voltage of 100 V is suddenly applied to a circuit consisting of a 100 Ω resistor, a 0.1 H inductor and a 100 μF capacitor in series. The capacitor is initially uncharged. Obtain the equation which shows how the capacitor voltage varies with time. **[$100 - 115.3\, e^{-500\,t} \sin (866\,t + \pi/3$ V]**

7. The voltage $v = 200 \sin 314t$ is suddenly applied at $t = 0$ to a circuit consisting of a 10 Ω resistor in series with a 0.1 H inductor. Deduce an equation showing how the current varies with time. **[$5.78\, e^{-100t} + 6.06 \sin (314\,t - 72°20')$]**

8. A 20 Ω resistor, a 0.01 H inductor and a 100 μF capacitor are connected in series. A d.c. voltage of 100 V is suddently applied to the circuit. Obtain the equation showing how the current through the circuit varies with time. Find the maximum current and the time at which it occurs. **[$10^4 e^{-100t}$; 3.67 A ; 0.001 second]**

9. A 4-μ F capacitor is initially charged to 300 V. It is discharged through a 100 mH inductance and a resistor in series :

(*a*) find the frequency of the discharge if the resistance is zero.

(*b*) how many cycles at the above frequency will occur before the discharge oscillation decays to 1/10 of its initial value if the resistance is 1 Ω.

(*c*) find the value of the resistance which would just prevent oscillations.

[(*a*) 796 Hz (*b*) 36.6 (*c*) 100 Ω]

Objective Tests – 22

1. Transient disturbance is produced in a circuit whenever

(*a*) it is suddently connected or disconnected from the supply
(*b*) it is shorted
(*c*) its applied voltage is changed suddenly
(*d*) all of the above.

2. There are no transients in pure resistive circuits because they

(*a*) offer high resistance
(*b*) obey Ohm's law
(*c*) have no stored energy
(*d*) are linear circuits.

3. Transient currents in electrical circuit are associated with

(*a*) inductors
(*b*) capacitors
(*c*) resisters
(*d*) both (*a*) and (*b*).

4. The transients which are produced due to sudden but energetic changes from one steady state of a circuit to another are called —— transients.

(*a*) initiation
(*b*) transition
(*c*) relaxation
(*d*) subsidence.

5. In an *R-L* circuit connected to an alternating sinusoidal voltage, size of transient current primarily depends on

(*a*) the instant in the voltage cycle at which circuit is closed
(*b*) the peak value of steady-state current
(*c*) the circuit impedance
(*d*) the voltage frequency.

6. Double-energy transients are produced in circuits consisting of

(*a*) two or more resistors
(*b*) resistance and inductance
(*c*) resistance and capacitance
(*d*) resistance, inductance and capacitance.

7. The transient current in a loss-free *L-C* circuit when excited from an ac source is a/an —— sine wave.

(*a*) over damped
(*b*) undamped
(*c*) under damped
(*d*) critically damped.

8. Transient current in an R–L–C circuit is oscilatory when

(*a*) $R = 0$ (*b*) $R > 2\sqrt{L/C}$

(*c*) $R < 2\sqrt{L/C}$ (*d*) $R = 2\sqrt{L/C}$

ANSWERS

1. *d* **2.** *c* **3.** *d* **4.** *b* **5.** *a* **6.** *d* **7.** *b* **8.** *c*

23 SYMMETRICAL COMPONENTS

23.1. Introduction

The method of symmetrical components was first proposed by C.L. Fortescue and has been found very useful in solving unbalanced polyphase circuits, for analytical determination of the

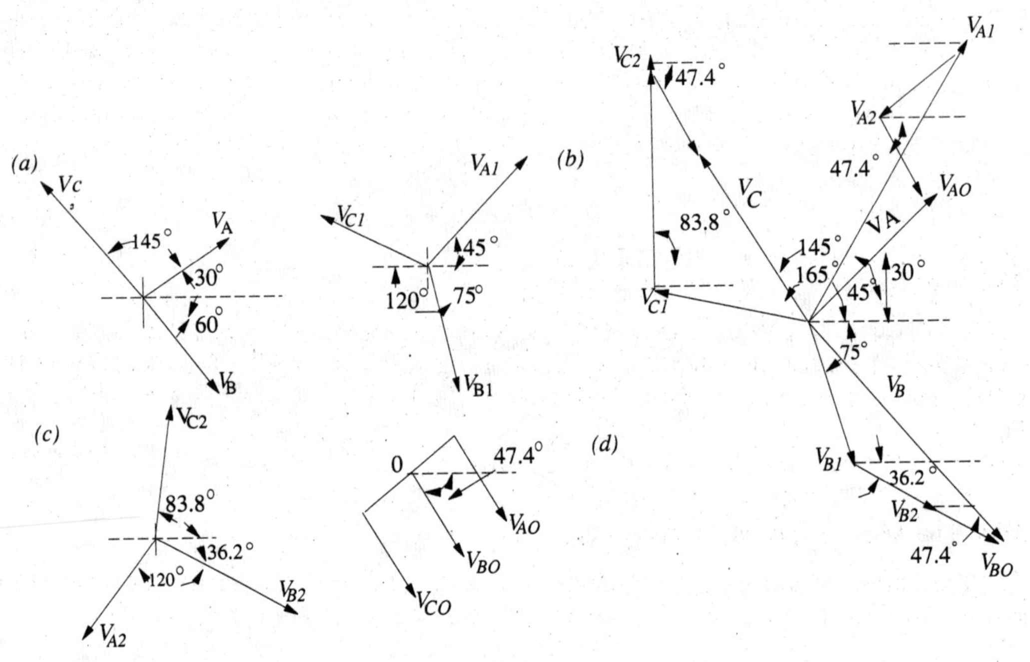

Fig. 23.1

Fig. 23.2

performance of polyphase electrical machinery when operated from a system of unbalanced voltages and for calculation of currents resulting from unbalanced faults. According to Fortescue's theorem, any unbalanced 3-phase system of vectors (whether representing voltages or currents) can be resolved into three *balanced* systems of vectors which are called its '*symmetrical components*'. In Fig. 23.1 (*a*) is shown a set of three unbalanced voltage vectors V_A, V_B and V_C having phase sequence $A \rightarrow B \rightarrow C$. These can be regarded as made up of the following symmetrical components :-

(*i*) A balanced system of 3-phase vectors V_{A1}, V_{B1} and V_{C1} having the phase sequence $A \rightarrow B \rightarrow C$ as the original set of three unbalanced vectors. These vectors constitute the positive-sequence components [Fig. 23.1 (*b*)].

(*ii*) A balanced system of 3-phase vectors V_{A2}, V_{B2} and V_{C2} having phase sequence $A \rightarrow C \rightarrow B$ which is opposite to that of the original unbalanced vectors. These vectors constitute the negative-sequence components [Fig. 23.1 (*c*)].

(*iii*) A system of three vectors V_{A0}, V_{BO} and V_{CO} which are equal in magnitude and are in phase with each other *i.e.* $V_{AO} = V_{BO} = V_{CO}$. These three co-phasal vectors form a uniphase system and are known as zero-sequence components [Fig. 23.1 (*d*)].

Hence, it means that an unbalanced 3-phase system of voltages or current can be regarded as due to the superposition of two symmetrical 3-phase systems having opposite phase sequences and a system of zero phase sequence *i.e.* ordinary single-phase current or voltage system. In Fig. 23.2, each of the original vectors has been reconstructed by the vector addition of its positive-sequence, negative – sequence and zero-sequence components. It is seen that

$$V_A = \overset{+ve}{V_{A1}} + \overset{-ve}{V_{A2}} + \overset{zero}{V_{A0}} \qquad ...(i)$$

$$V_B = V_{B1} + V_{B2} + V_{BO} \qquad ...(ii)$$

$$V_C = V_{C1} + V_{C2} + V_C\,0 \qquad ...(iii)$$

23.2. The Positive – sequence Components

As seen from above, the positive-sequence components have been designated as V_{A1}, V_{B1} and V_{C1}. The subscript 1 is meant to indicate that the vector belongs to the positive-sequence system. The letter refers to the original vector of which the positive-sequence vector is a component part.

These positive-sequence vectors are completely determined when the magnitude and phase of any one of these is known. Usually, these vectors are related to each other with the help of the operator a (for details, please refer to Art. 12.11). As seen from Fig. 23.1 (*b*).

$$\mathbf{V_{A1}} = V_{A1};\ \mathbf{V}_{B1} = a^2\mathbf{V}_{A1} = V_{A1}\angle -120°;\ \mathbf{V}_{C1} = a\mathbf{V}_{A1} == V_{A1}\angle 120°$$

23-3. The Negative – sequence Components

This system has a phase sequence of $A \to C \to B$. Since this system is also balanced, it is completely determined when the magnitude and phase of one of the vectors becomes known. The suffix 2 indicates that the vector belongs to the negative-sequence system. Obviously, as seen from Fig. from 23.1 (*c*).

$$\mathbf{V}_{A2} = V_{A2}; \mathbf{V}_{B2} = a\ V_{A2} = \mathbf{V}_{A2} = \angle 120°;\ \mathbf{V}_{C2} = a^2\mathbf{V}_{A2} = V_{A2}\angle -120°$$

23.4. The Zero – sequence Components

These three vectors are equal in magnitude and phase and hence form what is known as uniphase system. They are designated as V_{A0}, V_{B0} and V_{CO}. Since these are identical in magnitude

$$\therefore \qquad V_{AO} = V_{BO} = V_{CO}$$

23.5. Graphical Composition of Sequence Vectors

Fig. 23.2 shows how the original vector V_A has been obtained by the addition of V_{A1}, V_{A2} and V_{A0}. The same applies to other vectors V_B and V_C.

For simplicity, let us write V_{A1} as V_1, V_{A2} as V_2 and V_{A0} as V_0. Then

$$\mathbf{V}_A = V_1 + \mathbf{V}_2 + \mathbf{V}_0 \qquad ...(iv)$$

$$\mathbf{V}_B = a^2\mathbf{V}_1 + a\mathbf{V}_2 + \mathbf{V}_0 \qquad ...(v)$$

$$\mathbf{V}_C = a\mathbf{V}_1 + a^2\mathbf{V}_2 + \mathbf{V}_0 \qquad ...(vi)$$

23.6. Evaluation of $\mathbf{V}_{A1}$ or $\mathbf{V}_1$

The procedure for evaluating $\mathbf{V}_1$ is as follows :

Multiplying (v) by a and (vi) by a^2, we get

$$a\mathbf{V}_B = a^3\mathbf{V}_1 + a^2\mathbf{V}_2 + a\mathbf{V}_0; \qquad a^2\mathbf{V}_C = a^3\mathbf{V}_1 + a^4\mathbf{V}_2 + a^2\mathbf{V}_0$$

Now $a^3 = 1$ and $a^4 = a$, hence

$$a\mathbf{V}_B = \mathbf{V}_1 + a^2\mathbf{V}_2 + a\mathbf{V}_0 \qquad ...(vii)$$

$$a^2\mathbf{V}_C = \mathbf{V}_1 + a\mathbf{V}_2 + a^2\mathbf{V}_0 \qquad ...(viii)$$

Adding (iv), (vii) and ($viii$), we get

$$\mathbf{V}_A + a\mathbf{V}_B + a^2\mathbf{V}_C = 3\mathbf{V}_1 + \mathbf{V}_2(1 + a + a^2) + \mathbf{V}_0(1 + a + a^2) = 3\mathbf{V}_1$$

$$\therefore \quad \mathbf{V}_1 = \frac{1}{3}(\mathbf{V}_A + a\mathbf{V}_B + a^2\mathbf{V}_C) = \frac{1}{3}(\mathbf{V}_A + \mathbf{V}_B\angle 120° + \mathbf{V}_C\angle -120°)$$

$$= \frac{1}{3}\left[V_A + V_B\left(-\frac{1}{2} + j\frac{\sqrt{3}}{2}\right) + V_C\left(-\frac{1}{2} - j\frac{\sqrt{3}}{2}\right)\right]$$

This shows that, geometrically speaking, V_1 is a vector one-third as large as the vector obtained by the vector addition of the three original vectors V_A, $V_B\angle 120°$ and $V_C\angle -120°$.

23.7. Evaluation of $\mathbf{V}_{A2}$ or $\mathbf{V}_2$

Multiplying (vi) by a and (v) by a^2 and adding them to (iv) we get

$$a\mathbf{V}_C = a^2\mathbf{V}_1 + a^3\mathbf{V}_2 + a\mathbf{V}_0; \qquad a^2\mathbf{V}_B = a^4\mathbf{V}_1 + a^3\mathbf{V}_2 + a^2\mathbf{V}_0$$

$$\mathbf{V}_A + a^2\mathbf{V}_B + a\mathbf{V}_C = \mathbf{V}_1(1 + a + a^2) + 3\mathbf{V}_2 + \mathbf{V}_0(1 + a + a^2) = 3\mathbf{V}_2 \quad \text{Now, } 1 + a + a^2 = 0$$

$$\therefore \quad \mathbf{V}_2 = \frac{1}{3}(\mathbf{V}_A + a^2\mathbf{V}_B + a\mathbf{V}_C) = \frac{1}{3}(\mathbf{V}_A + \mathbf{V}_B\angle -120° + \mathbf{V}_C\angle 120°)$$

$$= \frac{1}{3}\left[\mathbf{V}_A + \mathbf{V}_B\left(-\frac{1}{2} - j\frac{\sqrt{3}}{2}\right) + \mathbf{V}_C\left(-\frac{1}{2} + j\frac{\sqrt{3}}{2}\right)\right]$$

23.8. Evaluation of $\mathbf{V}_{A0}$ or $\mathbf{V}_0$

Adding (iv), (v) and (vi), we get $\mathbf{V}_A + \mathbf{V}_B + \mathbf{V}_C = \mathbf{V}_1(1 + a + a^2) + \mathbf{V}_2(1 + a + a^2) + 3\mathbf{V}_0 = 3\mathbf{V}_0$

$$\therefore \quad \mathbf{V}_0 = \frac{1}{3}(\mathbf{V}_A + \mathbf{V}_B + \mathbf{V}_C)$$

It shows that V_0 is simply a vector one third as large as the vector obtained by adding the original vectors V_A, V_B and V_C.

To summarize the above results, we have

(i) $\mathbf{V}_1 = \frac{1}{3}(\mathbf{V}_A + a\mathbf{V}_B + a^2\mathbf{V}_C)$ (ii) $\mathbf{V}_2 = \frac{1}{3}(\mathbf{V}_A + a^2\mathbf{V}_B + a\mathbf{V}_C)$ (iii) $\mathbf{V}_0 = \frac{1}{3}(\mathbf{V}_A + \mathbf{V}_B + \mathbf{V}_C)$

Note. An unbalanced system of 3-phase currents can also be likewise resolved into its symmetrical components. Hence

$$\mathbf{I}_A = \mathbf{I}_1 + \mathbf{I}_2 + \mathbf{I}_0; \quad \mathbf{I}_B = a^2\mathbf{I}_1 + a\mathbf{I}_2 + \mathbf{I}_0; \quad \mathbf{I}_C = a\mathbf{I}_1 + a^2\mathbf{I}_2 + \mathbf{I}_0$$

Also, as before $\mathbf{I}_1 = \frac{1}{3}(\mathbf{I}_A + a\mathbf{I}_B + a^2\mathbf{I}_C)$; $\mathbf{I}_2 = \frac{1}{3}(\mathbf{I}_A + a^2\mathbf{I}_B + a\mathbf{I}_C)$; $\mathbf{I}_0 = \frac{1}{3}(\mathbf{I}_A + \mathbf{I}_B + \mathbf{I}_C)$ ...(ix)

It shows that I_0 is one-third of the neutral or earth-return current and is zero for an unearthed 3-wire system. It is seen from (*ix*) above that I_0 is zero if the vector sum of the original current vectors is zero. This fact can be used with advantage in making numerical calculations because the original system of vectors can then be reduced to two balanced 3-phase systems having opposite phase sequences.

Example 23.1. *Find out the positive, negative and zero-phase sequence components of the following set of three unbalanced voltage vectors :*

$$V_A = 10 \angle 30°;\ V_B = 30 \angle -60°;\quad V_C = 15 \angle 145°$$

Indicate on an approximate diagram how the original vectors and their different sequence components are located. **(Principles of Elect. Engg. –1, Jadavpur Univ. 1987)**

Solution. (*i*) **Positive-sequence vectors**

As seen from Art. 23.6

$$\mathbf{V}_1 = \frac{1}{3}(\mathbf{V}_A + a\mathbf{V}_B + a^2\mathbf{V}_C) = \frac{1}{3}(10 \angle 30° + a.\ 30 \angle{-60°} + a^2.\ 15\angle 145°)$$

$$= \frac{1}{3}(10 \angle 30° + 30 \angle 60° + 15 \angle 25°) = 12.42 + j\,12.43 = 17.6 \angle 45°$$

$\therefore \mathbf{V}_{A1} = 17.6 \angle 45°$; $\mathbf{V}_{B1} = 17.6 \angle 45° \times \angle{-120°} = 17.6 \angle{-75°}$

$\mathbf{V}_{C1} = 17.6 \angle 45° \times \angle 120° = 17.6 \angle 165°$

These are shown in Fig. 23.1 (*b*)

(*ii*) **Negative – sequence vectors**

As seen from Art. 23.7,

$$\mathbf{V}_2 = \frac{1}{3}(\mathbf{V}_A + a^2\mathbf{V}_B + a\mathbf{V}_C) = \frac{1}{3}(10 \angle 30° + a^2.30 \angle{-60°} + a.15 \angle 145°)$$

$$= \frac{1}{3}(10 \angle 30° + 30 \angle{-180^0} + 15\angle 265°) = -7.55 - j\,3.32 = 8.24 \angle{-156.2°}$$

$\mathbf{V}_{A2} = 8.24 \angle{-156.2°}$; $\mathbf{V}_{B2} = 8.24 \angle{-156.2°} \times \angle 120° = 8.24\angle -36.2°$

$\mathbf{V}_{C2} = 8.24 \angle{-156.2°} \times \angle -120° = 8.24\angle -276.2°$

These vectors are shown in Fig. 23.1 (c)

(*iii*) **Zero sequence vectors**

$$\mathbf{V}_0 = \tfrac{1}{3}(\mathbf{V}_A + \mathbf{V}_B + \mathbf{V}_C)$$

$$= \tfrac{1}{3}(10 \angle 30° + 30 \angle -60° + 15 \angle 145°) = 3.8 - j4.12 = 5.6 \angle -47.4°$$

These vectors are shown in Fig. 23.1 (*d*).

Example 23.2. *Explain how an unsymmetrical system of 3-phase currents can be resolved into 3 symmetrical component systems.*

Determine the values of the symmetrical components of a system of currents

$I_R = 0 + j\,120$ A ; $I_Y = 50 - j\,100$ A ; $I_B = -100 - j\,50$ A

Phase sequence is RYB. **(Elect. Engg.-I Bombay, Univ. 1986)**

Solution. $\mathbf{I}_R = 0 + j\,120 = 120 \angle 90°$

$\mathbf{I}_Y = 50 - j\,100 = 111.8 \angle{-63.5°}$; $\mathbf{I}_B = -100 - j50 = 111.8 \angle{-153.5°}$

(i) Positive – sequence Components

$$\mathbf{I}_1 = \frac{1}{3}(\mathbf{I}_R + a\mathbf{I}_y + a^2\mathbf{I}_B) = \frac{1}{3}\left[(0 + j120) + \left(-\frac{1}{2} + j\frac{\sqrt{3}}{2}\right)(50 - j100) + \left(-\frac{1}{2} - j\frac{\sqrt{3}}{2}\right)(-100 - j50)\right]$$

$= 22.8 + j\,108.3 = 110.7\angle 78.1°$ $\therefore$ $\mathbf{I}_{R1} = 110.7 \angle 78.1°$; $\mathbf{I}_{Y1} = 110.7\angle{-41.9°}$; $\mathbf{I}_{B1} = 110.7 \angle 198.1°$

(ii) Negative – sequence components

$$\mathbf{I}_2 = \frac{1}{3}(\mathbf{I}_R + a^2\mathbf{I}_Y + a\mathbf{I}_B) \quad = \frac{1}{3}(-18.3 + j65.1) = -6.1 + j21.7 = 22.5\angle 105.7°$$

$$\therefore \mathbf{I}_{R2} = 22.5\angle 105.7°;\ \mathbf{I}_{Y2} = 22.5\angle 225°;\ \mathbf{I}_{B2} = 22.5\angle -14.3°$$

(iii) Zero-sequencec component

$$\mathbf{I}_0 = \tfrac{1}{3}(\mathbf{I}_R + \mathbf{I}_Y + \mathbf{I}_B) = \tfrac{1}{3}[(0 + j120) + (50 - j100) + (-100 - j50)] = -16.7 - j10$$

As a check, it may be found that

$$\mathbf{I}_R = \mathbf{I}_{R1} + \mathbf{I}_{R2} + \mathbf{I}_0;\ \mathbf{I}_Y = \mathbf{I}_{Y1} + \mathbf{I}_{Y2} + \mathbf{I}_0;\ \mathbf{I}_{B1} + \mathbf{I}_{B2} + \mathbf{I}_0$$

Example 23.3. *In a 3-phase, 4-wire system, the currents in the R, Y and B lines under abnormal conditions of loading were as follows :*

$$I_R = 100\angle 30°;\ I_Y = 50\angle 300°;\ \mathbf{I}_B = 30\angle 180°$$

Calculate the positive, negative and zero-phase sequence currents in the R-line and the return current in the neutral conductor.

Solution. (*i*) The positive-sequence components of current in the *R*-line is

$$\mathbf{I}_1 = \frac{1}{3}(\mathbf{I}_R + a\mathbf{I}_Y + a^2\mathbf{I}_B)$$

Now

$$\mathbf{I}_R = 100\angle 30° = 50(\sqrt{3} + j)$$

$$\mathbf{I}_Y = 50\angle 300° = 50\left(\frac{1}{2} - j\frac{\sqrt{3}}{2}\right) = 25(1 - j\sqrt{3})$$

$$\mathbf{I}_B = 30\angle 180° = (-30 + j0)$$

$$\mathbf{I}_1 = \frac{1}{3}[50(\sqrt{3} + j) + 25(1 - j\sqrt{3}+)\left(-\frac{1}{2} + j\frac{\sqrt{3}}{2}\right)$$

$$+(-30)\left(-\frac{1}{2} - j\frac{\sqrt{3}}{2}\right)] = 58\angle 48.4°$$

(*ii*) The negative – sequence components of the current in the *R*-line is

$$\mathbf{I}_2 = \frac{1}{3}(\mathbf{I}_R + a^2\mathbf{I}_Y + a\mathbf{I}_B)$$

$$= \frac{1}{3}\left[50(\sqrt{3} + j) + 25(1 - j\sqrt{3})\left(-\frac{1}{2} - j\frac{\sqrt{3}}{2}\right) + (-30)\left(-\frac{1}{2} + j\frac{\sqrt{3}}{2}\right)\right] = 18.9\angle 24.9°$$

(*iii*) The zero – sequence component of current in the *R*-line is

$$\mathbf{I}_0 = \frac{1}{3}(\mathbf{I}_R + \mathbf{I}_Y + \mathbf{I}_B) = \frac{1}{3}[50(\sqrt{3} + j) + 25(1 - j\sqrt{3}) - 30] = 27.2\angle 4.7°$$

The neutral current is

$$\mathbf{I}_N = \mathbf{I}_R + \mathbf{I}_Y + \mathbf{I}_B = 3 \times \mathbf{I}_0 = 3 \times 27.2\angle 4.7° = 81.6\angle 4.7°$$

Example 23.4. *A 3-phase, 4-wire system supplies loads which are unequally distributed on the three phases. An analysis of the currents flowing in the direction of the loads in the R, Y and B lines shows that in the R-line, the positive phase sequence current is 200 ∠ 0° A and the negative phase sequence current is 100 ∠ 60°. The total observed current flowing back to the supply in the neutral conductor is 300 ∠300° A. Calculate the currents in phase and magnitude in the three lines.*

Assuming that the 3-phase supply voltages are symmetrical and that the power factor of the load on the R-phase is $\sqrt{3}/2$ leading, determine the power factor of the loads on the two other phases.

Solution. It is given that in *R*-phase [Fig. 23.3 (*a*)]

$\mathbf{I}_{R1} = 200\angle 0° = (200 + j0)$ A ; $\mathbf{I}_{R2} = 100\angle 60° = (50 + j\,86.6)$ A

$\mathbf{I}_{R0} = \frac{1}{3}\mathbf{I}_N = (300/3)\angle 300° = (50 - j\,86.6)$ A

$\mathbf{I}_R = \mathbf{I}_{R1} + \mathbf{I}_{R2} + \mathbf{I}_{R0} = (200 + j0) + (50 + j\,86.6) + (50 - j\,86.6) = (300 + j0) = 300\angle 0°$

Similarly, as seen from Fig. 23.3 (*b*) for the *Y*-phase

$\mathbf{I}_Y = \mathbf{I}_{Y1} + \mathbf{I}_{Y2} + \mathbf{I}_{Y0} = a^2\mathbf{I}_{R1} + a\mathbf{I}_{R2} + \mathbf{I}_{R0}$

$= 200\angle 0° - 120° + 100\angle 60° + 120°$

$+ 100\angle 300° = -100 - j173.2 - 100 + 50 - j86.6 = -150 - j259.8 = 300\angle 240°$ A

Similarly, as seen from Fig. 23-3 (*c*) for the *B*-phase

$\mathbf{I}_B = \mathbf{I}_{B1} + \mathbf{I}_{B2} + \mathbf{I}_{B0} = a\mathbf{I}_{R1} + a^2\mathbf{I}_{R2} + \mathbf{I}_{R0} = 200\angle 0° + 120° + 100\angle 60° - 120° + 100\angle 300° = 0$

Since the power factor of the *R*-phase is $\sqrt{3}/2$ leading, the current I_R leads the voltage V_R by 30° [Fig. 23.3(d)]

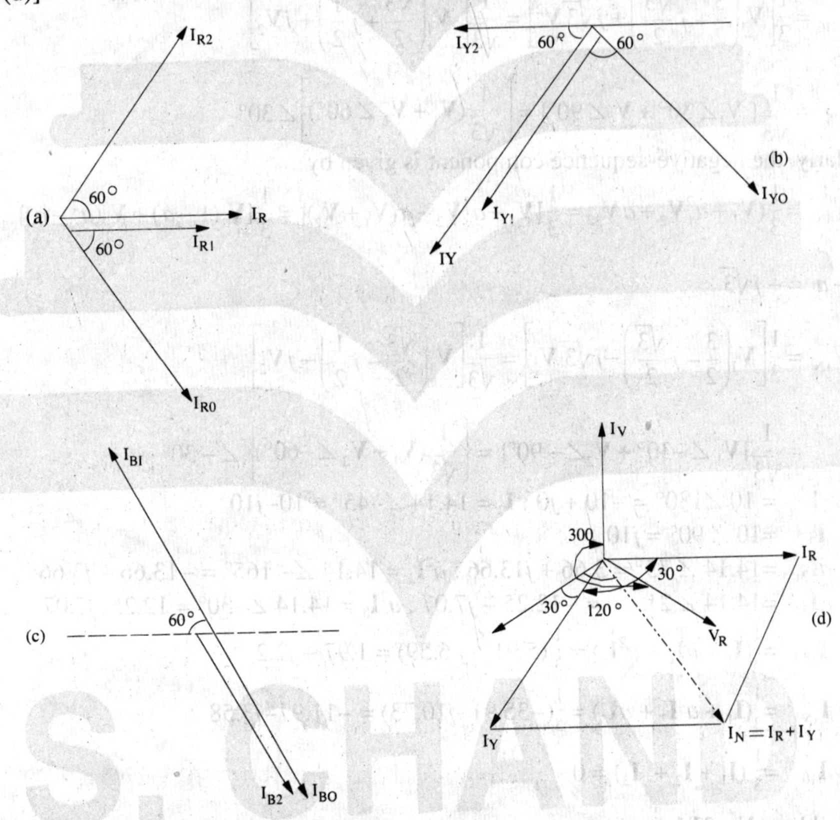

Fig. 23.3

(*d*). Now, phase angle of I_Y is 240° relative to I_R so that I_Y leads its voltage V_Y by 30°. Hence, power factor of *Y* phase is also $\sqrt{3}/2$ leading. The power factor of *B* line is indeterminate because the current in this line is zero.

Example 23.5. *Prove that in a 3-phase system if V_1, V_2 and V_3 are the three balanced voltages whose phasor sum is zero, the positive and negative sequence components can be expressed as*

$$V_{1p} = \left\{\frac{1}{\sqrt{3}}(V_1 + V_2\angle 60°)\right\}\angle 30° \; ; \; V_{1N} = \left\{\frac{1}{\sqrt{3}}(V_1 + V_2\angle -60°)\right\}\angle -30°$$

Phase sequence is 1-2-3.

A system of 3-phase currents is given as $I_1 = 10\angle 180°, I_2 = 14.14\angle -45°$ and $I_3 = 10\angle 90°$. Determine phasor expression for the sequence components of these currents. Phase sequence is 1-2-3. **(Elect. Engg-I, Bombay Univ. 1980)**

Solution. As seen from Art. 23.6

$$\mathbf{V}_{1P} = \frac{1}{3}(\mathbf{V}_1 + a\mathbf{V}_2 + a^2\mathbf{V}_3); \text{ Now, } \mathbf{V}_1 + \mathbf{V}_2 + \mathbf{V}_3 = 0 \quad \therefore \quad \mathbf{V}_3 = -(\mathbf{V}_1 + \mathbf{V}_2)$$

$$\mathbf{V}_{1P} = \frac{1}{3}[\mathbf{V}_1 + a\mathbf{V}_2 - a^2(\mathbf{V}_1 + \mathbf{V}_2)] = \frac{1}{3}[\mathbf{V}_1(1-a^2) + \mathbf{V}_2(a-a^2)]$$

Now, $1 - a^2 = \frac{2}{3} + j\frac{\sqrt{3}}{2}$ and $a - a^2 = j\sqrt{3}$

$$\therefore \quad \mathbf{V}_{1P} = \frac{1}{3}\left[\mathbf{V}_1\left(\frac{3}{2} + j\frac{\sqrt{3}}{2}\right) + j\sqrt{3}\,\mathbf{V}^2\right] = \frac{1}{\sqrt{3}}\left[\mathbf{V}_1\left(\frac{\sqrt{3}}{2} + j\frac{1}{2}\right) + j\mathbf{V}_2\right]$$

$$= \frac{1}{\sqrt{3}}[\mathbf{V}_1\angle 30° + \mathbf{V}_2\angle 90°] = \left[\frac{1}{\sqrt{3}}(\mathbf{V}_1 + \mathbf{V}_2\angle 60°)\right]\angle 30°$$

Similarly, the negative-sequence component is given by

$$\mathbf{V}_{1N} = \frac{1}{3}(\mathbf{V}_1 + a^2\mathbf{V}_2 + a\mathbf{V}_3) = \frac{1}{3}[\mathbf{V}_1 + a^2\mathbf{V}_2 - a(\mathbf{V}_1 + \mathbf{V}_2)] = \frac{1}{3}[\mathbf{V}_1(1-a) + \mathbf{V}_2(a^2 - a)]$$

Now, $a^2 - a = -j\sqrt{3}$

$$\therefore \quad \mathbf{V}_{1N} = \frac{1}{3}\left[\mathbf{V}_1\left(\frac{3}{2} - j\frac{\sqrt{3}}{2}\right) - j\sqrt{3}\,\mathbf{V}_2\right] = \frac{1}{\sqrt{3}}\left[\mathbf{V}_1\left(\frac{\sqrt{3}}{2} - j\frac{1}{2}\right) - j\mathbf{V}_2\right]$$

$$= \frac{1}{\sqrt{3}}[\mathbf{V}_1\angle -30° + \mathbf{V}_2\angle -90°] = \left\{\frac{1}{\sqrt{3}}(\mathbf{V}_1 + \mathbf{V}_2\angle -60°)\right\}\angle -30°$$

Now $\mathbf{I}_1 = 10\angle 180° = -10 + j0$; $\mathbf{I}_2 = 14.14\angle -45° = 10 - j10$

$\mathbf{I}_3 = 10\angle 90° = j10$

$a\mathbf{I}_2 = 14.14\angle 75° = 3.66 + j13.66$; $a^2\mathbf{I}_2 = 14.14\angle -165° = -13.66 - j3.66$

$a\mathbf{I}_3 = 14.14\angle 21 - ° = -12.25 - j7.07$; $a^2\mathbf{I}_3 = 14.14\angle -30° = 12.25 - j7.07$

$\mathbf{I}_{1P} = \frac{1}{3}(\mathbf{I}_1 + a\mathbf{I}_2 + a^2\mathbf{I}_3) = \frac{1}{3}(5.91 + j6.59) = 1.97 + j2.2$

$\mathbf{I}_{1N} = \frac{1}{3}(\mathbf{I}_1 + a^2\mathbf{I}_2 + a\mathbf{I}_3) = \frac{1}{3}(-35.91 - j10.73) = -11.97 - j3.58$

$\mathbf{I}_{10} = \frac{1}{3}(\mathbf{I}_1 + \mathbf{I}_2 + \mathbf{I}_3) = 0$

Tutorial Problems No. 23.1

1. The following currents were recorded in the *R*, *Y*, and *B* lines of a 3-phase system under abnormal conditions : $\mathbf{I}_R = 300\angle 300°$ A : $\mathbf{I}_Y = 500\angle 240°$ A ; $\mathbf{I}_B = 1{,}000\angle 60°$ A

Calculate the values of the positive, negative and zero phase-sequence components.

[$\mathbf{I}_1 = \mathbf{536}\angle\mathbf{-44°20'}$ A; $\mathbf{I}_2 = \mathbf{372}\angle\mathbf{171°}$ A ; $\mathbf{I}_0 = \mathbf{145}\angle\mathbf{23°20'}$]

2. Determine the symmetrical components of the three currents $I_0 = 10\angle 0°$; $\mathbf{I}_b = 100\angle 250°$ and $\mathbf{I}_C = 10\angle 110°$ A [$\mathbf{I}_1 = (\mathbf{39.45} + j\mathbf{5.22})$ A ; $\mathbf{I}_2 = (\mathbf{-20.24} + j\mathbf{22.98})$ A ; $\mathbf{I}_0 = (\mathbf{-9.21} - j\mathbf{28.19})$ A]

(Elect. Meas & Measuring Instru., Madras Univ. June 1976)

3. The three current vectors of a 3-phase, four-wire system have the following values ; $\mathbf{I}_A = 7 + j_0$, $\mathbf{I}_B = -12-j13$ and $\mathbf{I}_C = -2+j3$. Find the symmetrical components. The phase sequence is *A, B, C*.

[$\mathbf{I}_0 = -2.33-j3.33$; $\mathbf{I}_{A1} = (27.75 - j3.67)$; $\mathbf{I}_{B1} = (-17.05-j22.22)$; $\mathbf{I}_{C1} = (-10.7 + j25.9)$; $\mathbf{I}_{A2} = (0.25 + j13.67)$ A ; $\mathbf{I}_{B2} = (-11.97-j6.62)$ A ; $\mathbf{I}_{C2} = (11.73-j7.05)$ A]

23.9. Zero Sequence Components of Current and Voltage

Any circuit which allows the flow of positive-sequence currents will also allow the flow of negative-sequence currents because the two are similar. However, a fourth wire is necessary if zero-sequence components are to flow in the lines of the 3-phase system. It follows that the line currents of 3-phase 3-wire system can contain no zero sequence components whether it is delta - or star-connected. The zero sequence components of line-to-line voltages are non-existent regardless of the degree of imbalance in these voltages. It means that a set of unbalanced 3-phase, line-to-line voltages may be represented by a positive system and a negative system of balanced voltages. This fact is of considerable importance in the analysis of 3-phase rotating machinery. For example, the operation of an induction motor when supplied from an unbalanced system of 3-phase voltages, may be analysed on the basis of two balanced systems of voltages of opposite phase sequence.

Let us consider some typical 3-phase connections with reference to zero-sequence components of current and voltage.

(*a*) **Four-wire Star Connection.** Due to the presence of the fourth wire, the zero sequence currents may flow. The neutral wire carries only the zero-sequence current which is the sum of the zero-sequence currents in the three lines. Since the sum of line voltages is zero, there can be no zero sequence component of line voltages.

(*b*) **Three-wire Star Connection.** Since there is no fourth or return wire, zero-sequence components of current cannot flow. The absence of zero-sequence currents may be explained by considering that the impedance offered to these currents is infinite and that this impedance is situated between the star points of the generator and the load. If the two star points were joined by a neutral, only zero-sequence currents will flow through it so that only zero-sequence voltage can exist between the load and generator star points. Obviously, no zero-sequence component of voltage appears across the phase load.

(*c*) **Three-wire Delta Connection.** Due to the absence of fourth wire, zero-sequence components of currents cannot be fed into the delta-connected load. However, though line currents have to sum up to zero (whereas phase currents need not do so) it is possible to have a zero-sequence current circulating in the delta-connected load.

Similarly, individual phase voltages will generally possess zero-sequence components though components are absent in the line-to line voltage.*

23.10. Unbalanced Star Load Supplied from Unbalanced Three-phase Three-wire System

In this case, line voltages and load currents will consist of only positive and negative-sequence components (but no zero-sequence component). But load voltages will consist of positive, negative and zero-sequence components.

Let the line voltages be denoted by $\mathbf{V}_{RY}$, $\mathbf{V}_{YB}$ and $\mathbf{V}_{BR}$, line (and load) currents by $\mathbf{I}_R$, $\mathbf{I}_Y$ and $\mathbf{I}_B$, the load voltages by $\mathbf{V}_{RN}$, $\mathbf{V}_{YN}$ and $\mathbf{V}_{BN}$ and load impedances by $\mathbf{Z}_R$, $\mathbf{Z}_Y$, and $\mathbf{Z}_B$ (their values being the same for currents of any sequence).

Obviously, $\mathbf{V}_{RN} = \mathbf{I}_R\mathbf{Z}_R$; $\mathbf{V}_{YN} = \mathbf{I}_Y\mathbf{Z}_Y$ and $\mathbf{V}_{BN} = \mathbf{I}_B\mathbf{Z}_B$

If $\mathbf{V}_{RN1}$, $\mathbf{V}_{RN2}$ and $\mathbf{V}_0$ are the symmetrical components of $\mathbf{V}_{RN}$, then we have

$$\mathbf{V}_0 = \frac{1}{3}(\mathbf{V}_{RN} + \mathbf{V}_{YN} + \mathbf{V}_{BN}) = \frac{1}{3}(\mathbf{I}_R\mathbf{Z}_R + \mathbf{I}_Y\mathbf{Z}_Y + \mathbf{I}_B\mathbf{Z}_B)$$

*However, under balanced conditions, the phase voltages will possess no zero-sequence components.

$$= \frac{1}{3}[\mathbf{Z}_R(\mathbf{I}_{R1} + \mathbf{I}_{R2}) + \mathbf{Z}_Y(\mathbf{I}_{Y1} + \mathbf{I}_{Y2}) + \mathbf{Z}_B(\mathbf{I}_{B1} + \mathbf{I}_{B2})]$$

$$= \frac{1}{3}[\mathbf{Z}_R(\mathbf{I}_{R1} + \mathbf{I}_{R2}) + \mathbf{Z}_Y(a^2\mathbf{I}_{R1} + a\mathbf{I}_{R2}) + \mathbf{Z}_B(a\mathbf{I}_{R1} + a^2\mathbf{I}_{R2})]$$

$$= \mathbf{I}_{R1}.\frac{1}{3}(\mathbf{Z}_R + a^2\mathbf{Z}_Y + a\mathbf{Z}_B) + \mathbf{I}_{R2}.\frac{1}{3}(\mathbf{Z}_R + a\mathbf{Z}_Y + a^2\mathbf{Z}_B) = \mathbf{I}_{R1}\mathbf{Z}_{R2} + \mathbf{I}_{R2}\mathbf{Z}_{R1} \quad ...(i)$$

$$\mathbf{V}_{RN1} = \frac{1}{3}(\mathbf{V}_{RN} + a\mathbf{V}_{YN} + a^2\mathbf{V}_{BN}) = \frac{1}{3}(\mathbf{I}_R\mathbf{Z}_R + a\mathbf{I}_Y\mathbf{Z}_Y + a^2\mathbf{I}_B\mathbf{Z}_B)$$

$$= \frac{1}{3}[\mathbf{Z}_R(\mathbf{I}_{R1} + \mathbf{I}_{R2}) + a\mathbf{Z}_Y(\mathbf{I}_{Y1} + \mathbf{I}_{Y2}) + a^2\mathbf{Z}_B(\mathbf{I}_{B1} + \mathbf{I}_{B2})]$$

$$= \mathbf{I}_{R1}.\frac{1}{3}(\mathbf{Z}_R + \mathbf{Z}_Y + \mathbf{Z}_B) + \mathbf{I}_{R2}.\frac{1}{3}(\mathbf{Z}_R + a^2\mathbf{Z}_Y + a\mathbf{Z}_B) = \mathbf{I}_{R1}\mathbf{Z}_o + \mathbf{I}_{R2}\mathbf{Z}_{R2} \quad ...(ii)$$

$$\mathbf{V}_{RN2} = \mathbf{I}_{R2}\mathbf{Z}_0 + \mathbf{I}_{R1}\mathbf{Z}_{R1}$$

Similarly,

$$\mathbf{V}_{YN1} = \mathbf{I}_{Y1}\mathbf{Z}_0 + \mathbf{I}_{Y2}\mathbf{Z}_{Y2} = a^2(\mathbf{I}_{R1}\mathbf{Z}_0 + \mathbf{I}_{R2}\mathbf{Z}_{R2}) = a^2\mathbf{V}_{RN1}$$

$$\mathbf{V}_{BN1} = \mathbf{I}_{B1}\mathbf{Z}_0 + \mathbf{I}_{B2}\mathbf{Z}_{B2} = a(\mathbf{I}_{R1}\mathbf{Z}_0 + \mathbf{I}_{R2}\mathbf{Z}_{R2}) = a\mathbf{V}_{RN1}$$

$$\mathbf{V}_{YN2} = \mathbf{I}_{Y2}\mathbf{Z}_0 + \mathbf{I}_{Y1}\mathbf{Z}_{Y1} = a(\mathbf{I}_{R2}\mathbf{Z}_0 + \mathbf{I}_{R1}\mathbf{Z}_{R1}) = a\mathbf{V}_{RN2}$$

$$\mathbf{V}_{BN2} = \mathbf{I}_{B2}\mathbf{Z}_0 + \mathbf{I}_{B1}\mathbf{Z}_{B1} = a^2(\mathbf{I}_{R2} + \mathbf{Z}_0 + \mathbf{I}_{R1}\mathbf{Z}_{R1}) = a^2\mathbf{V}_{RN2}$$

Now, $\mathbf{V}_{RN1}$ and $\mathbf{V}_{RN2}$ may be determined from the relation between the line and phase voltages as given below :

$$\mathbf{V}_{RY} = \mathbf{V}_{RN} + \mathbf{V}_{NY} = \mathbf{V}_{RN} - \mathbf{V}_{YN} = \mathbf{V}_0 + \mathbf{V}_{RN1} + \mathbf{V}_{RN2} - (\mathbf{V}_0 + \mathbf{V}_{YN1} + \mathbf{V}_{YN2}]$$

$$= \mathbf{V}_{RN1}(1-a^2) + \mathbf{V}_{RN2}(1-a) = \frac{1}{2}\sqrt{3}[\mathbf{V}_{RN1}(\sqrt{3}+j1) + \mathbf{V}_{RN2}(\sqrt{3}-j1)] \quad ...(iii)$$

$$\mathbf{V}_{YB} = \mathbf{V}_{YN} + \mathbf{V}_{NB} = \mathbf{V}_{YN} - \mathbf{V}_{BN} = \mathbf{V}_0 + \mathbf{V}_{YN1} + \mathbf{V}_{YN2} - (\mathbf{V}_0 + \mathbf{V}_{BN1} + \mathbf{V}_{BN2})$$

$$= \mathbf{V}_{RN1}(a^2-a) - \mathbf{V}_{RN2}(a-a^2) = -j\sqrt{3}.\ \mathbf{V}_{RN1} + j\sqrt{3}\,\mathbf{V}_{RN2}$$

$$\therefore\ \mathbf{V}_{RN2} = \mathbf{V}_{RN1} - j\,\mathbf{V}_{YB}/\sqrt{3}$$

Substituting this value of $\mathbf{V}_{RN2}$ in Eq. (*iii*) above and simplifying, we have

$$\mathbf{V}_{RN1} = \frac{1}{3}[\mathbf{V}_{RY} + \frac{1}{2}\mathbf{V}_{YB}(1 + j\sqrt{3})] \quad ...(iv)$$

$$\mathbf{V}_{RN2} = \frac{1}{3}[\mathbf{V}_{RY} + \frac{1}{2}\mathbf{V}_{YB}(1 - j\sqrt{3})] \quad ...(v)$$

Having known $\mathbf{V}_{RN1}$ and $\mathbf{V}_{RN2}$, the currents $\mathbf{I}_{R1}$ and $\mathbf{I}_{R2}$ can be determined from the following equations :

$$\mathbf{I}_{R1} = \frac{(\mathbf{V}_{RN1}\mathbf{Z}_0 - \mathbf{V}_{RN2}\mathbf{Z}_{R2})}{\mathbf{Z}_0^2 - \mathbf{Z}_{R1}\mathbf{Z}_{R2}} \quad ...(vi)$$

$$\mathbf{I}_{R2} = \frac{(\mathbf{Z}_{RN2}\mathbf{Z}_0 - \mathbf{V}_{RN1}\mathbf{Z}_{R1})}{\mathbf{Z}_0^2 - \mathbf{Z}_{R1}\mathbf{Z}_{R2}} \quad ...(vii)$$

Alternatively, when $\mathbf{I}_{R1}$ becomes known, $\mathbf{I}_{R2}$ may be found from the relation

$$\mathbf{I}_{R2} = \frac{V_{RN2} - \mathbf{I}_{R1}\mathbf{Z}_{R1}}{\mathbf{Z}_0} \quad ...(viii)$$

The symmetrical components of the currents in other phases can be calculated from $\mathbf{I_{R1}}$ and $\mathbf{I_{R2}}$ by using the relations given in Art. 23.8.

The phase voltages may be calculated by any one of the two methods given below :

(*i*) directly by calculating the products $\mathbf{I_R Z_R}$, $\mathbf{I_Y Z_Y}$ and $\mathbf{I_B Z_B}$.

(*ii*) by first calculating the zero-sequence component $\mathbf{V_0}$ of the phase or load voltages and then adding this to the appropriate positive and negative sequence components.

For example, $\mathbf{V_{RN}} = \mathbf{V_0} + \mathbf{V_{RN1}} + \mathbf{V_{RN2}}$

23.11. Unbalanced Star Load Supplied from Balanced Three-phase, Three-wire System

It is a special case of the general case considered in Art. 23.10 above. In this case, the symmetrical components of the load voltages consist only of positive and zero-sequence components. This fact may be verified by substituting the value of $\mathbf{V_{YB}}$ in Eq. (*v*) of Art. 23.10. Now, in a balanced or symmetrical system of positive-phase sequence

$$\mathbf{V_{YB}} = \mathbf{V_{RY}}\left(-\frac{1}{2} - j\frac{\sqrt{3}}{2}\right)$$

Substituting this value in Eq. (*v*) above, we have

$$\mathbf{V_{RN2}} = \tfrac{1}{3}[\mathbf{V_{RY}} + \tfrac{1}{2}\,\mathbf{V_{YB}}\,(1 - j\sqrt{3})]$$

$$= \frac{1}{3}\left[\mathbf{V_{RY}} + \frac{1}{2}\cdot\mathbf{V_{RY}}\left(-\frac{1}{2} - j\frac{\sqrt{3}}{2}\right)(1 - j\sqrt{3})\right] = \frac{1}{3} - (\mathbf{V_{RY}} - \mathbf{V_{RY}}) = 0$$

Hence, substituting this value of $\mathbf{V_{RN2}}$ in Eq. (*iv*) and (*vi*) of Art. 23.10, we get

$$\mathbf{V_{RN1}} = j\mathbf{V_{YB}}\sqrt{3} = \frac{1}{2}\,\mathbf{V_{RY}} - j\frac{1}{2}\,\mathbf{V_{RY}}/\sqrt{3} \qquad \ldots(i)$$

$$\mathbf{I_{R1}} = \mathbf{V_{RN1}}\mathbf{Z_0}/(\mathbf{Z_0}^2 - \mathbf{Z_{R1}}\mathbf{Z_{R2}}) = \mathbf{V_{RN1}}\frac{\mathbf{Z_R} + \mathbf{Z_Y} + \mathbf{Z_B}}{\mathbf{Z_R Z_Y} + \mathbf{Z_Y Z_B} + \mathbf{Z_B Z_R}} \qquad \ldots(ii)$$

$$\mathbf{I_{R2}} = -\mathbf{I_{R1}}\cdot\mathbf{Z_{R1}}/\mathbf{Z_0} = -\mathbf{R1_0} = -\mathbf{I_{R1}}\frac{\mathbf{Z_R} + a\mathbf{Z_Y} + a^2\mathbf{Z_B}}{\mathbf{Z_R} + \mathbf{Z_Y} + \mathbf{Z_B}} \qquad \ldots(iii)$$

Example 23.6 illustrates the procedure for calculating the current in an unbalanced star-connected load when supplied from a symmetrical three-wire system.

Example 23.6. *A symmetrical 3-phase, 3-wire, 440-V system supplies an unbalanced Y-connected load of impedances $Z_R = 5\angle 30°\ \Omega$; $Z_Y = 10\angle 45°\ \Omega$, $Z_B = 10\angle 60°\ \Omega$. Phase sequence is $R \to Y \to B$. Calculate in the rectangular complex form the symmetrical components of the currents in the R-line.* **(Elect. Engg.-I, Bombay Univ. 1985)**

Solution. As shown in Fig. 23.4, the branch impedances are

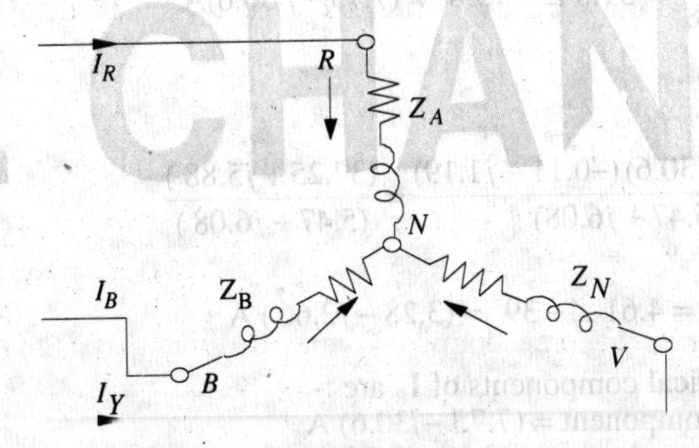

Fig. 23.4

$\mathbf{Z}_R = 5\angle 30° = (4.33 + j2.5)\ \Omega$

$\mathbf{Z}_Y = 10\angle 45° = (7.07 + j7.07)\ \Omega$

$\mathbf{Z}_B = 10\angle 60° = (5 + j8.66)\ \Omega$

$a\mathbf{Z}_y = 10\angle 165° = 10(-0.966 + j0.259) = -9.66 + j2.59)\ \Omega$

$a^2\mathbf{Z}_Y = 10\angle -75° = 10(0.259 - j0.966) = 2.59 - j9.66$

$a\mathbf{Z}_B = 10\angle 180° = (-10 + j\,0)$

$a^2\mathbf{Z}_B = 10\angle -60° = 10(0.5 - j0.866) = (5 - j8.66)$

The symmetrical component impedances required for calculation purposes are as follows :

$$\mathbf{Z}_0 = \frac{1}{3}(\mathbf{Z}_R + \mathbf{Z}_Y + \mathbf{Z}_B) = \frac{1}{3}(16.4 + j\,18.23) = (5.47 + j\,6.08)\ \Omega$$

$$\mathbf{Z}_{R1} = \frac{1}{3}(\mathbf{Z}_R + \mathbf{Z}_Y + a^2\mathbf{Z}_B)$$

$$= \frac{1}{3}(4.33 + j\,2.5 - 9.66 + j\,2.59 + 5 - j\,8.66)$$

$$= \frac{1}{3}(-0.33 - j\,3.57) = -0.11 - j\,1.19$$

$$\mathbf{Z}_{R2} = \frac{1}{3}(\mathbf{Z}_R + a^2\mathbf{Z}_Y + a\mathbf{Z}_B)$$

$$= \frac{1}{3}(4.33 + j2.5 + 2.59 - j\,9.66 - 10 + j0) = \frac{1}{3}(-3.08 - j\,7.16) = -1.03 - j\,2.55$$

Now

$$\mathbf{Z}_0^{\,2}\ \mathbf{Z}_{R1}\mathbf{Z}_{R2} = \frac{1}{3}(\mathbf{Z}_R\mathbf{Z}_Y + \mathbf{Z}_Y\mathbf{Z}_B + \mathbf{Z}_B\mathbf{Z}_R)$$

$$= \frac{1}{3}[50\angle 75° + 100\angle 105° + 50\angle 90°] = (-4.31 + j\,65)$$

Let $\mathbf{V}_{RY}$ be taken as the reference vectors so that $\mathbf{V}_{RY} = (440 + j0)$

Then, $\mathbf{V}_{RN1} = \frac{1}{2}\mathbf{V}_{RY} - j\cdot\frac{1}{2}\mathbf{V}_{RY}/\sqrt{3} = \frac{1}{2}(440 + j0) - j\frac{1}{2}(440 + j0)/\sqrt{3} = (220 - j\,127)$

Now, $$\mathbf{I}_{R_1} = \frac{\mathbf{V}_{RN1}\cdot\mathbf{Z}_0}{(\mathbf{Z}_0^{\,2} - \mathbf{Z}_{R_1}\mathbf{Z}_{R_2})} = \frac{(220 - j127)(5.47 + j6.08)}{(-4.31 + j65)} = \frac{1986 + j\,643}{(-4.31 + j\,65)}$$

$$= \frac{2088\angle 18°}{66.4\angle 93.8°} = 31.6\angle -75.8° \neq (7.75 - j\,30.6)\text{A}$$

$$\mathbf{I}_{R_2} = -\mathbf{I}_{R_2}\mathbf{Z}_{R_1}/\mathbf{Z}_0$$

$$= \frac{-(7.75 - j30.6)(-0.11 - j1.19)}{(5.47 + j6.08)} = \frac{(37.25 + j5.88)}{(5.47 + j6.08)}$$

$$= \frac{37.75\angle 9°}{8.18\angle 48°} = 4.61\angle -39° = (3.28 - j2.66)\text{ A}$$

Hence, the symmetrical components of $\mathbf{I}_R$ are :-

Positive-sequence component = (7.75 – j30.6) A

Negative-sequence component = (3.28–j2.66) A

Note. (*i*) $\mathbf{I}_R = \mathbf{I}_{R1} + \mathbf{I}_{R2}$

(*ii*) Symmetrical components of other currents are

$\mathbf{I}_Y = a^2\mathbf{I}_{B1}$ and $\mathbf{I}_{Y2} = a\mathbf{I}_{R2}$; $\mathbf{I}_{B1} = a\mathbf{I}_{R1}$ and $\mathbf{I}_{B2} = a^2\mathbf{I}_{R2}$

Example 23.7. *A balanced star-connected load takes 75 A from a balanced 3-phase, 4-wire supply. If the fuses in two of the supply lines are removed, find the symmetrical components of the line currents before and after the fuses are removed.*

Solution. The circuit is shown in Fig. 23.5.

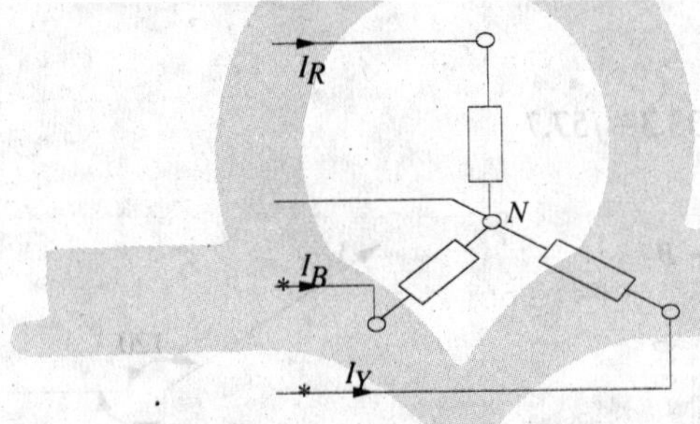

Fig. 23.5

Before fuses are removed

$\mathbf{I}_R = 75\ \angle 0°$; $\mathbf{I}_Y = 75\ \angle{-120°}$; $\mathbf{I}_B = 75\ \angle 120°$

$$\mathbf{I}_1 = \frac{1}{3}(\mathbf{I}_R + a\mathbf{I}_Y + a^2\mathbf{I}_B) = \frac{1}{3}(75\ \angle 0° + 75\ \angle 0° + 75\ \angle 360°) = 75\angle 0°\ \text{A}$$

$$\mathbf{I}_2 = \frac{1}{3}(\mathbf{I}_R + a^2\mathbf{I}_Y + a\mathbf{I}_B) = \frac{1}{3}(75\ \angle 0° + 75\ \angle 120° + 75\ \angle 240°) = 0$$

$$\mathbf{I}_0 = \frac{1}{3}(\mathbf{I}_R + \mathbf{I}_Y + \mathbf{I}_B) = \frac{1}{3}(75\ \angle 0° + 75\ \angle -120° + 75\ \angle 120°) = 0.$$

After fuses are removed

$\mathbf{I}_R = 75\ \angle 0°$; $\mathbf{I}_Y = \mathbf{I}_B = 0$

$$\mathbf{I}_1 = \frac{1}{3}(75\ \angle 0° + 0 + 0) = 25\ \angle 0°$$

$\mathbf{I}_2 = 25\ \angle 0°$, $\mathbf{I}_0 = 25\ \angle 0°$

Example 23.8. *Explain the terms : positive-, negative - and zero-sequence components of a 3-phase voltage system. A star-connected load consists of three equal resistors, each of 1 Ω resistance. When the load is connected to an unsymmetrical 3-phase supply, the line voltages are 200 V, 346 V and 400 V.*

Find the magnitude of the current in any one phase by the method of symmetrical components.

(Power Systems-II, A.M.I.E. 1989)

Solution. This question could be solved by using the following two methods :

(*a*) As shown in Fig. 23.6 (*a*), the line voltages form a closed right-angled triangle with an angle $= \tan^{-1}(346/200) = \tan^{-1}(1.732) = 60°$

Hence, if $\mathbf{V}_{AB} = 200\ \angle 0°$, then $\mathbf{V}_{BC} = 346\ \angle{-90°} = -j346$ and $\mathbf{V}_{CA} = 400\ \angle 120°$.

As seen from Eq. (*iv*) and (*v*) of Art. 23.10

$$\mathbf{V}_{AN1} = \frac{1}{3}\left[\mathbf{V}_{AB} + \frac{1}{2}\mathbf{V}_{BC}(1 + j\sqrt{3})\right]$$

$$= \frac{1}{3}\left[200 + \left(\frac{1}{2} + j\frac{\sqrt{3}}{2}\right)(-j346)\right] = 166.7 - j57.7$$

$$\mathbf{V}_{AN2} = \frac{1}{3}\left[\mathbf{V}_{AB} + \mathbf{V}_{BC}\left(\frac{1}{2} - \frac{\sqrt{3}}{2}\right)\right]$$

$$= \frac{1}{3}\left[200 + \left(\frac{1}{2} - j\frac{\sqrt{3}}{2}\right)(-j346)\right] = -33.3 - j57.7$$

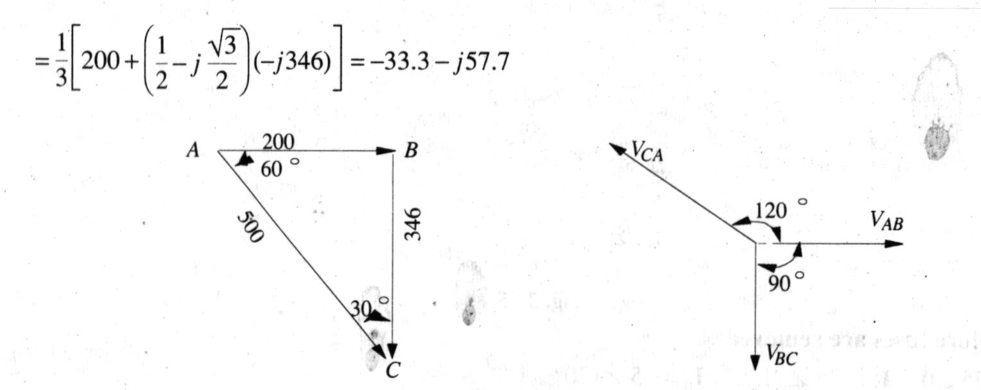

(*a*) **Fig. 23.6** (*b*)

Now, $\mathbf{Z}_A + \mathbf{Z}_B + \mathbf{Z}_C = 3$

$\mathbf{Z}_{A1} = \mathbf{Z}_A + a\mathbf{Z}_B + a^2\mathbf{Z}_C = 0$ (because $\mathbf{Z}_A = \mathbf{Z}_B = \mathbf{Z}_C$ and $1 + a + a^2 = 0$)

Similarly, $\mathbf{Z}_{A2} = \mathbf{Z}_A + a^2\mathbf{Z}_B + a\mathbf{Z}_C = 0$

Hence, from Eq. (*ii*) of Art. 23.10 we have $\mathbf{V}_{AN1} = \mathbf{I}_{A1}\mathbf{Z}_0 \quad \therefore \quad \mathbf{I}_{A1} = \mathbf{V}_{AN1}/\mathbf{Z}_0$

$$\text{Now, } \mathbf{Z}_0 = \frac{1}{3}(\mathbf{Z}_A + \mathbf{Z}_B + \mathbf{Z}_C) = 3/3 = 1\ \Omega \qquad \therefore \qquad \mathbf{I}_{A1} = (166.7 - j57.7)\,A$$

$$\mathbf{I}_{AS} = \mathbf{V}_{AN2}/\mathbf{Z}_0 = -33.3 - j57.7 \qquad \therefore \qquad \mathbf{I}_A = \mathbf{I}_{A1} + \mathbf{I}_{A2} \quad (\because \ \mathbf{I}_0 = 0)$$

$$\therefore \quad \mathbf{I}_A = (166.7 - j57.7) + (-33.3 - j57.7) = 133.4 - j115.4 = 176.4\angle -40.7° \text{ A}$$

(*b*) Using Millman's theorem and taking the line terminal *A* as reference point, the voltage between *A* and the neutral point *N* is

$$\mathbf{V}_{NA} = \frac{\mathbf{V}_{BA}\mathbf{Y}_B + \mathbf{V}_{CA}\mathbf{Y}_C}{\mathbf{Y}_A + \mathbf{Y}_B + \mathbf{Y}_C} = \frac{200\angle 180° \times 1 + 400\angle 120° \times 1}{3} = -133.3 + j115.3 = 176.4\angle 139.3°$$

$$\mathbf{V}_{AN} = 176.4\angle 139.3° - 180° = 176.4\angle -40.7° \quad \therefore \quad \mathbf{I}_A = \mathbf{I}_{AN}/\mathbf{Z} = 176.4\angle -40.7° \qquad \text{...as before}$$

Example 23.9. *Three equal impedances of (8 + j6) are connected in star across a 3-phase, 3-wire supply. The phase voltages are $V_A = (220 + j0)$, $V_B = (-j220)$ and $V_C = (-100 + j220)$ V. If there is no connection between the load neutral and the supply neutral, calculate the symmetrical components of A-phase current and the three line currents.*

Solution. Since there is no fourth wire, there is no zero-component current. Moreover, $\mathbf{I}_A + \mathbf{I}_B + \mathbf{I}_C = 0$. The symmetrical components of the *A*-phase voltages are

$$\mathbf{V}_0 = \frac{1}{3}(220 - j220 - 100 + j220) = 40\ \mathbf{V}^*$$

$$\mathbf{V}_{A1} = \frac{1}{3}[220 + (-0.5 + j0.866)(-j220) + (-0.5 - j0.866)(-100 + j220)]$$

$$= \frac{1}{3}[(660 + j86.6)] = (220 + j28.9)\ \text{V}$$

$$\mathbf{V}_{A2} = \frac{1}{3}[220 + (-0.5 - j0.866)(-j220) + (-0.5 + j0.866)(-100 + j220)]$$

$$= \frac{1}{3}(-120 - j86.6) = (-40 - j28.9)\ \mathbf{V}$$

The component currents in phase *A* are

$$\mathbf{I}_{A1} = \frac{\mathbf{V}_{A1}}{(8 + j6)} = \frac{220 + j28.9}{(8 + j6)} = (19.33 - j10.89)\ \text{A}$$

$$\mathbf{I}_{A2} = \frac{\mathbf{V}_{A2}}{(8 + j6)} = \frac{(-40 - j28.9)}{(8 + j6)} = (-4.93 - j0.09)\ \text{A}$$

$$\mathbf{I}_A = \mathbf{I}_{A1} + \mathbf{I}_{A2} = (14.4 - j10.8)\ \text{A};\ \mathbf{I}_B = \mathbf{I}_{B1} + \mathbf{B}_{B2} = a^2\mathbf{I}_{A1} + a\mathbf{I}_{A2}$$

$$\mathbf{I}_B = \left(-\frac{1}{2} - j\frac{\sqrt{3}}{2}\right)(19.93 - j10.89) + \left(-\frac{1}{2} - j\frac{\sqrt{3}}{2}\right)(-4.93 - j0.09)$$

Example 23.10. *Two equal impedance arms AB and BC are connected to the terminals A, B, C of a 3-phase supply as shown in Fig. 23.7. Each capacitor has a reactance of* $X = \sqrt{3}R$. *A high impedance voltmeter V is connected to the circuit at points P and Q as shown. If the supply line voltages* V_{AB}, V_{BC}, V_{CA} *are balanced, determine the reading of the voltmeter (a) when the phase sequence of the supply voltages is* $A \rightarrow B \rightarrow C$ *and (b) when the phase sequence is reversed. Hence, explain how this network could be employed to measure, respectively, the positive and negative phase sequence voltage components of an unbalanced 3-phase supply.*

Solution. The various currents and voltages are shown in Fig. 23.7.

Phase Sequence ABC

Taking $\mathbf{V}_{AB}$ as the reference vector, we have

$\mathbf{V}_{AB} = V(1 + j0)$; $\mathbf{V}_{BC} = a^2V$; $\mathbf{V}_{CA} = aV$

As seen from the diagram,

---

*However, it is not required in the problem.

$$\mathbf{I}_{AB} = \frac{\mathbf{V}_{AB}}{3R - jX} = \frac{V}{3R - j\sqrt{3}R} = \frac{V}{\sqrt{3}R(\sqrt{3}-j)}$$

$$\mathbf{I}_{BC} = \frac{\mathbf{V}_{BC}}{3R - jX} = \frac{a^2V}{\sqrt{3}R(\sqrt{3}-j)}$$

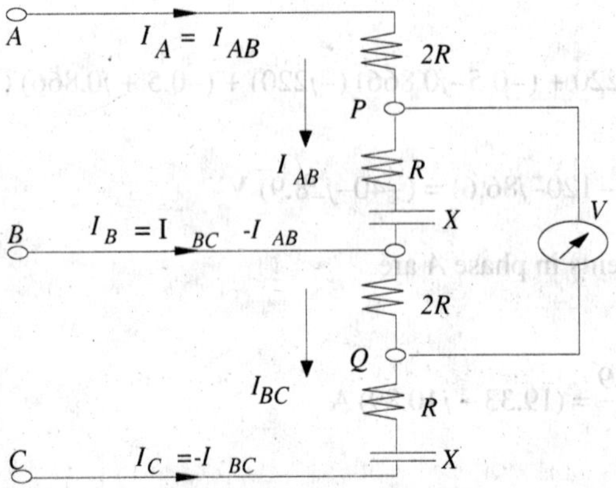

Fig. 23.7

Now, $\mathbf{V}_{PO} + \mathbf{V}_{OQ} = \mathbf{V}_{PQ}$

Also, $\mathbf{V}_{PQ} = \mathbf{I}_{AB}(R - jX) + \mathbf{I}_{BC}.2R$

$$= \frac{V}{\sqrt{3}R\,(\sqrt{3}-j)}.R(1 - j\sqrt{3}) + \frac{a^2V}{\sqrt{3}R(\sqrt{3}-j)}.2R$$

$$= \frac{V}{\sqrt{3}(\sqrt{3}-j)}.(1 - j\sqrt{3}+2a^2) \quad \left(\because \ a^2 = -\frac{1}{2} - j\frac{\sqrt{3}}{2}\right)$$

$$= \frac{V}{\sqrt{3}(\sqrt{3}-j)}.(-j2\sqrt{3}) \qquad \therefore \ V_{PQ} = \frac{V}{\sqrt{3}\times 2}\times 2\sqrt{2} = V$$

Hence, the voltmeter which is not phase sensitive will read the line voltage when phase sequence is $A \rightarrow B \rightarrow C$.

Phase Sequence ACB

In this case, $\mathbf{V}_{AB} = V(1 + j0)$; $\mathbf{V}_{BC} = aV$; $\mathbf{V}_{CA} = a^2V$

Also, $\mathbf{V}_{PQ} = V_{PO} + \mathbf{V}_{OQ}$

$$= \mathbf{I}_{AB}(R - jX) + \mathbf{I}_{BC}.2R \ . = \frac{V(R - jX)}{\sqrt{3}.R(\sqrt{3}-j)} + \frac{aV.2R}{\sqrt{3}.R(\sqrt{3}-j)}$$

$$= \frac{V}{\sqrt{3}(\sqrt{3}-j)}(1 - j\sqrt{3}+2a) = \frac{V}{\sqrt{3}(\sqrt{3}-j)}(1 - j\sqrt{3} - 1 + j\sqrt{3}) = 0$$

Hence, when the phase sequence is reversed, the voltmeter reads zero.

It can be proved that with sequence *ABC*, the voltmeter reads the positive sequence component ($\mathbf{V}_1$) and with phase sequence *ACB*, it reads the negative sequence component ($\mathbf{V}_2$). With phase sequence *ABC*

$$\mathbf{V}_{PQ} = \frac{\mathbf{V}_{AB}}{3R - j\sqrt{3}.R}(R - j\sqrt{3}R) + \frac{\mathbf{V}_{BC}}{3R - j\sqrt{3}R}2R = \frac{1}{3 - j\sqrt{3}}[(\mathbf{V}_1 + \mathbf{V}_2)(1 - j\sqrt{3}) + 2(a^2\mathbf{V}_1 + a\mathbf{V}_2)]$$

$$= \frac{1}{3 - j\sqrt{3}}[\mathbf{V}_1\, 1 - j\sqrt{3} + 2a^2) + \mathbf{V}_2(1 - j\sqrt{3} + 2a)] = \frac{1}{3 - j\sqrt{3}}\mathbf{V}_1(-j2\sqrt{3}) \quad \therefore \quad \mathbf{V}_{PQ} = \mathbf{V}_1$$

With phase sequence *ACB*

$$\mathbf{V}_{PQ} = \frac{1}{3 - j\sqrt{3}}[(\mathbf{V}_1 + \mathbf{V}_2)(1 - j\sqrt{3}) + 2(a\mathbf{V}_1 + a^2\mathbf{V}_2)]$$

$$= \frac{1}{3 - j\sqrt{3}}[\mathbf{V}_1(1 - j\sqrt{3} + 2a) + \mathbf{V}_2(1 - j\sqrt{3} + 2a^2)]$$

$$= \frac{1}{3 - j\sqrt{3}}[\mathbf{V}_2(-j2\sqrt{3})] \qquad \therefore \quad V_{PQ} = V_2$$

23.12. Measurement of Symmetrical Components of Circuits

The apparatus consists of two identical current transformers, two impedances of the same ohmic value, one being more inductive than the other to the extent that its phase angle is 60° greater and two identical ammeters A_1 greater and two identical ammeters A_1 and A_2 as shown in Fig. 23.8 (*a*). It can be shown that A_1 reads positive-sequence current only while A_2 reads negative-sequence current only. If the turn ratio of the current transformer is *K*, then keeping in mind that zero-sequence component is zero, we have

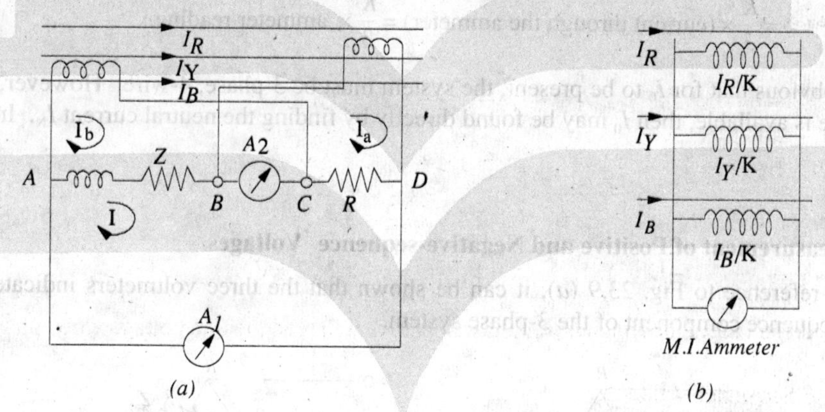

Fig. 23.8

$\mathbf{I}_a = \mathbf{I}_R/K = (\mathbf{I}_1 + \mathbf{I}_2)/K$

where $\mathbf{I}_1$ and $\mathbf{I}_2$ are the positive and negative-sequence components of the line current respectively. Similarly,

$$\mathbf{I}_b = \mathbf{I}_Y/K = (a^2\mathbf{I}_1 + a\mathbf{I}_2)/K$$

If the impedance of each ammeter is $R_A + jX_A$, then impedance between points *B* and *D* is

$$\mathbf{Z}_{BD} = (R + R_A + jX_A)$$

The value of **Z** is so chosen that

$$\mathbf{Z}_{AC} = \mathbf{Z} + R_A + jX_A = (R + R_A + jX_A)\angle 60° = \mathbf{Z}_{BD}\angle 60°$$

For finding the current read by A_1, imagine a break at point *X*. Thevenin voltage across the break *X* is

$$\mathbf{V}_{th} = \mathbf{I}_b\mathbf{Z}_{AC} + \mathbf{I}_a\mathbf{Z}_{BD} = \mathbf{I}_b\mathbf{Z}_{BD} = \mathbf{I}_b\mathbf{Z}_{BD}\angle 60° + \mathbf{I}a\mathbf{Z}_{BD} = (\mathbf{I}_b\angle 60° + \mathbf{I}_a)\,\mathbf{Z}_{BD}$$

Total impedance is series with this Thevenin voltage is

$$\mathbf{Z}_T = \mathbf{Z}_{AC} + \mathbf{Z}_{BD} = \mathbf{Z}_{BD}\angle 60° + \mathbf{Z}_{BD} = \left(\frac{3}{2} + j\frac{\sqrt{3}}{2}\right)\mathbf{Z}_{BD}$$

The current flowing normally through the wire in which a break has been imagined is

$$\mathbf{I} = \frac{\mathbf{V}_{th}}{\mathbf{Z}_T} = \frac{\mathbf{I}_b\angle 60° + \mathbf{I}_a}{(3/2) + j(\sqrt{3}/2)} = \frac{I}{K}\;\frac{(a^2\mathbf{I}_1 + a\mathbf{I}_2)\angle 60° + \mathbf{I}_1 + \mathbf{I}_2}{\sqrt{3}\angle 30°}$$

$$= \frac{1}{K}\;\frac{I_1(1\angle 300° + 1) + I_2(1\angle 180° + 1)}{\sqrt{3}\angle 30°} = \frac{I_1}{K}\angle -60°$$

It means that A_1 reads positive-sequence current only. The ammeter A_2 reads current which is

$$= \mathbf{I}_a + \mathbf{I}_b - \mathbf{I} = \frac{1}{K}(I_1 + I_2 + I_1\angle 240° + I_2\angle 120° - I_2\angle -60°)$$

$$= \frac{I_1}{K}\left(1 - \frac{1}{2} - j\frac{\sqrt{3}}{2} - \frac{1}{2} + j\frac{\sqrt{3}}{2}\right) + \frac{I_2}{K}(\mathrm{I} + 1\angle 120°) = \frac{I_2}{K}\angle 60°$$

In other words, A_2 reads negative-sequence current only.

Now, it will be shown that the reading of the moving-iron ammeter of Fig. 23-8(*b*) is proportional to the zero-sequence component.

$$\mathbf{I}_0 = \frac{1}{3}(\mathbf{I}_R + \mathbf{I}_Y + \mathbf{I}_B) = \frac{K}{3}\left(\frac{\mathbf{I}_R}{K} + \frac{\mathbf{I}_Y}{K} + \frac{\mathbf{I}_B}{K}\right)$$

$$= \frac{K}{3} \times \text{(current through the ammeter)} = \frac{K}{3} \times \text{ ammeter reading.}$$

It is obvious that for I_0 to be present, the system must be 3-phase, 4-wire. However, when the fourth wire is available, then I_0 may be found directly by finding the neutral current I_N. In that case $\mathbf{I}_0 = \frac{1}{3}I_N$.

23.13. Measurement of Positive and Negative-sequence *Voltages

With reference to Fig. 23.9 (*a*), it can be shown that the three voltmeters indicate only the positive-sequence component of the 3-phase system.

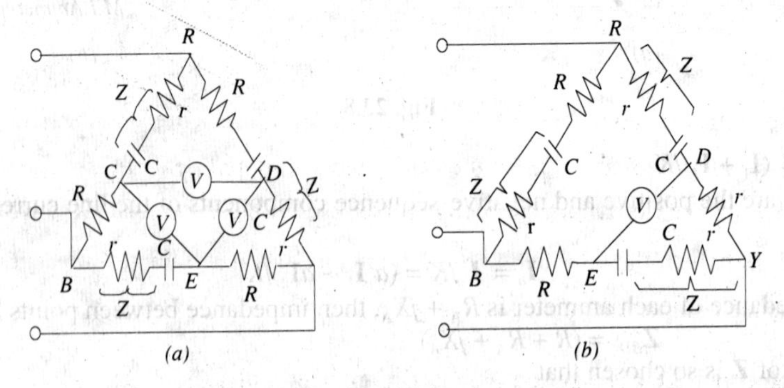

Fig. 23.9

*It is supposed to posses infinite impedance

The line-to-neutral voltage can be written (with reference to the red phase) as

$$\mathbf{V}_{RN} = \mathbf{V}_1 + \mathbf{V}_2 + \mathbf{V}_0 ; \mathbf{V}_{YN} = a^2\mathbf{V}_1 + a\mathbf{V}_2 + \mathbf{V}_0 ; \mathbf{V}_{BN} = a\mathbf{V}_1 + a^2\mathbf{V}_2 + \mathbf{V}_0$$

$$\mathbf{V}_{RY} = \mathbf{V}_{RN} + \mathbf{V}_{NY} = \mathbf{V}_{RN} - \mathbf{V}_{YN} = (1 - a^2)\,\mathbf{V}_1 + (1 - a)\,\mathbf{V}_2$$

$$\mathbf{V}_{YB} = \mathbf{V}_{YN} + \mathbf{V}_{NB} = \mathbf{V}_{YN} - \mathbf{V}_{BN} = (a^2 - a)\,\mathbf{V}_1 + (a - a^2)\,\mathbf{V}_2$$

$$\therefore \quad \mathbf{V}_{DY} = \frac{\mathbf{V}_{RY}\,(r + 1/j\omega C)}{(R + r + 1/j\omega C)} \qquad \mathbf{V}_{YE} = \frac{\mathbf{V}_{YB}.R}{(R + r + 1/j\omega C)}$$

$$\therefore \quad \mathbf{V}_{DE} = \frac{(r + 1/j\omega C)}{(R + r + 1/j\omega C)}\left[\mathbf{V}_{RY} + \frac{R}{(r + 1/j\omega C)}\mathbf{V}_{YB}\right]$$

The different elements of the bridge circuit are so chosen that

$$\frac{R}{r + j1/\omega C} = -a^2 = \frac{1}{2} + j\frac{\sqrt{3}}{2} \qquad \text{or} \quad R = \frac{r}{2} + \frac{\sqrt{3}}{2\omega C} + \frac{1}{j2\omega C} + j\frac{\sqrt{3}\,r}{2}$$

Equating the j-terms or quadrature terms on both sides, we have

$$0 = \frac{1}{j2\omega C} + j\frac{\sqrt{3}}{2} \qquad \therefore \quad \frac{1}{\omega C} = \sqrt{3}\,r \qquad \text{...}(i)$$

Similarly, equating the reference or real terms, we have

$$R = \frac{r}{2} + \frac{\sqrt{3}}{2\omega C} = \frac{r}{2} + \frac{3r}{2} = 2r \qquad \text{...}(ii)$$

$$\therefore \quad \frac{r + 1/j\omega C}{R + r + 1/j\omega C} = \frac{r - j\sqrt{3}r}{3r - j\sqrt{3}r} = \frac{r(1 - j\sqrt{3})}{r(3 - j\sqrt{3})} = \frac{1}{\sqrt{3}}\angle -30°$$

$$\text{and} \quad \mathbf{V}_{DE} = \frac{1\angle -30°}{\sqrt{3}}[(1 - a^2)\,\mathbf{V}_1 + (1 - a)\mathbf{V}_2 - a^2(a^2 - a)\,\mathbf{V}_1 - a^2(a - a^2)\mathbf{V}_2]$$

$$= \frac{1\angle -30°}{\sqrt{3}}(1 - a^2 - a^4 + a^3)\,\mathbf{V}_1 \quad [\because \; 1 - a = a^2(a - a^2)] = \sqrt{3}\,\mathbf{V}_1\angle -30°$$

Hence, the voltmeter* connected between points D and E measures $\sqrt{3}$ times the positive sequence component of the phase voltage. So do the other two voltmeters.

In Fig. 23.9 (*b*), the elements have been reversed. It can be shown that provided the same relation is maintained between the elements, the high impedance voltmeter measures $\sqrt{3}$ times the negative-sequence component of phase voltage.

23.14. Measurement of Zero-sequence Component of Voltage

The zero-sequence voltage is given by

---

*It is supposed to possess infinite impedance.

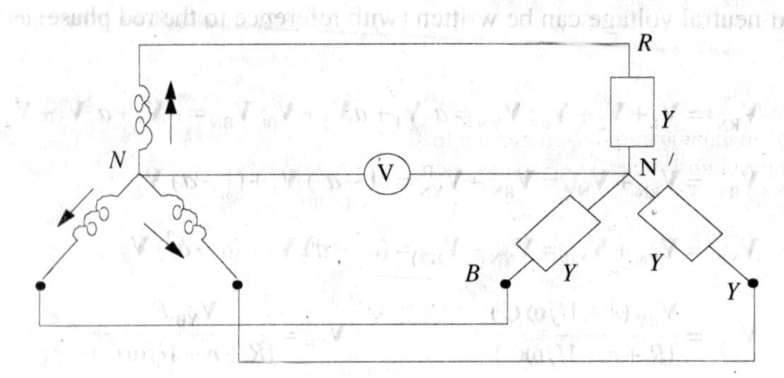

Fig. 23.10

$$\mathbf{V}_0 = \frac{1}{3}(\mathbf{V}_{RN} + \mathbf{V}_{YN} + \mathbf{V}_{BN})$$

Fig. 23.10 indicates one method of measuring $\mathbf{V}_0$. As seen

$$\mathbf{V}_{N'N} = \frac{\mathbf{V}_{RN}\mathbf{Y} + \mathbf{V}_{YN}\mathbf{Y} + \mathbf{V}_{BN}\,\mathbf{Y}}{3\mathbf{Y}}$$

$$= \frac{1}{3}(\mathbf{V}_{RN} + \mathbf{V}_{YN} + \mathbf{V}_{BN}) = \mathbf{V}_0$$

Hence, voltmeter connected between the neutral points measures the zero-sequence component of the voltage.

Tutorial Problem No. 23.2

1. Explain the essential features in the representation of an unsymmetrical three-phase system of voltages or currents by symmetrical components.

In a three-phase system, the three line currents are ; $\mathbf{I}_R = (30 + j50)$ A, $\mathbf{I}_Y = (15 - j45)$ A, $\mathbf{I}_B = (-40 + j70)$ A. Determine the values of the positive, negative and zero sequence components.

[52.2 A, 19.3 A, 25.1 A] (*London Univ.*)

2. The phase voltages of a three-phase, four-wire system are ; $\mathbf{V}_{RN} = (200 + j0)$ *V*, $\mathbf{V}_{YN} = (0 - j200)$ V, $\mathbf{V}_{BN} = (-100 + j\,100)$ *V*. Show that these voltages can be replaced by symmetrical components of positive, negative and zero sequence and calculate the magnitude of each component.

Sketch vector diagrams representing the positive, negative and zero-sequence components for each of the three phases.

[176 V, 38 V, 47.1 V] (*London Univ.*)

3. Explain, with the aid of a diagram of connections, a method of measuring the symmetrical components of the currents in an unbalanced 3-phase, 3-wire system.

If in such a system, the line currents, in amperes, are

$$\mathbf{I}_R = (10 - j2),\ \mathbf{I}_Y = (-2 - j4),\ \mathbf{I}_B = (-8 + j6)$$

Calculate their symmetrical components.

$[\mathbf{I}_{RO} = \mathbf{I}_{YO} = \mathbf{I}_{BO} = 0\ ;\ \mathbf{I}_{R1} = 7.89 + j\,0.732\ ;\ \mathbf{I}_{Y1} = a^2\mathbf{I}_{R1}\ ;\ \mathbf{I}_{B1} = a\mathbf{I}_{R1}$
$\mathbf{I}_{R2} = 2.113 - j\,2.732\ ;\ \mathbf{I}_{Y2} = a\mathbf{I}_{R2}\ ;\ \mathbf{I}_{B2} = a^2\mathbf{I}_{R2}]$

Objective Tests – 23

1. The method of symetrcial components is very useful for
(*a*) solving unbalanced polyphase circuits
(*b*) analysing the performance of 3-phase electrical machinery
(*c*) calculating currents resulting from unbalanced faults
(*d*) all of the above.

2. An unbalanced system of 3-phase voltages having *RYB* sequence actually consists of
(*a*) a positive-sequence component
(*b*) a negative-sequence component
(*c*) a zero-sequence component
(*d*) all of the above.

3. The zero-sequence component of the unbalanced 3-phase system of vectors V_A, V_B and V_C is —— of their vector sum.

(*a*) one-third
(*b*) one-half
(*c*) two-third
(*d*) one-fourth.

4. In the case of an unbalanced star-connected load supplied from an unbalanced 3-φ, 3 wire system, load currents will consist of
(*a*) positive-sequence components
(*b*) negative-sequence components
(*c*) zero-sequence components
(*d*) only (*a*) and (*b*).

ANSWERS

1. (*d*) **2.** (*d*) **3.** (*a*) **4.** (*d*)